Airplane Aerodynamics and Performance

Dr. Jan Roskam
Ackers Distinguished Professor of Aerospace Engineering
The University of Kansas, Lawrence

and

Dr. Chuan-Tau Edward Lan
Warren S. Bellows Distinguished Professor of Aerospace Engineering
The University of Kansas, Lawrence

1997

DARcorporation
Design, Analysis and Research Corporation
120 East 9th Street, Suite 2 • Lawrence, Kansas 66044, U.S.A.

PUBLISHED BY
Design, Analysis and Research Corporation (*DARcorporation*)
120 East Ninth Street, Suite 2
Lawrence, Kansas 66044
U.S.A.
Phone: (785) 832-0434
Fax: (785) 832-0524
e-mail: info@darcorp.com
http://www.darcorp.com

ISBN 1-884885-44-6

In all countries, sold and distributed by
Design, Analysis and Research Corporation
120 East Ninth Street, Suite 2
Lawrence, Kansas 66044
U.S.A.

> The information presented in this book have been included for their instructional value. They are not guaranteed for any particular purpose. The publisher does not offer any warranties or representations, nor does it accept any liabilities with respect to the information. It is sold with the understanding that the publisher is not engaged in rendering engineering or other professional services. If such services are required, the assistance of an appropriate professional should be sought.

Copyright © 1997 by Dr. Jan Roskam and Dr. C.T. Lan. All rights reserved
Printed in the United States of America
First Printing, 1997

Information in this document is subject to change without notice and does not represent a commitment on the part of *DARcorporation*. No part of this book may be reproduced, stored in a retrieval system or transmitted in any form or by any means, electronic or mechanical, photocopying, recording, or otherwise, without written permission of the publisher, *DARcorporation*.

TABLE OF CONTENTS

SYMBOLS AND ACRONYMS	xiii
INTRODUCTION	1
1. THE ATMOSPHERE	**3**
1.1 ATMOSPHERIC FUNDAMENTALS	3
1.2 THE INTERNATIONAL STANDARD ATMOSPHERE	4
1.2.1 Temperature Variation with Altitude	5
1.2.2 Pressure and Density Variation with Altitude	6
1.2.3 Barometric Altimeters	8
1.2.4 Viscosity	10
1.3 SUMMARY FOR CHAPTER 1	11
1.4 PROBLEMS FOR CHAPTER 1	11
1.5 REFERENCES FOR CHAPTER 1	12
2. BASIC AERODYNAMIC PRINCIPLES AND APPLICATIONS	**13**
2.1 THE CONTINUITY EQUATION	13
2.2 THE INCOMPRESSIBLE BERNOULLI EQUATION	15
2.3 COMPRESSIBILITY EFFECTS	17
2.3.1 The Isentropic Equation of State	17
2.3.2 The Speed of Sound	19
2.4 THE COMPRESSIBLE BERNOULLI EQUATION	20
2.5 MEASUREMENT OF AIRSPEED	21
2.5.1 Low–Speed Airspeed Indicators (Incompressible Flow)	21
2.5.2 High–Speed Airspeed Indicators (Compressible Flow)	22
2.5.3 Airspeed Corrections	26
2.5.6 Effect of the fuselage on wing aerodynamic center	56
2.6 THE KUTTA–JOULOWSKI THEOREM	32
2.7 THE LINEAR MOMENTUM PRINCIPLE	36
2.8 VISCOUS EFFECTS, THE BOUNDARY LAYER AND FLOW SEPARATION	37
2.8.1 The Laminar Boundary Layer	39
2.8.2 The Turbulent Boundary Layer	40
2.8.3 Flow Separation	42
2.9 SUMMARY FOR CHAPTER 2	46
2.9.1 The Continuity Equation	46

	2.9.2	The Bernoulli Equation	46
	2.9.3	The Kutta–Joulowski Theorem	47
	2.9.4	The Linear Momentum Principle	47
	2.9.5	Viscous Effects	47
2.10	PROBLEMS FOR CHAPTER 2	48
2.11	REFERENCES FOR CHAPTER 2	49

3. AIRFOIL THEORY   51

3.1	AIRFOIL GEOMETRY	51
3.2	AERODYNAMIC FORCES AND MOMENTS ON AN AIRFOIL	53
3.3	IMPORTANT AIRFOIL CHARACTERISTICS	56
	3.3.1	Lift Curve: c_l versus α	56
	3.3.2	Drag Polar: c_l versus c_d	57
	3.3.3	Pitching Moment Curve: c_m versus c_l or c_m versus α	59
3.4	AIRFOIL PRESSURE DISTRIBUTION	60
3.5	COMPRESSIBILITY EFFECTS	62
3.6	REYNOLDS NUMBER EFFECTS	68
3.7	DESIGN OF AIRFOILS	70
3.8	AIRFOIL MAXIMUM LIFT CHARACTERISTICS	77
	3.8.1	Geometric Factors Affecting Airfoil Maximum Lift at Low Speeds	78
		3.8.1.1 Thickness ratio	78
		3.8.1.2 Leading edge radius and leading edge shape	78
		3.8.1.3 Camber and location of maximum thickness	78
	3.8.2	Effect of Reynolds Number on Maximum Lift	80
	3.8.3	Effect of High Lift Devices on Airfoil Maximum Lift	80
		3.8.3.1 Trailing edge flaps	80
		3.8.3.2 Leading edge devices	86
		3.8.3.3 Boundary layer control	87
3.9	SUMMARY FOR CHAPTER 3	88
3.10	PROBLEMS FOR CHAPTER 3	92
3.11	REFERENCES FOR CHAPTER 3	93

4. WING THEORY   95

4.1	DEFINITION OF WING PROPERTIES	95
4.2	CIRCULATION, DOWNWASH, LIFT AND INDUCED DRAG	97
4.3	EVALUATION OF THE SPAN EFFICIENCY FACTOR e	104
4.4	AERODYNAMIC CENTER	108
4.5	WING STALL	109

	4.5.1	Airfoil Stall Characteristics	110
	4.5.2	Effect of Planform and Twist	110
		4.5.2.1 Taper ratio	110
		4.5.2.2 Aspect ratio	111
		4.5.2.3 Sweep angle	112
		4.5.2.4 Twist	114
	4.5.3	Stall Control Devices	115
		4.5.3.1 Twist or wash–out	115
		4.5.3.2 Variations in section shape	115
		4.5.3.3 Leading edge slats or slots near the tip	116
		4.5.3.4 Stall fences and snags	116
		4.5.3.5 Stall strips	117
		4.5.3.6 Vortex generators	117
4.6	COMPRESSIBILITY EFFECTS		119
4.7	HIGH LIFT DEVICES, SPOILERS, DIVE BRAKES, SPEED BRAKES		121
	4.7.1	Lift Induced by Partial Span Flaps Below the Stall	122
	4.7.2	Maximum Lift Coefficient with High Lift Devices	127
		4.2.7.1 Clean wing maximum lift coefficient	127
		4.2.7.2 Maximum lift increment due to high lift devices	128
		4.2.7.3 Examples of maximum lift increment due to high lift devices	129
	4.7.3	The Effect of Trim on Maximum Lift	129
	4.7.4	The Effect of Spoilers, Dive Brakes and Speed Brakes	132
4.8	SUMMARY FOR CHAPTER 4		133
4.9	PROBLEMS FOR CHAPTER 4		133
4.10	REFERENCES FOR CHAPTER 4		135

5. AIRPLANE DRAG   137

5.1	COMPLETE AIRPLANE DRAG POLARS		137
	5.1.1	Clean Airplane	137
	5.1.2	Effect of Flaps, Speed-Brakes and Landing Gear	145
5.2	UNDERSTANDING AIRPLANE DRAG CONTRIBUTIONS		146
	5.2.1	Wing Drag Coefficient, $C_{D_{wing}}$	148
		5.2.1.1 Wing zero lift drag coefficient, $C_{D_{0_{wing}}}$	148
		5.2.1.2 Wing drag coefficient due to lift, $C_{D_{L_{wing}}}$	154
		5.2.1.3 Wing drag coefficient due to compressibility, $C_{D_{wing_{wave}}}$	154
	5.2.2	Fuselage Drag Coefficient, $C_{D_{fuse}}$	157
		5.2.2.1 Fuselage zero lift drag coefficient, $C_{D_{0_f}}$	158

		5.2.2.2 Fuselage base drag coefficient, $C_{D_{f_b}}$	160
		5.2.2.3 Fuselage drag coefficient due to lift, $C_{D_{L_f}}$	161
	5.2.3	Empennage Drag Coefficient, $C_{D_{emp}}$	161
	5.2.4	Nacelle–Pylon Drag Coefficient, $C_{D_{np}}$	162
	5.2.5	Additional Drag Due to Integration Effects	163
		5.2.5.1 Wing planform shape effect on induced drag	163
		5.2.5.2 Wing tip shape effect on induced drag	163
		5.2.5.3 Wing twist effect on induced drag	165
		5.2.5.4 Wing sweep effect on induced drag	165
		5.2.5.5 Effect of tandem configurations on induced drag	165
		5.2.5.6 Effect of winglets on zero lift and on induced drag	166
	5.2.6	Flap Drag Coefficient, $C_{D_{flap}}$	167
	5.2.7	Landing Gear Drag Coefficient, $C_{D_{flap}}$	168
	5.2.8	Canopy/Windshield Drag Coefficient, $C_{D_{cw}}$	169
	5.2.9	Store Drag Coefficient, $C_{D_{store}}$	169
	5.2.10	Trim Drag Coefficient, $C_{D_{trim}}$	170
	5.2.11	Interference Drag Coefficient, $C_{D_{interf.}}$	172
	5.2.12	Miscellaneous Drag Coefficient, $C_{D_{misc}}$	173
	5.2.13	Drag Component Summary and Typical Examples	174
5.3	DETERMINATION OF DRAG IN THE WIND–TUNNEL		178
5.4	SIMPLIFIED METHOD FOR PREDICTING DRAG POLARS OF CLEAN AIRPLANES		188
5.5	SUMMARY FOR CHAPTER 5		194
5.6	PROBLEMS FOR CHAPTER 5		195
5.7	REFERENCES FOR CHAPTER 5		198

6. AIRPLANE PROPULSION SYSTEMS   201

6.1	PISTON ENGINES		208
	6.1.1	The Four–Stroke Piston Engine	209
	6.1.2	The Two–Stroke Piston Engine	210
	6.1.3	Geometric Layout of Piston Engines	212
	6.1.4	Power Output and Fuel Efficiency of Piston Engines	213
	6.1.5	Factors Affecting the Power Output of Piston Engines	214
	6.1.6	Performance Charts for Piston Engines	217

		6.1.7 Cooling and Installation of Piston Engines	221
		6.1.8 Scaling Laws for Piston Engines	222
6.2	JET ENGINES		225
		6.2.1 Basic Operation of Turbojets	225
		6.2.2 Characteristics of By–pass Engines and Turbofans	231
		6.2.3 Operational Characteristics of Turboprops	233
		6.2.4 Typical Mechanical Arrangements of Turbojets, Turbofans and Turboprops	235
		6.2.5 Thrust, Power and Fuel Consumption of Gas Turbine Engines	238
		6.2.6 Cooling Considerations for Turbine Engine Installations	249
		6.2.7 Noise Considerations for Gas Turbine Engines	252
		6.2.8 Thrust Reversal for Turbine Engines	255
		6.2.9 Scaling Laws for Turbine Engines	256
		6.2.10 Pulse Jets	258
		6.2.11 Ram Jets	259
6.3	ROCKET ENGINES		260
6.4	ELECTRIC MOTORS		261
6.5	SUMMARY FOR CHAPTER 6		262
6.6	PROBLEMS FOR CHAPTER 6		262
6.7	REFERENCES FOR CHAPTER 6		263

7. PROPELLER THEORY AND APPLICATIONS — 265

7.1	MOMENTUM THEORY		266
	7.1.1	Incompressible Theory for Positive Propeller Thrust	266
	7.1.2	Incompressible Theory for Negative Propeller Thrust	269
	7.1.3	Compressible Theory for Positive Propeller Thrust	273
7.2	SIMPLE BLADE–ELEMENT THEORY		275
7.3	COMBINED BLADE–ELEMENT THEORY AND MOMENTUM THEORY		277
7.4	ANALYSIS OF PROPELLER POWER LOSSES		282
	7.4.1	Axial and Rotational Momentum Losses	283
	7.4.2	Measured Losses and Some Interpretations	284
7.5	FACTORS AFFECTING PROPELLER PERFORMANCE		286
	7.5.1	Propeller Blade Angle	286
	7.5.2	Propeller Blade Geometry	287
	7.5.3	Propeller Blade Loading	288
	7.5.4	Propeller Shank Form	289
	7.5.5	Compressibility Effects	290
	7.5.6	Blockage, Scrubbing, Compressibility and Installation Effects on Propeller Efficiency	292
		7.5.6.1 Blockage effects	292
		7.5.6.2 Scrubbing drag effects	296
		7.5.6.3 Compressibility effects	296
	7.5.7	Free Propeller Efficiency and Installed Propeller Efficiency for Tractors and Pushers	298

7.6	PREDICTION OF PROPELLER PERFORMANCE		300
	7.6.1	Prediction of Static Thrust	300
	7.6.2	Prediction of In–Flight Thrust and Power	304
	7.6.3	Prediction of Negative Thrust	308
		7.6.3.1 Prediction of negative thrust or propeller thrust reversing	308
		7.6.3.2 Prediction of drag on stopped engines	311
	7.6.4	Prop–Fans and Ducted Propellers	311
		7.6.4.1 Prop–fans	311
		7.6.4.2 Ducted propellers	313
7.7	PROPELLER NOISE		315
	7.7.1	Sources and Characteristics of Propeller Noise	316
	7.7.2	Prediction Procedure for Propeller Noise	317
		7.7.2.1 Far–field noise	317
		7.7.2.2 Near–field noise	320
7.8	PROPELLER SELECTION		322
	7.8.1	Introduction to Propeller Selection	322
	7.8.2	Propeller Design Variables to be Selected	322
7.9	SUMMARY FOR CHAPTER 7		327
7.10	PROBLEMS FOR CHAPTER 7		327
7.11	REFERENCES FOR CHAPTER 7		328

8. FUNDAMENTALS OF FLIGHT PERFORMANCE 331

8.1	DEFINITION OF ANGLES AND AXIS SYSTEMS		331
8.2	STEADY, UN–POWERED FLIGHT		335
	8.2.1	Equations and Definitions	335
	8.2.2	Glide Angle, Rate–of–Descent and Speed	336
	8.2.3	Speed Polar or Hodograph	344
	8.2.4	Effect of Altitude, Weight and Wind	347
		8.2.4.1 Effect of altitude	347
		8.2.4.2 Effect of weight	348
		8.2.4.3 Speed polar in generalized coordinates and application in flight testing	349
		8.2.4.4 Effect of wind	350
	8.2.5	Unpowered Glide at High Speed	352
8.3	STEADY, POWERED FLIGHT		354
8.4	STEADY, LEVEL, POWERED FLIGHT		357
	8.4.1	Turbojet or Turbofan Driven Airplanes	357
	8.4.2	Propeller Driven Airplanes	363
8.5	SUMMARY FOR CHAPTER 8		369
8.6	PROBLEMS FOR CHAPTER 8		369
8.7	REFERENCES FOR CHAPTER 8		372

Table of Contents

9. CLIMB AND DRIFT–DOWN PERFORMANCE — 373

- 9.1 EQUATIONS OF MOTION — 374
- 9.2 CLIMB AND DRIFT–DOWN PERFORMANCE OF JET AIRPLANES — 379
  - 9.2.1 Climb Performance for Jet Airplanes — 379
    - 9.2.1.1 The general case — 379
    - 9.2.1.2 The case of parabolic drag and constant thrust — 383
  - 9.2.2 Climb Performance for Jet Airplanes at Steep Angles — 387
  - 9.2.3 Drift–Down Performance for Jet Airplanes — 388
- 9.3 CLIMB AND DRIFT–DOWN PERFORMANCE OF PROPELLER DRIVEN AIRPLANES — 389
  - 9.3.1 Climb Performance for Propeller Driven Airplanes — 389
    - 9.3.1.1 The general case — 389
    - 9.3.1.2 The case of parabolic drag and constant power — 399
  - 9.3.2 Climb Performance for Propeller Driven Airplanes at Steep Angles — 401
  - 9.3.3 Drift–Down Performance for Propeller Driven Airplanes — 403
- 9.4 METHODS FOR PREDICTING TIME–TO–CLIMB, TIME–TO–DRIFT–DOWN, AEO CEILINGS AND OEI CEILINGS — 404
  - 9.4.1 Method for Predicting Time-to-Climb Performance — 404
  - 9.4.2 Method for Predicting Time-to-Drift-Down Performance — 411
  - 9.4.3 Method for Predicting AEO and OEI Ceilings — 417
- 9.5 EFFECT OF FORWARD AND VERTICAL ACCELERATIONS ON CLIMB PERFORMANCE — 420
  - 9.5.1 Effect of Forward Acceleration — 420
    - 9.5.1.1 Climb at constant equivalent airspeed — 421
    - 9.5.1.2 Climb at constant Mach number — 423
  - 9.5.2 Effect of Vertical Acceleration — 424
- 9.6 EFFECT OF LANDING GEAR, FLAPS, STOPPED ENGINES, TRIM REQUIREMENTS, WEIGHT AND ALTITUDE ON CLIMB PERFORMANCE — 425
  - 9.6.1 Effect of Landing Gear and Flaps on Climb Performance — 425
  - 9.6.2 Effect of Stopped Engines and Trim on Climb Performance — 425
  - 9.6.3 Effect of Weight and Altitude on Climb Performance — 427
- 9.7 CLIMB PERFORMANCE REGULATIONS — 428
- 9.8 SUMMARY FOR CHAPTER 9 — 432
- 9.9 PROBLEMS FOR CHAPTER 9 — 432
- 9.10 REFERENCES FOR CHAPTER 9 — 434

10. TAKE–OFF AND LANDING — 435

- 10.1 THE TAKE–OFF PROCESS — 435
  - 10.1.1 Commercial Take–off Rules — 436
  - 10.1.2 Military Take–off Rules — 441
- 10.2 EQUATIONS OF MOTION DURING TAKE–OFF — 442

10.2.1	Equations of Motion During the Take–off Ground Roll	442
	10.2.1.1 Effect of wind on take–off	446
	10.2.1.2 Effect of speed on take–off thrust	447
	10.2.1.3 Effect of the ground on lift and drag	448
	10.2.1.4 Effect of thrust deflection on take–off	452
10.2.2	Equations of Motion During the Take–off Transition	452
10.2.3	Equations of Motion During the Climb–out to the Obstacle	453
10.3	PREDICTION OF THE TAKE–OFF DISTANCE	454
10.3.1	Accurate Method for Predicting the Take–off Distance	454
	10.3.1.1 Ground distance (accurate method)	454
	10.3.1.2 Transition distance (accurate method)	455
	10.3.1.3 Climb distance (accurate method)	455
10.3.2	Statistical Method for Predicting the Take–off Distance	456
	10.3.2.1 Statistical method for predicting the take–off distance of propeller driven, FAR 23 airplanes	456
	10.3.2.2 Statistical method for predicting the take–off distance of jet driven, FAR 25 airplanes	458
10.3.3	Approximate, Analytical Method for Predicting the Take–off Distance and the Time to Take–off	460
	10.3.3.1 Approximate method for calculating the ground distance component, S_{NGR}	460
	10.3.3.2 Approximate method for calculating the ground distance component, S_R	462
	10.3.3.3 Approximate method for calculating the air distance component, S_{TR}	464
	10.3.3.4 Approximate method for calculating the air distance component, S_{CL}	466
	10.3.3.5 Alternate method for calculating the total ground distance, S_G	466
	10.3.3.6 Method for calculating the time required to take–off, t_{TO}	466
10.3.4	Approximate, Analytical Method for Predicting the Balanced Fieldlength	472
10.3.5	Presentation of Take–off Performance Data	475
10.4	THE LANDING PROCESS	478
10.4.1	Commercial Landing Rules	479
10.4.2	Military Landing Rules	479
10.5	EQUATIONS OF MOTION DURING LANDING	480
10.5.1	Equations of Motion During the Landing Approach and the Descent from the Obstacle	480
10.5.2	Equations of Motion During the Landing Transition	481
10.5.3	Equations of Motion During the Landing Ground Roll	482
10.6	PREDICTION OF THE LANDING DISTANCE	487
10.6.1	Accurate Method for Predicting the Landing Distance	487
	10.6.1.1 Air distance covered during descent from the obstacle (accurate method)	487
	10.6.1.2 Air distance covered during the transition or flare (accurate method)	489

Table of Contents

	10.6.1.3 Ground distance covered with the nose–gear off the ground (accurate method)	491
	10.6.1.4 Ground distance covered with the nose–gear on the ground (accurate method)	493
10.6.2	Statistical Method for Predicting the Landing Distance	494
	10.6.2.1 Statistical method for predicting the landing distance of propeller driven, FAR 23 airplanes	494
	10.6.2.2 Statistical method for predicting the landing distance of jet driven, FAR 25 airplanes	494
10.6.3	Approximate, Analytical Method for Predicting the Landing Distance and the Time to Land	497
	10.6.3.1 Approximate method for calculating the air distance component, S_{LA}	497
	10.6.3.2 Approximate method for calculating the ground distance component, S_{LR}	498
	10.6.3.3 Approximate method for calculating the ground distance component, S_{LNGR}	498
	10.6.3.4 Method for calculating the time required to land, t_L	500
10.6.4	Presentation of Landing Distance Data	505
10.7	SUMMARY FOR CHAPTER 10	506
10.8	PROBLEMS FOR CHAPTER 10	506
10.9	REFERENCES FOR CHAPTER 10	507

11. RANGE, ENDURANCE AND PAYLOAD–RANGE — 509

11.1	PROPELLER DRIVEN AIRPLANES	509
11.1.1	Breguet Equations for Range and Endurance	510
11.1.2	Maximum Range and Endurance for Parabolic Drag Polars	513
	11.1.2.1 Maximum range for parabolic drag polars	513
	11.1.2.2 Sizing the wing for maximum range (Maximum C_L/C_D)	515
	11.1.2.3 Effect of wind on maximum range	516
	11.1.2.4 Maximum endurance for parabolic drag polars	517
	11.1.2.5 Sizing the wing for maximum endurance (Maximum C_L^3/C_D^2)	518
	11.1.2.6 Numerical example of a composite plot of C_D, C_L/C_D and C_L^3/C_D^2 versus C_L	519
11.1.3	Applications of the Range and Endurance Equations to a Propeller Driven Airplane With a Parabolic Drag Polar	521
11.1.4	Maximum Range and Endurance for Airplanes With Non–parabolic Drag Polars Including the Effect of Wind	523
11.1.5	Accurate Determination of Range and Endurance by Numerical Integration of Specific Range and Endurance	526

	11.1.5.1 Step–by–step procedure for determining range	527
	11.1.5.2 Step–by–step procedure for determining endurance (loiter)	533
11.2	JET AIRPLANES	535
	11.2.1 Breguet Equations for Range and Endurance	536
	11.2.1.1 Constant altitude cruise and endurance	536
	11.2.1.2 Constant speed cruise and endurance	539
	11.2.2 Maximum Range and Endurance for Parabolic Drag Polars	542
	11.2.2.1 Maximum range for parabolic drag polars (constant altitude)	542
	11.2.2.2 Maximum range for parabolic drag polars (constant Mach number)	543
	11.2.2.3 Sizing the wing for maximum range (Maximum C_L/C_D)	544
	11.2.2.4 Effect of wind on maximum range	544
	11.2.2.5 Maximum range for parabolic drag polars (constant Mach number)	544
	11.2.2.6 Sizing the wing for maximum endurance (Maximum C_L/C_D)	544
	11.1.2.7 Numerical example of a composite plot of C_D, C_L/C_D and $\sqrt{C_L}/C_D$ versus C_L	544
	11.2.3 Applications of the Range and Endurance Equations to a Jet Transport Airplane With a Parabolic Drag Polar	546
	11.2.4 Accurate Determination of Range and Endurance for Jet Airplanes With Non–parabolic Drag Polars by Use of Specific Range and Endurance	548
	11.2.4.1 Step–by–step procedure for determining range	549
	11.2.4.2 Step–by–step procedure for determining endurance (loiter)	551
	11.2.4.3 Suggested procedure to account for the effect of altitude, temperature and wind on range and endurance	553
11.3	THE PAYLOAD–RANGE DIAGRAM AND METHODS FOR PRESENTING USEFUL PAYLOAD–RANGE DATA	559
	11.3.1 The Payload–Range Diagram	559
	11.3.2 Method for Presenting Payload–Range Data to Account for Fieldlength Capability	561
11.4	REGULATIONS FOR RANGE AND ENDURANCE FUEL RESERVES	561
	11.4.1 Summary of FAR Parts 121 and 135 Requirements for Fuel Reserves	561
	11.4.2 Summary of Military Requirements for Fuel Reserves	564
11.5	SIZING OF WEIGHT OF AIRPLANES TO GIVEN MISSION PAYLOAD, RANGE AND ENDURANCE REQUIREMENTS	565
	11.5.1 Determination of the Regression Coefficients A and B With Numerical Examples	568
	11.5.2 Methods for Determining the Weight Fractions in Eqn (11.88)	569
	11.5.3 Three Solutions for Eqn (11.86)	572
	11.5.4 Example Application of Airplane Weight Sizing	573
11.6	SUMMARY FOR CHAPTER 11	576
11.7	PROBLEMS FOR CHAPTER 11	579
11.8	REFERENCES FOR CHAPTER 11	580

12. MANEUVERING AND THE FLIGHT ENVELOPE — 581

- 12.1 STALL SPEEDS AND MINIMUM SPEEDS — 581
- 12.2 BUFFET LIMITS — 586
- 12.3 LEVEL FLIGHT MAXIMUM SPEEDS AND CEILINGS — 592
  - 13.3.1 Level Flight Maximum Speeds — 592
  - 13.3.2 Level Flight Ceilings — 593
- 12.4 THE V–N DIAGRAM — 593
  - 12.4.1 FAR 23 — 593
    - 12.4.1.1 The 1–g stall speed, V_S — 594
    - 12.4.1.2 The design maneuvering speed, V_A — 595
    - 12.4.1.3 The design cruising speed, V_C — 595
    - 12.4.1.4 The design diving speed, V_D — 596
    - 12.4.1.5 The negative 1–g stall speed, $V_{S_{neg}}$ — 596
    - 12.4.1.6 The design limit load factors, $n_{lim_{pos}}$ and $n_{lim_{neg}}$ — 597
    - 12.4.1.7 The gust load load factor lines in Figure 12.12 — 597
  - 12.4.2 FAR 25 — 598
    - 12.4.2.1 The 1–g stall speed, V_{S_1} — 598
    - 12.4.2.2 The design maneuvering speed, V_A — 600
    - 12.4.2.3 The design speed for maximum gust intensity, V_B — 600
    - 12.4.2.4 The design cruising speed, V_C — 600
    - 12.4.2.5 The design diving speed, V_D — 600
    - 12.4.2.6 The negative 1–g stall speed, $V_{S_{1_{neg}}}$ — 600
    - 12.4.2.7 The design limit load factors, $n_{lim_{pos}}$ and $n_{lim_{neg}}$ — 600
    - 12.4.2.8 The gust load load factor lines in Figure 12.14 — 601
  - 12.4.3 Military Airplanes — 601
- 12.5 SYMMETRICAL PULL–UP: INSTANTANEOUS AND SUSTAINED — 603
  - 12.5.1 Instantaneous Pull–up — 603
  - 12.5.2 Sustained Pull–up — 604
- 12.6 STEADY, LEVEL AND COORDINATED TURNS: INSTANTANEOUS AND SUSTAINED — 606
  - 12.6.1 Instantaneous Turn — 607
  - 12.6.2 Sustained Turn — 607
- 12.7 SPIN — 613

12.8	ACCELERATED FLIGHT PERFORMANCE	616
	12.8.1 Application of Energy–State Approximation to Accelerated Flight Performance	617
	12.8.2 Energy Maneuverability	626
12.9	SUMMARY FOR CHAPTER 12	627
12.10	PROBLEMS FOR CHAPTER 12	628
12.11	REFERENCES FOR CHAPTER 12	628

APPENDIX A:	PROPERTIES OF THE U.S. STANDARD ATMOSPHERE	631
	TABLE A1 U.S. STANDARD ATMOSPHERE IN ENGLISH UNITS	631
	TABLE A2 U.S. STANDARD ATMOSPHERE IN METRIC UNITS	634
	TABLE A3 CONVERSION FACTORS	635

APPENDIX B:	ADVANCED AIRCRAFT DESIGN AND ANALYSIS SOFTWARE	637
B1	AAA: ADVANCED AIRCRAFT ANALYSIS	637
	B1.1 General Capabilities of the AAA Program	637
	B1.2 Brief Description of AAA Program Modules	638
	B1.2.1 Weight sizing module	638
	B1.2.2 Performance sizing module	638
	B1.2.3 Geometry module	639
	B1.2.4 High Lift Module	639
	B1.2.5 Drag polar module	639
	B1.2.6 Stability and control module	639
	B1.2.7 Weight and balance module	639
	B1.2.8 Installed thrust module	640
	B1.2.9 Performance analysis module	640
	B1.2.10 Stability and control derivatives module	640
	B1.2.11 Dynamics module	640
	B1.2.12 Control module	640
	B1.2.13 Cost analysis module	640
	B1.2.14 Database module	641
	B1.2.15 Help/setup module	641
B2	G.A.–CAD: GENERAL AVIATION COMPUTER AIDED DESIGN	641
B3	OTHER SOFTWARE PACKAGES	641
B4	REFERENCES	642

APPENDIX C:	PROPELLER PERFORMANCE CHARTS	643

APPENDIX D:	PROPELLER NOISE CHARTS	685

INDEX		699

SYMBOLS, SUBSCRIPTS AND ACRONYMS

Symbol	Description	Unit(s)
Regular		
a	Atmospheric lapse rate	$-0.00356616\ ^0F/ft$
a	Section or airfoil lift curve slope	1/rad or 1/deg
a_g	Acceleration on the ground	ft/sec^2, or m/sec^2
A	Propeller disk area	ft^2 or m^2
$A = b^2/S$	Wing aspect ratio	———
A_i	Cross sectional area of streamtube at section i	ft^2 or m^2
AF	Activity factor	———
b	Wing span	ft or m
B	Number of blades	———
BFL	Balanced fieldlength	ft or m
c	Section or airfoil chord	ft or m
$\bar{c}$	Mean geometric chord	ft or m
c_f	Skin friction coefficient	———
c_d	Airfoil or section drag coefficient	———
c_l	Airfoil or section lift coefficient	———
c_{l_α}	Airfoil or section lift curve slope	1/rad
$c_{l_{max}}$	Maximum section lift coefficient	———
c_m	Airfoil or section pitching moment coefficient	———

Symbols, Subscripts and Acronyms

Symbol	Description	Unit(s)

Regular (Continued)

Symbol	Description	Unit(s)
$\bar{c}_{m_0}$	Airfoil or section pitching moment coefficient for zero lift coefficient	———
c_p	Pressure coefficient	———
c_p	Specific fuel consumption	lbs/hr/hp
c_j	Specific fuel consumption	lbs/hr/lbs
c_r	Root chord	ft or m
c_t	Tip chord	ft or m
C	Constant in Eqn (2.26)	ft^2/sec^2
C_p	Specific heat at constant pressure	$ft^2/sec^2/{}^0R$
C_D	Drag coefficient	———
C_{D_b}	Base drag coefficient	———
C_{D_i}	Induced drag coefficient	———
C_{D_L}	Induced drag coefficient	———
C_{D_0}	Zero lift drag coefficient	———
C_f	Skin friction coefficient	———
C_L	Lift coefficient	———
$C_{L_{buffet}}$	Lift coefficient at buffet	———
C_{L_α}	Lift–curve–slope	1/rad or 1/deg
C_{L_i}	Design integrated lift coefficient	———
$C_{L_{max}}$	Maximum lift coefficient	———
C_m	Pitching moment coefficient	———
$C_N = C_R$	Normal force or resultant force coefficient	———
C_P	Propeller power coefficient	———
C_Q	Propeller torque coefficient	———
C_R	Resultant force coefficient	———

Symbol	Description	Unit(s)

Regular (Continued)

Symbol	Description	Unit(s)
C_T	Propeller thrust coefficient	———
C_v	Specific heat at constant volume	$ft^2/sec^2/{}^0R$
C.G.R.	Climb gradient	rad
d	Section or airfoil drag force	lbs or N
de	Change in energy	ft − lbs or N − m
dq	Change in heat	ft − lbs or N − m
d_T	Thrust moment arm	ft or m
dT	Change in temperature	${}^0R$
dv	Change in volume	ft^3 or m^3
D	Propeller diameter	ft or m
D_f	Friction force	lbs or N
e	Napierian logarithmic constant	2.7183
e	Oswald's efficiency factor	———
E	Endurance	hrs
f	Equivalent parasite area	ft^2 or m^2
F	Force	lbs or N
F_n	Net thrust	lbs or N
g	Acceleration of gravity	ft/sec^2, or m/sec^2
g_0	Acceleration of gravity at sea–level	$32.17\ ft/sec^2 = 9.806\ m/sec^2$

Symbol	Description	Unit(s)
Regular (Continued)		
h	Altitude	ft or m
Hg	Mercury	——
h_{flare}	Flare height	ft or m
$h_{INCL} = h_{TR}$	Height at end of transition	ft or m
h_{screen}	Screen or obstacle height	ft or m
$J = V/nD$	Propeller advance ratio	——
K_b	Flap span correction factor (p.123)	——
K_c	Flap chord correction factor (p.123)	——
K_Λ	Sweep correction factor (p.128)	——
K_g	Gust alleviation factor	——
l	Section or airfoil lift force	lbs or N
l	Length	ft or m
L	Characteristic length	ft or m
L	Lift	lbs or N
L'	Airfoil thickness location parameter: p.150	——
m	Section or airfoil pitching moment	ft − lbs or N − m
$\dot{m}$	Mass flow	slugs/sec or Kg/sec
M	Mach number	——
M^*	Critical Mach number for an unswept wing	——
M_{crit}	Critical Mach number for a swept wing	——
M_{dd}	Drag divergence Mach number	——

Symbols, Subscripts and Acronyms

Symbol	Description	Unit(s)

Regular (Continued)

Symbol	Description	Unit(s)
n	Propeller revolutions per second	1/sec
$n = L/W$	Load factor	——
n_{lim}	Design limit load factor	——
N	Normal force	lbs or N

Symbol	Description	Unit(s)
p	Atmospheric pressure	lbs/ft^2, or N/m^2
p_0	Atmospheric pressure at sea–level $2,116.2 \ lbs/ft^2 \ 0 = 1.013 \ N/m^2$	
p_a	Dry air pressure	lbs/ft^2, or N/m^2
p_t	Total pressure or stagnation pressure	lbs/ft^2, or N/m^2
p_v	Water vapor pressure	lbs/ft^2, or N/m^2
P	Power	$ft - lbs/sec$ or hp
P_{bl}	Power blade loading	$lbs/sec/ft$ or hp/ft^2
$P_c = P/\overline{q}AV$	Power diskloading coefficient	——
P_{av}	Available power	$ft - lbs/sec$ or hp
P_{shp}	Shaft horsepower	$ft - lbs/sec$ or hp

Symbol	Description	Unit(s)
$\overline{q} = \frac{1}{2}\varrho V^2$	Dynamic pressure	lbs/ft^2, or N/m^2
Q	Propeller torque	$ft - lbs$, or $N - m$

Symbol	Description	Unit(s)
r	Radial distance	ft or m
R	Propeller radius	ft or m
R	Range	nm or km
$\vec{R}$	Resultant aerodynamic force	lbs or N
R	Gas constant (for dry air: $R = 53.35 \ ft/^0R$, or $29.26 \ m/^0K$)	
RC or R.C.	Rate of climb	ft/sec or m/sec

Symbol	Description	Unit(s)
Regular (Continued)		
RD or R.D.	Rate of descent	ft/sec or m/sec
R_{flare}	Flare radius	ft or m
R_{loop}	Loop radius	ft or m
$R_{max_{glide}}$	Maximum glide distance	ft or m
R_t	Turn radius	ft or m
R_{TR}	Transition radius	ft or m
R_{LS}	Lifting surface correction factor	———
R_N or $R_e = \varrho VL/\mu$	Reynolds number	———
R_v	Gas constant for water vapor or steam	$R_v = 85.89$ ft/°R
R_{wf}	Wing–fuselage interference factor	———
S	Wing area	ft² or m²
SHP	Shaft horsepower	ft − lbs/sec or hp
S_{CL}	Take–off climb distance	ft or m
S_G	Take–off ground roll	ft or m
S_{NGR}	Take–off distance with nosegear on the ground	ft or m
S_L	Landing distance	ft or m
S_{LA}	Landing air distance	ft or m
S_{LG}	Landing ground roll	ft or m
S_{LNGR}	Landing distance with nosegear on the ground	ft or m
S_R	Take–off rotation distance	ft or m
S_{TR}	Take–off transition distance	ft or m
S_{LDES}	Landing distance to descend from the obstacle	ft or m
S_{LR}	Landing rotation distance	ft or m
S_{LTR}	Landing transition distance	ft or m

Symbol	Description	Unit(s)

Regular (Continued)

Symbol	Description	Unit(s)
S_{TO}	Take–off distance	ft or m
S_{TOBFL}	Take–off balanced fieldlength	ft or m
S_{TR}	Take–off transition distance	ft or m
S_{w_f}	Flapped wing area	ft^2 or m^2
S_{wet}	Wetted area	ft^2 or m^2
S.E.	Specific endurance	hrs/lbs
S.R.	Specific range	nm/lbs

Symbol	Description	Unit(s)
t	Section or airfoil thickness	ft or m
t	Time	sec or hr
T	Thrust	lbs or N
$\overline{T}$	Mean thrust during take–off run	lbs or N
T	Absolute temperature	^0R, or ^0K

$^0R = {^0F} + 459.7^0, {^0K} = {^0C} + 273.15^0$

Symbol	Description	Unit(s)
T_0	Absolute temperature at sea–level	

$59\ ^0F = 518.7\ ^0R = 15\ ^0C = 288.2\ ^0K$

Symbol	Description	Unit(s)
T.F.	Turbulence factor (p.185)	———
THP	Thrust horsepower	hp
TOP_{23}	Take–off parameter FAR 23	$lbs^2/ft^2 hp$
TOP_{25}	Take–off parameter FAR 25	———

Symbol	Description	Unit(s)
u	Flow velocity	ft/sec or m/sec or kts or km/hr
U	True airspeed	ft/sec or m/sec or kts or km/hr
U_{de}	Derived gust velocity	ft/sec or m/sec
U_1	Steady state speed	ft/sec or m/sec or kts or km/hr

Symbol	Description	Unit(s)
Regular (Continued)		
v	Specific volume	ft^3 or m^3
v	FFlow velocity	ft/sec or m/sec or kts or km/hr
V	True airspeed	ft/sec or m/sec or kts or km/hr
V_a	Speed of sound	ft/sec or m/sec or kts or km/hr
V_A	Approach speed	ft/sec or m/sec or kts or km/hr
V_A	Design maneuvering speed	ft/sec or m/sec or kts or km/hr
V_B	Design speed for maximum gust intensity	ft/sec or m/sec or kts or km/hr
V_C	Design cruising speed	ft/sec or m/sec or kts or km/hr
V_D	Design diving speed	ft/sec or m/sec or kts or km/hr
V_H	Design maximum level speed	ft/sec or m/sec or kts or km/hr
V_c	Calibrated airspeed	ft/sec or m/sec or kts or km/hr
V_e	Equivalent airspeed	ft/sec or m/sec or kts or km/hr
V_e	Exit or exhaust velocity	ft/sec or m/sec or kts or km/hr
V_i	Instrument indicated airspeed	ft/sec or m/sec or kts or km/hr
V_i	Flow volume per second at station i	ft^3/sec or m^3/sec
V_I	Indicated airspeed	ft/sec or m/sec or kts or km/hr
V_s	Stall speed	ft/sec or m/sec or kts or km/hr
V_{S_1}	1–g Stall speed	ft/sec or m/sec or kts or km/hr
V_j	Jet velocity	ft/sec or m/sec or kts or km/hr
V_θ	Tangential velocity	ft/sec or m/sec or kts or km/hr
V_1	Take–off decision speed	ft/sec or m/sec or kts or km/hr
V_2	Minimum safety speed	ft/sec or m/sec or kts or km/hr
V_2	Also: speed at the obstacle	ft/sec or m/sec or kts or km/hr

Symbol	Description	Unit(s)

Regular (Continued)

Symbol	Description	Unit(s)
V_{50}	Also: speed at the obstacle	ft/sec or m/sec or kts or km/hr
V_{EF}	Engine failure speed	ft/sec or m/sec or kts or km/hr
V_{flare}	Flare speed	ft/sec or m/sec or kts or km/hr
V_{INCL}	Initial climb–out speed	ft/sec or m/sec or kts or km/hr
V_{LOF}	Lift–off speed	ft/sec or m/sec or kts or km/hr
V_{MC}	Minimum control speed	ft/sec or m/sec or kts or km/hr
V_{MCG}	Minimum control speed on the ground	ft/sec or m/sec or kts or km/hr
V_{MU}	Minimum unstick speed	ft/sec or m/sec or kts or km/hr
V_R	Rotation speed	ft/sec or m/sec or kts or km/hr
V_S	Calibrated stall speed	ft/sec or m/sec or kts or km/hr
V_{TD}	Touchdown speed	ft/sec or m/sec or kts or km/hr
V_∞	Velocity at infinity	ft/sec or m/sec or kts or km/hr
V_w	Wind velocity	ft/sec or m/sec or kts or km/hr
V_g	Velocity relative to the ground	ft/sec or m/sec or kts or km/hr
w	Downwash velocity	ft/sec or m sec
$\dot{w}_f$ or $\dot{W}_f$	Fuel flow	lbs/sec
W	Airplane weight	lbs or N
W_{crew}	Crew eight	lbs or N
W_E	Empty weight	lbs or N
W_F	Fuel weight required for mission	lbs or N
$W_{F_{used}}$	Fuel weight used	lbs or N
$W_{F_{volumetric}}$	Maximum fuel weight due to fuel volume	lbs or N
W_{OWE}	Operating empty weight	lbs or N

Symbol | Description | Unit(s)

Regular (Continued)

Symbol	Description	Unit(s)
W_{PL}	Payload weight	lbs or N
W_{ramp}	Ramp weight	lbs or N
$W_{F_{sewu}}$	Fuel weight for engine start and warm-up	lbs or N
$W_{F_{taxi}}$	Fuel weight for taxi	lbs or N
W_{F_L}	Fuel weight for landing	lbs or N
$W_{F_{taxi/shut}}$	Fuel weight for taxi and shut-down	lbs or N
$W_{F_{TO}}$	Fuel weight for take-off	lbs or N
$W_{F_{CL}}$	Fuel weight for climb	lbs or N
$W_{F_{DE}}$	Fuel weight for descent	lbs or N
W_{tfo}	Trapped fuel and oil weight	lbs or N
W_{TO}	Take-off weight	lbs or N
$W_{TO_{max}}$	Maximum allowable take-off weight	lbs or N
$W_{CR_{begin}}$	Weight at start of cruise	lbs or N
$W_{CR_{end}}$	Weight at end of cruise	lbs or N
x_{ac}	Distance from l.e. to a.c.	ft or m
$\overline{x}_{ac}$	Aerodynamic center location in fraction of the chord	———

Greek

Symbol	Description	Unit(s)
α	Angle of attack	deg or rad
α^*	Angle of attack at end of linear range	deg or rad
α_0	Angle of attack for zero lift	deg or rad
α_i	Downwash induced angle of attack	deg or rad
$\alpha_{c_{l_{max}}}$	Angle of attack for maximum lift	deg or rad

Symbol	Description	Unit(s)
Greek (Continued)		
β	Geometric blade pitch angle	deg or rad
$\beta = \sqrt{M^2 - 1}$	Factor in Prandtl–Glauert transformation	———
γ	Ratio of specific heats of a gas	———
γ	Flight path angle	deg or rad
$\bar{\gamma}$	Negative of flight path angle	deg or rad
Γ	Sweep angle	deg or rad
Γ	Circulation	ft^2/sec or m^2/sec
$\delta = p/p_0$	Atmospheric pressure ratio	———
δ	Boundary layer thickness	ft or m
δ_{flap}	Flap deflection angle	deg or rad
Δy	Leading edge shape parameter	———
ΔV_i	Instrument error	ft/sec ^or^ m/sec
ΔV_p	Position error	ft/sec ^or^ m/sec
ε	Downwash angle	deg or rad
ε_T	Twist angle	deg or rad
ε_w	Wing twist angle	deg or rad
η	Spanwise station in fraction of b/2	———
η_p	Propeller efficiency	———
$\eta_{propulsive}$	Propulsive efficiency	———
η_t	Thermal efficiency	———
θ	Airplane pitch attitude angle (See Figure 1.6)	rad
$\theta = T/T_0$	Atmospheric temperature ratio	———
$\lambda = c_t/c_r$	Taper ratio	———
Λ	Sweep angle	deg or rad

Symbol	Description	Unit(s)

Greek (Continued)

Symbol	Description	Unit(s)
μ	Coefficient of viscosity	lbs-sec/ft^2
μ_g	Wheel-to-ground friction coefficient	———
$\mu_{g_{brake}}$	Ground braking friction coefficient	———
$\nu = \mu/\varrho$	Kinematic viscosity	ft^2/sec or m^2/sec
π	3.14	———
ϱ	Air density	slugs/ft^3, or Kg/m^3
ϱ_0	Air density at sea-level 0.002377 slugs/ft^3 = 1.225 Kg/m^3	
ϱ_t	Air density at stagnation	slugs/ft^3, or Kg/m^3
ϱ_a	Dry air density	slugs/ft^3, or Kg/m^3
ϱ_v	Water vapor density	slugs/ft^3, or Kg/m^3
$\sigma = Bc/\pi R$	Propeller solidity ratio	———
$\sigma = \varrho/\varrho_0$	Atmospheric density ratio	———
τ	Shear stress	lbs/ft^2, or N/m^2
Φ	Airplane bank angle (See Figure 1.6)	rad
Φ	Propeller blade helix angle (p.276)	rad
ϕ_T	Thrust line inclination angle w.r.t. YX-plane	rad
$\chi = (a_\infty \beta)/2\pi$	Lift-curve-slope ratio, see p.104	———
ψ	Heading angle	rad
$\dot{\psi}$	Turn rate	rad/sec

Subscripts

Note: A, S, b and $\bar{c}$ without a subscript indicates a wing property!

a	Air
ac or a.c. or A.C.	Aerodynamic center
AEO	All engines operating
av	Available

b	Body fixed

c	Canard
cg	Center of gravity
c/4 or 0.25	Relative to the quarter chord
c/2	Relative to the semi–chord
climb	Climb
CL	Climb
compr.	Compressible
cp or c.p. or C.P.	Center of pressure
crit.	Critical
cw	Canopy/windshield

dd	Drag divergence

e	Elevator
e or eff	Effective
emp	Empennage

f	Fuselage
flap	Flap
fus	Fuselage
free	Free (propeller)

g	Ground
gear	Gear

h	Horizontal
h	Horizontal tail
h	At altitude, h

Subscripts (Continued)

i	Item number i
incompr.	Incompressible
interf.	Interference

j	Jet

L	Landing
LA	Landing
LE or l.e.	Leading edge
LNGR	Landing nosegear on the ground
LR	Landing rotation
LTR	Loiter

max	Maximum
mg	Main gear, about or relative to main gear
min	Minimum
misc	Miscellaneous
m	Main gear
M	At some Mach number
M=0	At zero Mach number

n	Nosegear
neg	Negative
n–g	n–g Pull–up
NGR	Nosegear on the ground
np	Nacelle/pylon

OEI	One engine inoperative
OWE	Operating weight empty

p	Pylon, also: propeller
par	Parasite
PA	Powered Approach
pos	Positive

ref	Reference
reqd	Required

s	Standard
s.l.	Sea–level
slip	Slipstream

Subscripts (Continued)

s or static	Static
std	Standard
store	Store

trim	Trimmed
T	Thrust
TO	Takeoff

v	Vertical
v	Vertical tail

w or wing	Wing
wf	Wing+fuselage

x, y or z	In the x, y or z–direction

Acronyms

A/C	Air Conditioner
AEO	All Engines Operating
BMA	Best Mach number and Altitude
b.p.r. or B.P.R.	Bypass ratio
C.F.	Centrifugal force
C.G.R.	Climb Gradient
ESHP	Equivalent Shaft Horsepower
EPNdB	Equivalent Perceived Noise in dB
EPR	Engine Pressure Ratio
GAW	General Aviation Whitcomb
ICAO	International Civil Aviation Organization
MAP	Manifold Absolute Pressure
METO	Minimum Except Take–off
m.g.c.	Mean geometric chord
N1	Critical turbine RPM
OAT	Outside Air Temperature
OEI	One Engine Inoperative
S.F.	Shape factor (see Eqn 3.39)
SHP	Shaft Horsepower
RPM	Revolutions Per Minute
THP	Thrust Horsepower

Symbols, Subscripts and Acronyms

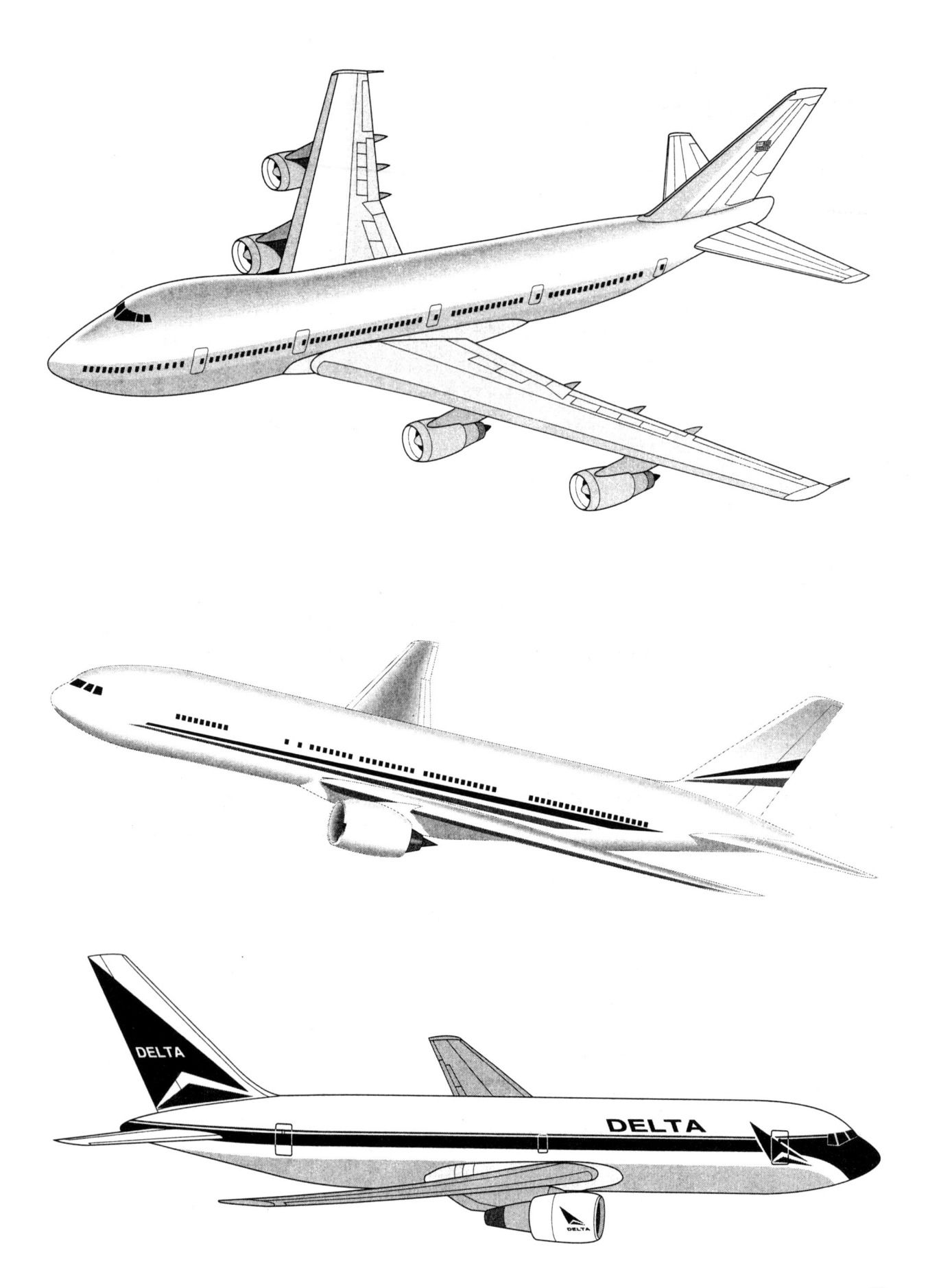

INTRODUCTION

This textbook covers those fundamentals of airplane aerodynamics, propulsion and performance which are required to predict the performance characteristics of fixed wing airplanes. The text is aimed at junior and senior level aeronautical engineering students. However, the authors have attempted to present the material in such a manner that practising engineers and engineering managers will find the material useful as well.

As part of airplane aerodynamics the following topics are covered:

Characteristics of the atmosphere are presented in Chapter 1. The basic principles of aerodynamics with applications are covered in Chapter 2. Fundamentals with applications of airfoil theory are given in Chapter 3.

Airfoils are integrated into wings, tails and/or canards to form lifting surfaces. Basic lifting surface theory (also known as wing theory) is developed in Chapter 4. Specifically this chapter contains a discussion of wing geometric parameters, circulation, downwash, lift and induced drag as well as aerodynamic center. Stall characteristics of wings and methods for controlling the stall characteristics are also discussed. The effects of compressibility on wing drag and on wing aerodynamic center are presented. Finally, high lift devices and speed (drag) brakes and spoilers are discussed.

Chapter 5 contains methods for estimating the drag characteristics (drag polars) of airplanes. Several example applications are given. This chapter also contains a basic discussion of how to reduce small scale windtunnel data to obtain full scale drag data.

As part of airplane propulsion the following topics are covered:

In Chapter 6 an overview of reciprocating, rotary and turbine engines is provided with example data for typical engines in each category.

Chapter 7 contains a discussion of propeller theory and application. Methods for estimating propeller performance (free and installed) and propeller noise are included.

As part of airplane performance the following topics are covered:

The fundamentals of airplane flight mechanics (i.e. the basic performance equations of motion) are discussed in Chapter 8.

Methods for calculating the speed, climb, descent and drift–down performance of airplanes are covered in Chapter 9. Applications to a range of airplanes are included. The effect of the airworthiness regulations on speed and climb performance is also presented.

In Chapter 10 methods for predicting the takeoff and landing fieldlength performance are given, including the effects of airworthiness regulations.

Methods for determining the range, endurance and payload–range diagram of airplanes are given in Chapter 11. Rules for determining fuel reserves are also summarized in this chapter. A method for rapid weight sizing of airplanes to meet given mission payload versus range and endurance requirements is also presented.

Chapter 12 contains methods for the prediction of maneuvering (pull–ups and turns) characteristics. The effect of specific excess power on climbs is also presented in this chapter. Several airplane characteristics at the edges of the flight envelope are also discussed.

Appendix A contains standard atmospheric characteristics in English and Metric units..

Appendix B contains information on the Advanced Aircraft Design and Analysis software.

Appendix C contains propeller performance charts

Appendix D contains propeller noise charts

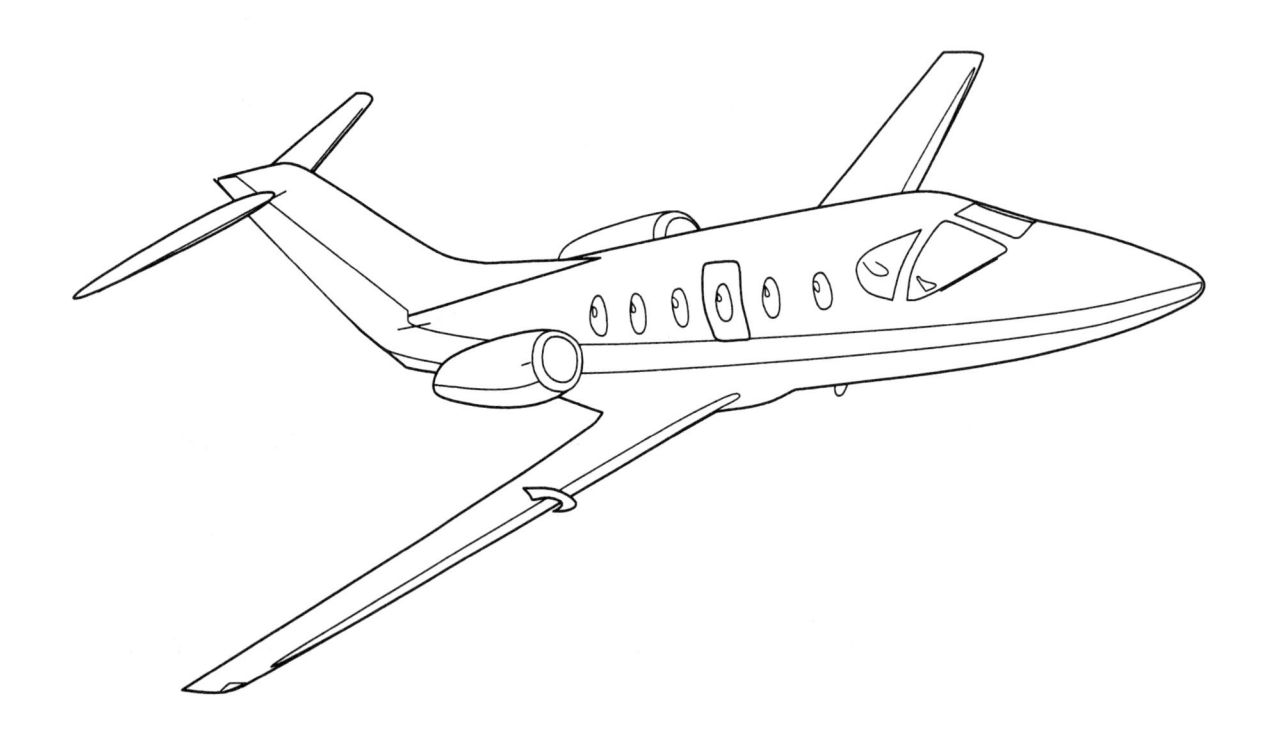

CHAPTER 1: THE ATMOSPHERE

The performance characteristics of airplanes and engines depend on the properties of the atmosphere in which they operate. In this chapter those properties of the atmosphere which are important to the determination of airplane and engine performance are presented. Applications to various airplane and engine performance scenarios are given.

1.1 ATMOSPHERIC FUNDAMENTALS

The atmosphere consists of a gaseous mixture (called the air) of approximately 78% nitrogen, 21% oxygen and 1% other gases, water vapor being one of those. The actual composition of the atmosphere varies with geographical locations and altitudes. However, in most applications to conventional aerodynamics, the atmosphere can be regarded as a homogeneous gas of uniform composition. The air may therefore be regarded as a gas which satisfies the perfect gas law:

$$p = \rho g R T \tag{1.1}$$

where:

p = atmospheric pressure in lbs/ft^2, or N/m^2

ρ = air density in $slugs/ft^3$, or kg/m^3

g = acceleration of gravity in ft/sec^2, or m/sec^2

R = gas constant (for dry air: R = 53.35 $ft/^0R$, or 29.26 $m/^0K$)

T = absolute temperature in ^0R, or ^0K ($^0R = {}^0F + 459.7^0$, $^0K = {}^0C + 273.15^0$)

Eqn (1.1) is also referred to as the equation of state. Note that when using Eqn (1.1) the temperature, T, must be the absolute temperature of the gas. When a significant amount of moisture (water vapor) is present in the atmosphere, the air density will change.

To determine how water vapor affects air density, consider a given volume filled with a dry air and water vapor mixture. Dalton's law of partial pressures (see Ref.1.1, p. 496) states that the observed pressure, p, of the mixture equals the sum of the dry air pressure, p_a and the water vapor pressure, p_v :

$$p = p_a + p_v \tag{1.2}$$

Also, since the total mass of air inside the volume is equal to the sum of the dry air mass and the water vapor mass, the observed density of the mixture is given by:

$$\varrho = \varrho_a + \varrho_v \tag{1.3}$$

In Eqn (1.3) it is also assumed that the distribution of masses of dry air and water vapor in the volume is uniform. By using the perfect gas law of Eqn (1.1) it is possible to show that the density of the mixture is given by:

$$\varrho = \frac{p - p_v}{gRT} + \frac{p_v}{gR_vT} = \frac{1}{gRT}\left[p - p_v(1 - \frac{R}{R_v})\right] \tag{1.4}$$

For water vapor or steam, $R_v = 85.89$ ft/^{0}R, (see Ref. 1.2, pages 4–23). It follows that $R < R_v$ so that the second term in Eqn (1.4) is negative. Therefore, the density of a mixture of dry air and water vapor is less than that of dry air. Normally the effect of water vapor on air density is small. To illustrate this fact, assume that the observed temperature and pressure of a mixture of dry air and water vapor are 90 degrees F and 2,116.2 lbs/ft2 respectively. If the relative humidity is 100%, the water vapor pressure can be found to be 100.6 lbs/ft2, (see Ref. 1.2, p. 4–80). The density of the mixture then follows from Eqn (1.4) as:

$$\varrho = \frac{1}{53.35 \times 32.17 \times 549.7}(2,116.2 - 100.6 \times 0.379) = 0.002203 \text{ slugs/ft}^3 \tag{1.5}$$

The corresponding dry air density is 0.002243 slugs/ft3. The density reduction in this case is therefore only 1.8%. Despite this small effect of water vapor on air density, water vapor does have a significant effect on engine performance and on supersonic aerodynamics. The effect of humidity (water vapor) on engine performance is discussed in Chapter 6. A discussion of the effect of water vapor on supersonic aerodynamics is beyond the scope of this text. The interested reader may wish to consult Ref. 1.3.

1.2 THE INTERNATIONAL STANDARD ATMOSPHERE

For purposes of airplane aerodynamics and performance calculations the atmosphere is divided into four regions as shown in Figure 1.1.

To provide a basis for comparing the performance characteristics of airplanes and to allow for the calibration of altimeters, it is desirable to have standard properties of the atmosphere which represent the so–called "average" conditions. Such standard properties have been established by the International Civil Aviation Organization (ICAO). These standard atmospheric characteristics are generally used by airplane and engine manufacturers around the world. In the U.S.A. there is a so–called U.S. Standard Atmosphere as well. The U.S. standard atmosphere is the same as the ICAO atmosphere for altitudes below 65,617 ft. This standard atmosphere is representative of the atmospheric characteristics in the region of the mid latitudes of the northern hemisphere.

The properties of the U.S. standard atmosphere are given in both English and Metric units in Appendix A.

Figure 1.1 The Four Regions of the Atmosphere

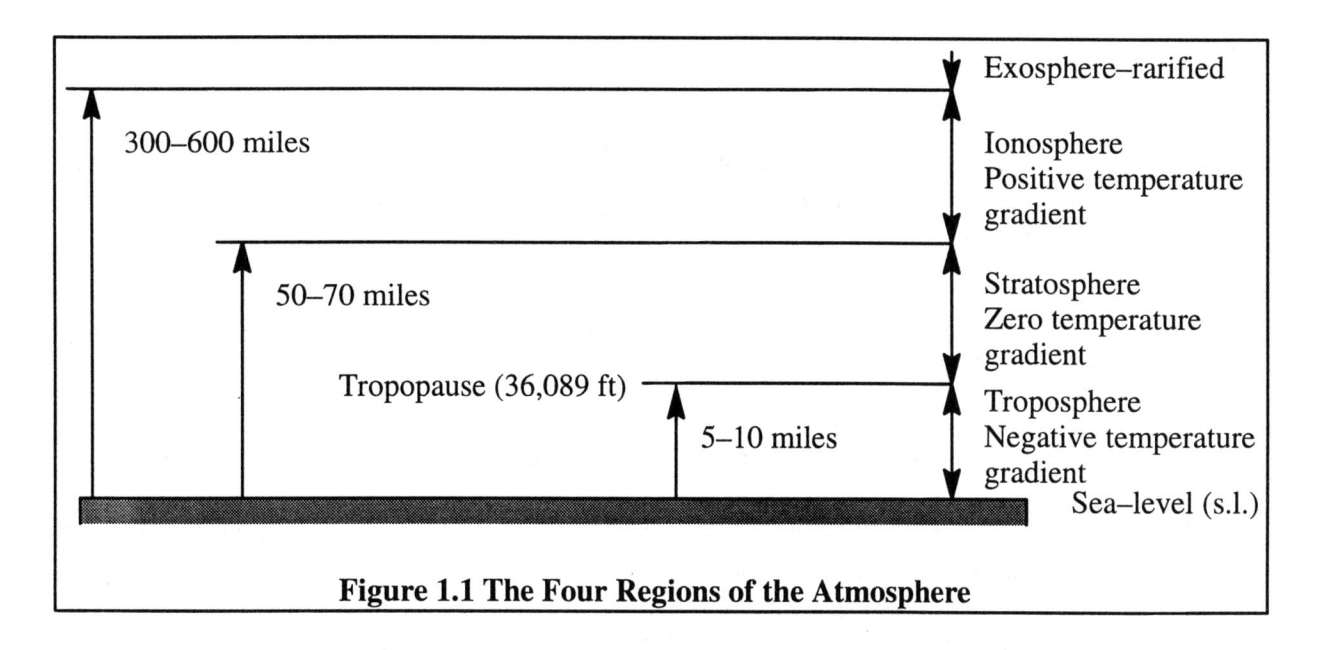

- 300–600 miles: Exosphere–rarified / Ionosphere — Positive temperature gradient
- 50–70 miles: Stratosphere — Zero temperature gradient
- Tropopause (36,089 ft)
- 5–10 miles: Troposphere — Negative temperature gradient
- Sea–level (s.l.)

According to the standard atmosphere the standard sea–level properties of the atmosphere are as follows:

$$g_0 = 32.17 \text{ ft/sec}^2 = 9.806 \text{ m/sec}^2 \tag{1.6a}$$

$$p_0 = 29.92 \text{ in Hg} = 2,116.2 \text{ lbs/ft}^2 = 1.013 \times 10^5 \text{ N/m}^2 \tag{1.6b}$$

$$T_0 = 59{}^0F = 518.7{}^0R = 15{}^0C = 288.2{}^0K \tag{1.6c}$$

$$\varrho_0 = 0.002377 \text{ slug/ft}^3 = 1.225 \text{ Kg/m}^3 \tag{1.6d}$$

For subsonic airplanes only the troposphere and the stratosphere are important.

1.2.1 TEMPERATURE VARIATION WITH ALTITUDE

In the standard atmosphere it is assumed that below an altitude of 36,089 ft, there is a constant drop of temperature of 0.00356616 deg. F per foot of altitude. This is referred to as the lapse rate of the atmosphere. Therefore, the temperature at any given altitude, h, can be written as:

$$T = T_1 + a(h - h_1) \tag{1.7}$$

where:

a = the lapse rate of the atmosphere = $-0.00356616 {}^0F/ft$

T_1 is the reference temperature at altitude h_1

h_1 is the reference altitude

At sea–level, $h_1 = 0$ and $T_1 = T_0$. Above an altitude of 36,089 ft in the stratosphere, the standard temperature is constant and roughly equal to –69.7 deg. F.

1.2.2 PRESSURE AND DENSITY VARIATION WITH ALTITUDE

To determine the variation of pressure and density with altitude consider the vertical force equilibrium of a small sample of air in the atmosphere. Figure 1.2 illustrates the forces acting on this cube of air. Because the three forces are in equilibrium in a quiescent atmosphere it follows:

$$p\,dx\,dy - (p + dp)dx\,dy - \rho g\,dx\,dy\,dh = 0 \qquad (1.8)$$

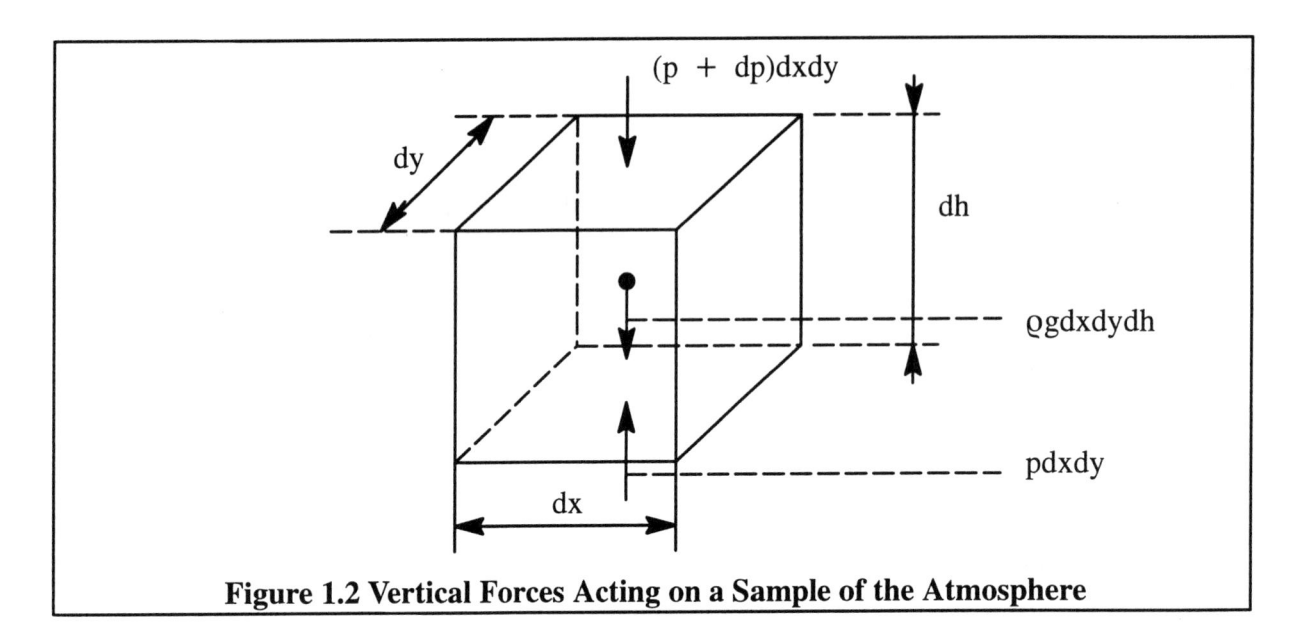

Figure 1.2 Vertical Forces Acting on a Sample of the Atmosphere

After simplification this yields:

$$dp = -\rho g\,dh \qquad (1.9)$$

Division of Eqn (1.9) by Eqn (1.1) yields:

$$\frac{dp}{p} = -\frac{dh}{RT} \qquad (1.10)$$

Because the temperature behavior in the troposphere differs from that in the stratosphere, two cases must be considered: from sea–level to 36,089 ft and from 36,089 to 65,617 ft.

Troposphere: sea–level to 36,089 ft

Differentiation of Eqn (1.7) results in:

$$dT = a\,dh \qquad (1.11)$$

Substitution of this result into Eqn (1.10) yields:

$$\frac{dp}{p} = -\frac{dT}{aRT} \qquad (1.12)$$

Eqn (1.12) can be integrated to provide the following relationship between the pressure at any altitude and the pressure at a reference altitude:

$$\frac{p}{p_1} = \left(\frac{T}{T_1}\right)^{-\frac{1}{aR}} = \left\{1 + \frac{a}{T_1}(h - h_1)\right\}^{-\frac{1}{aR}} \qquad (1.13)$$

From the perfect gas relation of Eqn (1.1) it is now possible to show that the relationship between the density at any altitude and the density at a reference altitude is as follows:

$$\frac{\varrho}{\varrho_1} = \left(\frac{p}{p_1}\right)\left(\frac{T}{T_1}\right) = \left(\frac{T}{T_1}\right)^{\left(-\frac{1}{aR} - 1\right)} \qquad (1.14)$$

Of particular interest are the relationships between temperature, pressure and density at sea–level to those at an arbitrary altitude:

$$\frac{T}{T_0} = \theta = 1 - \frac{ah}{T_0} = 1 - 6.875 \times 10^{-6} h \qquad (1.15)$$

$$\frac{p}{p_0} = \delta = \theta^{5.2561} \qquad (1.16)$$

$$\frac{\varrho}{\varrho_0} = \sigma = \theta^{4.2561} \qquad (1.17)$$

Stratosphere: 36,089 to 65,617 ft.

Because the temperature in the stratosphere is a constant –69.7 deg. F it is possible to integrate Eqn (1.10) directly to yield:

$$\ln\left(\frac{p}{p_{ref}}\right) = -\left(\frac{h - h_{ref}}{RT}\right) \qquad (1.18)$$

By taking the reference altitude to be 36,089 ft it is found that:

$$\frac{p}{p_{ref}} = e^{\left\{-\left(\frac{(h - 36,089)}{RT_{ref}}\right)\right\}} \qquad (1.19)$$

However, Eqns (1.15) through (1.17) show that at 36,089 ft:

$$\frac{T_{ref}}{T_0} = \theta = 0.75189 \quad \text{while}: T_{ref} = 390^0 R \qquad (1.20)$$

$$\frac{P_{ref}}{P_0} = \theta^{5.2561} = 0.2234 \tag{1.21}$$

$$\frac{\varrho_{ref}}{\varrho_0} = \theta^{4.2561} = 0.2971 \tag{1.22}$$

From Eqns (1.21) and (1.22) it can be found that:

$$\frac{p}{p_0} = 0.2234 \, e^{\left\{-\left(\frac{h - 36,089}{20,806.7}\right)\right\}} \tag{1.23}$$

$$\frac{\varrho}{\varrho_0} = 0.2971 \, e^{\left\{-\left(\frac{h - 36,089}{20,806.7}\right)\right\}} \tag{1.24}$$

Equations (1.15), (1.16), (1.17), (1.23) and (1.24) have been used to compute the standard atmospheric properties which are tabulated in Appendix A.

1.2.3 BAROMETRIC ALTIMETERS

Barometric altimeters are commonly used in airplanes to measure barometric altitude. Basically an altimeter is a pressure gauge which translates the measured pressure into an altitude reading which corresponds to that predicted by the standard atmosphere.

To calibrate altimeters the characteristics of the standard atmosphere are used. Reference 1.4 may be consulted for a fairly detailed treatment of various types of altimeter. The reader should also understand the difference between instruments which can determine height above ground, such as Radar Altimeters, Ground Proximity Warning Systems (GPWS) and Global Positioning System (GPS), and barometric sensors, such as barometric altimeters.

Figure 1.3 shows the faceplate of a typical barometric and radar altimeter.

The setting knob on the lower left of the barometric instrument is used to rotate the scale of the altimeter so that the instruments reads the correct altitude corresponding to local reference pressure conditions.

The radar altimeter does not require such a setting knob. Instead it is equipped with a test knob, to test it.

When an altimeter is "set" to a pressure of 29.92 "Hg (the standard atmospheric pressure at sea-level) the altitude it reads is defined as the "pressure altitude".

The altitude corresponding to a given density in the standard atmosphere is defined as the "density altitude". The altitude corresponding to a given temperature in the standard atmosphere is defined as the "temperature altitude".

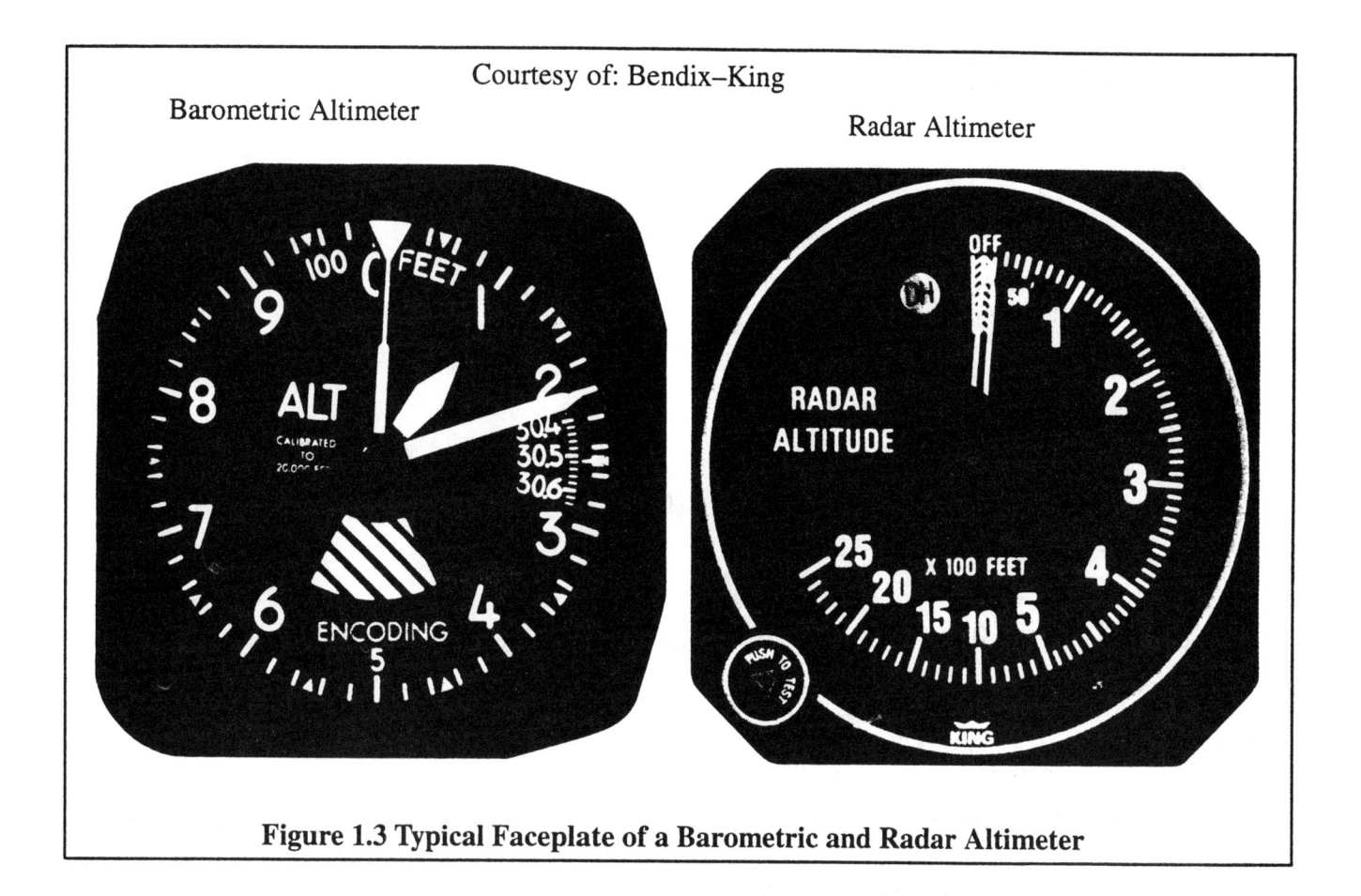

Figure 1.3 Typical Faceplate of a Barometric and Radar Altimeter

In an atmosphere with standard conditions all three altitudes are exactly the same. However, in a non–standard atmosphere all three altitudes will be different! The following examples are presented to enhance the reader's understanding of atmospheric characteristics.

Example 1.1 A standard altimeter indicates 15,000 ft when the ambient temperature is 35 deg. F. Calculate the density altitude and the temperature altitude.

Solution: From Table A1 at h = 15,000 ft, the standard temperature is 5.5 deg. F. Therefore, the atmosphere is not standard. Since the altimeter is a pressure gauge, it will read the correct pressure. From Table A1 the correct pressure at that altitude is 144x8.294 = 1,194 psf. The actual density therefore is:

$$\varrho = \frac{p}{gRT} = \frac{1,194}{32.17 \times 53.35 \times (35 + 459.7)} = 0.001406 \text{ slug/ft}^3$$

From Table A1 the standard altitude corresponding to this density is approximately 17,000 ft. A more accurate calculation of the density altitude can be done as follows. From Eqn (1.17):

$$\sigma = \frac{\varrho}{\varrho_0} = \frac{0.001406}{0.002377} = 0.5915$$

From Eqn (1.15):

$$\theta = 1 - 6.875 \times 10^{-6} h_{density} = \sigma^{\frac{1}{4.2561}}$$

Therefore:

$$h_{density} = \frac{1 - 0.88393}{6.875 \times 10^{-6}} = 16,883 \text{ ft}$$

This result is close to the 17,000 ft interpolated from Table A1 (Appendix A). To find the temperature altitude, Eqn (1.15) is used again:

$$\frac{T}{T_0} = \frac{494.7}{518.7} = 0.95373 = 1 - 6.875 \times 10^{-6} \times h_{temperature}$$

From this it is found that: $h_{temperature} = 6,730 \text{ ft}$. Note the large difference between temperature and density altitudes in this example.

During flight tests, when measurements are normally conducted under non–standard atmospheric conditions, engine performance data and airplane performance data are all transformed to what these data would have been under standard atmospheric conditions.

Example 1.2 Because density cannot be measured directly, the density is normally inferred by calculation from measurements of static, ambient temperature and static, ambient pressure. Calculate the density ratio, σ if measurements show that the altimeter reads 5,000 ft and the ambient, static air temperature is 80 deg. F.

Solution: From Eqn (1.20):

$$\theta = \frac{80 + 459.7}{518.7} = 1.0405$$

From Table A1 it is found that $\delta = 0.8320$. Therefore, from Eqn (1.1):

$$\sigma = \left(\frac{p}{p_0}\right)\left(\frac{T_0}{T}\right) = \frac{\delta}{\theta} = 0.7996$$

1.2.4 VISCOSITY

Another atmospheric property which is important in aerodynamics and performance is the viscosity of the air.

Because of the close relationship between viscosity and the behavior of the boundary layer in air flow around an airplane, a discussion of viscosity of the atmosphere is given in Chapter 2.

1.3　SUMMARY FOR CHAPTER 1

In this chapter the most important properties of the atmosphere were derived and discussed. It was shown that with the perfect gas law simple equations can be derived from which standard atmospheric conditions can be predicted. It was also shown that with the help of local measurements of temperature and pressure actual atmospheric conditions can be reconstructed. In this manner a meaningful comparison of airplane and engine performance based on in–flight measurements can always be arrived at.

1.4　PROBLEMS FOR CHAPTER 1

1.1　Calculate the pressure, density and temperature at 30,500 ft and at 61,500 ft in the standard atmosphere. Compare the results of the calculations with values interpolated from Table A1.

1.2　On a hot day, the measured temperature and pressure are 38 deg. C and 29.0 in. Hg respectively. Calculate the density ratio and the density.

1.3　A standard altimeter reads 14,000 ft when the ambient temperature is 35 deg. F. What is the density altitude?

1.4　At a certain altitude, a standard altimeter reads 10,000 ft. If the density altitude is 8,000 ft, find the true temperature at that altitude.

1.5　An airplane is fitted with an altimeter which is calibrated according to the standard atmosphere. On a certain day the pressure at sea–level is found to be 2,130 lbs/ft^2 and the measured temperature is 50 deg. F. The lapse rate of the temperature is –0.0039 deg. R per foot of altitude. If on this same day, the altimeter reads 15,000 ft, what is the true altitude of the airplane above sea–level?

1.6　An altimeter which is set to 29.92 in. Hg reads zero feet when the airplane is on the ground at an airport which is 1,500 ft above sea–level. The following data are taken during a climb if this airplane:

Pressure Altitude in ft	Temperature, T in deg. F	Pressure Altitude in ft	Temperature, T in deg. F
0	21	6,000	–1
1,000	17	7,000	–5
2,000	14	8,000	–9
3,000	10	9,000	–13
4,000	7	10,000	no reading
5,000	2		

If the altimeter reads 10,000 ft what will be the actual altitude of the airplane above sea–level? (Hint: Use an average lapse rate.)

1.5 REFERENCES FOR CHAPTER 1

1.1   Joos, G. and Freeman, I.M.; Theoretical Physics; Hafner Publishing Company, N.Y., 1950.

1.2   Baumeister, T. and Marks, L.S. (Editors); Standard Handbook for Mechanical Engineers; McGraw Hill Book Co., Seventh Edition, 1967.

1.3   Emmons, H.W. (Editor); Fundamentals of Gas Dynamics; page 526, Volume III in the series of High Speed Aerodynamics and Jet Propulsion; Princeton University Press, 1958.

1.4   McKinley, J.L. and Bent, R.D.; Basic Science for Aerospace Vehicles; McGraw Hill Book Co., Fourth Edition, N.Y., 1972.

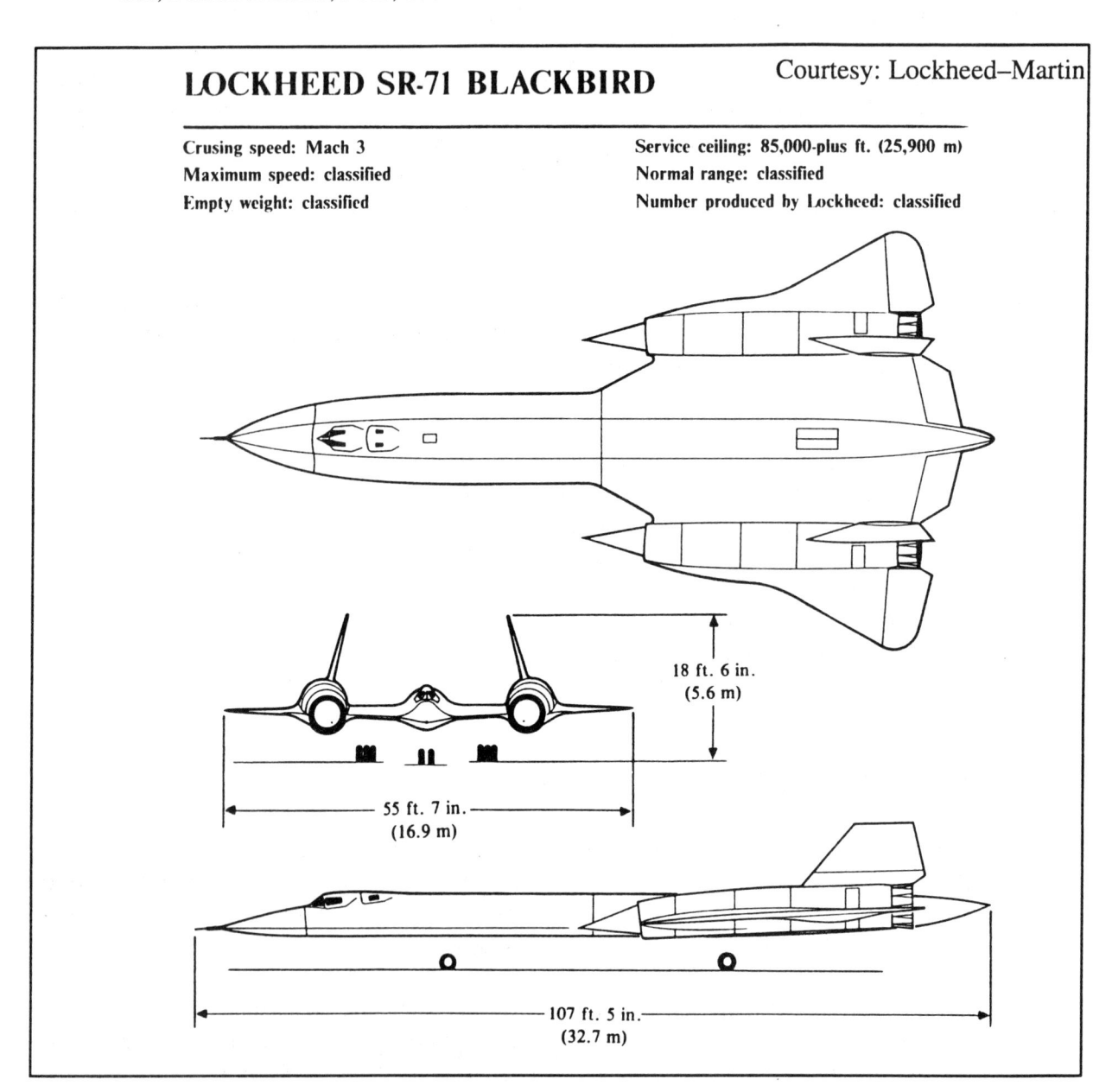

LOCKHEED SR-71 BLACKBIRD Courtesy: Lockheed–Martin

Crusing speed: Mach 3
Maximum speed: classified
Empty weight: classified

Service ceiling: 85,000-plus ft. (25,900 m)
Normal range: classified
Number produced by Lockheed: classified

18 ft. 6 in. (5.6 m)

55 ft. 7 in. (16.9 m)

107 ft. 5 in. (32.7 m)

CHAPTER 2: BASIC AERODYNAMIC PRINCIPLES AND APPLICATIONS

In this chapter several basic aerodynamic principles and their applications to airplane aerodynamics will be discussed. Throughout the discussions in this chapter the airflow will be considered steady. This means that all flow properties such as pressure, velocity, temperature and density are considered to be independent of time. Except in a small region close to the airplane surface, where viscosity is important, the airflow will also be assumed to be non–viscous or inviscid.

When an airplane model is placed in the middle of a windtunnel test section, the uniform airflow will be disturbed. This produces pressure forces on the model surface which in turn are responsible for lift and induced drag. In addition, there will be friction forces which cause friction drag. To explain the friction drag which acts on the model, the airflow must be considered to be viscous, at least in the region adjacent to the surface of the model. That region of the flow will be referred to as the boundary layer.

To understand and model these physical phenomena and others, some basic aerodynamic principles must be derived and discussed. Also, several applications of these principles will be outlined. The following basic aerodynamic principles are discussed:

    2.1   The continuity equation
    2.2   The incompressible Bernoulli equation
    2.3   Compressibility effects: the isentropic equation of state and an expression for the speed of sound
    2.4   The compressible Bernoulli equation
    2.5   Measurement of airspeed
    2.6   The Kutta–Joukowski theorem
    2.7   The linear momentum principle
    2.8   Viscous effects, boundary layer and flow separation

2.1 THE CONTINUITY EQUATION

Before discussing the continuity equation the concept of streamlines will be introduced. A streamline is a curve along which the tangent at any point on the curve always coincides with the flow velocity vector at that point. A group of streamlines adjacent to each other is referred to as a streamtube. Figure 2.1 shows an example of a streamtube.

From the definition of streamtube it is evident that the flow velocity normal to the exterior surface of the streamtube is always zero. Therefore, following the principle of conservation of mass, all mass passing through Section 1 in Figure 2.1 must also pass through Section 2.

Figure 2.1 Example of a Streamtube

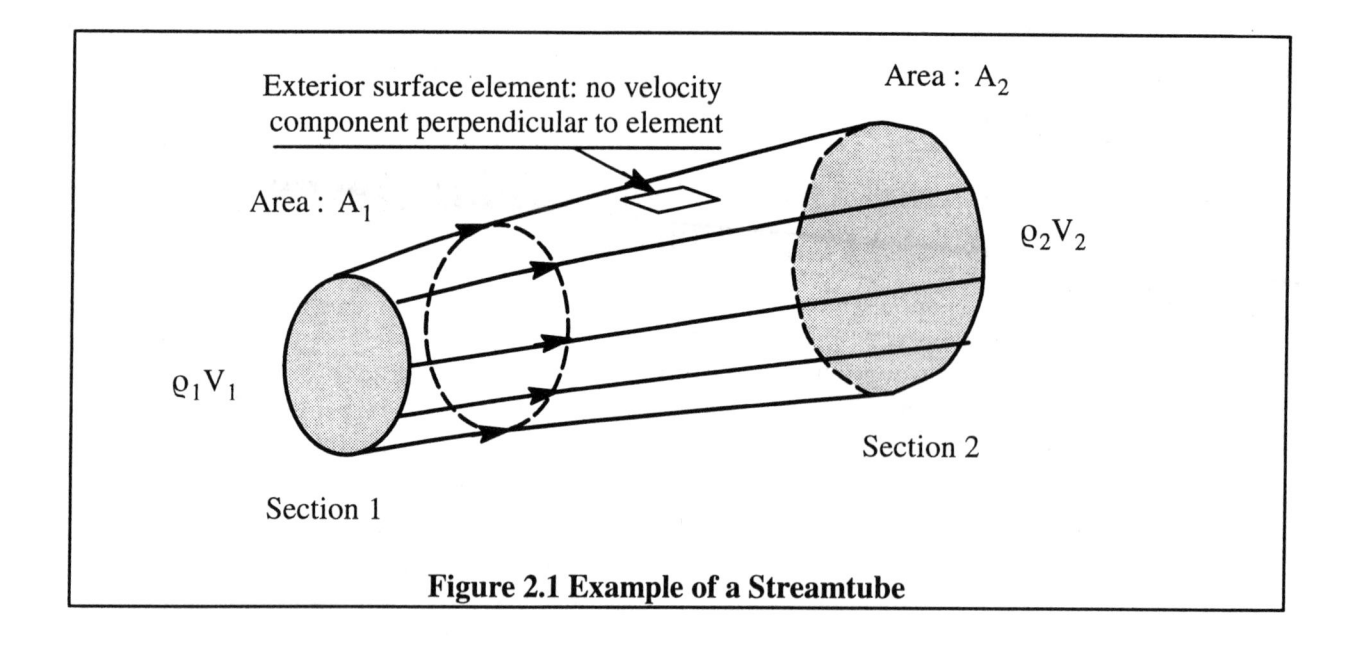

The cross sectional area of Section 1 is: A_1 and the local velocity of the flow is: V_1. The product of A_1 and V_1 is the flow volume per second, in ft3/sec. Therefore, the mass flow rate, $\dot{m}$ is given by:

$$\dot{m} = \varrho_1 A_1 V_1 = \varrho_2 A_2 V_2 \tag{2.1}$$

For constant mass flow, evidently:

$$\dot{m} = \varrho A V = \text{constant} \tag{2.2}$$

This equation is known as the streamtube **compressible continuity equation**.

<u>Definition:</u>   In incompressible flow, the density, ϱ remains constant.

If the airflow is to be considered incompressible, the density should remain constant throughout the entire streamtube. If that is the case, Eqn (2.1) can be written as:

$$A_1 V_1 = A_2 V_2 \tag{2.3}$$

Therefore, Eqn (2.2), in incompressible flow becomes:

$$AV = \text{constant} \tag{2.4}$$

This equation is known as the streamtube **incompressible continuity equation**.

The walls of a windtunnel can be thought of as the exterior surface of a streamtube. For a low speed windtunnel the airflow can be regarded as incompressible (a more precise criterion will be discussed in Section 2.5.2). In such a case, Eqn (2.4) shows that the airspeed can be increased by decreasing the cross–sectional area. This explains why the windtunnel test section has the smallest cross–sectional area to achieve higher test speeds. Figure 2.2 shows two examples of low speed wind tunnels.

Figure 2.2 Examples of Low Speed Wind Tunnels

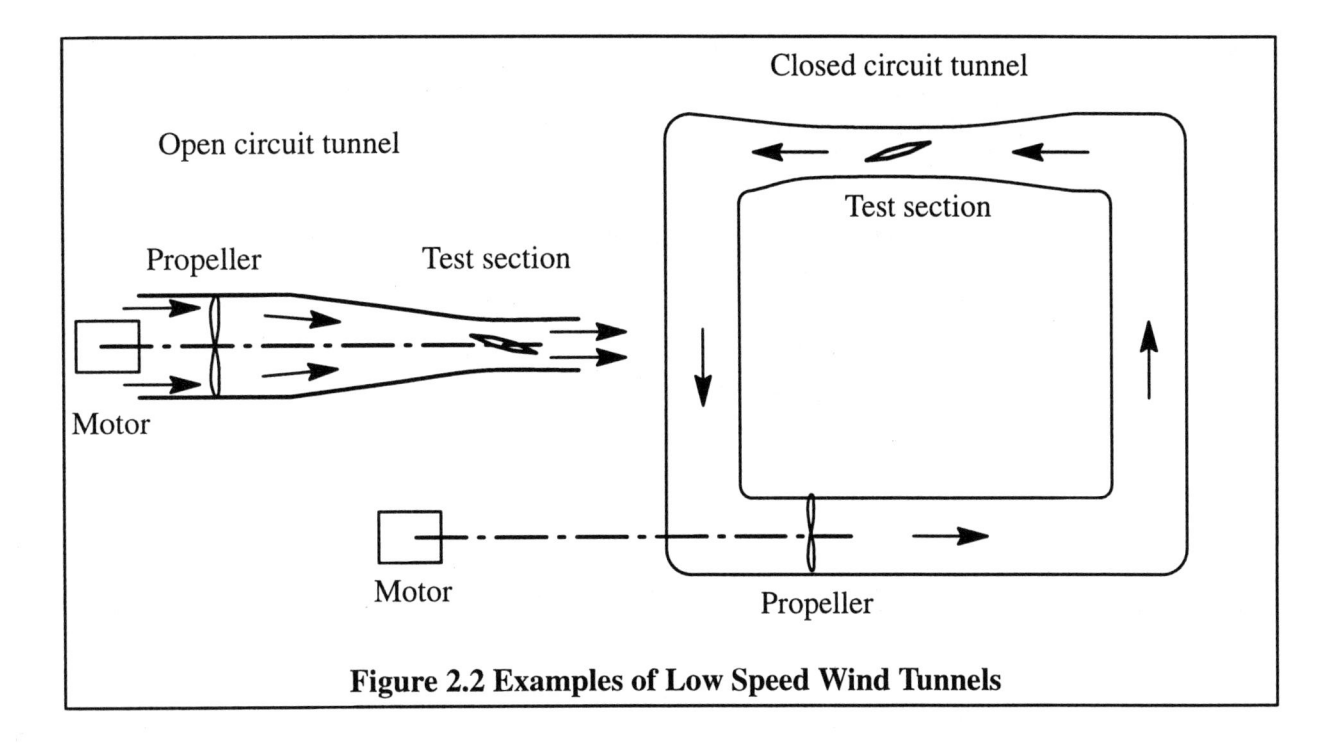

2.2 THE INCOMPRESSIBLE BERNOULLI EQUATION

Consider a streamtube with length, ds, while only pressure forces act on its surface. Such a streamtube is depicted in Figure 2.3.

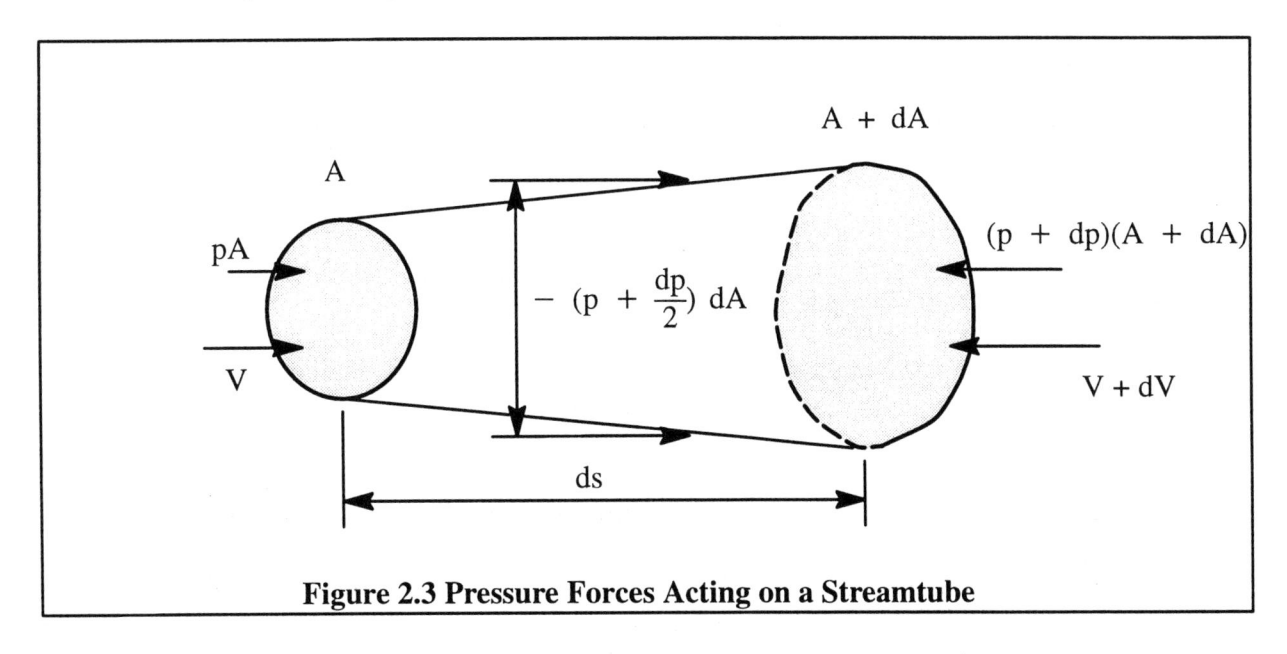

Figure 2.3 Pressure Forces Acting on a Streamtube

Applying Newton's Second Law to the streamtube segment results in:

$$m\frac{dV}{dt} = pA + (p + \frac{dp}{2})dA - (p + dp)(A + dA) \approx -Adp \qquad (2.5)$$

where: the product dpdA has been neglected because it is a second order term. Observe that the weight of the air mass has also been neglected in Eqn (2.5). For horizontal flow that certainly is acceptable. The left hand side of Eqn (2.5) can be cast in the form:

$$m\frac{dV}{dt} = m\frac{dV}{ds}\frac{ds}{dt} = \frac{m}{2}\frac{dV^2}{ds} \qquad (2.6)$$

It is now possible to rewrite Eqn (2.5) as:

$$\frac{Ads}{m}dp + \frac{1}{2}d(V^2) = 0 \qquad (2.7)$$

Since $m = \rho Ads$, Eqn (2.7) can be written as:

$$\frac{dp}{\rho} + \frac{1}{2}dV^2 = 0 \qquad (2.8)$$

Eqn (2.8) is referred to as Euler's equation of motion along a streamline. If the density, ρ remains constant within the streamtube (incompressible flow assumption) it is possible to integrate Eqn (2.8) directly and produce:

$$p + \frac{1}{2}\rho V^2 = \text{constant} \qquad (2.9)$$

This result is known as the **incompressible Bernoulli equation**. It is applicable along a streamline in incompressible flow.

The reader should realize that Eqn (2.7) states that the work done by the pressure forces equals the change in the kinetic energy of the air mass. Therefore, Eqn (2.9) is **not** applicable in situations where energy is added to the airflow. The latter is the case in jet engines and in propellers. An example application to a low speed windtunnel will now be presented.

Example 2.1: The windtunnel in Figure 2.4 has a test section at Station 2 of 4 ft by 4.25 ft. At Station 1 its cross section measures 13 ft by 13 ft. At some tunnel speed, the manometer reading is 28 inches. The manometer liquid has a specific gravity of 0.85*. Calculate the airspeed in the test section. Assume that the flow is incompressible and that standard sealevel conditions prevail.

Solution: From the manometer reading, the pressure difference between Stations 1 and 2 can be determined as:

$$\delta p = \frac{28}{12}\sin(30^0) \times 0.85 \times 62.4 = 62.4 \text{ lbs/ft}^2 \qquad (2.10)$$

Application of Bernoulli's Eqn (2.9) to Stations 1 and 2 yields:

---

* Specific gravity is defined as the ratio between the mass density of a substance to that of water. Water density is 1.94 slugs/ft3. The specific weight of water is therefore: $32.17 \times 1.94 = 62.4$ lbs/ft3.

Figure 2.4 Windtunnel Set-up for Example 2.1

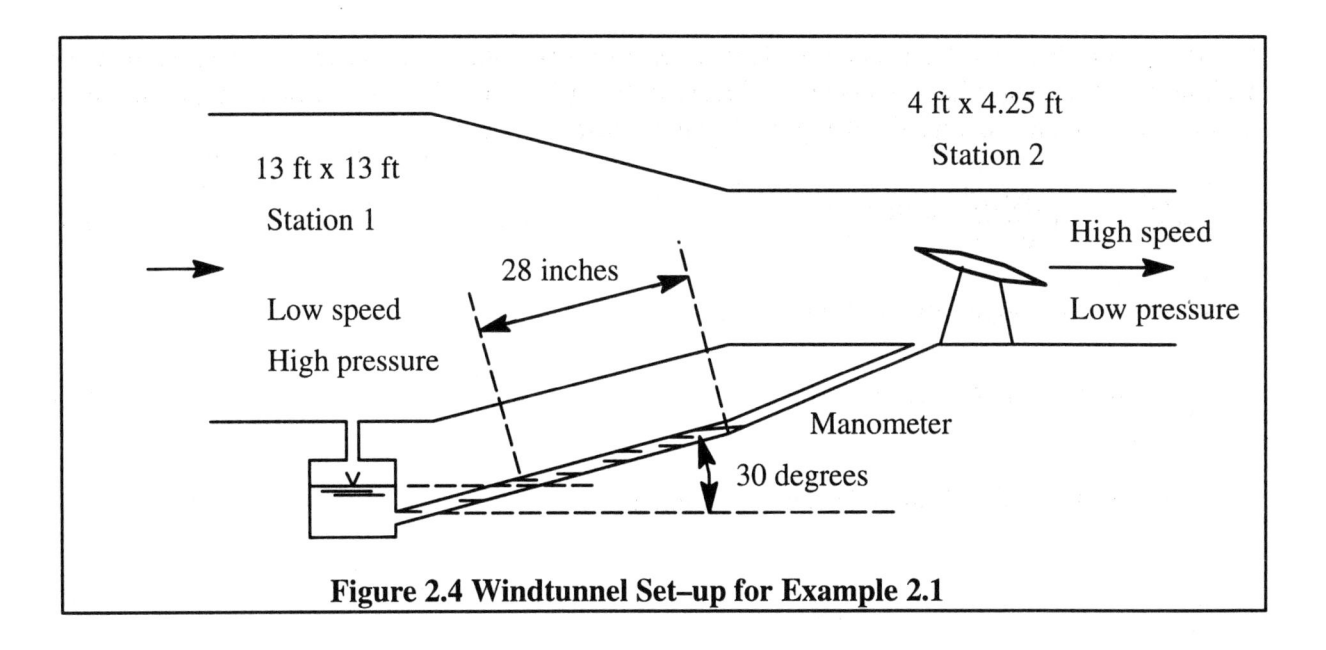

$$p_1 + \tfrac{1}{2}\varrho V_1^2 = p_2 + \tfrac{1}{2}\varrho V_2^2 \qquad (2.11)$$

From the continuity Eqn (2.4) it follows that:

$$A_1 V_1 = A_2 V_2 \qquad (2.12)$$

By eliminating V_1 from Eqns (2.11) and (2.12) it is found that:

$$p_1 - p_2 = \tfrac{1}{2}\varrho\left\{V_2^2 - \left(\frac{A_2}{A_1}\right)^2 V_2^2\right\} = \tfrac{1}{2}\varrho V_2^2\left\{1 - \left(\frac{A_2}{A_1}\right)^2\right\} \qquad (2.13)$$

From this it follows that:

$$V_2 = \sqrt{\frac{2(p_1 - p_2)/\varrho}{1 - \left(\frac{A_2}{A_1}\right)^2}} = \sqrt{\frac{2 \times 62.4/0.002377}{1 - \left(\frac{4 \times 3.25}{13 \times 13}\right)^2}} = 229 \text{ft/sec} \qquad (2.14)$$

2.3 COMPRESSIBILITY EFFECTS

When the airflow is regarded as compressible, the internal energy of the airflow must be considered. In this section, the so-called isentropic condition (equation of state) and an expression for the speed of sound will be derived.

2.3.1 THE ISENTROPIC EQUATION OF STATE

According to the first law of thermodynamics (Ref. 2.1, page 6), an amount of heat, dq, added to a unit mass of gas will result in a differential change in energy, de, and an expansion of volume, dv, of the gas. The quantity, v, is defined as the specific volume.

$$dq = de + pdv \tag{2.15}$$

In most applications of subsonic aerodynamics, the heat transfer within the flow can be neglected, except in a small region adjacent to the airplane surface, where viscosity and heat conduction may be important. For that reason, Eqn (2.15) can be approximated as:

$$de + pdv = 0 \tag{2.16}$$

Because air can be thought of as a perfect gas, the following two relationships apply in accordance with Ref. 2.1, page 12:

$$de = C_v dT \tag{2.17}$$

and:

$$C_v + gR = C_p \tag{2.18}$$

where:

C_v is the specific heat at constant volume in $ft^2/sec^2/{}^0R$

C_p is the specific heat at constant pressure in $ft^2/sec^2/{}^0R$

For a perfect gas, the equation of state, Eqn (1.1), can also be written as:

$$pv = gRT \tag{2.19}$$

Differentiation of this equation of state yields:

$$dT = \frac{pdv + vdp}{gR} \tag{2.20}$$

Substitution of Eqns (2.17) and (2.20) into Eqn (2.16) results in:

$$\frac{C_v}{gR}(pdv + vdp) + C_v \frac{dp}{p} = 0 \tag{2.21}$$

This result can be rewritten as:

$$(C_v + gR)\frac{dv}{v} + C_v \frac{dp}{p} = 0 \tag{2.22}$$

By using Eqn (2.18) and defining:

$$\gamma = C_p/C_v \text{ (also called the ratio of specific heats of the gas)} \tag{2.23}$$

it is found that Eqn (2.22) yields:

$$\gamma \frac{dv}{v} + \frac{dp}{p} = 0 \tag{2.24}$$

This equation can be integrated to produce:

$$pv^\gamma = \text{constant} \tag{2.25}$$

or:

$$\frac{p}{\varrho^\gamma} = \text{constant} \tag{2.26}$$

Basic Aerodynamic Principles and Applications

Equations (2.25) and (2.26) are known as the **isentropic conditions** or the **isentropic equations of state**.

For air at normal temperatures, a value of $\gamma = 1.4$ produces good results. For air at high temperatures, such as in an engine exhaust, $\gamma = 1.33$ is a better number.

The reader should keep in mind that Eqn (2.26) is valid along a streamline in airflow where compressibility is important. This tends to be the case at free stream Mach numbers above 0.30. The definition of Mach number is as follows:

$$M = \frac{V}{V_a} \qquad (2.27)$$

where: V is the true airspeed

V_a is the speed of sound

A discussion of the speed of sound is given in Sub–section 2.3.2.

2.3.2 THE SPEED OF SOUND

As a general rule, air is a compressible medium. When a disturbance producing an infinitesimal pressure change is generated at some point in the flow, this disturbance will be propagated throughout the air as a pressure wave travelling at the speed of sound. Knowing the magnitude of the speed of sound is important. If the flow velocity exceeds the propagation speed of disturbances, these disturbances will pile up to form strong waves, called shock waves. These shock waves in turn produce large changes in flow properties. One important consequence of all this is an increase in drag.

To derive an expression for the speed of sound, consider a one–dimensional duct with an infinitesimal pressure wave generated inside the gas at rest. If a coordinate system is chosen to be fixed with the disturbance, the gas velocity, V_a will be the propagation speed of the disturbance before being affected by it. The disturbance will then produce infinitesimal changes in density, pressure and velocity. These changes must satisfy both the continuity equation and Euler's equation of motion. Differentiating the continuity equation (2.2) for a constant area duct yields:

$$V_a d\varrho + \varrho dV_a = 0 \qquad (2.28)$$

The Euler equation of motion, Eqn (2.9) can be rewritten as:

$$dp + \varrho V_a dV_a = 0 \qquad (2.29)$$

Elimination of dV_a from Eqns (2.28) and (2.29) produces:

$$V_a^2 = \frac{dp}{d\varrho} \qquad (2.30)$$

Because the changes in the flow properties were assumed to be infinitesimal, the process can be assumed to be isentropic. Therefore, the isentropic equation of state, Eqn (2.26) must apply. Substituting Eqn (2.26) into Eqn (2.30) results in:

$$V_a = \sqrt{\frac{\gamma p}{\varrho}} = \sqrt{\gamma g R T} \qquad (2.31)$$

where the perfect gas relationship of Eqn (2.19) was used in the last step.

One important result of Eqn (2.31) is that the speed of sound varies with the square root of the absolute temperature. This implies that the speed of sound is constant in the stratosphere, where T is constant.

In the standard troposphere, in accordance with Eqns (1.15) and (2.31) the ambient speed of sound decreases with increasing altitude as follows:

$$V_a = \sqrt{\{\gamma g R T_0(1 - 6.875 \times 10^{-6} h)\}} = 1,116.39 \sqrt{(1 - 6.875 \times 10^{-6} h)} \text{ ft/sec} \quad (2.32)$$

where: h is measured in feet (ft) in the English system. In the metric system it can be shown that:

$$V_a = 340.3 \sqrt{(1 - 2.255 \times 10^{-5} h)} \text{ m/sec} \quad (2.33)$$

where: h is measured in meters (m). As an exercise the reader should check equations (2.31) and (2.32) against the numerical results listed in Tables A1 and A2.

2.4 THE COMPRESSIBLE BERNOULLI EQUATION

In the case of isentropic, compressible flow, Eqn (2.9) can be integrated in the following manner. By using Eqn (2.26) to eliminate the density, ϱ, in Eqn (2.9) to yield:

$$C^{1/\gamma} \frac{dp}{p^{1/\gamma}} + \frac{1}{2} dV^2 = 0 \quad (2.34)$$

where: C is the constant in Eqn (2.26). Eqn (2.34) can be integrated to produce:

$$\left(\frac{1}{1 - 1/\gamma}\right) C^{1/\gamma} p^{(1 - 1/\gamma)} + \frac{1}{2} V^2 = \text{constant} \quad (2.35)$$

With the help of Eqn (2.26) this result can be changed to:

$$\left(\frac{\gamma}{\gamma - 1}\right) \frac{p}{\varrho} + \frac{1}{2} V^2 = \text{constant} \quad (2.36)$$

Eqn (2.36) is known as the **compressible Bernoulli equation**. As stated before, this equation applies to isentropic flow. It is shown in Ref. 2.1, page 55, that Eqn (2.36) also applies to adiabatic flow in which no heat transfer occurs along the streamline. That implies that Eqn (2.36) also applies with shock waves present. In such a case, Eqn (2.36) can be interpreted as the conservation of energy across the shock waves (Ref. 2.1, page 56). For more details on shock waves the reader is encouraged to consult Ref. 2.1.

2.5 MEASUREMENT OF AIRSPEED

2.5.1 LOW-SPEED AIRSPEED INDICATORS (INCOMPRESSIBLE FLOW)

Consider a pitot static tube as shown in Figure 2.5.

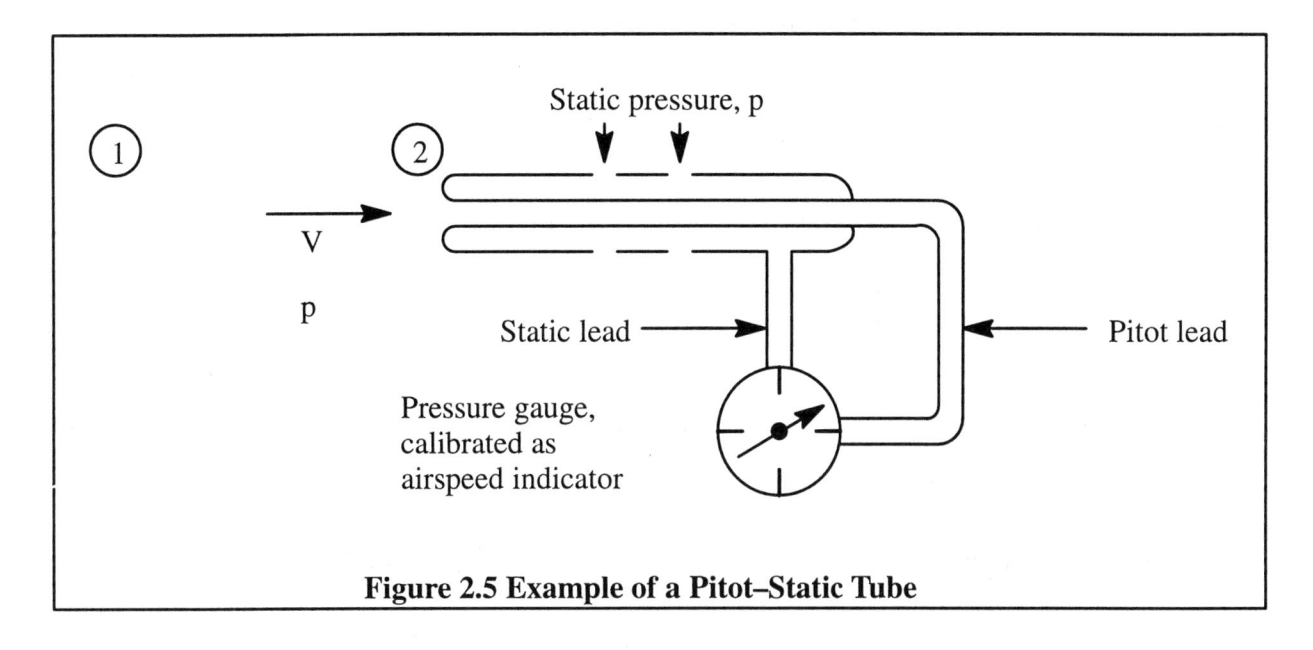

Figure 2.5 Example of a Pitot-Static Tube

Application of the incompressible Bernoulli equation (2.9), to a point 1 (located far away from the pitot tube) and to a point 2 (located at the inlet of the pitot tube) yields:

$$p + \frac{1}{2}\varrho V^2 = p_2 = p_t \tag{2.37}$$

where: p_t is the pressure at the stagnation point (zero speed). This pressure is also called the **stagnation pressure** or the **total pressure**. It follows that:

$$p_t - p = \frac{1}{2}\varrho V^2 = \bar{q} \tag{2.38}$$

where: $\bar{q}$ is defined as the **dynamic pressure**. Next, the density, ϱ, in Eqn (2.38) is replaced by the standard sealevel value. Following that the equation is solved for airspeed, called the equivalent airspeed, V_e:

$$V_e = \sqrt{\frac{2(p_t - p)}{\varrho_0}} \tag{2.39}$$

The pressure difference, $(p_t - p)$ can be sensed by the pitot tube shown in Figure 2.5. The airspeed indicator can be calibrated to indicate V_e with the help of Eqn (2.39). Therefore, the airspeed indicator essentially operates as a pressure gauge.

The true airspeed, V, at any altitude can be expressed as a function of V_e as follows:

$$V_e = \sqrt{\frac{2(p_t - p)}{\varrho}} = V_e\sqrt{\frac{\varrho_0}{\varrho}} = \frac{V_e}{\sqrt{\sigma}} \qquad (2.40)$$

The reader should observe that at sea–level, where $\sigma = 1.0$, the true airspeed and the equivalent airspeed are the same: $V = V_e$ as long as the flow is incompressible.

2.5.2 HIGH–SPEED AIRSPEED INDICATORS (COMPRESSIBLE FLOW)

The ratio of true airspeed to the ambient speed of sound at the same point is called the Mach number in accordance with Eqn (2.27). Whether airspeed is considered to be low or high, depends on the Mach number, M or M_∞.

Flight speeds where: $M_\infty < 1.0$ are referred to as **subsonic**.

Flight speeds where: $M_\infty > 1.0$ are referred to as **supersonic**.

Flight speeds in the range of roughly: $0.80 < M_\infty < 1.2$ are referred to as **transonic**.

The actual Mach range for which the flow over the surface of an airplane is transonic depends strongly on the configuration. Wing sweep angle and wing thickness ratio are some of the major parameters which determine this. Airplanes with no sweep and thick airfoils become transonic at much lower flight Mach numbers than airplanes with significant sweep and thin airfoils. For a detailed discussion of these effects, see Reference 2.2.

Roughly speaking, for flight Mach numbers above 0.30 the compressible Bernoulli equation should be used. In this case, Eqn (2.36) will be applied to two points in the flow field, points 1 and 2 in Figure 2.5. The following expression is obtained:

$$\left(\frac{\gamma}{\gamma - 1}\right)\frac{p}{\varrho} + \frac{1}{2}V^2 = \left(\frac{\gamma}{\gamma - 1}\right)\frac{p_t}{\varrho_t} \qquad (2.41)$$

Application of the isentropic equation of state, Eqn (2.26) to the same points 1 and 2 results in:

$$\frac{p}{\varrho^\gamma} = \frac{p_t}{\varrho_t^\gamma} \qquad (2.42)$$

Next, ϱ_t will be eliminated between Equations (2.41) and (2.42) to yield:

$$\frac{V^2}{2} = \left(\frac{\gamma}{\gamma - 1}\right)\frac{p}{\varrho}\left\{\left(\frac{p_t}{p}\right)^{\frac{\gamma-1}{\gamma}} - 1\right\} = \frac{V_a^2}{\gamma - 1}\left\{\left(\frac{p_t - p}{p} + 1\right)^{\frac{\gamma-1}{\gamma}} - 1\right\} \qquad (2.43)$$

or,

$$M^2 = \frac{2}{\gamma - 1}\left\{\left(\frac{p_t - p}{p} + 1\right)^{\frac{\gamma-1}{\gamma}} - 1\right\} \qquad (2.44)$$

From this equation it can be seen that when $(p_t - p)/p$ is measured, the Mach number can be calculated. An instrument which measures both $(p_t - p)$ and the static pressure, p, to indicate the Mach number through Eqn (2.44) is called a Mach meter. Note, that a conventional airspeed indicator will only measure $(p_t - p)$.

When, in the calibration of an airspeed indicator, $(p_t - p)/p$ is replaced by $(p_t - p)/p_0$ and V_a is replaced by V_{a_0}, the resulting airspeed is called the calibrated airspeed, V_c:

$$V_c^2 = \frac{2V_{a_0}^2}{\gamma - 1}\left\{\left(\frac{p_t - p}{p_0} + 1\right)^{\frac{\gamma-1}{\gamma}} - 1\right\} \qquad (2.45)$$

Eqns (2.44) and (2.45) can be used to establish a relationship between the true Mach number and the calibrated airspeed. Solving Eqn (2.45) for $(p_t - p)$ results in:

$$p_t - p = p_0\left[\left\{1 + \frac{\gamma - 1}{2}\left(\frac{V_c}{V_{a_0}}\right)^2\right\}^{\frac{\gamma}{\gamma - 1}} - 1\right] \qquad (2.46)$$

Next, Eqn (2.46) is substituted into Eqn (2.44) to yield the following relationship between M and V_c:

$$M^2 = \frac{2}{\gamma - 1}\left[\left[\frac{1}{\delta}\left[\left\{1 + \frac{\gamma - 1}{2}\left(\frac{V_c}{V_{a_0}}\right)^2\right\}^{\frac{\gamma}{\gamma - 1}} - 1\right] + 1\right]^{\frac{\gamma-1}{\gamma}} - 1\right] \qquad (2.47)$$

Figure 2.6 represents a plot of true Mach number versus calibrated airspeed for constant pressure altitudes. This plot was obtained with Eqn (2.47).

In certain applications it is convenient to solve Eqn (2.44) for the term $(p_t - p)$ as a function of M:

$$p_t - p = p\left[\left\{1 + \frac{(\gamma - 1)}{2}M^2\right\}^{\frac{\gamma}{\gamma - 1}} - 1\right] \qquad (2.48)$$

Observe that $(p_t - p)$ in incompressible flow is exactly equal to the dynamic pressure (See Eqn 2.38). However, in a compressible flow, the stagnation pressure is increased due to compressibility.

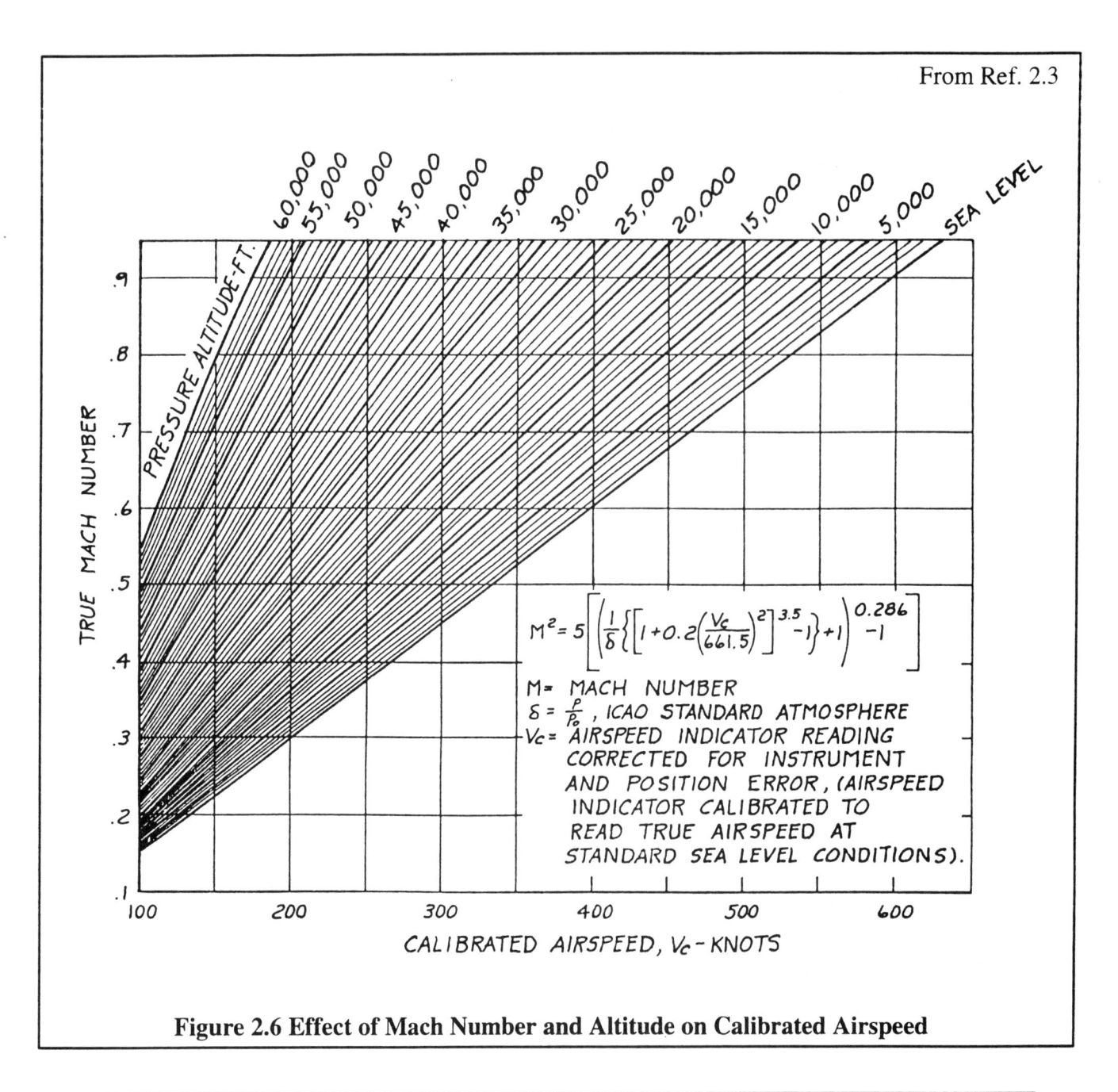

Figure 2.6 Effect of Mach Number and Altitude on Calibrated Airspeed

$$M^2 = 5\left[\left(\frac{1}{\delta}\left\{\left[1+0.2\left(\frac{V_c}{661.5}\right)^2\right]^{3.5}-1\right\}+1\right)^{0.286}-1\right]$$

M = MACH NUMBER
$\delta = \frac{P}{P_o}$, ICAO STANDARD ATMOSPHERE
V_c = AIRSPEED INDICATOR READING CORRECTED FOR INSTRUMENT AND POSITION ERROR, (AIRSPEED INDICATOR CALIBRATED TO READ TRUE AIRSPEED AT STANDARD SEA LEVEL CONDITIONS).

This can be shown with the help of Eqn (2.48). First, note that: $\{(\gamma - 1)M^2/2\} \approx 0.2M^2$ for air. Next, note that for small Mach numbers: $0.2M^2 \ll 1.0$. Therefore, Eqn (2.48) can be expanded with the binomial expansion formula to yield:

$$p_t - p \approx p\left\{1 + \left(\frac{\gamma}{\gamma-1}\right)\left(\frac{\gamma-1}{2}\right)M^2 + \left(\frac{\gamma}{\gamma-1}\right)\left(\frac{\gamma}{\gamma-1} - 1\right)\frac{1}{2}\frac{(\gamma-1)^2}{4}M^4 + \ldots - 1\right\} =$$

$$= \frac{p\gamma}{2}M^2\left(1 + \frac{1}{4}M^2 + \frac{1}{40}M^4 + \frac{1}{1600}M^6 + \ldots\right) \tag{2.49}$$

The reader is asked to show that:

$$\frac{p\gamma}{2}M^2 = \frac{1}{2}\rho V^2 = \bar{q} \tag{2.50}$$

Therefore:

$$p_t - p = \frac{1}{2}\rho V^2\left(1 + \frac{1}{4}M^2 + \frac{1}{40}M^4 + \frac{1}{1600}M^6 + \ldots\right) \tag{2.51}$$

It is seen that:

$$\text{at } M = 0.3 : \frac{p_t - p}{\bar{q}} \approx 1.0227$$
$$\text{at } M = 0.5 : \frac{p_t - p}{\bar{q}} \approx 1.064 \tag{2.52}$$

Therefore, the effect of compressibility on dynamic pressure is seen to be small at Mach numbers below 0.3. Above Mach numbers of about 0.5 the error made by neglecting compressibility is more than 6%. Compressibility effects should then be accounted for.

The reader is urged to memorize the following relationship which follows from Eqn (2.50):

$$\bar{q} = \frac{\gamma}{2} p_0 \frac{p}{p_0} M^2 = \frac{\gamma}{2} p_0 \delta M^2 = 1481.3 \, \delta M^2 \tag{2.53}$$

An example application of the effect of compressibility on an airspeed indicator, calibrated for incompressible flow, will now be discussed.

Example 2.2: An airspeed indicator on an airplane is calibrated in accordance with the incompressible flow assumption. While flying at sealevel, the indicator shows 500 mph. What is the true airspeed? Assume standard atmospheric conditions.

Solution: Because the airspeed indicator is a pressure gauge, it will measure the true pressure difference, $(p_t - p)$. Therefore:

$$p_t - p = \tfrac{1}{2}\varrho_0 V^2 = \tfrac{1}{2} \times 0.002377(500 \times 1.467)^2 = 639.4 \text{ psf}$$

The true Mach number can now be calculated from Eqn (2.44):

$$M^2 = \frac{2}{\gamma - 1}\left\{\left(\frac{p_t - p}{p} + 1\right)^{\frac{\gamma-1}{\gamma}} - 1\right\} = \frac{2}{0.4}\left\{\left(\frac{639.4}{2,116.2} + 1\right)^{\frac{0.4}{1.4}} - 1\right\} = 0.3917$$

Therefore, M = 0.6259 and the corresponding true airspeed is:

$$V = MV_a = 0.6259 \times 1,116.4 = 698.76 \text{ fps} = 476.3 \text{ mph}$$

It is seen, that this airspeed indicator overestimates the airspeed by about 5%.

2.5.3 AIRSPEED CORRECTIONS

Actual airspeed instruments aboard airplanes are calibrated to read true airspeed at all Mach numbers under standard sealevel conditions, provided no instrument or static pressure source errors present. To obtain airspeed from an imperfect airspeed instrument operating under non–standard conditions requires that the corresponding errors be accounted for. To this end, the following definitions for various airspeeds are used:

V_i is the **instrument indicated airspeed** which is <u>uncorrected for errors</u>. It <u>includes</u> the standard sealevel adiabatic compressible flow effect.

V_I (IAS) is the **indicated airspeed** <u>corrected for instrument error, ΔV_i</u>, only:

$$V_I = V_i + \Delta V_i \tag{2.54}$$

V_c (CAS) is the **calibrated airspeed**. It is equal to the indicator reading <u>corrected for position error due to incorrect static port location (ΔV_p) and also corrected for instrument error</u>:

$$V_c = V_I + \Delta V_p \tag{2.55}$$

V_e (EAS) is the **equivalent airspeed**. It is equal to the indicator reading <u>corrected for instrument error, for position error and for adiabatic compressible flow at the particular altitude</u>:

$$V_e = V_c + \Delta V_c \tag{2.56}$$

where: ΔV_c is called the compressibility correction.

V is the true airspeed. It is calculated from:

$$V = \frac{V_e}{\sqrt{\sigma}} \tag{2.57}$$

The various corrections will now be discussed.

The instrument error, ΔV_i can be determined during laboratory calibration tests.

The position error, ΔV_p arises from the location of the static pressure port on the airplane. Two examples of typical static port locations are shown in Figure 2.7. It can be conjectured from these examples that the static port locations will normally not provide for ambient (i.e. free stream) static pressures. The surface of the airplane causes distortions in the flow away from ambient conditions. To determine the error (position error) caused by the position of the static port the airplane is flown under a number of speed and altitude conditions in formation with a "pacer" airplane which has been accurately calibrated. By flying both airplanes together under stabilized speed conditions, the indicated readings of Mach number, airspeed and altitude are recorded for both airplanes. The indicated readings from the pacer airplane are first reduced to true values. The difference between the true and the indicated values for M and the test airplane is called: ΔM_p, so that:

$$M = M_I + \Delta M_p \tag{2.58}$$

The position error, ΔV_p can now be calculated from:

$$\Delta V_p = \frac{dV_c}{dM}\Delta M_p = \frac{\Delta M_p}{(dM/dV_c)} \tag{2.59}$$

where: (dM/dV_c) is found by differentiating Eqn (2.47). An alternative is to calculate (dM/dV_c) as follows:

$$dM/dV_c = \frac{dM}{dp}\frac{dp}{dV_c} \tag{2.60}$$

Differentiation of Eqn (2.48) results in:

$$\frac{dM}{dp} = -\left[\frac{1 + \frac{\gamma-1}{2}M^2}{\gamma M p}\right] \tag{2.61}$$

Differentiation of Eqn (2.46) yields:

$$\frac{dp}{dV_c} = -\gamma p_0 V_c \left\{1 + \frac{\gamma-1}{2}\left(\frac{V_c}{V_{a_0}}\right)^2\right\}^{\frac{1}{\gamma-1}} V_{a_0}^{\ 2} \tag{2.62}$$

Combining Eqns (2.61) and (2.62) produces:

$$\frac{dM}{dV_c} = \left[\frac{1}{\delta V_{a_0}^{\ 2} M}\right]\left(1 + \frac{\gamma-1}{2}M^2\right)\left\{1 + \frac{\gamma-1}{2}\left(\frac{V_c}{V_{a_0}}\right)^2\right\}^{\frac{1}{\gamma-1}} V_c \tag{2.63}$$

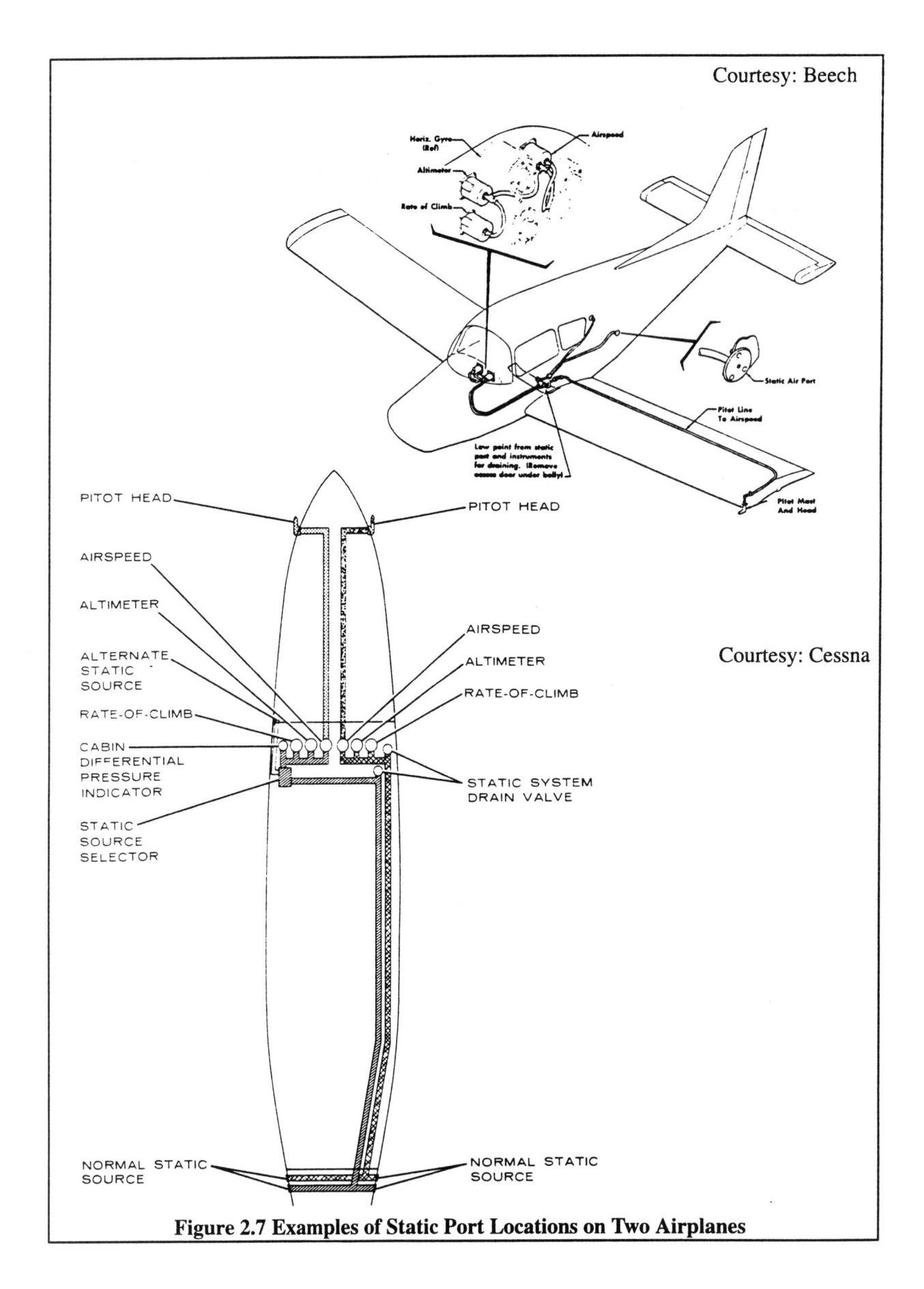

Figure 2.7 Examples of Static Port Locations on Two Airplanes

Eqn (2.63) has been used to generate Figure 2.8, where (dM/dV_c) is plotted against V_c for a range of pressure altitudes. These results should be used in conjunction with airspeed corrections on high subsonic airplanes.

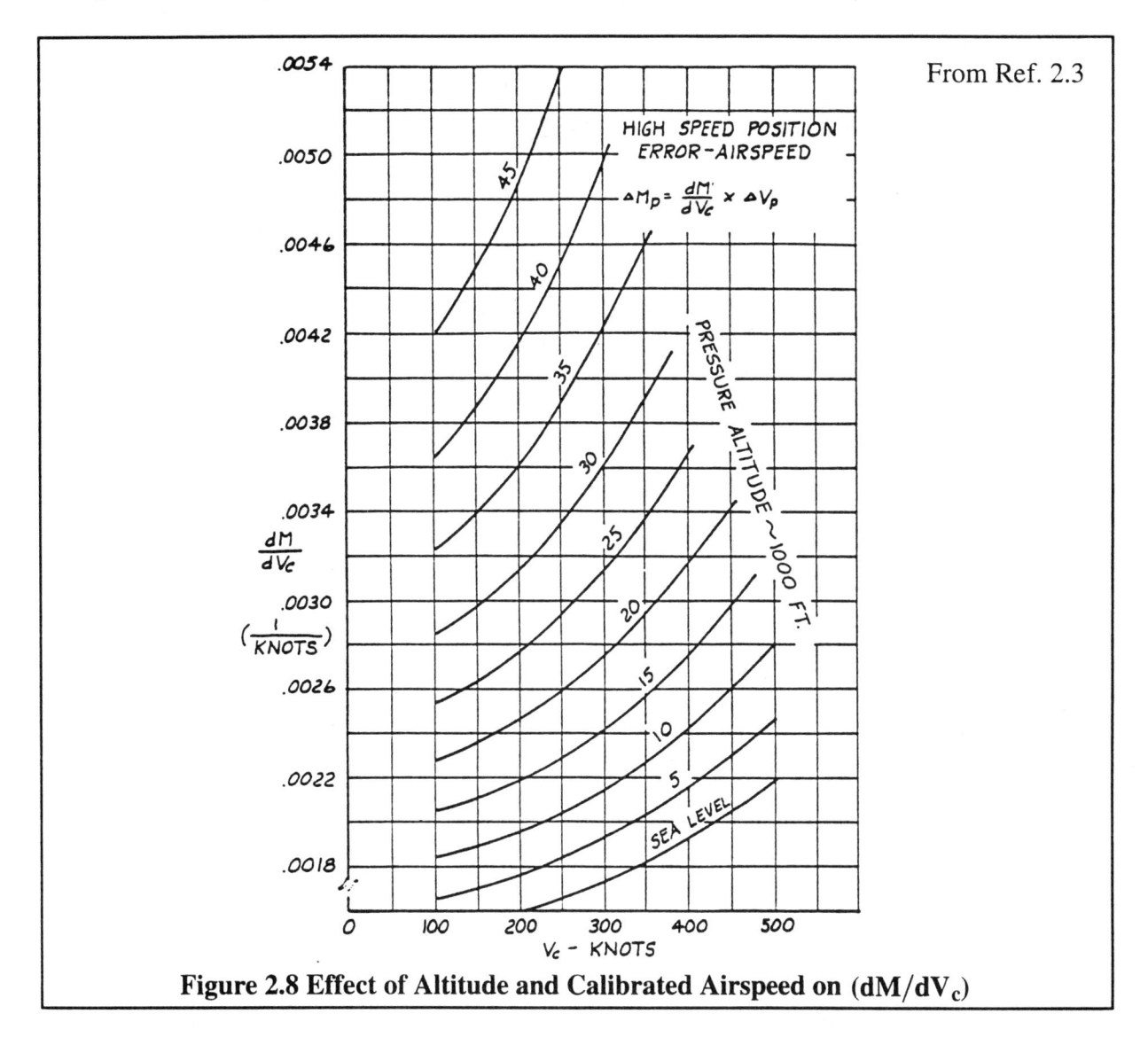

Figure 2.8 Effect of Altitude and Calibrated Airspeed on (dM/dV_c)

For low speed airplanes, a different procedure can be used. The pressure error, Δp caused by the local velocity at the static pressure port, (V_I) being different from the free stream velocity, V_c can be computed as follows:

$$\Delta p = \tfrac{1}{2}\varrho_0 V_c^2 - \tfrac{1}{2}\varrho_0 V_I^2 \qquad (2.64)$$

From this it follows that:

$$\frac{\Delta p}{\overline{q}} = \frac{V_c^2 - V_I^2}{V_c^2} = \frac{(V_I + \Delta V_p)^2 - V_I^2}{(V_I + \Delta V_p)^2} \approx \frac{2\Delta V_p}{V_I + 2\Delta V_p} \qquad (2.65)$$

Eqn (2.65) shows that $\Delta V_p/V_I$ is a function of $\Delta p/\bar{q}$ which depends on factors such as weight, flap deflection and other airplane configuration parameters. The quantity $\Delta p/\bar{q}$ must be determined from flight tests or from windtunnel tests for different flight conditions as well as for different airplane configurations.

Finally, the compressibility correction ΔV_c must be derived. To do this, rewrite Eqn (2.48) as:

$$p_t - p = p\left\{\left(1 + \frac{\gamma-1}{2\gamma}V^2\sigma\frac{\varrho_0}{p}\right)^{\frac{\gamma}{\gamma-1}} - 1\right\} \tag{2.66}$$

By equating Eqn (2.66) to Eqn (2.46) it follows that:

$$p\left\{\left(1 + \frac{\gamma-1}{2\gamma}V^2\sigma\frac{\varrho_0}{p}\right)^{\frac{\gamma}{\gamma-1}} - 1\right\} = p_0\left[\left\{1 + \frac{\gamma-1}{2}\left(\frac{V_c}{V_{a_0}}\right)^2\right\}^{\frac{\gamma}{\gamma-1}} - 1\right] \tag{2.67}$$

This equation can be solved for $V\sqrt{\sigma}$ to yield:

$$V\sqrt{\sigma} = V_e = \sqrt{\frac{2\gamma}{\gamma-1}\left(\frac{p}{\varrho_0}\right)} \sqrt{\left[\frac{\left\{1 + \frac{\gamma-1}{2}\left(\frac{V_c}{V_{a_0}}\right)^2\right\}^{\frac{\gamma}{\gamma-1}} - 1}{\delta} + 1\right]^{\frac{\gamma-1}{\gamma}} - 1} \tag{2.68}$$

The difference between V_e and V_c is ΔV_c. The compressibility correction, ΔV_c is plotted versus V_c for a range of altitudes and Mach numbers in the standard atmosphere in Figure 2.9.

Reference 2.3 contains a more detailed discussion of methods for in–flight measurement of speed and calibration of airspeed and Mach indicators.

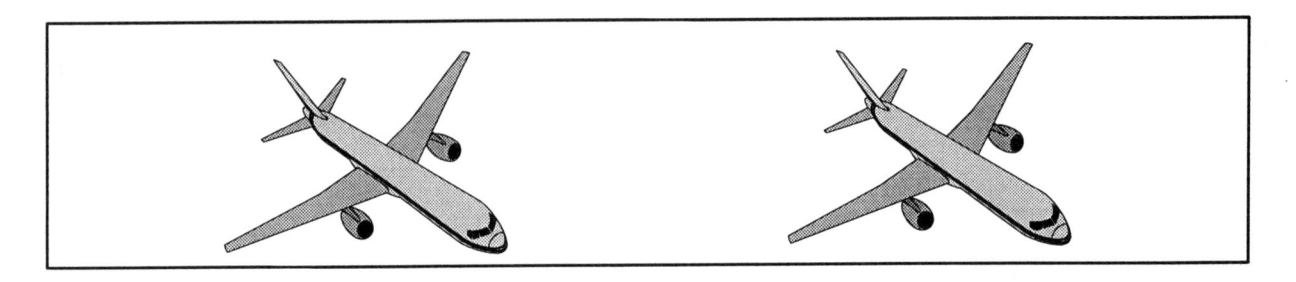

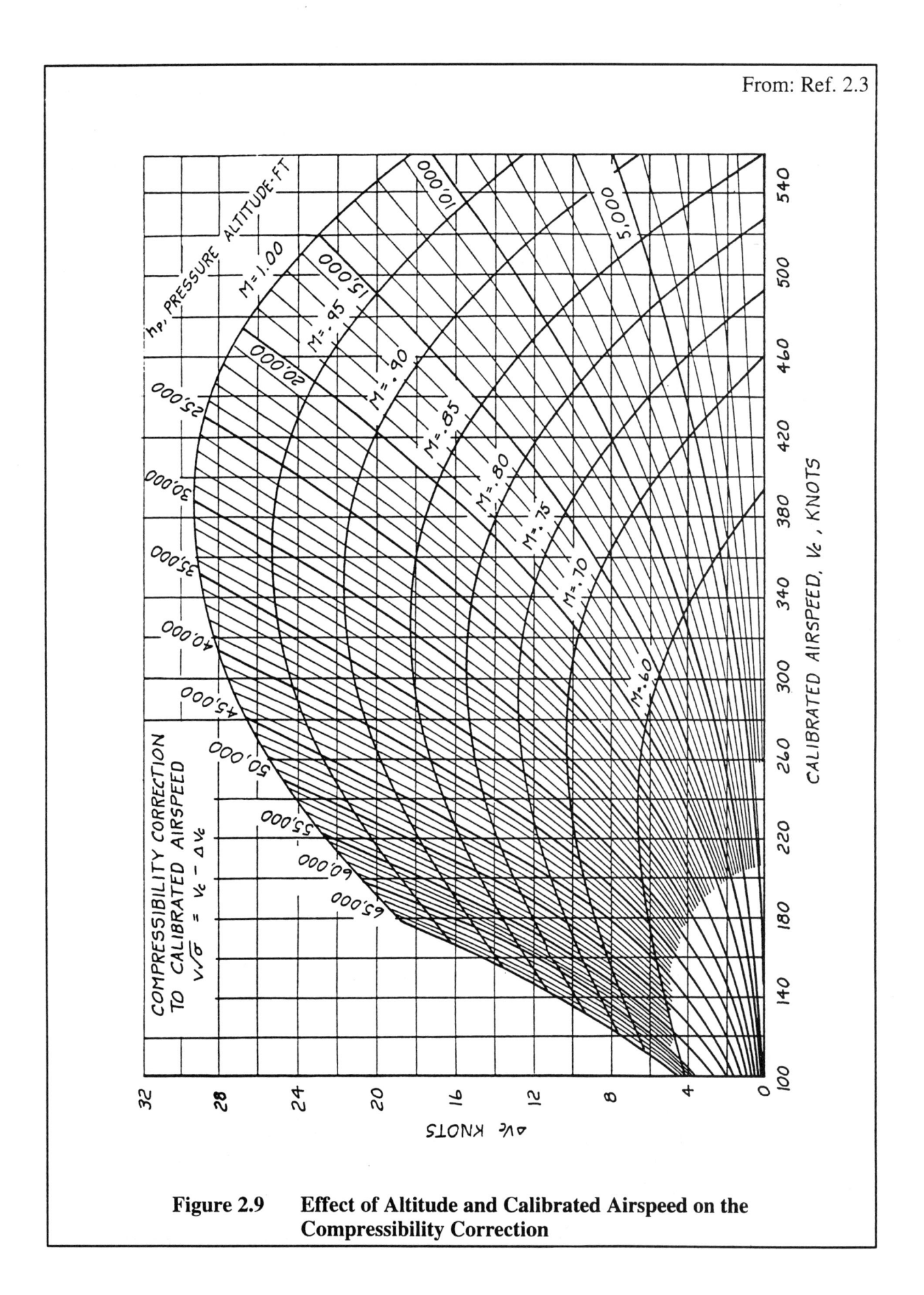

Figure 2.9 Effect of Altitude and Calibrated Airspeed on the Compressibility Correction

2.6 THE KUTTA–JOUKOWSKI THEOREM

The phenomenon of a wing producing lift can be explained with the concept of a vortex. A vortex produces a flow field of circular streamlines with induced velocity magnitude given by:

$$V_\theta = \frac{C}{r} \qquad (2.69)$$

where: C is a constant*. Figure 2.10 shows the relationship between a vortex and its induced velocity. The strength of a vortex is determined by its circulation, Γ. The circulation of a vortex is defined as:

$$\Gamma = \oint_l \vec{v} \cdot \vec{dl} \qquad (2.70)$$

where the integration is taken along a closed curve (l) as shown in Figure 2.10.

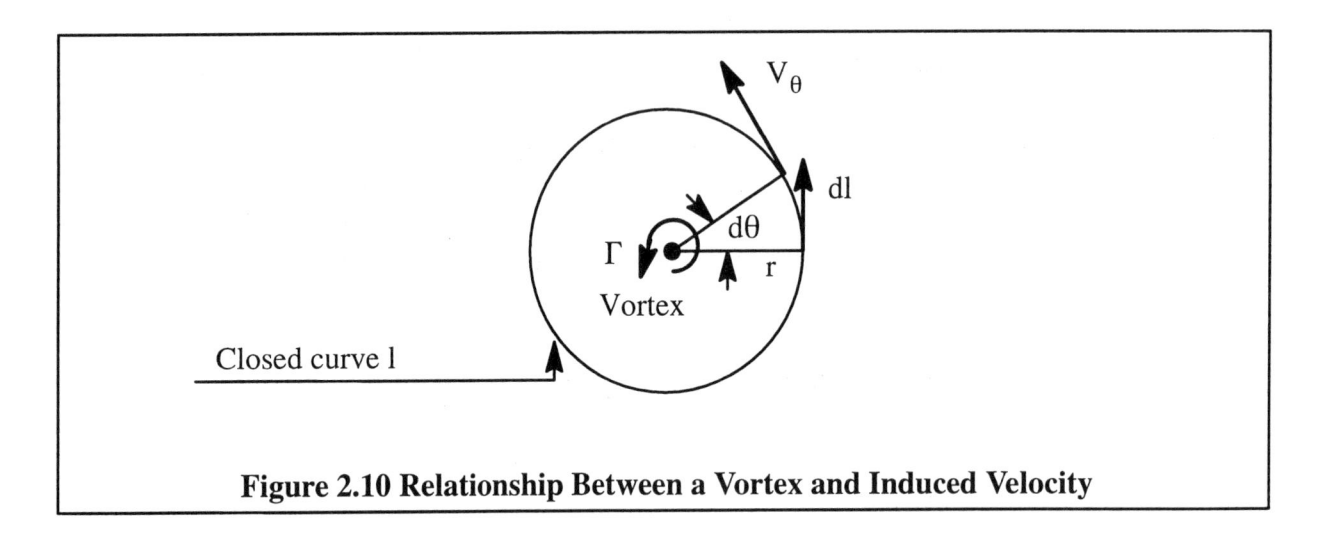

Figure 2.10 Relationship Between a Vortex and Induced Velocity

As an example of evaluating the integral of Eqn (2.70), consider a closed, circular path around the vortex, as shown in Figure 2.10. The circulation is then given by:

$$\Gamma = \int_0^{2\pi} V_\theta \, r \, d\theta = C \int_0^{2\pi} d\theta = 2\pi C \qquad (2.71)$$

It follows, that the constant, C, in Eqn (2.69) can be expressed in terms of the circulation through Eqn (2.70) so that Eqn (2.69) can in turn be written as:

$$V_\theta = \frac{\Gamma}{2\pi} \frac{1}{r} \qquad (2.72)$$

The flow around an airfoil (or wing section) can be represented by a vortex as shown in Figure 2.10. That this is the case can be visualized by considering a typical flow situation around an airfoil: see Fig. 2.11a. Such a flow situation can be seen to be analogous to the flow situation depicted in Fig. 2.11b where the idea of a vortex is superimposed.

---

* This equation is an analog of the so–called Biot and Savart Law. See Ref. 2.4, page 128.

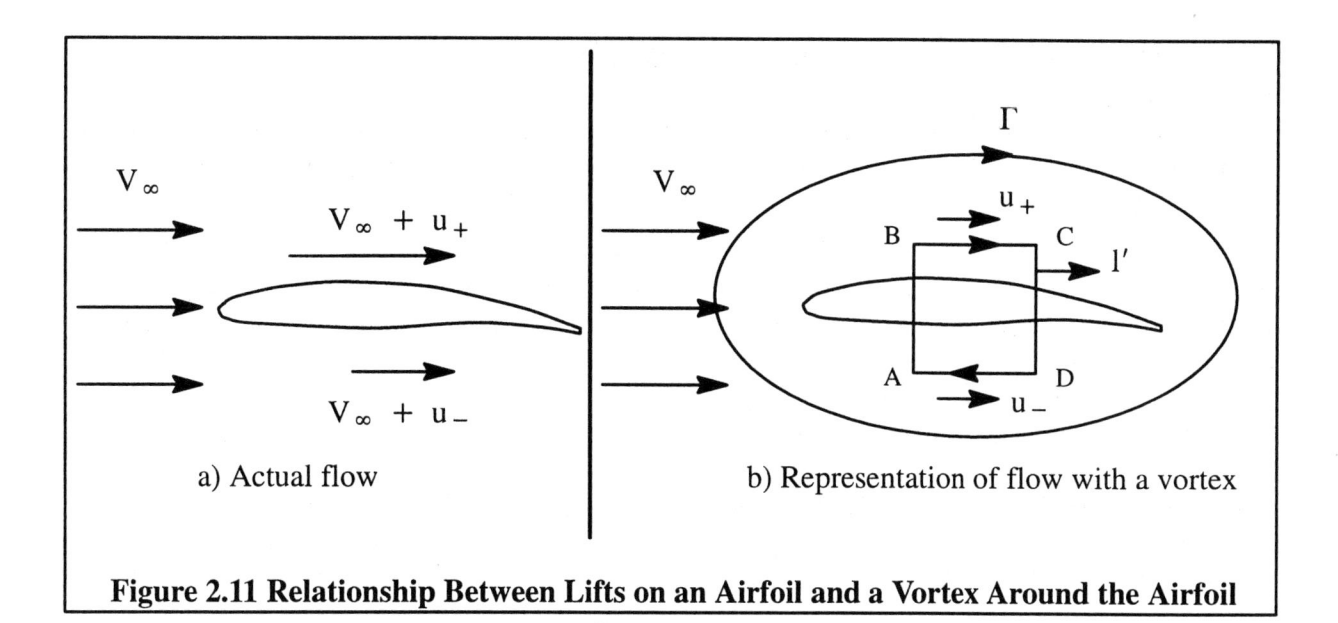

Figure 2.11 Relationship Between Lifts on an Airfoil and a Vortex Around the Airfoil

To see how a vortex is generated on a section of the wing, consider an airfoil in a uniform free stream as in Fig. 2.11a. To have an upward force (lift), the pressure acting on the upper surface must be lower than the pressure acting on the lower surface. According to Bernoulli's equation (Eqn 2.9), this implies that the tangential air velocity on the upper surface will be increased by an amount equal to u_+. At the same time, the tangential velocity of the air at the lower surface is changed by an amount equal to u_-. Assuming incompressible flow, the lift force over a length dx becomes:

$$dL = (p_- - p_+)dx \qquad (2.73)$$

However, according to Eqn (2.9):

$$p_+ + \tfrac{1}{2}\varrho V_+^2 = p_\infty + \tfrac{1}{2}\varrho V_\infty^2 \qquad (2.74a)$$

and:

$$p_- + \tfrac{1}{2}\varrho V_-^2 = p_\infty + \tfrac{1}{2}\varrho V_\infty^2 \qquad (2.74b)$$

By subtraction of these two equations it follows that:

$$(p_- - p_+) = \tfrac{1}{2}\varrho\{(V_\infty + u_+)^2 - (V_\infty + u_-)^2\} \approx \varrho V_\infty (u_+ - u_-) \qquad (2.75)$$

where the terms $(u_+^2 - u_-^2)$ has been neglected as a second order effect. It now follows from Eqns (2.73) and (2.75) that:

$$L = \varrho V_\infty \int (u_+ - u_-)dx \qquad (2.76)$$

Now, if a small closed curve, l' is chosen as shown in Fig. 2.11b, and the circulation, Γ, is calculated along it, the following result is obtained:

$$d\Gamma = (u_+ - u_-)dx \qquad (2.77)$$

Evidently, the contribution to the circulation measured over segments AB and CD in Fig. 2.11b is zero because of the zero velocity along these segments. From Eqns (2.76) and (2.77) the following relation is now obtained:

$$L = \rho V_\infty \Gamma \tag{2.78}$$

This equation is the well known **Kutta–Joukowski Theorem**.

According to the Kutta–Joukowski Theorem, the generation of lift on a wing can be explained by the existence of circulation in a uniform flow, and vice versa. This representation of lift force by circulation is convenient because it helps explain many flow characteristics around airplanes. As an example, the existence of wing tip vortices can be explained. It is well known by pilots that the tip vortices emanating from the wing tips of large, heavy airplanes can be quite strong. To make a rough estimate of the induced velocity in the vicinity of wing tip vortices, let W be the weight of the airplane and b be the span of the wing. Observe, that the lift calculated with Eqn (2.78) is the lift force per unit of span. According to Ref. 2.4, page 168, the total wing lift can be approximated as follows:

$$L = W \approx \rho V_\infty \Gamma(0.85b) \tag{2.79}$$

Solving for the circulation:

$$\Gamma = \frac{W}{\rho V_\infty (0.85b)} \tag{2.80}$$

Therefore, the circulation, Γ, around each wing tip can be assumed to have the value given by Eqn (2.80). As a consequence, at a point far downstream from the wing tip, the approximate magnitude of induced velocity will be given by Eqn (2.72):

$$V_\theta = \frac{W}{\rho V_\infty (0.85b)} \frac{1}{2\pi r} \tag{2.81}$$

The airplane lift coefficient in a level, steady state, straight line flight condition can be expressed as:

$$C_L = \frac{W}{\frac{1}{2}\rho V_\infty^2 S} \tag{2.82}$$

where: S is the wing area in ft^2.

After using Eqn (2.82) in Eqn (2.81) it is seen that:

$$V_\theta = \frac{C_L V_\infty S}{2\pi(0.85)b^2(r/b/2)} = \frac{C_L V_\infty}{5.341 \, A \, \bar{r}} \tag{2.83}$$

where: $A = b/S^2$ is called the wing aspect ratio

$\bar{r} = 2r/b$

Eqn (2.83) gives a prediction of the induced velocity due to one tip vortex only. Figure 2.12 shows a comparison of results predicted from Eqn (2.83) with experimental data from Ref. (2.6) and with estimates obtained from a theoretical model due to Betz and described in Ref. (2.7).

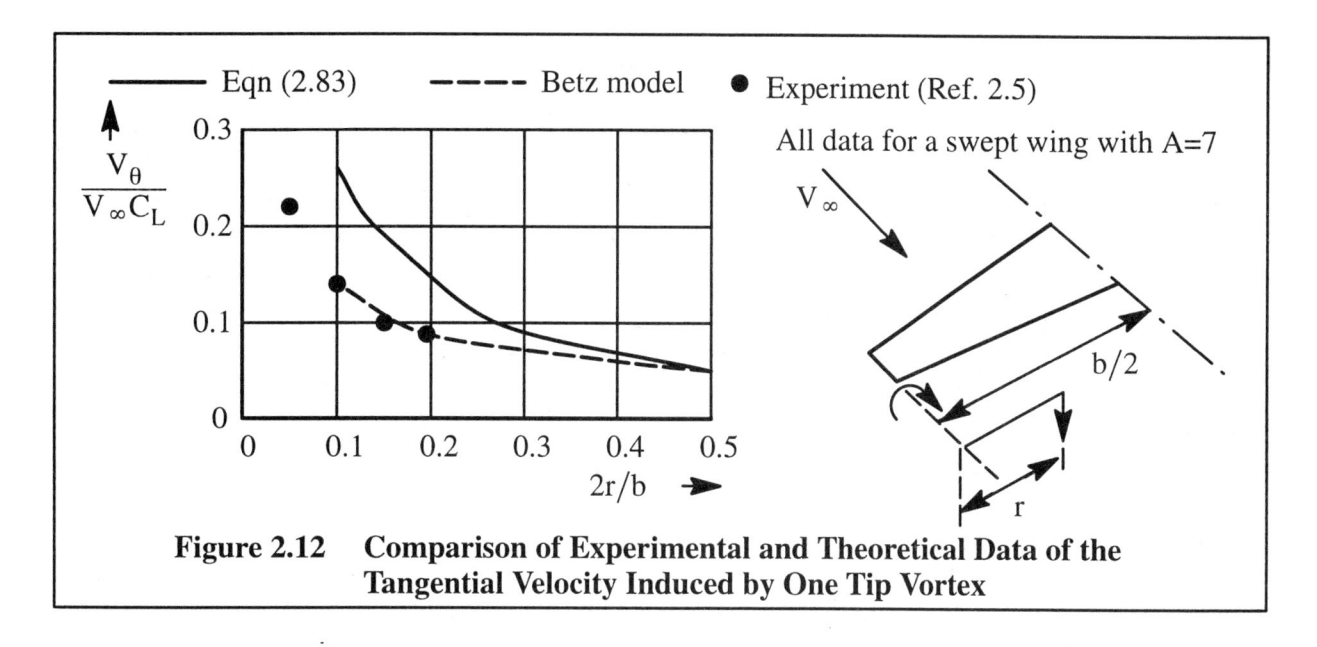

Figure 2.12 Comparison of Experimental and Theoretical Data of the Tangential Velocity Induced by One Tip Vortex

Observe, that Eqn (2.83) over predicts the tangential (or swirl) velocity of the tip vortex close to the core of the trailing vortex. For values of $2r/b$ in excess of 0.3 the predictions with Eqn (2.83) appear to be quite reasonable. The reason Eqn (2.83) is not accurate close to the core is that all trailing vorticity is assumed to be concentrated at the core in this model. Also, the detailed roll–up process of the tip vortex is glossed over by Eqn (2.83). In the Betz model these assumptions have been removed which results in a better prediction.

Magnitudes of V_θ values for three airplane types, as predicted with Eqn (2.83), are shown in Figure 2.13. The airplane data required to perform these predictions are given in Table 2.1. In all cases the flight condition is a takeoff climb.

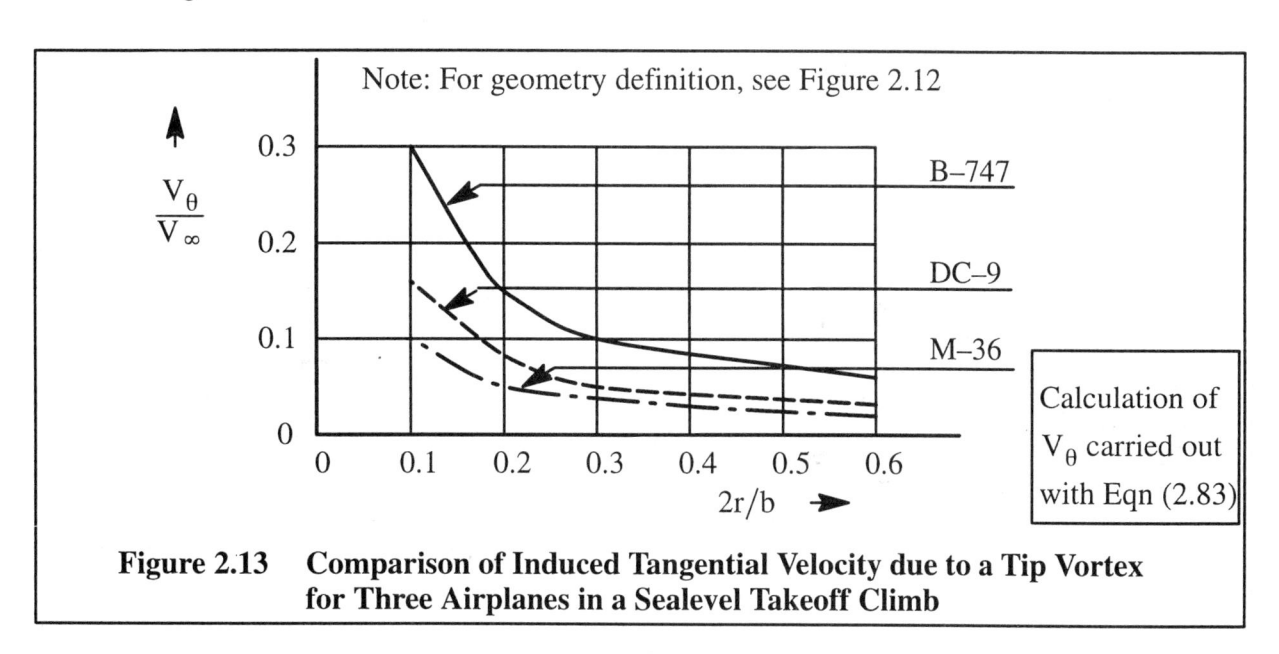

Figure 2.13 Comparison of Induced Tangential Velocity due to a Tip Vortex for Three Airplanes in a Sealevel Takeoff Climb

Table 2.1 Characteristics of Three Airplanes in Takeoff Climb at Sealevel						
Airplane Type	Weight (lbs)	Aspect Ratio (A)	Wing Area (ft²)	Climb Speed (kts)	Climb Speed (ft/sec)	$C_L = \frac{W}{qS}$
B–747	750,000	6.96	5,500	189	320	1.12
DC–9	140,000	9.62	1,279	198	334	0.83
M–36	18,000	5.74	253.3	250	422	0.34

These data are used to generate Figure 2.13

The results show that the tip vortices of a Boeing 747 will induce a tangential velocity which is twice as high as that of the DC-9. The reader should be aware of the fact that in a typical takeoff climb, the flaps will be deployed. Flaps will create their own vorticity. Such flap vortices have not been accounted for in Figure 2.13. One final comment. In reality the circulation strength of trailing vortices will gradually diminish as the distance behind the wing is increased. This effect has also not been accounted for.

2.7 THE LINEAR MOMENTUM PRINCIPLE

The fluid forces which act on an airplane can also be calculated by applying the linear momentum principle. According to Ref. 2.8, Chapter 3, the fluid forces acting on any control surface (cs) in steady flow can be computed from the following equation:

$$\vec{F} = \int_{cs} \rho \vec{V} \vec{V} \cdot d\vec{A} \tag{2.84}$$

The geometry associated with Eqn (2.84) is defined in Figure 2.14 for the case of a jet engine.

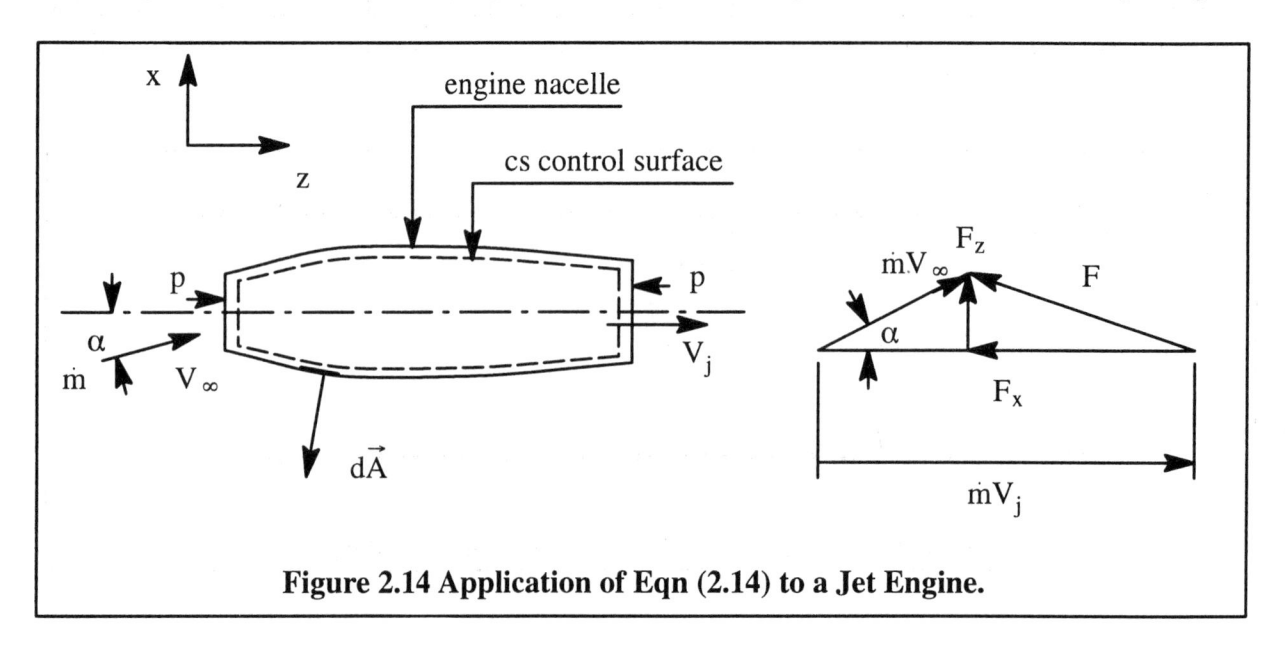

Figure 2.14 Application of Eqn (2.14) to a Jet Engine.

In Eqn (2.84), the vector $\vec{F}$ is the force (including the pressure forces) acting on the control surface, cs. The scalar $\vec{V} \cdot d\vec{A}$ is the volume flow rate. Note that $d\vec{A}$ is regarded as positive in an outward, normal direction from the surface. To illustrate the application of Eqn (2.84) to the jet engine nacelle example in Figure 2.14, let F_x and F_z be the force components acting on the engine in the x– and z– directions respectively. Therefore, $-F_x$ and $-F_z$ will be the corresponding force components acting on the control surface. It follows from Eqn (2.84) that, assuming that the mass flow rate, $\vec{V} \cdot d\vec{A}$, is nearly unchanged:

$$-F_x = (V_\infty \cos\alpha)(\dot{m}) + V_j(\dot{m}) = \dot{m}(V_j - V_\infty \cos\alpha) \tag{2.85}$$

and also that:

$$-F_z = (V_\infty \sin\alpha)(-\dot{m}) + 0 \tag{2.86}$$

Note that at the inlet: $\varrho\vec{V} \cdot d\vec{A} = -\dot{m}$ because $\vec{V}$ is in a direction opposite to $d\vec{A}$. Thus, the horizontal force component points to the negative x–direction and the normal force points upwards. The problem can also be solved graphically as shown on the right hand side of Figure 2.14. Note that $\dot{m}\vec{V}_\infty$ is drawn as a vector parallel to $\vec{V}_\infty$ and that $\dot{m}\vec{V}_j$ is parallel to $\vec{V}_j$.

More applications of the linear momentum principle are given in Chapters 4, 6 and 7.

2.8 VISCOUS EFFECTS, THE BOUNDARY LAYER AND FLOW SEPARATION

Sofar the airflow has been considered as non–viscous. In reality, within a thin layer, called the boundary layer, near the solid surface, viscosity does become important. The air particles will adhere to the surface as indicated in Figure 2.15.

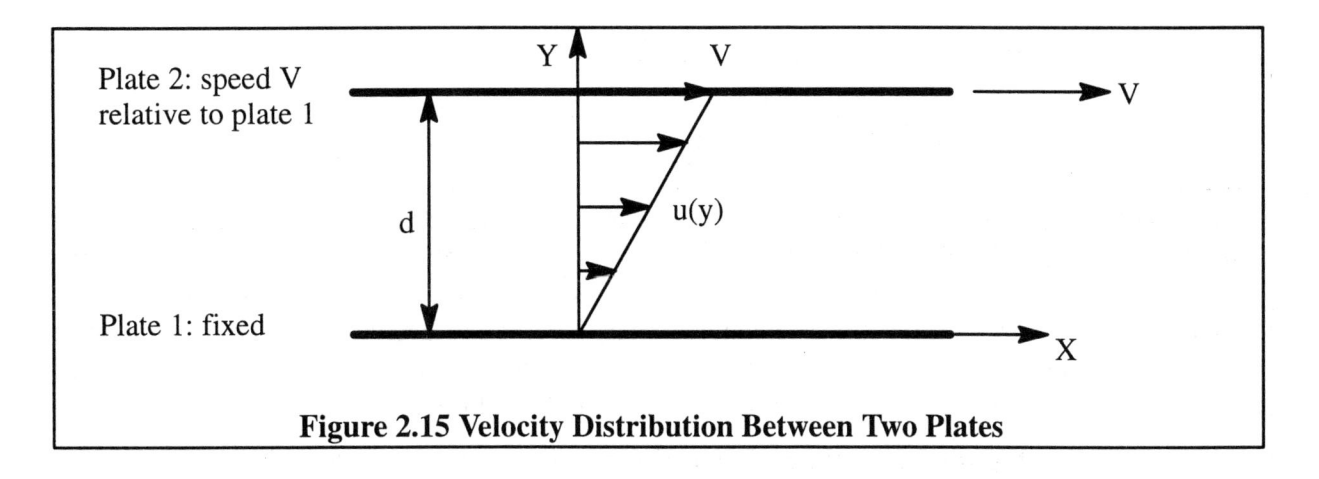

Figure 2.15 Velocity Distribution Between Two Plates

In Figure 2.15, two plates, a distance d apart, are shown such that one is moving with velocity V to the right while the other is held fixed. Due to viscosity, those air particles in contact with the surfaces will acquire the velocities of the surfaces. Therefore, the velocity distribution will be as shown in Figure 2.15. Experiments in Ref. 2.8 have shown that the force, F, needed to move the upper plate is directly proportional to the plate area, A, and the velocity, V, while being inversely proportional to the distance, d. Therefore:

$$F \propto \frac{AV}{d} \tag{2.87}$$

or

$$\tau = \frac{F}{A} = \mu \frac{du}{dy} \tag{2.88}$$

where: τ is the so-called shear stress in lbs/ft^2 or N/m^2. The proportionalty constant, μ is called the coefficient of viscosity. The units for μ are derived from Eqn (2.88) by a dimensional analysis:

$$\dim(\mu) = \frac{lbs/ft^2}{(ft/sec)ft} = lbs \frac{sec}{ft^2} \text{ or } N \frac{sec}{m^2} \tag{2.89}$$

For air, μ increases with temperature and can be calculated by the following approximate formula for the standard atmosphere, according to Ref. 2.8:

$$\mu = (10^{-10} \times 0.3170) \, T^{3/2} \left(\frac{734.7}{T + 216}\right) \text{ lbs } \frac{sec}{ft^2}, \text{ with T in } ^0R \tag{2.90}$$

or:

$$\mu = (1.458 \times 10^{-6}) T^{3/2} \left(\frac{1}{T + 110.4}\right) N \frac{sec}{m^2}, \text{ with T in } ^0K \tag{2.91}$$

Numerical values for μ in the standard atmosphere are given in Appendix A.

The type of air flow in the boundary layer depends on the smoothness of the flow approaching the body surface, the shape of the body, the surface roughness, the pressure gradient in the flow and the Reynolds number, R_N, of the flow. The Reynolds number is defined as follows:

$$R_N = \frac{\varrho V L}{\mu} = \frac{VL}{\nu} \tag{2.92}$$

where: $\nu = \frac{\mu}{\varrho}$ is the so-called kinematic viscosity.
L is the characteristic length of the body in the flow direction. As an example, usually, L is taken to be the chord length of an airfoil or the length of a fuselage.

At low Reynolds numbers, the flow in the boundary layer tends to be laminar (laminar boundary layer). However, above certain "transition Reynolds numbers", the flow becomes turbulent (turbulent boundary layer). The distinguishing characteristics of both types of boundary layers are displayed in Figure 2.16. Note that the turbulent boundary layer tends to have a more uniform velocity profile. However, in a turbulent boundary layer there is a lot of vertical (up and down) exchange of air particles. In a laminar boundary layer the air particles follow paths parallel to the plate.

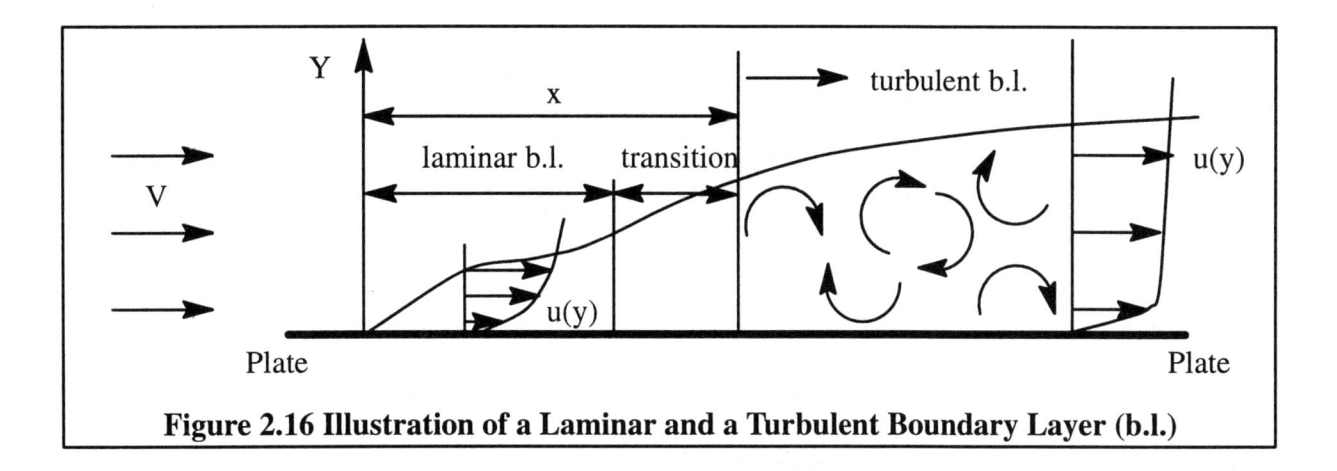

Figure 2.16 Illustration of a Laminar and a Turbulent Boundary Layer (b.l.)

According to Ref. 2.11, page 435, transition from laminar to turbulent takes place on a flat plate at a point, x, determined by:

$$\left(R_{N_x}\right)_{crit.} = \left(V \frac{x}{\nu}\right)_{crit.} \approx 3.5 \times 10^5 \text{ to } 10^6 \qquad (2.93)$$

In reality the transition is also affected by such factors as pressure gradient, surface curvature, free stream turbulence and surface roughness (Ref. 2.12). On an airfoil it has been found that the pressure gradient has an important influence on the location of the transition point. In general, the flow will remain laminar in a region of accelerating flow (i.e. with a favorable pressure gradient). In a positive pressure gradient, the boundary layer flow will be more unstable when disturbed and will transition to turbulent flow sooner. In the case of surface curvature, it has been found that the transition Reynolds number is almost unaffected by a convex surface. However, in a region of concave curvature the transition Reynolds number is reduced. In addition, the transition from laminar to turbulent boundary layers is hastened by an increase in free stream turbulence and by surface roughness. For practical applications to subsonic flow, the transition point can be approximately taken at the point of minimum pressure if the airfoil Reynolds number, $R_N = \frac{Vc}{\nu}$, exceeds 10^7 (See: Ref. 2.11, page 717).

2.8.1 THE LAMINAR BOUNDARY LAYER

As shown in Figure 2.16, the air velocity on the lower plate surface is zero and increases rapidly, away from the surface. Because the boundary layer velocity, u, will approach the external velocity, V, only gradually, a thickness defined at a location for which u = 0.99V is normally taken as the boundary layer thickness, δ. From theoretical calculations (Ref. 2.11, page 130), δ is given by:

$$\delta = \frac{5.2x}{\sqrt{R_{N_x}}} \qquad (2.94)$$

Eqn (2.94) shows that the boundary layer thickness grows parabolically in the downstream direction.

The "rubbing" of the boundary layer on the plate gives rise to friction forces, D_f, otherwise known as skin friction drag. The skin friction drag coefficient for **one side** of a flat plate in laminar flow is given by:

$$c_f = \frac{D_f}{\frac{1}{2}\varrho V^2 S} = \frac{1.328}{\sqrt{R_N}} \tag{2.95}$$

where: S is the area of one side of the plate only

R_N is based on the total plate length

2.8.2 THE TURBULENT BOUNDARY LAYER

When the flow is transitioned to turbulent flow, the boundary layer thickness will be increased. In fact, this phenomenon is often used to determine the location of the transition region. The turbulent boundary layer tends to have a more uniform velocity profile as shown in Figure 2.16. The thickness of the boundary layer can be determined from (Ref. 2.11, page 599):

$$\delta = \frac{0.37 x}{\left(R_{N_x}\right)^{1/5}} \tag{2.96}$$

The skin friction drag coefficient for a flat plate can be calculated from Schlichting's formula (Ref. 2.11, page 602):

$$c_f = \frac{0.455}{\left(\log_{10} R_N\right)^{2.58}} \tag{2.97}$$

In the transition region, the Prandtl–Schlichting skin friction formula (Ref. 2.11, page 602) may be used:

$$c_f = \frac{0.455}{\left(\log_{10} R_N\right)^{2.58}} - \frac{1700}{R_N} \tag{2.98}$$

The friction coefficient as given by Eqns (2.97) and (2.98) is for **one side** of a flat plate only. Figure 2.17a shows how the one–sided friction coefficient, c_f, varies with Reynolds number, R_N, in subsonic incompressible flow. Note the agreement between theory and experiment.

As it turns out, c_f is also a function of Mach number. Why this is the case, is discussed in Ref. 2.13. Since under compressible flow situations boundary layers are usually turbulent, Figure 2.17b shows how the one–sided, turbulent boundary layer skin friction coefficient, c_f varies with both Reynolds number and Mach number. Equation (2.99) fits the curves of Figure 2.17b:

$$c_f = \frac{0.455}{\left(\log_{10} R_N\right)^{2.58}\left(1 + 0.144 M^2\right)^{0.58}} \tag{2.99}$$

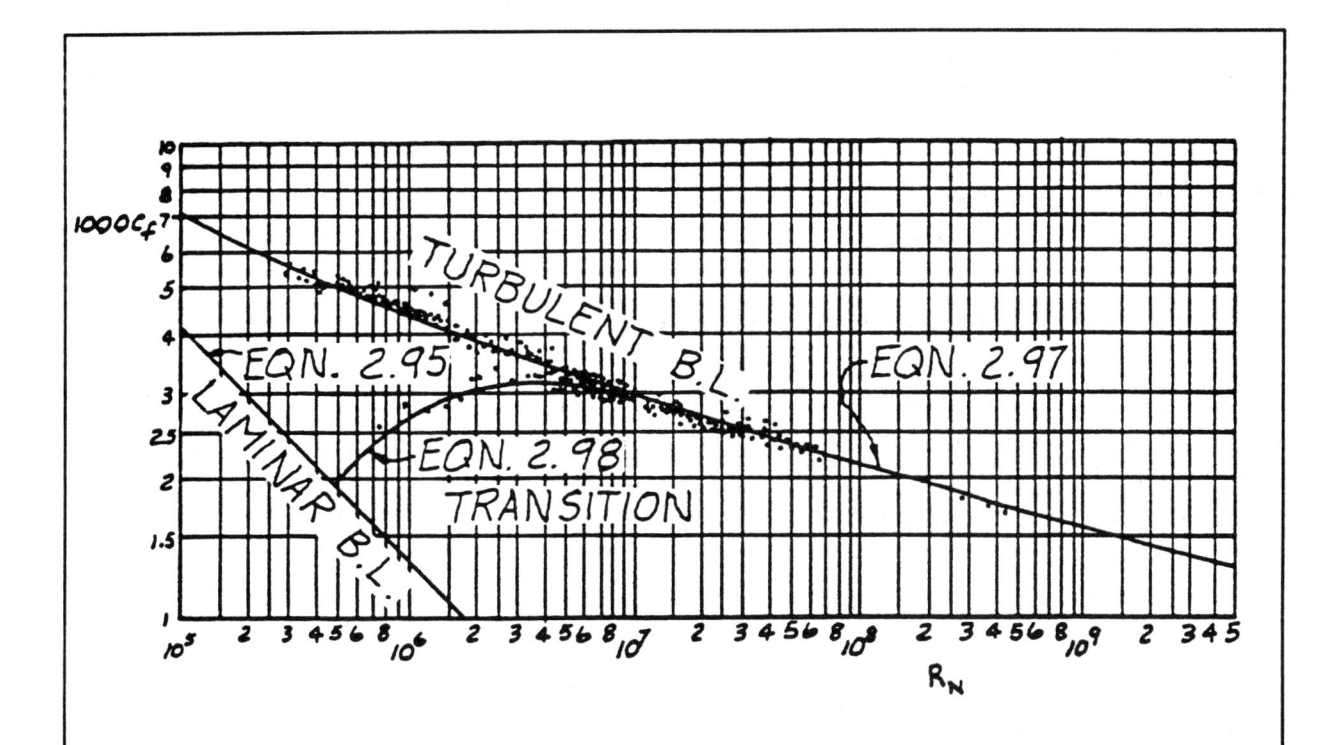

Figure 2.17a Variation of Flat–Plate Skin Friction Coefficient with Reynolds Number in Incompressible Flow

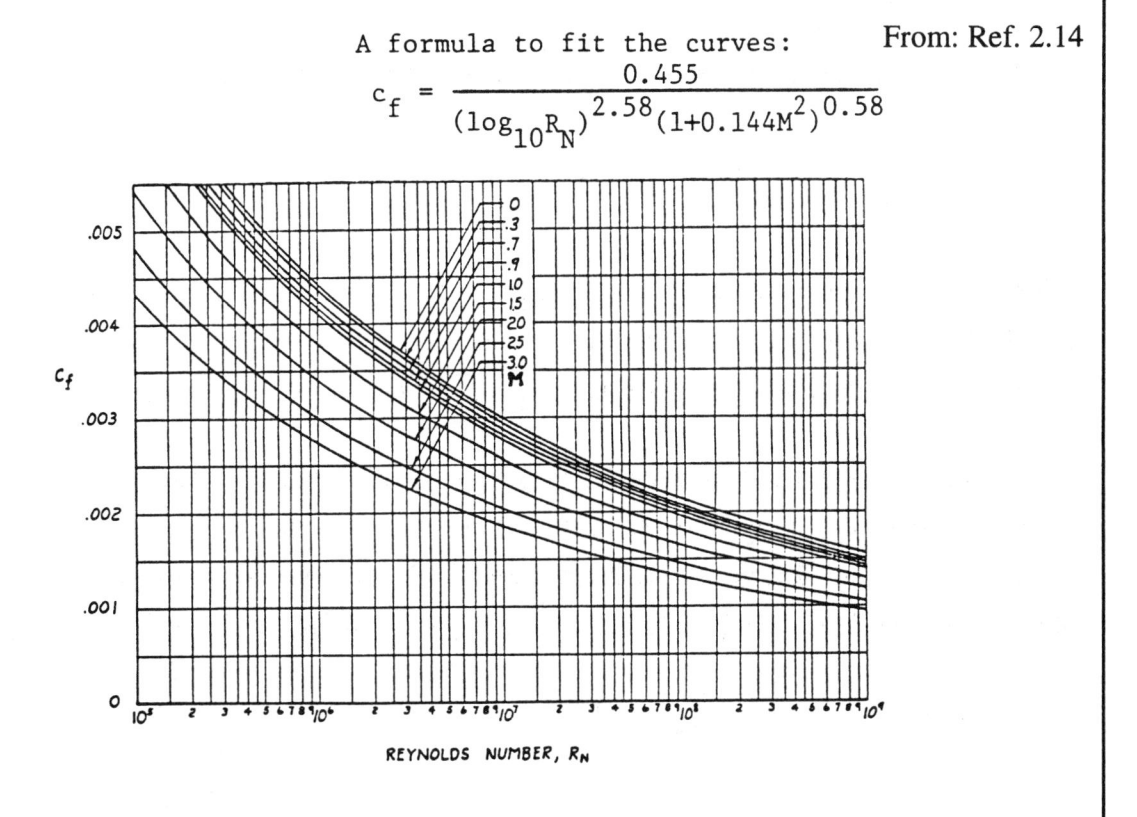

A formula to fit the curves: From: Ref. 2.14

$$c_f = \frac{0.455}{(\log_{10} R_N)^{2.58}(1+0.144 M^2)^{0.58}}$$

Figure 2.17b Variation of Flat–Plate Skin Friction Coefficient in a Turbulent Boundary Layer with Reynolds Number and Mach Number

2.8.3 FLOW SEPARATION

The flow over an airfoil is somewhat changed from that over a flat plate. The difference arises from the airfoil shape which makes the air velocity near the surface different from the free stream value. As shown in Figure 2.18, on the forward section of the airfoil, the air velocity increases downstream, so that the static pressure decreases in the same direction. Since the accelerating flow tends to assist the boundary layer to remain attached to the surface, the negative rate of change of pressure in the downstream direction is called a "favorable" pressure gradient.

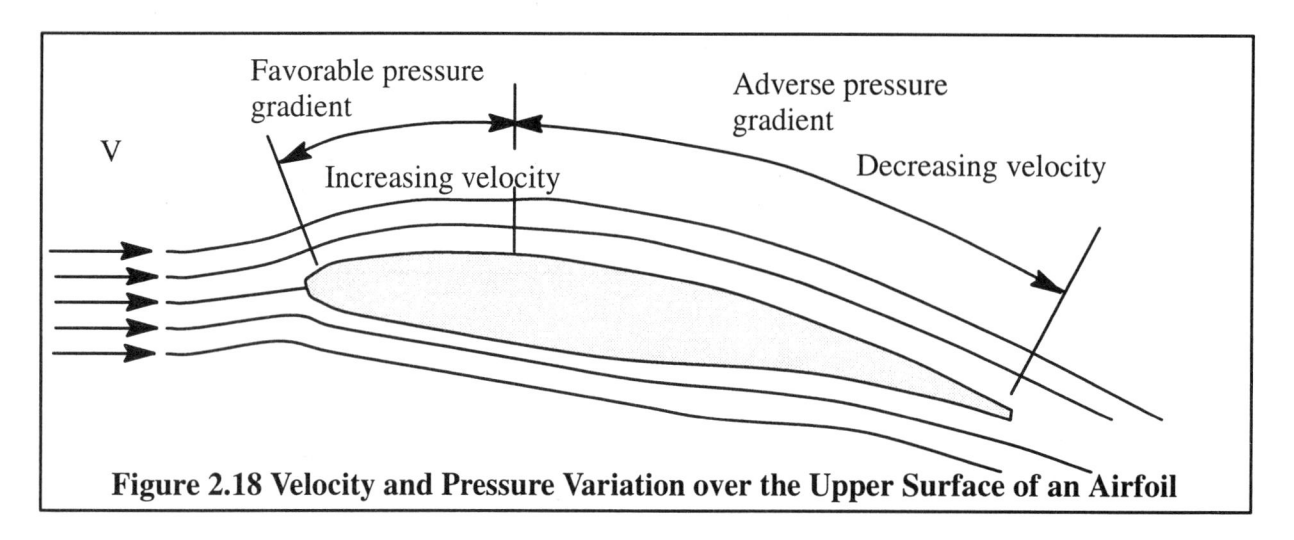

Figure 2.18 Velocity and Pressure Variation over the Upper Surface of an Airfoil

On the other hand, over the rear section of the airfoil, the opposite is true. The air velocity outside the boundary layer decreases and the pressure increases in the downstream direction. Therefore, the boundary layer flow must resist the increasing pressure while moving downstream. When air particles in the boundary layer do not have sufficient energy to reach the trailing edge, they will separate from the surface, creating a so–called wake. This is illustrated in Figure 2.19. When this happens, the resulting large difference in pressure distribution on fore and aft surfaces produces a large pressure drag. This pressure drag may be several times larger than the skin friction drag. Hence, the positive pressure gradient is often called an "adverse" pressure gradient.

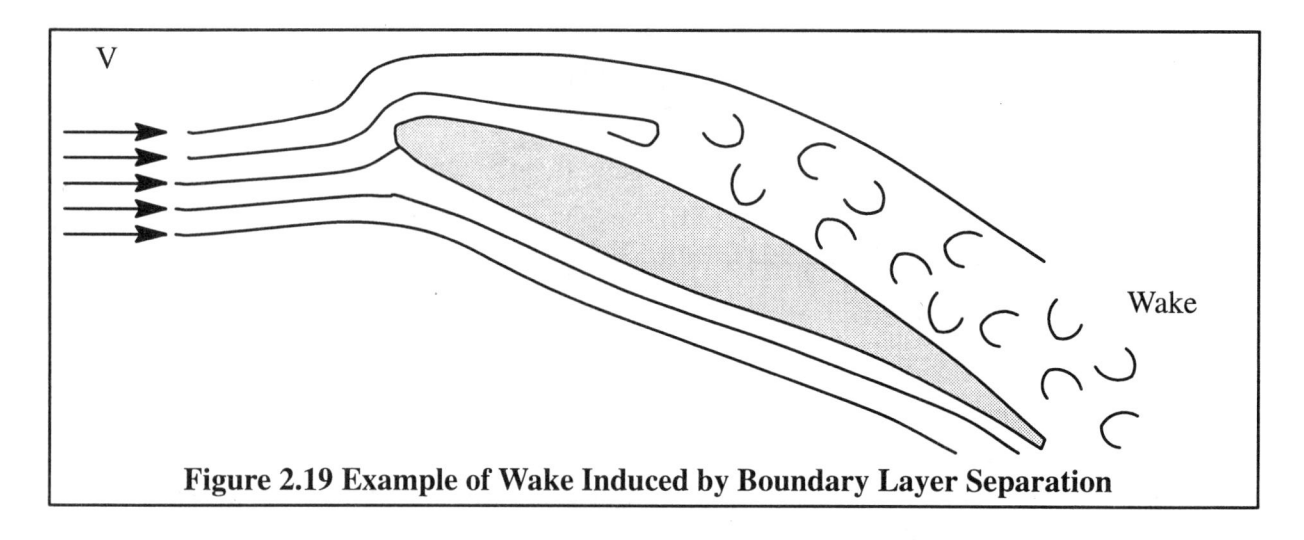

Figure 2.19 Example of Wake Induced by Boundary Layer Separation

When the boundary layer flow is turbulent, the air particles possess more energy to overcome the "adverse" pressure gradient over a larger distance. Therefore, flow separation can be delayed. This explains the abrupt drop in drag on spheres and circular cylinders when the transition Reynolds number (approximately 3×10^5) is reached, or when the boundary layer is changed artificially ("tripped") to turbulent flow by the use of so-called tripping wires (Ref. 2.11, page 434). In both cases, the turbulent boundary layer remains attached over a larger distance so that the wake is thinner. Figure 2.20 shows an example in the case of a sphere.

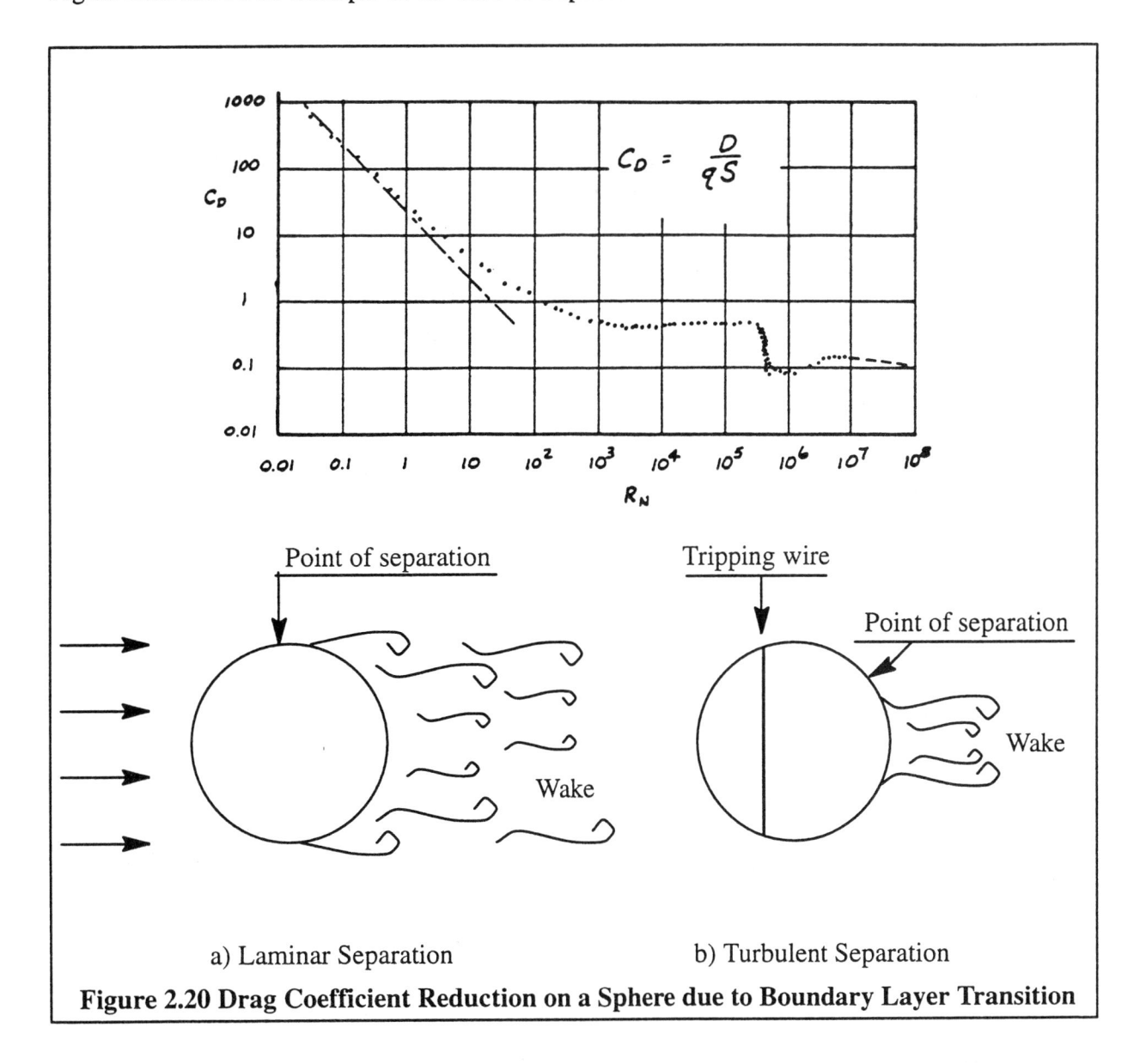

a) Laminar Separation   b) Turbulent Separation

Figure 2.20 Drag Coefficient Reduction on a Sphere due to Boundary Layer Transition

Finally, another point to be mentioned here is that the boundary layer possesses a lower energy level when compared to the external flow. Therefore, when jet engines are buried inside a fuselage (as in many fighter aircraft) a boundary layer bleed system must be used to remove the boundary layer air before it enters the intake. A simple version of such a bleed system is the so-called boundary layer splitter, an example of which is shown in Figure 2.21. Such a splitter prevents the engine from ingesting low energy air and thereby reducing its operating efficiency.

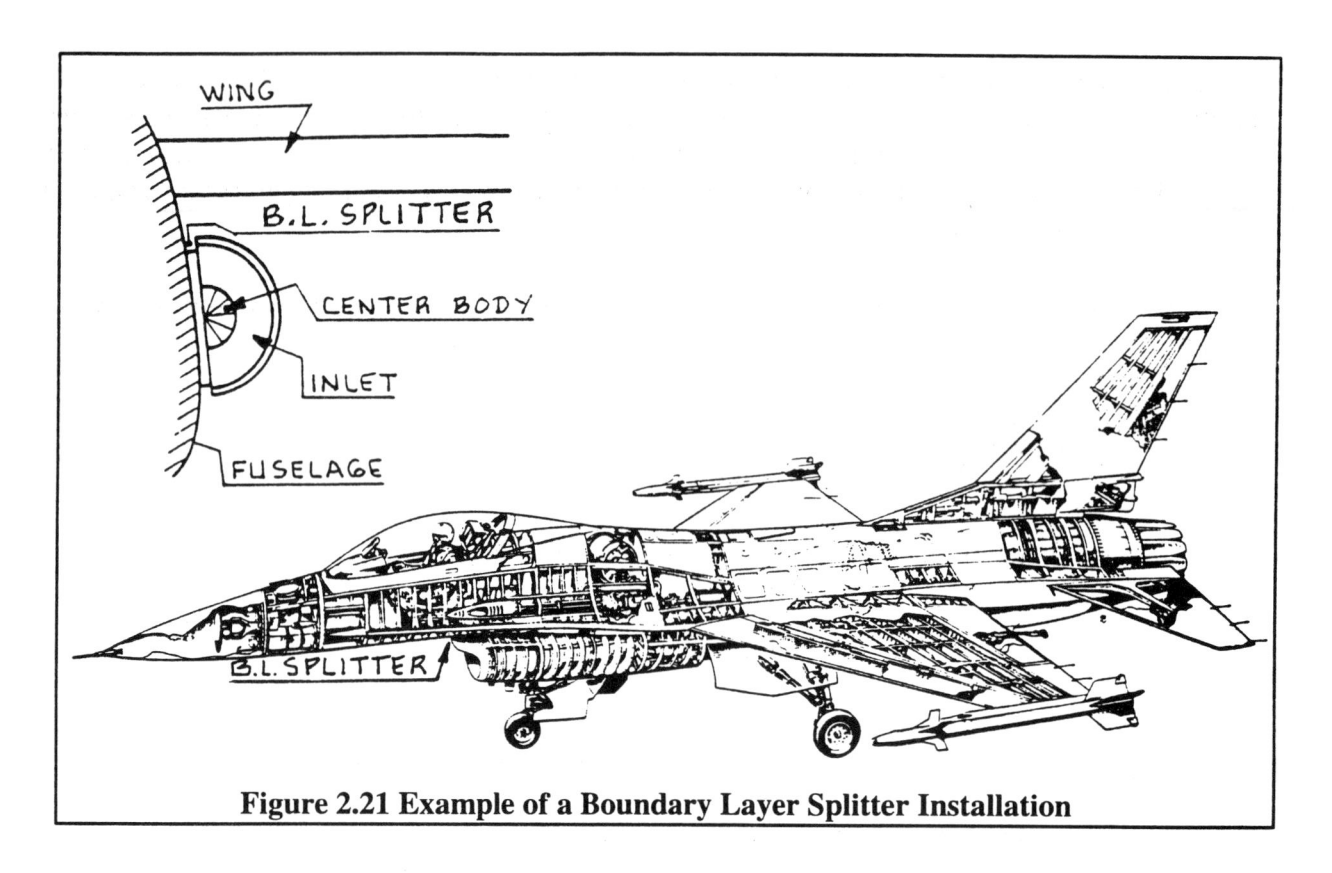

Figure 2.21 Example of a Boundary Layer Splitter Installation

The following example shows how the friction drag of a flat plate may be calculated.

Example 2.3: A flat plate of 10 ft span and 6 ft chord is placed in an air stream of 100 mph under standard sealevel conditions. If the transition Reynolds number is 10^6, calculate the total skin friction drag of the plate in lbs.

Solution: The location of the transition point is determined from: $10^6 = \frac{Vx}{\nu}$. Therefore:

$$x_{transition} = \frac{10^6 \nu}{V} = \frac{10^6 \times 1.572 \times 10^{-4}}{100 \times 1.467} = 1.072 \text{ ft}$$

The drag contribution of the first (and laminar) part of the plate (1.072 ft x 10 ft) may be computed from Eqn (2.95) by first computing the one-sided drag coefficient:

$$C_{D_{f_1}} = \frac{1.328}{\sqrt{R_N}} = \frac{1.328}{1,000} = 1.328 \times 10^{-3}$$

Next, the drag of the first part for the entire plate (both surfaces) is found from:

$$D_1 = 2C_D \tfrac{1}{2} \varrho V^2 S =$$

$$= 2 \times 1.328 \times 10^{-3} \times \tfrac{1}{2} \times 0.002377 \times (100 \times 1.467)^2 \times 1.072 \times 10 = 0.73 \text{ lbs}$$

To calculate the drag over the turbulent part of the plate the following procedure is used. First, the drag coefficient of the one–sided plate, assuming a completely turbulent boundary layer is determined. Then, the drag coefficient of the first part of the plate, assuming turbulent boundary layer, is calculated. Finally, these two contributions are subtracted, to given the drag of the second (and turbulent) part of the plate.

If the entire plate is assumed to be covered by a turbulent boundary layer, the drag coefficient follows from Eqn (2.97). First, the Reynolds number of the entire plate must be computed:

$$R_N = \frac{100 \times 1.467 \times 6}{1.572 \times 10^{-4}} = 5.6 \times 10^6$$

Next, the drag coefficient of the completely turbulent plate is found as:

$$C_{D_{f_2}} = \frac{0.455}{\left(\log_{10} R_N\right)^{2.58}} = 0.33 \times 10^{-2}$$

The drag of the entire plate, assuming completely turbulent flow then is:

$$D_2 = 2 \times 0.33 \times 10^{-2} \times \bar{q} \times 6 \times 10 = 10.13 \text{ lbs}$$

Similarly, the drag coefficient for the first part of the plate, assuming turbulent flow, is given by:

$$C_{D_{f_3}} = \frac{0.455}{\left(\log_{10} 10^6\right)^{2.58}} = 0.447 \times 10^{-2}$$

The reader will have noted that the transition Reynolds number had to be used to obtain this result. The drag of the entire first part of the plate, assuming turbulent flow then would have been:

$$D_3 = 2 \times 0.447 \times 10^{-2} \times \bar{q} \times 1.072 \times 10 = 2.45 \text{ lbs}$$

It now follows that the drag of the entire plate is found from:

$$D = D_2 - D_3 + D_1 = 10.13 - 2.45 + 0.73 = 8.41 \text{ lbs}$$

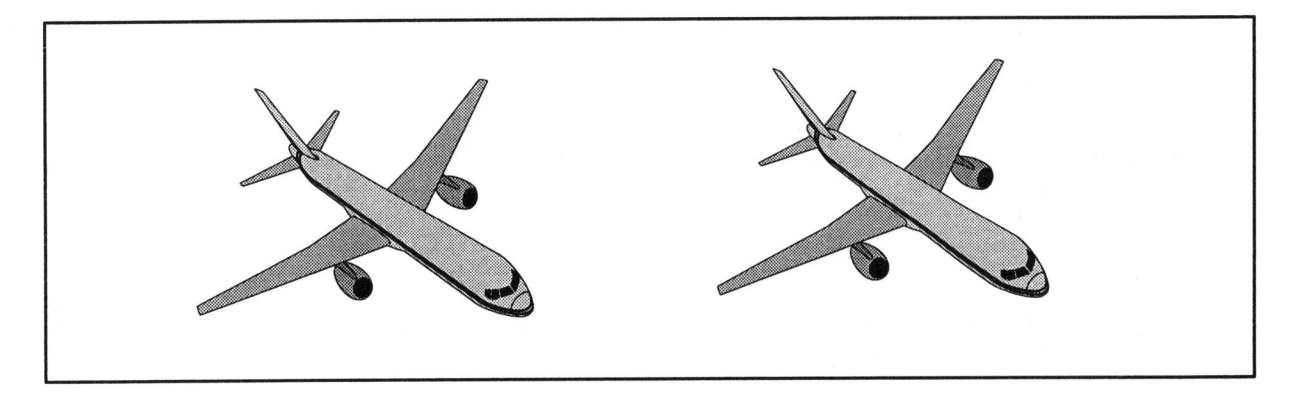

2.9 SUMMARY FOR CHAPTER 2

In this chapter various aerodynamic principles and formulas have been derived and discussed. For the convenience of the reader, the most important equations are summarized next. As a first approximation it is acceptable to assume that air flow in the subsonic speed range behave as an inviscid fluid. However, when calculating drag in subsonic flow this is not true. Even in subsonic flow, the effect of viscosity on the boundary layer and therefore on friction drag must be included.

2.9.1 THE CONTINUITY EQUATION: Conservation of mass along a streamtube, such as a windtunnel.

(1) For incompressible flow (approximately M < 0.3):

$$AV = \text{constant} \tag{2.4}$$

(2) For compressible flow (approximately M > 0.3):

$$\dot{m} = \varrho AV = \text{constant} \tag{2.2}$$

2.9.2 THE BERNOULLI EQUATION: Conservation of energy along a streamline.

(1) For incompressible flow (approximately M < 0.3):

$$p + \frac{1}{2}\varrho V^2 = \text{constant} \tag{2.9}$$

The isentropic equation of state, Eqn (2.26), is not needed.

<u>Application:</u> Definition of equivalent airspeed, V_e:

$$V_e = \sqrt{\frac{2(p_t - p)}{\varrho_0}} \tag{2.39}$$

(2) For compressible flow (approximately M > 0.3):

$$\left(\frac{\gamma}{\gamma - 1}\right)\frac{p}{\varrho} + \frac{1}{2}V^2 = \text{constant} \tag{2.36}$$

or:

$$M^2 = \frac{2}{\gamma - 1}\left\{\left(\frac{p_t - p}{p} + 1\right)^{\frac{\gamma - 1}{\gamma}} - 1\right\} \tag{2.44}$$

The isentropic equation of state can be used:

$$\frac{p}{\varrho^\gamma} = \text{constant} \tag{2.26}$$

Basic Aerodynamic Principles and Applications

The speed of sound is given by:

$$V_a = \sqrt{\frac{\gamma p}{\varrho}} = \sqrt{\gamma g R T} \tag{2.31}$$

Application: Definition of calibrated airspeed, V_c:

$$V_c^2 = \frac{2V_{a_0}^2}{\gamma - 1}\left\{\left(\frac{p_t - p}{p_0} + 1\right)^{\frac{\gamma-1}{\gamma}} - 1\right\} \tag{2.45}$$

Various airspeed corrections have been discussed in Section 2.5.3.

2.9.3 THE KUTTA–JOUKOWSKI THEOREM:
Relation between a force per unit span acting on a body and the circulation of a vortex:

$$L = \varrho V_\infty \Gamma \tag{2.78}$$

2.9.4 THE LINEAR MOMENTUM PRINCIPLE:
The calculation of forces acting on a body, if the air stream which passes it has a momentum change:

$$\vec{F} = \int_{cs} \varrho \vec{V}\vec{V} \cdot d\vec{A} \tag{2.84}$$

2.9.5 VISCOUS EFFECTS:
The effect of viscosity on the behavior and friction of laminar and turbulent boundary layers on one side of a flat plate:

For a laminar boundary layer:

$$c_f = \frac{D_f}{\frac{1}{2}\varrho V^2 S} = \frac{1.328}{\sqrt{R_N}} \tag{2.95}$$

For a turbulent boundary layer:

$$c_f = \frac{0.455}{\left(\log_{10} R_N\right)^{2.58}\left(1 + 0.144\, M^2\right)^{0.58}} \tag{2.97}$$

Application: The calculation of friction drag.

2.10 PROBLEMS FOR CHAPTER 2

2.1   Air having the standard sealevel density has a velocity of 100 fps at a section of a wind tunnel. At another section, having half as much cross sectional area as the first section, the flow velocity is 400 mph. What is the density at the second section?

2.2   A wind tunnel has a test section of 4 ft x 4 ft and a largest section of 15 ft x 15 ft. A manometer as shown in Figure 2.4 is used, with mercury as the manometer fluid. What will the manometer reading be if the test speed is 500 mph? Assume that the temperature at the test section is 80 deg. F. Also assume, that the pressure at the test section is the standard sealevel value. (Hint: compressibility cannot be neglected. Equate the density ratios obtained from the continuity and Bernoulli equations. The isentropic equation of state is also needed.)

2.3   A wind tunnel is as shown in Figure 2.4. If the highest attainable airspeed at the test section is 200 mph, what would the attainable test speed be if the cross sectional area at Station 2 is reduced to half of its original size? Assume that the incompressible flow rate is not changed.

2.4   A low speed airspeed indicator reads 200 mph when the airplane is flying at an altitude at which the altimeter reads 6,000 ft, while the ambient temperature is found to be 30 deg. F. Calculate the true airspeed.

2.5   A pitot static tube is used to measure the airspeed at the test section of a wind tunnel. If the pressure difference across the pitot static tube is 4 inches of water, what is the airspeed at the test section? If the ratio of the cross sectional area between the largest section and the test section is 100:1, what is the airspeed at the largest section? Assume incompressible flow and standard sealevel conditions.

2.6   Two flat plates, one having 6 ft span and 3 ft chord, the other having 9 ft span and 6 ft chord, are placed in different air streams. The free stream velocity for the smaller plate is 100 ft/sec. It is found that the total skin friction drag for the two plates is the same. Find the airspeed for the larger plate. Assume laminar flow at standard sealevel conditions.

2.7   An airplane is flying at a density altitude of 15,000 ft at an ambient temperature of –39 degrees F. If the wing chord is 6 ft and the equivalent airspeed is 200 kts, what is the overall Reynolds number of the wing?

2.8   Consider the stabilizer on a light airplane as a flat plate for the purpose of determining its skin friction drag. If the transition Reynolds number is 750,000, what is the total friction drag of a rectangular stabilizer having a span of 6 ft and a chord of 3 ft at a speed of 100 mph? Assume standard sealevel conditions.

2.9   In Example 2.3, compare the boundary layer thickness at a point 5 ft downstream of the leading edge if the boundary layer is completely laminar and turbulent respectively.

2.10   Verify Equations (2.32) and (2.33) by comparison with Table A1.

2.11 REFERENCES FOR CHAPTER 2

1.1   Liepmann, H.W. and Roshko, A.; Elements of Gasdynamics; John Wiley & Sons, Inc., 1963

2.2   Roskam, J.; Airplane Design, Parts II and III, DAR Corporation, 120 East 9th St., Suite 2, Lawrence, Kansas, 66044; 1989.

2.3   Anon.; Jet Transport Performance Methods; Boeing Document D6–1420, 1967, The Boeing Company, Seattle, WA.

2.4   Anon.; Naval Test Pilot School Flight Test Manual: Fixed Wing Performance, Theory and Flight Test Techniques; USNTPS–FTM No. 104, July, 1977, Patuxent River, MD.

2.5   Glauert, H.; The Elements of Aerofoil and Airscrew Theory; Cambridge University Press, Cambridge, U.K., 1959.

2.6   El–Ramly, Z. and Rainbird, W.J.; Flow Survey of the Vortex Wake Behind Wings; Journal of Aircraft, Volume 14, November 1977, pages 1102–1108.

2.7   Donaldson, C. duP. and Belanin, A.J.; Vortex Wakes of Conventional Aircraft; AGARD––AG–204, Advisory Group for Aerospace Research and Development of NATO, May 1975.

2.8   Streeter, V.L. and Wylie, E.B.; Fluid Mechanics; McGraw Hill Publishing Co., Seventh Edition, N.Y., N.Y., 1979.

2.9   Anon.; Pratt & Whitney Aircraft Group, Aeronautical Vest–Pocket Handbook; 16th Edition, June 1977.

2.10   U.S. Standard Atmosphere, 1976.

2.11   Schlichting, H.; Boundary Layer Theory; McGraw Hill Publishing Co., Sixth Edition, N.Y., N.Y., 1968.

2.12   Tani, I.; Boundary Layer Transition; Annual Review of Fluid Mechanics, Volume I, 1969, pages 169–196.

2.13   Hoerner, S.F.; Fluid Dynamic Drag; Chapter 15, Published by the author, 1965.

2.14   Hoak, D.E. et al; USAF Stability and Control Datcom.; USAF Flight Dynamics Laboratory, WPAFB, Ohio, 45433, April 1978.

CHAPTER 3: AIRFOIL THEORY

The objective of this chapter is to provide the reader with fundamental knowledge of and practical limitations to airfoil and associated aerodynamic behavior.

Basic geometric parameters used in describing and defining airfoils are presented in Section 3.1.

A discussion of fundamental aerodynamic force behavior of airfoils, including a corresponding non–dimensional parameter analysis is given in Section 3.2.

Section 3.3 covers the most important aspects of airfoil (= section) lift, drag and pitching moment behavior. Several important mathematical modelling aspects are also discussed.

The total force and moment coefficients (lift, drag and pitching moment) are the result of integrating the airfoil pressure distribution. Section 3.4 presents a discussion of pressure distributions.

The aerodynamic characteristics of airfoils (or sections) are strongly influenced by compressibility (i.e. Mach number) effects. Section 3.5 contains a discussion of the effect of compressibility on airfoil behavior.

The aerodynamic characteristics of airfoils (or sections) are also strongly influenced by viscous (i.e. Reynolds number) effects. Section 3.6 contains a discussion of the effect of Reynolds number on airfoil behavior.

With modern airfoil theory and high speed engineering work stations it has become possible to actually "design" airfoil shapes to meet certain prescribed lift, drag and pitching moment behaviors. A fundamental discussion of the effect of basic geometric airfoil parameters on airfoil performance is included in Section 3.7.

The maximum lift coefficient which airfoil sections can generate are of fundamental importance to wing and tail design. For that reason, a separate discussion on this topic is given in Section 3.8.

3.1 AIRFOIL GEOMETRY

An airfoil or section is a streamlined body which, when set at a suitable angle of attack, produces more lift than drag while also producing a manageable pitching moment. The NACA (National Advisory Committee for Aeronautics and predecessor of NASA*) has developed and tested many series of airfoils many of which are still being used today.

---

* NASA = National Aeronautics and Space Administration

Ref. 3.1 contains a systematic listing of geometric and aerodynamic data for the most important NACA airfoils. The reader should be familiar with the most important geometric features of airfoils. The following geometric definitions refer to Figure 3.1.

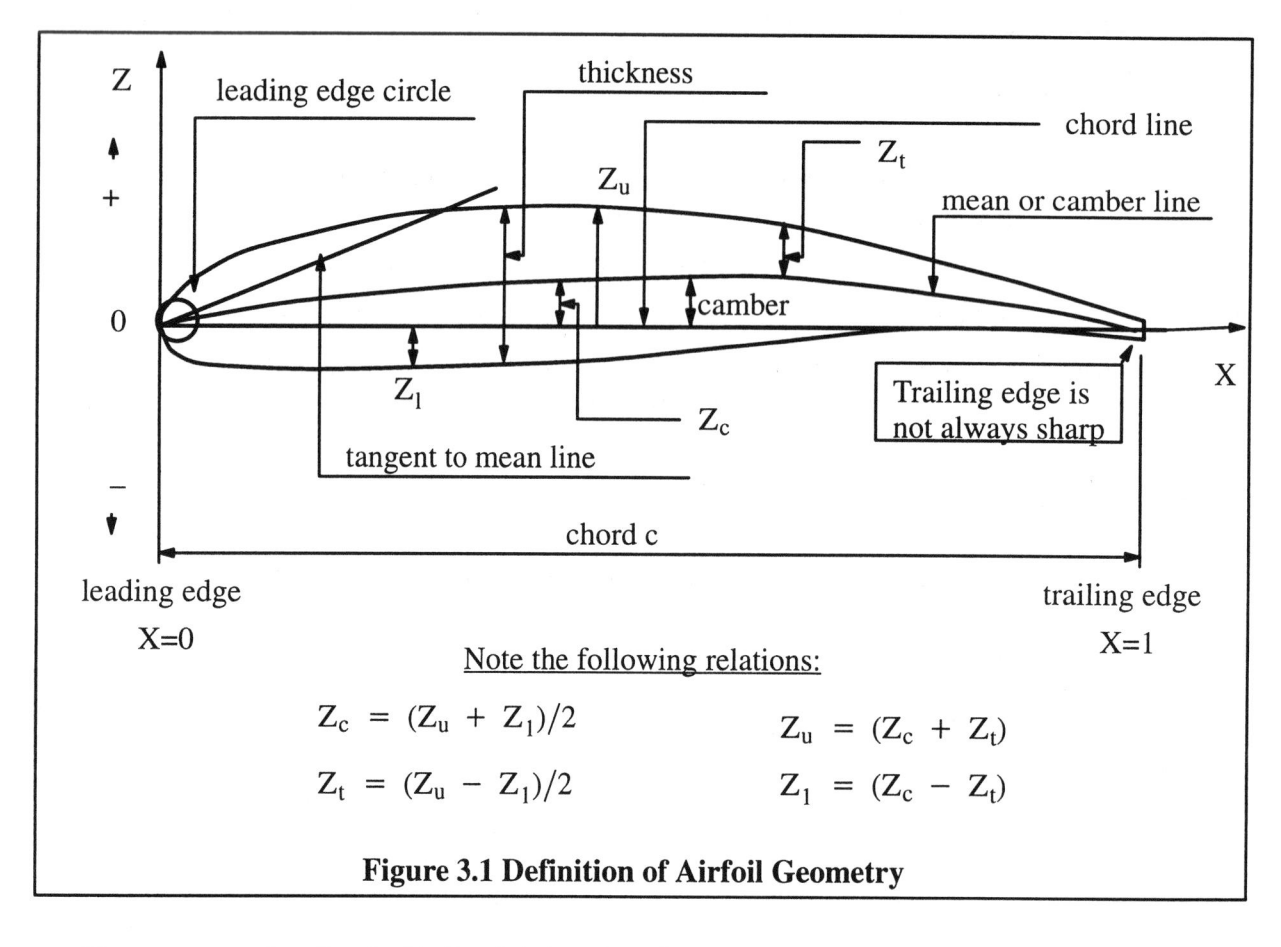

Note the following relations:

$$Z_c = (Z_u + Z_l)/2 \qquad Z_u = (Z_c + Z_t)$$
$$Z_t = (Z_u - Z_l)/2 \qquad Z_l = (Z_c - Z_t)$$

Figure 3.1 Definition of Airfoil Geometry

The <u>mean camber line</u> is the line joining the mid points between the upper and lower surfaces of an airfoil and measured perpendicular to the mean camber line. From an engineering accuracy viewpoint it is usually acceptable to define the camber line with measurements perpendicular to the chord line.

The <u>chord line</u> is the straight line which joins the end points of the mean camber line.

The <u>thickness</u> is the height of the airfoil measured normal to the chord line. The ratio of the maximum thickness to the chord length is called the thickness ratio.

The <u>camber</u> is the maximum distance of the mean line from the chord line.

The <u>leading edge radius</u> is the radius of a circle which is tangent to the upper and lower surfaces. The center of this circle is located on a tangent to the mean line drawn through the leading edge of this line.

3.2 AERODYNAMIC FORCES AND MOMENTS ON AN AIRFOIL

When an airfoil is placed in a moving stream of air, an aerodynamic force acting on the airfoil will be created. Experimentally, this force, F, has been found to depend on six variables:

- air velocity, V
- air density, ϱ
- characteristic area or size, S
- coefficient of viscosity, μ
- speed of sound, V_a
- angle of attack, α

To find the general form of this dependence a dimensional analysis can be used. Since there are six dimensional variables, (F, V, ϱ, S, μ and V_a), three dimensionless parameters can be found in accordance with Buckingham's π – theorem (See: Ref.3.2 p.183 and also Ref. 3.3). By using V, ϱ and S as the repeating variables, the first dimensionless parameter may be written as:

$$\pi_1 = V^a \varrho^b S^d F \tag{3.1}$$

In terms of the fundamental units: length, l, time, t and mass, m, Eqn (3.1) yields:

$$\text{Dim } \pi_1 = \left(\frac{l}{t}\right)^a \left(\frac{m}{l^3}\right)^b \left(l^2\right)^d \left(\frac{ml}{t^2}\right) \tag{3.2}$$

Because π_1 is a dimensionless quantity, equating equal powers of the units l, m and t produces:

$$\begin{aligned} \text{for l:} \quad & 0 = a - 3b + 2d + 1 \\ \text{for m:} \quad & 0 = b + 1 \\ \text{for t:} \quad & 0 = -a - 2 \end{aligned} \tag{3.3}$$

Solving these three equations for a, b and d it is found that: $a = -2$, $b = -1$ and $d = -1$. Therefore:

$$\pi_1 = \frac{F}{\varrho V^2 S} \tag{3.4}$$

Using the same process, the second dimensionless parameter may be written as:

$$\pi_2 = V^a \varrho^b S^d \mu \tag{3.5}$$

$$\text{Dim } \pi_2 = \left(\frac{l}{t}\right)^a \left(\frac{m}{l^3}\right)^b \left(l^2\right)^d \left(\frac{m}{lt}\right) \tag{3.6}$$

Equating powers of like variables results in:

for l : $0 = a - 3b + 2d - 1$

for m : $0 = b + 1$ (3.7)

for t : $0 = -a - 1$

Solving for a, b and d yields: a = –1, b = –1 and d = –1/2 and therefore:

$$\pi_2 = \frac{\mu}{\varrho V S^{1/2}} \tag{3.8}$$

The reader is asked to show that:

$$\pi_3 = \frac{V_a}{V} \tag{3.9}$$

In accordance with the π – theorem, the dimensionless relation which governs the aerodynamic forces can now be written as follows:

$$\frac{F}{\varrho V^2 S} = f'\left(\frac{\mu}{\varrho V S^{1/2}}, \frac{V_a}{V}, \alpha\right) \tag{3.10}$$

It is customary to replace $S^{1/2}$ (which has a linear dimension) by c, the chord of the airfoil. It can now be seen, that Eqn (3.10) contains two important dimensionless parameters:

$$M = \text{Mach number} = \frac{V}{V_a} \tag{3.11}$$

and

$$R_N = \text{Reynolds number} = \frac{\varrho V c}{\mu} \tag{3.12}$$

It is therefore possible to write:

$$F = \varrho V^2 S\, f(R_N, M, \alpha) \tag{3.13}$$

For airfoils, let S = c x 1, or the area per unit span. It is convenient to introduce the following dimensionless expression for the force component (i.e. the lift force) perpendicular to the free stream:

$$f(R_N, M, \alpha) = \frac{c_l}{2} \tag{3.14}$$

It is now possible to reduce Eqn (3.13) to its conventional form:

Airfoil Theory

$$l = c_l \frac{1}{2}\varrho V^2 c = c_l \bar{q} c \tag{3.15}$$

where: c_l is the so-called sectional lift coefficient

$\bar{q}$ is the dynamic pressure

In a similar manner, the force component in the free stream direction, i.e. the drag force can be written as:

$$d = c_d \frac{1}{2}\varrho V^2 c = c_d \bar{q} c \tag{3.16}$$

where: c_d is the so-called sectional drag coefficient

This derivation clearly establishes the fact, that the lift and drag coefficients, c_l and c_d are functions of α, R_N and M. By using a similar process a pitching moment coefficient, c_m can be defined so that the sectional pitching moment can be computed from:

$$m = c_m \frac{1}{2}\varrho V^2 c^2 = c_m \bar{q} c^2 \tag{3.17}$$

where: c_m is the so-called sectional pitching moment coefficient. It is defined as positive if the moment is nose-up.

Figure 3.2 shows how l, d, m and F act on a typical airfoil at some angle of attack, α. Note that l is the lift force in lbs/ft, d is the drag force in lbs/ft and m is the pitching moment in ft–lbs/ft. This so because the span was taken to be unity.

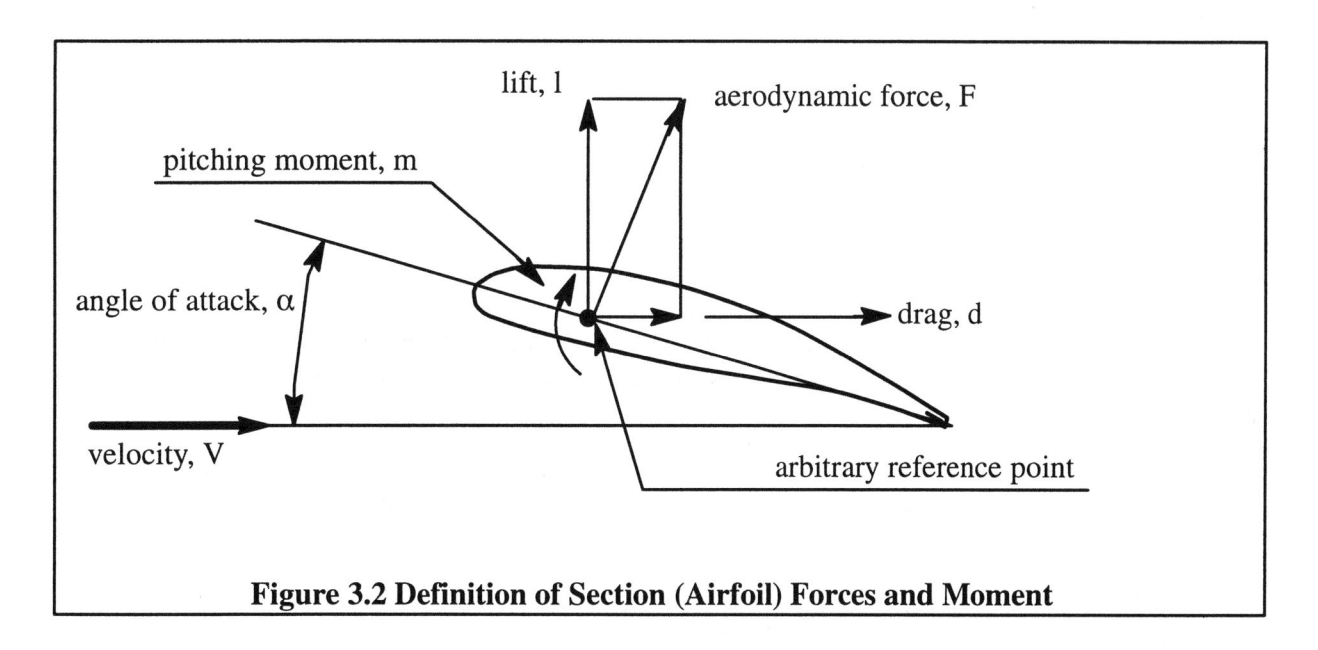

Figure 3.2 Definition of Section (Airfoil) Forces and Moment

3.3 IMPORTANT AIRFOIL CHARACTERISTICS

In the aerodynamic analysis of airplanes, the following airfoil relationships are of fundamental importance:

3.3.1 Lift Curve: c_l versus α

3.3.2 Drag Polar: c_l versus c_d

3.3.3 Pitching Moment Curve: c_m versus α or c_m versus c_l

Examples of these relationships will be discussed next.

3.3.1 LIFT CURVE: c_l VERSUS α

Typical curves of c_l versus α are shown in Figure 3.3a for two airfoils. A generic version of the c_l versus α relationship is shown in Figure 3.3b.

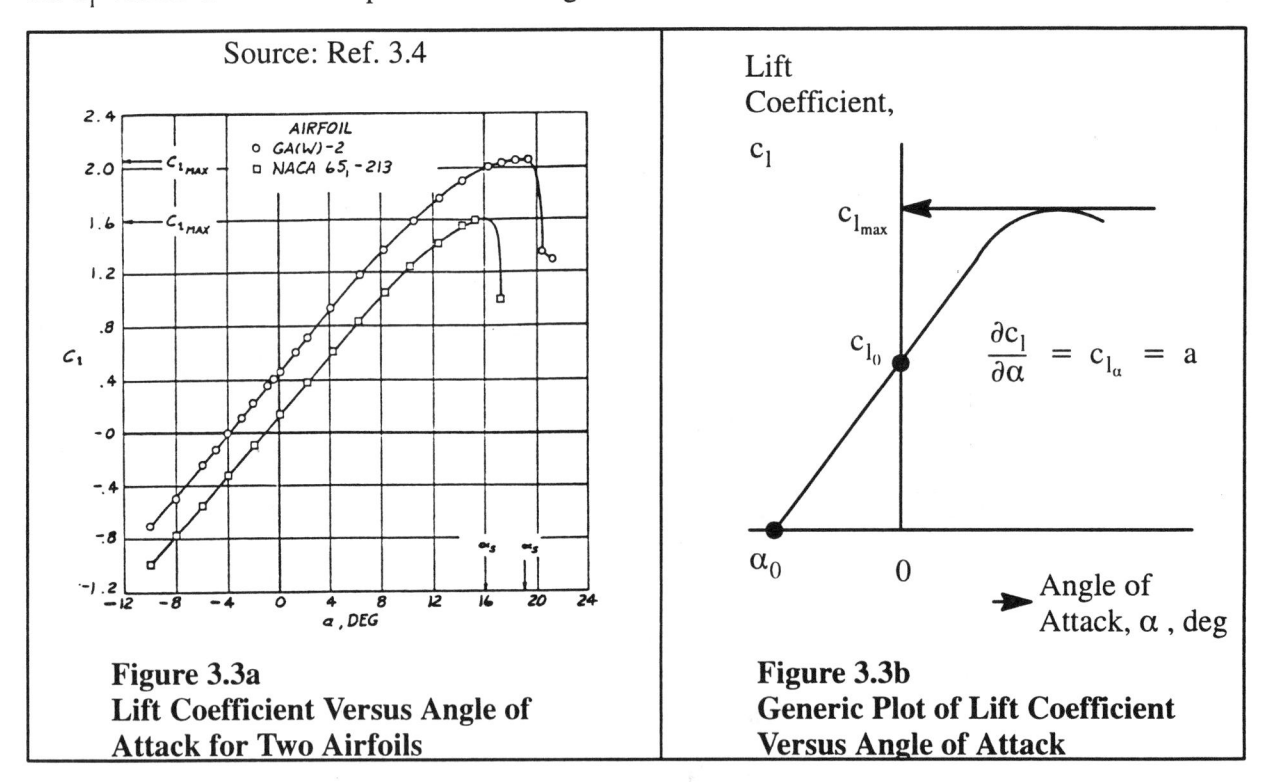

Figure 3.3a
Lift Coefficient Versus Angle of Attack for Two Airfoils

Figure 3.3b
Generic Plot of Lift Coefficient Versus Angle of Attack

The linear part of the lift curve can be mathematically represented by:
$$c_l = a(\alpha - \alpha_0) = c_{l_\alpha}(\alpha - \alpha_0) \tag{3.18}$$

where: a and c_{l_α} are different symbols for the so-called lift-curve-slope, in 1/rad. This quantity has a theoretical value of 2π per radian for very thin airfoils. Table 3.1 shows actual values of airfoil lift curve slope for various airfoils.

α_0 is the angle of attack for zero lift and is typically negative

It is seen in Figures 3.3 that a definite maximum value of c_l is reached in the nonlinear range of the curves. This maximum value is called $c_{l_{max}}$. Typical values for airfoil maximum lift coefficient are also given in Table 3.1. A more detailed discussion of airfoil $c_{l_{max}}$ and the associated so-called stall behavior is presented in Section 3.8.

3.3.2 DRAG POLAR: c_l VERSUS c_d

Typical curves of c_l versus c_d are shown in Figure 3.4a for two airfoils. The magnitude of c_d is frequently expressed in drag counts. One drag count is equivalent to $c_d = 0.0001$. A generic version of the c_l versus c_d relationship is shown in Figure 3.4b.

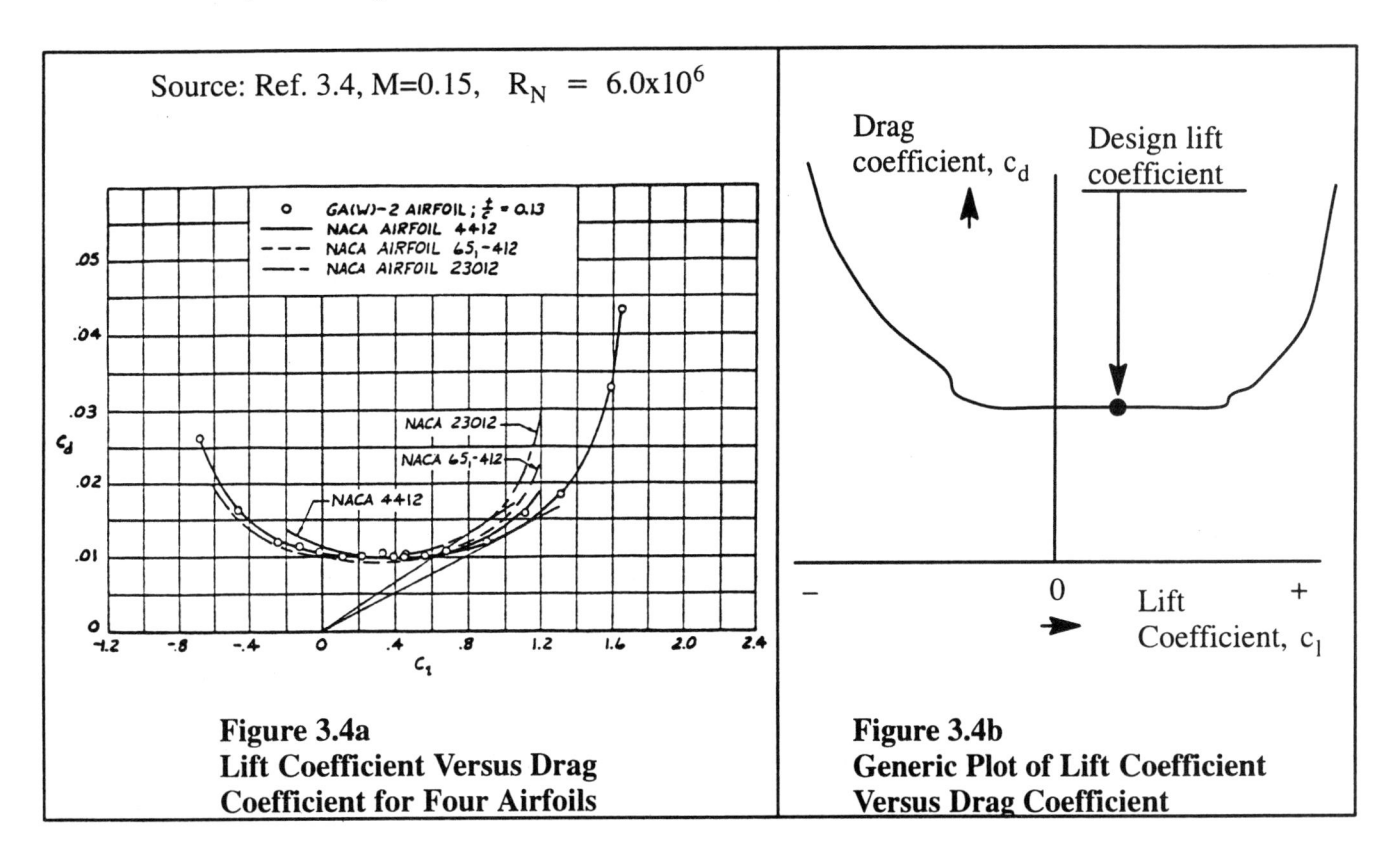

Figure 3.4a
Lift Coefficient Versus Drag Coefficient for Four Airfoils

Figure 3.4b
Generic Plot of Lift Coefficient Versus Drag Coefficient

A very important parameter which represents the aerodynamic efficiency of an airfoil is the so-called lift-to-drag ratio, c_l/c_d. The maximum value of this parameter is obtained by drawing the tangent from the origin to the c_l versus c_d curve. The lift coefficient at which this maximum value of c_l/c_d occurs is another important design parameter. How these parameters affect the performance of airplanes is discussed in Chapters 8–12.

Table 3.1 Experimental, Low Speed NACA Airfoil Data for Smooth Leading Edges

(Note: Data reproduced from Reference 3.1 for $R_N = 9 \times 10^6$)

Airfoil	α_0 (deg)	$\bar{c}_{m_0}$	c_{l_α} (1/deg)	$\bar{x}_{ac}$	$\alpha_{c_{l_{max}}}$ (deg)	$c_{l_{max}}$	α^* (deg)
0006	0	0	0.108	0.250	9.0	0.92	9.0
0009	0	0	0.109	0.250	13.4	1.32	11.4
1408	−0.8	−0.023	0.109	0.250	14.0	1.35	10.0
1410	−1.0	−0.020	0.108	0.247	14.3	1.50	11.0
1412	−1.1	−0.025	0.108	0.252	15.2	1.58	12.0
2412	−2.0	−0.047	0.105	0.247	16.8	1.68	9.5
2415	−2.0	−0.049	0.106	0.246	16.4	1.63	10.0
2418	−2.3	−0.050	0.103	0.241	14.0	1.47	10.0
2421	−1.8	−0.040	0.103	0.241	16.0	1.47	8.0
2424	−1.8	−0.040	0.098	0.231	16.0	1.29	8.4
23012	−1.4	−0.014	0.107	0.247	18.0	1.79	12.0
23015	−1.0	−0.007	0.107	0.243	18.0	1.72	10.0
23018	−1.2	−0.005	0.104	0.243	16.0	1.60	11.8
23021	−1.2	0	0.103	0.238	15.0	1.50	10.3
23024	−0.8	0	0.097	0.231	15.0	1.40	9.7
64–006	0	0	0.109	0.256	9.0	0.80	7.2
64–009	0	0	0.110	0.262	11.0	1.17	10.0
64_1–012	0	0	0.111	0.262	14.5	1.45	11.0
64_1–212	−1.3	−0.027	0.113	0.262	15.0	1.55	11.0
64_1–412	−2.6	−0.065	0.112	0.267	15.0	1.67	8.0
64–206	−1.0	−0.040	0.110	0.253	12.0	1.03	8.0
64–209	−1.5	−0.040	0.107	0.261	13.0	1.40	8.9
64–210	−1.6	−0.040	0.110	0.258	14.0	1.45	10.8
64A010	0	0	0.110	0.253	12.0	1.23	10.0
64A210	−1.5	−0.040	0.105	0.251	13.0	1.44	10.0
64A410	−3.0	−0.080	0.100	0.254	15.0	1.61	10.0
64_1A212	−2.0	−0.040	0.100	0.252	14.0	1.54	11.0
64_2A215	−2.0	−0.040	0.095	0.252	15.0	1.50	12.0

Note: For definition of symbols, see the list of Symbols

3.3.3 PITCHING MOMENT CURVE: c_m VERSUS c_l or c_m VERSUS α

Figure 3.5a shows typical airfoil data for c_m versus c_l and c_m versus α.

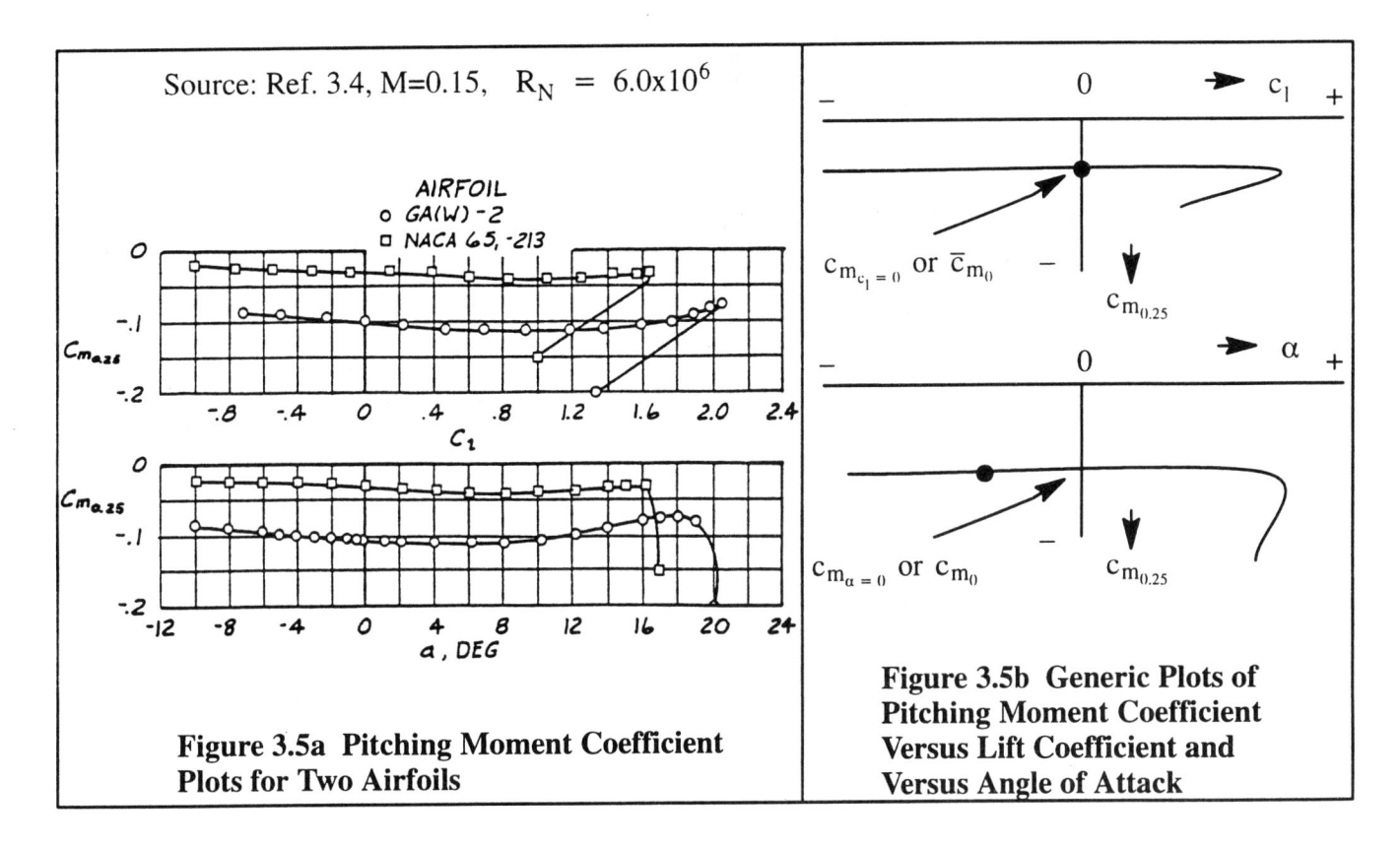

Figure 3.5a Pitching Moment Coefficient Plots for Two Airfoils

Figure 3.5b Generic Plots of Pitching Moment Coefficient Versus Lift Coefficient and Versus Angle of Attack

The magnitude of the pitching moment coefficient, c_m depends on the location of the moment reference center. This moment reference center is normally identified in a subscript to c_m.

In Figure 3.5a the moment reference center is the quarter chord point, identified in the subscript as 0.25c or simply 0.25. Generic plots of c_m versus c_l and c_m versus α are shown in Figure 3.5b. Numerical values for the parameter c_{m_0} are given in Table 3.1 for several types of airfoil.

A very important reference point on an airfoil is its so-called aerodynamic center or a.c. The aerodynamic center is defined as that point about which the variation of the pitching moment coefficient with angle of attack is zero. To find the a.c., assume that in some experimental set-up the moment reference center was selected to be a distance x from the leading edge. Figure 3.6 shows the corresponding geometry. Neglecting the moment contribution due to drag it is seen that:

$$c_{m_{ac}}\bar{q}c^2 = c_{m_x}\bar{q}c^2 + c_l\bar{q}c(x_{ac} - x) \tag{3.19}$$

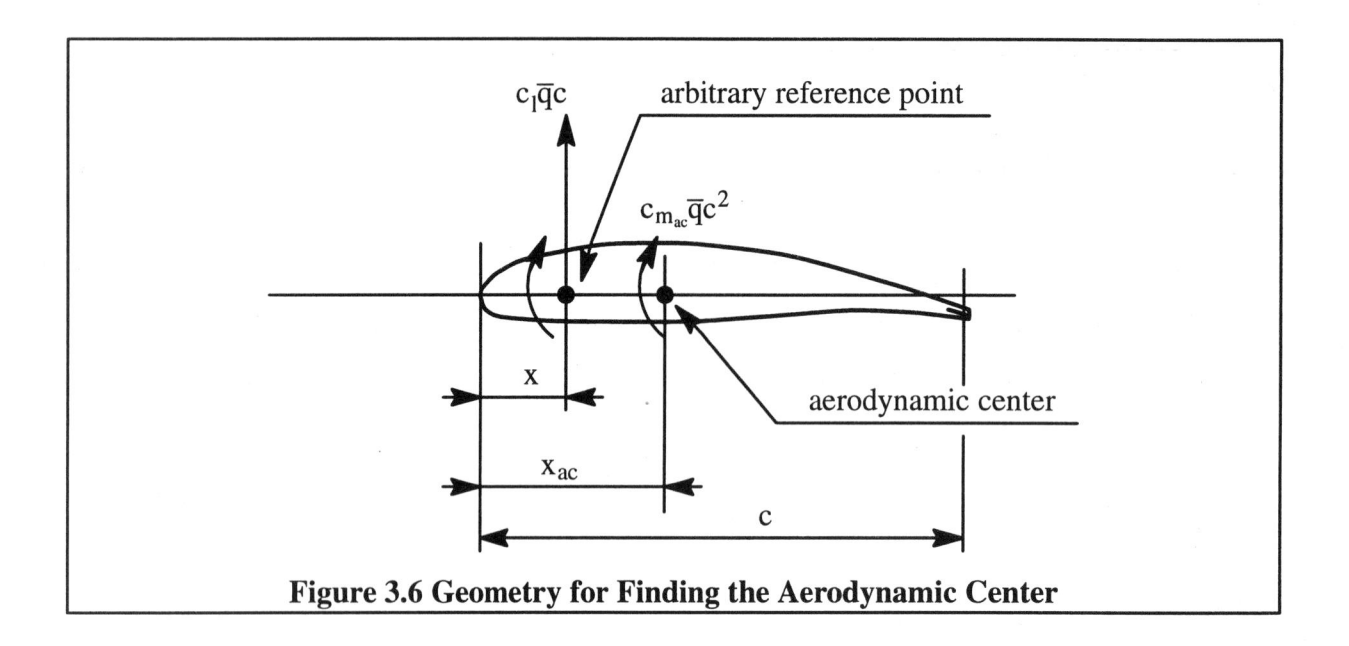

Figure 3.6 Geometry for Finding the Aerodynamic Center

$$c_{m_{ac}} = c_{m_x} + c_l\left(\frac{x_{ac} - x}{c}\right) \tag{3.20}$$

By definition, $c_{m_{ac}}$ is independent of the angle of attack, α, and therefore:

$$\frac{\partial c_{m_{ac}}}{\partial \alpha} = 0 = \frac{\partial c_{m_x}}{\partial c_l} + \frac{\partial c_l}{\partial \alpha}\left(\frac{x_{ac} - x}{c}\right) \tag{3.21}$$

From this it follows that:

$$\frac{x_{ac}}{c} = \frac{x}{c} - \frac{\partial c_{m_x}}{\partial c_l} \tag{3.22}$$

Using experimental data of c_{m_x} versus c_l it is therefore possible to compute the location of the aerodynamic center, x_{ac}. From experimental data taken at low subsonic Mach numbers it is normally found that the aerodynamic center is at the quarter chord point: $\frac{x_{ac}}{c} \approx 0.25$.

3.4 AIRFOIL PRESSURE DISTRIBUTION

The pressure distribution over an airfoil is important for load calculations and for control surface hinge moment calculations. The pressure distribution is normally expressed in terms of the so-called pressure coefficient, c_p, which is defined as:

$$c_p = \frac{p - p_\infty}{\bar{q}_\infty} \tag{3.23}$$

Airfoil Theory

At low speed, use of the incompressible Bernoulli equation (Eqn (2.10), allows Eqn (3.23) to be cast in the following form:

$$c_p = 1 - \left(\frac{V}{V_\infty}\right)^2 \qquad (3.24)$$

Eqn (3.24) shows that in incompressible flow $c_p = 0$ at the stagnation point, where V=0. A typical example of a pressure distribution over an airfoil at low speed is shown in Figure 3.7.

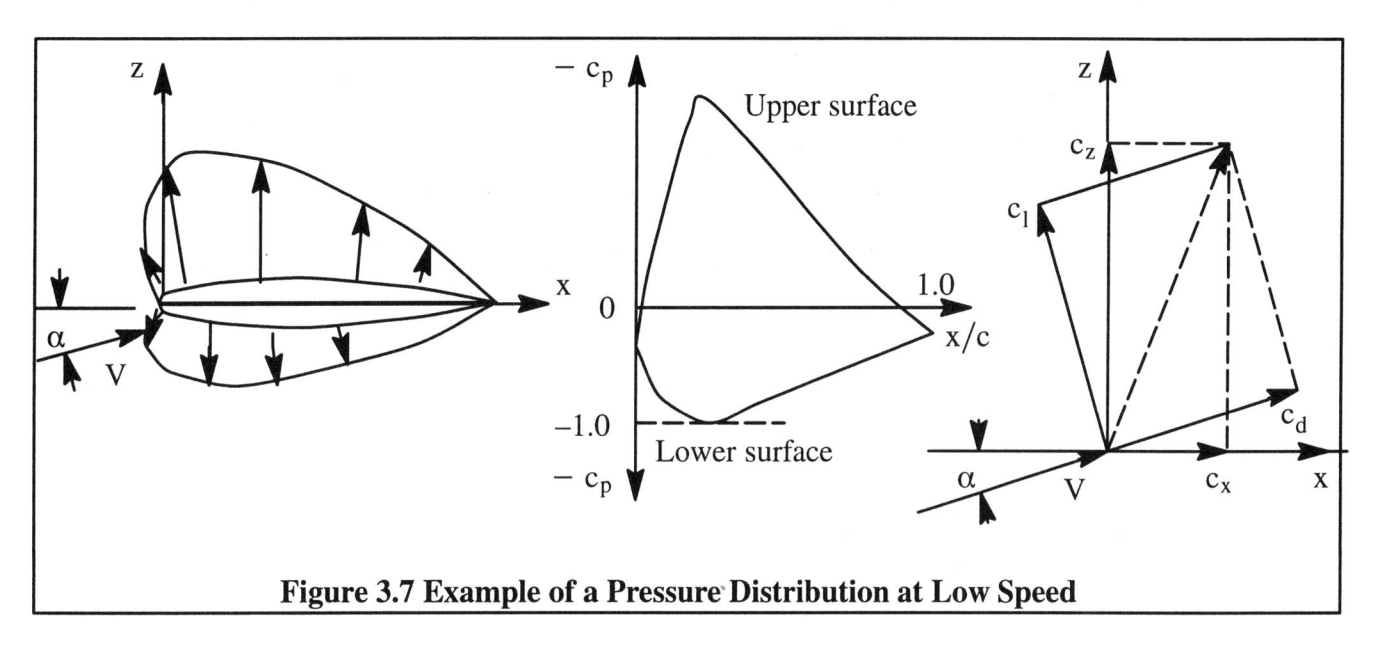

Figure 3.7 Example of a Pressure Distribution at Low Speed

The effect of angle of attack and of control surface deflection on the pressure distribution over an airfoil is shown in Figure 3.8. Note the peak pressure near the leading edge as a result of a change in angle of attack. Also note the peak pressure near the hinge line as a result of a control surface deflection.

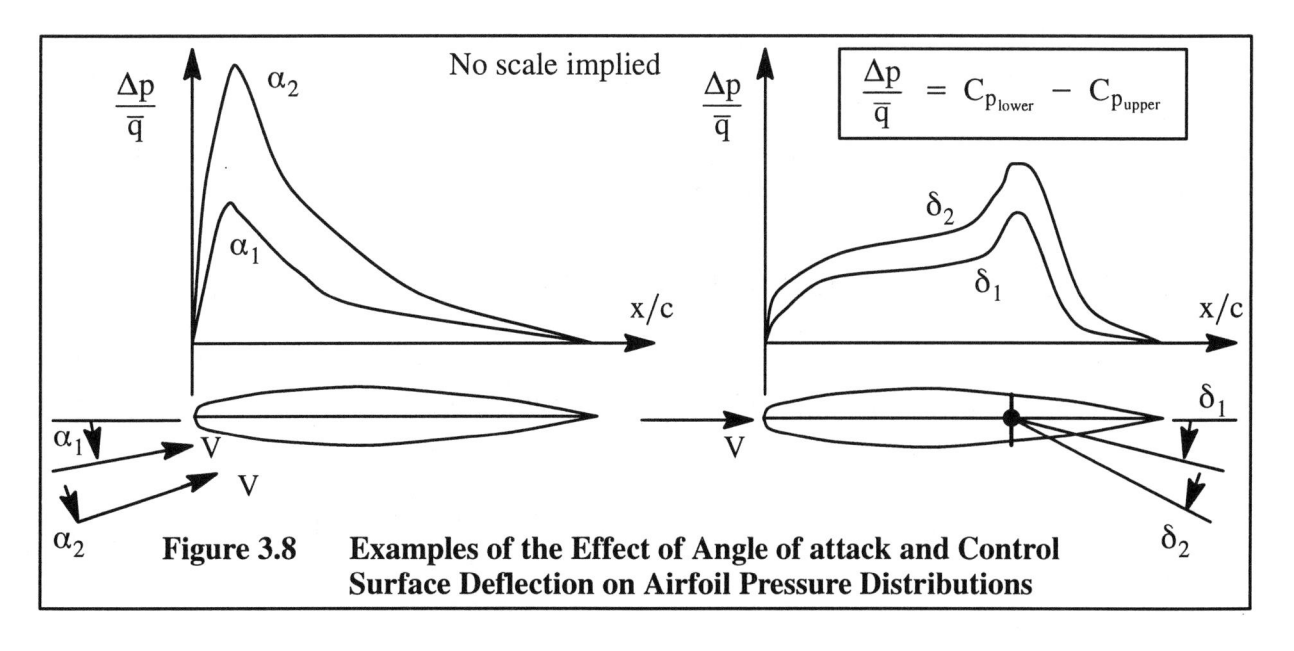

Figure 3.8   Examples of the Effect of Angle of attack and Control Surface Deflection on Airfoil Pressure Distributions

By integrating the pressure distributions, c_l and c_d can be obtained. The force coefficients c_z and c_x now follow from:

$$c_z = \frac{Z}{qc} = \int_0^1 \left(c_{p_{lower}} - c_{p_{upper}}\right) d\left(\frac{x}{c}\right) \qquad (3.25a)$$

$$c_x = \frac{X}{qc} = \int_0^1 \left(c_{p_{lower}} - c_{p_{upper}}\right) d\left(\frac{z}{c}\right) \qquad (3.25b)$$

From these equations it is possible to compute c_l and c_d as:

$$c_l = c_z \cos\alpha - c_x \sin\alpha \qquad (3.26a)$$

$$c_d = c_z \sin\alpha + c_x \cos\alpha \qquad (3.26b)$$

The vectorial relationship between c_l and c_d on the one hand and c_z and c_x on the other hand is also illustrated in Figure 3.7.

Finally, the pitching moment coefficient relative to the leading edge may be computed from:

$$c_{m_{leading\ edge}} = \int_0^1 \left(c_{p_{lower}} - c_{p_{upper}}\right) \frac{x}{c} d\left(\frac{x}{c}\right) \qquad (3.27)$$

3.5 COMPRESSIBILITY EFFECTS

Due to the variation of local velocity over the curved surface of an airfoil, the local Mach number of the flow can vary considerably from that of the free stream. By increasing the free stream Mach number at a given angle of attack, eventually a free stream Mach number is reached for which sonic speed occurs somewhere on the airfoil surface. That free stream Mach number from which this first occurs is defined as the **critical Mach number**.

The pressure coefficient at some point on an airfoil, where sonic velocity is first reached, is called the critical pressure coefficient. In the following an expression will be derived from which the critical pressure coefficient can be determined.

From the compressible Bernoulli equation (2.36) it follows that:

$$\frac{V^2}{2} - \frac{V_\infty^2}{2} = -\left(\frac{1}{\gamma-1}\right)\left(\frac{\gamma p}{\varrho} - \frac{\gamma p_\infty}{\varrho_\infty}\right) = \left\{\frac{\gamma p_\infty}{\varrho_\infty(\gamma-1)}\right\}\left(\frac{p\varrho_\infty}{\varrho p_\infty} - 1\right) =$$

$$= -\left(\frac{\gamma}{\gamma-1}\right)\frac{p_\infty}{\varrho_\infty}\left\{\frac{p}{p_\infty}\left(\frac{p}{p_\infty}\right)^{\frac{-1}{\gamma}} - 1\right\} =$$

$$= -\left(\frac{\gamma}{\gamma-1}\right)\frac{p_\infty}{\varrho_\infty}\left\{\left(\frac{p}{p_\infty}\right)^{\frac{\gamma-1}{\gamma}} - 1\right\} = \left(\frac{V_{a_\infty}^2}{\gamma-1}\right)\left\{\left(\frac{p}{p_\infty}\right)^{\frac{\gamma-1}{\gamma}} - 1\right\} \qquad (3.28)$$

From Eqn (3.23) it follows that:

$$\frac{p}{p_\infty} = 1 + \frac{c_p \bar{q}_\infty}{p_\infty} = 1 + \frac{\gamma c_p M_\infty^2}{2} \qquad (3.29)$$

Substitution into Eqn (3.28) and rearranging now yields:

$$V^2 = V_\infty^2 - \left(\frac{2V_{a_\infty}^2}{\gamma-1}\right)\left\{\left(1 + \frac{\gamma c_p M_\infty^2}{2}\right)^{\frac{\gamma-1}{\gamma}} - 1\right\} \qquad (3.30)$$

At the critical Mach number, $V = V_a$, the speed of sound. With the help of Eqn (2.31), the latter may be expressed as:

$$V_a^2 = \frac{\gamma p}{\varrho} = \left(\frac{\gamma p_\infty}{\varrho_\infty}\right)\frac{p\varrho_\infty}{p_\infty \varrho} = V_{a_\infty}^2\left(\frac{p}{p_\infty}\right)^{\frac{\gamma-1}{\gamma}} = V_{a_\infty}^2\left(1 + \frac{\gamma c_p M_\infty^2}{2}\right)^{\frac{\gamma-1}{\gamma}} \qquad (3.31)$$

At the critical condition, Eqns (3.30) and (3.31) are equated to each other and solved for the critical pressure coefficient. The result is:

$$c_{p_{crit}} = \frac{2}{\gamma M_\infty^2}\left[\left\{\frac{M_\infty^2(\gamma-1) + 2}{\gamma+1}\right\}^{\frac{\gamma}{\gamma-1}} - 1\right] \qquad (3.32)$$

Eqn (3.32) represents the value of the pressure coefficient, c_p, at that point on the airfoil where sonic velocity is first reached for a given free stream Mach number, M_∞. This expression can also be used to determine the critical Mach number of an airfoil at a given angle of attack, α. To do this, assume that at $M_\infty = 0$, the most negative pressure coefficient on the airfoil is given by c_{p_0}. According to Glauert (Ref. 3.1, page 256), the pressure coefficient at the same point and for the same α but for another M_∞ is approximately given by:

$$c_p = \frac{c_{p_0}}{\sqrt{1 - M_\infty^2}} \qquad (3.33)$$

where the factor, $\sqrt{1 - M_\infty^2}$, is known as the Prandtl–Glauert Transformation Factor. A more accurate formula was derived by Von Kármán and Tsien (Ref.3.1, page 258) as:

$$c_p = \frac{c_{p_0}}{\sqrt{1 - M_\infty^2} + \dfrac{C_{p_0}}{2\left(\sqrt{1 - M_\infty^2} + 1\right)}} \tag{3.34}$$

By plotting Eqn (3.32) and either Eqn (3.33) or (3.34) as a function of M_∞ the critical Mach number is found at the intersection of the curves. Figure 3.9 illustrates this.

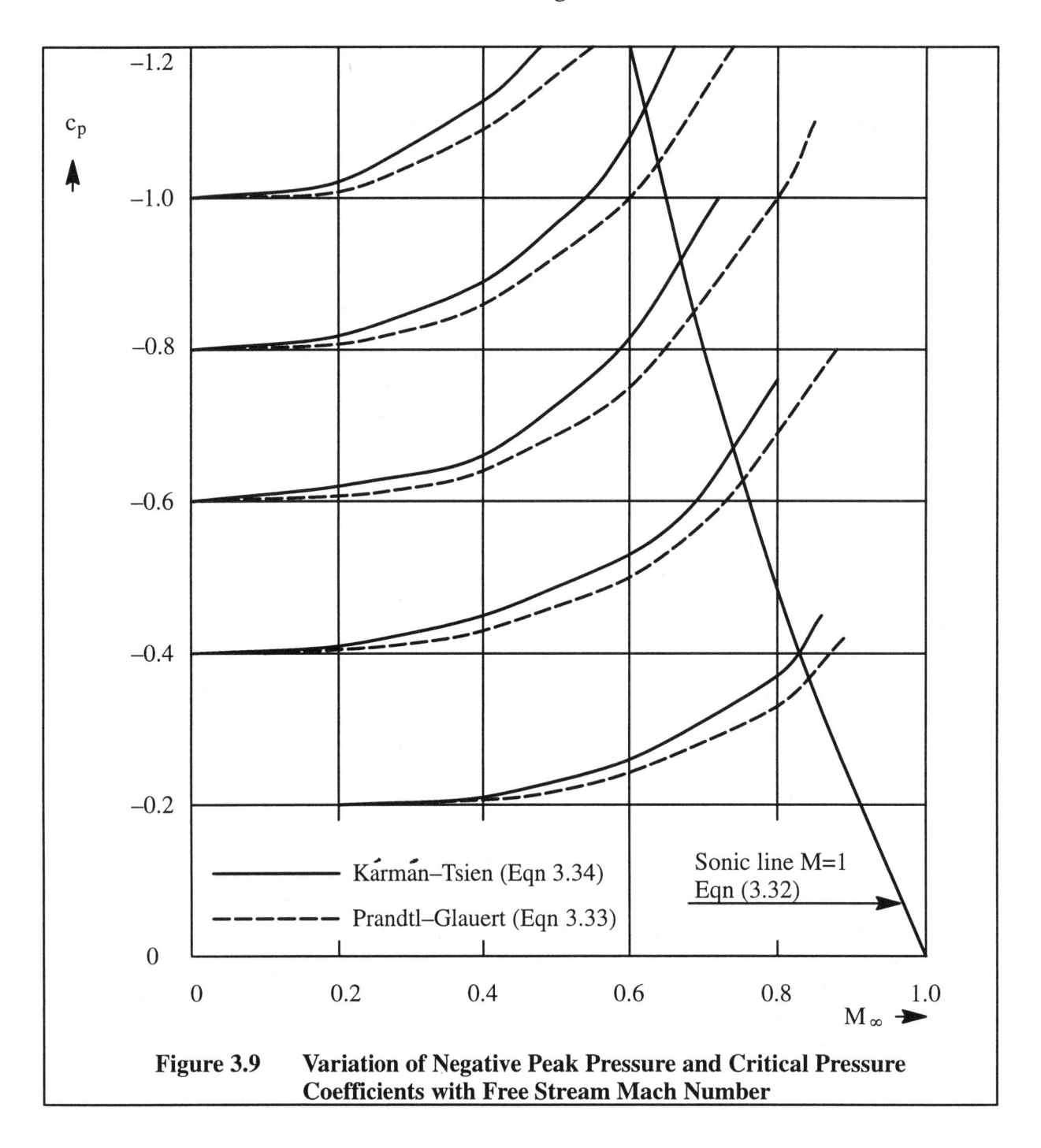

Figure 3.9   **Variation of Negative Peak Pressure and Critical Pressure Coefficients with Free Stream Mach Number**

Because the lift coefficient of an airfoil can be found by integrating c_p around the airfoil, it may be seen that, from the Prandtl–Glauert formula (Eqn (3.33):

$$c_l = \frac{c_{l_0}}{\sqrt{1 - M_\infty^2}} \qquad (3.35)$$

By the same line of reasoning the airfoil lift curve slope, c_{l_α} can be estimated from:

$$c_{l_\alpha} = \frac{c_{l_{\alpha_{M=0}}}}{\sqrt{1 - M_\infty^2}} \quad \text{or}: \quad a = \frac{a_0}{\sqrt{1 - M_\infty^2}} \qquad (3.36)$$

The Kármán–Tsien and Prandtl–Glauert equations are valid only for subsonic flow conditions below the critical Mach number, M_{crit}. Because of the increase in lift curve slope, the lift–versus–angle–of–attack curves of airfoils at different subsonic Mach numbers are as shown in Figure 3.10. The maximum value of the lift coefficient can, at first, either increase or decrease with Mach number. However, eventually, the maximum lift coefficient will decrease with increasing Mach number. The reason is shock–induced boundary layer separation.

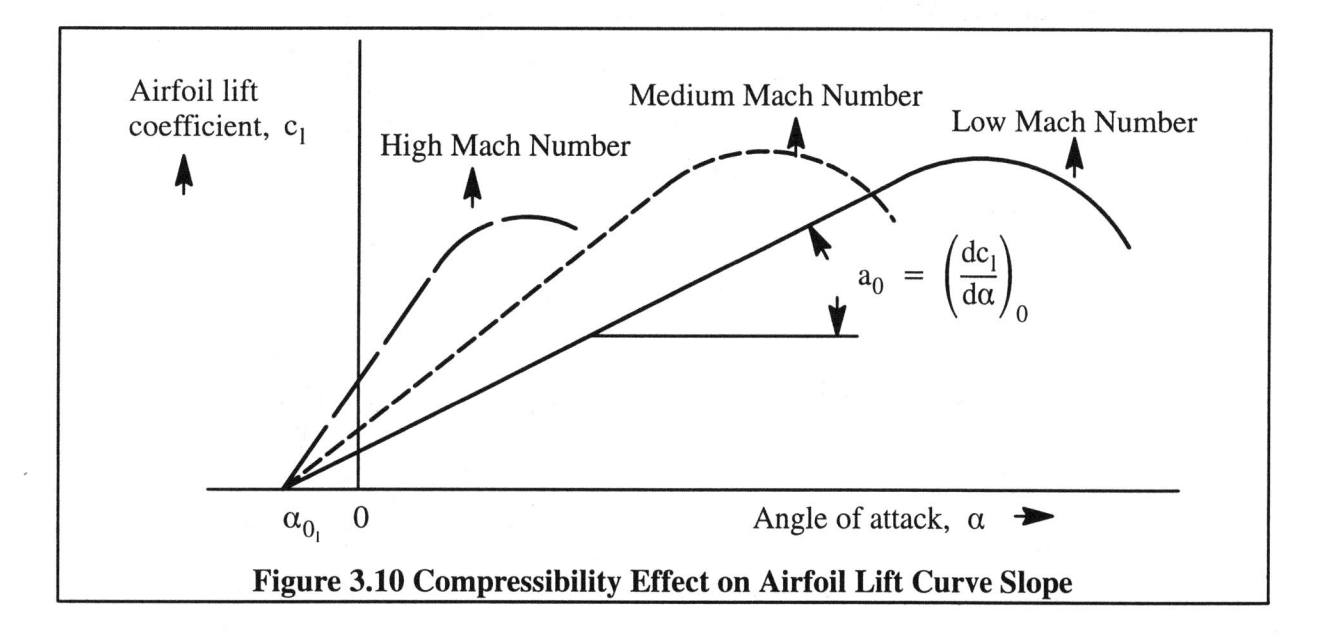

Figure 3.10 Compressibility Effect on Airfoil Lift Curve Slope

With further increase in free stream Mach number beyond the critical value, a supersonic region on the upper surface will appear. Eventually, a strong enough shock wave will terminate the supersonic region and may even separate the boundary layer. This is illustrated in Figure 3.11. Due to shock induced flow separation, the drag coefficient will increase sharply. Figure 3.12 shows the drag rise of an entire airplane due to these phenomena. Observe that the zero–lift drag coefficient rises rapidly in the high subsonic Mach number range. This drag rise is referred to as drag divergence. There are two definitions used to characterize the free stream Mach number where the drag rise begins: one due to Douglas, the other due to Boeing.

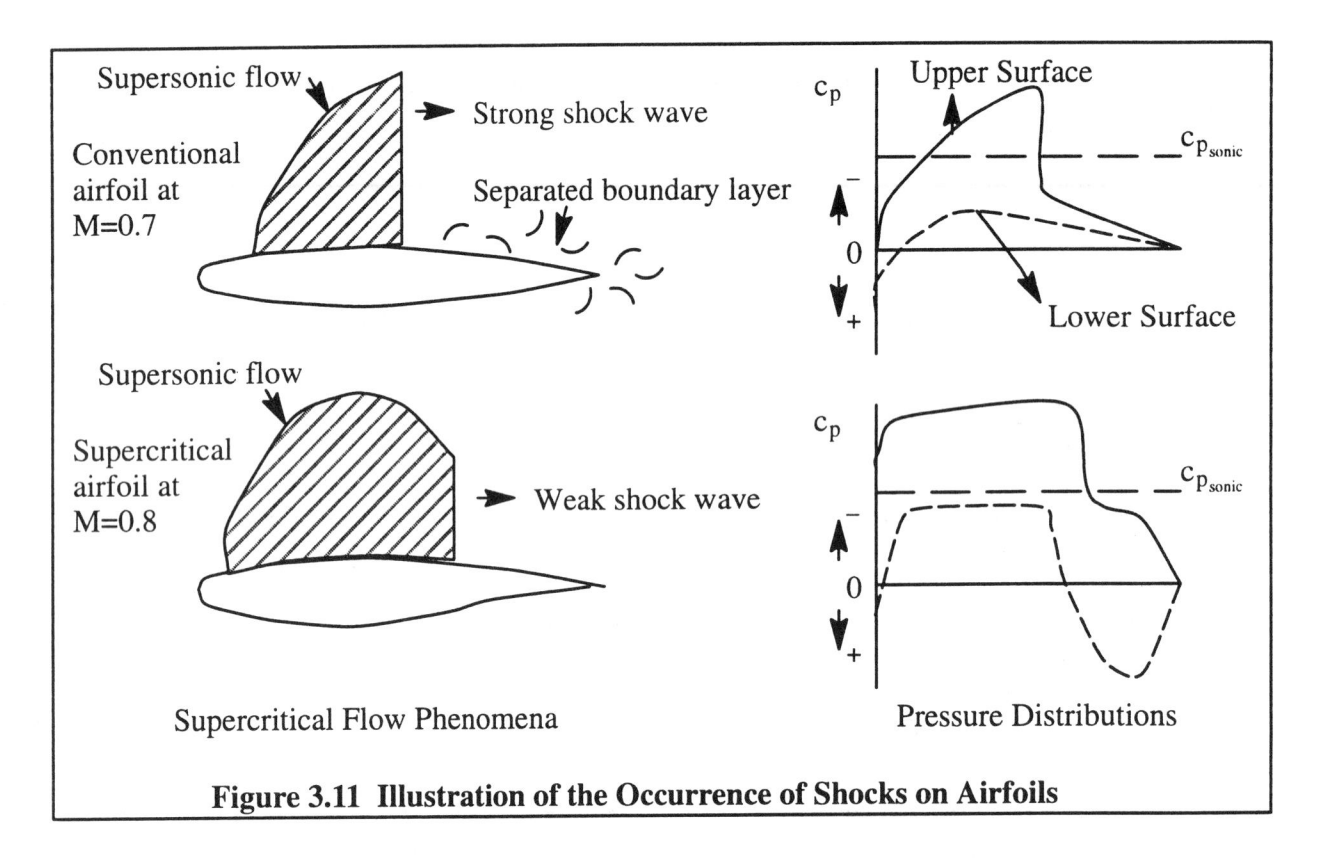

Airfoil Theory

Figure 3.11 Illustration of the Occurrence of Shocks on Airfoils

In the Douglas definition, the free stream Mach number at which $\frac{\partial c_d}{\partial M} = 0.1$ is defined as the drag divergence Mach number, M_{dd}. In the Boeing definition, the free stream Mach number for which the drag rise first reaches a value of $\Delta c_d = 0.0020$ above the incompressible level, is defined as the drag divergence Mach number. Both definitions are also illustrated in Figure 3.12. In practice, these definitions produce very similar results.

Airfoils designed for transonic cruise applications should have as high a drag divergence Mach number as possible. Examples of actual dragrise data on some airfoils are shown in Figure 3.13.

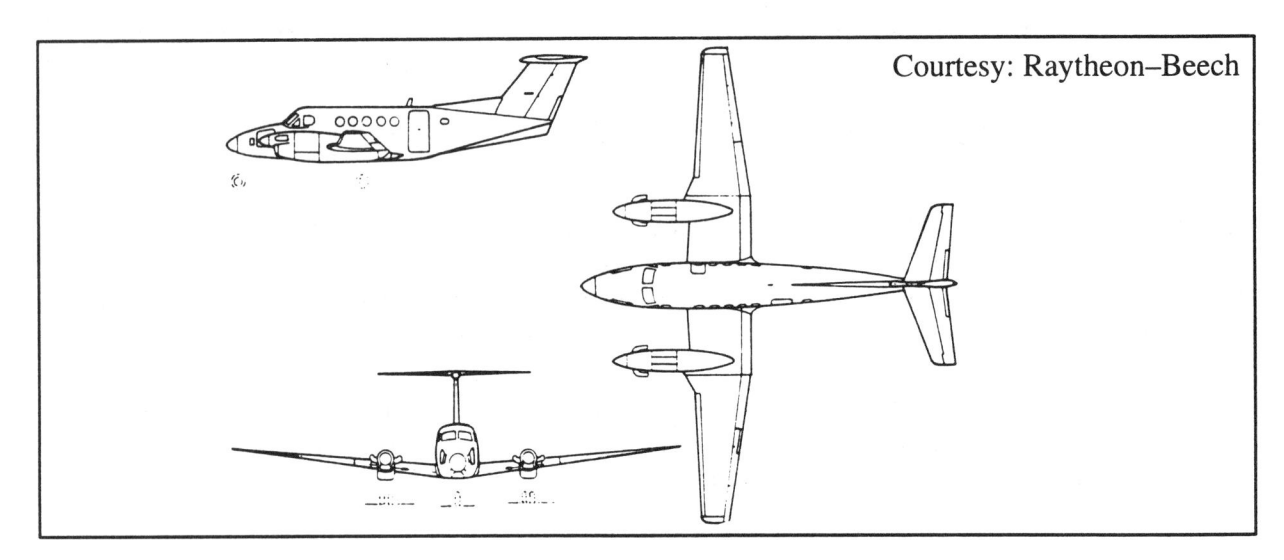

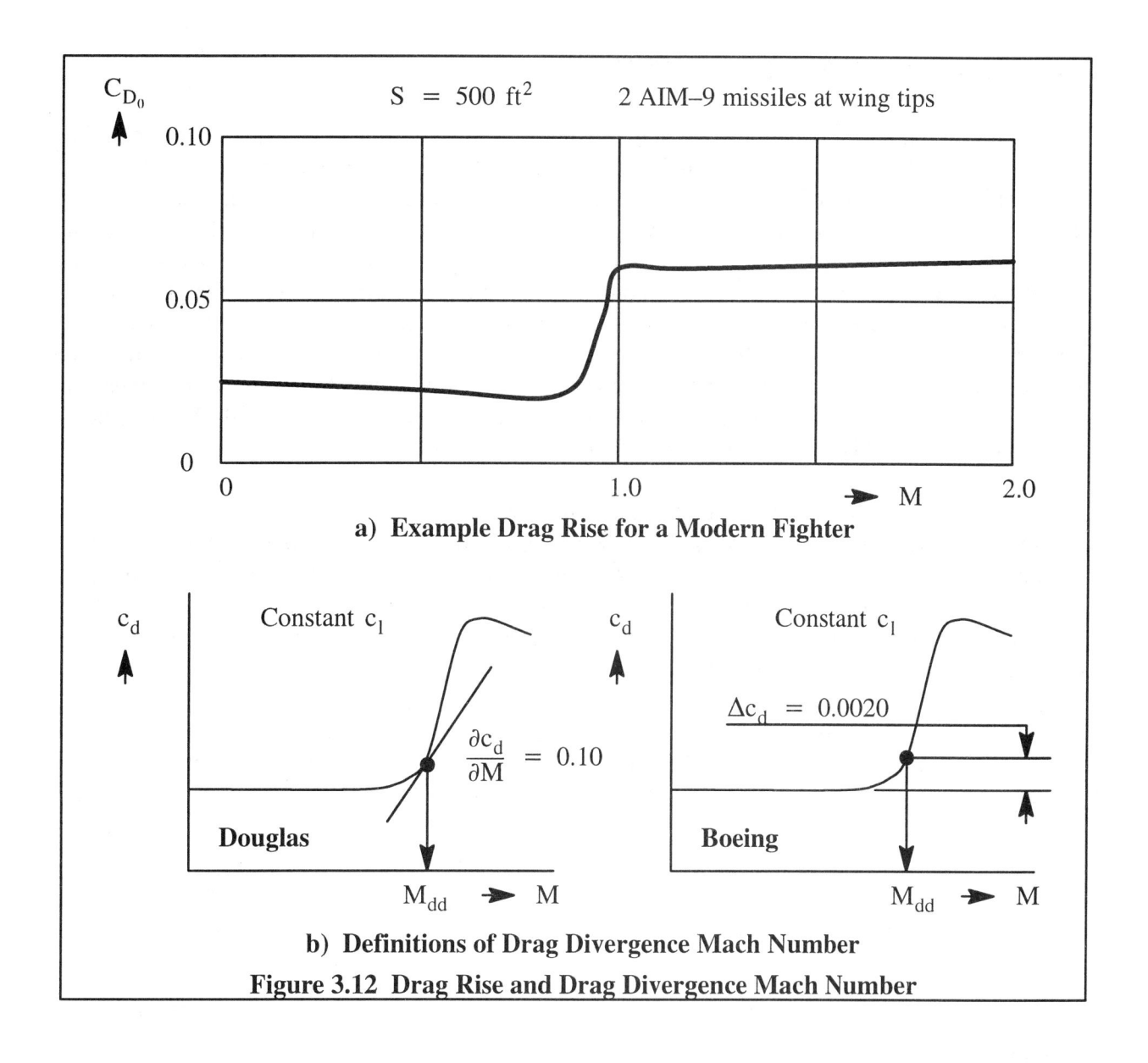

Airfoil Theory

a) Example Drag Rise for a Modern Fighter

$S = 500 \text{ ft}^2$ 2 AIM–9 missiles at wing tips

Douglas: $\frac{\partial c_d}{\partial M} = 0.10$

Boeing: $\Delta c_d = 0.0020$

b) Definitions of Drag Divergence Mach Number

Figure 3.12 Drag Rise and Drag Divergence Mach Number

$c_l = 0.7$ $t/c = 0.10$

Conventional
Early Supercritical
Refined Supercritical

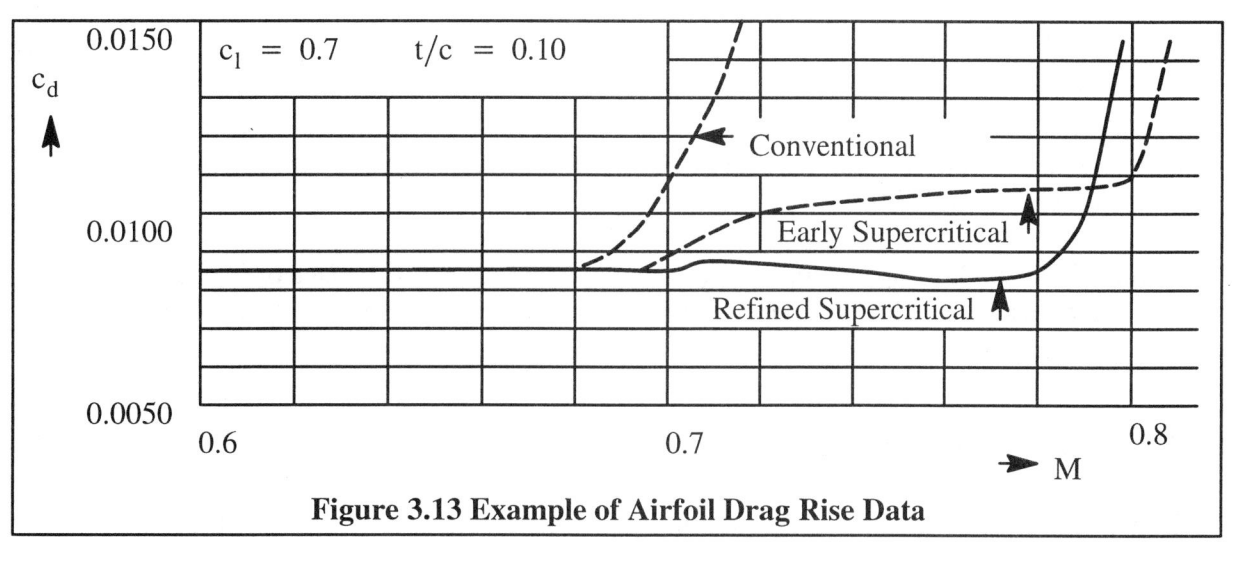

Figure 3.13 Example of Airfoil Drag Rise Data

3.6 REYNOLDS NUMBER EFFECTS

Due to the large difference in Reynolds number between windtunnel and full scale flight, the boundary layer characteristics of the model will not correctly simulate those of the full scale airplane. This will create some obvious differences in the aerodynamic forces. Whereas the boundary layer on a full scale wing is frequently mostly turbulent, extensive regions of laminar flow may exist on a windtunnel model, particularly at low model Reynolds numbers. The most noticeable effect of Reynolds number is on the drag level and on the maximum lift coefficient. Therefore, it is essential to simulate the full scale fully turbulent condition on the model surface before the model data can be extrapolated to full scale flight. The standard practice is to use a so–called trip–strip located at 5% to 10% of the chord. Such a trip–strip consists of a narrow band of distributed roughness of known grit size. Figure 3.14 illustrates an example trip–strip on a windtunnel model. As a practical guide in determining the location and the grit–size of the trip–strip, Reference 3.5 can be used.

Experiments in subsonic flow indicate that if narrow, sparsely distributed, bands of roughness are used, the measured minimum drag is nearly unchanged for any roughness height which is less than the boundary layer thickness (Ref. 3.6). This implies that a proper choice of roughness grit can provide essentially a zero drag penalty due to roughness itself as long as the transition occurs. For the transition to occur, the roughness size should be such that the windtunnel Reynolds number, based on the roughness height, is at least 600.

Once the boundary layer characteristics have been correctly simulated, the Reynolds number correction for drag can be applied as follows. The minimum profile drag coefficient can be expressed as:

$$c_{d_{min}} = (S.F.) \times 2c_f \tag{3.37}$$

where: c_f is the skin friction coefficient of a flat plate with the same transition location as the airfoil.
   S.F. is the shape factor which represents the thickness and the associated viscosity effects.

For a fully turbulent boundary layer, c_f may be calculated from Schlichting's formula, given in Eqn (2.97) as:

$$c_f = \frac{0.455}{\left(\log_{10} R_N\right)^{2.58}} \tag{3.38}$$

Note that this formula is valid only for low Mach numbers. A formula which applies at all Mach numbers is given as Eqn (2.99).

Several forms of the shape factor are in existence. Hoerner (Ref. 3.6, page 6–6) has shown, from a collection of early data, that the shape factor for sections with maximum thickness at 30% chord can be represented empirically by:

$$S.F. = 1 + 2\left(\frac{t}{c}\right) + 60\left(\frac{t}{c}\right)^4 \tag{3.39}$$

It therefore follows that the full scale minimum profile drag coefficient can be obtained from:

Courtesy: Rinaldo Piaggio

Figure 3.14 Example of a Wind–tunnel Model with a Trip Strip

$$c_{d_{min}}(\text{full scale}) = c_{d_{min}}(\text{model}) - \Delta c_{d_{min}} \tag{3.40}$$

where:

$$\Delta c_{d_{min}} = 2(S.F.)\{c_f(\text{model}) - c_f(\text{full scale})\} \tag{3.41}$$

There still do not exist accurate theoretical methods for correcting the maximum lift coefficient from tunnel data to full scale. Jacobs and Sherman have shown some experimental results for the Reynolds number effect on $c_{l_{max}}$ for several NACA airfoils (Ref. 3.8). Additional data can be found in Ref. 3.9. Most aircraft manufacturers have their own in–house correction procedures for extrapolating tunnel $c_{l_{max}}$ data to full scale. Such procedures are then based upon their experience obtained in comparing model and airplane data.

3.7 DESIGN OF AIRFOILS

Because lifting surfaces (such as wings, tails, canards and pylons) can be thought of as spanwise arrangements of airfoils, the basic characteristics of airfoils have a major effect on the behavior of lifting surfaces. It is therefore important to be aware of those airfoil characteristics which have the potential of being 'driving' factors in airplane lift, drag, and pitching moment.

To design an airfoil for any specific requirement involving lift, drag or pitching moment, several effects of airfoil geometry on airfoil aerodynamics should be understood. It has been found that the most important geometric parameters are:

1) maximum thickness ratio, $(t/c)_{max}$

2) shape of the mean line (also referred to as camber). If the mean line is a straight line, the airfoil is said to be symmetrical.

3) leading edge shape or Δy parameter and leading edge radius (l.e.r.)

4) trailing edge angle, ϕ_{TE}

Figure 3.15 provides a geometric interpretation for these parameters, most of which were also defined in Figure 3.1. The reader should consult Ref. 3.1 for a detailed discussion of airfoil parameters and airfoil characteristics. Ref. 3.1 also contains a large body of experimental data on a variety of NACA airfoils (NACA = National Advisory Committee on Aeronautics, predecessor of NASA, the National Aeronautics and Space Administration). In addition, this reference contains explanations for the numerical designations used with all NACA airfoils. Table 3.2 defines the most important NACA airfoil designations.

It is noted that the NACA 6–series airfoils were designed to have mean camber lines which produce a near uniform chordwise loading from the leading edge to a point x/c = a, and a linearly decreasing load from this point to the trailing edge. Any time this condition is met, the corresponding a–value is given after the airfoil designation.

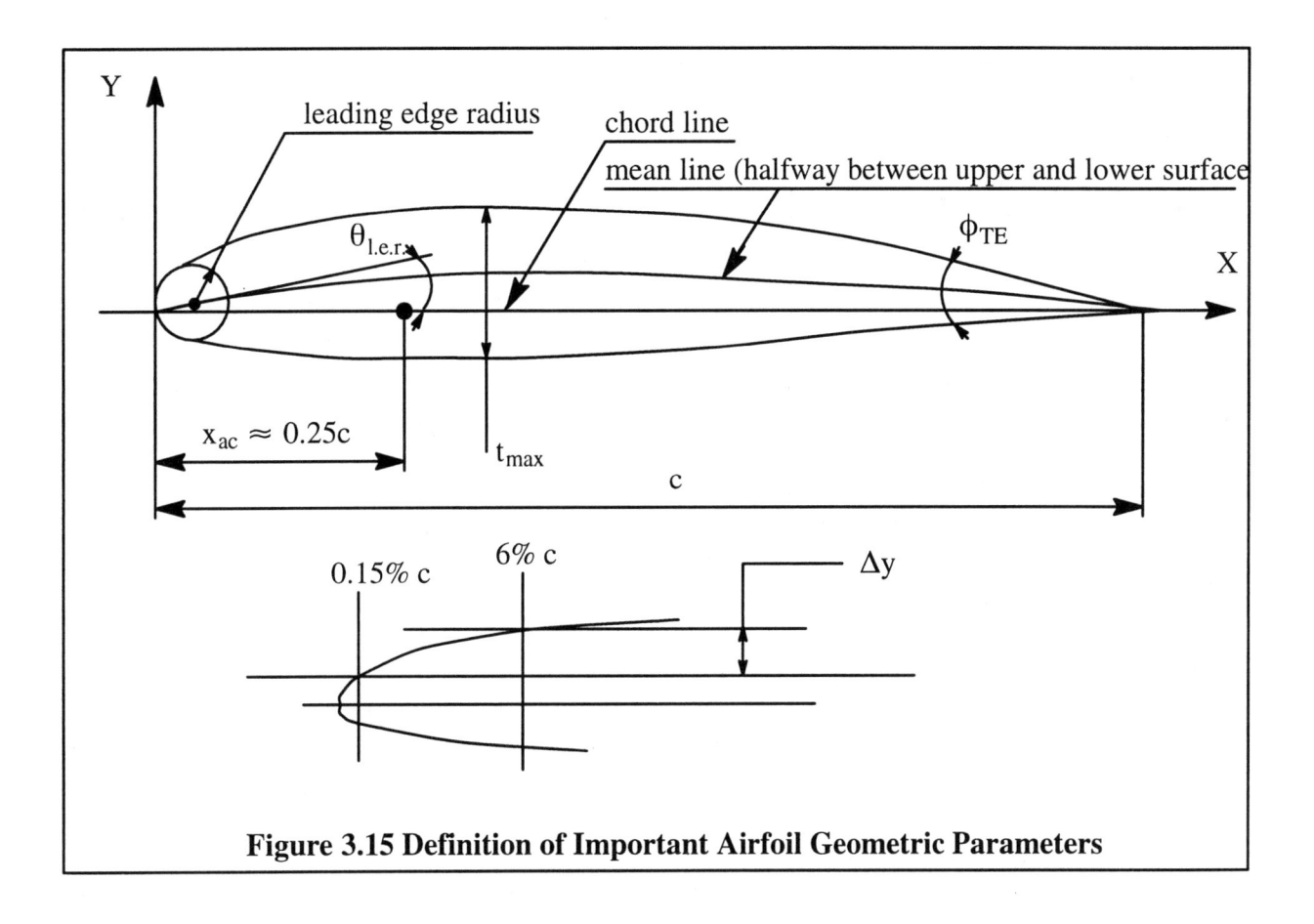

Figure 3.15 Definition of Important Airfoil Geometric Parameters

Examples are: NACA 66(215)–216 with a = 0.6 and NACA 65(318)–217 with a = 0.5. These two examples represent airfoils with thickness distributions obtained by linearly increasing or decreasing the ordinates of the originally derived distribution. In the last example, the airfoil has a 17% thickness ratio. Its ordinates were derived from the airfoil with 18% thickness distribution. The digit 3 represents in tenths, one half of the extent of the low drag range. When this digit is omitted, it implies that the low drag range is less than 0.1.

Since the late 1950's NASA has engaged in the design of airfoils for transonic transport and fighter applications. These so–called supercritical airfoils have a higher M_{dd} value than the conventional NACA 6–series airfoils as illustrated in Figure 3.13. These supercritical airfoils are characterized by very little camber in the forward portion. On the other hand, the rearward portion is severely cambered. Figure 3.16 presents an example of a supercritical airfoil.

During the course of these recent airfoil research activities, new airfoils for lower speed applications have also been derived. Examples are the low–speed airfoils, such as LS(1)–0417 and LS(1)–0413, the medium speed airfoils, such as MS(1)–0313 and natural laminar flow airfoils, such as NLF(1)–416. The LS(1)–0417 airfoil is also known as the GA(W)–1 airfoil (W stands for Whitcomb) and the LS(1)–0413 airfoil is also known as the GA(W)–2 airfoil. Figure 3.17 shows a comparison of older NACA airfoils with the GA(W)–2 airfoil. Figure 3.17 also shows a comparison of the camber and thickness distributions for the GA(W)–1 airfoil with those for the NACA 65_3–018 airfoil. Several key design features of the 17% thick GA(W)–1 airfoil are:

Table 3.2 Examples of NACA Airfoil Designations

4–digit airfoils   **Example: NACA 4412**

4   camber: 0.04c

4   position of the camber at 0.4c from the leading edge (L.E.)

12   maximum thickness ratio: 0.12c

5–digit airfoils   **Example: NACA 23015**

2   camber: 0.02c

     the design lift coefficient is 0.15 times the first digit for this series

30   position of the camber at (0.30/2) = 0.15c from the leading edge (L.E.)

15   maximum thickness ratio: 0.15c

6–series airfoils   **Example: NACA 65_3–421**

6   series designation

5   minimum pressure occurs at 0.5c

3   the drag coefficient is near its minimum value over a range of
     lift coefficients of 0.3 above and below the design lift coefficient

4   design lift coefficient is 0.4

21   maximum thickness ratio: 0.21c

7–series airfoils   **Example: NACA 747A315**

7   series designation

4   favorable pressure gradient on the upper surface from the L.E.
     to 0.4c at the design lift coefficient

7   favorable pressure gradient on the lower surface from the L.E.
     to 0.7c at the design lift coefficient

A   a serial letter to distinguish different sections having the same numerical
     designation but different mean line or different thickness distribution

3   design lift coefficient is 0.3

15   maximum thickness ratio: 0.15c

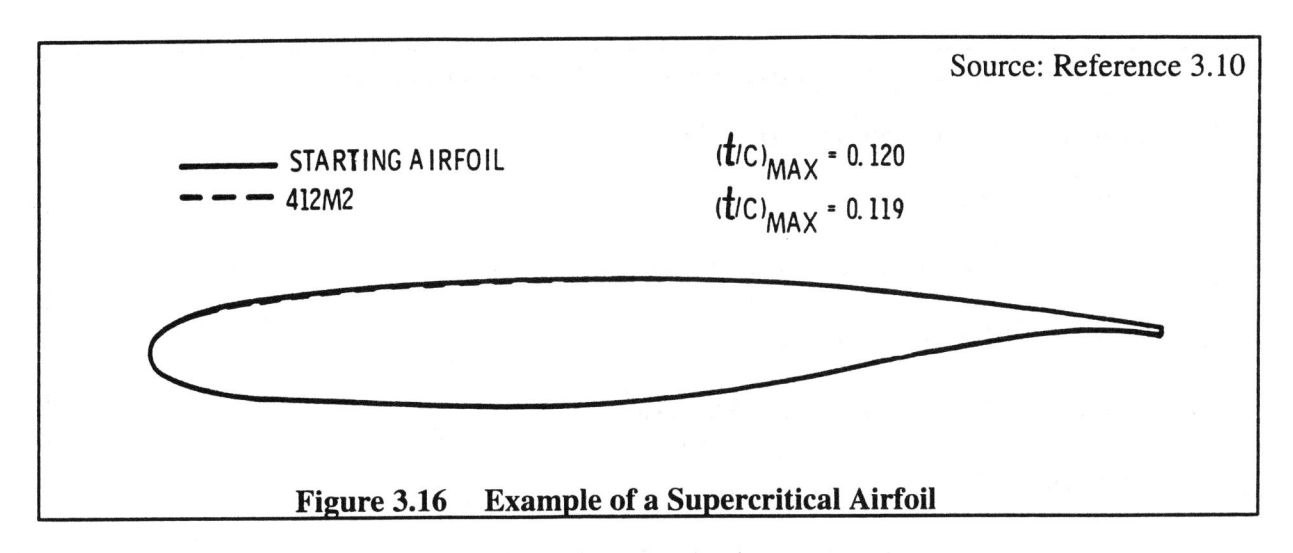

Airfoil Theory

Figure 3.16 Example of a Supercritical Airfoil

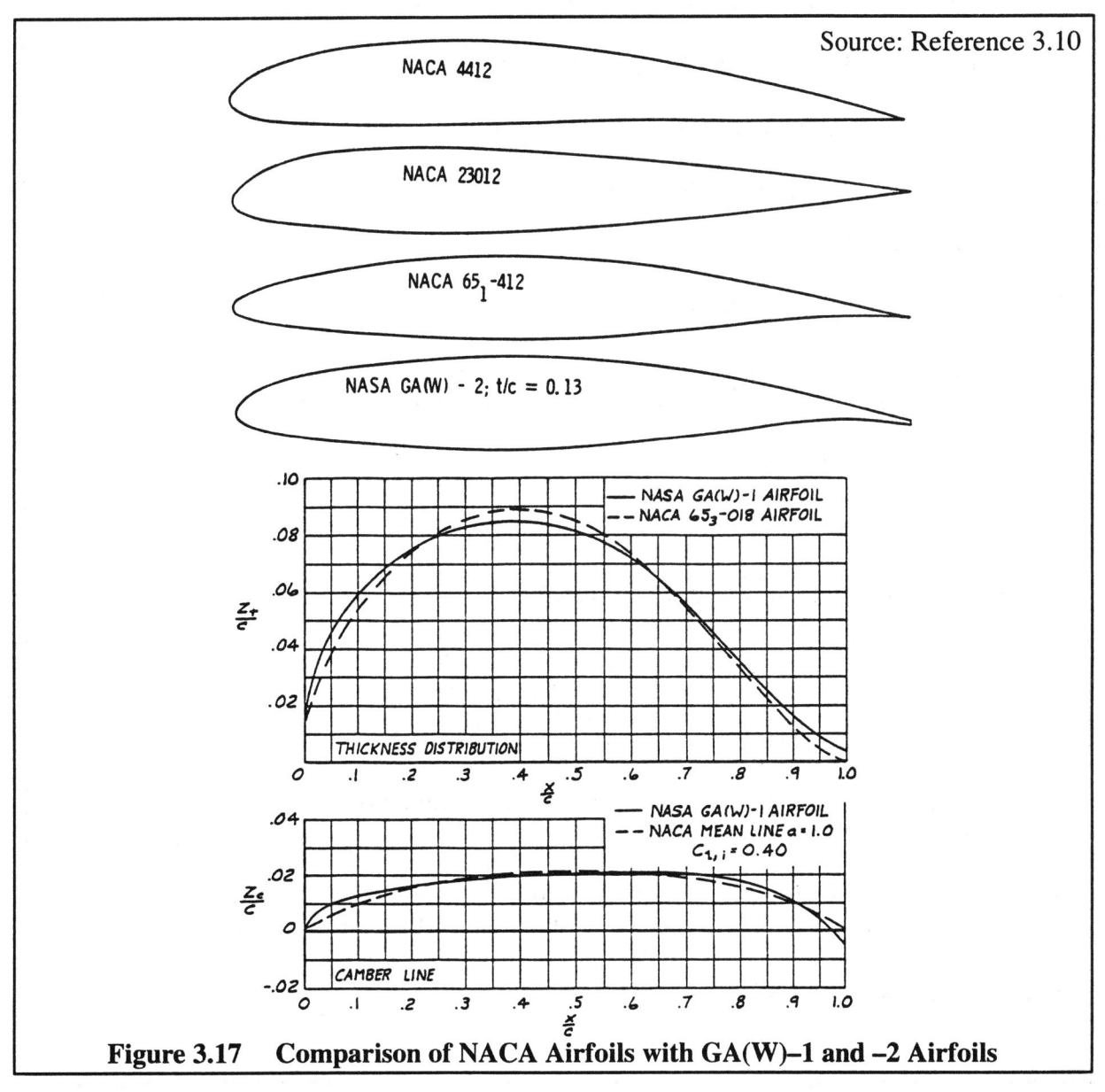

Source: Reference 3.10

STARTING AIRFOIL $(t/c)_{MAX} = 0.120$
412M2 $(t/c)_{MAX} = 0.119$

Figure 3.17 Comparison of NACA Airfoils with GA(W)–1 and –2 Airfoils

Source: Reference 3.10

a) A large upper surface leading edge radius (0.06c) was used to alleviate the peak negative pressure coefficients and therefore delay airfoil stall to a higher angle of attack.

b) The airfoil was contoured to provide an approximate uniform chordwise load distribution near the design lift coefficient of 0.4.

c) A blunt trailing edge was provided with the upper and lower surface slopes approximately equal to moderate the upper surface pressure recovery and thus postpone the stall.

Test results in References 3.11 (for GA(W)–1) and 3.4 (for GA(W)–2) show that the section maximum lift coefficient, $c_{l_{max}}$ of this type airfoils is about 30% greater than that of a typical older NACA 6–series airfoil. This is achieved with a section lift–to–drag ratio, c_l/c_d at $c_l = 0.9$ which is about 50% greater! Figures 3.18a and 3.18b show some example data. In Figure 3.18b, the so–called NACA standard roughness is a large wrap–around roughness as compared with the narrow strip roughness strip now used as the NASA standard.

In selecting an airfoil for an airplane lifting surface (wing, canard, horizontal or vertical tail) the following considerations are important:

1) Drag (for example, one may wish to obtain the highest possible cruise speed)

2) Lift–to–drag ratio at values of the lift coefficient which are important to the airplane (for example, one may wish to design for a given climb rate with one engine inoperative)

3) Thickness (for example, one may wish to design the wing for a low structural weight)

4) Thickness distribution (for example, one may wish to design for a large internal fuel volume)

5) Stall characteristics (for example, one may wish to design for gentle stall characteristics)

6) Drag rise behavior (for example, one may wish to design for a high drag divergence Mach number. This item is closely linked to item 1).

It is clear from these six items that airfoil design and/or airfoil selection will have to be done with a number of compromises in mind to achieve an acceptable overall result. Table 3.3 lists a number of practical airfoil applications.

Part VI of Reference 3.10 may be consulted for rapid, empirical methods to predict section lift, drag and pitching moment characteristics from the basic geometric parameters seen in Figure 3.14.

Airfoil Theory

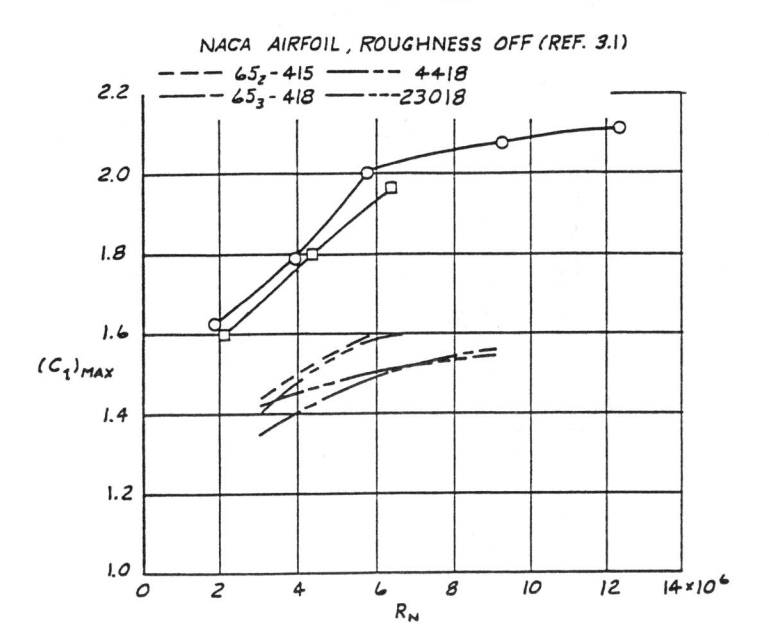

Source: Reference 3.10

a) Variation of Section Maximum Lift Coefficient with Reynolds Number

Source: Reference 3.10

$R_N = 6 \times 10^6$ and $M = 0.2$

b) Variation of Section Drag Coefficient with Section Lift Coefficient

Figure 3.18 Comparison of Aerodynamic Characteristics of Some NACA and NASA Airfoils

Table 3.3 Examples of Airfoil Applications			
Airplane Type	Wing	Horiz. Tail	Vert. Tail
Beech Bonanza	NACA 23016.5 (mod) at root NACA 23012 (mod) at tip -3^0 twist	NACA 0009	NACA 0009
Beech Queen Air B80	NACA 23020 (mod) at root NACA 23012 (mod) at tip -3^0 54' twist	NACA 0009	NACA 0009
Beech Skipper	NASA GA(W)–1	NACA 0009	NACA 0009
Beech Duchess	NACA 63_2A415 (mod) at root	NACA 0009	NACA 0009
Cessna 210 Centurion	NACA 64_2A215 at root NACA 64_1A412 at tip -2^0 twist	NACA 0009	NACA 0009
Cessna T–37	NACA 2418 at root NACA 2412 at tip	NACA 0012	NACA 0012
Cessna 337 Skymaster	NACA 2412 at root NACA 2409 at tip -2^0 twist	NACA 0009	NACA 0009
Cessna 500 Citation	NACA 23014 (mod) at root NACA 23012 at tip -3^0 twist	NACA 0009	NACA 0009
Piper PA–23 Aztec	USA*35–B (mod) t/c = 14% -2.5^0 twist	NACA 0009	NACA 0009
Piper PA–31T Cheyenne	NACA $63_2$415 at root NACA 63 A212 at tip -2.5^0 twist	NACA 0009	NACA 0009
Lockheed 1329–25 Jetstar	NACA 63A112 at root NACA 63A309 at tip -2^0 twist	NACA 0009	NACA 0009
LTV A–7 Corsair	NACA 65A007	Not available	Not available
Northrop F–5A	NACA 65A004.8 (mod)	Not available	Not available
Boeing 747	Boeing proprietary airfoils. t/c=13.44% inboard t/c=7.8% mid–span t/c=8% tip	Not available	Not available

3.8 AIRFOIL MAXIMUM LIFT CHARACTERISTICS

The maximum lift characteristics of an airfoil as well as the associated stall behavior are of great importance to airplane performance.

Whenever the airflow around an airfoil separates, stall is said to have started. From a $c_l - \alpha$ viewpoint there are two types of stall: gradual and abrupt. Figure 3.19 shows examples of each type.

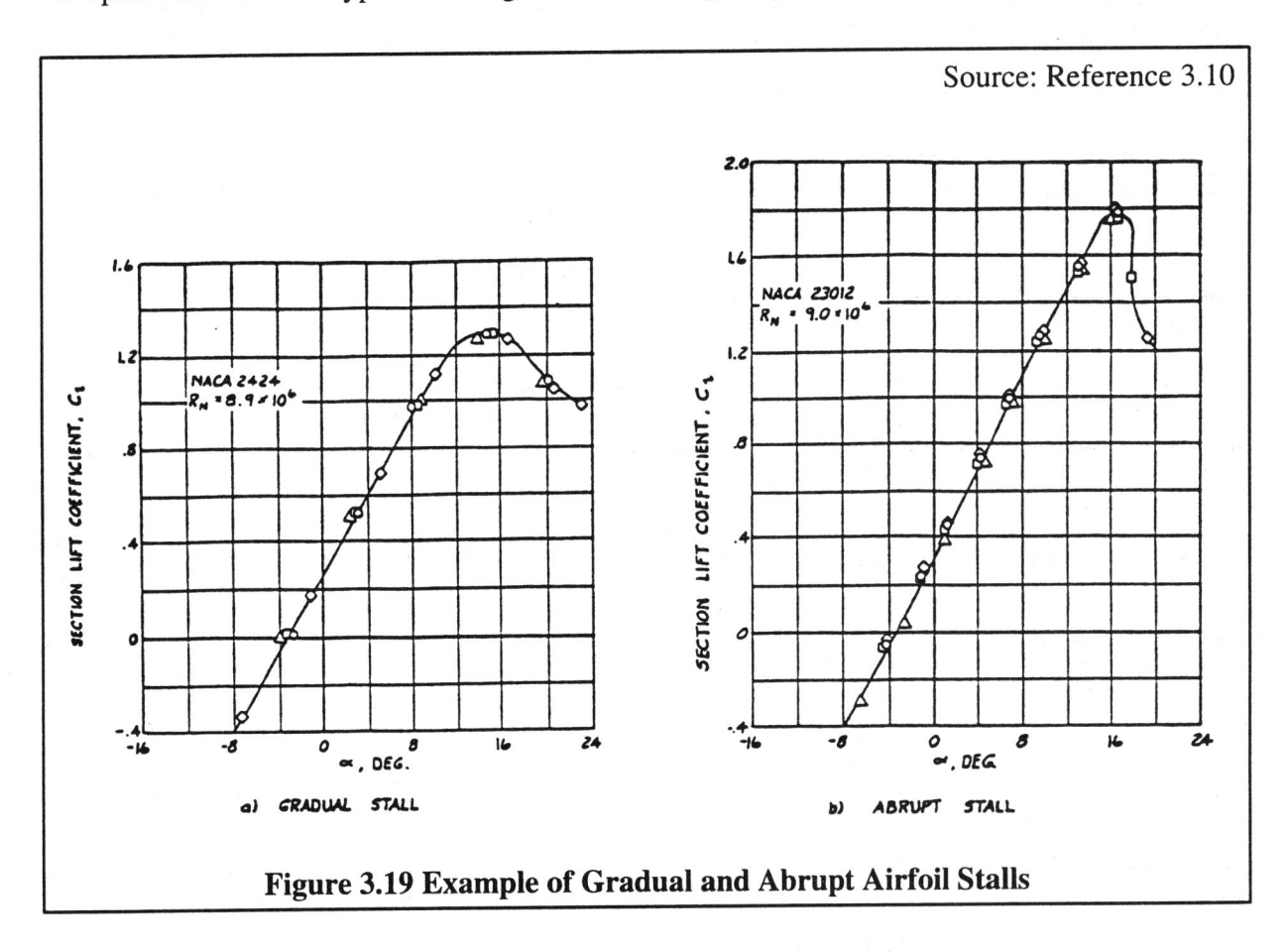

Figure 3.19 Example of Gradual and Abrupt Airfoil Stalls

The first type of stall is characterized by a gradual stall followed by a shallow drop–off of the section lift coefficient. This type of stall frequently occurs on airfoils with moderate or thick sections.

The second type of stall is characterized by an abrupt drop–off of the section lift coefficient. It is often associated with thin airfoil sections.

The main airfoil design features which affect section stall and therefore the maximum lift coefficient are:

    a) thickness ratio   b) leading edge radius
    c) camber   d) location of maximum thickness

These four factors are discussed in Sub–section 3.8.1.

3.8.1 GEOMETRIC FACTORS AFFECTING AIRFOIL MAXIMUM LIFT AT LOW SPEEDS

3.8.1.1 Thickness Ratio

Figure 3.20 shows how airfoil $c_{l_{max}}$ is affected by airfoil thickness ratio, t/c. It is shown in Sub–sub–section 3.8.1.2, that for a given thickness ratio, $c_{l_{max}}$ depends strongly on the leading edge radius and on the leading edge shape. Figure 3.20 also shows that the modern LS series of airfoils have considerably higher values of $c_{l_{max}}$ than conventional NACA airfoils. For the NACA airfoils, a thickness ratio of around 13% will generally produce the highest possible section lift coefficient. For the LS series of airfoils the highest vale of $c_{l_{max}}$ occurs at a thickness ratio of about 15%.

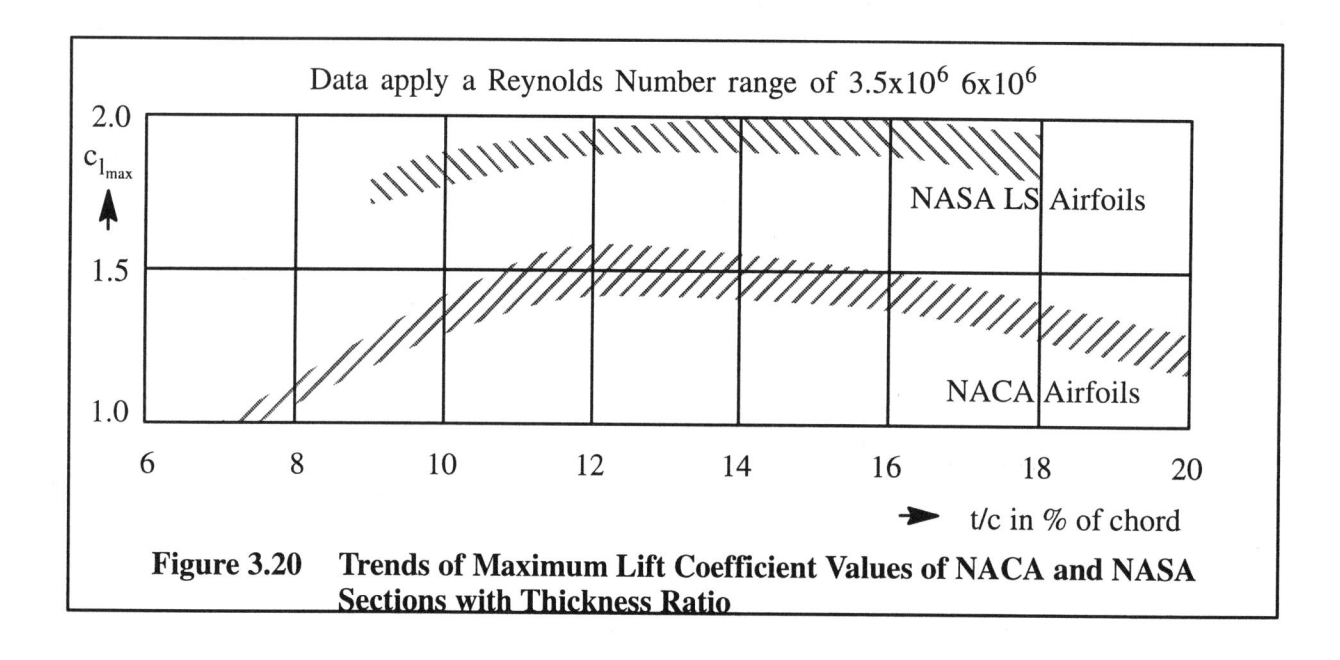

Figure 3.20 Trends of Maximum Lift Coefficient Values of NACA and NASA Sections with Thickness Ratio

3.8.1.2 Leading Edge Radius and Leading Edge Shape

The effect of leading edge radius and leading edge shape is more or less reflected in a geometric parameter, called: $\frac{z_5}{t}$, where z_5 is the local thickness of the airfoil at 5% chord and t is the maximum thickness of the airfoil. Figure 3.21 shows the effect of $\frac{z_5}{t}$ on section $c_{l_{max}}$ for NACA symmetrical airfoils of different thickness ratios. A large value of $\frac{z_5}{t}$ indicates a large leading edge radius. It is seen that large leading edge radii are beneficial in producing large values of $c_{l_{max}}$ at low speeds.

Figure 3.21 Variation of Maximum Lift Coefficient with Geometry of NACA Symmetrical Airfoils at a Reynolds Number of 6x10⁶

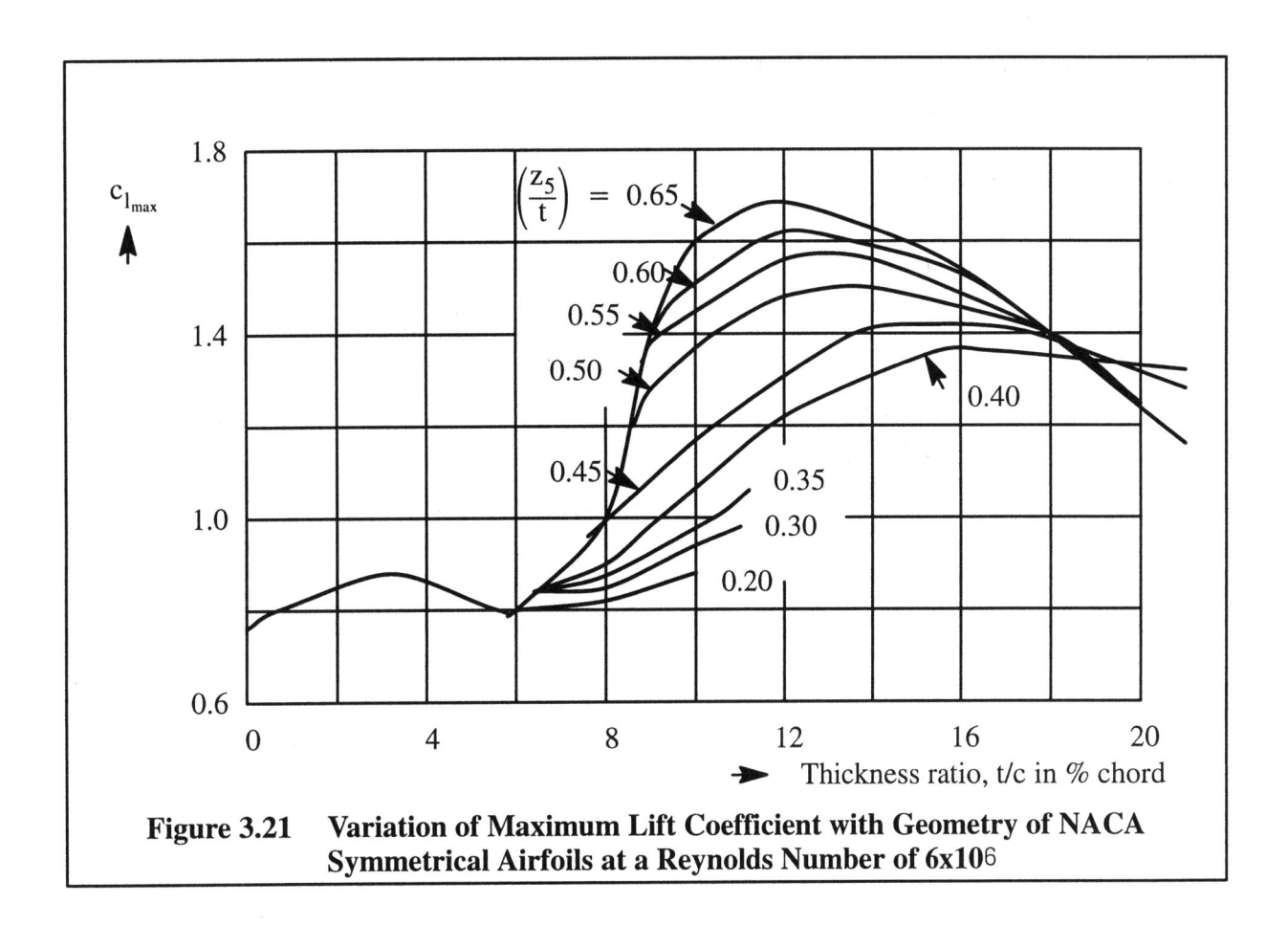

3.8.1.3 Camber and Location of Maximum Thickness

Experimental data show that the maximum lift coefficient of a cambered section depends not only on the amount of camber and camber line shape, but also on the thickness and nose radius of the section on which it is used.

In general, the addition of camber is always beneficial to $c_{l_{max}}$ and the benefit grows with increasing camber. The increment to maximum lift due to camber is least for sections with relatively large leading edge radii (i.e. the benefit of camber grows with reduction of the parameter $\frac{z_5}{t}$; and camber is more effective on thin sections than on thick sections.

In addition, a forward position of maximum camber produces higher values of $c_{l_{max}}$. For example, the NACA 23012 airfoil (with 2% maximum camber at 0.15 chord) has a $c_{l_{max}}$ of 1.79 as compared with 1.67 for NACA 4412 (with 4% camber at 0.4 chord but the same thickness distribution) at a Reynolds number of 9x10⁶.

3.8.2 EFFECT OF REYNOLDS NUMBER ON MAXIMUM LIFT

For airfoils of moderate thickness ratio, there is a significant increase in $c_{l_{max}}$ with increasing Reynolds Number as shown in Figure 3.18a. On the other hand, the effect of Reynolds number for thin airfoils is relatively insignificant. In general, these Reynolds number effects are less for cambered than for symmetrical sections. At low Reynolds Number, the effect of camber is more significant. The opposite is true for Reynolds numbers greater than 6x10⁶., where camber looses some of its effect.

3.8.3 EFFECT OF HIGH LIFT DEVICES ON AIRFOIL MAXIMUM LIFT

The airfoil maximum lift coefficient can be significantly increased by using high lift devices. In the following, effects of trailing edge flaps, leading edge devices and boundary layer control (BLC) on $c_{l_{max}}$ will be discussed.

3.8.3.1 Trailing Edge Flaps

The following five types of trailing edge flaps will be discussed:

a) Plain flaps   b) Split flaps   c) Slotted Flaps
d) Fowler flaps   e) Double slotted flaps

a) Plain flaps

Plain trailing edge flaps are formed by hinging the rear–most part of a wing section about a point within the contour, as shown in Figure 3.22a. By definition, a downward deflection is positive, an upward deflection is negative. The main effect of flap deflection is an increase in the effective camber of the airfoil. The resulting change in lift curve is illustrated in Figure 3.23. It should be noted that the data in Figure 3.23 are based on the flaps–up airfoil chord. It is seen that the flap makes the angle of attack for zero lift much more negative without significantly affecting the lift–curve–slope. However, note that the stall angle of attack is reduced. Although not shown, the drag coefficient is increased and the lift–to–drag ratio is reduced. The pitching moment of the section becomes more negative with the flap down.

For most commonly used wing sections, the flap deflection angle can be increased to about 15 degrees without flow separation. As a general rule, the $c_{l_{max}}$ of the section will increase up to deflections of 60 to 70 degrees for flap–to–chord ratios of up to 0.3. At those high flap deflections, the drag will be very much larger. For example data on airfoils with plain flaps, the reader should consult Refs 3.1, 3.8 and 3.9. If there is a gap between the leading edge of the flap and the cove in which the flap rotates, significant loss of lift can occur. This is because the high pressure air from the lower surfaces will "leak" through this gap toward the negative pressure region on the upper surface. It has been observed that a gap of 1/300th of the chord results in a loss of 0.35 in $c_{l_{max}}$ (Ref. 3.12).

Many single engine, light airplanes use plain flaps.

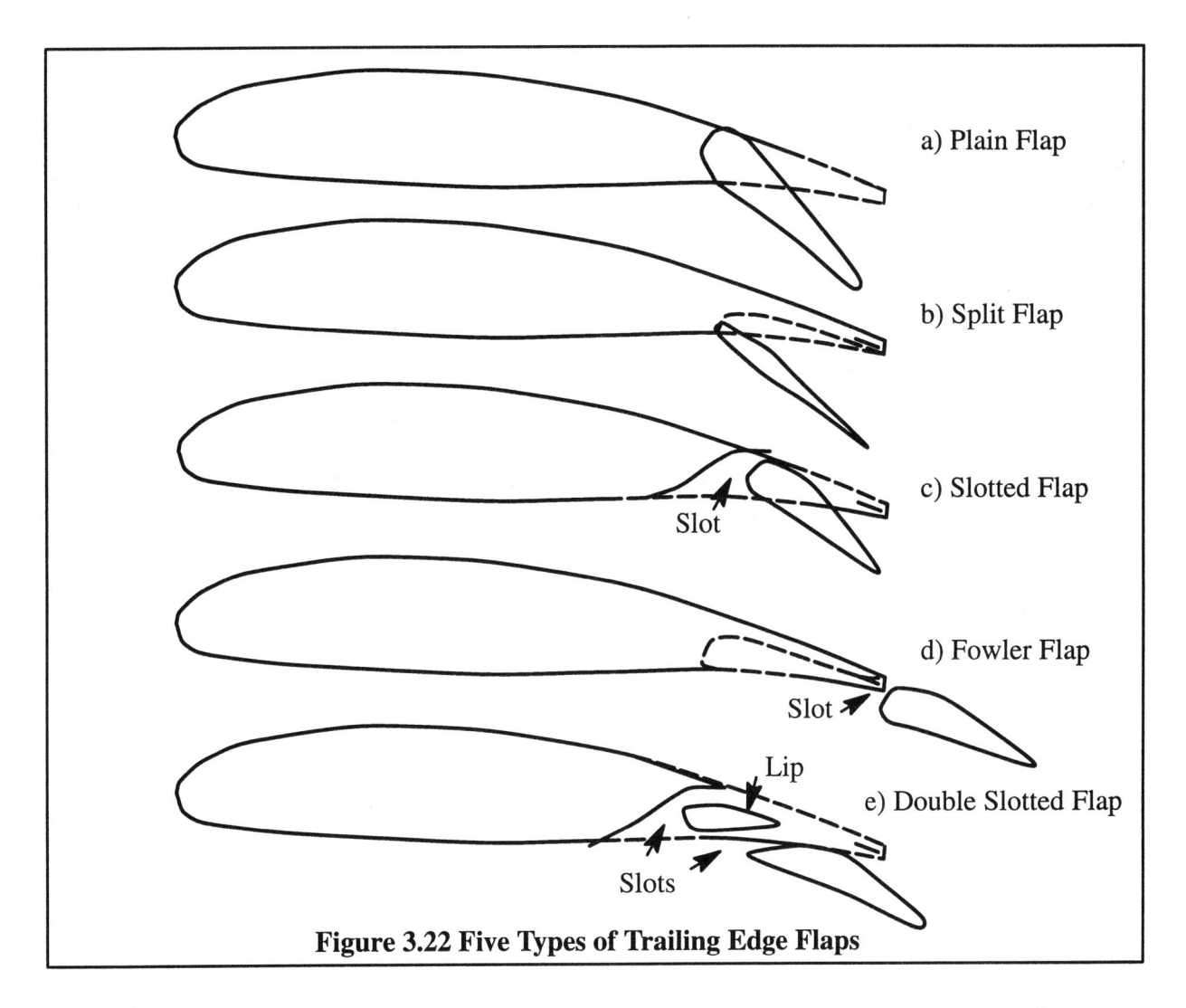

Airfoil Theory

Figure 3.22 Five Types of Trailing Edge Flaps

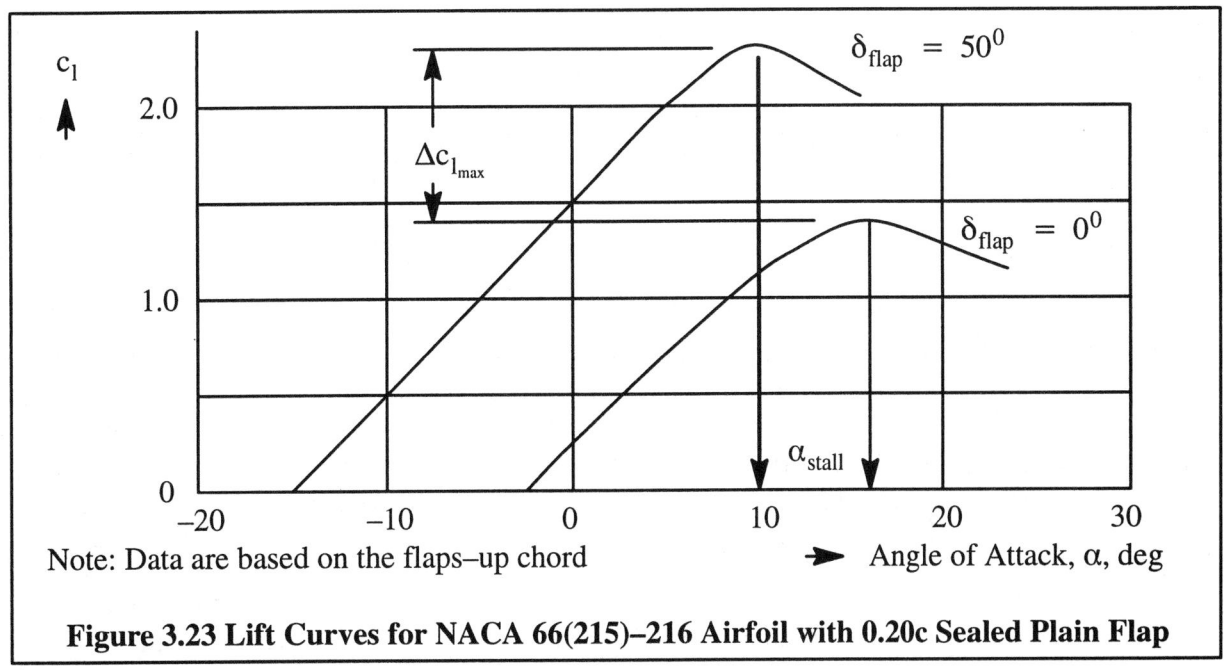

a) Plain Flap
b) Split Flap
c) Slotted Flap
d) Fowler Flap
e) Double Slotted Flap

Figure 3.23 Lift Curves for NACA 66(215)–216 Airfoil with 0.20c Sealed Plain Flap

Note: Data are based on the flaps–up chord

b) Split flaps

The usual split flap is formed by deflecting the aft portion of the lower surface about a hinge point on the surface at the forward edge of the deflected portion. Figure 3.22b shows a typical split flap. Split flaps derive their effectiveness from the large increase in camber produced. Because the upper surface is less cambered than the lower surface with the flap deflected, the separation effects on the upper surface will be less marked than those associated with a plain flap. Therefore, the flap performance at high angle of attack is improved. The lift–curve slope with split flaps is higher and the stall angle of attack is somewhat lower than that for the flaps–up airfoil, but higher than that for the plain flap. The angle of attack for zero lift is reduced, but not by quite so much as is the case with a plain flap. The increment in $c_{l_{max}}$ is larger than with a plain flap. However the drag associated with a split flap is very much higher than that associated with a plain flap because of the large wake.

For airfoils with normal thickness ratios, high benefits in $c_{l_{max}}$ can be obtained by using 20% to 25% chord split flaps with deflection angles of 60 to 70 degrees. Figure 3.24 provides some data based on Reference 3.1. The Douglas DC–3 already used split flaps.

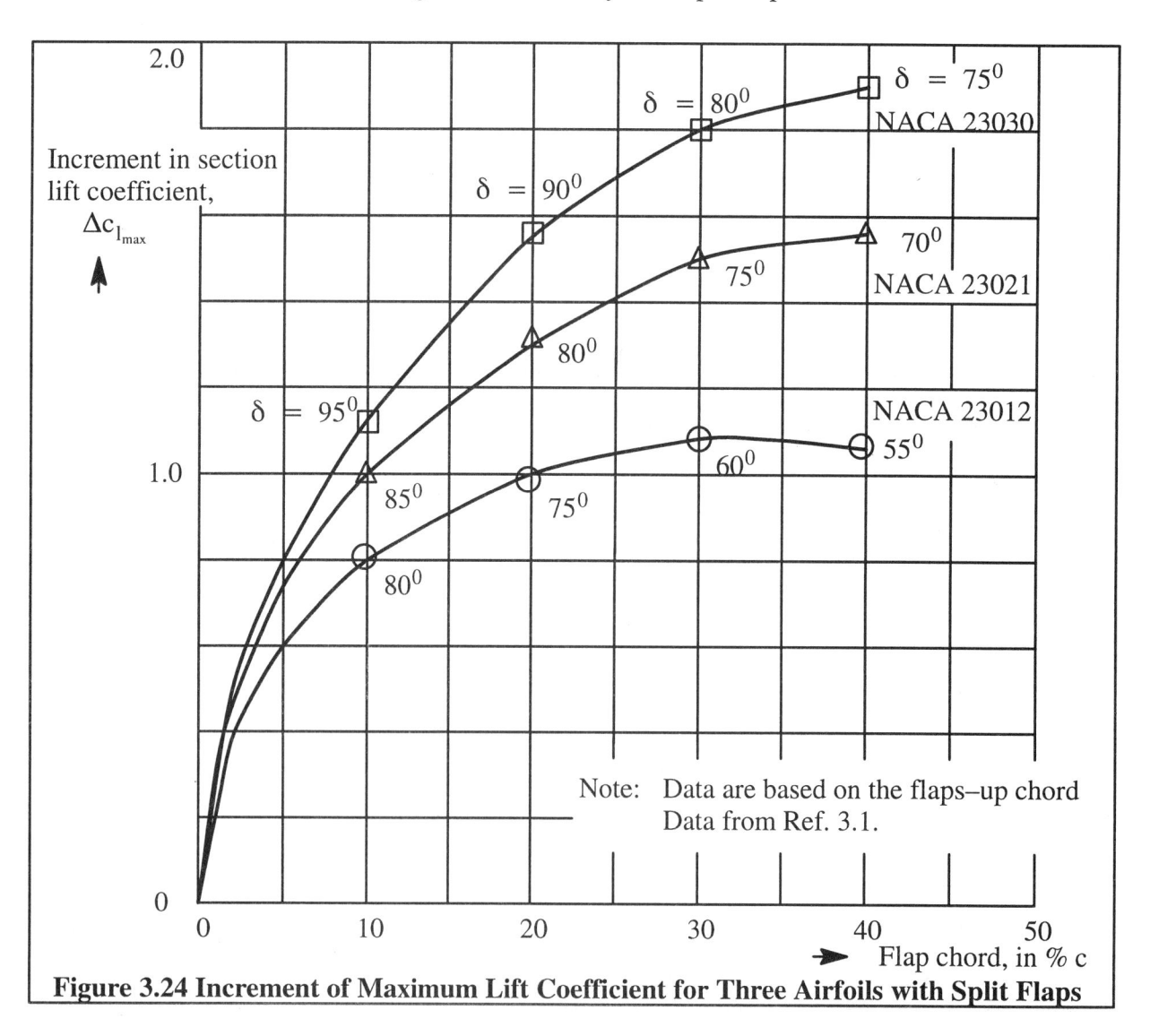

Figure 3.24 Increment of Maximum Lift Coefficient for Three Airfoils with Split Flaps

c) Slotted flaps

Slotted flaps provide one or more slots between the main portion of the wing section and the deflected flap. Figure 3.22c provides an example of a single slotted flap. The slot(s) duct high energy air from the lower surface to the upper surface and direct this air in such a manner as to delay flow separation over the flap by providing boundary layer control. The high lift is again derived from the increase in camber which is characteristic for a slotted flap arrangement. The increment in $c_{l_{max}}$ is much higher than that associated with either a split or a plain flap while the drag is much lower, due to the boundary layer control effect. However, the pitching moment associated with slotted flaps tends to be high and negative. This does have a slight depressing effect on the trimmed maximum lift coefficient of an airplane.

Common flap–chord ratios used with slotted flaps range from 0.25 to 0.30. The aerodynamic characteristics of slotted flaps are very sensitive to the details of the slot entry and lip design. Typical incremental values for section maximum lift coefficients which can be attained with slotted flaps are shown in Figure 3.25 in comparison with those attained by split flaps.

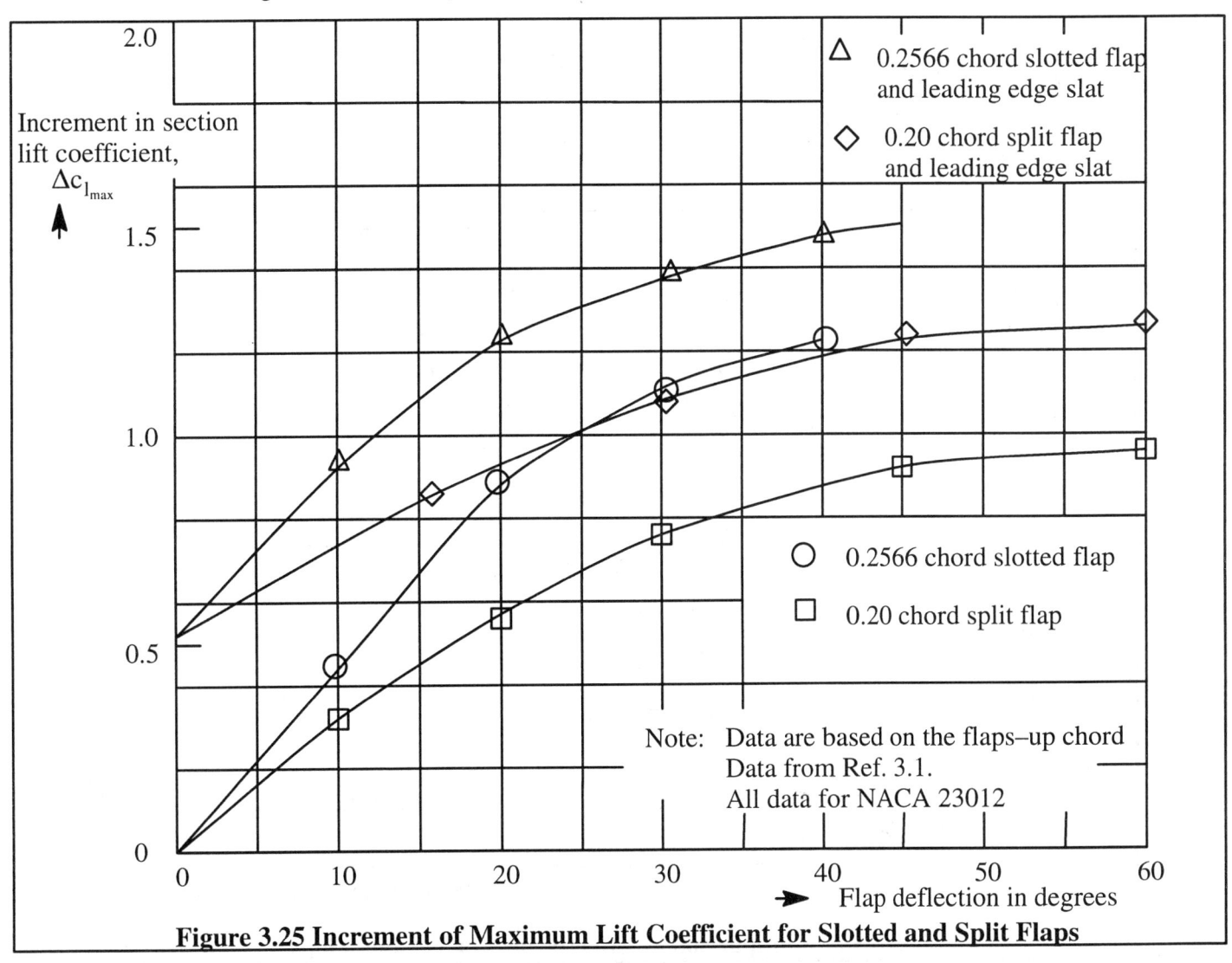

Figure 3.25 Increment of Maximum Lift Coefficient for Slotted and Split Flaps

d) Fowler flaps

The Fowler flap (see Figure 3.22d) employs the same principle as the slotted flap, but it also moves backward while deflecting downward. The backward motion increases the effective wing area. Fowler flaps were already used by Lockheed on the piston powered Electra airliners before WWII.

The relative effectiveness of the flap types a) through d) in Figure 4.22 is given in Figure 3.26. An important experimental result is that the incremental value of $c_{l_{max}}$, called $\Delta c_{l_{max}}$ is essentially independent of Reynolds number. Also, flaps have been found to have a rather insignificant effect on the derivative dc_m/dc_l before stall and therefore on the location of the aerodynamic center. Table 3.4 lists typical applications for various types of flaps.

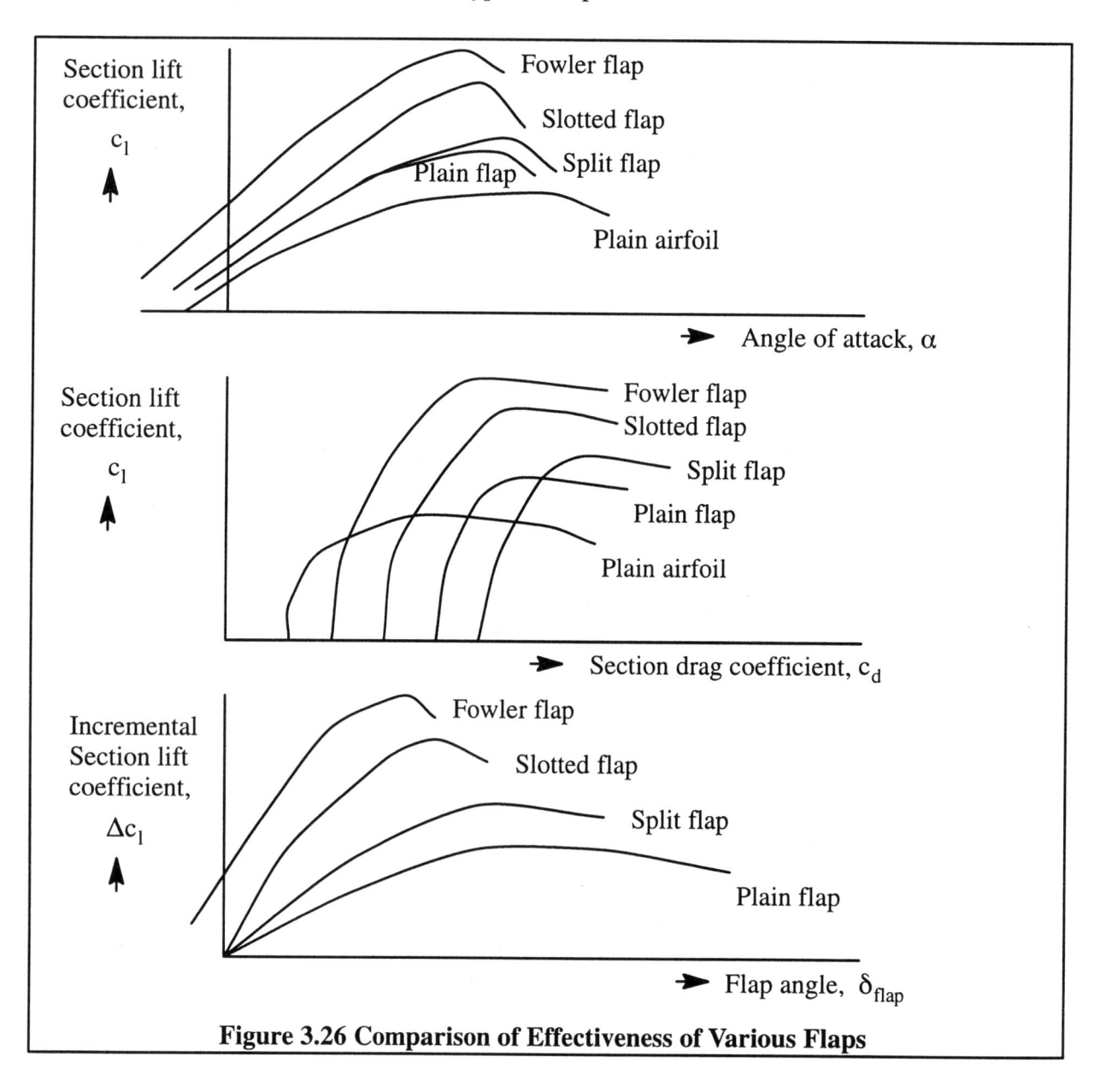

Figure 3.26 Comparison of Effectiveness of Various Flaps

Table 3.4 Examples of Use of Leading and Trailing Edge Devices

Airplane Type	Trailing Edge Device Type	Leading Edge Device
Cessna 414A (Chancellor)	Split flap	None
Cessna 208 Caravan	Single slotted flaps	None
Cessna 550 Citation II	Single slotted flaps	None
Boeing 707	Fowler flaps	Krueger flaps
Boeing 737	Triple slotted Fowler flaps	Krueger flaps and slats
Boeing 747	Triple slotted Fowler flaps	Krueger flaps and variable camber slats
Boeing 767	Single and double slotted flaps	Slats
Raytheon–Beech 1900D	Single slotted flaps	None
Raytheon–Beech Starship	Single slotted Fowler flaps	None
Raytheon–Beech 400A	Single and double slotted Fowler flaps	None
Bombardier–Learjet 35/36	Single slotted flaps	None
Lockheed–Martin C–130	Fowler flaps	None
Lockheed–Martin F–16	Plain flaperon	Leading edge flap
McDonnell–Douglas C–17	Double slotted flaps	Slats
McDonnell–Douglas MD–87	Double slotted flaps	Slats
McDonnell–Douglas MD–11	Double slotted flaps	Slats
Mooney Ovation	Single slotted flaps	None
Northrop–Grumman E–2C	Fowler flaps and drooped ailerons	None
Fokker F–50	Single slotted flaps	None
Fokker F–100	Double slotted Fowler flaps	None
Pilatus PC–7	Split flaps	None
Pilatus PC–12	Fowler flaps	None
AVRO RJ100	Fowler flaps	None

e) Double slotted flaps

Figure 3.22e shows an example of a double slotted flap. It is basically an enhancement of the single slotted flaps. Combinations of Fowler flaps and slotted flaps (called double and triple slotted Fowler flaps) have been used on several modern jet transports. The Boeing 727, 737 and 747 are examples.

3.8.3.2 Leading Edge Devices

There are three main types of leading edge devices:

a) and b) Slats   c) Slots   d) through g) Leading edge flaps

Sketches of these leading edge devices are shown in Figure 3.27.

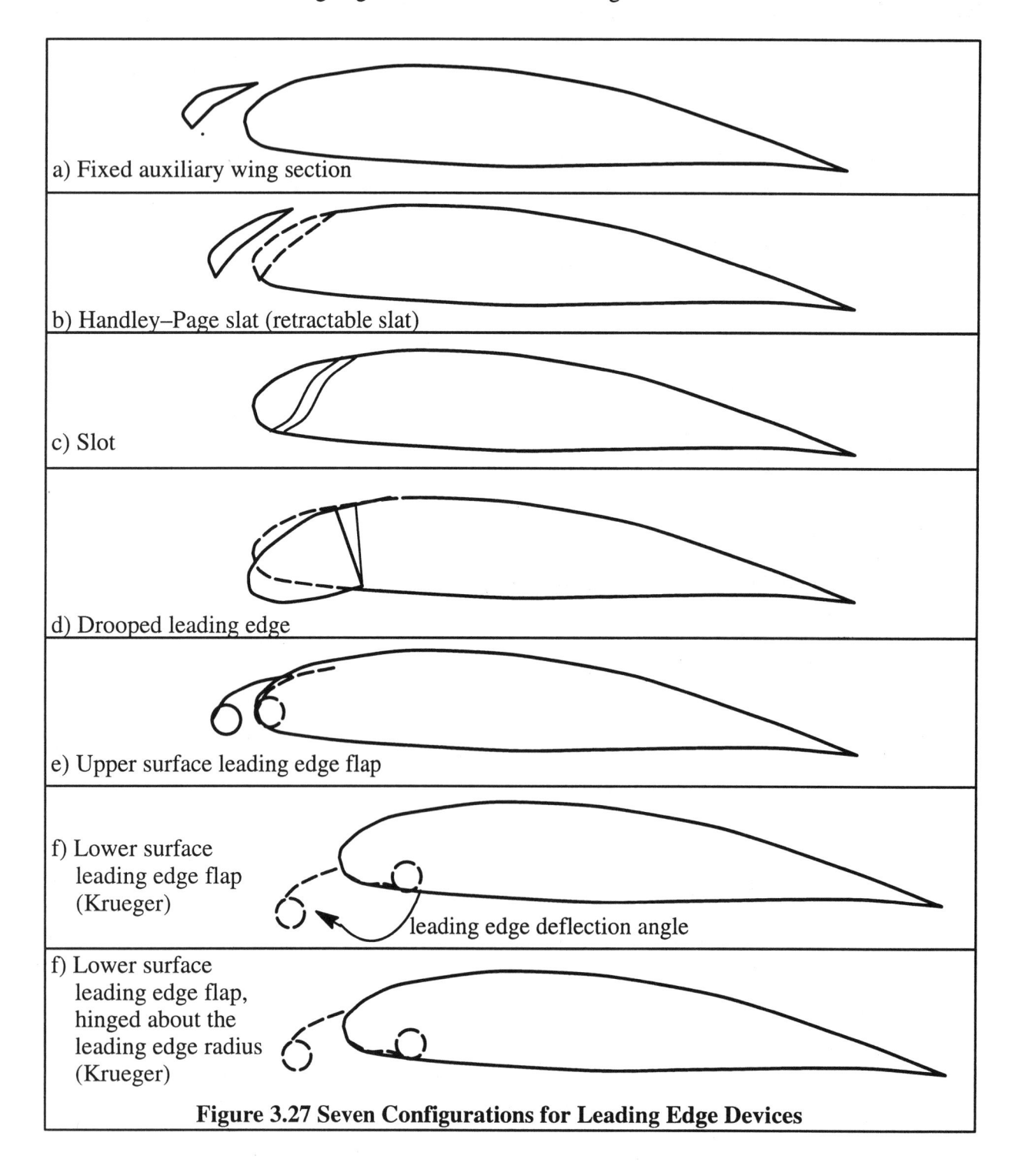

a) Fixed auxiliary wing section

b) Handley–Page slat (retractable slat)

c) Slot

d) Drooped leading edge

e) Upper surface leading edge flap

f) Lower surface leading edge flap (Krueger)

leading edge deflection angle

f) Lower surface leading edge flap, hinged about the leading edge radius (Krueger)

Figure 3.27 Seven Configurations for Leading Edge Devices

a) and b) Slats

Leading edge slats (see Figures 3.27a and b) consist of airfoils mounted ahead of the leading edge of the wing such as to assist in turning the air around the leading edge at high angles of attack and therefore delay leading edge stalling. These slats may either be fixed or retractable. The use of slats may increase the maximum lift coefficient, $c_{l_{max}}$, by as much as 0.50. The effect of slats was shown in Figure 3.25. The Handley–Page slat has been used on airplanes since WWI. Modern applications include the North American F–86 and Sabreliner.

c) Slots

When the slot (see Figure 4.27c) is located near the leading edge, this leading edge device differs only in detail from the leading edge slat. Additional slots may be introduced at various chordwise stations. The effectiveness of the slot derives from its boundary layer control (BLC) effect. At low angle of attack, the minimum section profile drag may be greatly increased by the presence of such slots. Early applications include the Globe Swift (outboard wing) and the horizontal tail of the Cessna Cardinal.

d) through g) Leading Edge Flaps

A leading edge flap may be formed by bending down the forward portion of the wing section, to form a droop: see Figure 4.27d. Other types of leading edge flaps are formed by extending a surface downward and forward from the vicinity of the leading edge (Krueger flap): see Figures 4.27e through g. Although these flaps are not as powerful as trailing edge flaps, they tend to be mechanically simpler. They can also be very effective when combined with the use of trailing edge flaps. Leading edge flaps reduce the severity of the pressure peak ordinarily associated with with high angles of attack and thereby help in delaying separation.

Some test results on a NACA 64_1–012 airfoil with leading edge flaps and trailing edge split flaps are presented in Figure 3.28. The flap angle, indicated in the figure, is measured clockwise from the airfoil lower surface as shown in Figure 3.27. It is seen that the upper surface leading edge flaps appear to be more effective in increasing section $c_{l_{max}}$.

3.8.3.3 Boundary Layer Control

Higher maximum lift coefficients can also be achieved by boundary layer control (BLC). BLC may involve injecting high speed air (or gas) parallel to the wall (called "blowing'), or removing the low energy boundary layer by "suction", or both. Blowing is done to re–energize the boundary flow to delay separation, while suction is equivalent to eliminating the low–energy shear layer. It is shown in Figure 3.29, that with the slot internal pressure 12 times higher than the free stream dynamic pressure and a slot width of 0.667% of the chord, the value of $c_{l_{max}}$ can be enhanced from 1.5 without blowing to 2.9 with blowing.

Suction effect can be provided by using a suitable suction pump to suck the shear layer through

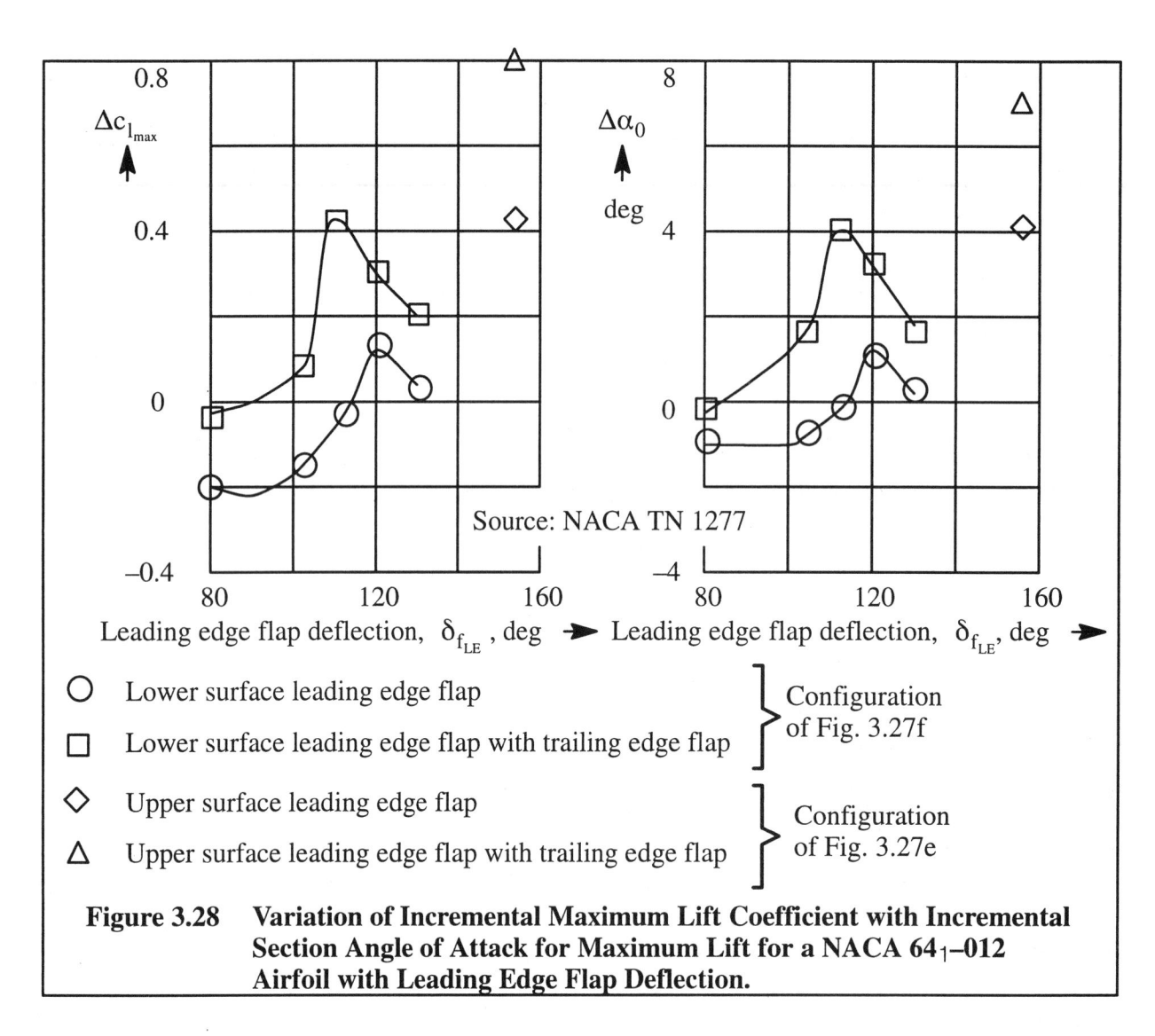

Figure 3.28 Variation of Incremental Maximum Lift Coefficient with Incremental Section Angle of Attack for Maximum Lift for a NACA 64_1–012 Airfoil with Leading Edge Flap Deflection.

Source: NACA TN 1277

○ Lower surface leading edge flap
□ Lower surface leading edge flap with trailing edge flap
} Configuration of Fig. 3.27f

◇ Upper surface leading edge flap
△ Upper surface leading edge flap with trailing edge flap
} Configuration of Fig. 3.27e

a porous airfoil skin. This was successfully applied to a F–86F airplane in 1953. This type of boundary layer control can not only increase the lift, but also delay transition to reduce the skin friction drag. This is merely one of many possible concepts in so–called boundary layer control. Other concepts use suction achieved by ejector blowing, as shown in Figure 3.30. This principle has been used in the design of a NASA Augmentor Wing STOL airplane.

It should be noted that, when the blowing slot is at the flap hinge and the blowing air momentum is high, the device is frequently referred to as a "blown flap". In this case, not only can the flow separation over the flap surface be delayed, but also the circulation, and hence the lift, can be increased. For more detail on the effect of blowing, References 3.13 – 3.15 may be consulted.

3.9 SUMMARY FOR CHAPTER 3

In this chapter several important airfoil characteristics were described and corresponding experimental evidence was shown.

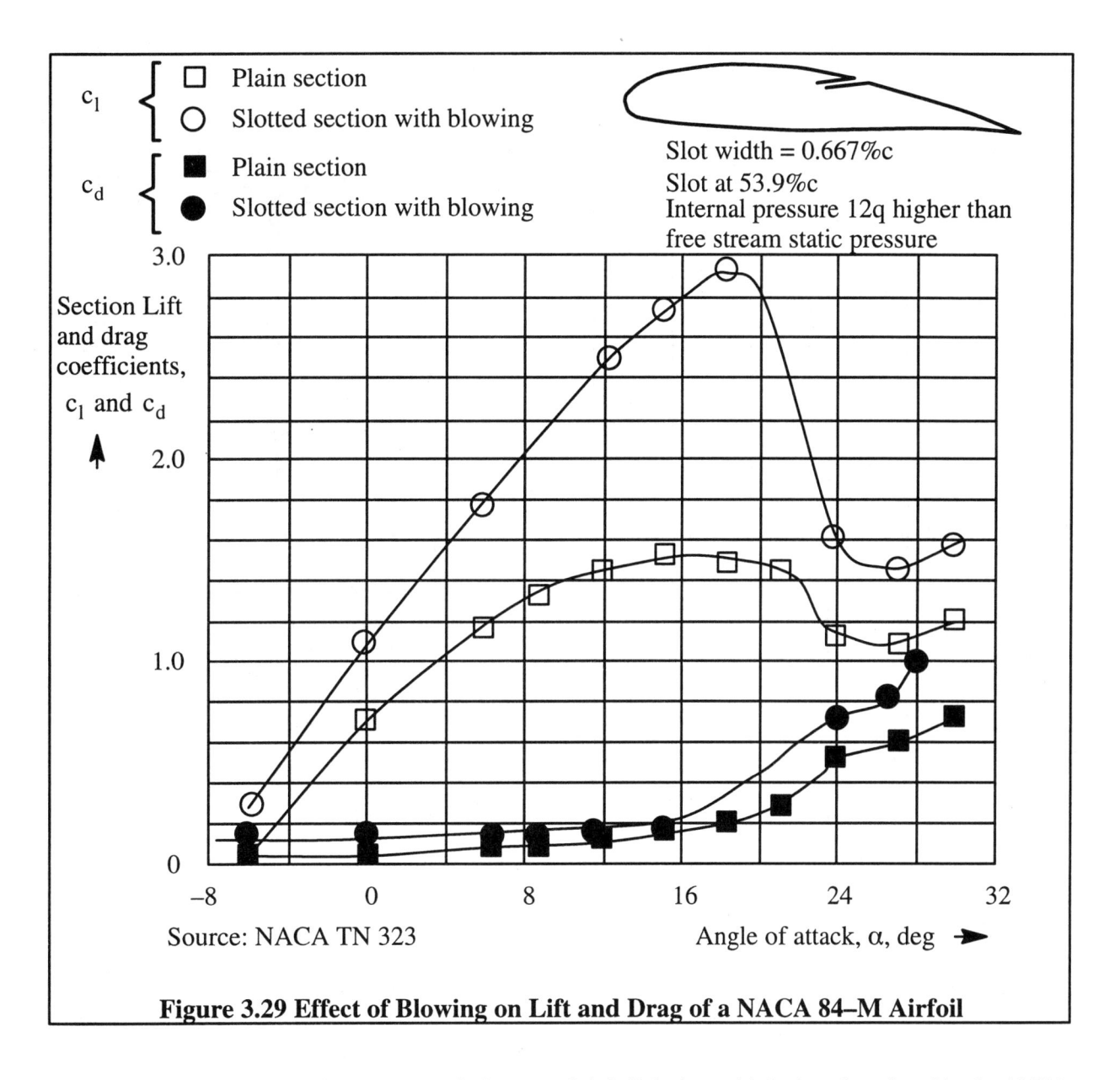

Figure 3.29 Effect of Blowing on Lift and Drag of a NACA 84–M Airfoil

Based on the computational state–of–the–art of airfoil design which already existed in the 1970's (See Refs 3.16 – 3.19), it is recommended that for most new airplane designs, airfoils be specifically tailored to the mission requirements of that airplane. The tailoring procedure should balance the requirements for good performance (i.e. low drag) with requirements for good stability and control (i.e. benign stall, low negative pitching moments and low sensitivity to rain and ice conditions).

An indication of the excellent agreement between experimental and theoretical determination of the aerodynamic characteristics of airfoils is shown in Figure 3.31.

The reader may well ask the following question. Is it possible to control the camber of a wing in a continuous manner? This is in fact what birds do. It is this thought which was behind the so–called NASA mission adaptive wing which was flight tested on a F–111 fighter. See Ref. 3.20.

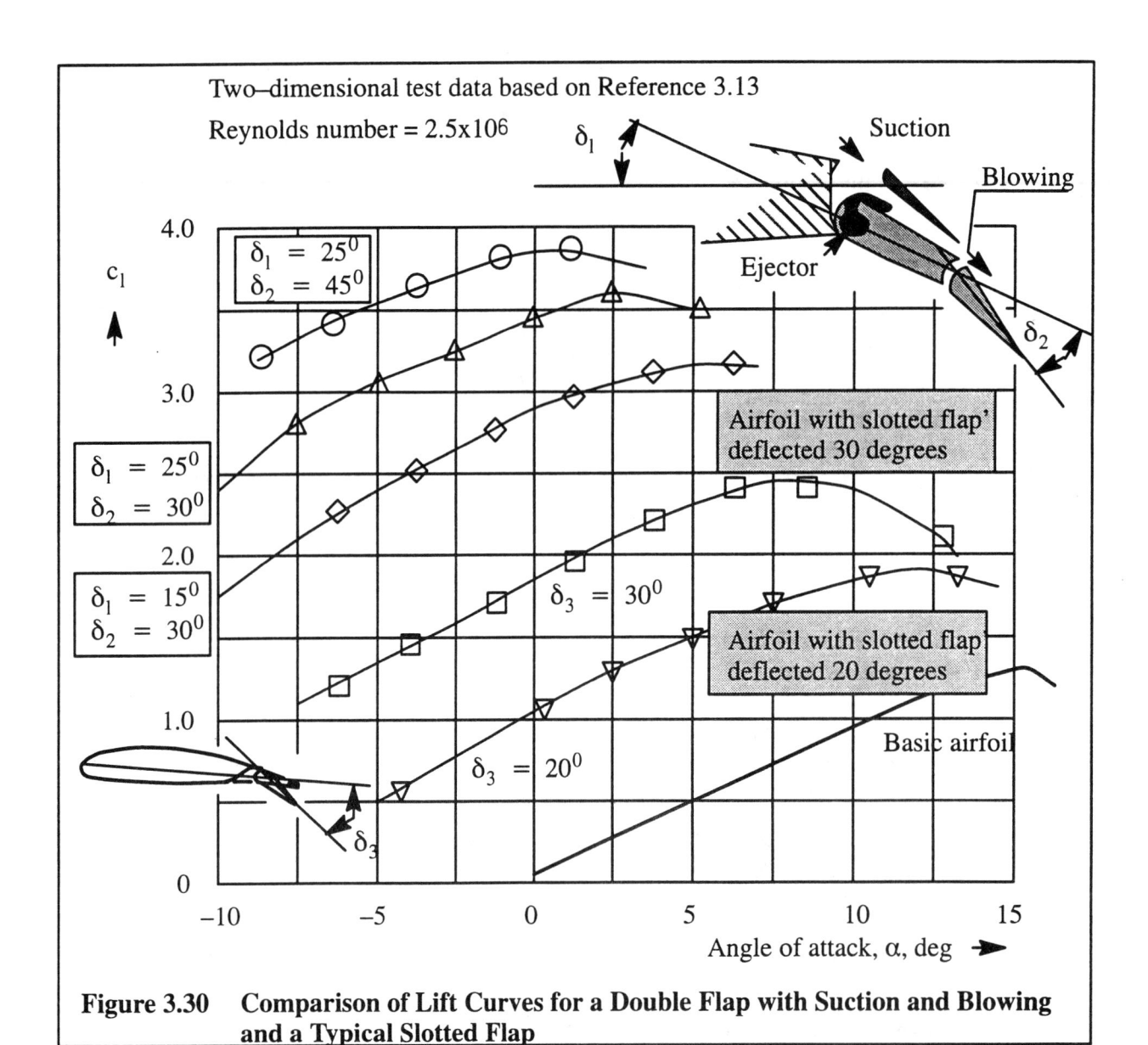

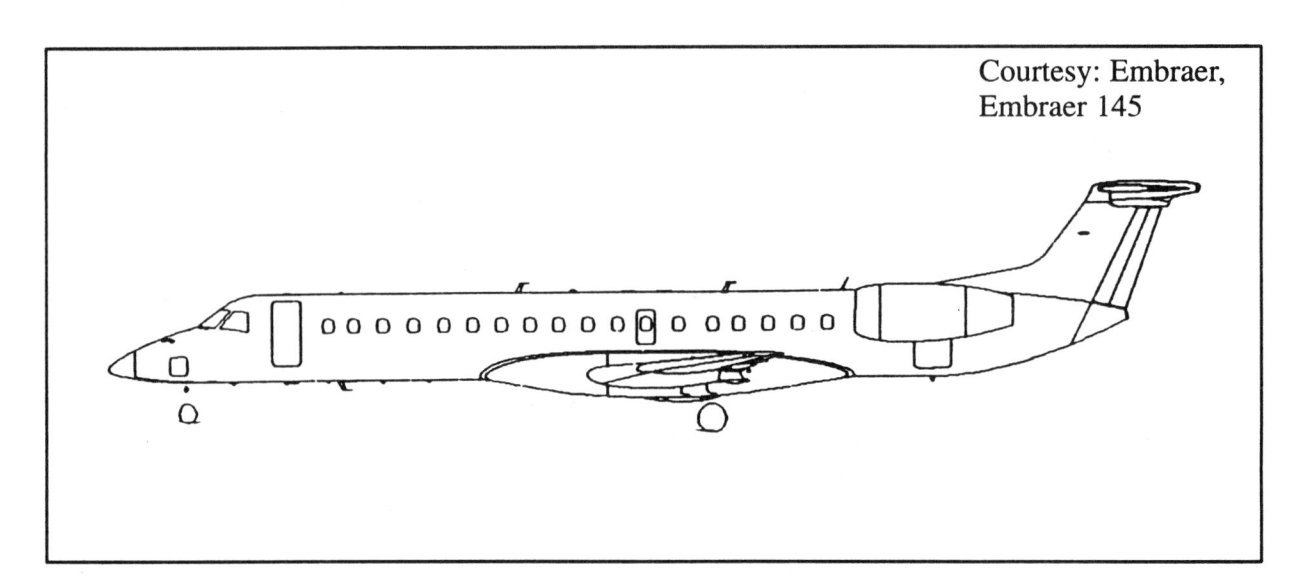

Figure 3.30 Comparison of Lift Curves for a Double Flap with Suction and Blowing and a Typical Slotted Flap

Courtesy: Embraer, Embraer 145

Airfoil Theory

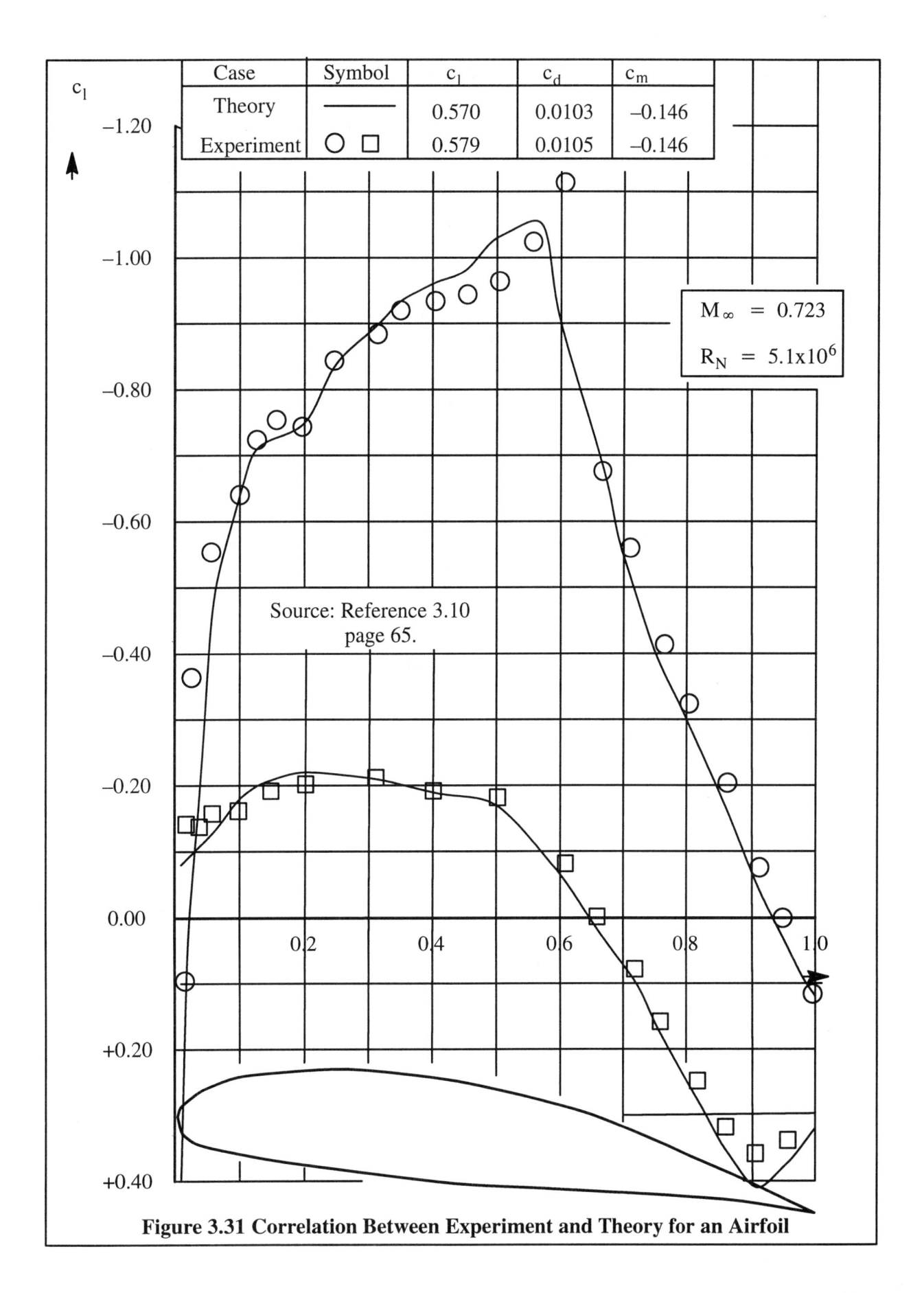

Case	Symbol	c_l	c_d	c_m
Theory	——	0.570	0.0103	−0.146
Experiment	○ □	0.579	0.0105	−0.146

$M_\infty = 0.723$

$R_N = 5.1 \times 10^6$

Source: Reference 3.10 page 65.

Figure 3.31 Correlation Between Experiment and Theory for an Airfoil

3.10 PROBLEMS FOR CHAPTER 3

3.1 An airfoil with thickness ratio of 17% has the following geometrical and aerodynamic characteristics:

x/c	(z/c) upper	(z/c) lower	C_p upper at $M_\infty = 0.0$
0.0	0.0	0.0	0.0
0.005	0.02035	−0.01444	−0.35
0.0125	0.03069	−0.02052	−0.43
0.025	0.04165	−0.02691	−0.48
0.05	0.05600	−0.03569	−0.70
0.075	0.06561	−0.04209	−0.76
0.100	0.07309	−0.04700	−0.77
0.150	0.08413	−0.05426	−0.75
0.20	0.09209	−0.05926	−0.75
0.25	0.09778	−0.06265	−0.73
0.30	0.10169	−0.06448	−0.73
0.35	0.10409	−0.06517	−0.73
0.40	0.10500	−0.06483	−0.725
0.45	0.10456	−0.06344	−0.723
0.50	0.10269	−0.06091	−0.72
0.55	0.09917	−0.05683	−0.71
0.60	0.09374	−0.05061	−0.66
0.65	0.08604	−0.04265	−0.588
0.70	0.07639	−0.03383	−0.50
0.75	0.06517	−0.02461	−0.35
0.80	0.05291	−0.01587	−0.26
0.85	0.03983	−0.00852	−0.16
0.90	0.02639	−0.00352	−0.05
0.95	0.01287	−0.00257	−0.05
1.00	−0.00074	−0.00783	+0.10

Derive an airfoil with a thickness ratio of 13% with the same camber.

3.2 For the airfoil of Problem 3.2, calculate and plot the pressure distribution at $M_\infty = 0.0$ and at $M_\infty = 0.6$ by using the Prandtl–Glauert transformation.

3.3 For the airfoil of Problem 3.1, find the critical Mach number.

3.4 An airfoil has a lift curve slope of 6.3 per radian and an angle of attack for zero lift of −2 degrees. At what angle of attack will the airfoil develop a lift of 140 lbs/ft at 100 mph under standard sea–level conditions? Assume that c = 8 ft.

3.5 The test results of a NACA 23012 airfoil model at $R_N = 3 \times 10^6$ show the following:

α in degrees	c_l	
0	0.15	$c_{d_{min}} = 0.0065$
9	1.20	

Determine the lift curve slope per degree and per radian.

3.6 For the data of Problem 3.5 determine $c_{d_{min}}$ if $R_N = 20 \times 10^6$.

3.7 For the airfoil of Problem 3.1, determine the pressure coefficient at x/c = 0.10 when the airspeed is 450 mph under standard sea–level conditions. Hint: use the Prandtl–Glauert and Kárman–Tsien formulas.

3.8 The pressure at a point on an airplane flying at 250 mph at sea–level is 8 psi absolute. Find the pressure coefficient at this point.

3.9 Prove that Eqn (3.9) is correct.

3.10 Use Jane's All The World's Aircraft (Ref. 3.21) to identify 10 airplanes for each of the six type of trailing edge flaps discussed in Chapter 3.

3.11 REFERENCES FOR CHAPTER 3

3.1 Abbott, I.A. and Von Doenhoff, A.E.; Theory of Wing Sections; Dover, 1959.

3.2 Vernard, J.K.; Elementary Fluid Mechanics; J. Wiley & Sons; 1957.

3.3 Buckingham, E.; Model Experiments and the Focus of Empirical Equations; Trans. ASME, Vol. 37, p.263, 1915.

3.4 McGhee, R.J., Beasley, W.D. and Somers, D.M.; Low Speed Aerodynamic Characteristics of a 13–Percent–Thick Airfoil Section Designed for General Aviation Applications; NASA TMX–72697, May, 1975.

3.5 Braslow, A. and Knox, E.; Simplified Method for Determination of Critical Height of Dis–tributed Roughness Particles for Boundary Layer Transition at Mach Numbers from 0.0 to 5.0; NACA TN 4363, September, 1958.

3.6 Braslow, A.L., Hicks, R.M. and Harris, R.V., Jr.; Use of Grit–Type Boundary Layer Transition Trips on Wind Tunnel Models; NASA TN D–3579, September, 1966.

3.7 Hoerner, S.F.; Fluid Dynamic Drag; published by the author, 1965.

3.8 Jacobs, E.N. and Sherman, A.; Airfoil Section Characteristics as Affected by Variations of the Reynolds Number; NACA Technical report 586, 1937.

3.9 Hoerner, S.F. and Borst, H.V.; Fluid Dynamic Lift; published by L.A. Hoerner, 1975.

3.10 Anon.; Advanced Technology Airfoil research, Volume I; NASA Conference Publication 2045, Part I; Langley Research Center, Hampton, VA, March 7–9, 1978.

3.11 McGhee, R.J. and Beasley, W.D.; Low Speed Aerodynamic Characteristics of a 17–Percent–Thick Airfoil Section Designed for General Aviation Applications; NASA TN D–7428, December, 1973.

3.12 Nonweiler, T.; Maximum Lift for Symmetrical Wings; Aircraft Engineering, Vol. 27, No. 311, p. 3–8, January 1955.

3.13 Poisson–Quinton, Ph. and Lepage, L.; Survey of French Research on the Control of Boundary Layer and Circulation; in Boundary Layer and Flow Control; ed. by G.V. Lachmann, Vol. 1; Pergamon Press, 1961.

3.14 Kohlman, D.L.; Introduction to STOL Airplanes; Iowa State University Press, 1979.

3.15 McCormick, Jr., B.W.; Aerodynamics of V/STOL Flight; Academic Press, 1967.

3.16 Bauer, F., Garabedian, P. and Korn, D.; A Theory of Supercritical Wing Sections, with Computer Programs and Examples; Lecture Notes in Economic and Mathematical Systems, Vol. 6, Springer Verlag, New York, 1972.

3.17 Freuler, R.J. and Gregorek, G.M.; An Evaluation of Four Single Element Airfoil Analytic Methods; Proceedings of Advanced Technology Airfoil Research Conference, NASA Publication 2045, March, 1978.

3.18 Stevens, W.A., Goradia, S.H. and Braden, J.A.; Mathematical Model for Two–Dimensional Multi–Component Airfoils in Viscous Flows, NASA CR——1843, July, 1971.

3.19 Bauer, F., Garabedian, P., Korn, D. and Jameson, A,; Supercritical Wing Sections II, A Handbook, Lecture Notes in Economics and Mathematical Systems, Vol. 108, Springer Verlag, New York, 1975.

3.20 DeCamp, R.W. and Hardy, R,; Mission Adaptive Wing research Programme; Aircraft Engineering, January 1981, pp.10–11.

3.21 Jackson, P.; Jane's All the World's Aircraft; Jane's Information Group Ltd, 163 Brighton Road, Coulsdon, Surrey CR5 2NH, U.K., 1996–1997

CHAPTER 4: WING THEORY

The purpose of this chapter is to familiarize the reader with several key aspects of wing (or planform) theory and applications.

After a discussion of basic planform geometric parameters in Section 4.1, a discussion of circulation, downwash, lift and induced drag is given in Section 4.2.

From an airplane design viewpoint, span efficiency, aerodynamic center location, stall behavior, high speed characteristics and flaps are of prime importance. Sections 4.3 through 4.7 provide introductions to these subjects.

The material presented in this chapter is aimed at wings (or planforms). The reader will recognize that tail surfaces, canard surfaces as well as many other types of surfaces found in airplanes are also planforms. Therefore, the material in this chapter can be applied directly to such other surfaces.

4.1 DEFINITION OF WING PROPERTIES

Figure 4.1 shows a typical straight, tapered wing planform. The reader is encouraged to memorize the geometric properties shown in this figure.

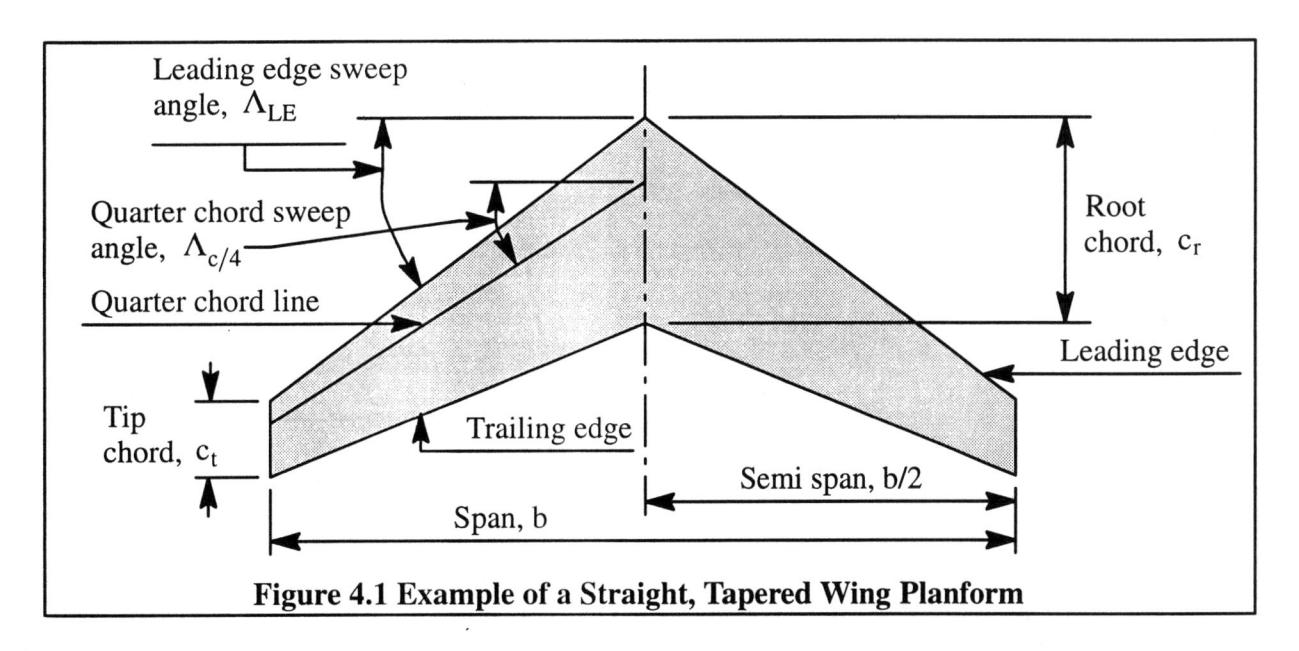

Figure 4.1 Example of a Straight, Tapered Wing Planform

The wing area, S is defined as the shaded area in Figure 4.1. In general, S is defined as the area of the wing planform, projected onto a plane of reference which is usually the wing root chord plane.

It is seen from Figure 4.1 that S may be determined from:

$$S = \frac{b}{2}(c_r + c_t) \tag{4.1}$$

In addition to wing area, other important parameters are the so-called wing aspect ratio, A and the taper ratio, λ, which are defined as:

$$A = \frac{b^2}{S} \tag{4.2}$$

and

$$\lambda = \frac{c_t}{c_r} \tag{4.3}$$

The wing sweep angle, Λ, is also of major importance. The sweep angle is normally measured either relative to the leading edge (Λ_{LE}) or relative to the quarter chord line ($\Lambda_{c/4}$).

To define lift and drag coefficients, the wing area, S, is required. To define a pitching moment coefficient it is necessary to use S in combination with a characteristic length. Normally, the so-called mean geometric chord (m.g.c.) of the wing is used for this characteristic length. The mean geometric chord of a wing is defined as:

$$m.g.c. = \bar{c} = \frac{2}{S} \int_0^{b/2} c^2 dy \tag{4.4}$$

The reader is asked to show, that for straight tapered wings the m.g.c. becomes:

$$m.g.c. = \frac{2}{3} c_r \left(\frac{\lambda^2 + \lambda + 1}{\lambda + 1} \right) \tag{4.5}$$

The simple geometric construction shown in Figure 4.2 can be used to quickly locate the m.g.c. for a straight, tapered wing.

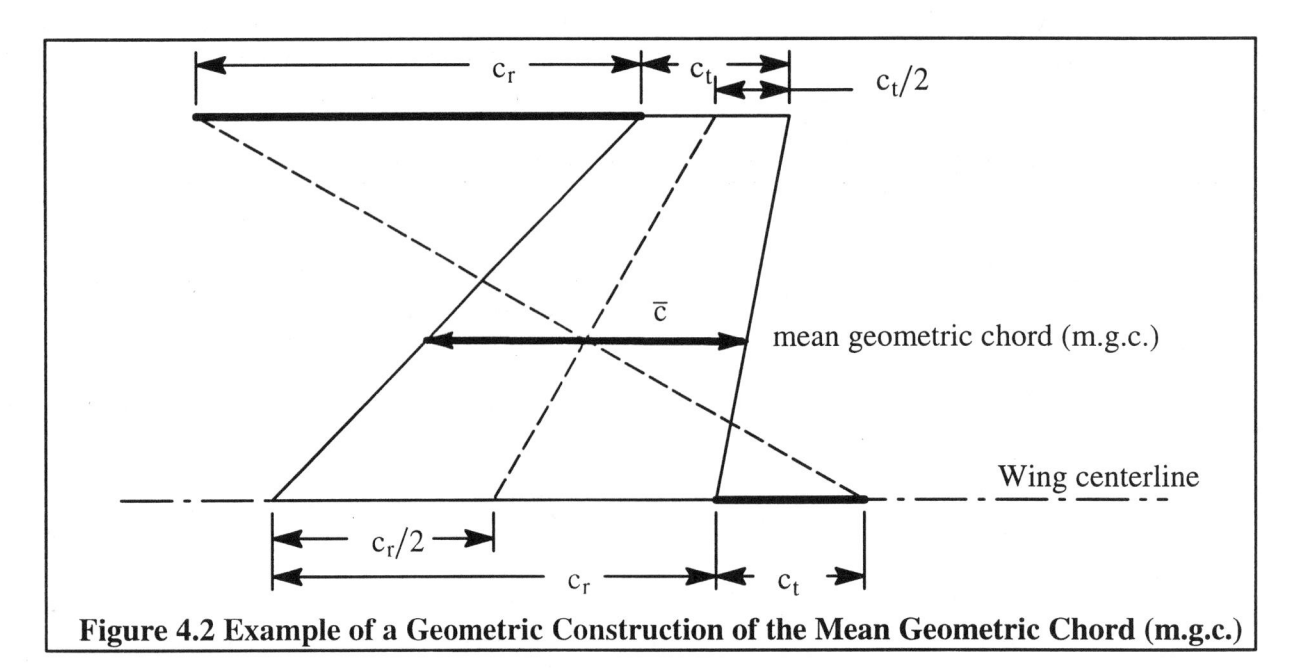

Figure 4.2 Example of a Geometric Construction of the Mean Geometric Chord (m.g.c.)

In Chapter 3 the equations for airfoil lift, drag and pitching moment coefficients were given as Eqns (3.15) through (3.17). By analogy the equations for wing (planform) lift, drag and pitching moment coefficients are as follows:

$$L = C_L \bar{q} S \tag{4.6}$$

$$D = C_D \bar{q} S \tag{4.7}$$

$$M = C_m \bar{q} S \bar{c} \tag{4.8}$$

4.2 CIRCULATION, DOWNWASH, LIFT AND INDUCED DRAG

The circulation about a wing is produced by the difference in pressure between the upper and the lower surfaces of a wing. This pressure difference induces a flow from the lower surface toward the upper surface around the leading edge and around the tips. The strength of this flow is measured in terms of circulation. The circulation can be represented by means of a bound vortex distribution on the wing and its associated trailing vortices. Figure 4.3 illustrates these ideas.

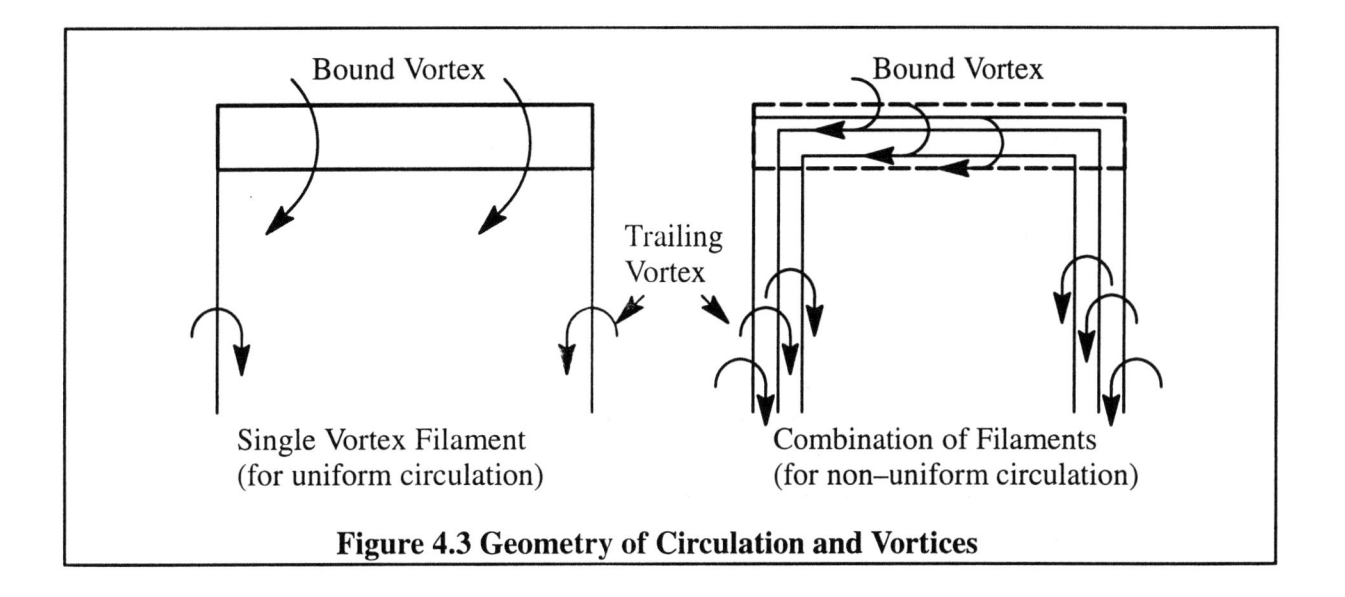

Figure 4.3 Geometry of Circulation and Vortices

The net effect of the trailing vortices is to produce a downwash velocity distribution, w, on the wing. This downwash is shown in Figure 4.4. From Figure 4.4 it may be seen that:

$$\tan \alpha_i \approx \alpha_i = \frac{w}{V} \tag{4.9}$$

In accordance with Eqn (2.78), the total force on the wing is given by:

$$F = \int_{-b/2}^{b/2} \varrho \, V \, \Gamma \, dy \tag{4.10}$$

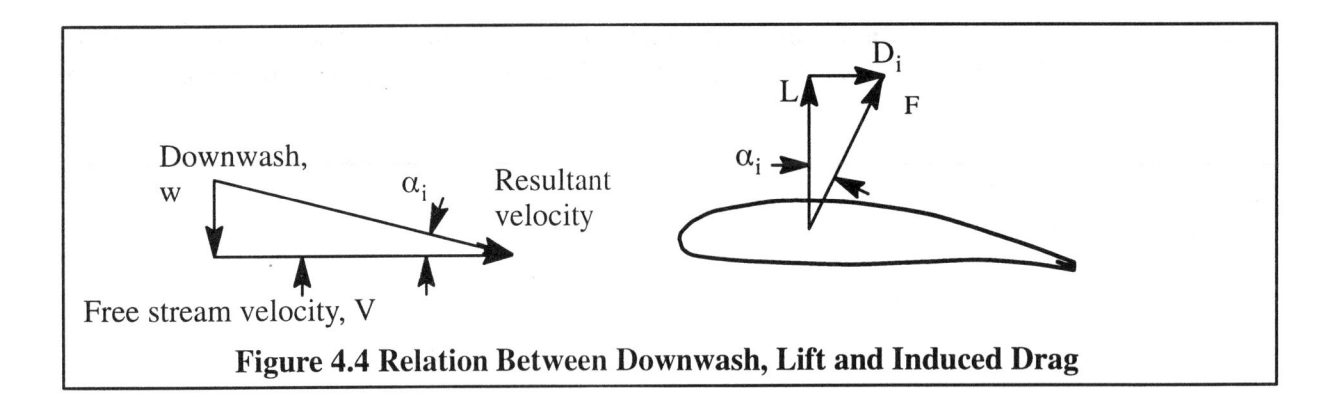

Figure 4.4 Relation Between Downwash, Lift and Induced Drag

Since α_i is small, $F \approx L$ and therefore:

$$D_i = L \tan \alpha_i \approx L \frac{w}{V} \approx L \alpha_i \qquad (4.11)$$

Let w_1 be the downwash far behind the wing. Next, consider a stream tube of cross sectional area, S' behind the wing. Figure 4.5 shows the wing with the stream tube.

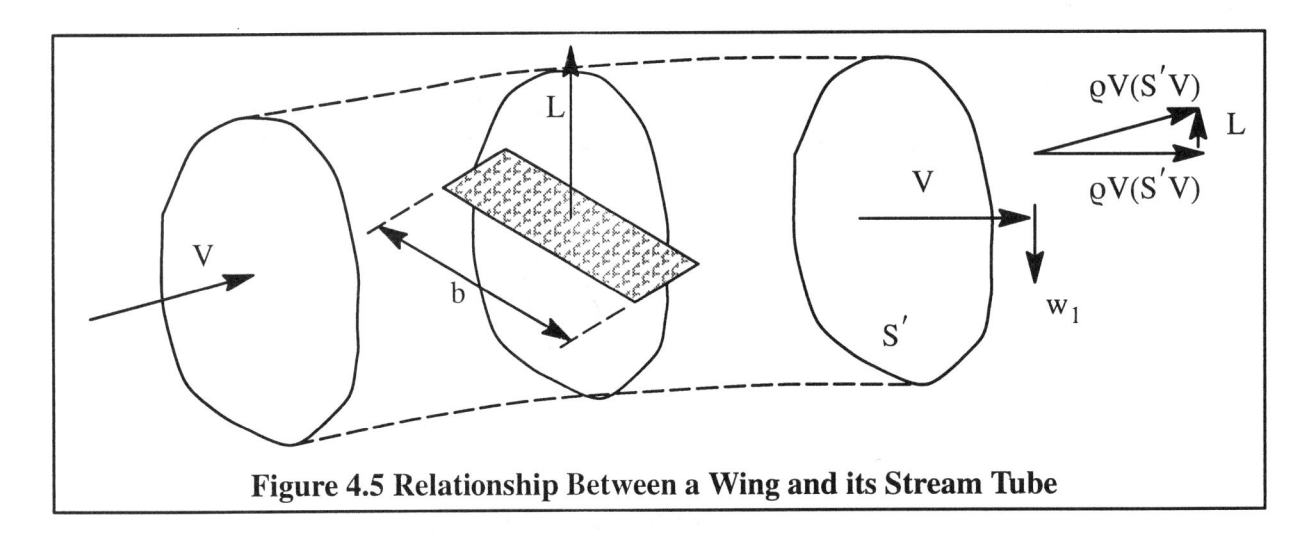

Figure 4.5 Relationship Between a Wing and its Stream Tube

Two methods for calculating lift will now be considered:

- the momentum method
- the energy method

- The Momentum Method

According to the linear momentum principle, assuming uniform downwash over S':

$$L = \varrho V(S'V) \frac{w_1}{V} = \varrho w_1 (S'V) \qquad (4.12)$$

● The Energy Method

The work done on the air mass per unit time equals the kinetic energy increase per unit time. Therefore:

$$D_i V = \varrho S' V \frac{w_1^2}{2} \tag{4.13}$$

Dividing by V yields:

$$D_i = \varrho S' \frac{w_1^2}{2} \tag{4.14}$$

Recalling Eqn (4.11):

$$L = \frac{D_i}{\alpha_i} = \frac{\varrho S' \frac{w_1^2}{2}}{w/V} \tag{4.15}$$

From this it follows that:

$$\varrho w_1 (S'V) = \varrho S' \frac{V w_1^2}{2w} \tag{4.16}$$

From this, it is now seen that:

$$2w = w_1 \tag{4.17}$$

This result means that the induced downwash far behind the wing is twice that of the downwash on the wing itself. The area, S' may be written as: $S' = \frac{\pi b^2}{4}$. Thus, it is seen that Eqn (4.15) may be written as:

$$L = \varrho(2w)\frac{\pi b^2}{4} V \tag{4.18}$$

This is just another way of writing Eqn (4.12). Now solving for the downwash, w:

$$w = \frac{2L}{\varrho \pi b^2 V} = \frac{C_L \bar{q} S V}{\pi b^2 \bar{q}} = \frac{C_L S V}{\pi b^2} = \frac{C_L V}{\pi A} \tag{4.19}$$

The angle, α_i may now be written as:

$$\alpha_i = \frac{w}{V} = \frac{C_L}{\pi A} \tag{4.20}$$

Also, from Eqn (4.11) it follows that:

$$D_i = L\alpha_i = C_L \bar{q} S \frac{C_L}{\pi A} \tag{4.21}$$

The induced drag coefficient can therefore be written as:

$$C_{D_i} = \frac{D_i}{\bar{q}S} = \frac{C_L^2}{\pi A} \tag{4.22}$$

These equations are only valid for wings with uniform downwash distribution. The latter can be achieved only if the span loading of the wing is elliptical. It has been shown that in such a case the induced drag coefficient is a minimum. It is shown in Ref. 4.1, page 142, that this condition can be achieved by using an elliptical planform. When, as is normally the case, the downwash distribution is not uniform, a correction factor "e" is introduced to yield:

$$C_{D_i} = \frac{C_L^2}{\pi A e} \tag{4.23}$$

Therefore, the induced angle, α_i can be rewritten as:

$$\alpha_i = \frac{C_L}{\pi A e} \tag{4.24}$$

where: e is the span efficiency factor, also known as Oswald's efficiency factor. Experimentally it has been found that e ranges from 0.85 to 0.95 for a wing by itself. A discussion on how to evaluate "e" from tunnel data is given in Section 4.3.

The factor Ae is frequently referred to as the effective aspect ratio: A_{eff}. The total drag coefficient for a wing can therefore be written as:

$$C_D = C_{D_0} + \frac{C_L^2}{\pi A_{eff}} \tag{4.25}$$

where: C_{D_0} is the lift independent sum of skin friction and pressure drag.

The factor Ae can be used to determine the lift curve slope of one wing from knowledge of the lift curve slope of another wing. To show this, assume that two wings have high but different aspect ratios. Also assume that both wings use the same airfoil. According to Prandtl's Lifting Line Theory (Ref. 4.1 page 137) if these wings are placed at the same effective angle of attack, $\alpha_a - \alpha_i$, their lift coefficient, C_L will be the same. The angle, $\alpha_a = \alpha - \alpha_0$ is called the absolute angle of attack and α_0 is the angle of attack for zero lift. Therefore:

$$\alpha_{a_1} - \frac{C_L}{\pi (Ae)_1} = \alpha_{a_2} - \frac{C_L}{\pi (Ae)_2} \tag{4.26}$$

or:

$$\alpha_{a_1} = \alpha_{a_2} + \frac{C_L}{\pi}\left\{\frac{1}{(Ae)_1} - \frac{1}{(Ae)_2}\right\} \tag{4.27}$$

Because the lift curve slope, a, is related to the lift coefficient by:

$$C_L = a\alpha_a \text{ (with } \alpha \text{ in radians)} \tag{4.28}$$

it follows that:

$$\frac{C_L}{a_1} = \frac{C_L}{a_2} + \frac{C_L}{\pi}\left\{\frac{1}{(Ae)_1} - \frac{1}{(Ae)_2}\right\} \tag{4.29}$$

From the latter the lift curve slope of one wing follows from that for the other wing from:

$$a_1 = \frac{a_2}{1 + \frac{a_2}{\pi}\left\{\frac{1}{(Ae)_1} - \frac{1}{(Ae)_2}\right\}} \quad \text{Note well: } a_2 \text{ is measured in } 1/\text{rad} \tag{4.30}$$

Again, this equation allows an estimate of the lift curve slope, a_1 of one wing with effective aspect ratio, $(Ae)_1$ if the corresponding, but different, properties of the other wing are known. This result should be used only for high aspect ratio wings. Generally for A>5 the relation works well. Figure 4.6 shows the effect of aspect ratio of wings with the same angle of attack for zero lift. Note that wings of varying aspect ratio tend to have approximately the same angle of attack for zero lift. The lift curve slope is seen to decrease significantly with increasing aspect ratio. That effect agrees with Eqn (4.30).

For the same wing area and for the same airfoils, the wing zero lift drag is also essentially independent of aspect ratio. Eqn (4.25) also suggests this. Application of Eqn (4.23) to two wings which differ only in aspect ratio yields:

$$C_{D_1} = C_{D_2} + \frac{C_L^2}{\pi}\left\{\frac{1}{(Ae)_1} - \frac{1}{(Ae)_2}\right\} \tag{4.31}$$

The drag polars in Figure 4.6 show that, although the drag coefficients at zero lift are essentially independent of aspect ratio, marked reductions in the drag coefficient, C_D, occur as the aspect ratio is increased. This is particularly true at high values of lift coefficient, C_L.

Equations 4.27 and 4.31 were used to reduce the data of Figure 4.6 to a wing with a common aspect ratio of 5. The data then take the form shown in Figure 4.7. Observe that, for all practical purposes, the data have collapsed onto one curve. It can therefore be concluded, that the characteristics of a wing of one aspect ratio may be predicted with considerable accuracy from similar data on a wing of different aspect ratio.

Wing Theory

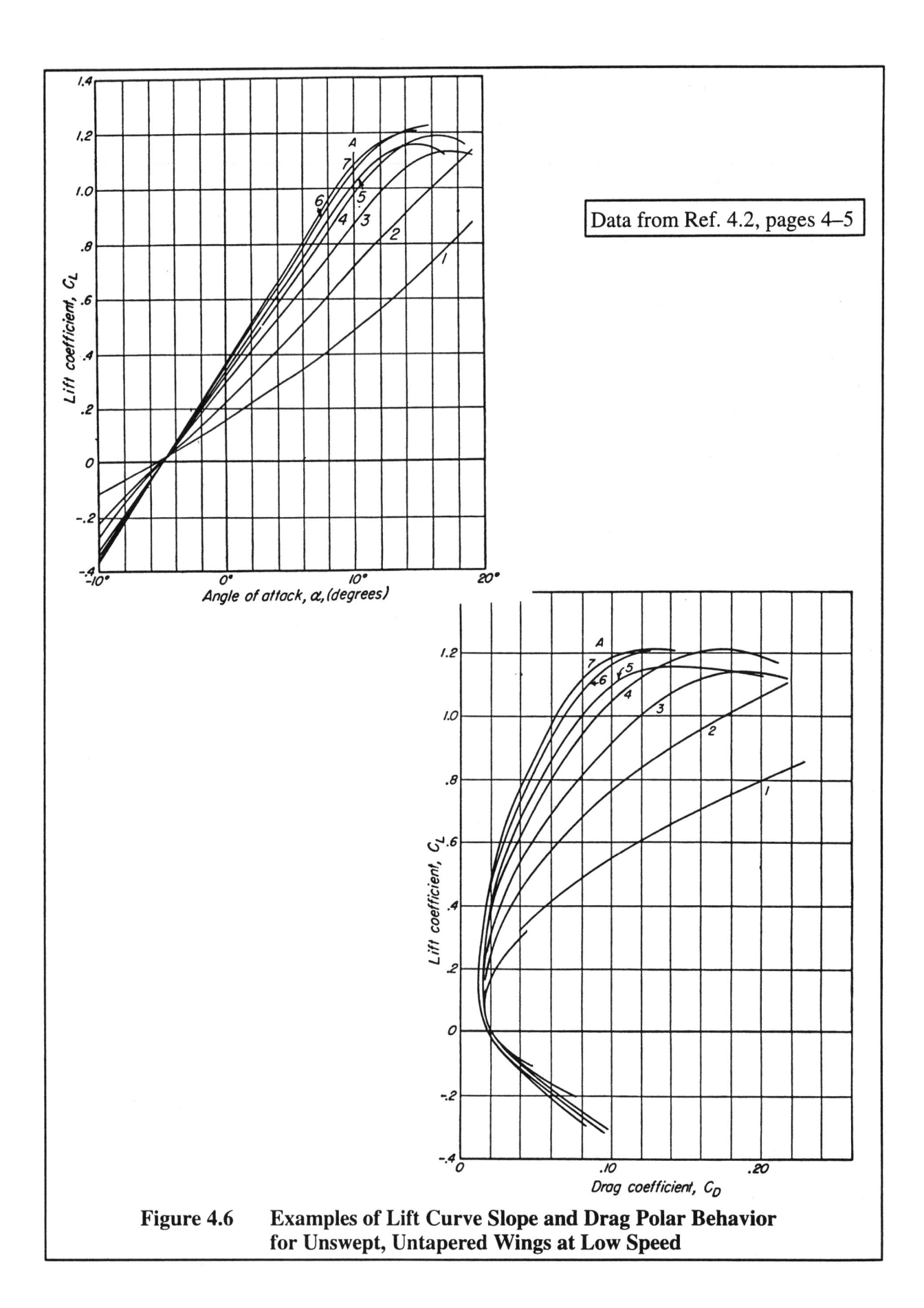

Data from Ref. 4.2, pages 4–5

Figure 4.6 Examples of Lift Curve Slope and Drag Polar Behavior for Unswept, Untapered Wings at Low Speed

Wing Theory

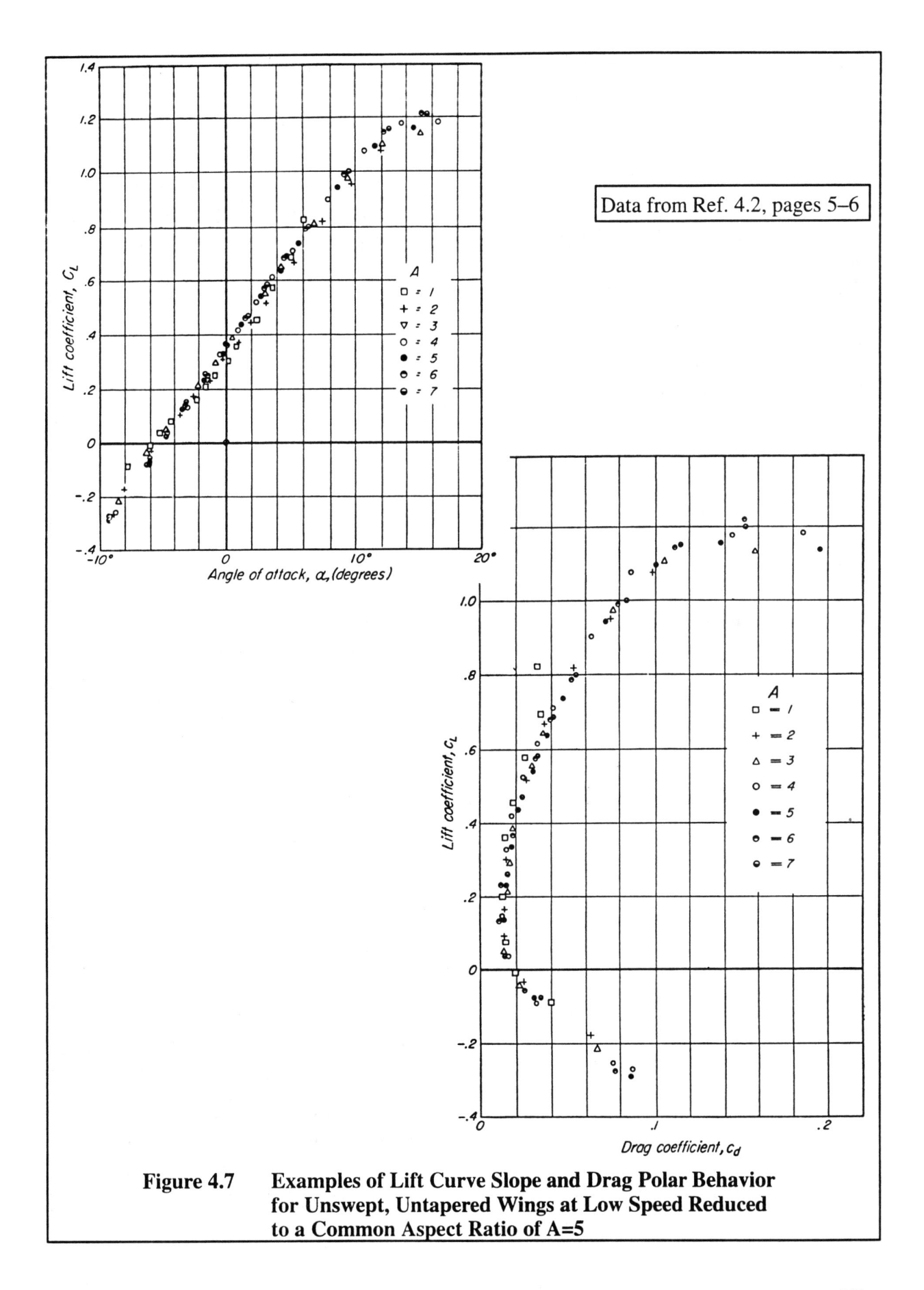

Data from Ref. 4.2, pages 5–6

Figure 4.7 Examples of Lift Curve Slope and Drag Polar Behavior for Unswept, Untapered Wings at Low Speed Reduced to a Common Aspect Ratio of A=5

Eqn (4.30) can be used to determine the lift curve slope of a wing 1 when the characteristics of a wing 2 with the same airfoil section are known. If wing 2 has an infinite aspect ratio (i.e. is the equivalent of a two–dimensional airfoil section), Eqn (4.30) can be reduced to:

$$a_1 = \frac{a_\infty}{1 + \frac{a_\infty}{\pi A e}} \qquad (4.32)$$

The reader should be careful in using Eqn (4.32). It has been found to produce good results only for wings with aspect ratios larger than about 5 and with sweep angles below about 15 degrees. An improved version of Eqn(4.32) has been developed by Lowry and Polhamus (See Reference 4.3) for arbitrary aspect ratios and sweep angles in subsonic flow:

$$a = \frac{2\pi A}{\left\{2 + \sqrt{\frac{A^2 \beta^2}{\varkappa^2}\left(1 + \frac{\tan^2 \Gamma_{c/2}}{\beta^2}\right) + 4}\right\}} \qquad (4.33)$$

where: $\beta^2 = 1 - M_\infty^2$ \qquad (4.33a)

and: $\varkappa = \frac{a_\infty}{2\pi/\beta}$

The wing lift curve slope, a, is also given the symbol: C_{L_α}.

In Eqn (4.33), the term $2\pi/\beta$ is the theoretical, sectional lift curve slope obtained by the Prandtl–Glauert transformation of Eqn (3.34). It follows that $\varkappa$ is the ratio of experimental airfoil lift curve slope at a subsonic Mach number to its theoretical value. It can be shown that if A is large and is taken as the effective aspect ratio, A_{eff}, while $\Lambda_{c/2}$ is small, and $a_\infty = 2\pi$, Eqn (4.32) can be reduced to Eqn (4.31) for $M_\infty = 0$.

More accurate results can be obtained through the so–called lifting surface theories (see, for example, Refs 4.4 through 4.6). Lifting surface theories can calculate not only the lift curve slope of any arbitrary wing, but also the induced drag, pitching moment and load distribution.

4.3 EVALUATION OF THE SPAN EFFICIENCY FACTOR e

To expedite performance calculations and to simplify reduction of flight test data, it is convenient to represent the lift–drag coefficient curve by the following equation:

$$C_D = C_L + C_1 C_L + C_2 C_L^2 + \ldots \ldots \qquad (4.34)$$

For wings with A > 3 and Λ < 30 deg and not incorporating airfoils with wide, low drag buckets, the following parabolic drag polar equation provides a reasonable curve fit for lift coefficient values ranging from 0 to 1:

$$C_D = C_{D_0} + \frac{C_L^2}{\pi A e} \qquad (4.35)$$

Wing Theory

The factor e in Eqn (4.35) is called the Oswald efficiency factor. This factor can be enhanced by features such as tip–tanks, winglets and proper wing–fuselage blending (i.e. fairing). The basic form of Eqn (4.35) has been shown to be applicable at subsonic and supersonic Mach numbers. However, at high Mach numbers, e has been shown to be dependent on Mach number. This fact is illustrated in Figures 4.8 and 4.9 for wing–body combinations and for a supersonic fighter. The vertical axis is the derivative of the drag coefficient with respect to the square of the lift coefficient:

$$\frac{dC_D}{dC_L^2} = \frac{1}{\pi A e} \tag{4.36}$$

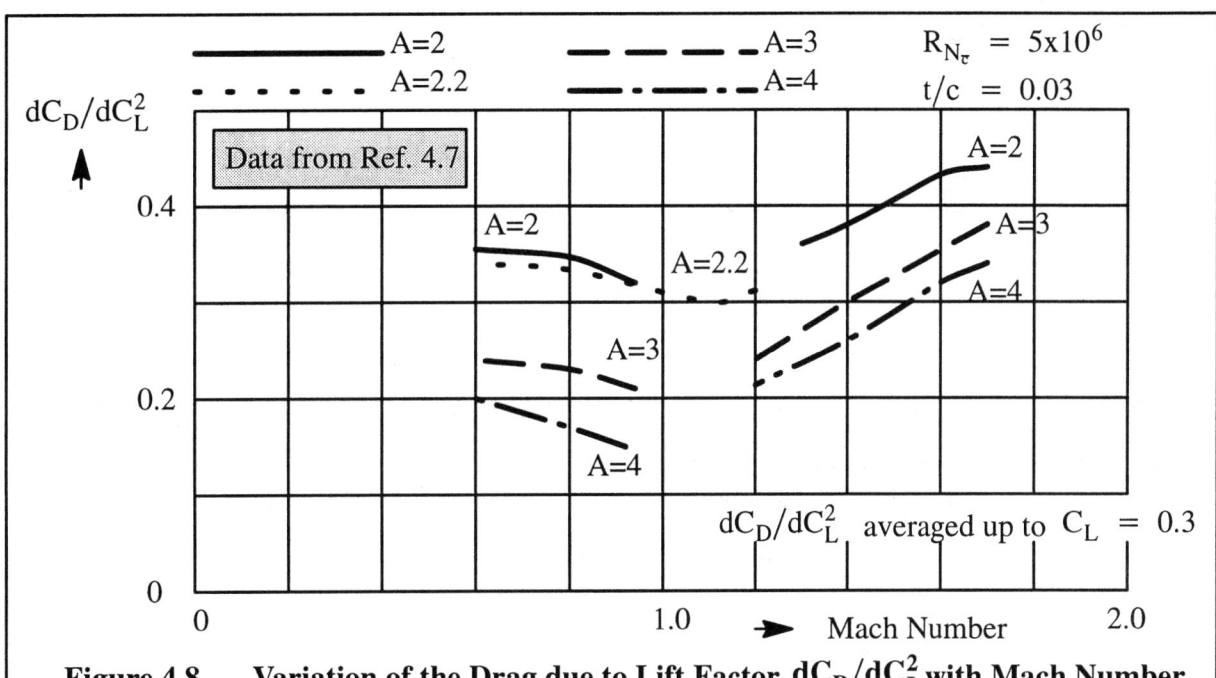

Figure 4.8 Variation of the Drag due to Lift Factor, dC_D/dC_L^2 with Mach Number for Delta Wing Plus Body Combinations

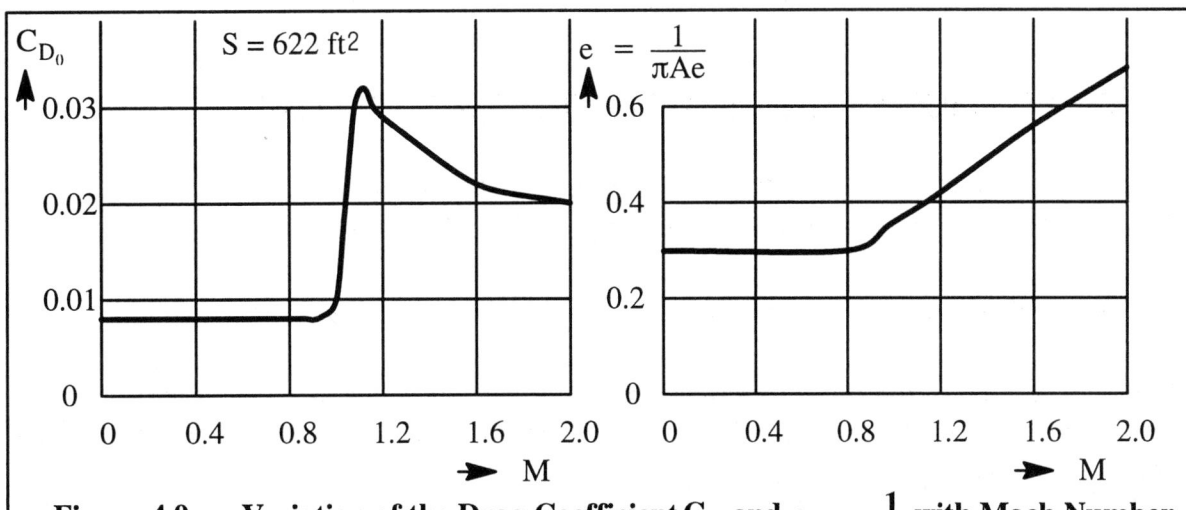

Figure 4.9 Variation of the Drag Coefficient, C_D and $e = \frac{1}{\pi A e}$ with Mach Number for a Supersonic Fighter Configuration

At supersonic speeds, C_{D_0} will increase with Mach Number because of the addition of wave drag. This sharp drag rise is also shown in Figure 4.9.

Ideally, if C_D of Eqn (4.35) is plotted versus C_L^2, a straight line should be obtained, with a slope of $1/\pi Ae$. This has been shown to be the case only for Reynolds numbers higher than about 5×10^6. Figure 4.10 backs this up with some test data.

Figure 4.10 Example of the dependence of dC_D/dC_L^2 on Reynolds Number

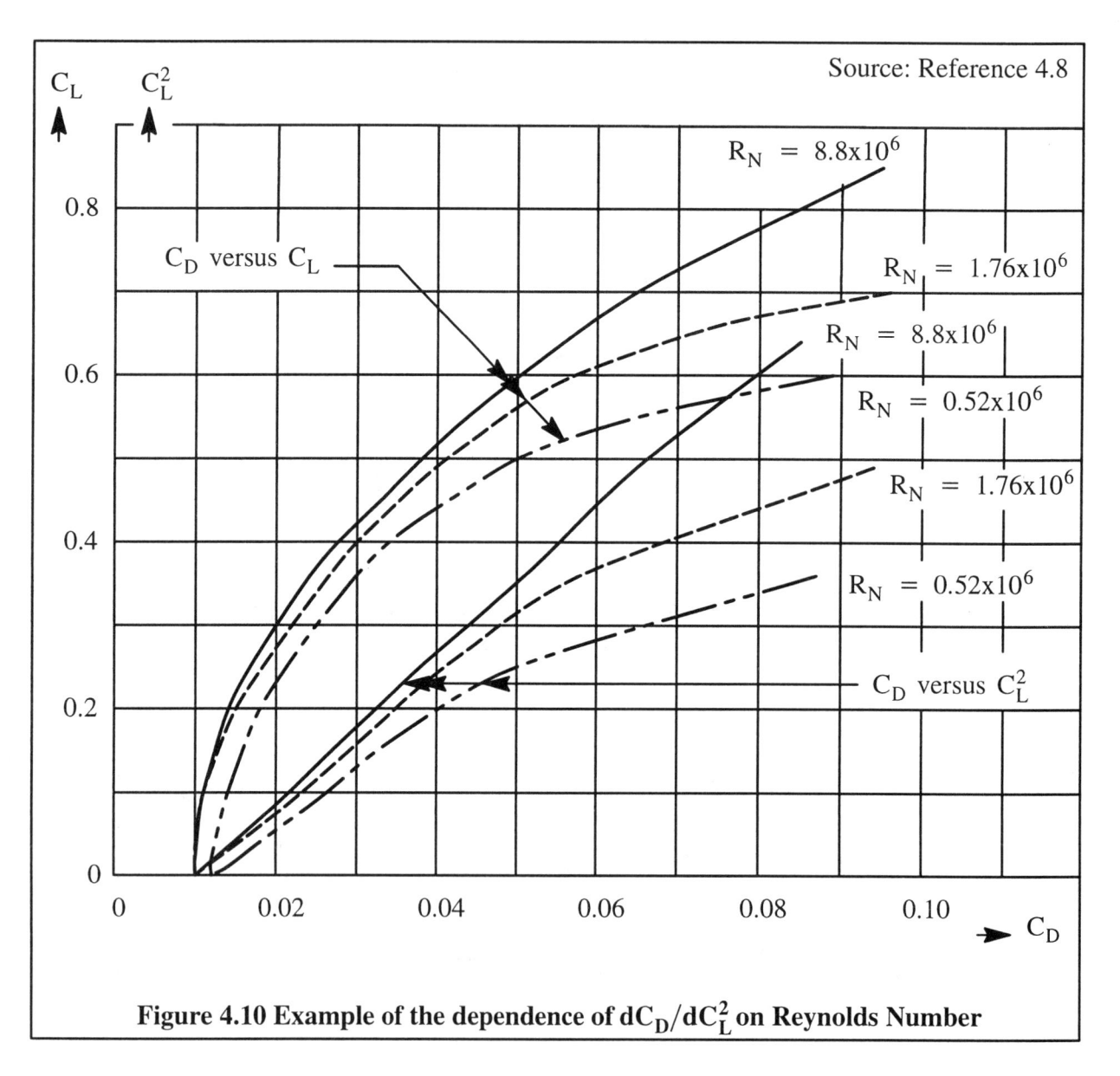

Table 4.1 lists typical in flight Reynolds numbers associated with the wing, horizontal tail, vertical tail and fuselages of a number of airplanes.

Table 4.1 Example of Typical Flight Reynolds Numbers of Airplanes					
		Reynolds Number			
		All Dimensions in Feet			
Airplane Type	Flight Condition	Wing	Hor. Tail	Vert. Tail	Fuselage
Cessna Stationair 7		$\bar{c}_w = 5.2$	$\bar{c}_h = 3.7$	$\bar{c}_v = 3.3$	$l_{fus} = 28.3$
	Max. Cruise Speed at 10,000 ft: $V_{CR} = 170$ mph	6.4×10^6	4.6×10^6	4.1×10^6	34.8×10^6
	Stall Speed (flaps down) at Sealevel: $V_S = 67$ mph	3.2×10^6	2.3×10^6	2.1×10^6	17.4×10^6
Learjet Model 36		$\bar{c}_w = 7.0$	$\bar{c}_h = 4.1$	$\bar{c}_v = 7.9$	$l_{fus} = 45.9$
	Max. Cruise Speed at 35,000 ft: $V_{CR} = 534$ mph	13.5×10^6	7.9×10^6	15.2×10^6	88.5×10^6
	Stall Speed (flaps down) at Sealevel: $V_S = 106$ mph	6.9×10^6	4.0×10^6	7.8×10^6	45.2×10^6
Boeing 727-200		$\bar{c}_w = 16.8$	$\bar{c}_h = 12.8$	$\bar{c}_v = 22.2$	$l_{fus} = 130.4$
	Max. Cruise Speed at 25,000 ft: $V_{CR} = 599$ mph	49.0×10^6	37.3×10^6	64.8×10^6	378.1×10^6
	Stall Speed (flaps down) at Sealevel: $V_S = 122$ mph	19.1×10^6	14.6×10^6	25.2×10^6	148.0×10^6
Boeing 747-200B		$\bar{c}_w = 30.8$	$\bar{c}_h = 24.6$	$\bar{c}_v = 27.0$	$l_{fus} = 229.0$
	Max. Cruise Speed at 30,000 ft: $V_{CR} = 608$ mph	77.8×10^6	62.1×10^6	68.3×10^6	578.0×10^6
	Stall Speed (flaps down) at Sealevel: $V_S = 116$ mph	32.8×10^6	26.2×10^6	28.8×10^6	244.0×10^6

4.4 AERODYNAMIC CENTER

The aerodynamic center (a.c.) of a wing is defined in the same way as that of an airfoil section (see Chapter 3). The aerodynamic center of a wing is defined as that point about which the variation of pitching moment coefficient, C_m is invariant with angle of attack, α. This definition is contrasted with that of the center of pressure (c.p.). The center of pressure is that point at which the pitching moment coefficient is zero. It follows, that the location of the c.p. varies with angle of attack.

To determine the a.c. from experimental data, assume that the moment center for the data is at a distance, x from the leading edge of the mean geometric chord of the wing: see Figure 4.11. Taking moments about the aerodynamic center it follows that:

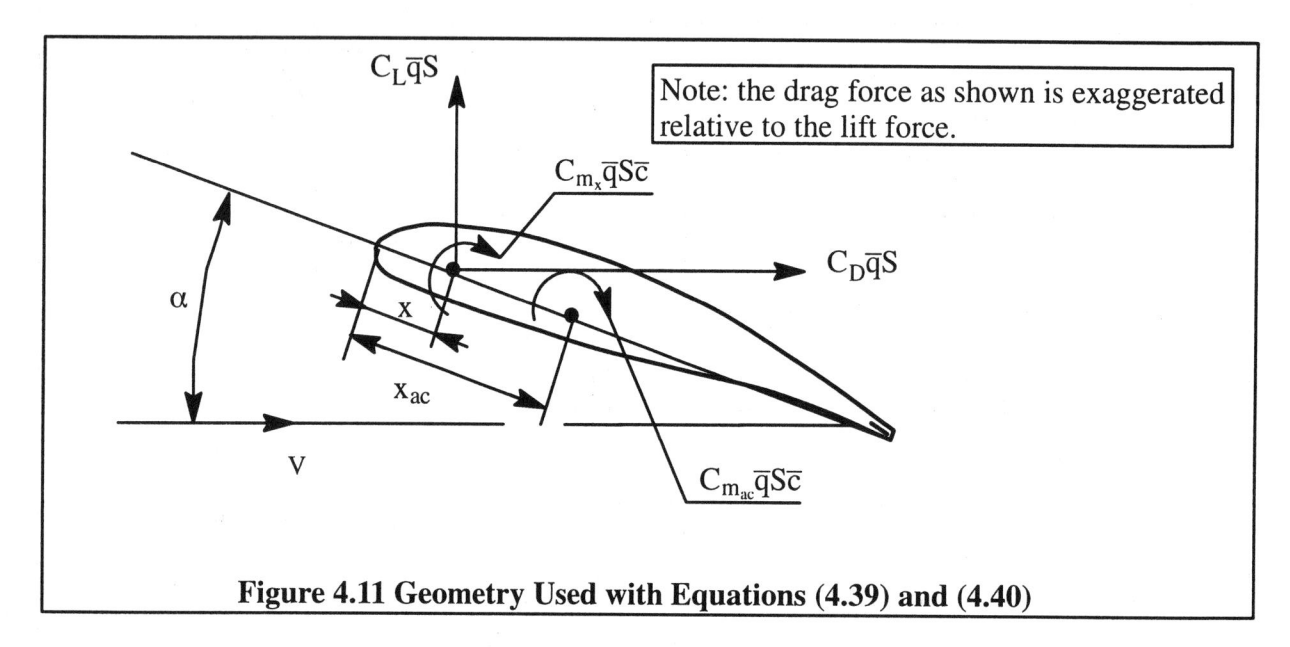

Figure 4.11 Geometry Used with Equations (4.39) and (4.40)

$$C_{m_{ac}} \bar{q} S \bar{c} = C_{m_x} \bar{q} S \bar{c} + C_L \bar{q} S (x_{ac} - x) \cos\alpha + C_D \bar{q} S (x_{ac} - x) \sin\alpha \qquad (4.37)$$

Solving for x_{ac}, it is found that:

$$\frac{x_{ac}}{\bar{c}} = \frac{x}{c} - \frac{C_{m_x} - C_{m_{ac}}}{C_L \cos\alpha + C_D \sin\alpha} \qquad (4.38)$$

As long as the angle of attack is small, this can be written as:

$$\frac{x_{ac}}{\bar{c}} \approx \frac{x}{c} - \frac{C_{m_x} - C_{m_{ac}}}{C_L} \qquad (4.39)$$

At zero lift, the pressure distribution on the wing appear as a pure moment. Since a pure moment may be transferred to any location without changing its magnitude, and since $C_{m_{ac}}$ is independent of angle of attack, the zero lift pitching moment coefficient, $\bar{C}_{m_0}$ must be equal to the pitching moment coefficient about the aerodynamic center, $C_{m_{ac}}$. Therefore:

$$C_{m_{ac}} = \overline{C}_{m_0} \qquad (4.40)$$

Figure 4.12 presents some experimental data showing typical a.c. locations for several wings. Empirical methods for determining a.c. locations and pitching moments of arbitrary wings may be found in References 4.9 and 4.10.

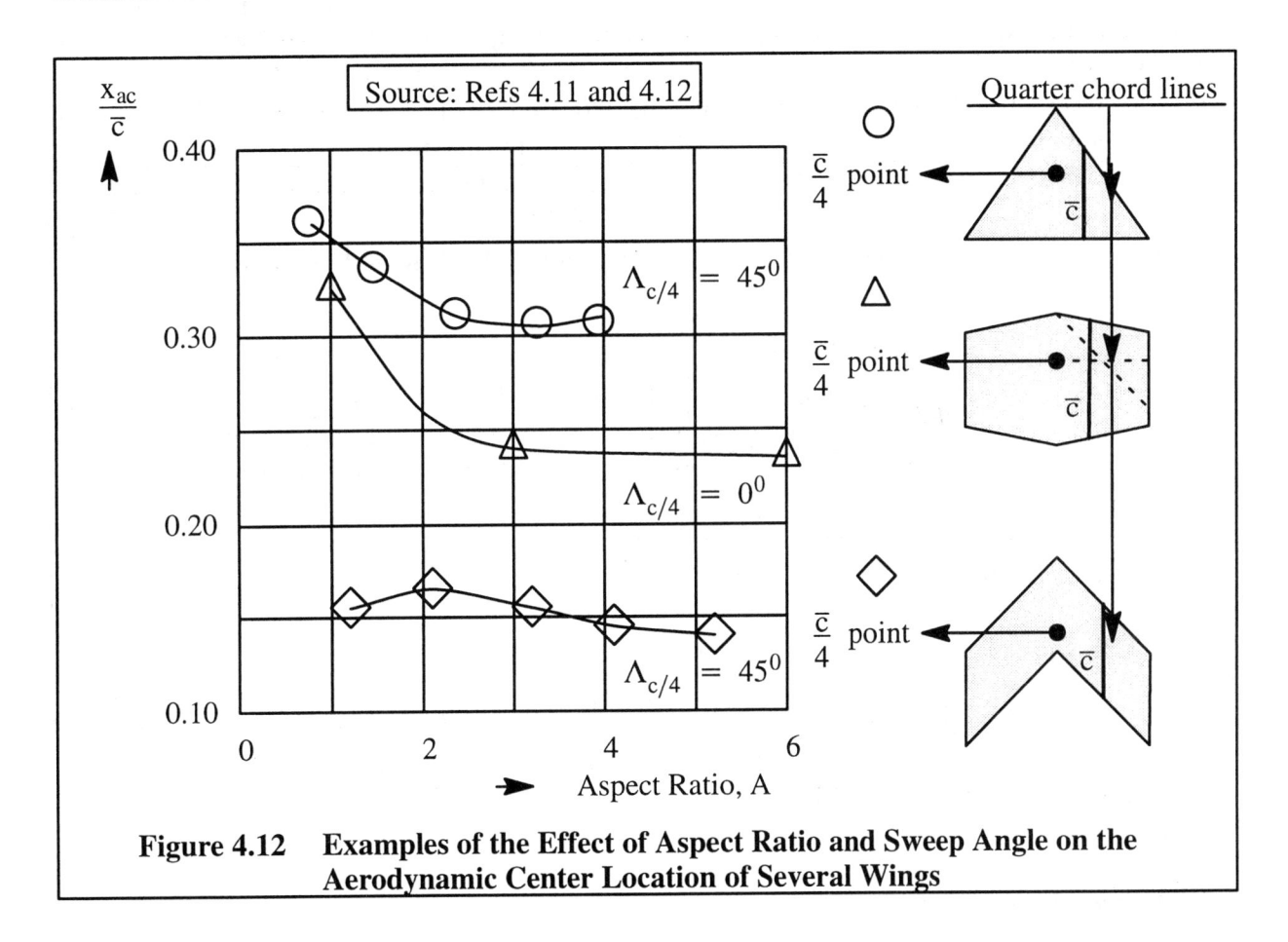

Figure 4.12 Examples of the Effect of Aspect Ratio and Sweep Angle on the Aerodynamic Center Location of Several Wings

4.5 WING STALL

Wing stall is caused by flow separation. How flow separation progresses in a chordwise and spanwise manner depends on the following items:

4.5.1 Airfoil stall characteristics

4.5.2 Planform geometry and twist

4.5.3 Stall control devices

The stall behavior of wings (or more general of lifting surfaces) is important for the following reasons:

1. In FAR* 23 airplanes, the stall speed at maximum weight may not be more than 61 knots (for W<6,000 lbs unless certain crash safety provisions have been incorporated into the design). Also, the bank angle must not exceed 15 degrees between the onset of the stall with the wings level and the completion of the recovery.

2. In FAR 25 airplanes, an airplane must not reach a bank angle of more than 20 degrees between the onset of the stall with the wings level and the completion of the recovery.

3. It is generally of great interest for either the performance of an airplane or for reasons of stability and control, to achieve the highest possible value for the maximum lift coefficient on a wing. This is true within certain constraints involving mission requirements and cost considerations.

For these three reasons, the stall characteristics of wings are of great interest to designers.

4.5.1 AIRFOIL STALL CHARACTERISTICS

Airfoil stall behavior and the factors which affect it were discussed in Chapter 3. The reader should realize, that sudden airfoil stall behavior does not necessarily imply sudden wing stall behavior. The effects of planform design can significantly modify any airfoil tendency to rapid stall. These planform effects are discussed in Sub–section 4.5.2.

4.5.2 EFFECT OF PLANFORM AND TWIST

The following planform effects are important in affecting the stall behavior of a wing (or lifting surface):

4.5.2.1	Taper ratio	4.5.2.3	Sweep angle
4.5.2.2	Aspect Ratio	4.5.2.4	Twist and camber

4.5.2.1 Taper Ratio

A wing with a rectangular planform (taper ratio of 1.0) has a larger downwash angle at the tip than at the root. Therefore, the effective angle of attack at the tip is reduced compared to that at the root. Therefore, the tip will tend to stall later than the root. However, as shown in Section 4.2 a rectangular wing planform is also aerodynamically inefficient. This is because the spanwise load distribution is far from elliptical, which is needed to minimize induced drag. To reduce induced drag a planform is tapered, to approximate the ideal, elliptical span load distribution. The result of taper is a smaller tip chord. That in turn results in a lower tip Reynolds number as well as a lower tip induced downwash angle. Both effects lower the angle of attack at which stall occurs and therefore the tip may stall before the root. This is undesirable from a viewpoint of lateral stability and lateral

---

*   FAR: Federal Aviation Regulations

controllability as the stall is approached. To counteract these tendencies, twist is applied to many wings. Figure 4.13 illustrates the effect of wing taper on the spanwise load distribution. It is seen that decreasing the taper ratio will increase the loading at the tip, which in turn promotes tip stall. This problem can be solved with twist as shown in Sub–sub section 4.5.2.4.

Another problem with a rectangular wing is that it is also structurally inefficient: there is a lot of area outboard, which supports very little lift. Taper helps solve this problem as well.

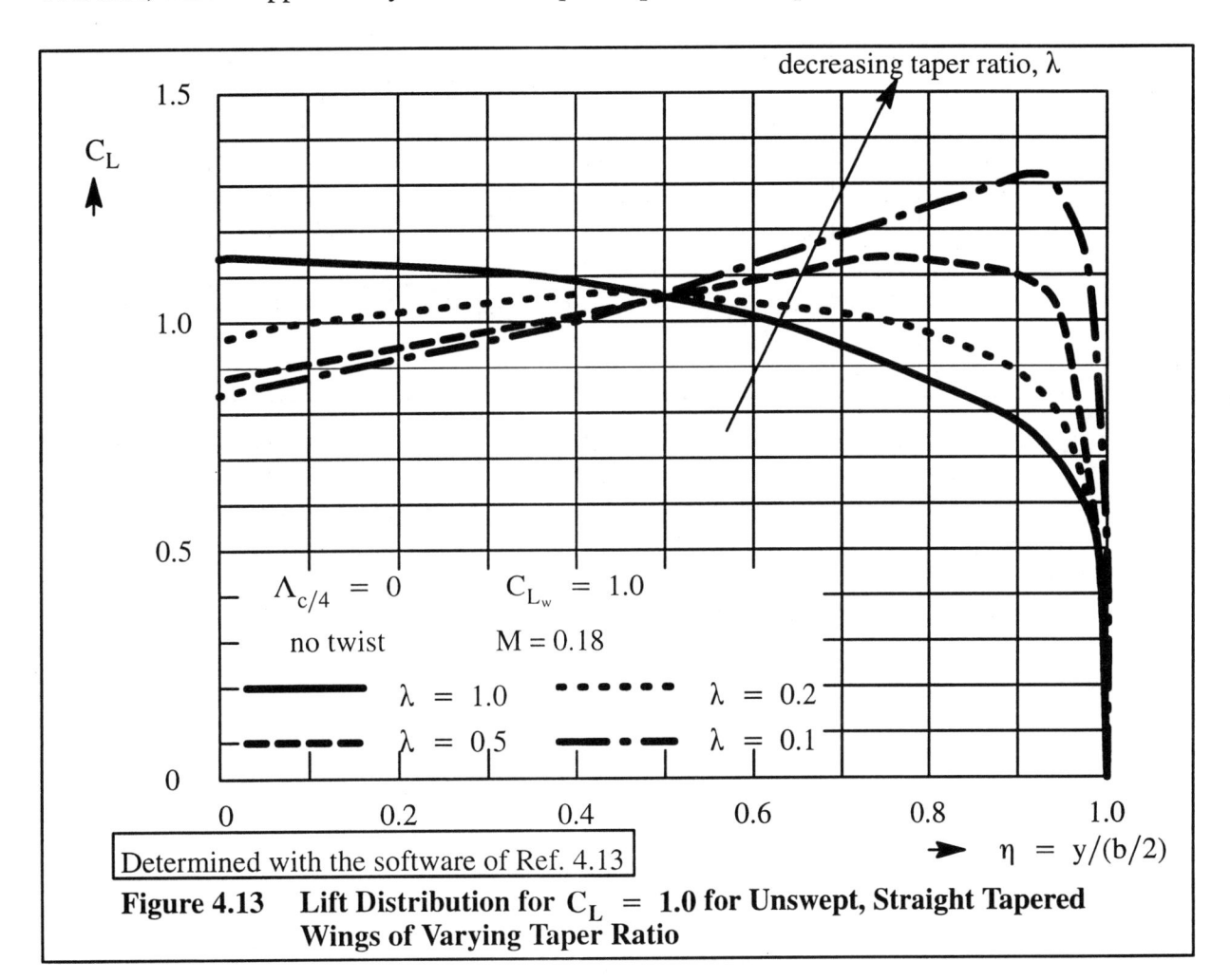

Figure 4.13 Lift Distribution for $C_L = 1.0$ for Unswept, Straight Tapered Wings of Varying Taper Ratio

4.5.2.2 Aspect Ratio

The following discussion applies to wings with very low sweep angles. As the wing aspect ratio increases, the wing behaves more and more like an airfoil. That is, its flow characteristics are more and more 2–dimensional. An exception is always the region at the wing tip. Therefore, it can be expected that the maximum lift coefficient, $C_{L_{max}}$ will increase with increasing aspect ratio, up to a number corresponding to section maximum lift coefficient, $c_{l_{max}}$. The increase is slight as can be seen from the experimental data in Figure 4.14 (Ref. 4.14, p. 16–3) for sweep angles around zero: see the shaded box.

Chapter 4

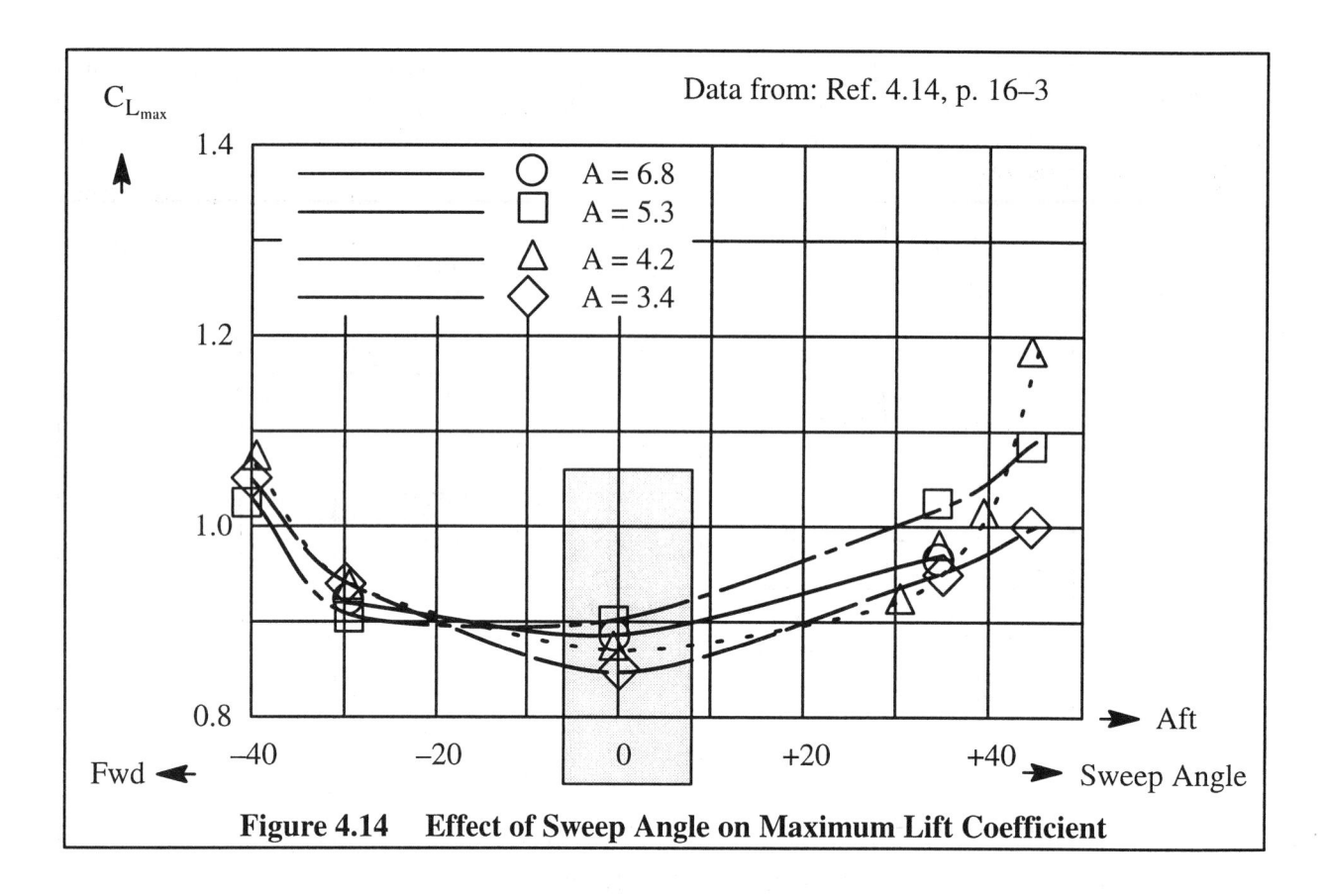

Figure 4.14   Effect of Sweep Angle on Maximum Lift Coefficient

4.5.2.3 Sweep Angle

On most aft swept wing airplanes the wing tips are located behind the center of gravity. Therefore, any loss of lift at the wing tips causes the center of pressure to move forward. This in turn will cause the airplane nose to come up. This pitch–up tendency can cause the angle of attack of the airplane to increase even further. That can result in a loss of control. The reader is asked to visualize why a forward swept wing airplane would exhibit a pitch–down tendency in a similar situation.

In addition, an aft swept wing will tend to have tip stall because of the tendency toward outboard, spanwise flow, causing the boundary layer to thicken as it approaches the tips. A swept forward wing, for the same reason, would tend toward root stall.

The maximum lift coefficient (defined as the maximum value reached by the lift coefficient as angle of attack is increased) can actually increase with increasing sweep angle. This is shown in Figure 4.14. However, the accompanying variation of pitching moment with angle of attack can lead to serious pitch control difficulties because of the tendency toward pitch–up as shown in the insert of Figure 4.15. Whether or not pitch–up occurs depends not only on the combination of aspect ratio and sweep angle, but also on airfoil type, twist and taper ratio. Figure 4.15 shows a boundary between stable and unstable pitching moment behavior as aspect ratio and sweep angle are varied. For these reasons a range of useful lift coefficients is defined as that range within which control problems are "manageable". Under this rather loose definition it can be shown that the maximum, useful lift coefficient actually decreases with increasing sweep angle. An example is shown in Figure 4.16 which is based on Reference 4.14, page 16–6.

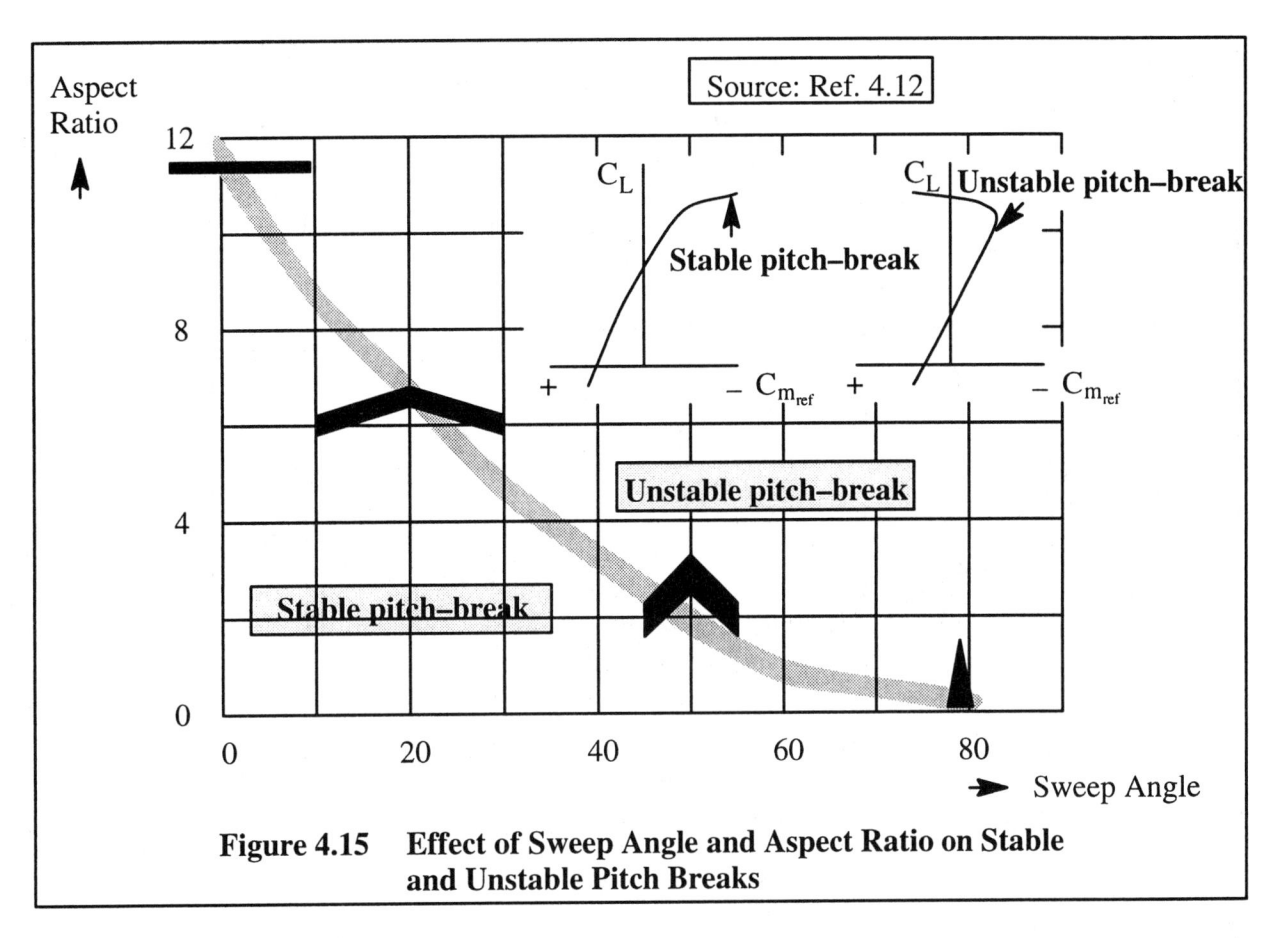

Figure 4.15 Effect of Sweep Angle and Aspect Ratio on Stable and Unstable Pitch Breaks

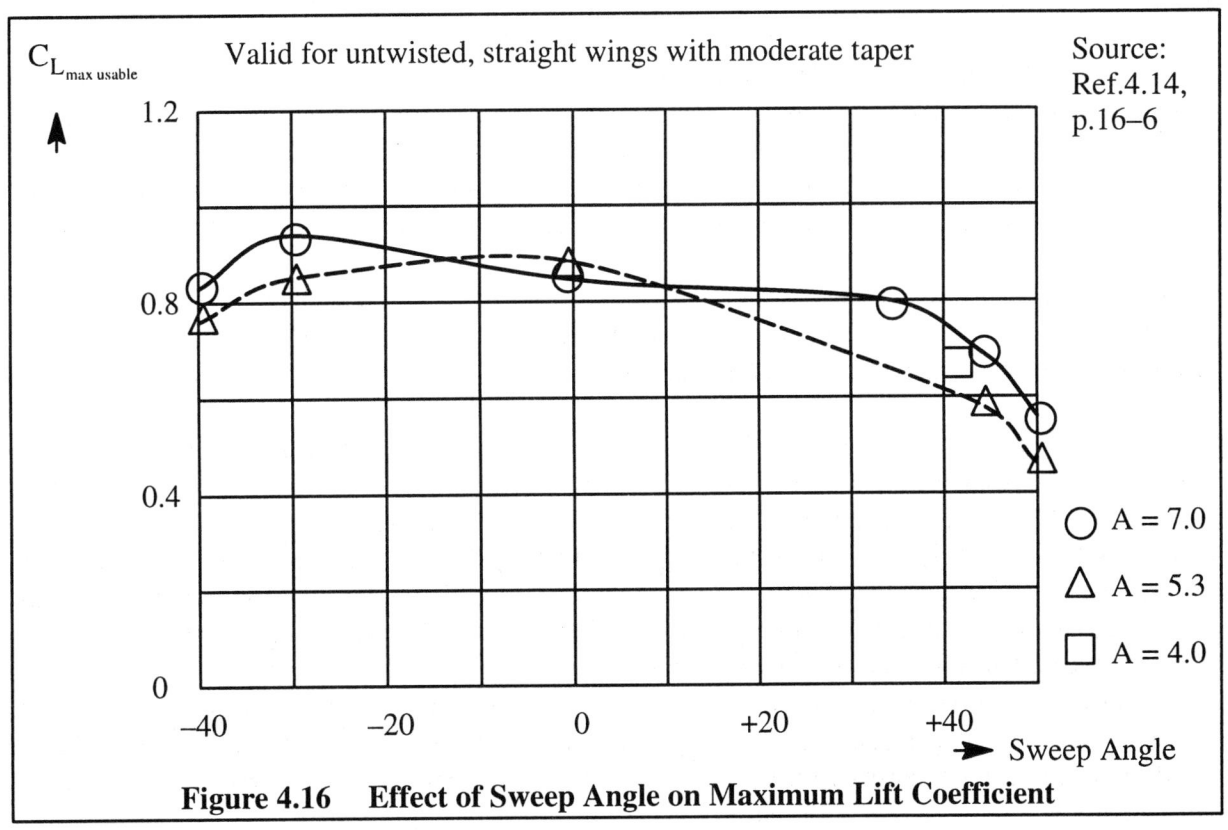

Figure 4.16 Effect of Sweep Angle on Maximum Lift Coefficient

The trend is for the useful wing $C_{L_{max}}$ to decrease with increasing aft sweep in the moderate sweep angle range of +/–25 degrees. Initially, this decrease follows the cosine–rule of Ref. 4.15, page 339):

$$C_{L_{max}}(\Gamma) = \{C_{L_{max}}(\Gamma = 0)\}\cos\Gamma \tag{4.41}$$

For higher sweep angles the maximum lift coefficient falls off rapidly with increasing sweep angles (fore and aft!).

4.5.2.4 Twist

If the angles of attack of spanwise sections of a wing are not equal, the wing is said to have twist. If the angle of attack at the tip is less than that at the root the wing is said to have wash–out or negative twist. With wash–out the wing tip will be at a lower angle of attack than the root thus delaying tip stall. Figure 4.17 illustrates how twist influences the spanwise load distribution. Note that the load is concentrated further inboard with wash–out (negative twist).

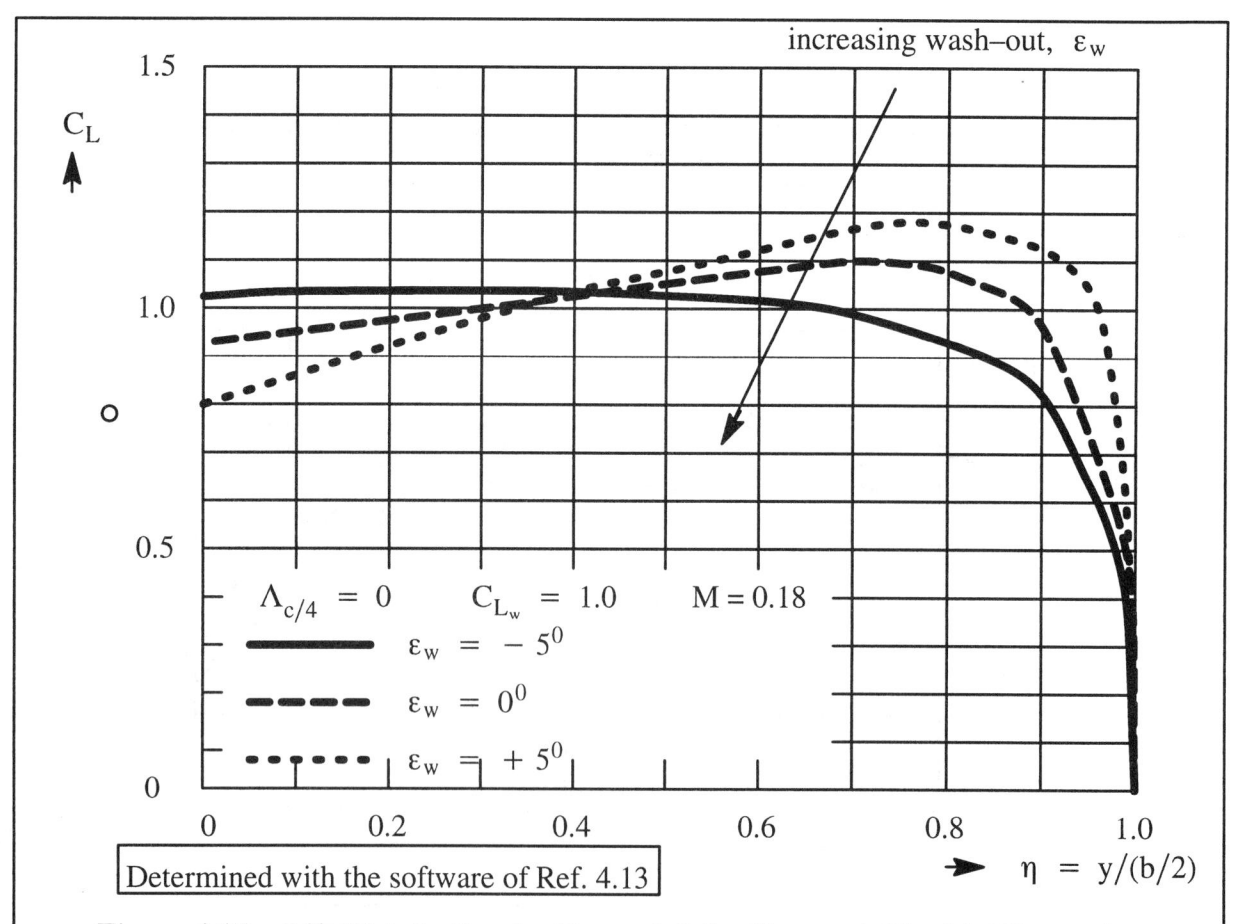

Figure 4.17   Lift Distribution for $C_L = 1.0$ for Unswept, Straight Tapered Wings with Three Twist Angles

4.5.3 STALL CONTROL DEVICES

In the following a number of devices for delaying tip stall are enumerated.

4.5.3.1 Twist or Wash–out

The effectiveness of washout in reducing tip stall was discussed in Sub–sub–section 4.5.2.4. Examples of the numerical magnitude of twist (or wash–out) on several airplanes are provided in Table 4.2.

Table 4.2 Examples of Washout in Several Airplanes

Airplane Type	Wing Incidence Angle in Degrees		Twist or Wash–out in degrees
	At Root	At Tip	
Cessna Stationair 6	+ 1.5	– 1.5	3.0
Cessna 310	+ 2.5	– 0.5	3.0
Cessna Titan	+ 2.0	– 1.0	3.0
Cessna Citation I	+ 2.5	– 0.5	3.0
Beechcraft T–34C	+4.0	+1.0	3.0
Beechcraft 55 Baron	+4.0	0.0	4.0
Beechcraft Queenair	+3.9	0.0	3.9
Beechcraft Kingair	+4.8	0.0	4.8
Beechcraft T–1A Jayhawk	+3.0	–3.3	6.6
Gulfstream IV	+3.5	–2.0	5.5
Northrop–Grumman E–2C Hawkeye	+4.0	+1.0	3.0
Piper PA–28–161 Warrior	+2.0	–1.0	3.0
Piper Cheyenne	+1.5	–1.0	2.5
Piper Tomahawk	+2.0	0.0	2.0
Fokker F–50	+3.5	+1.5	2.0

4.5.3.2 Variations in Section Shape

Many airplanes have wings with spanwise varying airfoil sections. A frequently used feature which accomplishes the same as twist is to change camber in the spanwise direction. This is sometimes referred to as aerodynamic twist.

4.5.3.3 Leading Edge Slats or Slots Near the Tip

It was shown in Chapter 3 that leading edge slats or slots can significantly enhance the value of $c_{l_{max}}$ of an airfoil. Such devices are and have been used over part of the outboard span of a wing to delay tip stall. As seen in Figure 3.??, the angle of attack at which the section stalls is increased by the use of slats or slots.

Examples of slatted wings are the North American F–86 and Sabreliner airplanes. A fixed slot is used on the Globe Swift. Movable slats are used on the DC–9 and Boeing 727 and 737.

4.5.3.4 Stall Fences and Snags

Stall fences are used to prevent the boundary layer from drifting outboard toward the tips. Boundary layers on swept wings tend to do this because of the spanwise pressure gradient of a swept wing. Similar results can be achieved with a leading edge snag. Such snags tend to create a vortex which act like a boundary layer fence. Examples of a boundary layer fence and a leading edge snag are shown in Figure 4.18.

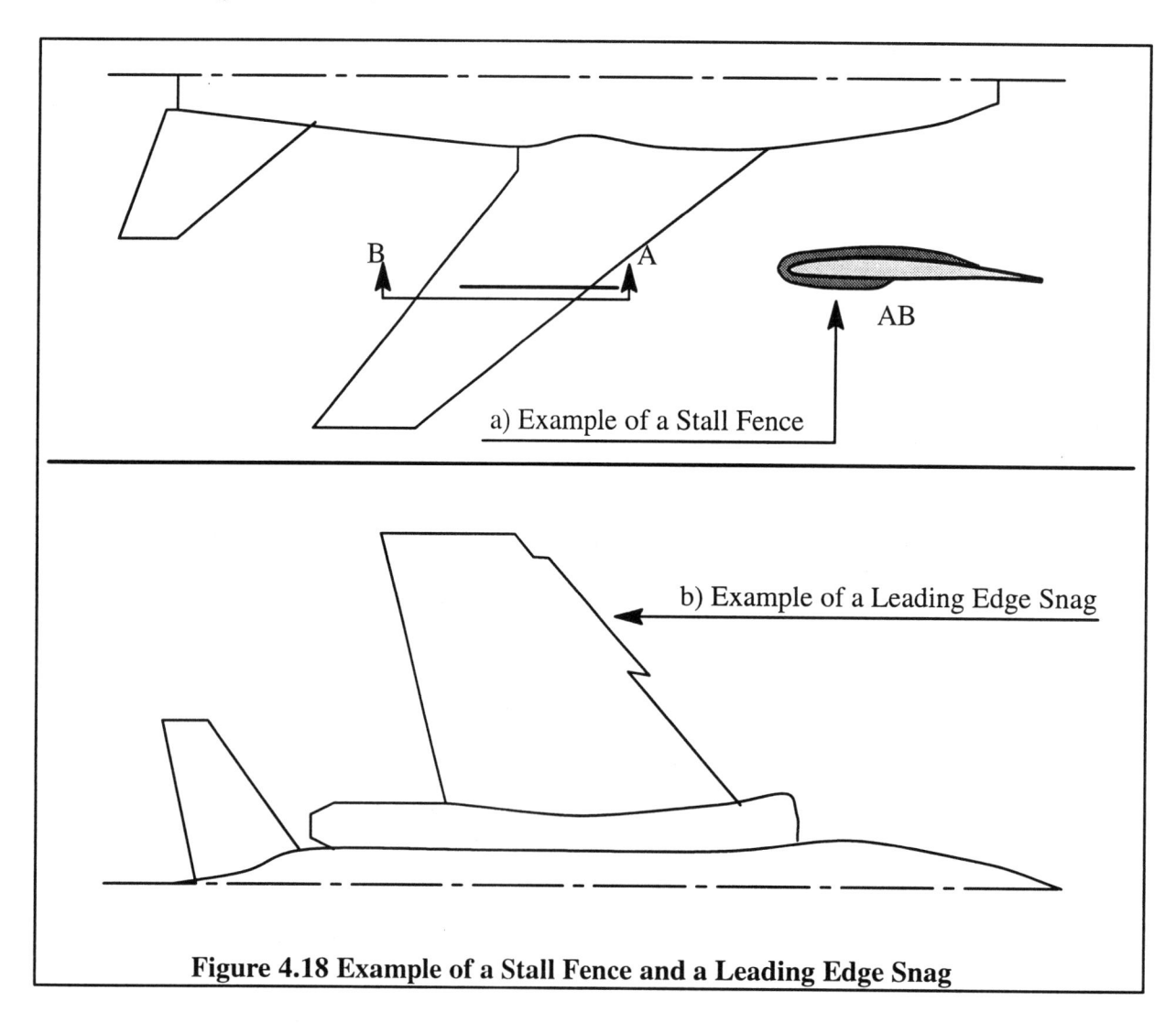

Figure 4.18 Example of a Stall Fence and a Leading Edge Snag

4.5.3.5 Stall Strips

Stall strips are usually angular devices installed at the leading edge and extending over a limited span of the wing. Their purpose is to actually induce stall at a given angle of attack. Stall strips are often added to airplanes during flight test to correct certain unsatisfactory stall behaviors. An example of a stall strip is shown in Figure 4.19.

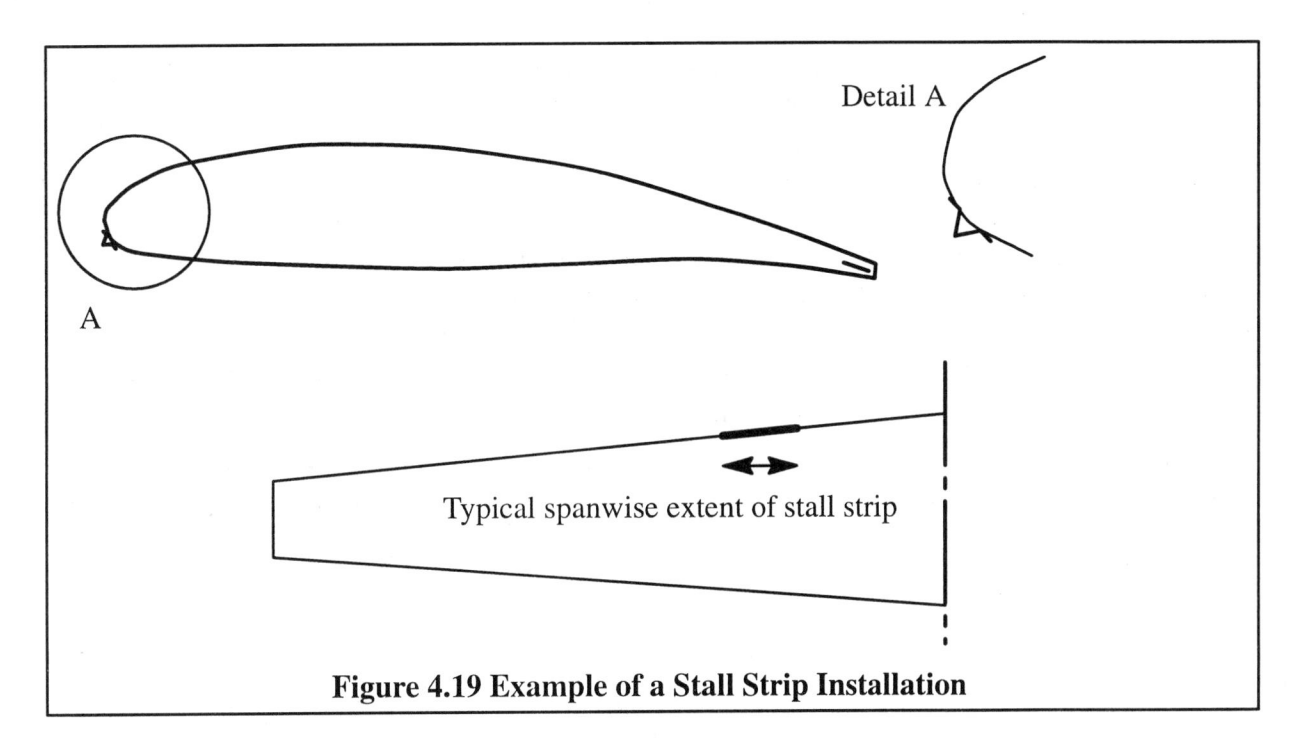

Figure 4.19 Example of a Stall Strip Installation

4.5.3.6 Vortex Generators

Vortex generators (v.g.'s) are very small, low aspect ratio wings (they look like jet engine turbine blades) places vertically at some local angle of attack on the wing, fuselage or tail surfaces of airplanes. The span of these v.g.'s is typically selected to be just outside the local edge of the boundary layer. These vortex generators will produce lift and therefore tip vortices near the edge of the boundary layer. The vortices will mix with the high energy fluid just outside the boundary layer and therefore raise the kinetic energy level of the flow inside the boundary layer. This increase in energy level allows the boundary layer to advance further into an adverse pressure gradient before separating.

Figure 4.20 shows an example of a vortex generator on a lifting surface. Figure 4.21 shows the effect of a vortex generator on the lift of a GA(W)–1 airfoil.

Vortex generators are used in many different sizes and shapes. They are typically added to an airplane after flight tests have uncovered certain flow separation problems. For that reason they are often referred to as "aerodynamic afterthoughts". The precise number and orientation of v.g.'s is normally determined in a series of sequential flight trials. Even though v.g.'s are beneficial in delaying local wing stall, they can generate measurable increases in cruise drag. Most of today's jet transports have large numbers of vortex generators on wings, tails and nacelles.

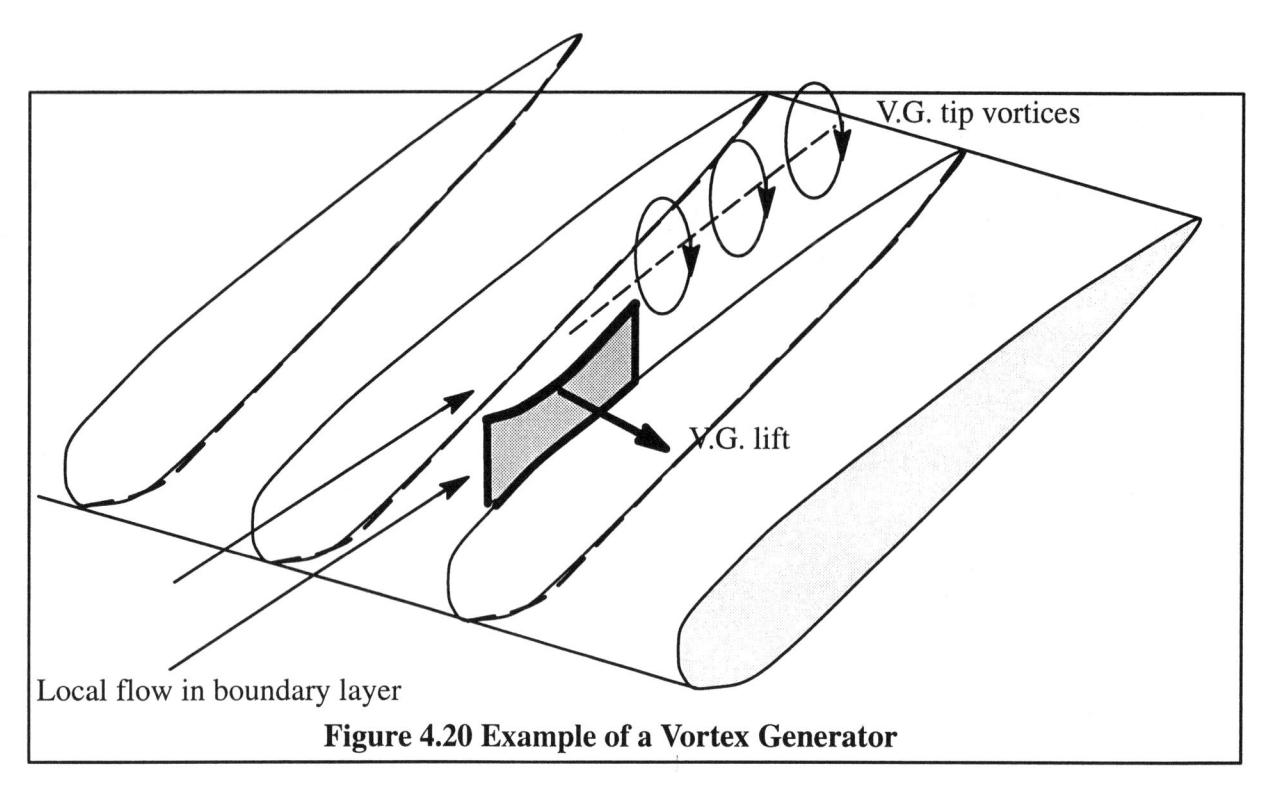

Figure 4.20 Example of a Vortex Generator

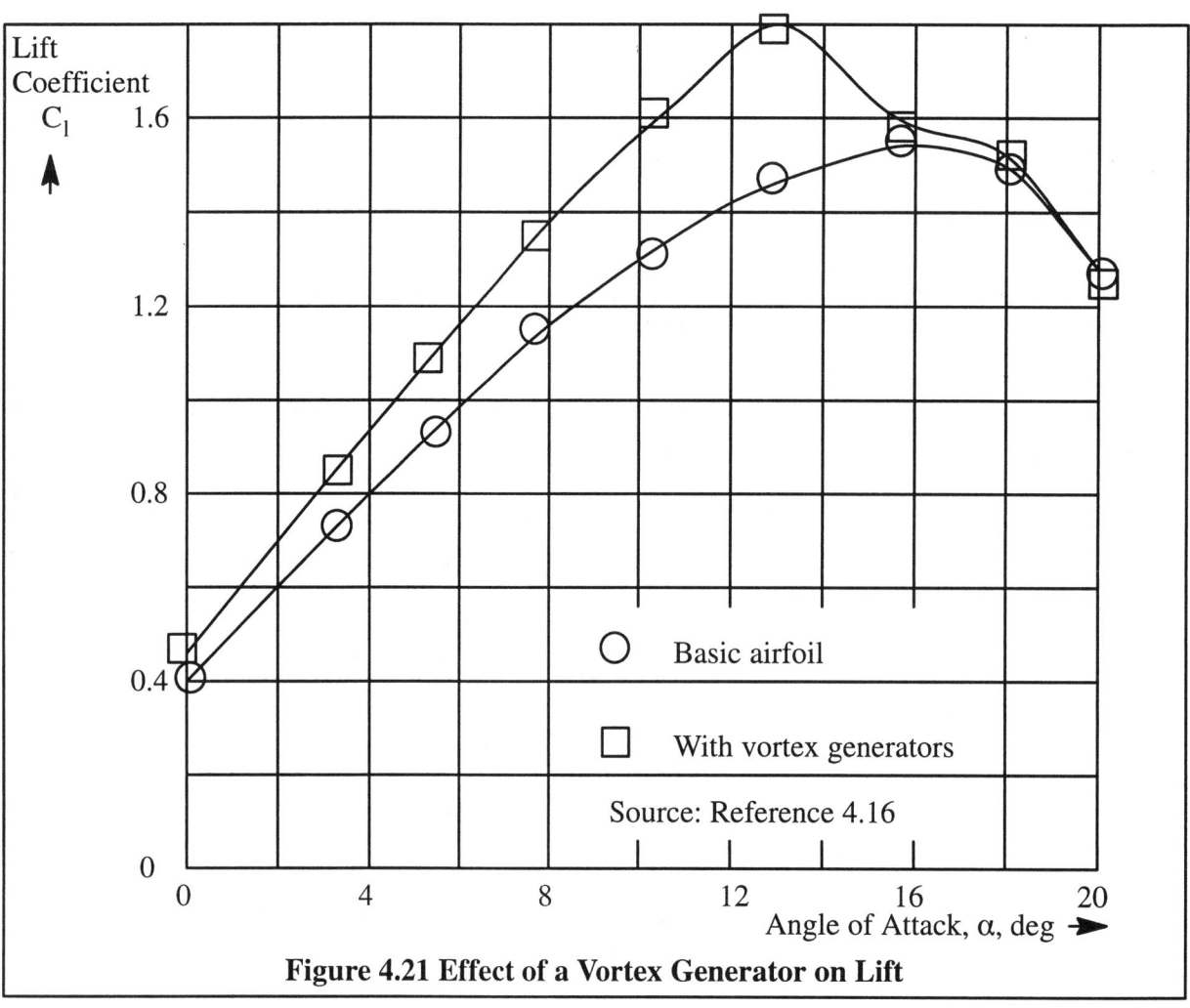

Figure 4.21 Effect of a Vortex Generator on Lift

4.6 COMPRESSIBILITY EFFECTS

It was already shown in Section 3.5 that compressibility effects can have serious effects on airfoil drag, lift and pitching moment. It can therefore be expected that the same holds true for wings.

Compressibility effects on airfoils can be delayed by thickness and camber tailoring (See Section 3.7). On wings it is possible to delay the effects of compressibility not only by tailoring thickness and camber, but also by tailoring the sweep angle. Figure 4.22 shows two examples.

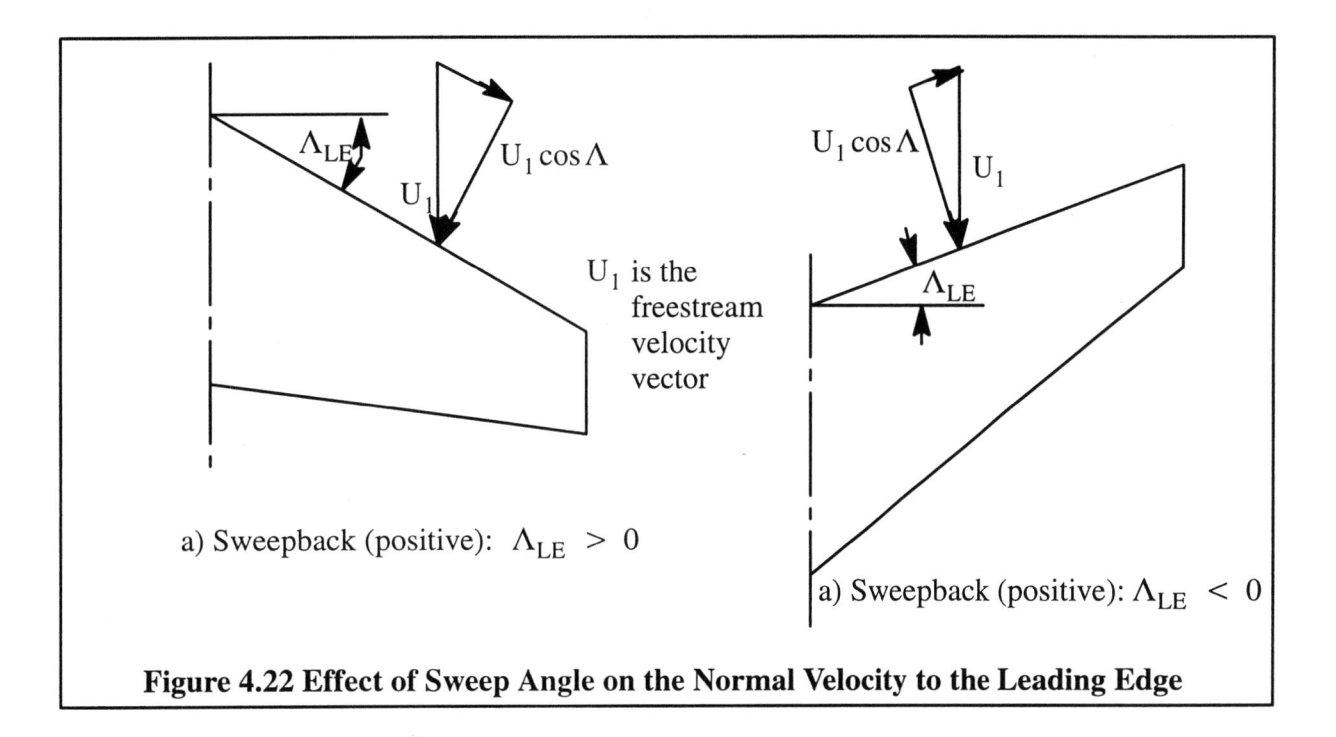

Figure 4.22 Effect of Sweep Angle on the Normal Velocity to the Leading Edge

a) Sweepback (positive): $\Lambda_{LE} > 0$

a) Sweepback (positive): $\Lambda_{LE} < 0$

As seen in Figure 4.22, the free-stream velocity vector can be resolved into components normal and parallel to the leading edge. The normal component is responsible for the aerodynamic characteristics, and its associated Mach number is: $M_\infty \cos \Lambda_{LE}$. It follows that if the critical Mach number for the same wing, but unswept, is denoted by M^*, then the critical Mach number for the swept wing would be given by the relation:

$$M_{crit} \cos \Lambda = M^* \qquad (4.42)$$

or,

$$M_{crit} = M^*/\cos \Lambda \qquad (4.43)$$

This result can also be interpreted as follows. In a flow with M_∞, the sectional characteristics would correspond to an effective Mach number given by: $M_{eff.} = M_\infty \cos \Lambda_{LE}$. For example, if $M_\infty = 0.9$ (which may exceed the drag divergence Mach number for an unswept wing), the effective

Mach number for a swept wing with $\Lambda_{LE} = 30^0$ is:

$$M_{eff} = 0.9\cos 30^0 = 0.779 \tag{4.44}$$

This could be below the drag divergence Mach number, M_{dd}. Thus, it is possible with a highly swept wing to delay to high values of M_∞ the troublesome aerodynamic characteristics associated with transonic flow. This argument is exactly valid only in two–dimensional flow. The 3–D effects near the tip and the root tend to make such a relationship too optimistic. A useful empirical formula for an approximate value of M_{crit} is given in Ref. 4.17, page 103 as:

$$M_{crit} = M^* \sqrt{(\sec \Lambda_{LE})} \tag{4.45}$$

Experiments show that the use of sweepback not only increases M_{dd}, but also reduces the rate at which the drag coefficient rises in the transonic region. This tendency is illustrated in Figure 4.23.

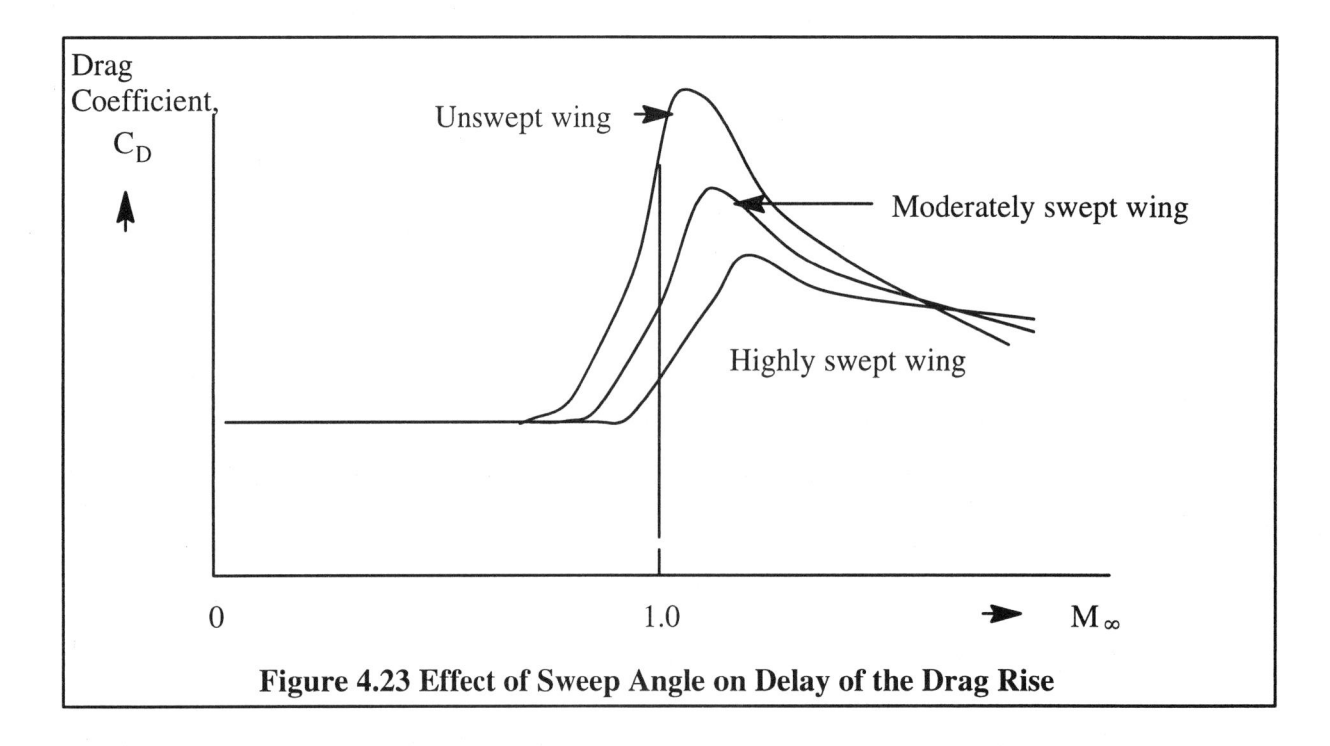

Figure 4.23 Effect of Sweep Angle on Delay of the Drag Rise

Experiments leading up to the X–29 program have shown that forward sweep offers roughly a leading edge sweep benefit of approximately five degrees over aft sweep. The reason is that what counts is the sweep angle of the so–called shock–sweep line. For the same leading edge sweep angle magnitude, a forward swept wing tends to have a five degree higher shock–sweep angle than an aft swept wing.

Experiments also indicate that, other things being equal, a reduction in aspect ratio gives a rise in M_{dd}, and therefore helps to reduce transonic drag (Ref. 4.18, page 15–16). This reduction in transonic drag more than compensates for the accompanying increase in induced drag.

For low aspect ratio wings edge vortex separation becomes important. Leading edge vortex separation is illustrated in Figure 4.24. The vortex creates low pressure on the upper surface and thereby produces (vortex) lift. The vortex separation can also occur along the tip chord and along the wing or body mounted strakes (Lexes!). In transonic flow, the interaction of vortex flow with shock waves can create a very complex flow field.

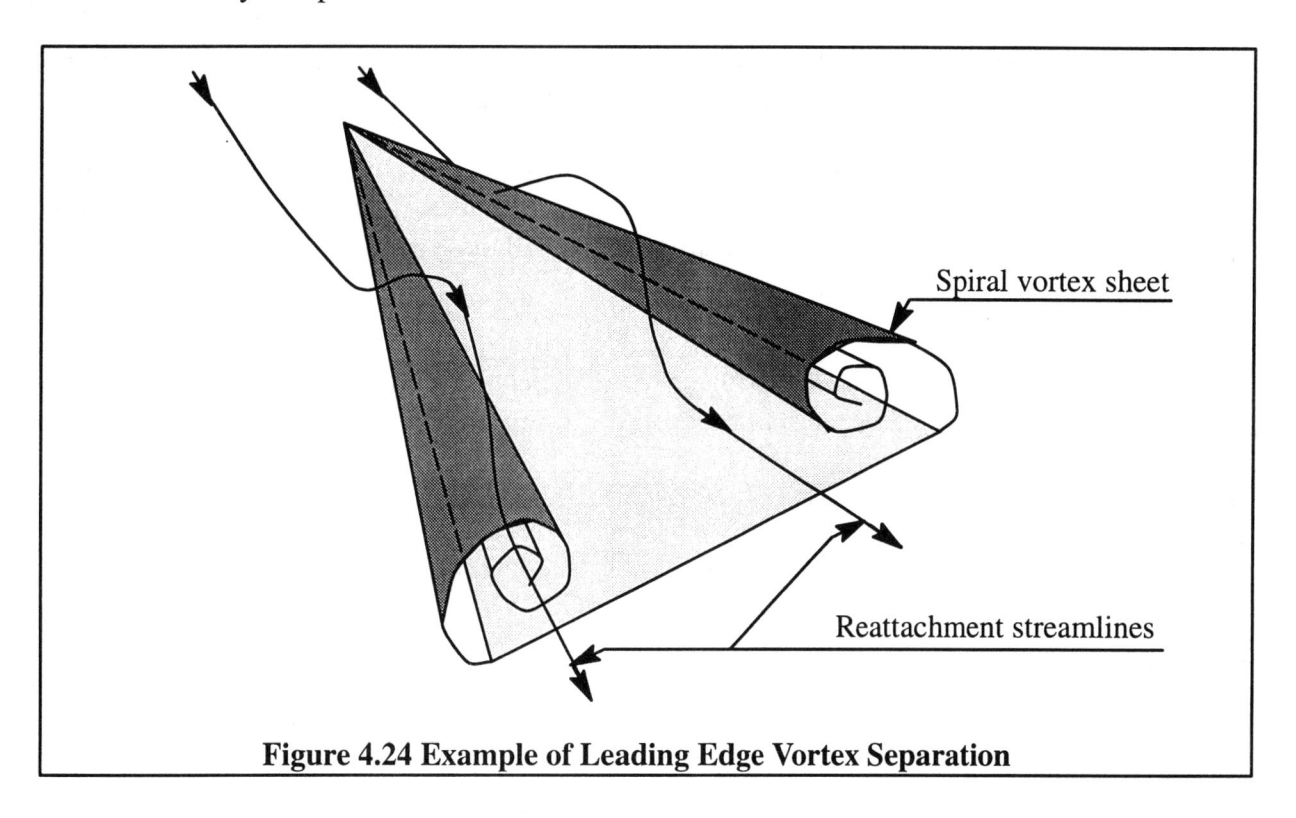

Figure 4.24 Example of Leading Edge Vortex Separation

4.7 HIGH LIFT DEVICES, SPOILERS, DIVE BRAKES, SPEED BRAKES

From the definition of lift coefficient, the speed in level flight, with L = W, is given by:

$$V = \sqrt{\frac{2L}{\varrho C_L S}} = \sqrt{\frac{2W}{\varrho C_L S}} \tag{4.46}$$

If the maximum lift coefficient, $C_{L_{max}}$ is used in Eqn (4.46), the airplane stall speed is obtained as:

$$V_s = \sqrt{\frac{2}{\varrho C_{L_{max}}}\left(\frac{W}{S}\right)} \tag{4.47}$$

The ratio, W/S, in Eqn (4.47) is referred to as the wing loading of the airplane.

Because takeoff and landing speeds depend on the stall speed (as shown in Chapter 10), it is necessary to reduce the stall speed, whenever short runways are to be used. From Eqn (4.47) it is seen that the stall speed can be reduced if either $C_{L_{max}}$ is increased, or W/S is decreased, i.e. high lift or low wing loading. Wing loadings of modern airplanes range from around 10 psf for small, low performance airplanes to 160 psf for large, high performance airplanes. Devices which increase airplane lift are referred to as high lift devices. Such devices come in two categories: those which change the airfoil geometry and those which control the boundary layer (BLC). In practice, these devices are frequently referred to as trailing edge and leading edge flaps for the first category and BLC devices for the second category.

The sectional (i.e. airfoil or 2D) characteristics of high lift devices were discussed in Section 3.8. In the following, some three–dimensional (3D) effects will be discussed.

Although significant progress in computational aerodynamics has been made toward the theoretical prediction of high lift effects, the current state–of–the–art is still not satisfactory. Therefore, only semi–empirical methods will be discussed. In all these methods, knowledge of the sectional characteristics, $c_{l_{max}}$ is generally the starting point for any three–dimensional calculations. These three–dimensional calculations take the form of corrections to two–dimensional data as a result of partial span effects, fuselage interference, etc. A method for estimating the incremental lift due to trailing edge flaps below the stall angle will be discussed first.

4.7.1. LIFT INDUCED BY PARTIAL SPAN FLAPS BELOW STALL

Figure 4.25 shows calculated results for a swept wing with a partial span flap deflected by 10 degrees at an angle of attack of 10 degrees.

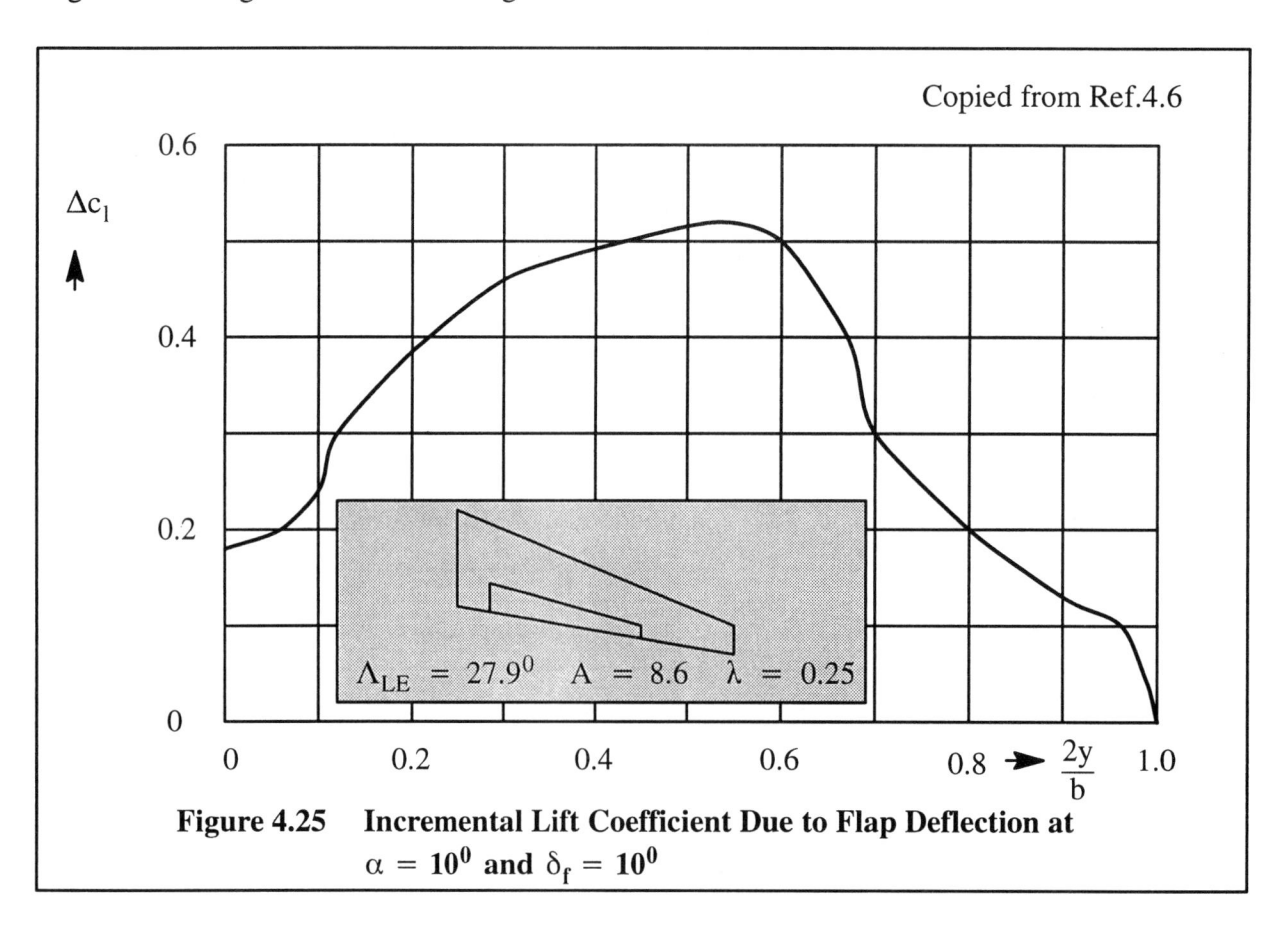

Figure 4.25 Incremental Lift Coefficient Due to Flap Deflection at $\alpha = 10^0$ and $\delta_f = 10^0$

Wing Theory

It is seen that the incremental sectional lift coefficient due to the flap, Δc_l, varies considerably inside the flap region. It is also seen that Δc_l maintains significant values even outside the flap region. This so–called flap–lift carry–over is due to three–dimensional effects. Even though theoretical methods for calculating flap induced lift distributions are sufficiently accurate at low angles of attack, at high angles of attack experimental data should be used.

When sectional data are available, the method by Lowry and Polhamus (Reference 4.3) can be used to estimate the total incremental lift coefficient due to the flap, ΔC_L. Because of its importance in the cycle of airplane design and airplane performance analysis it will be summarized next. According to Lowry/Polhamus it is possible to write:

$$\Delta C_L = \frac{\partial C_L}{\partial \delta_f} \delta_f K_b = C_{L_{\delta_f}} \delta_f K_b \qquad (4.48)$$

where: δ_f is the flap deflection angle in radians

K_b is the flap span factor defined as the ratio of partial–span–flap lift coefficient to full–span lift coefficient

The three–dimensional effect due to the flaps, mentioned before, is accounted for through this flap span factor, K_b. Equation (4.48) can be re–written as follows:

$$\Delta C_L = \frac{C_{L_{\delta_f}}}{c_{l_{\delta_f}}} (c_{l_{\delta_f}} \delta_f) K_b = \frac{C_{L_{\delta_f}}/C_{L_\alpha}}{c_{l_{\delta_f}}/c_{l_\alpha}} \frac{C_{L_\alpha}}{c_{l_\alpha}} \Delta c_{l_f} K_b \qquad (4.49)$$

where: Δc_{l_f} is the sectional lift coefficient increment due to the flaps.

The so–called flap–chord factor, K_c is defined as follows:

$$K_c = \frac{C_{L_\delta}/C_{L_\alpha}}{c_{l_\delta}/c_{l_\alpha}} \qquad (4.50)$$

Including this flap–chord factor, Eqn (4.49) can be written as:

$$\Delta C_L = \Delta c_{l_f} \left(\frac{C_{L_\alpha}}{c_{l_\alpha}}\right) K_c K_b \qquad (4.51)$$

The ratio of three–dimensional over two–dimensional lift curve slopes can be estimated from Eqn (4.33) with $a_\infty = 2\pi$:

$$\frac{C_{L_\alpha}}{c_{l_\alpha}} = \frac{A}{\frac{a_\infty}{\pi} + \sqrt{A^2\left(1 + \frac{\tan^2 \Lambda_{c/2}}{\beta^2}\right) + \left(\frac{a_\infty}{\pi}\right)^2}} \qquad (4.52)$$

where: β is defined in Eqn (4.33a).

The reader will observe that Eqn (4.52) is a variation of Eqn (4.33). The flap–span factors K_b and K_c may be found from Figures 4.26 and 4.27 respectively. The reader should carefully note, that the flap–span factor, K_b, as defined in Figure 4.27 applies to a flap which runs from the centerline of the wing to the outboard flap span station, $b_f/b/2$. Figure 4.28 shows how K_b should be determined for a flap which runs from a partial span inboard station to a partial span outboard station.

The following example illustrates an application of this method:

Example: For the wing in Figure 4.25, with a partial span flap deflected at 10 degrees, find ΔC_L at M=0.2. The wing has an aspect ratio of A=8.5, a taper ratio of λ=0.253 and a semi–chord sweep angle of $\Lambda_{c/2} = 21.3^0$. The inboard and outboard flap stations of the flap are $b_f/b/2$ =0.110 and 0.655 of the semi–span respectively.

Solution: Eqn (4.51) will be used to determine ΔC_L. For a wing with a taper ratio of λ=0.253 and $b_f/b/2$ =0.110 and 0.655, values for K_b are found from Figure 4.27 as 0.80 and 0.16 respectively. With a flap chord ratio of $c_f/c = 0.3$ it is seen from the insert in Fig. 4.26 that $(\alpha_\delta)_{c_l} = 0.66$. Hence, from Figure 4.26 it follows that for an aspect ratio of 8.5, $K_c = 1.025$. The ratio of lift–curve slopes in three–dimensional and two–dimensional flow can be estimated from Eqn (4.52), while assuming that $a_\infty = 2\pi$:

$$\frac{C_{L_\alpha}}{c_{l_\alpha}} = \frac{8.5}{2 + \sqrt{8.5^2\left(1 + \frac{\tan^2 21.3}{(1 - 0.2^2)}\right) + 4}} = 0.748 \qquad (4.53)$$

At this point, if experimental data were available for the incremental, sectional lift coefficient due to flaps, Δc_{l_f}, Eqn (4.51) could be used to determine ΔC_L. For the present purpose, such experimental data are assumed not to be available. Therefore, the following course is taken, using the theoretical results of Ref.. 4.6. According to this reference, the following equations can be used to estimate Δc_{l_f}:

$$\Delta c_{l_f} = 2\delta_f (\tau + \sin\tau) \qquad (4.54)$$

where: $\tau = \cos^{-1}(2x_f - 1)$ \qquad (4.55)

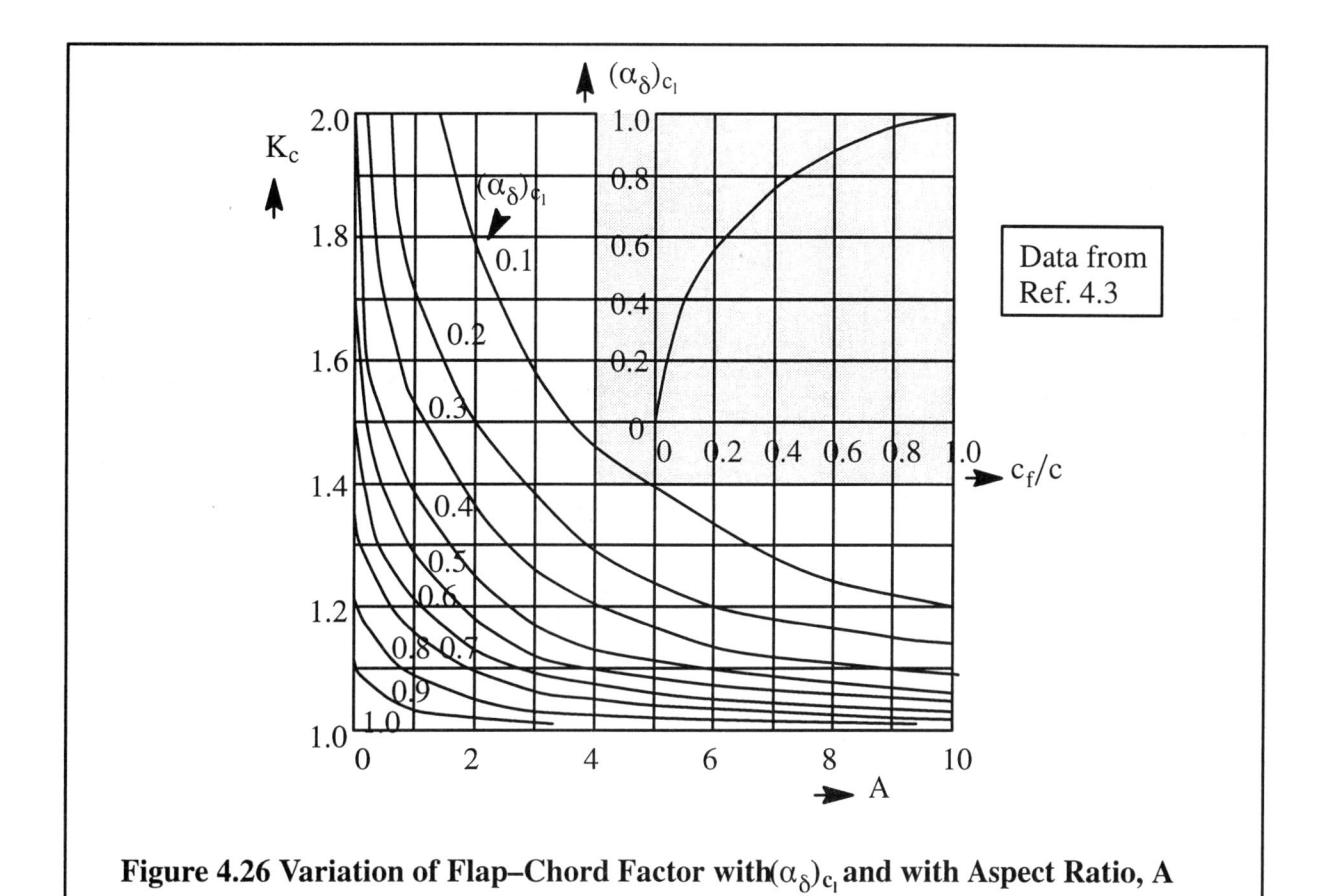

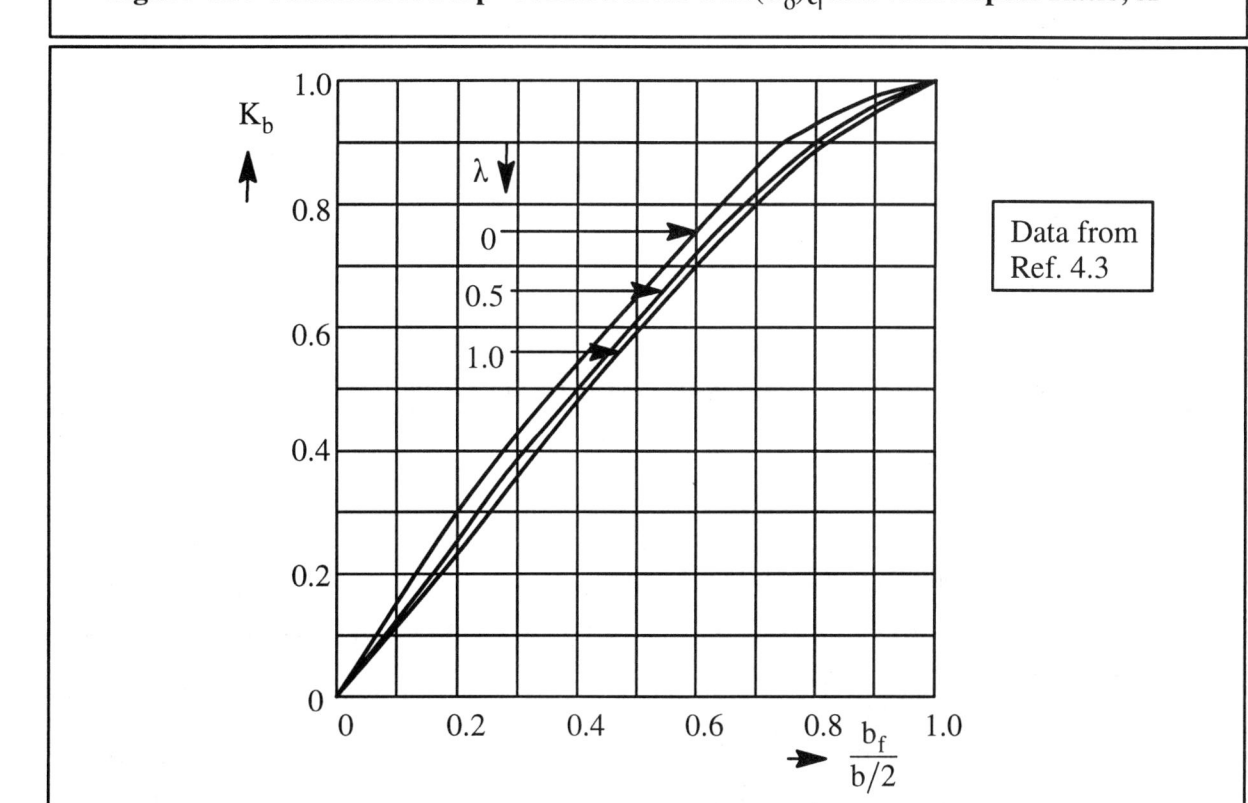

Figure 4.26 Variation of Flap–Chord Factor with $(\alpha_\delta)_{c_l}$ and with Aspect Ratio, A

Figure 4.27 Variation of Flap–Span Factor with Flap Span for Inboard Flaps

Figure 4.28 Definition of the Flap Span Factor, K_b for Other Than Inboard Flaps

NOTE: $\dfrac{S_{wf}}{S} = \dfrac{\text{Area ABCD}}{\text{Area EFGH}}$

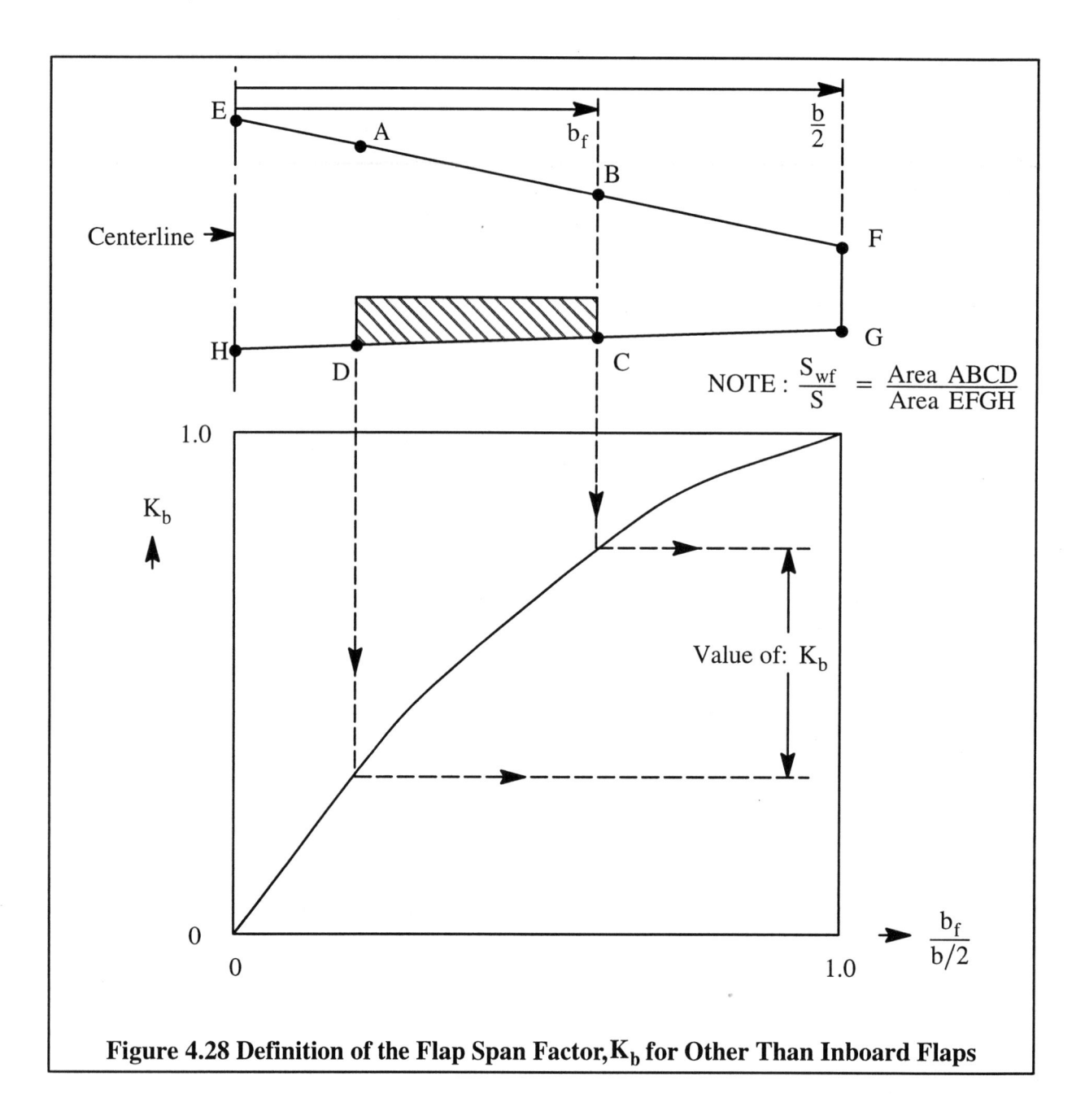

and: $x_f = (1 - c_f/c)$ \hfill (4.56)

It therefore follows that from Eqn (4.55):

$\tau = \cos^{-1}(2 \times 0.7 - 1) = 66.42^0 = 1.16$ rad. From Eqn (4.54) it is now found that: $\Delta c_{l_f} = 2 \times \dfrac{10 \times \pi}{180}(1.16 + \sin 66.42^0) = 0.725$. Finally, from Eqn (4.51) it follows that: $\Delta C_L = 0.725 \times 0.748 \times 1.025 \times (0.8 - 0.16) = 0.356$. This value was found to be comparable to that of 0.333 found with the lifting surface method of Reference 4.6.

4.7.2. MAXIMUM LIFT COEFFICIENT WITH HIGH LIFT DEVICES

A simple, yet acceptable, method for estimating the maximum lift coefficient of a wing with high lift devices is to add the maximum lift coefficient increments due to leading edge and trailing edge devices to the maximum lift coefficient of the clean wing. This is essentially the method presented in Reference 4.19. First, a method for estimating the maximum lift coefficient for a clean wing will be discussed.

4.7.2.1 Clean Wing Maximum Lift Coefficient

It has been found that for moderate to high aspect ratio wings with moderate sweep angles, the maximum lift, is to a large extent determined by the maximum sectional lift. A pragmatic assumption in estimating $C_{L_{max}}$ for a wing is that the complete wing will attain its maximum lift when any spanwise section first reaches $c_{l_{max}}$. The reason is, that once a section is stalled, the resulting flow separation will spread quickly to neighboring sections as the angle of attack is increased. As a result the wing will loose part of its lifting capability.

For such a wing the maximum lift coefficient can be estimated with the following procedure.

1.) Assume that the spanwise variation of sectional lift coefficient, $c_{l_{max}}(y)$ is known. This is plotted as shown in Figure 4.29.

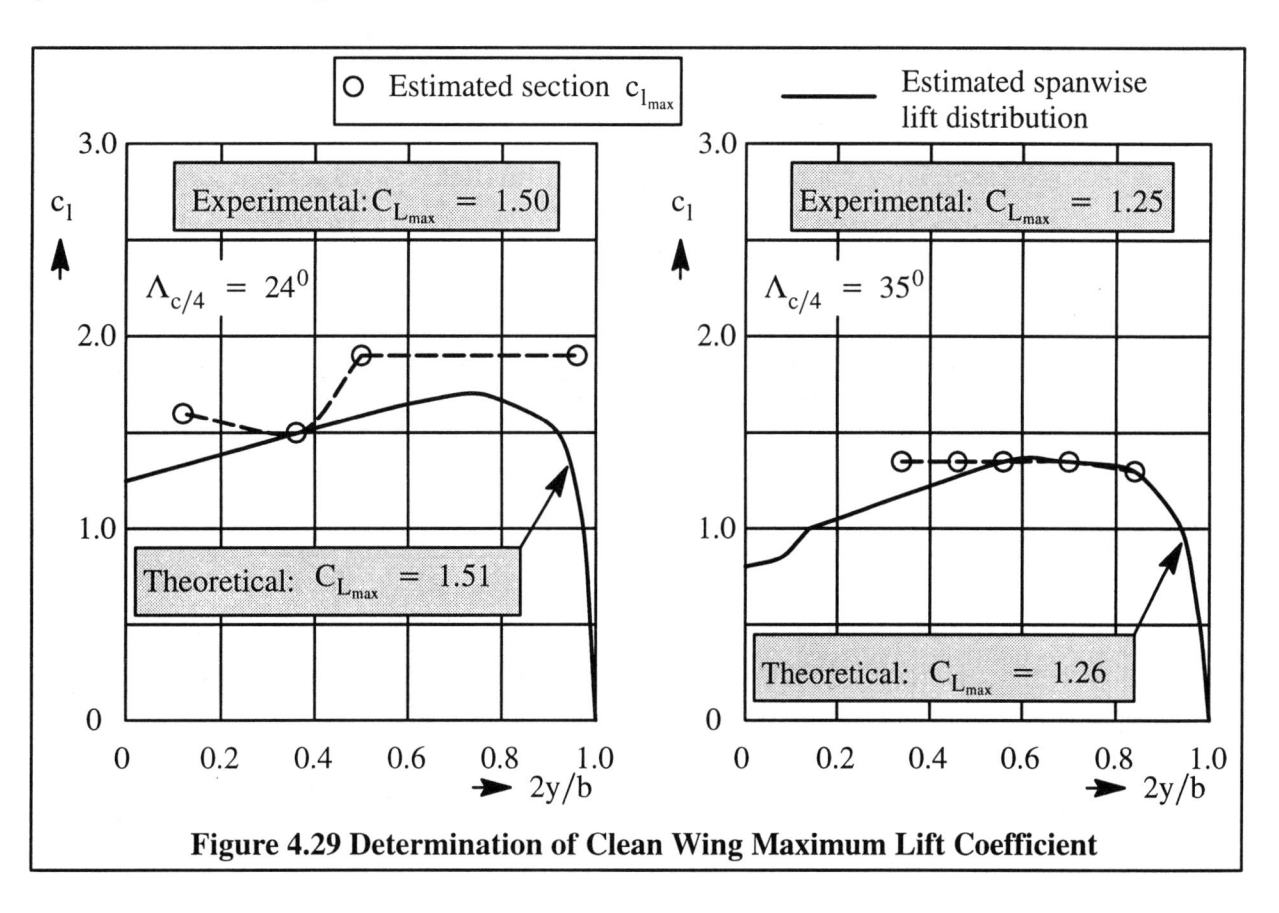

Figure 4.29 Determination of Clean Wing Maximum Lift Coefficient

2.) Use a lifting surface method to predict the spanwise lift distribution for a range of angles of attack. An example is the method of Ref. 4.6. At each angle of attack, the integrated spanwise lift distribution will result in a value of wing lift coefficient. The wing maximum lift coefficient is that calculated value which corresponds to the one for which a local lift coefficient first reaches the spanwise line of maximum section lift coefficients.

3.) This approach is illustrated in Figure 4.29 for two wings. As can be seen, the accuracy is good for both cases.

The reader should realize that, as the sweep angle is increased (above about 40 deg.) and the aspect ratio is reduced (below about 4.5) the three–dimensional induction effects become important. When edge separated vortex flow is present (see Section 4.6), wing stall can be delayed to larger angles of attack. In such cases, the flow characteristics are quite three–dimensional. Refs 4.9 and 4.20 present a method for estimating $C_{L_{max}}$ for this type of wing.

4.7.2.2 Maximum Lift Increment Due to High Lift Devices

For efficient trailing edge flaps, the sectional value of flap maximum lift coefficient increment can be used to calculate the three–dimensional value of a wing, by the following empirical relation:

$$(\Delta C_{L_{max}})_{flaps} = (\Delta c_{l_{max}}) \left(\frac{S_{w_f}}{S}\right) K_\Lambda \quad (4.57)$$

where: $\Delta c_{l_{max}}$ is the sectional value of maximum lift coefficient increment due to trailing edge flaps

S_{w_f}/S is the ratio of flapped wing area to the wing reference area. See Fig.4.28.

K_Λ is a factor which has been empirically derived (Ref.4.19) to account for the effect of sweep:

$$K_\Lambda = \left(1 - 0.08 \cos^2 \Lambda_{c/4}\right) \cos^{3/4} \Lambda_{c/4} \quad (4.58)$$

For leading edge devices, the effect of sweep is more pronounced. According to Ref. 4.21, p.549, the incremental value of $C_{L_{max}}$ due to leading edge high lift devices can be written as:

$$(\Delta C_{L_{max}})_{l.e.\ devices} = (\Delta c_{l_{max}}) \left(\frac{S_{w_f}}{S}\right) \cos^2 \Lambda_{c/4} \quad (4.59)$$

It must be realized, that in most practical situations the actual increment of $C_{L_{max}}$ due to any type of flap will be reduced by flow around flap tracks, flap supports, interference with nacelles, unforeseen leakage paths and aeroelastic deformation.

4.7.2.3 Examples of Maximum Lift Increment Due to High Lift Devices

Table 4.3 was prepared to provide the reader with some insight into the ranges of numerical values associated with clean and flapped configurations. The first twelve configurations apply to pre–1957 high lift technology and the data were taken from Ref. 4.22. The last three represent more recent technology. Although not reflected in Table 4.3, the Airbus 320 achieves a trimmed maximum lift coefficient of 3.2. This airplane uses also drooped ailerons, leading edge slats and double slotted trailing edge flaps. The word "trimmed" is used to indicate that the lift coefficient quoted is being achieved with zero total pitching moment on the airplane. The effect of trim is discussed next.

4.7.3 THE EFFECT OF TRIM ON MAXIMUM LIFT

Deployment of high lift devices is usually accompanied by a significant change in pitching moment. Trailing edge flaps tend to generate negative (nose down) pitching moments, while leading edge devices tend to produce positive (nose up) pitching moments. These pitching moments must be balanced. In conventional airplane configurations (tail aft) this balance is achieved with the horizontal tail. In a canard airplane the balance is obtained with the canard surface. It is important to be able to predict the effect of this balance requirement on the so–called trimmed value of maximum lift coefficient, $C_{L_{max_{trim}}}$ which can be achieved. A detailed discussion of these trim effects on conventional, pure canard and three–surface airplanes, see Ref. 4.10, pages 344–353) and Ref. 4.23, pages 234–237. In the present text, only the trim effect for a tail aft configuration will be examined.

Figure 4.30 shows the geometry needed to discuss trim for a conventional airplane in the flaps–up configuration.

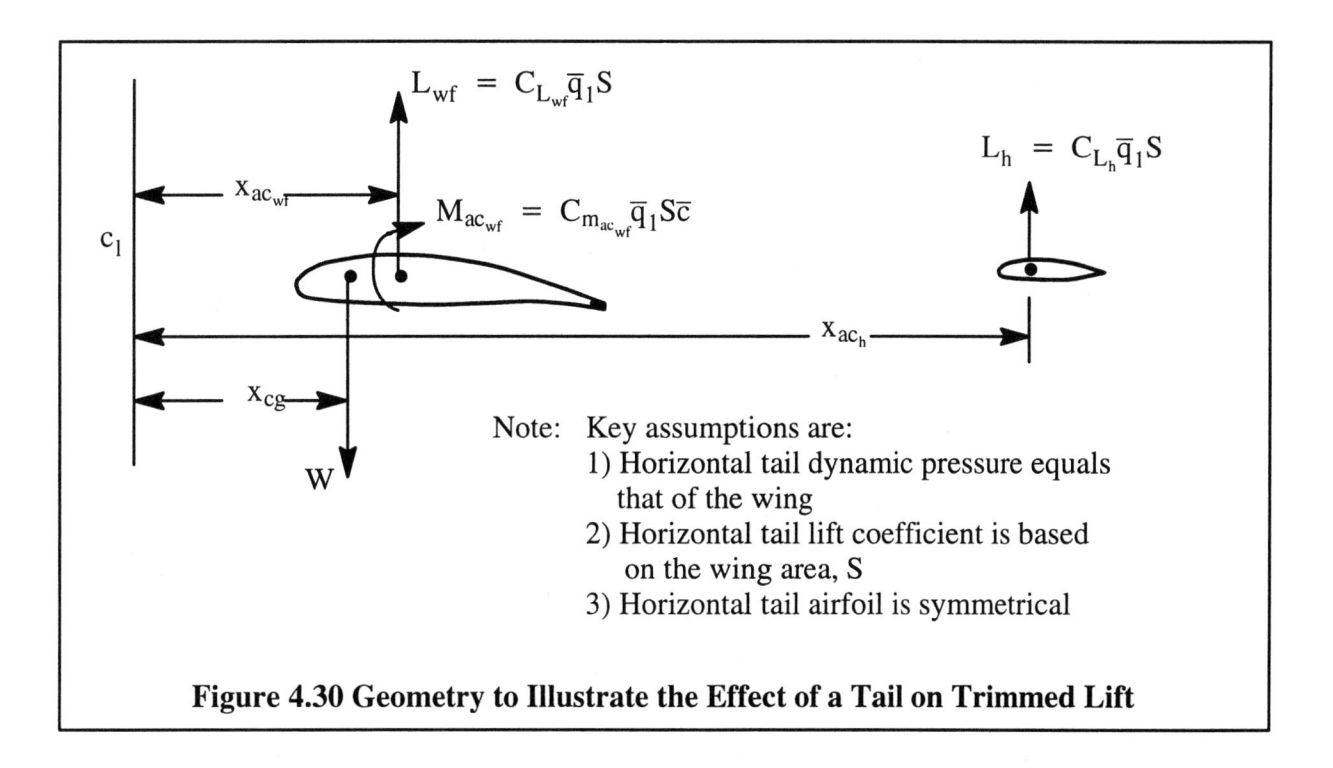

Figure 4.30 Geometry to Illustrate the Effect of a Tail on Trimmed Lift

$L_{wf} = C_{L_{wf}} \bar{q}_1 S$

$L_h = C_{L_h} \bar{q}_1 S$

$M_{ac_{wf}} = C_{m_{ac_{wf}}} \bar{q}_1 S \bar{c}$

Note: Key assumptions are:
1) Horizontal tail dynamic pressure equals that of the wing
2) Horizontal tail lift coefficient is based on the wing area, S
3) Horizontal tail airfoil is symmetrical

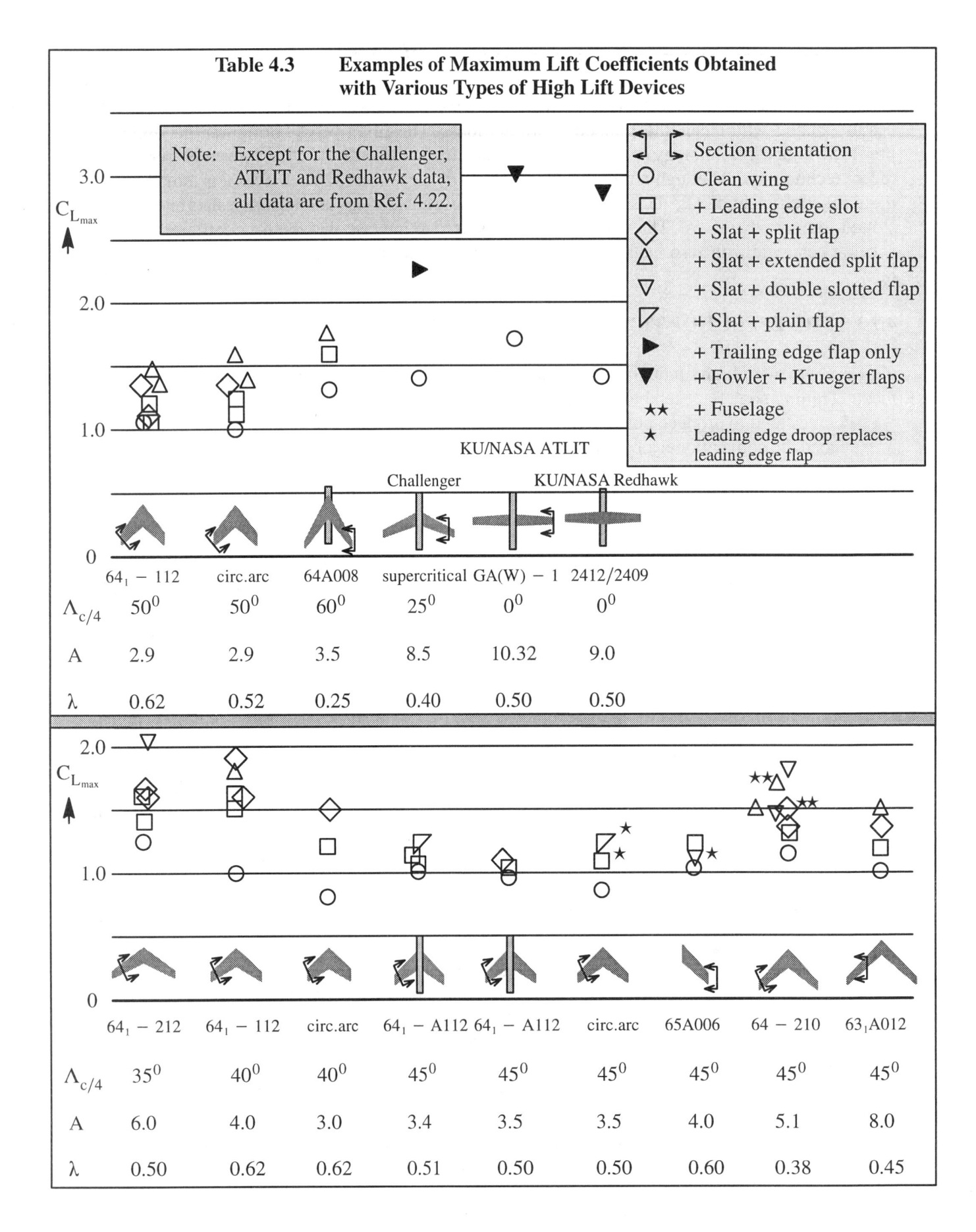

Table 4.3   Examples of Maximum Lift Coefficients Obtained with Various Types of High Lift Devices

Note: Except for the Challenger, ATLIT and Redhawk data, all data are from Ref. 4.22.

Legend:
- ⌐ ⌐ Section orientation
- ○ Clean wing
- □ + Leading edge slot
- ◇ + Slat + split flap
- △ + Slat + extended split flap
- ▽ + Slat + double slotted flap
- ▱ + Slat + plain flap
- ▶ + Trailing edge flap only
- ▼ + Fowler + Krueger flaps
- ★★ + Fuselage
- ★ Leading edge droop replaces leading edge flap

				Challenger	KU/NASA ATLIT	KU/NASA Redhawk
Section	64_1-112	circ.arc	64A008	supercritical	GA(W)-1	2412/2409
$\Lambda_{c/4}$	$50°$	$50°$	$60°$	$25°$	$0°$	$0°$
A	2.9	2.9	3.5	8.5	10.32	9.0
λ	0.62	0.52	0.25	0.40	0.50	0.50

Section	64_1-212	64_1-112	circ.arc	64_1-A112	64_1-A112	circ.arc	65A006	64-210	63_1A012
$\Lambda_{c/4}$	$35°$	$40°$	$40°$	$45°$	$45°$	$45°$	$45°$	$45°$	$45°$
A	6.0	4.0	3.0	3.4	3.5	3.5	4.0	5.1	8.0
λ	0.50	0.62	0.62	0.51	0.50	0.50	0.60	0.38	0.45

In an equilibrium flight condition, airplane weight, W, must be balanced by the total lift on the airplane, L. This total lift, L, is in turn equal to the sum of wing and tail lift:

$$L = W = L_{wf} + L_h \tag{4.60}$$

In terms of the trimmed lift coefficient this can be written as:

$$C_{L_{trim}} = C_{L_{wf}} + C_{L_h} \tag{4.61}$$

The condition for pitching moment equilibrium about the center of gravity yields:

$$-C_{L_{wf}}\bar{q}_1 S(x_{ac_{wf}} - x_{cg}) + C_{m_{ac_{wf}}}\bar{q}_1 S\bar{c} - C_{L_h}\bar{q}_1 S(x_h - x_{cg}) = 0 \tag{4.62}$$

This equation can be solved for the tail lift coefficient as follows:

$$C_{L_h} = -C_{L_{wf}}\frac{(x_{ac_{wf}} - x_{cg})}{(x_h - x_{cg})} + C_{m_{ac_{wf}}}\frac{\bar{c}}{(x_h - x_{cg})} \tag{4.63}$$

By substituting C_{L_h} back into Eqn (4.61) the trimmed airplane lift coefficient is obtained as:

$$C_{L_{trim}} = C_{L_{wf}}\left\{1 - \frac{(x_{ac_{wf}} - x_{cg})}{(x_h - x_{cg})}\right\} + C_{m_{ac_{wf}}}\frac{\bar{c}}{(x_h - x_{cg})} \tag{4.64}$$

The term $(x_{ac_{wf}} - x_{cg})/(x_h - x_{cg})$ tends to be negligible when compared to 1.0 for most conventional airplanes. The term $C_{m_{ac_{wf}}}$ is nearly always negative. The greater the amount of camber used in a wing airfoil, the larger negative $C_{m_{ac_{wf}}}$ will be. Therefore, the trimmed lift coefficient will be less than the wing lift coefficient. This results in a lift and induced drag penalty.

The reader should recognize the fact, that flow field changes due to flap deployment can in fact alter the angle of attack of a horizontal tail. On airplanes like the Cessna 172 the change in down-wash over the tail, as a result of flap deployment, produces a nose up pitching moment which more than negates the nose down moment produced by the flaps. In the Cessna 172 this results in the pilot having to push on the control wheel when the flaps are lowered!

With canard configurations, a similar analysis will show that the trimmed lift coefficient will be larger than the wing lift coefficient. Thus it would appear that a canard configuration has an advantage over a conventional configuration. However, if the down-wash from the canard has a significant effect on the wing, this conclusion may not be be correct. Therefore, the reader is urged to always perform a detailed analysis, before jumping to a conclusion about the pros and cons of various types of configurations.

4.7.4 THE EFFECT OF SPOILERS, DIVE BRAKES AND SPEED BRAKES

Spoilers are normally wing mounted, plate–like devices which are used to destroy a part of the lift and/or to create extra drag. Spoilers are used in flight to help an airplane slow down more rapidly and/or to adjust the descent rate. They are also used as lift dumpers and drag creators, once on the runway, right after touch–down. Figure 4.31 shows examples of wing mounted spoilers.

In many airplanes, wing mounted spoilers are also used for part of the roll control requirements. The four outboard spoilers on the 747 are an example. It is also possible, to use spoilers for all of the roll control requirements of an airplane. If that is done, the entire trailing edge of the wing is available for high lift devices. That technology was used in the ATLIT (Ref. 4.24) and Redhawk airplanes. The resulting maximum lift capability is shown in Table 4.3. Several Robertson STOL conversions also use this approach (Ref. 4.25).

Dive brakes and speed brakes (these words are used interchangeably) are normally wing or fuselage mounted, plate–like devices which are used to increase drag. Dive brakes are used in fighter airplanes to allow a rapid slow–down in level flight or to arrest a speed–up in a dive. They are also used to help slow the airplane down immediately following touch–down. Figure 4.31 also shows an example of a fuselage mounted speed brake.

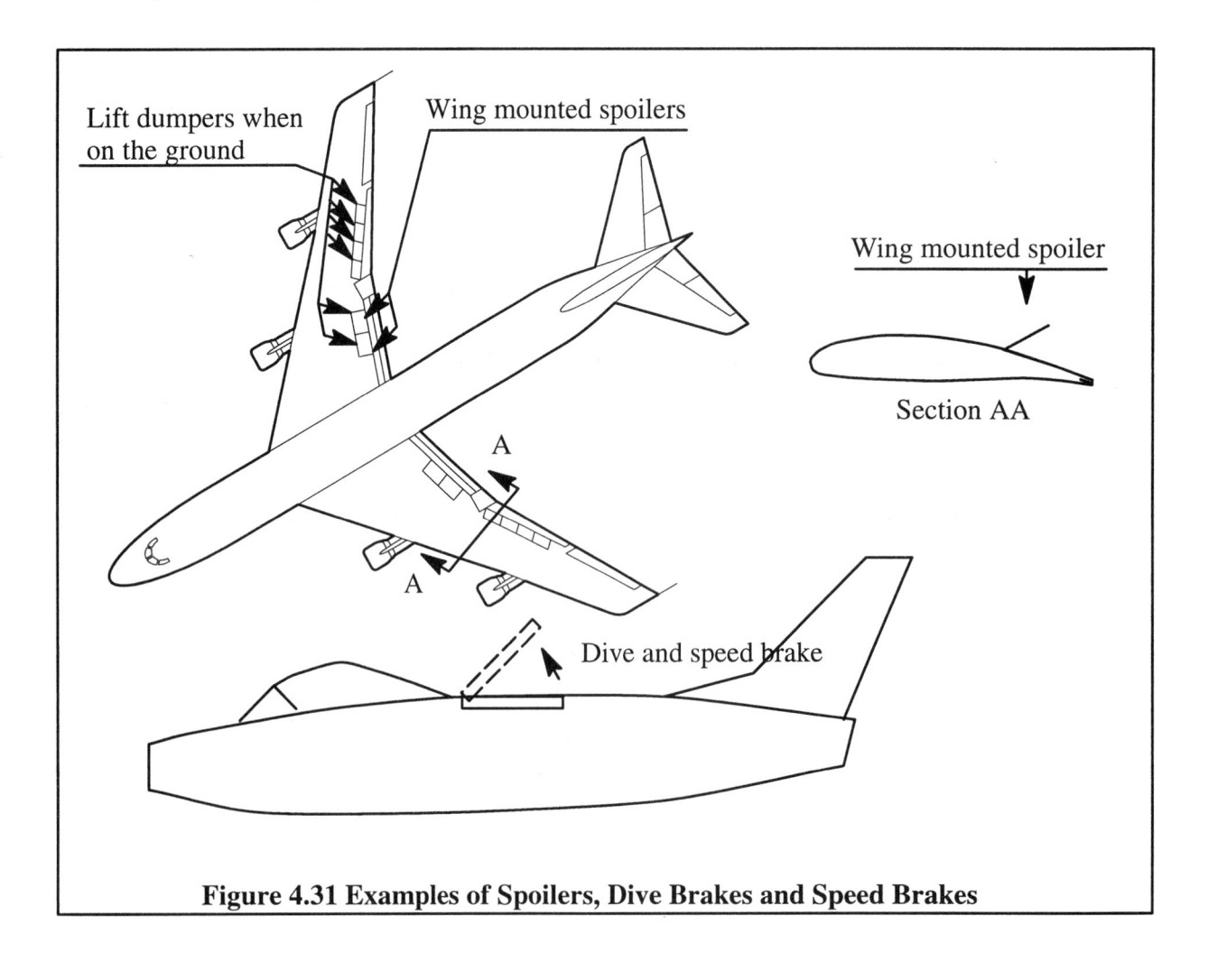

Figure 4.31 Examples of Spoilers, Dive Brakes and Speed Brakes

4.8 SUMMARY FOR CHAPTER 4

In this chapter, the reader has been given an insight into the lift, drag and pitching moment behavior of airplane lifting surfaces. It is shown, that the section (or airfoil) characteristics discussed in Chapter 3, form an essential ingredient in the prediction of lifting surface characteristics. Both empirical and theoretical methods were presented.

4.9 PROBLEMS FOR CHAPTER 4

4.1 A straight, tapered wing with a span of 30 ft has leading and trailing edge sweep angles of 45 deg. and 15 deg. respectively. The wing area is 280 sq.ft. Determine the magnitude of: the root chord, tip chord and the mean geometric chord. Also, show a dimensioned drawing of the planform with the location of the m.g.c. clearly marked.

4.2 Test results obtained on a NACA 23012 airfoil show the following data:

α^0	c_l
0	0.15
9	1.20

If this airfoil is used in the design of an elliptical wing with aspect ratio, A=7.0, determine the wing lift curve slope at low Mach numbers.

4.3 A rectangular wing model of 40 in. by 5 in. has the following characteristics which were determined in a windtunnel test:

e=0.87 $C_{L_\alpha} = a = 0.09$ 1/deg $\alpha_0 = -3^0$

If a full scale, rectangular wing of 42 ft by 6 ft is constructed, what lift will this wing develop at $\alpha = 5^0$ and 120 mph under standard sea–level conditions? Assume that e=0.87 for the full scale wing as well.

4.4 The following data are obtained from a windtunnel test of a wing with A=10.32 for an advanced twin engine light airplane:

α^0	C_L	C_D
–2	0.0634	0.0060
0	0.2598	0.0083
2	0.4545	0.0141
4	0.6483	0.0231
6	0.8412	0.0349
8	1.0327	0.0495
10	1.2217	0.0668
12	1.4059	0.0865

Determine the drag polar, using Eqn (4.36). What is the maximum lift–to–drag ratio and for what value of the lift coefficient does it occur?

4.5 The following data were obtained from a windtunnel test of a wing:

$$\text{at } C_L = 0, \; C_{m_{\bar{c}/4}} = -0.126$$

$$\text{at } C_L = 0.6, \; C_{m_{\bar{c}/5}} = -0.15$$

Find the aerodynamic center location for this wing, assuming that the angles of attack are small.

4.6 An airplane with a weight of 5,000 lbs has a wing area of 250 sq.ft. If the airplane lift–curve slope is 6.0 per radian and the angle of attack for zero lift is –2 deg., calculate the angle of attack (in degrees) of this airplane in level flight at a speed of 200 mph under standard sea–level conditions.

4.7 How fast must the wing of problem 4.4 travel at an angle of attack of 4 degrees to produce a lift of 1.87 KN at 9,000 ft altitude? Assume standard atmospheric conditions.

4.8 A low speed airplane has the following characteristics:

Airfoil: NACA 23012, with a value of a_∞ per the data in Problem 4.2.

$A = 9.0$ and $\lambda = 0.5$

$\Lambda_{c/2} = 0$ deg. and $b = 31.4$ ft.

The wing is equipped with a partial span flap as shown in Figure 4.32. During takeoff, the angle of attack is set at 8 degrees. Determine the flap angle required to generate a total lift coefficient of 1.34. Hint: use Eqns (4.32) and (4.52).

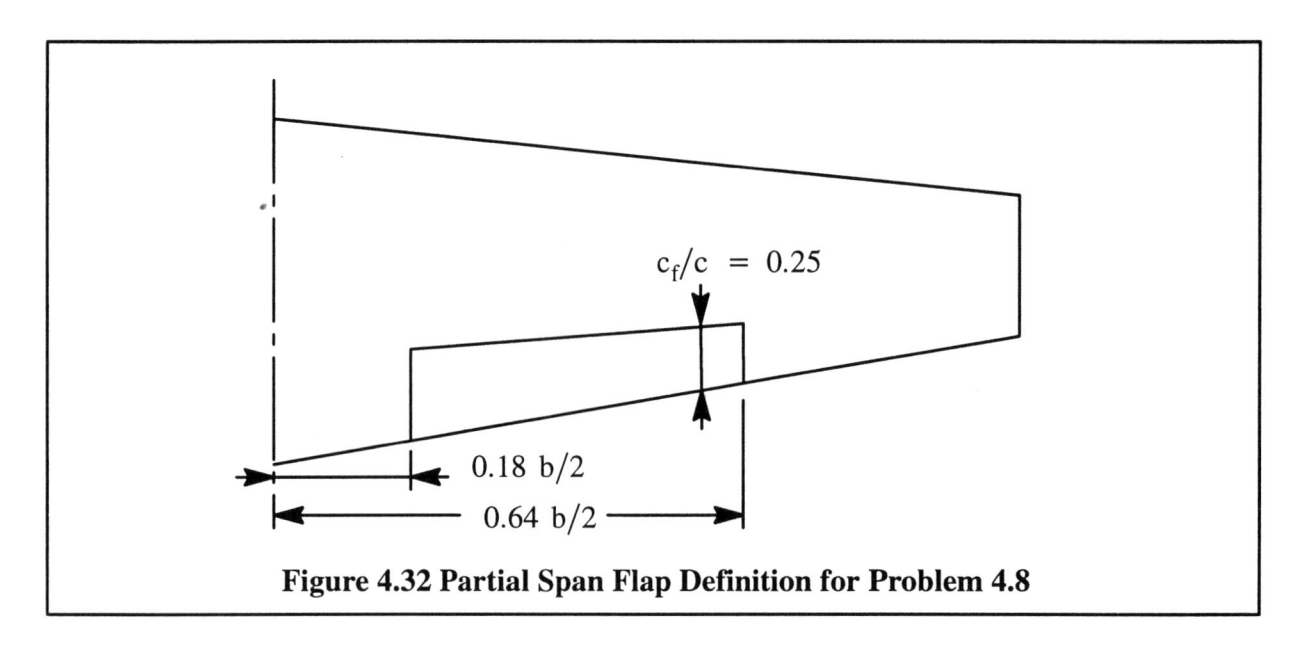

Figure 4.32 Partial Span Flap Definition for Problem 4.8

4.10 REFERENCES FOR CHAPTER 4

4.1 Glauert, H.; The Elements of Aerofoil and Airscrew Theory; Cambridge University Press, United Kingdom, 1959.

4.2 Abbott. I.A. and Von Doenhoff, A.E.; Theory of Wing Sections; Dover Publications, 1959.

4.3 Lowry, J.G. and Polhamus, E.C.; A method for Predicting Lift Increments due to Flap Deflection at Low Angles of Attack in Incompressible Flow; NACA TN 3911, 1957.

4.4 Giesing, J.P.; Lifting Surface Theory for Wing–Fuselage Combinations; Report DAC–67212, Vol. 1, McDonnell–Douglas Corporation, August 1, 1968.

4.5 Margason, R.J. and Lamar, J.E.; Vortex–Lattice Fortran Program for Estimating Subsonic Aerodynamic Characteristics of Complex Planforms; NASA TN D–6142, February, 1971.

4.6 Lan, C.E.; A Quasi–Vortex–Lattice Method in Thin Wing Theory; Journal of Aircraft, Vol. 11, September 1974, pp. 518–527.

4.7 Osborne, R.S. and Kelly, T.C.; A Note on the Drag Due to Lift of Delta Wings at Mach Numbers up to 2.0; NASA TN D–545, 1960.

4.8 Relf, E.F.; Note on the Lift Slope and Some Other Properties of Delta and Swept–Back Wings; Aeronautical Research Council of Great Britain, R&M No. 3111, 1959.

4.9 Hoak, D.E. et al; USAF Stability and Control Datcom; Air Force Flight Dynamics Laboratory, Wright–Patterson Air Force Base, Ohio, Revised April 1978.

4.10 Roskam, J.; Airplane Design, Part VI, Preliminary Calculation of Aerodynamic, Thrust and Power Characteristics; 1990, DARCorporation, 120 East 9th Street, Suite 2, Lawrence, KS, 66044.

4.11 Truckenbrodt, E.; Experimentelle Und Theoretische Untersuchungen An Symmetrisch Angestromten Pfeil Und Delta Flugeln; Zeitschrift Fur Flugwissenschaften, Band 2; 1953, pp. 185–201.

4.12 Shortal, J.A. and Maggin, B.; Effect of Sweepback and Aspect Ratio on Longitudinal Stability Characteristics of Wings at Low Speeds; NACA TN 1093, July 1946.

4.13 Anon.; Advanced Aircraft Analysis Software Manual, Version 1.7, 1996; DARCorporation, 120 East 9th Street, Suite 2, Lawrence, KS, 66044.

4.14 Hoerner, S.F. and Borst, H.V.; Fluid Dynamic Lift; Published by the author, 1975.

4.15　Wimpenny, J.C.; The Design and Application of High Lift Devices; International Congress on Subsonic Aeronautics, Annals of the New York Academy of Sciences, Vol. 154, Art. 2, November 1968.

4.16　Wentz, W.H. and Seetharam, H.C.; Development of a Fowler Flap System for a High Performance General Aviation Airfoil; NASA CR–2443, December 1974.

4.17　Perkins, C.D. and Hage, R.E.; Airplane Performance, Stability and Control, John Wiley & Sons, New York, 1949.

4.18　Hoerner, S.F.; Fluid Dynamic Drag; Published by the author, 1965.

4.19　Callaghan, J.G.; Aerodynamic Prediction Methods for Aircraft at Low Speeds with Mechanical High Lift Devices; Paper No. 2 in Prediction Methods for Aircraft Aerodynamic Characteristics, AGARD–LS–67, 1974.

4.20　Nicolai, L.M.; Fundamentals of Aircraft Design; University of Dayton, 1975.

4.21　Torenbeek, E.; Synthesis of Subsonic Airplane Design; Kluwer–Boston, Hingham, Maine, 1982.

4.22　Furlong, G.C. and McHugh, J.G.; A Summary and Analysis of the Low Speed Longitudinal Characteristics of Swept Wings at High Reynolds Number; NACA TR 1339, 1957.

4.23　Roskam, J.; Airplane Flight Dynamics and Automatic Flight Controls, Parts I and II; 1995, DARCorporation, 120 East 9th Street, Suite 2, Lawrence, KS, 66044.

4.24　Holmes, B.J.; Flight Evaluation of an Advanced Technology Light Twin–Engine Airplane (ATLIT); NASA CR–2832, July 1977.

4.25　Taylor, J.W.R.; Jane's All The World's Aircraft; 1980–1981, Jane's Publishing Company, London, England.

CHAPTER 5: AIRPLANE DRAG

The objective of this chapter is to provide the reader with fundamental insight into the most important contributions to airplane drag. Mathematical and experimental techniques used in representing and predicting airplane drag will be presented.

In Section 5.1 the complete drag behavior of an airplane is discussed. The individual contributions to airplane drag by its many components is discussed in Section 5.2. Methods for measuring drag in wind–tunnels are discussed in Section 5.3.

A simple method for estimating airplane drag polars is given in Section 5.4. For detailed methods the reader is encouraged to consult References 5.1, 5.2 and 5.3. A user–friendly computer program, based on Reference 5.1 is described in Appendix B.

5.1 COMPLETE AIRPLANE DRAG POLARS

5.1.1 CLEAN AIRPLANE

A clean airplane is defined as an airplane in its cruise configuration. For most (but not all) airplanes this means flaps up and gear up.

For many conventional airplanes it has been found possible to represent the relationship between drag and lift coefficients by the following parabolic form:

$$C_D = C_{D_0} + \frac{C_L^2}{\pi A e} \tag{5.1}$$

where: C_{D_0} is the zero–lift drag coefficient

e is Oswald's efficiency factor. A value of e=1 indicates an elliptical lift distribution.

$\frac{C_L^2}{\pi A e} = C_{D_i}$ is the induced drag coefficient

In a practical situation, magnitudes for C_{D_0} and e can be found with experimental, theoretical and/or empirical methods. Figure 5.1 shows an example of a theoretical drag polar in accordance with Eqn (5.1). Point A is the point at which the lift–to–drag ratio is a maximum. The reader is asked to verify that the following relationships hold at point A:

$$C_D = 2C_{D_0} \text{ at } C_L \text{ for } (C_L/C_D)_{max} \tag{5.2}$$

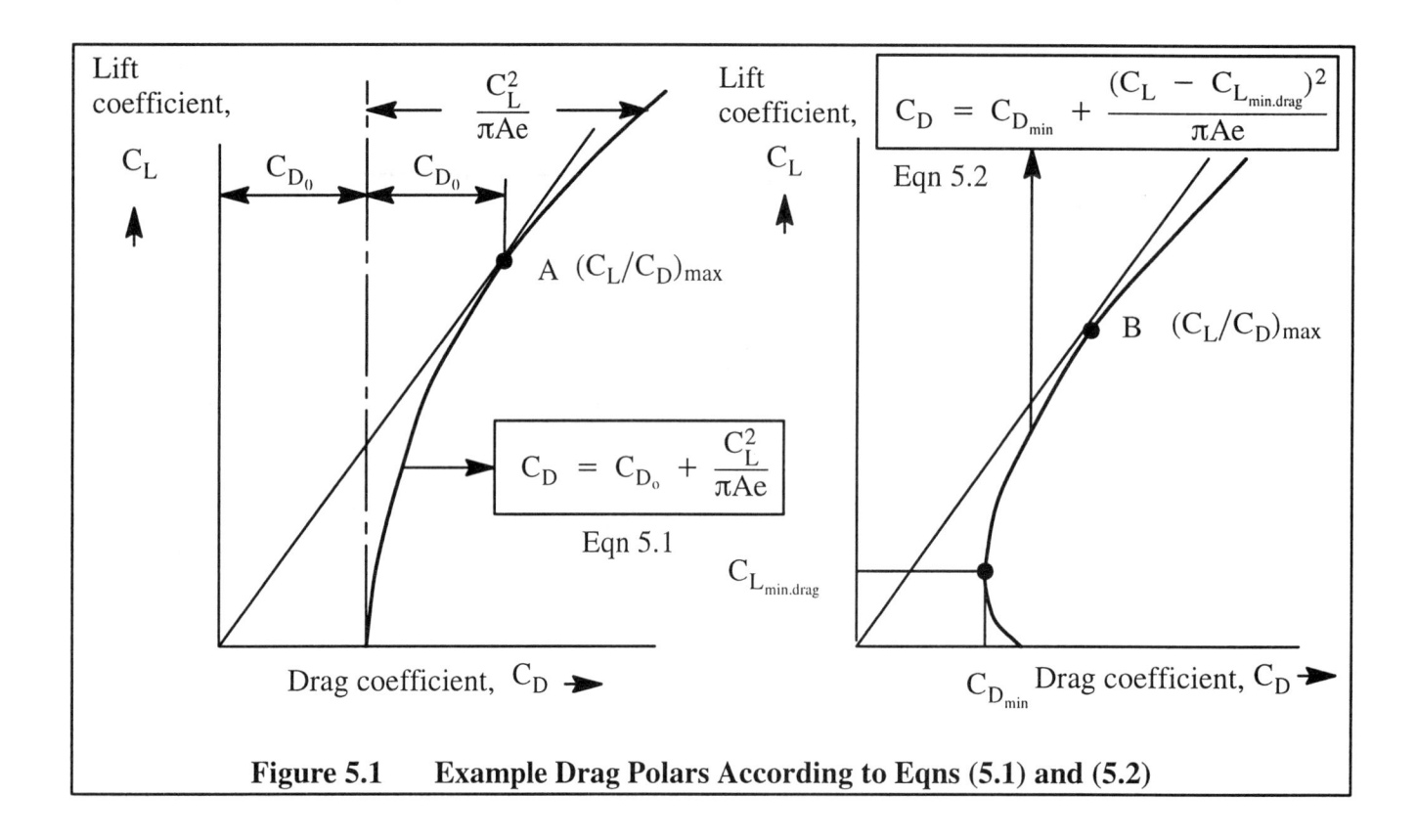

Airplane Drag

Figure 5.1   Example Drag Polars According to Eqns (5.1) and (5.2)

and
$$C_L \text{ for } (C_L/C_D)_{max} = \sqrt{\pi A e C_{D_0}} \qquad (5.3)$$

If the drag coefficient, C_D, is plotted versus the square of the lift coefficient, C_L^2, a straight line should be the result. Figure 5.2 presents a comparison of actual and theoretical values of C_D plotted against C_L^2 for two different airplanes. It is clear that the parabolic form of the drag polar is reasonable only for a limited range of lift coefficients. One reason for this is, that at higher values of the lift coefficient the flow will usually begin to separate from an airplane. This results in an increase in drag.

It is observed from Figure 5.2 that the drag levels of the two airplanes are very different. This is partly due to the difference in the propulsion system: jet airplanes tend to be 'cleaner' than piston/propeller airplanes. Another reason is better aerodynamic design as a result of advances in aerodynamics technology. A final reason is the manufacturing difference between the two airplanes: the AT–6 used round head riveting, while the S–211 is flush riveted.

Observe also from Figure 5.2, that $C_{D_0} \neq C_{D_{min}}$. The reason for this is the rather highly cambered airfoil used in the wing of the S–211. Whenever significant camber is used in an airplane, a better representation for the drag polar is:

Figure 5.2 Comparison of Actual with Parabolic Approximations for Drag Polars

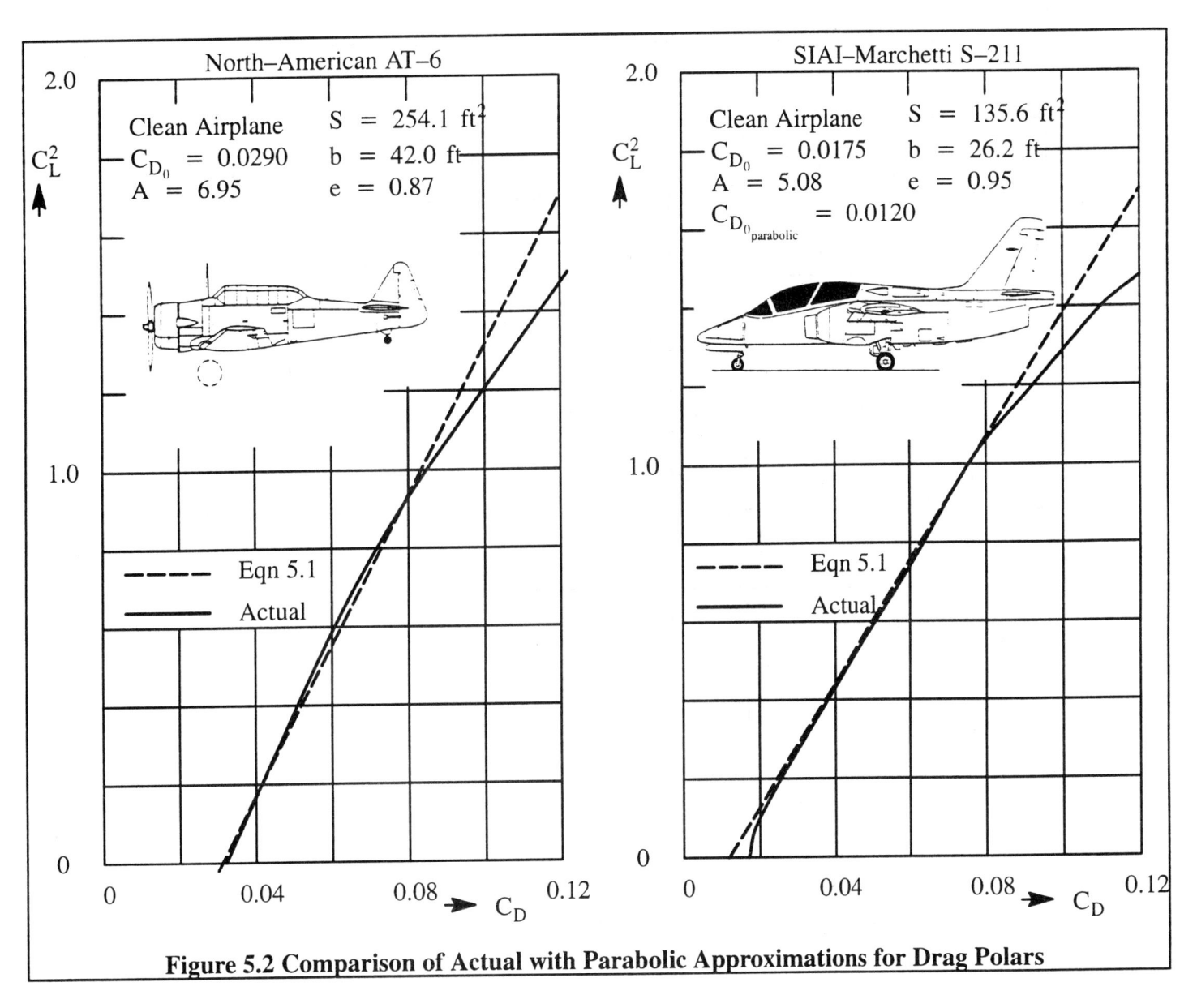

$$C_D = C_{D_{min}} + \frac{(C_L - C_{L_{min.drag}})^2}{\pi Ae} \tag{5.4}$$

where: $C_{D_{min}}$ is the minimum drag coefficient

$C_{L_{min.drag}}$ is the lift coefficient at the minimum drag coefficient

At high subsonic speeds, whenever shock waves occur and/or whenever shock induced boundary layer separation occurs, the drag rises much more rapidly with lift coefficient, than indicated by either of the parabolic approximations given by Eqns (5.1) or (5.4). Figure 5.3 shows the drag rise behavior of a subsonic jet transport and of a subsonic jet trainer as a function of Mach number. In Chapter 3, page 62, the so-called critical Mach number was introduced as that free stream Mach number for which sonic velocity is first reached somewhere on an airfoil. The same definition is used for wings, for bodies and also for entire airplanes.

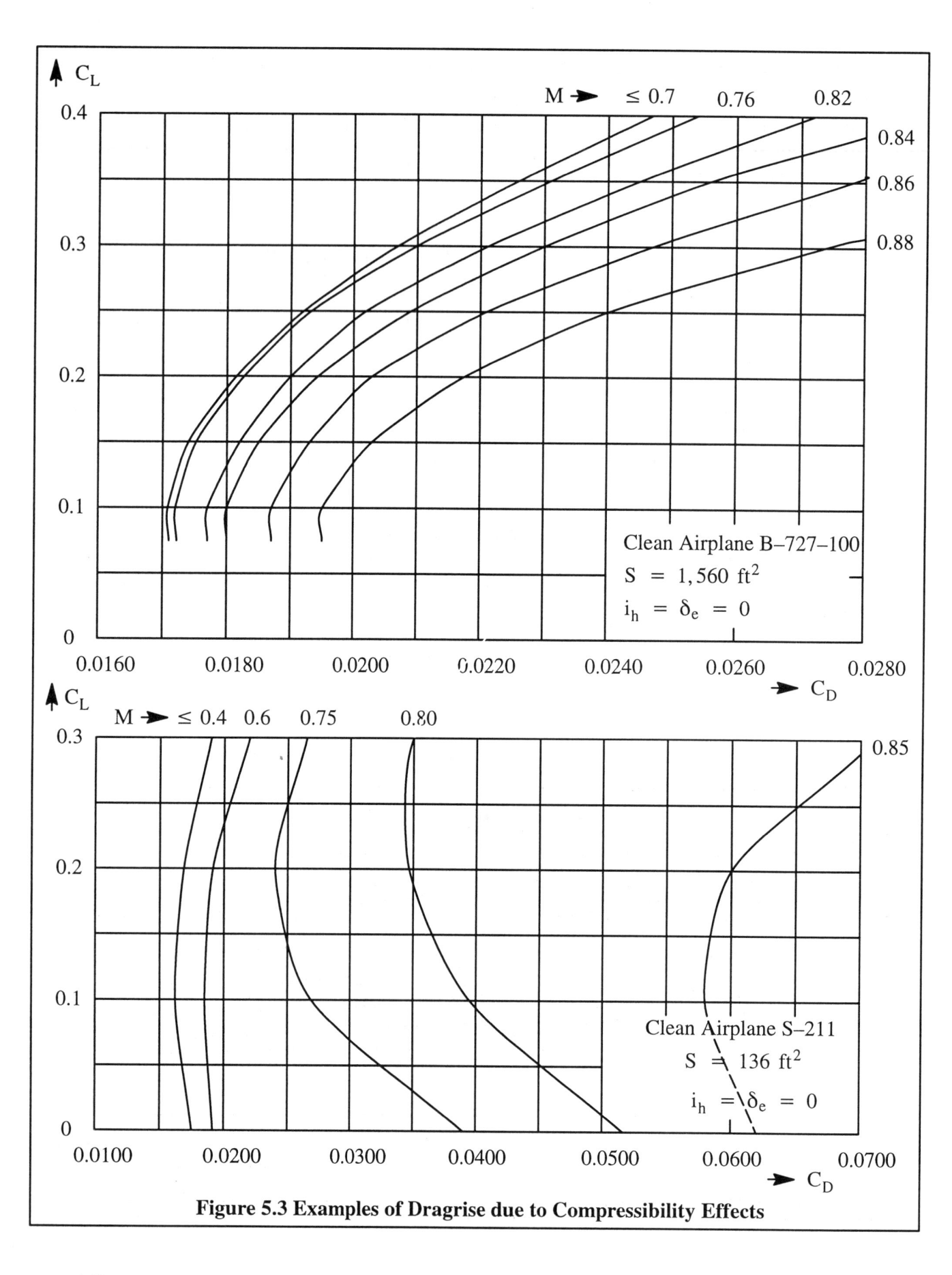

Figure 5.3 Examples of Dragrise due to Compressibility Effects

Airplane Drag

Of greater practical interest is the so-called drag divergence Mach number, M_{dd}. This was also defined in Chapter 3 for an airfoil. Two definitions were given: the Boeing and Douglas definition respectively. These definitions are repeated next, as applied to the entire airplane.

a) Boeing Definition

M_{dd} is that free stream Mach number for which the drag due to compressibility first reaches 20 drag counts ($\Delta C_D = 0.0020$) above the incompressible level.

b) Douglas Definition

M_{dd} is that free stream Mach number for which the slope of the dragrise, $\delta C_D/\delta M$, first reaches the value 0.10.

These definitions are most easily applied when the drag rise behavior of airplanes is represented in a cross-plot of drag coefficient, at constant lift coefficient, versus Mach number. The reader is asked to apply these definitions to the dragrise behavior at different lift coefficients of the B-727-100 and the S-211 of Figure 5.3 and determine how closely they agree.

According to Chapter 4 (Figure 4.22), both critical Mach number and drag divergence Mach number depend strongly on the sweep angle. As it turns out, they also depend on the thickness ratio of lifting surfaces. These effects are illustrated in Figure 5.4 for conventional, non-super-critical airfoils.

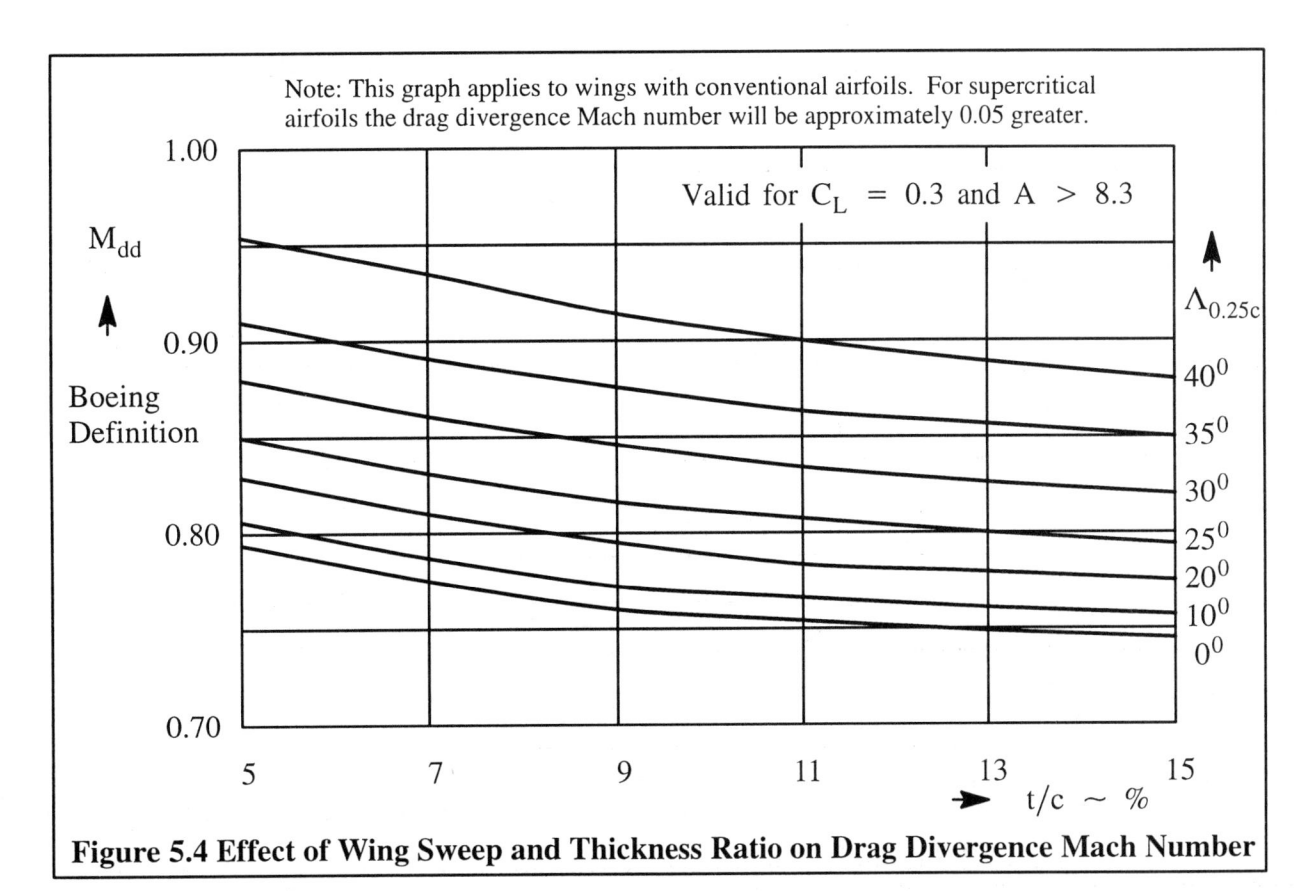

Figure 5.4 Effect of Wing Sweep and Thickness Ratio on Drag Divergence Mach Number

It has also been pointed out in Chapter 4, that at high Mach numbers, the occurrence of shock waves and shock induced separation will cause changes in lift behavior. In fact, if the free stream Mach number becomes too high, a phenomenon, called buffet, will occur. At the buffet boundary, lift behaves in an oscillatory manner. This can cause severe stability and control problems. Since the effect of Mach number on lift and drag occurs simultaneously it is their interaction which must be understood. Aerodynamic design details, such as airfoil design, thickness ratio variation with span, twist and design and location of the horizontal tail all can play an important role. An early example of the effect of Mach number on lift and drag characteristics of a complete airplane is shown in Figure 5.5.

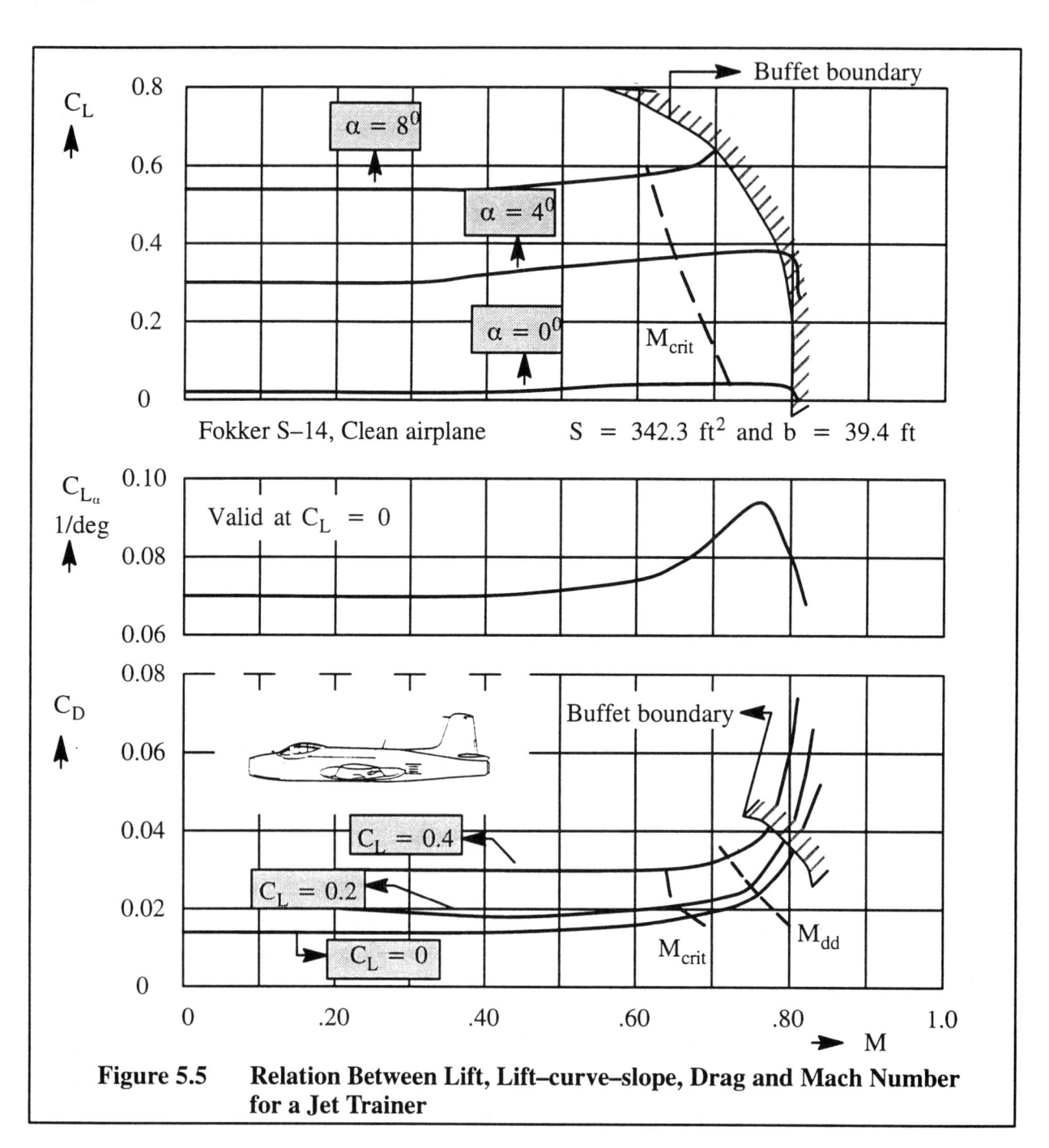

Figure 5.5 Relation Between Lift, Lift–curve–slope, Drag and Mach Number for a Jet Trainer

Airplane Drag

For airplanes at supersonic speeds, the drag behavior is also strongly dependent on the Mach number as well as on the lift coefficient. Figures 5.6 show an example of this for an early supersonic fighter. For such airplanes, the drag polar is sometimes written as:

$$C_D = C_{D_0} + kC_L^2 \tag{5.5}$$

For the airplane of Figures 5.6, values for C_{D_0}, k and $C_{L_{\alpha_{at\,C_L=0}}}$ are plotted in Figure 5.7.

In modern fighters, leading and trailing edge flaps as well as the stabilizer, are often used to trim the airplane at higher lift coefficients. The drag polars of such airplanes therefore include a schedule of surface deflections. This schedule is normally automatically programmed in the control laws of the automatic flight control system.

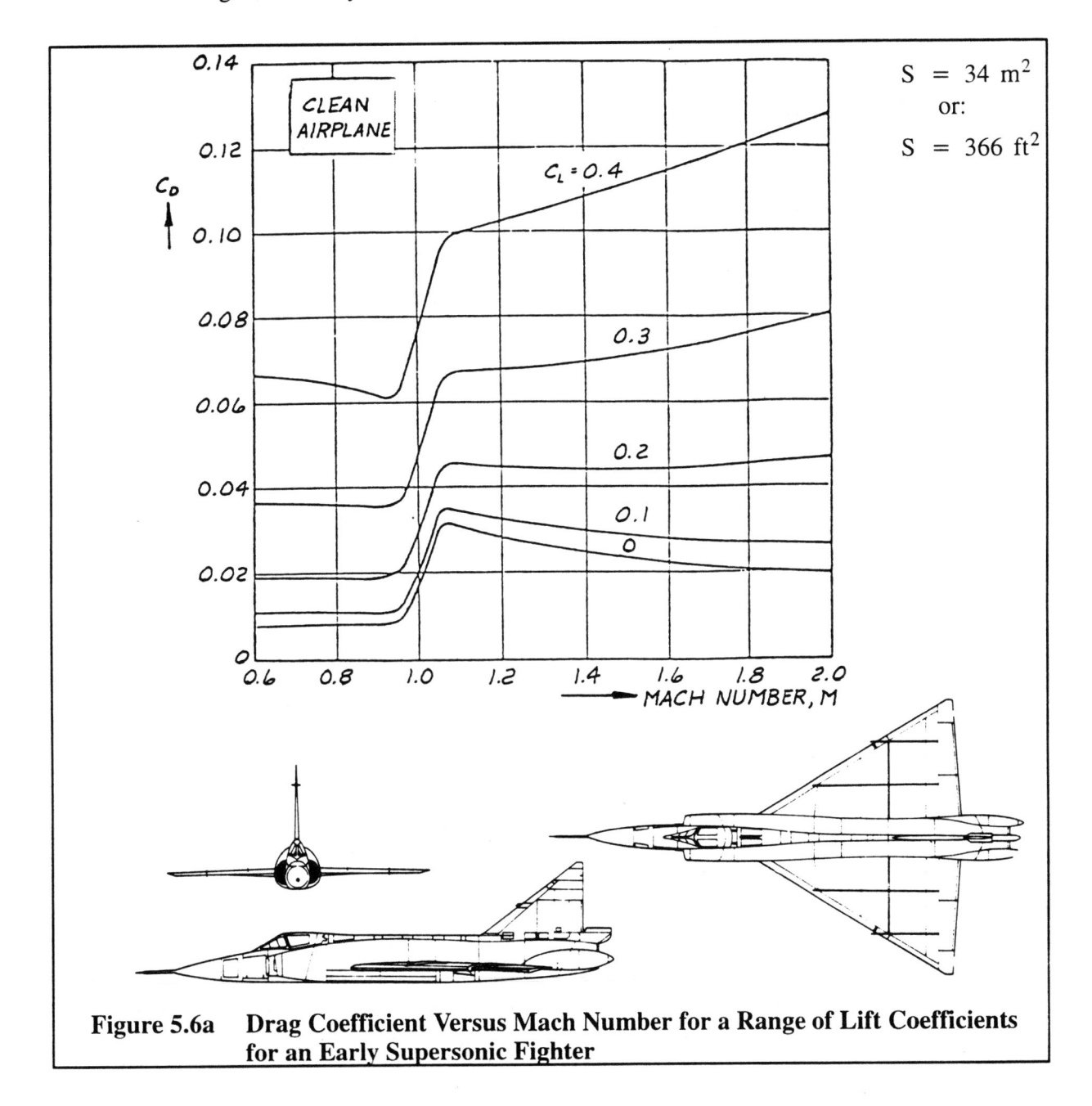

Figure 5.6a Drag Coefficient Versus Mach Number for a Range of Lift Coefficients for an Early Supersonic Fighter

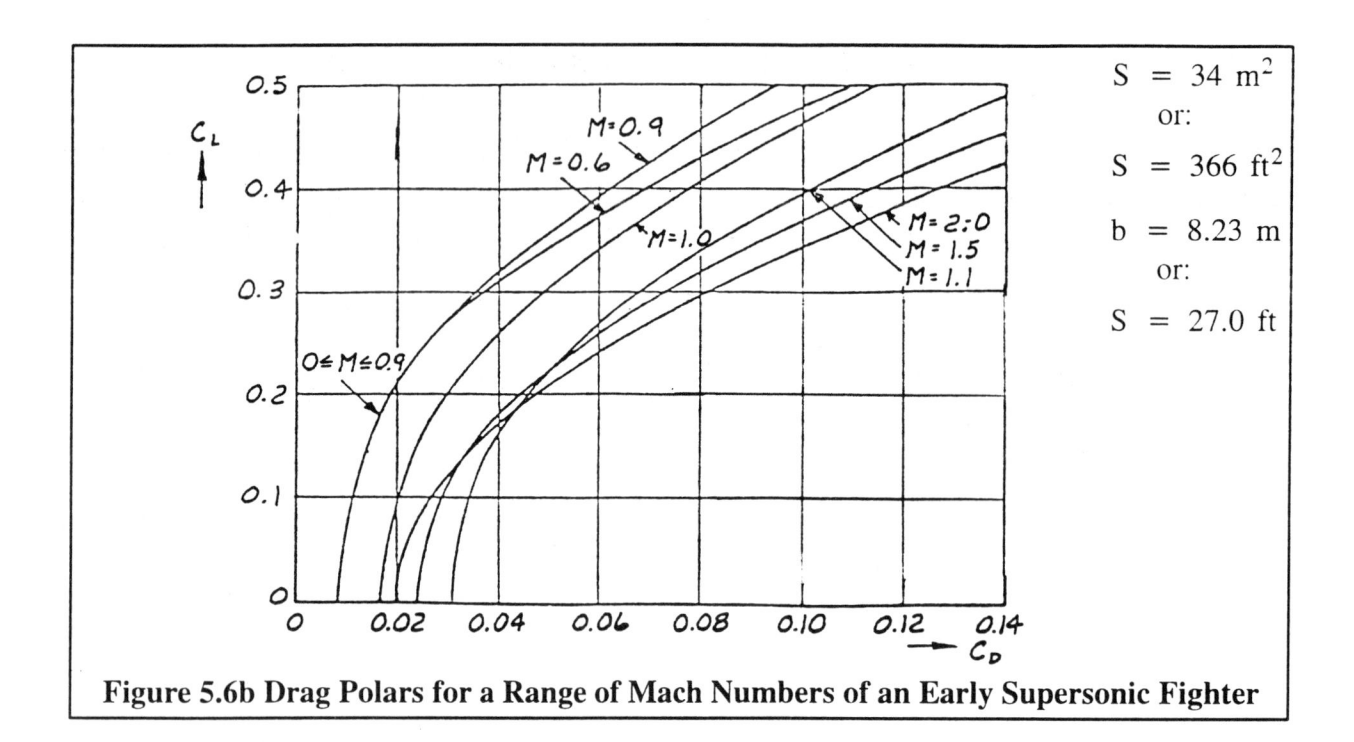

Airplane Drag

Figure 5.6b Drag Polars for a Range of Mach Numbers of an Early Supersonic Fighter

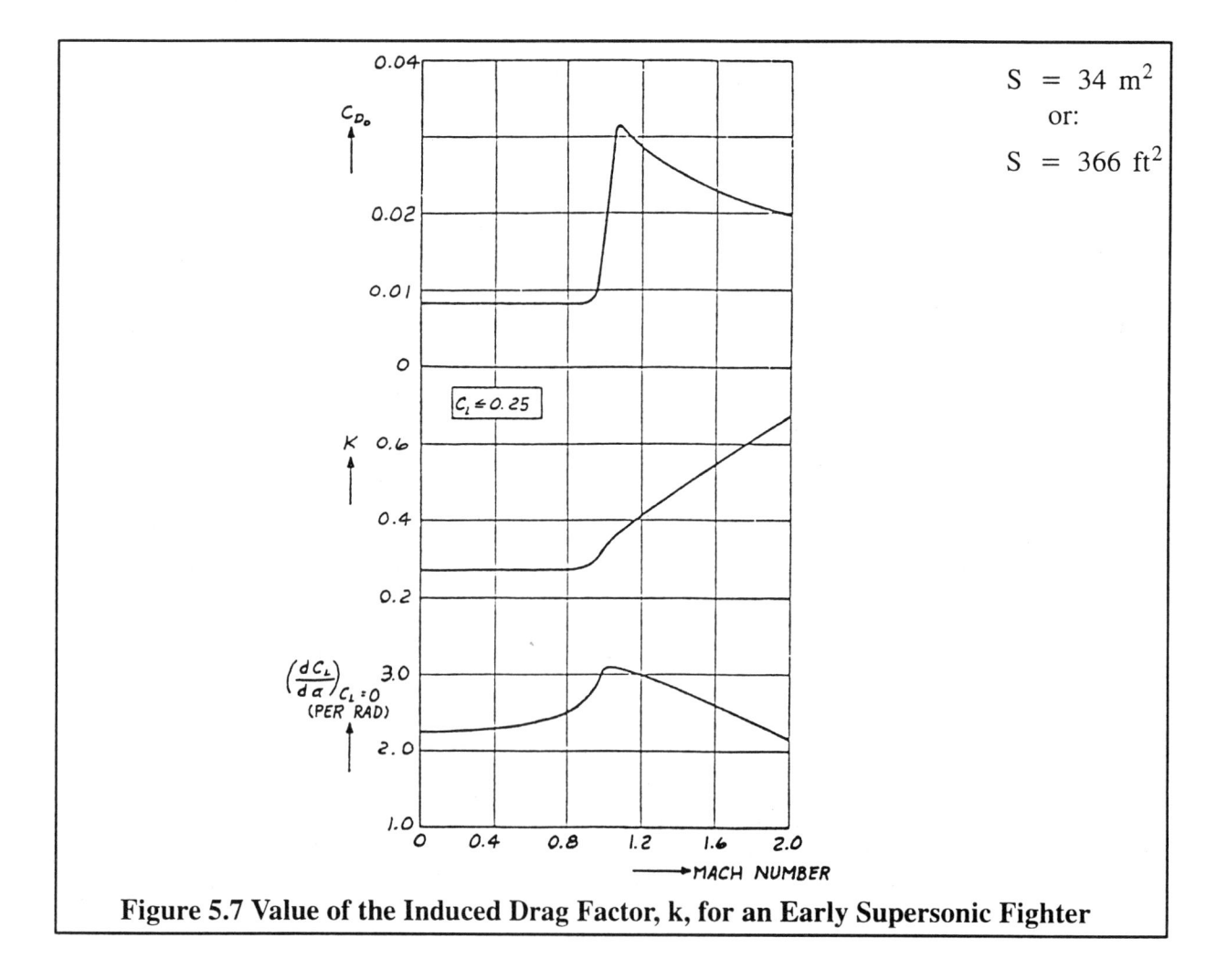

S = 34 m^2
or:
S = 366 ft^2
b = 8.23 m
or:
S = 27.0 ft

Figure 5.7 Value of the Induced Drag Factor, k, for an Early Supersonic Fighter

S = 34 m^2
or:
S = 366 ft^2

5.1.2 EFFECT OF FLAPS, SPEED–BRAKES AND LANDING GEAR

The discussion in Sub–section 5.1.1 has dealt only with the so–called 'clean' airplane. Whenever flaps, speed–brakes or landing gear are deployed, the drag of airplanes increases significantly. Figure 5.8 shows an example of these drag effects for the case of a small jet trainer airplane.

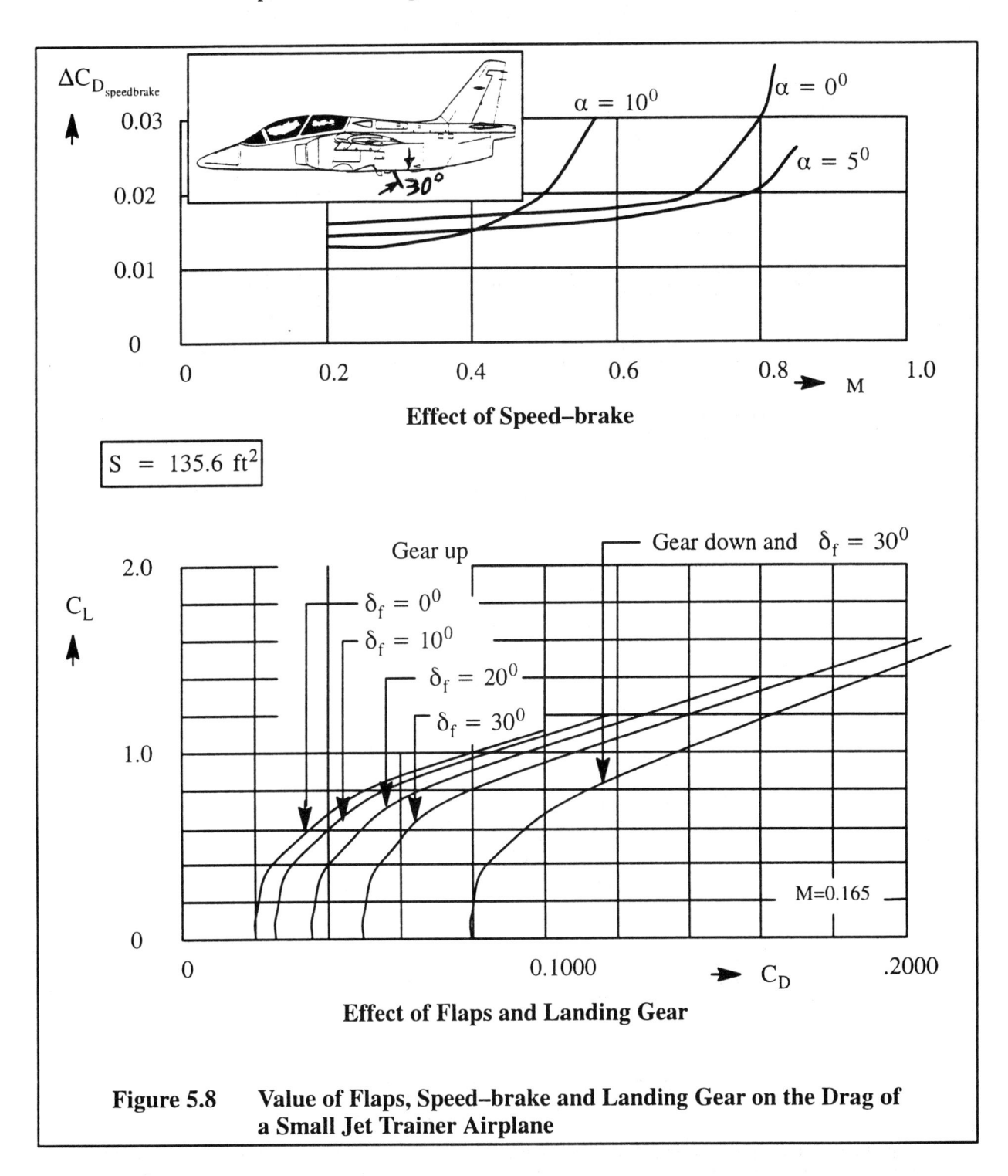

Effect of Speed–brake

$S = 135.6 \text{ ft}^2$

Effect of Flaps and Landing Gear

Figure 5.8 Value of Flaps, Speed–brake and Landing Gear on the Drag of a Small Jet Trainer Airplane

5.2 UNDERSTANDING AIRPLANE DRAG CONTRIBUTIONS

To understand how and which components of an airplane contribute to the total drag of an airplane, the drag polar is frequently split into the following components:

a) parasite drag, defined as that drag which is not dependent on the production of lift. This drag contribution has been identified in Eqns (5.1) and (5.4), as C_{D_0} and $C_{D_{min}}$.

b) induced drag, defined as that drag which directly depends on the production of lift. This drag contribution was identified in Eqn (5.1) as C_{D_i}.

In reality it has been found that such a simple split is difficult to achieve. The main reason is that in many airplanes parts of the parasite drag can become dependent on lift. Another reason is, that at high Mach numbers, compressibility drag will become very significant. Figure 5.9 shows typical zero–lift (parasite) and induced drag breakdowns for a transport and for a fighter airplane.

The relative importance of parasite and induced drag varies with flight condition and with airplane type. Note the excellent lift–to–drag ratio of the jet transport (L/D=15.8) in cruise and compare this to that of the fighter (L/D=6.1 in subsonic cruise and L/D=1.7 in supersonic cruise). One reason for the very low lift–to–drag ratio in supersonic cruise is the occurrence of wave drag due to compressibility.

Experienced aerodynamicists know, that any drag build–up method is somewhat arbitrary and relies on certain bookkeeping assumptions. For example, the internal drag of jet inlets is most often accounted for as a reduction in installed thrust and not as an increase in drag. Most airplane manufacturers have developed their own (often proprietary) procedures for estimating airplane drag.

To help in understanding the basics of drag component and drag source build–up, the following equation is suggested in Reference 5.1:

$$C_D = C_{D_{wing}} + C_{D_{fus}} + C_{D_{emp}} + C_{D_{np}} + C_{D_{flap}} + C_{D_{gear}} + C_{D_{cw}} + \qquad (5.6)$$
$$+ C_{D_{store}} + C_{D_{trim}} + C_{D_{interf.}} + C_{D_{misc}}$$

where: $C_{D_{wing}}$ is the wing drag coefficient, see Sub–section 5.2.1

$C_{D_{fuse}}$ is the fuselage drag coefficient, see Sub–section 5.2.2

$C_{D_{emp}}$ is the empennage drag coefficient, see Sub–section 5.2.3

$C_{D_{np}}$ is the nacelle/pylon drag coefficient, see Sub–section 5.2.4

When the wing, fuselage, empennage and nacelles are integrated, several additional factors which influence airplane drag must be accounted for. Some of these are discussed in Sub–section 5.2.5

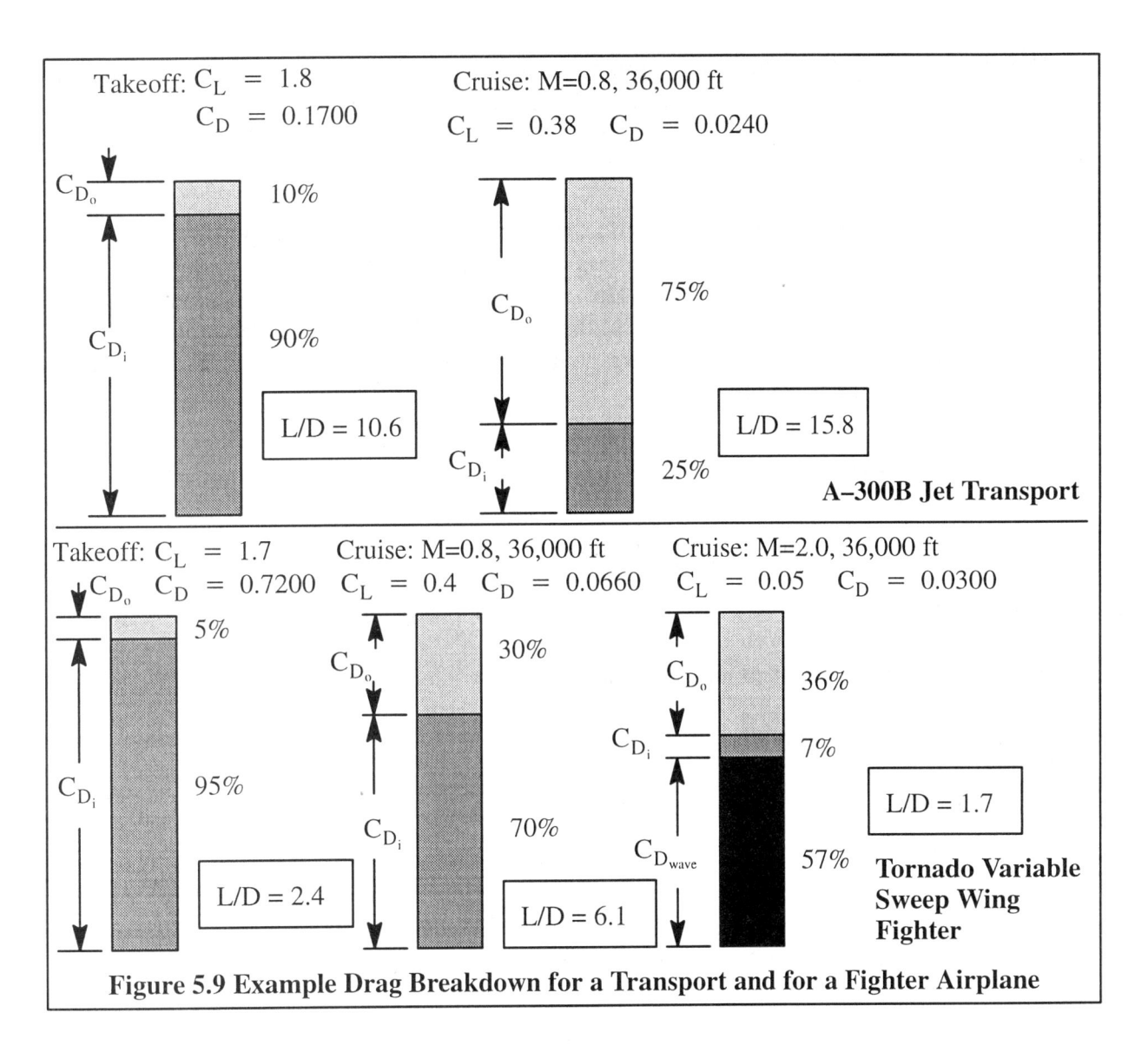

Airplane Drag

Figure 5.9 Example Drag Breakdown for a Transport and for a Fighter Airplane

A-300B Jet Transport:
- Takeoff: $C_L = 1.8$, $C_D = 0.1700$; C_{D_o} 10%, C_{D_i} 90%; L/D = 10.6
- Cruise: M=0.8, 36,000 ft; $C_L = 0.38$, $C_D = 0.0240$; C_{D_o} 75%, C_{D_i} 25%; L/D = 15.8

Tornado Variable Sweep Wing Fighter:
- Takeoff: $C_L = 1.7$, $C_D = 0.7200$; C_{D_o} 5%, C_{D_i} 95%; L/D = 2.4
- Cruise: M=0.8, 36,000 ft; $C_L = 0.4$, $C_D = 0.0660$; C_{D_o} 30%, C_{D_i} 70%; L/D = 6.1
- Cruise: M=2.0, 36,000 ft; $C_L = 0.05$, $C_D = 0.0300$; C_{D_o} 36%, C_{D_i} 7%, $C_{D_{wave}}$ 57%; L/D = 1.7

$C_{D_{flap}}$ is the flap drag coefficient, see Sub-section 5.2.6

$C_{D_{gear}}$ is the landing gear drag coefficient, see Sub-section 5.2.7

$C_{D_{cw}}$ is the canopy/windshield drag coefficient, see Sub-section 5.2.8

$C_{D_{store}}$ is the store(s) drag coefficient, see Sub-section 5.2.9

$C_{D_{trim}}$ is the trim drag coefficient, see Sub-section 5.2.10

$C_{D_{interf.}}$ is the interference drag coefficient, see Sub-section 5.2.11

$C_{D_{misc}}$ is the miscellaneous drag coefficient, see Sub-section 5.2.12

The computer program described in Appendix B allows a rapid calculation of airplane component drag and complete airplane drag for arbitrary configurations and Mach numbers from low subsonic to supersonic.

Airplane Drag

Before discussing the various drag components it is instructive to consider Figure 5.10 which shows typical drag breakdowns as determined for five different airplanes. It is seen, that the wing and fuselage constitute the largest drag contributions in these airplanes. This is to a significant extent due to the large amount of wetted area of these components.

It should be noted here that various manufacturers of business jets and transports have developed very accurate predictive techniques for the drag of their designs. Accuracies of plus or minus 1% in the cruise configuration are not unusual. However, because of the very close interaction between drag and installed thrust and the intricacies of the bookkeeping decisions made in their prediction, many manufacturers find it necessary to engage in extensive drag reduction programs after a new airplane has entered the flight test phase.

5.2.1 WING DRAG COEFFICIENT, $C_{D_{wing}}$

Drag contributions due to the wing consist of:

- friction drag
- profile or thickness drag
- drag due to lift, also called induced drag
- interference drag
- compressibility or wave drag

The wing drag coefficient may be determined from:

$$C_{D_{wing}} = C_{D_{0_w}} + C_{D_{L_w}} \tag{5.7}$$

where: $C_{D_{0_w}}$ is the zero–lift drag coefficient of the wing

$C_{D_{L_w}}$ is the drag coefficient due to lift of the wing (induced drag)

Wing zero–lift drag is discussed in Sub–sub–section 5.2.1.1, while wing drag due to lift is covered in Sub–sub–section 5.2.1.2. The effect of compressibility is briefly discussed in Sub–sub–section 5.2.1.3.

5.2.1.1 Wing Zero–Lift Drag Coefficient, $C_{D_{0_w}}$

The subsonic, wing zero–lift drag coefficient may be empirically determined from:

$$C_{D_{0_w}} = R_{wf} R_{LS} C_{f_w} \left\{ 1 + L'(t/c) + 100(t/c)^4 \right\} \frac{S_{wet_w}}{S} \tag{5.8}$$

where: R_{wf} is the so-called wing–fuselage interference factor. This factor be estimated from Figure 5.11.

R_{LS} is a lifting surface correction factor which depends on the sweep angle of the locus of maximum airfoil thickness on the wing, $\Lambda_{(t/c)_{max}}$. See: Figure 5.12.

Airplane Drag

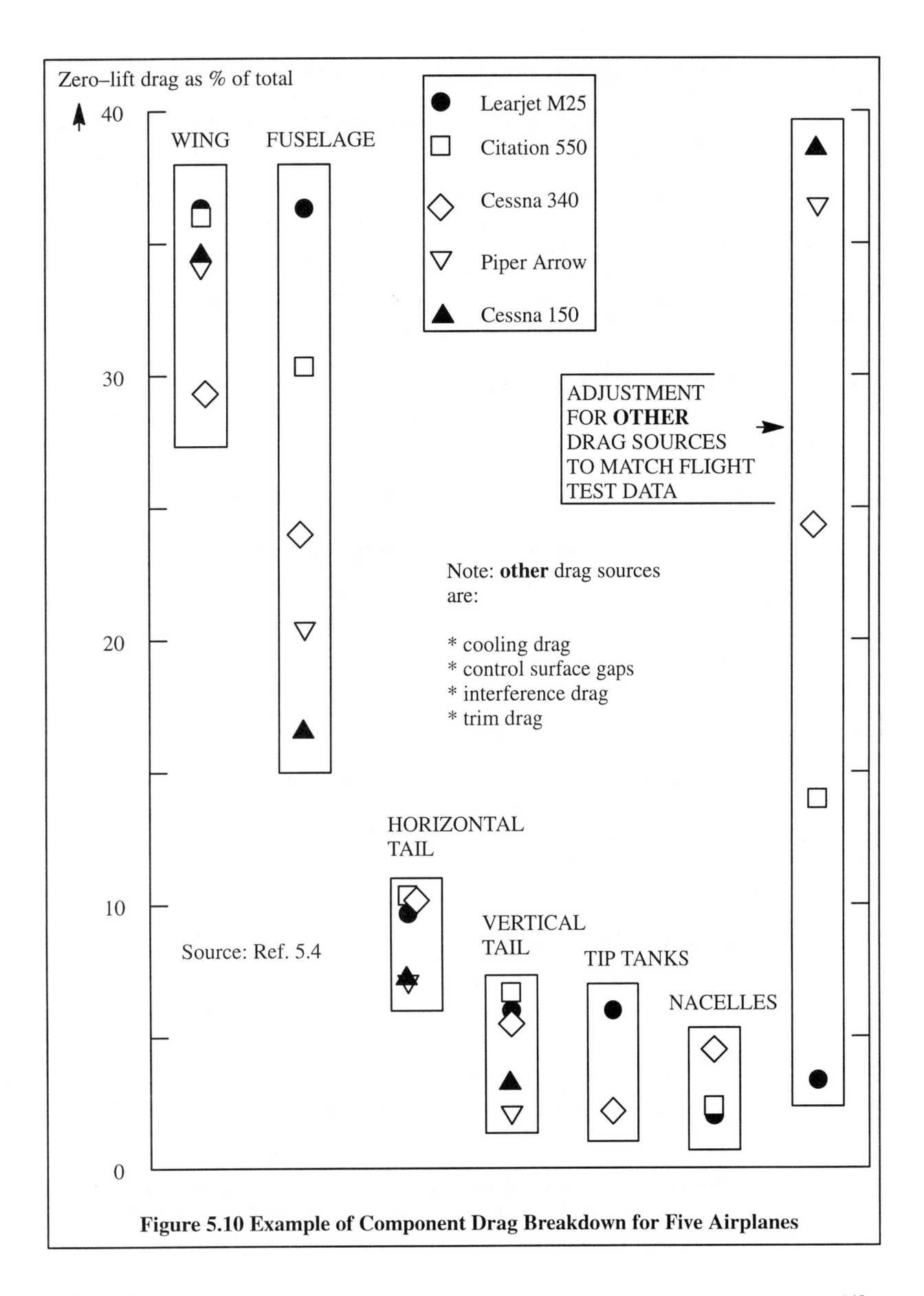

Zero–lift drag as % of total

Legend:
- ● Learjet M25
- □ Citation 550
- ◇ Cessna 340
- ▽ Piper Arrow
- ▲ Cessna 150

ADJUSTMENT FOR **OTHER** DRAG SOURCES TO MATCH FLIGHT TEST DATA →

Note: **other** drag sources are:

* cooling drag
* control surface gaps
* interference drag
* trim drag

Source: Ref. 5.4

Categories shown on chart: WING, FUSELAGE, HORIZONTAL TAIL, VERTICAL TAIL, TIP TANKS, NACELLES

Figure 5.10 Example of Component Drag Breakdown for Five Airplanes

C_{f_w} is the turbulent flat plate friction coefficient of the wing. The general, turbulent, flat plate friction coefficient, C_f is shown in Figure 5.13* as a function of Mach number and the Reynolds number, see Eqn (2.97). For the wing Reynolds number use:

$$R_{N_w} = \frac{\varrho U_1 \bar{c}_{w_e}}{\mu} \tag{5.9}$$

where: $\bar{c}_{w_{exposed}}$ is the mean geometric chord of the exposed (wetted) part of the wing. Figure 5.14 shows how the exposed wing area is defined.

μ is the coefficient of viscosity of air: see Appendix A.

L' is the airfoil thickness location parameter, found from Figure 5.15.

t/c is the thickness ratio of the wing, defined at the m.g.c of the exposed wing.

S_{wet_w} is the wetted area of the wing, see: Figure 5.14.

S is the wing area, also called the wing reference area.

Note Well: the friction drag prediction given here is valid only for smooth surfaces. For surfaces with roughness due to manufacturing and/or various types of paint, additional friction may result. Reference 5.1 contains a detailed method to account for such roughness effects.

The thickness ratio term, t/c, in Eqn (5.8) accounts for the profile drag. Because a wing is nearly always attached to a fuselage, additional drag due to interference between these components will result. Sub–section 5.2.10 contains a physical explanation of interference drag.

Clearly, the wetted area, S_{wet_w} plays an important role in determining wing drag. The wetted area of a wing depends on how much of the wing is exposed. The difference between wing reference, or planform, area and wing exposed area is shown in Figure 5.14. For most cases the wetted area of a wing can be determined from:

$$S_{wet_w} = K_w S_{exposed} \tag{5.10}$$

where: $S_{exposed}$ is the exposed planform area of the wing as defined in Figure 5.13

K_w is the so–called surface area factor which may be estimated from:

$$K_w = 1.9767 + 0.5333\left(\frac{t}{c}\right) \text{ for : } \frac{t}{c} \geq 0.05 \tag{5.11}$$

$$K_w = 2.0 \text{ for : } \frac{t}{c} < 0.05$$

For most wings it is acceptable to use for t/c the value associated with the mean geometric chord.

---

* It should be noted that Figure 5.13 was obtained by converting incompressible values into compressible values through Eckert's method of Ref.5.5. The required physical quantities, such as density and viscosity, are evaluated at a chosen reference temperature given by Eckert.

Figure 5.11 Wing–Fuselage Interference Factor

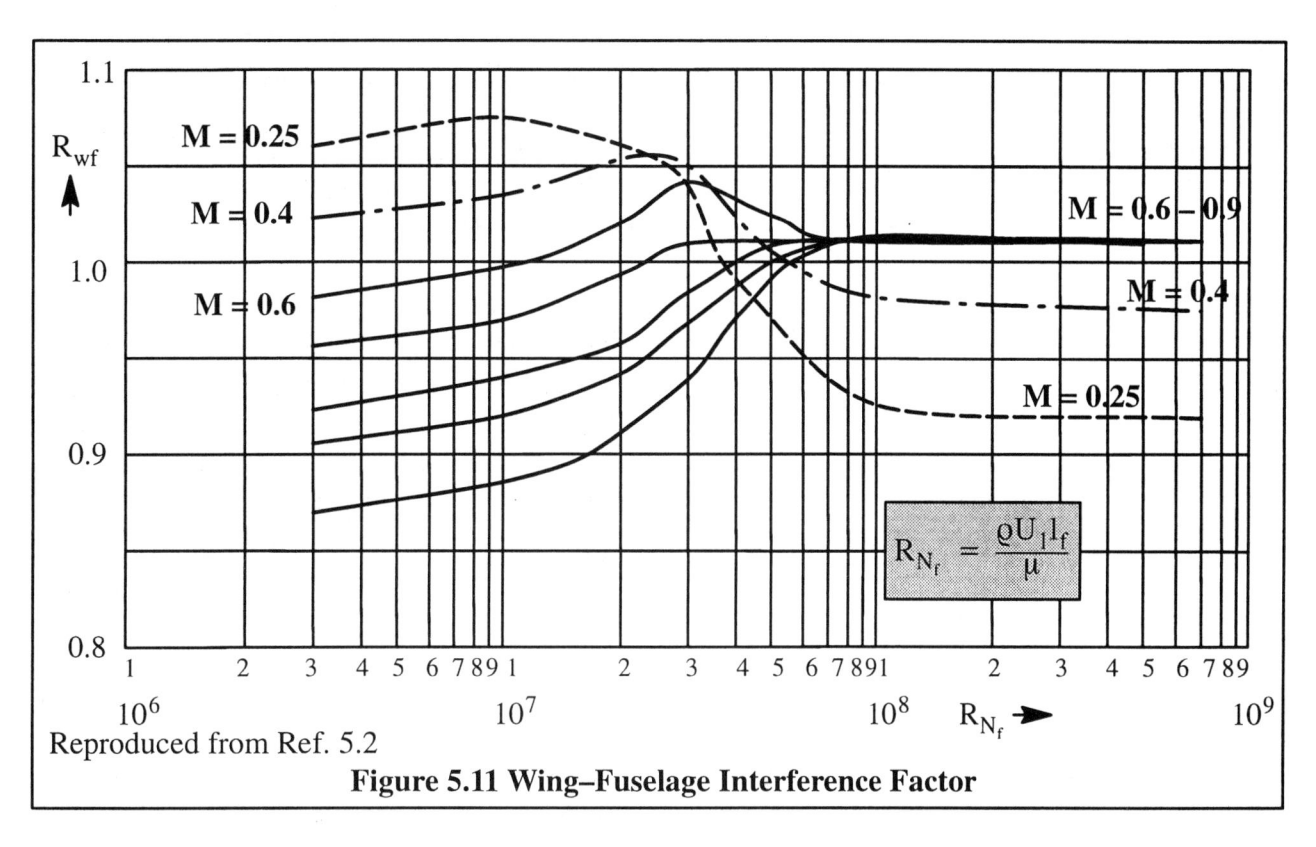

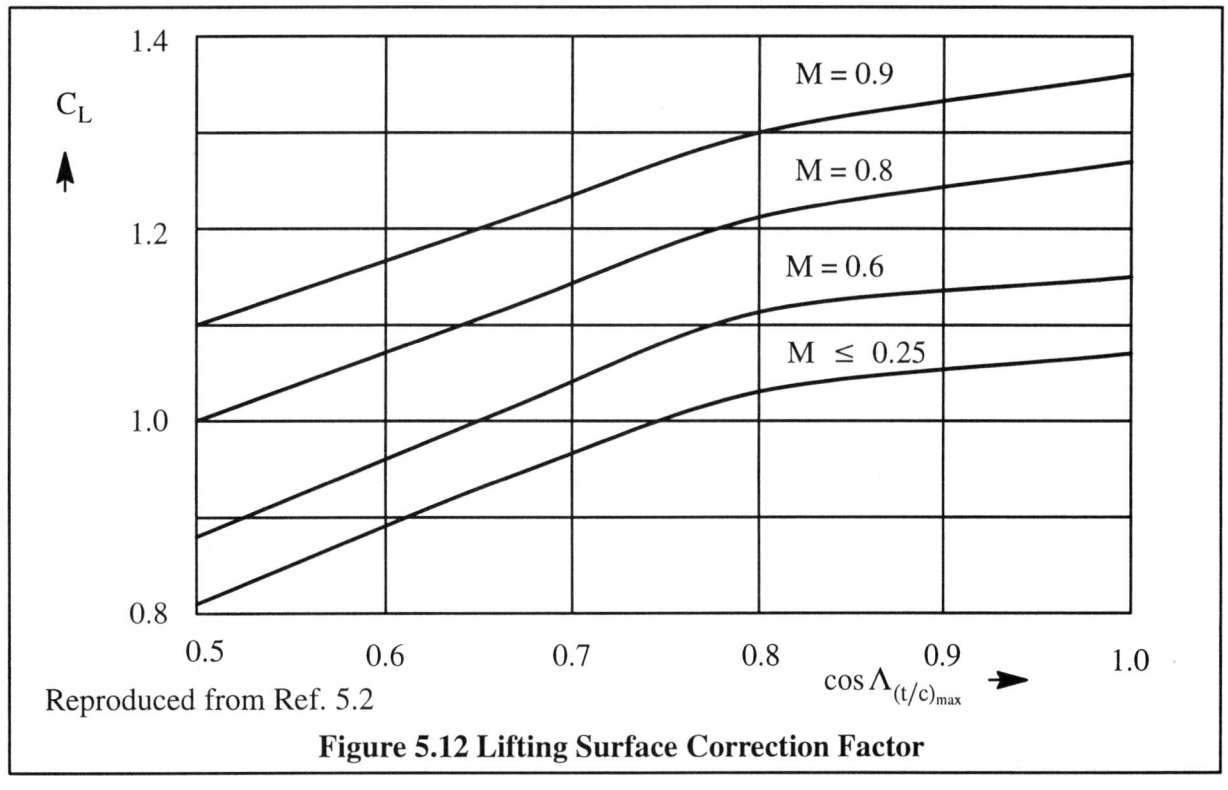

$$R_{N_f} = \frac{\rho U_1 l_f}{\mu}$$

Reproduced from Ref. 5.2

Figure 5.12 Lifting Surface Correction Factor

Reproduced from Ref. 5.2

In designing a new airplane, the selection of the wing area, S has important repercussions to cruise performance, fieldlength performance and, because most wings are used to store fuel, to range performance. References 5.6 and 5.7 contain methods for determining the wing area.

$$C_f = \frac{0.455}{(\log_{10} R_N)^{2.58}} (1 + 0.144 M^2)^{0.58}$$

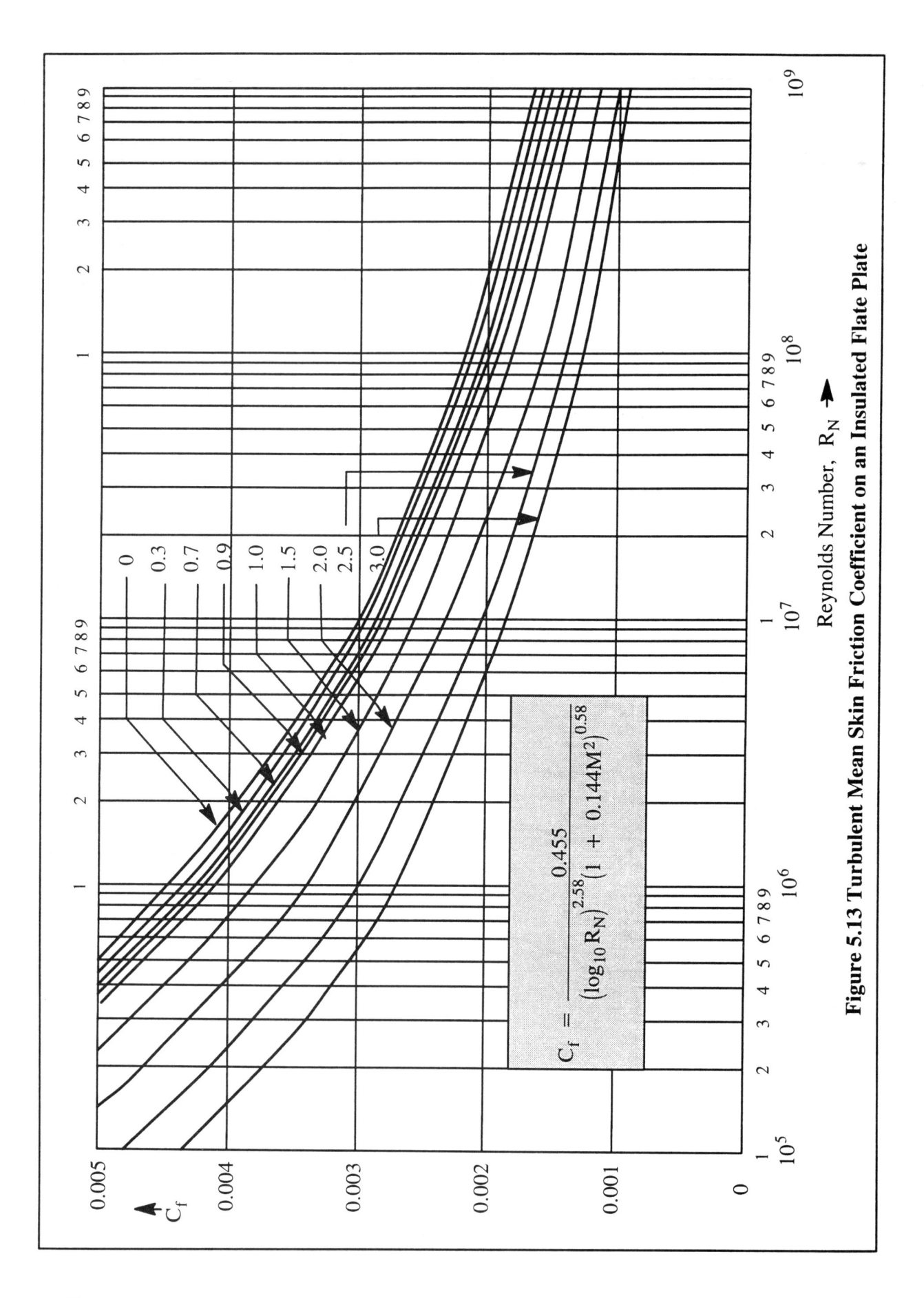

Figure 5.13 Turbulent Mean Skin Friction Coefficient on an Insulated Flate Plate

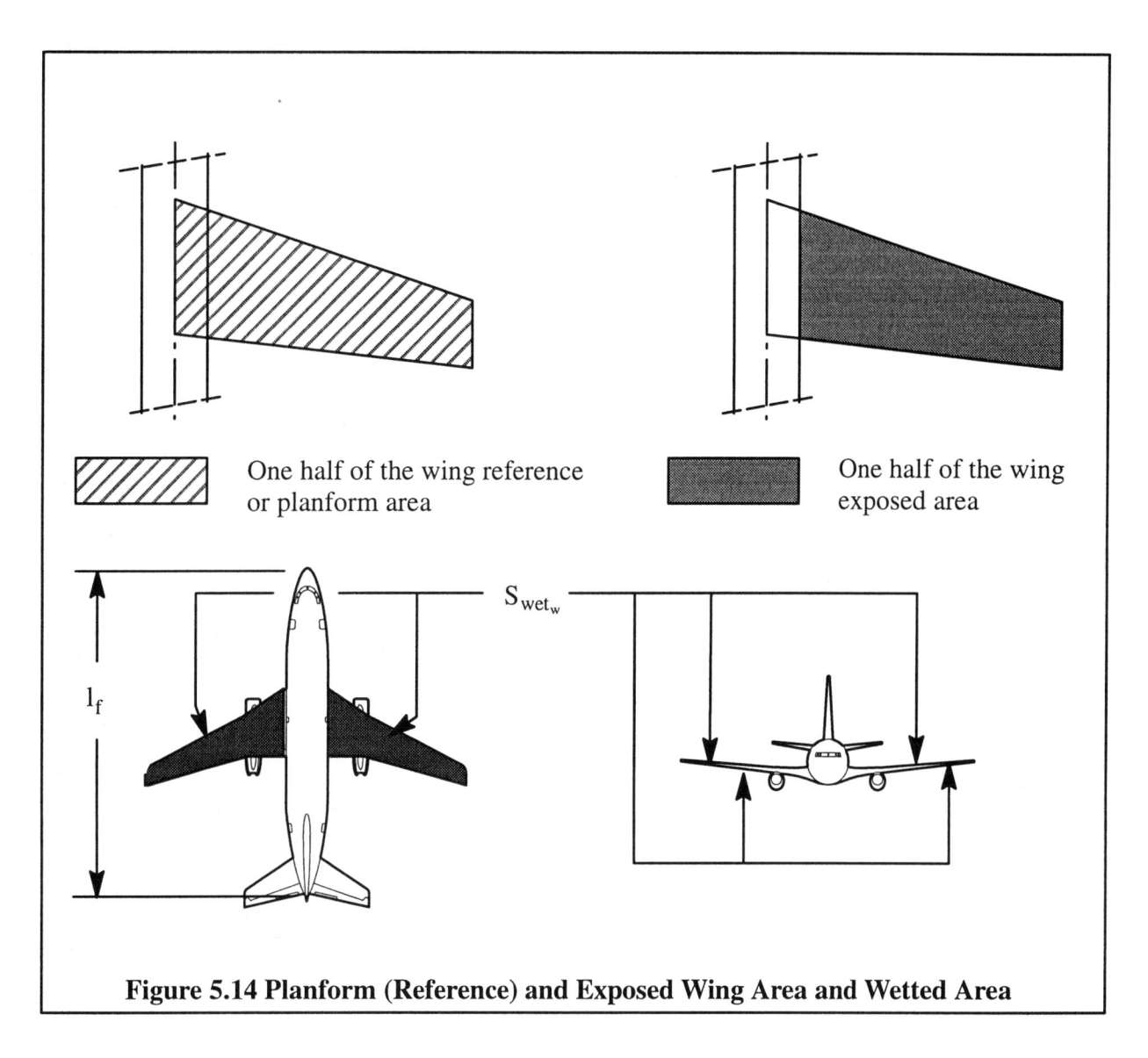

Figure 5.14 Planform (Reference) and Exposed Wing Area and Wetted Area

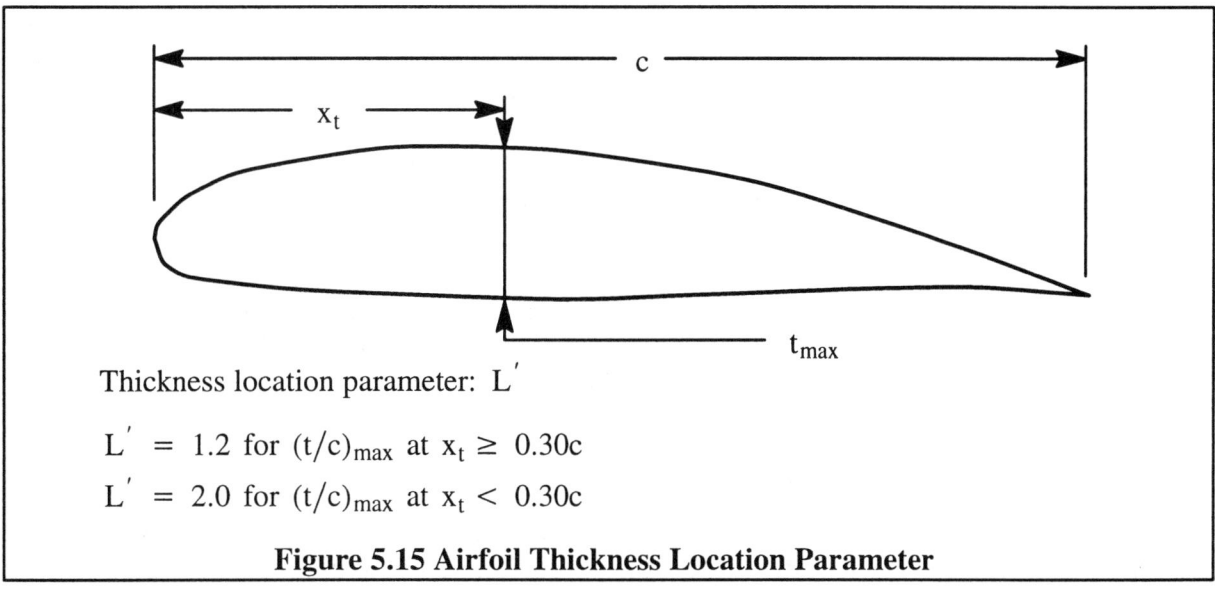

Thickness location parameter: L'

$L' = 1.2$ for $(t/c)_{max}$ at $x_t \geq 0.30c$

$L' = 2.0$ for $(t/c)_{max}$ at $x_t < 0.30c$

Figure 5.15 Airfoil Thickness Location Parameter

5.2.1.2 Wing Drag Coefficient Due To Lift, $C_{D_{L_w}}$

The wing drag coefficient due to lift may be determined from:

$$C_{D_{L_w}} = \frac{C_{L_w}^2}{\pi A e} \tag{5.12}$$

where: C_{L_w} is the wing lift coefficient. The wing lift coefficient differs from the airplane lift coefficient, C_L, because of trim considerations. Reference 5.1 contains a method for computing C_{L_w}. For inherently stable, tail–aft airplanes, C_{L_w} may be obtained from:

$$C_{L_w} = 1.05 \, C_L \tag{5.13}$$

e is Oswald's efficiency factor. A detailed method for estimating e is given in Ref. 5.1. In most cases, values for e range from 0.75 to 0.85.

Note Well: Eqn. (5.12) does not specifically account for wing twist. Reference 5.1 contains a method which does account for a linear twist distribution.

5.2.1.3 Wing Drag Coefficient Due To Compressibility, $C_{D_{wing_{wave}}}$

This drag increment includes the variation of profile drag with Mach number, shock wave drag and drag from shock induced separation.

Within a boundary layer, the effect of compressibility is to increase the temperature. Therefore, the density of the air in the boundary layer is reduced so that the skin friction drag in compressible flow will be lower. However, because of a simultaneous increase in viscosity, the decrease in laminar skin friction drag is to a large extent eliminated. On the other hand, the turbulent boundary layer skin friction drag has been found to decrease with Mach number in subsonic flow, in accordance with Ref. 5.3, page 15–9:

$$C_{f_{compr.}} = C_{f_{incompr.}} (1 - 0.09 M^2) \tag{5.14}$$

The interference drag of wing–fuselage–nacelle combinations is usually increased due to compressibility. This increase normally outweighs any decrease in friction drag. One example of the variation of compressibility drag with Mach number and with lift coefficient is shown in Figure 5.16.

Reducing the transonic drag due to compressibility effects is the equivalent to increasing the drag divergence Mach number. One way to accomplish this is to use the so–called transonic area ruling method. Designers apply area ruling by relying on the following area ruling concept:

<u>Area Rule Concept:</u>

The flow about a low aspect ratio wing–fuselage combination at transonic and supersonic speeds is similar to the flow about a body of revolution having the same distribution of cross–sectional area.

Figure 5.16 Effect of Mach Number and Lift Coefficient on Compressibility Drag of a Business Jet

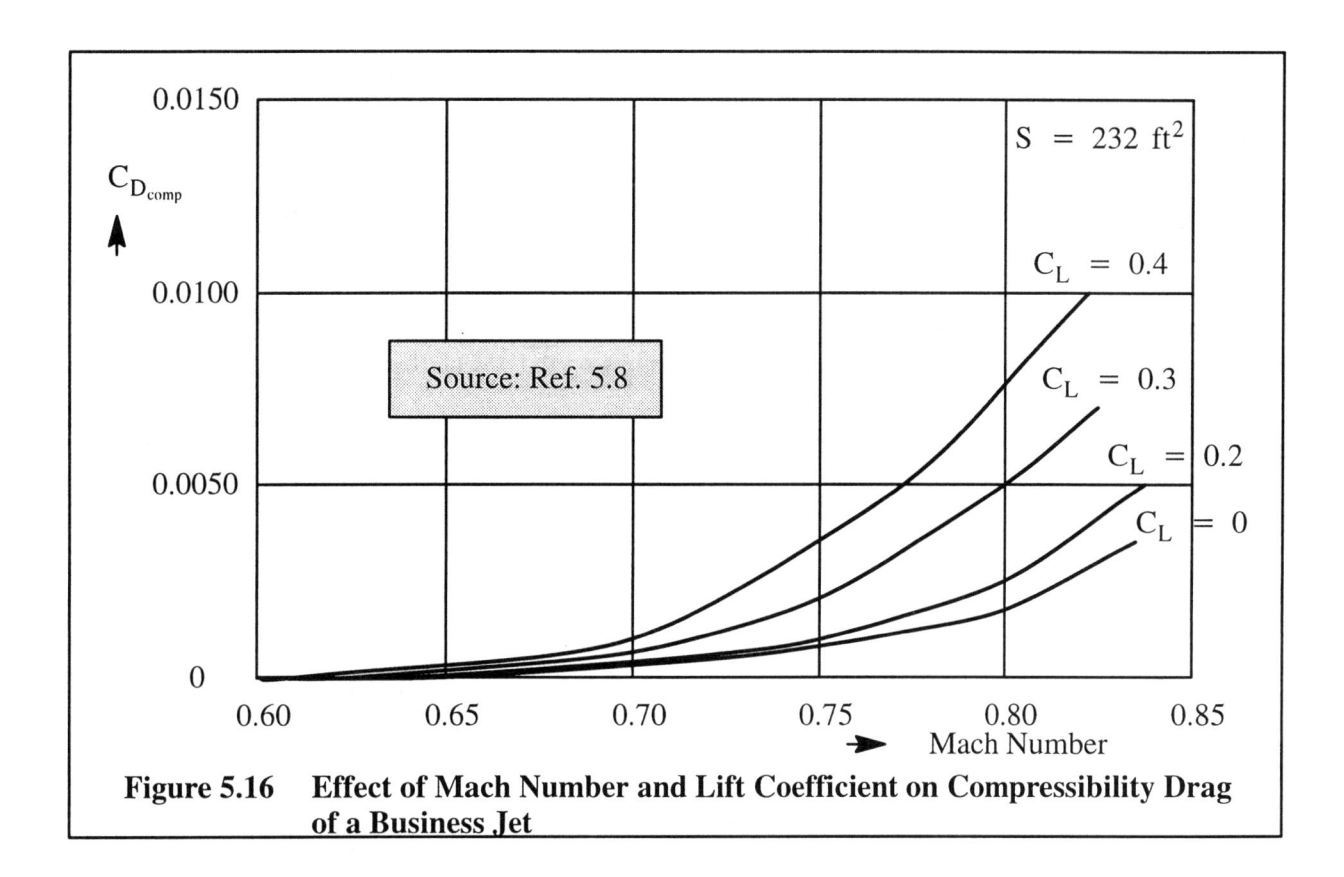

According to this rule, it can be assumed that at large distances from the body the disturbances in the flow are independent of the arrangement of the components and are only a function of the cross–sectional area distribution. This means that the drag of a wing–fuselage combination can be calculated as though the combination were a body of revolution with equivalent cross sections.

This area rule concept suggests the most desirable way to arrange the various components of an airplane (fuselage, canard, canopy, wing, nacelles and vertical (and/or) horizontal tail) for minimum wave drag at a particular value of the free stream Mach number, M_∞. What is needed is a smooth equivalent body of revolution. The theoretical body of revolution with minimum wave drag is the so–called Sears–Haack body which is described in Refs 5.1 and 5.9. Such a body is characterized by the following mathematical shape:

$$\left(\frac{r}{r_{max}}\right)^2 = \left\{1 - \left(\frac{2x}{l_f}\right)^2\right\}^{3/2} \quad \text{for} \quad -\frac{l_f}{2} \le x \le \frac{l_f}{2} \tag{5.15}$$

The volume and wave drag of an equivalent body shaped according to Eqn (5.15) are:

$$\text{Volume} = \frac{3}{16}\pi l_f A_{max} \quad \text{for equivalent body of revolution} \tag{5.16}$$

$$C_{D_{wave}} = \frac{9}{2}\frac{\pi}{l_f^2} A_{max} \quad \text{for equivalent body of revolution} \tag{5.17}$$

where: A_{max} is the maximum cross sectional area of the equivalent body of revolution

Figure 5.17 shows an example of the actual drag rise of an early fighter airplane with that of its equivalent body of revolution. Note the similarity in drag rise due to compressibility between these two cases.

The practical consequence of this area ruling concept is that the cross sectional area of the fuselage must be reduced in the region of the other components. Figure 5.18 shows an example cross sectional area distribution with a smooth equivalent body of revolution (matched to fit the maximum cross sectional area) superimposed. The shaded areas would require a change in local fuselage cross section to match the actual cross sectional area distribution to the one dictated by the smooth body.

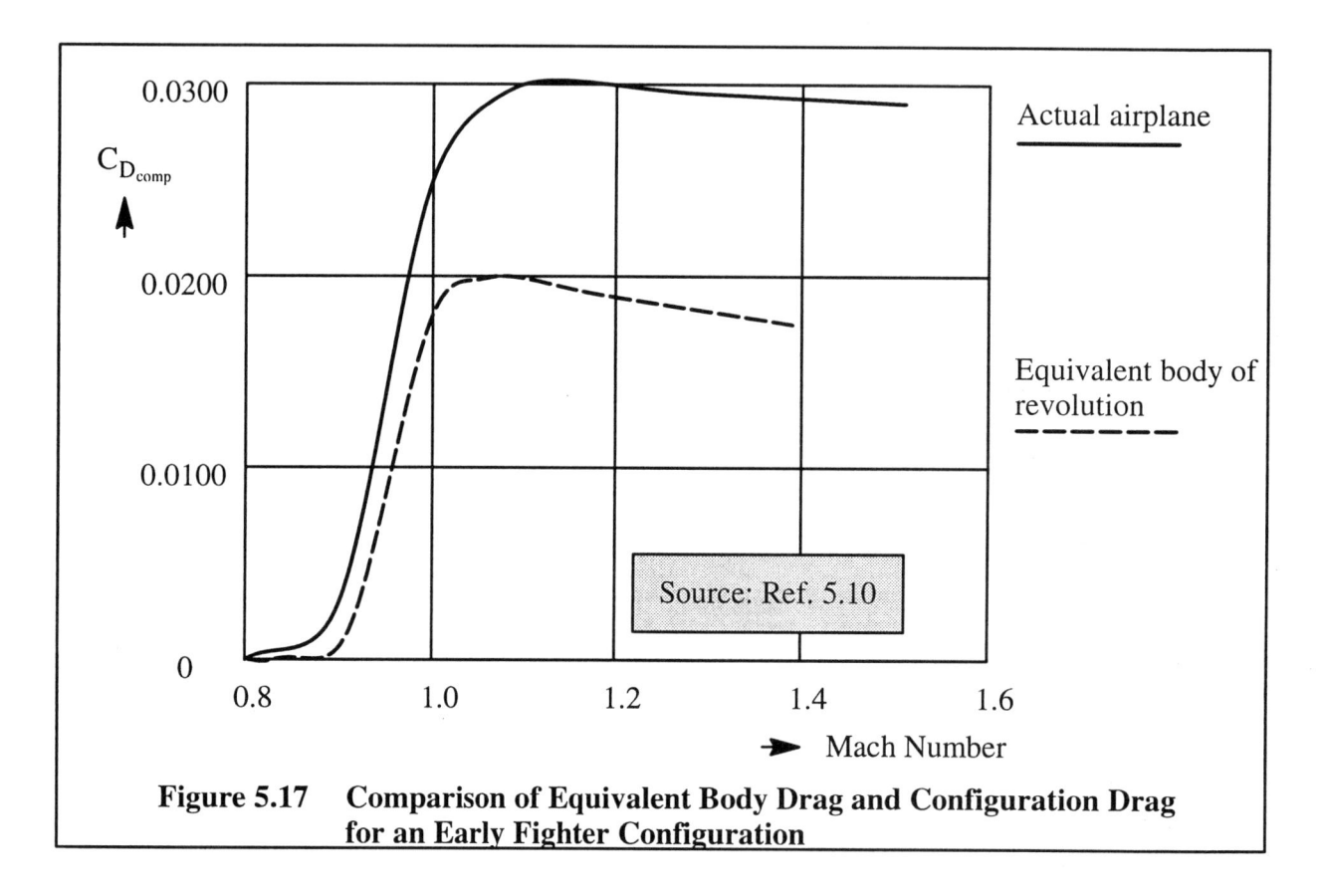

Figure 5.17 Comparison of Equivalent Body Drag and Configuration Drag for an Early Fighter Configuration

Note well: In the case of flow–through engine nacelles, the cross sectional area of the stream tube of the engine should be subtracted from the total nacelle cross sectional area!!!!

When placing bodies on wings (such as a tip tank) or close to a fuselage (such as an engine nacelle) a lot of attention must be paid to transonic effects on interference drag. Local area ruling is often applied. Examples are the NF–5 tip–tank/wing intersection, the DC–10 center–engine/vertical–tail intersection and the Cessna Citation X nacelle/fuselage intersection.

It was already shown, that wing thickness ratio and wing sweep angle have an important effect on drag due to compressibility. The effects of thickness ratio on the critical Mach number (and thus on drag due to compressibility was shown in Figure 5.4. The effect of wing sweep was discussed in Section 4.6. Detailed methods for predicting the effect of Mach number on wing zero–lift drag and on wing drag due to lift are beyond the scope of this text. See Reference 5.1 for details.

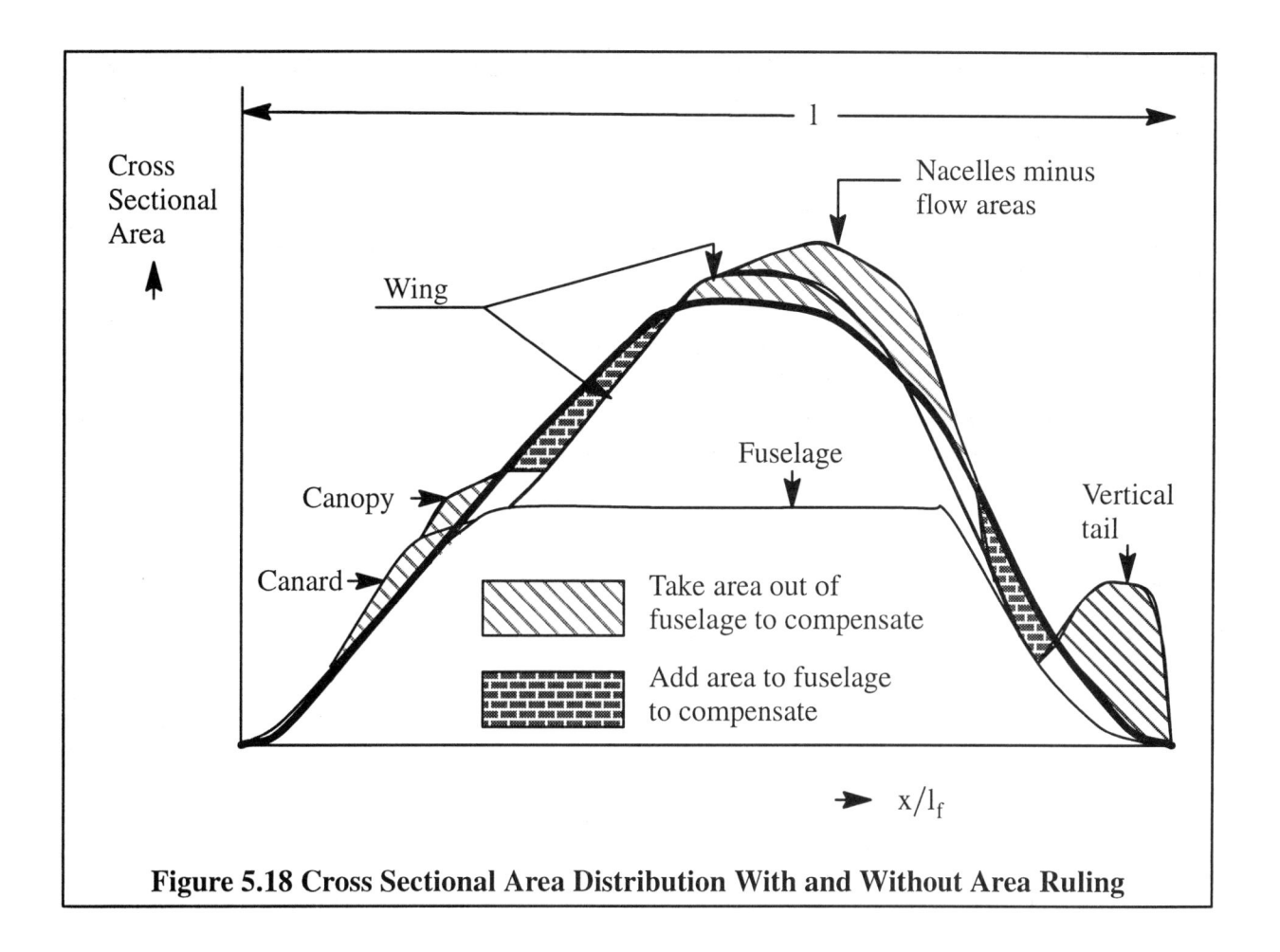

Figure 5.18 Cross Sectional Area Distribution With and Without Area Ruling

5.2.2 FUSELAGE DRAG COEFFICIENT, $C_{D_{fuse}}$

Drag contributions due to the fuselage consist of:

- friction drag
- form and base drag
- drag due to lift, also called induced drag
- interference drag
- compressibility or wave drag

The fuselage drag coefficient may be determined from:

$$C_{D_f} = C_{D_{0_f}} + C_{D_{b_f}} + C_{D_{L_f}} \tag{5.18}$$

where: $C_{D_{0_f}}$ is the zero–lift drag coefficient of the fuselage

$C_{D_{b_f}}$ is the base drag coefficient of the fuselage

$C_{D_{L_f}}$ is the drag coefficient due to lift of the fuselage (induced drag)

Fuselage zero–lift drag is covered in Sub–sub–section 5.2.2.1. The base drag of the fuselage is discussed in Sub–sub–section 5.2.2.2. Fuselage induced drag, at normal flight angles of attack, tends to be very small. A detailed method for its prediction is found in Ref. 5.1. The fuselage can significantly influence the wing drag due to lift. That effect is discussed in Sub–section 5.2.5.

When a fuselage is integrated with a wing (and with nacelles and the empennage), extra drag, called: interference drag, is produced. Several aspects of interference drag are discussed in Sub–section 5.2.10.

Finally, compressibility effects due to an isolated fuselage tend to be small, as long as the fuselage has a reasonable fineness ratio (length over diameter). When a fuselage is integrated with a wing (and with nacelles and the empennage) the compressibility drag becomes a strong function of the progression of cross–sectional area. That effect is also discussed in Sub–section 5.2.5.

5.2.2.1 Fuselage Zero–Lift Drag Coefficient, $C_{D_{0_f}}$

The subsonic, fuselage zero–lift drag coefficient may be empirically determined from:

$$C_{D_{0_f}} = R_{wf} C_{f_f} \left\{ 1 + \frac{60}{(l_f/d_f)^3} + 0.0025(l_f/d_f) \right\} \frac{S_{wet_f}}{S} \tag{5.19}$$

where: R_{wf} is the wing–fuselage interference factor. This factor was also included in the prediction of wing zero–lift drag: see Figure 5.11.

C_{f_f} is the turbulent flat plate skin–friction coefficient of the fuselage. The general, turbulent, flat plate friction coefficient, C_f is shown in Figure 5.13 as a function of Mach number and the Reynolds number. For the fuselage Reynolds number use:

$$R_{N_f} = \frac{\varrho U_1 l_f}{\mu} \tag{5.20}$$

where: l_f is the fuselage length as defined in Figure 5.14.

l_f/d_f is the fuselage fineness ratio. Table 5.1 contains typical values for several airplanes. Definitions for l_f and for d_f are also given in Table 5.1.

S_{wet_f} is the wetted area of the fuselage. Table 5.1 gives examples of the ratio of fuselage wetted area to wing area for several airplanes.

The wetted area of a fuselage is determined by a designer's decision to employ a certain size fuselage cross–section. For example, for a given number of passengers, the designer must decide how much headroom, how much leg room, how many seats abreast, how much elbow room and how much aisle room needs to be provided. Clearly, the greater the "creature comfort", the greater the wetted area and, therefore, the higher the friction drag.

Table 5.1 Examples of Fuselage Fineness Ratios and Wetted Areas				
Type	l_f/d_f	S (ft^2)	S_{wet_f} (ft^2)	S_{wet_f}/S
Cessna 210	5.02	175	319	1.82
Cessna 207	5.69	174	425	2.44
Cessna 185	5.15	176	292	1.68
Cessna 310	5.40	179	306	1.71
Cessna 414	5.52	195.7	488	2.49
Beech Sierra	5.22	146	332	2.27
Beech Bonanza ('58)	4.98	181	323	1.78
Beech Baron	5.69	199.2	362	1.82
Beech Duke	5.59	212.9	586	2.28
Beech King Air	6.06	294	652	2.22
Piper Navajo	5.97	229	502	2.19
Piper Seneca	5.68	206.5	356	1.72
Learjet M24	8.80	232	502	2.16
Shorts SD3–30	7.43	453	1,543	3.41
Fokker F–28–4000	8.59	850	2,454	2.89
Boeing 757–200	11.7	1,951	5,601	2.87

5.2.2.2 Fuselage Base Drag Coefficient, $C_{D_{f_b}}$

When a fuselage is truncated a so–called base area is created. Figure 5.19 shows a typical example of a fuselage with base area.

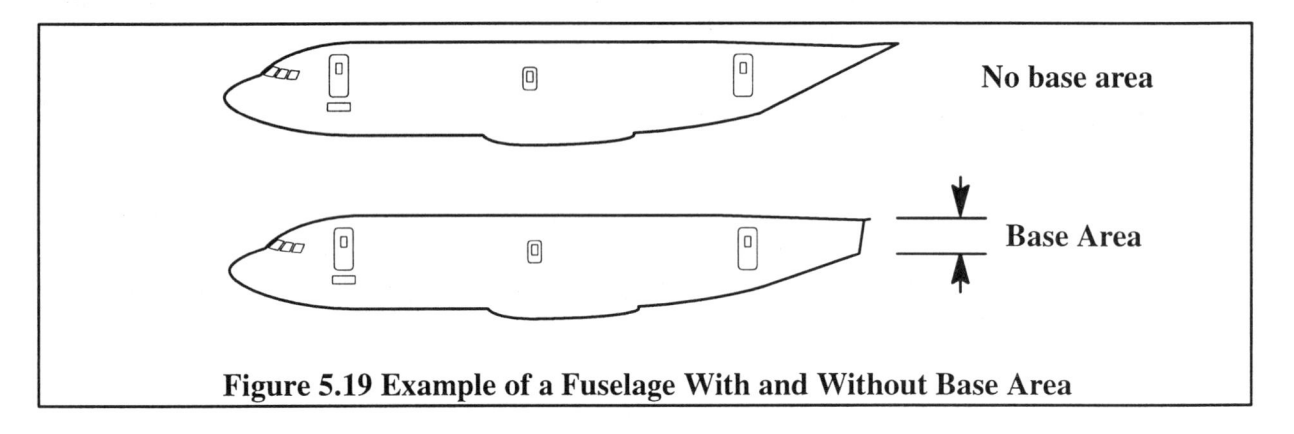

Figure 5.19 Example of a Fuselage With and Without Base Area

As a result of this base area, additional drag (called base drag) is incurred. Frequently, fuselage base area is caused by the exhaust of a jet engine or an A.P.U. (Auxiliary Power unit). With the engine running, a high velocity exhaust will emanate from the base. That negates the effect of the base area. Therefore, with the engine running, the base area may be assumed to be zero!

When a fuselage is not properly streamlined, flow separation may occur. This in effect causes a certain amount of base area to appear and results in base drag. Reference 5.1 contains methods for estimating the base drag coefficient. Figure 5.20 shows examples of base drag coefficients.

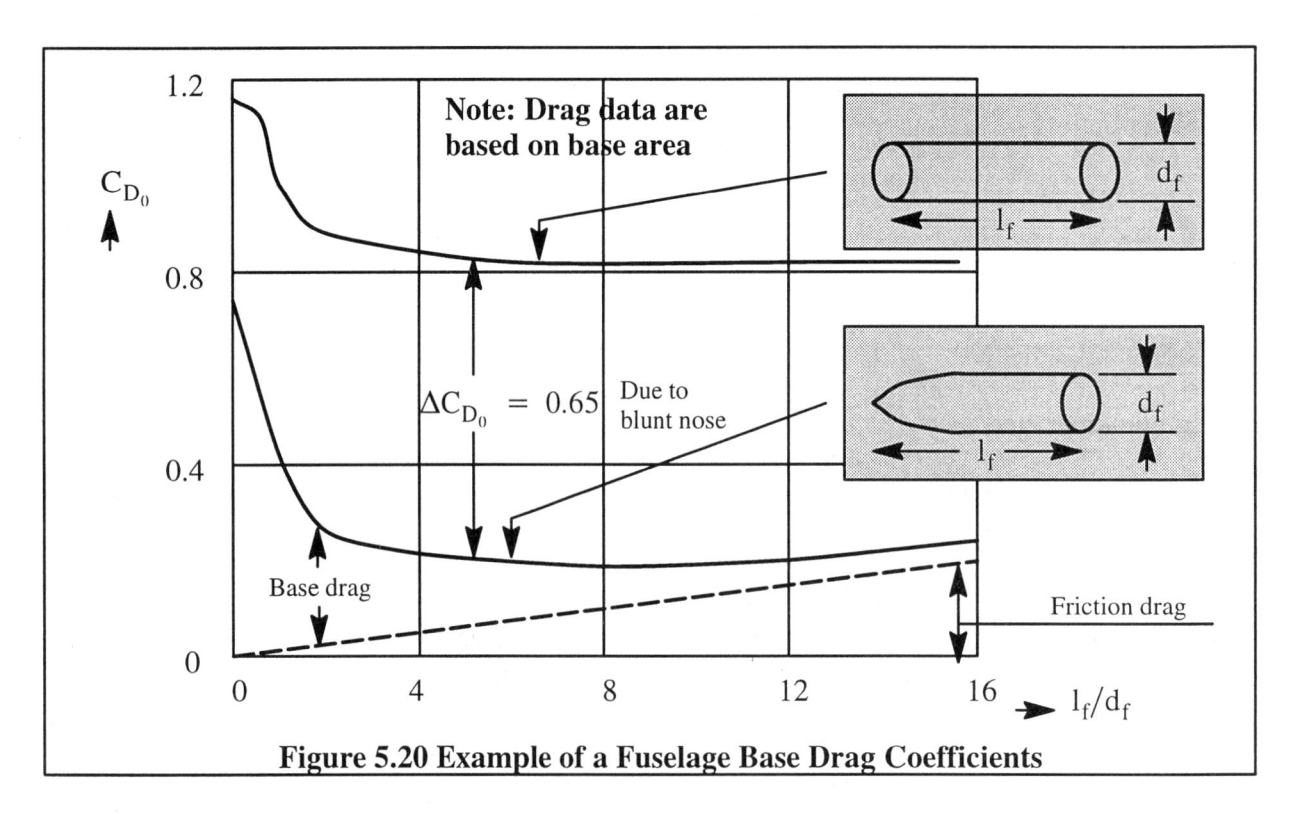

Figure 5.20 Example of a Fuselage Base Drag Coefficients

5.2.2.3 Fuselage Drag Coefficient Due To Lift, $C_{D_{L_f}}$

In the normal, take–off, cruise and landing angle of attack range the induced drag due a fuselage tends to be quite negligible. Exceptions are highly integrated wing–fuselage configurations, such as Burnelli type airplanes. A detailed treatment of fuselage induced drag is beyond the scope of this text but may be found in References 5.1 and 5.2.

5.2.3 EMPENNAGE DRAG COEFFICIENT, $C_{D_{emp}}$

The empennage (i.e. tail surfaces) of an airplane normally consist of the horizontal tail and the vertical tail. In some instances a canard is used instead of, or in addition to a horizontal tail. The total drag due to all these surfaces is referred to as the empennage drag. Drag contributions due to the empennage consist of:

- friction drag
- profile or thickness drag
- drag due to lift, also called induced drag
- interference drag
- compressibility or wave drag

The empennage drag coefficient may be determined from:

$$C_{D_{emp}} = C_{D_{0_{ht}}} + C_{D_{0_{vt}}} + C_{D_{0_c}} \qquad (5.21)$$

where: $C_{D_{0_{ht}}}$ is the zero–lift drag coefficient of the horizontal tail

$C_{D_{0_{vt}}}$ is the zero–lift drag coefficient of the vertical tail

$C_{D_{0_c}}$ is the zero–lift drag coefficient of the canard

The zero–lift contributions for these empennage surfaces may be computed with Eqn (5.8) suitably modified to account for differences in Reynolds number, thickness and wetted area.

There also are induced drag contributions due to these empennage surfaces. This is accounted for as so–called trim drag in Sub–section 5.2.10.

The effect of compressibility on empennage drag is not discussed in this text. It was already shown, that surface thickness ratio and sweep angle have an important effect on drag due to compressibility. The effect of thickness ratio on drag due to compressibility was shown in Figure 3.13. The effect of sweep was discussed in Section 4.6.

Detailed methods for predicting the effect of Mach number on wing zero–lift drag and on wing drag due to lift are beyond the scope of this text. Such methods are contained in Reference 5.1.

Interference effects are briefly discussed in sub–sub–section 5.5.5.11.

5.2.4 NACELLE–PYLON DRAG COEFFICIENT, $C_{D_{np}}$

The incremental drag due to nacelles and/or nacelle/pylon combinations for jet or propeller installations consists of:

- friction drag
- form and base drag
- drag due to lift, also called induced drag
- interference drag
- compressibility or wave drag

Normally, a nacelle/pylon combination is split into its two components: the actual nacelle and the pylon which is used to attach the nacelle to a fuselage or to a wing. In most single engine light airplanes, there is no pylon. A typical exception is the Lake –4 amphibious airplane. Figure 5.21 provides illustrations of these cases.

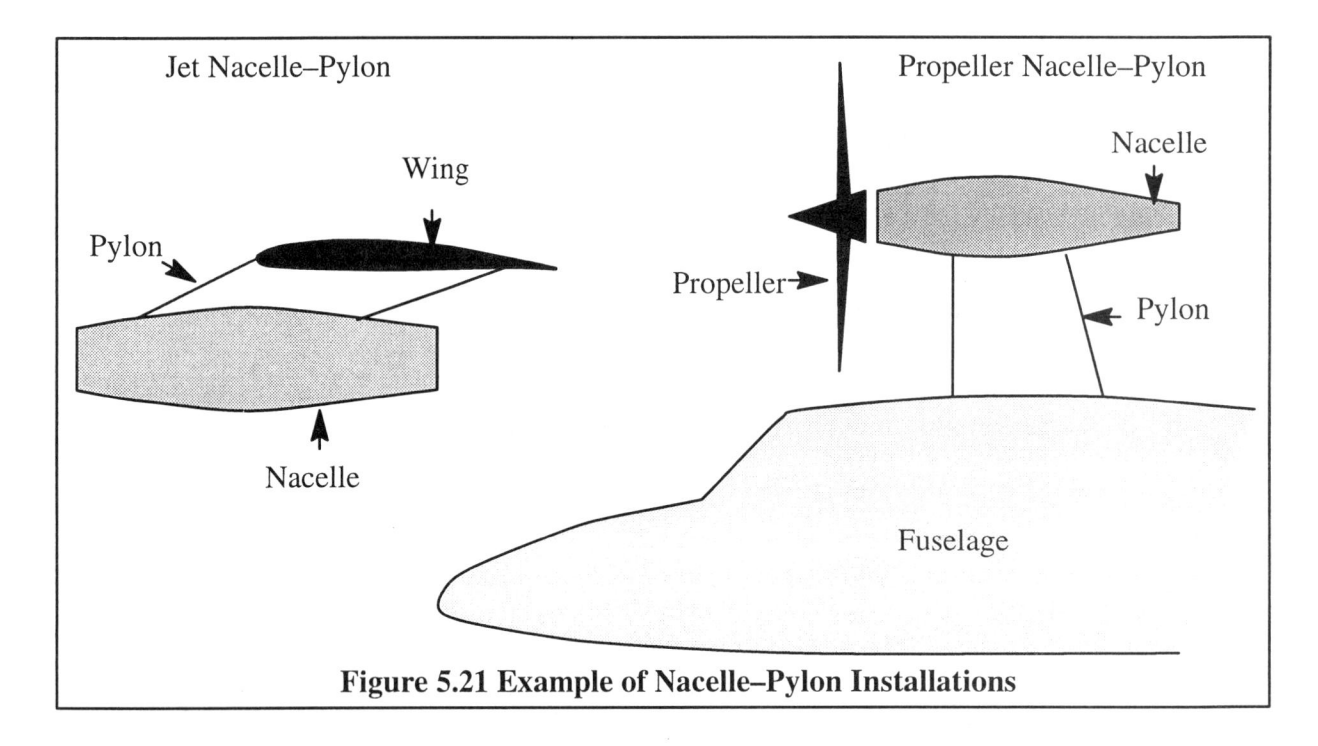

Figure 5.21 Example of Nacelle–Pylon Installations

For the nacelle itself, the first three drag contributions can be adequately estimated using the same methodology used for determining the fuselage drag. Care must be used in determining the wetted area of the nacelle. Only the external area of the nacelle is counted as wetted area. Any drag caused by internal flow through the nacelle is normally accounted for against the installed thrust. This, of course, is a bookkeeping decision.

For the pylon, the first three drag contributions can be adequately estimated using the same methodology used for determining the wing drag.

To avoid penalties due to compressibility, the pylon sweep angle and the pylon thickness ratio are subject to the same considerations as those for a wing or tail. In most cases, by using local area ruling on the nacelle–pylon–wing or nacelle–pylon–fuselage ensemble, it is possible to avoid wave drag penalties in the transonic speed regime.

For a discussion of interference drag, the reader is referred to sub–sub–section 5.2.5.11.

5.2.5 ADDITIONAL DRAG DUE TO INTEGRATION EFFECTS

When the wing, fuselage, empennage and nacelles are integrated, several additional factors which may influence airplane drag must be accounted for. Many of these factors involve drag due to lift (or induced drag). Some of these are discussed here.

5.2.5.1 Wing Planform Shape Effect on Induced Drag

As might be expected, the planform shape of the wing has a significant effect on induced drag. It was seen in Chapter 4, that the induced drag of a wing with elliptical planform and moderate to high aspect ratio, is theoretically the minimum possible. Pure trapezoidal planforms are a good approximation to elliptical planforms and are usually easier to produce. Mixed trapezoidal and rectangular planforms (the latter are called Hershey–bar wings) incur a modest induced drag penalty. The reader may keep the following numbers in mind:

For **rectangular planforms**, the additional induced drag penalty, according to the method of Ref. 5.3, Chapter 7, may be estimated from:

$$\Delta\left(\frac{dC_{D_i}}{dC_L^2}\right) \approx 0.003 \tag{5.22}$$

By using Eqn (4.24) this implies a change in the induced drag factor 1/e of:

$$\Delta\left(\frac{1}{e}\right) \approx (0.003)\,\pi A \tag{5.23}$$

For **straight tapered (trapezoidal) planforms**, with taper ratios ranging from 0.3 to 0.4 the additional induced drag may be estimated from:

$$C_{D_{i_{straight-taper}}} \approx (1.01 \text{ To } 1.02) \times C_{D_{i_{elliptical}}} \tag{5.24}$$

Figure 5.22 shows examples of these three planform shapes. A mixed planform, is also shown.

5.2.5.2 Wing Tip Shape Effect on Induced Drag

Tip vortices may roll up and get around the lateral edges of a wing. This has the effect of reducing the effective span of the wing. Figure 5.23 illustrates these effects.

Experiments show that wings with square or sharp tips have the widest effective span. When tip–tanks are used, favorable effects can be derived from this effective increase in span and also from

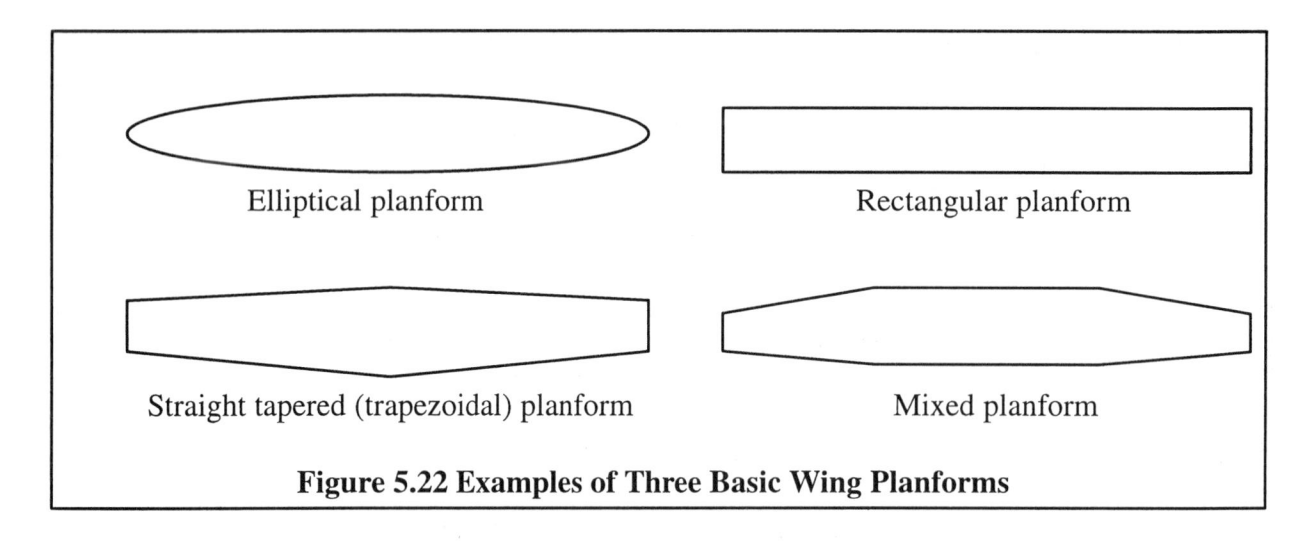

Figure 5.22 Examples of Three Basic Wing Planforms

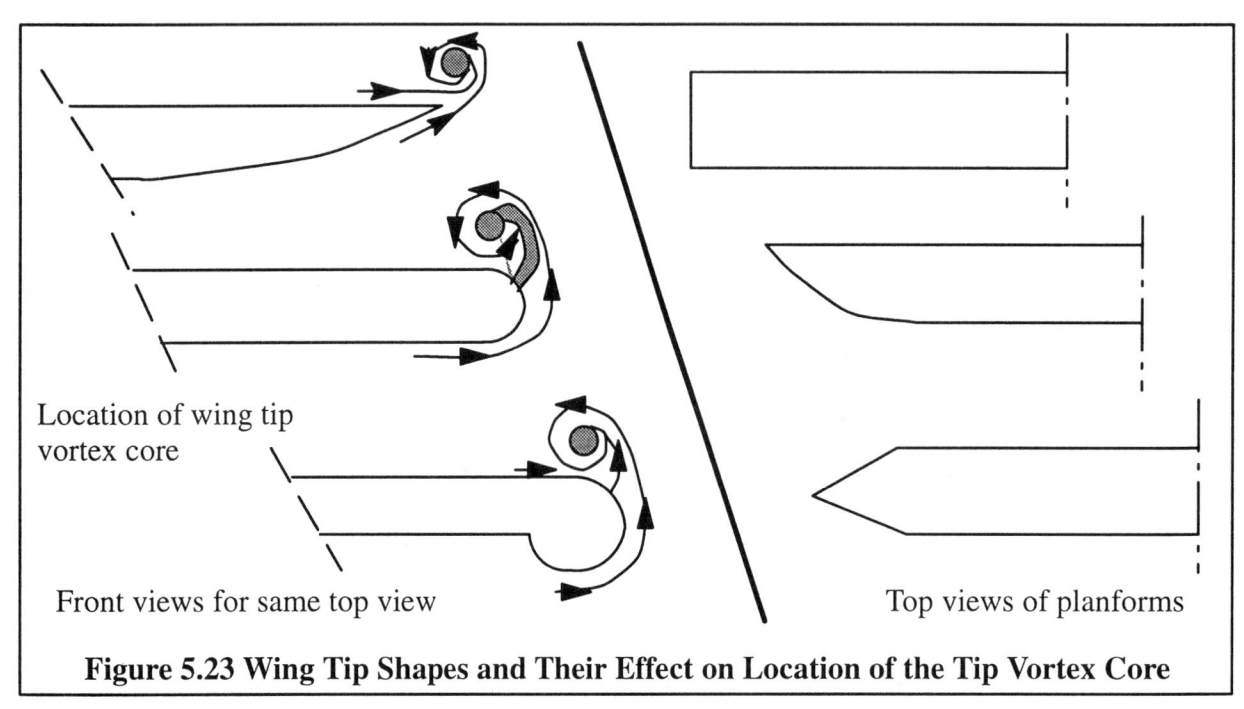

- Elliptical planform
- Rectangular planform
- Straight tapered (trapezoidal) planform
- Mixed planform

Location of wing tip vortex core

Front views for same top view

Top views of planforms

Figure 5.23 Wing Tip Shapes and Their Effect on Location of the Tip Vortex Core

the increase in wing area and from the so–called end–plate effects. However, the tip vortices will tend to move somewhat inboard which in turn reduces the effective span. For most practical purposes it is acceptable to assume that, in case of tip tanks, half of the tip tank diameter can be added to the actual wing span to obtain the aerodynamically effective span.

Work done by Van Dam et al (Ref. 5.11) shows that at low speeds and for moderate high aspect ratio wings tip planform shapes similar to the one shown in Figure 5.24 can reduce the induced drag by a significant amount when compared to conventional tip shapes. This can be of great importance in engine–out climb cases, where the lift coefficient (and therefore the induced drag) is high.

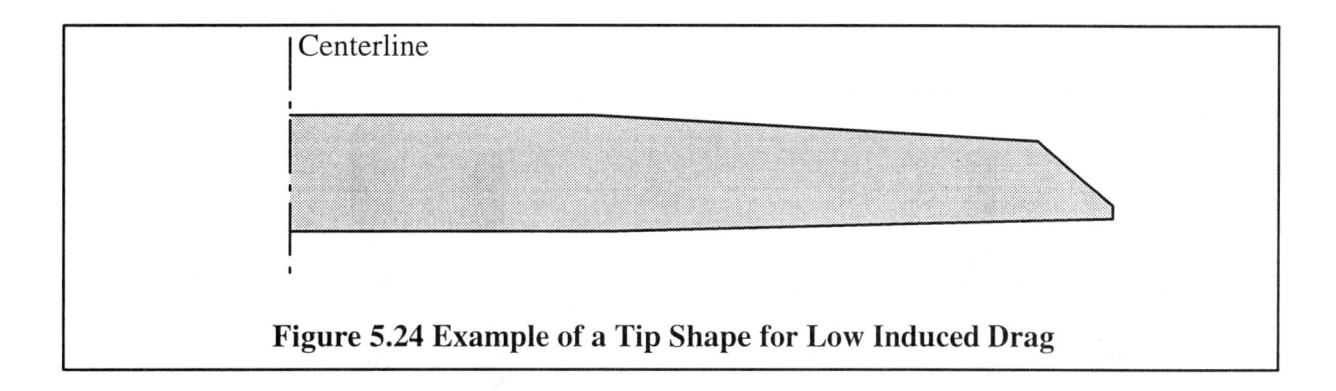

Figure 5.24 Example of a Tip Shape for Low Induced Drag

5.2.5.3 Wing Twist Effect on Induced Drag

Washout of a wing, also referred to as twist, changes the spanwise lift distribution of a wing. This has been found to increase induced drag. The effect may be estimated from Eqn (4.8) in Ref. 5.1 or Ref. 5.3, page 7–7. Because slightly forward swept wings tend to have favorable stall characteristics, no twist needs to be incorporated. This result is a small savings in induced drag.

5.2.5.4 Wing Sweep Effect on Induced Drag

For swept aft wings, the lift tends to be more concentrated toward the tip. By decreasing the taper ratio, the lift concentration near the tip can be reduced. However, reduction in taper ratio also results in a reduction in Reynolds number. This in turn usually results in a reduction of local maximum lift coefficient. Therefore, small taper ratio wings tend toward tip stall. As a result, swept aft wings with realistic taper ratios tend to have lift distributions which are fairly far removed from the ideal elliptical shape. As a result the induced drag is increased. It should be noted that this is not the case with a forward swept wing. Experiments indicate that the induced drag of aft swept wings is proportional to the inverse of the cosine of the sweep angle:

$$\frac{dC_{D_i}}{dC_L^2} \approx \frac{1}{\cos \Lambda_{c/4}} \tag{5.25}$$

5.2.5.5 Effect of Tandem Configurations on Induced Drag

Figure 5.25 shows typical wing arrangements referred to as a tandem configurations.

Whether or not such a configuration should be called a tandem depends on two factors: 1) the relative size of the rear surface compared to the front surface, and 2) whether or not the rear surface carries positive lift. Wing theory shows that the induced drag of a tandem configuration is equal to that of a single wing with the combined lift distribution. This result follows from Munk's Stagger Theorem (see: Ref. 5.12).

The horizontal tail of a conventional airplane, or the front surface of a canard airplane can be thought of as a tandem configuration if the tail or the canard is large relative to the wing. Because of the requirement for static longitudinal stability in most conventional airplanes (See Ref. 5.13) a tail normally carries a down load to trim the airplane, while a canard normally carries an up load to trim the airplane. The resulting induced drag can be substantial as shown in Section 5.2.6.

Figure 5.25 Examples of Tandem Configurations

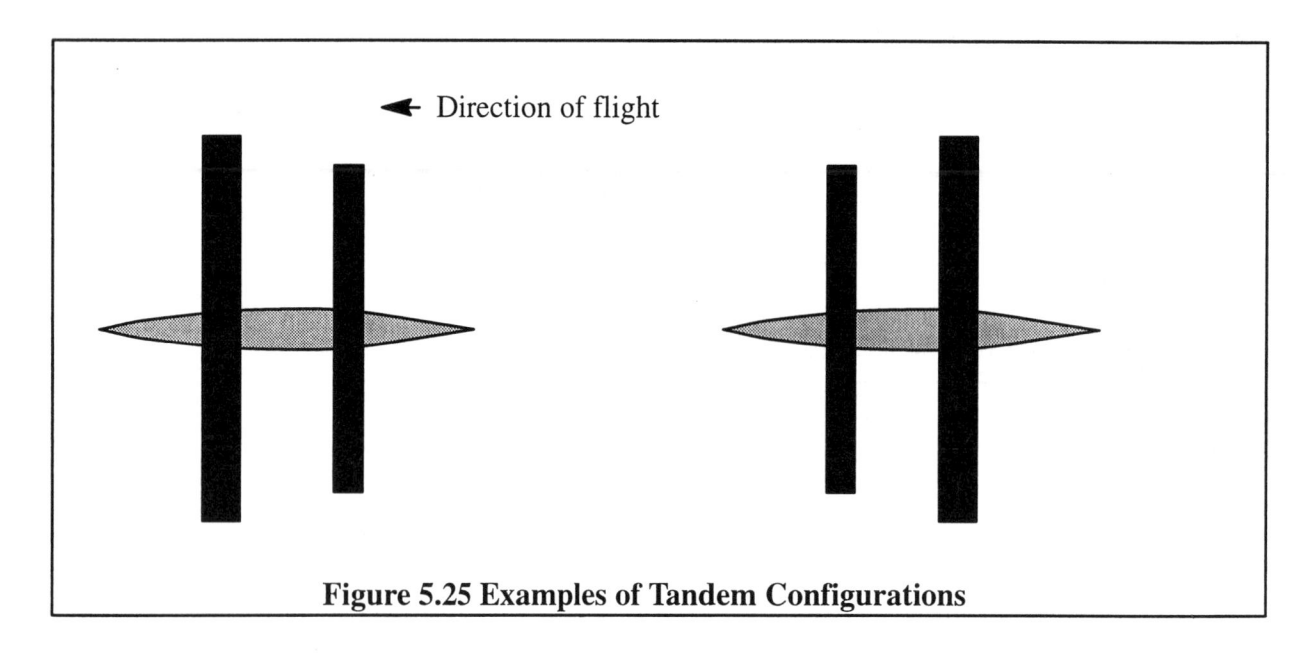

5.2.5.6 Effect of Winglets on Zero Lift and on Induced Drag

Winglets are small, nearly vertical lifting surfaces, mounted rearward and/or downward relative to the wing tips. Examples of winglets on several modern airplanes are shown in Figure 5.26.

The total velocity of the air relative to the inside surface of a winglet consists of two components: the increased velocity over the inward surface (due to local angle of attack) and the high velocity over the forward portion of a wing tip. To avoid that the total velocity approaches the local speed of sound (which would result in additional compressibility effects), winglets are placed rearward on wing tips.

To explain how winglets work, refer to the insert sketch in Figure 5.26. The wing tip circulation produces a large sidewash component, even at low airplane angles of attack. The resultant lift component is perpendicular to the local flow vector. It is seen that because of the forward orientation of this winglet lift an effective thrust force is produced. This thrust force effectively reduces the airplane induced drag. Some readers may recognize the fact, that a winglet acts similar to the sail on a sailboat while tacking into the wind! Therefore, some companies refer to winglets as tip–sails. An example is the Beech Starship.

A question which is always asked is: are winglets more effective than increases in span? This question is tricky to answer because of the effects of wing root bending and torsion, flutter, trade between induced drag and friction drag and weight. Based on studies performed in References 5.11, 5.14 and 5.15 the only general conclusion is that it depends on the mission and on the configuration of an airplane whether a winglet wins out over a wing tip extension.

One practical problem with wing tip extensions is the limited gate parking spaces which are available at some airports. This is precisely the reason why the Boeing 747–400 has winglets rather than added span. Finally, in some cases management may perceive the very "looks" of winglets to be sufficiently attractive, to use them for that reason.

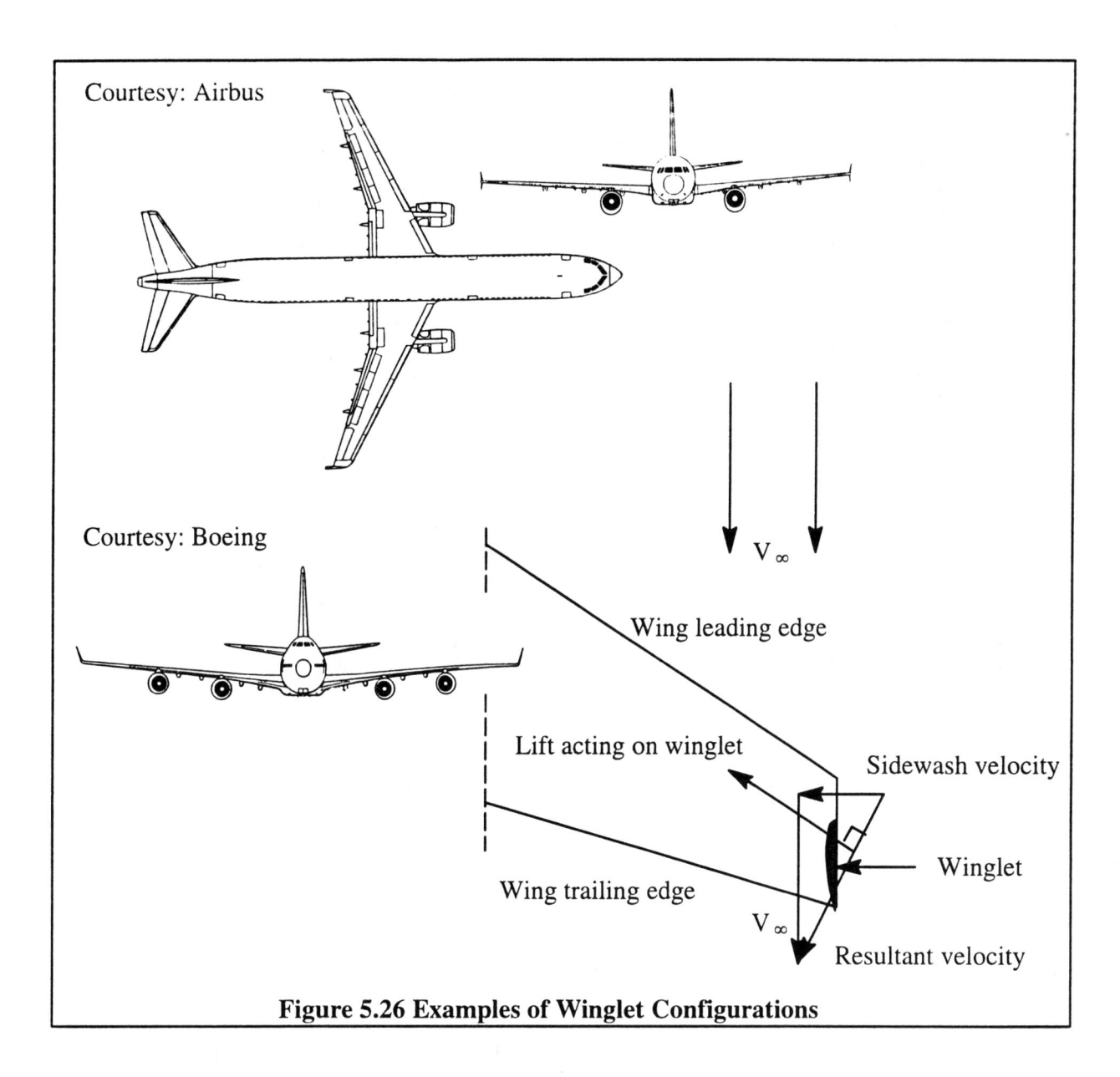

Figure 5.26 Examples of Winglet Configurations

5.2.6 FLAP DRAG COEFFICIENT, $C_{D_{flap}}$

Drag contributions due to flaps consist of:

- parasite drag
- drag due to lift, also called induced drag
- interference drag

The flap drag coefficient may be determined from:

$$C_{D_{flap}} = C_{D_{0_{flap_{par}}}} + C_{D_{L_{flap}}} + C_{D_{0_{flap_{interf.}}}} \tag{5.26}$$

where: $C_{D_{0_{flap_{par}}}}$ is the parasite (zero–lift) drag coefficient due to the flap

$C_{D_{L_{flap}}}$ is the drag coefficient due to lift of the flap (induced drag)

$C_{D_{0_{flap_{interf.}}}}$ is the drag coefficient due to interference between the flap and the wing/fuselage

The three–dimensional flow field due to leading edge and trailing edge flaps is very difficult to model. For that reason, it is customary to estimate the parasite drag contribution due to flaps from two–dimensional data. In preliminary design it is acceptable to use:

$$C_{D_{0_{flap_{par}}}} = \Delta c_{d_p} \left(\frac{S_{w_f}}{S}\right) \cos \Lambda_{c/4} \qquad (5.27)$$

where: $C_{D_{L_{flap}}}$ is the two–dimensional profile drag increment due to the flap(s). This quantity may be estimated with the methods of References 5.1 and 5.2.

Deflection of partial span flaps changes the shape of the spanwise load distribution. The additional induced drag due to the flaps is a function of the incremental lift coefficient due to the flap and is a function of the flap deflection angle itself. Also, the spanwise distribution of the section zero–lift angle of attack is changed by the deflected flap. For these reasons, the induced drag due to the flap(s) will have a component which is proportional to the square of the incremental lift coefficient. This component can be calculated by using lifting surface theories, see Ref. 4.6.

According to Ref. 5.1 the induced drag coefficient due to flaps may be estimated from:

$$C_{D_{L_{flap}}} = K^2 \left(\Delta C_{L_{flap}}\right)^2 \cos \Lambda_{c/4} \qquad (5.28)$$

where: K is an empirical factor given in Ref. 5.1

$\Delta C_{L_{flap}}$ is the incremental lift coefficient due to the flap at a given angle of attack.

Interference drag due to the flaps arises from a very complicated flow field in the vicinity of the flaps and neighboring lifting surfaces and bodies. According to Ref. 5.1, p. 88, the following empirical relations may be used to estimate the interference drag for two types of flap:

$$C_{D_{0_{flap_{interf.}}}} = -0.15 \Delta C_{D_p} \text{ for split flaps} \qquad (5.29)$$

and

$$C_{D_{0_{flap_{interf.}}}} = +0.40 \Delta C_{D_p} \text{ for slotted flaps} \qquad (5.30)$$

Examples of the effect of flaps on airplane drag are shown in Figure 5.9 and 9.11.

5.2.7 LANDING GEAR DRAG COEFFICIENT, $C_{D_{gear}}$

Landing gears cause extra drag during takeoff and landing for airplanes with retractable gears and in all flight conditions for airplanes with fixed gears. The magnitude of the landing gear drag coefficient is proportional to the frontal area of the gear (struts, shock absorbers and tires) as presented to the oncoming flow. Because of this, gear drag can be a function of angle of attack. Methods for estimating gear drag are contained in Refs 5.1, 5.2 and 5.16.

It should be noted, that in some airplanes the gear drag increment can be a function of the flap deflection. The reason for this is the change in downwash at the gear caused by the flaps. An example of this is shown in Figure 5.27.

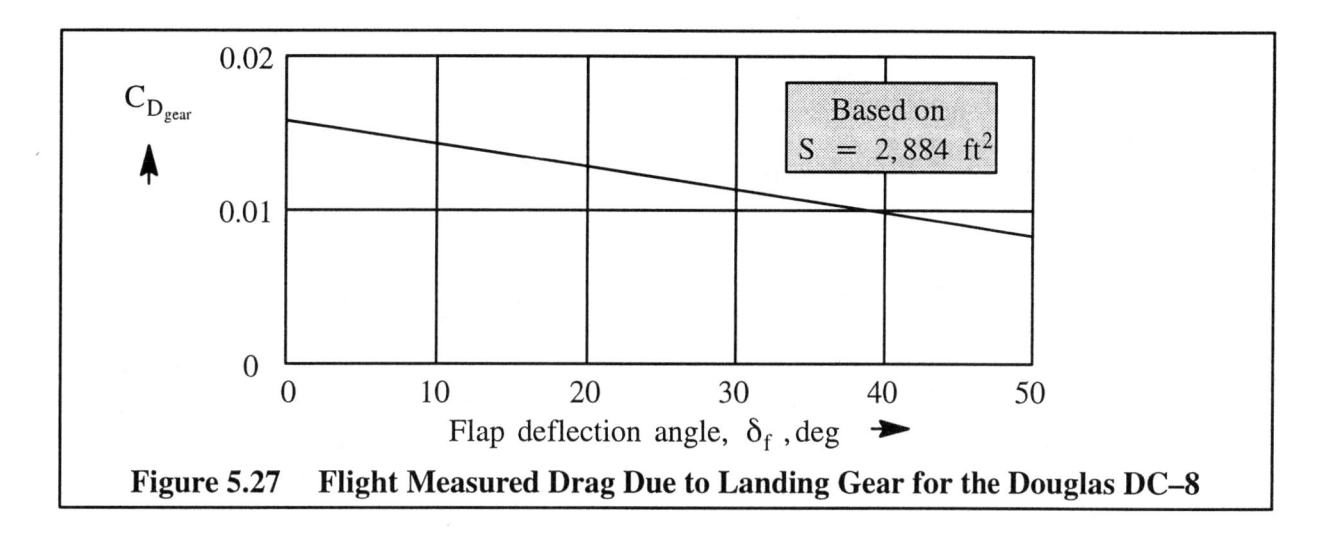

Figure 5.27 Flight Measured Drag Due to Landing Gear for the Douglas DC-8

5.2.8 CANOPY/WINDSHIELD DRAG COEFFICIENT, $C_{D_{cw}}$

The drag coefficient due to canopy or windshield depends strongly on the cross-sectional shape, the stream-wise shape, the fineness ratio (i.e. length divided by height for canopies) and the Mach number. References 5.1 and 5.3 contain data for estimating this drag increment. The multitude of shapes in use suggests that there is no unanimity on the "best" shape for these items.

5.2.9 STORE DRAG COEFFICIENT, $C_{D_{store}}$

The store drag coefficient depends on the size and number of stores carried externally by an airplane. For an arbitrary number of stores this may be determined from:

$$C_{D_{store}} = \sum_{i=0}^{i=n} \left\{ \left(K_{store_i} \right) \left(C_{D_{store_i}} \right) \right\} \tag{5.31}$$

where: K_{store_i} is the store interference factor. For semi-submerged stores, K=0.7 while for external stores, K=1.3. The reader should also consult Ref. 5.17. for geometric data on military stores as well as for suggestions how to integrate them into an airplane.

$C_{D_{store_i}}$ is the store drag coefficient for an isolated store. This may be computed with the fuselage zero-lift drag coefficient method of Sub-section 5.2.2.

Airplane Drag

5.2.10 TRIM DRAG COEFFICIENT, $C_{D_{trim}}$

Trim drag is defined as that extra drag required to produce a condition of zero moments on the airplane. Normally, for an airplane under symmetrical power, trim drag is caused by loads on the tail and/or on the canard to assure zero pitching moment. These aerodynamic trim loads cause an increase in induced drag which is called the trim drag. For an airplane with one engine out, trim drag is also increased by the requirement to produce lift (i.e. side force) on the vertical tail and to deflect the lateral controls, to assure zero yawing and rolling moments.

Figure 5.28 shows four possible layout cases. It is seen that for a conventional airplane, the trim load on the horizontal tail can be zero, up or down, depending on how the airplane is balanced (i.e. where the center of gravity is in relation to the tail–off aerodynamic center). For a canard airplane the load on the canard will always be up under the condition assumed in Figure 5.28.

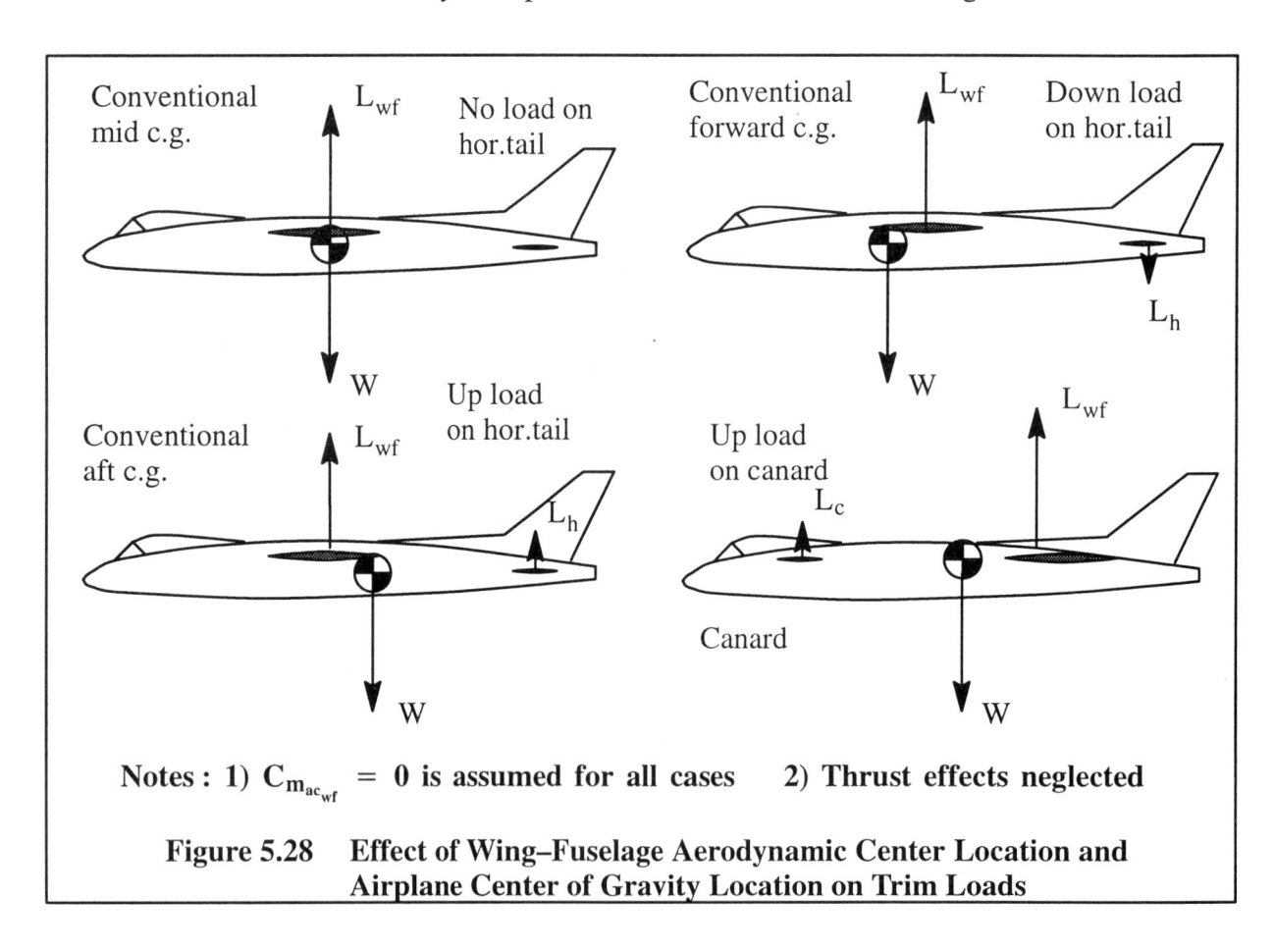

Notes: 1) $C_{m_{ac_{wf}}} = 0$ is assumed for all cases   2) Thrust effects neglected

Figure 5.28   Effect of Wing–Fuselage Aerodynamic Center Location and Airplane Center of Gravity Location on Trim Loads

For the conventional case the trim drag for an up or down load on the horizontal tail would be determined from:

$$C_{D_{0_{trim}}} = \frac{(C_{L_h})^2}{\pi A_h e_h} \frac{\bar{q}_h}{\bar{q}} \frac{S_h}{S} \tag{5.32}$$

For a canard configuration the trim drag may be computed in a similar manner.

Trim drag does depend also on: a) fuselage camber shape   and   b) thrust line location.

In conventional, inherently stable airplanes, the camber shape of the fuselage can have a significant effect on the zero–lift pitching moment coefficient of an airplane. An extreme example of a fuselage shape which takes advantage of this effect is the Lockheed Constellation: see Figure 5.29.

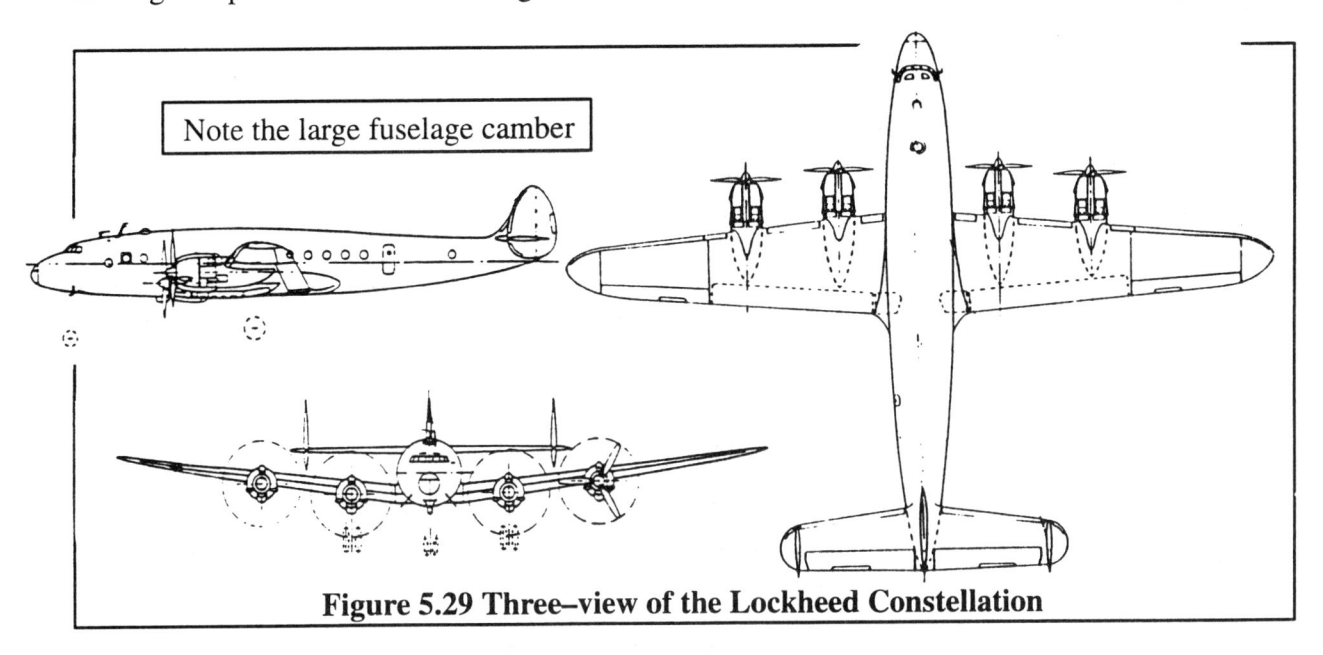

Note the large fuselage camber

Figure 5.29 Three–view of the Lockheed Constellation

The location of the thrust–line relative to the center of gravity can have a large effect on the zero lift pitching moment coefficient (and thus on trim drag) of an airplane. How to estimate the effect of thrust–line location on $\overline{C}_{m_0}$ is illustrated in Figure 5.30. Whether or not this effect increases or decreases trim drag depends on whether the thrust line is located above or below the c.g., on the level of stability (or lack thereof) and on the airplane configuration.

The additional zero–lift pitching moment coefficient due to thrust–line offset is:

$$\Delta \overline{C}_{m_0} = \frac{T \, d_T}{\overline{q}_1 S \overline{c}}$$

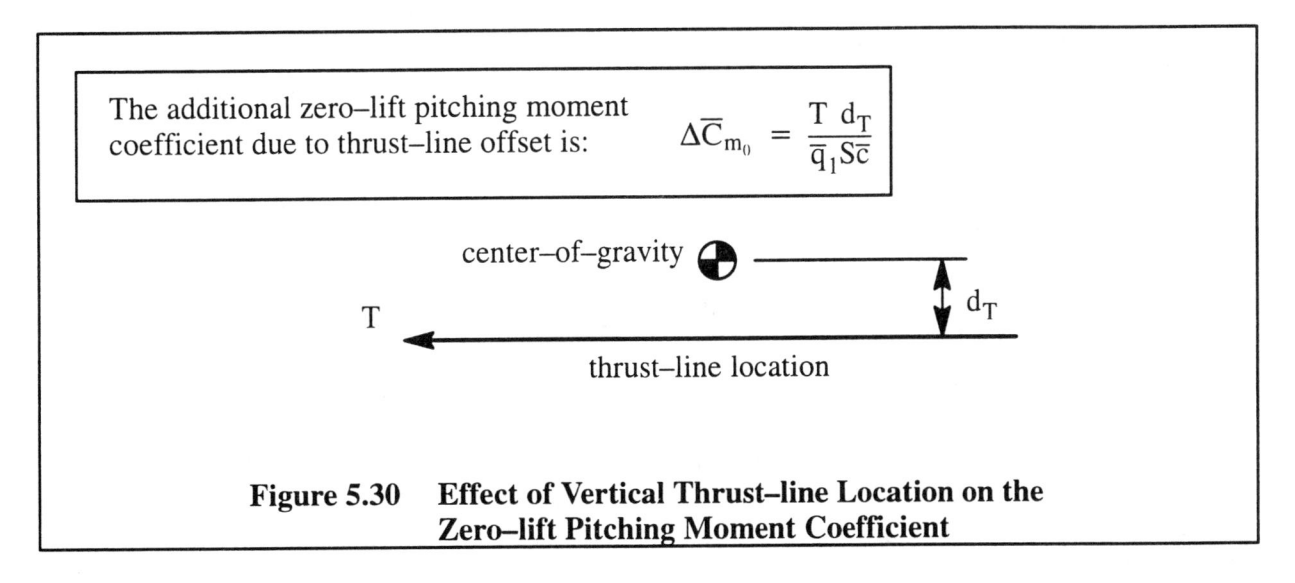

Figure 5.30   Effect of Vertical Thrust–line Location on the Zero–lift Pitching Moment Coefficient

Typically, trim drag under symmetrical power can range from 0.5% to 5% of total airplane (untrimmed) drag, depending on the location of the center of gravity.

5.2.11 INTERFERENCE DRAG COEFFICIENT, $C_{D_{interf.}}$

Interference drag arises from changes in the flow pattern which accompanies the placing of two bodies (or surfaces) in close proximity. Ref. 5.3 (Chapter 8) presents many examples of mutual interference between various components of airplanes.

Whenever two streamlined bodies are placed side–by–side, the average velocity near the surface of each body is always increased. The adverse pressure gradient which is associated with the decrease in velocity over the aft portion of bodies is responsible for a considerable increase in the drag coefficient.

If the streamlined bodies are placed in tandem, the drag of the rear body can be increased because it is immersed in the low energy wake from the front body. Thus, the total drag of two bodies placed close together will nearly always be higher than the sum of the individual drags of the isolated bodies. An example of these trends is shown in Figure 5.31 for two struts.

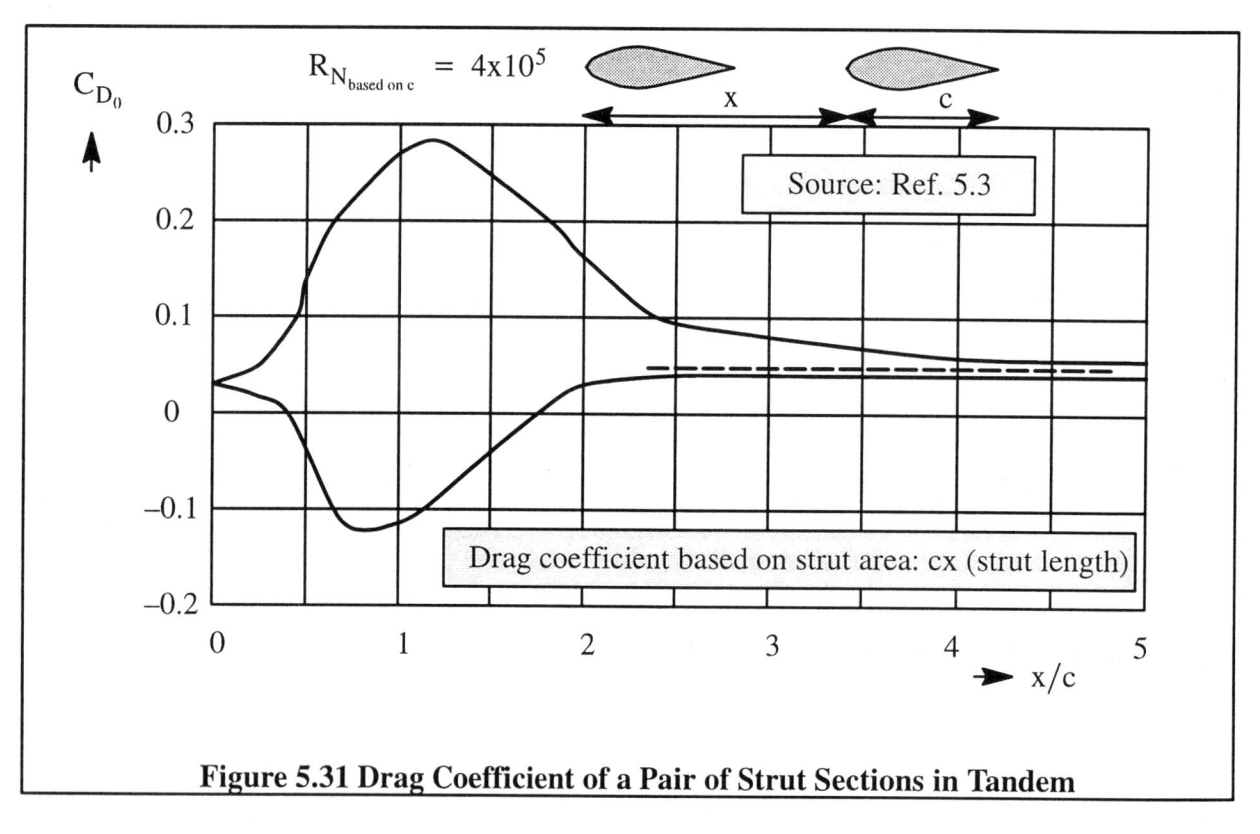

Figure 5.31 Drag Coefficient of a Pair of Strut Sections in Tandem

In many airplane applications, small bodies are attached to wings and fuselages. Examples are antennas, optical and/or radar sensors, missiles, bombs and other stores. The interference drag caused by such attachments can be reduced by placing the body in a region of favorable pressure gradients of the main body (wing or fuselage). In many cases, missiles can be placed in so–called conformal manner with a significant reduction in the installed drag. Some examples may be found in Ref. 5.17.

Finally, interference drag is also caused by the integration of wings with a fuselage as well as the integration of nacelles with a wing or with a fuselage. In these cases, interference drag consists of two types:

* a parasitic type, corresponding to boundary layer and pressure losses

and

* an induced type, caused by changes in the lift distribution.

The parasitic (viscous) interference drag of a wing attached to a fuselage is smallest with the wing near the nose of the fuselage, such as on the DH Mosquito bomber of WWII. This interference drag reaches a maximum for wing locations farther aft on a fuselage, particularly if the wing is mounted behind the location for maximum width of a fuselage. The thickness of the boundary layer which increases along the fuselage and produces reductions in local dynamic pressure, is responsible for this result.

The interference drag of tail surfaces tends to be smaller for tails installed at the rear end of a fuselage than for tails installed further forward.

The parasitic (viscous) interference drag of engine nacelles can be explained in a similar manner. A rough, but simple, rule at nearly zero lift is that wing–nacelle interference drag is approximately equal to the section drag of a wing area, twice as large as the wing portion covered by the nacelle.

The reader should note, that at the intersections of wings and fuselages as well as of wings and nacelles, the wetted area of the wing is in fact reduced. That factor should also be accounted for.

The induced interference drag of wing–fuselage and wing–nacelle combinations occurs mainly because of the change in spanwise lift distribution. The final lift distribution with the fuselage or nacelle added to the wing tends to be less elliptical which increases the induced drag.

To reduce total drag it is necessary for the designer to find the appropriate geometric layout of all airplane components such that the total interference drag is kept to a minimum. Also, properly designed transition sections (also known as fairings or fillets) can reduce interference drag in a significant manner.

5.2.12 MISCELLANEOUS DRAG COEFFICIENT, $C_{D_{misc}}$

The miscellaneous drag coefficient, $C_{D_{misc}}$ is used to account for those extra drag items which cannot be conveniently accounted for in the eleven drag contributions discussed in Sub–sections 5.2.1 through 5.2.11. Typical of such additional drag items are:

* protuberances
* inlet spillage drag
* cooling air drag,
* antenna drag

* struts not accounted for under interference drag
* drag associated with cabin–heating air inlets
* nozzle and/or exhaust system integration drag
* gaps, surface roughness

References 5.1 and 5.3 contain more information on these topics.

5.2.13 DRAG COMPONENT SUMMARY AND TYPICAL EXAMPLES

Figure 5.32 shows two types of total cruise drag breakdowns for a business jet airplane: a) a component breakdown and b) a causal breakdown. Table 5.2 shows an actual drag breakdown for a business jet in cruise.

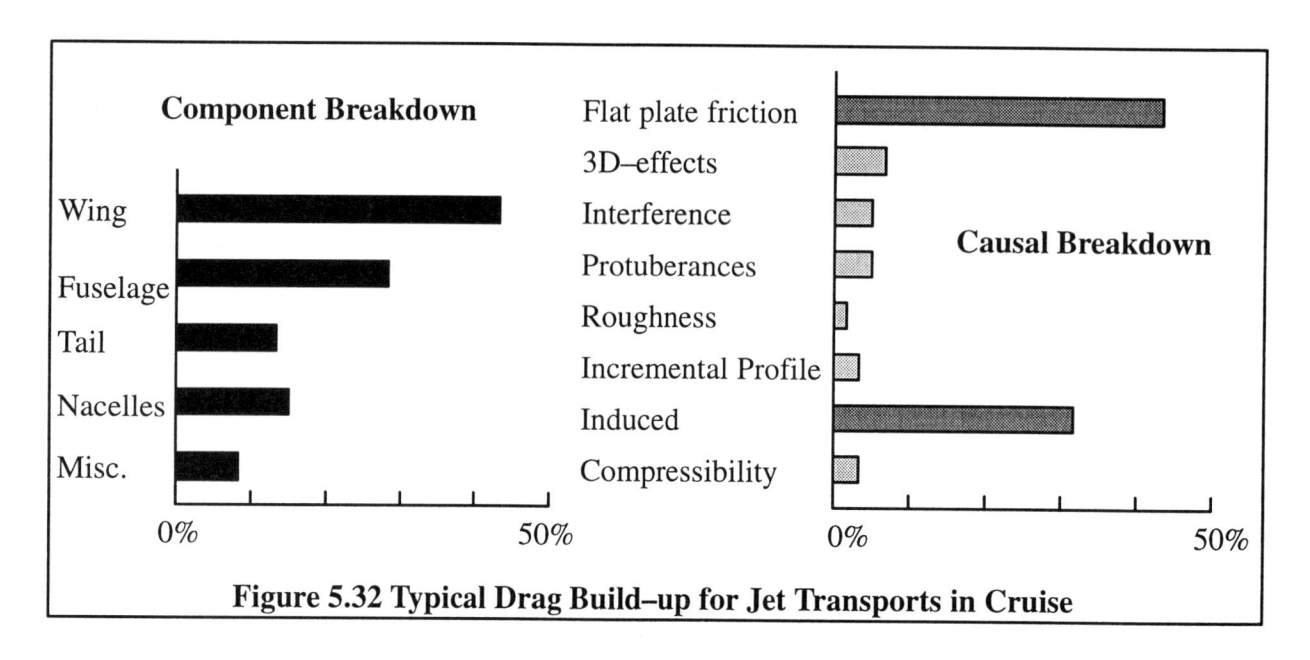

Figure 5.32 Typical Drag Build-up for Jet Transports in Cruise

It is clear, that skin friction and induced drag are the largest causal contributors to cruise drag. The reader is reminded, that friction drag is driven by the wetted area. A large fuselage, providing excellent creature comfort, also causes large wetted area and thus large friction drag. The induced drag is driven by airplane weight, flight condition and wing planform design.

From a pragmatic viewpoint, once the key design decisions have been made which "nail down" fuselage size and wing design, the designer can reduce drag only by paying attention to the other details. Doing so can pay significant dividends in drag reduction as illustrated by the examples in Figures 5.33a and 5.33b which are taken from Reference 5.18.

References 5.3 and 5.18 are recommended for anyone interested in pursuing drag reductions on existing or new airplanes.

A word of caution: More often than not, reducing drag items such as those mentioned in Figure 5.33 results in an increase in manufacturing cost! Whether or not such an increase is warranted depends on trade-offs between manufacturing cost increases and operation cost decreases. The manufacturer must keep in mind that operators are willing to pay for drag decreases in higher acquisition cost only if this results in higher profit potential.

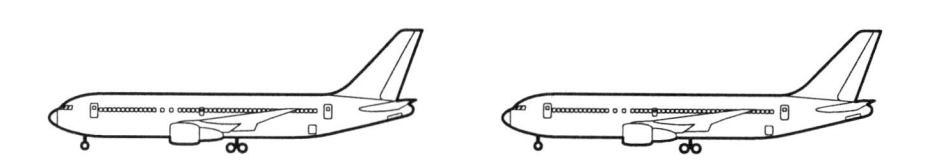

Table 5.2 Cruise Drag Breakdown for the Learjet Model 25 (Ref. 5.8)

$M = 0.75$	$C_L = 0.336$	$C_D = 0.0338$
Source	ΔC_D	% of Total

Causal Drag Breakdown:

Source	ΔC_D	% of Total
Profile drag (skin friction)	0.0180	53.25
Profile drag variation with lift	0.0007	2.07
Interference drag	0.0031	9.17
Roughness and gap drag	0.0015	4.44
Induced drag	0.0072	21.30
Compressibility drag	0.0028	8.28
Trim drag	0.0005	1.48
Total drag	0.0338	100.00

Profile Drag Breakdown: $\Delta C_{D_{profile\ friction}} = 0.0180$

Source	$\Delta C_{D_{profile\ friction}}$	% of Total
Wing	0.0053	29.57
Fuselage	0.0063	34.95
Tip tanks	0.0021	11.83
Tip tank fins	0.0001	0.54
Necelles	0.0012	6.45
Pylons	0.0003	1.61
Horizontal tail	0.0016	9.14
Vertical tail	0.0011	5.91
Total profile drag (friction)	0.0180	100.00

Courtesy: Bombardier Learjet

Airplane Drag

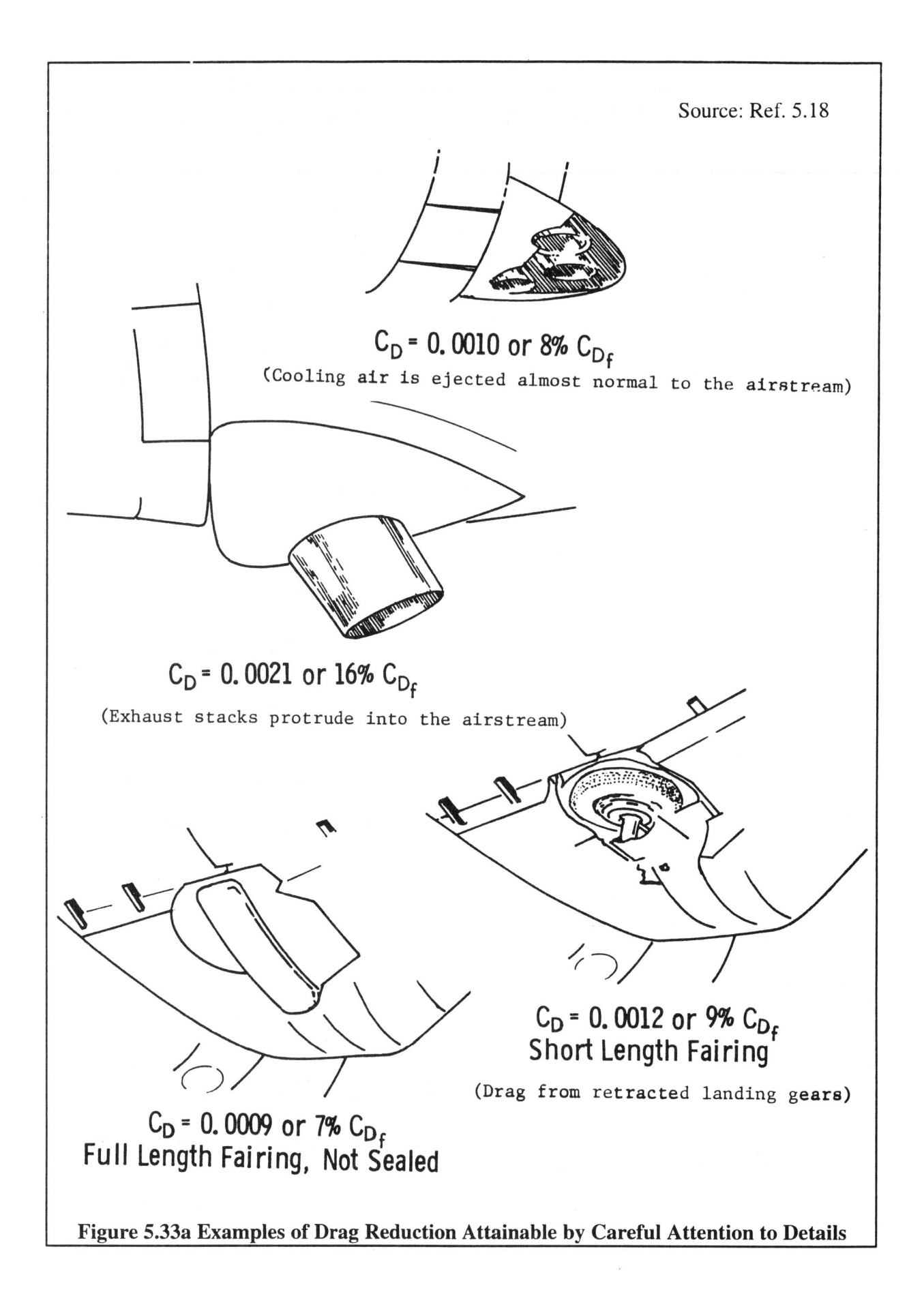

Source: Ref. 5.18

$C_D = 0.0010$ or 8% C_{D_f}
(Cooling air is ejected almost normal to the airstream)

$C_D = 0.0021$ or 16% C_{D_f}
(Exhaust stacks protrude into the airstream)

$C_D = 0.0012$ or 9% C_{D_f}
Short Length Fairing
(Drag from retracted landing gears)

$C_D = 0.0009$ or 7% C_{D_f}
Full Length Fairing, Not Sealed

Figure 5.33a Examples of Drag Reduction Attainable by Careful Attention to Details

Airplane Drag

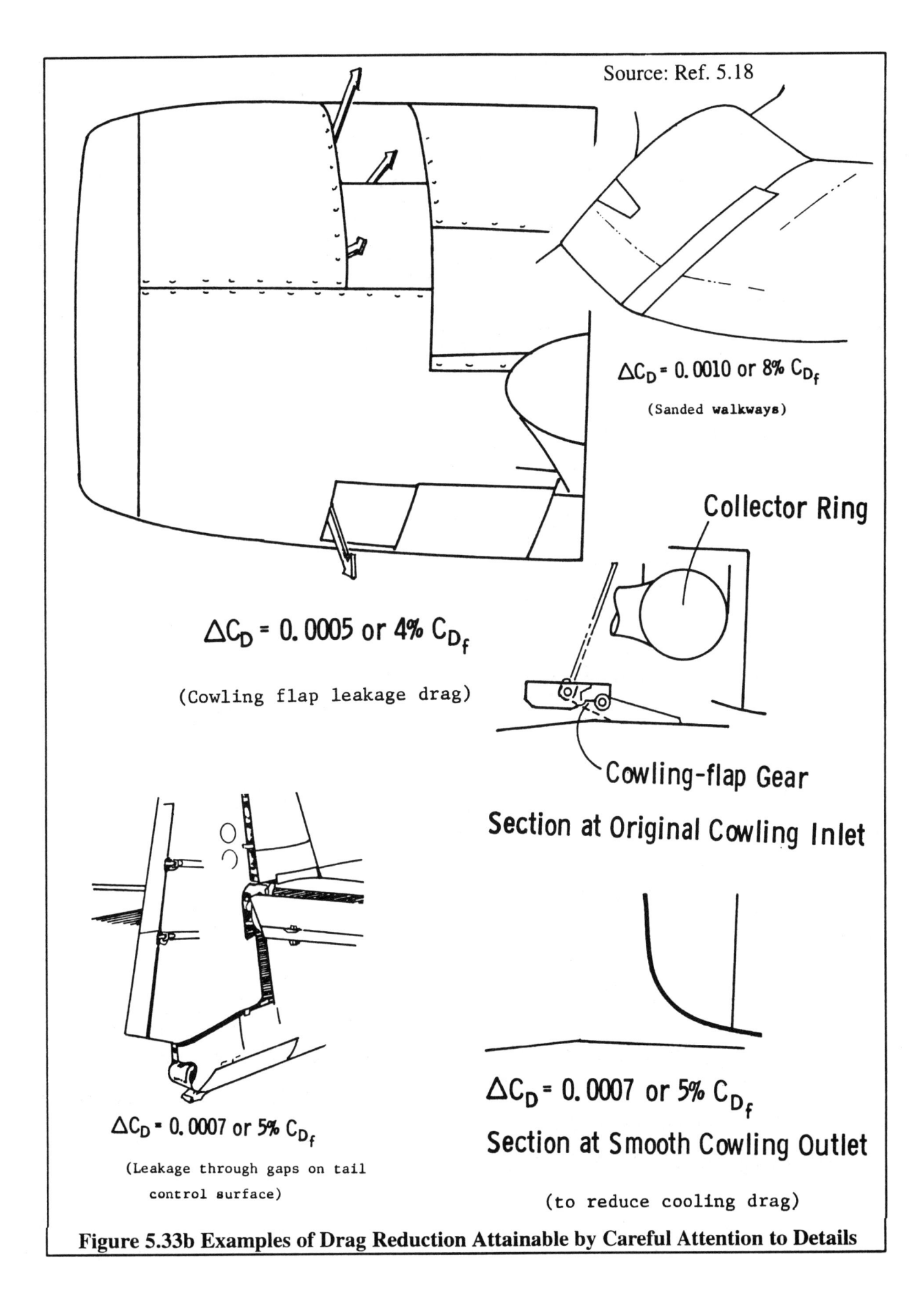

Source: Ref. 5.18

$\Delta C_D = 0.0010$ or 8% C_{D_f}

(Sanded walkways)

$\Delta C_D = 0.0005$ or 4% C_{D_f}

(Cowling flap leakage drag)

Collector Ring

Cowling-flap Gear

Section at Original Cowling Inlet

$\Delta C_D = 0.0007$ or 5% C_{D_f}

(Leakage through gaps on tail control surface)

$\Delta C_D = 0.0007$ or 5% C_{D_f}

Section at Smooth Cowling Outlet

(to reduce cooling drag)

Figure 5.33b Examples of Drag Reduction Attainable by Careful Attention to Details

Chapter 5

5.3 DETERMINATION OF DRAG IN THE WIND–TUNNEL

Despite the increasing importance of computational aerodynamics, wind–tunnels remain indispensable and cost effective tools for predicting the full–scale aerodynamic characteristics of airplanes, particularly in the case of drag. Wind–tunnels also provide a means to realistically evaluate the component drag build–up of an airplane.

Figures 5.34 shows an example of the size of a typical windtunnel model for the SIAI–Marchetti S–211 (Northrop–Grumman JPATS) jet trainer. Figures 5.35 and 5.36 are included to give the reader some idea of how a windtunnel model is broken down into various components. This allows the testing of a gradual build–up of a configuration to assess the drag contribution of the various components. Figure 5.37 shows the complete model installed in the Boeing Transonic Windtunnel.

The type of installation in Figure 5.37 is referred to as a "sting–type" installation. Another frequently used installation is the so–called "strut–type" installation. Figure 5.38 shows an example.

Table 5.3 summarizes several pros and cons of these two installations

With each of these installations a means of measuring the three forces and three moments which act on a model must be provided. Figure 5.39 shows two examples of internal strain gauge balances which are typically used with "sting–type" installations. Figure 5.40 shows an example of an external strain gauge balance which is typically used with "strut–type" installations.

References 5.19 and 5.20 contain detailed discussions of subsonic and supersonic windtunnel testing procedures and the corresponding data reduction methods.

Primarily because of differences in Reynolds number and the associated effect on the boundary layer, it is necessary to analytically manipulate windtunnel data before full scale drag predictions can be made. Also, the effect of the weight of the windtunnel model (in all its components) must be accounted for. As a general rule, to predict full scale flight data from windtunnel model data, the following three steps must be taken:

Step 1)   Tare Corrections. Windtunnel data must be corrected for the weight of the model. These corrections are referred to as "tares".

Step 2)   Flow–field Corrections. Windtunnel data must be corrected for differences between the flow–field in the tunnel and the flow–field in flight. These differences arise because of wall–to–model interference effects in the test section which change the effective angle of attack of a model.

Step 3)   Reynolds Number and Boundary Layer Corrections. The windtunnel model data, as corrected under 1) and 2) must be extrapolated to the full scale flight regime. This extrapolation usually consists of accounting for differences in Reynolds number and boundary layer characteristics.

Figure 5.34 Example of a Windtunnel Model of the S–211 Jet Trainer Airplane (Courtesy of SIAI–Marchetti, Sesto–Calende, Italy)

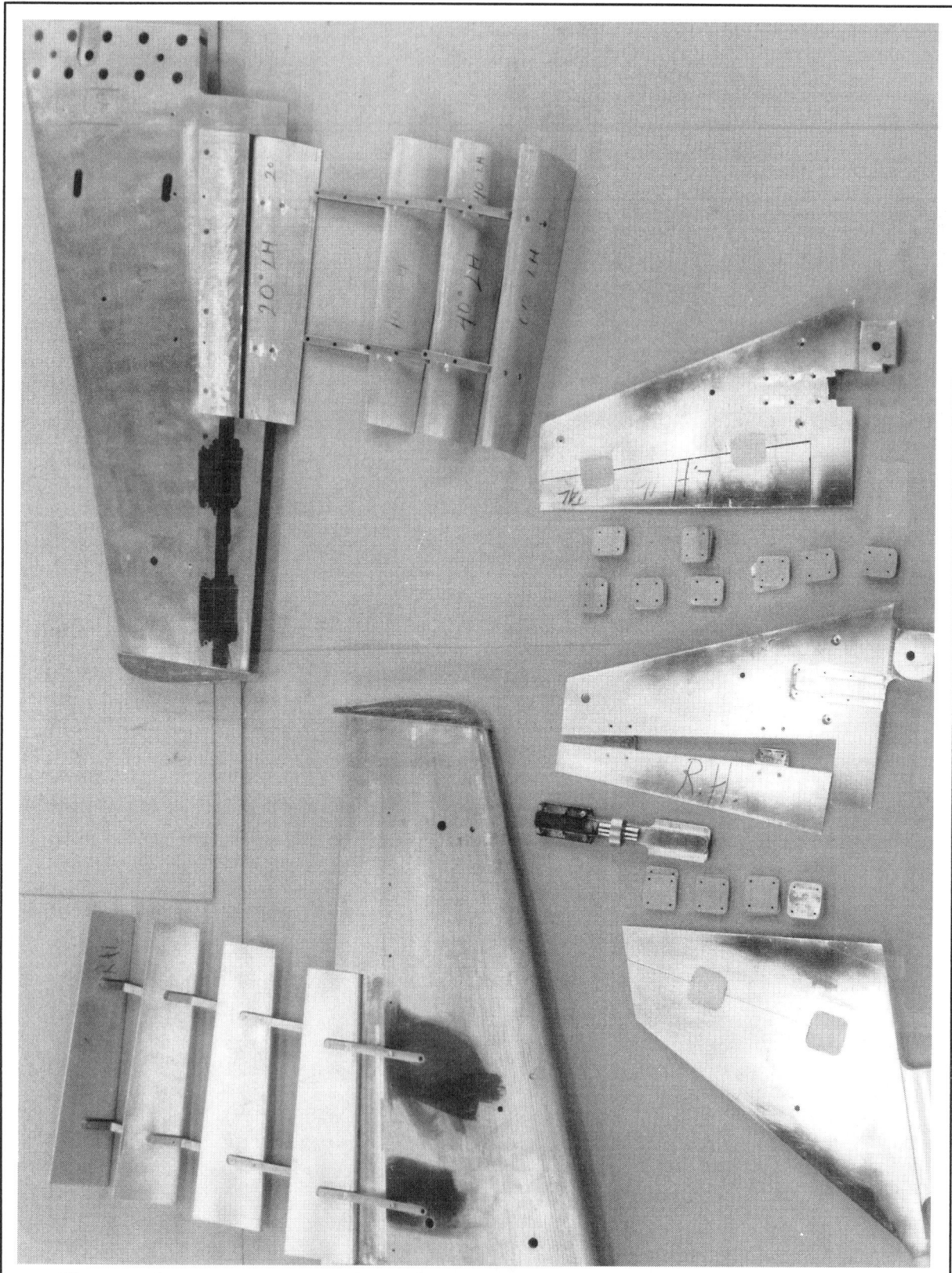

Figure 5.35 Example of a Wing, Wing Flap and Empennage Breakdown for a Windtunnel Model of the S–211 Jet Trainer Airplane
(Courtesy of SIAI–Marchetti, Sesto–Calende, Italy)

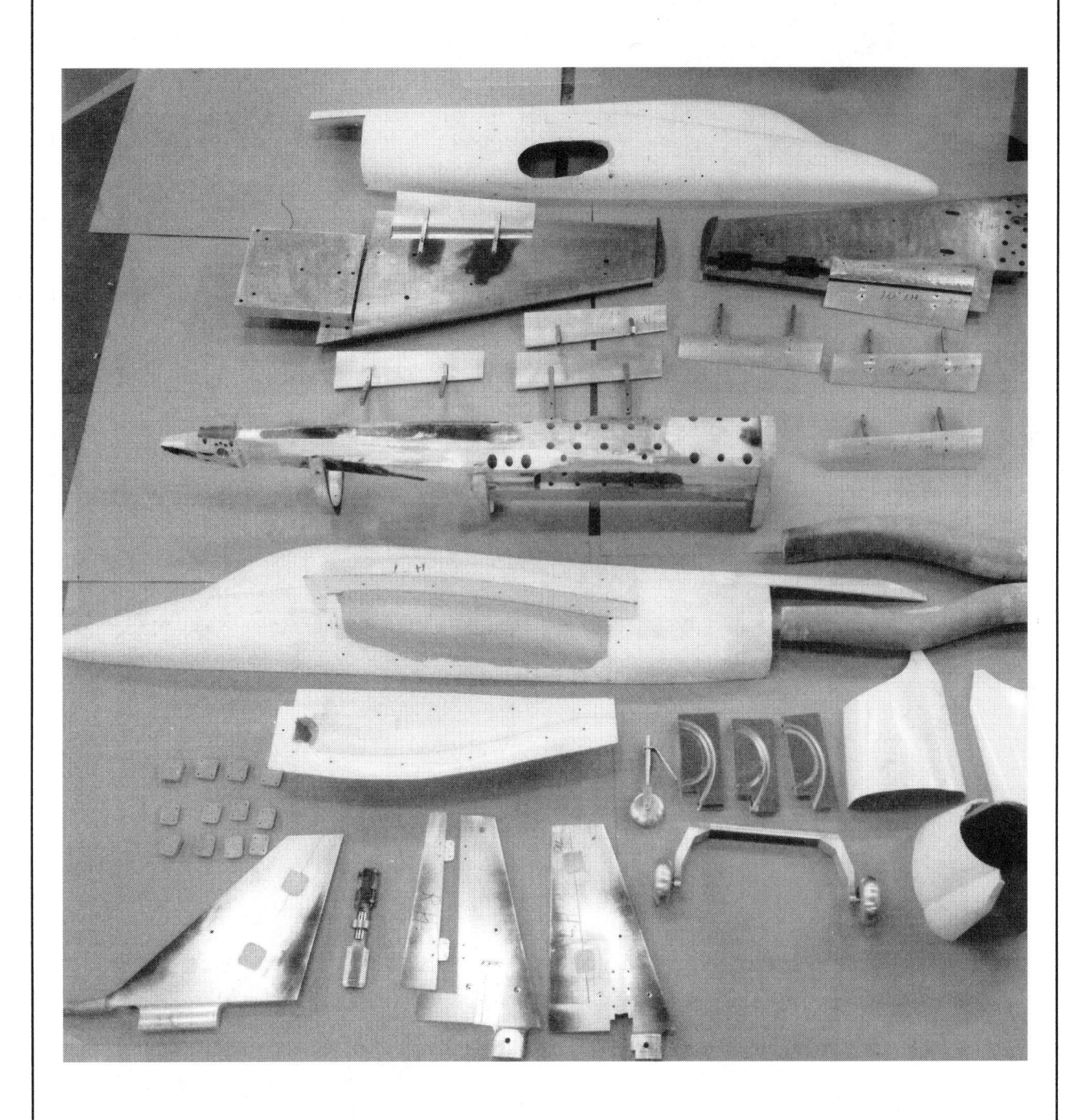

Figure 5.36 Example of Component Breakdown for a Windtunnel Model of the S–211 Jet Trainer Airplane
(Courtesy of SIAI–Marchetti, Sesto–Calende, Italy)

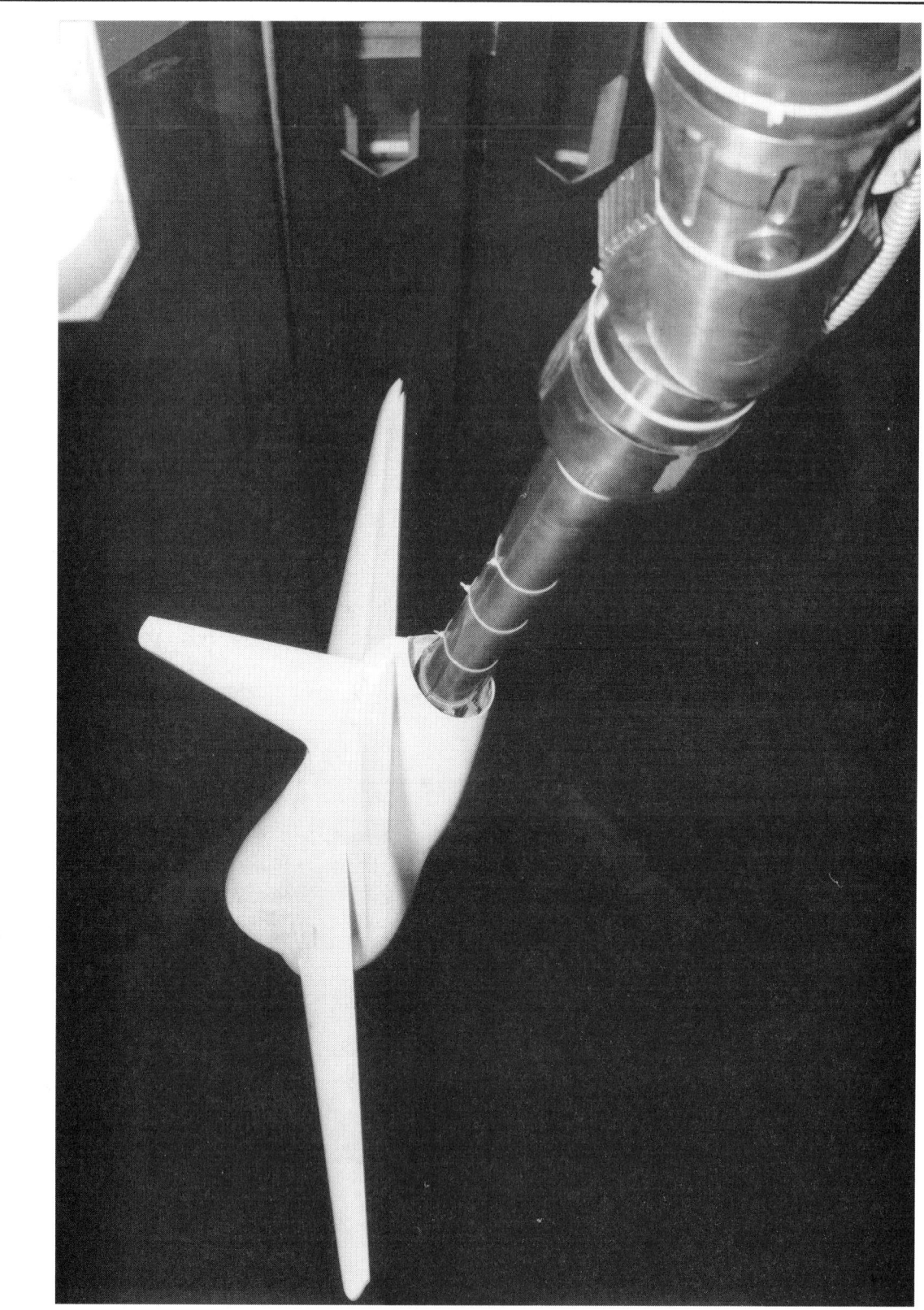

Figure 5.37 Example of the S–211 Windtunnel Model Installed in the Boeing Transonic Windtunnel
(Courtesy of SIAI–Marchetti, Sesto–Calende, Italy)

Airplane Drag

Figure 5.38 Example of a Strut Type Windtunnel Model Installation
(Courtesy of Gates–Learjet Corporation)

Table 5.3 Summary of Pros and Cons of Sting–Type and Strut–Type Wind–tunnel Model Installations

	Sting–Type (Figure 5.37)	Strut–Type (Figure 5.38)
PROS	• No strut interference This is particularly important for high speed and sideslip testing • Internal balance	• No model distortions are required
CONS	• Aft end model distortions are usually required to accommodate the sting • Possible fouling between model and sting at high angles of attack	• Limited to low speed testing • External balance required

Courtesy: SIAI Marchetti, Sesto Calende, Italy

Figure 5.39 Examples of Internal Strain Gauge Balances

Figure 5.40 Example of an External Strain Gauge Balance Installation

To make the necessary tare corrections (Step 1), the weight of each windtunnel model component must be determined so that the actual model weight in all its configurations is known. Tare corrections are discussed in detail in Ref. 5.19.

To make the flow–field corrections (Step 2), three types of corrections are normally required:

a) Tunnel wall corrections to account for the effect of the limited size of the test section. This includes corrections due to flow blockage which comes about because the model cross section, vertical to the test section, may not be small compared to the cross sectional area of the test section. This becomes a problem at high angles of attack.

b) Corrections to account for the effects of the model suspension system. Examples of such effects are: aerodynamic interference between the model and the suspension system and aeroelastic effects due to the model deforming the suspension system.

c) Corrections to account for imperfections in the tunnel flow. Examples of such imperfections could be: turbulence caused by roughness in the tunnel walls, by turning vanes and by non–uniform velocity distribution due to rotational flow.

In these Step 2 corrections, the effect of turbulence produced in the tunnel by the tunnel propellers, tunnel guide vanes, screens etc. should be included. One way to indicate the degree of turbulence in a windtunnel is by the use of the so–called turbulence factor (T.F.).

It is known that the critical Reynolds number of a sphere tested in free air is about 385,000. At this Reynolds number the laminar separation point and the boundary layer transition point coincide. The drag coefficient (based on maximum cross–sectional area) at this critical Reynolds number is 0.30. If the sphere is now tested in the windtunnel, the critical Reynolds number at which the drag coefficient, C_D is 0.30 (based on maximum cross sectional area) will be denoted by: R_{N_C}. The turbulence factor of the tunnel is then defined as:

$$\text{T.F.} = \frac{385,000}{R_{N_C}} \qquad (5.33)$$

The effective Reynolds number, R_{N_e}, under any testing condition will be:

$$R_{N_e} = (\text{T.F.}) \times R_N \qquad (5.34)$$

The Step 3 corrections are necessary because of the usual large differences in Reynolds number between that of the windtunnel model (small scale) and that of the airplane in flight (full scale). The reader is reminded of Figures 2.17b and 5.11 which illustrate the effect of Reynolds number on the skin friction coefficient.

The difference between small scale and full scale Reynolds number affects primarily the friction drag coefficient and the maximum lift coefficient. This is so because of the difference in boundary layer characteristics between small scale and full scale. While the boundary layer on a large airplane is usually fully turbulent, large regions of laminar flow may exist on the model. Because it is essential to simulate the full scale boundary layer development accurately for drag prediction, the standard practice is to solve this problem by using so–called trip strips. These trip strips consist of a narrow band of carborandum grit.

On lifting surfaces (such as wing, tail, canard) the trip strip is usually placed at about 5% to 10% of the local chord, On bodies (such as fuselage, nacelles and stores) the trip strip location is usually at about 5% of the length of the body.

To be precise, the trip strip location should be such that: $R_{N_K} \geq 600$, where R_{N_K} is based on the roughness height, K, the velocity at the top of the roughness, U_K, and the kinematic viscosity at the top of the roughness, ν_K (see Ref. 5.21).

The method of artificially fixing transition on the tunnel model, even with the rough method of locating the trip strip as outlined above, has been found to produce good results in sub–critical flow. Below the critical Mach number, M_{crit}, the full scale wing pressure distribution is in good agreement with small scale pressure distributions measured on models equipped with trip strips. However, for supercritical flow, $M > M_{crit}$, the problem of simulating high Reynolds number at small scale becomes much more complicated. This is because the boundary layer thickness must be correctly simulated at the location of the shock wave. If the trip strip is located too far forward, the boundary layer will be unrealistically thick at the shock. This places the shock in the wrong position or leads to premature shock–induced separation. Therefore, in supercritical flow, the trip strip must be located further aft than the normal position to simulate properly the shock characteristics. The correct trip strip location may be calculated by theoretical methods or can be determined empirically. The reader may wish to consult References 5.21 and 5.22.

If access is available to a cryogenic test facility (such as the National Transonic Test Facility at NASA Langley) small scale tests at closer to full scale Reynolds number is possible.

The upshot of all this is, that after obtaining the windtunnel data, corrections must be applied to account for these Reynolds number effects. One method of doing this is to adjust the measured small scale values of C_{D_0} or $C_{D_{min}}$ according to the following equation:

$$\left(\frac{C_{D_{min}}}{C_f}\right)_{windtunnel} = \left(\frac{C_{D_{min}}}{C_f}\right)_{fullscale} \tag{5.35}$$

Another way of applying the Reynolds number corrections is to calculate $C_{D_{min}}$ as:

$$\left(C_{D_{min}}\right)_{fullscale} = \left(C_{D_{min}}\right)_{windtunnel} - \Delta C_{D_{min}} \tag{5.36}$$

where: $\Delta C_{D_{min}}$ is the decrease in turbulent friction due to the difference in Reynolds numbers. Figure 5.11 can be used to find the appropriate values for C_f or $\Delta C_{D_{min}}$. Any additional adjustment

Airplane Drag

to the drag polar shape can be done empirically by comparing with known drag polar shapes of existing similar airplanes.

Eventually, any methods used to account for Reynolds number effects must be checked against flight test data.

Detailed discussions of other corrections are beyond the scope of this text. However, these can be found in Reference 5.19. The accuracy of extrapolating windtunnel data to full scale results is discussed in Reference 5.23.

Because of the great importance of the effects of the turbulence factor and Reynolds number on skin friction drag, the following example problem is presented.

Example Problem:

The drag of a wing model is measured at a test Reynolds number of 3×10^6 and a turbulence factor of 2.0. The measured drag coefficient, based on wing area, is 0.0082. Find the equivalent free–air drag coefficient if the full scale Reynolds number is 9×10^6. Assume that the Mach number is close to zero.

Solution:

The effective Reynolds number is given by Eqn (5.27) as: $R_N = 2.0 \times 3 \times 10^6 = 6 \times 10^6$. Using the flat–plate solution of Eqn (2.97) for turbulent flow, the drag coefficient correction can be obtained as:

$$\Delta C_D = 2.0 \left\{ \frac{0.455}{\left(\log_{10} 6 \times 10^6\right)^{2.58}} - \frac{0.455}{\left(\log_{10} 3 \times 10^6\right)^{2.58}} \right\} = -0.0008$$

Comment 1: this correction is to account for the effect of the tunnel turbulence on the test Reynolds number.

Comment 2: the factor 2.0 accounts for the fact that the wing has two sides which are exposed to friction drag

Hence, the equivalent free–air drag coefficient of the model in the tunnel is:

$$C_{D_{(R_N = 6 \times 10^6)}} = 0.0082 - 0.0008 = 0.0074$$

Using Figure 5.11 to correct for the difference between the full scale Reynolds number and the effective test Reynolds number it is seen that:

$$\Delta C_{D_{min}} = 0.00325_{\text{at } R_N = 6 \times 10^6} - 0.00305_{\text{at } R_N = 9 \times 10^6} = 0.0002$$

The predicted full scale drag coefficient of the wing is therefore:

$$C_{D_{full\ scale}} = 0.0074 - 0.0002 = 0.0072$$

5.4 SIMPLIFIED METHOD FOR PREDICTING DRAG POLARS OF CLEAN AIRPLANES

The build–up methods for drag polar prediction of new designs as outlined in Sections 5.2 and 5.3 will produce good results. When done by hand, a large expenditure of engineering man–hours is required. By using the software of Reference 5.24 significant savings in man–hours can be obtained. However, even with the use of software the amount of time spent in predicting airplane drag is not insignificant. For that reason, in preliminary design, a faster method has been developed which has been found to yield adequate results.

This faster method is based on the so–called "equivalent–flat–plate" drag of airplanes. The method assumes that the drag polar is parabolic:

$$C_D = C_{D_0} + \frac{C_L^2}{\pi A e} \tag{5.37}$$

In this method the zero lift drag coefficient, C_{D_0}, may be determined from:

$$C_{D_0} = \frac{f}{S_{ref}} \tag{5.38}$$

where: f is the equivalent parasite area of the airplane.

For clean airplanes (flaps up, gears up, cooling flaps closed, speed brakes retracted) the equivalent parasite area, f, may be determined from Figures 5.41, 5.42 and 5.43.

As seen from Figures 5.41–5.43, to determine the equivalent parasite area, 'f', it is necessary to know the airplane wetted area, S_{wet} and the equivalent skin friction coefficient, C_f of the airplane. How to do this is explained next.

First, the airplane wetted area, S_{wet} may be calculated from:

$$S_{wet} = S_{wet_{fuselage}} + S_{wet_{wing}} + S_{wet_{empennage}} + S_{wet_{nacelles}} - S_{wet_{intersections}} \tag{5.39}$$

This equation may have to be adjusted for airplanes with different configurations. Values for the wetted area of any airplane component can be determined from any reasonable three–view. For fuselages, nacelles and stores the wetted area may be obtained from integration of a perimeter plot as shown in Figure 5.44. Corrections to the wetted area may have to be made to prevent accounting twice for the intersections of such items as wing on fuselage.

Second, the equivalent skin friction coefficient, C_f depends on the type and accuracy of the manufacturing technology (smoothness) used. Specific airplane points in Figures 5.41–5.43 were entered to provide the reader with a yardstick against which to judge any given new airplane design.

Airplane Drag

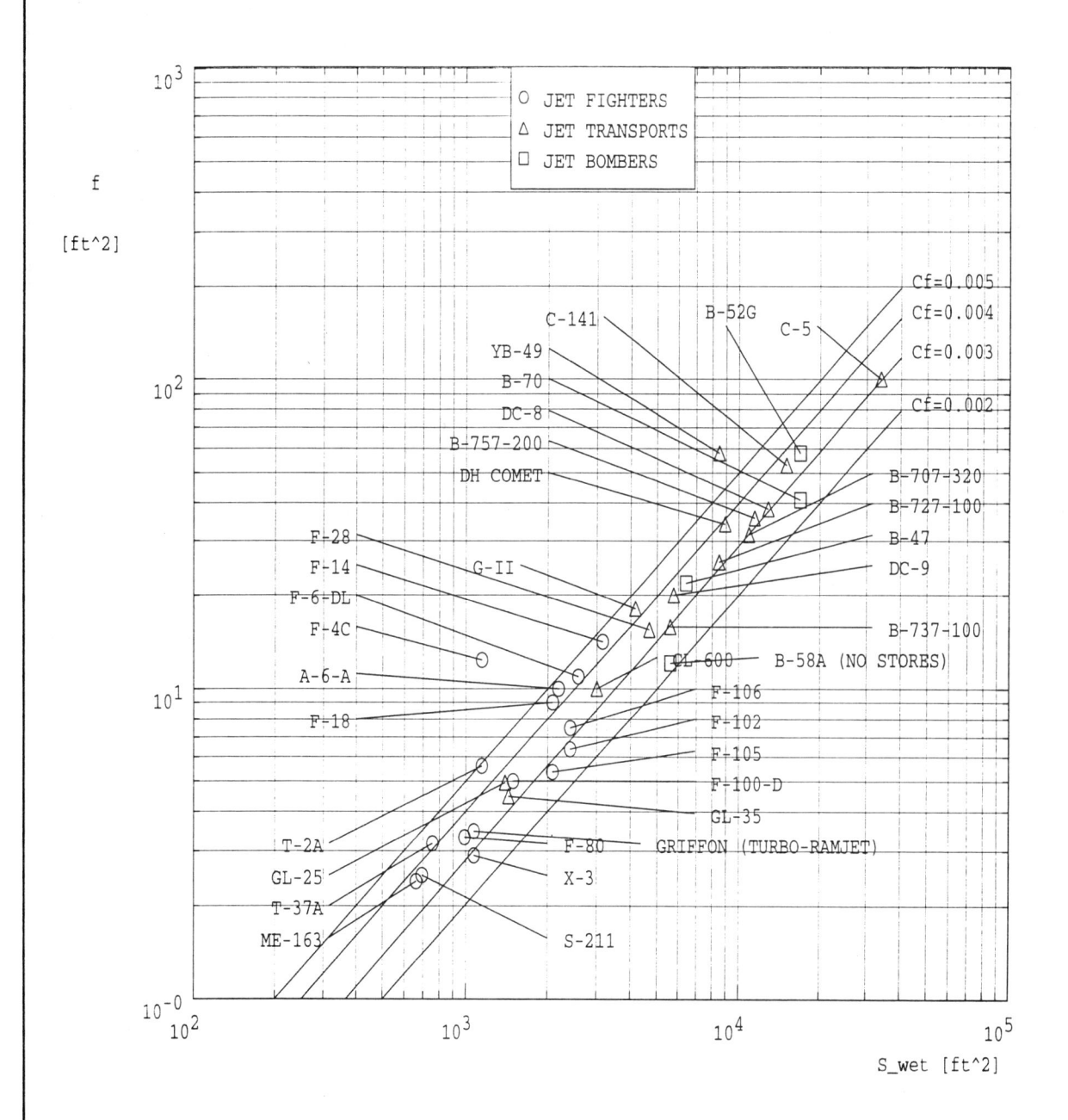

> Note: This figure appears as a help file in the software of Reference 5.24

[Figure: log-log plot of parasite area f [ft^2] (y-axis, 10^0 to 10^3) versus wetted area S_wet [ft^2] (x-axis, 10^2 to 10^5) with data points for jet fighters (○), jet transports (△), and jet bombers (□). Reference lines shown for $C_f = 0.005, 0.004, 0.003, 0.002$.

Aircraft labeled include: C-141, B-52G, C-5, YB-49, B-70, DC-8, B-757-200, DH COMET, B-707-320, B-727-100, B-47, DC-9, F-28, F-14, G-II, F-6-DL, F-4C, B-737-100, A-6-A, CL-600, B-58A (NO STORES), F-18, F-106, F-102, F-105, F-100-D, CL-35, T-2A, F-80, GRIFFON (TURBO-RAMJET), GL-25, X-3, T-37A, ME-163, S-211.]

> The parameter C_f represents the turbulent boundary layer, equivalent skin friction coefficient of the airplane

Figure 5.41 Parasite Area Versus Wetted Area for Jet Fighters, Jet Bombers and Jet Transports

Airplane Drag

> Note: This figure appears as a help file in the software of Reference 5.24

The parameter C_f represents the turbulent boundary layer, equivalent skin friction coefficient of the airplane

Figure 5.42 Parasite Area Versus Wetted Area for Multi-engine Propeller Driven Airplanes

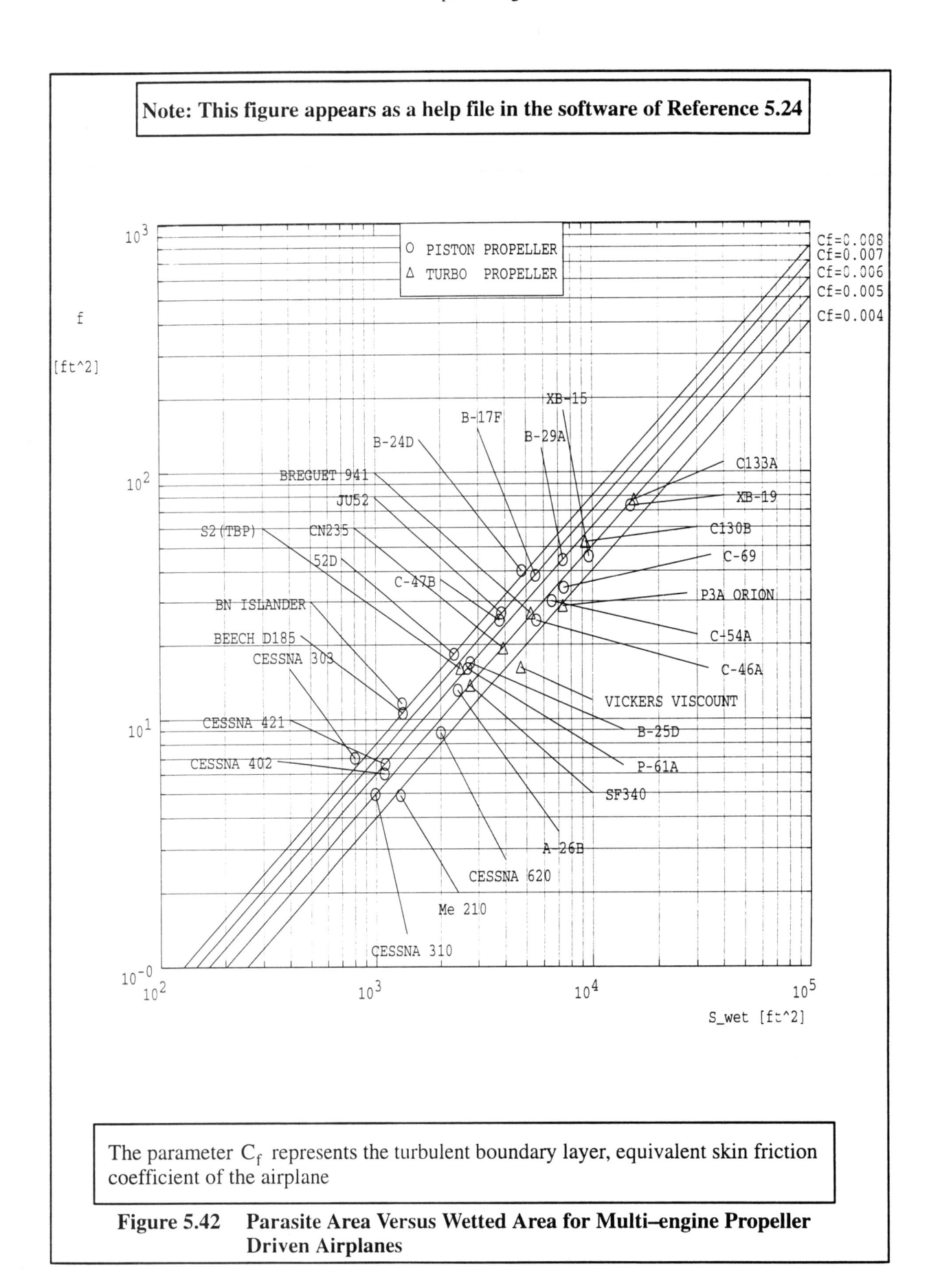

Airplane Drag

> Note: This figure appears as a help file in the software of Reference 5.24

[Plot: f [ft^2] vs S_{wet} [ft^2], log-log scale from 10^2 to 10^4 on x-axis and 10^0 to 10^3 on y-axis. Diagonal lines show C_f = 0.015, 0.010, 0.008, 0.006, 0.005, 0.004, 0.003. Symbols: ○ RETRACTABLE GEAR, △ FIXED GEAR.

Data points labeled: FIESELER STORCH, CESSNA T188C SPRAY BAR ON, T188C SPRAY BAR OFF, PT-18, PT-13, CESSNA L-5, Me-109, YPT-16, PIPER CUB, CESSNA-152, TAYLORCRAFT, THORPE T-18, KR-1, VARIEZE, CESSNA-210, BEECH-35, TYPICAL SAILPLANE, P-47D, P-35, NDN FIRECRACKER, P-40 E,F, P-36A, CESSNA-180, P-51B, P-51F.]

> The parameter C_f represents the turbulent boundary layer, equivalent skin friction coefficient of the airplane

Figure 5.43 Parasite Area Versus Wetted Area for Single–engine Propeller Driven Airplanes

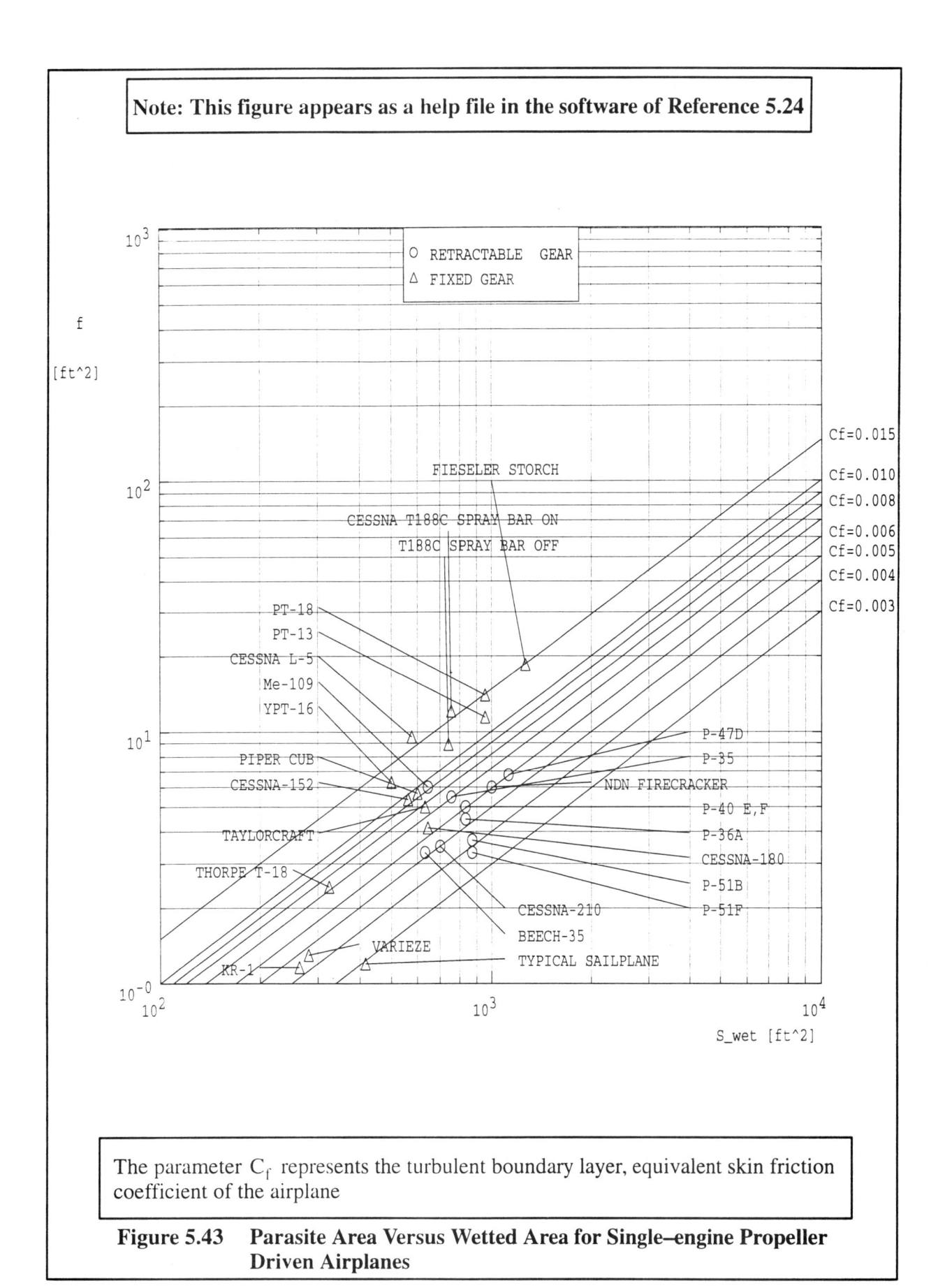

The reader should not be too optimistic about the level of aerodynamic smoothness which can be obtained. If that were feasible, existing manufacturers would have done so.

The value of Oswald's efficiency factor 'e' varies from airplane to airplane, depending on wing planform design, on wing–fuselage and on wing–nacelle integration. Example numbers for C_{D_0} and 'e' are given in Table 5.4.

Reference 5.25 contains yet another empirical method for predicting airplane drag. The latter method was derived by analyzing flight test data of many military airplanes, mostly trainers and fighters.

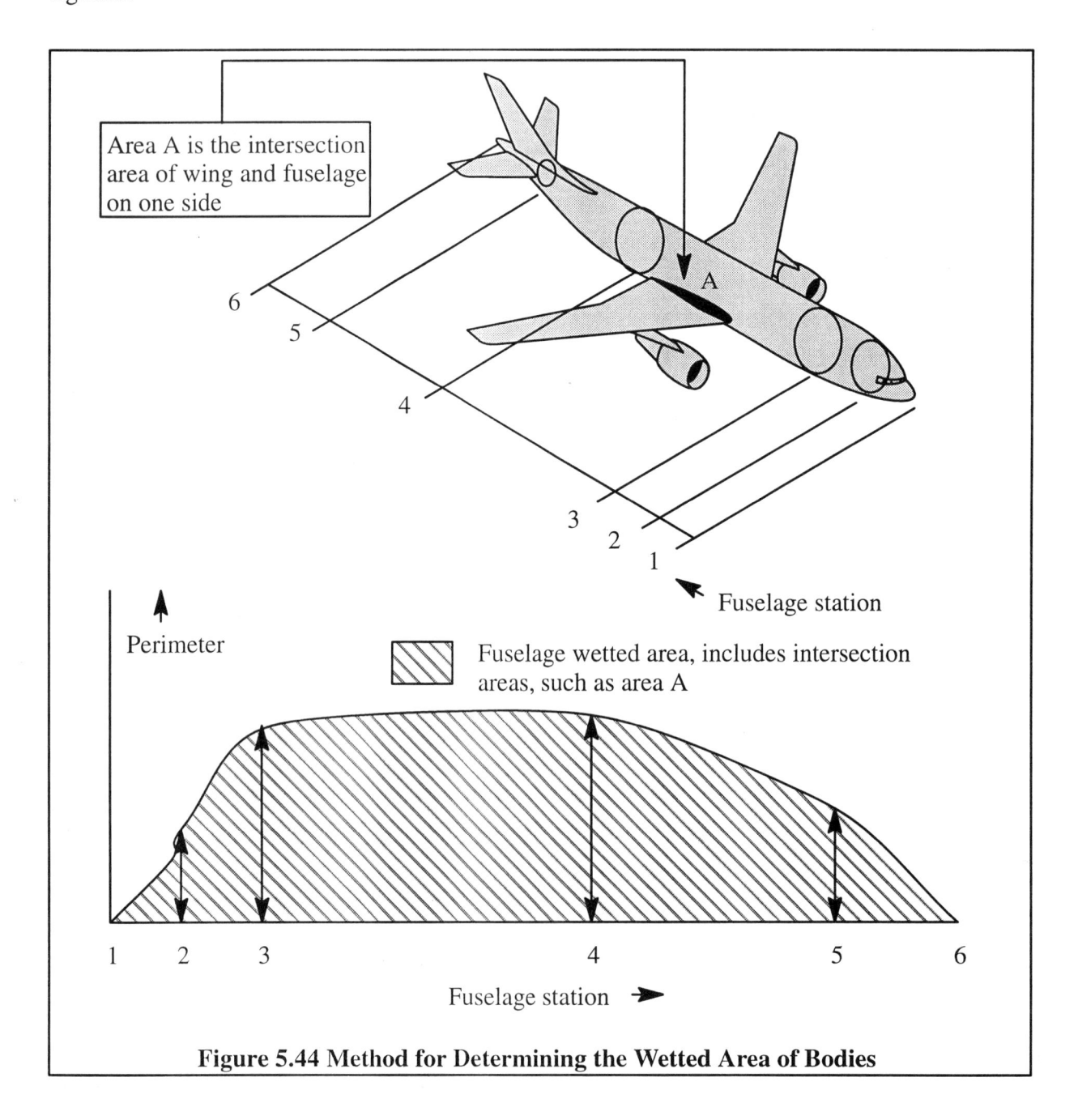

Figure 5.44 Method for Determining the Wetted Area of Bodies

Table 5.4 Summary of Drag Data for Various Airplanes

Type	Wing Area, S	Aspect Ratio	$\frac{S_{wet}}{S}$	Drag Polar	e	$\left(\frac{C_L}{C_D}\right)_{max}$ @ C_L
	ft^2			$C_{D_0} + \frac{C_L^2}{\pi A e}$		
C-150	160	7.0	?	$0.0327 + 0.0592 C_L^2$	0.77	11.3 @ 0.74
C-172	174	7.5	3.7	$0.0281 + 0.0552 C_L^2$	0.77	12.7 @ 0.71
C-180	174	7.5	?	$0.0246 + 0.0572 C_L^2$	0.75	13.3 @ 0.66
C-182	174	7.5	4.0	$0.0293 + 0.0506 C_L^2$	0.84	13.0 @ 0.75
C-185	174	7.5	?	$0.0207 + 0.0494 C_L^2$	0.86	15.6 @ 0.65
C-310	175	7.3	4.6	$0.0263 + 0.0596 C_L^2$	0.73	12.6 @ 0.66
Skyrocket	183	6.7	?	$0.0163 + 0.0579 C_L^2$	0.82	16.3 @ 0.53
Saab 340	450	11.0	?	$0.0285 + 0.0362 C_L^2$	0.80	15.6 @ 0.89
DC 9-30	1,001	6.8	6.5	$0.0211 + 0.0450 C_L^2$	0.81	16.7 @ 0.50
B 707-320	3,050	7.1	5.0	$0.0131 + 0.0650 C_L^2$	0.70	19.6 @ 0.45
A-340	3,908	9.5	?	$0.0165 + 0.0435 C_L^2$	0.77	18.5 @ 0.60
B 767	3,050	8.0	?	$0.0135 + 0.0592 C_L^2$	0.67	17.2 @ 0.50
C-17	3,800	7.2	?	$0.0175 + 0.0510 C_L^2$	0.87	16.4 @ 0.55
Learjet M25	232	5.0	5.6	$0.0260 + 0.0078 C_L^2$	0.82	10.9 @ 0.58
G-II	800	6.0	?	$0.0230 + 0.0057 C_L^2$	0.93	14.0 @ 0.63

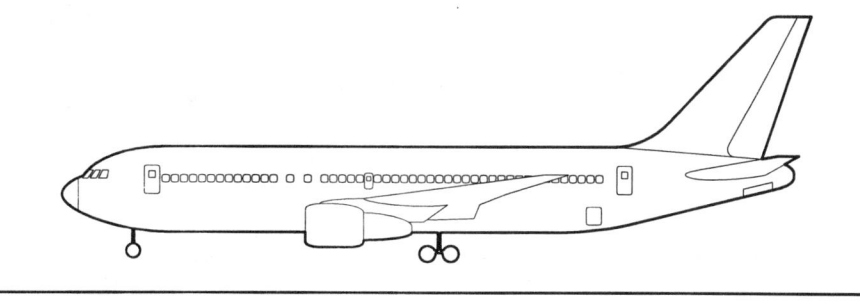

5.5 SUMMARY FOR CHAPTER 5

In this chapter various physical sources for airplane total drag, examples of airplane drag breakdown and drag prediction methods (theoretical, empirical and experimental) were presented. For quick drag prediction methods, the very user–friendly software of Ref. 5.24 as described in Appendix B is highly recommended.

The importance of airplane drag in predicting the operating economy of airplanes cannot be overemphasized. Airplanes flying today have mostly turbulent boundary layers. Exceptions are the Bellanca Skyrocket (not in production) and the Piaggio P–180 Avanti. In these airplanes significant laminar flow runs exist, primarily on the lifting surfaces. Figure 5.45 shows the very large drag reduction potential which still exists. That potential will be realized only if economical ways can be found to maintain laminar flow at high sweep angles and at high Mach numbers. Whether or not this will come to pass will depend not only on the development of appropriate technology but also on the actual cost of fuel. As always, airplane designers must carefully balance the cost of ownership versus the cost of operating an airplane.

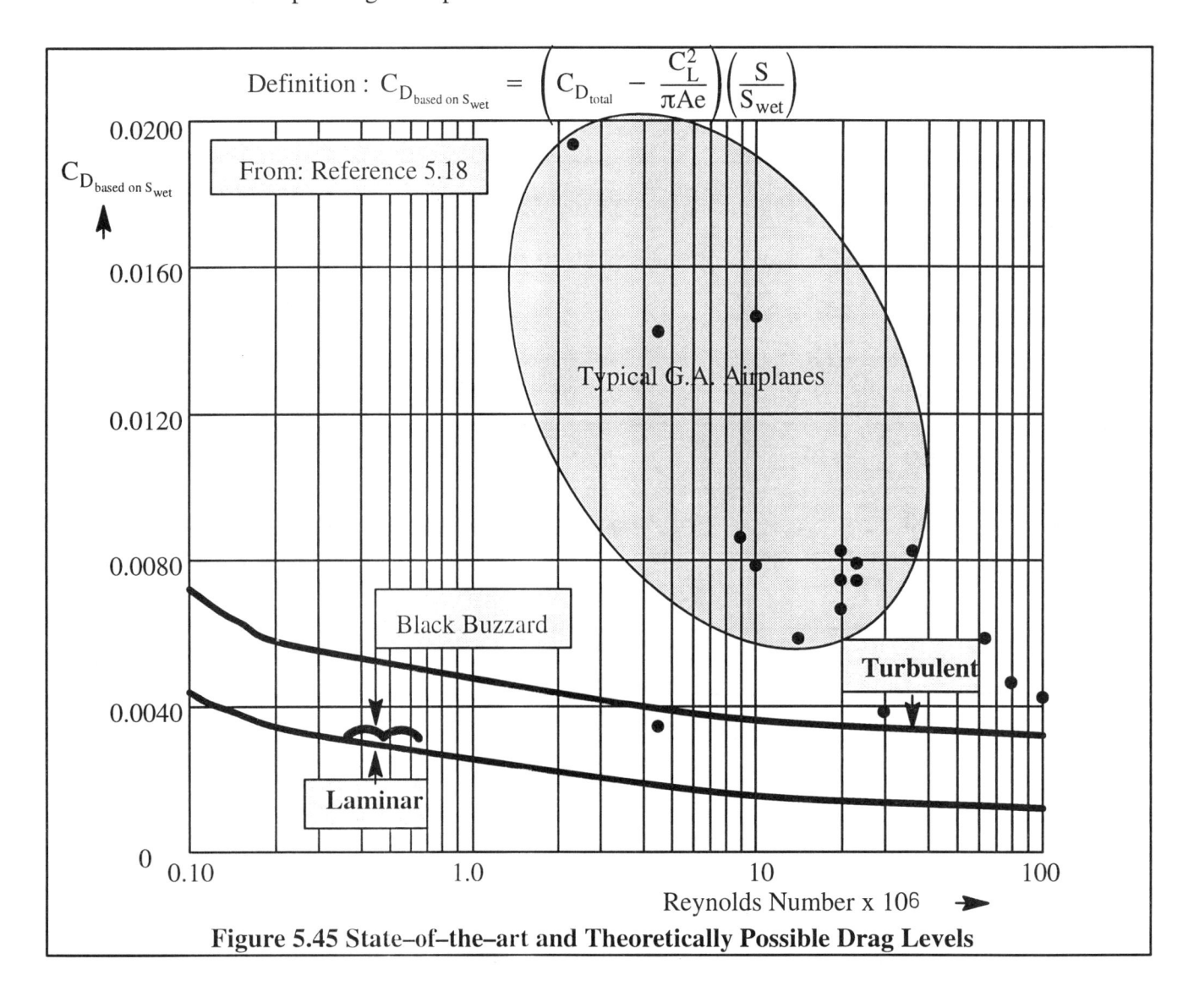

$$\text{Definition}: C_{D_{based\ on\ S_{wet}}} = \left(C_{D_{total}} - \frac{C_L^2}{\pi Ae}\right)\left(\frac{S}{S_{wet}}\right)$$

Figure 5.45 State–of–the–art and Theoretically Possible Drag Levels

5.6 PROBLEMS FOR CHAPTER 5

5.1 Prove that Eqn (5.2) is correct.

5.2 Rederive Eqns (5.23) through 5.25) for a three–surface airplane.

5.3 Show that for Eqn (5.3) the following relations hold:

$$C_{D_{at\,(C_L/C_D)max}} = 2C_{D_{min}} + \frac{2C_{L_{min.drag}}}{\pi Ae} - \frac{2C_{L_{min.drag}}}{\pi Ae}\sqrt{\pi AeC_{D_{min}} + C_{L_{min.drag}}^2}$$

and

$$C_{L_{at\,(C_L/C_D)max}} = \sqrt{\pi AeC_{D_{min}} + C_{L_{min.drag}}^2}$$

5.4 Show that for two airplanes having the same wing parasite drag coefficient and the same span efficiency factor, the total wing drag of one airplane can be related to that of the other airplane (called the reference airplane) at the same lift as follows:

$$D_w = D_{0_{wr}}\frac{S}{S_r} + (D_{wr} - D_{0_{wr}})\frac{S_r A_r}{S A}$$

where the subscript 'r' indicates the reference airplane.

5.5 Let: $r = \frac{D_w}{D_{wr}}$ and $P = \frac{D_{0_{wr}}}{D_{wr}}$.

1) Show that for rectangular wings the equation in Problem 5.4 can be written as:

$$r = P\frac{bc}{b_r c_r} + (1-P)\frac{b_r^2}{b^2}$$

2) Given are the following characteristics for a typical light airplane:

$W = 2,800$ lbs $S = 174$ ft^2 $A = 7.4$ $C_{D_{o_w}} = 0.0090$

$C_{D_{fhv}} = 0.0170$ $e = 0.75$ cruise altitude $= 8,000$ ft

Determine P for this airplane at the cruise altitude.
Note: fvh = fuselage+vertical tail+ horizontal tail.

3) If D_{total} represents the total drag of the airplane, then:

$$D_{total} = rD_{wr} + C_{D_{o_{fvh}}}\bar{q}S$$

and

$$D_{total_r} = C_{D_r}\bar{q}S_r = \left(C_{D_{0_w}} + C_{D_{0_{fvh}}} + \frac{C_L^r}{\pi A_r e}\right)\bar{q}S_r$$

Assume that $b = b_r$, but $c \neq c_r$. Calculate and plot D_{total}/D_{total_r} versus c/c_r for V=100 mph and for V=200 mph at 8,000 ft altitude for c/c_r ranging from 0.4 to 1.0.

5.6 In Problem 5.4, now assume that $c = c_r$. Show that if the wing span is reduced there will be a reduction in drag only if $P > \frac{2}{3}$.

5.7 Assume that the pitching moment characteristics of a propeller driven airplane are as shown in Figure 5.46. If the elevator deflection will generate an incremental pitching moment coefficient given by: $\Delta C_m = C_{m_{\delta_e}}\delta_e$, calculate and plot the elevator angle, δ_e, to trim versus the thrust coefficient for zero flaps and for flaps down. Assume that $C_{m_{\delta_e}} = -0.02$ 1/deg and $\alpha = 6$ degrees.

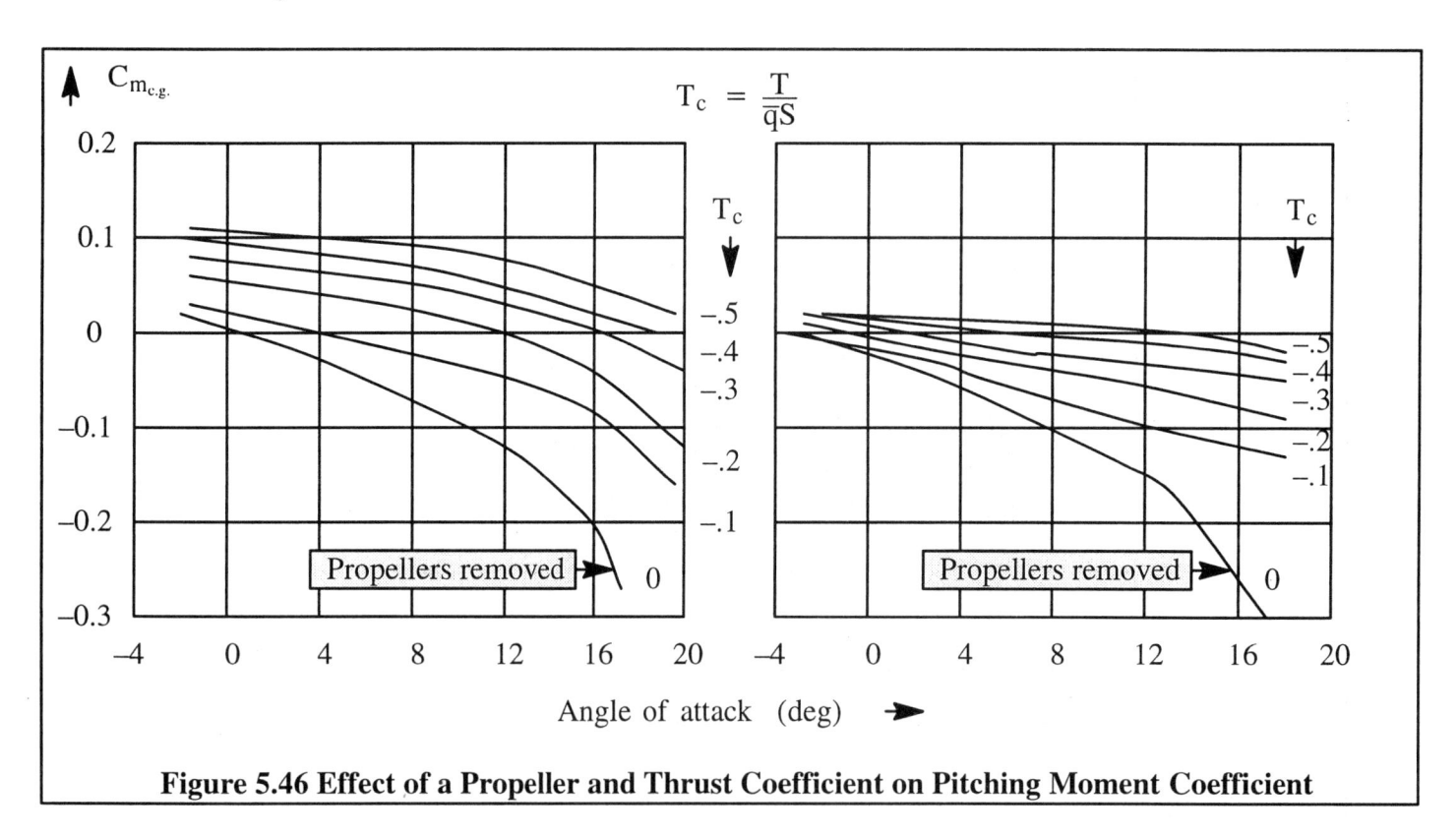

Figure 5.46 Effect of a Propeller and Thrust Coefficient on Pitching Moment Coefficient

5.8 Use the method of Section 5.4 to estimate the drag polar of the following airplanes:

a) Boeing 777   b) Airbus A–330   c) EMB–145
d) Beech/Raytheon 1900   e) EMB Brasilia   f) ATR–42

Hint: determine the wetted areas of these airplanes from Jane's All The World's Aircraft.

5.9 Windtunnel test results for a small jet trainer are shown in Figure 5.47. Determine the drag divergence Mach number (M_{DD}) as a function of lift coefficient, C_L, for the Boeing and Douglas definitions respectively.

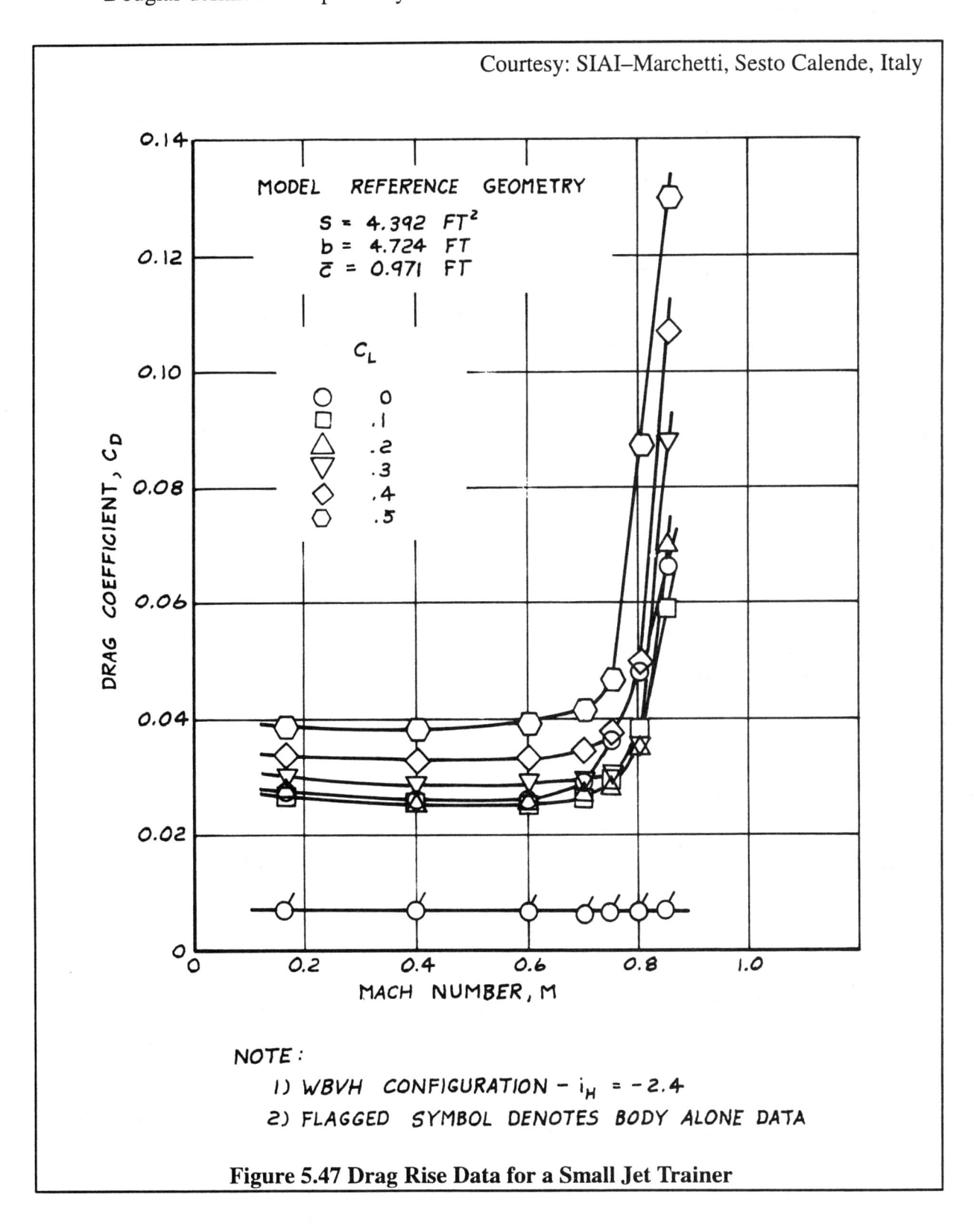

Figure 5.47 Drag Rise Data for a Small Jet Trainer

5.7 REFERENCES FOR CHAPTER 5

5.1 Roskam, J.; Airplane Design, Part VI, Preliminary Calculation of Aerodynamic, Thrust and Power Characteristics; DARCorporation, 120 East Ninth Street, Suite 2, Lawrence, Kansas 66044.

5.2 Hoak, D.E.; USAF Stability and Control Datcom; April 1978, Flight Control Division, Air Force Flight Dynamics Laboratory, Wright–Patterson Air Force Base, Dayton, Ohio.

5.3 Hoerner, S.F.; Fluid Dynamic Drag; Hoerner Fluid Dynamics, Brick Town, N.J., 1965.

5.4 Anderson, S.B.; General Overview of Drag; Page 11 in Proceedings of the NASA–Industry–University General Aviation Drag Reduction Workshop; Edited by Jan Roskam, The University of Kansas, Lawrence, Kansas, 1975.

5.5 Eckert, E.R.G.; Engineering Relations for Friction and Heat Transfer to Surfaces in High Velocity Flow; Journal of the Aeronautical Sciences, Vol. 22, 1955, pp. 585–587.

5.6 Roskam, J.; Airplane Design, Part I, Preliminary Configuration Design and Integration of the Propulsion System; DARCorporation, 120 East Ninth Street, Suite 2, Lawrence, Kansas 66044.

5.7 Roskam, J.; Airplane Design, Part II, Preliminary Sizing of Airplanes; DARCorporation, 120 East Ninth Street, Suite 2, Lawrence, Kansas 66044.

5.8 Ross, R. and Neal, R.D.; Learjet Model 25 Drag Analysis; Page 353 in Proceedings of the NASA–Industry–University General Aviation Drag Reduction Workshop; Edited by Jan Roskam, The University of Kansas, Lawrence, Kansas, 1975.

5.9 Ashley, H. and Landahl, M.T.; Aerodynamics of Wings and Bodies; Addison–Wesley, 1965.

5.10 Nelson, R.L. and Welsh, C.J.; Some Examples of the Application of the Transonic and Supersonic Area Rules to the Prediction of Wave Drag; NACA TN D–446, 1960.

5.11 Van Dam, C.P., Holmes, B.J. and Pitts, C.; Effects of Winglets on Performance and Handling Qualities of General Aviation Aircraft; AIAA Paper 80–???, Presented at the AIAA Aircraft Systems and Technology Meeting, Anaheim, CA, August 4, 1980.

5.12 Munk, M.K.; The Minimum Induced Drag of Aerofoils; NACA Report No. 121, 1921.

5.13 Roskam, J.; Airplane Flight Dynamics and Automatic Flight Controls, Part I; DAR Corporation, 120 East Ninth Street, Suite 2, Lawrence, Kansas 66044, 1995.

5.14 Flechner, S.G. and Jacobs, P.F.; Experimental Results of Winglets on First, Second and Third Generation Jet Transports; NASA TM 72674, May 1978.

5.15 Whitcomb, R.T.; A Design Approach and Selected Wind–Tunnel Results at High Subsonic Speeds for Wing–tip Mounted Winglets; NASA TN D–8260, July 1976.

5.16 Corning, G.; Supersonic and Subsonic CTOL and VTOL Airplane Design; Published by the author, Box 14, College Park, Maryland, 1976.

5.17 Roskam, J.; Airplane Design, Part IV, Layout Design of Landing Gear and Systems; DARCorporation, 120 East Ninth Street, Suite 2, Lawrence, Kansas 66044.

5.18 Proceedings of the NASA–Industry–University General Aviation Drag Reduction Workshop; Edited by Jan Roskam, The University of Kansas, Lawrence, Kansas, 1975.

5.19 Rae, W.H., Jr. and Pope, A.; Low–Speed Wind Tunnel Testing; John Wiley and Sons, 1984.

5.20 Pope, A. and Goin, K.L.; High–Speed Wind Tunnel testing; John Wiley and Sons, 1965.

5.21 Paterson, J.H., MacWilkinson, D.G. and Blackerley, W.T.; A Survey of Drag Prediction Techniques Applicable to Subsonic and Transonic Aircraft Design; Paper No. 1 in AGARD CP–124, Aerodynamic Drag, 1973.

5.22 Bowes, G.M.; Aircraft Lift and Drag Prediction and Measurement; Paper in AGARD LS–67, Prediction Methods for Aircraft Aerodynamic Characteristics, 1974.

5.23 Brown, C.E. and Chen, C.F.; An Analysis of Performance Estimation Methods for Aircraft; NASA CR–921, November 1967.

5.24 Anon.; Advanced Aircraft Analysis Software; DARCorporation, 120 East Ninth Street, Suite 2, Lawrence, Kansas 66044, 1996 (See also Appendix B).

5.25 Feagin, R.C. and Morrison, Jr., W.D.; Delta Method, an Empirical Drag Buildup Technique; NASA CR–151971, December 1978.

Courtesy: Embraer

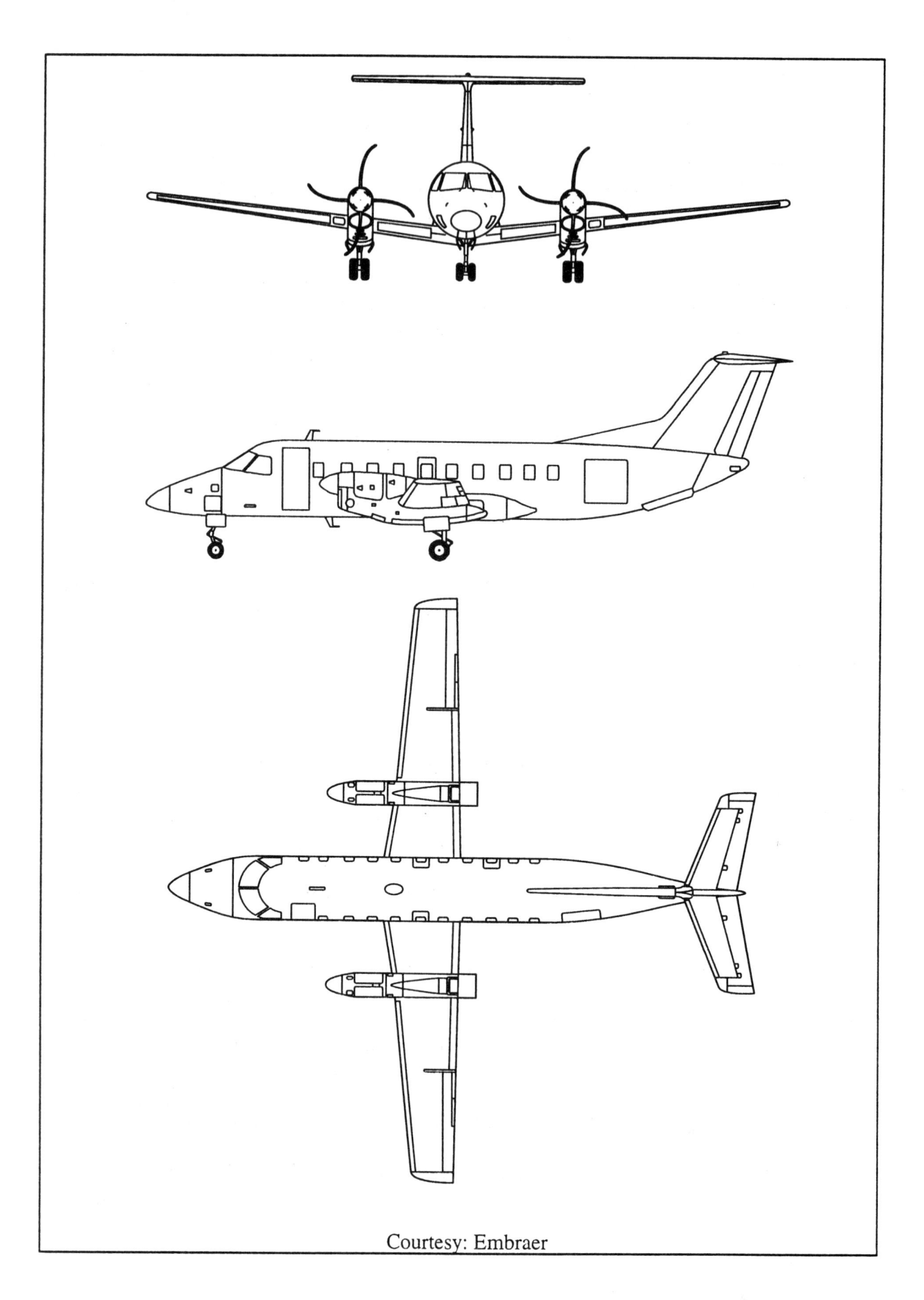

CHAPTER 6: AIRPLANE PROPULSION SYSTEMS

The objective of this chapter is to provide the reader with fundamental insight into the most important propulsion systems used in airplanes. The prevailing types of engines used in airplanes are: piston (reciprocating or rotary) engines, turbojets, turbofans and turboprops. Reasons for this are mostly: acquisition cost, operational cost, fuel availability, installed weight and drag of the integrated propulsion system.

Broadly speaking, airplane propulsion systems can be classified into four categories:

6.1 Piston engines   6.2 Jet engines
6.3 Rocket engines   6.4 Electric motors

These four engine types are discussed in the corresponding sections. Operating cycles, thermodynamic operating principles and typical engine arrangements are also presented only for piston engines and jet engines. A brief overview of the contents of each section is given next.

6.1 Piston engines

Examples are: Piston engines with Otto or Diesel cycles which normally drive propellers.

The invention of the modern four–stroke piston gasoline engine is usually attributed to August Otto and Eugen Langen who built such an engine in 1876. Otto cycle engines require an electric spark to initiate combustion. In Diesel engines ignition is triggered by the high temperature associated with compression of a gas. The Diesel engine was developed by Rudolf Diesel in 1897. The first practical aeronautical application of the piston engine was in the Wright Flyer of 1903. Diesel engines were used extensively on German airplanes during WWII. Figure 6.1 shows six example applications of piston engines in airplanes.

Several important characteristics of four–stroke and two–stroke piston engines are discussed in Sub–sections 6.1.1 and 6.1.2 respectively. Types of layouts used for these engines are summarized in Sub–section 6.1.3. Examples of how engine shaft horsepower and s.f.c. (specific fuel consumption) vary with altitude and engine r.p.m. (rotations per minute) are given in Sub–section 6.1.4. Various factors which affect engine power are discussed in Sub–section 6.1.5.

Performance charts (or power diagrams) and their use are reviewed in Sub–section 6.1.6.

Some cooling and cooling drag considerations are given in Sub–section 6.1.7. Finally, approximate scaling laws involving the variation of engine size, and weight with required power output are provided in Sub–section 6.1.8.

6.2 Jet engines

Examples are: turbojets, turbofans, un–ducted fans, turboprops, pulse–jets and ram–jets.

The turbojet engine was first developed along separate lines in Germany (Hans Pabst von Ohain) and in England (Sir Frank Whittle) before and during WWII. The first operational jet fighters using turbojets were the Messerschmitt 162 and the Gloster Meteor. Figures 6.2a and 6.2b each show six example applications of jet engines in airplanes.

The basic operation of turbojets is covered in Sub–section 6.2.1. Characteristics of By–pass engines, turbofans and turboprops are discussed in Sub–sections 6.2.2 and 6.2.3 respectively. Sub–section 6.2.4 shows typical mechanical arrangements used in turbine engines.

Thrust, power and fuel consumption trends with altitude and speed for these engines are discussed in Sub–section 6.2.5.

Cooling and noise considerations are presented in Sub–sections 6.2.6 and 6.2.7 respectively. Example thrust reversers are shown in Sub–section 6.2.8. Approximate scaling laws involving the variation of engine size, and weight with required thrust output are provided in Sub–section 6.2.9. Pulse–jets and ram–jets are briefly discussed in Subsections 6.2.10 and 6.2.11.

6.3 Rocket engines

Several early high speed airplanes such as the Bell X–1 series and the Messerschmitt 163 used liquid rocket engines. Solid rocket engines (in airplane applications) have been limited to JATO (Jet Assisted Take–off) applications. The DeHavilland Comet used JATO rockets to assist in takeoff from very high altitude fields. Figure 6.3 shows four example applications of rocket engines.

Because of the very high specific fuel consumption, rocket engines are unlikely to be applied to vehicles which must fly at Mach numbers of less than roughly 4. Above that, in hypersonic applications certain forms of rocket engines will probably be used. For these reasons, rocket engines have not been given much emphasis in this text.

6.4 Electric motors

Although still a rarity, electric (engines) motors are being used to drive propellers in some applications. Despite the energy storage problems associated with electrical systems, they are attractive because of their quiet operation and small installation profile. Figure 6.4 shows two example applications of electric motors to airplanes. Because of the paucity of applications at the time of publication, these powerplants have also not been given much emphasis.

The reader is encouraged to find the names corresponding to the airplanes in Figures 6.1 – 6.4.

It is important to be aware of typical speed–altitude operating ranges of the various forms of aircraft propulsion. Figure 6.5 shows typical areas of application for these propulsion systems in relation to airplane flight envelopes.

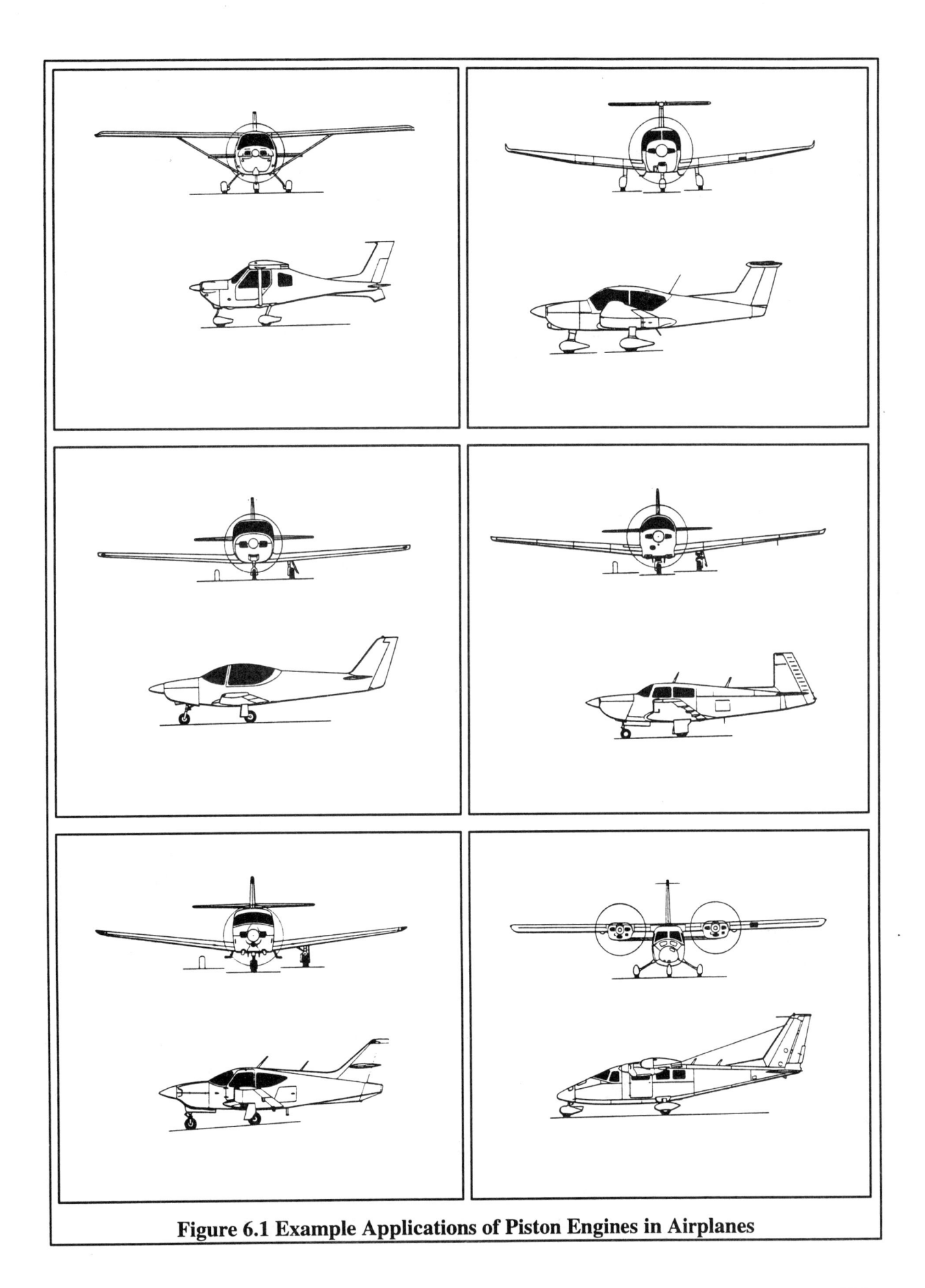

Figure 6.1 Example Applications of Piston Engines in Airplanes

Figure 6.2a Example Applications of Jet Engines in Airplanes

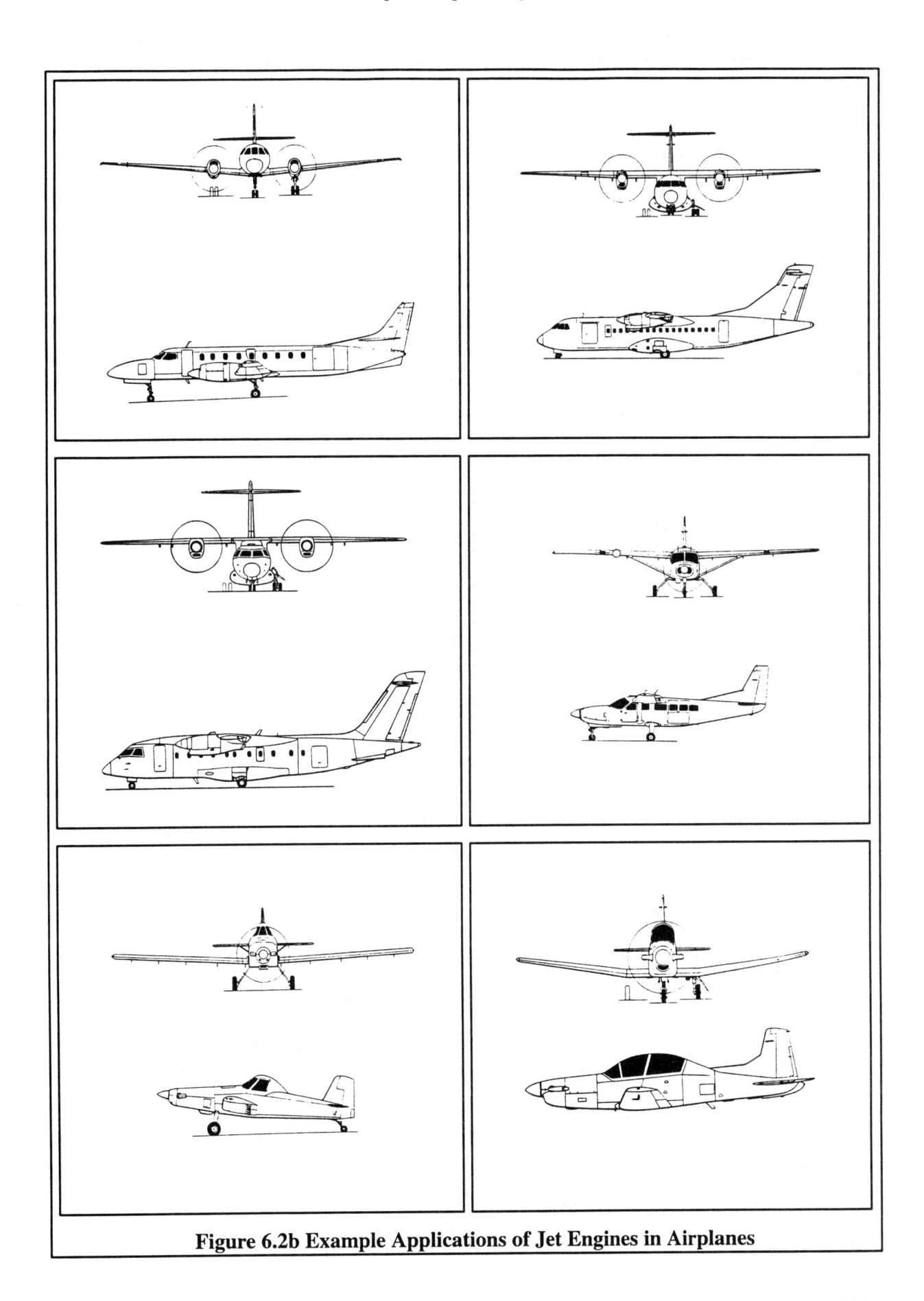

Figure 6.2b Example Applications of Jet Engines in Airplanes

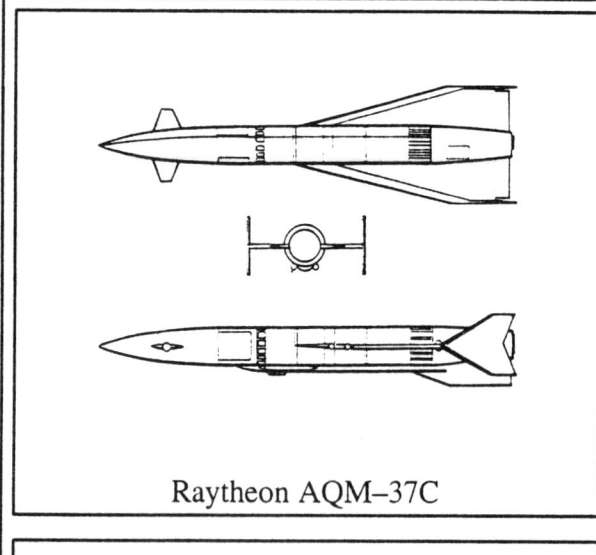

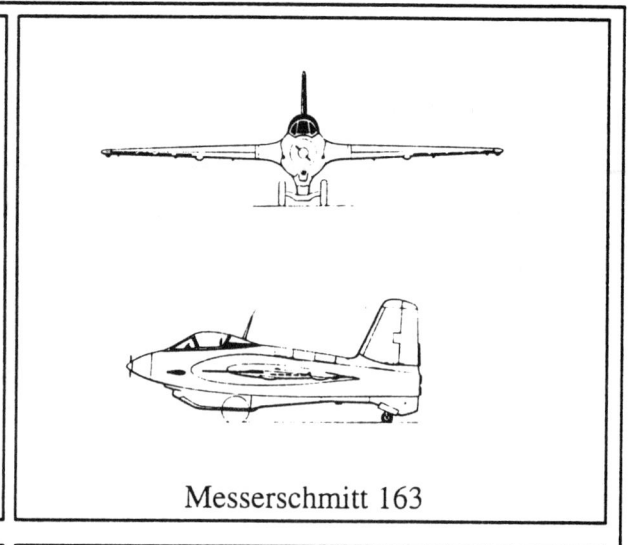

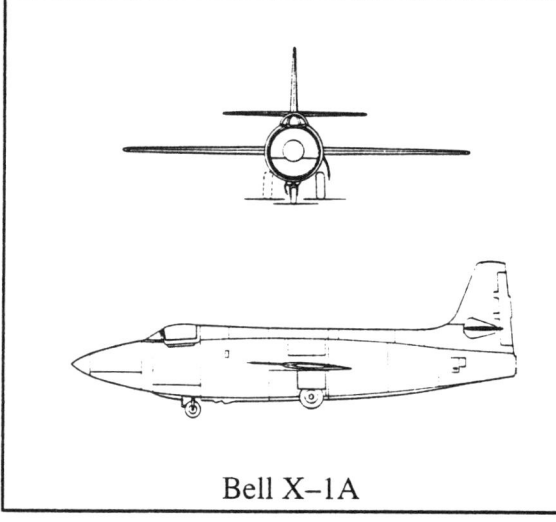

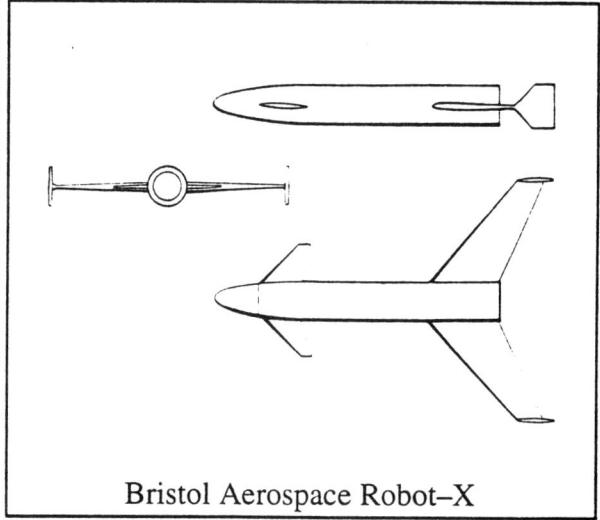

Raytheon AQM–37C	Messerschmitt 163
Bell X–1A	Bristol Aerospace Robot–X

Figure 6.3 Example Applications of Rocket Engines in Airplanes

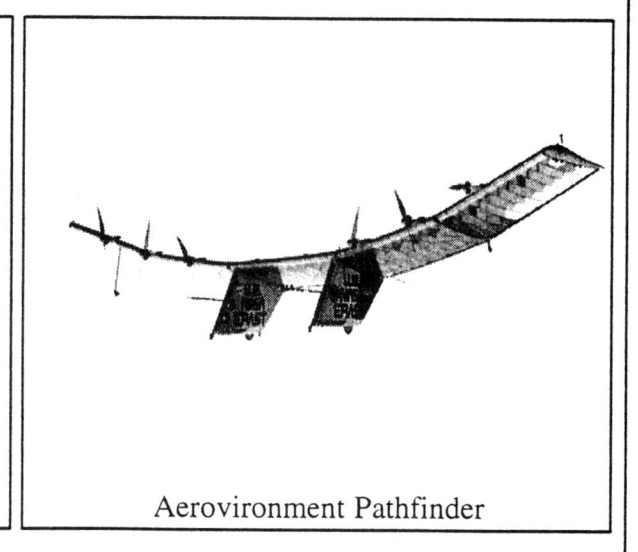

Aerovironment FQM-151A Pointer	Aerovironment Pathfinder

Figure 6.4 Example Applications of Electric Motors in Airplanes

Airplane Propulsion Systems

Note: The flight envelopes are intended to show approximate envelopes only.

Figure 6.5 Aircraft Propulsion Systems in Relation to Flight Envelopes

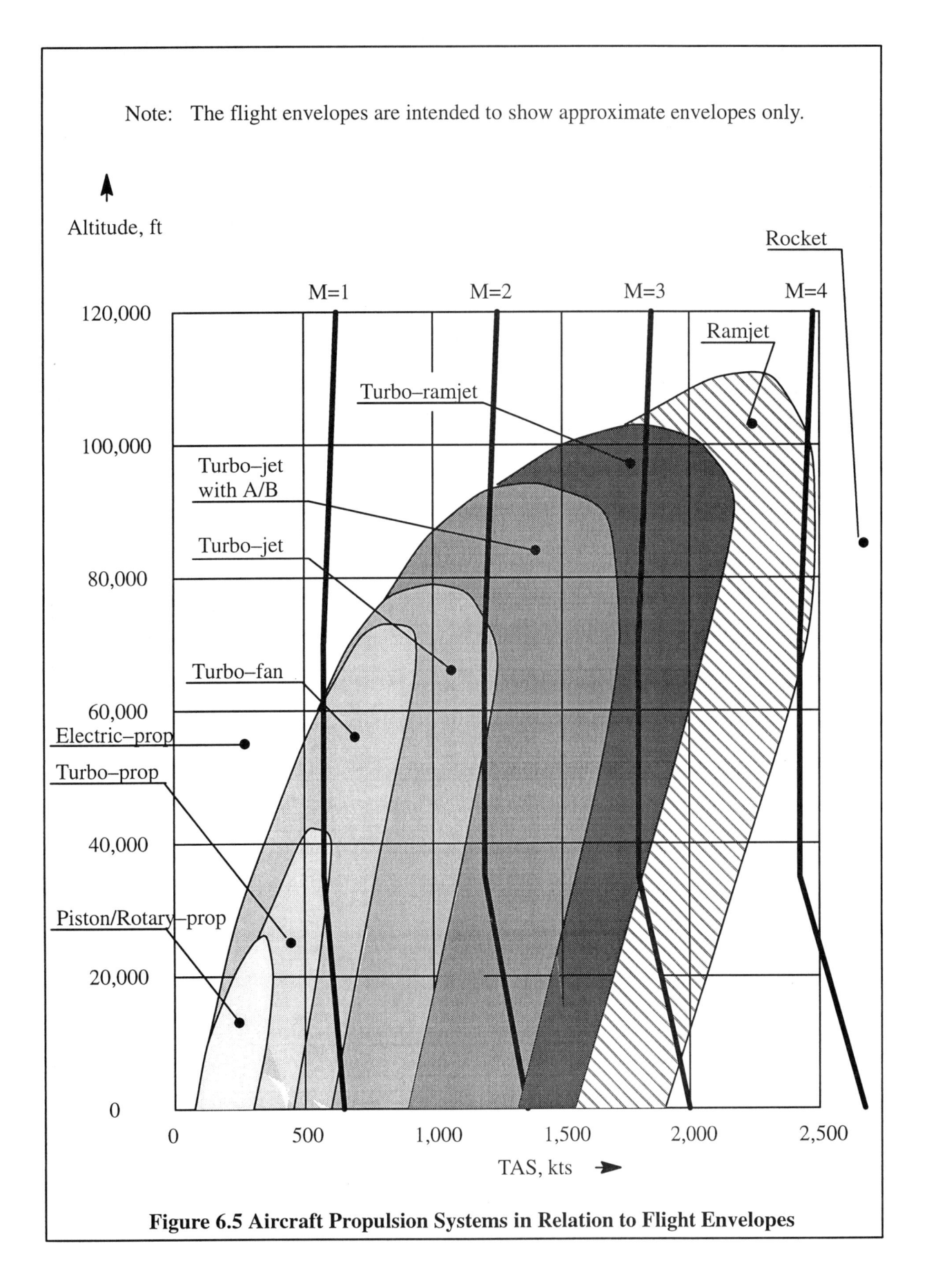

Chapter 6 207

6.1 PISTON ENGINES

In piston engines, the combustion process is intermittent as opposed to continuous. Piston engines are normally configured as four–stroke, two stroke or rotary engines using either spark ignition or compression ignition. The latter are more commonly referred to as Diesel engines. References 6.1–6.4 cover piston engines in some detail.

Figure 6.6 shows a conceptual cross–section of a four-cylinder piston engine which drives a propeller directly from the crankshaft. This is referred to as direct drive as opposed to geared drive, where the crankshaft would drive the propeller via a gear box.

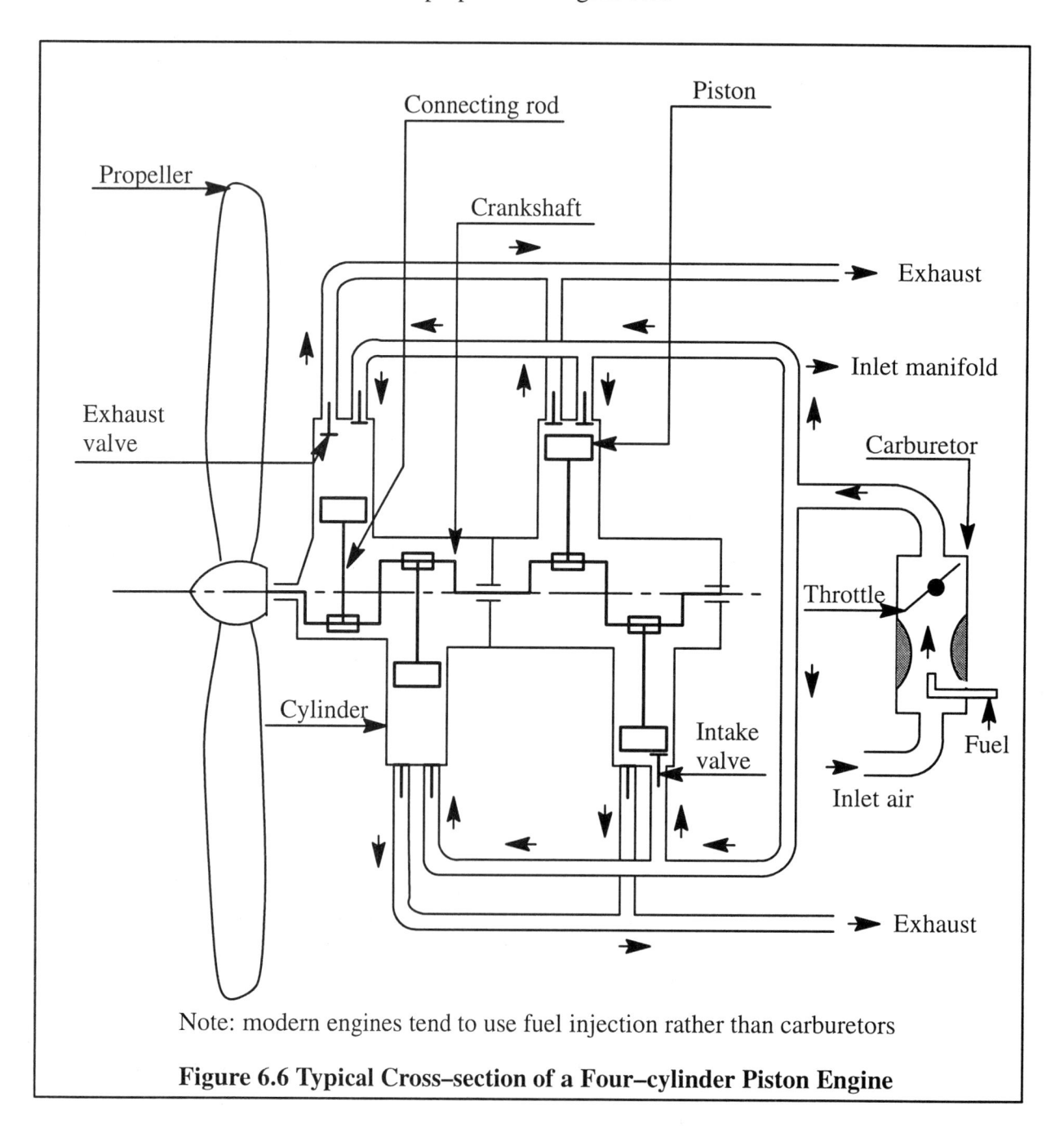

Note: modern engines tend to use fuel injection rather than carburetors

Figure 6.6 Typical Cross–section of a Four–cylinder Piston Engine

The four pistons are connected to the crankshaft with a connecting rod. Each piston moves back and forth inside a cylinder with two valves: an inlet valve and an exhaust valve. The inlet valve opens and closes as dictated by a so-called cam-shaft (not shown) which is driven off the crankshaft. When the inlet valve opens it admits a fuel-air mixture which in turn comes from the inlet manifold. The inlet manifold is fed by a carburetor which is controlled by a throttle. In modern engines the fuel is injected directly into the cylinder without passing through a carburetor (fuel injection).

Next, the basic operating principles of four-stroke versus two-stroke engines will be discussed.

6.1.1 THE FOUR-STROKE PISTON ENGINE

Figure 6.7 shows a schematic diagram of one cylinder of a four-stroke piston engine which is the most common type of piston engine for airplanes. Four different strokes are required in each cylinder for this engine to provide power to the crankshaft.

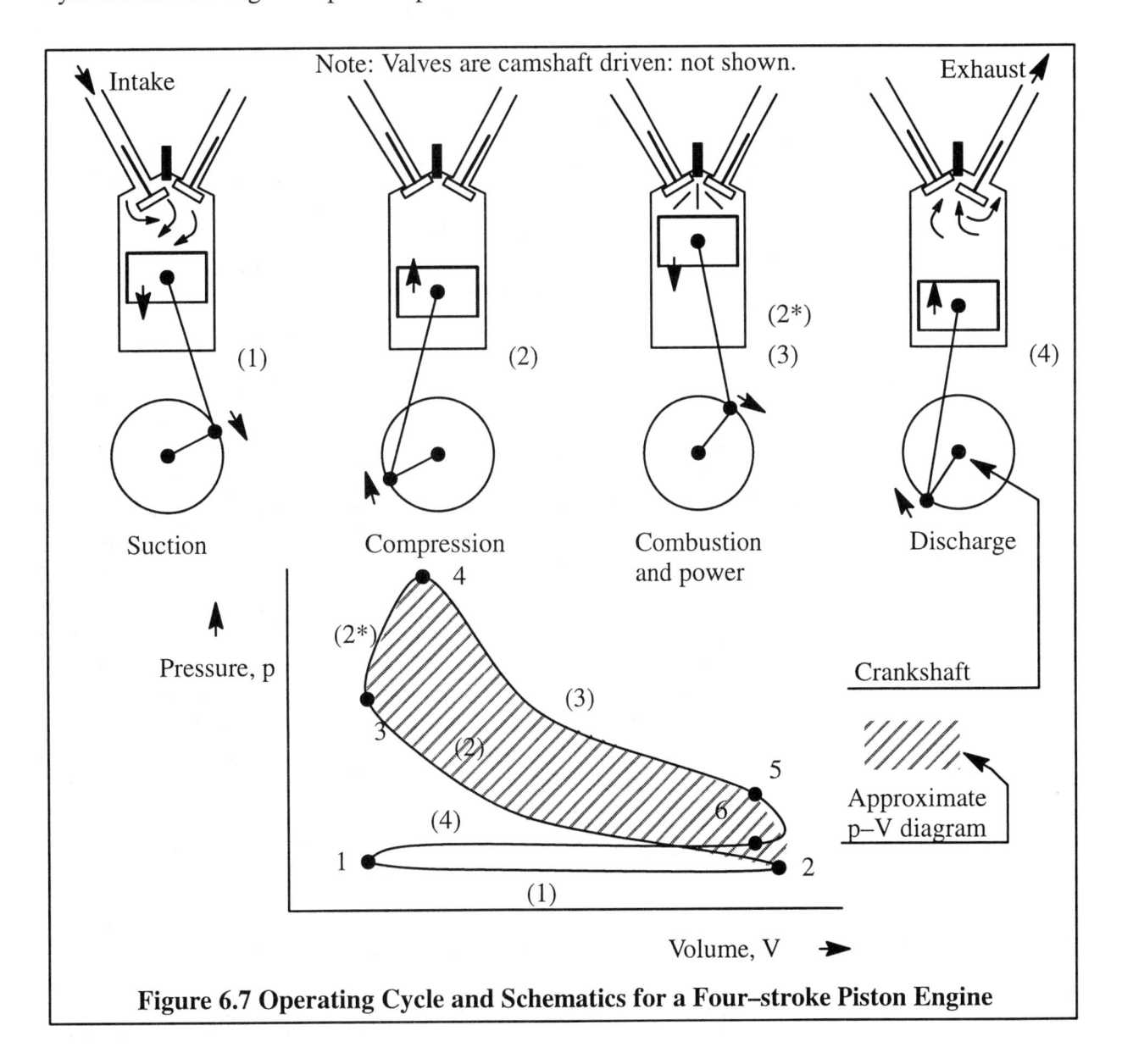

Figure 6.7 Operating Cycle and Schematics for a Four-stroke Piston Engine

In a four–stroke engine, during the suction (or intake) stroke (1) the piston moves from the top of the cylinder toward the bottom. A fuel–air mixture is sucked into the cylinder through the intake valve which is in the open position. Note from the p–V diagram in Figure 6.3, that the pressure remains roughly constant but the volume increases in going from point 1 to point 2. During the compression stroke (2) the piston moves from the bottom to the top of the cylinder. Both valves remain closed and as a result, the pressure greatly increases to that at point 3. Then, the fuel–air mixture is ignited and combustion takes place at roughly constant volume (2*). Therefore, the pressure increases to that at point 4. Next, the burning mixture expands (roughly) isentropically. As a consequence, the piston is forced to move downward during the combustion and power stroke (3) to the point labelled 5. At that point the exhaust valve opens and the pressure drops to that at point 6 which is roughly equivalent to that at point 2. Finally, during the discharge (exhaust) stroke (4), the piston moves upward and drives the combustion products out of the cylinder through the open exhaust valve. During this stroke, the volume decreases from point 6 to point 1 and the pressure stays roughly the same.

Because of the proximity of points 6 and 2, the p–V cycle is often simplified to that of the shaded area in Figure 6.7.

Observe that in a four–stroke engine the crankshaft turns twice over 360 degrees during the four strokes. In most engines, the crankshaft drives a so–called camshaft (not shown in Figures 6.6 or 6.7) which in turn actuates the valves. For a more detailed description of the four–stroke piston engine see References 6.1–6.4.

6.1.2 THE TWO–STROKE PISTON ENGINE

Figure 6.8 shows a schematic of a two–stroke piston engine. First, consider the compression event, item a). In this case, the piston is moving up and compresses a fuel–air mixture above the piston. At the same time a low pressure is created in the crankcase below the piston. This low pressure opens the spring–loaded valve and admits a new fuel–air mixture. When the piston is at the top the valve closes and the fuel–air mixture on top of the piston is ignited. This results in what is called the ignition and power event which is considered next. During this event the piston is driven down. As the piston moves down, because of the relative location of the cylinder exhaust port, it is opened. This allows most of the burned mixture to escape through the exhaust in a process known as scavenging. As the piston continues down, the inlet port is opened next and the new fuel–air mixture is sucked into the cylinder. Together, these actions are called the exhaust and intake event, which is labelled c) in Figure 6.8.

Both the four–stroke and two–stroke engines types discussed so–far use spark ignition to achieve combustion. Another method of achieving ignition is to use compression ignition. This is used in Diesel engines. In Diesel engines compression of the fuel–air mixture is large enough to cause a rise in temperature which automatically triggers ignition. Diesel engines were used on a number of German airplanes during WWII. A major advantage of such engines is their ability to burn many different types of fuel, including certain oils. Despite many advantages, Diesel engines have not found a significant niche in aviation. A good summary of potential Diesel engine applications is given in Reference 6.5.

Figure 6.8 Operating Cycle and Schematics for a Two-stroke Piston Engine

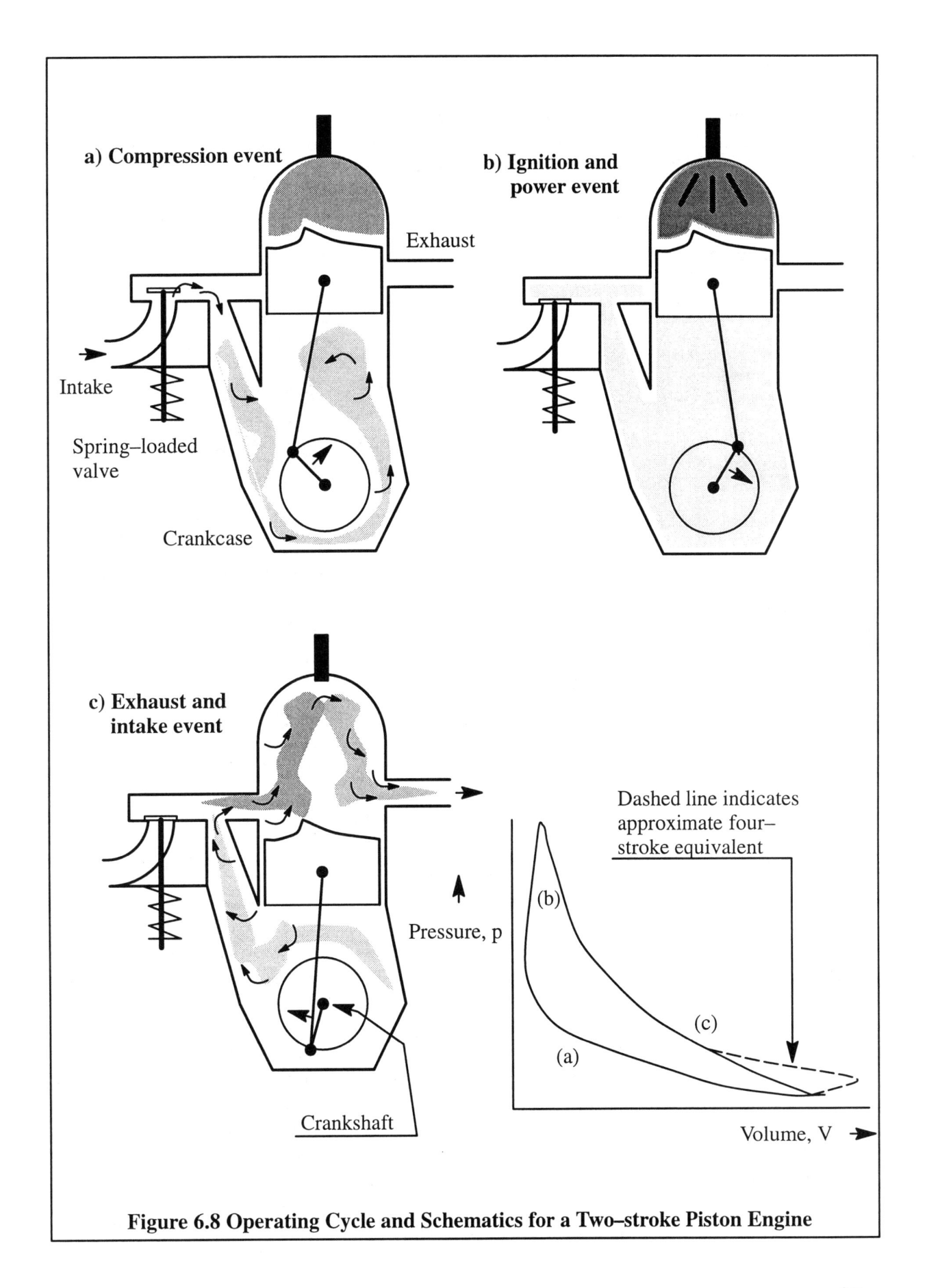

6.1.3 GEOMETRIC LAYOUT OF PISTON ENGINES

Piston engines can be configured in many different geometric layouts. Examples are: horizontally opposed cylinder, radial engines, in–line (upright and inverted), V and W. Figure 6.9 illustrates four of these forms with current example applications.

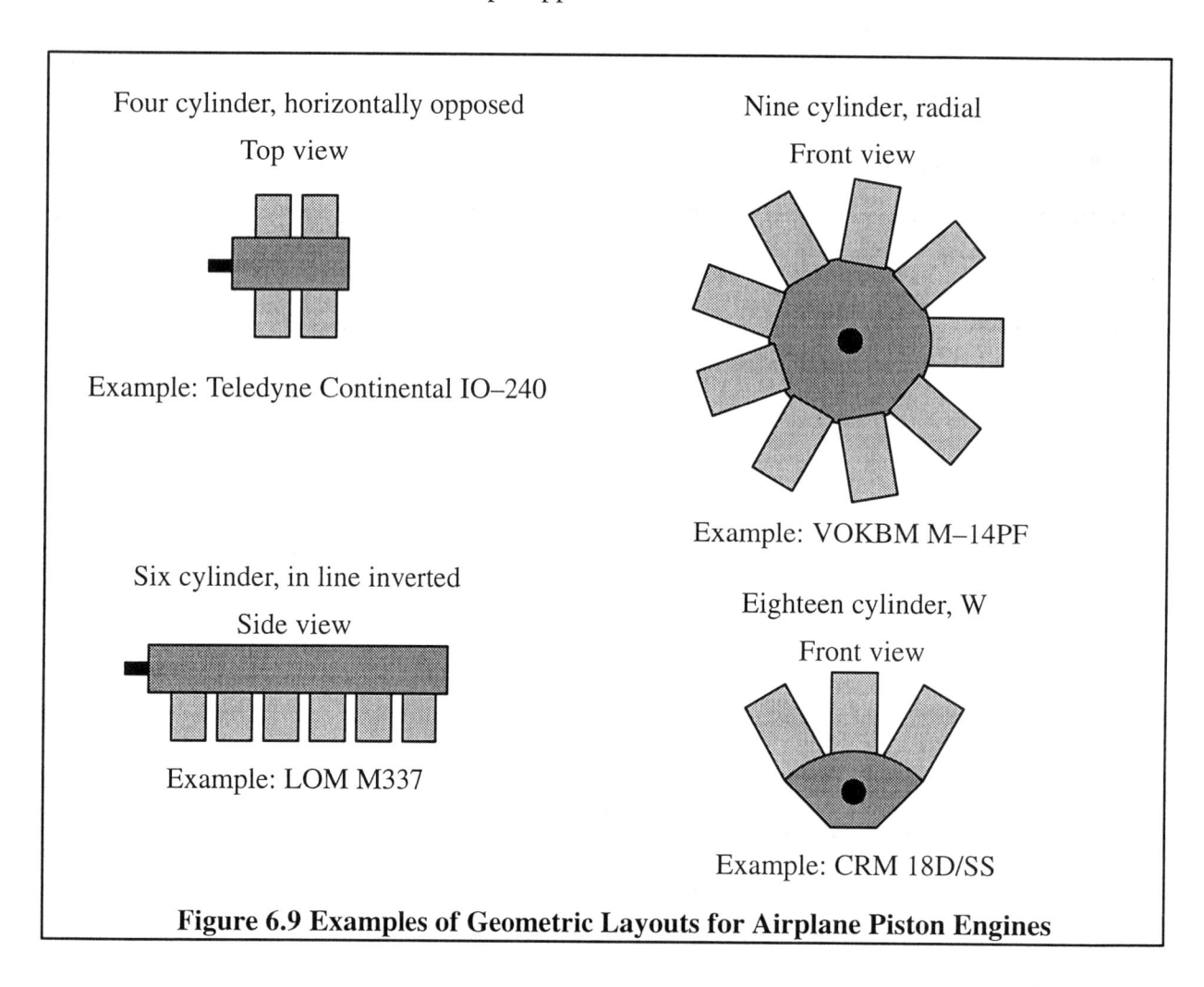

Figure 6.9 Examples of Geometric Layouts for Airplane Piston Engines

Four cylinder, horizontally opposed — Top view — Example: Teledyne Continental IO–240

Nine cylinder, radial — Front view — Example: VOKBM M–14PF

Six cylinder, in line inverted — Side view — Example: LOM M337

Eighteen cylinder, W — Front view — Example: CRM 18D/SS

The horizontally opposed cylinder form is the most widely used for light airplanes. In the past, radial engines and in–line engines were popular in larger and in high performance airplanes. Radial engines are still used in many agricultural airplane applications. In principle, any geometric layout can be used for either four–stroke, two–stroke, spark ignition or compression ignition.

Piston engines are limited to relatively low performance airplanes: flight Mach numbers up to around 0.6. By turbo–charging they can be operated at considerable altitudes albeit at a significant penalty in weight and complexity. At low flight speeds a major advantage of the piston engine is its low specific fuel consumption. A disadvantage is the relatively low mean time between overhaul (MTBO) which is typically of the order of several hundreds of hours. Turbine engines on the other hand have MTBO values of thousands of hours.

6.1.4 POWER OUTPUT AND FUEL EFFICIENCY OF PISTON ENGINES

In the following some important terminology used with piston engines is introduced. In addition, formulas from which power available and fuel efficiency can be determined will be given.

The compression ratio of a piston engine is defined as the ratio of the cylinder volume with the piston at the bottom to that with the piston at the top. Compression ratios range from around 6 in spark ignition engines to around 16 in compression ignition engines.

The power delivered to the output shaft (crankshaft) is referred to as the brake–horse–power (BHP), also called the shaft–horse–power (SHP), P_{shp}. If the output shaft drives a propeller which provides a thrust, T at a speed, V then the product TV is defined as the power available, P_{av}. The ratio of power available to shaft–horse–power is called the propeller efficiency, η_p:

$$\eta_p = \frac{TV}{SHP} = \frac{P_{av}}{P_{shp}} \tag{6.1}$$

Typical values for propeller efficiency range from 0.8 to 0.9 although lower and higher efficiencies are possible. How the efficiency of a propeller is determined by its design characteristics (diameter, blade geometry, number of blades, blade angle and r.p.m.) is discussed in Chapter 7.

The SHP of an engine depends on a number of factors as illustrated in the following equation:

$$ISHP = MEP \; S_{piston} \frac{l_{piston}}{12} \frac{RPM}{2 \times 60 \times 550} N_{piston} \tag{6.2}$$

where: ISHP is the so-called indicated shaft horsepower
MEP is the mean effective pressure across a piston head in p.s.i. per cycle
S_{piston} is the cross sectional area of the piston in in^2
l_{piston} is the stroke of the piston in inches
RPM stands for the rotations per minute of the crankshaft
N_{piston} is the number of cylinders (pistons) in the engine

Because of various mechanical losses, the actual shaft horsepower, SHP is less than the indicated shaft–horsepower, ISHP. It is customary to rewrite Eqn (6.2) as:

$$SHP = \eta_{mech} ISHP = \frac{\eta_{mech} MEP \; V_{cyl} \; RPM \; N_{piston}}{792,000} \tag{6.3}$$

where: η_{mech} is the mechanical efficiency in converting MEP work to crankshaft work

$V_{cyl} = S_{piston} l_{piston}$ is the useful volume of each cylinder

Airplane Propulsion Systems

The thermal input, T.I. (in BTU/hr) to the engine from the fuel flow can be obtained from:

$$\text{T.I.} = \dot{w}_f \, \text{H.V.} \tag{6.4}$$

where: $\dot{w}_f$ is the fuel flow in lbs/hour
H.V. is the heating value of the fuel in BTU/lbs

Noting that one BTU equals 778 ft–lbs/sec, the mechanical output, M.O. of the engine can be written as:

$$\text{M.O.} = \frac{\text{SHP} \times 550 \times 60 \times 60}{778} = 2{,}545(\text{SHP}) \quad \text{in BTU/hr} \tag{6.5}$$

The thermal efficiency of the engine can now be expressed as:

$$\eta_{th} = \frac{2{,}545 \, (\text{SHP})}{\dot{w}_f \, (\text{H.V.})} = \frac{2{,}545}{\left\{\left(\frac{\dot{w}_f}{\text{SHP}}\right)(\text{H.V.})\right\}} = \frac{2{,}545}{(\text{s.f.c.})(\text{H.V.})} \tag{6.6}$$

where: s.f.c. $= \dfrac{\dot{w}_f}{(\text{SHP})}$ is the specific fuel consumption in lbs of fuel per shaft–horse–power per hour

The overall propulsion system efficiency is therefore:

$$\eta_{overall} = \eta_{th} \, \eta_p \tag{6.7}$$

If the propeller efficiency is, say 87% and the thermal efficiency is, say 30%, the overall efficiency is about 26%.

6.1.5 FACTORS AFFECTING THE POWER OUTPUT OF PISTON ENGINES

Several factors influence the amount of power developed by piston engines. The most important factors are discussed in the following.

1) Heat release per pound of air.

When the heat release is high, the temperature and the pressure of the combustion products in each cylinder will be higher also. The magnitude of heat release depends on the fuel heating value (H.V.) and on the fuel–air ratio (F/A). If F/A is high, the mixture is said to be too rich and combustion may not be complete. If F/A is too low (lean) combustion may not take place. The effect of the fuel–air ratio on a typical p–V cycle is shown in Figure 6.10a.

2) Charge per stroke

The quantity of air (mass) introduced into the cylinder controls the quantity of heat released for a given amount of fuel. The quantity of air depends on the intake pressure also called the manifold

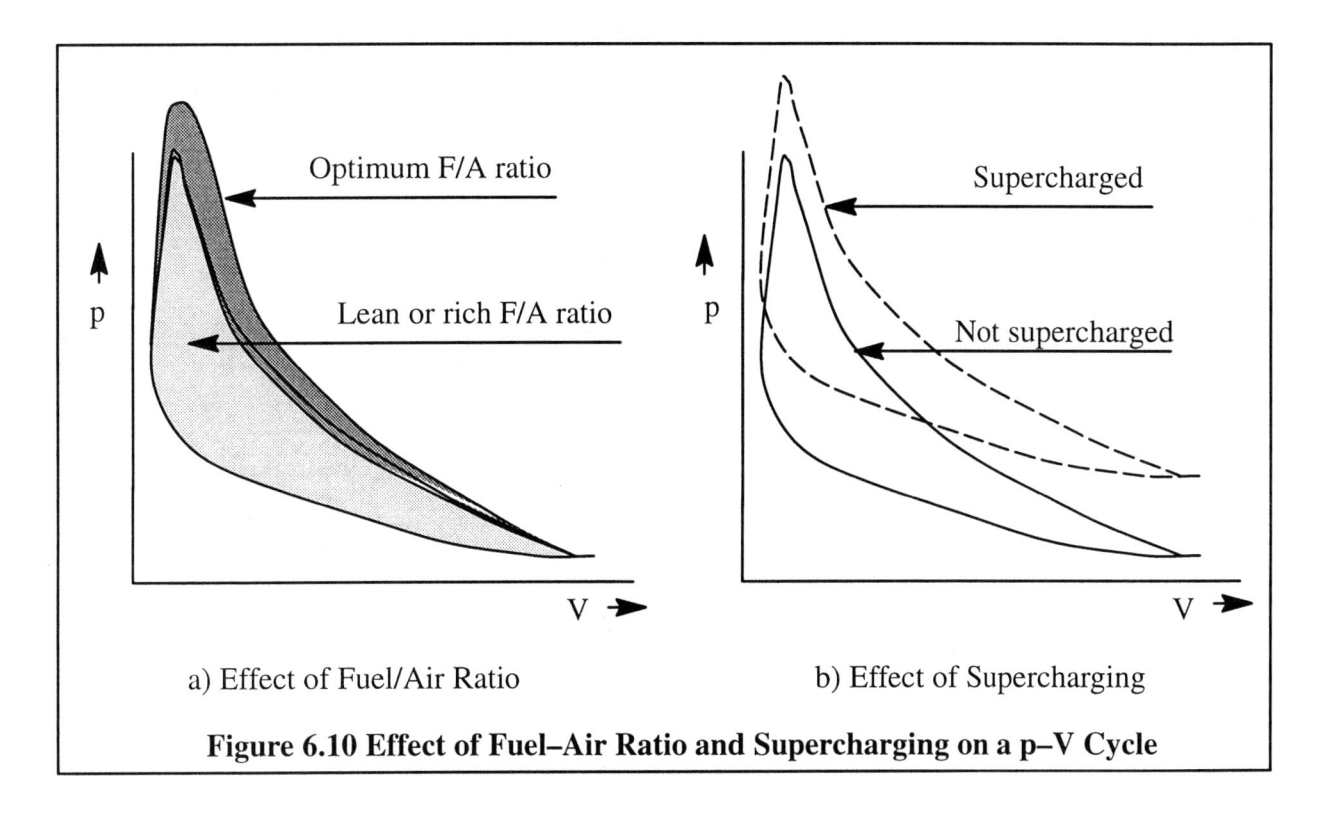

Figure 6.10 Effect of Fuel–Air Ratio and Supercharging on a p–V Cycle

a) Effect of Fuel/Air Ratio

b) Effect of Supercharging

absolute pressure or MAP. The charge per stroke decreases with altitude because of the changes in the atmosphere. The charge per stroke can be increased at any altitude by supercharging. With supercharging the air is compressed by an air–pump (compressor) before being admitted into the cylinders. The effect of supercharging on the p–V diagram is shown in Figure 6.10b.

3) Maximum permissible RPM

As shown by Eqn (6.3), increasing the RPM will increase the power output. Every engine has an upper limit for the RPM because of inertial or structural limitations. Typical maximum RPM values for piston engines are 2,200 to 3,500.

4) Effect of altitude

Complete burning of the fuel in a given F/A mixture requires a sufficient quantity of air. At higher altitudes the density of air is less. Therefore, at any given throttle setting the power output decreases with altitude. It should be noted that the throttle setting determines the power level via the fuel control system. If the throttle is gradually opened at higher altitudes such as to maintain manifold absolute pressure (MAP), more power can be obtained because the reduction in back pressure will tend to increase the airflow and thus the power. Figure 6.11 illustrates the effect of altitude on engine power. An empirical formula for relating power at altitude to power at sea–level is:

$$\frac{SHP_h}{SHP_{h=0}} \approx \left\{(1.132)\frac{\varrho_h}{\varrho_{h=0}} - 0.132\right\} \quad \text{Applies to non} - \text{supercharged engines only} \quad (6.8)$$

where: ϱ_h is the air density in slugs/ft^3 at altitude, h

Airplane Propulsion Systems

Figure 6.11 Effect of Altitude and Supercharging on Engine Power

$$\frac{SHP_h}{SHP_{h=0}} \approx \left\{ (1.132)\frac{\varrho_h}{\varrho_{h=0}} - 0.132 \right\}$$

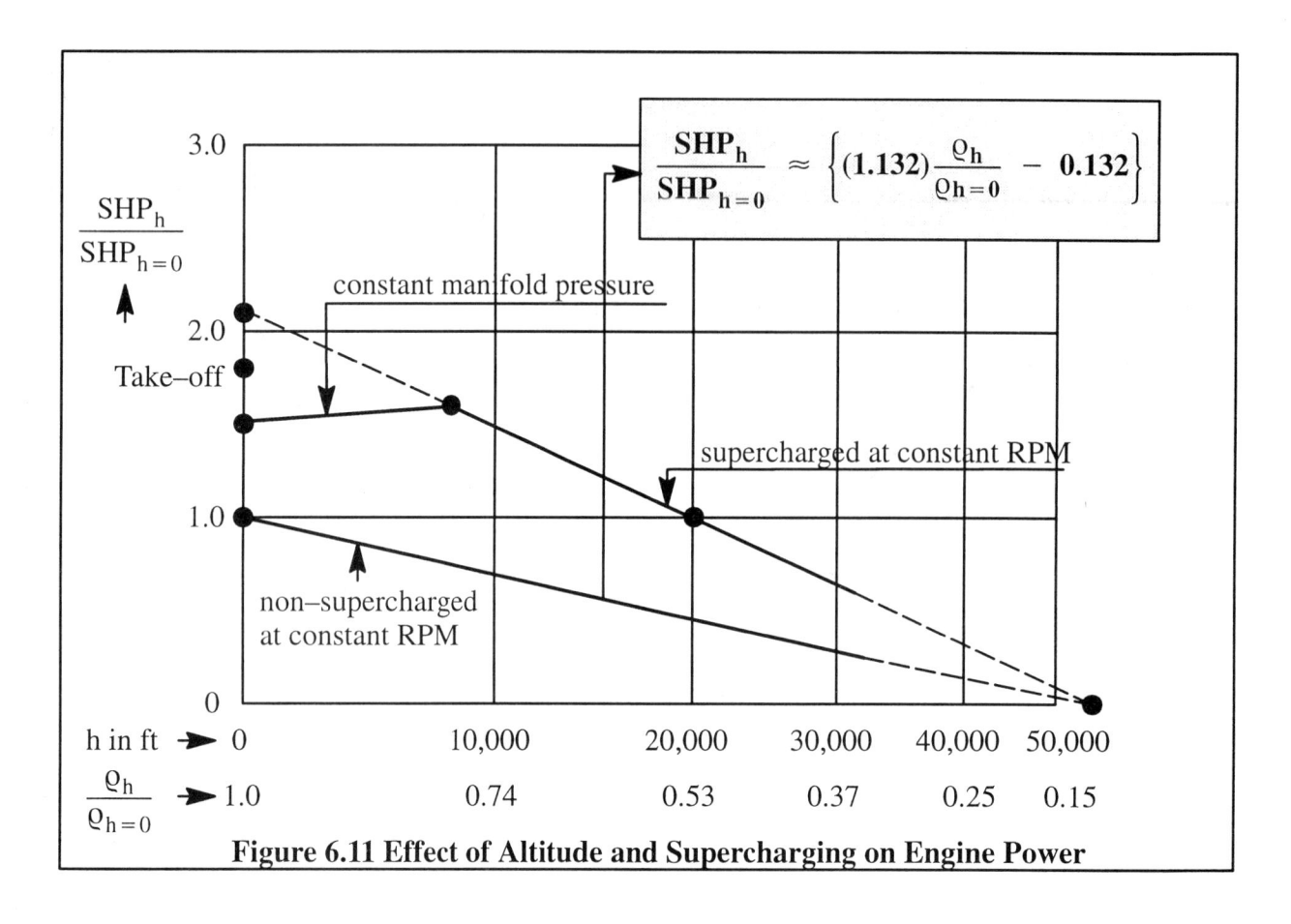

h in ft →	0	10,000	20,000	30,000	40,000	50,000
$\frac{\varrho_h}{\varrho_{h=0}}$ →	1.0	0.74	0.53	0.37	0.25	0.15

5) Effect of Air Temperature

$$\frac{SHP_T}{SHP_{T_s}} \approx \sqrt{\frac{460 + T_s}{460 + T}} \qquad \text{Applies to non}-\text{supercharged engines only} \qquad (6.9)$$

where: SHP_T is the shaft horsepower at the actual air temperature, T

SHP_{T_s} is the shaft horsepower at the standard air temperature, T_s

Evidently at higher air temperatures (hot day) the SHP output of the engine will decrease.

6) Supercharging

In supercharged engines a compressor (usually a centrifugal compressor) driven by either the crankshaft or by a turbine (turbo–super–charger) will increase the MAP and therefore the power output as well. The power used to drive the supercharger directly from the crankshaft (via a gearing system) can amount to 6%–10% of the total power. At low altitudes, the power output may in fact be less than that with a non–supercharged engine. If the power used to drive the supercharger is obtained via a turbine, driven by the exhaust gases, there is no power loss at low altitude. The engine is then said to be flat–rated (i.e. constant SHP up to some altitude).

The effect of supercharging on the p–V diagram is shown in Figure 6.10b. The effect of altitude on the power output of a supercharged engine is also shown in Figure 6.11. In the example, the engine delivers sea–level power at 20,000 ft altitude! A schematic example of a supercharged engine is given in Figure 6.12. High performance general aviation airplanes use supercharged engines.

7) Compounding

If the turbine which is driven by the exhaust gases is also connected to the engine crankshaft so that it provides power to the crankshaft, the arrangement is referred to as compounding. Compound engines were used on the Lockheed L–1049C and DC–7C transports of the early fifties. With the advent of the much simpler (and at high speeds more efficient) turbojets and turbofans, compound engines are no longer a viable alternative.

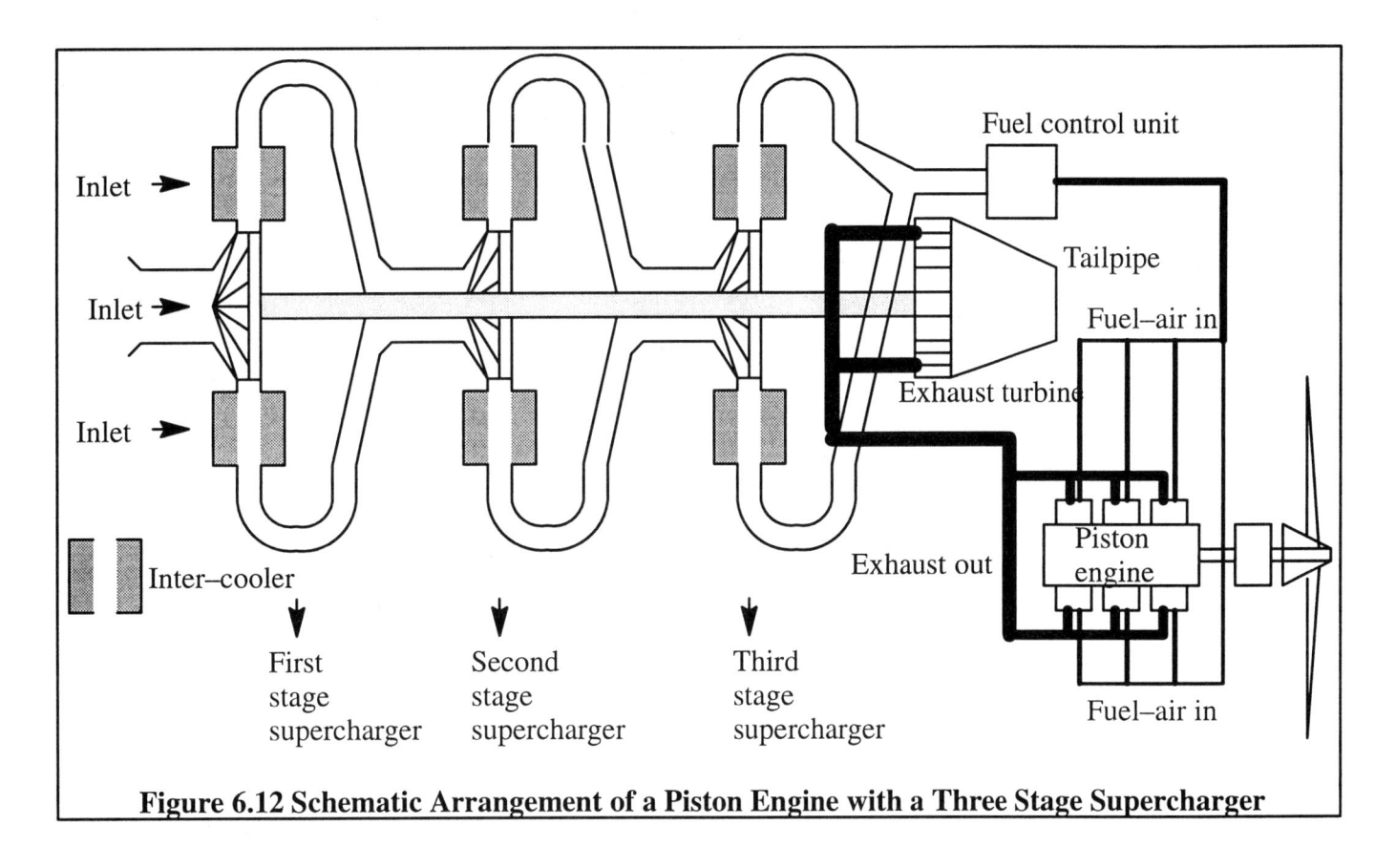

Figure 6.12 Schematic Arrangement of a Piston Engine with a Three Stage Supercharger

6.1.6 PERFORMANCE CHARTS FOR PISTON ENGINES

Piston engine performance is normally presented in the form of a so–called power diagram or performance chart. In these charts the brake–horse–power (BHP) delivered at the shaft is plotted versus engine RPM, altitude and MAP (Absolute Manifold Pressure in in. Hg.). The reader may assume that the notation SHP used so–far and the notation BHP are synonymous.

The power output of piston engines is classified in terms of power ratings. Typically these ratings are as follows:

1) Take–off power: the maximum power allowed during take–off.

2) Military power: the maximum power allowed for a limited period of time. This rating is used in military applications only.

3) Maximum continuous power (or normal rated power). This rating may be used for maximum climb performance and for maximum level speed. This power level is also referred to as METO: maximum except take–off.

4) Cruise power. Typically there are two cruise power ratings:

   a) Performance cruise: 75% of take–off power at 90% of maximum RPM

   b) Economy cruise: 65% of take–off power.

To determine the actual power available in a given flight condition, a numerical example will be given using the performance diagram for a 200 bhp piston engine as shown in Figure 6.13. The corresponding engine fuel flow data are given in Figure 6.14.

Example: Find the BHP available and the fuel flow required at 2,300 RPM and 22 in. Hg MAP for a pressure altitude of 2,500 ft. and an air temperature of 20 degrees Fahrenheit above standard atmospheric temperature. Assume the mixture is set at that for best power.

Solution:

Step 1: Locate point A on the full throttle RPM (2,300) and 22 in. Hg MAP line on the altitude performance side of the diagram in Figure 6.13.

Step 2: At sea–level for the given MAP and engine RPM, locate point B and transfer it to point C.

Step 3: Connect points C and A with a straight line.

Step 4: Find point D at the required pressure altitude of 2,500 ft

Step 5: Project point D on the vertical BHP scale and read the available BHP as 126 hp. This would be the BHP available if the air temperature were standard.

Step 6: Correct the available BHP for the effect of temperature deviation with Eqn (6.9). At 2,500 ft the standard temperature is 50.2 deg. F. Under the prevailing conditions the actual temperature is 50.2 + 20 = 70.2. The actual BHP is therefore:

$$SHP_T = 126 \sqrt{\frac{460 + 50.2}{460 + 70.2}} = 126 \times 0.96 = 121.2 \text{ hp}$$

Step 7: To find the fuel consumption in lbs/hr read the fuel flow chart in Figure 6.14 for 121.2 bhp at 2,300 RPM at the best power setting for the mixture. The resulting fuel flow is 59 lbs/hr.

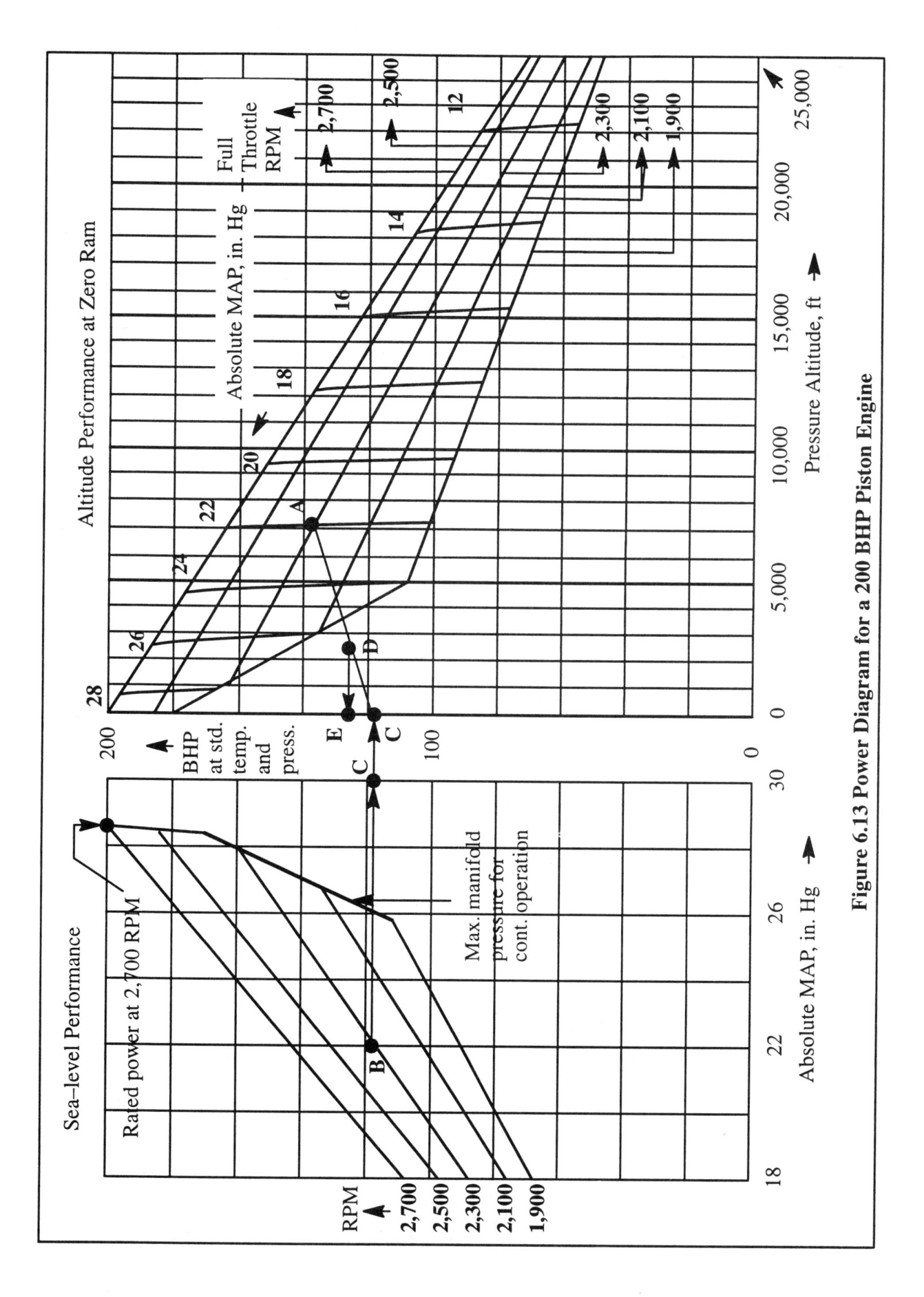

Figure 6.13 Power Diagram for a 200 BHP Piston Engine

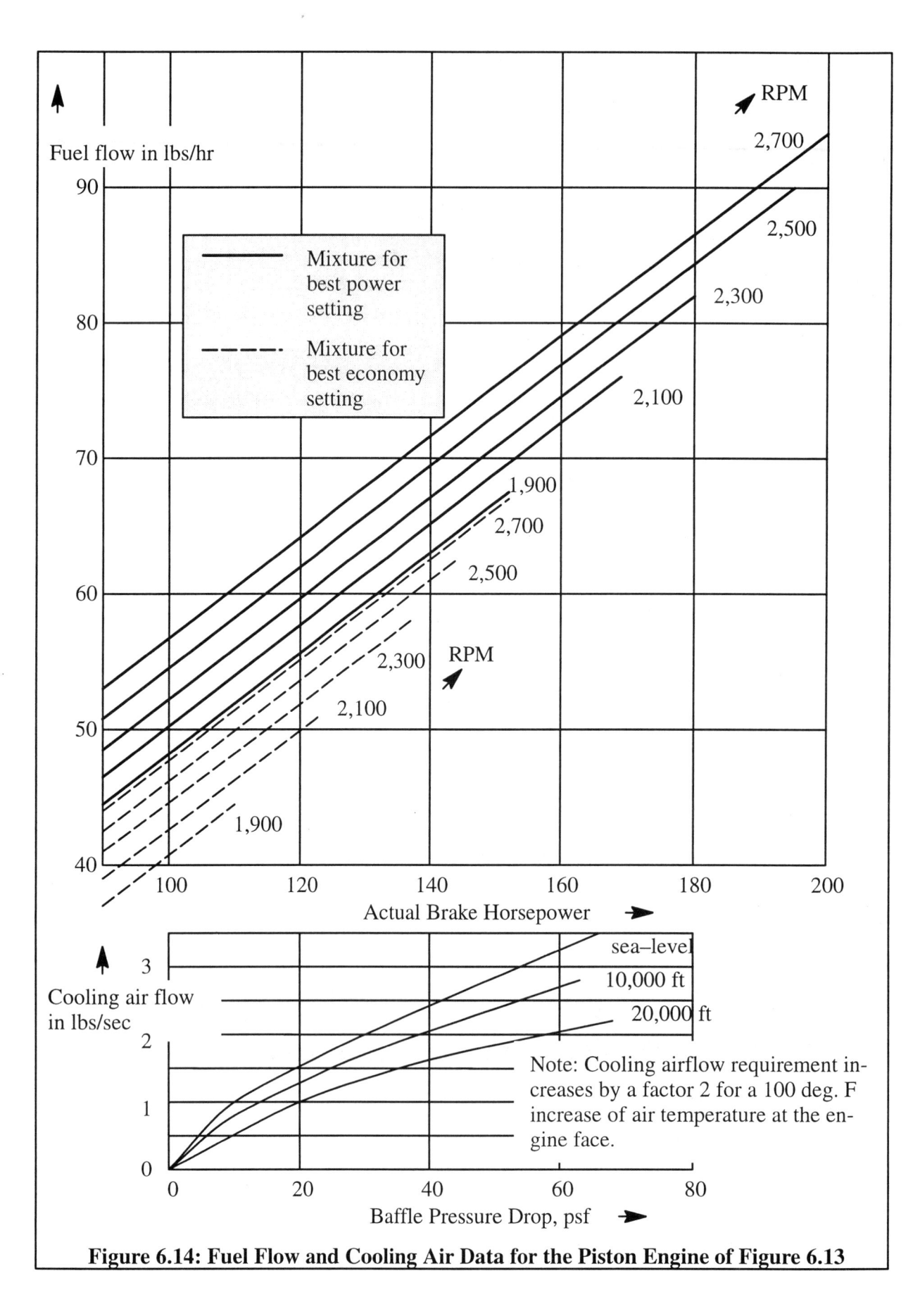

Figure 6.14: Fuel Flow and Cooling Air Data for the Piston Engine of Figure 6.13

6.1.7 COOLING AND INSTALLATION OF PISTON ENGINES

Because of the heat released in the combustion process all piston engines have to be cooled to ensure acceptable engine life and maintenance costs. Two methods of cooling are being used: air cooling and liquid cooling.

In air–cooled engines, ram air is ducted around the cylinder heads for cooling. After passing by the cylinder heads the air is ducted overboard. This air cooling process can create a significant amount of drag. The Townsend ring and NACA cowls of Figure 6.15 were developed to manage the flow of cooling air and reduce the cooling drag of earlier un–cowled, radial engine installations.

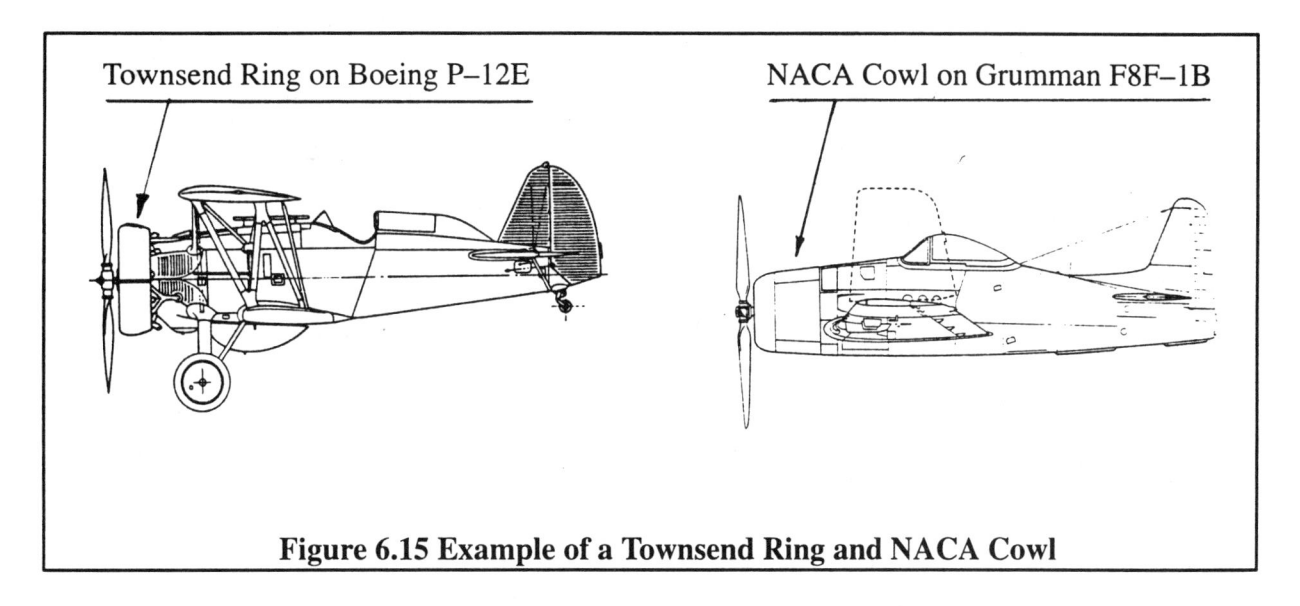

Figure 6.15 Example of a Townsend Ring and NACA Cowl

References 6.6–6.8 provide detailed data on cooling drag associated with air cooled installations. The ducting of cooling air through engine compartments with horizontally opposed engines for tractors and pushers is discussed in Ref. 6.9. Figure 6.16 shows two example arrangements.

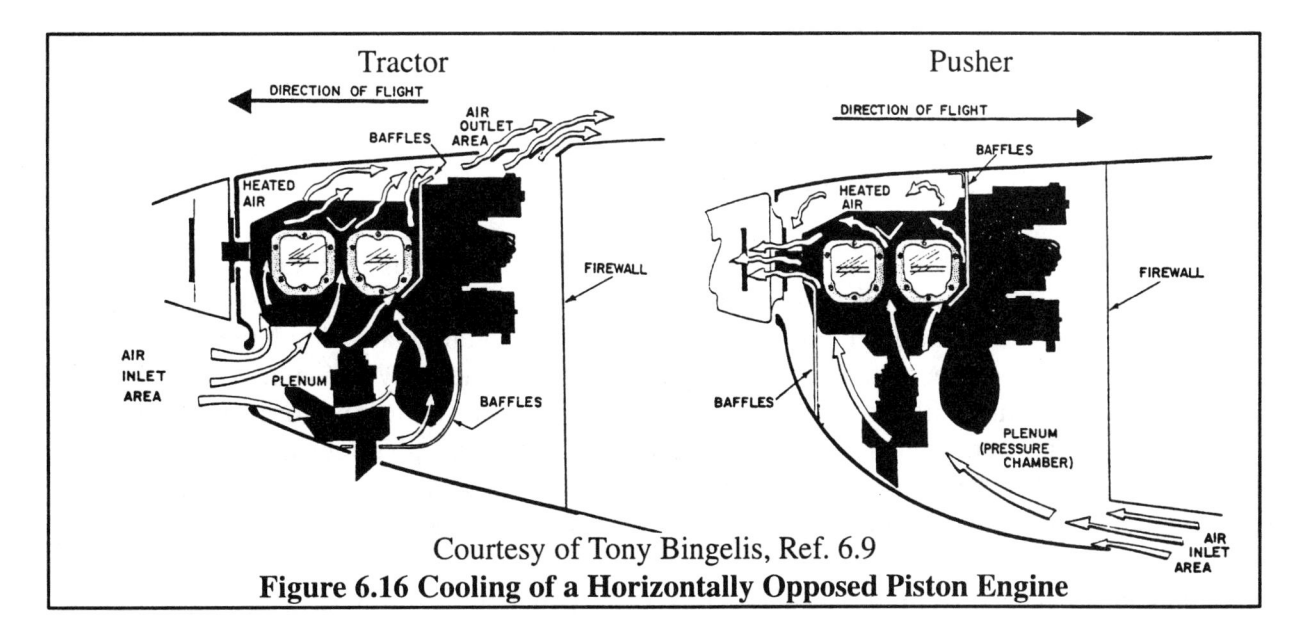

Courtesy of Tony Bingelis, Ref. 6.9
Figure 6.16 Cooling of a Horizontally Opposed Piston Engine

To ensure that the cooling air flows all around the cylinder heads, extensive use is made of so-called baffle plates. Figure 6.17a shows an example of a baffle plate arrangement for a two-row radial engine.

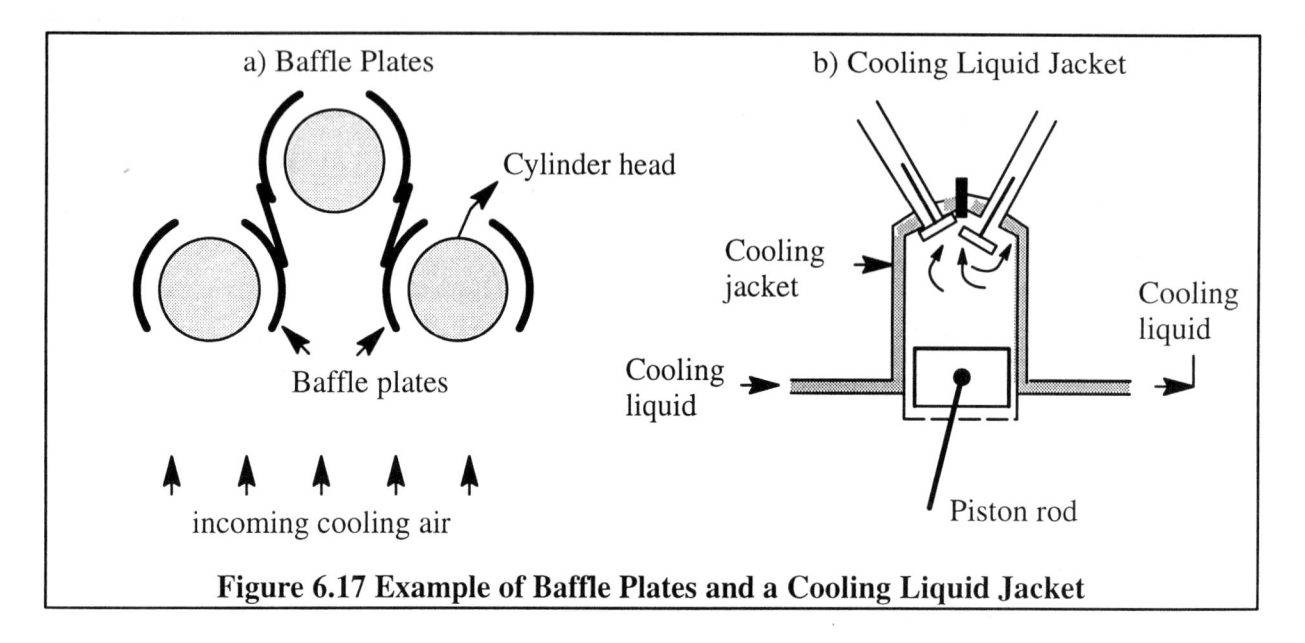

Figure 6.17 Example of Baffle Plates and a Cooling Liquid Jacket

In liquid cooled engines, a liquid is used to cool the critical engine components. An example of a liquid cooling arrangement for one cylinder is shown in Figure 6.17b. The cooling liquid is pumped around the hot engine parts and then sent to a radiator. Cooling air is ducted through the radiator to cool the cooling liquid. A good discussion with data of radiator installations is given in Chapter 9 of Reference 6.6. By properly designing such an installation very little drag penalties have to be paid. In the P-51 example it is said that a net thrust was in fact produced. One must of course take account also of added weight and complexity. Added weight shows up in induced drag and added complexity in maintenance and repair cost.

Methods for estimating cooling drag for air-cooled and liquid cooled engines are contained in Ref. 6.6. A more recent method for estimating cooling airflow requirements, duct sizes and cooling drag is given in Ref. 6.10. An example of specific engine cooling airflow requirements was given in Figure 6.14.

6.1.8 SCALING LAWS FOR PISTON ENGINES

The authors were unable to find quick scaling rules which can be used for the geometric sizing of conventional, gasoline powered piston engines. Such rules would be helpful in establishing the required engine envelope during preliminary design, so that a preliminary installation diagram can be prepared. A study of potential, modern airplane Diesel engines may be found in Ref. 6.5. This reference also contains scaling rules for predicting Diesel engine size and performance parameters for applications different than those discussed in Ref. 6.5. Even though no certified aviation Diesel engines are on the market at the time this book was published, they are very attractive powerplants due to their low s.f.c. and their ability to utilize a variety of low cost fuels.

Some of the scaling rules found in Ref. 6.5 are presented in the following.

For engine weight:
$$W_{wet} = 15.17(SHP_{t.o.})^{0.610} \qquad (6.10)$$

where: W_{wet} is the weight in lbs of the engine, including accessories and fluids in the lines

$SHP_{t.o.}$ is the shaft horsepower required for take-off

For engine specific fuel consumption at sea-level and take-off:
$$s.f.c_{t.o.} = 0.541(SHP_{t.o.})^{-0.062} \qquad (6.11)$$

For engine specific fuel consumption at 15,000 ft and maximum cruise power:
$$s.f.c_{t.o.} = 0.454(SHP_{t.o.})^{-0.055} \qquad (6.12a)$$

For engine specific fuel consumption at 25,000 ft and maximum cruise power:
$$s.f.c_{t.o.} = 0.525(SHP_{t.o.})^{-0.079} \qquad (6.12b)$$

For engine geometric size:

Consider Figure 6.18 for the dimensions used in the scaling rules for geometric size.

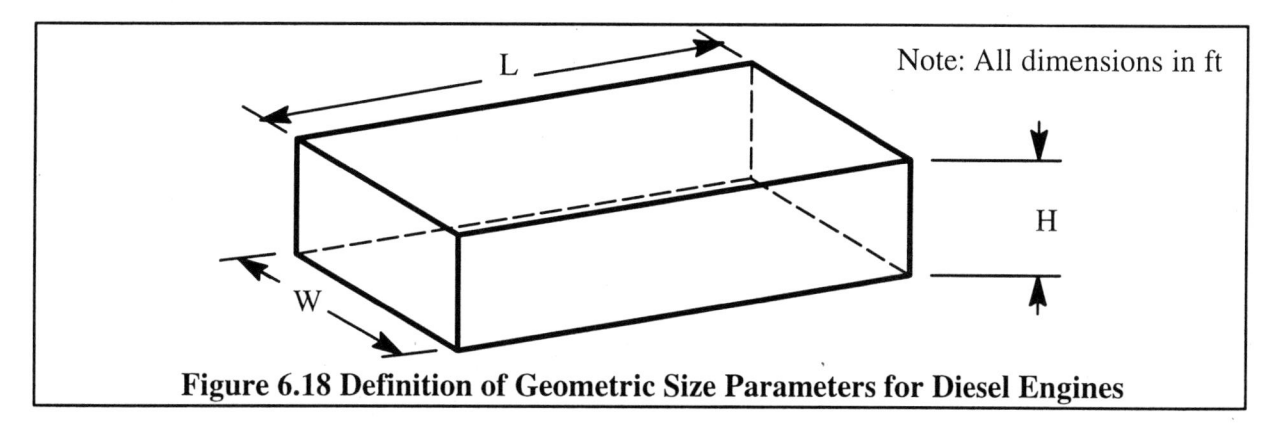

Figure 6.18 Definition of Geometric Size Parameters for Diesel Engines

For engine length, L, of 6 and 8 cylinder engines:
$$L = 1.56(SHP_{t.o.})^{0.15} \quad \text{in ft} \qquad (6.13)$$

For engine width, W, of 6-cylinder engines:
$$W = 1.5 + 0.00134(SHP_{t.o.}) \quad \text{in ft} \qquad (6.14)$$

For engine width, W, of 8-cylinder engines:
$$W = 1.9 + 0.00042(SHP_{t.o.}) \quad \text{in ft} \qquad (6.15)$$

For engine height, H, of 6-cylinder engines:

$$W = 1.7 + 0.00156(SHP_{t.o.}) \quad \text{in ft} \tag{6.16}$$

For engine height, H, of 8-cylinder engines:

$$W = 1.8 + 0.00081(SHP_{t.o.}) \quad \text{in ft} \tag{6.17}$$

A plot of engine weight versus SHP for conventional, gasoline powered piston engines is given in Figure 6.19. This plot can be used for rapidly estimating dry weight of piston engines ranging from 40 – 1,000 hp.

For a set of empirical formulas to estimate reciprocating engine weight and required cylinder volume (which relates to the geometric engine size) may be found in Ref, 6.11 on pages 109 and 110.

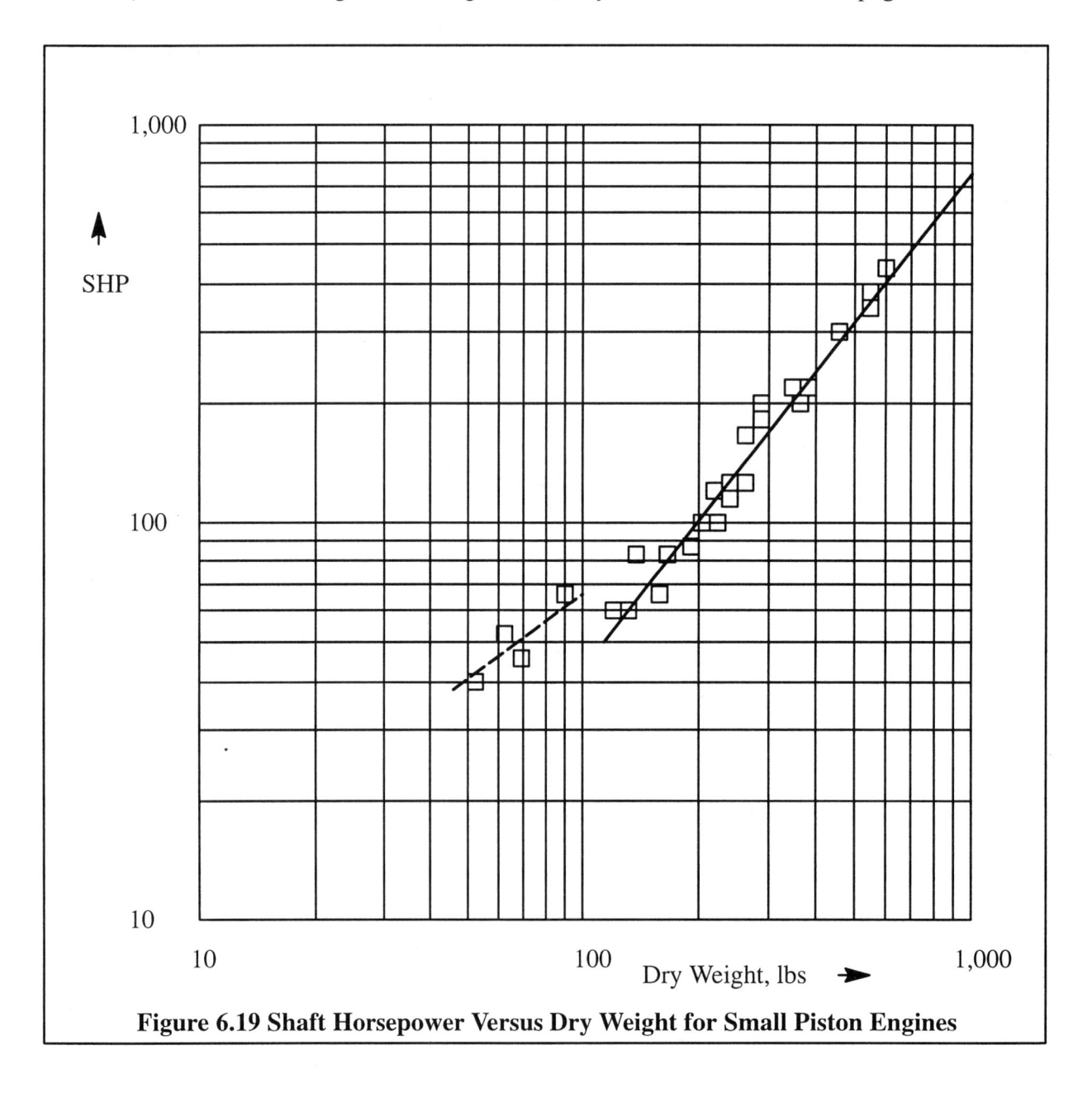

Figure 6.19 Shaft Horsepower Versus Dry Weight for Small Piston Engines

6.2 JET ENGINES

Examples of jet engines are: turbojets, turbofans, un–ducted fans, turboprops, ram–jets and pulse–jets. Some basic operational characteristics of turbojets are discussed in Sub–section 6.2.1. Examples of how engine thrust and s.f.c. vary with altitude and Mach number are given. The effect of afterburners is also discussed. In Sub–section 6.2.2 by–pass engines and turbofans are briefly discussed. Several aspects of turboprops are covered in Sub–section 6.2.3.

Examples of mechanical arrangements (layouts) of turbojet, turbo–fan, turbo–prop and prop–fan engines are shown in Sub–section 6.2.4. Examples of thrust, power and fuel consumption of several actual engines are presented in Sub–section 6.2.5. Installation examples and cooling considerations are presented in Sub–section 6.2.6. Noise considerations associated with gas turbine engines are presented in Sub–section 6.2.7. Thrust reversal is a required operational feature for most jet engines. A brief discussion of thrust reversers is given in Sub–section 6.2.8. In Sub–section 6.2.9, approximate scaling laws involving the variation of engine size, and weight with required thrust output are given. Pulse–jets and ram–jets are briefly discussed in Sub–sections 6.2.10 and 6.2.11 respectively.

6.2.1 BASIC OPERATION OF TURBOJETS

The basic operation of a turbojet engine can be explained with Figure 6.20.

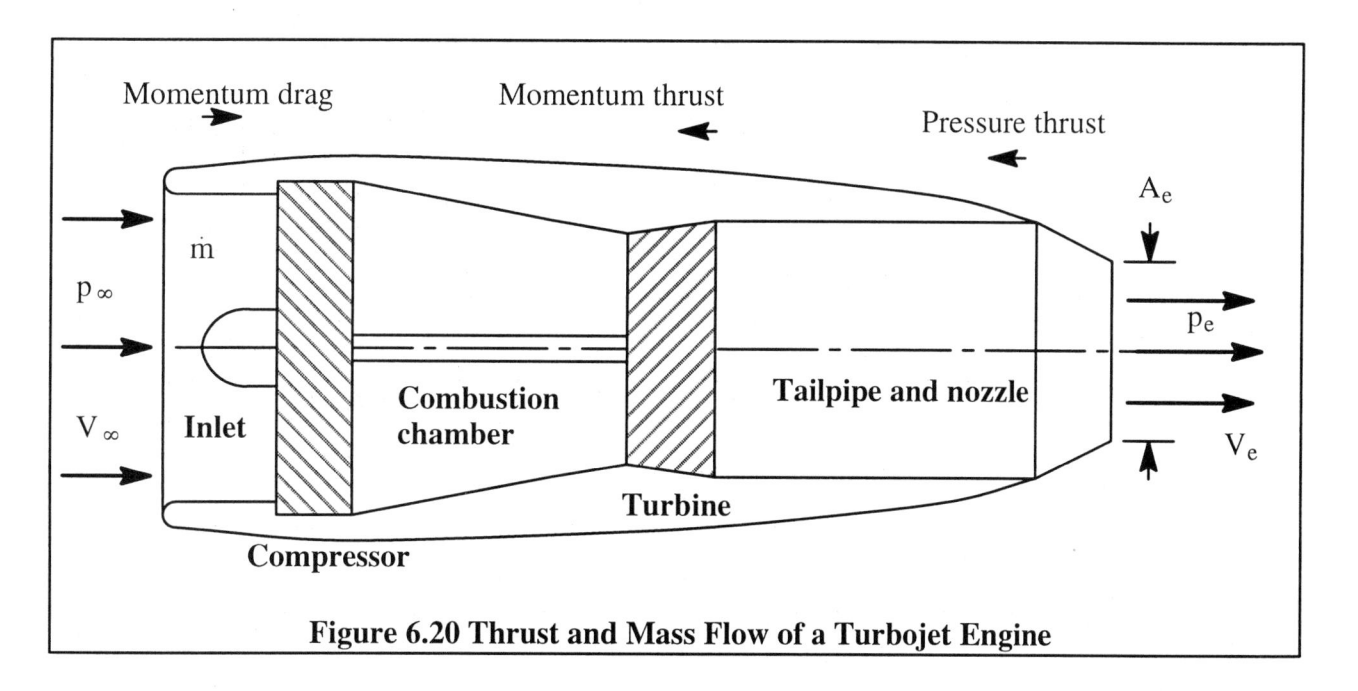

Figure 6.20 Thrust and Mass Flow of a Turbojet Engine

The air entering the inlet duct is slowed down and enters the compressor at a local Mach number of around 0.4 or less. Next, the air is compressed and ducted into the combustion chamber where part of the air (approximately 25%) is used in a continuous combustion process. The heated mixture of air and combustion products enters the turbine and imparts on it the energy needed to drive the compressor. The remaining, still high energy, air is expanded through a nozzle which increases its kinetic energy. In a p–V (pressure–volume) diagram this cycle (referred to as the Brayton–cycle) can be represented as in Figure 6.21. This cycle will be traced in the following.

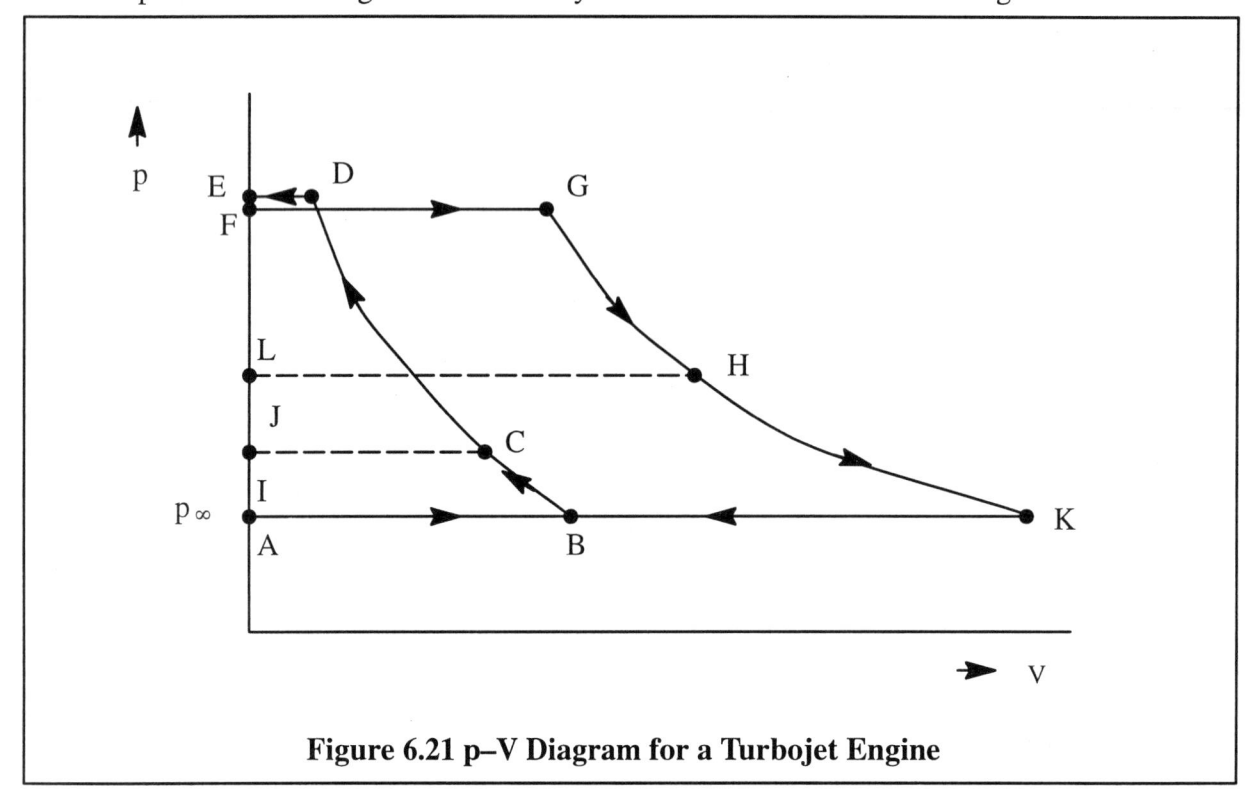

Figure 6.21 p–V Diagram for a Turbojet Engine

The air is slowed in the inlet toward the compressor along BC. Then, the air is compressed by the compressor along CD and subsequently discharged into the combustion chamber along DE. As a result of losses the combustion chamber delivers the gas mixture to the turbine at a pressure corresponding to point F along FG. The heated mixture now expands through the turbine and drives the turbine along GH. The mixture then expands through the nozzle to form a high velocity jet along HK. In Figure 6.21 the pressure at point K is assumed to be p_∞. In reality this is not the case and it is normally higher: $p_e > p_\infty$. This is because in subsonic flow the nozzle is normally choked as V_e is greater than sonic speed. The work developed by the turbine is represented by area AFGK while the work expended by the compressor is represented by area JCDE. The remaining area ABCJ represents the compression work recovered from the kinetic energy of the incoming air. It is referred to as ram.

Using the principle of conservation of linear momentum the **net** thrust, F_n, which is the propulsive force actually imparted to the airplane, may be seen from Figure 6.20 to be:

$$F_n = \dot{m}(V_e - V_\infty) + (p_e - p_\infty)A_e \tag{6.18}$$

The term: $\dot{m}V_\infty$ is commonly referred to as the momentum drag. The term: $\dot{m}V_e$ is referred to as the momentum thrust. The term: $(p_e - p_\infty)A_e$ is called the pressure thrust.

The actual mass flow at the exit, $\dot{m}_e$ differs from that at the inlet, $\dot{m}$ by about 2% as a result of fuel being added in the combustion chamber. In deriving Eqn (6.18) this has been neglected.

The propulsive efficiency of a jet engine can be expressed as:

$$\eta_{propulsive} = \frac{\text{work done on the airplane}}{\text{energy imparted to the engine airflow}} \quad (6.19)$$

The work done on the airplane is the product of net thrust and airplane speed. The energy imparted to the airflow is the work done plus the energy wasted in the exhaust. The latter can be written as: $0.5\dot{m}(V_e - V_\infty)^2$. Therefore, Eqn (6.19) can also be expressed as:

$$\eta_{propulsive} = \frac{V_\infty \{\dot{m}(V_e - V_\infty) + (p_e - p_\infty)A_e\}}{V_\infty \{\dot{m}(V_e - V_\infty) + (p_e - p_\infty)A_e\} + 0.5\dot{m}(V_e - V_\infty)^2} \quad (6.20)$$

If it is assumed that the nozzle fully expands the exhaust gasses to atmospheric pressure, the pressure terms in Eqn (6.20) are negligible. Therefore:

$$\eta_{propulsive} = \frac{V_\infty \{\dot{m}(V_e - V_\infty)\}}{V_\infty \{\dot{m}(V_e - V_\infty)\} + 0.5\dot{m}(V_e - V_\infty)^2} = \frac{2V_\infty}{V_\infty + V_e} \quad (6.21)$$

It is seen from Eqn (6.21) that at low flight speeds: $V_\infty \ll V_e$, and the efficiency will be low. At high flight speeds, the efficiency of the gas turbine engine improves.

As an example, consider a jet airplane flying at 350 mph with a jet exhaust velocity of 1,100 mph. The propulsive efficiency, according to Eqn (6.21) is: $2 \times 350/(350 + 1100) = 0.48$.

Next, consider the same airplane flying at a speed of 550 mph and the same jet exhaust velocity. Now the propulsive efficiency is: $2 \times 550/(550 + 1100) = 0.67$ which is considerably more.

A typical example of how propulsive efficiency of a pure turbojet engine varies with speed at constant altitude is given in Figure 6.22a. A comparison with other types of gas turbine engines is also shown. Note that the pure jet engine, from an efficiency point of view, does not compare well with a turboprop until rather high flight speeds. The reason for the high propulsive efficiency of a propeller at low speeds is explained in Chapter 7.

From Eqn (6.18) it is seen that if the exit velocity remains constant (independent of airplane speed) the net thrust will decrease linearly with air–speed. However, due to the so–called 'ram ratio' effect the actual thrust decrease with air–speed is much less severe. This ram ratio is defined as the ratio between the total air pressure at the compressor entry to the static air pressure at the inlet entry.

As a result of the ram ratio, the mass flow rate of air delivered to the compressor actually increases with increasing air–speed. In addition, the jet exit velocity also tends to increase somewhat with increasing air–speed. For these reasons the thrust does not decrease quite so dramatic with speed. A typical trend is shown in Figure 6.23.

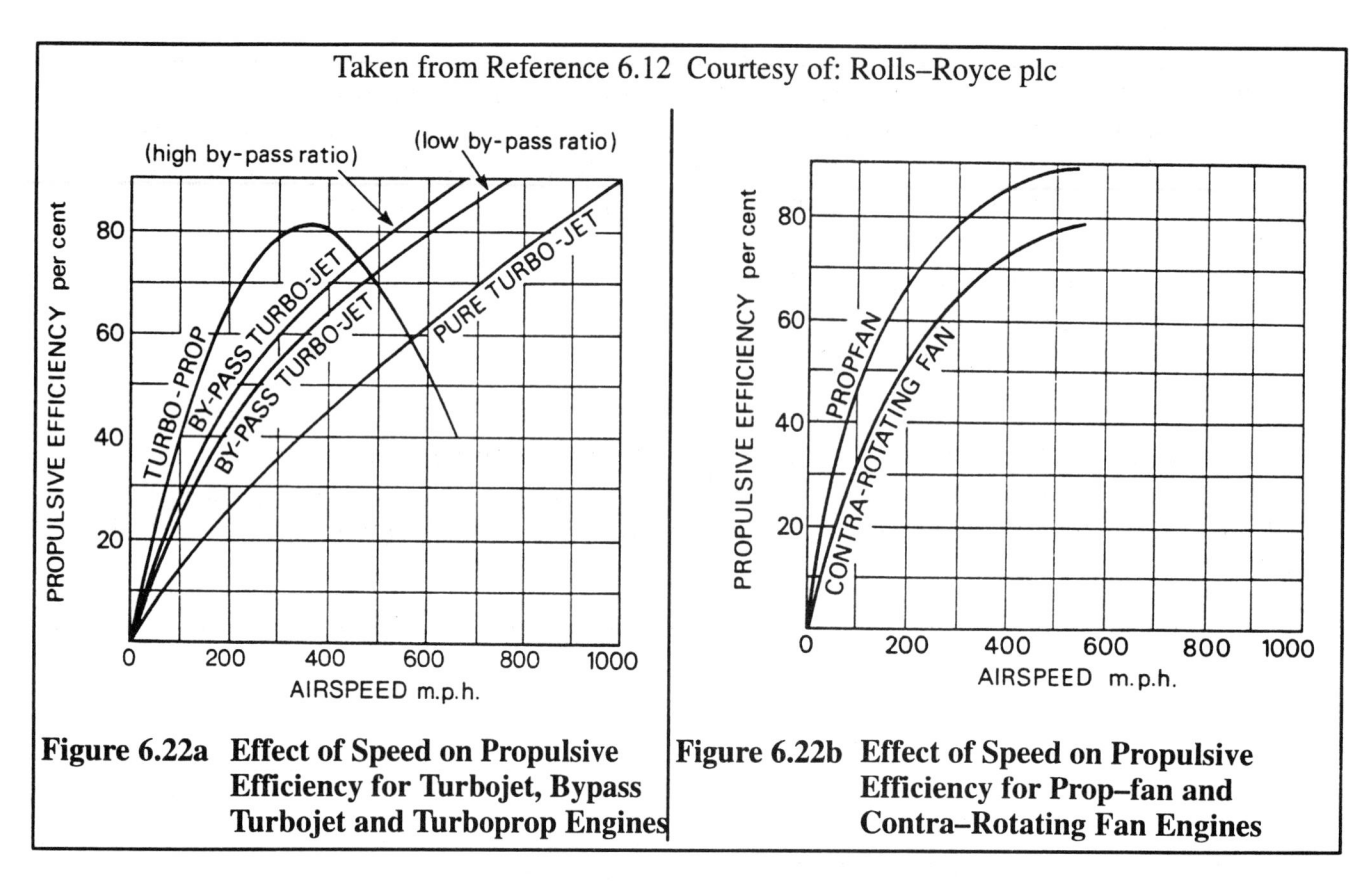

Figure 6.22a Effect of Speed on Propulsive Efficiency for Turbojet, Bypass Turbojet and Turboprop Engines

Figure 6.22b Effect of Speed on Propulsive Efficiency for Prop–fan and Contra–Rotating Fan Engines

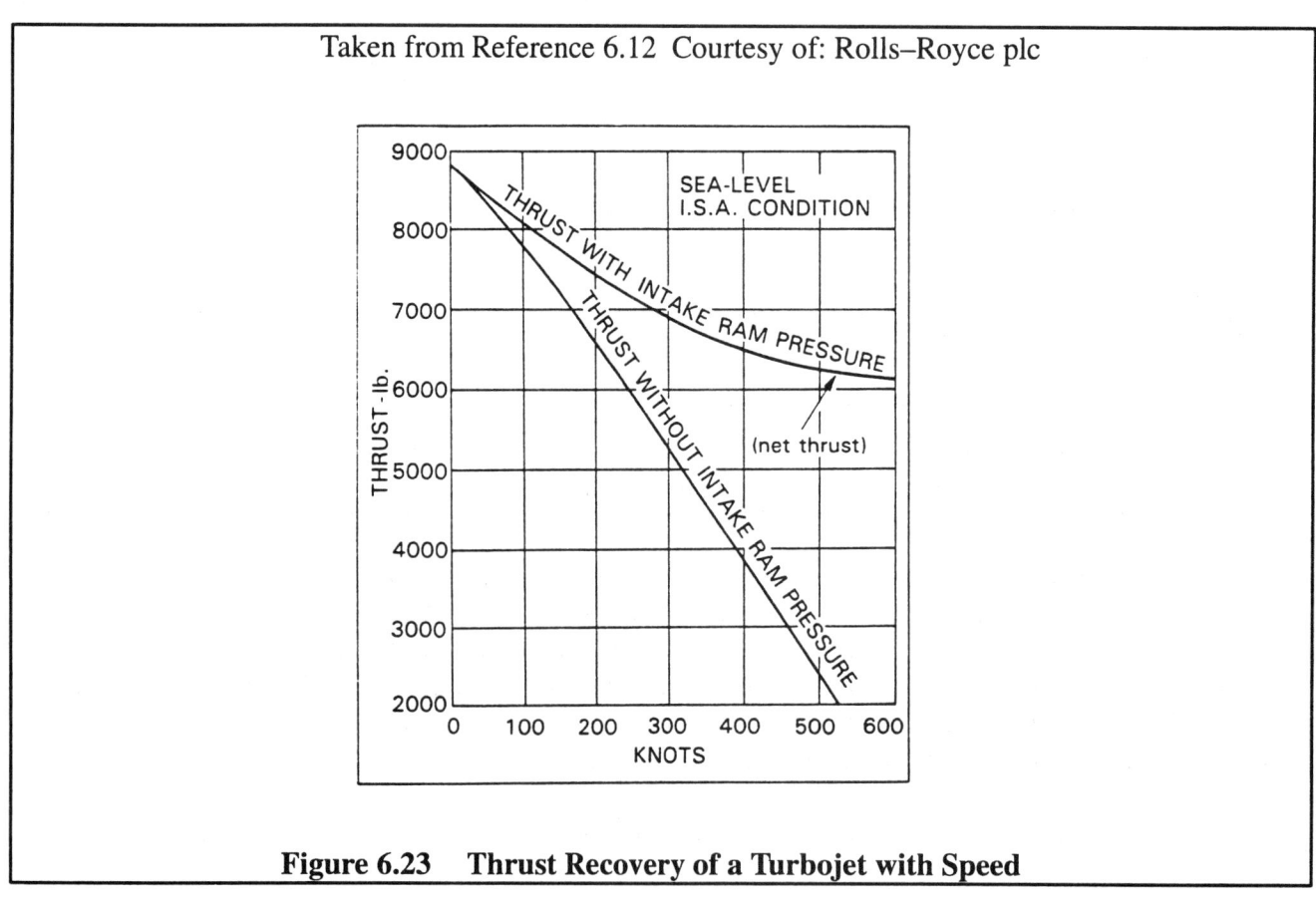

Figure 6.23 Thrust Recovery of a Turbojet with Speed

As seen, when air–speed is increased, the mass flow into the engine increases. As a result of the increased mass flow, more fuel is required to sustain combustion and therefore the fuel consumption in terms of lbs/hr increases with increasing speed. To enable a consistent comparison of one gas turbine type versus another the idea of thrust specific fuel consumption has been introduced:

$$\text{t.s.f.c.} = \frac{\text{fuel consumption in lbs/hr}}{\text{thrust output in lbs}} = \frac{\text{lbs/hr}}{\text{lbs}} \tag{6.22}$$

Because the fuel consumption increases with speed while the thrust decreases with speed, the thrust specific fuel consumption or t.s.f.c. will also increase with speed. Example trends are shown in Figure 6.24. The nonlinearity in the fuel consumption is caused by the fact that as speed increases the stresses on certain engine components increase to a point where the fuel flow must be artificially restricted.

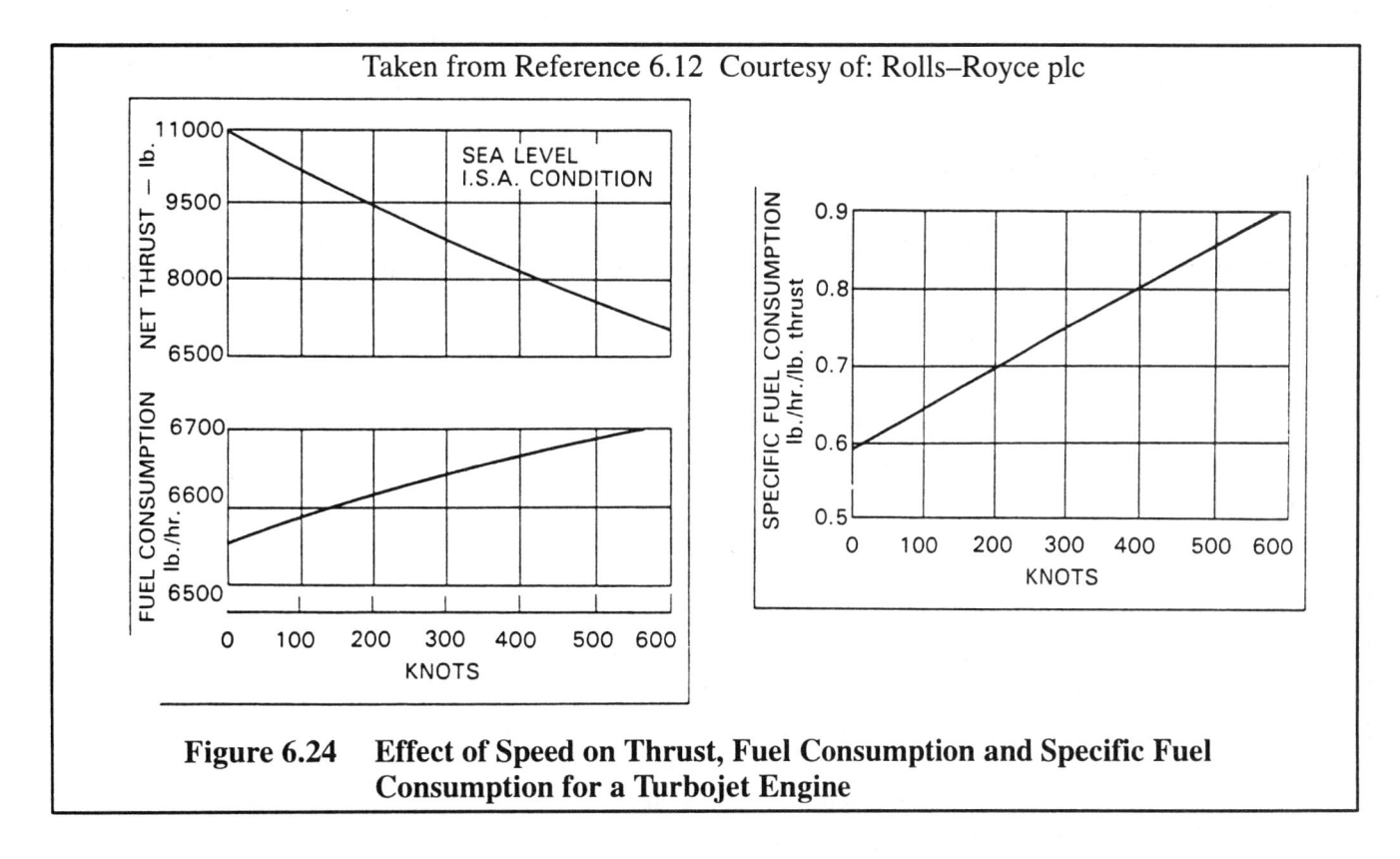

Figure 6.24 Effect of Speed on Thrust, Fuel Consumption and Specific Fuel Consumption for a Turbojet Engine

The altitude at which an airplane flies affects the thrust in two unrelated ways. First, because air density decreases with increasing altitude, the mass flow rate (at constant speed) is reduced with increasing altitude. This causes the thrust to decrease with altitude. Second, as altitude increases, the air temperature decreases. This has the effect of increasing the air density. The net result is that thrust still decreases with altitude but at a slower rate. The reader is referred to Refs 6.13 and 6.14 for the physics and the mathematics which are involved.

A typical relation between thrust, fuel flow, s.f.c. and altitude is shown in Figure 6.25.

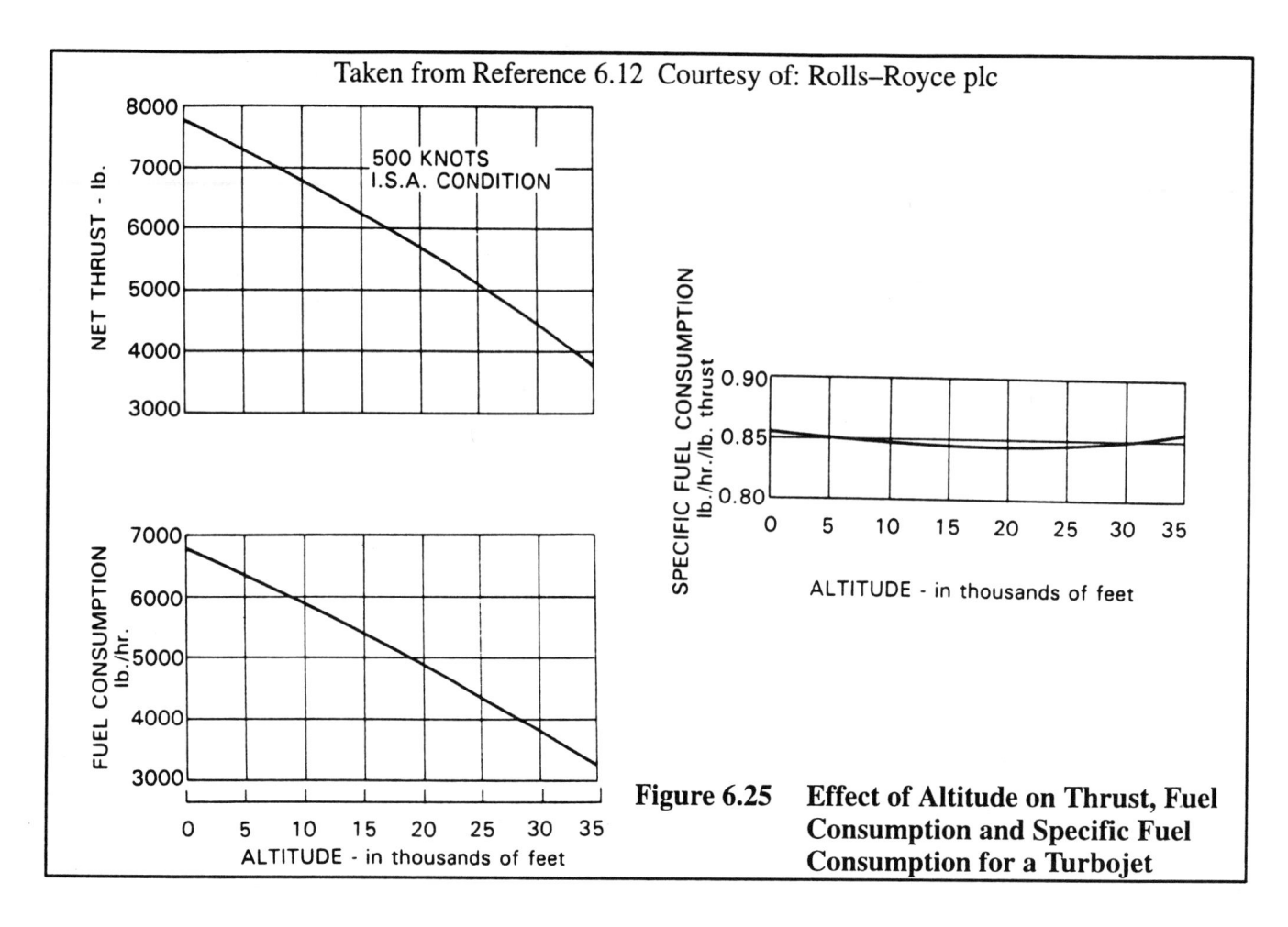

Figure 6.25 Effect of Altitude on Thrust, Fuel Consumption and Specific Fuel Consumption for a Turbojet

Taken from Reference 6.12 Courtesy of: Rolls–Royce plc

To attain higher propulsive efficiencies than are possible with a straight turbojet the idea of by-pass was introduced. In that case, not all the air passes through the 'hot' part of the engine. Examples of layouts of by–pass engines are shown in Sub–section 6.2.4.

Particularly in military applications, turbojet engines and even by–pass turbojet engines (defined in Sub–section 6.2.2) require the addition of an afterburner to enhance thrust for limited periods of time. An afterburner is basically a duct through which the exhaust gasses pass. Because only about 25% of the air passing through a turbojet is used for combustion, the gas mixture which emanates from the last turbine stage still contains enough air to sustain combustion in the afterburner duct. Figure 6.26 shows an example of an afterburner arrangement. Depending on the engine, the installed thrust can be increased by about 50% through the use of an afterburner. A drawback is the increased fuel consumption: the s.f.c with afterburner on increases usually by at least a factor 2.

Another drawback of an afterburner is the fact that the basic engine thrust with the afterburner not operating (but present) is less than the thrust produced by the engine without the afterburner installed. As can be expected, the weight of the engine + afterburner is much higher than that of the basic engine.

An afterburner can be thought of as a ramjet engine added to the turbojet. Ramjets are briefly discussed in Sub–section 6.2.10.

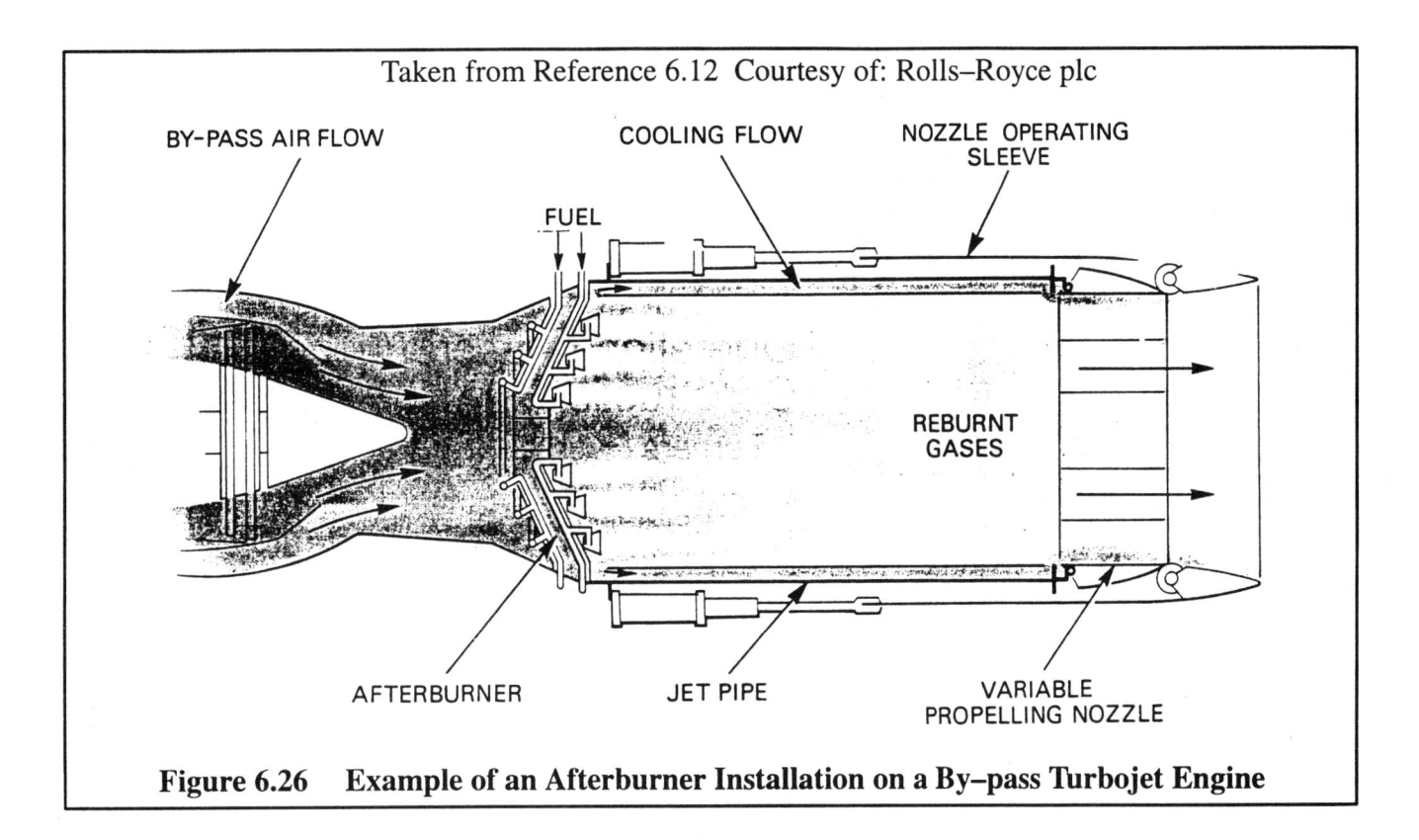

Figure 6.26 Example of an Afterburner Installation on a By-pass Turbojet Engine

6.2.2 CHARACTERISTICS OF BY-PASS ENGINES AND TURBOFANS

Whenever not all of the air entering a jet engine passes through the hot core, the engine is referred to as a by-pass engine. The by-pass ratio of an engine has the following definition:

$$\text{by} - \text{pass ratio} = \text{b.p.r.} = \frac{\text{mass flow rate through the fan}}{\text{mass flow rate through the turbine}} \tag{6.23}$$

When the by-pass ratio is low, it is referred to as a by-pass turbojet. When the by-pass ratio is high, such engines are generally referred to as turbo-fan engines. The boundary between by-pass turbojets and turbofans is not easily definable nor does this really matter.

Trends of propulsive efficiency with speed for low and high by-pass ratio with speed were shown in Figure 6.22a. Early by-pass engines (PW JT8D used in the Boeing 727, 737 and DC-9) had low by-pass ratios which ranged from slightly above 1.0 to 2.0. Modern high by-pass engines, such as turbofans have by-pass ratios of around 4.0 for the Garrett TFE 731 (used in several Learjet, Cessna and other business jets) to 6.0 for the PW 2000 (used in the Boeing 757 and McDD C-17A). Future arrangements such as the prop-fan and unducted fan shown in Sub-section 6.2.4 may have by-pass ratios as high as 15.0.

A typical schematic layout of a turbo-fan engine is shown in Figure 6.27. In this example the engine is configured as a so-called two-spool arrangement.

By analogy of Eqn (6.21) the propulsive efficiency of a by-pass arrangement can be written as:

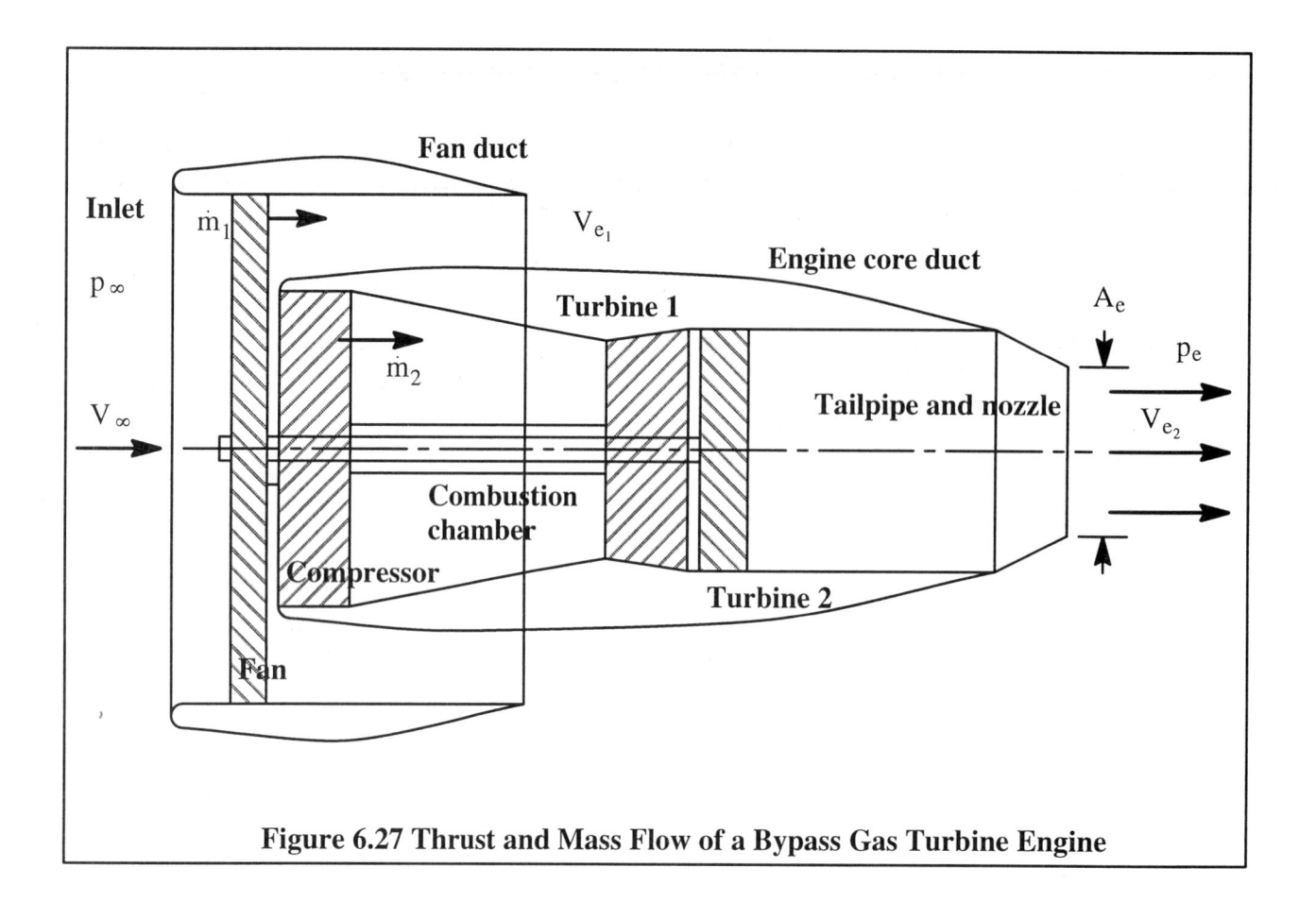

Figure 6.27 Thrust and Mass Flow of a Bypass Gas Turbine Engine

$$\eta_{propulsive} = \qquad (6.24)$$

$$= \frac{V_\infty \{\dot{m}_1(V_{e_1} - V_\infty)\} + V_\infty \{\dot{m}_2(V_{e_2} - V_\infty)\}}{V_\infty \{\dot{m}_1(V_{e_1} - V_\infty)\} + 0.5\dot{m}_1(V_{e_1} - V_\infty)^2 + V_\infty \{\dot{m}_2(V_{e_2} - V_\infty)\} + 0.5\dot{m}_2(V_{e_2} - V_\infty)^2}$$

As an exercise, consider the following case, not untypical for a modern low by–pass engine:

$V_\infty = 583$ mph $\qquad \dot{m}_1 = 15.3$ slugs/sec $\qquad \dot{m}_2 = 3.1$ slugs/sec

$V_{e_1} = 781$ mph $\qquad V_{e_2} = 812$ mph

Substitution of these numbers into Eqn (6.18) yields a propulsive efficiency of approximately 85%. Therefore, by increasing the by–pass ratio, much higher efficiencies can be obtained by gas turbine engines. By using a so–called rear–mounted fan, by–pass ratios of 15 can probably be achieved, further improving the propulsive efficiency. An example of a mechanical arrangement of such a powerplant is shown in Sub–section 6.2.4.

A major advantage of any by–pass engine is the fact that a mixing of external, cooler air with the internal, hotter air can be arranged which in turn reduces the noise of the engine. This aspect is discussed briefly in Sub–section 6.2.7.

6.2.3 OPERATIONAL CHARACTERISTICS OF TURBOPROPS

A turboprop engine is basically a turbojet designed to drive a propeller. The propeller is driven by a gearbox which receives its power directly from a shaft driven by the hot core. In modern turboprops multiple spool arrangements are often used as shown in Sub-section 6.2.4.

For purposes of this discussion the turboprop arrangement of Figure 6.28 will be used.

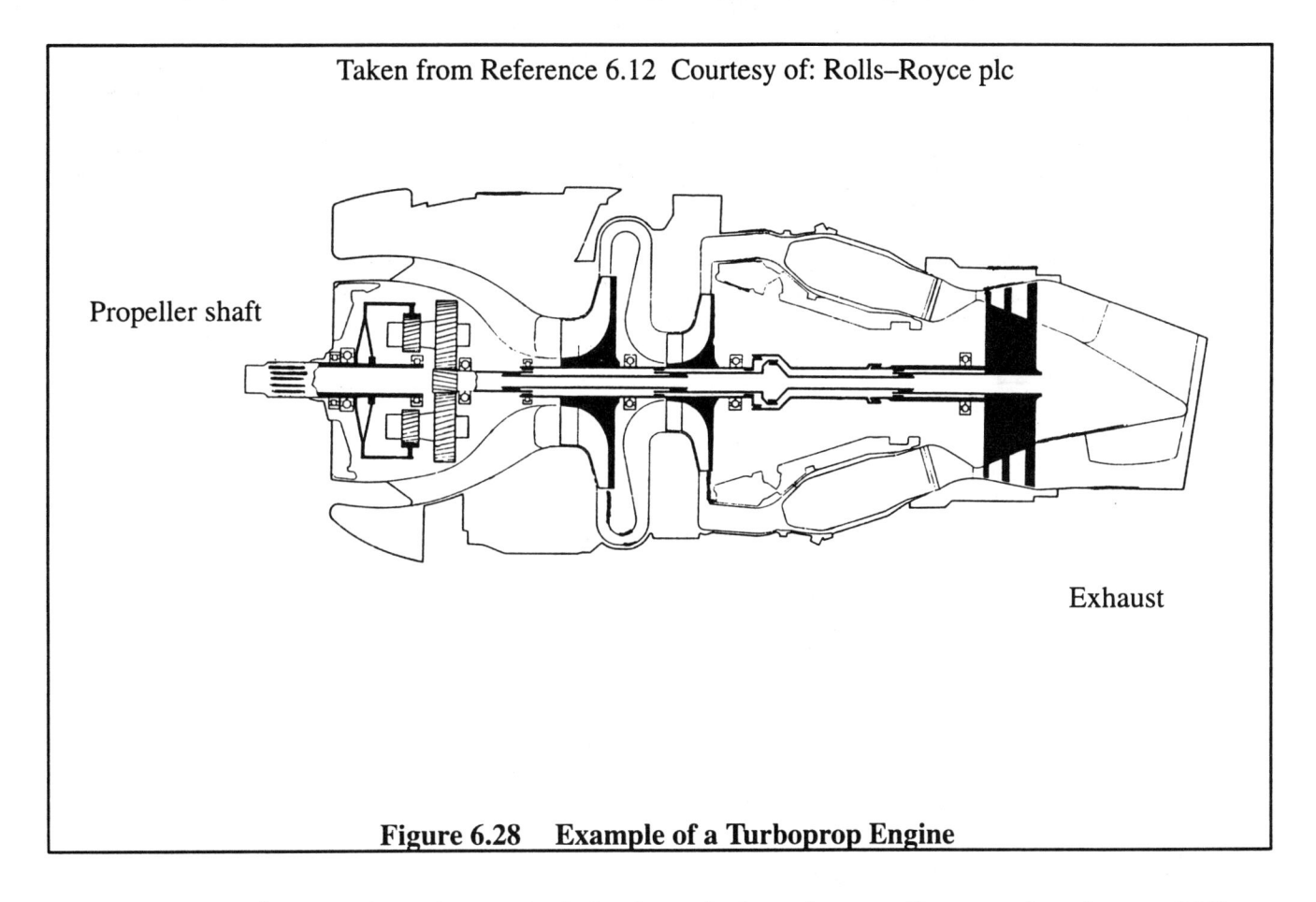

Figure 6.28 Example of a Turboprop Engine

The thrust output of a turboprop is derived mostly from the propeller even though up to 20% of the effective thrust may come from the jet exhaust. A major advantage of the turboprop is its high efficiency, particularly at moderate flight speeds. Because of the presence of a reduction gear box and a propeller, the propulsive efficiency of a turboprop is written as:

$$\eta_{propulsive} = \eta_t \, \eta_{mech} \, \eta_p \tag{6.25}$$

where: η_t is the turbine/compressor efficiency

η_{mech} is the mechanical efficiency of the reduction gear box

η_p is the propeller efficiency

The total thrust horsepower of a turboprop is given by:

$$\text{THP} = \eta_p \text{SHP} + \frac{F_{net} V_\infty}{550} \tag{6.26}$$

where: F_{net} is the net, jet exhaust thrust

SHP is the shaft–horsepower delivered to the propeller

The total thrust horsepower, THP can be used to define the concept of equivalent shaft horsepower, ESHP, as:

$$\text{ESHP} = \text{SHP} + \frac{F_{net} V_\infty}{550 \, \eta_p} \tag{6.27}$$

The specific fuel consumption of a turbo–prop is usually given in lbs of fuel per ESHP. An example of how SHP, net thrust and fuel consumption vary with altitude for a typical turboprop engine is shown in Figure 6.29.

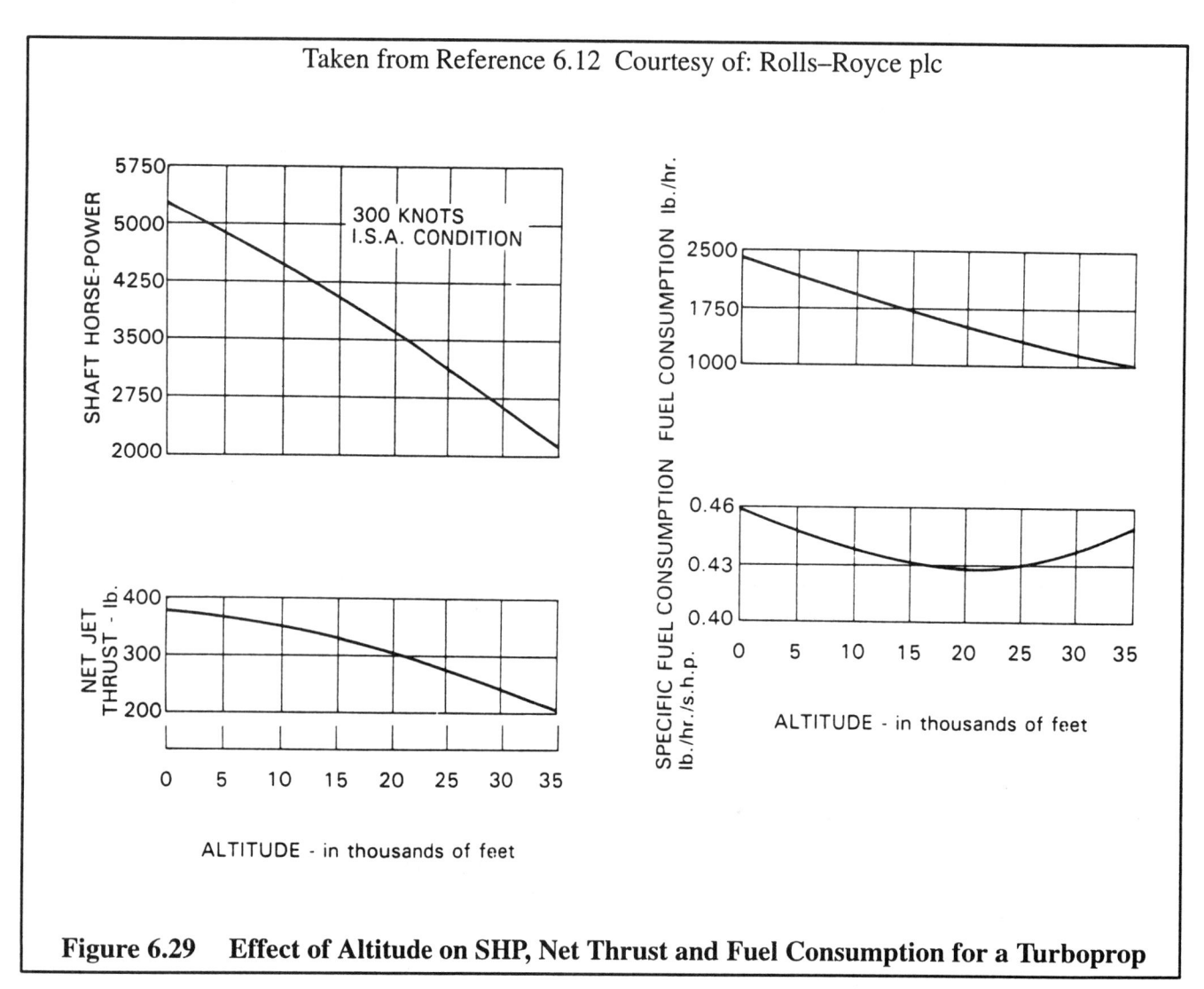

Figure 6.29   Effect of Altitude on SHP, Net Thrust and Fuel Consumption for a Turboprop

6.2.4 TYPICAL MECHANICAL ARRANGEMENTS OF TURBOJETS, TURBOFANS AND TURBOPROPS

Examples of the layout or the mechanical arrangement of several types of gas turbine engines are given in Figures 6.30 and 6.31. A discussion of some of these layouts follows.

Figure 6.30a shows a straight turbojet engine with a double entry, centrifugal compressor and a single turbine driving it. Several early turbojet engines were of this type.

Because of the poor efficiency of a jet engine at low speeds, using the jet engine to drive a propeller (turboprop) is an attractive alternative for airplanes in moderate speed regimes. Figure 6.30b shows an example of a turboprop layout with a single entry, two–stage, centrifugal compressor driven by three turbine wheels. Note that the layouts of Figure 6.30a and 6.30b both have a single spool. The shaft forward of the first compressor stage drives a gear box which in turn drives a propeller (not shown).

An alternate way to arrange a straight turbojet is to use an axial compressor. Figure 6.30c shows an arrangement of a seventeen stage, axial compressor driven by a three stage turbine, all still on one spool.

A more efficient way to design a gas turbine engine is to use two or more spools. Figure 6.30d shows an arrangement for a turbo–propeller engine with two spools. Gas turbines are very attractive in certain helicopter applications as well as in ground vehicle and stationary powerplant applications. In the latter cases these engines are referred to as turbo–shaft engines. In a helicopter, the output shaft of the turbo–shaft engine is used to drive the rotor. An example of such an engine, is given in Figure 6.30e. Note that this engine has three spools: a free turbine which drives the last centrifugal compressor stage, a second turbine which drives a four–stage axial flow compressor and a two–stage turbine which drives the gear box which in turn drives the rotor (not shown).

It was already shown that by creating a bypass a gas turbine engine can be made more efficient. Figure 6.31a shows an example of an early bypass turbojet engine. This example has a twin–spool design. The first two–stage turbine drives a twelve stage axial flow compressor. The second two–stage turbine drives a four stage fan/compressor. Some of the air from this stage is by–passed around the core engine.

Figure 6.31b shows an example of a modern high bypass ratio fan engine with a triple spool arrangement. A single stage turbine drives a six–stage axial flow compressor (low pressure). Another single stage turbine drives another six–stage axial flow compressor (high pressure). Finally, a three stage turbine drives a large front fan. This fan can be thought of as a multi-bladed, ducted propeller.

It was mentioned in Sub–section 6.2.2 that very high propulsive efficiencies can be achieved with higher by–pass ratios. Examples of very high by–pass engine concepts are shown in Figures 6.31c and 6.31d. Figure 6.31c shows a concept of an un–ducted, contra–rotating prop–fan engine. Figure 6.19d illustrates a concept of a very high by–pass ratio contra–rotating, ducted fan. Whether or not such engines will materialize depends on trades between fuel cost, engine complexity, engine acquisition cost and engine maintenance cost.

Airplane Propulsion Systems

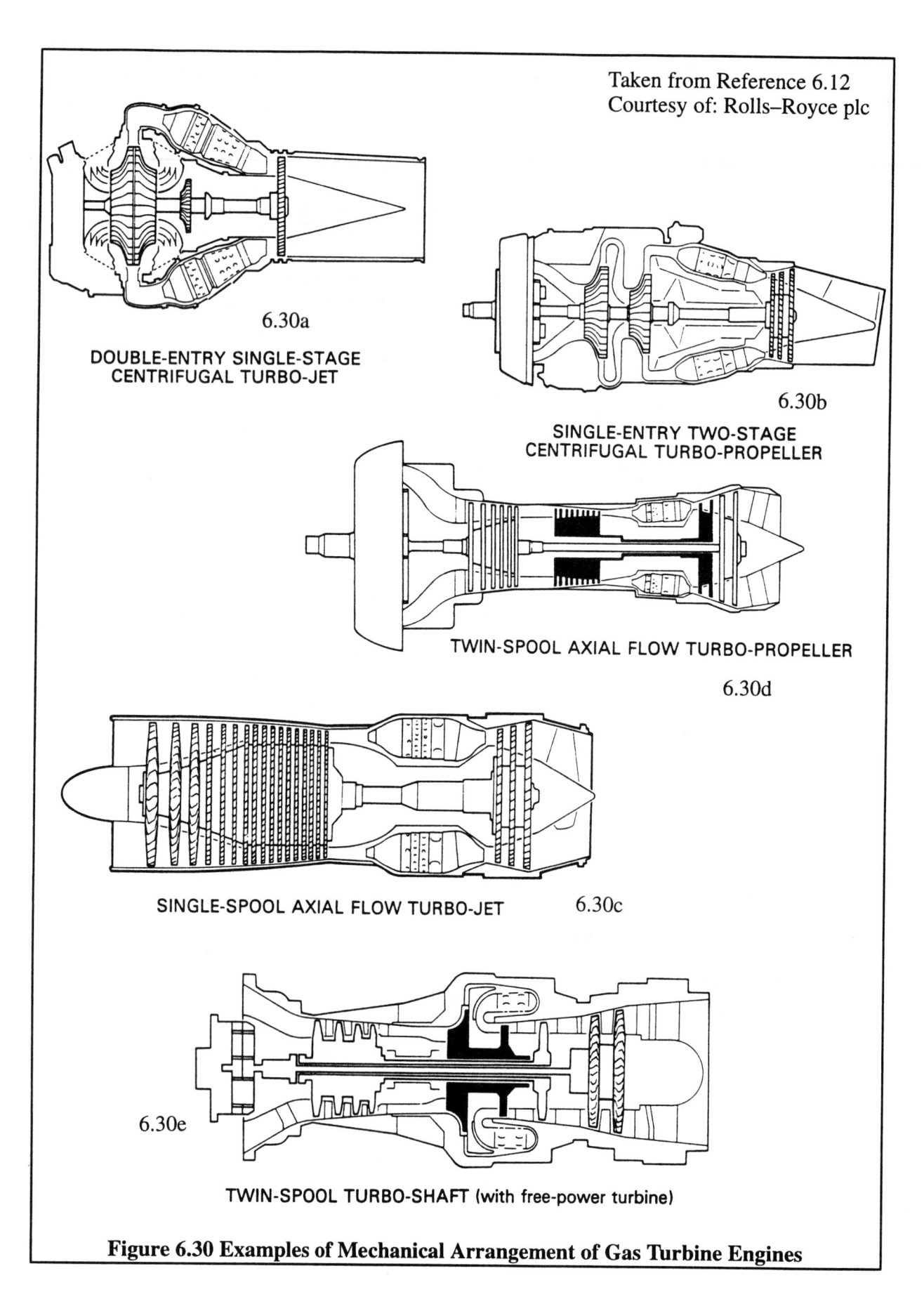

Taken from Reference 6.12
Courtesy of: Rolls–Royce plc

6.30a
DOUBLE-ENTRY SINGLE-STAGE CENTRIFUGAL TURBO-JET

6.30b
SINGLE-ENTRY TWO-STAGE CENTRIFUGAL TURBO-PROPELLER

TWIN-SPOOL AXIAL FLOW TURBO-PROPELLER
6.30d

SINGLE-SPOOL AXIAL FLOW TURBO-JET 6.30c

6.30e
TWIN-SPOOL TURBO-SHAFT (with free-power turbine)

Figure 6.30 Examples of Mechanical Arrangement of Gas Turbine Engines

Airplane Propulsion Systems

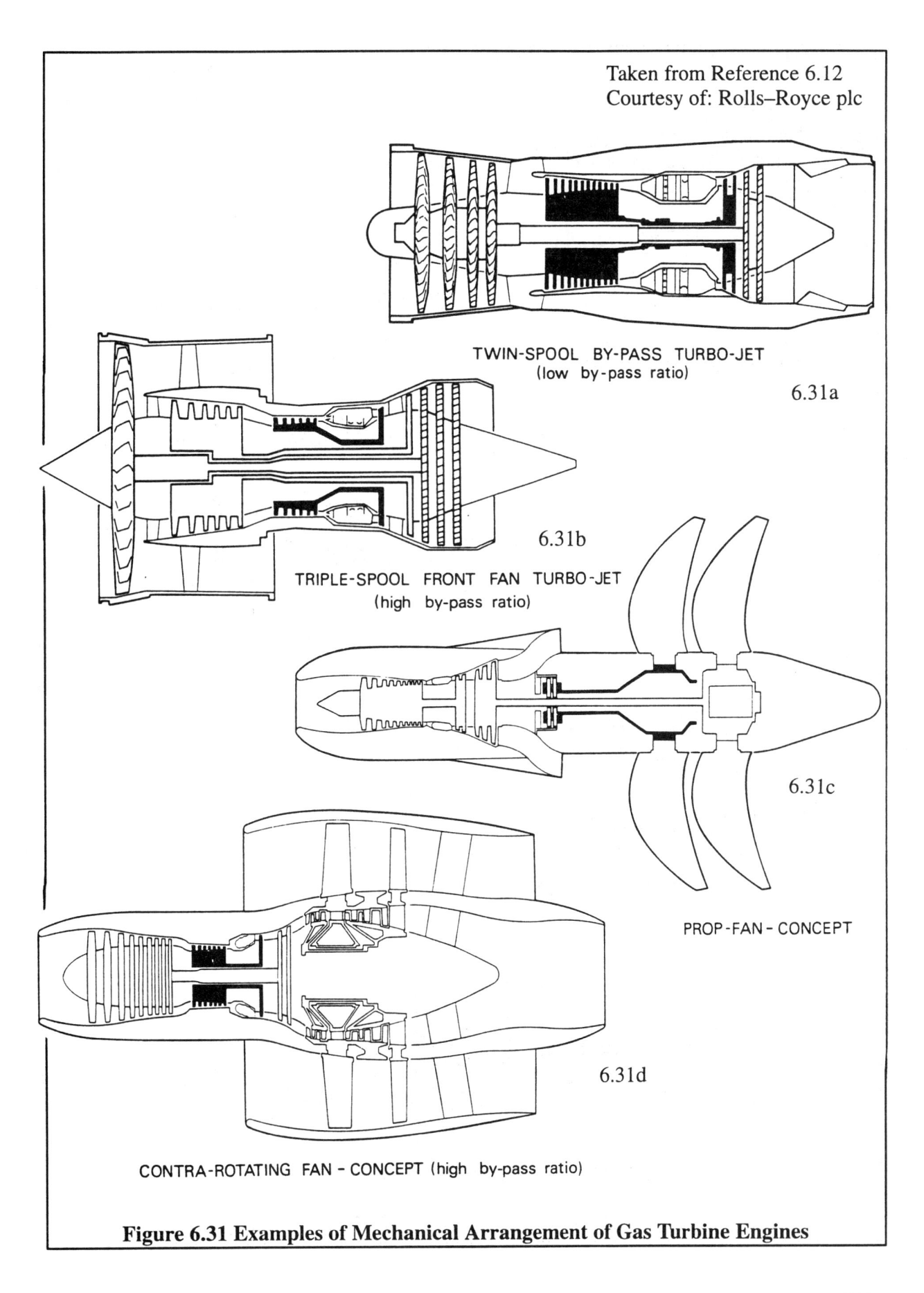

Taken from Reference 6.12
Courtesy of: Rolls–Royce plc

TWIN-SPOOL BY-PASS TURBO-JET
(low by-pass ratio)

6.31a

TRIPLE-SPOOL FRONT FAN TURBO-JET
(high by-pass ratio)

6.31b

PROP-FAN-CONCEPT

6.31c

CONTRA-ROTATING FAN - CONCEPT (high by-pass ratio)

6.31d

Figure 6.31 Examples of Mechanical Arrangement of Gas Turbine Engines

6.2.5 THRUST, POWER AND FUEL CONSUMPTION OF GAS TURBINE ENGINES

In this Sub–section, the basic parameters which determine the thrust of a jet engine will be investigated. Three dimensionless parameters will be derived. These dimensionless parameters are used to reduce engine performance data from one set of atmospheric conditions to a standard set for presentation in engine performance charts or plots. Finally, several actual examples will be given of the variation of thrust, fuel consumption and airflow with altitude. Tabulated data on engine weight, thrust and s.f.c. are also included.

It has been found that the net thrust output of a turbojet engine, F_n, depends strongly on the following five variables:

* air speed, V (in ft/sec)
* ambient air temperature, T in deg R
* ambient air pressure, p (in lbs/sqft)
* engine RPM, N (in r.p.m.)
* engine diameter, D (in ft)

Therefore, there are six physical parameters and three basic units: m in slugs, l in ft and t in sec. In accordance with the Buckingham π–Theorem of Chapter 3, there should be $6 - 3 = 3$ dimensionless parameters which define the relationship between these physical parameters. Using F_{net}, D and N as the repeating variables, the first dimensionless parameter may be written as:

$$\pi_1 = F_{net}^a D^b N^c V \tag{6.28}$$

In terms of the units this yields:

$$0 = \left(\frac{ml}{t^2}\right)^a (l)^b \left(\frac{1}{t}\right)^c \left(\frac{l}{t}\right) \tag{6.29}$$

The following algebraic equations define the relationship between the powers a, b and c:

$$\text{l:} \quad 0 = a + b + 1 \qquad b = -1 \tag{6.30}$$

$$\text{m:} \quad 0 = a \qquad a = 0 \tag{6.31}$$

$$\text{t:} \quad 0 = -2a - c - 1 \qquad c = -1 \tag{6.32}$$

As a result, the first dimensionless parameter is:

$$\pi_1 = \frac{V}{ND} \tag{6.33}$$

Observing that the temperature, according to the kinetic theory of gasses, is proportional to the square of velocity, the unit of T is considered to be the square of velocity. It is now left up to the reader to show that the other two parameters are:

$$\pi_2 = \frac{\sqrt{T}}{ND} \tag{6.34}$$

$$\pi_3 = \frac{pD^2}{F} \tag{6.35}$$

The dimensionless parameters π_1 and π_2 can be combined to create another dimensionless parameter, i.e. $V/\sqrt{T}$. The relationship between π_1, π_2 and π_3 can now be written as:

$$\frac{F_{net}}{pD^2} = f\left(\frac{V}{\sqrt{T}}, \frac{ND}{\sqrt{T}}\right) \tag{6.36}$$

Three observations are in order:

1) For any given engine, the diameter, D, is fixed.
2) It is conventional to refer engine data to standard sea–level conditions.
3) $V/\sqrt{T}$ can be replaced by the Mach number, M.

As a result, the conventional format for Eqn (6.36) becomes:

$$\frac{F_{net}}{\delta} = f\left(M, \frac{N}{\sqrt{\theta}}\right) \tag{6.37}$$

In Eqn (6.37) the terms F_{net}/δ and $N/\sqrt{\theta}$ are referred to as the corrected thrust and the corrected RPM respectively. An example variation of F_{net}/δ with $N/\sqrt{\theta}$ for a small turbo–fan engine is shown in Figure 6.32a.

The air mass flow rate, $\dot{m}_a$ (in slugs/sec), through a gas turbine engine is also corrected with these parameters:

$$\frac{\dot{m}_a\sqrt{T}}{\delta} = f\left(M, \frac{N}{\sqrt{\theta}}\right) \tag{6.38}$$

To illustrate the use of these corrections the following example is offered.

Example: An engine is operating in air with p = 14.5 psi and T = 300 F. Under these conditions the following measurements are available:

$F_{net} = 6,100$ lbs, $\dot{W}_a = 107$ lbs/sec and $N = 100\%$ rpm

Correct these observed data to standard sea–level conditions.

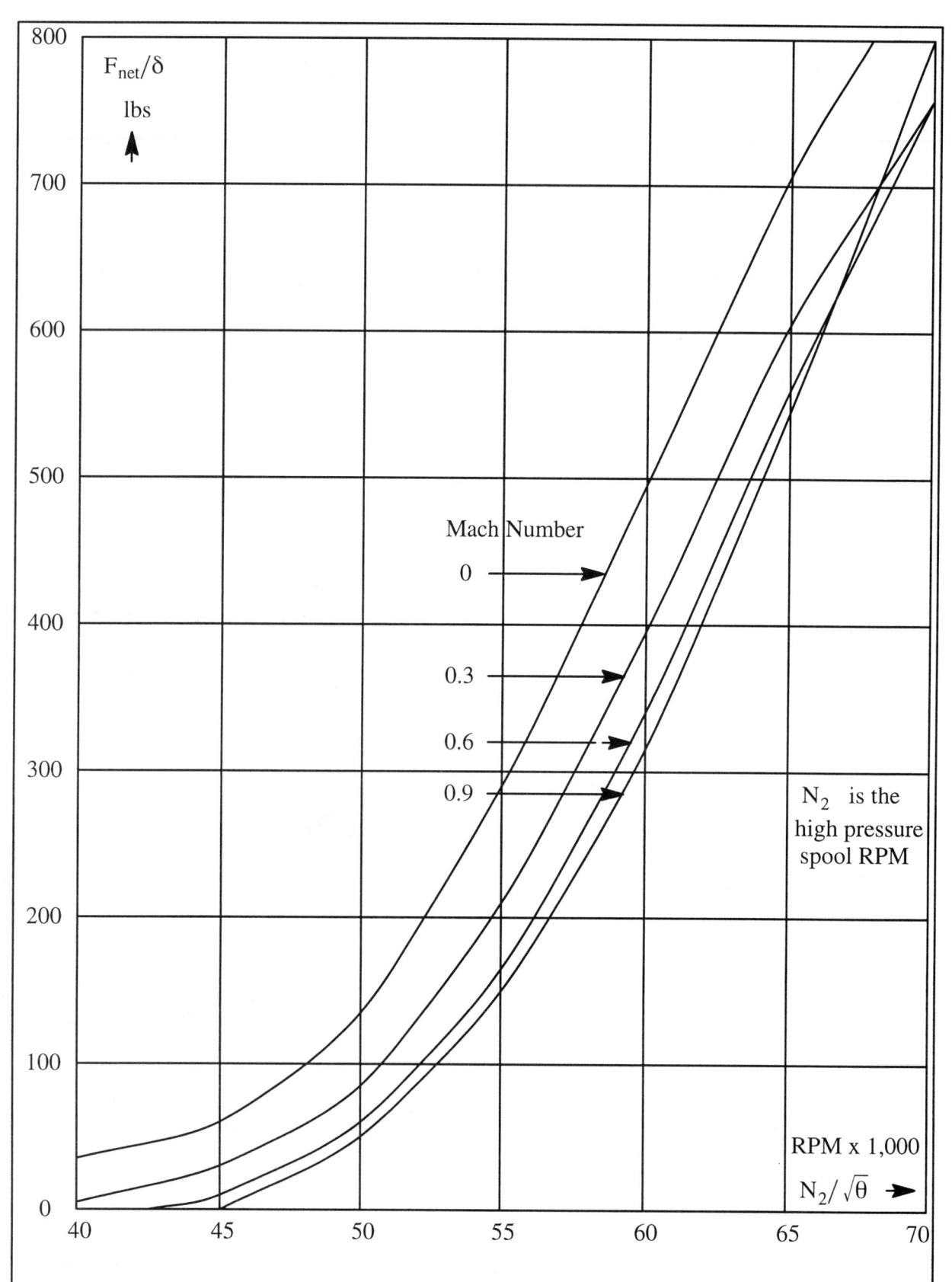

Figure 6.32a   Effect of Mach Number on the Relation between Corrected Net Trust and Corrected RPM of the High Pressure Spool

Solution: The pressure and temperature parameters are:

$$\delta = 14.5/14.7 = 0.986, \quad \theta = (460 + 30)/518.7 = 0.945 \text{ and } \sqrt{\theta} = 0.972$$

Therefore:

$$(F_{net})_{std} = \frac{F_{net}}{\delta} = \frac{6,100}{0.987} = 6,180 \text{ lbs}$$

$$(\dot{W}_a)_{std} = \frac{\dot{W}_{net}\sqrt{T}}{\delta} = \frac{107 \times 0.972}{0.986} = 105.5 \text{ lbs/sec}$$

$$(N)_{std} = \frac{N}{\sqrt{T}} = \frac{100}{0.972} = 103\%$$

It is of interest to see which components of an engine contribute to thrust or drag. Figure 6.32b shows an example thrust distribution for a single spool, axial flow turbojet.

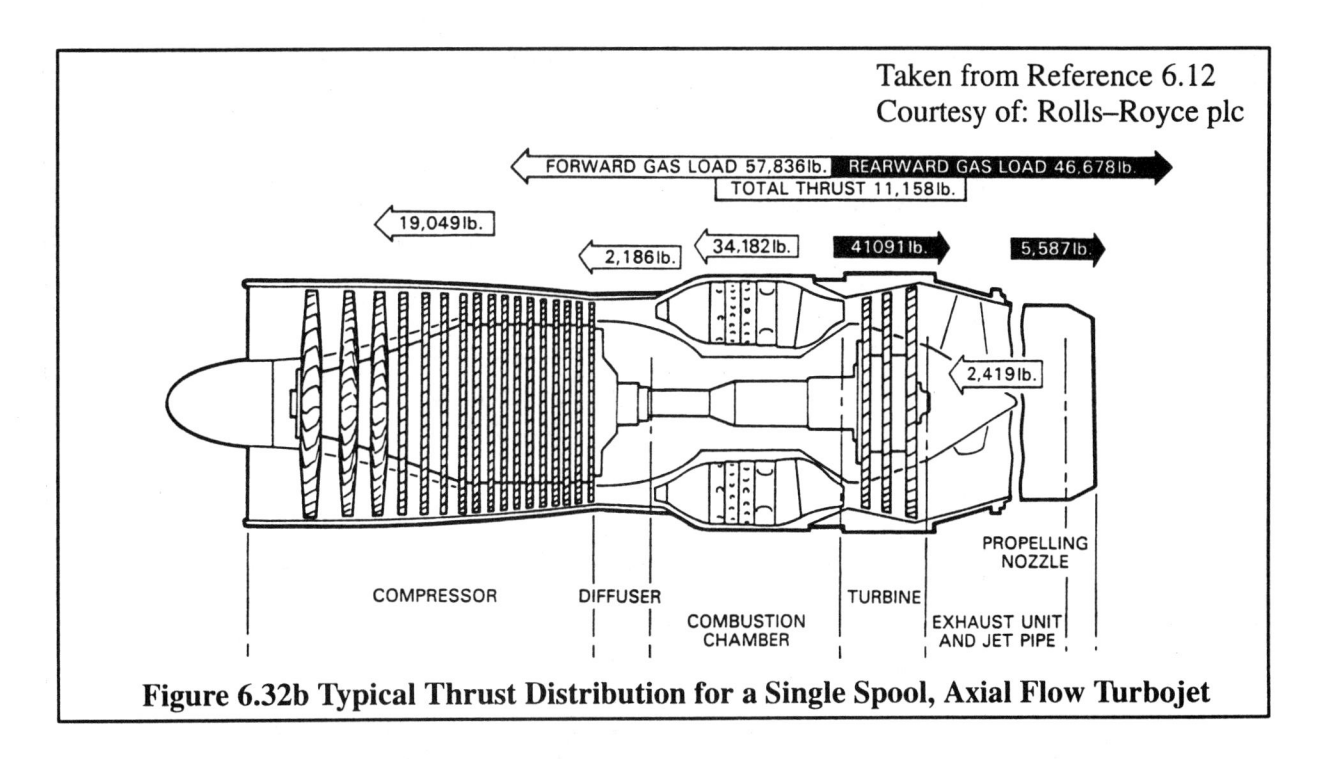

Figure 6.32b Typical Thrust Distribution for a Single Spool, Axial Flow Turbojet

Example variations of engine thrust and s.f.c. with altitude, Mach number and thrust rating are given in Figures 6.33 and 6.34 for a small and large turbo–fan respectively.

For a turbo–propeller engine, example variations of SHP and fuel flow with altitude and power rating are given in Figure 6.35.

Finally, Tables 6.1, 6.2 and 6.3 provide some data on sea–level and cruise altitude performance of a wide range of engines. The effect of engine thrust and SHP on engine dry weight is plotted logarithmically in Figure 6.36a and 6.36b respectively.

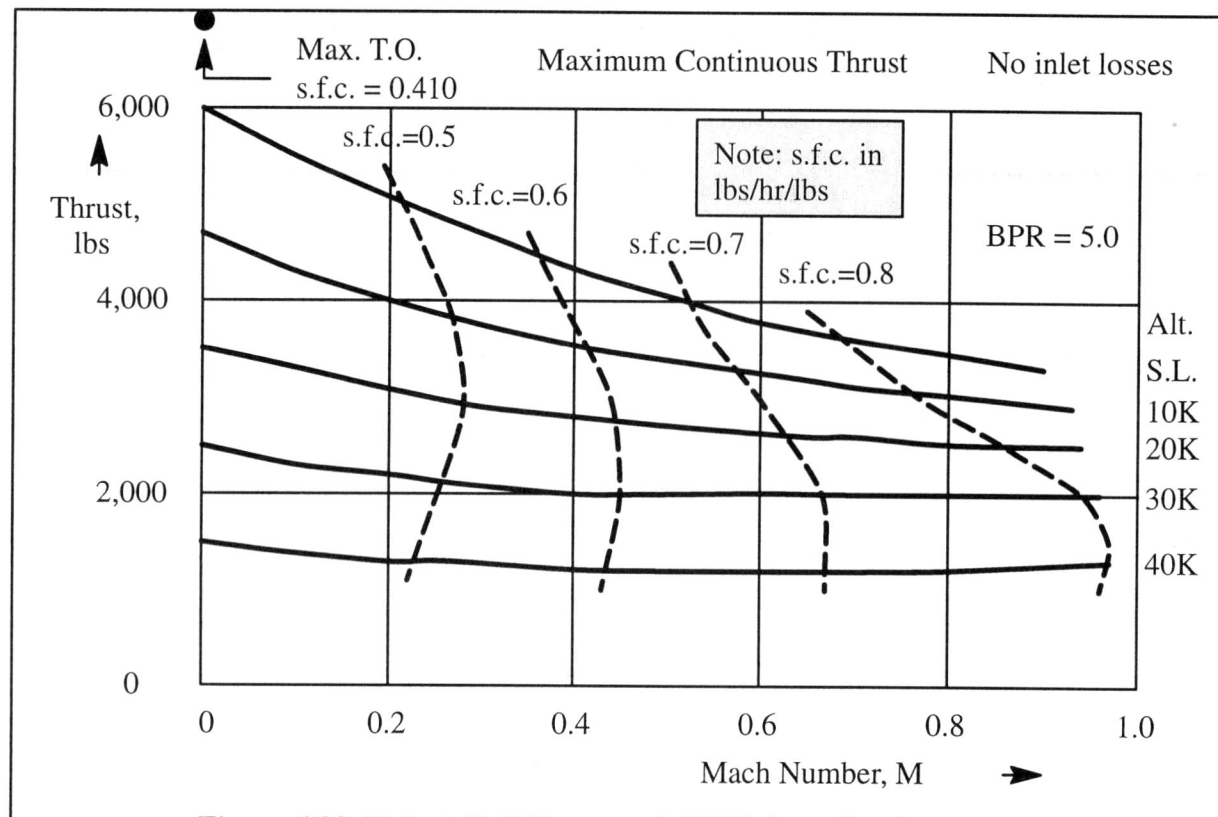

Figure 6.33 Uninstalled Thrust and S.F.C. for a Small Turbofan

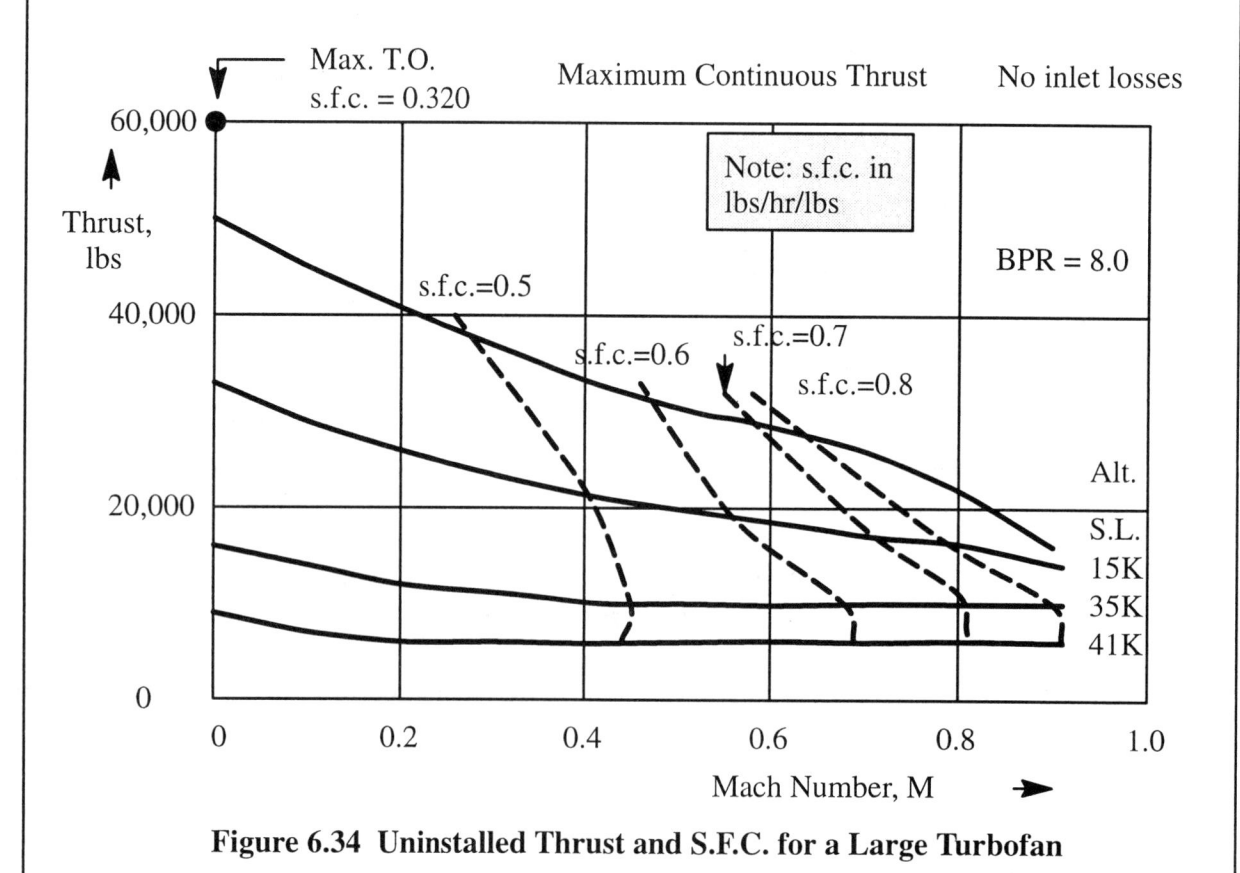

Figure 6.34 Uninstalled Thrust and S.F.C. for a Large Turbofan

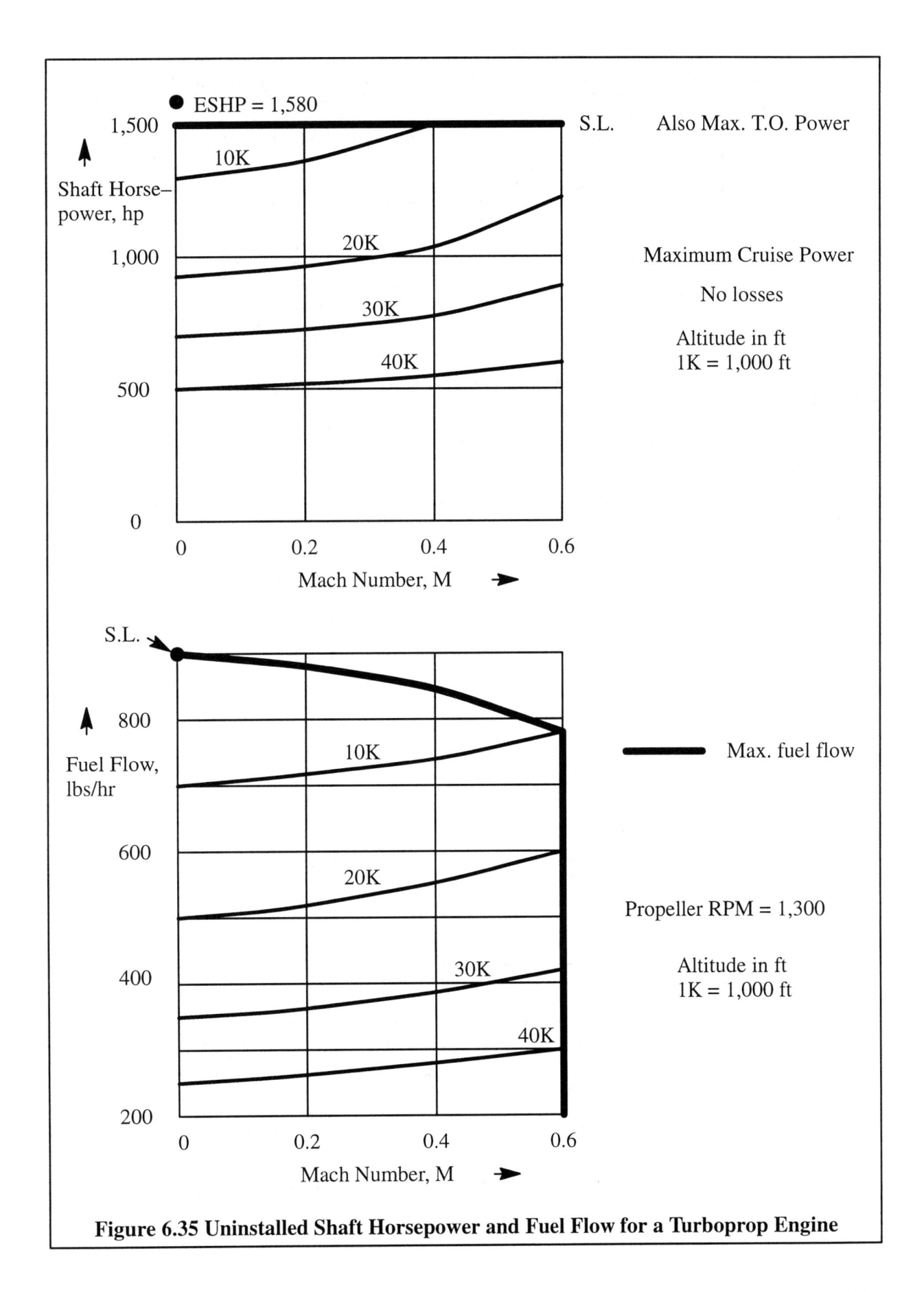

Figure 6.35 Uninstalled Shaft Horsepower and Fuel Flow for a Turboprop Engine

Table 6.1 Weight and Performance Data for Turbojet and Turbofan Engines

Manuf./Type/ Config./B.P.R.	Weight (lbs)	Thrust at take–off s.l.s. (lbs)	s.f.c. at take–off s.l.s. (lbs/hr/lbs)	Massflow take–off s.l.s. (lbs/sec)	Alt./M	Thrust at altitude/M (lbs)	s.f.c. at altitude/M (lbs/hr/lbs)
(1)/CF6–6D/ TBF/5.72	7,896	40,000	0.346	1,303	35K/0.8	7,160 (80% max)	0.616
(1)/CF6–32/ TBF/4.90	7,140	36,500	0.357	1,104	35K/0.8	6,630 (80% max)	0.609
(1)/CF6–50/ TBF/4.26	8,731	51,000	0.390	1,450	35K/0.8	8,720 (80% max)	0.628
(1)/CF6–80/ TBF/4.66	8,435	48,000	0.344	1,433	35K/0.8	8,260 (80% max)	0.592
(1)/CF34/ TBF/6.30	1,580	8,650	0.362	332	40K/0.8	1,420 (max)	0.728
(1)/CFM56–2/ TBF/6.00	4,610	24,000	???	817	35K/0.8	5,188 (max)	0.661
(1)/CJ610–5/ TBJ/1.00	402	2,950	0.980	44	36K/0.8	870 (max)	1.150
(1)/CF700/ TBF/1.93	725	4,200	0.660	126	36K/0.8	1,060 (max)	0.980
(1)/J79–17/ TBJ/0	3,873	17,820	1.980	170	35K/0.9	2,600 (max)	0.980
					35K/2.0	18,600 (max)	2.070
(1)/TF34–400/ TBF/6.2	1,478	9,275	0.363	338	36K/0.8	1,896 (intermediate)	0.682
(1)/F404–402/ TBF/0.27	2,282	17,700	???	146	???	???	???
(2)/CFE738/ TBF/5.3	1,325	5,725	0.372	210	40K/0.8	1,464 (max)	0.640
(3)/535E4/ TBF/4.3	7,264	42,000	???	1,150	35K/0.8	8,700 (max)	0.598
(3)/RB211/ TBF/4.3	9,814	60,600	???	1,604	35K/0.85	11,813 (max)	0.570
(3)/Trent800/ TBF/???	13,133	90,000	???	???	35K/0.83	13,000 (max)	0.557

Type: (1) = G.E.
(2) = G.E./Allied Signal
(3) = Rolls Royce

Manuf./Type/Config./B.P.R. = Manufacturer/Type/Configuration/By–pass Ratio

Table 6.2 Weight and Performance Data for Turbojet and Turbofan Engines							
Manuf./Type/ Config./B.P.R.	Weight (lbs)	Thrust at take–off s.l.s. (lbs)	s.f.c. at take–off s.l.s. (lbs/hr/lbs)	Massflow take–off s.l.s. (lbs/sec)	Alt./M	Thrust at altitude/M (lbs)	s.f.c. at altitude/M (lbs/hr/lbs)
(4)/TFE731-2/ TBF/2.66	743	3,500	???	113	40K/0.8	755 (max)	0.815
(4)/TFE731-5/ TBF/3.48	890	4,600	???	143	40K/0.8	1,000 (max)	0.760
(4)/ALF502L/ TBF/5.0	1,311	7,500	0.428	???	???	???	???
(5)/JT8D-219/ TBF/1.77	4,612	21,000	???	488	35K/0.8	5,250 (max)	0.737
(5)/PW4000/ TBF/4.85	9,400	56,000	???	1,705	35K/0.8	???	0.537
(5)/PW2000/ TBF/6.00	7,300	38,250	???	1,340	35K/0.8	???	0.563
(5)/PW300/ TBF/4.3	993	4,679	0.388	???	40K/0.8	1,155 (max)	0.681
(6)/CFM56-2A2 TBF/6.0	4,820	24,000	???	817	35K/0.8	5,188 (max)	0.661
(7)/IAEV2500/ TBF/4.6	5,224	30,000	???	848	35K/0.8	5,752 (max)	0.575
(8)/Larzac 04/ TBF/1.13	639	2,966	0.710	63	35K/0.8	772 (max)	???
(9)/FJ44/ TBF/3.28	447	1,900	0.475	???	36K/0.7	506 (max. cont.)	0.758
(10)/Adour 871/ TBF/0.80	1,330	5,900	0.740	???	39K/0.8	???	0.955

Type: (4) = Allied Signal

(5) = Pratt & Whitney

(6) = CFM International (G.E./SNECMA)

(7) = International Aero Engines

(8) = Turbomeca

(9) = Williams/Rolls Royce

(10) = Rolls Royce/Turbomeca

Manuf./Type/Config./B.P.R. = Manufacturer/Type/Configuration/By–pass Ratio

Table 6.3 Weight and Performance Data for Turboprop Engines					
Manufacturer/ Type	Weight (lbs)	ESHP/SHP/Prop RPM at maximum take–off setting, s.l.s.* (hp)	s.f.c. at take–off s.l.s. (lbs/hr/eshp)	ESHP/SHP/Prop RPM at maximum cruise setting, s.l.s.* (hp)	Massflow take–off s.l.s. (lbs/sec)
(1)/PT6A–11	328	528/500/2,200	0.647	528/500/2,200	???
(1)/PT6A–27	328	715/680/2,200	0.602	715/680/2,200	6.8
(1)/PT6A–34	331	783/750/2,200	0.595	783/750/2,200	???
(1)/PT6A–65B	481	1,174/1,100/1,700	0.536	1,174/1,100/1,700	9.5
(1)/PW118	861	1,892/1,800/1,300	0.498	1,593/1,513/1,300	???
(1)/PW123B	992	2,626/2,500/1,200	0.463	2,136/2,030/1,200	???
(1)/PW115	841	1,580/1,500/1,300	0.544	1,580/1,500/1,300	???
(2)/TPE331–1	336	705/665/???	0.605	690/650/???	6.2
(2)/TPE331–10	380	???/1,000/???	0.560	???	???
(3)/CT7–5	676	1,685/1,630/???	0.466	1,465/1,417/???	???
(3)/CT64–820–4	1,145	3,133/3,133/1,160	0.486	2,745/2,745/1,015	???
(4)/Dart RDa7 Mk535–2/TS1637	1,369	????/2,080/????	0.760	????/1,835/????	23.5
(5)/250–B17C	195	420/420/????	0.657	369/369/????	3.45
(5)/250–C30	240	650/650/????	0.592	557/557/????	5.6
(5)/T56–A–15	1,825	4,910/4,591/1,106	0.501	4,365/4,061/1,106	32.4

Type: (1) = Pratt & Whitney, Canada   (3) = General Electric

(2) = Allied Signal   (4) = Rolls Royce

(5) = Allison

* In calculating ESHP from SHP it is usually assumed, that 2.5 lbs of axial jet thrust is equivalent to 1 hp.

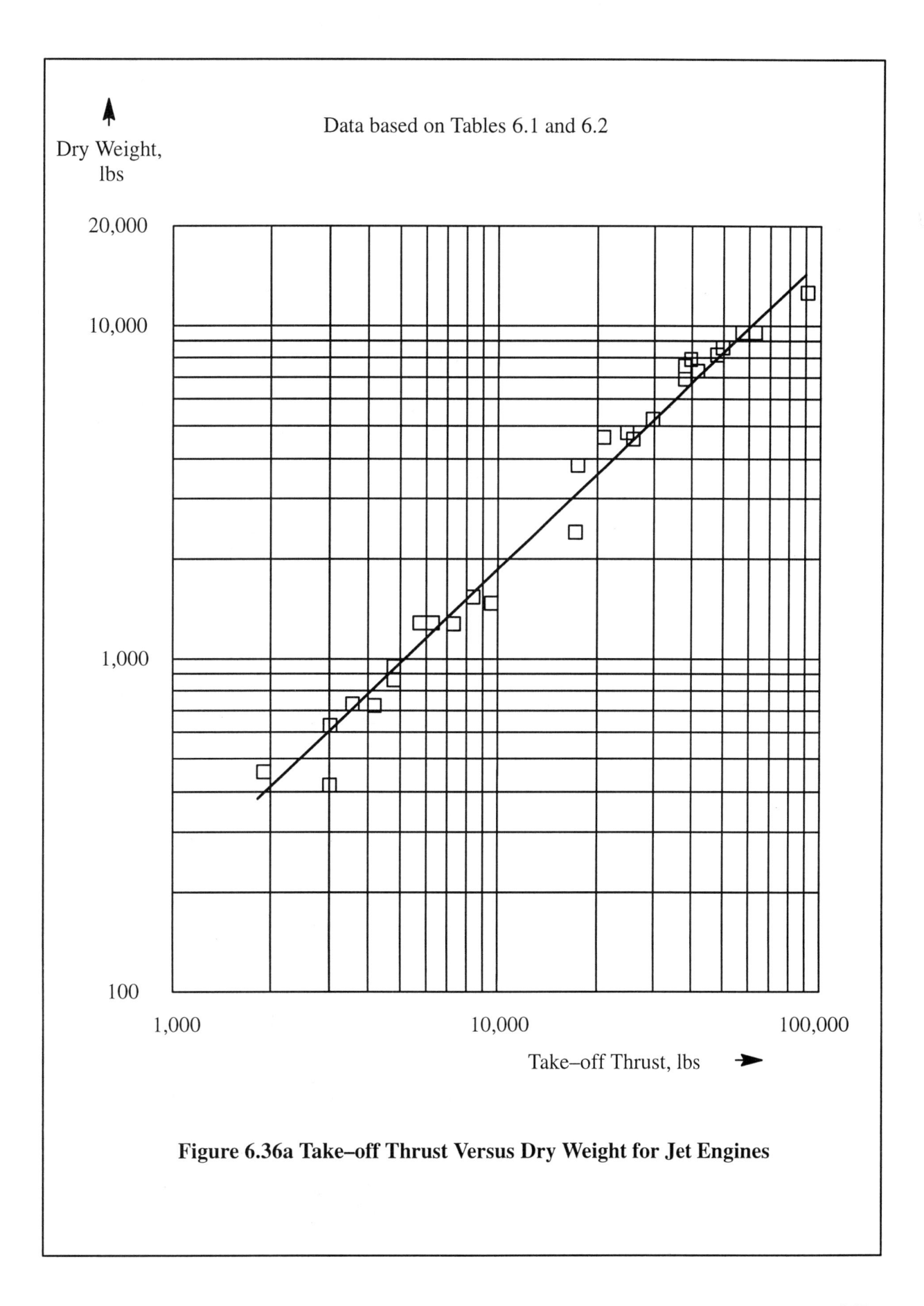

Figure 6.36a Take-off Thrust Versus Dry Weight for Jet Engines

Data based on Table 6.3

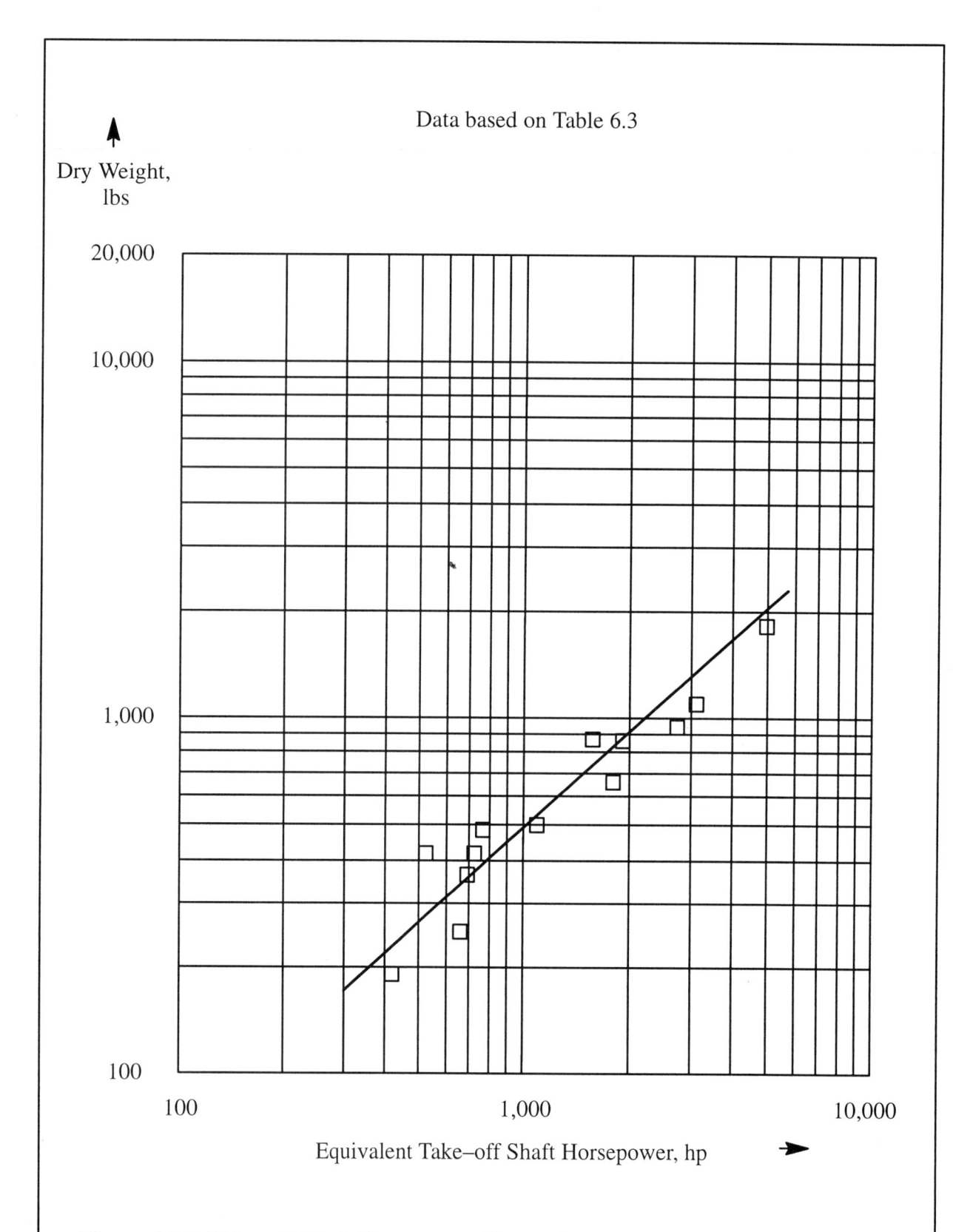

Figure 6.36b Take–off Shaft Horsepower Versus Dry Weight for Turboprop Engines

6.2.6 COOLING CONSIDERATIONS FOR TURBINE ENGINE INSTALLATIONS

Various components of jet engines need to be cooled to ensure safe and efficient operation. This is particularly true for the combustor, for the turbine blades and for various accessories. Because of the direct relationship between turbine inlet temperature and engine operating efficiency, much development emphasis has been given to combustor and turbine blade materials and designs which can tolerate such high temperatures. In fact, many of today's turbine engines operate at turbine inlet temperatures which are above the melting point of the materials used in the turbine blades. Adequate cooling techniques are a must. Figure 6.37 shows an example of how the cooling of turbine blades has changed from the 60's to today.

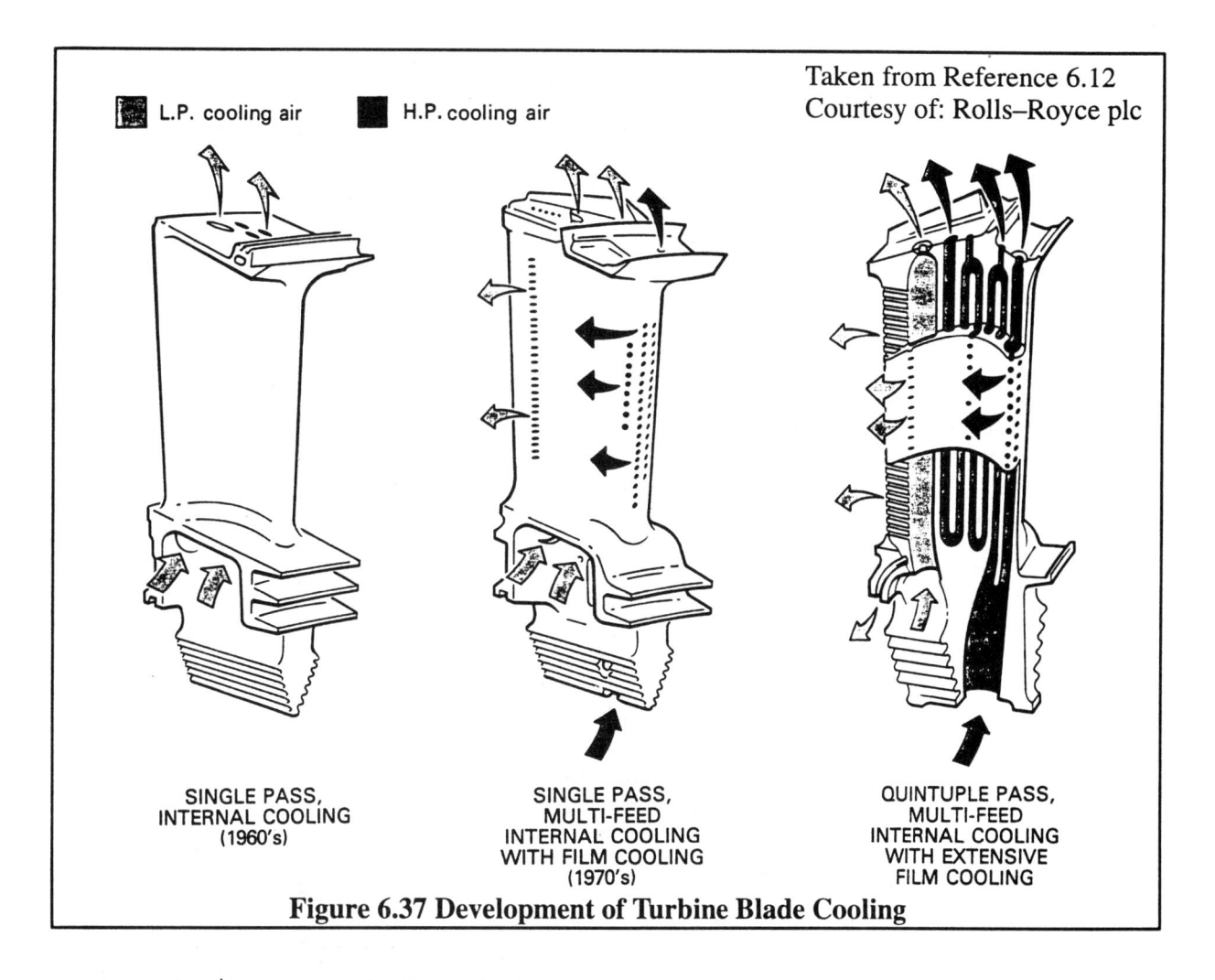

Figure 6.37 Development of Turbine Blade Cooling

Several engine accessories, in particular the engine generator, must also be cooled. In flight, this is done by ducting outside air from special cooling air intakes toward the accessories. During ground operation this does not work and low pressure air is tapped from the compressor and ducted to the accessories. This would hurt the efficiency of the engine in flight and therefore a valving system is used to switch from external air to compressor air and vice–versa. A schematic of an arrangement for cooling the engine generator is shown in Figure 6.38. The effect of the cooling air requirements on drag and installed thrust must be accounted for: methods for doing this are given in Ref. 6.15.

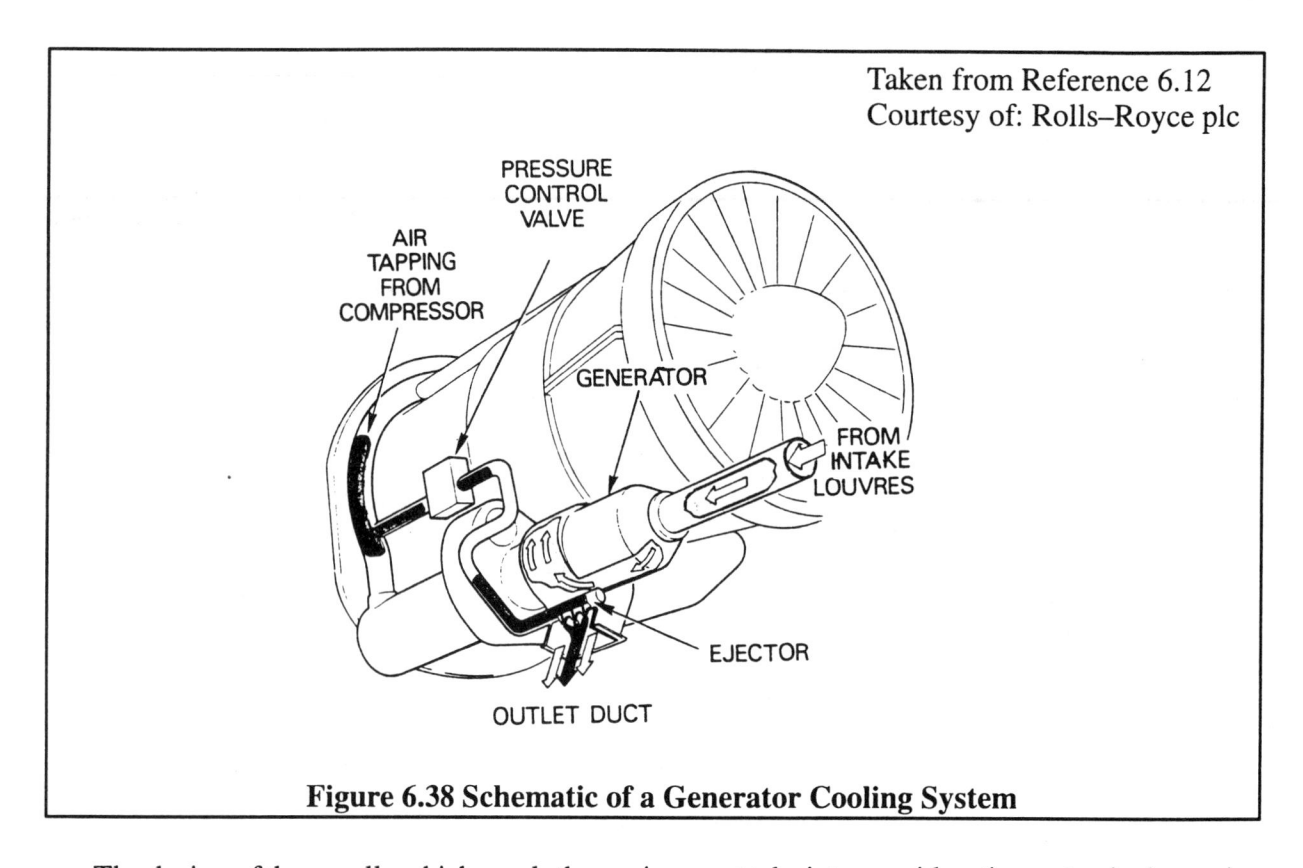

Figure 6.38 Schematic of a Generator Cooling System

The design of the nacelle which cowls the engine must take into consideration not only the engine envelope and its cooling and accessibility requirements, but also the envelope and location of the accessory drive system. Figure 6.39 shows some examples of accessory drive installations.

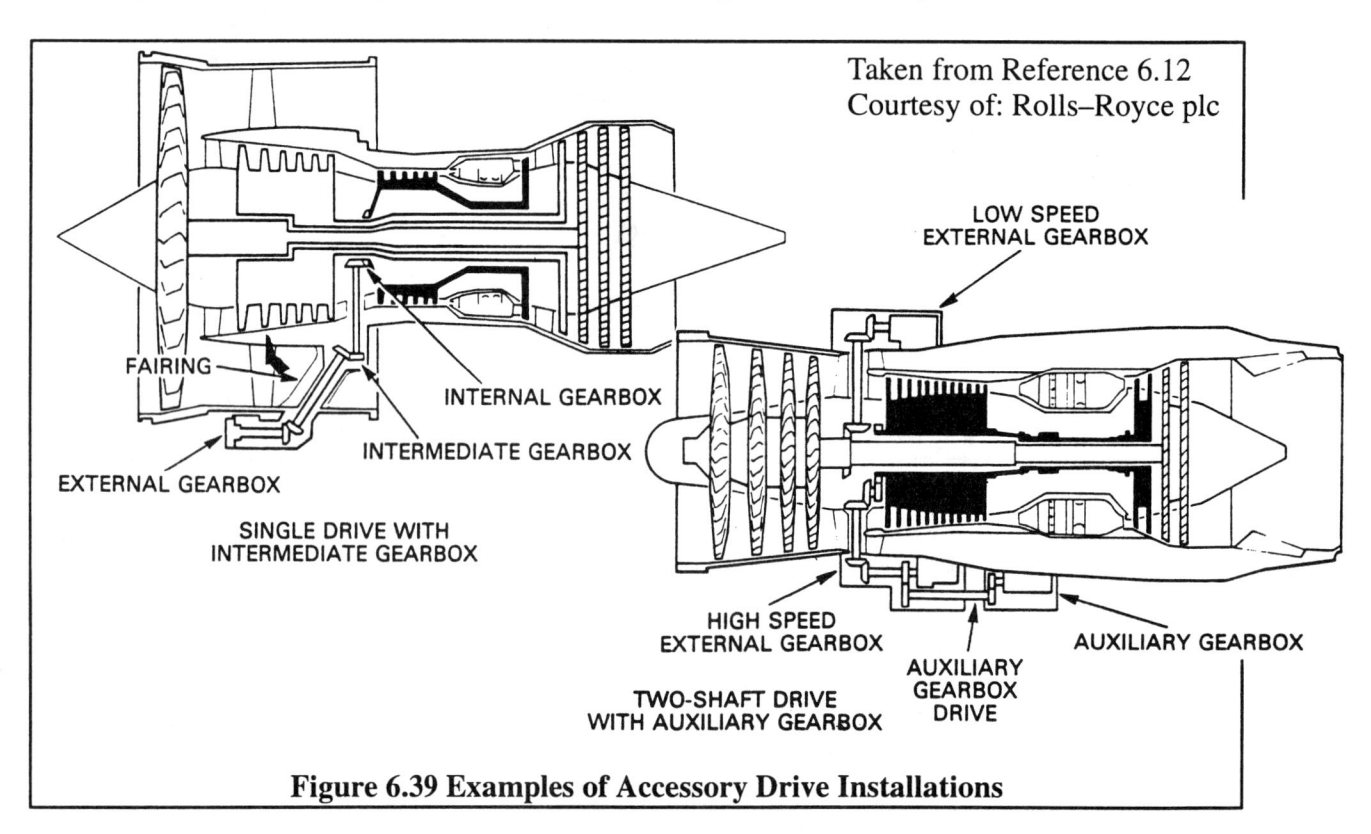

Figure 6.39 Examples of Accessory Drive Installations

It should be expected that of large power outputs are required for the accessory drive system, they will tend to require a significant amount of volume. This in turn will affect the size and shape of the nacelle which in turn affects the weight and drag of the airplane.

Also, structural provisions must be made to mount the engine to the airframe. These structural provisions must take into account the weight and the thrust output of the engine. This also requires additional volume in the nacelle. Example engine installations are shown in Figure 6.40.

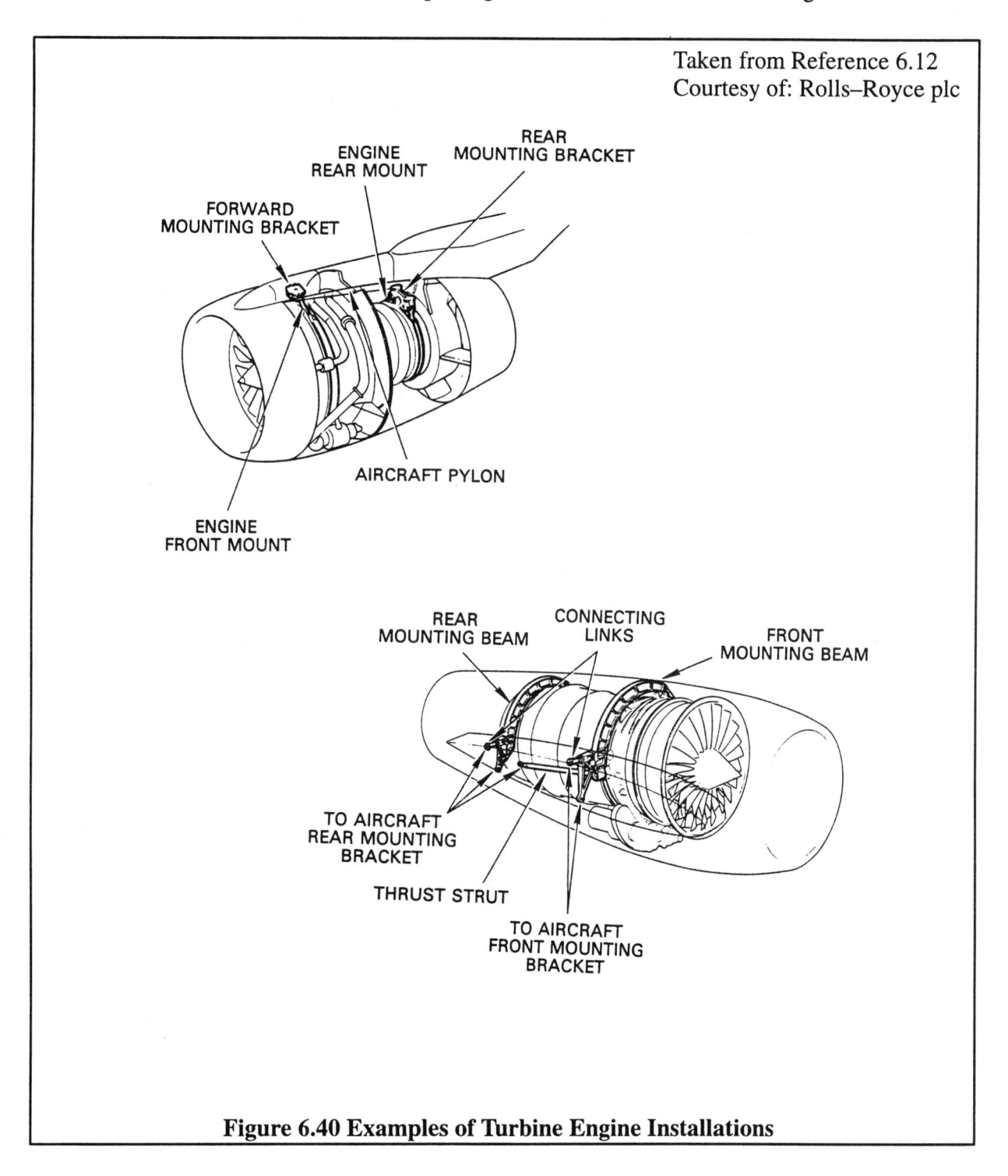

Figure 6.40 Examples of Turbine Engine Installations

6.2.7 NOISE CONSIDERATIONS FOR GAS TURBINE ENGINES

As a result of serious public complaints, airplane noise has become an important design and development consideration since the late 1960's. Maximum allowable noise limits were set with the introduction of FAR 36, also in the late 60's. There have been three stages in the setting of noise certification standards for airplanes: Stages 1, 2 and 3. The requirements under Stages 1 and 2 allow higher noise levels for old technology airplanes. Conditions under which these levels are still acceptable are outlined in FAR 36. The higher noise levels are allowed in view of the fact that operating economics dictate that many older (noisy) airplanes will operate well into the next century.

Since 1975, all newly designed airplanes must meet the FAR 36, Stage 3 requirements. A graphical representation of the allowable noise levels as they apply to turbojet driven transports is given in Figure 6.41.

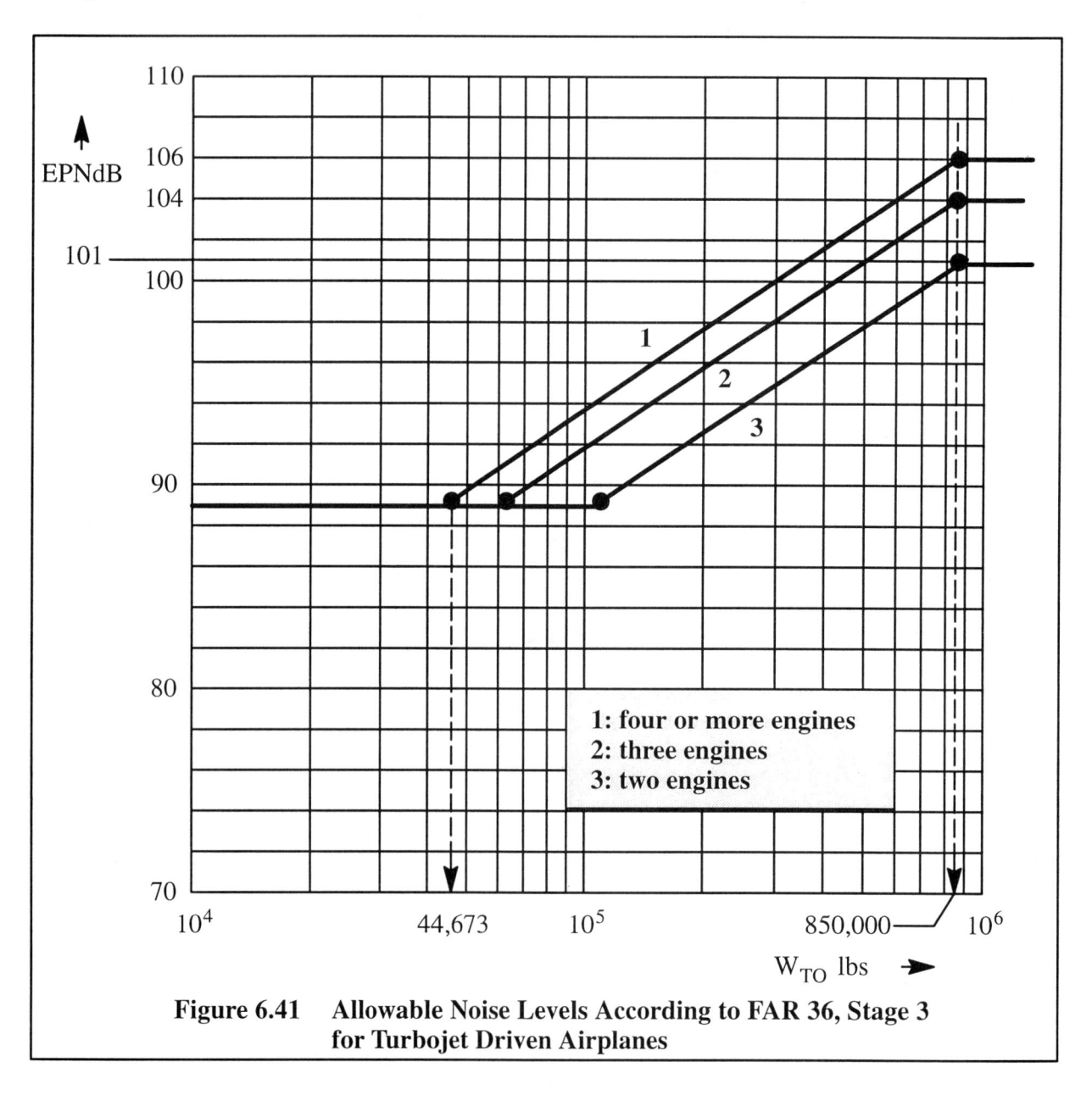

Figure 6.41 Allowable Noise Levels According to FAR 36, Stage 3 for Turbojet Driven Airplanes

It will be noted from Figure 6.41 that the allowable noise levels are stated in EPNdB (Equivalent Perceived Noise in dB) for turbojet driven airplanes. FAR 36 also specifies how noise must be measured and analyzed during take–off, fly–over or approach to landing to arrive at numerical values for actual airplane EPNdB levels. The required noise measurements must be taken at precisely defined locations relative to the runway. Figure 6.42 shows what these locations are for turbojets..

Human perception of noise is complicated by the fact that noise level, tone and duration all play a role. A detailed discussion of noise, its perception and its contributory sources is considered to be beyond the scope of this text. The reader is referred to FAR 36. The latter contains a detailed specification of what constitutes noise, how it is to be measured for purposes of aircraft certification and how it is to be analyzed to obtain EPNdB and dB(A) data.

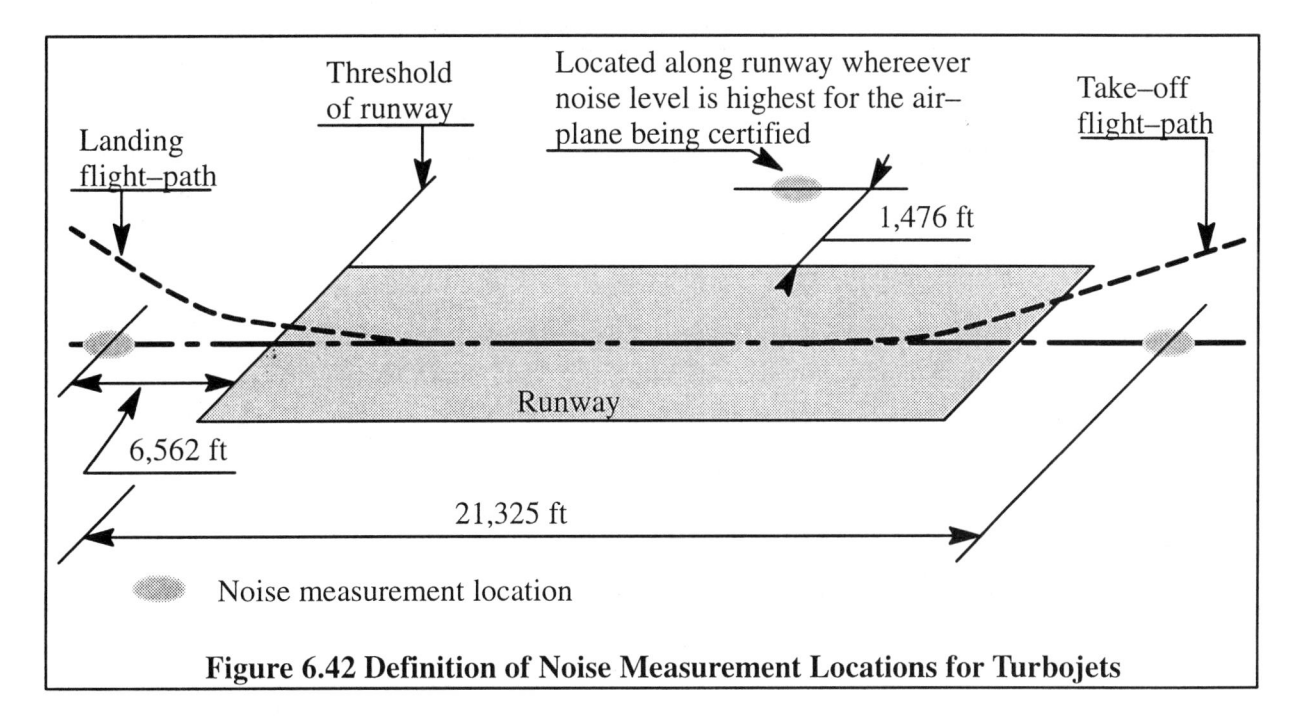

Figure 6.42 Definition of Noise Measurement Locations for Turbojets

Sources of noise which contribute significantly to the perceived noise of an airplane during take–off, fly–over or approach to landing are:

* Engine noise   * Propeller noise   * Airframe noise

In the case of airplanes equipped with gas turbine engines, the engine noise constitutes the primary source of noise. Chapter 7 contains some information on propeller noise as well as references on that subject. For more details on engine noise the reader may consult Refs 6.14, 6.16 and – 6.17. In this Sub–section only some very basic considerations of engine noise are included.

The primary sources of engine noise are qualitatively depicted in Figure 6.43 for a turbojet and a turbo–fan respectively. Note that turbine noise, jet noise, compressor noise and fan noise all play individual roles. Many design details of these engine components contribute their own noise levels, and tones during the time it takes an airplane to pass over (or by) a person.

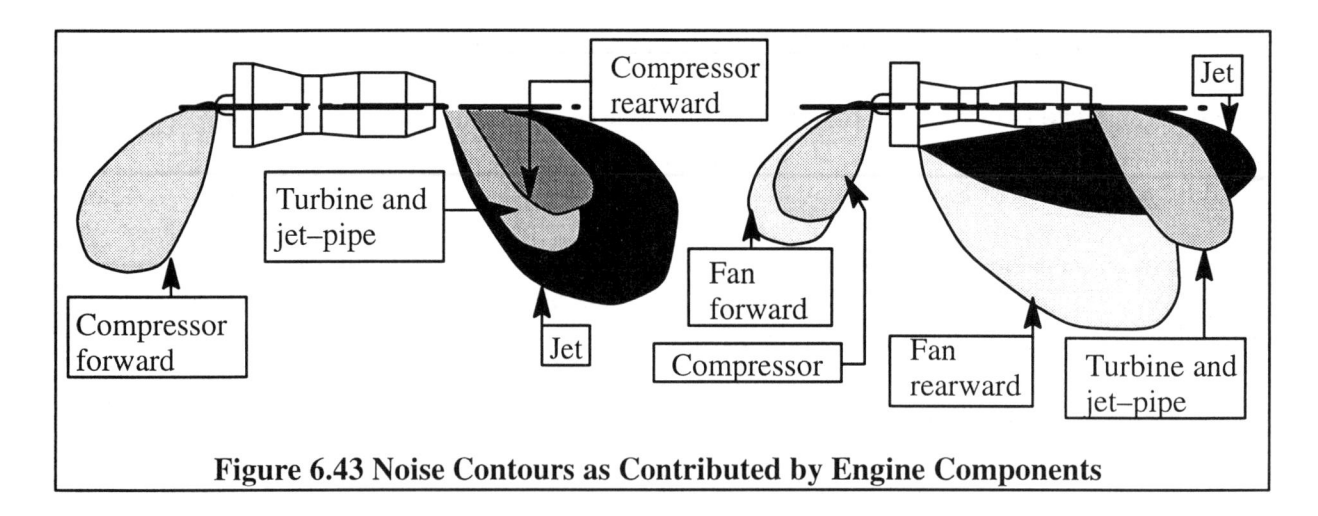

Figure 6.43 Noise Contours as Contributed by Engine Components

The jet noise can be reduced by careful mixing of external air or fan air with the exhaust gases. In many engines this is still being achieved with noise suppressors. Combustor noise and machinery noise such as that produced by fans, compressors and turbines can be reduced by the application of various types of liners. Reference 6.11 contains some examples of these techniques. All these methods do increase weight and also tend to increase fuel consumption.

It has been found that when the bypass ratio exceeds roughly 5.0 the internal machinery noise emanating from the engine (fan, compressor and turbine) dominates over the jet noise. Figure 6.44 shows the trends of noise reduction achieved with noise suppression techniques and high bypass ratio engines. Apparently, an engine noise floor is being approached.

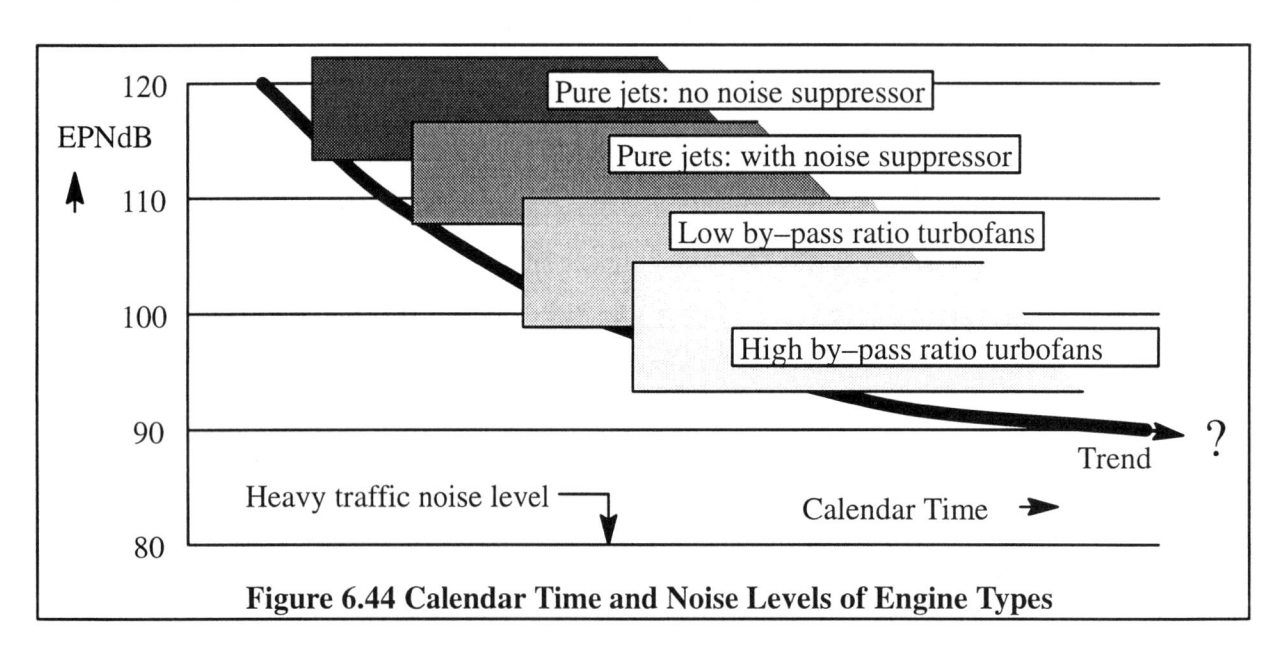

Figure 6.44 Calendar Time and Noise Levels of Engine Types

6.2.8 THRUST REVERSAL FOR TURBINE ENGINES

Reversing the thrust of an airplane during the ground part of the landing process has become an operational necessity. To reverse thrust in the case of a low by-pass ratio engine requires that the hot exhaust gases be reversed. This is usually accomplished with a set of clamshell doors or with bucket doors. Examples of both are shown in Figure 6.45.

Taken from Reference 6.12
Courtesy of: Rolls-Royce plc

CLAMSHELL DOORS IN FORWARD THRUST POSITION

CLAMSHELL DOORS IN REVERSE THRUST POSITION

ACTUATOR EXTENDED AND BUCKET DOORS IN FORWARD THRUST POSITION

ACTUATOR AND BUCKET DOORS IN REVERSE THRUST POSITION

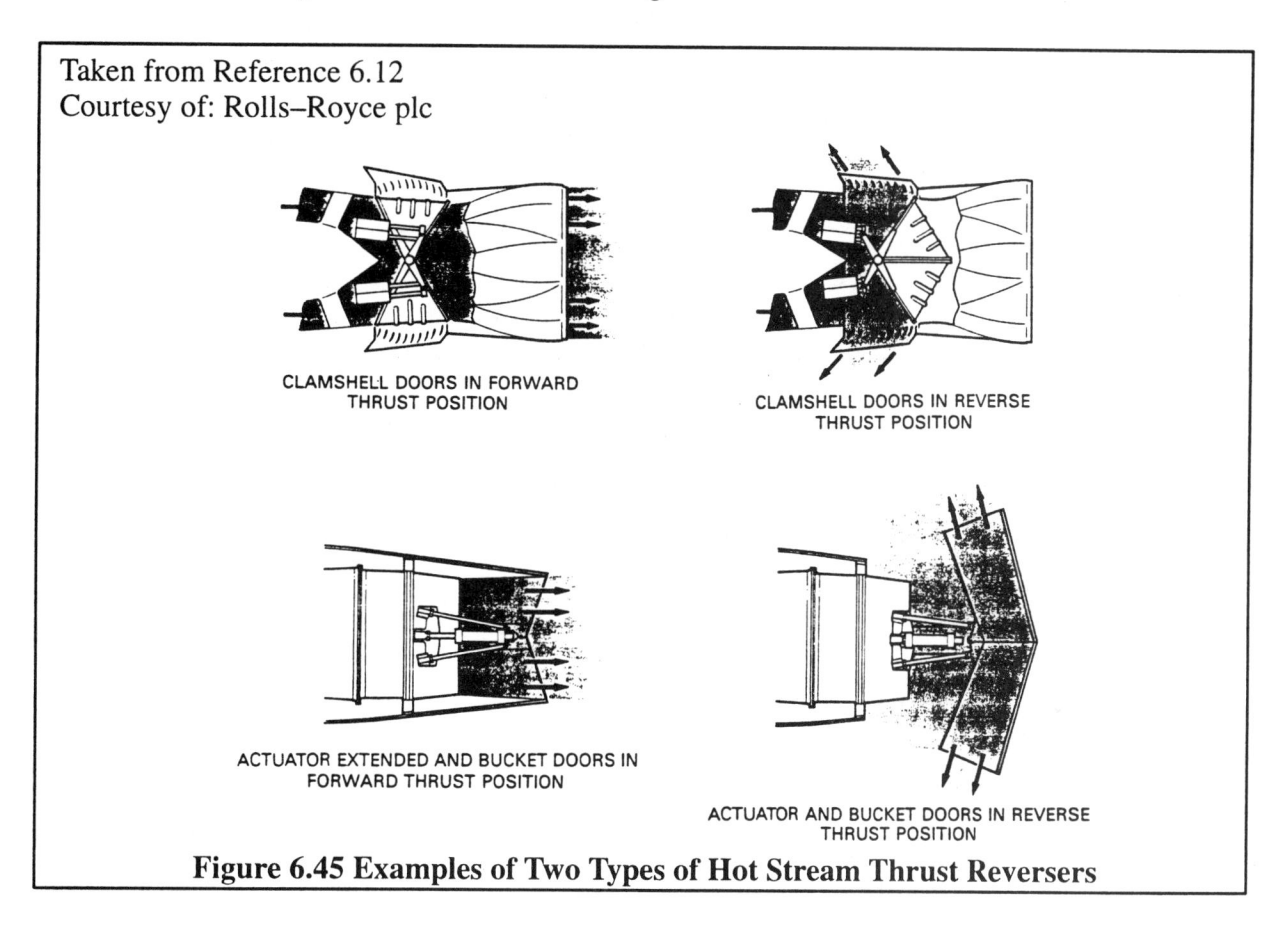

Figure 6.45 Examples of Two Types of Hot Stream Thrust Reversers

In the case of high bypass ratio engines the cold air-stream can be reversed. An example of how that can be done is shown in Figure 6.46.

Taken from Reference 6.12
Courtesy of: Rolls-Royce plc

COLD STREAM REVERSER IN FORWARD THRUST POSITION

COLD STREAM REVERSER IN REVERSE THRUST POSITION

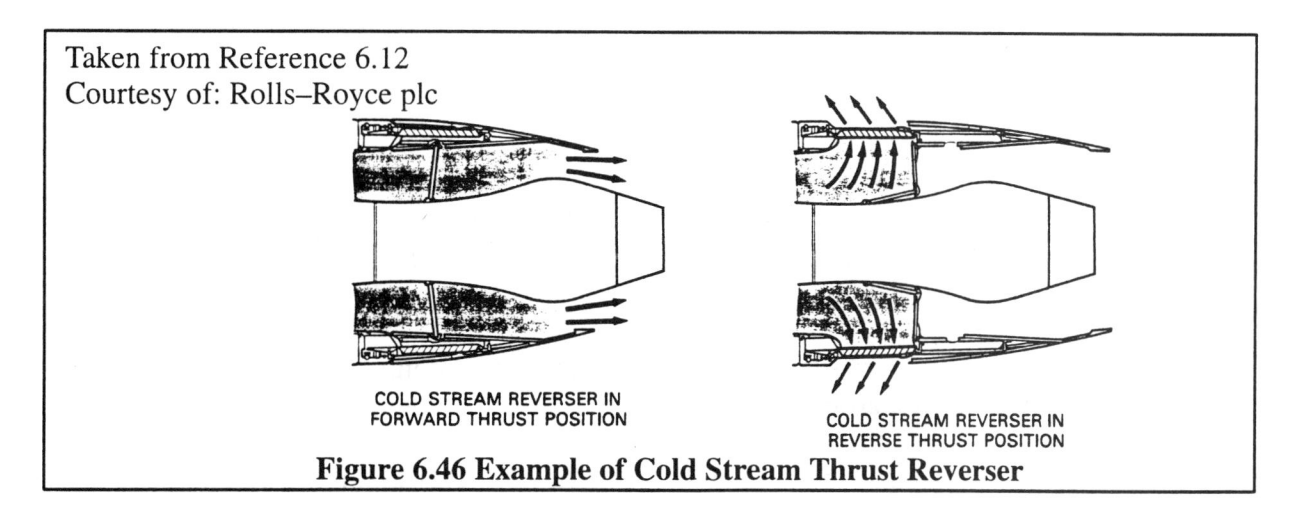

Figure 6.46 Example of Cold Stream Thrust Reverser

6.2.9 SCALING LAWS FOR TURBINE ENGINES

When analyzing a new airplane design it is frequently not possible to use an already existing engine. In such instances so–called "rubberized" engine data are required which relate the most important size and the weight characteristics of the engine to the required thrust output for the new design. Assume that Figure 6.47 represents the installation drawing of an existing engine for which thrust and s.f.c. data are available. It is also assumed that the dry weight of that engine is known.

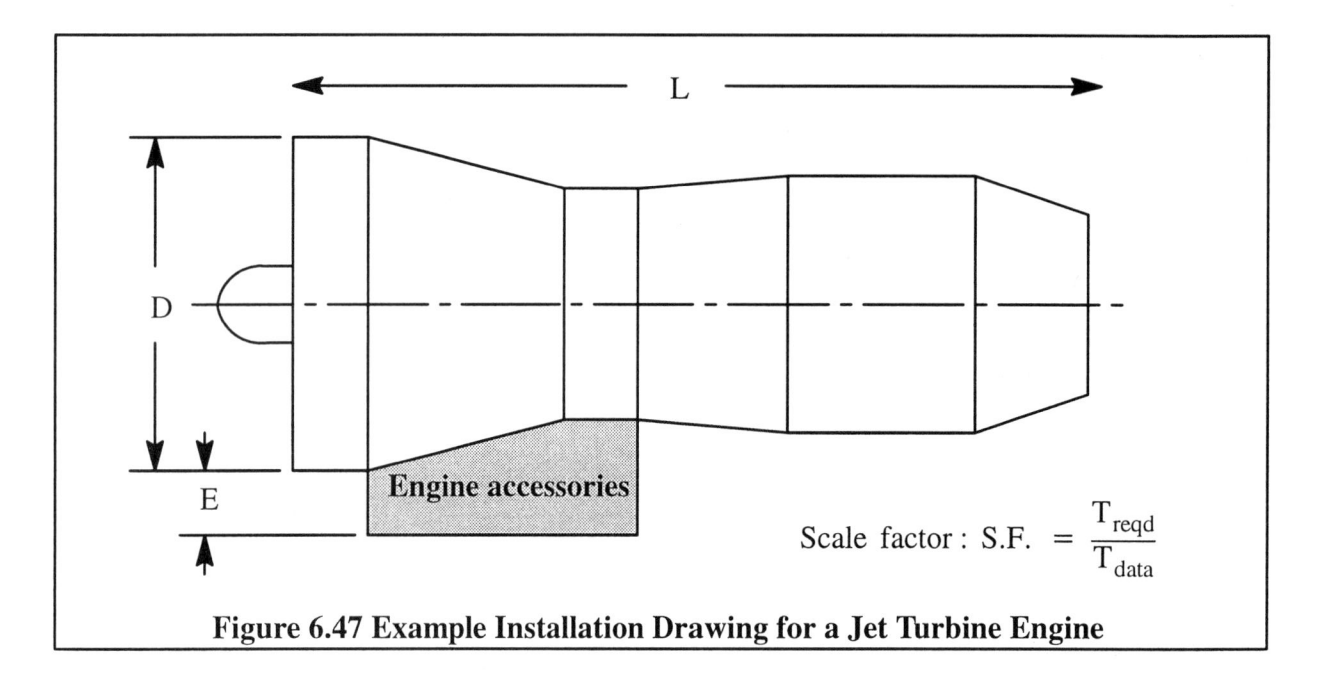

Scale factor : S.F. $= \dfrac{T_{reqd}}{T_{data}}$

Figure 6.47 Example Installation Drawing for a Jet Turbine Engine

Assume, that in some flight condition, the required thrust for a new engine design is: T_{reqd}. Also assume, that the thrust available, for the same flight condition, of an actual engine for which data are available is: T_{data}. In such a case, the following scaling rules (taken from Ref. 6.18) can be used to estimate the length, diameter and weight of the new engine:

$$L_{reqd} = L_{data}(S.F.)^{0.4} \qquad (6.39)$$

$$D_{reqd} = D_{data}(S.F.)^{0.5} \qquad (6.40)$$

$$W_{e_{reqd}} = W_{e_{data}}(S.F.)^{1.1} \qquad (6.41)$$

where: S.F. is the scale factor defined in Figure 6.47.

Observe the engine accessories box underneath the engine. For most engines that is where the accessory box is located. A reasonable assumption for the dimension E is: E = 0.25D.

It is also possible to estimate the s.f.c. at take–off, the s.f.c. at cruise, the cruise thrust, length, diameter and weight for a new engine when the design maximum take–off thrust is known. The following equations were also taken from Ref. 6.18.

a) For non–after–burning engines:

$$\text{s.f.c}_{\text{take-off}} = (0.67)e^{(-0.12\,\text{BPR})} \tag{6.42}$$

$$\text{s.f.c}_{\text{cruise}} = (0.88)e^{(-0.05\,\text{BPR})} \tag{6.43}$$

$$T_{\text{cruise}} = (0.60)(T_{TO})^{0.9} e^{(0.02\,\text{BPR})} \tag{6.44}$$

$$L = (2.22)(T_{TO})^{0.4}(M_{\text{max}})^{0.2} \tag{6.45}$$

$$D = (0.393)(T_{TO})^{0.5} e^{(0.04\,\text{BPR})} \tag{6.46}$$

$$W_e = (0.084)(T_{TO})^{1.1} e^{(-0.045\,\text{BPR})} \tag{6.47}$$

b) For after–burning engines:

$$\text{s.f.c}_{\text{take-off}} = (2.1)e^{(-0.12\,\text{BPR})} \tag{6.48}$$

$$\text{s.f.c}_{\text{cruise}} = (1.04)e^{(-0.186\,\text{BPR})} \tag{6.49}$$

$$T_{\text{cruise}} = (1.60)(T_{TO})^{0.74} e^{(0.023\,\text{BPR})} \tag{6.50}$$

$$L = (3.06)(T_{TO})^{0.4}(M_{\text{max}})^{0.2} \tag{6.51}$$

$$D = (0.288)(T_{TO})^{0.5} e^{(0.04\,\text{BPR})} \tag{6.52}$$

$$W_e = (0.063)(T_{TO})^{1.1}(M_{\text{max}})^{0.25} e^{(-0.81\,\text{BPR})} \tag{6.53}$$

where: e = 2.7183

BPR is the engine by–pass rato

T_{TO} is the maximum take–off thrust

M_{max} is the maximum level flight Mach number

For methods to size new engines and to predict their performance Ref. 6.13 is recommended.

6.2.10 PULSE–JETS

Strictly speaking the pulse–jet is an intermittent combustion engine. However, because its propulsive force comes directly from the momentum changes in a gas flow it is included as a jet engine. There are two types of pulse–jets: valve–type and acoustic–type. Figure 6.48 shows a cross section of a pulse–jet with the shutter valves in two positions: open and closed.

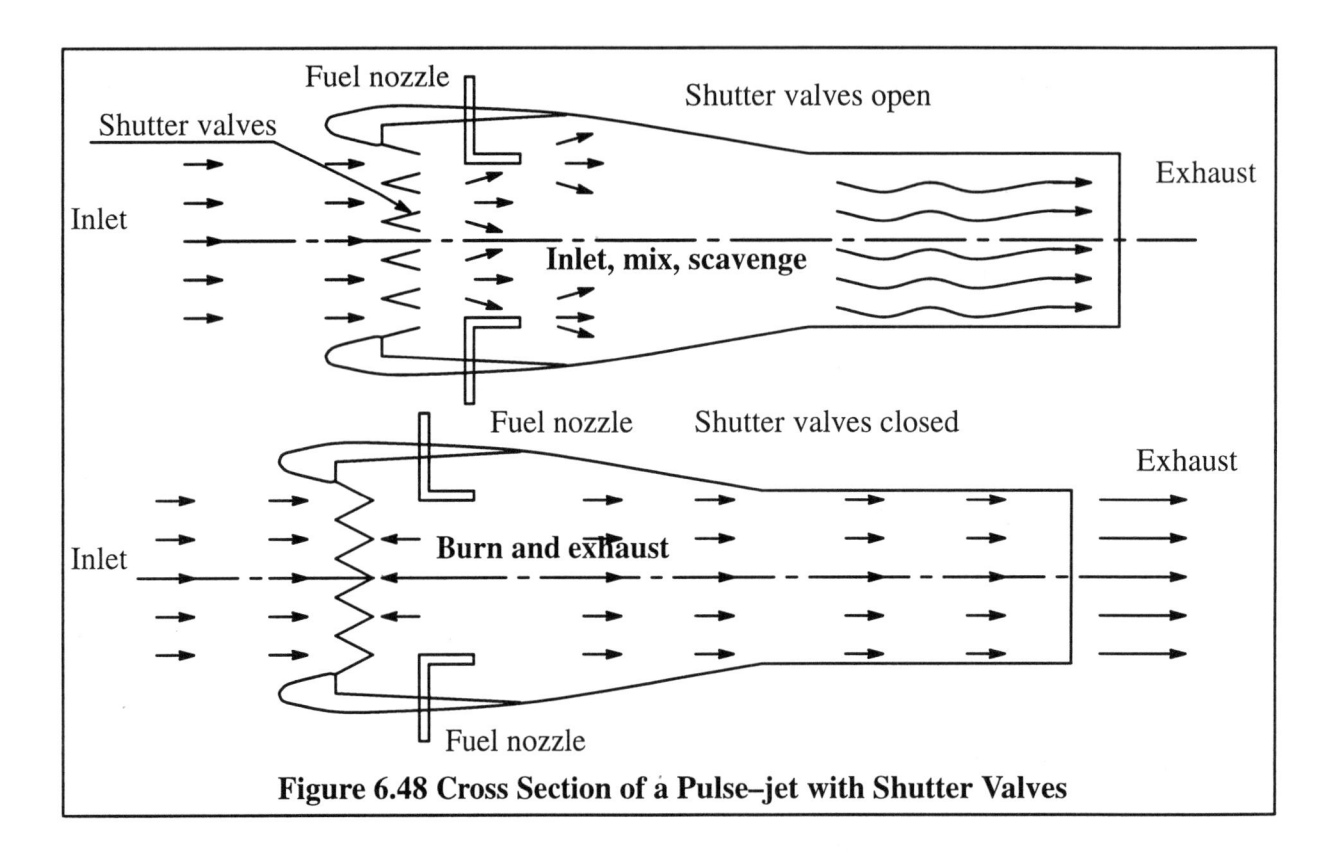

Figure 6.48 Cross Section of a Pulse–jet with Shutter Valves

With the valves open a new charge of air is admitted. The air mixes with the fuel. Next, an explosion takes place which shuts the valves and forces the combustion products out the exhaust. Therefore, a pulsating thrust of a given frequency is created.

The valves are simple leaf–spring type of shutters. Except for these valves, the pulse–jet does not have any moving parts. Pulse–jets are therefore very simple and cheap to construct. Their fuel consumption is very high and the accompanying noise is unacceptable by modern standards. This type of pulse–jet was employed on the German WWII V–1 flying bombs built by Fieseler Aircraft.

Figure 6.49 shows a cross section of an acoustic–type pulse–jet. In this type of pulse–jet the combustion process creates two shock wave fronts, one travelling down the long upper tube, the other down the short lower tube. By properly "tuning" of the system a stable, resonating combustion process can be achieved which yields a considerable thrust. This type of pulse–jet does not contain any moving parts at all. Fuel consumption is very high. Again, the noise level is unacceptable by modern standards. The French engine manufacturer SNECMA developed these pulse–jets in the late fourties for use on drones. One application was the Dutch AT–21 target drone built by Aviolanda Aircraft from 1954–1958.

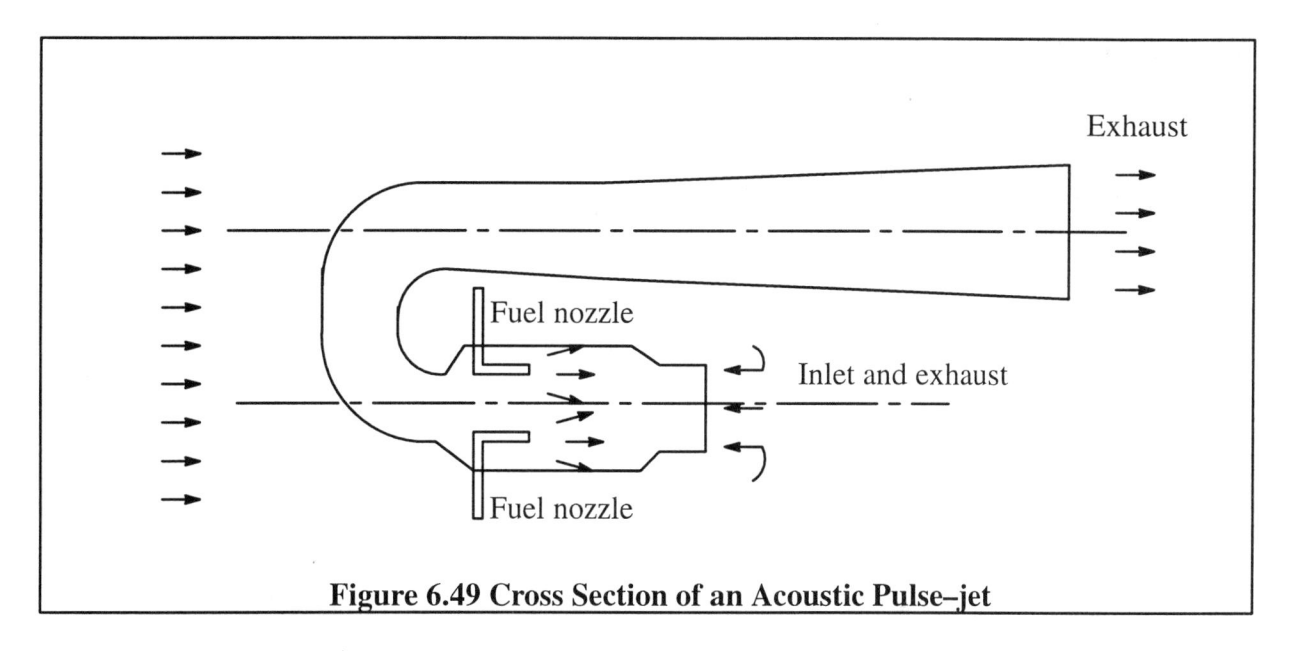

Figure 6.49 Cross Section of an Acoustic Pulse-jet

6.2.11 RAMJETS

Figure 6.50 gives an example schematic of a ramjet engine.

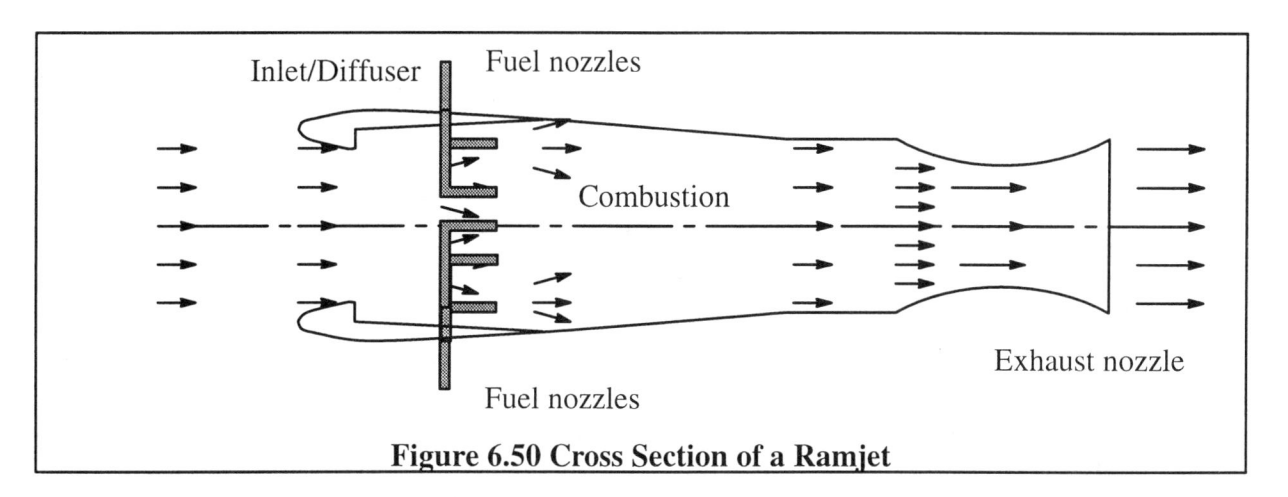

Figure 6.50 Cross Section of a Ramjet

A ramjet operates as follows: as the air enters the inlet it is compressed adiabatically (approximately) which causes its temperature to increase and its velocity to decrease. The air is further heated by the combustion. Also, as a result of the injected fuel, the mass of the mixture increase in the range of 5%–10%. The high temperature gases expand in the nozzle and leave at high velocity. The thrust of a ramjet is equal to the net rate of change of momentum of the gases passing through the engine minus the external drag of the engine. Ramjets also are very simple to build but they do have high fuel consumption. Ramjets do not deliver net thrust at low speeds. In other words, some minimum forward speed is required to obtain net thrust. Ramjets have been used in a variety of missiles. The Boeing Bomarc of the fifties is one example. Various high speed forms of ramjets, known as air–turbo ramjets and scram–jets are finding applications in very high supersonic and hypersonic applications. References 6.14 and 6.19 contain some speculative information about some of these applications.

6.3 ROCKET ENGINES

At very high altitudes, where air–breathing engines are no longer practical, the rocket engine offers a viable alternative method of aircraft propulsion. Fundamentally, there are two types of rockets: liquid fuel and solid fuel rockets.

An example arrangement for a liquid fuel rocket engine is given in Figure 6.51

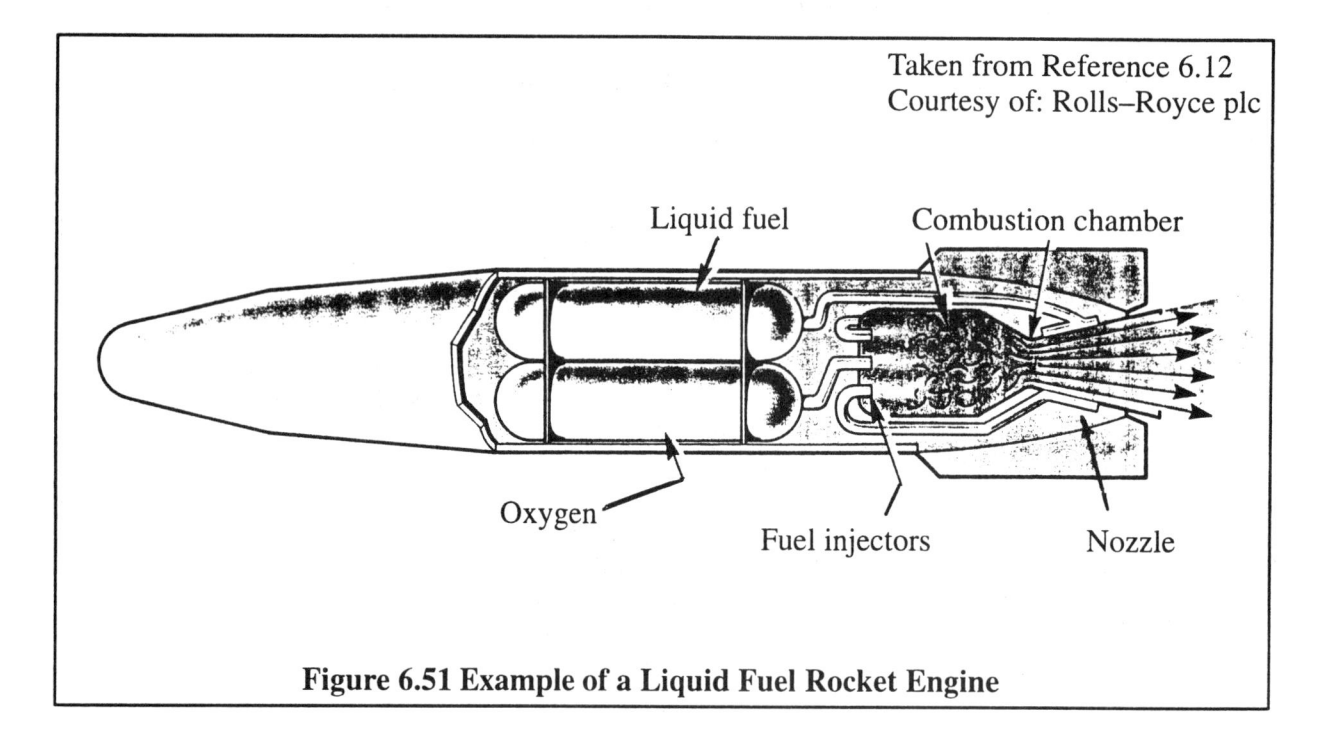

Figure 6.51 Example of a Liquid Fuel Rocket Engine

Liquid oxygen and fuel are pumped via a system of lines and valves into a combustion chamber. Once ignited, combustion is continuous until either oxygen or fuel supply is shut off. In most instances the fuel is used to cool the nozzle before being pumped into the combustion chamber. The injectors must be designed so that the oxidizer and the fuel are properly mixed. The combustion process typically yields very high temperatures and pressures. Typical velocities in the exhaust nozzle range from 8,000 to 13,000 ft/sec. The supersonic nature of the exhaust is usually visible in the form of diamond type shock patterns.

Because there is only gas flow coming from the nozzle, the thrust of a rocket engine is given by:

$$F_{net} = \dot{m}_{o+f} V_e = \frac{\dot{W}_f}{g} V_e \qquad (6.54)$$

where: $\dot{m}_{o+f}$ is the mass flow rate of fuel plus oxydizer (i.e. total mass flow rate through the nozzle, in slugs/sec

V_e is the exit velocity at the nozzle

$\dot{W}_{o+f}$ is the flow rate of fuel plus oxydizer in lbs/sec

The efficiency of a rocket engine is often given in terms of the so-called specific impulse, I_{sp} which is defined as:

$$I_{sp} = \frac{F_{net}}{\dot{W}_{o+f}} = \frac{V_e}{g} \tag{6.55}$$

Chemical rockets can achieve specific impulse values of around 400–500 sec.

An example arrangement for a solid fuel rocket engine is given in Figure 6.52

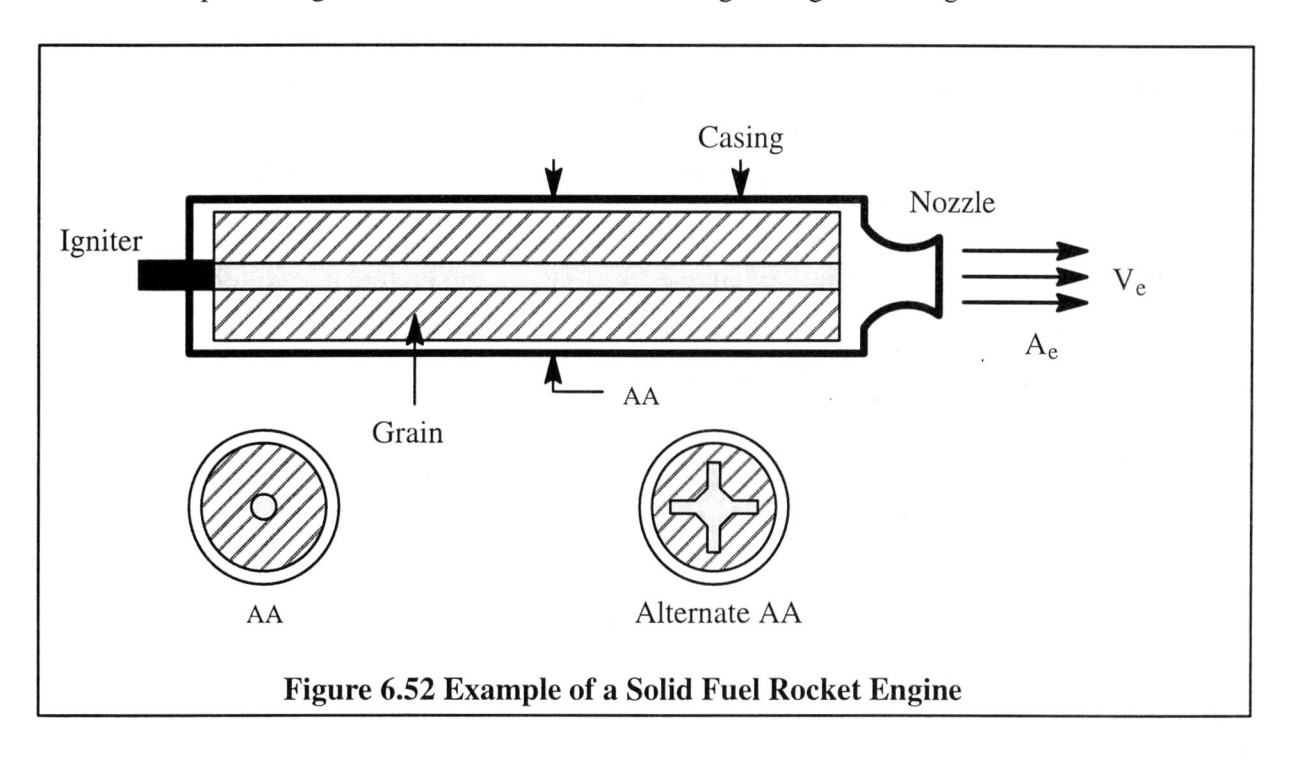

Figure 6.52 Example of a Solid Fuel Rocket Engine

Solid fuel rocket engines are basically very simple: they consist of a casing, a nozzle, the grain and an igniter. The grain is a mixture of fuel and oxydizer. The cross sectional shape of the grain is important. Burning tends to take place perpendicular to the inner grain surface. Therefore, to achieve high mass flow rates the burning surface area should be as large as possible. Two grain cross sections are shown in Figure 6.52 Other cross sections are also being used.

6.4 ELECTRIC MOTORS

A significant advantage of electric motors is their rather vibration–free and quiet operation. A disadvantage has been their high weight per unit power output. With the advent of rare earth magnets, electric motors are now considerably lighter than two decades ago. For that reason, they are a candidate for propulsion of certain classes of vehicles. An example is the solar–cell/electric–motor/propeller system used in the Aerovironment Pathfinder shown in Figure 6.4. A major problem with electric motor propulsion still is the weight associated with the energy storage (i.e. battery) and/or energy generation system.

6.5 SUMMARY FOR CHAPTER 6

In this chapter the reader has been provided with some basic insights into the operation and performance characteristics of aircraft engines. Most of the emphasis is on piston engines and turbine engines because these types are used in the great majority of airplanes. The fundamental thermodynamic cycle characteristics of these engine types are briefly discussed.

Trends of the effect of altitude and speed on the thrust and power output of these engines are given. Example data for engine weight and fuel consumption are also presented. Rough scaling rules for determining engine weight, size and performance for a given required output in thrust or power are included.

Cooling requirements are discussed for piston and turbine engines. for A brief discussion of noise characteristics of jet engines is also provided with a summary of the history of regulatory noise limitations.

For a more detailed study of aircraft propulsion systems References 6.1–6.4, 6.12 and 6.13 are recommended. Example installations of engines in airplanes can be found in Reference 6.20.

6.6 PROBLEMS FOR CHAPTER 6

6.1 A 9–cylinder engine with a bore (i.e. diameter of the piston head) of 5 inches and a stroke of 6 inches runs at 2,200 RPM. The indicated mean effective pressure is 200 psi. If the brake horsepower is 530 hp, calculate the mechanical efficiency.

6.2 An aircraft engine develops 260 BHP at 2,300 RPM and burns 21 gallons of fuel per hour. Calculate the brake thermal efficiency if the fuel has a heating value of 20,000 BTU per lbs. The unit weight of the fuel (gasoline) may be taken as 6 lbs/gallon.

6.3 If a 14–cylinder engine with a bore of 5.5 inches and a stroke of 6 inches develops 1,100 hp at a takeoff RPM of 2,300, find the BMEP in hp.

6.4 For the piston engine of Figures 6.13 and 6.14 determine the available horsepower and fuel flow at 2,500 RPM, a MAP of 22 in. Hg. and 5,000 ft altitude with an atmospheric temperature of 30 degrees F. above standard. Assume a mixture setting for best economy. What is the specific fuel consumption in lbs/hr/shp in this case?

6.5 A turbojet engine is operated at standard 20,000 ft altitude conditions with an exit pressure of 9 psi and an exit area of 180 in^2. If $\gamma = 1.33$ and the Mach number at the exit is 1.0, what is the gross thrust? What percentage of the gross thrust is due to the pressure thrust?

6.6 A liquid fuel rocket consumes 300 lbs/sec of propellants. The gas exit velocity is found to be 6,500 ft/sec. Calculate the specific impulse.

6.7 REFERENCES FOR CHAPTER 6

6.1   Taylor, C.F.; The Internal Combustion Engine in Theory and Practice (Volumes I and II); M.I.T. Press, Cambridge, Massachusetts, 1966.

6.2   McKinley, J.L. and Bent, R.D.; Powerplants for Aerospace Vehicles, Third Edition; McGraw–Hill Book Co., 1965.

6.3   Smith, H.; Aircraft Piston Engines; McGraw–Hill Book Co., 1981.

6.4   Rogowski, A.R.; Elements of Internal–Combustion Engines; McGraw–Hill Book Co., 1953.

6.5   Brouwers, A.P.; Lightweight Diesel Engine Designs for Commuter Type Aircraft; Teledyne Continental Motors, Contractors Report No. 995, NASA CR–165470, 1981

6.6   Hoerner, S.F.; Fluid Dynamic Drag; Library of Congress Catalog Card No. 57–13009; Published by the author.

6.7   Dearborn, C.H. and Silverstein, A.; Drag Analysis of Single–Engine Military Airplanes Tested in the NACA Full–Scale Wind Tunnel; NACA Wartime Report ACR, October 1940.

6.8   Lange, R.H.; A Summary of Drag Results From Recent Langley Full–Scale Tunnel Tests of Army and Navy Airplanes; NACA ACR No. L5A30, 1945.

6.9   Bingelis, T.; Firewall Forward; Published by the author, 8509 Greenflint Lane, Austin, Texas 78759; 1984.

6.10 Monts, F.; The Development of Reciprocating Engine Installation Data for General Aviation Aircraft; SAE Paper 730325, presented at the Business Aircraft Meeting, Wichita, Kansas, April 3–6, 1973.

6.11 Torenbeek, E.; Synthesis of Subsonic Airplane Design; Delft University Press, Martinus Nijhoff Publishers, 1982, Delft, The Netherlands.

6.12 Anon.; The Jet Engine; Rolls–Royce plc, 1986, Derby, England.

6.13 Mattingly, J.D., Heiser, W.H. and Daley, D.H.; Aircraft Engine Design; AIAA Education Series, 1987.

6.14 Kerrebrock, J. l.; Aircraft Engines and Gas Turbines; The MIT Press, Cambridge, Massachusetts, 1977.

6.15 Roskam, J.; Airplane Design, Part VI; Preliminary Calculation of Aerodynamic, Thrust and Power Characteristics; Design, Analysis and Research Corporation, 120 East Ninth Street, Lawrence, Kansas, 66044.

6.16   Anon.; Aircraft Engine Noise Reduction; NASA SP–311, 1972.

6.17   Lighthill, M.J.; On Sound Generated Aerodynamically, I: General Theory; Proceedings of the Royal Society, Series A, Volume 211, 1952, pp 564–587.

6.18   Raymer, D.P.; Aircraft Design: A Conceptual Approach; AIAA Education Series, 1989.

6.19   Sweetman, Bill; Aurora, The Pentagon's Secret Hypersonic Spyplane; Motorbooks International, Wisconsin, 1993.

6.20   Roskam, J.; Airplane Design, Part III; Layout Design of Cockpit, Fuselage, Wing and Empennage; Design, Analysis and Research Corporation, 120 East Ninth Street, Lawrence, Kansas, 66044.

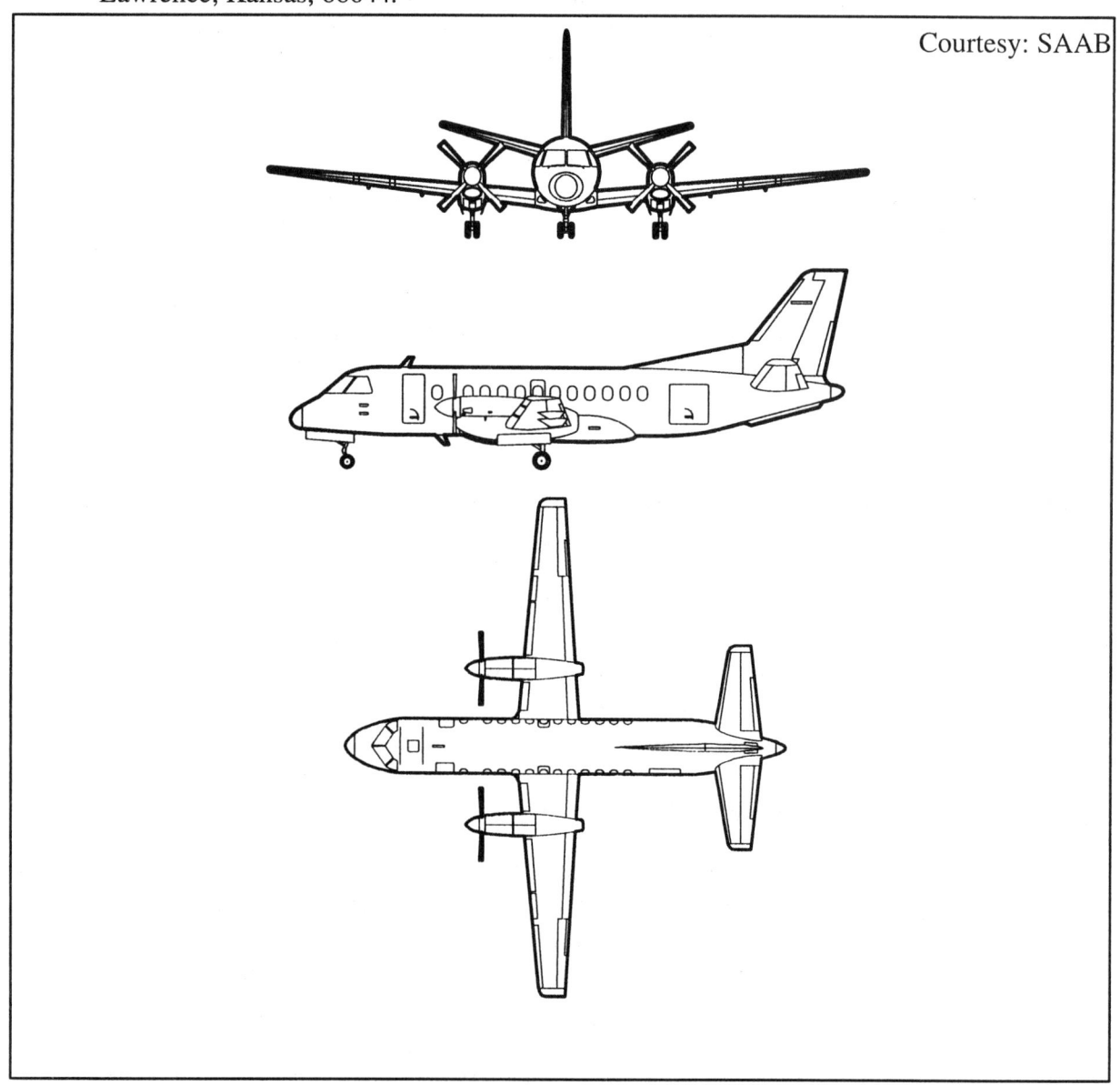

Courtesy: SAAB

CHAPTER 7: PROPELLER THEORY AND APPLICATIONS

The objective of this chapter is to provide the reader with fundamental insight into the theory and operation of airplane propellers. Once the required thrust and/or power for a propeller driven airplane are known the next step in the design process is to select or design an appropriate propeller.

A propeller delivers thrust by imparting a relatively small velocity increase to a relatively large air–mass. A turbojet does just the opposite: it imparts a relatively large velocity change to a relatively small air–mass. Turbo–fans, prop–fans and ducted propellers are somewhere in between these two extremes.

The simplest theory which describes the operation of a propeller is the so–called momentum theory. It is discussed in Section 7.1. A problem with momentum theory is that it lacks sufficient detail to allow for the geometric design of a propeller.

One theory which can be used to design propeller blades is the so–called blade–element theory. In this theory details of propeller airfoil geometry and other aerodynamic characteristics can be accounted for. Fundamental aspects of this theory are presented in Section 7.2. However, this theory does not account for the downwash produced by thrust generating propeller blades and the effect of this downwash on blade thrust.

A more complete theory is the combined blade–element and momentum theory, presented in Section 7.3.

Various sources of propeller losses are analyzed and discussed in Section 7.4 and factors which affect propeller performance are outlined in Section 7.5. Propellers can affect the drag polars of an airplane, to the point where a significant difference between power–on and power–off drag polars can exist. That effect is also discussed in Section 7.5.

A discussion of free–propeller efficiency versus installed propeller efficiency (also called propulsive efficiency) is also given in Section 7.5.

In Section 7.6 the use of so–called propeller charts in predicting propeller performance is presented. Step–by–step procedures for estimating propeller efficiency and propeller thrust (static and in–flight) are described. Inverse applications which allow the geometry of a propeller to be established for given performance objectives are also presented. Because of their bulk, the charts themselves are contained in Appendix C.

Propeller noise and the potential of propeller applications for high subsonic flight speeds is discussed in Section 7.7.

7.1 MOMENTUM THEORY

Propellers are used for generating positive as well as negative thrust. In the latter case the propeller is said to operate in reverse. An incompressible theory for positive and negative thrust generation is discussed in Sub–sections 7.1.1 and 7.1.2 respectively. When generating thrust at higher subsonic speeds compressibility effects must be accounted for. That is discussed in Sub–section 7.1.3

7.1.1 INCOMPRESSIBLE THEORY FOR POSITIVE PROPELLER THRUST

In the following theory it will be assumed that the flow through the propeller disk is incompressible and irrotational. The physical propeller disk is replaced by a thin "actuator disk" which is thought to consist of an infinite number of blades. The actuator disk is assumed to be uniformly loaded. The inflow and outflow are also assumed to be uniform. Figure 7.1 illustrates this idealized situation.

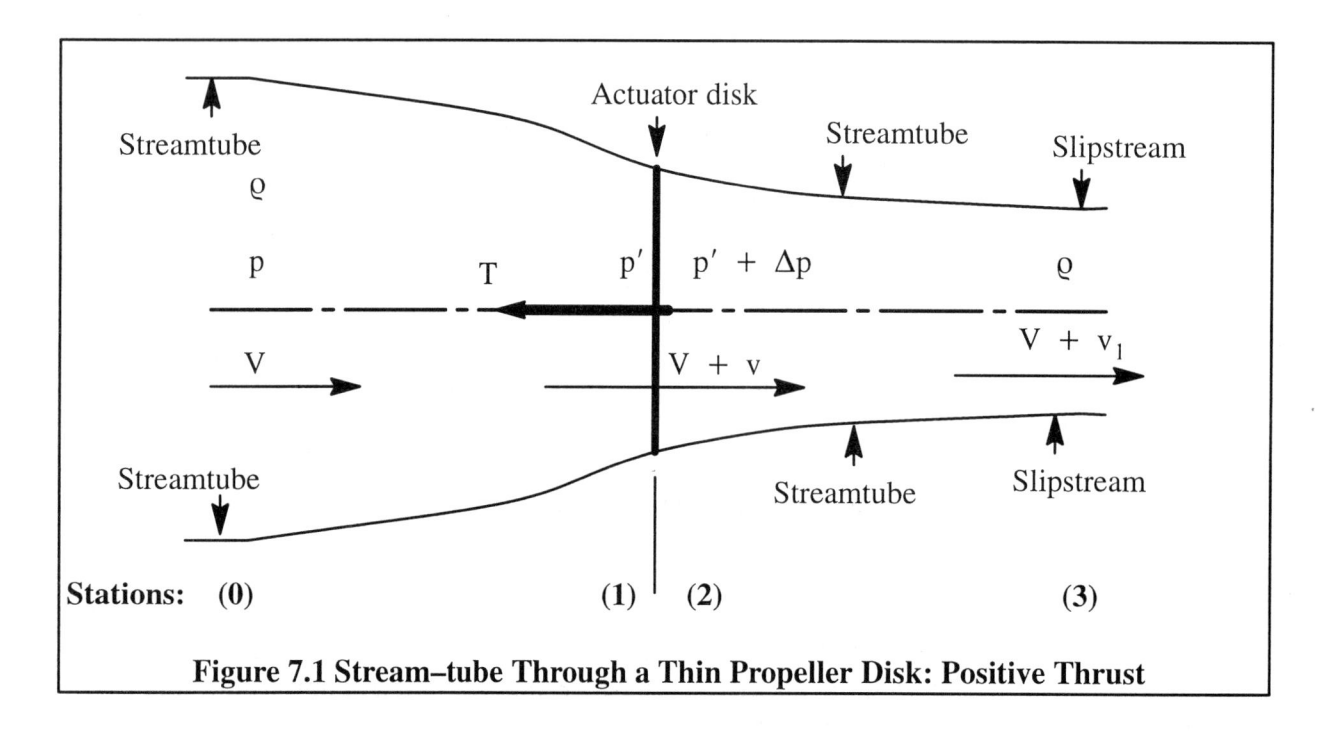

Figure 7.1 Stream–tube Through a Thin Propeller Disk: Positive Thrust

Far in front of the disk the static pressure and speed conditions are given by p and V respectively. At the disk, the velocity of the air is assumed to be V + v. The velocity cannot change in a discontinuous fashion because to do so would require an infinite acceleration. The pressure in front of the disk is assumed to be p' while the pressure behind the disk is assumed to be p' + Δp. The pressure increase, Δp is caused by the propeller which adds energy to the flow. The incompressible Bernoulli equation will now be applied to the flow. However, because the Bernoulli equation applies only to constant energy flow it has to be applied in two stages: first to the flow ahead of the propeller and next to the flow behind the propeller. The result is:

$$p + 0.5\varrho V^2 = H = p' + 0.5\varrho(V + v)^2 \quad \text{ahead of the disk} \tag{7.1}$$

and

$$p + 0.5\varrho(V + v_1)^2 = H_1 = p' + \Delta p + 0.5\varrho(V + v)^2 \quad \text{behind the disk} \qquad (7.2)$$

where: H is the total pressure ahead of the disk

H_1 is the total pressure behind the disk

v_1 is the induced velocity increase far downstream

Solving for Δp by subtracting Eqns (7.1) and (7.2) from each other results in:

$$\Delta p = H_1 - H = p + 0.5\varrho(V + v_1)^2 - (p + 0.5\varrho V^2) = \varrho\left(V + \frac{v_1}{2}\right)v_1 \qquad (7.3)$$

If A represents the propeller disk area, then the thrust produced by the propeller is:

$$T = A\Delta p = A\varrho\left(V + \frac{v_1}{2}\right)v_1 \qquad (7.4)$$

According to the linear momentum principle of fluid mechanics, see Eqn (2.84), the following relationship must apply:

$$\Sigma\vec{F} = \int_S \varrho\vec{V}\left(\vec{V} \cdot d\vec{A}\right) \qquad (7.5)$$

where: $\Sigma\vec{F}$ is the total force acting on the fluid in a control volume bounded by stations (0) and (3) in Figure 7.1.

If T is the fluid force acting on the propeller then $\Sigma\vec{F}$ must act in the opposite direction to the mass flow. It therefore follows that Eqn (7.5) is reduced to:

$$T = \varrho V\{-(V + v)A\} + \varrho(V + v_1)\{(V + v)A\} = \varrho A(V + v)v_1 \qquad (7.6)$$

When Equations (7.4) and (7.6) are equated it is seen that:

$$v_1 = 2v \qquad (7.7)$$

Therefore, the incompressible momentum theory indicates that the induced velocity far downstream is twice that at the disk. The reader may observe from Eqn (4.16) that this conclusion is similar to that relative to downwash induced by a wing as discussed in Chapter 4, Section 4.2.

To determine the propeller efficiency it is noted that the power output (i.e. the useful power) is given by TV, while the power input is equal to the thrust caused by air pressure on the propeller disk and multiplied by the air velocity through the disk, or: $T(V+v)$. Therefore, the propeller efficiency can be written as:

$$\eta_p = \frac{TV}{T(V+v)} = \frac{V}{V+v} \qquad (7.8)$$

The efficiency of Eqn (7.8) is the ideal, free propeller efficiency. The actual efficiency will be lower because of energy losses due to slipstream rotation, blade profile drag, non–uniform flow, compressibility effects and propeller blockage due to a fuselage and/or a nacelle. None of these losses have been accounted for in this simple theory. It is seen from Eqn (7.8) that it is desirable to keep the induced velocity, v, low if the propeller efficiency is to be high. Combining Eqns (7.4) and (7.7) produces:

$$T = 2\rho A(V+v)v \qquad (7.9)$$

For a given flight velocity, V, and thrust, T, Eqn (7.9) can be solved for the induced velocity:

$$v = -\frac{V}{2} + \sqrt{\left(\frac{V}{2}\right)^2 + \frac{T}{2A\rho}} \qquad (7.10)$$

It follows from Eqn (7.10) that to keep the induced velocity small (in turn to keep the efficiency high) the quantity T/A must be kept low. The quantity T/A is referred to as the propeller disk loading. A low disk loading, for a given T, requires a large diameter propeller. Large diameter propellers are undesirable for several reasons:

1) Large diameters tend to cause high tip speeds which leads to compressibility effects

2) Large diameters tend to cause ground clearance problems

3) Large diameters tend to introduce structural problems with the propeller blades (high stress)

Under static conditions (V=0) the induced velocity as expressed by Eqn (7.10) can also be written as:

$$v_{static} = v_s = \sqrt{\frac{T}{2A\rho}} \qquad (7.11)$$

By introducing this static induced velocity back into Eqn (7.10) it becomes:

$$\frac{v}{v_s} = -\frac{V}{2v_s} + \sqrt{\left(\frac{V}{2v_s}\right)^2 + 1} \qquad (7.12)$$

Note that only positive values of v/v_s have any physical significance. From Eqn (7.12) a plot of v/v_s versus V/v_s can be generated for positive v/v_s, see: Figure 7.2.

It may be seen that, for a given amount of thrust, the induced velocity decreases as the flight speed is increased. This implies higher efficiency so long as compressibility effects (to be discussed in Sub–section 7.1.3) are not important. The part of Figure 7.2 which represents negative thrust will be discussed in Sub–section 7.1.2.

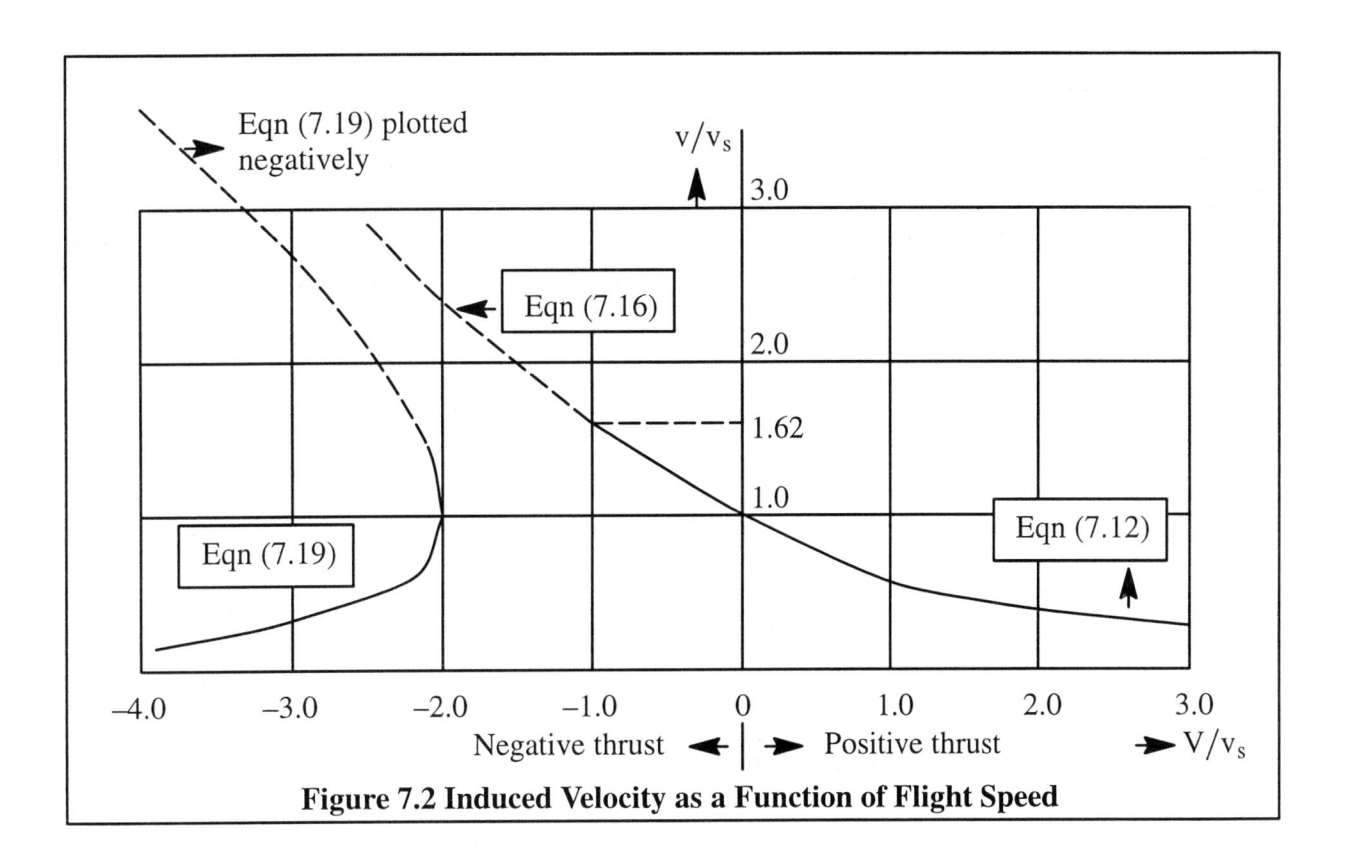

Figure 7.2 Induced Velocity as a Function of Flight Speed

7.1.2 INCOMPRESSIBLE THEORY FOR NEGATIVE PROPELLER THRUST

First, apply the continuity equation to stations (2) and (3) in Figure 7.1:

$$A_3(V + 2v) = A(V + v) \tag{7.13}$$

where: A_3 is the cross-sectional area of the slip-stream at station (3)

Eqn (7.13) can also be written as:

$$\frac{R_3}{R} = \sqrt{(1 + \frac{V}{v})/(2 + \frac{V}{v})} \tag{7.14}$$

where: R is the propeller radius

R_3 is the radius of the slipstream at station (3)

Figure 7.3 shows a plot of Eqn (7.14). It is of interest to study what happens as V/v changes from positive to negative.

Note, that under static conditions (V=0), $R_3/R = \sqrt{1/2} = 0.707$. It is also seen that as V/v becomes negative, R_3/R decreases.

When V/v approaches the value −1, R_3/R approaches zero. In that case, the mass flow rate

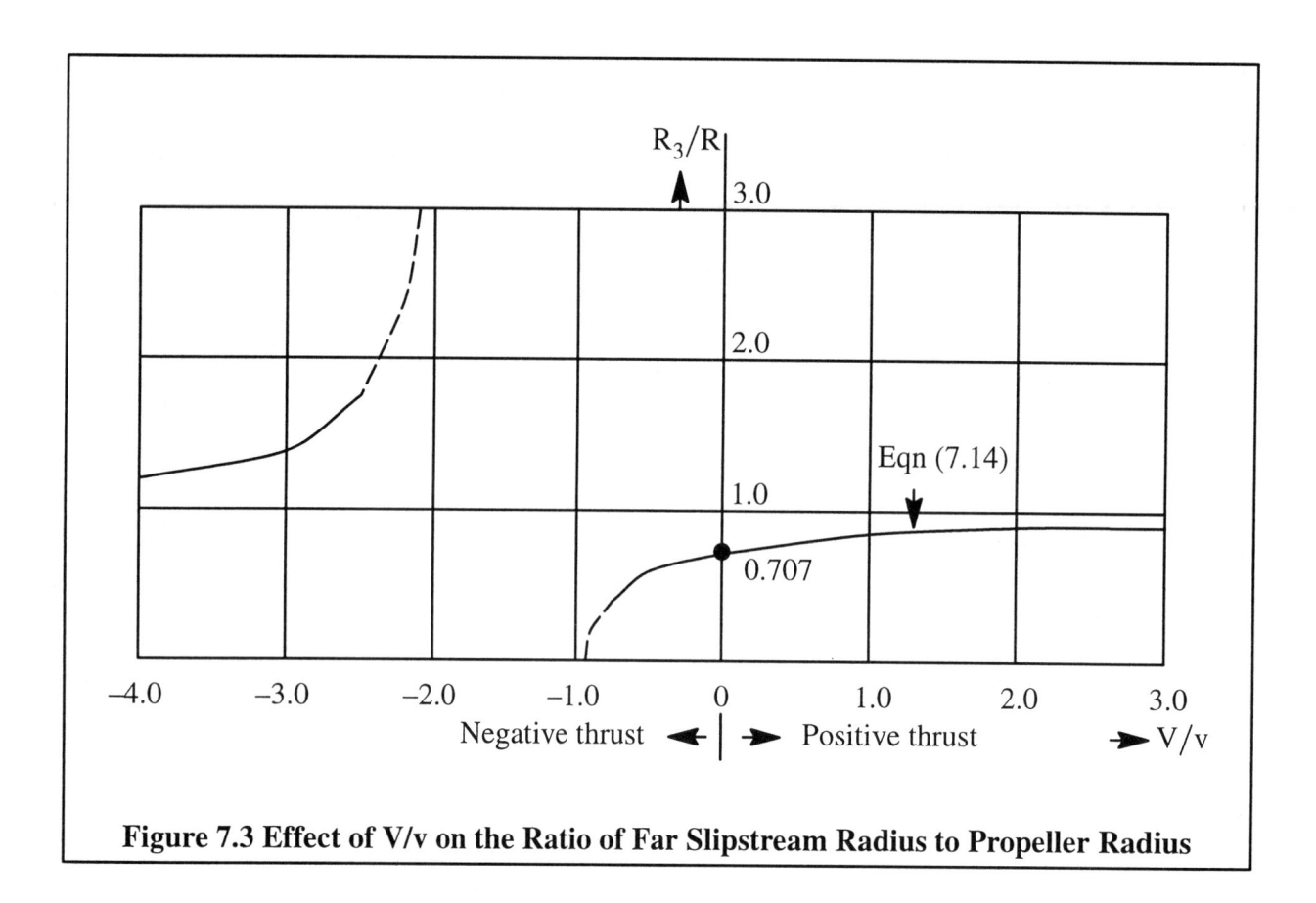

Figure 7.3 Effect of V/v on the Ratio of Far Slipstream Radius to Propeller Radius

through the propeller, $A(V + v)$, becomes zero. This indicates that the concept of simple momentum theory becomes meaningless.

When $-2.0 < V/v < -1.0$, R_3/R is imaginary. Again, the present flow model is not applicable.

When $-1.0 < V/v < 0.0$ the flow situation is as depicted in Figure 7.4. This situation is referred to as reversed, or negative thrust.

By applying Eqn (7.5) to Figure 7.4 it is found that:

$$-T = -\varrho A\{(-V + v)V\} + \varrho A(-V + v)\{-(-V + 2v)\} = 2\varrho A(-V + v)v \tag{7.15}$$

This expression could also have been obtained by replacing V by $-v$ in Eqn (7.9). Solving for v yields:

$$v = \frac{V}{2} + \sqrt{\left(\frac{V}{2}\right)^2 + \frac{T}{2A\varrho}} \tag{7.16}$$

Eqn (7.16) is plotted in Figure 7.2 as an extension of Eqn (7.12).

When $V/v < -2.0$ (left side curve in Figure 7.3) the flow situation is as shown in Figure 7.5. This case is applicable to negative thrust (i.e. drag) production in a high speed dive.

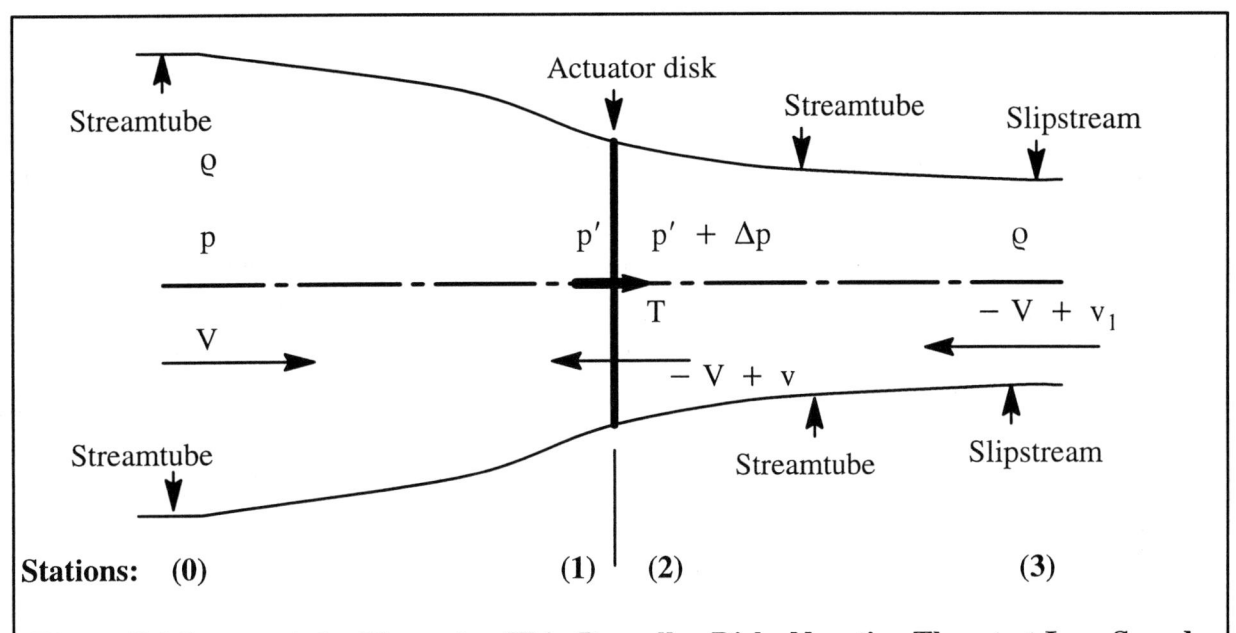

Figure 7.4 Stream-tube Through a Thin Propeller Disk: Negative Thrust at Low Speeds

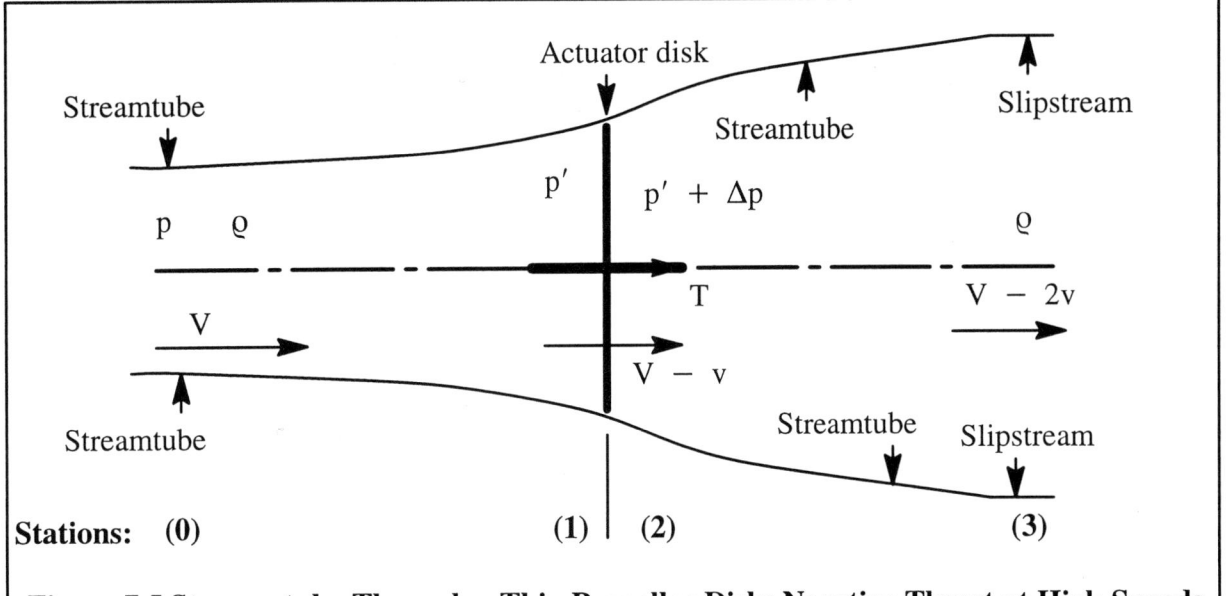

Figure 7.5 Stream-tube Through a Thin Propeller Disk: Negative Thrust at High Speeds

The momentum equation, for the case of Figure 7.5, ($V/v < -2.0$) now yields:

$$-T = -\varrho A\{(V-v)V\} + \varrho A(V-v)\{(V-2v)\} = 2\varrho A(V-v)v \qquad (7.17)$$

From this it follows that:

$$v = \frac{V}{2} - \sqrt{\left(\frac{V}{2}\right)^2 - \frac{T}{2A\varrho}} \qquad (7.18)$$

Using the definition of static induced velocity (from Eqn 7.11) this can be written as:

$$\frac{v}{v_s} = \frac{V}{2v_s} - \sqrt{\left(\frac{V}{2v_s}\right)^2 - 1} \qquad (7.19)$$

Eqn (7.19) is also plotted in Figure 7.2, but negatively, to distinguish it from the positive thrust case. Note that in deriving Eqn (7.19) appropriate signs for T, V and v have been used in Eqn (7.17). Therefore, only positive values of V/v_s should be used in Eqn (7.19). The present situation is similar to that of a helicopter in vertical descent. The expressions for induced velocity in the case of negative thrust will be used in Sub–section 7.6.6 which deals with thrust reversing.

An example application will now be presented.

Example 7.1:

A propeller with a diameter of 10 ft produces a positive thrust of 1,000 lbs at 100 mph under standard sea–level conditions. Determine:

a) The maximum attainable efficiency using the incompressible momentum theory.

b) The pressure difference between atmospheric pressure and the pressure immediately in front of the propeller.

Solution:

a) From Eqn (7.10) the induced velocity is given by:

$$v = \frac{-100 \times 1.467}{2} + \sqrt{\frac{(100 \times 1.467)^2}{4} + \frac{4 \times 1,000}{2 \times 0.002377 \times 3.14 \times 10^2}} = 16.42 \text{ fps}$$

The maximum attainable efficiency follows from Eqn (7.8) as:

$$\eta_p = \frac{100 \times 1.467}{100 \times 1.467 + 16.42} = 89.9\%$$

b) From Eqn (7.1) it is found that the pressure difference between atmospheric pressure and the pressure immediately in front of the propeller is:

$$p - p' = \frac{0.002377}{2}\left\{(100 \times 1.467 + 16.42)^2 - (100 \times 1.467)^2\right\} = 6.05 \text{psf}$$

7.1.3 COMPRESSIBLE THEORY FOR POSITIVE PROPELLER THRUST

A compressible momentum theory is useful to predict the general trend of the characteristics of high speed propellers. To develop such a theory the compressible continuity and Bernoulli equations must be used and solved numerically. Presentation of this theory, an example of which is given in Reference 7.1, is beyond the scope of this text.

However, it is of interest to present some example results from this theory. Figure 7.6 shows these example results as obtained from Reference 7.1. Note that the subscripts refer to the same propeller flow stations as used in Figure 7.1.

Observe that the pressure variation through the propeller disk is larger for compressible flow than for in–compressible flow. This implies that in compressible flow the pressure gradient on the blade boundary layer is more adverse so that interference effects between the propeller and a nacelle, body or fuselage will also be increased.

As the air passes through the propeller disk, its velocity is decreased by compressibility*. Therefore, the blade camber designed with incompressible theory may not be proper in compressible flow.

Examining the Mach number variation in Figure 7.6, it is seen that the Mach number immediately behind the propeller (M_2) is always less than the free–stream Mach number.

At the same time, the Mach number immediately ahead of the propeller (M_1) is always larger. This implies that a body placed ahead of the propeller (i.e. a pusher configuration) will suffer greater compressibility effects than a tractor configuration, other things being equal.

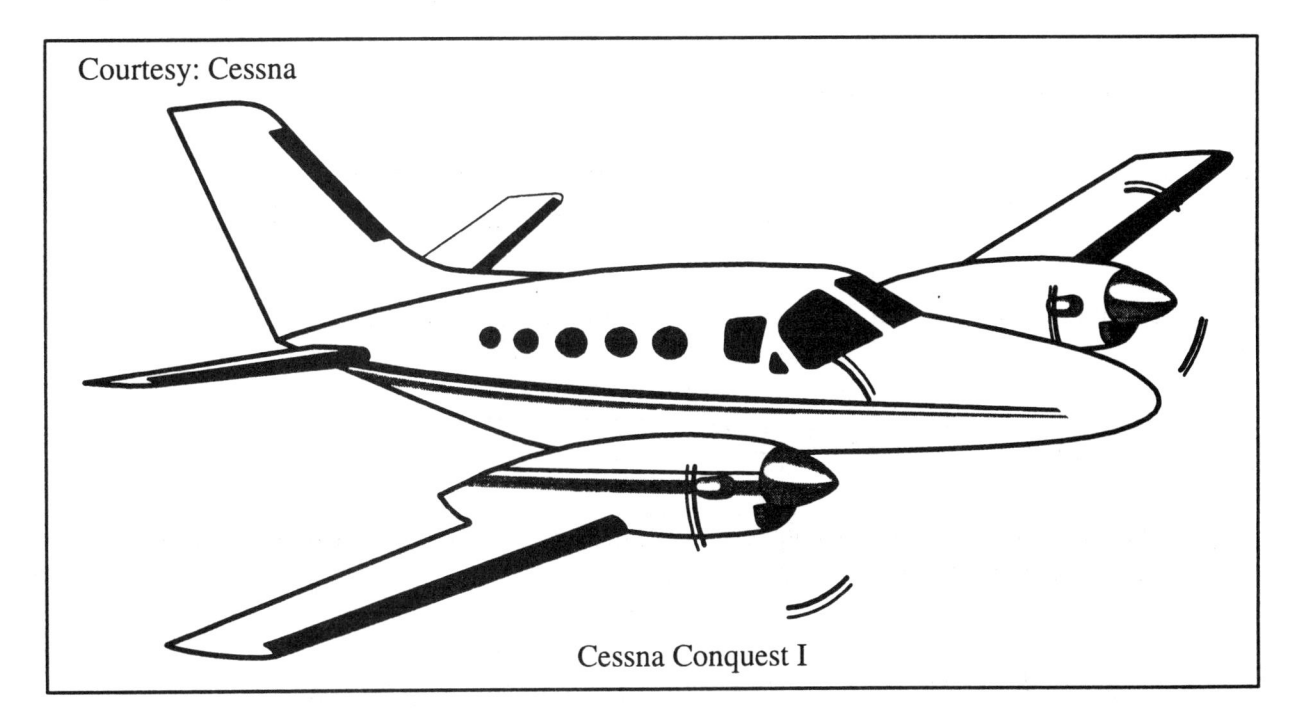

Courtesy: Cessna

Cessna Conquest I

---

*It is noted that similar effects are found in jet engine compressors.

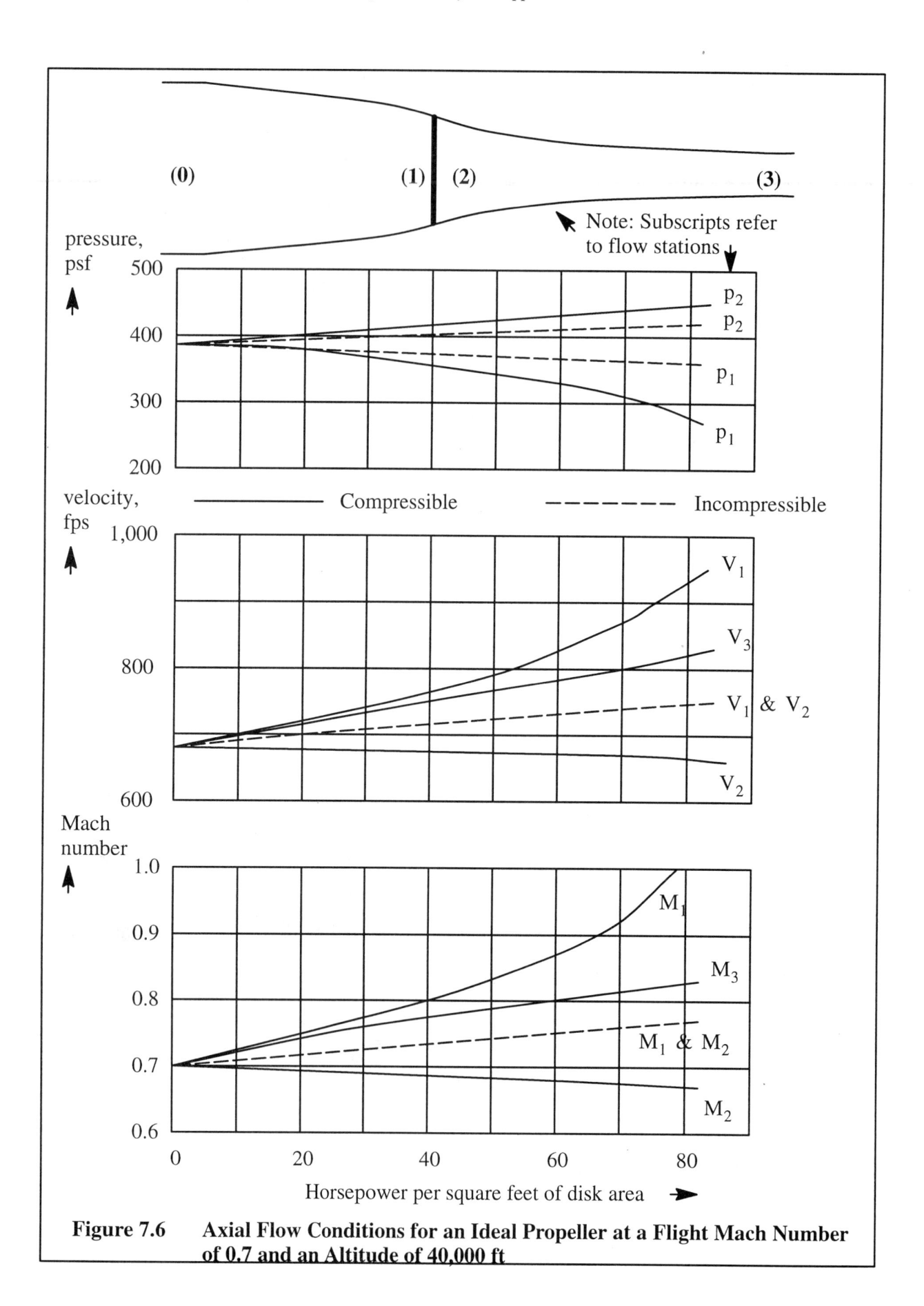

Propeller Theory and Applications

Figure 7.6 Axial Flow Conditions for an Ideal Propeller at a Flight Mach Number of 0.7 and an Altitude of 40,000 ft

274   Chapter 7

7.2 SIMPLE BLADE–ELEMENT THEORY

In this theory the local aerodynamic forces acting on a blade element at a distance r from the propeller spin axis will be considered. Figure 7.7 shows the blade element and the forces.

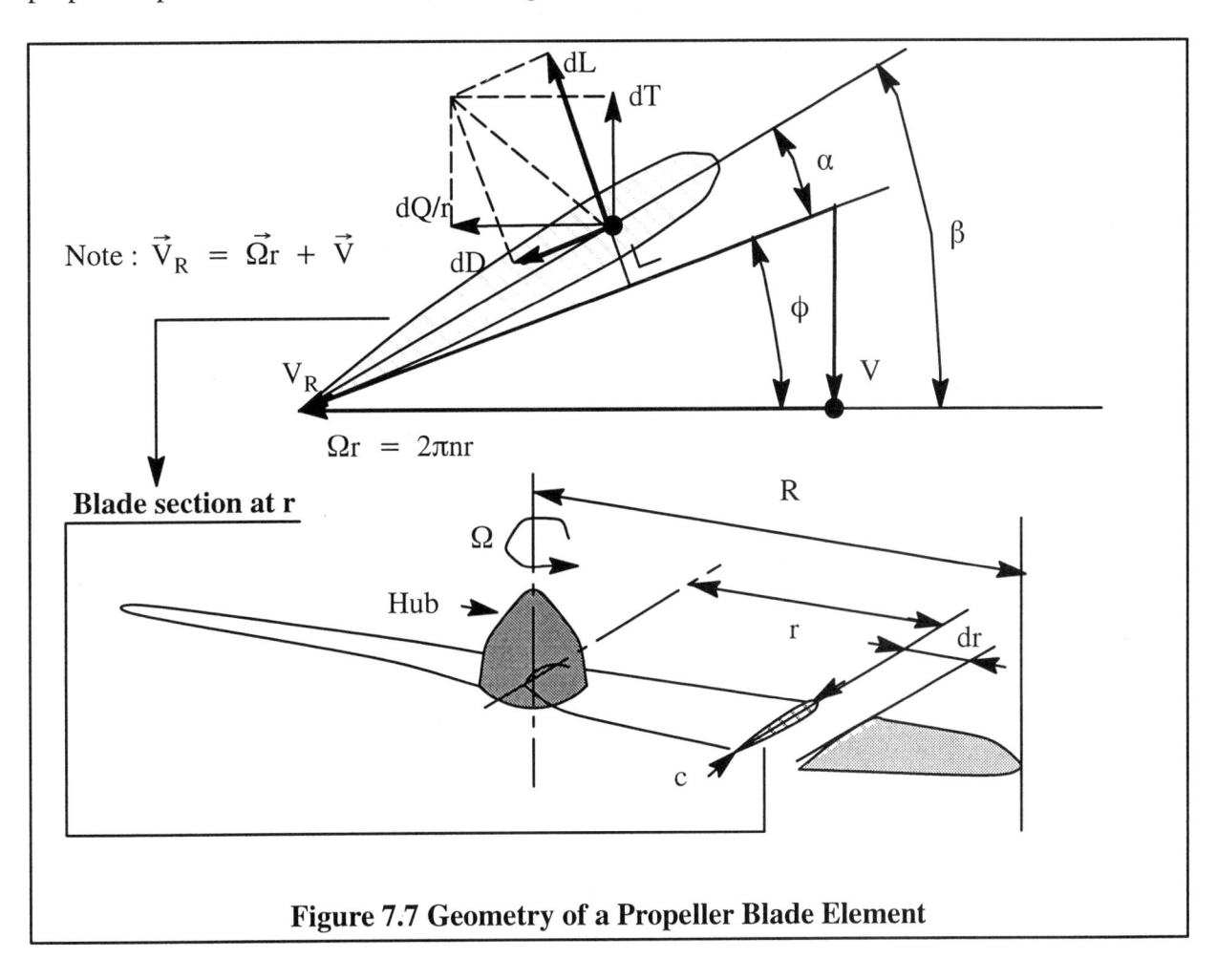

Figure 7.7 Geometry of a Propeller Blade Element

Observe that the blade element at radius r is moving with a spin velocity $2\pi nr$, where n is the propeller rotational velocity in cycles per second. In addition, the propeller has a forward (i.e. airplane) velocity, V.

The lift and drag forces acting on the element are dL and dD respectively. Note that dL acts perpendicular to the vectorial sum of V and $2\pi nr$, while dD acts along but opposite to that vector sum.

The force dT is the actual propeller thrust of the blade element. The force perpendicular to dT is the force which produces the propeller torque, Q (in ft–lbs). For that reason it is labeled dQ/r.

The angle, β, is called the geometric pitch angle of the blade element.

The airplane speed, V, is also called the propeller advance velocity.

The speed due to rotation, Ωr, is much less at the hub of the propeller than at the tip.

Chapter 7

The advance velocity is the same across the blade. Therefore, if the propeller blade were not twisted, the local angle of attack of a blade element near the hub is much less than that at the tip and could even become negative. For that reason it is necessary to increase the geometric pitch angle, β, near the hub such as to maintain an efficient angle of attack. That is why propeller blades appear to be so severely twisted near the hub.

Note from Figure 7.7 that the blade element thrust, dT, is composed of a lift and a drag component:

$$dT = dL\cos\phi - dD\sin\phi = 0.5\,\varrho\,V_R^2\,c\,dr\,(c_l\cos\phi - c_d\sin\phi) \tag{7.20}$$

where: ϕ is the so-called helix angle

c is the blade chord at radius r

The propeller torque is caused by the lift and drag components in the plane of rotation:

$$dQ = (dL\sin\phi + dD\cos\phi)r = 0.5\,\varrho\,V_R^2\,c\,r\,dr\,(c_l\sin\phi + c_d\cos\phi) \tag{7.21}$$

If η is the blade-element efficiency, then:

$$\eta = \frac{VdT}{\Omega dQ} = \frac{V}{\Omega r}\frac{(c_l\cos\phi - c_d\sin\phi)}{(c_l\sin\phi + c_d\cos\phi)} = \tan\phi\frac{(c_l\cos\phi - c_d\sin\phi)}{(c_l\sin\phi + c_d\cos\phi)} \tag{7.22}$$

Before Eqns (7.20) through (7.22) can be integrated over the blade itself and over the entire propeller (number of blades), to determine the total thrust, T, and the total torque, Q, the sectional aerodynamic characteristics of each blade element must be known. However, these characteristics can be computed accurately only if the induced velocity due to lift production on each blade element is accounted for. This induced velocity alters the local blade angles of attack.

The model of Figure 7.7 does not account for the induced velocity distribution. One assumption which can be made is that actual $c_l - c_d$ section data can be used without accounting for the spanwise induced effects. Although such a method can give fairly good results, a better theory which combines blade-element and momentum theory, is available and will be discussed in Section 7.3.

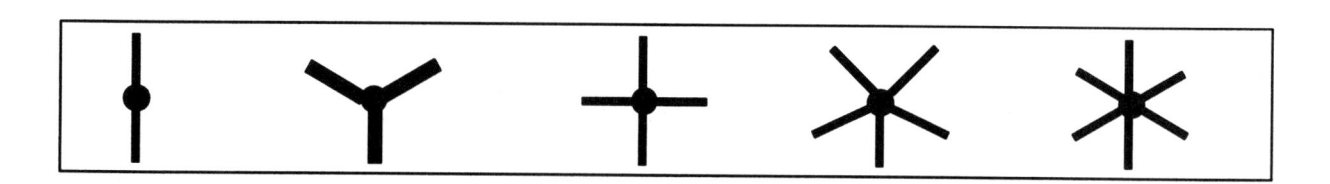

7.3 COMBINED BLADE–ELEMENT THEORY AND MOMENTUM THEORY

In this theory, the induced flow due the production of lift on a propeller blade is accounted for. In momentum theory, thrust can be expressed in terms of induced velocity. If the number of blades is B, the total elemental thrust from simple blade theory (neglecting the section drag term) is:

$$BdT = BdL\cos\phi_0 = B\ c_l\ 0.5\ \varrho\ V_{R0}^2\ c\ dr\ \cos\phi_0 \tag{7.23}$$

where: ϕ_0 is defined in Figure 7.8

V_{R_0} is also defined in Figure 7.8

Note from Figure 7.8 that the induced velocity component in the thrust direction is: $V_i\cos\phi_0$. Therefore, according to momentum theory, the total elemental thrust may also be written as:

$$BdT = \varrho(2\pi r dr)(V + V_i\cos\phi_0)(2V_i\cos\phi_0) \tag{7.24}$$

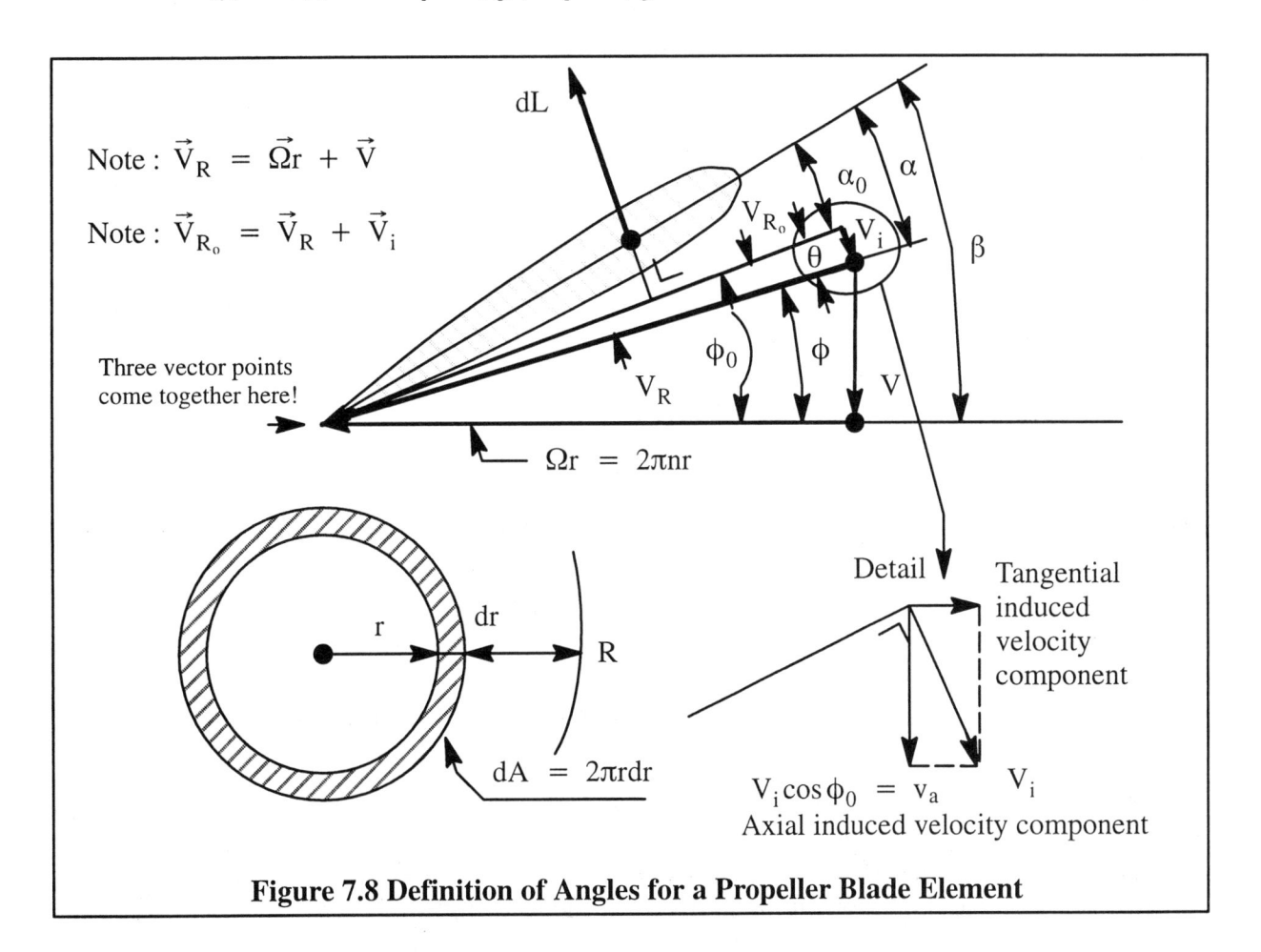

Figure 7.8 Definition of Angles for a Propeller Blade Element

By equating Eqns (7.23) and (7.24) it can be found that:

$$V_i = \frac{B\, c_l\, c\, (V_{R_0})^2}{8\pi r\, (V + V_i \cos\phi_0)} \tag{7.25}$$

To simplify Eqn (7.25) it will be assumed that the induced angle, θ of Figure 7.8, is small so that:

$$\sin\phi_0 = \sin(\phi + \theta) \approx \sin\phi + \theta\cos\phi \approx \frac{V + V_i \cos\phi_0}{V_{R_0}} \tag{7.26}$$

Furthermore, it is seen that:

$$\tan\theta = V_i/V_{R_0} \approx \theta \tag{7.27}$$

and:

$$c_l = a_0 \alpha_0 = a_0(\beta - \phi - \theta) \tag{7.28}$$

where: a_0 is the section lift curve slope of the blade airfoil.

Using Eqns (7.26) through (7.28) it is possible to rewrite Eqn (7.25) in terms of the angle, θ, as follows:

$$\theta \approx B \left(\frac{c}{8\pi r}\right) \frac{a_0(\beta - \phi - \theta)}{(\sin\phi + \theta\cos\phi)} \tag{7.29}$$

In a general case, the blade chord, c, will vary across the propeller blade radius, r. For many propellers, a good approximation can be obtained by assuming a constant blade chord.

The solidity ratio of a propeller, σ, is defined as the ratio of the total blade area to that of the disk. For a propeller of constant blade chord, c, this yields:

$$\sigma = \frac{B\, c\, R}{\pi R^2} = \frac{B\, c}{\pi R} \tag{7.30}$$

The blade station, r, is non–dimensionalized as:

$$x = \frac{r}{R} \tag{7.31}$$

$x = r/R$. With x and σ it is possible to show that Eqn (7.29) can be written as:

$$\theta^2 \cos\phi + \theta\left(\sin\phi + \frac{\sigma a_0}{8x}\right) - \frac{\sigma a_0}{8x}(\beta - \phi) = 0 \tag{7.32}$$

This equation has the following solution for the induced angle, θ:

$$\theta = \frac{1}{2\cos\phi}\left\{-\left(\sin\phi + \frac{a_0\sigma}{8x}\right) + \sqrt{\left(\sin\phi + \frac{a_0\sigma}{8x}\right)^2 + 4\cos\phi\frac{a_0\sigma}{8x}(\beta - \phi)}\right\} \quad (7.33)$$

Eqn (7.33) can be used to compute the induced velocity in the following manner. First, the following assumptions are introduced:

$$\cos\phi \approx 1.0 \qquad \sin\phi \approx V/V_R \approx V/xV_t \qquad V_R \approx xV_t \quad (7.34)$$

where: V_t stands for the propeller tip speed

Observe from Figure 7.8:

$$V_i/V_R = \tan\theta \quad \text{and thus:} \quad V_i \approx \theta V_R \quad \text{or:} \quad V_i \approx \theta x V_t \quad (7.35)$$

Therefore, by introducing assumptions (7.35) into Eqn (7.33), the induced velocity becomes:

$$V_i = V_t\left\{-\left(\frac{V}{2V_t} + \frac{a_0\sigma}{16}\right) + \sqrt{\left(\frac{V}{2V_t} + \frac{a_0\sigma}{16}\right)^2 + \frac{a_0\sigma\beta x}{8} - \frac{a_0\sigma V}{8V_t}}\right\} \quad (7.36)$$

As a side comment, Equation (7.36) is often used to calculate the induced velocity on helicopter rotors in vertical climbing flight: see Ref. 7.2. In such flight conditions the rotor (i.e. propeller) thrust is very high. For conditions of relatively low thrust, Eqn (7.32) can be solved by neglecting the θ^2-term. This yields:

$$\theta \approx \frac{\beta - \phi}{1 + \frac{8x\sin\phi}{\sigma a_0}} \quad (7.37)$$

From Figures (7.7) and (7.8) it is seen that:

$$V_{R_0} = V_R\cos\theta = \left(\frac{2\pi nr}{\cos\phi}\right)\cos\theta \quad (7.38)$$

Using Figure 7.8, the elemental thrust of a B-bladed propeller is seen to be:

$$dT = B(dL\cos\phi_0 - dD\sin\phi_0) = B0.5\varrho(V_{R_0})^2 cdr(c_l\cos\phi_0 - c_d\sin\phi_0) =$$
$$= B\varrho\left(\frac{2\pi^2 n^2 r^2}{\cos^2\phi}\cos^2\theta\right)cdr(c_l\cos\phi_0 - c_d\sin\phi_0) \quad (7.39)$$

Similarly, the elemental torque is found as:

$$dQ = B(dL\sin\phi_0 + dD\cos\phi_0) = B\varrho\left(\frac{2\pi^2 n^2 r^3}{\cos^2\phi}\cos^2\theta\right)cdr(c_l\sin\phi_0 + c_d\cos\phi_0) \quad (7.40)$$

To simplify these equations, the following functions of ϕ and θ are introduced:

$$\lambda_T = \left(\frac{\cos^2\theta}{\cos^2\phi}\right)(c_l\cos\phi_0 - c_d\sin\phi_0) \qquad (7.41)$$

$$\lambda_Q = \left(\frac{\cos^2\theta}{\cos^2\phi}\right)(c_l\sin\phi_0 + c_d\cos\phi_0) \qquad (7.42)$$

This, while also using: $x = r/R$, simplifies Eqns (7.39) and (7.40) to:

$$dT = 2B\rho\pi^2n^2r^2cdr\lambda_T = 2B\rho\pi^2n^2x^2R^3c\lambda_T dx \qquad (7.43)$$

and:

$$dQ = 2B\rho\pi^2n^2r^3cdr\lambda_Q = 2B\rho\pi^2n^2x^3R^4c\lambda_Q dx \qquad (7.44)$$

The so-called propeller thrust and torque coefficients are respectively **defined** as follows:

$$C_T = \frac{T}{\rho n^2 D^4} \qquad (7.45)$$

$$C_Q = \frac{Q}{\rho n^2 D^5} \qquad (7.46)$$

Setting $D=2R$, introducing $Bc = \sigma\pi R$ from Eqn (7.30), and differentiating Eqns (7.45) and (7.46) with respect to x yields:

$$\frac{dC_T}{dx} = \frac{dT/dx}{16\rho n^2 R^4} = \frac{2\rho\pi^3 n^2 R^4 x^2 \sigma\lambda_T}{16\rho n^2 R^4} = 3.88 x^2 \sigma\lambda_T \qquad (7.47)$$

and:

$$\frac{dC_Q}{dx} = \frac{dQ/dx}{32\rho n^2 R^5} = \frac{2\rho\pi^3 n^2 R^5 x^3 \sigma\lambda_Q}{32\rho n^2 R^5} = 1.94 x^3 \sigma\lambda_Q \qquad (7.48)$$

To determine the magnitudes of C_T and C_Q for a given propeller design, the following step-by-step procedure is suggested:

Step 1: Determine the helix angle, ϕ, from: $\tan\phi = \dfrac{V}{2\pi nr}$

Step 2: Determine the propeller solidity ratio, σ from: $\sigma = \dfrac{Bc}{\pi R}$

Step 3: Select a blade station: $x = r/R$

Step 4: Determine the induced angle, θ, from Eqn (7.37): $\theta \approx \dfrac{\beta - \phi}{1 + \dfrac{8x\sin\phi}{\sigma a_0}}$ a_0 in rad

Note: if the propeller airfoil is cambered, the angle θ should be computed relative to the airfoil zero-lift line and not the airfoil chord-line!

Step 5: Determine the section angle of attack from: $\alpha_0 = \beta - \phi - \theta$

Step 6: Calculate the section lift coefficient from: $c_l = a_0 \alpha_0$

Step 7: Determine the section drag coefficient, c_d from c_l versus c_d data for the airfoil being used at station x=r/R

Step 8: Calculate the angle, ϕ_0 from: $\phi_0 = \phi + \theta$

Step 9: Compute the coefficients: λ_T and λ_Q from Eqns (7.41) and (7.42)

Step 10: Determine the derivatives dC_T/dx and dC_Q/dx from Eqns (7.47) and (7.48)
Observe, that if a cambered airfoil is used, the angle β must be measured from the zero–lift line of such an airfoil!

Step 11: Repeat Steps 1–10 for a sufficient number of blade stations, x, and plot the results versus x. An example is shown in Figure 7.9.

Step 12: Numerically integrate the results of Step 11 over the propeller to obtain C_T and C_Q

It may be seen from Figure 7.9 that a major part of the propeller thrust and torque is derived from the outer parts of the propeller blades. That is one reason for keeping the blade radius close to the hub as small as practical from a structural viewpoint.

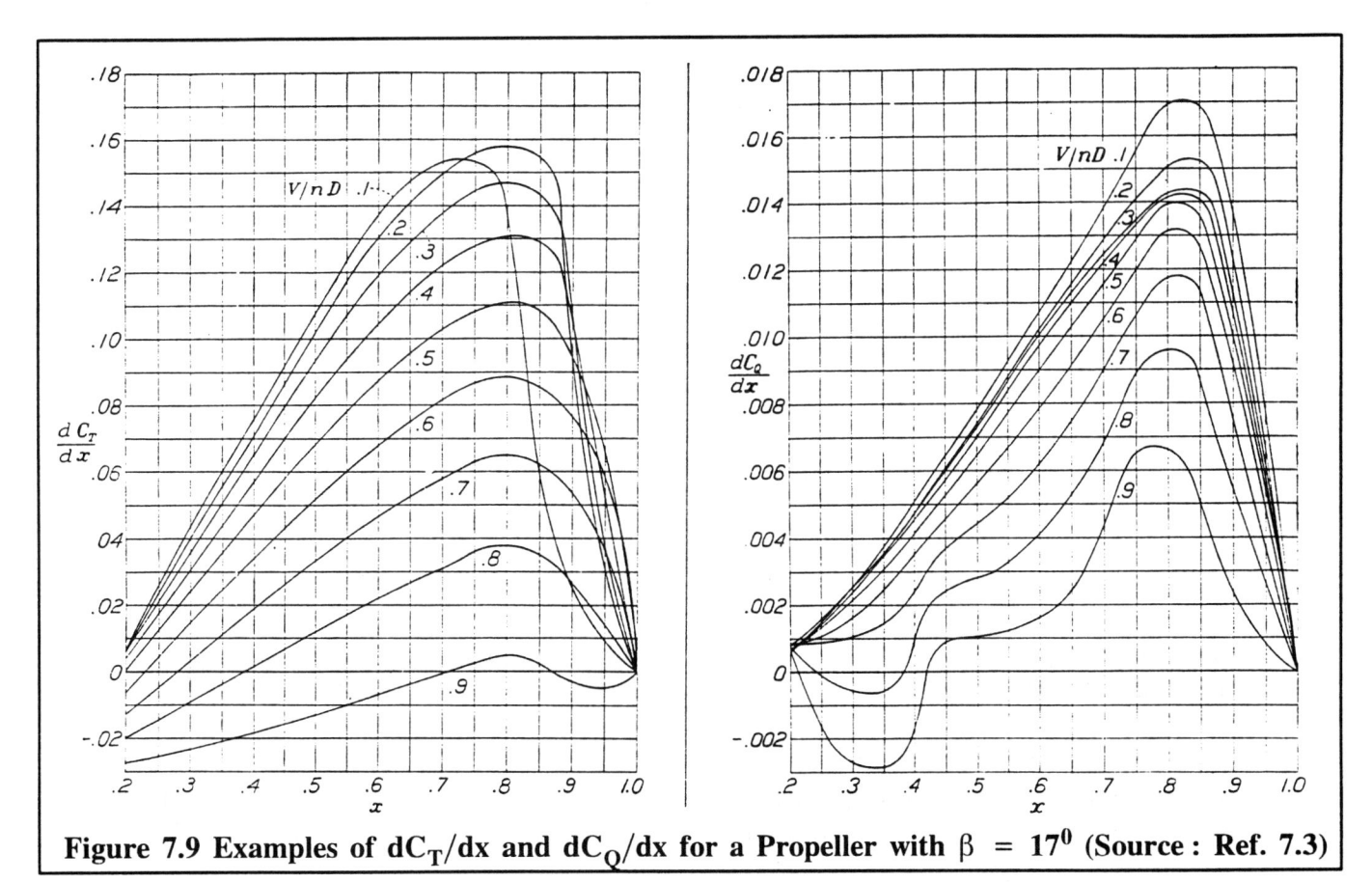

Figure 7.9 Examples of dC_T/dx and dC_Q/dx for a Propeller with $\beta = 17^0$ (Source: Ref. 7.3)

7.4 ANALYSIS OF PROPELLER POWER LOSSES

When carrying out thrust or power calculations for propeller driven airplanes it is necessary to account for various power losses which occur. In doing so, a number of symbols related to propeller performance are used repeatedly.

The reader is advised to commit the following definitions a) through e) to memory:

a) Propeller thrust coefficient (already defined in Eqn (7.45):

$$C_T = \frac{T}{\varrho n^2 D^4} \tag{7.49}$$

b) Propeller torque coefficient (already defined in Eqn (7.46):

$$C_Q = \frac{Q}{\varrho n^2 D^5} \tag{7.50}$$

c) Propeller power coefficient:

$$C_P = \frac{P}{\varrho n^3 D^5} \tag{7.51}$$

where: P is the power delivered by the engine to the propeller.

d) Propeller advance ratio:

$$J = \frac{V}{nD} \tag{7.52}$$

e) Propeller efficiency:

$$\eta_p = \frac{TV}{P} = J \frac{C_T}{C_P} \tag{7.53}$$

Observe, that power, P, can also be written as:

$$P = 2\pi n Q \tag{7.54}$$

For this reason, the power and torque coefficients, C_P and C_Q, are related by:

$$C_P = 2\pi C_Q \tag{7.55}$$

For a free propeller (without a nacelle behind it) there are three types of propeller losses: axial momentum losses, rotational momentum losses and losses due to blade profile drag.

To provide an insight into these three losses, an analysis based on Reference 7.3 will be presented in Sub–section 7.4.1 for axial and rotational momentum losses. A discussion of measured propeller losses, which include the losses due to blade profile drag, is presented in Sub–section 7.4.2.

7.4.1 AXIAL AND ROTATIONAL MOMENTUM LOSSES

The energy lost as a result of the axial flow velocity, $V_i \cos\phi_0 = v_a$, may be expressed as:

$$E_a = \int v_a dT = \varrho n^2 D^4 V \int_0^1 a \frac{dC_T}{dx} dx \tag{7.56}$$

where: $a = v_a/V$

Call the average rotational velocity in the propeller slip–stream, ω. The energy lost per unit time due to this rotational velocity, ω, may be written as:

$$E_r = \int \omega dQ = (2\pi n)\varrho n^2 D^5 \int_0^1 a' \frac{dC_Q}{dx} dx \tag{7.57}$$

where: $a' = \omega/2\pi n$

If both Eqns (7.56) and (7.57) are divided by the power input, P, the following non–dimensional forms are obtained:

$$\frac{E_a}{P} = \frac{J}{C_P} \int_0^1 a \frac{dC_T}{dx} dx \tag{7.58}$$

and:

$$\frac{E_r}{P} = \frac{2\pi}{C_P} \int_0^1 a' \frac{dC_Q}{dx} dx \tag{7.59}$$

Next, assume that the thrust and torque distributions over the blade are known from some experiment and also assume that the quantities a and a' can be determined. In that case Eqns (7.58) and (7.59) can be used to compute the axial and rotational momentum power losses.

A method for estimating both a and a', for a **two–bladed** propeller, will next be presented.

According to the momentum theory represented by Eqn (7.24) the elemental thrust is:

$$dT = \varrho(2\pi r dr)(V + v_a)2v_a \tag{7.60}$$

where the second factor 2 accounts for the two blades.

By using Eqns (7.52), (7.49) x=r/R and $a = v_a/V$ it can be shown that:

$$dC_T/dx = \pi J^2 x(1 + a)a \tag{7.61}$$

Propeller Theory and Applications

The ratio of the axial component of induced velocity to the forward speed, $a = v_a/V$ may be solved from Eqn (7.61) as:

$$a = \frac{-1 + \sqrt{\left(1 + 4\frac{dC_T/dx}{\pi J^2 x}\right)}}{2} \tag{7.62}$$

The minus sign in front of the square root in Eqn (7.62) is left off for physical realism.

The elemental torque is obtained by taking the axial mass flow through the ring $2\pi r dr$ and multiplying it by the rotational velocity, ωr times 2 to account for the two blades:

$$dQ = \rho(2\pi r dr)(V + v_a)2(\omega r)r \tag{7.63}$$

By using Eqns (7.52), (7.50), $x = r/R$ and $a' = \omega/2\pi n$ it can be shown that:

$$dC_Q/dx = \frac{\pi^2}{2} J x^3 (1 + a) a' \tag{7.64}$$

From this it follows that:

$$a' = \left(\frac{dC_Q}{dx}\right)\left(\frac{2}{\pi^2 J x^3 (1 + a)}\right) \tag{7.65}$$

It is left as an exercise for the reader to derive similar methods for propellers with more than two blades. See Problem 7.13.

7.4.2 MEASURED LOSSES AND SOME INTERPRETATIONS

By applying Eqns (7.58) and (7.59) to experimental propeller data, the power losses can be computed. Typical results of doing this for a **two-bladed** propeller are shown in Figure 7.10, where the so-called power disk-loading coefficient, P_c is defined as:

$$P_c = \frac{P}{\bar{q}AV} \tag{7.66}$$

A detailed analysis in Ref. 7.3 indicates that the dimensionless axial loss, E_a/P (Eqn 7.51) is roughly proportional to P_c while E_r/P is roughly proportional to C_Q/J. Therefore, if the engine power is doubled, the loss due to axial velocity is also doubled.

The reduction in efficiency due to rotational velocity is seen to be small in Figure 7.10, amounting to only about 1%. The reader must keep in mind, that the data in Figure 7.10 apply only to a two-bladed propeller with a relatively low solidity ratio. The rotational losses will increase for high blade angle settings and for higher solidity ratios.

According to Reference 7.3, the rotational losses can be almost completely recovered by using contra-rotating propellers albeit at a price of increased complexity.

The reader will note from Figure 7.9 that in the propeller hub area (x approaching 0) large losses can occur. Such losses can be lowered by using a properly designed spinner.

The so-called "other" losses in Figure 7.10 are due to errors in the computation of the losses due to axial and rotational velocity and also due to losses caused by the profile drag of the propeller blades.

In Section 7.5 several factors which influence the performance of propellers are discussed.

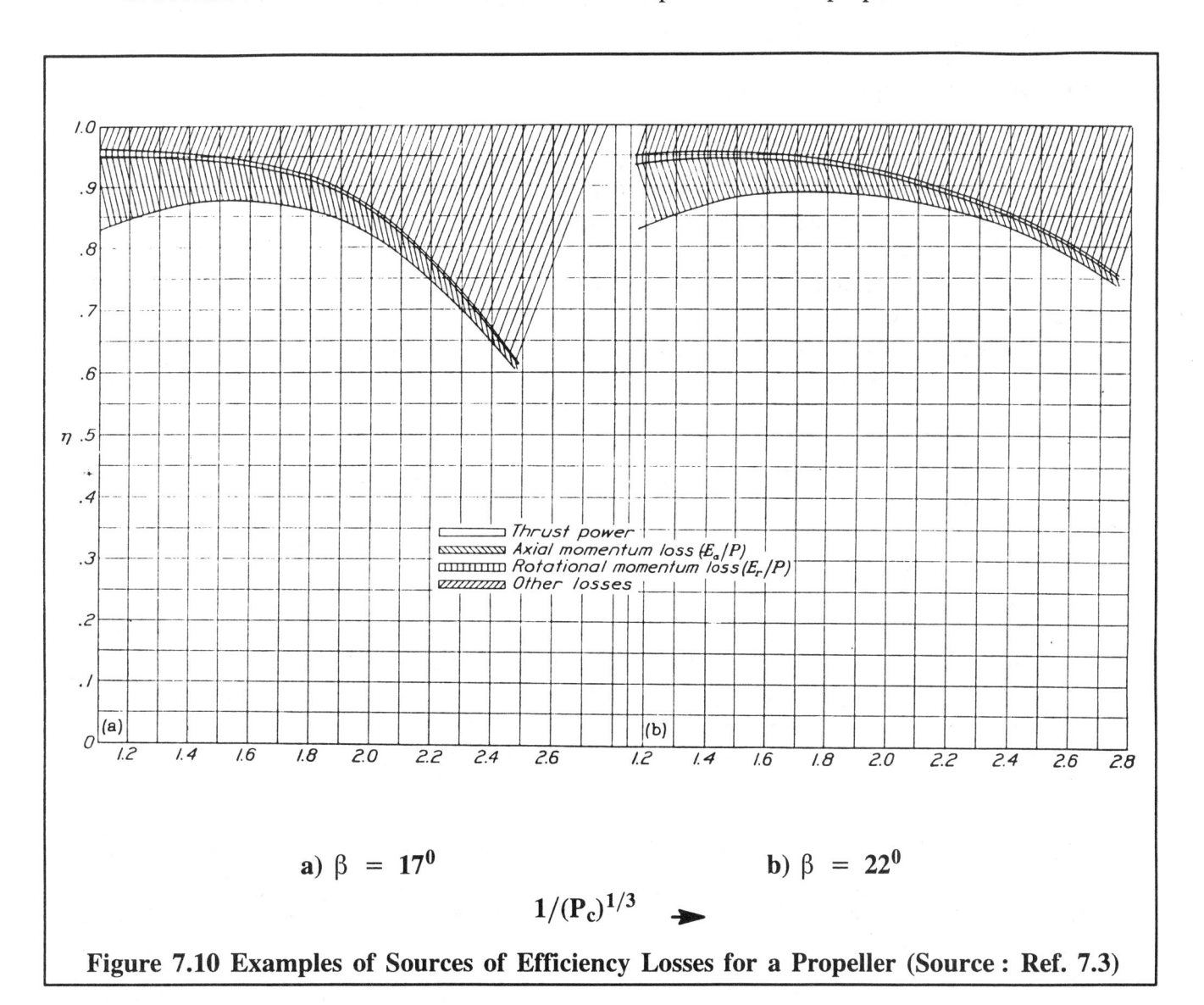

a) $\beta = 17^0$ b) $\beta = 22^0$

$1/(P_c)^{1/3}$ ➤

Figure 7.10 Examples of Sources of Efficiency Losses for a Propeller (Source : Ref. 7.3)

7.5 FACTORS AFFECTING PROPELLER PERFORMANCE

In this Section a discussion is presented of several factors which have been found to influence propeller performance. These factors are:

7.5.1   Propeller Blade Angle   7.5.2   Propeller Blade Geometry

7.5.3   Propeller Blade Loading   7.5.4   Propeller Shank Form

7.5.5   Compressibility Effects   7.5.6   Blockage and Installation Effects

A more detailed review of most of these factors is given in References 7.4 through 7.7.

A discussion of free propeller efficiency versus installed propeller efficiency (also called propulsive efficiency) is given in Sub–section 7.5.7.

7.5.1   PROPELLER BLADE ANGLE

Referring to Figures 7.7 or 7.8, the angle, β, is called the geometric pitch angle of the propeller blade element. If this angle is fixed, the propeller is referred to as a **fixed pitch propeller**. If this angle is variable, the propeller is referred to as a **variable pitch propeller**.

The geometric pitch angle of a propeller varies along the radius for reasons noted in Section 7.2. For reference purposes and a given blade design, the geometric pitch angle of a propeller is typically defined at the 3/4 radius. The ideal, isolated propeller efficiency is strongly dependent on this blade pitch angle. This can be seen from the constant efficiency lines in the propeller charts given in Appendix C: efficiency variations from 40 % to 85% are not unusual.

Fixed pitch propellers have the advantage of simplicity. However, their efficiency varies widely over the range of flight conditions and engine r.p.m. When designing a fixed pitch propeller, two choices normally present themselves: design the pitch angle for best propeller performance (i.e. thrust output) at take–off, or design the pitch angle for best propeller performance (i.e. highest efficiency) in cruise.

If the pitch angle is selected for best take–off thrust, the cruise performance will suffer. By the same token, if the pitch angle is selected for best cruise thrust, the take–off performance will suffer. Some intermediate position can, of course, also be taken. There also are ground–adjustable propellers. These are fixed–pitch propellers, the pitch angle of which can be altered to another fixed angle while on the ground.

There are two types of variable pitch propellers: constant speed propellers and adjustable blade angle propellers. In a constant speed propeller a mechanism automatically adjusts the propeller pitch angle to keep the propeller RPM constant. In an adjustable blade propeller the selection of in–flight pitch angle is left to the pilot.

At the time of publication, most airplanes use either fixed pitch or constant speed propellers.

7.5.2 PROPELLER BLADE GEOMETRY

The distribution of the propeller blade chord from the hub to the tip affects the ability of a propeller to absorb power. If a propeller absorbs a torque, Q, in ft–lbs at a rotational velocity, n, in cycles per second, the absorbed power is equal to:

$$P = 2\pi n Q = 2\pi n \int_0^{1.0R} dQ \approx 2\pi n \int_{0.15R}^{1.0R} dQ \tag{7.67}$$

where: dQ may be found from Eqn (7.40) or from (7.44).
the lower integration limit is changed in recognition of the fact that the propeller hub region does not contribute much to the torque.

Using Eqn (7.44) this yields:

$$P \approx 2\pi n \int_{0.15R}^{1.0R} dQ = 4B\varrho\pi^3 n^3 \int_{0.15R}^{1.0R} x^3 R^4 c \lambda_Q dx \tag{7.68}$$

At this point a simplification is introduced: the coefficient, λ_Q, defined by Eqn (7.42), may be thought of as an average propeller blade drag coefficient, $\bar{c}_d$. Introduction of this average drag coefficient into Eqn (7.68) yields:

$$P \approx 4B\varrho\pi^3 n^3 \bar{c}_d \int_{0.15R}^{1.0R} x^3 R^4 c\, dx \tag{7.69}$$

At this point the so-called blade activity factor, AF is introduced with the following definition:

$$AF = \frac{10^5}{D^5} \int_{0.15R}^{1.0R} cr^3 dr = \frac{10^5}{16} \int_{0.15R}^{1.0R} \left(\frac{c}{D}\right) x^3 dx \tag{7.70}$$

It should be noted, that the factor 10^5 is used only for the purpose of giving the activity factor a convenient numerical value. Using this definition of AF yields for Eqn (7.69):

$$P \approx 4B\varrho\pi^3 n^3 \bar{c}_d \left(\frac{D}{10}\right)^5 AF \tag{7.71}$$

Typical propeller activity factors range from about 70 to about 200. As an example, the blade activity factor for the Lockheed C–130 is AF = 162. Table 7.1 shows additional examples of blade activity factors used in several airplanes.

Experiments in Ref. 7.4 showed that with the same activity factor, AF, and advance ratio, J, be–

Table 7.1 Examples of Propeller Blade Activity Factors, AF			
Airplane or propeller design	Number of blades, B	Blade Activity Factor, AF	Integrated design lift coefficient, $C_{L_{int.}}$
Cessna 310	2	90	RAF–6
Hamilton Standard 6903A with WAC R–3350 compound engine	3	103	0.50 (NACA–16 series
Hamilton Standard 6921A with WAC R–3350 compound engine	3	132	0.586 (NACA–16 series
Hamilton Standard 4D15A3 with turboprop engine	3	200	0.427
Cessna 404	3	107	0.55
Cessna 337 (front prop)	2	112	0.51
Cessna 337 (rear prop)	2	105	0.57
Cessna 180K	2	100	0.50
Cessna S172E	2	104	0.53
Cessna 150M	2	77	0.60
Data obtained by courtesy of McCauley Accessory Division of Cessna Aircraft Company			

low the J value for maximum efficiency, tapered blades with broad roots and narrow tips can attain higher constant speed efficiencies than constant chord blades. These results were obtained for power coefficients, C_P, above 0.1. For power coefficients well below 0.1 constant chord blades were found to yield slightly higher efficiencies at all advance ratios.

7.5.3 PROPELLER BLADE LOADING

The blade loading of a propeller is defined as:

$$P_{bl} = \frac{4P}{\pi BD^2} \qquad (7.72)$$

Table 7.2 shows typical take–off propeller blade loadings for several airplane types. Specific data for 60 airplanes may be found in Ref. (Design, Pt II). The blade loading can be used for the preliminary sizing of the diameter of the propeller for a given airplane. By selecting a representative number for the blade loading, knowing the take–off horsepower and also selecting the number of blades typical for the application, Eqn (7.72) can be used to solve for the required propeller diameter.

According to Ref. 7.5, high blade loadings tend to have an adverse effect on propeller efficiency, particularly at low advance ratios, i.e. J < 2.0. The experimental data in Ref. 7.5 indicate that improvements in efficiencies may be obtained by increasing the solidity ratio of a propeller. This in turn can be achieved by increasing the number of blades. It can be seen from Eqn (7.72) that increasing the number of blades will decrease the blade loading. Ref. 7.5 also indicates that increasing the number of blades does not result in significant efficiency penalties at larger advance ratios. However, increasing the number of blades will increase propeller complexity, cost and weight.

Table 7.2 Typical Propeller Blade Loadings for Various Airplane Types					
Airplane Type	Propeller Type	Number of Blades	Power Loading per Blade in shp/ft^2		
			High	Average	Low
Homebuilts	Fixed pitch	2	3.2	2.4	1.0
FAR 23 Single Engine Types	Constant speed, some fixed pitch	2 or 3	3.9	3.2	2.0
Agricultural Types	Constant speed	2, 3 or 4	5.9	3.8	2.2
Military Trainers	Constant speed	2 or 3	6.2	4.4	2.8
FAR 23 Twin Engine Types	Constant speed	2 or 3	4.8	3.7	2.8
Regional Turboprops	Constant speed	3 or 4	5.2	4.6	3.4
Data from Reference 7.6					

7.5.4 PROPELLER SHANK FORM

The shape of a propeller blade changes significantly toward the root because of the need to attach the blade to a hub. The area of the blade close to the hub is referred to as the shank of the propeller blade. The shank area is indicated for two propellers in Figure 7.11.

The blade and the hub can be one solid unit as shown in Figure 7.11a for a typical fixed pitch, 2–bladed McCauley propeller.

In the case of constant speed propellers, the blade must be able to rotate relative to the hub. An example of a constant speed, 4–bladed Hartzell propeller is also shown in Figure 7.11b.

Experiments reported in Ref. 7.5 have shown that the following conclusions apply to the shank:

1) for constant speed propellers faired shanks given better efficiencies thank round shanks.

2) The design pitch angle at the shank should not exceed 90 degrees.

3) The shank profile should have as large a maximum lift coefficient as possible. For that reason, thin shanks are not desirable.

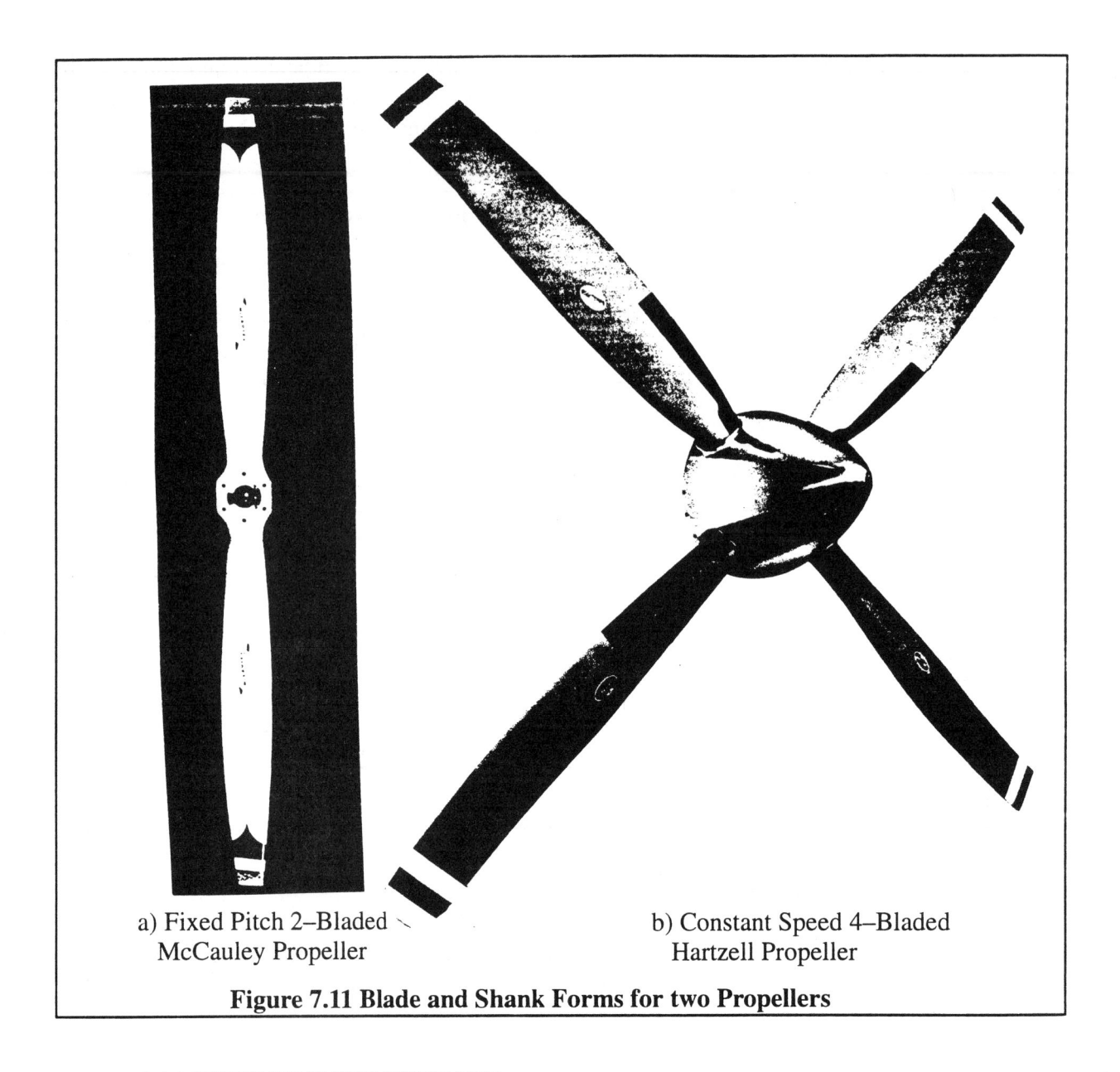

a) Fixed Pitch 2–Bladed McCauley Propeller

b) Constant Speed 4–Bladed Hartzell Propeller

Figure 7.11 Blade and Shank Forms for two Propellers

7.5.5 COMPRESSIBILITY EFFECTS

The combined effect of propeller tip speed and the forward speed of the airplane may subject the propeller tip region to compressibility effects. Neglecting the propeller induced velocity, it is seen from Figure 7.8 that the following relationship holds between forward speed, U, propeller rotational velocity, n (in rps) and the tip speed, U_{tip}:

$$U_{tip} = \sqrt{(\pi Dn)^2 + U^2} \tag{7.73}$$

In terms of the tip Mach number this becomes:

$$M_{tip} = \sqrt{\left(\frac{\pi Dn}{a}\right)^2 + M^2} \tag{7.74}$$

Even though a propeller suffers relatively small efficiency losses, even when the tip operates slightly above M = 1.0 (because most of the thrust is derived from more inward blade sections) it will produce unacceptable noise. To reduce noise to acceptable levels, the tip speed should be kept below roughly 800 ft/sec which is well below the speed of sound in sea–level standard conditions (a = 1,116 ft/sec). This produces a tip Mach number of 0.72.

The reader will observe that if a maximum tip Mach number is selected and the forward speed and the propeller RPM are known, the diameter of the propeller is frozen by Eqn. (7.74) and yields:

$$D = \sqrt{\frac{a^2}{\pi^2 n^2}\left(M_{tip}^2 - M^2\right)} \tag{7.75}$$

This method is also used as a way to select a propeller diameter during the early design stages of a new airplane. Observe, that at low takeoff speeds, M will be much smaller than M_{tip}.

If noise is not a design consideration and tip–losses due to compressibility are to be avoided, thin blade sections and high solidity ratios have been found to be the favorable design solution. An example of the effect of compressibility on propeller efficiency for a propeller with an 8% thick section is given in Figure 7.12 (from Ref. 7.7).

A method to account for the effect of compressibility on efficiency is given in Section 7.5.5.

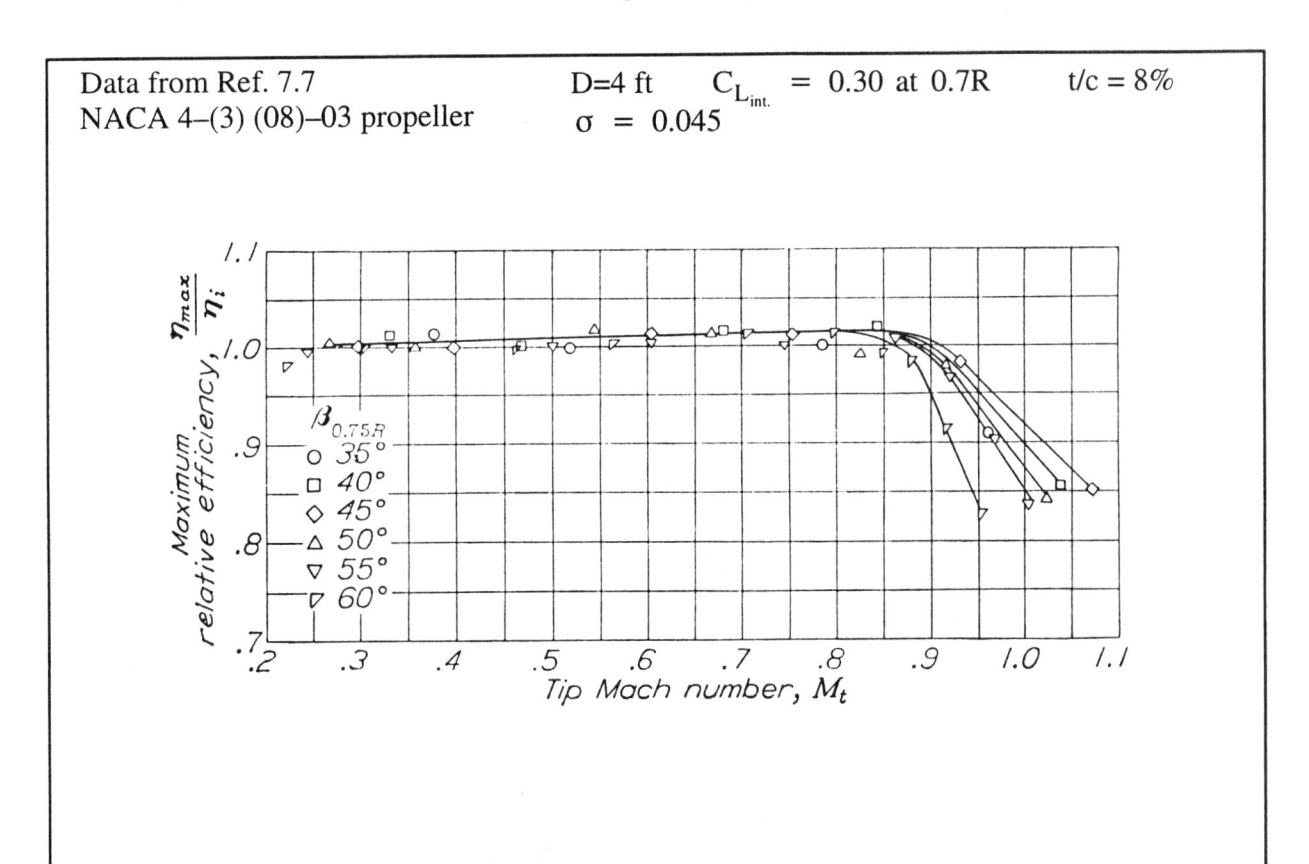

Figure 7.12 Effect of Compressibility on Relative Maximum Efficiency for a Propeller

7.5.6 BLOCKAGE, SCRUBBING, COMPRESSIBILITY AND INSTALLATION EFFECTS ON PROPELLER EFFICIENCY

The shape of the fuselage and/or nacelle and the type of air inlet installed on the nacelle behind or forward of a propeller does affect the installed efficiency of the propeller. These effects are referred to as blockage effects and are discussed in Sub–sub–section 7.5.6.1.

The slipstream created by a propeller can increase the drag of the airplane: it is referred to as scrubbing drag. This increased drag can also be interpreted as a reduction in the installed thrust or the installed efficiency of a propeller. It is discussed in Sub–sub–section 7.5.6.2.

As illustrated in Figure 7.12, the propeller efficiency will decline appreciably when the tip Mach number exceeds about 0.90. This effect is referred to as the compressibility effect. It is also influenced by blade camber, i.e. the design integrated lift coefficient. This effect is discussed in Sub–sub–section 7.5.6.3.

The consequence of these three effects, together referred to as installation effects, is to lower the installed propeller efficiency, relative to the efficiency of a free propeller. The following equation is suggested to account for these effects:

$$\eta_{p_{installed}} = F_{blockage} F_{scrubbing} F_{compressibility} \eta_{p_{free\ for\ J_{effective}}} \qquad (7.76)$$

where: $\eta_{p_{installed}}$ is the installed efficiency of the propeller accounting for blockage, installation and compressibility effects

$F_{blockage}$ is a factor which accounts for the type of inlet on the nacelle: It is discussed in 7.5.6.1.

$F_{scrubbing}$ is a factor which accounts for the scrubbing drag of those airplane components located in the slipstream. It is discussed in 7.5.6.2.

$F_{compressibility}$ is a factor which accounts for blade tip compressibility effects. It is discussed in 7.5.6.3.

$\eta_{p_{free\ for\ J_{effective}}}$ is the free propeller efficiency determined with an effective advance ratio, $J_{effective}$ as discussed also in 7.5.6.1

7.5.6.1 Blockage Effects

The effect of body nose shape (or nacelle shape) on the installed efficiency of a propeller (as opposed to a free propeller) may be thought of as consisting of two parts:

a) a change in the local velocity magnitude (usually a reduction) felt by that area of the propeller closest to the hub. This effect is strongly influenced by the ratio of body diameter to propeller diameter.

b) a change in the propeller blade angle of attack caused by the change in flow angularity for those propeller sections closest to the hub.

A method to account for effect a) is suggested by Torenbeek in Ref.7.8. The method suggests to use an effective advance ratio defined by:

$$J_{effective} = (1 - h)J \tag{7.77}$$

where: h accounts for the retardation of the airflow through the propeller disk, caused by the body behind it. An empirical relation to determine h is:

$$h = 0.329 \frac{S_{body}}{D^2} \tag{7.78}$$

where: S_{body} is the cross sectional area of the body behind the propeller disk

D is the propeller diameter.

The effective advance ratio is then to be used instead of the geometric advance ratio in the calculation of the free propeller efficiency, corrected for blockage effect a): $\eta_{p_{free\ for\ J_{effective}}}$.

A method to account for effect b) is based on Ref. 7.9. The factor, $F_{blockage}$ in Eqn (7.76) is determined from Figure 7.13. Observe, that the blockage factor depends on the type of inlet on the nacelle: annular or scoop. Figure 7.14 shows an example of each.

Both blockage effects can be reduced by streamlining of the body (or nacelle). Examples of good and bad streamlining are given in Figure 7.15.

Extending the shaft of a propeller can help in obtaining good streamlining. An example is shown in Figure 7.16. Bear in mind, that shaft extensions longer than about 8 inches result in vibrations which can only be reduced only with the installation of an additional bearing which adds weight.

Propeller spinners have also been shown to increase the installed efficiency of a propeller. Examples of spinners are shown in Figure 7.15 and 7.16.

The effect of spinners on efficiency becomes negligible if a propeller is placed ahead of a large, blunt body (Ref. 7.10). Examples are the large diameter radial engines of WWII. Most of the propellers installed forward of these engines did not have spinners: they would not have been effective.

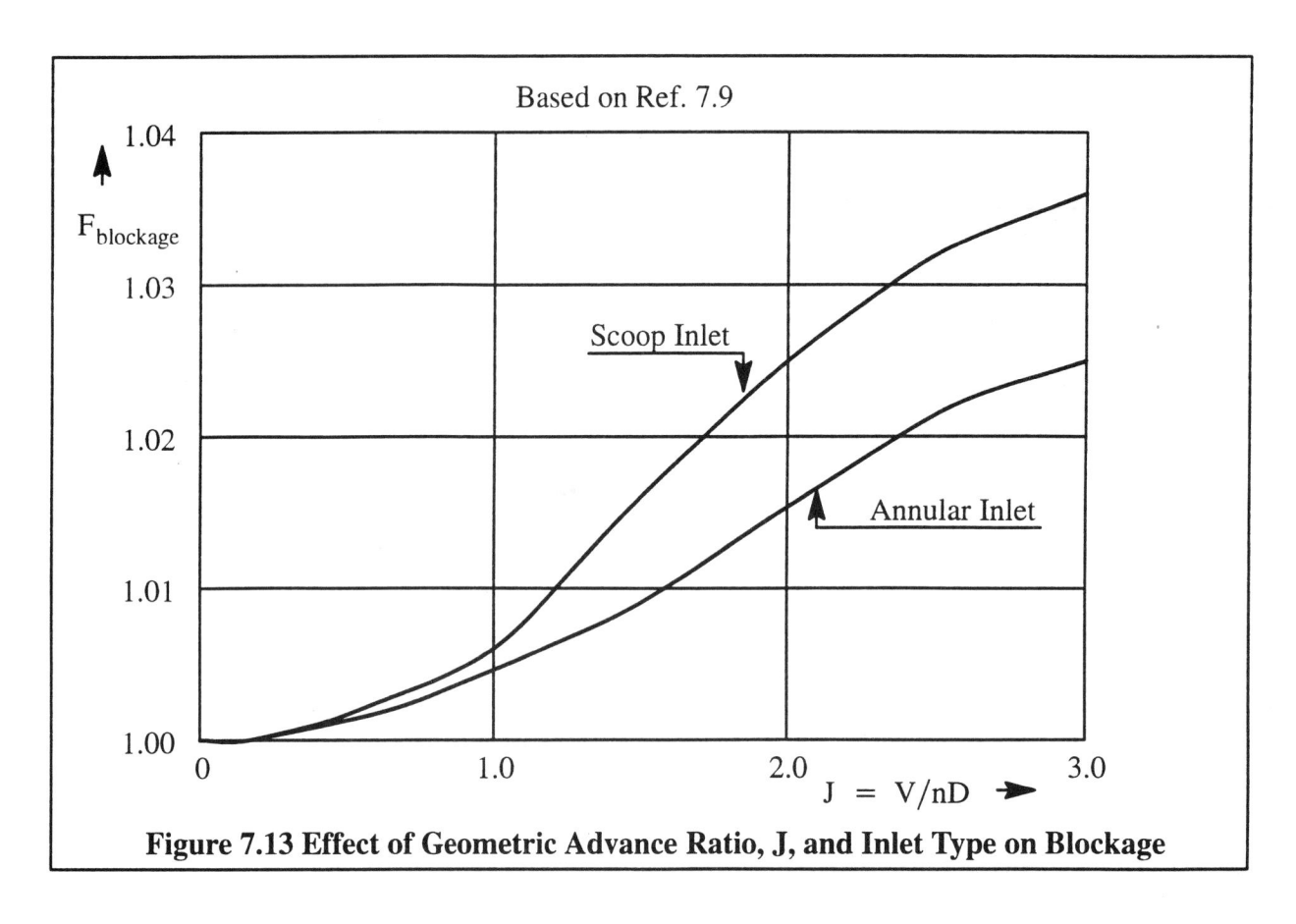

Figure 7.13 Effect of Geometric Advance Ratio, J, and Inlet Type on Blockage

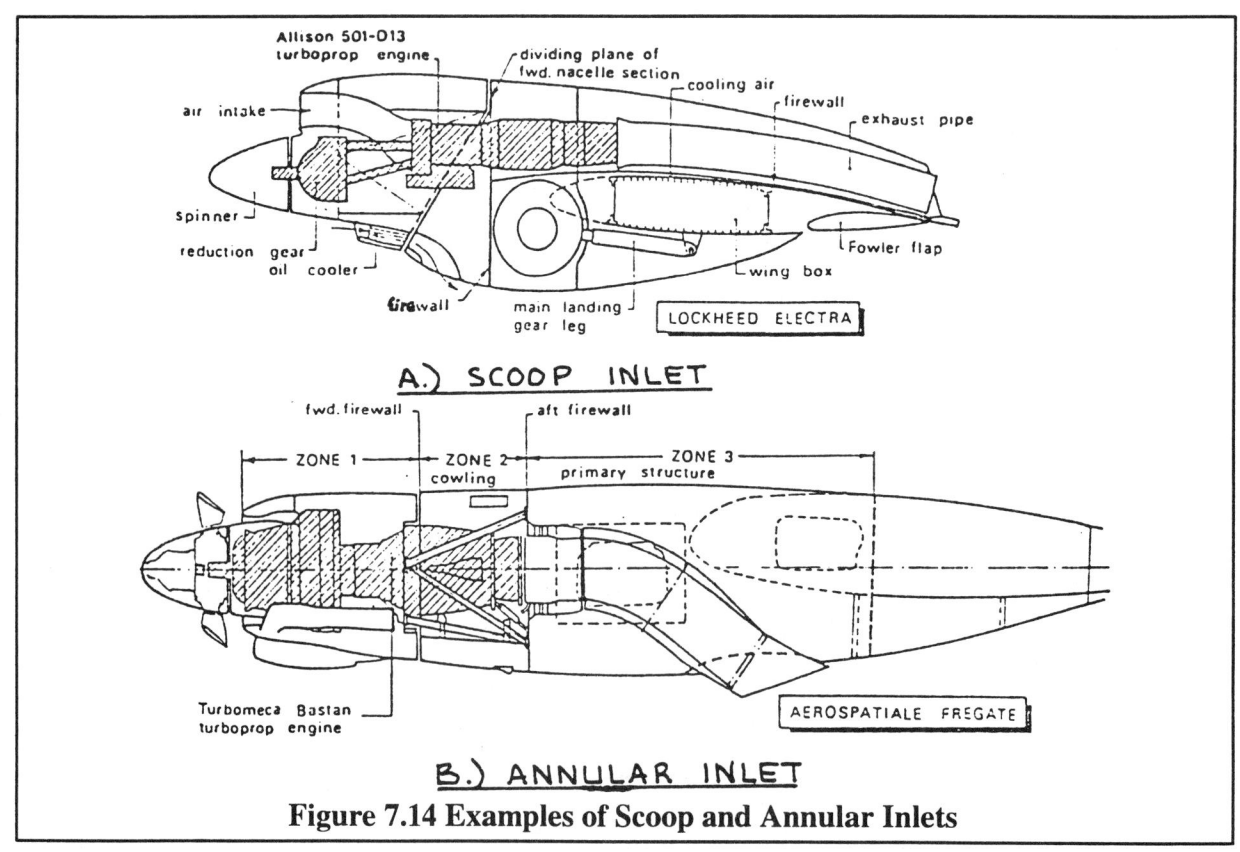

Figure 7.14 Examples of Scoop and Annular Inlets

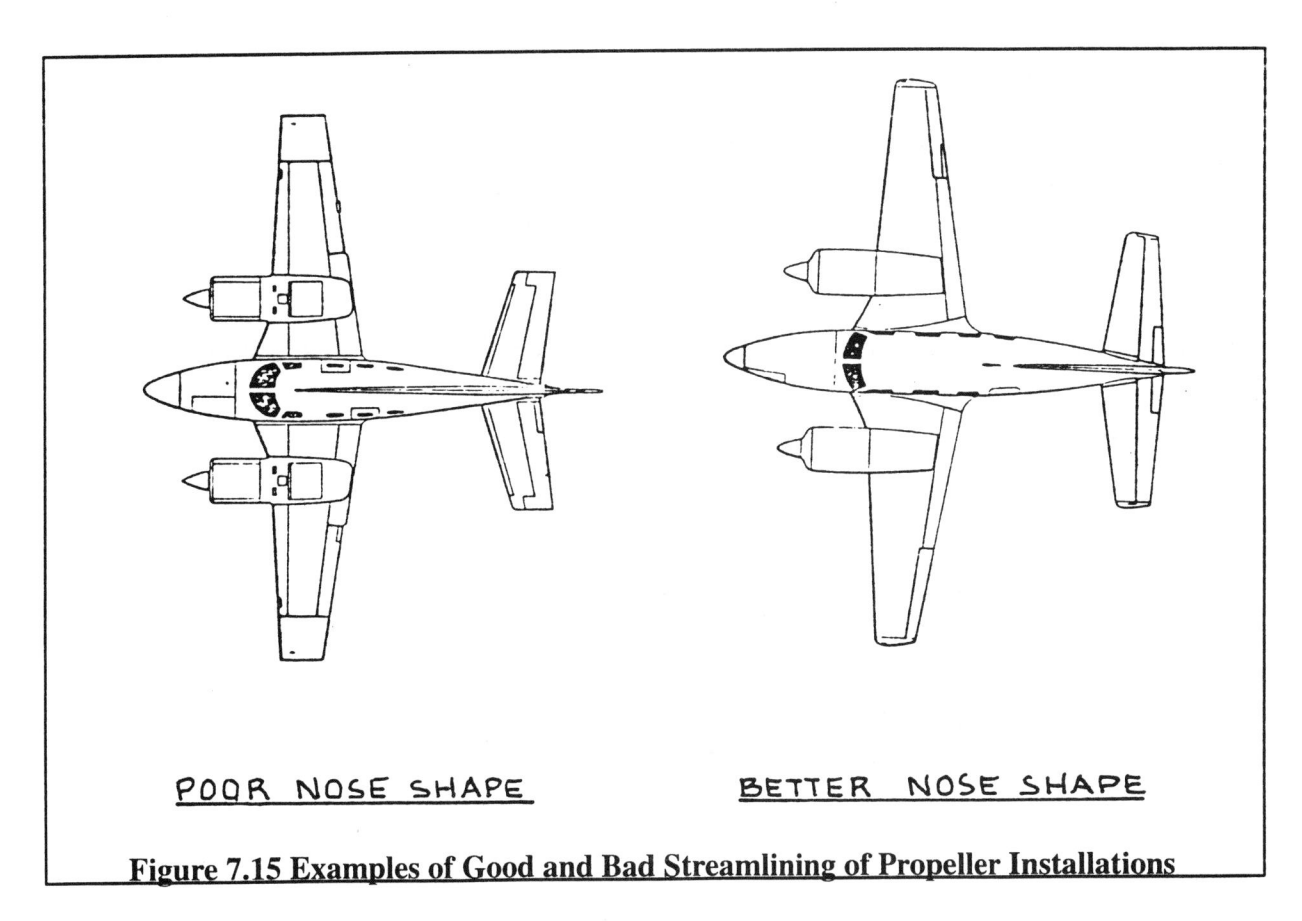

POOR NOSE SHAPE   BETTER NOSE SHAPE

Figure 7.15 Examples of Good and Bad Streamlining of Propeller Installations

Courtesy: Tony Bingelis

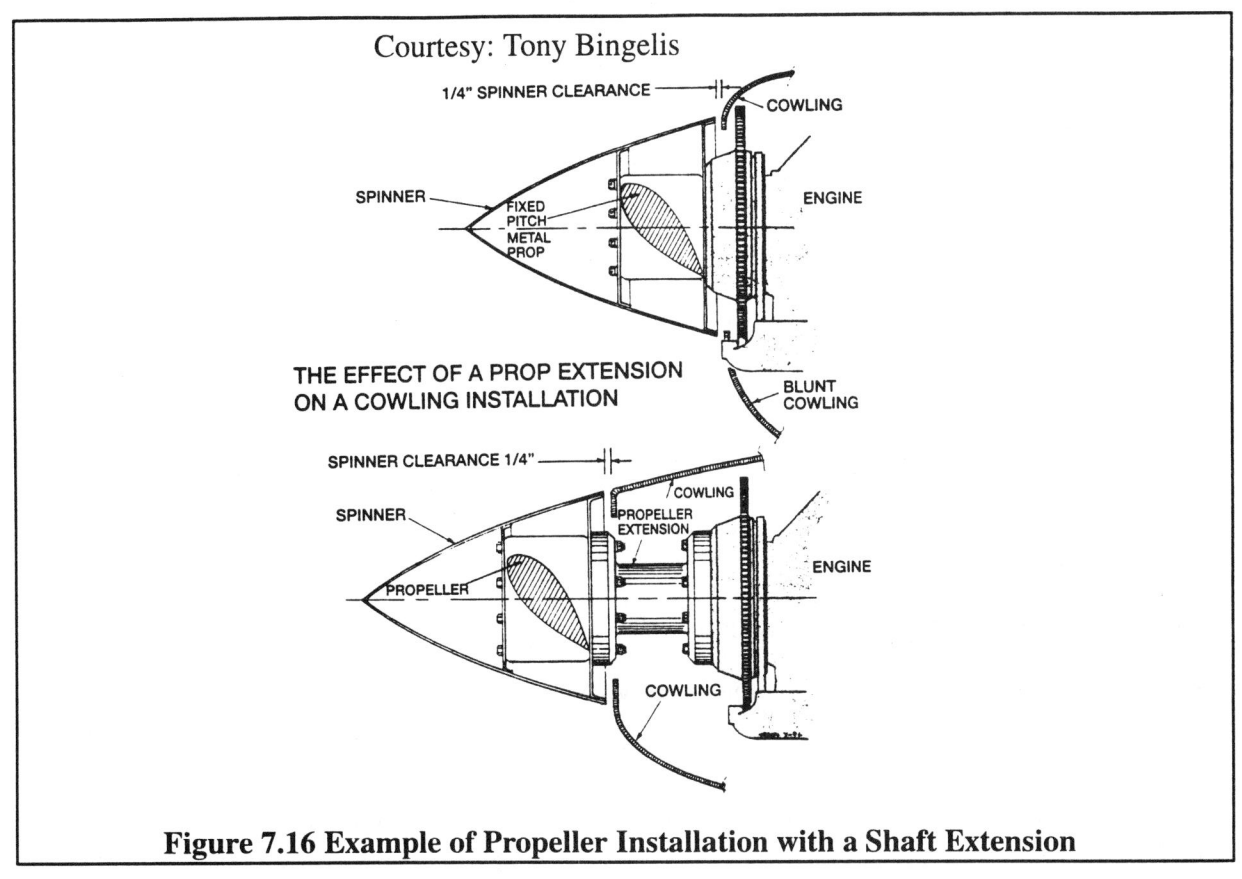

THE EFFECT OF A PROP EXTENSION ON A COWLING INSTALLATION

Figure 7.16 Example of Propeller Installation with a Shaft Extension

7.5.6.2 Scrubbing Drag Effects

The part of the body, nacelle, wing or other airplane component which is directly located in the slipstream will experience a change in drag, called the scrubbing drag. This effect can be 'counted' as a change (increase) in airplane drag or as a change (decrease) in the installed thrust of the propeller.

The scrubbing drag effect can be accounted for by a change in the propeller efficiency through the scrubbing factor, $F_{scrubbing}$, in Eqn (7.76). According to Ref. 7.8 (Torenbeek), based on data from DeHavilland Propellers, this scrubbing factor may be estimated from:

$$F_{scrubbing} = 1 - 1.558 \left(\frac{\sigma f_{slip}}{D^2} \right) \qquad (7.79)$$

where: σ is the atmospheric density ratio
  D is the propeller diameter, in ft
  f_{slip} is the parasite drag area of the those aircraft components which are immersed in the propeller slipstream. This parasite area may be estimated from:

$$f_{slip} \approx 0.0040(S_{wet\ slipstream}) \qquad (7.80)$$

where: 0.0040 is an average skin friction coefficient

  $S_{wet\ slipstream}$ is the wetted area of those aircraft components which are immersed in the propeller slipstream, in ft²

7.5.6.3 Compressibility Effects

The effects of compressibility on the efficiency of a propeller can be accounted for with a compressibility factor, $F_{compressibility}$ which may be determined as follows.

First, the incremental Mach number, $\Delta M_{blade\ camber}$ is determined from Figure 7.17 at the appropriate value of the design integrated lift coefficient, C_{L_i}.

Second, the compressibility correction factor, $F_{compressibility}$ is read from Figure 7.18 at the geometric advance ratio, $J = U/nD$ and at the effective Mach number: $M_{eff} = M + \Delta M_{blade\ camber}$, where M is the free stream Mach number of the airplane.

If needed, the blade tip Mach number may be estimated from Eqn (7.74).

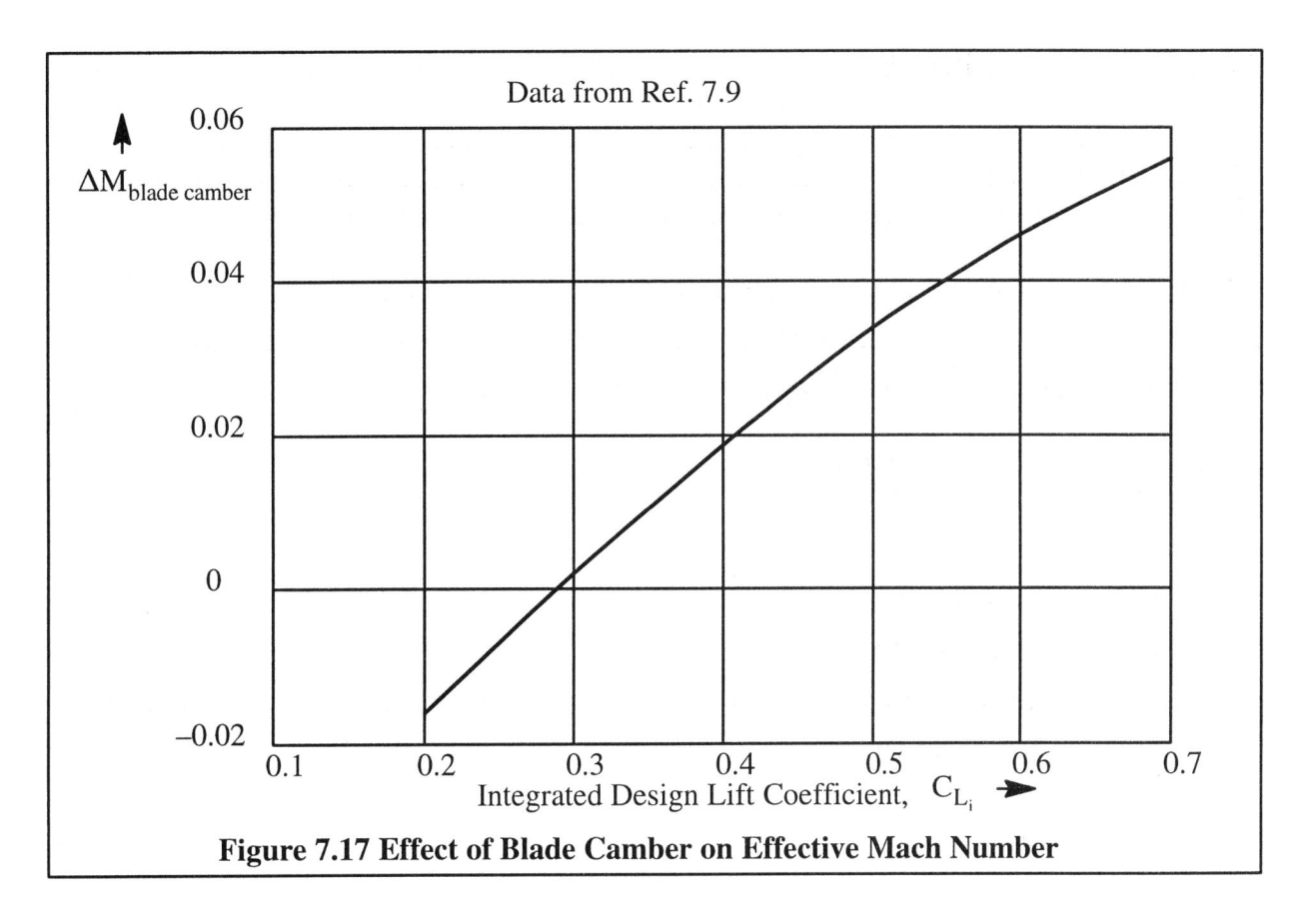

Figure 7.17 Effect of Blade Camber on Effective Mach Number

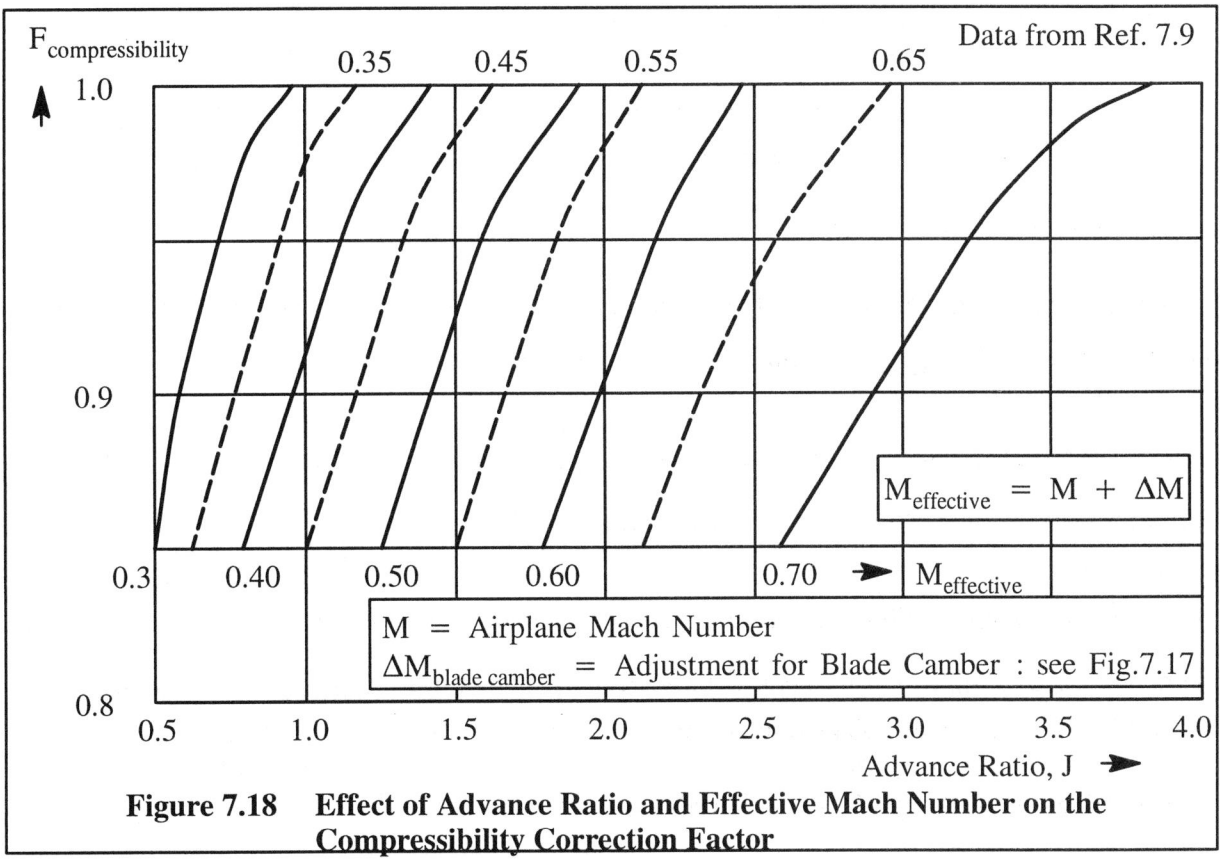

Figure 7.18 Effect of Advance Ratio and Effective Mach Number on the Compressibility Correction Factor

$M_{effective} = M + \Delta M$

M = Airplane Mach Number
$\Delta M_{blade\ camber}$ = Adjustment for Blade Camber : see Fig. 7.17

7.5.7 FREE PROPELLER EFFICIENCY AND INSTALLED PROPELLER EFFICIENCY FOR TRACTORS AND PUSHERS

Because there exists a lot of confusion about the use and meaning of the terms installed propeller efficiency and propulsive efficiency, their relationship is summarized next.

The installed efficiency of a propeller is also referred to as the propulsive efficiency:

$$\eta_{propulsive} = \frac{T_{installed} V}{SHP_{installed}} = \eta_{p_{installed}} \qquad (7.81)$$

Note, that the installed propeller efficiency, $\eta_{p_{installed}}$, is obtained from the free propeller efficiency, $\eta_{p_{free\ for\ J_{effective}}}$, by means of Eqn (7.76).

The installed shaft–horsepower, $SHP_{installed}$, is obtained by subtracting power requirements to drive various systems: fuel pumps, hydraulic pumps, generators, etc. from the engine SHP listed by the manufacturer.

The reader should carefully consult engine manufacturers data: in many cases the engine manufacturer in stating engine SHP has already subtracted the power required to drive systems essential for engine operation. These effects are referred to as engine installation effects. Ref. 7.11 may be consulted for detailed methods to estimate engine installation losses.

Propellers can be installed as a tractor or as a pusher. Figure 7.19 shows an example of each.

Everything discussed so–far applies, strictly speaking, to tractor propeller installations.

In a pusher configuration the scrubbing drag of the slipstream is basically non–existent. However, the propeller will be operating in the wake of the body ahead of it: nacelle or fuselage. Depending on the shaping of the body ahead of the propeller, the installed efficiency can be reduced. If care is taken in streamlining the body ahead of the propeller (such as was done in the case of the example in Figure 7.18 it may be assumed that the installed efficiency of the rear propeller is the same as that of the front propeller. Therefore, the method used to calculate $\eta_{p_{installed}}$ for a pusher propeller is identical to that for the tractor.

Figure 7.20 shows how the installed efficiency of a tractor compares with that of a pusher as a function of the ratio of body diameter to propeller diameter. Note, that if this ratio is about 0.5 or less the installed efficiencies of tractor and pusher propellers are the same.

When a pusher propeller is installed immediately behind a lifting surface, vorticity shed off the trailing edge can cause propeller blade excitation which may result in fatigue problems. In such cases it is recommended to locate the propeller plane well behind the trailing edge of the lifting surface: a distance of at least one half of the lifting surface chord should be adequate.

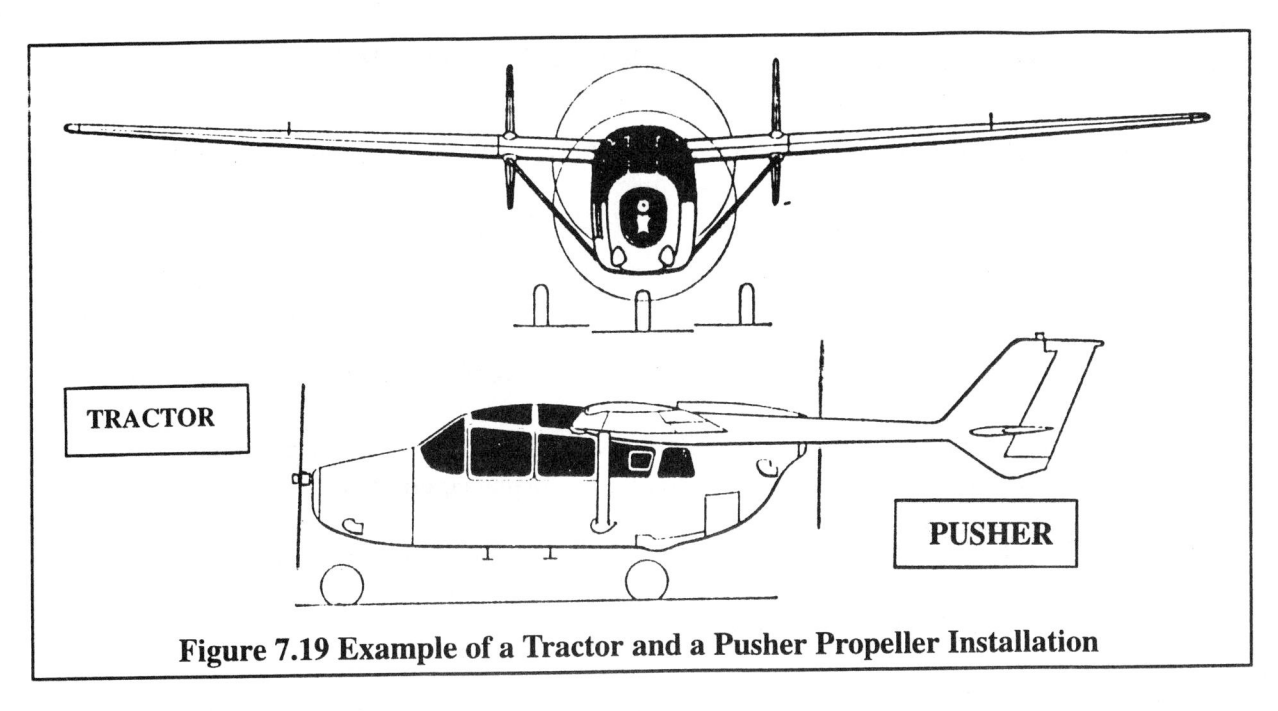

Figure 7.19 Example of a Tractor and a Pusher Propeller Installation

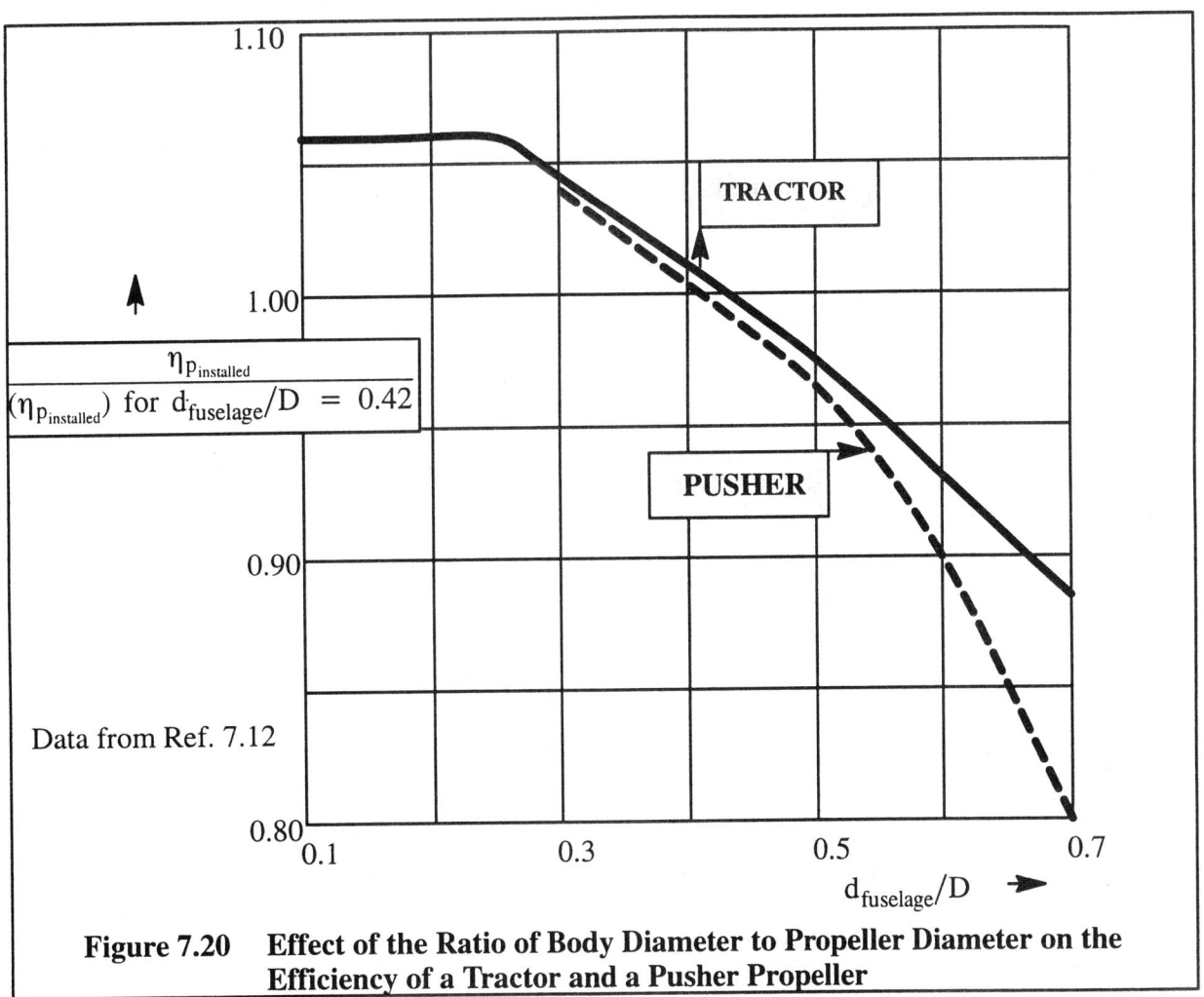

Figure 7.20 Effect of the Ratio of Body Diameter to Propeller Diameter on the Efficiency of a Tractor and a Pusher Propeller

7.6 PREDICTION OF PROPELLER PERFORMANCE

In this Section, methods for predicting propeller performance are presented as follows:

7.6.1   Prediction of Static Thrust   7.6.2   Prediction of In–flight Thrust and Power

7.6.3   Prediction of Negative Thrust

Section 7.6.4 contains a very brief discussion of prop–fans and ducted propellers

In predicting the performance of propellers, use is made of so–called propeller performance charts. References 7.13 and 7.14 present propeller performance charts for the following cases:

a) 3– and 4– bladed propellers for light to medium transports in Ref. 7.13

b) 2–bladed propellers for light, general aviation airplanes in Ref. 7.14.

Additional propeller performance charts are included in Ref. 7.15. During the 80's and 90's several computerized methods for estimating propeller performance have become available. One good example is given in Ref. 7.16. That method also includes a method from estimating propeller noise. A large family of propeller performance charts and a method for computing propeller performance was generated by Hamilton–Standard in the 1950's: see Reference 7.9. Copies of these charts are included in Appendix C as Charts C1 through C40. The methods presented in this Section are based on Reference 7.9.

7.6.1 PREDICTION OF STATIC THRUST

Since propeller thrust is derived from the lift generated by the propeller blades, the average lifting capability of a propeller blade is important. For that reason the so–called integrated design lift coefficient, C_{L_i} is introduced. By definition:

$$C_{L_i} = 4 \int_{0.15}^{1.0} (c_{l_d}) x^3 dx \tag{7.82}$$

where: c_{l_d} is the section design lift coefficient of the blade

The blade airfoil is normally selected so that the section drag is lowest at the section design lift coefficient. This then determines the camber of the section. Typical numerical values for integrated design lift coefficients are: $C_{L_i} \approx 0.35 - 0.60$. As a general rule, lower values of C_{L_i} lead to good cruise performance while higher values of C_{L_i} lead to good low speed performance.

A step–by–step procedure for predicting the static thrust of propellers will now be presented.

The procedure is designed for 3– and 4–bladed propellers. For 2–bladed propellers and for propellers with more than four blades an extrapolation process is used, based on data for 3–bladed and 4–bladed propellers.

Step 1: Determine the following input information:

    a) Altitude

    b) Engine shaft horsepower actually delivered to the propeller, SHP

    c) Propeller rpm, N and rps, n

    d) Blade activity factor, AF

    e) Integrated design lift coefficient, C_{L_i}

    f) Number of blades, B

Step 2: Calculate the propeller power coefficient:

$$C_p = \frac{550(SHP)}{\varrho n^3 D^5} \qquad (7.83)$$

where: SHP is the shaft horsepower, actually delivered by the engine to the propeller

        ϱ is the atmospheric density at the selected altitude, in slugs/ft3

        n is the propeller rotational speed in revolutions per second, rps

        D is the propeller diameter, in ft

Step 3: For three–bladed and for four–bladed propellers, select one of the propeller Charts C1 through C8 depending on the number of blades, activity factor and integrated design lift coefficient.

For two–bladed propellers (B = 2) and for propellers with more than four blades (B > 4) proceed to Step 5.

Step 4: From the appropriate chart, determine the ratio C_T/C_P. Proceed to Step 6.

Step 5: For propellers with B = 2 and/or B > 4 determine the appropriate C_T/C_P values for three–bladed and for four–bladed propellers at the values of C_P, AF and C_{L_i} for the B–bladed propeller at hand.

For a two–bladed propeller, calculate C_T/C_P from:

$$\left(\frac{C_T}{C_P}\right)_{B=2} = \left(\frac{C_T}{C_P}\right)_{B=3} - \left\{\left(\frac{C_T}{C_P}\right)_{B=4} - \left(\frac{C_T}{C_P}\right)_{B=3}\right\} \qquad (7.84)$$

For a B–bladed propeller with B > 4, calculate C_T/C_P from:

$$\left(\frac{C_T}{C_P}\right)_B = \left(\frac{C_T}{C_P}\right)_{B=4} + \left\{\left(\frac{C_T}{C_P}\right)_{B=4} - \left(\frac{C_T}{C_P}\right)_{B=3}\right\}(B - 4) \qquad (7.85)$$

Step 6: Calculate the static propeller thrust from:

$$T_{static} = 33,000\left(\frac{C_T}{C_P}\right)\left(\frac{SHP}{ND}\right) \quad , \text{where N is in rpm} \qquad (7.86)$$

This equation follows from Eqns (7.49) and (7.51).

Two example applications will now be presented: Example 7.2 for a three–bladed propeller and Example 7.3 for a two–bladed propeller.

Example 7.2: A three–bladed propeller has the following characteristics:

N = 2,800 rpm   D = 6.7 ft C_{L_i} = 0.50   AF = 120

The engine shaft horsepower delivered to the propeller is SHP = 400 hp.

Calculate the static thrust at sea–level

Solution:

Step 1: The following input information is available:

a) Altitude: sea–level   b) SHP = 400 hp
c) N = 2,800 rpm   d) AF = 120
e) C_{L_i} = 0.50   f) B = 3

Step 2:

$$C_p = \frac{550 \times 400}{0.002377 \times (2,800/60)^3 \times 6.7^5} = 0.0673$$

Step 3: Select Chart C2 from Appendix C.

Step 4: From Chart C2 and for C_{L_i} = 0.50 it is seen that: C_T/C_P = 2.32

Step 5: Not applicable, since B = 3

Step 6: From Eqn (7.85) find the static thrust as:

$$T_{static} = 33,000 \times 2.32 \times \left(\frac{400}{2,800 \times 6.7}\right) = 1,632 \text{ lbs}$$

This represents the static thrust delivered by a free propeller. The effect of nacelle blockage on static thrust is often negligible. Therefore, this answer is assumed to apply to the installed propeller.

Example 7.3: A two–bladed propeller has the following characteristics:

   $N = 2{,}800$ rpm $D = 6.7$ ft $C_{L_i} = 0.50$ $B = 2$

The engine shaft horsepower delivered to the propeller is SHP = 400 hp.

Calculate the static thrust

Solution:

Step 1: The following input information is available:

   a) Altitude: sea–level   b) SHP = 400 hp

   c) $N = 2{,}800$ rpm   d) AF = 120

   e) $C_{L_i} = 0.50$ f) $B = 2$

   g) $D = 6.7$ ft

Step 2:
$$C_p = \frac{550 \times 400}{0.002378 \times (2{,}800/60)^3 \times 6.7^5} = 0.0673$$

Steps 3: Not applicable to a two–bladed propeller.

Steps 4: Not applicable to a two–bladed propeller.

Step 5: Select charts C2 and C4 for 3–bladed and 4–bladed propellers with the same AF and integrated design lift coefficient. Read off the following values:

From Chart C2: $C_T/C_P = 2.32$ From Chart C4: $C_T/C_P = 2.42$

With Eqn (7.83):

$$\left(\frac{C_T}{C_P}\right)_{B=2} = 2.32 - (2.42 - 2.32) = 2.3$$

Finally, with Eqn (7.85):

$$T_{static} = 33{,}000\left(\frac{C_T}{C_P}\right)\left(\frac{SHP}{ND}\right) = 33{,}000 \times 2.3 \frac{400}{2{,}800 \times 6.7} = 1{,}618 \text{ lbs}$$

7.6.2 PREDICTION OF IN-FLIGHT THRUST AND POWER

A step–by–step procedure for predicting the in–flight thrust of propellers will be presented. The procedure is applies to 3– and 4–bladed propellers. For 2–bladed propellers and for propellers with more than four blades an extrapolation process is used, based on 3–bladed and 4–bladed propellers.

Step 1: Determine the following input information:

    a) Altitude and speed, U
    b) Engine shaft horsepower actually delivered to the propeller, SHP
    c) Propeller rpm, N and rps, n
    d) Blade activity factor, AF
    e) Integrated design lift coefficient, C_{L_i}
    f) Number of blades, B

Step 2: a) Calculate the propeller advance ratio, J, from Eqn (7.52):

$$J = \frac{V}{nD} \qquad (7.87)$$

b) Calculate the effective propeller advance ratio from Eqn (7.76):

$$J_{effective} = (1 - h)J \qquad (7.88)$$

Step 3: Calculate the propeller power coefficient, C_P from Eqn (7.83):

$$C_P = \frac{550 \times SHP}{\varrho n^3 D^5} \qquad (7.89)$$

Step 4: For three–bladed and for four–bladed propellers, select one of the propeller charts C9 through C40 depending on the number of blades, activity factor and integrated design lift coefficient.

For two–bladed propellers (B = 2) and for propellers with more than four blades (B > 4), proceed to Step 9.

Step 5: From the appropriate chart in Step 4 determine the free propeller efficiency for the power coefficient determined in Step 3 **and** for the effective advance ratio found in Step 2b:

$\eta_{P_{free\ for\ J_{efffective}}}$. Note from the propeller chart that the free propeller efficiency also implies a certain blade angle.

Step 6: Determine the installed propeller efficiency from Eqn (7.76):

$$\eta_{P_{installed}} = F_{blockage} F_{scrubbing} F_{compressibility} \eta_{P_{free\ for\ J_{effective}}} \qquad (7.90)$$

Step 7: Calculate the installed propeller thrust from Eqn (7.81):

$$T_{installed} = \frac{\eta_{P_{installed}} SHP_{installed}}{U} \qquad (7.91)$$

Step 8: Calculate the available thrust horsepower from:

$$\text{THP}_{av} = \frac{T_{installed} U}{550} \qquad (7.92)$$

Step 9: For propellers with B = 2 and/or B > 4 determine the appropriate $\eta_{p_{free}}$ values for three–bladed and for four–bladed propellers at the values of C_P, AF and C_{L_i} for the B–bladed propeller at hand. This step yields: $\eta_{p_{free_{B=3}}}$ and $\eta_{p_{free_{B=4}}}$.

Note: $\eta_{p_{free}}$ should really be: $\eta_{p_{free\ for\ J_{efffective}}}$ if the effective J differs significantly from J.

Step 10: For a two–bladed propeller (B=2), calculate $\eta_{p_{free_{B=2}}}$ from:

$$\eta_{p_{free_{B=2}}} = \eta_{p_{free_{B=3}}} + \left(\eta_{p_{B=3}} - \eta_{p_{B=4}}\right) \qquad (7.93)$$

For a propeller with more than 4 blades (B>4) calculate $\eta_{p_{free_{B=B}}}$ from:

$$\eta_{p_{free_{B=B}}} = \eta_{p_{free_{B=4}}} + \left(\eta_{p_{B=4}} - \eta_{p_{B=3}}\right)(B - 4) \qquad (7.94)$$

Step 11: Complete Steps 6–9 for the B–bladed propeller

Step 12: To compute installed propeller thrust and installed propeller power for a range of speeds, repeat this procedure for as many speeds, U, as required.

An example application will now be presented.

Example 7.4: A three–bladed propeller has the following characteristics:

N = 2,800 rpm   D = 6.7 ft C_{L_i} = 0.50   AF = 120

The engine shaft horsepower delivered to the propeller is SHP = 400 hp. The propeller is installed forward of a body such that the following installation characteristics prevail: h=0.20 and $F_{blockage}$ = 0.85. The factors h and $F_{blockage}$ are defined in Eqns (7.78) and (7.76) respectively. Assume that the following additional correction factors have been determined for this airplane: $F_{scrubbing}$ = 0.95 and $F_{compressibility}$ = 0.99. Calculate and plot the available thrust and available thrust horsepower versus speed at sea–level for: U=50 kts, U= 100 kts and U=200 kts.

Solution:

Step 1: The following input information is available:

a) Altitude: sea–level   b) SHP = 400 hp

c) N = 2,800 rpm
d) AF = 120
e) C_{L_i} = 0.50
f) B = 3
g) D = 6.7 ft
h) h = 0.20
i) $F_{blockage}$ = 0.85
j) $F_{scrubbing}$ = 0.95
k) $F_{compressibility}$ = 0.99

Step 2: Calculate the effective propeller advance ratio from Eqns (7.86) and (7.87):

$$J_{effective} = (1 - h)\frac{U}{nD}$$

For U=50 kts: $J_{effective} = (1 - 0.20)\frac{50 \times 1.688}{(2,800/60) \times 6.7} = 0.22$

For U=100 kts: $J_{effective} = (1 - 0.20)\frac{100 \times 1.688}{(2,800/60) \times 6.7} = 0.43$

For U=200 kts: $J_{effective} = (1 - 0.20)\frac{200 \times 1.688}{(2,800/60) \times 6.7} = 0.86$

Step 3: Calculate the propeller power coefficient, C_P from Eqn (7.88):

$$C_P = \frac{550 \times 400}{0.002378(2,800/60)^3 6.7^5} = 0.067$$

Step 4: The current propeller has an activity factor of AF=120. There are no charts for this activity factor. Charts C15 and C19 apply to the current 3-bladed propeller except for their activity factors of 100 and 140 respectively.

Step 5: The following free propeller efficiencies are found:

			$\eta_{P_{free}}$ for $J_{effective}$		
			Chart C15	Chart C19	Interpolation
U, kts	$J_{effective}$	C_P	AF=100	AF=140	AF=120
50	0.22	0.067	0.41	0.43	0.42
100	0.43	0.067	0.67	0.67	0.67
200	0.86	0.067	0.87	0.83	0.85

Step 6: The installed propeller efficiency is found from Eqn (7.89) as:

U, kts $\eta_{P_{installed}}$

50        $\eta_{P_{installed}} = 0.85 \times 0.95 \times 0.99 \times 0.42 = 0.34$

100        $\eta_{P_{installed}} = 0.85 \times 0.95 \times 0.99 \times 0.67 = 0.54$

200        $\eta_{P_{installed}} = 0.85 \times 0.95 \times 0.99 \times 0.85 = 0.68$

Step 7: The installed thrust is found with Eqn (7.90) as:

U, kts	$T_{installed}$
50	$T_{installed} = \dfrac{0.34 \times 400 \times 550}{50 \times 1.688} = 886$ lbs
100	$T_{installed} = \dfrac{0.54 \times 400 \times 550}{100 \times 1.688} = 704$ lbs
200	$T_{installed} = \dfrac{0.68 \times 400 \times 550}{200 \times 1.688} = 443$ lbs

Step 8: The available thrust horsepower is found with Eqn (7.91) as:

U, kts	THP_{av}
50	$THP_{av} = \dfrac{886 \times 50 \times 1.688}{550} = 136$ hp
100	$THP_{av} = \dfrac{704 \times 100 \times 1.688}{550} = 216$ hp
200	$THP_{av} = \dfrac{443 \times 200 \times 1.688}{550} = 272$ hp

Figure 7.21 shows a plot of the results of Steps 7 and 8.

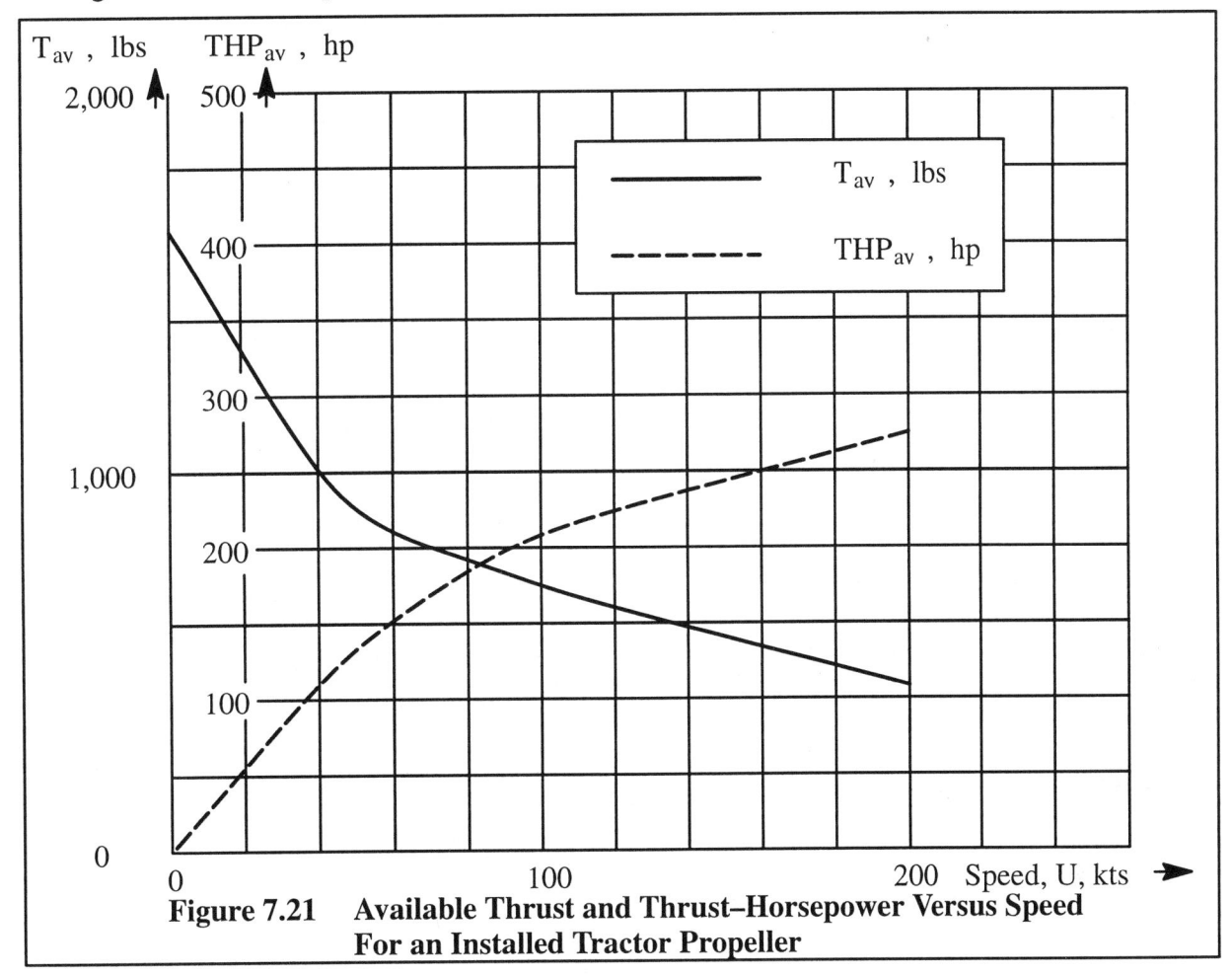

Figure 7.21 Available Thrust and Thrust–Horsepower Versus Speed For an Installed Tractor Propeller

7.6.3 PREDICTION OF NEGATIVE THRUST

Propellers can produce negative thrust in the following situations:

a) The blade angles are reversed and power is applied to produce negative or braking thrust. This also referred to as thrust reversing.

b) An engine is inoperative and the propeller is stopped or wind–milling and as a result produces drag. This drag can be viewed as a negative thrust.

These cases will be discussed in Sub–sub–sections 7.6.3.1 and 7.6.3.2 respectively.

7.6.3.1 Prediction of Negative Thrust or Propeller Thrust Reversing

When the propeller blades are set at a negative pitch angle and power is applied to the engine, negative thrust will be produced. This negative thrust is used to help slow the airplane during the groundrun following touchdown.

A method for estimating the 'braking' thrust of a propeller is presented in Reference 7.17. A simple, approximate method will be presented here. This approximate method is based on the assumption that the absolute magnitude of lift generated by a propeller blade depends on the absolute magnitude of the blade angle of attack. This idea is worked out in the following.

Propeller thrust can be regarded as proportional to the lift acting on the blades because the drag coefficient in Eqn (7.18) is usually much smaller that the lift coefficient. According to Eqn (7.28) and Figure 7.17, which shows a blade section at a negative pitch angle, the blade section angle of attack can be written as:

$$\alpha = (-\beta + \theta - \phi - \alpha_0) \tag{7.95}$$

where: α_0 is the blade section angle of attack for zero lift.
β is called the geometric pitch angle of the blade element
ϕ is the so–called helix angle
θ, is the induced angle

The angles β, ϕ and θ were defined in Sections 7.2 and 7.3.

The section lift coefficient now follows from:
$$c_l = a_0(-\beta + \theta - \phi - \alpha_0) = -a_0\{\beta - (\theta - \phi) + \alpha_0\} \tag{7.96}$$

It is seen from Figure 7.22 that:

$$\theta - \phi = \frac{V_i - V}{\Omega r} \tag{7.97}$$

Figure 7.22 Definition of Angles for a Propeller Blade Element at Negative Pitch

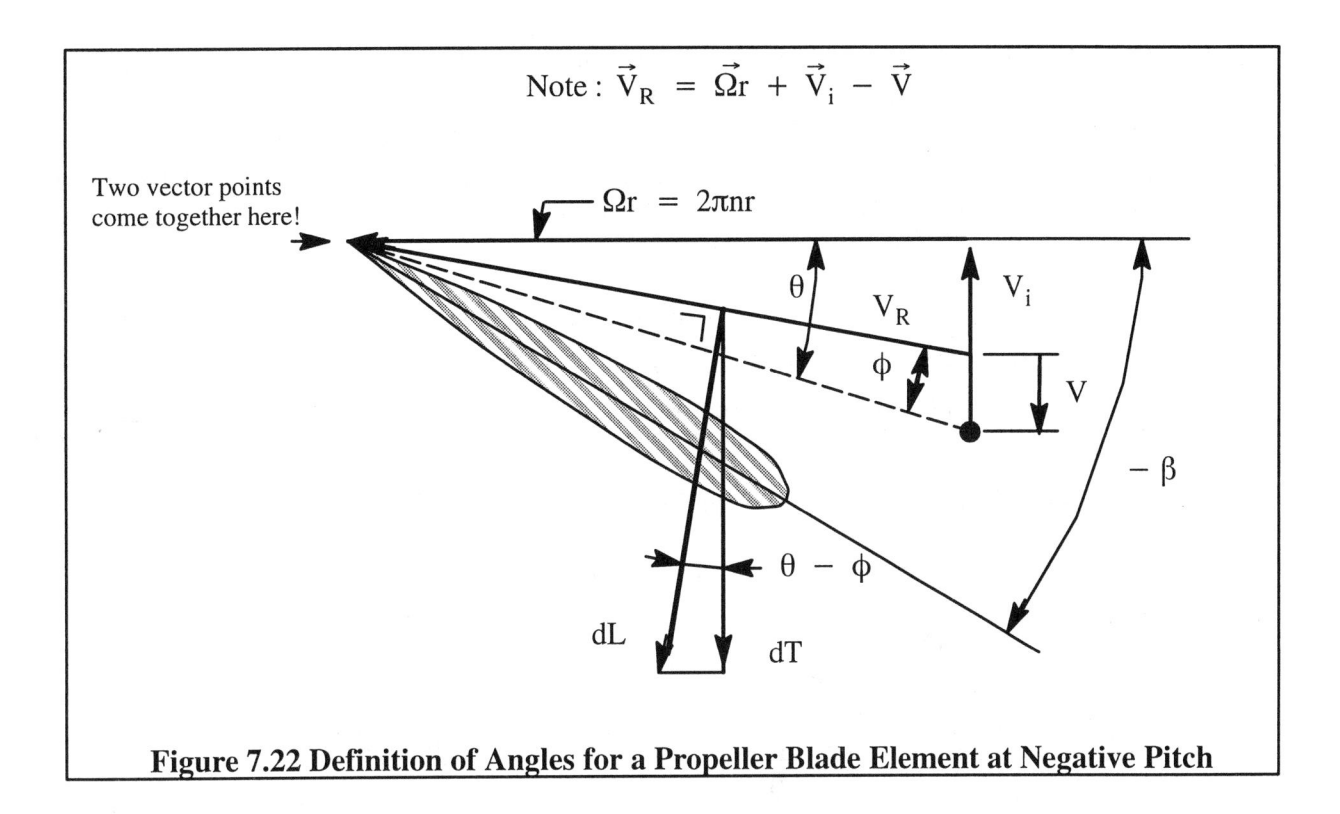

With Eqn (7.18) this now yields:

$$\beta - \phi = \frac{V_i - V}{\Omega r} = \frac{1}{\Omega r}\left\{\frac{V}{2} + \sqrt{\left(\frac{V}{2}\right)^2 + \frac{T}{2A\varrho}} - V\right\} = \frac{v_c}{\Omega r} = \phi_c \qquad (7.98)$$

where: v_c is the induced velocity for positive thrust, see also Eqn (7.10).

Eqn (7.95) may now be written as:

$$c_l = a_0(-\beta - \phi_c - \alpha_0 + 2\alpha_0) = -a_0\{(\beta + 2\alpha_0) - \phi_c - \alpha_0\} \qquad (7.99)$$

Eqn (7.99) suggests that negative thrust can be calculated approximately from propeller charts for positive thrust but using an equivalent pitch angle of: $(\beta + 2\alpha_0)$. If blade airfoil data for zero-lift angle of attack are not available, this quantity may be approximated by:

$$\alpha_0 \approx -7.3\, c_{l_{design}} \text{ in deg} \qquad (7.100)$$

At a given rated shaft horsepower, approved for braking operation, the magnitude of the negative pitch angle needed to absorb that power must be known for this method to be applicable. If it is not known, it may be assumed to be equal to the pitch angle needed for producing positive thrust at the same horsepower.

An application of this approximate method is now presented.

Chapter 7

Example 7.5: A 3–bladed, free propeller of 15.8 ft diameter and a blade activity factor of 100 is set at a negative pitch angle at a speed of 87 kts at sea–level. The propeller is driven by a piston engine, rated at 1,440 shp at 2,400 rpm. The propeller is geared to the engine crankshaft with a gearbox with a gearing ratio of 0.40. If the design integrated lift coefficient is 0.30 calculate the braking thrust.

Solution:

Step 1: Having no blade airfoil data available, the zero lift angle of attack of the blade is estimated with Eqn (7.100) as: $\alpha_0 \approx -7.3\, c_{l_{design}} = -7.3 \times 0.3 = -2.2$ deg.

Step 2: The propeller rps, n, is found as: n = 2,400 x 0.4 / 60 = 16 rps

The power coefficient is found from Eqn (7.89) as:

$$C_P = \frac{550 \times 1,440}{0.002378 \times 16^3 \times 15.8^5} = 0.083$$

The advance ratio, J, is found from Eqn (7.87) as: J = U/nD, or
J = (87x1.688)/(16x15.8) = 0.58

Step 3: Chart C14 applies to this propeller. According to Chart C14 with J=0.58 and $C_P = 0.083$ the equivalent positive pitch angle is 21.5 degrees. Therefore, the equivalent pitch angle for braking is: $(\beta + 2\alpha_0) = 21.5 - 2 \times 2.2 = 17.1$ degrees, or, actually, –17.1 degrees.

Step 4: To find the braking (reversing) thrust, the propeller efficiency is needed. From Chart C14, with a pitch angle of 17.1 degrees and J = 0.58, the corresponding efficiency is: $\eta_{p_{free}} = 0.81$. The reversing thrust now follows from Eqn (7.81) as:

$$T_{brake} = -\frac{\eta_P 550 SHP}{U} = -\frac{0.81 \times 550 \times 1,440}{87 \times 1.688} = -4,368 \text{ lbs}$$

This example is the one given in Ref. 7.17 with as results: –18 degrees for the pitch angle and –4,730 lbs for the thrust. The agreement is fairly good.

The reader should remember that this approximate method applies only at relatively low speeds.

At high speeds and low positive pitch angles, negative thrust (or drag) may occur because the angle ϕ in Eqn (7.28) may be larger than β. This situation may be associated with high speed diving or with inoperative engines. The possibility of using negative thrust to reduce the dive speed of an airplane will not be considered here. Interested readers should consult Ref. 7.18 as an example.

7.6.3.2 Prediction of Drag on Stopped Engines

When an engine which drives a propeller becomes inoperative, one of two scenarios must be accounted for:

Scenario 1: the propeller stops rotating

If a fixed pitch propeller stops rotating there will be a significant amount of flow separation from the leading and trailing edges of the propeller blades which act like flat plates set almost normal to the flow. This results in extra drag which can be estimated from the following incremental drag coefficient:

$$\Delta C_{D_{\text{stopped fixed pitch propeller}}} = (0.10 + \cos^2 \beta_{0.7}) \frac{S_{\text{blade planform}}}{S} \qquad (7.101)$$

If a variable pitch propeller stops, the blades can be rotated (feathered) in such a way that the blades are more or less aligned with the flow. In that case, assuming no flow separation, the extra drag is caused by friction. The incremental drag coefficient can now be estimated from:

$$\Delta C_{D_{\text{stopped variable pitch propeller}}} = c_{f_{\text{propeller}}} B \frac{2 S_{\text{blade planform}}}{S} \qquad (7.102)$$

Scenario 2: the propeller is windmilling

The windmilling rpm of the propeller depends on engine friction. If the torque due to engine friction is known from the engine manufacturer, the windmilling drag can be obtained from References 7.18 and 7.19 for Clark Y and RAF 6 blades. The procedure for estimating this drag is as follows:

1) Determine the windmilling rpm by finding the engine speed at which the power input at the propeller is equal to that required to drive the engine against the known friction torque.

2) With the known rpm, the power input and propeller pitch angle can be obtained.

3) The negative thrust (equals drag) of the propeller can now be found with the method of Sub–sub–section 7.6.3.1.

7.6.4 PROP–FANS AND DUCTED PROPELLERS

7.6.4.1 Prop–fans

The prop–fan, an advanced form of the turbo–prop, has been under consideration since 1976. The main purpose of the concept is to maintain a high propulsive efficiency at subsonic Mach numbers of around 0.8 while also retaining the good fuel efficiency of the turbo–prop. A typical prop–fan configuration is shown in Figure 7.23.

Figure 7.23 Example of a Prop–fan Configuration

The projected propulsive efficiency of the prop–fan is compared with turbo–props and high bypass ratio turbofans in Figure 7.24. It is seen that at a Mach number of around 0.8 the prop–fan is about 15% more efficient than current high bypass ratio turbo–fans. At lower speeds the efficiency of the prop–fan is considerably better.

The higher propulsive efficiency of the prop–fan results in lower overall fuel consumption. The better efficiency of the prop–fan is due to the sweeping of the blades and the area ruling of the spinner. Typical blade sweep angles used in prop–fans are 45 degrees near the tip. The number of blades in a prop–fan is typically 8 to 10 (as opposed to 4–6 for a trubo–prop). Coupled with a smaller blade diameter than a turbo–prop, the disk power loading in cruise is increased from around 10–15 SHP/D^2 for typical turbo–props to around 38 SHP/D^2 for typical prop–fans. SHP stands for the absorbed horsepower and D for the blade tip diameter in feet. Lower disk power loadings cause a large increase in propeller weight and gearbox weight with only a modest increase in propulsive efficiency. The high disk power loadings of the prop–fan produce lower overall fuel consumption.

However, the superior propulsive efficiency of the prop–fan has to be balanced against the greater mechanical complexity (blade angles must be variable!), potentially higher vibration levels and the problem of blade separation. Noise levels of the prop–fan have been shown to be quite acceptable during takeoff and landing. Part reason for the acceptable noise is the very high blade sweep near the tip. Reference 7.20 contains a discussion of prop–fan characteristics.

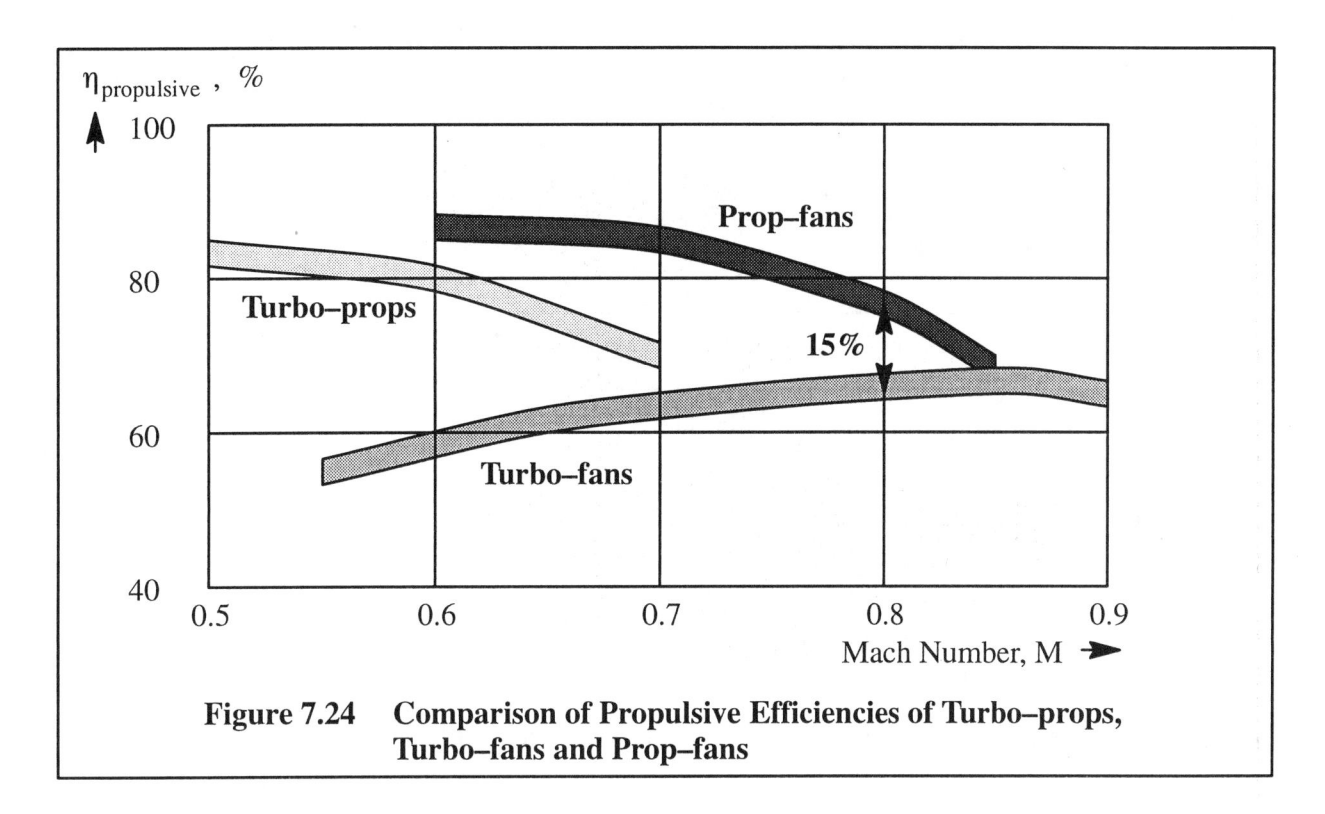

Figure 7.24 Comparison of Propulsive Efficiencies of Turbo–props, Turbo–fans and Prop–fans

Reasons why the prop–fan has not found its way into aircraft are the associated high cost, the relatively low price of fuel (compared with the fuel prices during the fuel crisis of the early 70's which triggered the early prop–fan activity) and the fact that the efficiencies of conventional propellers and turbo–fans have also been improving.

7.6.4.2 Ducted propellers

A ducted propeller consists of a duct, inside of which a propeller rotates.

Advantages of ducted propellers are: lower noise, uniform loading along the blade span and elimination of the propeller swirl (i.e. rotational) losses associated with a conventional, un–ducted propeller.

Disadvantages of the ducted propeller are: larger installed weight and added wetted area drag due to the external flow around the duct and the internal scrubbing drag. Figure 7.25 shows a schematic of a ducted propeller. The stator blades located behind the propeller (i.e. fan blades) serve two purposes: duct attachment and flow straightener to eliminate most of the rotational losses caused by the propeller blades.

An example of a ducted propeller was flown by Dowty–Rotol of England in a Pilatus/Britten/Norman Islander airframe in the late 70's. It used a piston engine to drive the fan. Example performance characteristics of a typical Dowty–Rotol propulsor are given in Table 7.3

Figure 7.25 Schematic of a Ducted Propeller

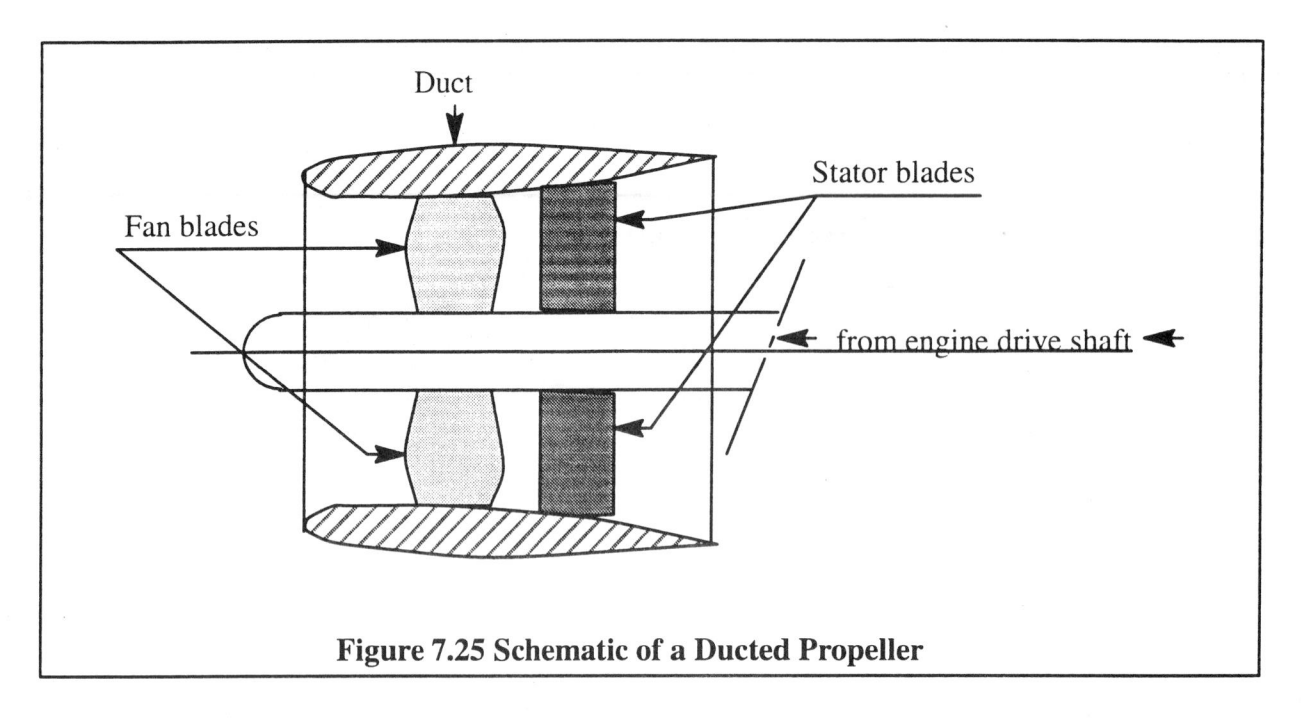

Table 7.3 Example Performance Particulars of a Small Dowty–Rotol Ducted Propeller

Maximum diameter	Hub tip ratio	Rotor blades	Stator blades
3.0 ft	0.45:1	7	8
Fan tip speed	Space chord ratio at root	Space chord ratio at tip	Fan pressure ratio
525 ft/sec	1.19:1	3.18:1	1.026:1
Fan mass flow	Sea–level thrust at 70 kts	Horsepower	Weight
100 lbs/sec	373 lbs	165 (shaft)	Not known

Except in airplanes like the Edgley Optica and the Rhein Flugzeugbau Fan–trainer (of which only a few were built) the ducted propeller still awaits a successful production application.

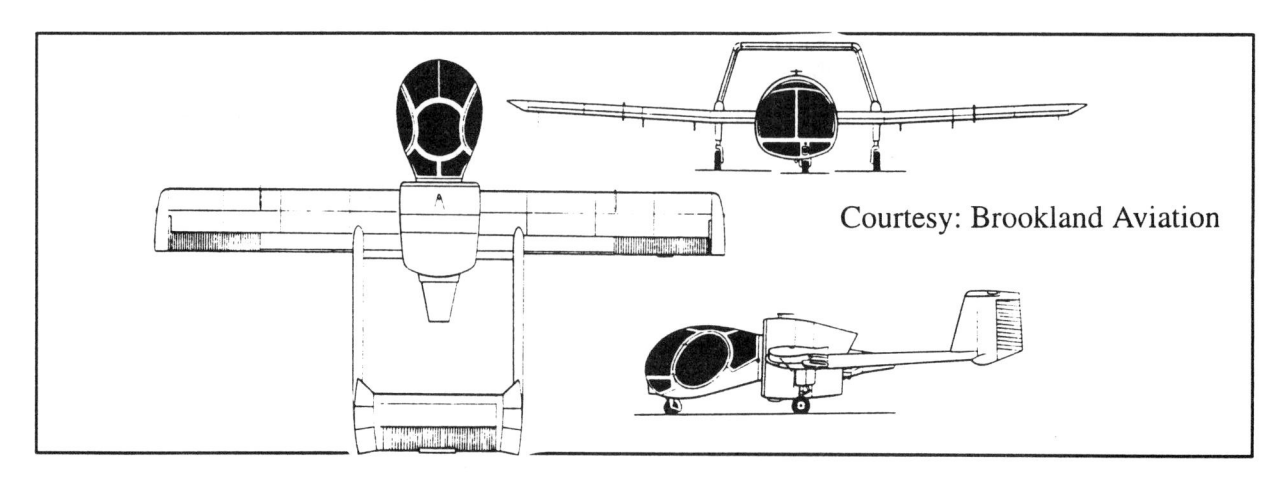

Courtesy: Brookland Aviation

7.7 PROPELLER NOISE

Since 1975, all newly designed airplanes must meet the FAR 36, Stage 3 requirements of Reference 7.21. A graphical representation of the allowable noise levels as they apply to small propeller driven airplanes is given in Figures 7.26.

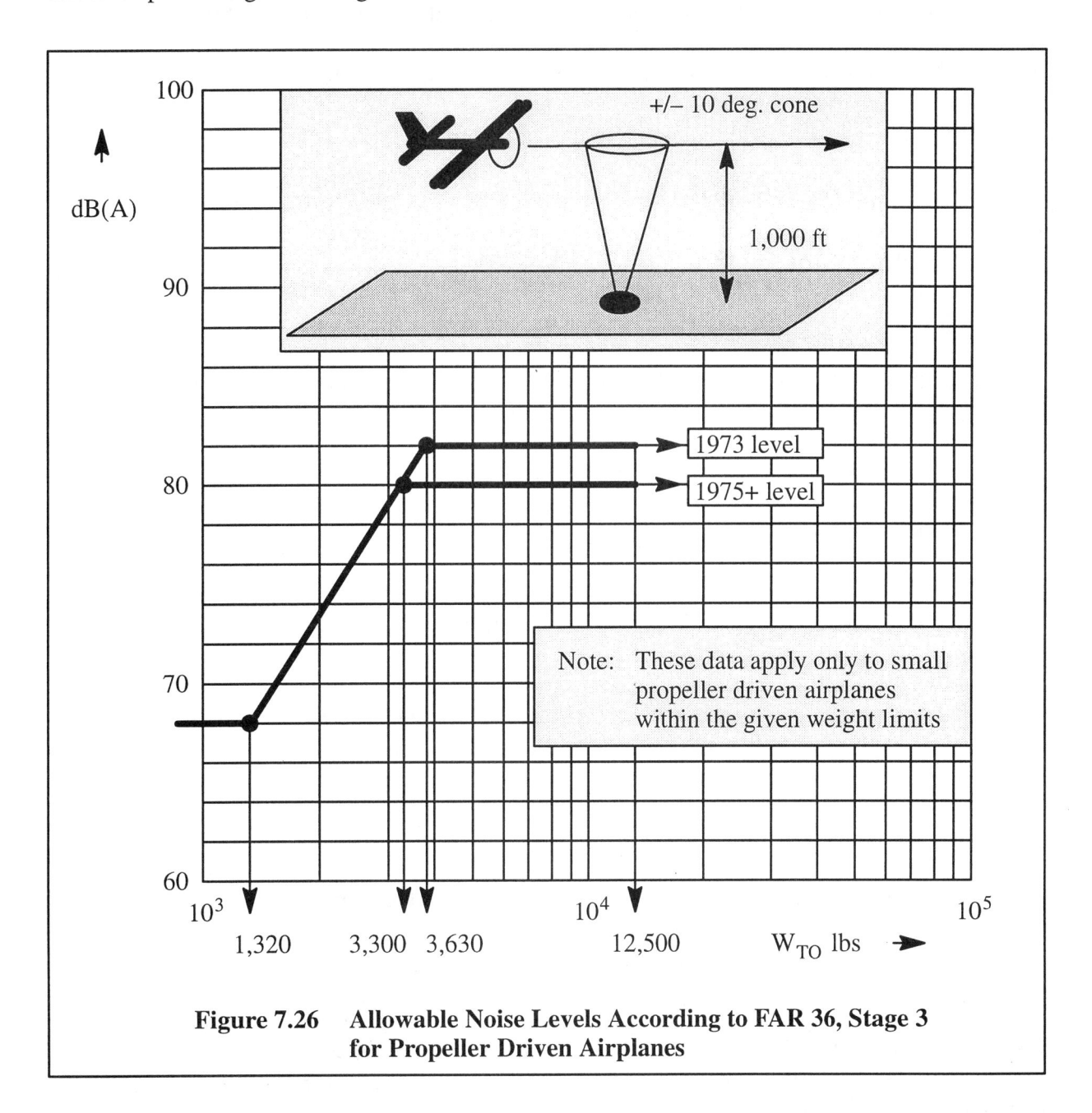

Figure 7.26 Allowable Noise Levels According to FAR 36, Stage 3 for Propeller Driven Airplanes

Note from Figure 7.26 that the allowable noise levels are stated in terms of dB(A) (i.e. Average Decibels) for propeller driven airplanes.

FAR 36 also specifies how noise must be measured and analyzed during take–off, fly–over or approach to landing to arrive at numerical values for actual airplane dB(A) levels. The required noise measurements must be taken at a defined locations relative to the ground. For small, propeller driven airplanes the measurement location is given in Figure 7.18. The noise level in dB(A) is that read through a so–called "A" filter. Characteristics for such a filter are described in Reference 7.21.

Sources and characteristics of propeller noise are discussed in Sub–section 7.7.1. An empirical procedure for estimating the noise of a propeller driven airplane is described in Sub–section 7.7.2.

7.7.1 SOURCES AND CHARACTERISTICS OF PROPELLER NOISE

Noise generated by propeller driven airplanes arises from several sources:

a) engine: exhaust noise, turbine noise, compressor noise, crankshaft/valve noise, gearbox noise

b) aerodynamic noise

c) propeller noise

In most propeller driven airplanes the propeller generated noise tends to dominate. For that reason, only propeller noise will be discussed.

Propeller noise consists of two components: rotational and vortex noise.

1) Rotational noise

This describes all sound which has discrete frequencies occurring at harmonics of the blade passage frequency which is defined as the product of the number of blades, B and the rotational frequency in rad/sec. This noise is generated because of the oscillatory pressure field acting on the air at a fixed point near the propeller disk. This noise is mainly associated with the production of thrust and torque. At high propeller speeds, blade thickness noise may also become important. The rotational noise level is maximum in the plane of rotation and increases with the power absorbed by the propeller, with increased propeller diameter, with fewer blades and with increased tip–speed.

2) Vortex noise

Vortex noise, or broadband noise, describes the sound produced by the unsteady pressure field associated with vortices shed from the trailing edge and tip of the blades as well as those associated with turbulence effects in the air stream itself.

The vortex shedding is a function of flow velocity. For a rotating blade, the sectional velocity varies across the blade span, resulting in a broad band of shedding frequencies. Since the noise level is proportional to the sixth power of the section velocity, it follows that the frequencies associated with the tip section tend to be of the greatest amplitude. Therefore, to reduce vortex noise, the tip speed should be reduced. To make up for lost thrust, the blade area can be increased. Obviously, increasing the blade area will reduce the propeller efficiency. It also has an effect on the structural design and therefore on the weight of the propeller.

The perceived noise from a given propeller is attenuated through a number of factors: distance from the source and energy losses.

Because sound pressure level is inversely proportional to the square of the distance from the noise source to the observer, it falls of roughly by 20log2 = 6 dB for every doubling of that distance.

Energy losses can be due to heat conduction and radiation, viscosity, diffusion, humidity, etc.

Other factors which affect noise attenuation are wind gradients and turbulence of the atmosphere. In fact, attenuation measured up–wind may exceed that measured down–wind by as much as 25 to 30 dB according to Ref. 7.22.

7.7.2 PREDICTION PROCEDURE FOR PROPELLER NOISE

Several computational methods for prediction of propeller noise are available: Ref. 7.16 is one example. It is also feasible to design a propeller for minimum noise subject to certain imposed aerodynamic constraints, as shown in Ref. 7.23.

In the following, a step–by–step, empirical procedure developed by Hamilton Standard Division of United Technologies Corporation and based on available test data, will be presented. The procedure is based on Ref. 7.9, on Appendix B of Ref. 7.22 and on Ref. 7.24. This procedure is judged to be adequate for conducting preliminary design studies. Since it is based on test data, engine noise is assumed to be included.

The step–by–step procedure deals with the prediction of far–field noise and near–field noise in Sub–sub–sections 7.7.2.1 and 7.7.2.2 respectively. The propeller noise charts which are referred to in 7.7.2.1 and 7.7.2.2 are included in Appendix D as Charts D1 through D13.

7.7.2.1 Far–Field Noise

Far–field noise is that noise at a distance greater than one propeller diameter away from the propeller tip. It can be estimated by using propeller noise charts D1 through D8 in Appendix D of this text by following the step–by–step procedure outlined next.

Step 1: Calculate the rotational tip Mach number from:

$$M_{rotation} = \frac{\pi n D}{V_a} \tag{7.103}$$

where: n is the propeller rotational velocity in rps (rotations per second)

V_a is the speed of sound in ft/sec

Step 2: Based on the power input to the propeller and its rotational tip speed, determine the far field, partial noise level, FL1, from Chart D1 in Appendix D.

Step 3: Find the far field, partial noise level, FL2, based on blade count and propeller diameter from Chart D2 in Appendix D.

Step 4: To account for atmospheric absorption and spherical spreading, find the near field, partial noise level, FL3, from Chart D3 in Appendix D.

Step 5: Read the correction, DI, for the directivity pattern from Chart D4 in Appendix D, where the azimuth angle, $\theta = 0$ degrees, is on the propeller axis in the forward direction as shown in Chart D4.

Step 6: The following corrections, NC, apply depending on the number of propellers:

For 1 propeller: NC = 0 dB
For 2 propellers: NC = 3 dB
For 4 propellers   NC = 6 dB

For airplanes with 3 or more than 4 propellers it is suggested to interpolate or extrapolate as required.

Step 7: The overall sound pressure level, OSPL, may be estimated from:

$$\text{OSPL} = \text{FL1} + \text{FL2} + \text{FL3} + \text{DI} + \text{NC} \quad \text{in dB} \tag{7.104}$$

Step 8: The perceived noise level, PNL, may be obtained from:

$$\text{PNL} = \text{OSPL} + \Delta\text{PNL} \quad \text{in dB} \tag{7.105}$$

where: ΔPNL may be found from Charts D5 through D8. It is noted that the helical tip Mach number is found from:

$$M_{tip} = M_\infty \sqrt{1 + \left(\frac{\pi}{J}\right)^2} \tag{7.106}$$

Step 9: To determine the dB(A) weighted sound, calculate:

$$\text{dB(A)} = \text{PNL} - 14 \quad \text{in dB} \tag{7.107}$$

The dB(A) weighted sound should then be judged against Figure 7.26 to determine the FAR 36 certifiability of the proposed propeller installation.

If the noise level is required in EPNdB, proceed as follows:

For take-off:
$$\text{EPNdB} = \text{PNL} - 4 \quad \text{in dB} \tag{7.108}$$

For approach:
$$\text{EPNdB} = \text{PNL} - 2 \quad \text{in dB} \tag{7.109}$$

Next, this procedure is illustrated with an example.

Propeller Theory and Applications

Example 7.6: The following data apply to a large, propeller driven airplane:

Propeller diameter:,	D	= 13.5 ft
Number of blades:	B	= 4
Power input to the propeller:	SHP	= 3,260 hp
RPM	RPM	= 1,020 rpm
Number of propellers		= 4
Airplane speed:	U	= 180 kts
Ambient temperature	T	= 86 deg. F

Determine the predicted noise level at a distance of 584 ft and an azimuth angle of 105 degrees.

Solution:

Step 1: $n = 1{,}020$ rpm $= 17$ rps

From Eqn (2.31): $V_a = \sqrt{\gamma gRT} = \sqrt{1.4 \times 32.2 \times 53.35 \times 546} = 1{,}146$ ft/sec

From Eqn (7.103): $M_{rot} = \dfrac{\pi \times 17 \times 13.5}{1{,}146} = 0.63$

Step 2: From Chart D1: FL1 = 96 dB

Step 3: From Chart D2: FL2 = – 2 dB

Step 4: From Chart D3: FL3 = – 1 dB

Step 5: From Chart D4: DI = +0.5 dB

Step 6: For the case of four propellers: NC = +6 dB

Step 7: From Eqn (7.104): OSPL = 96 – 2 – 1 + 0.5 + 6 = 99.5 dB

Step 8: $J = \dfrac{V}{nD} = \dfrac{180 \times 1.688}{17 \times 13.5} = 1.32$

From Eqn (7.106): $M_{tip} = \dfrac{180 \times 1.688}{1{,}146} \sqrt{1 + \left(\dfrac{\pi}{1.32}\right)^2} = 0.68$

From Chart D7: ΔPNL = + 1 dB

Step 9: From Eqn (7.107): dB(A) = 110.5 – 14 = 86.5 dB

7.7.2.2 Near–Field Noise

Near field noise is that noise at a distance of less than one propeller diameter of the propeller tip. Near–field noise data are required as an input for assessing acoustic fatigue potential of the structure of an airplane. It is also a required input for assessing the cabin or cockpit interior noise. Near–field noise can be estimated by using propeller noise charts D9 through D13 found in Appendix D of this text and by following the step–by–step procedure outlined next.

Step 1: Determine the helical tip Mach number from Eqn (7.106)

Step 2: Determine the near field, partial noise level, NL1, based on power absorbed by the propeller and on its diameter from Chart D9 in Appendix D.

Step 3: Find the correction for the number of blades, BC from Table 7.3:

Table 7.3 Blade Correction Term, BC, for Multi–bladed Propellers

Number of blades, B =	Blade Correction Term, BC in dB
2	+6
3	+2.5
4	0
6	–3.5
8	–6

Step 4: Find the near field, partial noise level, NL2, due to tip speed and tip distance, from Chart D10 in Appendix D. Note that Y is the distance from the propeller tip to the fuselage skin, measured in the plane of the propeller disk.

Step 5: Find the near field, partial noise level correction, XC, due to the fore and aft distance of the propeller from the point on the fuselage skin where the noise is to be determined from Chart D11 in Appendix D. Note that X is counted positive forward of the propeller disk.

Step 6: Obtain the reflection correction, RC, from Chart D12 in Appendix D.

Step 7: Calculate the overall, near–field sound pressure level, NFOSPL from:

$$\text{NFOSPL} = \text{NL1} + \text{BC} + \text{NL2} + \text{XC} + \text{RC} \quad \text{in dB} \qquad (7.110)$$

Step 8: To account for the harmonic distribution of the sound pressure level over the various harmonics caused by a rotating propeller, a correction, ΔNFPNL must be applied. This correction may be read from Chart D13 in Appendix D. The helical tip Mach number follows from Eqn (7.106).

Step 9: Calculate the perceived near–field noise level for each harmonic from:

$$\text{NFPNL} = \text{NFOSPL} + \Delta\text{NFPNL} \quad \text{in dB} \qquad (7.111)$$

An example application will now be presented.

Propeller Theory and Applications

Example 7.7: The following data apply to a small, single engine, propeller driven airplane:

Propeller diameter:,	D	= 9 ft
Number of blades:	B	= 3
Power input to the propeller:	SHP	= 300 hp
RPM	RPM	= 1,584 rpm
Number of propellers		= 1
Airplane speed:	U	= 125 kts
Ambient temperature	T	= 60 deg. F

Determine the predicted noise level associated with the fundamental and first harmonics on the fuselage skin at Y = 1.25 ft and X = 0 ft.

Solution:

Step 1: $n = 1{,}584/60 = 26.4$ rps

From Eqn (2.31): $V_a = \sqrt{\gamma g R T} = \sqrt{1.4 \times 32.2 \times 53.35 \times 520} = 1{,}118$ ft/sec

From Eqn (7.103): $M_{rot} = \dfrac{\pi \times 26.4 \times 9.0}{1{,}118} = 0.67$

Step 2: From Chart D9: NL1 = 130 dB

Step 3: From Table 7.3: BC = 2.5 dB

Step 4: With Y/D = 1.25/9 = 0.14, Chart D10 yields: NL2 = –0.5 dB

Step 5: From Chart D11: XC = 0 dB

Step 6: Assume that the fuselage cross section is circular, Chart D12 yields: RC = + 4 dB

Step 7: From Eqn (7.110): NFOSPL = 130 + 2.5 – 0.5 + 0 + 4 = 136 dB

Step 8: $M_\infty = 125 \times 1.688/1{,}118 = 0.19$

$J = V/nD = (125 \times 1.688)/(26.4 \times 9) = 0.89$

From Eqn (7.106):

$$M_{tip} = M_\infty \sqrt{1 + \left(\dfrac{\pi}{J}\right)^2} \; 0.19 \sqrt{1 + \left(\dfrac{3.14}{0.89}\right)^2} = 0.70$$

From Chart D13: fundamental harmonic: ΔNFPNL = –2.2 dB
second harmonic: ΔNFPNL = –8.0 dB

Step 9: From Eqn (7.111): fundamental harmonic: NFPNL = 136 – 2.2 = 133.8 dB
second harmonic: NFPNL = 136 – 8.0 = 128.0 dB

7.8 PROPELLER SELECTION

7.8.1 INTRODUCTION TO PROPELLER SELECTION

For any given airplane, a propeller is selected to satisfy the following **five** performance conditions:

* Take–off performance
* Climb performance
* Cruise performance
* Maximum speed capability
* FAR 36 noise requirements

Chapters 8 through 12 deal with the prediction of performance characteristics of propeller driven airplanes. Once the combination of airframe and engine type has been selected, the available shaft horsepower and the RPM of the propeller shaft are fixed. Within these constraints a propeller is then selected (or designed) as a compromise between these five conditions.

Other propeller selection (or design) considerations are the following **six** factors:

* weight
* complexity
* blade stress
* vibrations
* cost
* propeller diameter.

The very diameter of the propeller can be an important consideration: large diameter propellers tend to require long landing gears in low wing airplanes.

The reader should be aware of the fact that propellers with different numbers of blades (2, 3, 4, 5, 6 and more) can all satisfy the five performance conditions listed before. Therefore, anyone of the last six factors could well influence the final selection.

7.8.2 PROPELLER DESIGN VARIABLES TO BE SELECTED

The following fundamental propeller design variables must be specified when selecting or designing a propeller:

1) Diameter, D
2) Activity factor, AF and number of blades, B
3) Blade airfoil section and design lift coefficient

Some airframers prefer to design their own propellers. Most will provide a propeller manufacturer with the design specifications for items 1–3 and sufficient selection criteria so that a compromise between the **eleven** factors of Sub–section 7.8.1 can be reached.

The selection of the fundamental propeller design variables will be briefly discussed.

Propeller Theory and Applications

1) Propeller diameter, D

Everything else being the same, the propeller diameter should be as large as possible for good propulsive efficiency. Limitations such as ground clearance, noise, compressibility and blade stress levels may dictate an upper bound. Noise, compressibility and blade stress levels are all related to the tip speed. A reasonable starting point is to specify the allowable magnitude of the propeller tip speed in terms of the tip Mach number. For low noise, a helical tip Mach number limit could be 0.72 as discussed in Sub–section 7.5.4. According to Eqn (7.75) the propeller diameter then follows from:

$$D = \sqrt{\frac{a^2}{\pi^2 n^2}\left(M_{tip}^2 - M^2\right)} \qquad (7.112)$$

2) Activity factor, AF and number of blades, B

The activity factor, AF and the number of blades, B are closely related to the amount of power which can be absorbed by a propeller. This may be seen from Eqns (7.70) and (7.71) which are repeated here as Eqns (7.113) and (7.114):

$$AF = \frac{10^5}{D^5} \int_{0.15R}^{1.0R} cr^3 dr = \frac{10^5}{16} \int_{0.15R}^{1.0R} \left(\frac{c}{D}\right) x^3 dx \qquad (7.113)$$

$$P \approx 4B\varrho\pi^3 n^3 \bar{c}_d \left(\frac{D}{10}\right)^5 AF \qquad (7.114)$$

Typical propeller activity factors range from about 70 to about 200. For a given amount of power, P, and for a given propeller diameter, D, the number of blades, B and the activity factor, AF cannot be arbitrarily selected. The reason is, as seen from Eqn (7.113), the activity factor depends on the blade chord, c. If a large blade chord is selected, the number of blades is restricted because the circumference puts an upper limit on the number of blades. A recommended procedure is to plot the propeller efficiency versus the activity factor (staying within the normal range of activity factors of 70–200) and determine the best activity factor.

Other considerations are: weight, complexity of the variable pitch mechanism and blade flutter.

3) Blade airfoil section and design lift coefficient

Blade airfoil section and design integrated lift coefficient are related as may be seen from Eqn (7.82) repeated here as Eqn (7.115):

$$C_{L_i} = 4 \int_{0.15}^{1.0} (c_{l_d}) x^3 dx \qquad (7.115)$$

where: c_{l_d} is the section design lift coefficient of the blade

The section design lift coefficient depends on the blade airfoil camber and thickness distribution.

Thick airfoils tend to have lower critical Mach numbers, but higher lift coefficients at lower speeds. If take–off and landing performance are of the utmost performance, thick blade airfoils with high camber will be useful. For high speed applications, thin blade airfoils with low camber are more appropriate. Typical design integrated lift coefficients range from 0.35 to 0.60. To avoid compressibility penalties, the selected design integrated lift coefficient should not exceed the value derived from Figure 7.27.

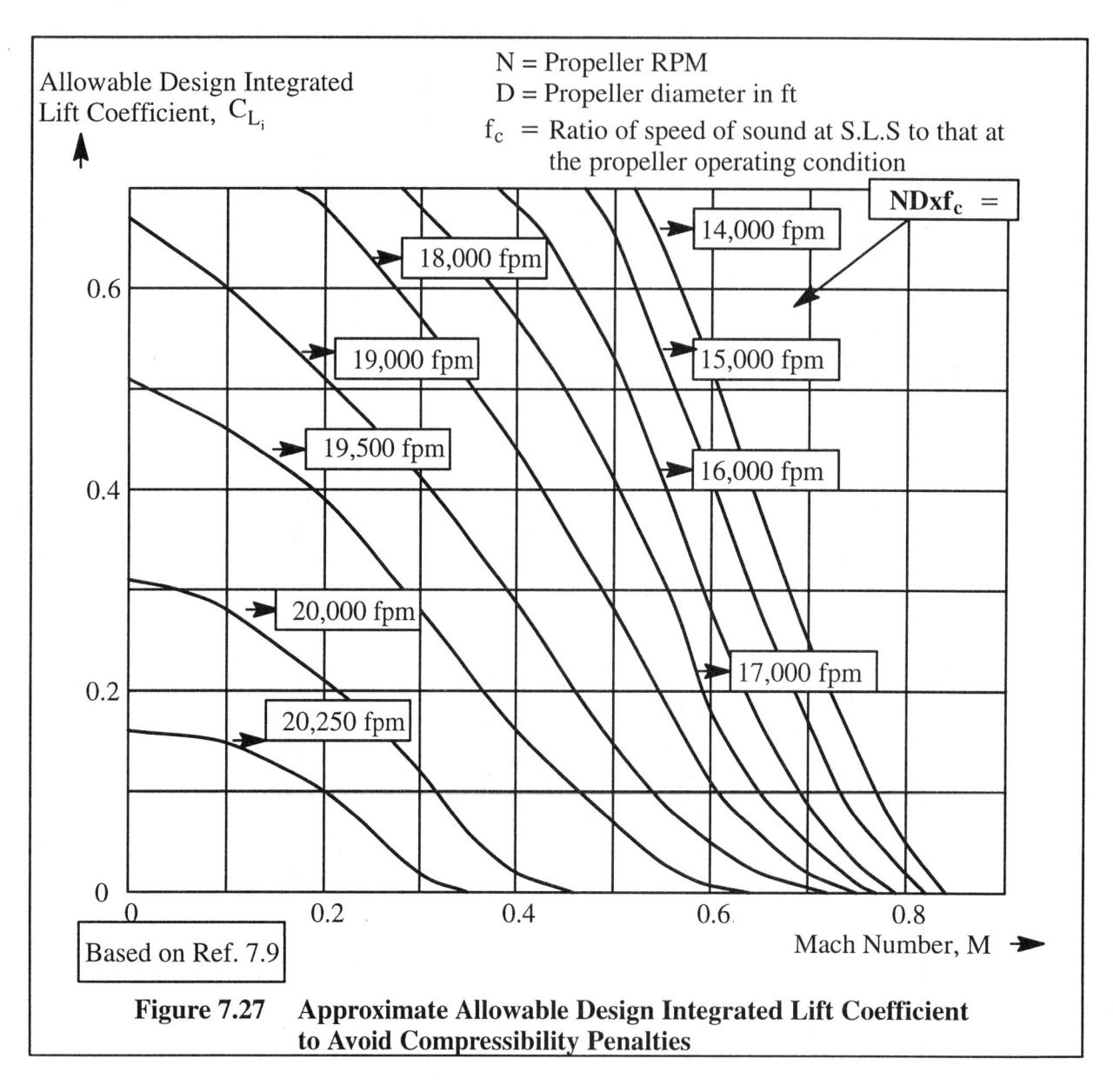

Figure 7.27   Approximate Allowable Design Integrated Lift Coefficient to Avoid Compressibility Penalties

The most commonly used blade airfoil sections are the RAF–6, Clark Y and NACA–16 series of airfoils. These airfoils can be compared by considering Figures 7.28 and 7.29. As a general rule, the RAF–6 section has high camber and offers good take–off thrust. The Clark Y section has moderate camber and low minimum drag. The NACA–16 sections are designed for high speed applications and are not normally used with engines below about 700 shp. In several modern applications new blade airfoils are specifically derived to suit a given purpose: better lift, better lift–to–drag ratio and better blade stiffness can all be criteria used in the design of new propeller blade airfoils.

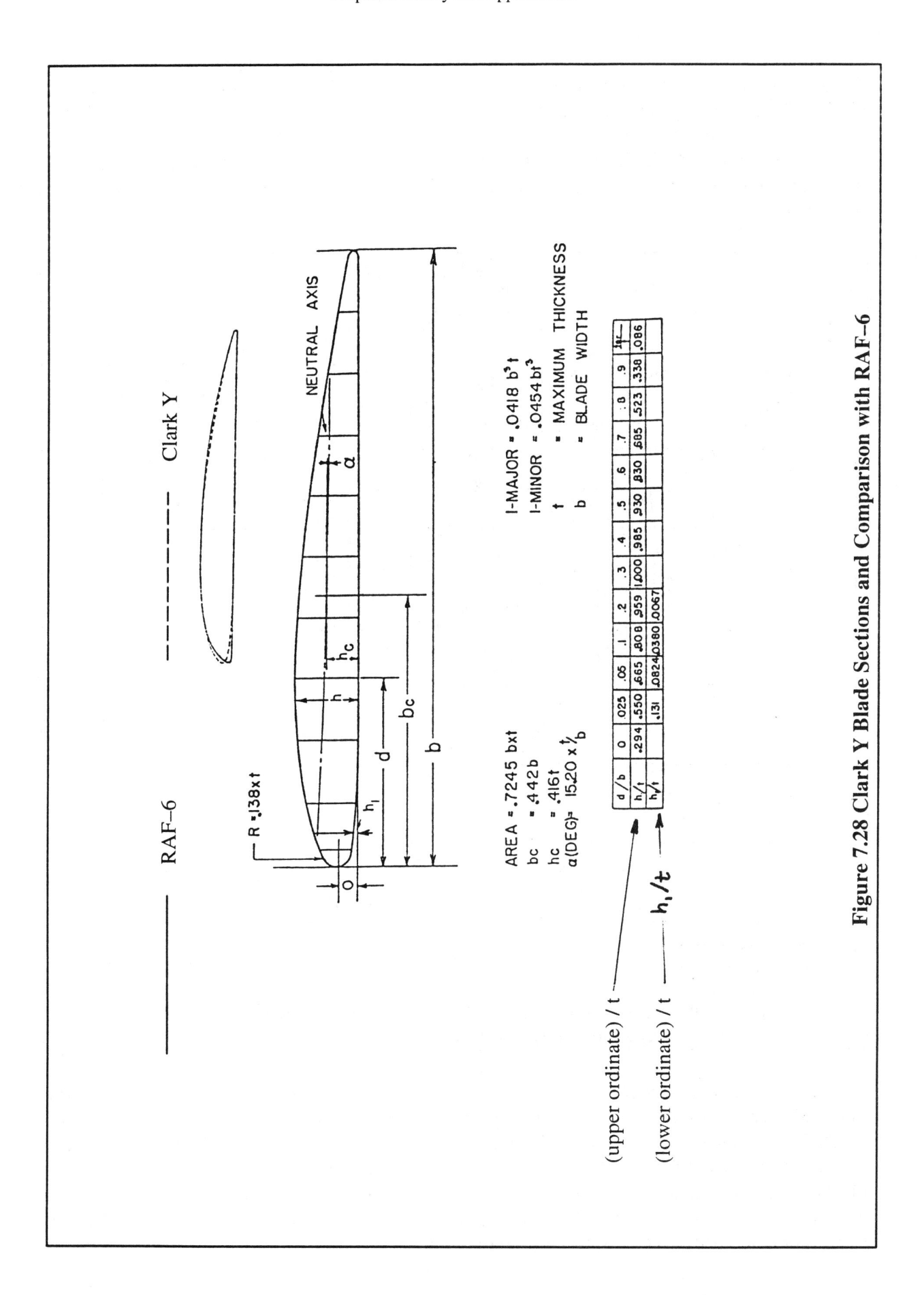

Figure 7.28 Clark Y Blade Sections and Comparison with RAF-6

AREA = .7245 b×t
bc = .442 b
hc = .416 t
α(DEG) = 15.20 × t/b

I-MAJOR = .0418 $b^3 t$
I-MINOR = .0454 bt^3
t = MAXIMUM THICKNESS
b = BLADE WIDTH

R = 1.38×t

d/b	0	.025	.05	.1	.2	.3	.4	.5	.6	.7	.8	.9	1.0
h/t	.294	.550	.665	.808	.959	1.000	.985	.930	.830	.605	.523	.338	.086
h_1/t	.131	.0824	.0380	.0067									

(upper ordinate) / t ⟶ h/t
(lower ordinate) / t ⟶ h_1/t

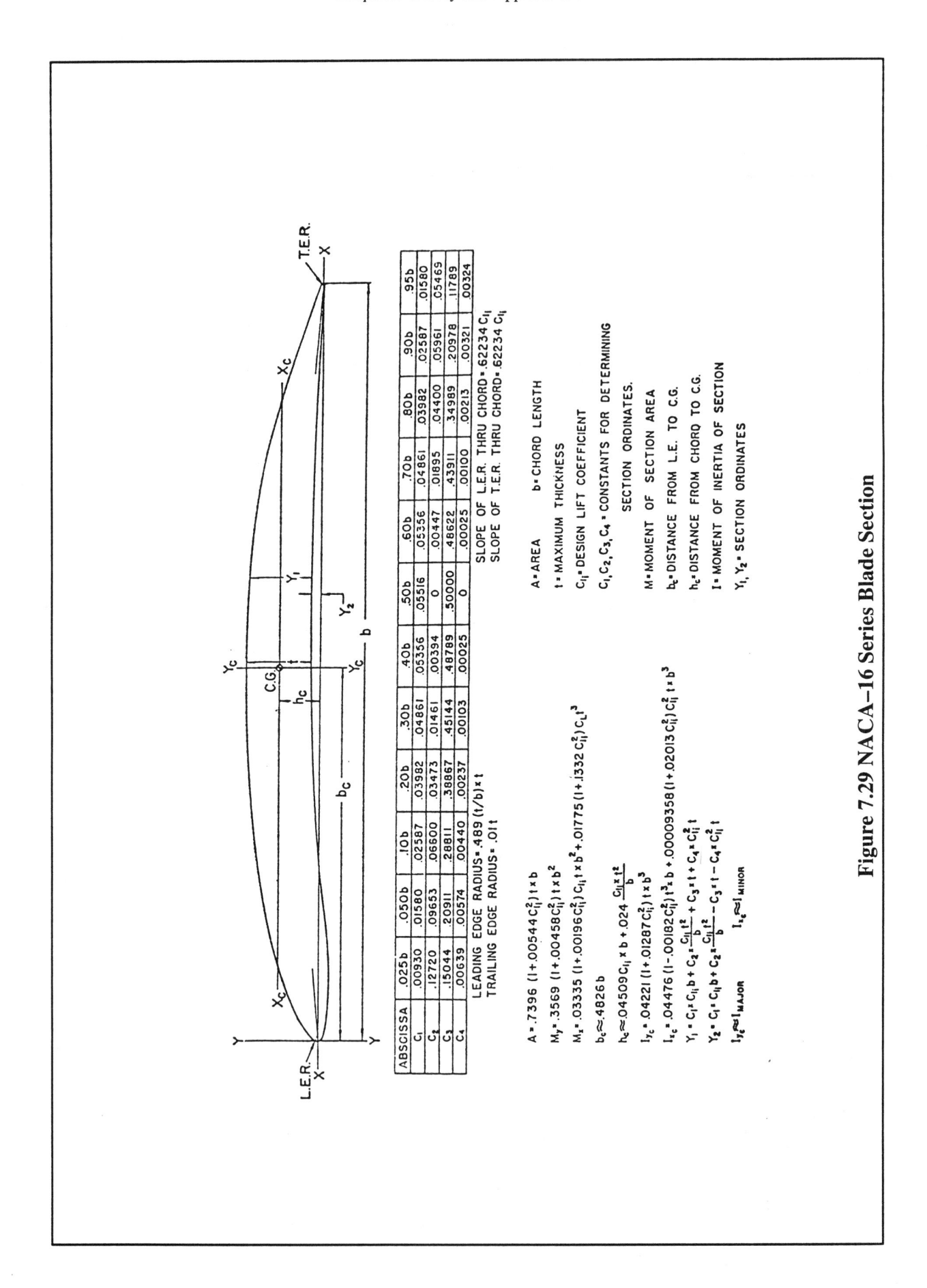

ABSCISSA	.025b	.050b	.10b	.20b	.30b	.40b	.50b	.60b	.70b	.80b	.90b	.95b
C_1	.00930	.01580	.02587	.03982	.04861	.05356	.05516	.05356	.04861	.03982	.02587	.01580
C_2	.12720	.09653	.06600	.03473	.01461	.00394	0	.00447	.01895	.04400	.05961	.05469
C_3	.15044	.20911	.28811	.38867	.45144	.48789	.50000	.48622	.43911	.34989	.20978	.11789
C_4	.00639	.00574	.00440	.00237	.00103	.00025	0	.00025	.00100	.00213	.00321	.00324

LEADING EDGE RADIUS = .489 $(t/b) \times t$
TRAILING EDGE RADIUS = .01t

SLOPE OF L.E.R. THRU CHORD = .62234 C_{li}
SLOPE OF T.E.R. THRU CHORD = .62234 C_{li}

A = AREA b = CHORD LENGTH
t = MAXIMUM THICKNESS
C_{li} = DESIGN LIFT COEFFICIENT
C_1, C_2, C_3, C_4 = CONSTANTS FOR DETERMINING SECTION ORDINATES.
M = MOMENT OF SECTION AREA
b_c = DISTANCE FROM L.E. TO C.G.
h_c = DISTANCE FROM CHORD TO C.G.
I = MOMENT OF INERTIA OF SECTION
Y_1, Y_2 = SECTION ORDINATES

$A = .7396 (1 + .00544 C_{li}^2) t \times b$

$M_y = .3569 (1 + .00458 C_{li}^2) t \times b^2$

$M_x = .03335 (1 + .00196 C_{li}^2) C_{li} t \times b^2 + .01775 (1 + .1332 C_{li}^2) C_{li} t^3$

$b_c \approx .4826 b$

$h_c \approx .04509 C_{li} \times b + .024 \dfrac{C_{li} \times t^2}{b}$

$I_{y_c} = .04221 (1 + .01287 C_{li}^2) t \times b^3$

$I_{x_c} = .04476 (1 - .00182 C_{li}^2) t^3 \times b + .00009358 (1 + .02013 C_{li}^2) C_{li}^2 t \times b^3$

$Y_1 = C_1 t C_{li} b + C_2 = \dfrac{C_{li} t^2}{b} + C_3 t t + C_4 \times C_{li}^2 t$

$Y_2 = C_1 t C_{li} b + C_2 = \dfrac{C_{li} t^2}{b} - C_3 t t - C_4 \times C_{li}^2 t$

$I_{x_c} \approx I_{MAJOR}$ $I_{y_c} \approx I_{MINOR}$

Figure 7.29 NACA–16 Series Blade Section

7.9 SUMMARY FOR CHAPTER 7

In this chapter several fundamental aspects of propeller theory and its practical application to propeller performance prediction problems were covered. Momentum theory, blade element theory and a combined blade–element and momentum theory were presented.

Methods for predicting the static thrust and the variation of thrust with forward flight speed of a free propeller were included.

In the real world, propellers are installed such that various extra losses accrue. Methods for predicting the efficiency of a propeller, including various installation losses, were also presented. Extensive use has been made of the Hamilton Standard propeller charts of Appendix C.

Fundamentals of far–field propeller noise analysis and prediction were covered in relation to the FAR 36 requirements. In addition, a method for predicting near–field propeller noise was given so that a preliminary estimate of propeller induced fatigue can be made.

7.10 PROBLEMS FOR CHAPTER 7

7.1 A propeller with a diameter of 15 ft produces a thrust of 8,000 lbs while moving at a speed of 400 kts at an altitude where the density is 0.0020 slugs/ft^3. Using the momentum theory, determine: a) the induced flow velocity through the disk and b) the final wake velocity.

7.2 Determine the diameter of the circular slipstream far behind a 10 ft diameter propeller which produces a thrust of 2,500 lbs at a speed of 200 kts under standard sea–level conditions.

7.3 For the propeller of Problem 7.1, calculate the minimum power which must be supplied to the propeller, according to the momentum theory.

7.4 A propeller operates at a disk loading of 10 lbs/ft^2 at a flight speed of 250 kts under standard sea–level conditions. Calculate the propeller efficiency according to momentum theory.

7.5 An airplane has a wing area, S, and a drag coefficient, C_D while flying at a speed, V, in an atmospheric density, ϱ. The airplane has a propeller with a disk area, A. Show, that in un–accelerated flight, the slipstream velocity in the propeller, V_s, is given by:

$$V_s = \sqrt{1 + \frac{S}{A}C_D}\text{, according to momentum theory.}$$

7.6 Determine the induced angle, θ, under the following conditions: number of blades, B=4; propeller diameter, D=10 ft; blade station, r=4 ft from the centerline; forward airplane velocity, U=200 mph; blade angle, β=25 deg.; section lift–curve slope, $a_0 = 0.10$ deg.$^{-1}$; rotational speed of the propeller, N=2,000 rpm.

7.7   The thrust and torque at 4 ft radius of a 2–bladed propeller are 145 lbs/ft and 175 ft–lbs/ft, respectively, under static conditions at sea–level. Determine the slipstream rotation in radians per sec. immediately behind the propeller disk at a radius of 4 ft.

7.8   A 4–bladed propeller is to drive an airplane at 275 mph at sea–level. The propeller RPM is 1,200 rpm. If the blade element at 4 ft radius has an absolute angle of attack of 6 degrees and the thrust grading is 190 lbs/ft per blade, find the blade chord at the 4 ft radius. Assume $a_0 = 0.10$ deg.$^{-1}$.

7.9   A propeller has the following blade chord distribution:

x/R	0.2	0.3	0.4	0.5	0.6	0.7	0.8	0.9	1.0
100c/D	10.0	9.3	8.6	7.9	7.3	6.6	5.9	5.2	0

Calculate the activity factor.

7.10   A propeller driven airplane has two engines, each developing a SHP=780 hp at 10,000 ft altitude in the standard atmosphere. The three–bladed propeller has a diameter of 9.0 ft, an activity factor of 140 per blade, an integrated design lift coefficient of 0.5 and a rotational speed of 2,100 rpm. Calculate and plot the thrust and efficiency in a speed range of 100 to 300 mph. Neglect blockage and compressibility effects.

7.11   A single engine airplane has the following characteristics:
3–bladed propeller with AF = 140; C_{L_i} = 0.30 ; D = 6.7 ft; N = 2,800 rpm; SHP = 285 hp.
Calculate and plot the take–off thrust versus speed in a range of 0 to 80 mph.

7.12   Calculate the far–field noise level of the propeller of Problem 7.11 at a speed of 80 mph, at a distance of 1,000 ft from the propeller with an azimuth of 90 degrees.

7.13   Derive a method for determining a and a' for a 3–bladed propeller: see Sub–section 7.4.1 for a method which applies to a 2–bladed propeller.

7.11   REFERENCES FOR CHAPTER 7

7.1   Vogeley, A.W.; Axial–Momentum Theory for Propellers in Compressible Flow; NACA TN 2164, 1951.

7.2   Stepniewski, W.A.; Basic Aerodynamics and Performance of the Helicopter; AGARD LS–63 on Helicopter Aerodynamics and Dynamics, March 1973.

7.3   Stickle, G.W. and Crigler, J.L.; Propeller Analysis from Experimental Data; NACA TR 712, 1941.

7.4 Reid, E.G.; The Influence of Blade–Width Distribution on Propeller Characteristics; NACA TN 1834, 1949.

7.5 Reid, E.G.; Studies of Blade Shank Form and Pitch Distribution for Constant–Speed Propellers; NACA TN 947, 1945.

7.6 Roskam, J.; Airplane Design, Part II: Preliminary Configuration Design and Integration of the Propulsion System, 1989; DARCorporation, 120 East Ninth Street, Suite 2, Lawrence, KS 66044.

7.7 Stack, J. et al; Investigation of the NACA 4–(3) (08)–03 and NACA 4–(3) (08)–045 Two Blade Propellers at Forward Mach Numbers to 0.725 to Determine the Effects of Compressibility and Solidity on Performance; NACA TR 999, 1950.

7.8 Torenbeek, E.; Synthesis of Subsonic Airplane Design; Kluwer Boston Inc., Hingham, Maine, 1982.

7.9 Anon.; Generalized Method of Propeller Performance Estimation; Hamilton Standard Report PDB 6101 A, United Aircraft Corporation, June 1963.

7.10 Stickle, G.W. et al; Effect of Body Nose Shape on the Propulsive Efficiency of a Propeller; NACA TR 725, 1941.

7.11 Roskam, J.; Airplane Design, Part VI: Preliminary Calculation of Aerodynamic, Thrust and Power Characteristics, 1990; DARCorporation, 120 East Ninth Street, Suite 2, Lawrence, KS 66044.

7.12 Wood, K.D.; Technical Aerodynamics; Published by the author, 1955 (out of print).

7.13 Gilman, J., Jr.; Propeller–Performance Charts for Transport Airplanes; NACA TN 2966, 1953.

7.14 Crigler, J.L. and Jaquis, R.E.; Propeller–Efficiency Charts for Light Airplanes; NACA TN 1338, 1947.

7.15 Gray, W.H. and Mastrocola, N.; Representative Operating Charts of Propellers Tested in the NACA 20–Foot Propeller Research Tunnel; NACA Wartime Report L–286, 1943.

7.16 Worobel, R.; Computer Program User's Manual for Advanced General Aviation Propeller Study; NASA CR–2066, May 1972.

7.17 Anon.; Approximate Estimation of Braking Thrust of Propellers; Engineering Sciences Data Unit, ESDU ED1/2, December 1952.

7.18 Hartman, E.P. and Biermann, D.; The Negative Thrust and Torque of Several Full–Scale Propellers and Their Application to Various Flight Problems; NACA TR 641, 1938.

7.19 Anon.; Approximate Estimation of the Drag of Windmilling Propellers; Engineering Sciences Data Unit, ESDU ED1/1, April 1962.

7.20 Dugan, J.F., Gatzen, B.S. and Adamson, W.M.; Prop–Fan Propulsion–Its Status and Potential; Society of Automotive Engineers, SAE Paper No. 780995, November 1978.

7.21 Anon.; Federal Aviation Regulations, Part 36, Noise Standards: Aircraft Type and Airworthiness Certification; Department of Transportation, June 1974.

7.22 Marte, J.E. and Kurtz, D.W.; A Review of Aerodynamic Noise from Propellers, Rotors, and Lift Fans; Technical Report 32–1462, Jet propulsion Laboratory, California Institute of Technology, January 1, 1970.

7.23 Woan, C.J.; Propeller Study, Part II: The Design of Propellers for Minimum Noise; NASA CR–155005, July 1977.

7.24 Anon.; Prediction Procedure for Near–Field and Far–Field Propeller Noise; Aerospace Information Report 1407, Society of Automotive Engineers, May 1977.

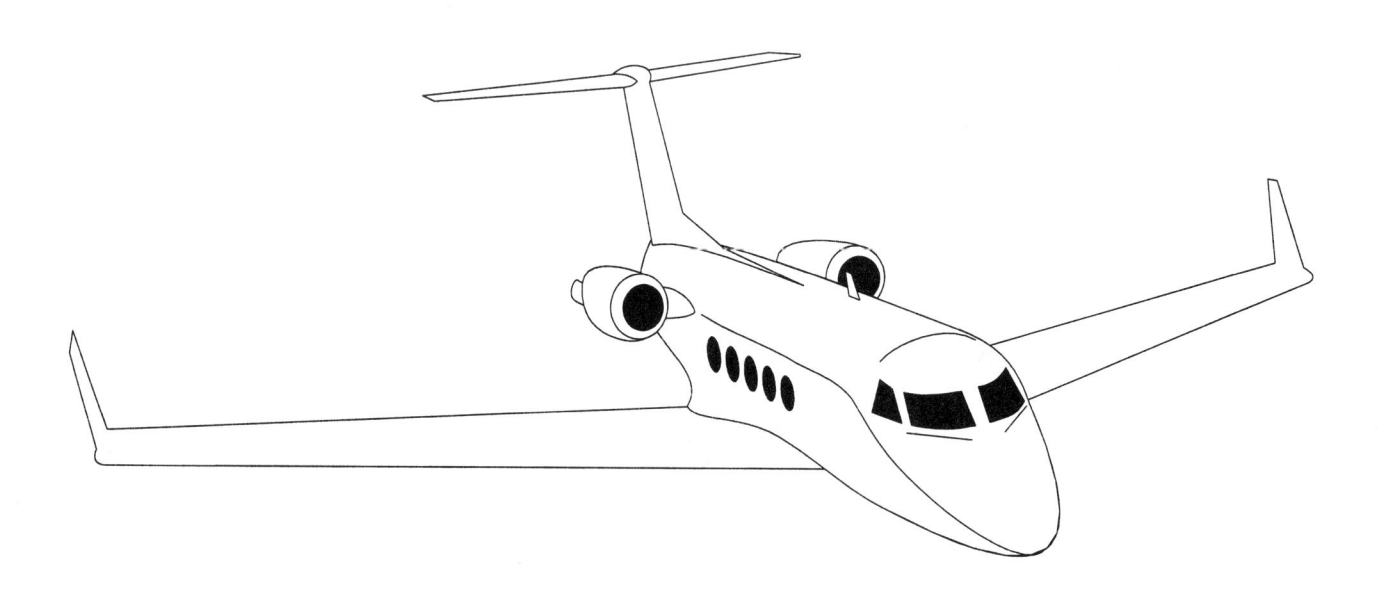

CHAPTER 8: FUNDAMENTALS OF FLIGHT PERFORMANCE

In this chapter, the fundamentals needed for calculating in–flight performance characteristics of airplanes will be derived and discussed. Before doing this certain conventions for axes and angles must be introduced. That is accomplished in Section 8.1.

Next, the fundamental aspects of in–flight airplane performance characteristics will be presented in the corresponding sections:

8.2 Steady, un–powered flight   8.3 Steady, powered, flight

8.4 Steady, powered, level flight

The methods in this chapter are concerned primarily with the level flight performance of airplanes. Methods to calculate climb performance, fieldlength performance, cruise and endurance performance and maneuvering performance are presented in Chapters 9 through 12 respectively.

8.1 DEFINITION OF ANGLES AND AXIS SYSTEMS

Figure 8.1 shows how the most important angles and axis systems are defined.

The reader should carefully note the meaning of several key words in the title of Figure 8.1:

* The word steady implies that no accelerations act on the airplane.
  Exceptions to this will be noted in Chapter 12.

* The word symmetrical implies that the flight takes place at zero sideslip.

* The words straight line imply, that only straight line flight paths are assumed.
  Exceptions to this will be noted in Chapter 12.

The meaning of the symbols in Figure 8.1 should be clearly understood:

U_1 is the steady speed of the airplane relative to the air

X_b is the body fixed X–axis which points through the c.g. The selection of the orientation of this axis relative to the airframe is arbitrary. In transport airplane the X–axis is usually selected to be parallel to the cabin floor.

Z_b is the body fixed Z–axis which points through the c.g. and is perpendicular to X_b

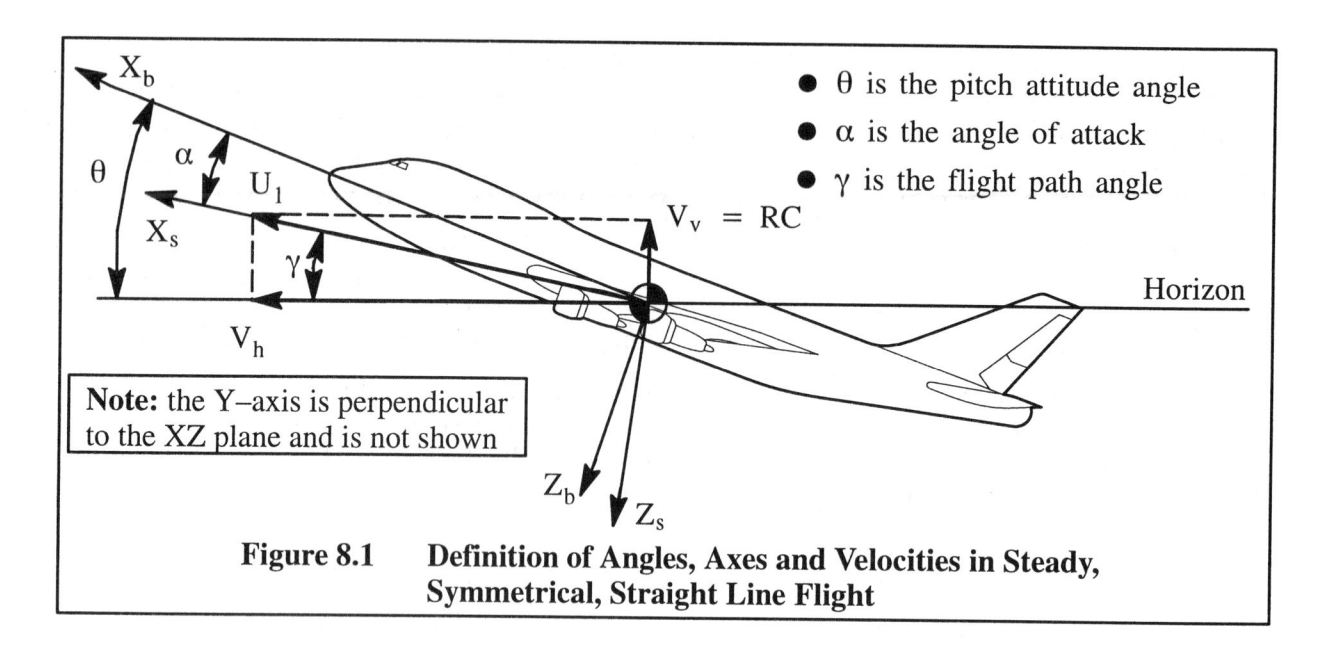

Figure 8.1 Definition of Angles, Axes and Velocities in Steady, Symmetrical, Straight Line Flight

- θ is the pitch attitude angle
- α is the angle of attack
- γ is the flight path angle

Note: the Y–axis is perpendicular to the XZ plane and is not shown

X_s is the stability X–axis which points through the c.g. and points along the steady state velocity vector, U_1

Z_s is the stability z–axis which points through the c.g. and is perpendicular to X_s

The following angles are of key importance to in–flight performance:

α is the airplane angle of attack, defined as the angle between X_b and U_1

θ is the airplane pitch attitude angle, defined as the angle between X_b and the horizon

γ is the flight path angle, defined as the angle between U_1 and the horizon

A very important relationship, which is always satisfied and should be memorized is:

$$\theta = \alpha + \gamma \tag{8.1}$$

It should be noted, that the angles α, γ, and θ as shown in Figure 8.1, are positive.

The following velocity components are of key importance to in–flight performance:

$V_v = RC$ is the vertical velocity component, also called the rate–of–climb, RC. If the rate of climb is negative, it is called the rate–of–descent, RD. Note that if RC is negative, the flight path angle, γ, is also negative.

V_h is the horizontal velocity component.

The following relationship between speed, U_1, flight path angle, γ, and rate–of–climb, RC, must also be kept in mind:

$$RC = U_1 \sin \gamma \tag{8.2}$$

A fundamental assumption which will be made in all performance discussions in this and subsequent chapters is that the airplane is **trimmed** about all axes X, Y and Z in Figure 8.1. This means that the total pitching moment, rolling moment and yawing moment which act on the airplane are assumed to be zero. The drag effect of this was discussed in Chapter 5, Sub–section 5.2.10.

It will also be assumed that airplane does not undergo any acceleration in the Y–axis direction. The net result of all this is that only equations in the X and Z axis directions are needed to describe the performance characteristics of an airplane.

Finally, the airplane is assumed to have no angular rates, with exceptions noted in Chapter 12. From a Newtonian viewpoint, the airplane is therefore treated as a point–mass model on which only forces (and no moments) can act. For a more general treatment of the equations of motion, the reader is referred to Chapter 1 of Reference 8.1.

Figure 8.2 provides a definition of all forces which act on the airplane in steady, un–accelerated flight. Observe especially the thrust force at an inclination angle, ϕ_T. For most airplanes this angle is very small. Obviously, when thrust vectoring is used, it is not small. Note the body–fixed axis and the stability axis notations. Remember that stability axes are also body–fixed axes.

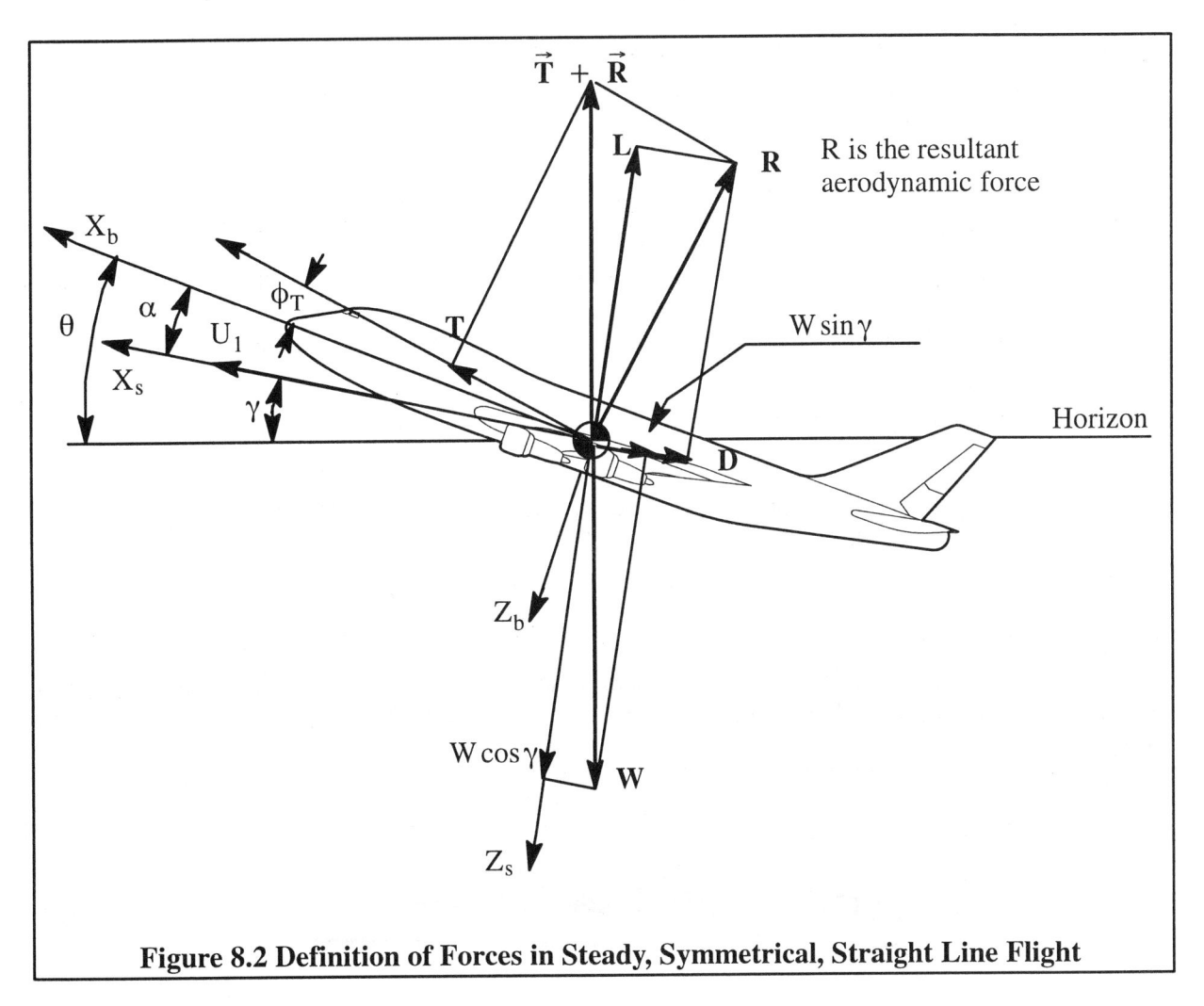

Figure 8.2 Definition of Forces in Steady, Symmetrical, Straight Line Flight

It can be seen from Figure 8.2 that a condition for in-flight equilibrium is:

$$\vec{R} + \vec{T} + \vec{W} = 0 \tag{8.3}$$

For analytical purposes Eqn (8.3) is normally split into its two component forms in the stability axis system:

$$T\cos(\alpha + \phi_T) - D - W\sin\gamma = 0 \tag{8.4}$$

and

$$T\sin(\alpha + \phi_T) + L - W\cos\gamma = 0 \tag{8.5}$$

Substitution of the standard form for expressing lift, L and drag, D yields:

$$T\cos(\alpha + \phi_T) - C_D \bar{q} S - W\sin\gamma = 0 \tag{8.6}$$

and

$$T\sin(\alpha + \phi_T) + C_L \bar{q} S - W\cos\gamma = 0 \tag{8.7}$$

Eqns (8.6) and (8.7) have seven variables which together completely define a steady state flight condition. These variables are:

W, h (altitude through ϱ), α, ϕ_T, V, γ, and T

In nearly all real world situations, the variables W, ϕ_T and h (altitude) will be pre-selected. That leaves four variables in Eqns (8.6) and (8.7). Of these four variables, two must be arbitrarily selected, before the other two can be solved for from Eqns (8.6) and (8.7). Example cases of practical interest are:

* Level flight ($\gamma = 0$) at a given speed, V. The magnitudes of required thrust, T, and the angle of attack, α, follow from Eqns (8.6) and (8.7).

* Flight at a given thrust, T, and any desired flight path angle, γ,. The variables speed, V, and angle of attack, α, follow from Eqns (8.6) and (8.7).

It is instructive to examine first the case of un-powered flight. As will be seen (and totally expected), in un-powered flight, maintaining altitude at constant speed is not possible. This yields the important glider design problem of finding that aerodynamic design which results in the lowest possible rate-of-descent, RD. Un-powered flight is discussed in Section 8.2 while powered flight is discussed in Section 8.3.

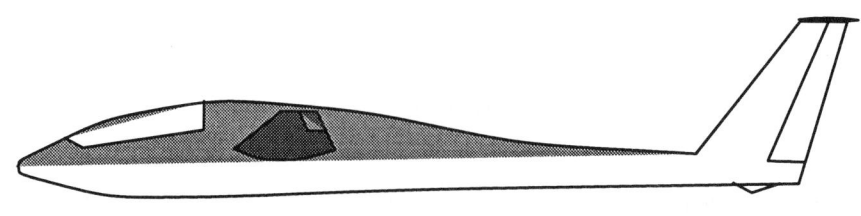

8.2 STEADY, UN–POWERED FLIGHT

8.2.1 EQUATIONS AND DEFINITIONS

Figure 8.3 shows the forces which act on an airplane in un–powered flight. The angles shown in Figure 8.3 have been exaggerated for clarity. Since T=0 in this case, Eqn (8.3) yields:

$$\vec{R} + \vec{W} = 0 \tag{8.8}$$

Since $\vec{R}$ is the resultant of lift, $\vec{L}$, and drag, $\vec{D}$, it is possible to write:

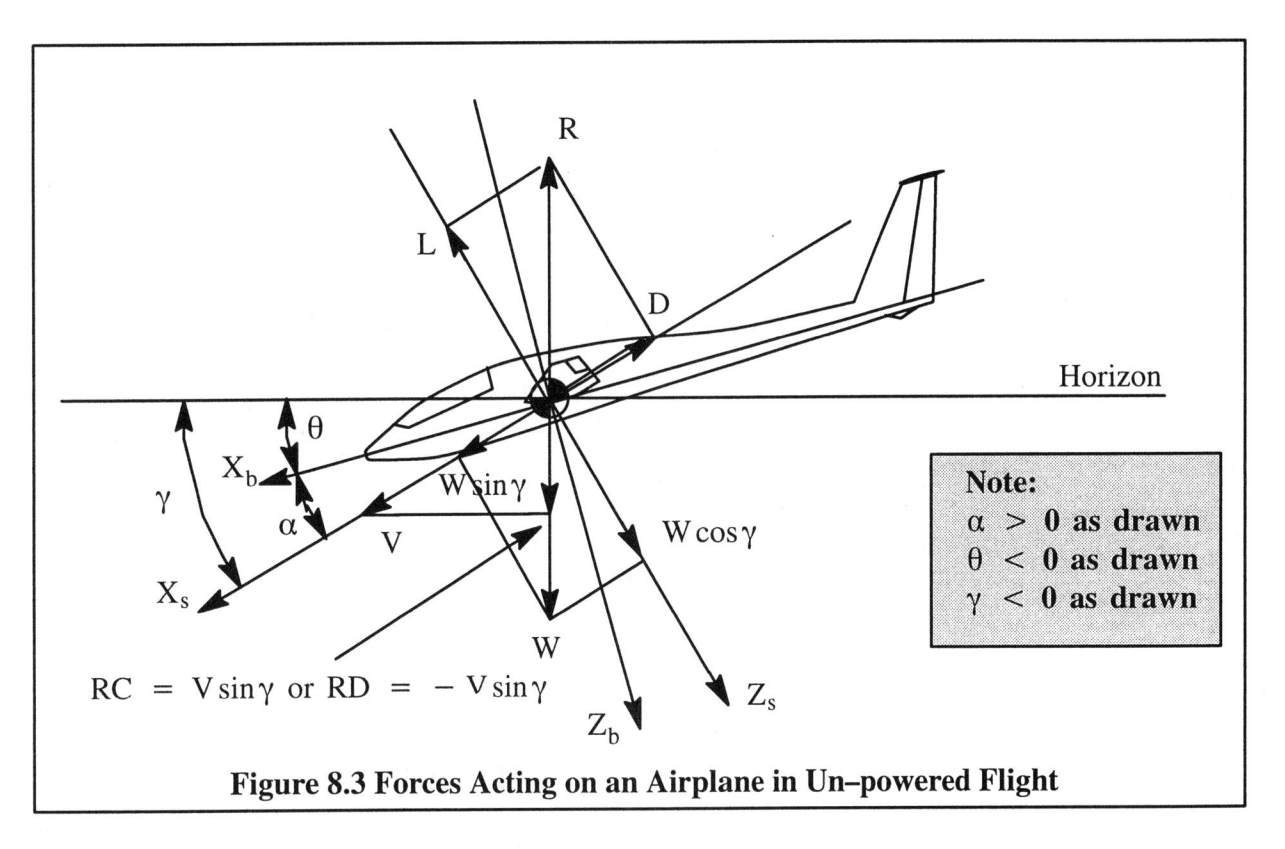

Figure 8.3 Forces Acting on an Airplane in Un–powered Flight

Note:
$\alpha > 0$ as drawn
$\theta < 0$ as drawn
$\gamma < 0$ as drawn

$RC = V\sin\gamma$ or $RD = -V\sin\gamma$

$$C_R \bar{q} S = \bar{q} S \sqrt{\left(C_L^2 + C_D^2\right)} = W \tag{8.9}$$

where: C_R is the resultant force coefficient.

The vector relationship of Eqn (8.8) can be written in component form as:

$$-D - W\sin\gamma = 0 \tag{8.10}$$

and

$$L - W\cos\gamma = 0 \tag{8.11}$$

The reader should observe, that Eqns (8.10) and (8.11) can also be obtained from Eqns (8.6) and (8.7) by simply substituting: T=0.

Because the flight path angle, γ, is negative as shown in Figure 8.3, it is useful to introduce:

$$\bar{\gamma} = -\gamma \tag{8.12}$$

With this construct and upon introducing the usual drag and lift coefficients Eqns (8.10) and (8.11) can be re-written as:

$$C_D \bar{q} S = W \sin \bar{\gamma} \tag{8.13}$$

and

$$C_L \bar{q} S = W \cos \bar{\gamma} \tag{8.14}$$

Observe, that there are five variables in these two equations:

W, h (altitude through ϱ), α, V, and γ

Since (in the absence of thermals) equilibrium flight can only be maintained in descending flight, altitude will in fact vary. However, since the objective of any gliding flight is to descend as slowly as possible, altitude will not change rapidly. Therefore, selecting altitude, although mathematically speaking fictitious, is a reasonable approximation. By selecting weight and altitude, three variables remain, one of which must be selected.

It is physically evident, that to stay aloft as long as possible in a glider, the numerical value of RD should be minimized. To cover the longest track (distance) along the ground, the numerical magnitude of the flight path angle, $\bar{\gamma}$ (glide angle), should be minimized. Both cases are important in the design of gliders and in the conduct of power-off flight in powered airplanes.

The effect of aerodynamic design on the glide angle, $\bar{\gamma}$, the rate-of-descent, RD, and air-speed, V, are discussed in Sub-section 8.2.2.

8.2.2 GLIDE ANGLE, RATE-OF-DESCENT AND SPEED

An expression for the glide angle, $\bar{\gamma}$, can be recovered from Eqn (8.13) and (8.14) by division:

$$\tan \bar{\gamma} = \frac{C_D}{C_L} \tag{8.15}$$

The speed in the glide, V, can be solved from Eqn (8.9) as:

$$V = \sqrt{\frac{2W}{\varrho S C_R}} \tag{8.16}$$

The reader is asked to show, by using Eqn (8.14) that this can be written as:

$$V = \sqrt{\frac{2W}{\rho S C_L} \cos \bar{\gamma}} \qquad (8.17)$$

From Eqn (8.15) it follows that **the smallest glide angle (i.e. the longest track along the ground) is obtained when flying at the maximum possible lift–to–drag ratio, C_L/C_D**. If the drag polar is assumed to be parabolic {See Eqn (5.1)} it is possible to show that the maximum lift–to–drag ratio, C_L/C_D is:

$$\left(\frac{C_L}{C_D}\right)_{max} = \frac{1}{2}\sqrt{\frac{\pi A e}{C_{D_0}}} \qquad (8.18)$$

This maximum lift–to–drag ratio occurs at a lift coefficient given by:

$$C_{L_{(L/D)max}} = \sqrt{\pi A e C_{D_0}} \qquad (8.19)$$

Therefore, design features which are conducive to shallow glides are: high aspect ratio and low zero–lift drag coefficient. For a given weight and altitude, Eqn (8.17) can be used to determine the speed at which flight should be conducted to achieve the condition reflected by Eqn (8.18). The reader is asked to show that this speed (for small $\bar{\gamma}$) is obtained from:

$$V = \sqrt{\frac{2W}{\rho S \sqrt{\pi A e C_{D_0}}}} \qquad (8.20)$$

The rate–of–descent, RD, follows from:

$$RD = V \sin \bar{\gamma} = V \frac{C_D}{C_L} \cos \bar{\gamma} = \sqrt{\frac{W}{S} \frac{2}{\rho} \frac{C_D^2}{C_L^3} \cos^3 \bar{\gamma}} \qquad (8.21)$$

As a general rule, the angle $\bar{\gamma}$ will be sufficiently small* to use the approximation:

$$\cos \bar{\gamma} \approx 1.0 \qquad (8.22)$$

When this approximation is used in Eqn (8.17), the speed in a "shallow" glide can be written as:

$$V = \sqrt{\frac{2W}{\rho S C_L}} \qquad (8.23)$$

The corresponding rate–of–descent (Eqn 8.21) can be written as:

---

* The reader should observe that even if the relatively large value of $\bar{\gamma} = 18^0$ is chosen, the error in $\cos \bar{\gamma}$ will be only 5%, while the error in $\cos^{3/2} \bar{\gamma}$ will be only 7%.

$$RD = V\sin\overline{\gamma} = V\frac{C_D}{C_L} = \sqrt{\frac{W}{S}\frac{2}{\varrho}\frac{C_D^2}{C_L^3}} \tag{8.24}$$

From Eqn (8.24) it follows that **the minimum rate–of–descent (i.e. the longest time aloft) occurs when the flight is conducted at the maximum possible value of C_L^3/C_D^2**.

The maximum value of C_L^3/C_D^2 can be obtained by setting its derivative with respect to C_L equal to zero. The reader should verify that for parabolic drag polars the results are:

$$\left(C_L^3/C_D^2\right)_{max} = \frac{3\pi Ae}{16}\sqrt{\frac{3\pi Ae}{C_{D_0}}} \tag{8.25}$$

and

$$C_{L_{\left(C_L^3/C_D^2\right)_{max}}} = \sqrt{3\pi AeC_{D_0}} \tag{8.26}$$

Therefore, design features which are conducive to low descent rates are: high aspect ratio and low zero–lift drag coefficient.

Clearly, the aerodynamic ratios C_L/C_D and C_L^3/C_D^2 play a very important role in the performance of gliders. The same can be said about the glide performance of airplanes with power–off.

Figure 8.4 shows a drag polar plot of a glider, with C_L/C_D and C_L^3/C_D^2 superimposed as functions of the lift coefficient.

For a given weight, W, wing area, S and altitude, h (this determines the density), by selecting a range of values for angle of attack, it is possible to compute: C_L/C_D, C_L^3/C_D^2, V and RD. The corresponding performance data are shown in Table 8.1.

The glider selected for this example is a relatively poor one by modern standards. Its lowest flight path angles are high compared to those achievable with modern gliders. Trimmed lift–to–drag ratios of 40.0 to 50.0 have already been achieved, yielding descent angles as low as 1.2 degrees! Despite this, the reader is asked to verify that the approximations of Eqns (8.22) – (8.24) are quite good!

The performance results of Table 8.1 are plotted in Figure 8.5. The reader is asked to check out the significance of the points labeled A, B and C in Figures 8.4 and 8.5. These very important points are discussed next.

Figure 8.4 Aerodynamic Characteristics of a Glider

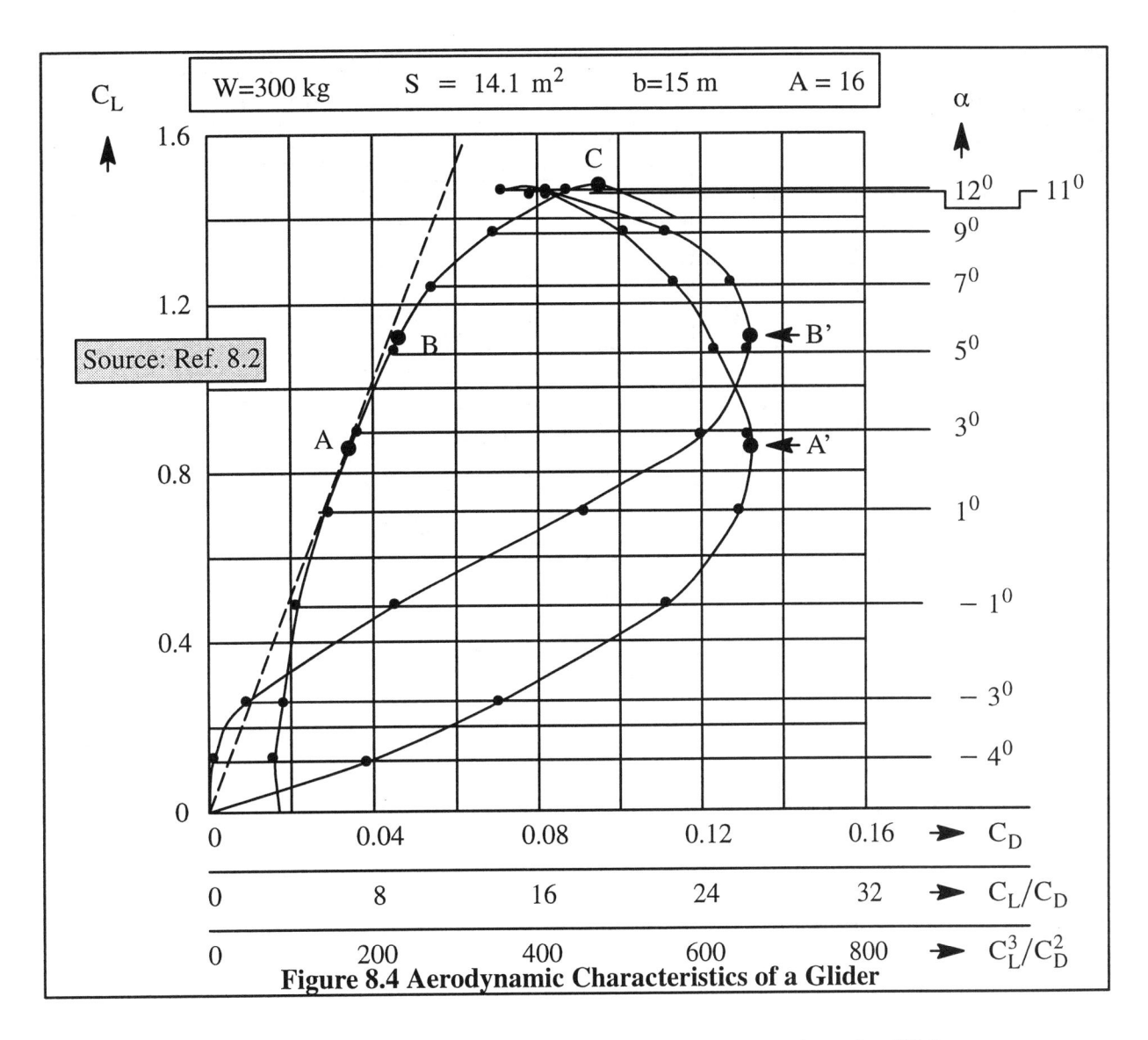

Table 8.1 Calculation of Performance Characteristics of a Glider

α deg.	C_L	C_D	C_L/C_D	C_L^3/C_D^2	$\tan\bar{\gamma}$	$\bar{\gamma}$ deg.	V m/sec	V km/hr	RD m/sec
12	1.47	0.0950	15.5	352	0.0645	3.7	15.2	54.7	0.97
11	1.46	0.0865	16.9	415	0.0592	3.4	15.3	55.0	0.90
9	1.36	0.0675	20.2	553	0.0495	2.8	15.8	56.8	0.78
7	1.23	0.0535	22.9	641	0.0437	2.5	16.7	60.0	0.73
5	1.08	0.0440	24.5	644	0.0408	2.3	17.8	64.0	0.72
3	0.90	0.0350	25.7	595	0.0389	2.2	19.4	69.7	0.76
1	0.70	0.0275	25.4	453	0.0394	2.3	22.1	79.5	0.87
−1	0.49	0.0220	22.0	225	0.0455	2.6	26.5	95.2	1.20
−3	0.25	0.0180	13.9	48	0.0719	4.1	36.5	131.0	2.62
−4	0.12	0.0160	7.5	6.8	0.1333	7.6	53.2	191.0	7.03

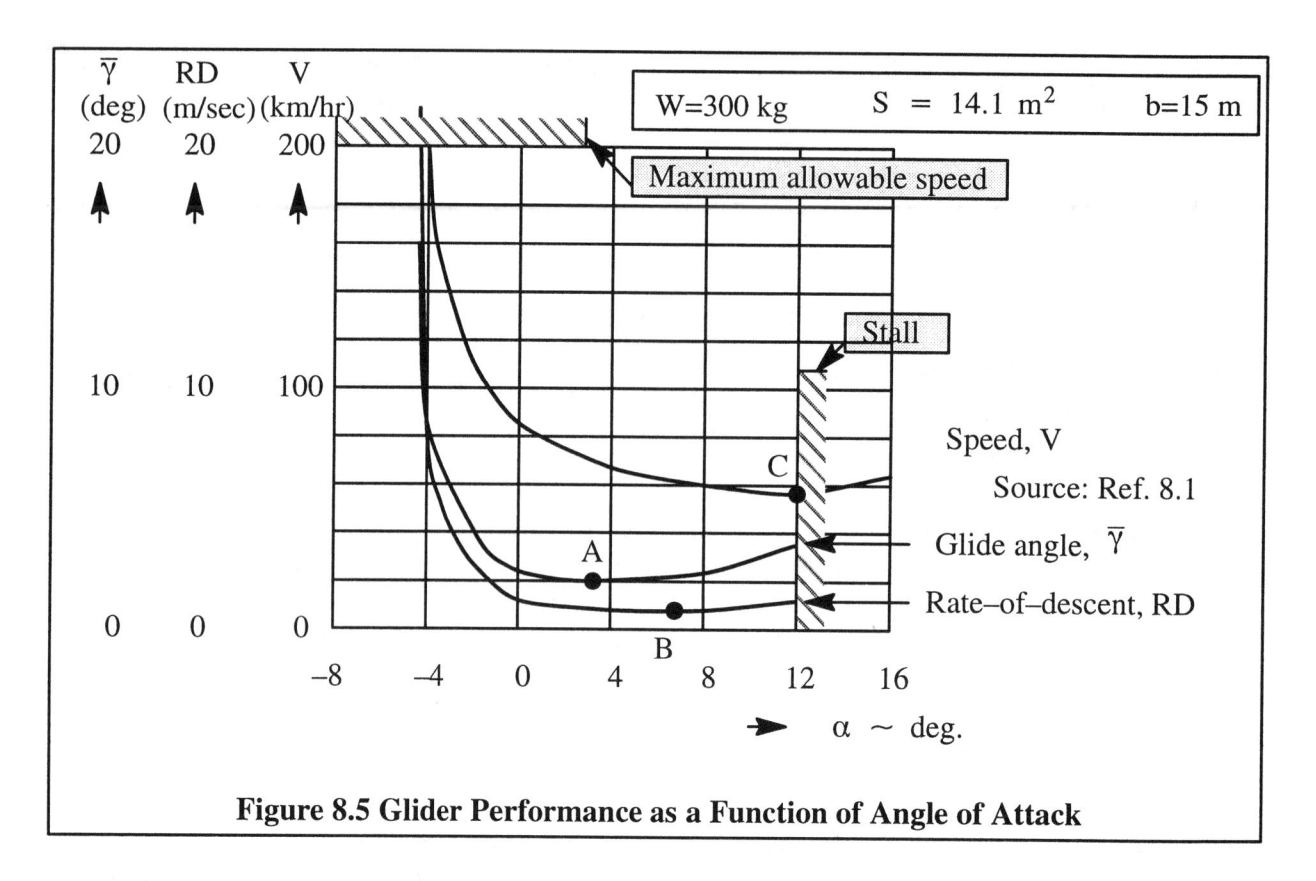

Figure 8.5 Glider Performance as a Function of Angle of Attack

Point A: Minimum glide path angle

Point A in Figure 8.4 represents the point of maximum lift-to-drag ratio, C_L/C_D. It may be seen from Eqn (8.15) that this corresponds to the lowest glide (or descent) angle. That this is indeed the case may also be seen from point A in Figure 8.5. Note that the descent angle minimum is a "weak" minimum.

When flown at this lowest glide angle, the airplane will cover the longest distance along the ground (ground track), before contacting the ground.

The minimum glide angle can be recovered from Eqns (8.15) and (8.18) as:

$$\tan \overline{\gamma}_{min} = \frac{1}{(C_L/C_D)_{max}} = 2\sqrt{\frac{C_{D_0}}{\pi Ae}} \tag{8.27}$$

The maximum distance covered in a steady glide is found from:

$$R_{max_{glide}} = \frac{h}{\tan \overline{\gamma}_{min}} = \frac{h}{2\sqrt{\frac{C_{D_0}}{\pi Ae}}} \tag{8.28}$$

where: h   is the altitude at which the glide is commenced

The speed at which this maximum performance glide (from a range covered viewpoint) must be flown can be determined with the help of Eqns (8.23) and (8.19) as:

$$V_{\text{max glide range}} = \sqrt{\frac{2W}{\varrho S \sqrt{\pi A e C_{D_0}}}} \qquad (8.29)$$

The reader is asked to show, that the drag polar of Figure 8.4 can be approximated by a parabolic drag polar with A=15.6, e=0.82 and $C_{D_0} = 0.0150$. The speed for best glide performance, using sea–level density data can then be computed from Eqn (8.29) as 20.9 m/sec. This agrees closely with the data of Table 8.1 which use the actual drag polar.

Modern gliders have maximum lift–to–drag ratios of around 45. When starting one mile high, they can cover a distance of 45 miles before contacting the ground. Note, that even single–engine propeller driven airplanes in a power–off condition, with a maximum lift–to–drag ratio of around 10 will be able to cover a distance of 10 miles before contacting the ground if the engine fails at one mile high.

Point B: Minimum rate–of–descent

Point B in Figure 8.4 represents the point where C_L^3/C_D^2 is a maximum. As seen from Eqn (8.24), the rate–of–descent, RD, is a minimum at this point. This may be verified by looking at Point B in Figure 8.5. As was the case with the descent angle, the rate–of–descent minimum is also a "weak" minimum.

When flown at maximum C_L^3/C_D^2, the airplane will stay aloft the longest time before contacting the ground.

The minimum rate–of–descent can be recovered from Eqns (8.24) and (8.25) as:

$$RD_{\min} = \sqrt{\frac{W}{S}\frac{1}{\varrho}\frac{1}{\left(C_L^3/C_D^2\right)_{\max}}} = \sqrt{\frac{W}{S\varrho}\frac{10.67}{\pi A e}\sqrt{\frac{C_{D_0}}{3\pi A e}}} \qquad (8.30)$$

The maximum time aloft follows from:

$$t_{\text{maximum aloft}} = \frac{h}{RD_{\min}} = \frac{h}{\sqrt{\dfrac{W}{S\varrho}\dfrac{10.67}{\pi A e}\sqrt{\dfrac{C_{D_0}}{3\pi A e}}}} \qquad (8.31)$$

For the glider data in Table 8.1, using 5,000 ft as the starting altitude, the estimated maximum time aloft is roughly 18 minutes. However, this estimate uses the density of 5,000 ft in Eqn (8.31). This is not a correct way of performing this calculation because density increases with decreasing altitude. The actual maximum time aloft will therefore be somewhat less. The reader is asked in Problem 8.5 to determine a better estimate by accounting for the change in density with altitude.

Point C: Minimum speed or stall speed

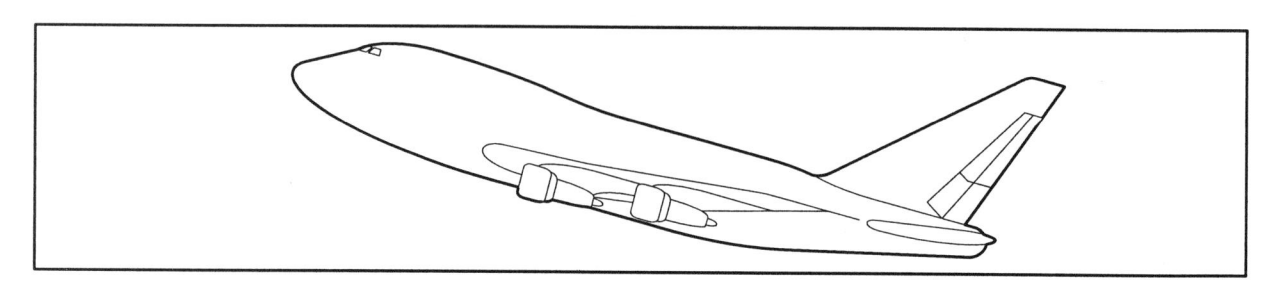

Point C in Figure 8.4 is the point of maximum lift coefficient. This point corresponds to the minimum flight speed (i.e. stall speed) as may be verified by looking at Point C in Figure 8.5. Again, another "weak" minimum is encountered.

The minimum flight speed may be recovered from Eqn (8.23) by setting the lift coefficient equal to the maximum lift coefficient:

$$V_{min} = V_s = \sqrt{\frac{2W}{\varrho S C_{L_{max}}}} \qquad (8.32)$$

It may be seen from Eqn (8.32) that to achieve a low stall speed, the wing loading, W/S, should be low and/or the maximum lift coefficient, $C_{L_{max}}$ should be high. Table 8.2 shows typical maximum lift coefficients for a range of airplanes with the flaps up and with the flaps retracted.

Some comments on the practical meaning of stall speed are in order. The stall speed according to Eqn (8.32) is also referred to as the 1g–stall–speed. In flight testing, the measured minimum airspeed is always lower than this 1g–stall–speed. The regulations of FAR 23 and 25 require that the in–flight determination of V_s be done with a deceleration of no more than 1 knot per second:

$$dV/dt \leq 1.0 \text{ knot/sec} \qquad (8.33)$$

Figure 8.6 shows the difference between the stall speed as obtained from a stalling maneuver and the 1–g stall speed as obtained from Eqn (8.32). Note that the in–flight minimum speed occurs at a negative value of load–factor.

In modern high performance airplanes it is possible, that the minimum speed is determined by control limitations. That can lead to minimum speeds which are higher than the 1–g stall speed, which corresponds to the maximum lift coefficient. This is an important issue because typical approach and takeoff speeds are referred to the minimum speed by certain factors. Examples are the approach speed which is defined as 1.3 times the stall (or minimum) speed and the take–off speed which is defined as 1.2 times the stall (or minimum) speed. For purposes of preliminary design it is customary to use the 1–g stall speed as defined by Eqn (8.32) as the minimum speed, unless control problems can be expected. In such cases, the minimum control speed must be established and used.

Table 8.2 Typical Values of Maximum Lift Coefficients for Airplanes

Airplane type	S	A	$\Lambda_{c/4}$	W_{TO}	$(W/S)_{TO}$	$C_{L_{max}}$ flaps − up	$C_{L_{max}}$ flaps − down
	ft^2		deg.	lbs	lbs/ft^2		
Transports							
Boeing 727–200	1,700	6.9	32	185,000	109	0.95	2.5
Boeing 737–200	980	8.8	25	116,000	118	??	3.2
McDD DC–10	3,861	6.8	35	455,000	118	??	2.5
Beech 1900D	310	10.9	0	16,950	54.7	1.6	2.3
Boeing 777	4,605	8.7	31.6	506,000	110	??	2.9
SAAB 2000	450	11	0	29,000	64.4	1.7	2.5
Trainers							
Cessna 152	157	6.7	0	1,670	10.6	1.3	1.7
Piper Tomahawk	125	9.3	0	1,670	13.4	1.7	1.8
Grumm.–Am. AA–1	101	6.0	0	1,600	15.8	1.6	1.7
Pilatus PC–9	325	6.3	0	4,960	15.3	0.7	0.9
SIAI-M S–211	136	5.6	15.5	6,063	44.7	??	2.4
Gulfstr.–Am. GA–7	184	7.4	0	3,800	20.7	1.2	1.5
Business Airplanes							
Learjet M36A	253	5.7	13	18,000	71	??	2.1
Dassault Falcon 20	440	6.4	30	28,700	65	??	2.7
Cessna 525	323	8.4	0	14,100	44	??	1.7
Piaggio P–180	172	12.3	0	11,550	67.2	1.7	2.3
Pilatus BN2T	325	7.4	0	7,000	21.5	2.3	3.1
Fighters							
Northrop F–5A	170	3.8	24	21,000	124	??	2.0
Dassault Mirage III–E375		1.9	61	30,000	80	??	1.0
Sukhoi 27B	667	3.5	40	49,600	74.4	??	1.9
FAMA IA 63	168	6.0	0	8,377	49.8	1.7	??
Gliders (Egrett is powered)							
Schweizer SGS-1-35	104	23.3	0	930	8.9	3.4	No flaps
Fournier RF–9	194	16.6	0	1,642	8.5	1.7	No flaps
Caproni A–21S	174	25.6	0	1,419	8.2	2.1	No flaps
Grob Egrett II	427	27.5	0	10,362	24.3	1.6	2.0

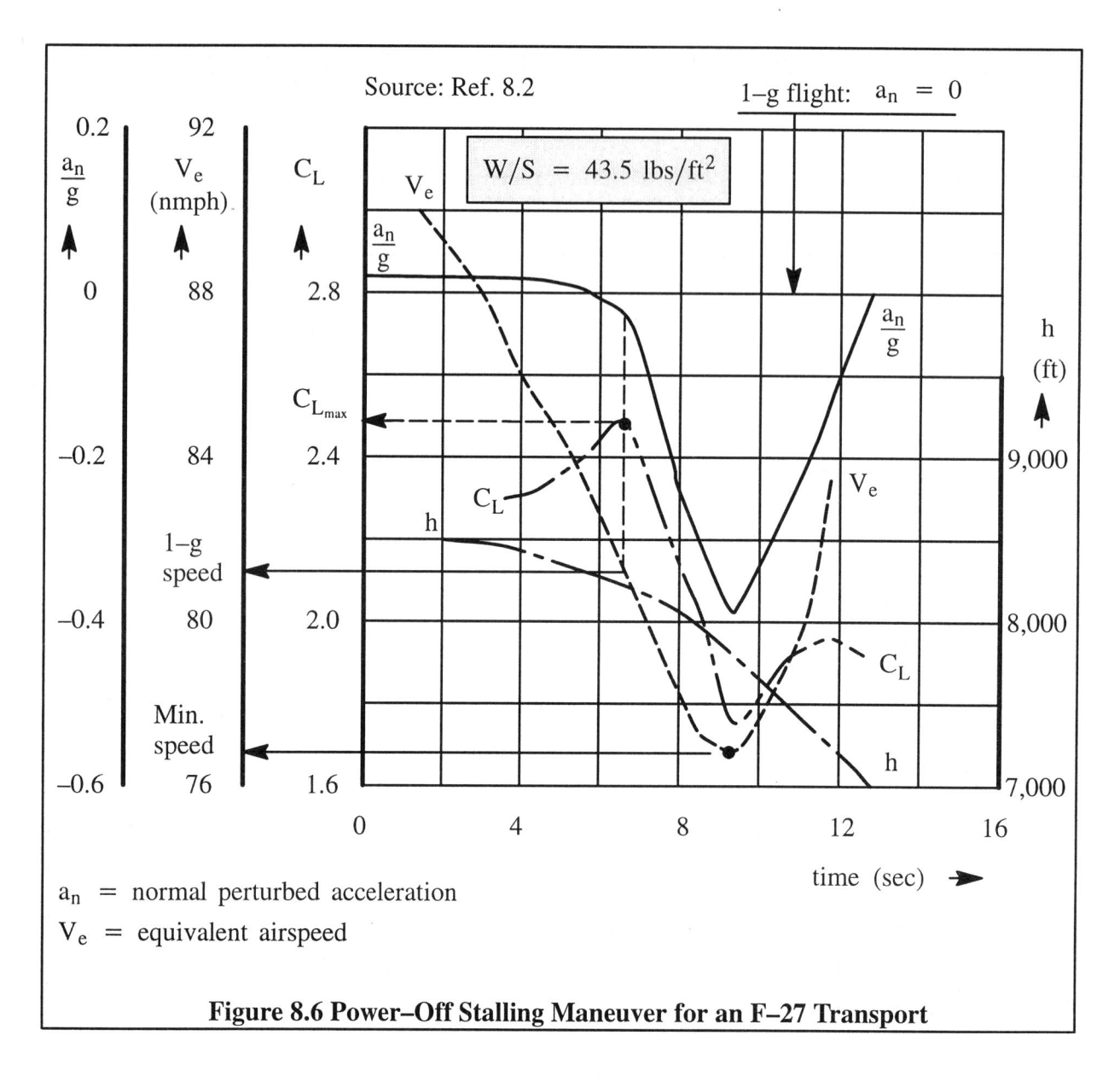

Figure 8.6 Power–Off Stalling Maneuver for an F–27 Transport

8.2.3 SPEED POLAR OR HODOGRAPH

The true air–speed, $\vec{V}$, of an airplane in a glide has a horizontal and a vertical component. The vertical component was called the rate–of–descent, RD. A graph which relates the horizontal component of velocity to the rate–of–descent RD is referred to as a speed polar or hodograph. A conceptual hodograph is shown in Figure 8.7.

If the flight path angle, $\bar{\gamma}$ is small (see footnote on page 337), the horizontal component of velocity, $V_h = V\cos\bar{\gamma}$ may be replaced (i.e. approximated) by: V.

It is clear from Eqns (8.15) through (8.32) that there exist explicit mathematical relationships between a hodograph and the drag polar of an airplane.

Figure 8.7 Hodograph or Speed Polar

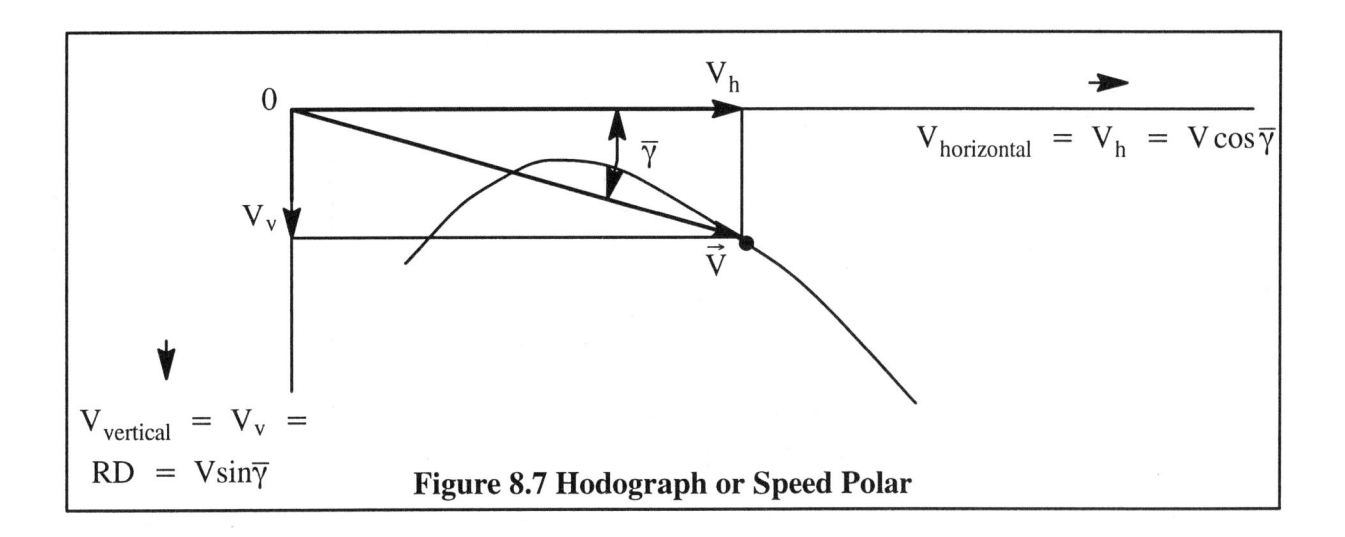

Figure 8.8 shows six points on a hodograph and on a drag polar which each have particular significance. The significance of these six points is summarized in Table 8.3. The reader is encouraged to study the relationship between the six points on the drag polar and the corresponding six points on the hodograph. The following comments are in order.

Note that the minimum air–speed, V_{min}, (point 1) occurs below the stall speed, V_s, (point 2)! The reason is that: $C_{R_{min}} < C_{R_{at\,C_{L_{max}}}}$ as seen from Figure 8.8.

Observe that the sustained, vertical dive speed corresponds to point 6 in Figure 8.8. The flight path angle, $\bar{\gamma}$, is 90 degrees. The condition of equilibrium then is: $D = W$. The speed in this flight condition is referred to as a terminal dive. For the glider of Figure 8.4, with W=300 kg and $C_{D_0} = 0.0170$, the dive speed at sea–level may be determined as: $V_D = 509$ km/hr. The maximum, aerodynamically attainable air–speed occurs at point 5. Here, with the drag polar of Figure 8.4, it is found that $C_{R_{min}} = 0.0167$. This yields: $V_{max} = 514$ km/hr.

In reality, maximum allowable airspeeds are usually determined by structural integrity considerations. Examples are: maximum allowable load factor in a pull–up from a dive, flutter or divergence.

For the glider in this example, the maximum allowable speed, from a load factor point of view, turns out to be: $V_{max} = 200$ km/hr. With the help of Figure 8.5 it may be seen that this speed corresponds to a flight path angle of only $\bar{\gamma} = 8.3$ degrees. Clearly, a sustained vertical dive in a glider is not an advisable maneuver!

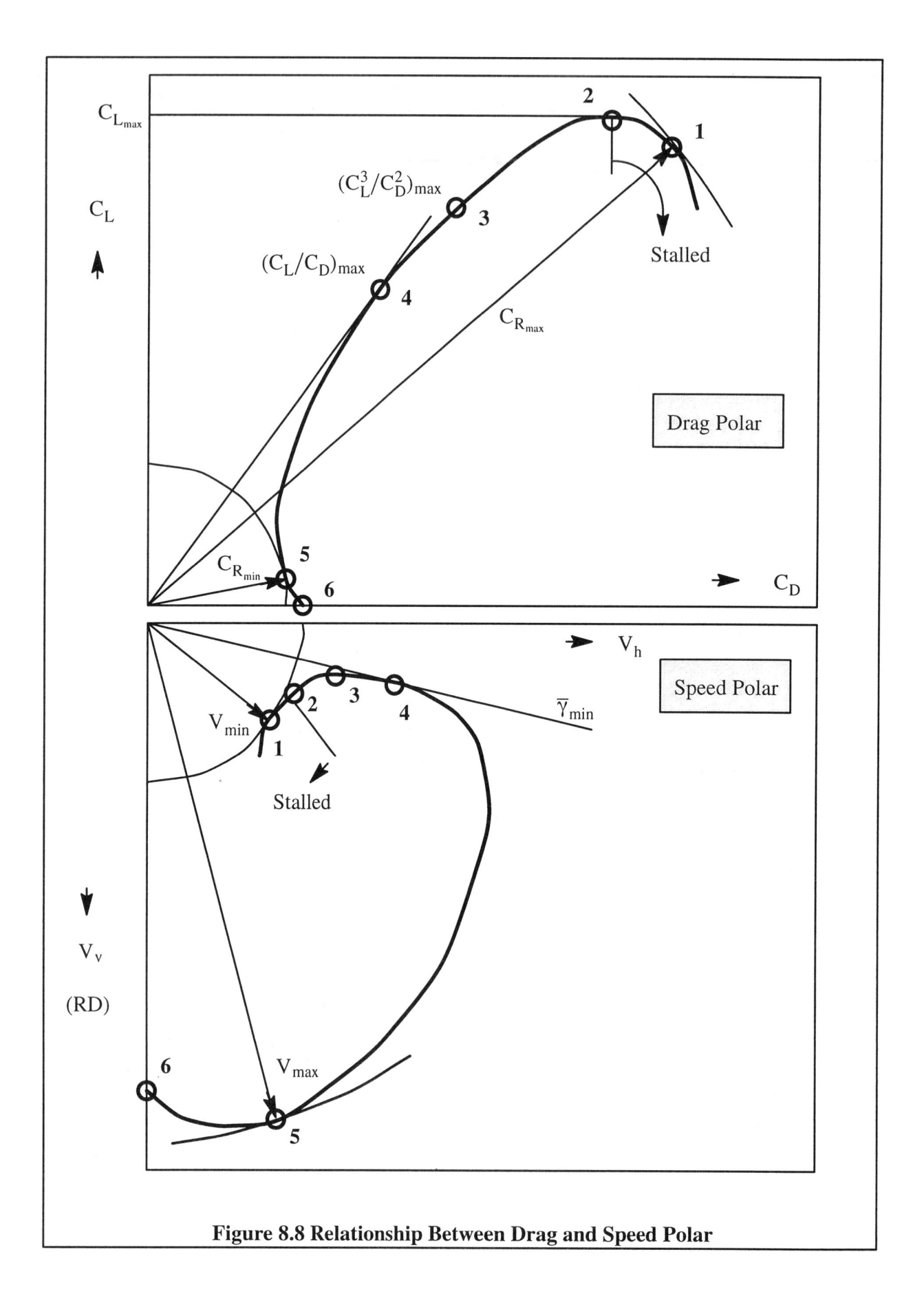

Figure 8.8 Relationship Between Drag and Speed Polar

Table 8.3	Summary of Important Points on the Speed Polar (Hodograph) and on the Drag Polar		
Point	Drag Polar	Speed Polar	Equation Number
1	Maximum resultant force coefficient $C_{R_{max}}$	Minimum air-speed V_{min}	(8.16)
2	Maximum lift coefficient $C_{L_{max}}$	1-g Stall speed V_s	(8.27)
3	$(C_L^3/C_D^2)_{max}$	Minimum rate of descent RD_{min}	(8.24)
4	$(C_L/C_D)_{max}$	Minimum glide path angle $\bar{\gamma}_{min}$	(8.21)
5	Minimum resultant force coefficient $C_{R_{min}}$	Maximum dive speed	(8.16)
6	Zero lift $C_L = 0$	Vertical dive $\bar{\gamma} = 90^0$	Solve V from Eqn (8.13) $\bar{\gamma} = 90^0$ $C_D = C_{D_0}$

8.2.4 EFFECT OF ALTITUDE, WEIGHT AND WIND

Another interesting application of the speed polar of Figure 8.8 is that it lends itself readily to a discussion of the effect of altitude, weight and wind on airplane performance during a glide. These effects are discussed in 8.2.4.1 through 8.2.4.4 respectively.

8.2.4.1 Effect of Altitude

As long as compressibility is not important, the drag polar does not change with flight speed (Mach number effects are negligible). For a glider at low speeds this condition is certainly satisfied.

The glide path angle, $\bar{\gamma}$, depends only on C_L/C_D as seen from Eqn (8.15). Since C_L/C_D itself depends on the angle of attack, the flight path angle depends only on angle of attack. A consequence of this, since $\theta = \alpha + \gamma$, is that the pitch attitude angle is also determined solely by the angle of attack. Altitude is therefore not a factor in any of these angles. However, the speed associated with a glide at a given altitude and angle of attack does depend on altitude as seen from Eqns (8.16) and (8.17).

The effect of altitude on the speed polar of the glider of Figure 8.4 is shown in Figure 8.9. Because points on the speed polar with the same angle of attack must have the same flight path angle, all such points lie on a straight line through the origin.

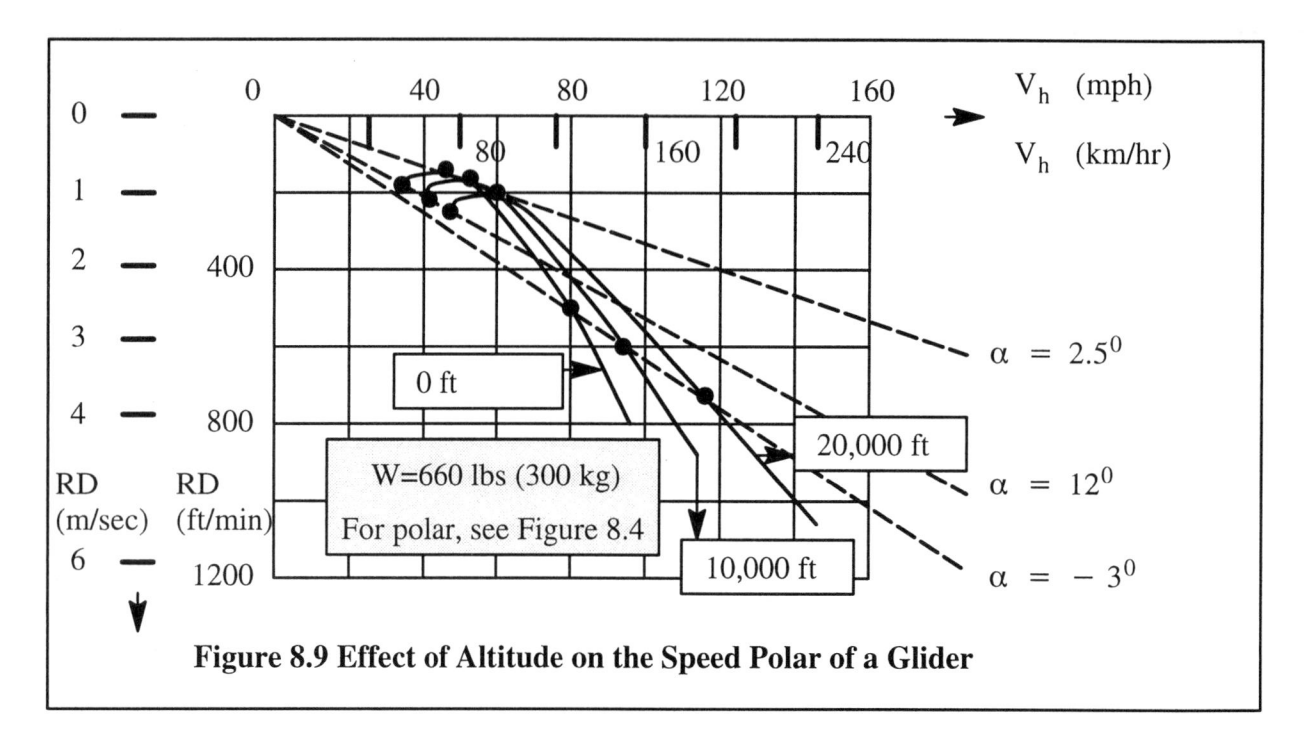

Figure 8.9 Effect of Altitude on the Speed Polar of a Glider

From Eqns (8.16) and (8.20) it follows that for two flight conditions at the same angle of attack, but at different altitudes h_1 and h_2 with densities ϱ_1 and ϱ_2 the following holds:

$$\frac{V_2}{V_1} = \frac{RD_2}{RD_1} = \sqrt{\frac{\varrho_1}{\varrho_2}} \qquad (8.34)$$

The reader is asked to show, that at constant wing loading, flight conditions for minimum air-speed (or minimum glide angle, or minimum descent rate) occur at the same equivalent air-speed at all altitudes. To a pilot this property is extremely useful: the air-speed indicator in the cockpit displays equivalent air-speed!

8.2.4.2 Effect of Weight

The effect of airplane weight on the speed polar can be determined as follows. For identical angle of attack values it is seen from Eqn (8.17) that:

$$\frac{V_2}{V_1} = \frac{RD_2}{RD_1} = \sqrt{\frac{W_2}{W_1}} \qquad (8.35)$$

An interesting conclusion, also derivable from Eqn (8.15), is that the minimum glide path angle is independent of weight. However, the equivalent air-speed for a given value of angle of attack, does depend on weight.

8.2.4.3 Speed Polar in Generalized Coordinates and Application in Flight Testing

The reader should observe from Eqn (8.17) and (8.18) that another way to write these is:

$$\frac{V\sqrt{\varrho}}{\sqrt{W}} = f_1(\alpha) \tag{8.36}$$

and

$$\frac{RD\sqrt{\varrho}}{\sqrt{W}} = f_2(\alpha) \tag{8.37}$$

It may be seen that for any given value of angle of attack, α, the quantities $V\sqrt{\varrho}/\sqrt{W}$ and $RD\sqrt{\varrho}/\sqrt{W}$ are constant. By elimination of α from Eqns (8.36) and (8.37) it will be found that:

$$\frac{RD\sqrt{\varrho}}{\sqrt{W}} = f\left(\frac{V\sqrt{\varrho}}{\sqrt{W}}\right) \tag{8.38}$$

It is possible to refer the quantities ϱ and W to reference values, such as ϱ_0 and W_0, the density at sea–level and any reference weight respectively. In that case Eqn (8.38) can be rewritten as:

$$\frac{RD\sqrt{\varrho/\varrho_0}}{\sqrt{W/W_0}} = f\left[\frac{V\sqrt{\varrho/\varrho_0}}{\sqrt{W/W_0}}\right] \tag{8.39}$$

In this form the speed polar is referred to as the generalized speed polar. An example of such a generalized speed polar is shown in Figure 8.10. It is very useful in reducing flight test data. In any flight test it is always difficult to run the test at the same conditions of atmospheric density and airplane weight. With the help of Eqn (8.39) it is possible to collapse all flight test data of speed and rate of descent into one form as evident from Figure 8.10.

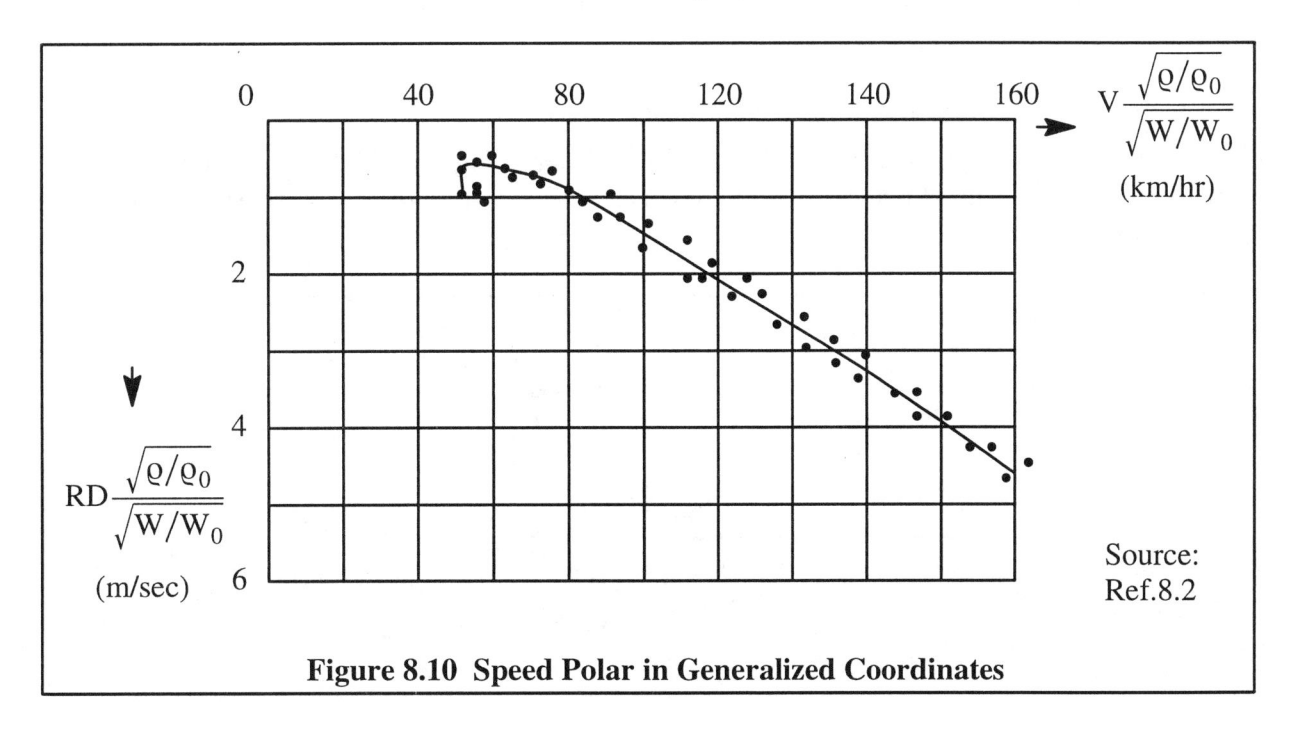

Figure 8.10 Speed Polar in Generalized Coordinates

Source: Ref.8.2

8.2.4.4 Effect of Wind

The effect of wind on airplane speed relative to the ground can be visualized with the help of Figure 8.11. The airplane in Figure 8.11 should be thought of as flying inside a very large air mass (box A) while the air mass (box A with everything in it) is moving relative to the ground with a speed, called the wind speed. The airplane ground speed therefore is the vectorial sum of the air–speed of the airplane and the wind speed of the air mass relative to the ground.

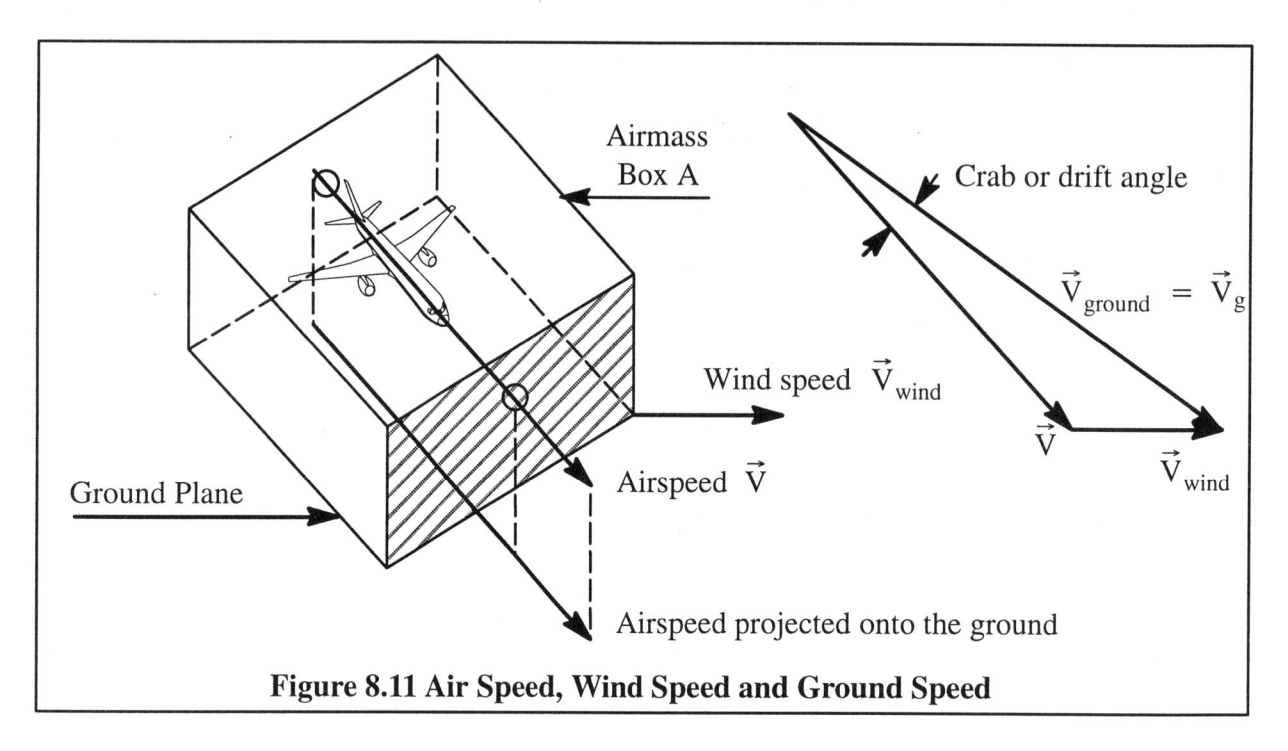

Figure 8.11 Air Speed, Wind Speed and Ground Speed

It is seen from the vector diagram in Figure 8.11 that:

$$\vec{V}_{ground} = \vec{V}_g = \vec{V} + \vec{V}_{wind} \tag{8.40}$$

The wind speed, $\vec{V}_{wind}$ is assumed to be constant for the air mass inside box A, but oriented in an arbitrary direction. The reader may have observed that in Figure 8.11 all speed vectors are assumed to be in a horizontal plane. In reality this is not always the case. A more general situation is shown in Figure 8.12.

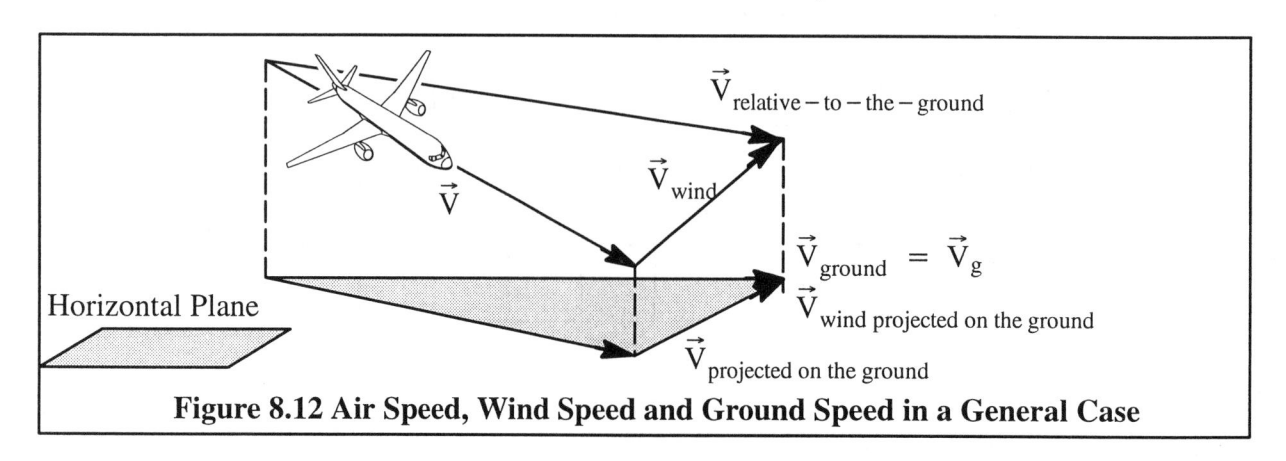

Figure 8.12 Air Speed, Wind Speed and Ground Speed in a General Case

In the following, the wind speed vector, $\vec{V}_{wind}$ is assumed to be parallel to the plane of symmetry of the airplane. In that case $\vec{V}_{wind}$ has a vertical component, $\vec{V}_{updraft}$, and a horizontal component, $\vec{V}_{headwind}$, also in the plane of symmetry.

The effect of these wind components on the speed polar of an airplane is shown in Figure 8.13.

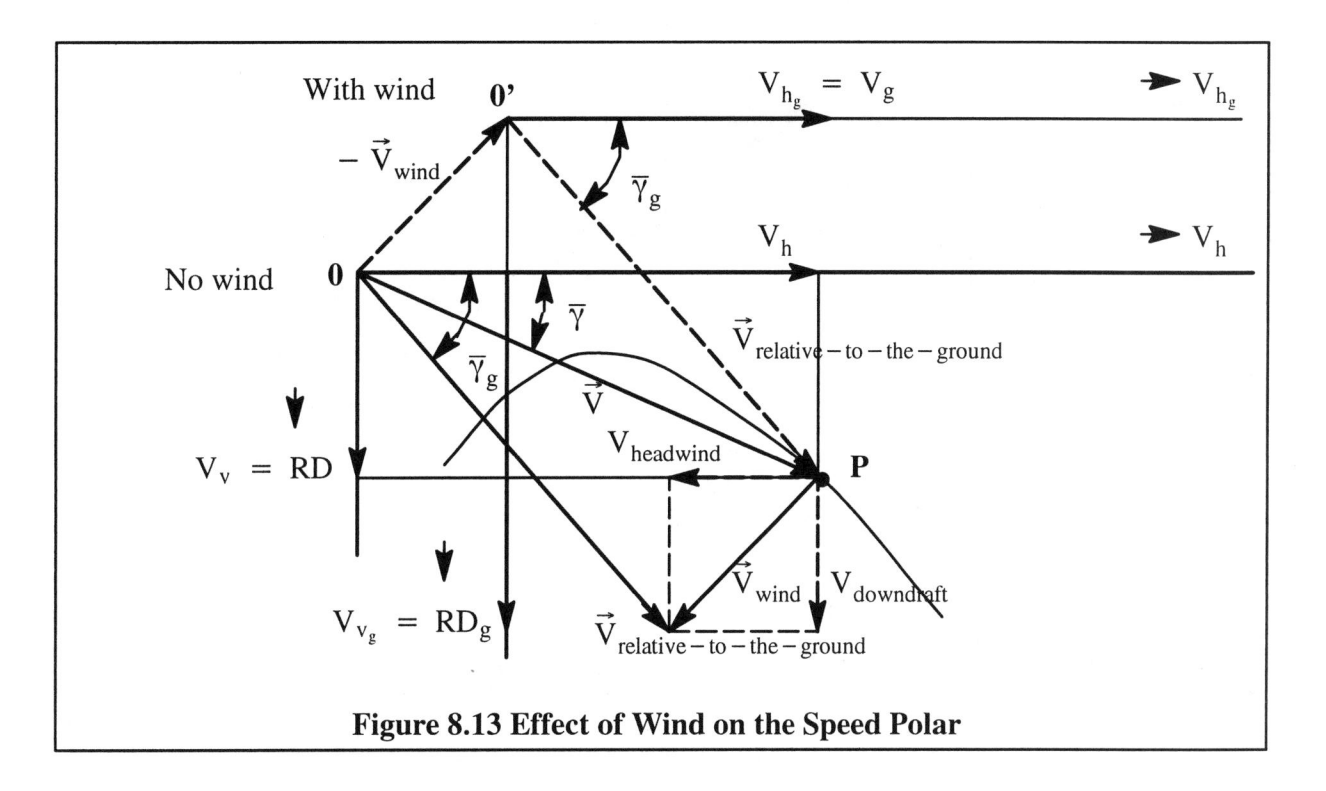

Figure 8.13 Effect of Wind on the Speed Polar

The total velocity relative to the ground, $\vec{V}_{relative-to-the-ground}$, in Figure 8.13 is seen to be the vector O'P if the airplane experiences a wind, $\vec{V}_{wind}$. The ground velocity, V_g, also known as ground–track velocity, is the horizontal component of $\vec{V}_{relative-to-the-ground}$. The wind case shown in Figure 8.13 is for a combined head–wind and down–draft. The origin of the speed polar, O', with wind, is obtained from the old origin, O, no wind, by:

1. For head–wind only: Move O along the positive V_h axis.

2. For up draft only: Move O along the positive RD axis.

3. For tail–wind only: Move O along the negative V_h axis.

4. For down draft only: Move O along the negative RD axis.

The angle, $\bar{\gamma}_g$, is called the actual glide path angle. Figure 8.14 summarizes the effect of wind in accordance with items 1 through 4.

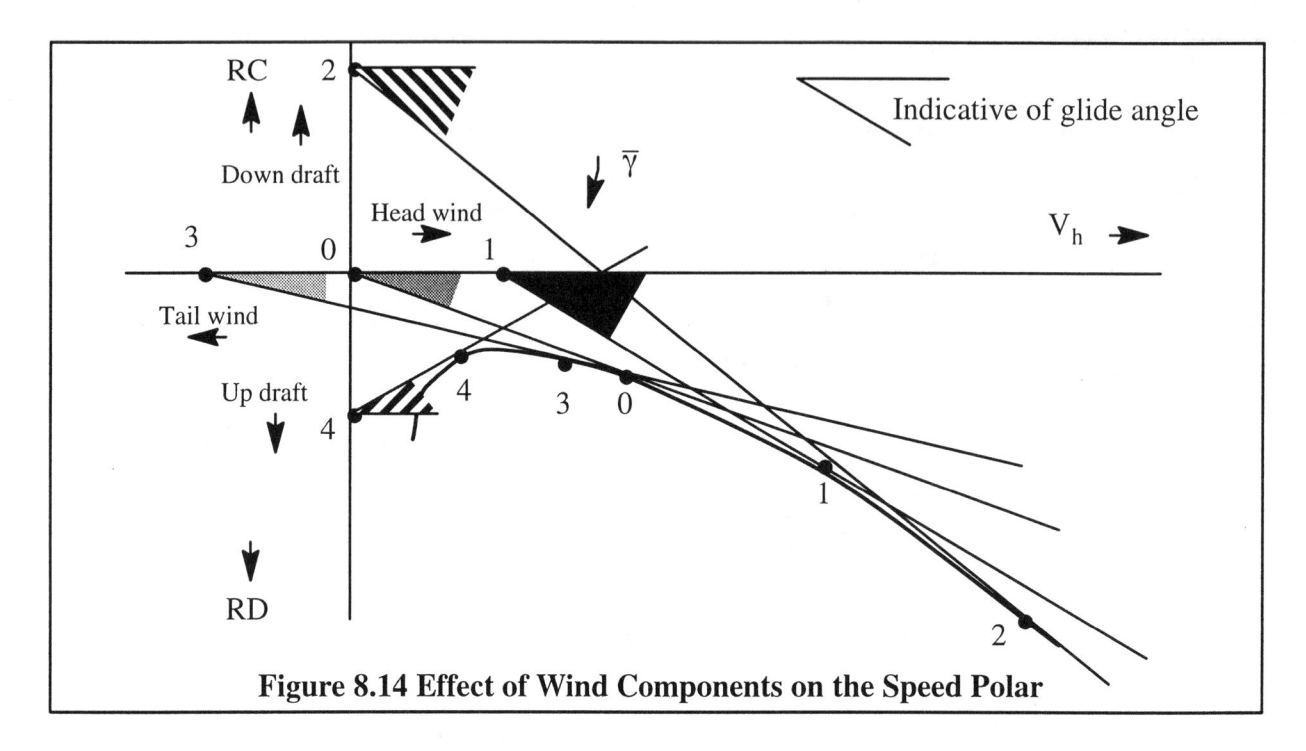

Figure 8.14 Effect of Wind Components on the Speed Polar

Clearly, the minimum glide path angle does depend strongly on the wind. It is seen that head winds will steepen the glide path angle, tail winds will make them more shallow. Similarly, up drafts will make glide angles more shallow (they can even make them positive, as shown!) while down drafts can make for very steep glide angles. Pilots must adjust for this when flying gliders or airplanes with the power off to attain the best possible glide performance.

8.2.5 UN–POWERED GLIDE AT HIGH SPEED

It may be seen from Eqn (8.17) that when the wing loading, W/S and the flight path angle, $\bar{\gamma}$, increase in magnitude, the corresponding value of speed will also increase. As the speed increases, so does the Mach number. At some value of Mach number, compressibility effects can no longer be neglected. When that is the case, Eqn (8.15) must be changed from:

$$\tan\bar{\gamma} = \frac{C_D}{C_L} = f(\alpha) \tag{8.41}$$

to

$$\tan\bar{\gamma} = \frac{C_D}{C_L} = f(\alpha, M) \tag{8.42}$$

As a consequence, the flight condition (flight path angle) is no longer governed uniquely by the angle of attack, but also by the Mach number. A graphical solution for determining the flight path angle in this case will now be presented.. The procedure to find this graphical solution is as follows:

Step 1: Select a wing loading, W/S
Step 2: Select a flight Mach number and the corresponding drag polar
Step 3: Select an altitude. This determines the magnitude of the pressure ratio, δ

Step 4: Calculate C_R from Eqn (8.43):

$$C_R = \frac{W/S}{1481.3 \, \delta \, M^2} \tag{8.43}$$

Step 5: Draw the circle with radius, C_R and find the intersect point with the drag polar at the given Mach number.
Step 6: Determine the flight path angle as shown in Figure 8.15.

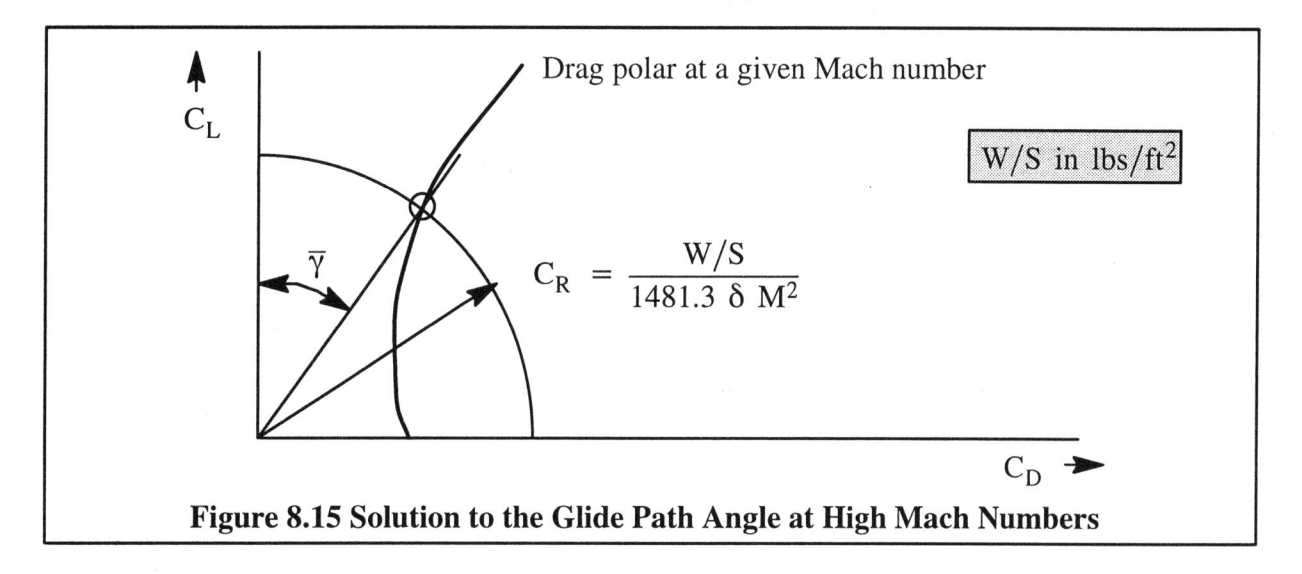

Figure 8.15 Solution to the Glide Path Angle at High Mach Numbers

A special case would be a vertical, terminal dive. In that case the drag of the airplane equals the weight. This condition is reflected by Eqn (8.44):

$$C_{D_0} = \frac{W/S}{1481.3 \, \delta \, M^2} \tag{8.44}$$

A graphical solution for the terminal Mach number at a given altitude is suggested in Figure 8.16. The flight path angle in this case is 90 degrees!

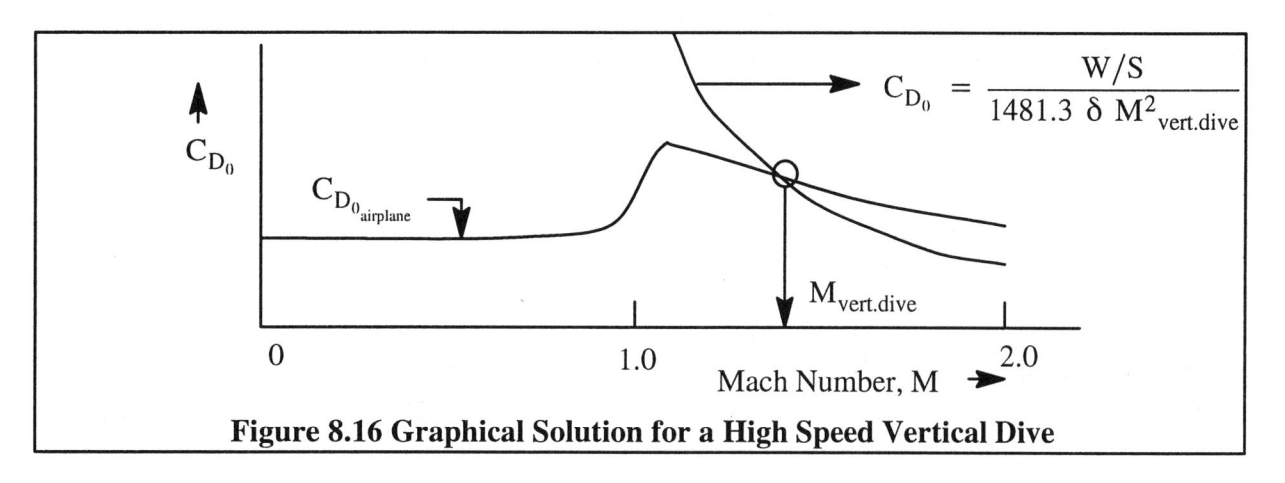

Figure 8.16 Graphical Solution for a High Speed Vertical Dive

8.3 STEADY, POWERED FLIGHT

In powered flight the engine(s) is (are) assumed to be operating. Depending on the amount of power (or thrust) used and depending on the flight condition the airplane can be in a steady state or in an accelerating flight condition. In this chapter only steady state, straight line flight conditions will be considered. The case of accelerated flight is covered in Chapters 9 and 12. Also the power (or thrust) will be assumed to be symmetrical.

For steady, symmetrical powered flight, the equations of motion of the airplane are copied from Eqns (8.4) and (8.6):

$$T\cos(\alpha + \phi_T) - D - W\sin\gamma = 0 \qquad (8.45)$$

and

$$T\sin(\alpha + \phi_T) + L - W\cos\gamma = 0 \qquad (8.46)$$

These equations are written in the stability axis system as pointed out in Figure 8.2.

In the case of conventional airplanes the angle $(\alpha + \phi_T)$, i.e. sum of angle of attack and thrust inclination angle, is sufficiently small so that $\sin(\alpha + \phi_T) \approx 0$ and $\cos(\alpha + \phi_T) \approx 1.0$ are reasonable approximations*. Therefore, Eqns (8.45) and (8.46) can be written as:

$$T - D - W\sin\gamma = 0 \qquad (8.47)$$

and

$$L - W\cos\gamma = 0 \qquad (8.48)$$

The equilibrium condition reflected by Eqns (8.47) and (8.48) is depicted in Figure 8.17a for climbing flight and in Figure 8.17b for level flight. Note, that for the level flight case: $\gamma = 0$ and this reduces Eqns (8.47) and (8.48) to:

$$T - D = 0 \qquad (8.49)$$

and

$$L - W = 0 \qquad (8.50)$$

It was shown in Chapter 6, that for jet driven airplanes the thrust is roughly constant with speed, but variable with altitude and throttle setting. For a given speed, altitude and throttle setting the available thrust is called: $T_{available} = T_{av}$.

For a jet airplane, Eqn (8.49) implies that in level flight:

---

* For certain types of V/STOL and vectored thrust airplanes this approximation is not acceptable.

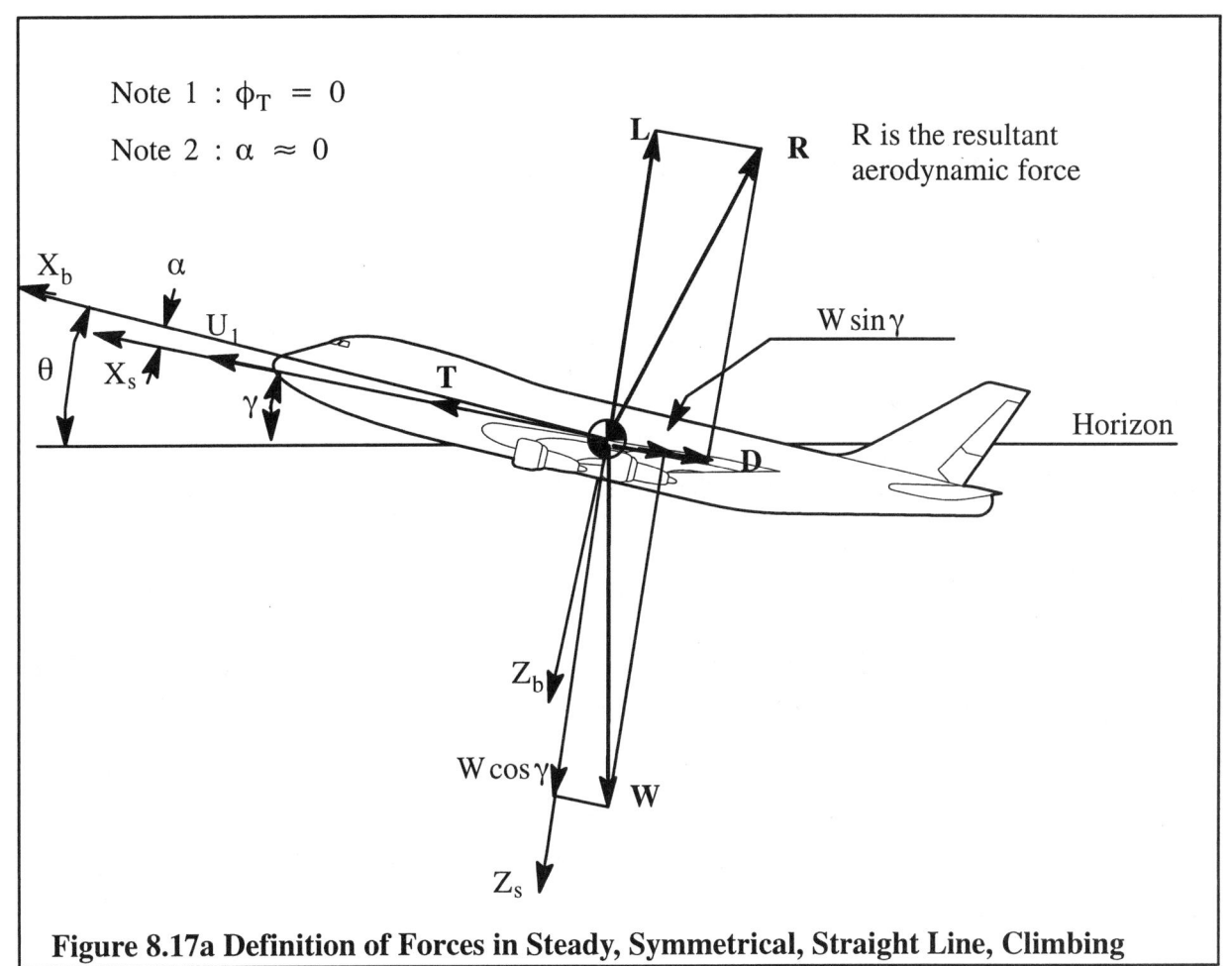

Figure 8.17a Definition of Forces in Steady, Symmetrical, Straight Line, Climbing Flight

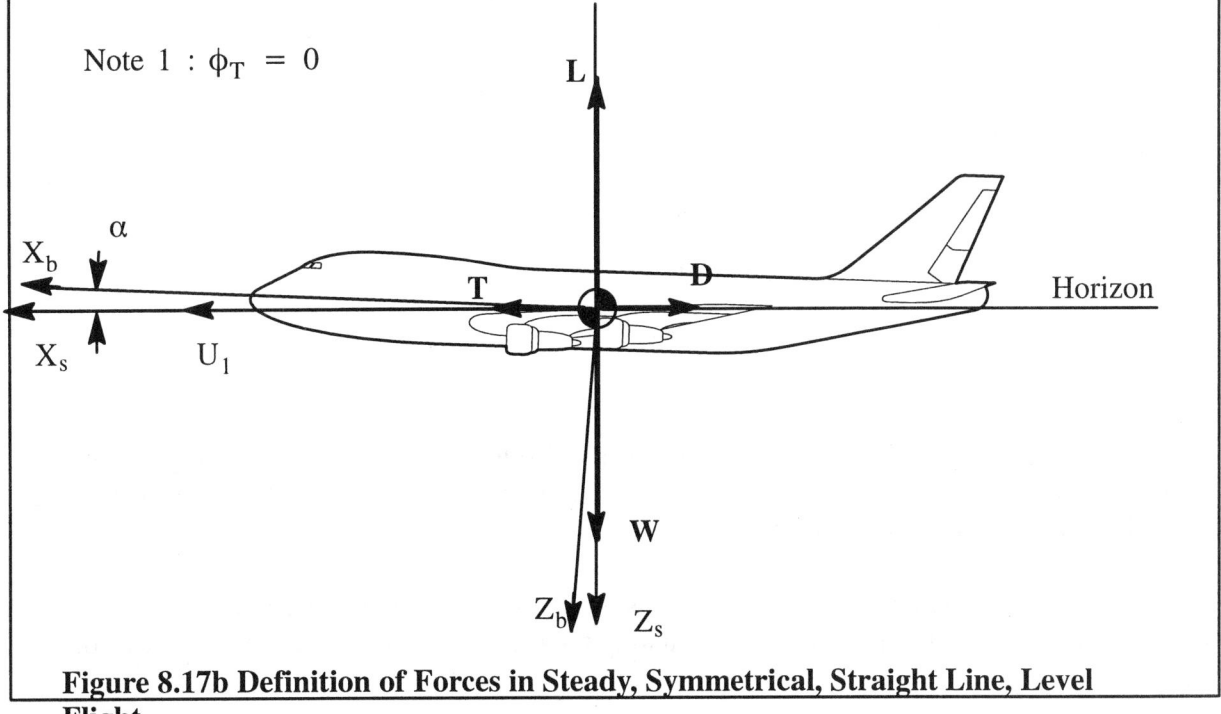

Note 1 : $\phi_T = 0$
Note 2 : $\alpha \approx 0$

R is the resultant aerodynamic force

Figure 8.17b Definition of Forces in Steady, Symmetrical, Straight Line, Level Flight

Note 1 : $\phi_T = 0$

$$T = D \quad \text{or:} \quad D = T_{reqd} \tag{8.52}$$

Since, according to Eqn (8.50), L = W, it follows by division that:

$$T_{reqd} = \frac{W}{L/D} = \frac{W}{C_L/C_D} \tag{8.53}$$

The lift–to–drag ratio, C_L/C_D follows from the drag polar, once the lift coefficient is determined from Eqn (8.50): $C_L = W/\overline{q}S$.

For a jet airplane, Eqn (8.47) implies that for climbing flight:

$$T = D + W\sin\gamma \quad \text{or:} \quad T_{av} = T_{reqd} + T_{climb} \tag{8.54}$$

Clearly, the thrust required to maintain constant speed climb consists of the sum of the thrust required to maintain level flight and the thrust required to keep the flight path inclined.

It was shown in Chapter 6, that for propeller driven airplanes the power available is roughly constant with speed. It has been found useful to express Eqns (8.47) and (8.49) in terms of power rather than in terms of thrust. This is done by multiplying these equations by the speed, U_1. For Eqn (8.47) this yields:

$$TU_1 = DU_1 + WU_1\sin\gamma \tag{8.55}$$

where: TU_1 = Power Available = P_{av}

DU_1 = Power Required = P_{reqd}

$WU_1\sin\gamma$ = Climb Power = P_{climb}

Note that, since: $U_1\sin\gamma$ = Rate of climb = RC it follows that:

$$P_{climb} = WU_1\sin\gamma = W(RC) \tag{8.56}$$

Clearly, according to Eqn (8.55) the power available to maintain an airplane in a steady climb is equal to the sum of the power required to maintain level flight and the power required to climb at a given flight path angle.

Because of the importance of level flight, this flight condition will be discussed in detail in Section 8.4. The subject of climbing flight, including the effect of forward acceleration, will be discussed separately in Chapter 9.

8.4 STEADY, LEVEL, POWERED FLIGHT

For steady, level flight, Eqns (8.49) and (8.50) apply, and may be rewritten as:

$$T = D = C_D \bar{q} S \qquad (8.57)$$

and

$$L = W = C_L \bar{q} S \qquad (8.58)$$

For a given weight, W a given altitude (which defines density) and a given airplane configuration, the variables in Eqns (8.57) and (8.58) are: α, V and T. One of these variables must be arbitrarily selected before a solution can be found. For example, if the speed, V is selected, the equations uniquely determine the angle of attack and the thrust required in that flight condition. Two cases will be considered:

      8.4.1 Turbojet or Turbofan Driven Airplanes

      8.4.2 Propeller Driven Airplanes

8.4.1 TURBOJET OR TURBOFAN DRIVEN AIRPLANES

This case will be considered first, for a parabolic drag polar and a linear relationship between C_L and α and second for a general drag polar and $C_L - \alpha$ curve.

1) Parabolic drag polar and linear C_L vs α

With a parabolic drag polar, Eqn (8.57) can be written as:

$$T = D = C_D \bar{q} S = \left(C_{D_0} + \frac{C_L^2}{\pi A e} \right) \bar{q} S = D_0 + D_i \qquad (8.59)$$

The symbols D_0 and D_i represent the zero–lift drag and the induced drag respectively. The reader is asked to show that with the usual definition of wing aspect ratio, $A = b^2/S$, Eqn (8.59) can be rewritten as:

$$T_{reqd} = \left(C_{D_0} \bar{q} S + \frac{(W/b)^2}{\pi e \bar{q}} \right) \qquad (8.60)$$

By also introducing the definition of equivalent parasite area, f, of Eqn (5.38) it follows that:

$$T_{reqd} = \left(f \bar{q} + \frac{(W/b)^2}{\pi e \bar{q}} \right) \qquad (8.61)$$

This result has several important implications:

1) The thrust required for level flight consists of two terms: one proportional to dynamic pressure, the other proportional to the inverse of dynamic pressure.

2) To minimize the first term (zero lift drag term) it is important to design the airplane with as small a parasite area as possible. A discussion of parasite area is given in Chapter 5.

3) To minimize the second term (induced drag term) the so–called span loading, W/b, must be made as small as possible.

Figure 8.18 shows a graphical interpretation of these results.

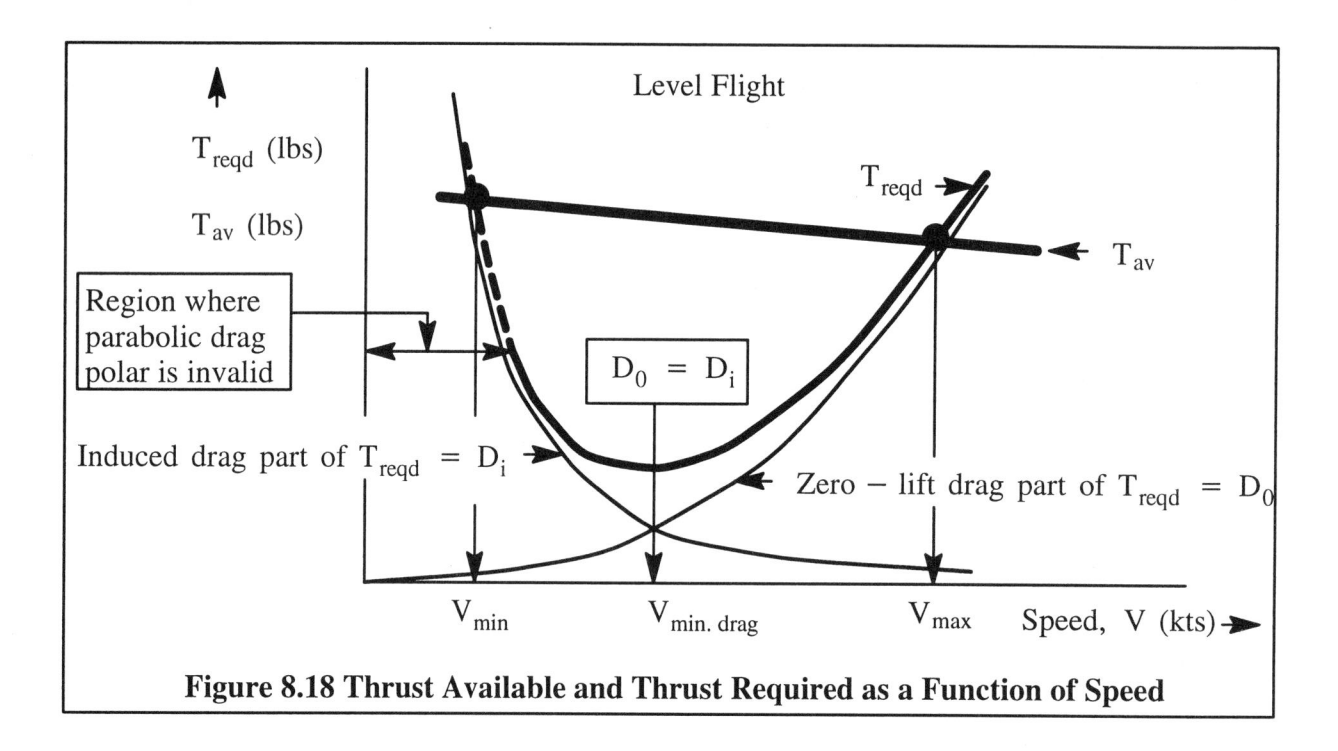

Figure 8.18 Thrust Available and Thrust Required as a Function of Speed

Note that for a given level of available thrust, T_{av}, there are two solutions to the level flight speed: a lower one, called V_{min} and a higher one, called V_{max}. In Reference 8.1, Chapter 5, it is shown that an airplane exhibits speed stability at V_{max} but is speed–unstable at V_{min}. At the latter speed the airplane is said to be flying at "the back of the thrust required curve". In both cases, the angle of attack follows from the lift coefficient of Eqn (8.58) and the lift–curve, C_L vs α. If the latter is linear it can be expressed as:

$$C_L = C_{L_0} + C_{L_\alpha} \alpha \tag{8.62}$$

where: C_{L_0} is the lift coefficient for zero angle of attack

C_{L_α} is the airplane lift–curve slope

Methods for computing C_{L_0} and C_{L_α} are found in Reference 8.3.

Combining Eqns (8.62) and (8.58) yields the following solution for α:

$$\alpha = \frac{C_L - C_{L_0}}{C_{L_\alpha}} = \frac{\frac{W}{\bar{q}S} - C_{L_0}}{C_{L_\alpha}} \qquad (8.63)$$

where: $\bar{q}$ represents the dynamic pressure at any speed where the thrust available equals the thrust required.

Note in Figure 8.18 that there is a speed for which the thrust required (or drag) is a minimum. The condition for minimum thrust required (or minimum drag) occurs with the airplane flying at its maximum lift–to–drag ratio, as may be seen from combining Eqns (8.57) and (8.58):

$$T_{reqd} = D = \frac{W}{L/D} = \frac{W}{C_L/C_D} \qquad (8.64)$$

This minimum drag condition can be investigated further by imposing the condition:

$$\frac{\partial T_{reqd}}{\partial V} = \frac{\partial D}{\partial V} = 0 \qquad (8.65)$$

By combining this condition with Eqn (9.59) it is found that:

$$V_{min.\ drag} = V_{min.thrust\ reqd} = \sqrt{\frac{\left(\frac{2}{\varrho}\frac{W}{S}\right)}{\sqrt{C_{D_0}\pi Ae}}} \qquad (8.66)$$

Substitution of this result into Eqn (8.59) it can be shown that at the speed for minimum drag:

$$D_0 = D_i = W\sqrt{\frac{C_{D_0}}{\pi Ae}} \qquad (8.67)$$

and therefore:

$$D_{min} = 2D_0 = 2W\sqrt{\frac{C_{D_0}}{\pi Ae}} = 2\frac{W}{b}\sqrt{\frac{f}{\pi e}} \qquad (8.68)$$

The thrust available is normally controlled by the pilot (or by the auto–pilot) through the engine throttles. The amount of thrust that can be commanded depends on the engine ratings and on the flight condition. Typically, the situation depicted in Figure 8.19 prevails. By adjusting the throttle a pilot can adjust the level of available thrust within some range. Therefore, any level flight speed between V_{min} and V_{max} can be obtained.

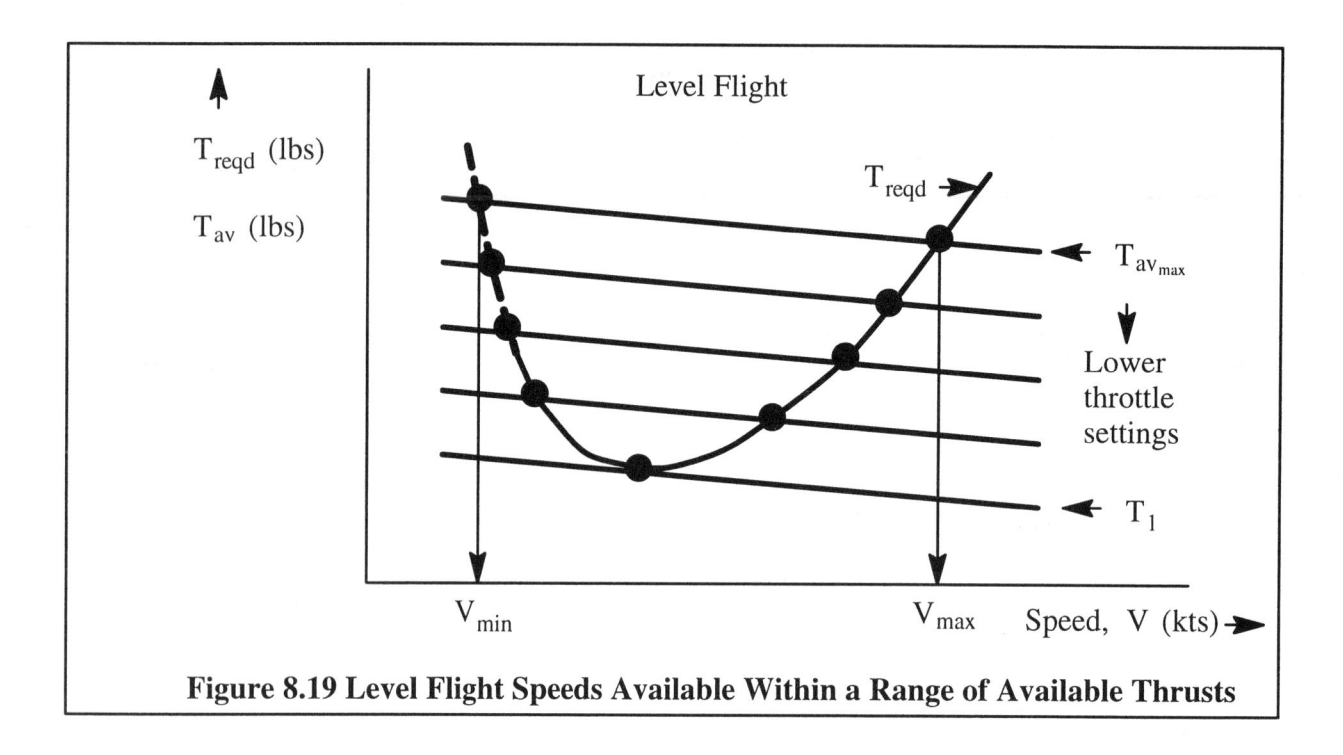

Figure 8.19 Level Flight Speeds Available Within a Range of Available Thrusts

For thrust settings below the line labelled, T_1 (= where thrust available is tangent to the thrust required line) level flight cannot be maintained and the airplane will start to drift down (i.e. glide with some thrust). The drift–down performance of airplanes is important in determining the altitude at which safe level flight can be maintained with one or more engines failed.

For ETOPS (Extended range Twin engine Operations) approval a detailed definition of the drift–down capabilities of twin passenger jets is required. Drift–down performance is discussed in detail in Chapter 9.

2) General drag polar and C_L vs α curve

Consider an early fighter airplane configuration with the general drag characteristics given in Figure 8.20. The insert in Figure 8.20 also gives the Mach number dependent lift coefficient versus angle of attack relationship.

Assume that the question is to find the level flight speeds at sea–level for an airplane weight of W=18,302 lbs, a wing–loading of W/S=50.2 lbs/ft² and an available thrust of 15,000 lbs without afterburner and 22,000 lbs with afterburner (in reality, thrust will vary with Mach number).

One procedure for finding the solution of this problem is to tabulate a range of speeds (Mach numbers) and use Eqn (8.58) to calculate the level flight lift coefficient. That information is used with Figure 8.20 to determine the drag coefficient and the angle of attack. Eqn (8.57) is then used to calculate the required thrust. Table 8.4 shows the results.

By plotting the results as shown in Figure 8.21 it is seen that the solutions for equilibrium speeds and angle of attack are:

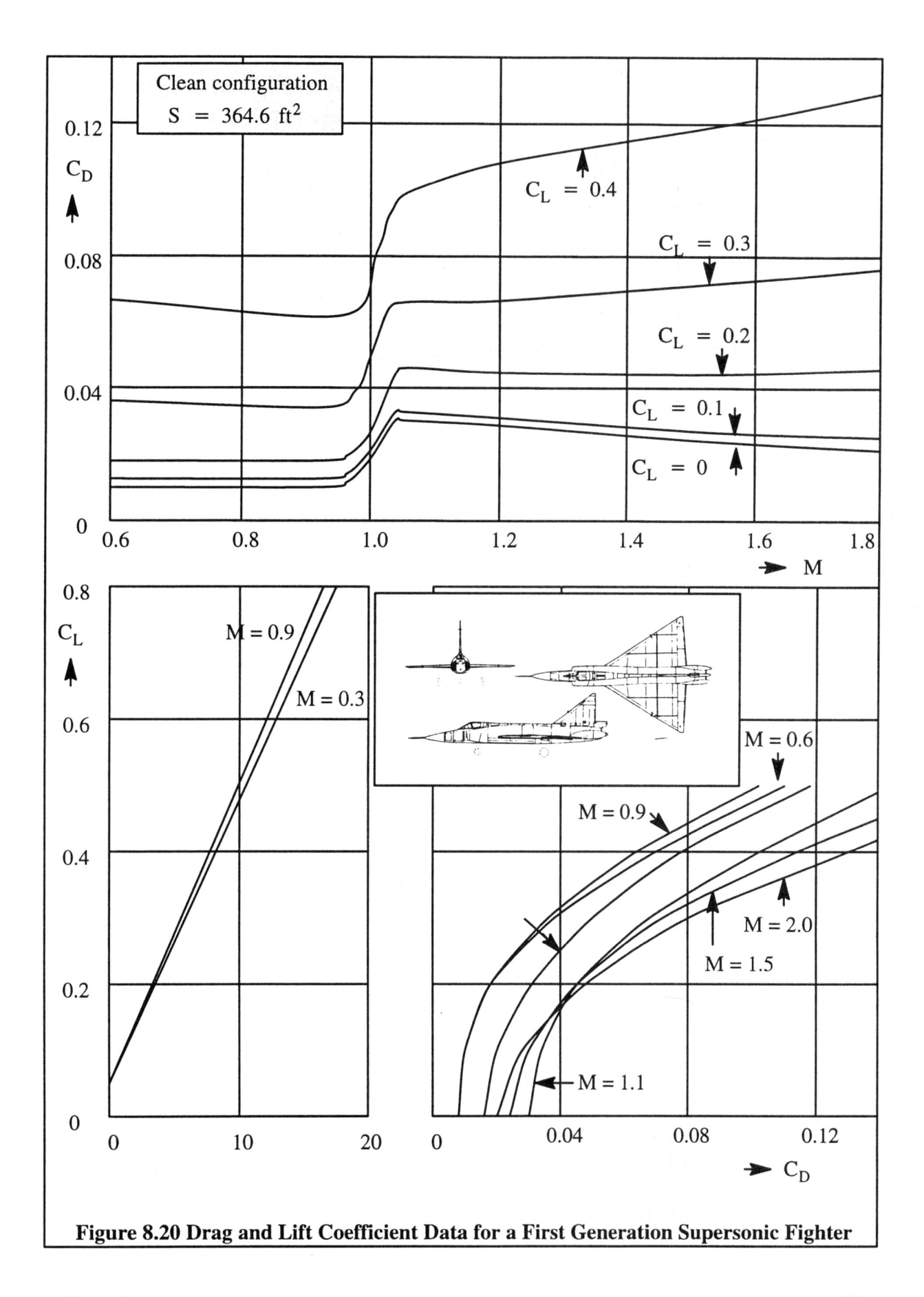

Figure 8.20 Drag and Lift Coefficient Data for a First Generation Supersonic Fighter

Fundamentals of Flight Performance

Table 8.4 Calculation of Required Thrust and Angle of Attack for the Fighter of Figure 8.20

M	V ft/sec	$\bar{q}$ lbs/ft²	C_L	α deg	C_D	D = T_{reqd} lbs
0.22	245.6	71.7	0.70	15.5	0.216	5,647
0.3	335.0	133.4	0.376	7.0	0.0595	2,894
0.4	446.6	237.1	0.212	3.5	0.0203	1,755
0.6	669.9	533.6	0.094	1.5	0.0103	2,004
0.8	893.2	948.6	0.0529	0	0.0088	3,044
0.9	1,004.9	1,200.7	0.0418	−0.3	0.0086	3,765
1.0	1,116.5	1,482.2	0.0339	−0.5	0.0172	9,295
1.1	1,228.2	1,793.6	0.0280	−0.7	0.0300	19,618
1.2	1,339.8	2,134.3	0.0235	−0.9	0.0280	21,789
1.4	1,563.1	2,905.1	0.0173	−1.1	0.0255	27,010
1.6	1,786.4	3,794.4	0.0132	−1.3	0.0240	33,203
1.8	2,009.7	4,802.2	0.0104	−1.5	0.0220	38,519

S = 364.6 ft²   Sea–level   W = 18,302 lbs

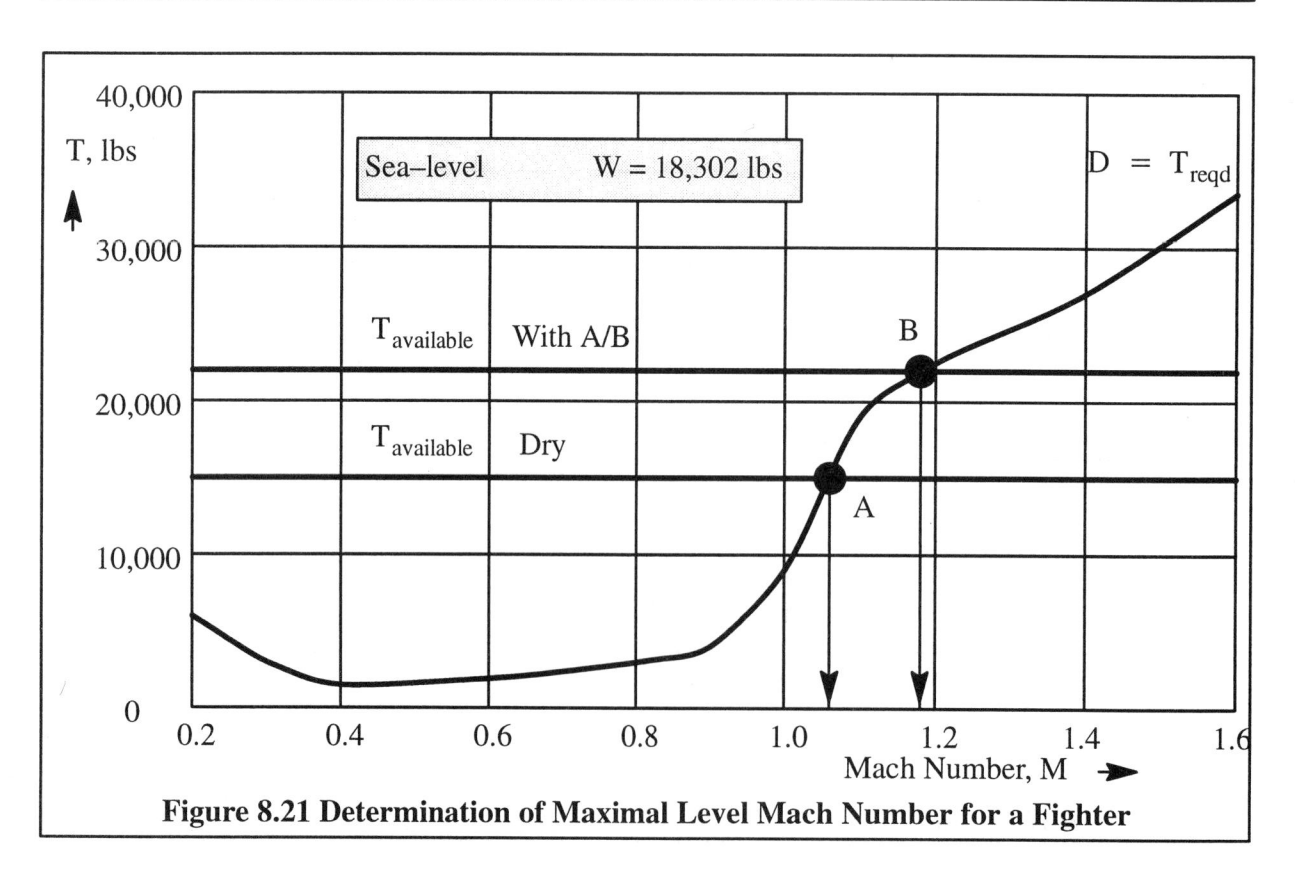

Figure 8.21 Determination of Maximal Level Mach Number for a Fighter

362   Chapter 8

At Point A, M = 1.06   At Point B, M = 1.18

The reader will observe that the thrust required curve is strongly influenced by the behavior of the drag polars as a function of Mach number. Note also, that for this particular airplane, because of the steep drag–rise, the afterburner does not significantly increase the maximum level speed attainable.

8.4.2 PROPELLER DRIVEN AIRPLANES

This case will be considered first, for a parabolic drag polar and a linear relationship between C_L and α and second for a general drag polar and $C_L - \alpha$ curve.

1) Parabolic drag polar and linear C_L vs α

Using a parabolic drag polar the power required for level speed can be written as:

$$P_{reqd} = DV = C_D \bar{q} S V = \left(C_{D_0} + \frac{C_L^2}{\pi A e}\right) \bar{q} S V = C_{D_0} \frac{1}{2} \rho V^3 S + \frac{W^2}{\pi A e \frac{1}{2} \rho V S} = P_{reqd_0} + P_{reqd_i} \tag{8.69}$$

where: P_{reqd_0} is the power required to overcome the zero–lift drag

P_{reqd_i} is the power required to overcome the induced drag

By using $C_{D_0} = \frac{f}{S}$ and $A = b^2/S$ it is possible to rewrite Eqn (8.69) as:

$$P_{reqd} = f \frac{1}{2} \rho V^3 + \frac{2(W/b)^2}{\pi e \rho V} = P_{reqd_0} + P_{reqd_i} \tag{8.70}$$

This result has several important implications:

1) The power required for level flight consists of two terms: one proportional to the third power of air–speed, the other proportional to the inverse of air–speed.

2) To minimize the first term (zero lift drag term) it is important to design the airplane with as small a parasite area as possible. A discussion of parasite area, f, is given in Chapter 5.

3) To minimize the second term (induced power term) the span loading, W/b, must be made as small as possible.

Figure 8.22 shows a graphical interpretation of these results.

Note that for a given level of available power, P_{av}, there are two solutions to the level flight speed: a lower one, called V_{min}, and a higher one, called V_{max}. The angle of attack at each level flight speed follows from the lift coefficient of Eqn (8.58) and the lift–curve, C_L vs α. If the latter is linear the solution for angle of attack is as given by Eqn (8.63).

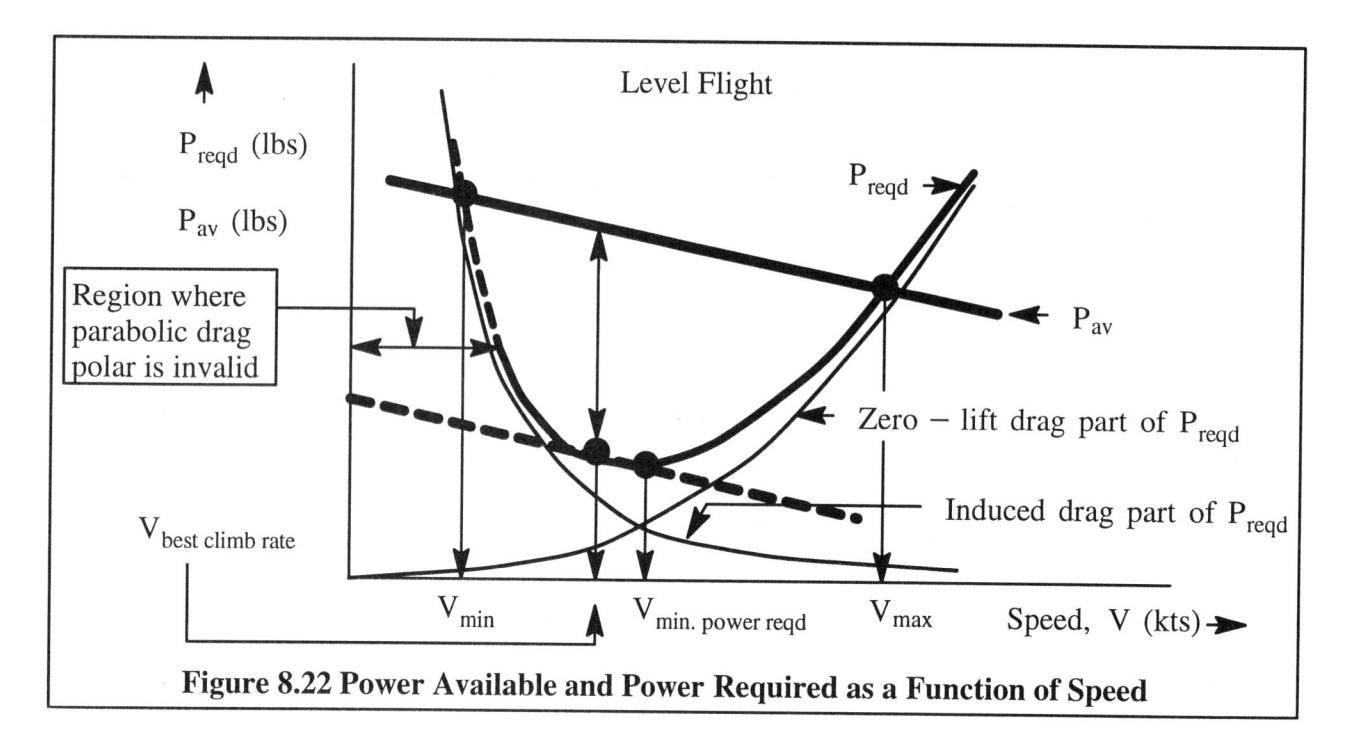

Figure 8.22 Power Available and Power Required as a Function of Speed

The power required for level flight, following Eqn (8.55) may also be written as:

$$P_{reqd} = DV = \frac{WV}{C_L/C_D} = W\sqrt{\frac{W}{S}\frac{2}{\varrho}\frac{C_D^2}{C_L^3}} \quad (8.71)$$

The minimum power required to fly at level speed therefore occurs when C_L^3/C_D^2 is a maximum. This minimum required power condition can be investigated further by imposing the condition:

$$\frac{\partial P_{reqd}}{\partial V} = 0 \quad (8.72)$$

With a parabolic drag polar it can be shown that the speed at which this minimum occurs is:

$$V_{min.\ power\ reqd} = \sqrt{\frac{W}{S}\frac{2}{\varrho}\frac{1}{\sqrt{3C_{D_0}\pi Ae}}} \quad (8.73)$$

By substituting this result back into Eqn (8.71) it is found that:

$$P_{reqd\ minimum} = \frac{4}{3}W\sqrt{\frac{W}{S}\frac{2}{\varrho}\sqrt{\frac{3C_{D_0}}{(\pi Ae)^3}}} \quad (8.74)$$

By using $C_{D_0} = \frac{f}{S}$ and $A = b^2/S$ it is possible to rewrite Eqn (8.74) as:

$$P_{\text{reqd minimum}} = \frac{4}{3}\sqrt{\left(\frac{W}{b}\right)^3 \frac{2}{\varrho}} \sqrt{\frac{f}{(\pi e)^3}} \tag{8.74}$$

This equation again illustrates the importance of low span loading and low equivalent parasite area in minimizing power required for level flight.

It is of interest to compare the speeds for minimum drag, Eqn (8.66), with that for minimum power, Eqn (8.73). The result is:

$$\frac{V_{\text{min. thrust reqd}}}{V_{\text{min. power reqd}}} = 3^{1/4} = 1.32 \tag{8.75}$$

It is also of interest to compare the minimum power required for level flight with the power required for minimum drag (thrust required). The power required for minimum drag is found by combining Eqn (8.66) with Eqn (8.68). The result is:

$$P_{\text{reqd minimum drag}} = 2W \sqrt{\frac{W}{S}\frac{2}{\varrho}} \sqrt{\frac{C_{D_0}}{(\pi A e)^3}} \tag{8.76}$$

The ratio of the two powers required is found by dividing Eqn (8.76) by Eqn (8.74) to yield:

$$\frac{P_{\text{reqd. min. drag}}}{P_{\text{min. reqd}}} = \left(\frac{1}{2}\right) 27^{1/4} = 1.14 \tag{8.77}$$

The following conclusions are presented for airplanes with parabolic drag polars:

1) The speed for minimum thrust required (or speed for minimum drag) is 32% greater than the speed for minimum power required for level flight.

2) The power required for minimum drag is 14% greater than the minimum power required for level flight.

2) General drag polar and C_L vs α curve

Consider an early turbo–propeller driven airplane configuration with the general drag characteristics given in Figure 8.23. The lower graph in Figure 8.23 also gives the lift coefficient versus angle of attack relationship.

Assume that the question is to find the maximum and minimum level flight speeds at sea–level for an airplane weight of W = 36,000 lbs, a wing–loading of W/S = 44.6 lbs/ft² and a total available thrust–horsepower of 2,100 hp in the speed range of 200–300 ft/sec. Table 8.5 shows the calculation results in terms of the drag (i.e. thrust required) and power required.

Fundamentals of Flight Performance

$S = 807$ ft^2 $A = 12$

$C_{L_{max}} = 1.45$

Figure 8.23 Drag and Lift Coefficient Versus Angle of Attack Data for a Twin Turboprop Regional Transport

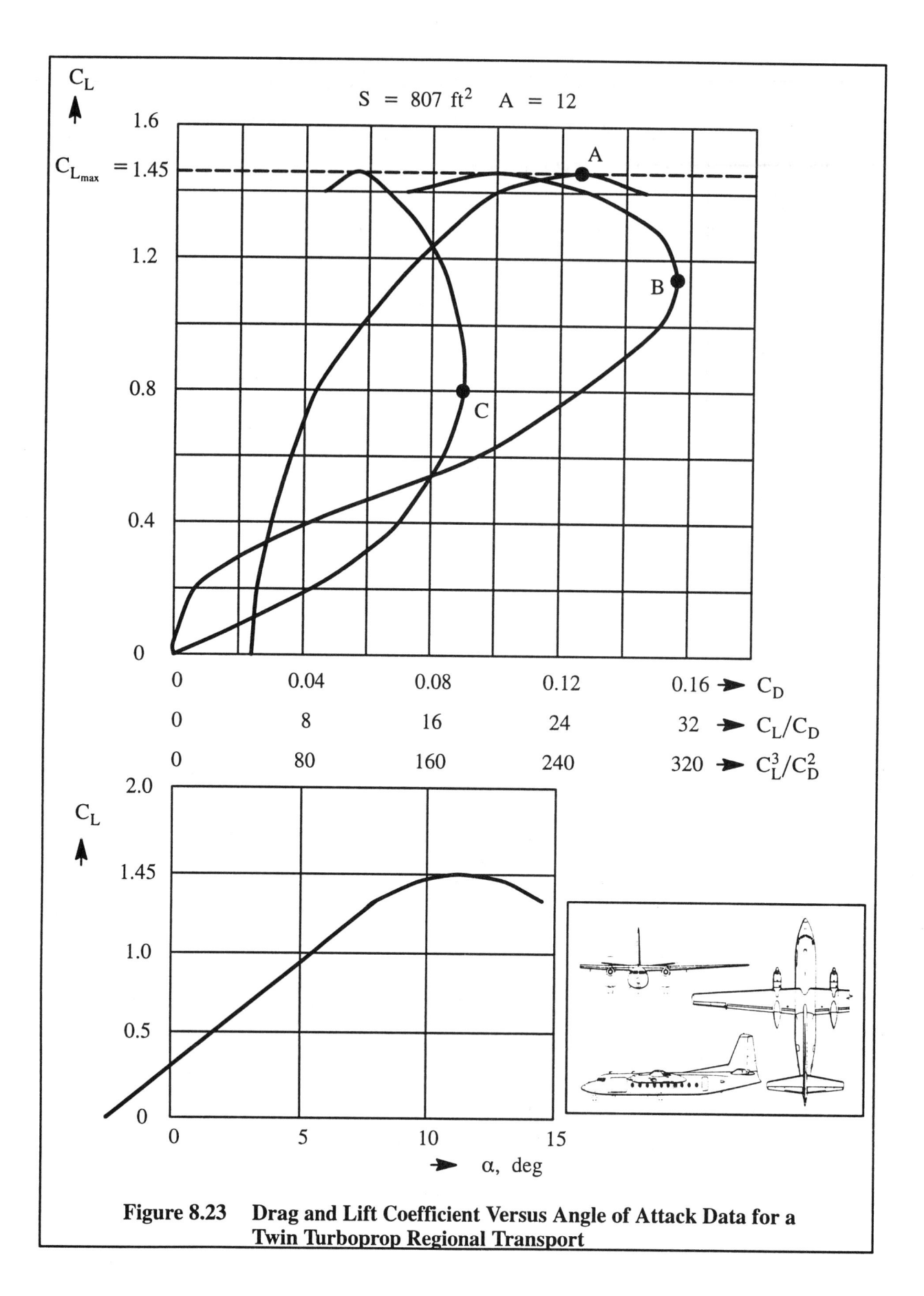

Table 8.5 Calculation of Drag and Power Required for the Transport of Figure 8.23
W = 36,000lbs   W/S = 44.6 lbs/ft²   Sealevel

V, ft/sec	V, kts	C_L	C_D Fig. 8.23	C_L/C_D Fig. 8.23	α, deg Fig. 8.23	D, lbs	P_{reqd}, hp
161	95.3	1.45	0.125	11.6	11.5	3,103	908
170	100.7	1.30	0.085	15.3	7.5	2,353	727
180	106.6	1.16	0.070	16.6	6.5	2,169	710
200	118.5	0.94	0.053	17.7	5.1	2,034	740
250	148.1	0.60	0.036	16.7	2.4	2,156	980
300	177.7	0.42	0.032	13.1	1.2	2,748	1,499
350	207.3	0.31	0.027	11.5	0.1	3,130	1,992
400	237.0	0.23	0.025	9.2	−0.7	3,913	2,846

The results of Table 8.5 are plotted in Figure 8.24. Note that at the stall (minimum air–speed), the airplane still has adequate horsepower to climb! The maximum speed which can be achieved at sea–level is 355 ft/sec, or 210 kts.

It is of interest to study points A,B and C on the drag characteristics plots in Figure 8.23 in comparison to the similarly labeled points in Figure 8.24.

First, consider Points A, B and C in Figure 8.23:

Point A represents the maximum lift coefficient, $C_{L_{max}}$.

Point B is the point where C_L^3/C_D^2 is maximum.

At Point C, C_L/C_D has a maximum.

Next, consider Points A, B and C in Figure 8.24:

Point A represents the minimum speed, $V_{minimum}$, which can be achieved. Note that the amount of power required to fly level at the minimum speed is less than what is available at the maximum lift coefficient. This is definitely not always the case.

Point B is the point where the power required for level flight is a minimum. Evidently this occurs when C_L^3/C_D^2 is at its maximum.

Point C is the point where the thrust required to maintain level flight (or the drag) is a minimum. Evidently this occurs when C_L/C_D has a maximum. Observe, that the tangent from the origin to the power required curve has a minimum also at point C.

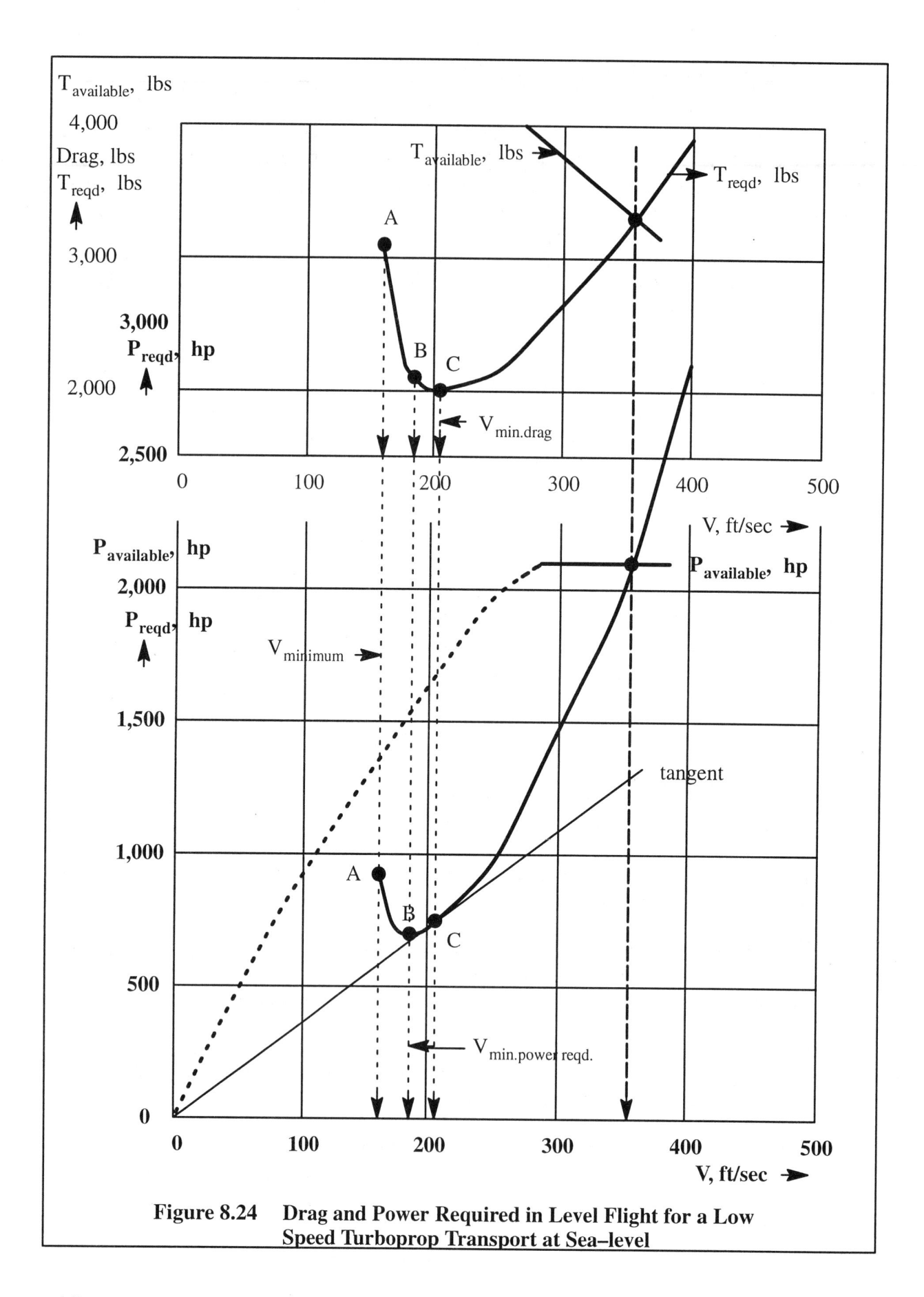

Figure 8.24 Drag and Power Required in Level Flight for a Low Speed Turboprop Transport at Sea-level

8.5 SUMMARY FOR CHAPTER 8

In this chapter the fundamentals required to determine the steady, level flight performance characteristics of airplanes have been developed for airplanes with parabolic drag polars as well as for the more general case of arbitrary drag polars. Theory and applications were presented separately for jet powered and for propeller driven airplanes.

Most of the fundamental issues raised in this chapter will be seen to re–occur in the discussion of airplane performance characteristics in other flight situations, particularly those covered in Chapters 9 and 10.

8.6 PROBLEMS FOR CHAPTER 8

8.1 Verify step–by–step that Equations (8.25) and (8.26) are correct.

8.2 Verify step–by–step that Equations (8.29), (8.30) and (8.31) are correct.

8.3 The drag polar characteristics of a high altitude reconnaissance airplane and a jet trainer are given in Figure 8.25. Calculate and plot their glide characteristics in a manner similar to what was done in Table 8.1 for Figures 8.4 and 8.5. Compare the:

1) Minimum glide path angle, the speed for minimum glide path angle, the rate of descent and the horizontal distance covered in a steady glide

2) The speed for minimum rate of descent, the glide path angle, the rate of descent and the duration during the steady glide

Assume that the altitude is a constant 5,000 ft in the standard atmosphere. The wing loading of airplane a) is 25.7 lbs/ft^2 and that of airplane b) is 59.1 lbs/ft^2.

8.4 Repeat problem 8.3 for the case where the starting altitude is 10,000 ft instead of 5,000 ft and still keep the altitude constant.

8.5 Repeat problem 8.4 for the case where the effect of varying altitude must be accounted for.

8.6 The drag polar of a light twin airplane in its clean configuration is:

$$C_D = 0.0358 + 0.0405\, C_L^2$$

The weight of the airplane is 4,200 lbs and its wing area is 155 ft^2. Using analytical procedures, calculate the following under standard sea–level conditions:

1) the maximum lift–to–drag ratio   2) the speed for minimum drag
3) the minimum power required   4) the speed for minimum power required

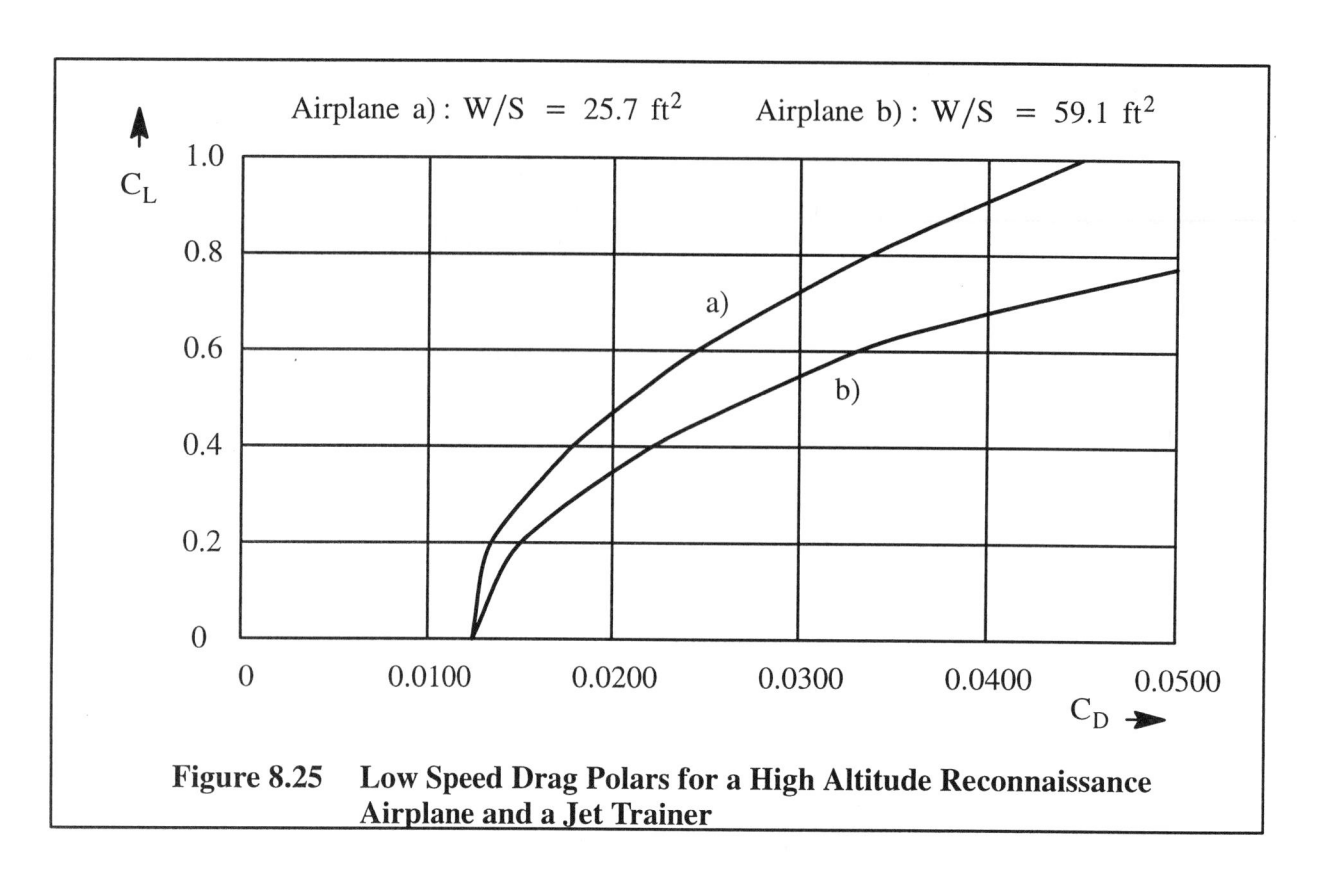

Airplane a): W/S = 25.7 ft²   Airplane b): W/S = 59.1 ft²

Figure 8.25   Low Speed Drag Polars for a High Altitude Reconnaissance Airplane and a Jet Trainer

8.7   A glider weighs 800 lbs and has a wing loading of 12 lbs/ft². Its drag polar is:

$$C_D = 0.0100 + 0.0220\, C_L^2$$

Assume that the glider is launched at 1,500 ft in still air and over level ground. Assume standard atmospheric conditions. Calculate the following:

a) the greatest distance it can cover along the ground

b) the speed for the greatest distance it can cover along the ground

c) the longest duration it can stay airborne

d) the speed for the longest duration it can stay airborne

Assume that the effect of changing density can be neglected.

8.8)   Verify that Eqns (8.66) and (8.73) are correct.

8.9)   The maximum, trimmed lift coefficient of the airplane in problem 8.6 with the flaps down is found to be 3.0. When approaching to land with power off, the airplane is flown at 15% above the stall speed. Calculate the increment in zero lift drag coefficient, (C_{D_o}), which must be provided by a combination of flaps and speed brakes if a glide path angle of 20 degrees is desired. What is the rate of descent? Assume standard, sea–level conditions.

8.10 For airplane a) in problem 8.3, assume that its zero lift drag coefficient, (C_{D_o}), is increased by 0.0500 by extending speed brakes when flying at 100 kts at 1,000 ft altitude.

Calculate the glide path angle and the rate of descent. Use standard atmospheric conditions.

8.11 The lift–drag characteristics of a Cessna Cardinal type airplane are given in Figure 8.26.

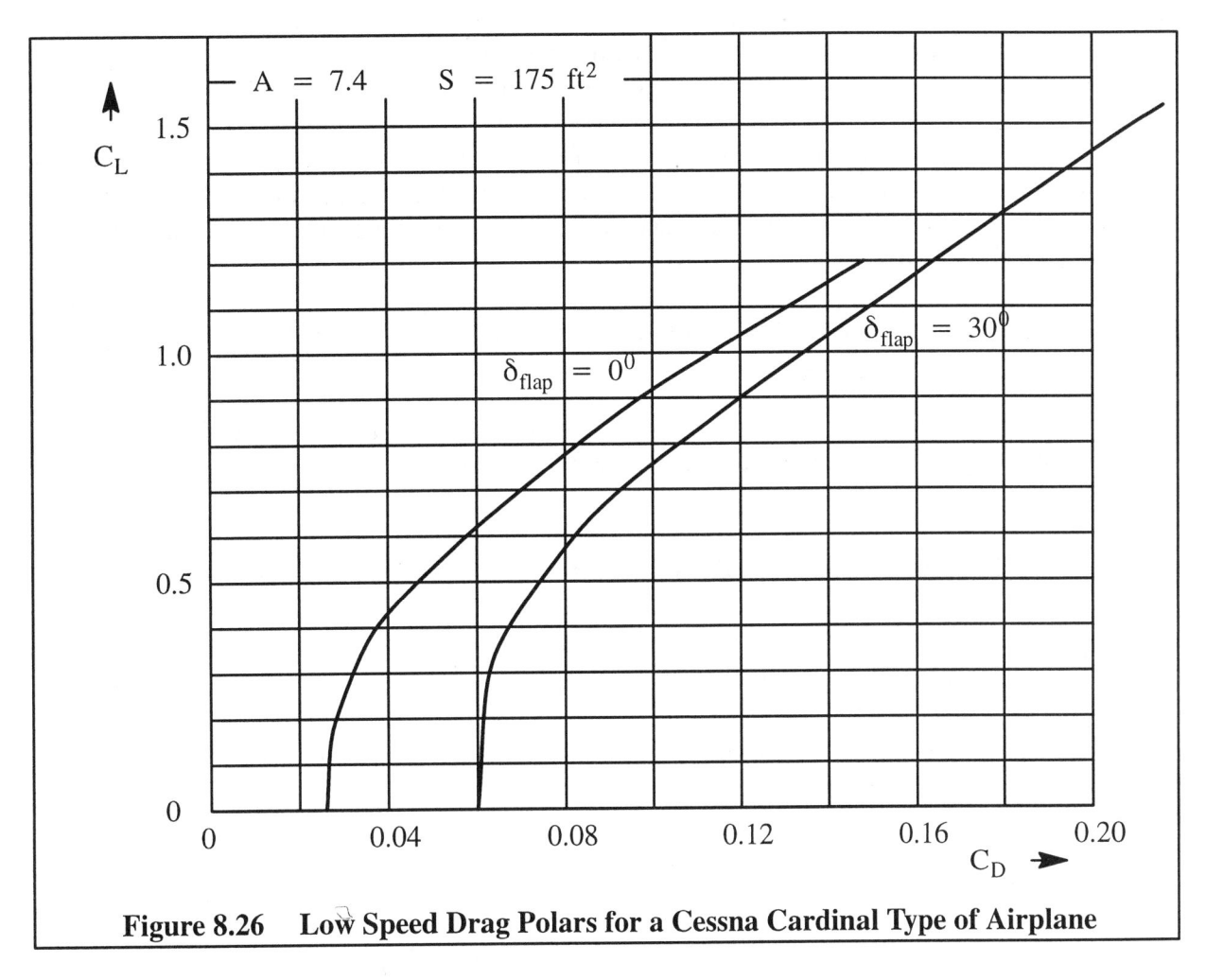

Figure 8.26 Low Speed Drag Polars for a Cessna Cardinal Type of Airplane

Assume that at the maximum level speed the available thrust horsepower is 150 hp. Calculate and plot the thrust and the power required curves. Find the following:

a) the minimum flight speed

b) the speed for minimum drag

c) the speed for minimum power required

d) the maximum level flight speed

Assume a weight of 2,500 lbs and standard sea–level conditions. Work this problem for zero degrees flap deflection and for 30 degrees flap deflection.

8.12 A jet powered airplane has the following drag polar:

$$C_D = 0.0200 + 0.045 \, C_L^2$$

The wing area is S = 200 ft². The available thrust is 3,500 lbs at 5,000 ft in a standard atmosphere.

   a) Using analytical procedures only, find the maximum level speed.
   b) Find the maximum level speed using a plot of thrust available and thrust required versus speed. Check the answer with a).
   c) What is the lift–to–drag ratio of the airplane at its maximum speed?

8.13 Assume that the airplane of problem 8.12 is equipped with a rocket engine capable of delivering a maximum thrust of 6,000 lbs at 60,000 ft. The incremental drag coefficient due to compressibility may be assumed to be 0.0020. Calculate the maximum speed in level flight. What is the corresponding Mach number? What is the lift–to–drag ratio?

8.14 An airplane with a weight of W lbs and a wing area of S ft² has a drag polar given by:

$$C_D = C_{D_0} + C_L^2/\pi Ae$$

The maximum available thrust is T which may be considered independent of speed at a given altitude. Derive an expression for the maximum level flight speed at that given altitude.

8.7 REFERENCES FOR CHAPTER 8

8.1 Roskam, J.; Airplane Flight Dynamics and Automatic Flight Controls, Part I; Design, Analysis and Research Corporation, 120 East 9th Street, Suite 2, Lawrence, KS 66044; 1995.

8.2 Wittenberg, H.; Notes on Flight Mechanics (in Dutch), Delft University of Technology, Delft, The Netherlands, 1972.

8.3 Roskam, J.; Airplane Design, Part VI, Preliminary Calculation of Aerodynamic, Thrust and Power Characteristics; Design, Analysis and Research Corporation, 120 East 9th Street, Suite 2, Lawrence, KS 66044; 1989.

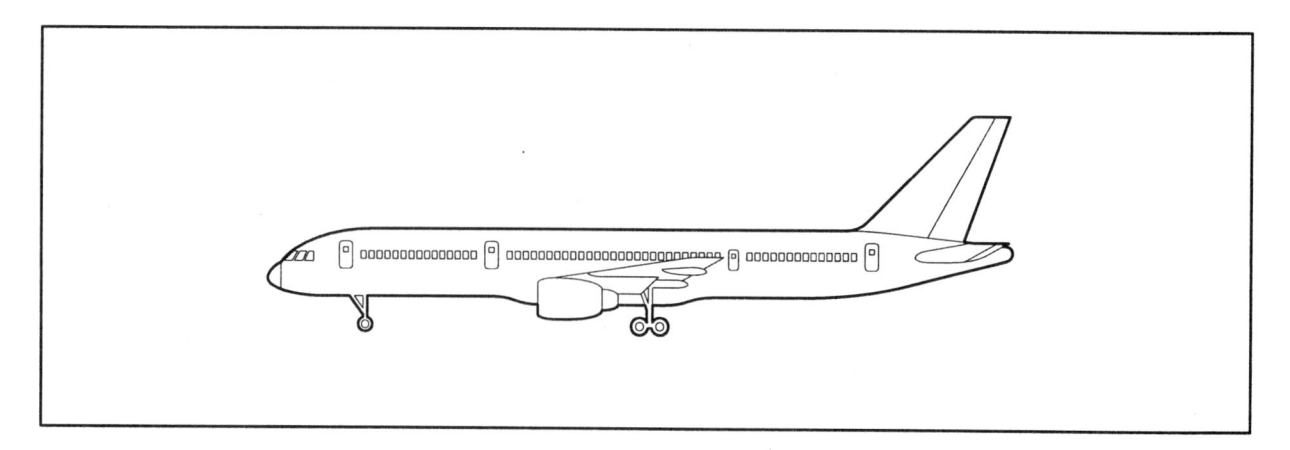

CHAPTER 9: CLIMB AND DRIFT–DOWN PERFORMANCE

In this chapter methods for determining the climb and drift–down performance of airplanes are presented. Climb and drift–down performance is of great importance in assuring:

1) Rapid climb to cruise or combat altitudes for civil and military airplanes respectively
2) Rapid climb–outs away from obstacles for agricultural and patrol airplanes
3) Capability to climb with one engine out during take–off and landing flight phases
4) Drift–down performance following an engine shut–down at cruise altitude and the ability to clear mountain ranges in such conditions
5) Climb reserves at the operational ceiling of commercial airplanes and at the combat ceiling of military airplanes

The equations of motion for a general climb situation are developed in Section 9.1. The general climb performance characteristics of jet and propeller driven airplanes are presented in Sections 9.2 and 9.3 respectively for shallow flight path angles.

Most climbs are indeed conducted at relatively shallow flight path angles: 15 degrees or less. At shallow flight path angles certain simplifications in the equations of motion are acceptable. That is no longer the case for flight path angles above (roughly) 15 degrees. Sections 9.2 and 9.3 also deal with the general case of steep flight path angles.

Once methods for determining the climb and drift–down performance are available, times to climb to altitude, operational ceilings and drift–down times can be evaluated. Methods for doing that are presented in Section 9.4.

In certain types of military airplanes the effect of forward and vertical accelerations can no longer be neglected. Acceleration effects on climb performance are discussed in Section 9.5.

Landing gears, flaps, speed brakes, stopped engines, trim requirements, etc., all affect the climb capabilities of an airplane. Also, the weight and altitude of the airplane (wing–loading) affect climb performance. These factors are highlighted in Section 9.6.

The climb performance of airplanes is subjected to airworthiness requirements which stipulate minimum required climb rates and/or climb gradients (angles). These requirements establish adequate levels of safety with all engines operating (AEO) as well as with one engine inoperative (OEI). The climb performance requirements of airplanes (regulations) are summarized in Section 9.7.

9.1 EQUATIONS OF MOTION

It will be assumed that the airplane center of gravity moves in a plane, vertical to the surface of the earth. It is also assumed that the airplane is in complete moment trim: no net rolling, pitching or yawing moments exist. Furthermore, no initial angular rates exist. The only accelerations allowed for are those along the flight path and perpendicular to the flight path. For a detailed treatment of the airplane equations of motion, see Chapter 1 of Reference 9.1.

Figure 9.1 shows the forces which act on the airplane in an accelerated climb condition.

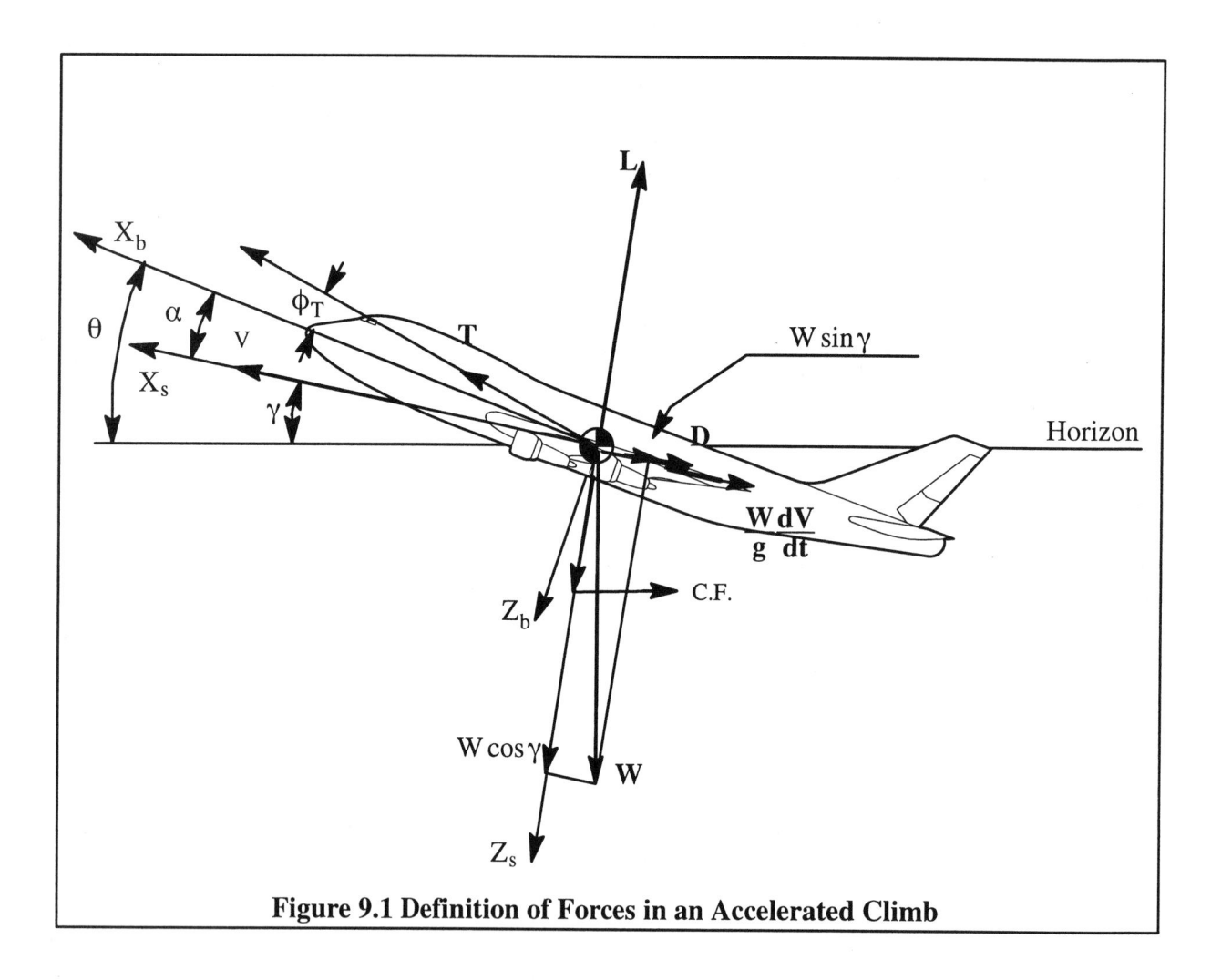

Figure 9.1 Definition of Forces in an Accelerated Climb

The equations of motion along the flight path and perpendicular to the flight path are:

$$T\cos(\alpha + \phi_T) - D - W\sin\gamma = \frac{W}{g}\frac{dV}{dt} \tag{9.1}$$

$$T\sin(\alpha + \phi_T) + L - W\cos\gamma = \text{C.F.} = \frac{W}{g}V\dot\gamma \tag{9.2}$$

The general case of accelerated flight performance will be taken up in Chapter 12. Most climbs are conducted with a straight line flight path. In that case the centrifugal acceleration is not present and the equations reduce to:

$$T\cos(\alpha + \phi_T) - D - W\sin\gamma = \frac{W}{g}\frac{dV}{dt} \tag{9.3}$$

and

$$T\sin(\alpha + \phi_T) + L - W\cos\gamma = 0 \tag{9.4}$$

At this point the additional assumption of small angles is made: the thrust inclination angle, ϕ_T, the angle of attack, α and the flight path angle, γ, are all assumed to be sufficiently small so that:

$$\sin(\alpha + \phi_T) \approx 0 \quad \cos(\alpha + \phi_T) \approx 1.0 \quad \sin\gamma \approx \gamma \quad \cos\gamma \approx 0 \tag{9.5}$$

With these additional assumptions, Eqn (9.3) and (9.4) become:

$$T - D - \frac{W}{g}\frac{dV}{dt} = W\sin\gamma \tag{9.6}$$

and

$$L \approx W \tag{9.7}$$

The rate–of–climb, R.C. of the airplane is defined as:

$$\text{R.C.} = \frac{dh}{dt} = V\sin\gamma \approx V\gamma \tag{9.8}$$

With the help of Eqn (9.6) the R.C. can be written as:

$$\text{R.C.} = \frac{(T-D)V}{W} = \frac{V}{g}\frac{dV}{dh}\frac{dh}{dt} = \frac{V}{g}\frac{dV}{dh}\text{R.C.} \tag{9.9}$$

Solving for the rate–of–climb:

$$\text{R.C.} = \frac{\frac{(T-D)V}{W}}{1 + \frac{V}{g}\frac{dV}{dh}} \tag{9.10}$$

The term $(V/g)(dV/dh)$ is called the acceleration factor. The effect of this acceleration factor on climb performance will be discussed in Section 9.5. As long as the climb takes place at a constant true air–speed, V, the following will hold: $dV/dh = 0$. In that case:

$$\text{R.C.} = \frac{(T - D)V}{W} = \frac{(T_{av} - T_{reqd})V}{W} \qquad (9.11)$$

The reader should compare this result with Eqn (8.54). Equation (9.11) applies directly to jet airplanes where the engine output is considered in terms of thrust. In the case of propeller driven airplanes the engine output is considered in terms of power. Therefore, for propeller driven airplanes Eqn (9.11) is written as:

$$\text{R.C.} = \frac{(T - D)V}{W} = \frac{(P_{av} - P_{reqd})}{W} \qquad (9.12)$$

The reader should compare this result with Eqn (8.55).

The climb performance of airplanes can also be defined in terms of the so-called climb gradient, C.G.R. Expressed as a corollary to Eqn (9.11) it is defined for jet airplanes as:

$$\text{C.G.R.} = \sin\gamma = \frac{\text{R.C.}}{V} = \frac{(T - D)}{W} = \frac{(T_{av} - T_{reqd})}{W} \qquad (9.11a)$$

Expressed as a corollary to Eqn (9.12) it is defined for propeller driven airplanes as:

$$\text{C.G.R.} = \sin\gamma = \frac{\text{R.C.}}{V} = \frac{(T - D)V}{WV} = \frac{(P_{av} - P_{reqd})}{WV} \qquad (9.12a)$$

The airworthiness requirements of FAR 23/25 (Ref. 9.2) and the military requirements of References 9.3 and 9.4 use R.C. and C.G.R. specifications, depending on the type of airplane and the flight condition. Section 9.7 contains a summary of these requirements.

Applications of Eqns (9.11) and (9.12) to jet airplanes and propeller driven airplanes are discussed in Sections 9.2 and 9.3 respectively.

Eqns (9.11) and (9.12) represent the case of an airplane in a straight line, constant speed climb. The forces which act on the airplane are shown in Figure 9.2. The reader should observe that the only difference between Figures 9.1 and 9.2 is the absence of the acceleration forces. The angles ϕ_T, α and γ are assumed to be sufficiently small so that condition (9.5) still applies.

When an airplane is in its design cruise flight condition and one engine becomes inoperative (OEI) the airplane will have to descend to a lower altitude where level flight on the remaining engine(s) again is possible. In the transition between AEO high altitude cruise and OEI lower altitude cruise the airplane is said to be in a drift-down, or power-on glide flight condition. Figure 9.3 shows the forces which act on the airplane in such a case. The equations of motion in a steady drift-down case are:

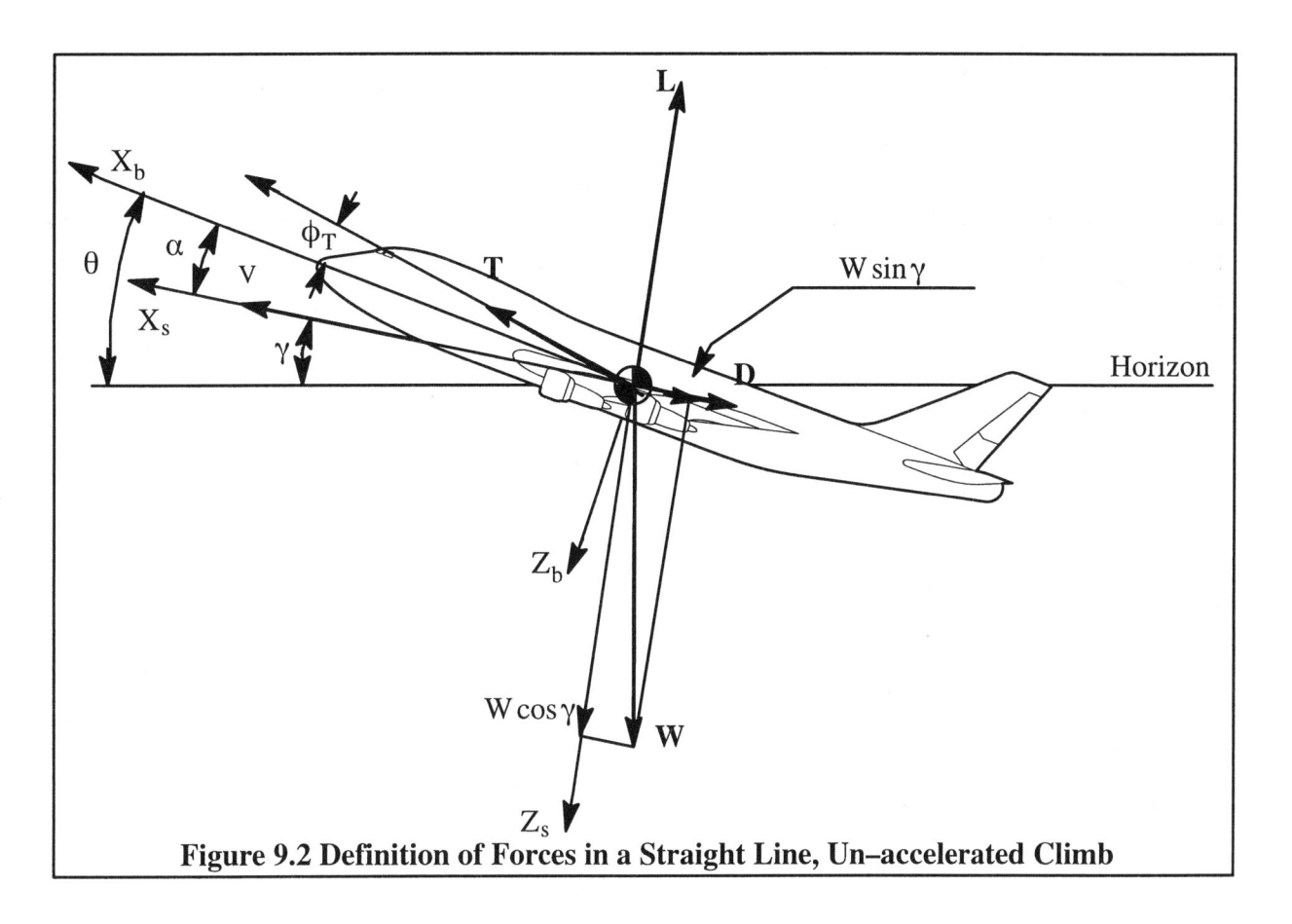

Climb and Drift-down Performance

Figure 9.2 Definition of Forces in a Straight Line, Un-accelerated Climb

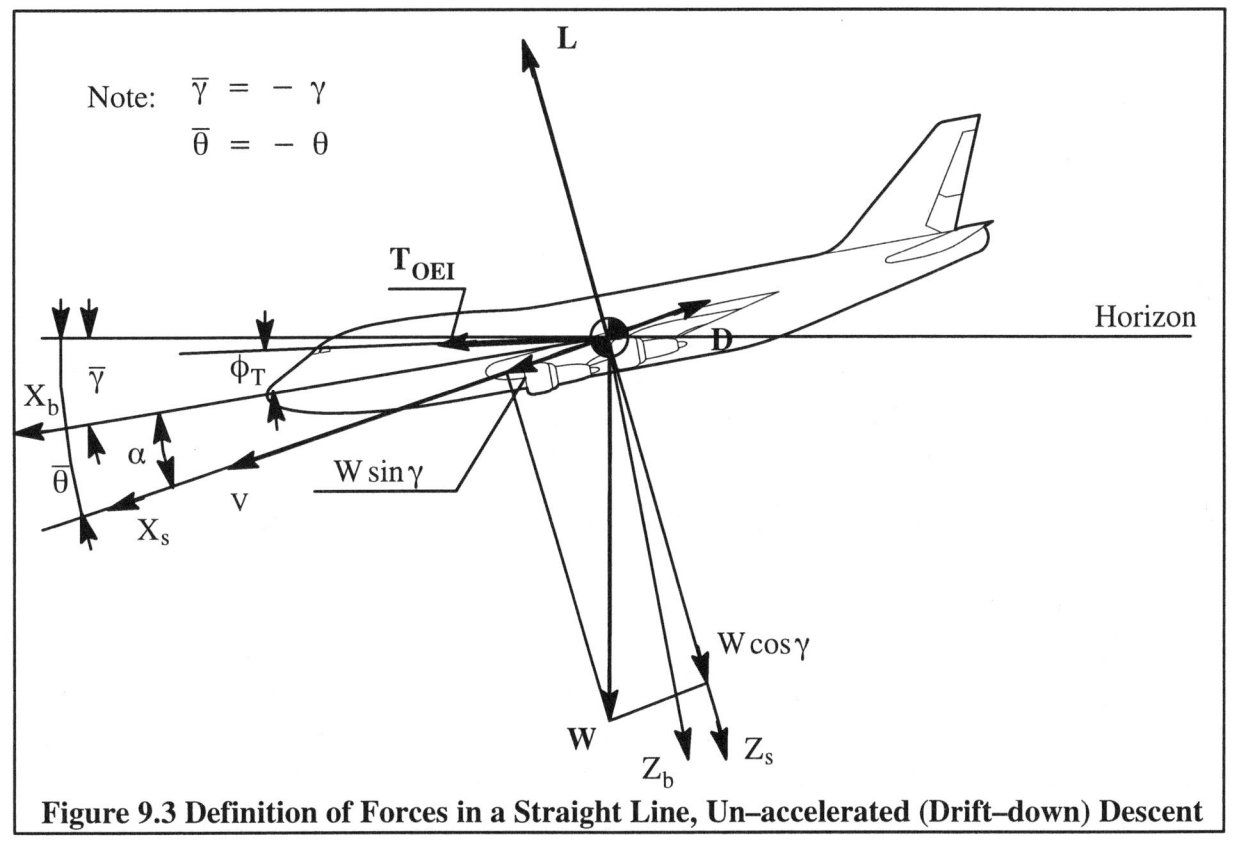

Note: $\bar{\gamma} = -\gamma$
$\bar{\theta} = -\theta$

Figure 9.3 Definition of Forces in a Straight Line, Un-accelerated (Drift-down) Descent

Chapter 9 377

$$T\cos(\alpha + \phi_T) - D + W\sin\overline{\gamma} = 0 \tag{9.13}$$

$$T\sin(\alpha + \phi_T) + L + W\cos\overline{\gamma} = 0 \tag{9.14}$$

For small angles, these equations may be re–written as:

$$T - D + W\overline{\gamma} = 0 \tag{9.15}$$

$$L = W \tag{9.16}$$

The rate of descent, R.D., as seen from Figure 9.3, is:

$$R.D. = -\frac{dh}{dt} = -V\sin\gamma \approx -V\gamma \tag{9.17}$$

With the help of Eqn (9.15) the R.D. can be written as:

$$R.D. = \frac{(D - T)V}{W} \tag{9.18}$$

For jet driven airplanes this yields:

$$R.D. = \frac{(D_{OEI} - T_{OEI})V}{W} = \frac{(T_{reqd_{OEI}} - T_{av_{OEI}})V}{W} \tag{9.19}$$

For propeller driven airplanes this yields:

$$R.D. = \frac{(D_{OEI} - T_{OEI})V}{W} = \frac{(P_{reqd_{OEI}} - P_{av_{OEI}})}{W} \tag{9.20}$$

Applications of Eqns (9.19) and (9.20) to jet airplanes and propeller driven airplanes are discussed in Sections 9.2 and 9.3 respectively.

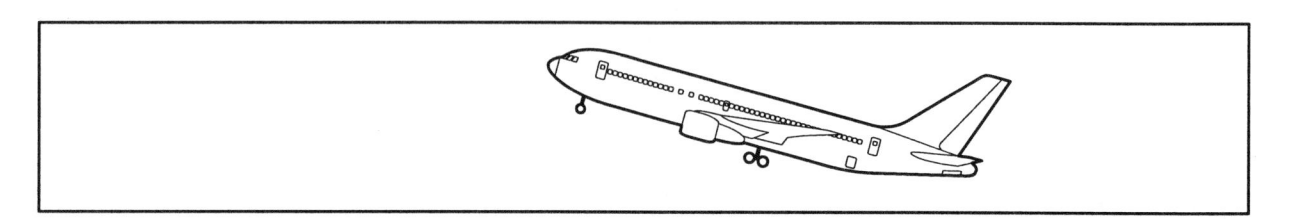

9.2 CLIMB AND DRIFT–DOWN PERFORMANCE OF JET AIRPLANES

For a given altitude, speed and engine power setting, a given jet engine will deliver a certain amount of installed thrust, T_{av}. Methods for determining the installed thrust from engine manufacturers un–installed thrust data are given in Ref. 9.5.

9.2.1 CLIMB PERFORMANCE FOR JET AIRPLANES

9.2.1.1 The General Case

The rate of climb, R.C. of a jet airplane follows from Eqn (9.11):

$$\text{R.C.} = \frac{(T - D)V}{W} = \frac{(T_{av} - T_{reqd})V}{W} \tag{9.21}$$

For a given amount of available thrust, the rate of climb clearly depends on the thrust required for level flight. The latter depends on weight, altitude and drag characteristics. For these reasons it is instructive to examine how the required thrust varies with these three parameters. The fundamental equation for thrust required for level flight was derived as Eqn (8.64):

$$T_{reqd} = D = \frac{W}{L/D} = \frac{W}{C_L/C_D} \tag{9.22}$$

The minimum thrust required for level flight occurs when the airplane is flown at the maximum value of L/D. A useful way to investigate the effect of both weight and altitude on the thrust required for level flight of jet airplanes is to write Eqn (9.22) as:

$$T_{reqd} = D = C_D(\tfrac{1}{2}\varrho V^2)S = C_D(1,481.3\delta M^2)S \tag{9.23}$$

This can be rewritten as:

$$\left(\frac{T_{reqd}}{\delta}\right) = C_D(1,481.3M^2)S \tag{9.24}$$

For L = W the following, similar relation can be written:

$$\left(\frac{W}{\delta}\right) = C_L(1,481.3M^2)S \tag{9.25}$$

An example application will now be discussed. This example deals with a large turbo–fan jet transport with the drag characteristics shown in Figure 9.4.

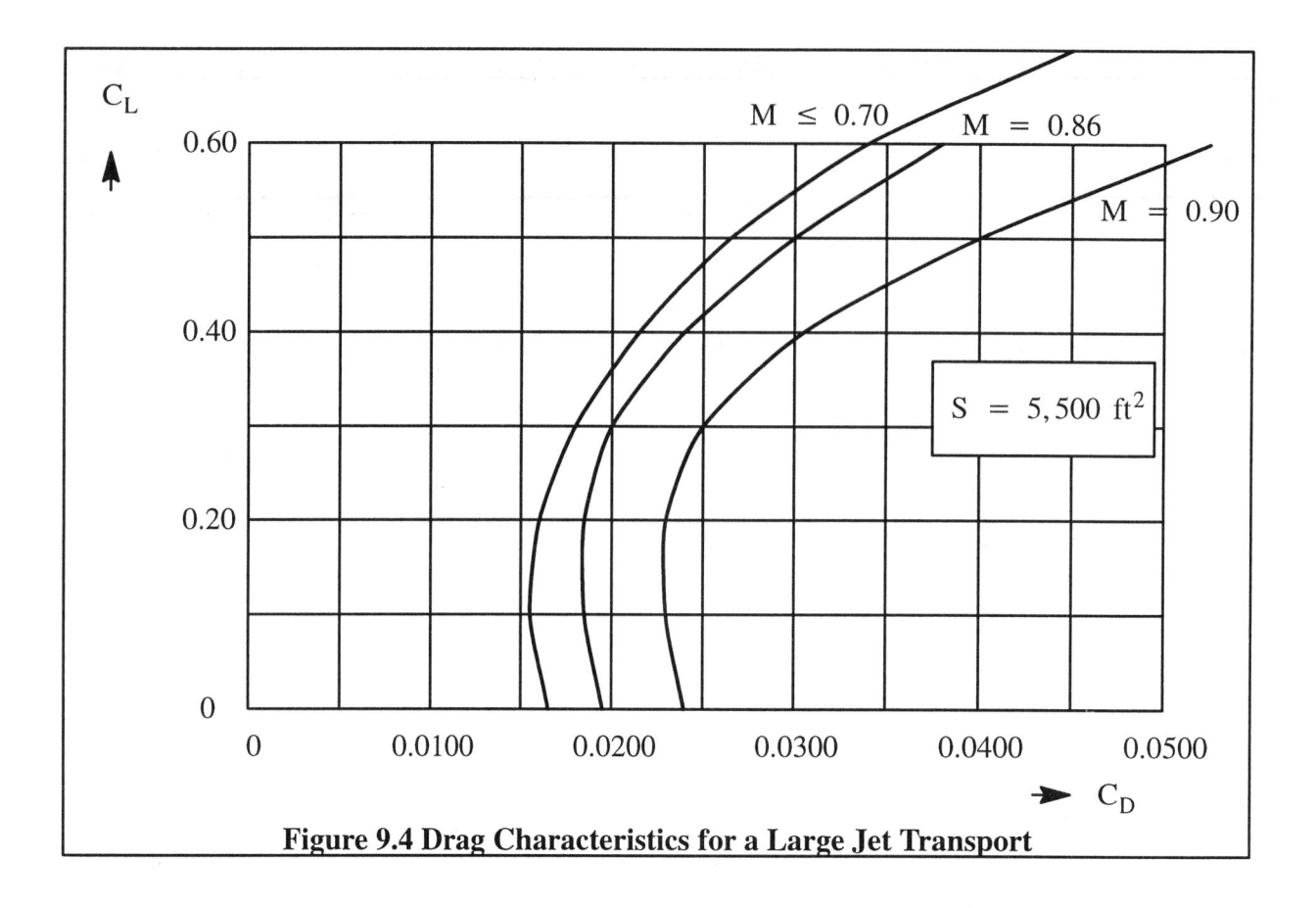

Figure 9.4 Drag Characteristics for a Large Jet Transport

The required thrust, divided by the atmospheric pressure ratio, δ, T_{reqd}/δ, for level flight is now calculated versus the weight divided by δ, W/δ, with the help of a table such as Table 9.1. For the example at hand the calculation results are shown in Table 9.2 and the graphical results are shown in Figure 9.5. Also shown in Figure 9.5 is the available (installed) thrust, divided by the atmospheric pressure ratio, δ, T_{av}/δ. The available thrust is shown for two thrust settings: maximum and 80% of maximum continuous thrust. The shaded rectangles in Table 9.2 represent flight conditions for which the lift coefficient was outside the range of drag data in Figure 9.5.

Points 1, 2 and 3 show the effect of increasing weight at constant altitude on the maximum speed (Mach number) attainable at a given thrust setting. Points 3, 4, 5 and 6 show the effect of altitude (at constant weight) on the maximum speed (Mach number) attainable at a given thrust setting. Observe, that with a weight of 637,000 lbs at 36,000 ft altitude, level flight is no longer possible: the available thrust is not sufficient.

All data required to compute the climb performance of the example jet transport are now at hand. Eqn (9.21) will be slightly modified to fit the T_{reqd}/δ and T_{av}/δ data format. To that end, the speed of the airplane, U is rewritten as:

Table 9.1 Calculation of Thrust Required for Level Flight

W/δ \ M		0.2	0.4	0.6	0.8	→ etc.
$(W/\delta)_1$	$C_L = \dfrac{W/\delta}{(1,481.3 M^2)S}$ → obtain C_D from drag polar → $\left(\dfrac{T_{reqd}}{\delta}\right) = C_D(1,481.3 M^2)S$ →					
$(W/\delta)_1$ ↓ etc.						

Table 9.2 Example Calculations of Thrust Required in Level Flight for a Jet Transport

Alt.	W/δ lbs		M→ 0.40	0.50	0.60	0.70	0.80	0.90
Sea	637,000	C_L	0.49	0.31	0.22	0.16	0.12	0.10
		C_D	0.0260	0.0185	0.0170	0.0160	0.0175	0.0230
		T_{reqd}/δ	33,892	37,681	49,861	63,874	91,248	151,781
10K	926,141	C_L	0.71	0.45	0.32	0.23	0.17	0.15
		C_D	0.0450	0.0235	0.0190	0.0175	0.0180	0.0230
		T_{reqd}/δ	58,659	47,865	55,727	69,862	93,855	151,781
20K	1,385,084	C_L		0.67	0.48	0.35	0.26	0.22
		C_D		0.0420	0.025	0.0195	0.0185	0.0235
		T_{reqd}/δ		85,545	73,324	77,846	96,462	155,081
30K	2,141,177	C_L			0.73	0.54	0.41	0.33
		C_D			0.0475	0.0290	0.0235	0.0265
		T_{reqd}/δ			139,316	115,771	122,533	174,879
36K	2,831,111	C_L				0.71	0.54	0.43
		C_D				0.0450	0.0320	0.0340
		T_{reqd}/δ				179,645	166,854	224,373

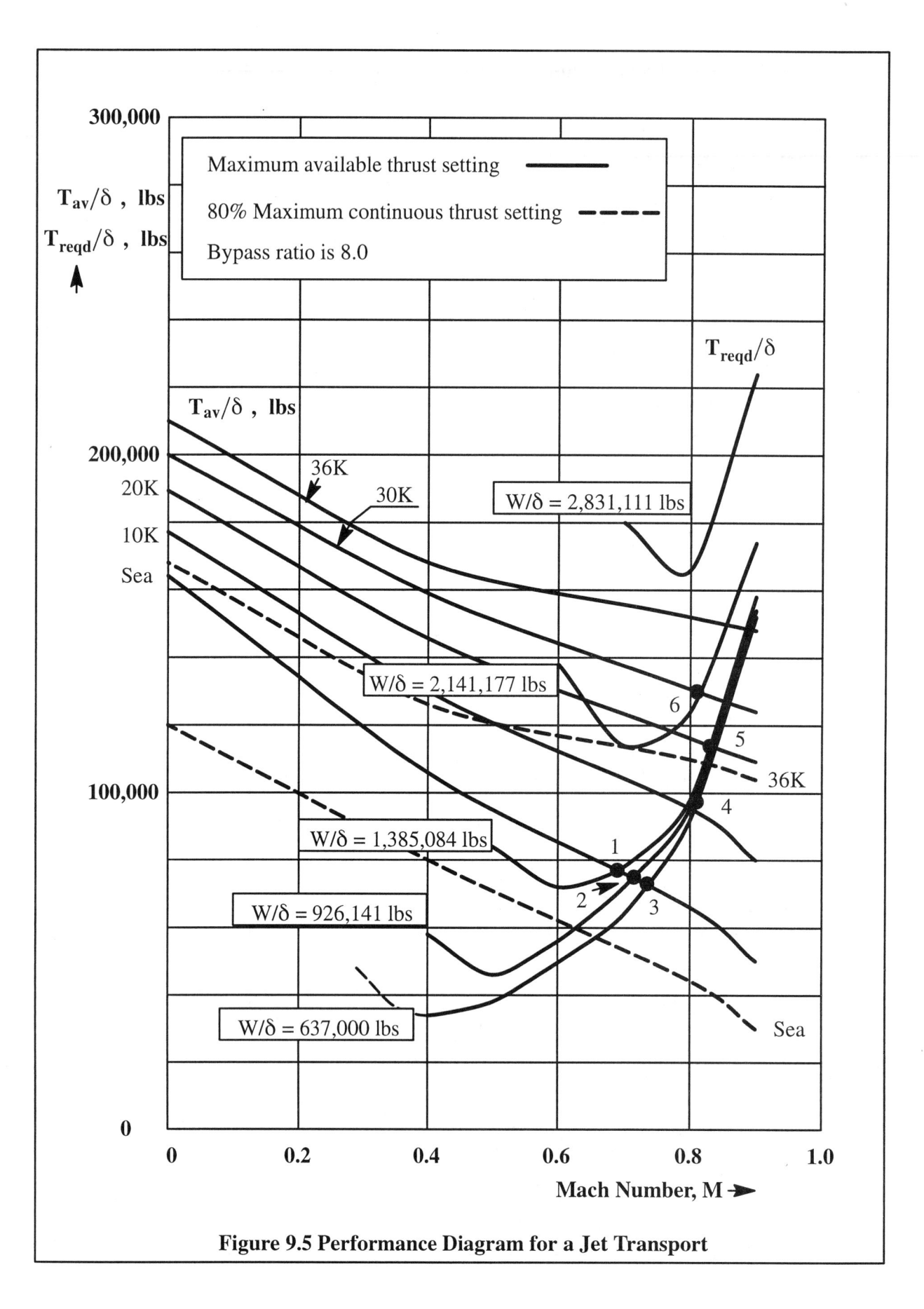

Figure 9.5 Performance Diagram for a Jet Transport

$$V = MV_a = M\frac{V_a}{V_{a_0}}V_{a_0} = M\sqrt{\theta}V_{a_0} \qquad (9.26)$$

where: V_{a_0} is the speed of sound at sea-level

θ is the temperature ratio at altitude

Upon division of the numerator and denominator of Eqn (9.21) by the atmospheric pressure ratio, δ, it is found that:

$$\text{R.C.} = V_{a_0}\sqrt{\theta}M\left(\frac{T_{av}/\delta - T_{reqd}/\delta}{W/\delta}\right) \qquad (9.27)$$

It follows that the maximum climb rate for a jet airplane occurs when the product of the flight Mach number, M, and the excess thrust is a maximum.

For the example jet transport the climb rates in ft/min. are determined with the help of Eqn (9.27) and Figure 9.5. The results are tabulated in Table 9.3 and plotted in Figure 9.6. The climb rates shown are typical for transport airplanes.

It is of interest to observe from Figure 9.5 that the Mach number for best climb rate increases with increasing altitude. **Thus, if an airplane is to climb at the best climb rate from sea–level to high altitude it must continually accelerate along its flight path!** To properly account for this requires a correction to the rate–of–climb. This correction will be discussed in Section 9.5.

9.2.1.2 The Case Of Parabolic Drag and Constant Thrust

For airplanes with straight turbojets or turbofans with moderate bypass ratios the thrust at a constant altitude is essentially independent of speed*. As shown in Figure 9.4, even high bypass ratio turbofans, at high altitude and high Mach number behave approximately like constant thrust machines. For these reasons it makes sense to study the case of a jet airplane with a parabolic drag polar and constant (speed independent) thrust. Using Eqn (9.21) with the condition L = W yields:

$$\text{R.C.} = \left(\frac{T_{av} - D}{W}\right)V = \sqrt{\frac{2W}{\varrho S}}(C_L)^{-1/2}\left(\frac{T_{av}}{W} - \frac{C_D}{C_L}\right) =$$

$$= \sqrt{\frac{2W}{\varrho S}}\left\{\left(\frac{T_{av}}{W}\right)(C_L)^{-1/2} - \left[\frac{C_{D_0} + \frac{C_L^2}{\pi Ae}}{(C_L)^{3/2}}\right]\right\} \qquad (9.28)$$

---

* The reason why high bypass ratio turbofans behave differently is because they are more like propellers (ducted propellers to be specific) and propeller thrust does decrease with speed for a given shaft horsepower input. See Chapters 6 and 7.

Climb and Drift–down Performance

The maximum rate of climb occurs at a lift coefficient which can be recovered from the condition: $\partial RC/\partial C_L = 0$. The reader is asked to show that after differentiation and simplification this optimum condition is:

$$\frac{C_L^2}{\pi Ae} + \frac{T_{av}}{W}C_L - 3C_{D_0} = 0 \qquad (9.29)$$

Therefore, the lift coefficient for best rate of climb is found from:

$$C_{L_{\text{best rate of climb}}} = -\left(\frac{T_{av}}{W}\right)\frac{\pi Ae}{2} + \sqrt{\left\{\left(\frac{T_{av}}{W}\right)\frac{\pi Ae}{2}\right\}^2 + 3C_{D_0}\pi Ae} \qquad (9.30)$$

For the jet transport drag polar of Figure 9.4 at Mach numbers around 0.7 it is found that:

$$C_D \approx 0.0150 + 0.048C_L^2 \quad \text{so that}: C_{D_0} = 0.0150 \text{ and } 1/\pi Ae = 0.048 \qquad (9.31)$$

Assume, using Figure 9.5, that at 30,000 ft the available thrust–to–weight ratio is a constant 0.064 over a Mach number range from 0.6 to 0.8. The lift coefficient for best rate of climb as found from Eqn (9.30) is: 0.51. The corresponding flight Mach number is: M = 0.70. Figure 9.5 suggests that the actual best rate–of–climb speed occurs around M = 0.72. It is seen that the parabolic approximation is rather good in this case.

The flight path angle, γ, follows from:

$$\gamma = RC/V = \frac{T_{av} - D}{W} \qquad (9.32)$$

Clearly, for constant thrust, the best climb angle occurs at the best lift–to–drag ratio. For the case of a parabolic drag polar it was seen in Chapter, 5, Eqns (5.2) and (5.3) that this happens when:

$$C_{D_i} = C_{D_0} \quad \text{or for}: \quad \frac{C_L^2}{\pi Ae} = C_{D_0} \qquad (9.33)$$

From this it follows that, for constant thrust and a parabolic drag polar:

$$C_{L_{\text{best climb }\angle}} = \sqrt{\frac{C_{D_0}}{\pi Ae}} \qquad (9.34)$$

For the airplane represented by the data in Figures 9.4 and 9.5 it is found that:

Table 9.3 Example Calculations of Climb Rate Versus Mach Number for a Jet Transport

Alt.	W/δ lbs	M	$T_{av}/\delta - T_{reqd}/\delta$	$\dfrac{T_{av}/\delta - T_{reqd}/\delta}{W/\delta}$	$a_0 M \sqrt{\theta}$ ft/sec	RC, ft/min
Sea	637,000	0.35	76,000	0.119	390.6	2,789
		0.40	74,000	0.116	446.4	3,107
		0.50	56,000	0.088	558.0	2,946
		0.60	36,000	0.565	669.6	2,270
		0.70	16,000	0.0251	781.2	1,176
		0.74	0	0	825.8	0
10K	926,141	0.40	73,000	0.0788	430.8	2,037
		0.50	75,000	0.0810	538.5	2,617
		0.60	56,000	0.0605	646.2	2,342
		0.70	34,000	0.0367	753.9	1,660
		0.72	0	0	775.4	0
20K	1,385,084	0.50	54,000	0.0390	518.3	1,213
		0.60	56,000	0.0404	621.9	1,507
		0.70	46,000	0.0332	725.6	1,445
		0.80	19,000	0.0137	829.2	682
10K		0.83	0	0	860.3	0
30K	2,141,177	0.60	6,000	0.0028	596.7	100
		0.70	24,000	0.0112	696.1	468
		0.80	6,000	0.0028	795.6	134
		0.82	0	0	815.5	0
36K	2,831,111	0.70	−26,000	−0.0092	677.8	−374
		0.78	−11,000	−0.0039	755.3	−177
		0.80	−15,000	−0.0053	774.7	−246
		0.90	−70,000	−0.2470	871.5	−1,292

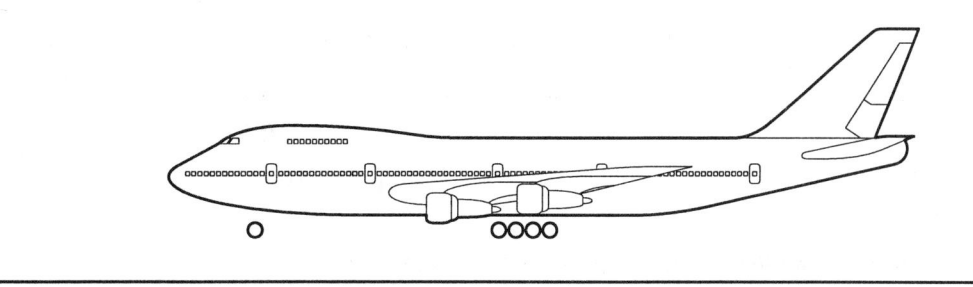

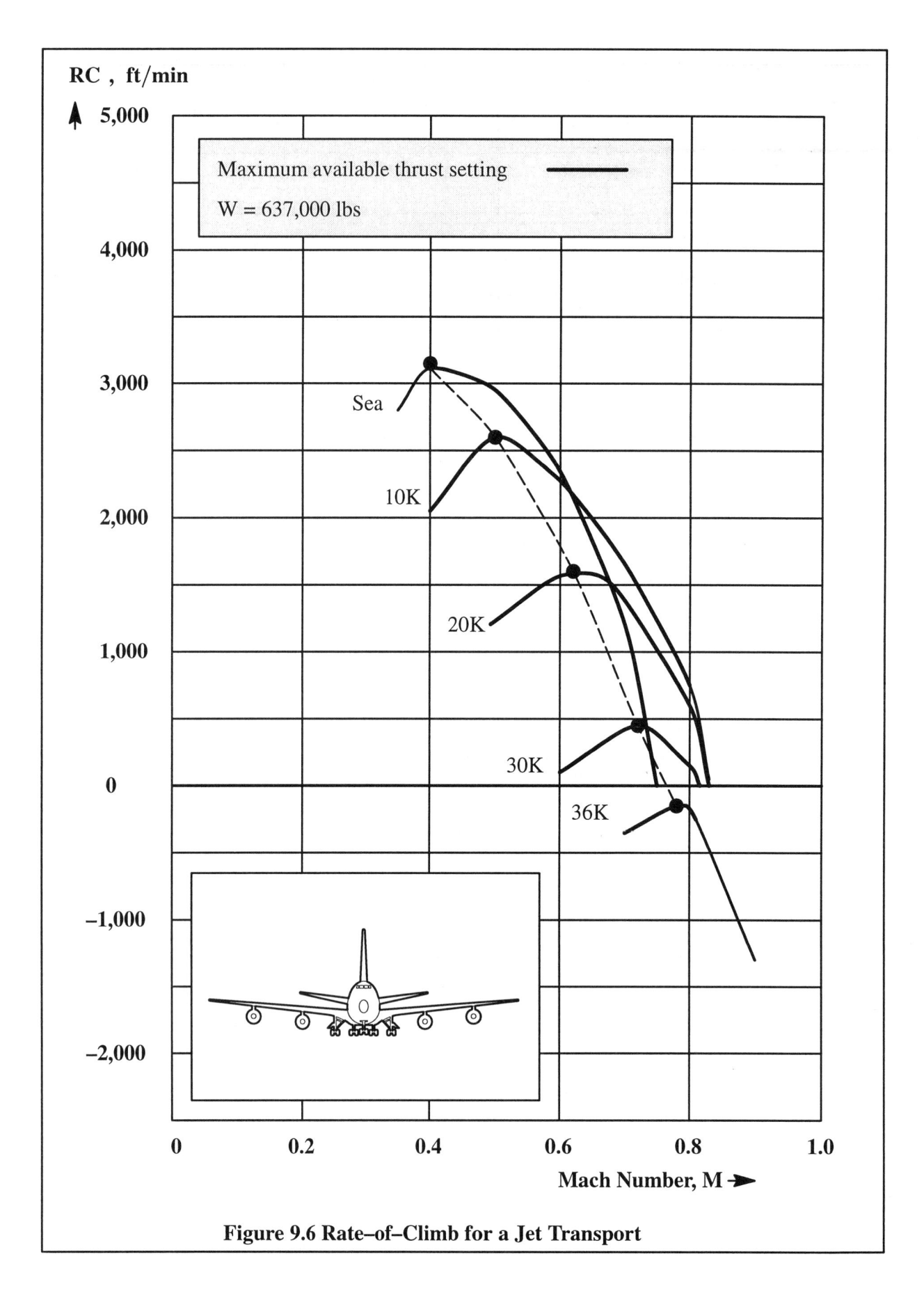

Climb and Drift–down Performance

Figure 9.6 Rate-of-Climb for a Jet Transport

Climb and Drift-down Performance

$$C_{L_{\text{best climb }\gamma}} = \sqrt{C_{D_0}\pi Ae} = \sqrt{\frac{0.0160}{0.050}} = 0.57$$

At sea–level this corresponds to a Mach number of 0.37. The corresponding climb angle is found from Figure 9.5 as:

$$\gamma = RC/V = \frac{T_{av} - D}{W} = \frac{112,000 - 33,000}{637,000} = 0.124 \text{ rad} = 7.1 \text{ deg.}$$

9.2.2 CLIMB PERFORMANCE FOR JET AIRPLANES AT STEEP ANGLES

Equations (9.6) and (9.7) were derived for climbing flight at small flight path angles. For large flight path angles these equations become:

$$T - D = W\sin\gamma \tag{9.35}$$

and

$$L = W\cos\gamma \tag{9.36}$$

By dividing drag into lift it is seen that:

$$\frac{L}{D} = \frac{C_L}{C_D} = \frac{W\cos\gamma}{T - W\sin\gamma} \tag{9.37}$$

Replacing $\cos\gamma$ by $\sqrt{1 - \sin^2\gamma}$ and solving for $\sin\gamma$ it follows that:

$$\sin\gamma = \frac{C_L^2 T}{W(C_L^2 + C_D^2)} - \sqrt{\left\{\frac{C_L^2 T}{W(C_L^2 + C_D^2)}\right\}^2 - \frac{C_L^2 T^2 - C_D^2 W^2}{W^2(C_L^2 + C_D^2)}} \tag{9.38}$$

It is seen from this equation that the climb angle is a function of C_L/C_D and T/W. Evidently, whenever the thrust–to–weight ratio is large, the climb angle will be large.

In the general case of a non–parabolic drag polar and an available thrust which varies with speed, Eqns (9.35) and (9.36) can be solved iteratively. Such an iteration might proceed as follows:

1) Assume that the weight, W, the altitude, h, the flight speed, V and the trust–to–weight ratio, T/W are given.
2) Assume first, that the flight path angle is zero. Eqn (9.36) is then used to calculate the lift coefficient, C_L.

3) From the drag polar (non–parabolic), read C_D for the computed C_L.

4) Compute the flight path angle, γ from Eqn (9.35).

5) Substitute γ back into Eqn (9.36) and repeat the process until the flight path angle has converged. A convergence accuracy of 0.1 degrees should be adequate.

9.2.3 DRIFT–DOWN PERFORMANCE FOR JET AIRPLANES

Whenever a multi–engine airplane is operated at its normal cruise altitude with all engines operating (AEO) it will still be able to climb when the throttles are advanced to their most forward position. However, if one engine becomes inoperative (OEI) the airplane may no longer be able to climb or maintain the AEO cruise altitude: the thrust required (including the drag due to the inoperative engine and the drag caused by re–trimming the airplane) may be greater than the thrust available with OEI. In that case the airplane is said to be in a drift–down mode. The airplane must then be flown to the OEI cruise altitude.

The thrust required to fly level with OEI will be called: $T_{reqd_{OEI}}$. Because of the increased drag with OEI it should be expected that: $T_{reqd_{OEI}} > T_{reqd_{AEO}}$. The thrust available with OEI will be called: $T_{av_{OEI}}$. It should be expected that: $T_{reqd_{OEI}} < T_{av_{AEO}}$. Modification of Eqn (9.11) yields for the rate–of–descent, RD:

$$RD = \frac{(D_{OEI} - T_{av_{OEI}})V}{W} = \frac{(T_{reqd_{OEI}} - T_{av_{OEI}})V}{W} \qquad (9.39)$$

Using notation similar to that employed in Eqn (9.27) this rate–of–descent can be rewritten as:

$$RD = V_{a_0}\sqrt{\theta}\,M\left(\frac{T_{reqd_{OEI}}/\delta - T_{av_{OEI}}/\delta}{W/\delta}\right) \qquad (9.40)$$

Of concern are the altitude at which OEI level flight is possible, the time required to drift down to that altitude and the steepness of the flight path during the drift–down process.

The OEI cruise altitude is defined as that altitude for which the rate–of–climb (RC) with OEI is at least 500 ft/min. A more detailed discussion of drift–down performance is given in Section 9.4.

9.3 CLIMB AND DRIFT–DOWN PERFORMANCE OF PROPELLER DRIVEN AIRPLANES

For a given altitude, speed and engine power setting, a given engine (piston or turbo–prop) will deliver a certain amount of shaft horsepower to the propeller shaft, SHP_{av}. The propeller, when free of the airplane (free propeller) would covert this shaft horsepower into available thrust horsepower, called, $THP_{av_{free}}$, with a free propeller efficiency of $\eta_{p_{free}}$:

$$THP_{av_{free\,\alpha}} = \eta_{p_{free}} SHP_{av} \qquad (9.41)$$

The free propeller suffers three types of losses as pointed out in Chapter 7: rotational slipstream losses, tangential slipstream losses and blade profile drag losses. Together, these losses are accounted for through the free propeller efficiency of $\eta_{p_{free}}$. However, most propellers operate in front of or behind a body, fuselage or nacelle. This results in additional installation losses (blockage effects and slipstream effects) which cause the actual propeller efficiency (also called propulsive efficiency), η_p to be less than that of the free propeller. In Chapter 7 these effects were lumped in to the installation efficiency, $\eta_{p_{installed}}$. As a result, the actual available thrust horsepower is:

$$THP_{av} = \eta_{p_{free}} \eta_{p_{installed}} SHP_{av} = \eta_{propulsive} SHP_{av} \qquad (9.42)$$

It will be assumed in this chapter that the actually available thrust horsepower, THP_{av} is known. For methods to compute the various propeller efficiencies the reader is referred to Chapter 7.

9.3.1 CLIMB PERFORMANCE OF PROPELLER DRIVEN AIRPLANES

9.3.1.1 The General Case

The rate of climb, R.C. of a propeller driven airplane follows from Eqn (9.12):

$$R.C. = \frac{(T_{av} - D)V}{W} = \frac{(THP_{av} - THP_{reqd})}{W} \qquad (9.43)$$

The available power depends on the available shaft horsepower and on the propulsive efficiency as shown in Eqn (9.42). The reader is referred to Chapter 7 for methods to predict the free and installed propeller efficiencies which together define the propulsive efficiency.

As seen from Eqn (9.43), for a given amount of available power, the rate of climb clearly depends on the power required for level flight. For a given power setting, the power available depends on altitude (after accounting for installation and power extraction losses). Figure 9.7 shows an example of available shaft horsepower for different altitudes for two types of engine: a normally aspirated

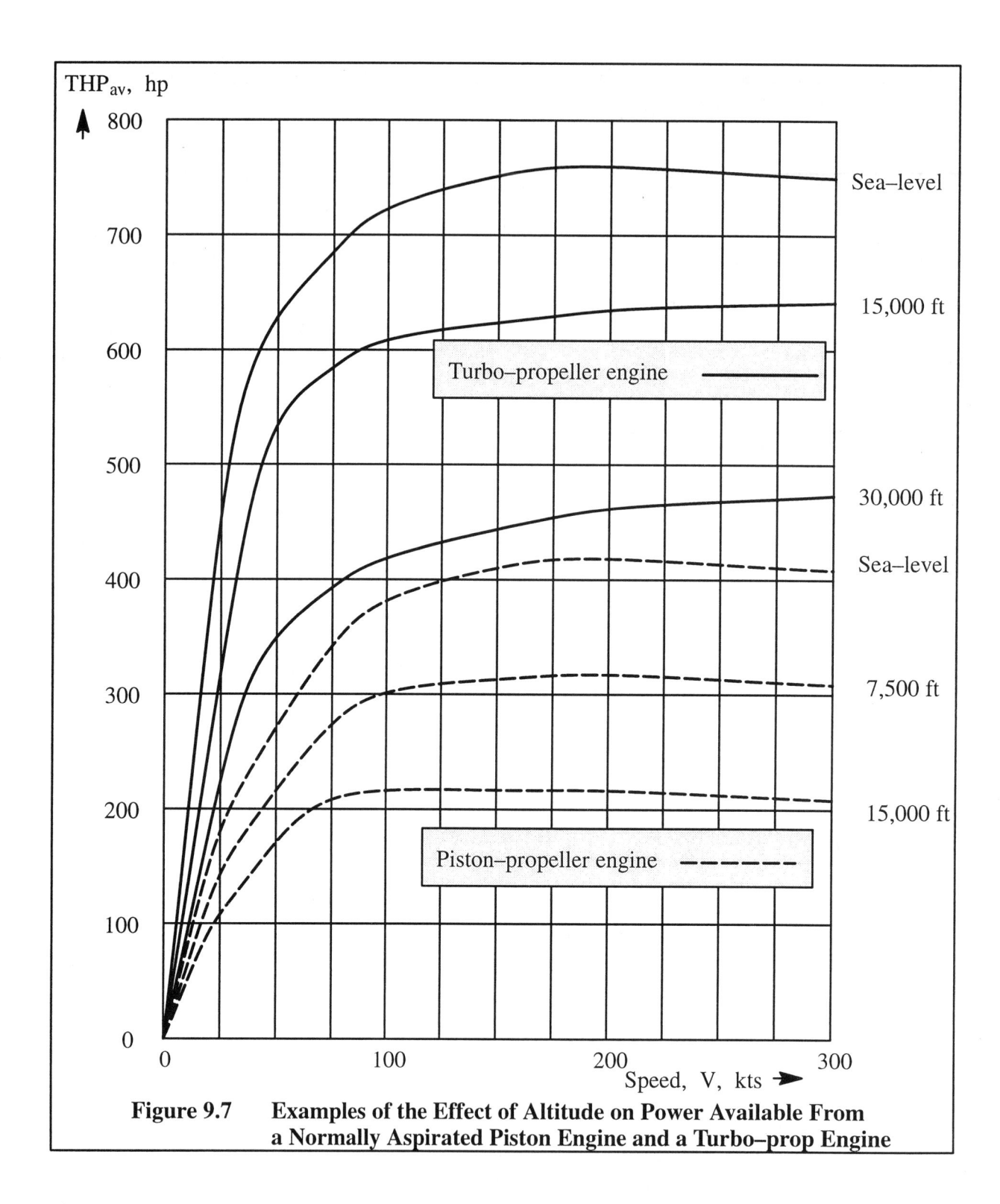

Figure 9.7 Examples of the Effect of Altitude on Power Available From a Normally Aspirated Piston Engine and a Turbo–prop Engine

piston engine and a turbo–prop engine. Observe that the available shaft horsepower is essentially speed independent over a wide speed range. The available power, as can be expected from the very definition of power, becomes zero at zero air–speed.

The fundamental equations for thrust and power required for level flight were derived before as Eqns (8.64) and (8.71):

$$T_{reqd} = D = \frac{W}{L/D} = \frac{W}{C_L/C_D} \qquad (9.44)$$

$$P_{reqd} = \frac{DV}{550} = \frac{WV}{550(C_L/C_D)} = \frac{W}{550}\sqrt{\frac{W}{S}\frac{2}{\varrho}\frac{C_D^2}{C_L^3}} \qquad (9.45)$$

The power required, as seen from Eqn (9.45), depends on altitude, weight, drag characteristics and wing area. It is instructive to examine how the required power varies with these four parameters. This is done in the following.

Effect of Altitude and Weight on Power Required

The speed for level flight at sea–level at a given weight and lift coefficient is:

$$V_{s.l.} = \sqrt{\frac{2W_{ref}}{\varrho_0 S C_L}} \qquad (9.46)$$

The thrust required for level flight at sea–level at a given weight and lift coefficient is:

$$T_{reqd_{s.l.}} = \frac{W_{ref}}{C_L/C_D} \qquad (9.47)$$

where: C_L/C_D is taken at the lift coefficient of Eqn (9.46).

The thrust horsepower required for level flight at sea–level at a given weight and lift coefficient is:

$$THP_{reqd_{s.l.}} = \frac{1}{550}\left(\sqrt{\frac{2W_{ref}^3}{\varrho_0 S}}\right)\frac{C_D}{C_L^{3/2}} \qquad (9.48)$$

where: $C_D/C_L^{3/2}$ is taken at the lift coefficient of Eqn (9.46).

For flight at the same lift coefficient, the effect of weight and altitude on the speed follows from Eqn (9.46):

$$V = V_{s.l.}\sqrt{\frac{W}{W_{ref}}\frac{1}{\sigma}} \qquad (9.49)$$

For flight at the same lift coefficient, the effect of weight on the thrust required follows from Eqn (9.47):

Climb and Drift–down Performance

$$T_{reqd} = T_{reqd_{s.l.}} \frac{W}{W_{ref}} \qquad (9.50)$$

For flight at the same lift coefficient, the effect of weight and altitude on the power required follows from Eqn (9.48):

$$THP_{reqd} = THP_{reqd_{s.l.}} \sqrt{\left(\frac{W}{W_{ref}}\right)^3 \frac{1}{\sigma}} \qquad (9.51)$$

The effect of altitude on the power required of an example turbo–propeller airplane is shown in Figure 9.8. Two observations are in order:

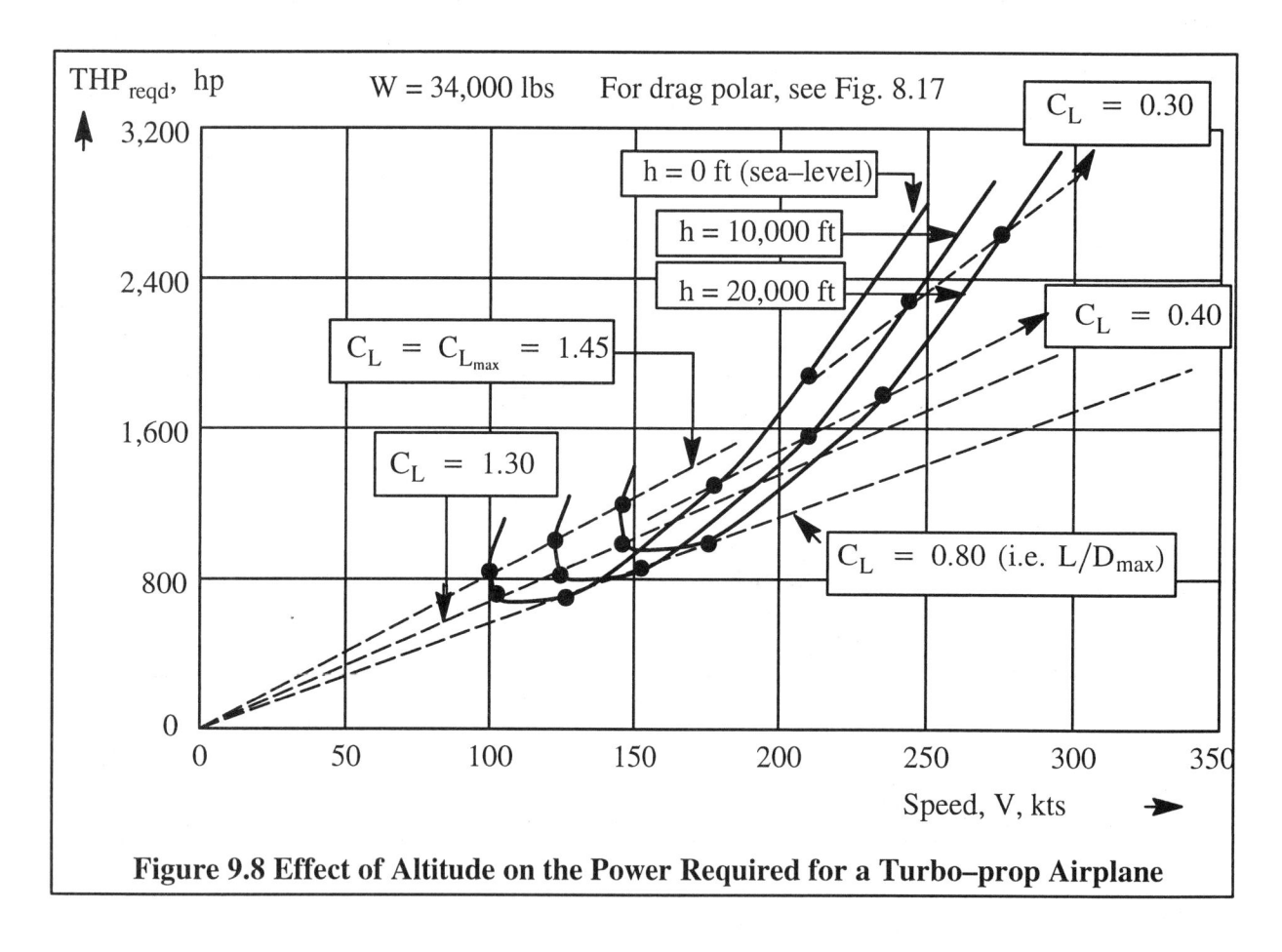

Figure 9.8 Effect of Altitude on the Power Required for a Turbo–prop Airplane

Observation 1: at higher altitude and at higher speeds, the power required is less than that required at sea–level. A physical explanation can be given as follows. At higher altitude the lift coefficient required for level flight will increase. This causes the induced drag to increase. However, in the higher speed regimes the lift coefficients will still be relatively small. Because the $C_L - C_D$ curve is relatively flat in the range of low lift coefficients, , the drag does not increase very rapidly with altitude. On the other hand, the power required is also proportional to atmospheric density. The latter decreases rapidly with increasing altitude. The overall effect is for the power required for level flight to decrease with increasing altitude: see Figure 9.8.

Observation 2: the speed for maximum rate–of–climb increases with altitude. This requires the airplane to accelerate along its flight path during a climb. In turn, this need to accelerate requires that a correction to the rate–of–climb must be made. Section 9.5 contains a discussion of this correction.

The effect of weight on the power required of an example turbo–propeller airplane is shown in Figure 9.9.

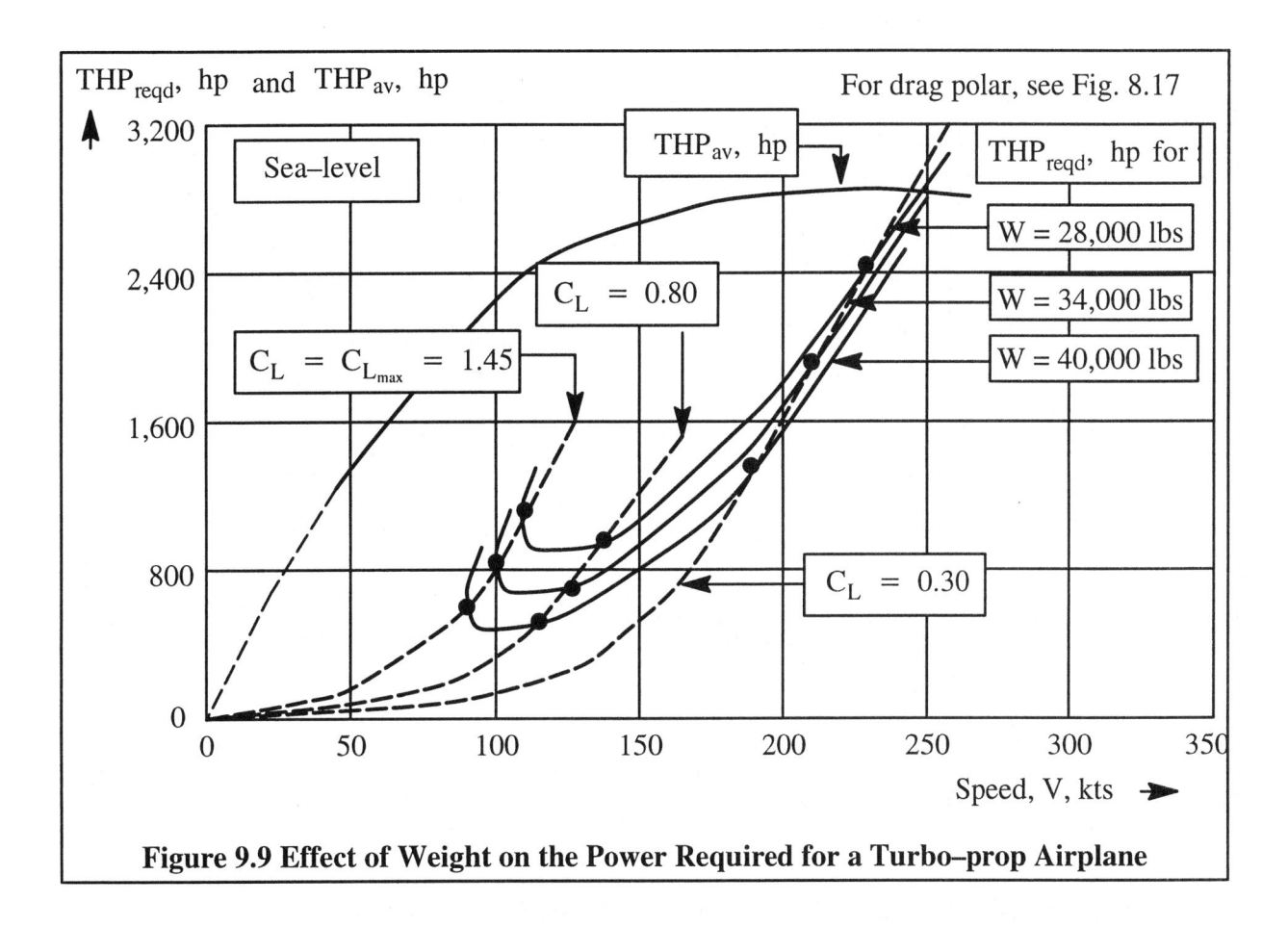

Figure 9.9 Effect of Weight on the Power Required for a Turbo–prop Airplane

Effect of Wing Area on Power Required

The wing loading of an airplane can be varied in two different ways:

a) For a given wing area, the weight can be varied. That case was discussed in the foregoing. The drag polar of the airplane does not change in this case.

b) For a given weight, the wing area can be changed. This is an exercise frequently considered during the preliminary design stages of an airplane when it is not yet certain how much wing area should be used. If the wing area for a given airplane design is changed, the drag polar changes. This effect will be illustrated for the case of a parabolic drag polar:

$$C_D = C_{D_0} + \frac{C_L^2}{\pi A e} \qquad (9.52)$$

This drag polar is now written explicitly in terms of the wing contribution to zero-lift drag and the contribution to zero-lift drag of the rest of the airplane:

$$C_D = \left[C_{D_{0_{wing}}} + \frac{\sum_1^n C_{D_n} S_n}{S} \right] + \frac{C_L^2}{\pi A e} \qquad (9.53)$$

where: $C_{D_{0_{wing}}}$ is the wing zero lift drag coefficient

$\sum_1^n C_{D_n} S_n$ is the equivalent flat plate area of the rest of the airplane

By now varying S it is possible to account for the effect of changing wing area on the drag polar of an airplane. Figure 9.10 illustrates this effect for a twin turbo-propeller transport airplane.

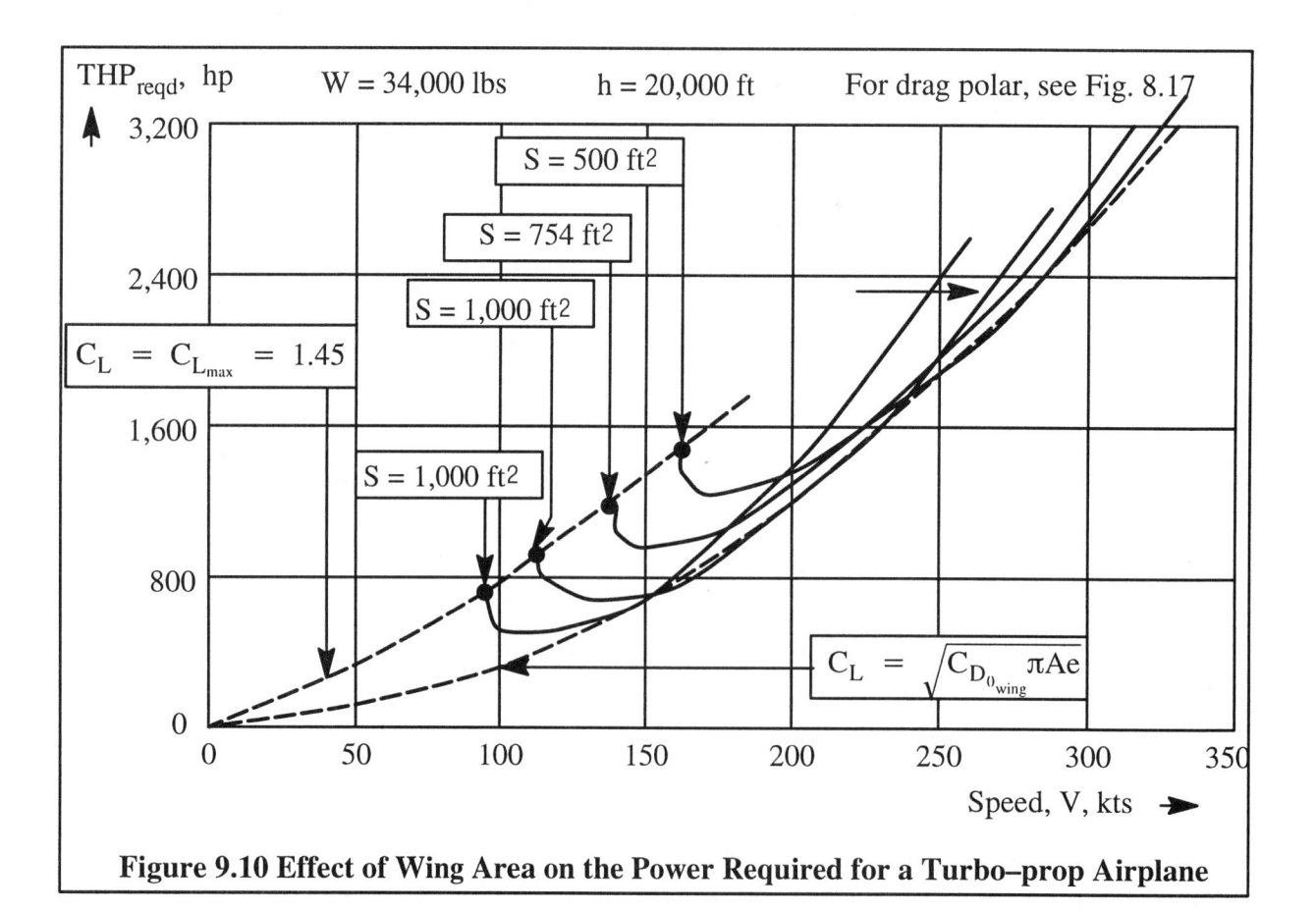

Figure 9.10 Effect of Wing Area on the Power Required for a Turbo-prop Airplane

Observe that at low speeds the power required for level flight increases for decreasing wing area (i.e. increasing wing loading, W/S). This is because of the induced drag increase associated with the high lift coefficients required for flight at lower speeds.

At higher speeds, the power required for level flight decreases with decreasing wing area (i.e. increasing wing loading). This is because the zero lift drag is less for smaller wing areas. Also observe a rather wide 'in–between' range of airspeeds, where the trends are variable.

Effect of Drag Characteristics on Power Required

It is seen from Eqns (9.47) and (9.48) that thrust and power required both depend on the relationship between drag coefficient and lift coefficient, in other words on the drag polar. The drag polar for a given airplane can vary significantly with the configuration of the airplane and with the operating state of the engines.

Flaps up or down, landing gear up or down and AEO or OEI can have a significant effect on the drag characteristics of an airplane as shown in Figure 9.11 for an example turbo–propeller airplane. Note that at the same lift coefficient the drag increases significantly as a result of flap deployment, landing gear deployment or OEI.

Observe from Figure 9.11 that the zero lift drag of the example turbo-propeller airplane increases **by a factor of five** from clean, AEO to flaps full down, gear down and OEI. This will affect not only the power required for level flight, but, for a given amount of power available, the rate–of–climb capability of the airplane. This effect will be further discussed in Section 7.4.

To illustrate the application of the methods discussed, a numerical example will now be given.

Example 9.1: A twin engine, propeller driven airplane has the following characteristics:

$$W = 4,600 \text{ lbs} \qquad S = 175 \text{ ft}^2 \qquad C_{L_{max}} = 1.31$$

$$C_D = 0.0293 + 0.0557 C_L^2 \text{ see Fig.9.12}$$

The task is to compute the power required versus speed and the rate–of–climb versus speed for three altitudes: sea–level, 7,500 ft and 15,000 ft. The power available data are given in Figure 9.13.

Solution: Sea–level: The relationship between speed and lift coefficient is found from Eqn (9.15):

$$C_L = \frac{W}{\frac{1}{2}\varrho V^2 S} = \frac{4,600}{0.5 \times 0.002377 \times 1.688^2 V^2 S} = \frac{7,759}{V^2} \quad \text{with V in knots}$$

The lift coefficient should not be greater than 1.31, the stated maximum lift coefficient. This puts a lower limit on the speed, V. Table 9.4 shows the results of calculating the power required with Eqns (9.44) and (9.45) at sea–level. Table 9.5 shows the results of calculating the power required at 7,500 ft and 15,000 ft by using Eqns (9.49) and (9.51). Results of these calculations are also plotted in Figure 9.13.

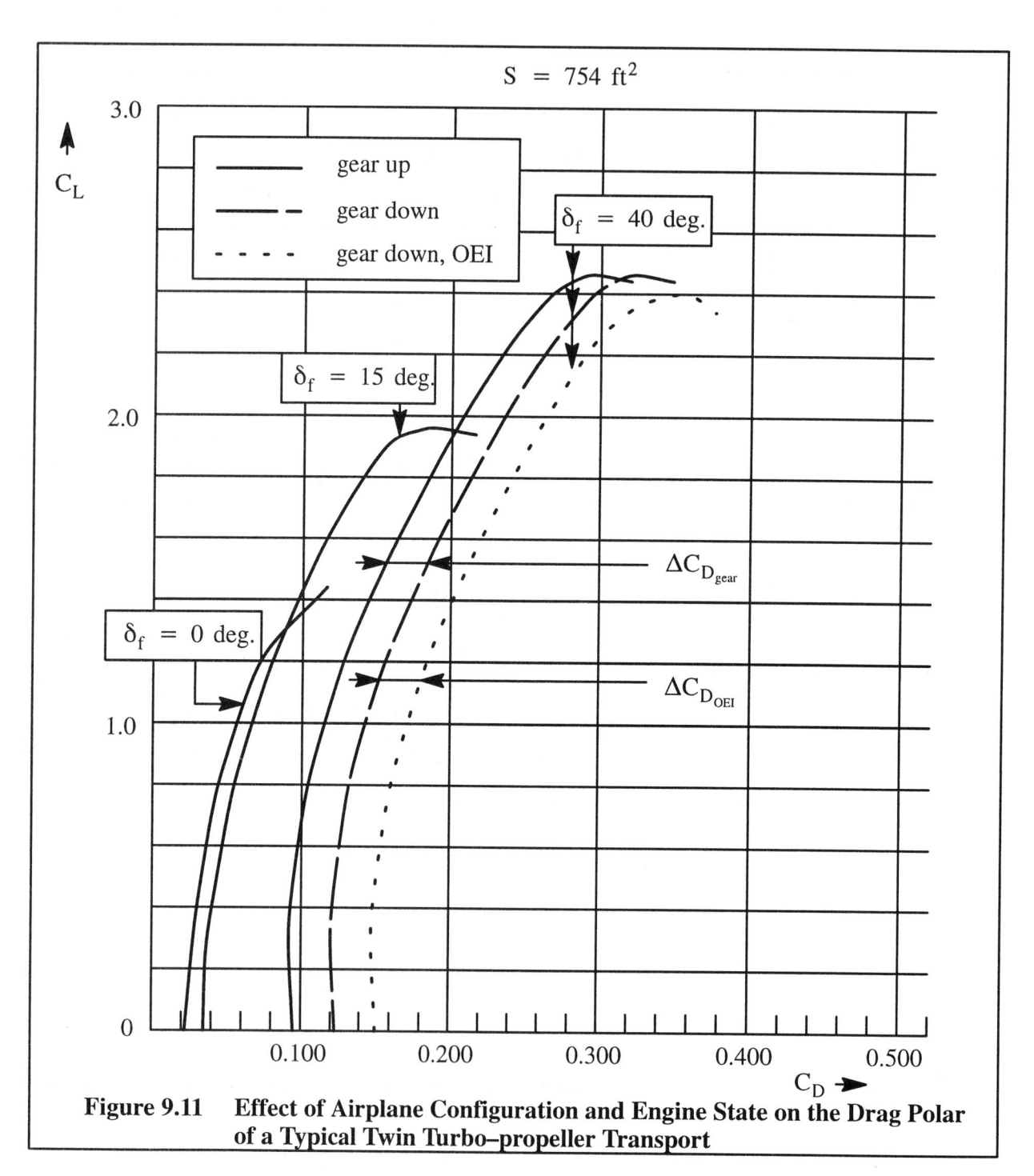

Figure 9.11 Effect of Airplane Configuration and Engine State on the Drag Polar of a Typical Twin Turbo–propeller Transport

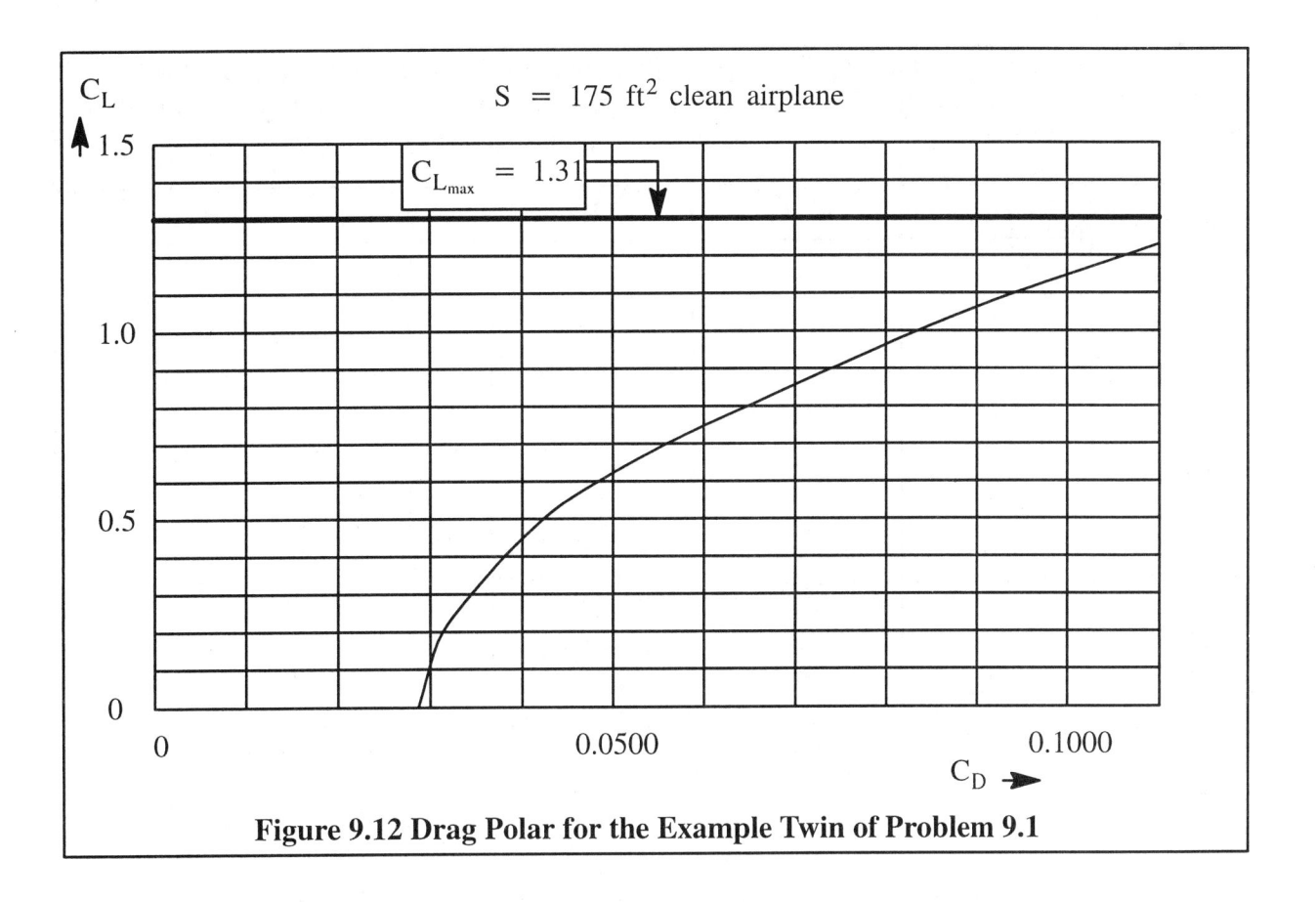

Figure 9.12 Drag Polar for the Example Twin of Problem 9.1

$S = 175$ ft² clean airplane

$C_{L_{max}} = 1.31$

Table 9.4 Calculation of Power Required for Level Flight at Sea-level for the Twin with Drag Characteristics of Figure 9.12.

V, kts	C_L	C_D From Fig. 9.12	$T_{reqd} = W\dfrac{C_D}{C_L}$ (lbs)	$THP_{reqd} = \dfrac{T_{reqd}V}{550} \times 1.688$
80	1.21	0.1108	421	103
90	0.96	0.0806	386	107
100	0.77	0.0623	372	114
120	0.54	0.0455	388	143
140	0.40	0.0382	439	189
160	0.30	0.0343	526	258
180	0.24	0.0325	623	344
200	0.19	0.0313	758	465

$$C_L = \frac{2W}{0.002377 \times V^2 \times 1.688^2 \times 175} = \frac{7{,}759}{V^2} \text{ with V in kts}$$

Climb and Drift–down Performance

Table 9.5 Calculation of Power Required for Level Flight at 7,500 ft and at 15,000 ft for the Twin with Drag Characteristics of Figure 9.12.

Sea–level		7,500 ft, $\sigma = 0.7981$		15,000 ft, $\sigma = 0.6292$	
$V_{s.l.}$, kts	$THP_{reqd_{s.l.}}$	V, kts	THP_{reqd}	V, kts	THP_{reqd}
80	103	90	115	101	130
90	107	101	120	113	135
100	114	112	128	126	144
120	143	134	160	151	180
140	189	157	211	177	238
160	258	179	289	202	325
180	344	201	385	227	434
200	465	224	520	252	586

$$V = V_{s.l.}\sqrt{\frac{W}{W_{ref}}\frac{1}{\sigma}} \qquad \frac{W}{W_{ref}} = 1.0 \qquad THP_{reqd} = THP_{reqd_{s.l.}}\sqrt{\left(\frac{W}{W_{ref}}\right)^3 \frac{1}{\sigma}}$$

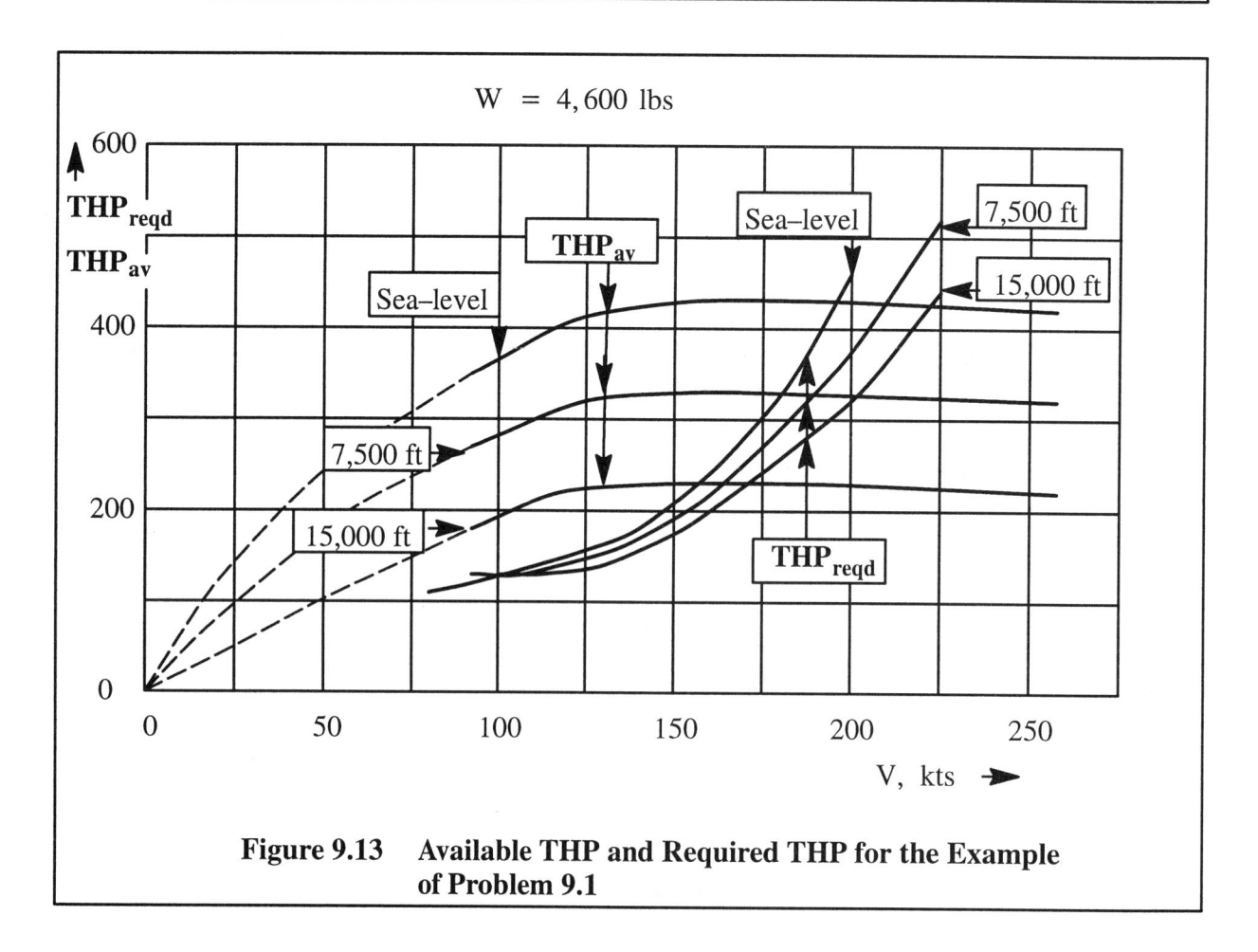

Figure 9.13 Available THP and Required THP for the Example of Problem 9.1

9.3.1.2 The Case Of Parabolic Drag and Constant Power

It is seen from Figure 9.13 that in the "normal" speed range the power available is roughly constant with speed. If the drag polar is assumed to be parabolic certain simplifications can be made.

From Eqns (9.43) and (9.45) it can be concluded that the best rate of climb occurs when the ratio $C_D/C_L^{3/2}$ is a maximum. With a parabolic drag polar this condition is satisfied when:

$$\frac{\partial}{\partial C_L}\left[\frac{C_L^{3/2}}{C_D}\right] = \frac{\partial}{\partial C_L}\left[\frac{C_L^{3/2}}{C_{D_0} + C_L^2/\pi Ae}\right] = \frac{\frac{3}{2}C_L^{1/2}}{C_{D_0} + \frac{C_L^2}{\pi Ae}} +$$

$$- \frac{C_L^{3/2}\frac{2C_L}{\pi Ae}}{\left(C_{D_0} + \frac{C_L^2}{\pi Ae}\right)^2} = 0 \qquad (9.54)$$

or when:

$$\frac{3}{2}C_L^{1/2}\left(C_{D_0} + \frac{C_L^2}{\pi Ae}\right) - C_L^{3/2}\frac{2C_L}{\pi Ae} = 0 \qquad (9.55)$$

This implies that:

$$C_{D_0} = \frac{C_L^2}{3\pi Ae} = \frac{1}{3}C_{D_i} \qquad (9.56)$$

The lift coefficient for the best rate of climb occurs at:

$$C_{L_{best\ climb\ rate}} = \sqrt{3C_{D_0}\pi Ae} \qquad (9.57)$$

The following procedure can be used to estimate the maximum rate of climb for a propeller driven airplane when the drag polar is parabolic and the power available is constant with speed:

Step 1: Use Eqn (9.57) to estimate $C_{L_{best\ climb\ rate}}$

Step 2: Find the speed for best rate of climb from: $V = \sqrt{\dfrac{2W}{\varrho S C_{L_{best\ climb\ rate}}}}$

Step 3: Calculate the drag coefficient from: $C_D = C_{D_0} + C_{L_{best\ climb\ rate}}^{\ 2}/\pi Ae$

Step 4: Calculate the drag from: $W\dfrac{C_D}{C_{L_{best\ climb\ rate}}}$

Step 5: Calculate the power required from: $THP_{reqd} = DxV$

Step 6: Determine the maximum climb rate from: $RC_{max} = (THP_{av} - THP_{reqd})/W$

It is not quite as easy to find an analytical expression for the maximum climb angle. For that reason a graphical procedure is normally used. In Problem 9.4 the reader is asked to develop an analytical method.

An example application of the Step 1–6 procedure for finding the maximum climb rate of a turbo–prop airplane will now be presented.

Example 9.2: A twin engine, turbo–propeller driven airplane has the following characteristics:

$W = 36,000$ lbs $\qquad S = 450$ ft^2 $\qquad C_{L_{max}} = 1.4$

$C_D = 0.0200 + 0.05C_L^2$

The task is to equip this airplane with two turbo–propeller engines with sufficient available thrust–horse–power so that the maximum speed is 400 kts at sea–level. At this speed, the compressibility drag increment is assumed to be 0.0015. The maximum available THP established on this basis may be assumed to be constant with speed. Calculate the maximum rate–of–climb, the speed at which this occurs and also determine the corresponding climb angle.

Solution: At the maximum speed of 400 kts at sea–level the lift coefficient is found as:

$$C_L = \frac{W}{qS} = \frac{36,000}{0.5 \times 0.002377 \times 400^2 \times 1.688^2 \times 450} = 0.148$$

The drag coefficient then is: $C_D = 0.0200 + 0.05 \times 0.148^2 + 0.0015 = 0.0226$

The drag is equal to: $D = 0.0226 \times 0.5 \times 0.002377 \times 400^2 \times 1.688^2 \times 450 = 5,513$ lbs

The required thrust horsepower is: $THP_{reqd} = \dfrac{5,513 \times 400 \times 1.688}{550} = 6,768$ hp

The latter now is the available thrust horse power for the maximum climb rate since the power available was assumed to be constant with speed. At the maximum climb rate, with Eqn (9.57):

$$C_{L_{best\ climb\ rate}} = \sqrt{3C_{D_0}\pi Ae} = \sqrt{3 \times (0.0200)/0.05} = 1.095$$

The speed for maximum rate of climb now follows from:

$$V = \sqrt{\frac{2 \times 36,000}{0.002377 \times 1.095 \times 450}} = 248 \text{ fps} = 147 \text{ kts}$$

At this speed, the drag coefficient is found from: $C_D = 0.0200 + 0.05 \times 1.095^2 = 0.0800$

Note that the compressibility drag increment is now left out because the flight speed is low.

The corresponding drag is: $D = W\dfrac{C_D}{C_L} = \dfrac{36,000}{1.095/0.0800} = 36,000/13.7 = 2,628$ lbs

The corresponding required horsepower is: $THP_{reqd} = 2,628 \times 248/550 = 1,185$ hp

The maximum rate-of-climb now follows from Step 6 as:

$$RC_{max} = (THP_{av} - THP_{reqd})/W = \frac{(6,768 - 1,185) \times 550}{36,000} = 85.3 \text{ ft/sec} = 5,116 \text{ ft/min}$$

The corresponding climb angle follows from: $\gamma = 85.3/248 = 0.34$ rad $= 19.5$ deg.

The reader will observe that this is no longer a small climb angle: it is larger than 15 degrees! Therefore, the small angle equations on which the solution to this problem is based may be questionable. Sub-section 9.3.2 deals with this question.

9.3.2 CLIMB PERFORMANCE FOR PROPELLER DRIVEN AIRPLANES AT STEEP ANGLES

For climbing flight at large flight path angles Eqns (9.35) and (9.36) were used for jet airplanes. They will be used here for a propeller driven airplane.

$$T - D = W\sin\gamma \tag{9.58}$$

and

$$L = W\cos\gamma \tag{9.59}$$

Eqn (9.58) is modified in terms of power:

$$TV - DV = WV\sin\gamma \quad \text{or:} \quad DV = TV - WV\sin\gamma \tag{9.60}$$

Dividing Eqn (9.59) by Eqn (9.60) yields:

$$\frac{L}{DV} = \frac{W\cos\gamma}{TV - WV\sin\gamma} = \frac{W\cos\gamma}{THP_{av} - WV\sin\gamma} \tag{9.61}$$

This can be rewritten as:

Climb and Drift–down Performance

$$\frac{C_L}{C_D} = \frac{WV\sqrt{1-\sin^2\gamma}}{THP_{av} - WV\sin\gamma} \quad (9.62)$$

By rearrangement this results in the following quadratic equation for the flight path angle:

$$\sin^2\gamma\left\{1 + \left(\frac{C_L}{C_D}\right)^2\right\} - 2\frac{THP_{av}}{WV}\left(\frac{C_L}{C_D}\right)^2\sin\gamma + \left(\frac{THP_{av}}{WV}\right)^2\left(\frac{C_L}{C_D}\right)^2 - 1 = 0 \quad (9.63)$$

The flight path angle then follows from:

$$\sin\gamma = \left(\frac{THP_{av}}{WV}\right)\frac{C_L^2}{C_L^2 + C_D^2} +$$

$$- \sqrt{\left(\frac{C_L^2}{C_L^2 + C_D^2}\frac{THP_{av}}{WV}\right)^2 - \left\{\frac{C_L^2}{C_L^2 + C_D^2}\left(\frac{THP_{av}}{WV}\right)^2\right\} + \frac{C_D^2}{C_L^2 + C_D^2}} \quad (9.64)$$

Note that the minus sign is required (and not the plus sign) in front of the square root.

The procedure for finding the flight path angle in a practical case is an iterative one:

Step 1: Determine the available thrust–horse–power, THP_{av} and assume that it is constant with speed, V.

Step 2: Select a specific value of speed, U

Step 3: Solve Eqn (9.59) for the lift coefficient, C_L while assuming that $\gamma = 0$.

Step 4: Find the lift-to-drag ratio, C_L/C_D.

Step 5: Use Eqn (9.64) to solve for γ.

Step 6: Solve Eqn (9.59) for the lift coefficient, C_L while assuming that $\gamma = \gamma$.

Step 7: Find the new lift-to-drag ratio, C_L/C_D.

Step 8: Use Eqn (9.64) to solve for γ.

Repeat this iteration until the flight path angle is within 0.1 degrees.

For the case of Example 9.2 it is seen that Eqn (9.64) yields:

$$\sin\gamma = \left(\frac{6,768}{36,000\times248/550}\right)\left(\frac{1.095^2}{1.095^2 + 0.0800^2}\right) +$$

$$-\sqrt{\left(\frac{1.095^2}{1.095^2 + 0.0800^2}\right)^2\left(\frac{6,768}{36,000\times248/550}\right)^2 - \frac{1.095^2}{1.095^2 + 0.0800^2}\left(\frac{6,768}{36,000\times248/550}\right)^2 + \frac{0.0800^2}{1.095 + 0.0800^2}}$$

or: $\sin\gamma = 0.9947\times0.4169 - \sqrt{0.9947^2\times0.4169^2 - 0.9947\times0.4169^2 + 0.0053} = 0.35$

This yields: γ = 20.4 deg. This is not very different from what was found with the small angle theory in Sub–section 9.3.2. The reason is that this angle is still not very steep.

9.3.3 DRIFT–DOWN PERFORMANCE FOR PROPELLER DRIVEN AIRPLANES

Whenever a multi–engine airplane is operated at its normal cruise altitude with all engines operating (AEO) it will still be able to climb when the throttles are advanced to their most forward position. However, if one engine becomes inoperative (OEI) the airplane may no longer be able to climb or maintain the AEO cruise altitude: the power required (including the drag due to the inoperative engine and the drag caused by re–trimming the airplane) may be greater than the power available with OEI. In that case the airplane is said to be in a drift–down mode. The airplane must then be flown to the OEI cruise altitude. For small propeller driven twins with normally aspirating piston engines the single engine ceiling may be considerably less than the terrain height over which the flight is being conducted. Pilots had better be aware of this.

The power required to fly level with OEI will be called: $THP_{reqd_{OEI}}$. Because of the increased drag with OEI it should be expected that: $THP_{reqd_{OEI}} > THP_{reqd_{AEO}}$. The power available with OEI will be called: $THP_{av_{OEI}}$. It should be expected that: $THP_{av_{OEI}} < THP_{reqd_{OEI}}$. The rate–of–descent, RD, is found by a modification of Eqn (9.11):

$$RD = \frac{-(D_{OEI} - T_{av_{OEI}})V}{W} = -\frac{(THP_{reqd_{OEI}} - THP_{av_{OEI}})}{W} \quad (9.65)$$

Of concern are the altitude at which OEI level flight is possible, the time required to drift down to that altitude and the steepness of the flight path during the drift–down process.

The OEI cruise altitude is defined as that altitude for which the rate–of–climb (RC) with OEI is at least 500 ft/min. A more detailed discussion of drift–down performance is given in Section 9.4.

9.4 METHODS FOR PREDICTING TIME–TO–CLIMB, TIME–TO––DRIFT–DOWN, AEO CEILINGS AND OEI CEILINGS

In this Section the following performance aspects will be presented:

9.4.1 Method for Predicting Time–to–Climb Performance
9.4.2 Method for Predicting Time–to–Drift–down Performance
9.4.3 Method for Predicting AEO and OEI Ceilings

9.4.1 METHOD FOR PREDICTING TIME–TO–CLIMB PERFORMANCE

The rate–of–climb, R.C. of an airplane was defined in Eqn (9.8) as:

$$\text{R.C.} = \frac{dh}{dt} \tag{9.66}$$

The time required to climb from one altitude to another can therefore be evaluated from:

$$t_2 - t_1 = \int_{h_1}^{h_2} \frac{dh}{\text{R.C.}} \tag{9.67}$$

The horizontal distance covered during the climb may be estimated from:

$$ds = \int_{h_1}^{h_2} V \cos\gamma \, dt \approx V_{ave}(t_2 - t_1) = R_{CL} \tag{9.68}$$

The weight of the fuel consumed during the climb may be estimated from:

$$W_{F_{CL}} = \int_{t_1}^{t_2} \dot{W}_{F_{CL}} \, dt = \int_{h_1}^{h_2} \frac{\dot{W}_{F_{CL}}}{\text{R.C.}} dh \tag{9.69}$$

The integrals in Eqns (9.67) through (9.69) are normally evaluated numerically. There are several reasons for this:

a) The rate–of–climb, R.C. depends on airplane configuration, weight, altitude thrust or power setting, all of which may vary during the climb

b) The speed during the climb is in general not a constant.

c) The fuel flow rate, $\dot{W}_f$, depends on airplane configuration, weight, altitude thrust or power setting, all of which may vary during the climb.

The functional relationships needed to perform the integrations explicitly are normally so complicated that explicit integration becomes un–practical. Numerical integration can be accomplished in the following four steps:

Step 1: Determine the maximum rate–of–climb for a range of altitudes and for a range of weights, all at a given thrust or power setting. Which thrust or power setting should be used depends on the type of engine and on any installation limitations* which are associated with that engine. These calculations may be done:

> at maximum continuous thrust or power
> at maximum climb thrust or power
> at any other thrust or power required

These calculations are normally done for standard atmospheric conditions with increments of +/– 5 degrees C.

The maximum rate–of–climb may be determined with the method implied by Eqn (9.27) for jets and Eqn (9.43) for propeller driven airplanes.

Step 2: Plot the maximum rate–of–climb versus altitude and weight for a given thrust or power setting at a given atmospheric temperature.

An example graph is given in Figure 9.14. Note the three ceiling definitions.

Step 3: Eqns (9.67) through (9.69) can be evaluated numerically with a table or spreadsheet. An example tabulation is given in Table 9.6.

Notes with Step 3: The calculation starts by dividing the expected airplane ceiling into incremental ranges of altitude. The following incremental ranges are recommended:

for fighters:	2,000 ft
for jet transports:	4,000 ft
for propeller driven transports:	5,000 ft

---

* Typical installation limitations may involve the use of air–conditioning packs, anti– or de–icing systems and other systems which consume a significant amount of power.

Climb and Drift–down Performance

Figure 9.14 Example of Maximum Rate-of-climb Versus Altitude and Weight

Turbo–prop transport for a range of weights, given power setting, given temperature

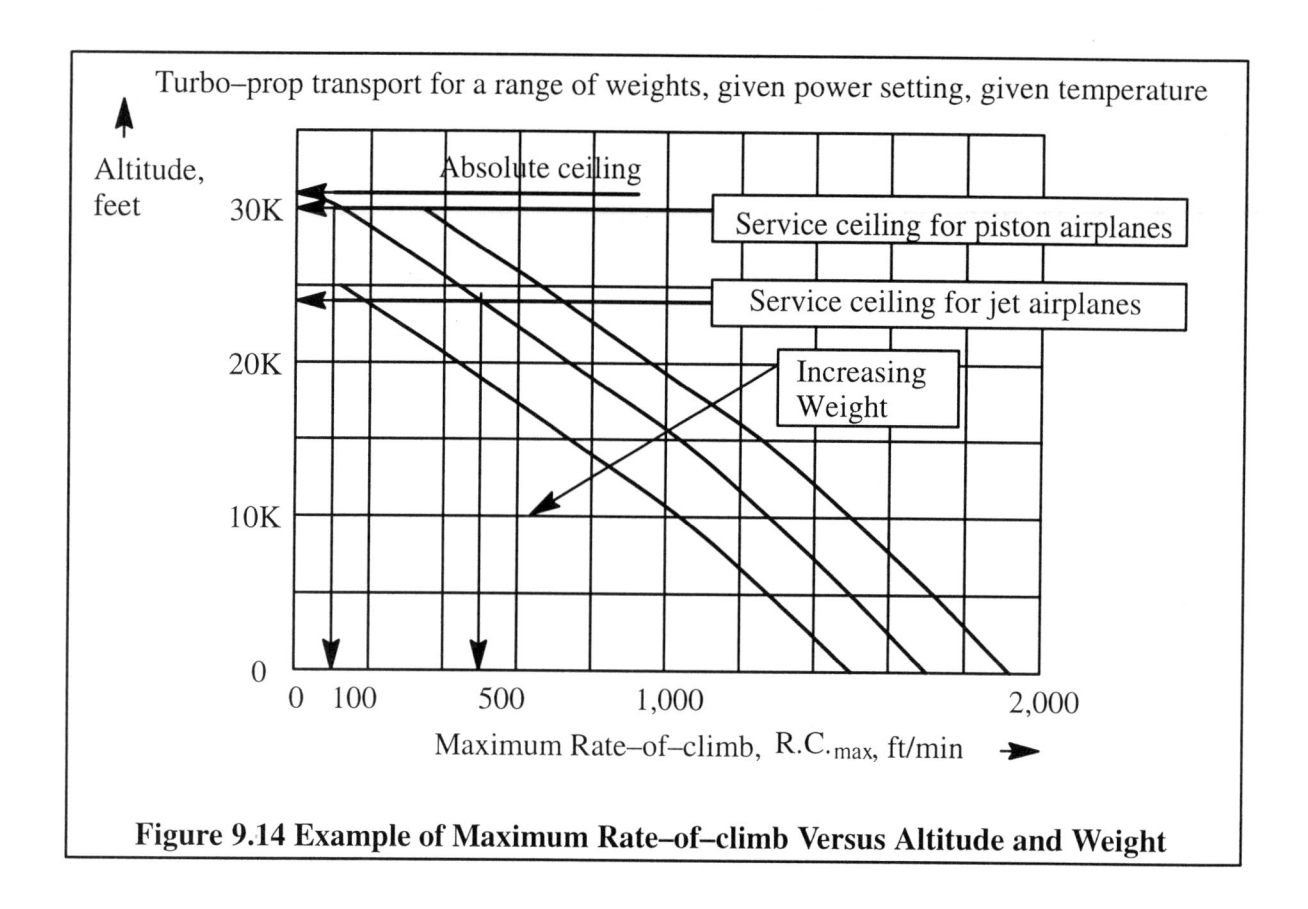

(Chart: Altitude, feet vs. Maximum Rate-of-climb, R.C.$_{max}$, ft/min; showing Absolute ceiling, Service ceiling for piston airplanes, Service ceiling for jet airplanes, and Increasing Weight curves)

Table 9.6 Numerical Integration for Time–to–climb, Distance Covered and Fuel Used

Range of Press. Alt.	Average Press. Alt.	Increment in Press. Alt.	Increment in True Alt.	Average Weight	Average R.C.	Time Increment	Total time
(1)	(2)	(3)	(4)	(5)	(6)	(7)	(8)
Select h_{1_p} to h_{2_p}	Compute $\dfrac{h_{1_p} + h_{2_p}}{2}$	$\Delta h_p = h_{1_p} - h_{2_p}$	$\Delta h_{true} = \Delta h_p \times \dfrac{\text{Abs. T}}{\text{Abs.Std, T}}$	Assume W_{ave}	R.C.$_{ave}$ See Fig. 9.15	$\Delta t = \dfrac{\Delta h_{true}}{\text{R.C.}_{ave}}$	$\Sigma \Delta t$

(9)	(10)	(11)	(12)	(13)	(14)
Average flight speed V_{ave} From climb performance data	Incremental horizontal distance covered $\Delta S = V_{ave} \times \Delta t$	Horizontal distance covered $\Sigma \Delta S$	Average fuel flow rate $\dot{W}_f$ From engine data	Incremental fuel used $\Delta W_f = \dot{W}_f \times \Delta t$	Total fuel used in climb $\Sigma \Delta W_f$

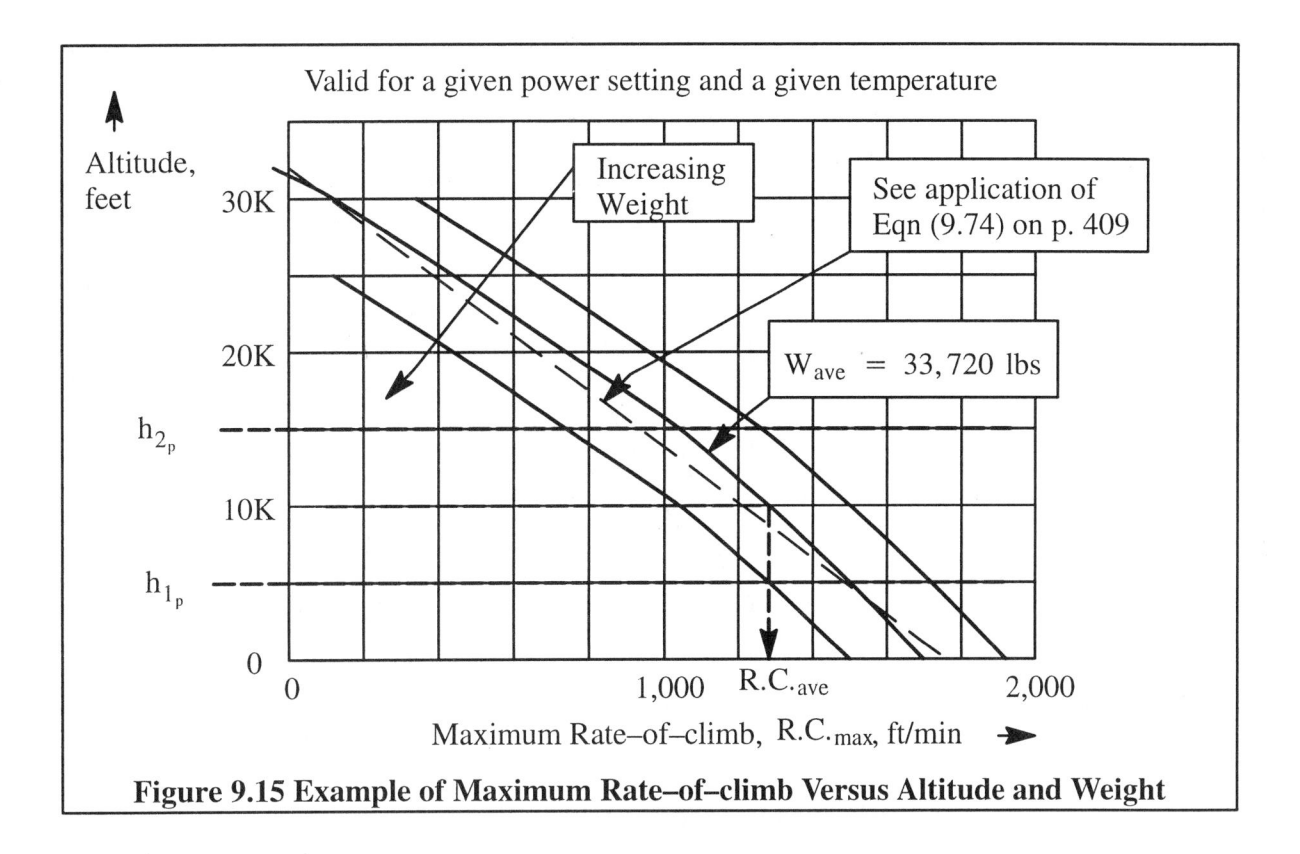

Figure 9.15 Example of Maximum Rate–of–climb Versus Altitude and Weight

The calculations should be done for a range of airplane weights at the beginning of the climb. The selected range of weights depends on the airplane type. For a B–747, the weight difference between take–off and landing on a maximum payload–range mission is about 40%. In such a case weight increments of 5% should be accounted for in climb calculations.

The R.C. calculated for each step is the average R.C. for the average weight of the airplane in climbing from h_1 to h_2.

After the fuel weight used in climbing from one altitude to another has been determined, the average weight assumed for the climb calculations may have to be iterated.

Step 4: Plot the variations of distance, fuel used and time–to–climb for the range of altitudes and airplane weights. An un–scaled example of such plots is shown in Figure 9.16.

A numerical example of the calculation of the time to climb to altitude is now presented for the case of the turbo–prop airplane at a constant weight of 33,720 lbs. The maximum rate–of–climb data are those of Figure 9.15. The example is given in Table 9.7.

Table 9.7 shows that the time required to climb to 30,000 ft (which is approximately the service ceiling for this airplane) is just under 44 minutes.

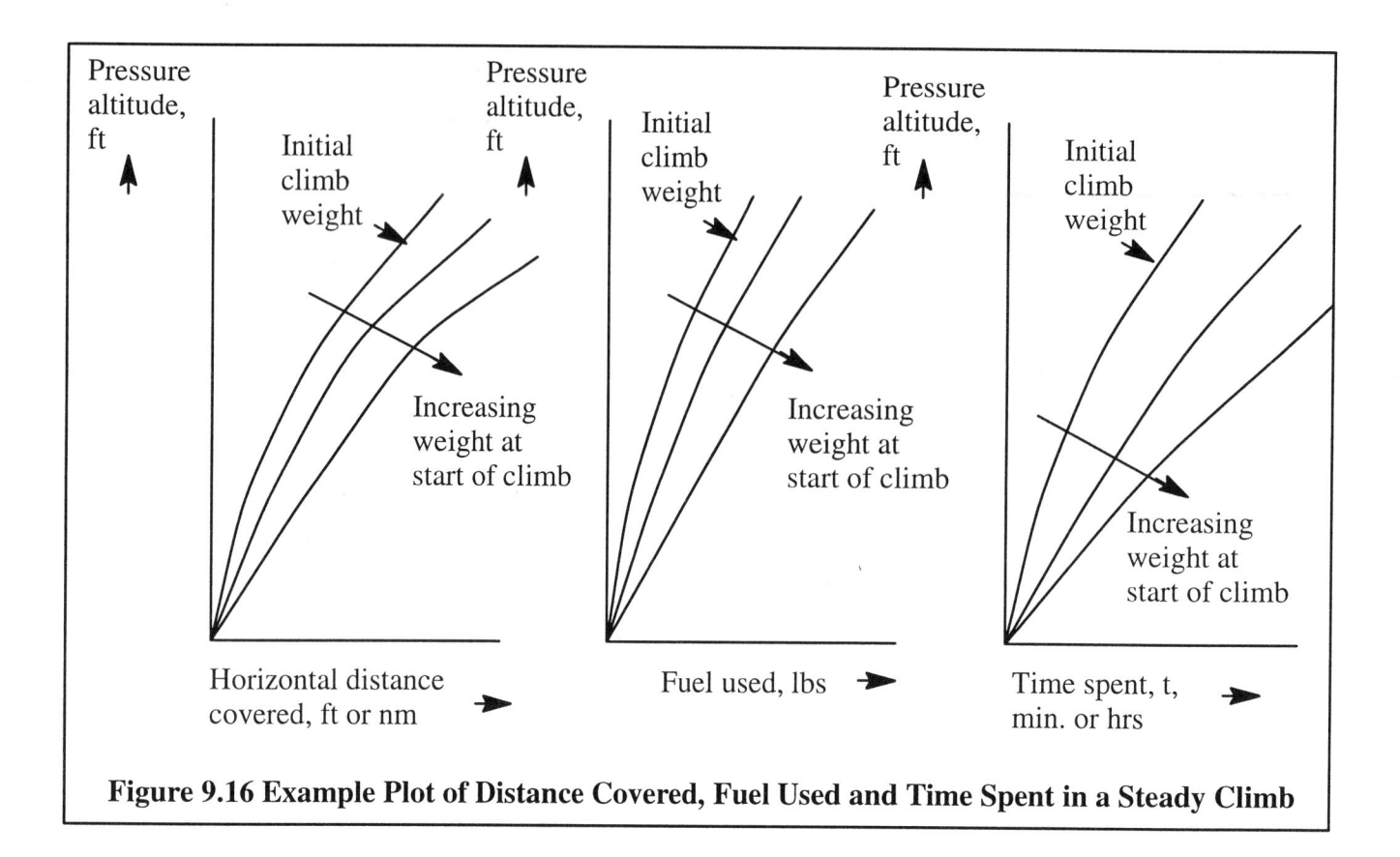

Figure 9.16 Example Plot of Distance Covered, Fuel Used and Time Spent in a Steady Climb

Table 9.7 Numerical Determination of the Time-to-Climb to 30,000 ft					
Altitude, ft	Maximum Rate-of-climb, ft/min $R.C._{max}$	Average Rate-of-climb, ft/min $R.C._{ave}$	Time increment, min. Δt_i	Total time spent, min. $t = \Sigma \Delta t_i$
0	1,700			0
		1,600	3.13	
5,000	1,500			3.13
		1,390	3.60	
10,000	1,280			6.73
		1,155	4.33	
15,000	1,030			11.06
		885	5.65	
20,000	740			16.71
		585	8.55	
25,000	430			25.26
		275	18.18	
30,000	120			43.44
	↓ From Fig. 9.15			↓ Time to climb to 30,000 ft

It is seen from Figure 9.14 that the maximum rate–of–climb, when plotted against altitude can be represented by sequential straight line segments, or in some cases, by one straight line. For a given altitude interval the R.C. at any altitude, h, may be written as:

$$R.C. = R.C._{h_1} - k(h - h_1) \qquad (9.70)$$

where the slope, k, may be determined from:

$$k = \frac{R.C._{h_2} - R.C._{h_1}}{h_1 - h_2} \qquad (9.71)$$

In such a case, the time–to–climb can evaluated directly from Eqn (9.67) as:

$$t_2 - t_1 = \frac{1}{k}\ln\left\{\frac{R.C._{h_1}}{R.C._{h_1} - k(h_2 - h_1)}\right\} = \frac{1}{k}\ln\left(\frac{R.C._{h_1}}{R.C._{h_2}}\right) \qquad (9.72)$$

In the following it is assumed that the entire plot of maximum R.C. versus altitude can be represented by a straight line, as shown in Figure 9.17.

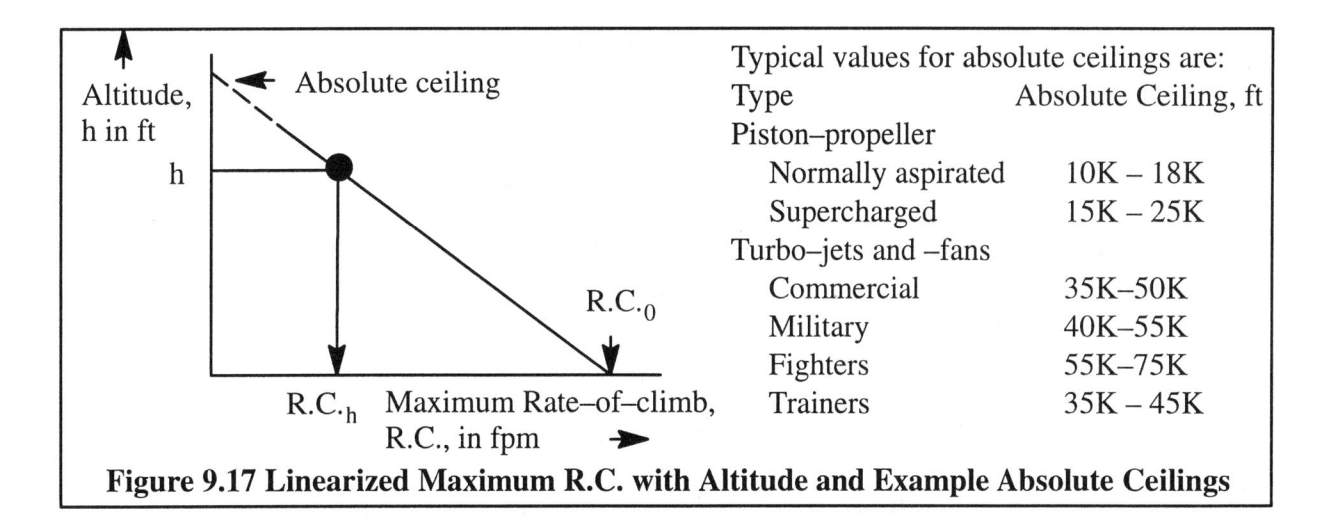

Typical values for absolute ceilings are:

Type	Absolute Ceiling, ft
Piston–propeller	
Normally aspirated	10K – 18K
Supercharged	15K – 25K
Turbo–jets and –fans	
Commercial	35K–50K
Military	40K–55K
Fighters	55K–75K
Trainers	35K – 45K

Figure 9.17 Linearized Maximum R.C. with Altitude and Example Absolute Ceilings

In this case, the climb performance can be modelled as:

$$R.C. = R.C._0(1 - h/h_{absolute}) \qquad (9.73)$$

The time–to–climb to a given altitude, h, for a given absolute ceiling, $h_{absolute}$, and for a given sea–level maximum rate–of–climb, $R.C._0$, may be determined from:

$$t_{climb} = \frac{h_{absolute}}{R.C._0} \ln\left(\frac{1}{1 - h/h_{absolute}}\right) \qquad (9.74)$$

In preliminary design, Eqn (9.74) can be used to determine the required, maximum sea–level rate–of–climb, $R.C._0$ required to satisfy a given time–to–climb to altitude requirement.

Eqn (9.74) can be applied to the case of the turbo–prop transport of Figure 9.15. The straight line approximation, drawn in Fig. 9.15 suggests:

$R.C._0 = 1,850$ ft/min and $H_{absolute} = 32,000$ ft

Eqn (9.74) yields for the time–to–climb to 30,000 ft: 48 minutes. This is not too far from the numerically determined time of 44 minutes.

Another useful graphical representation of the ability of an airplane to reach a given altitude for a given weight, is to plot:

   a) the thrust–limited altitude as a function of weight for various thrust settings
   b) the buffet–limited altitude for various thrust settings

Figure 9.18, based on the Fall 1996 issue of Boeing Airliner magazine shows an example.

The maximum certified altitude (MCA) of the 747–400 is 45,100 ft. The MCA depends on the ability of the fuselage structure to withstand the pressure differential between the cabin pressure and the prevailing outside air pressure. As seen in Figure 9.18, the 747–400 with PW4000 engines can reach that altitude only at very low weights.

The thrust–limited altitude is plotted for the following two cases:

   1) with a 100 ft/min residual climb capability
   2) with a 300 ft/min residual climb capability

The buffet–limited altitude is plotted for the following two cases:

   1) with a 0.2g margin to initial buffet (this allows 34 degrees of bank in a level turn)
   2) with a 0.3g margin to initial buffet (this allows 40 degrees of bank in a level turn)

Initial buffet is defined as a condition of a peak–to–peak 0.1g accelerometer reading at the pilot's seat track. It is normally caused by a shock induced flow separation somewhere on the wing (high speed buffet) or a stall induced flow separation somewhere on the wing (low speed buffet).

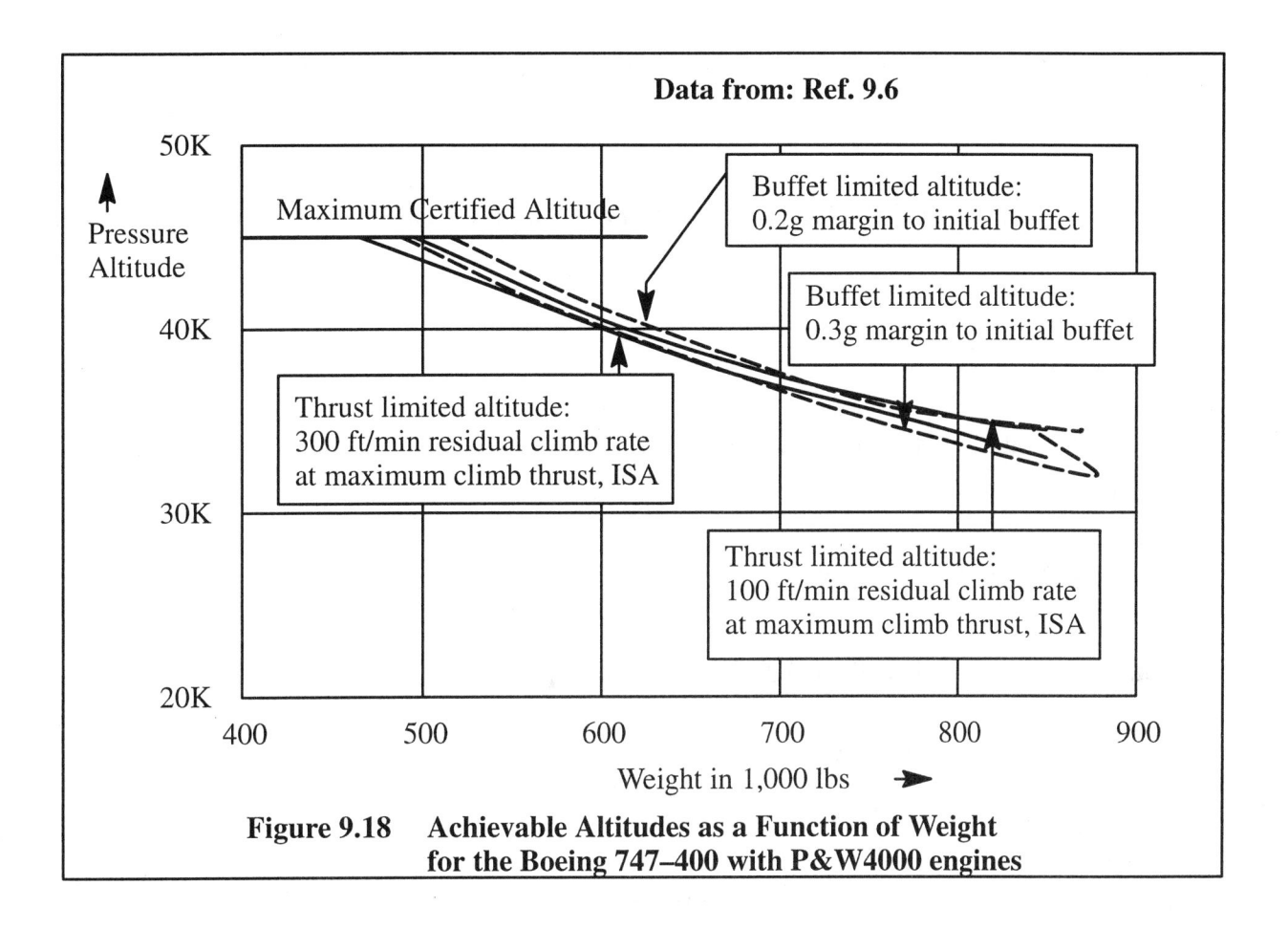

Figure 9.18 Achievable Altitudes as a Function of Weight for the Boeing 747-400 with P&W4000 engines

9.4.2 METHOD FOR PREDICTING TIME-TO-DRIFT-DOWN PERFORMANCE

When one engine becomes inoperative (OEI) the performance of an airplane is affected in two ways:

a) The amount of available thrust (or power) is reduced

b) The drag of the airplane is increased because of the extra drag on the stopped engines and the trim drag required to keep the airplane on a straight line flight path

As a result, when an airplane is operating at its maximum cruise altitude, it will have to descend (i.e. drift down) to a lower altitude. Particularly for transoceanic twins, operating under ETOPS (Extended Twin Operations) rules, the time to drift-down, the distance covered and the remaining range capability of the airplane on one engine and with a higher drag level all at the OEI cruise ceiling can become important issues.

The rate-of-descent (R.D.) of an airplane with OEI may be determined with Eqns (9.40) and (9.65) for jet and for propeller driven airplanes respectively.

The rate–of–descent, R.D. of an airplane was defined in Eqn (9.17) as:

$$\text{R.D.} = -\frac{dh}{dt} \tag{9.75}$$

The time required to drift–down (descend) from one altitude to another follows from:

$$t_2 - t_1 = \int_{h_1}^{h_2} \frac{-dh}{\text{R.D.}} \tag{9.76}$$

The horizontal distance covered during the drift–down maneuver may be estimated from:

$$ds = \int_{h_1}^{h_2} V\cos\gamma\, dt \approx V_{ave}(t_2 - t_1) \tag{9.77}$$

The weight of the fuel consumed during the descent may be estimated from:

$$W_f = \int_{t_1}^{t_2} \dot{W}_f\, dt = \int_{h_1}^{h_2} \frac{-\dot{W}_f}{\text{R.D.}}\, dh \tag{9.78}$$

The integrals in Eqns (9.75) through (9.78) are normally evaluated numerically. There are several reasons for this:

a) The rate–of–descent, R.D. depends on airplane configuration, weight, altitude thrust or power setting, added drag, all of which may vary during the climb
b) The speed during the climb is in general not a constant.
c) The fuel flow rate, $\dot{W}_f$, depends on airplane configuration, weight, altitude thrust or power setting, added drag, all of which may vary during the climb.

The functional relationships needed to perform the integrations explicitly are normally so complicated that explicit integration becomes un–practical. Numerical integration can be accomplished in the following four steps:

Step 1: Determine the minimum rate–of–descent for a range of altitudes and weights, at a given thrust or power setting for the OEI condition. The thrust or power setting to be used depends on the type of engine and on any installation limitations (See foot–note on p. 405.) which are associated with that engine. The calculations may be done:
  at maximum continuous thrust or power
  at maximum climb thrust or power
  at any other thrust or power required

These calculations are normally done for standard atmospheric conditions with increments of +/– 5 degrees C. The minimum rate–of–descent may be determined with the method implied by Eqn (9.40) for jets and Eqn (9.65) for propeller airplanes.

Step 2: Plot the minimum rate–of–descent versus altitude and weight for a given thrust or power setting at a given atmospheric temperature.

Figure 9.19 shows the thrust available and thrust required data for a range of altitudes for a twin engined business jet with a weight of 16,000 lbs. Observe that the thrust available with one engine inoperative is roughly half that with all engines operating.

Also observe that the thrust required with one engine inoperative is somewhat **higher**

than that required for all engines operating. The reason is item b) on page 411.

For the case of Figure 9.19 it is seen that the airplane must descend for altitudes higher than 20,000 ft. Table 9.8 contains the minimum descent rate calculations based on the data on the right side of Figure 9.19. The data are plotted in Figure 9.20.

The descent calculations are continued until the airplane reaches its so–called OEI cruise altitude. The OEI cruise altitude is defined as that altitude for which (at the given weight) the airplane can climb at R.C. = 100 ft/min.

For the airplane of Figure 9.19 at a weight of 16,000 lbs that altitude is established by interpolation in Figure 9.20: the OEI ceiling is 25,200 ft.

Step 3: Eqns (9.76) through (9.78) can be evaluated numerically with a table or spreadsheet. An example tabulation is also given in Table 9.8. The calculation starts by dividing the expected airplane ceiling into incremental ranges of altitude. The following incremental ranges are recommended:

for fighters:	2,000 ft
for jet transports:	4,000 ft
for propeller driven transports:	5,000 ft

It is seen in Figure 9.20 that the R.D. versus altitude is essentially a straight line. With Eqn (9.76), this would theoretically result in an infinite time to descend. In practice the airplane will be flown with a rate of descent not less than about 100 ft/min. The descent time from 40,000 ft to the OEI ceiling may be evaluated by numerical integration. The results are also shown in Table 9.8. The descent time to the OEI ceiling (also called the drift–down time) is estimated to be 45.9 min.

Climb and Drift-down Performance

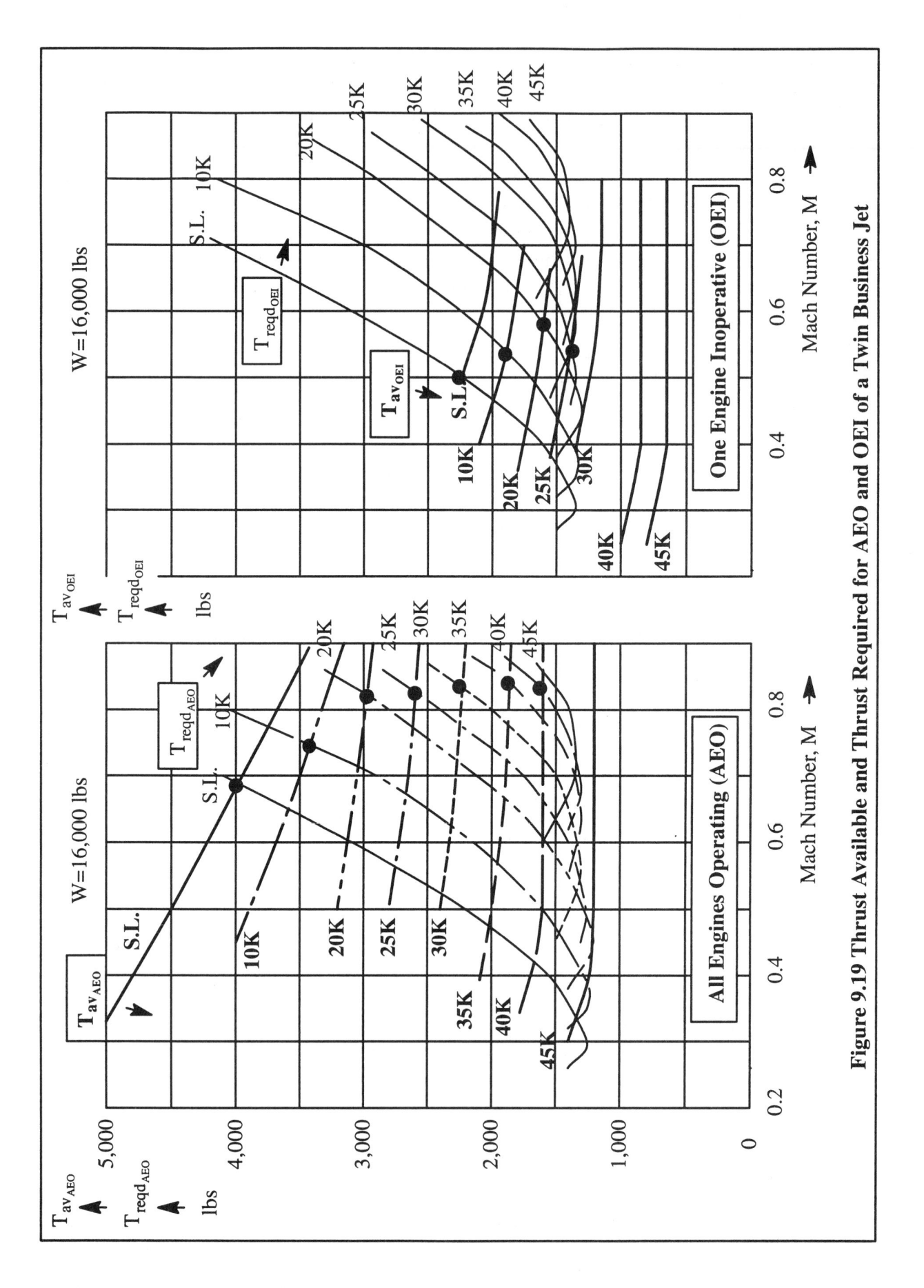

Figure 9.19 Thrust Available and Thrust Required for AEO and OEI of a Twin Business Jet

Table 9.8 Numerical Determination of the Time–to–Descend to OEI Cruise Altitude

Altitude, ft	$T_{reqd_{OEI}} - T_{av_{OEI}}$ lbs	M	V ft/sec	Minimum Rate –of–descent, ft/min	
40,000	500	0.70	678	1,271	W = 16,000 lbs Standard atmosphere
35,000	333	0.63	613	765	These data are plotted in Figure 9.20
30,000	167	0.57	567	355	
25,000	−67	0.51	518	−130	

Altitude, ft	Altitude Increment, ft	Average Rate–of–descent, R.D.$_{ave}$ ft/min	Time increment, min. Δt_i	Total time spent, min. $t = \Sigma \Delta t_i$
40,000				
	4,000	1,100	3.6	3.6
36,000				
	4,000	700	5.7	9.3
32,000				
	4,400	350	12.6	21.9
27,600				
	2,400	100	24.0	45.9
25,200				
		↓		↓
		From Fig. 9.20		Time to drift–down to OEI ceiling

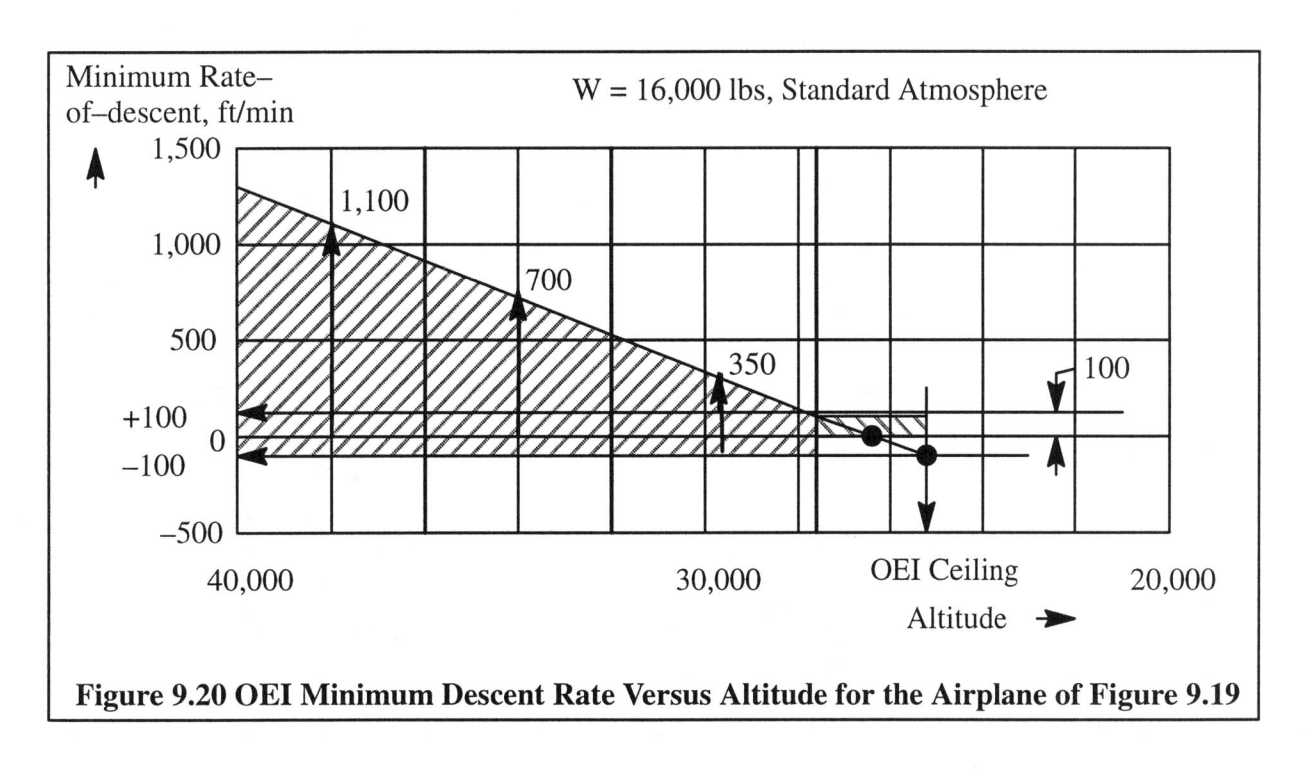

Figure 9.20 OEI Minimum Descent Rate Versus Altitude for the Airplane of Figure 9.19

Chapter 9

The horizontal distance covered depends on the average speed which, according to Table 9.8, is roughly 598 ft/sec. The horizontal distance covered is given by:
45.9 x 60 x 598 / 6,076 = 271 n.m.

Notes with Step 3: The distance covered during the OEI drift–down is an important parameter in flight track planning for over–water flights, where so–called "wet foot–prints" must be avoided. A wet foot–print (meaning landing in the water, following flame–out of the remaining engine due to fuel starvation) occurs if an airplane, when suffering an engine out condition during a prolonged over–water flight, cannot reach a suitable diversion airport. The operational importance of the wet foot–print idea is discussed in detail in Reference 9.8.

Note that there is a problem with these calculations. The minimum rate of descent occurs at a lower Mach number as the airplane descends. This requires the pilot to slow the airplane down. Low speed buffet considerations may inhibit this. Also, to ease pilot workload in such a descent, it would be preferred to fly at a constant Mach number. Modern flight management systems (FMC) can be programmed to fly a drift–down pattern automatically.

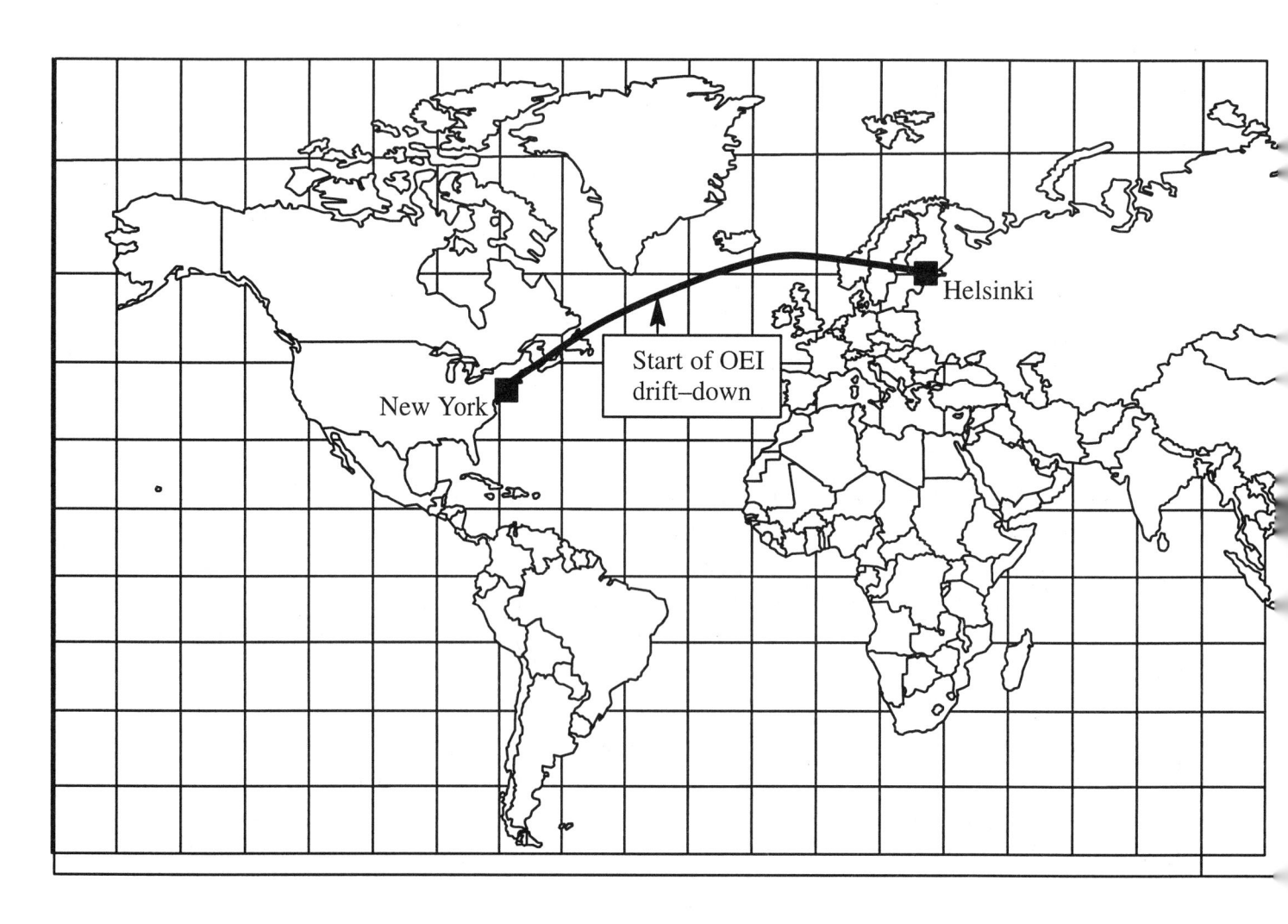

9.4.3 METHOD FOR PREDICTING AEO AND OEI CEILINGS

Several previous discussions have already mentioned various definitions of airplane ceilings. It is useful to review these definitions in one place.

Absolute Ceiling

The absolute ceiling of an airplane is that altitude at which the rate–of–climb, R.C. = 0.

For a given weight, the rate of climb depends on the air–speed and on the thrust (or power) level selected by the pilot. In the case of the absolute ceiling at a given weight, the thrust level is assumed to be the maximum continuous thrust (or power) allowable in civil operations and the maximum military thrust (or power) in military operations. The speed is that at which the thrust required (or power required) becomes tangent to the thrust available (or power available) curves.

Figure 9.21 shows a graphical example illustrating how the absolute ceiling comes about.

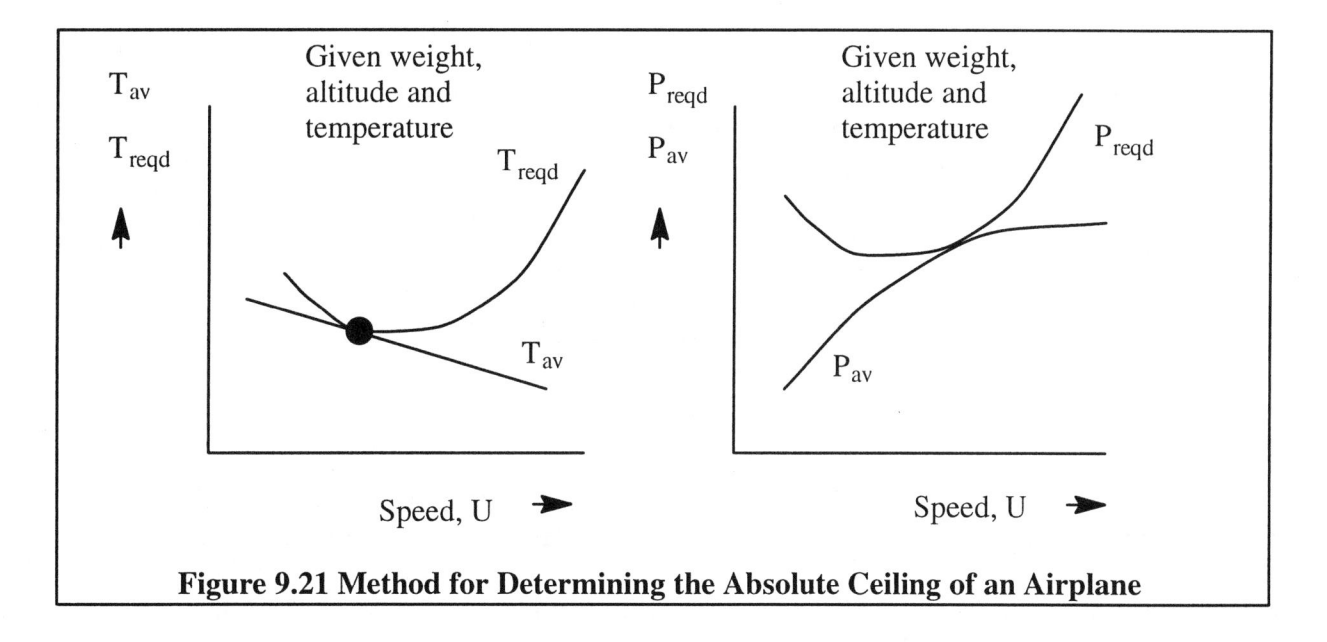

Figure 9.21 Method for Determining the Absolute Ceiling of an Airplane

A more direct graphical method for determining the absolute ceiling is to plot the maximum available climb rate versus altitude as shown in Figure 9.17. Extending Figure 9.17 to negative climb rates (i.e. rates of descent) results in a combination of Figures 9.17 and 9.20 and is shown in Figure 9.22.

This definition of absolute ceiling applies to AEO and OEI operations. Obviously with OEI the available thrust (or power) will be less than that for AEO. Also, with OEI the drag of the airplane will be higher than that with AEO. Figure 9.19 provides an illustration of the difference between AEO and OEI characteristics.

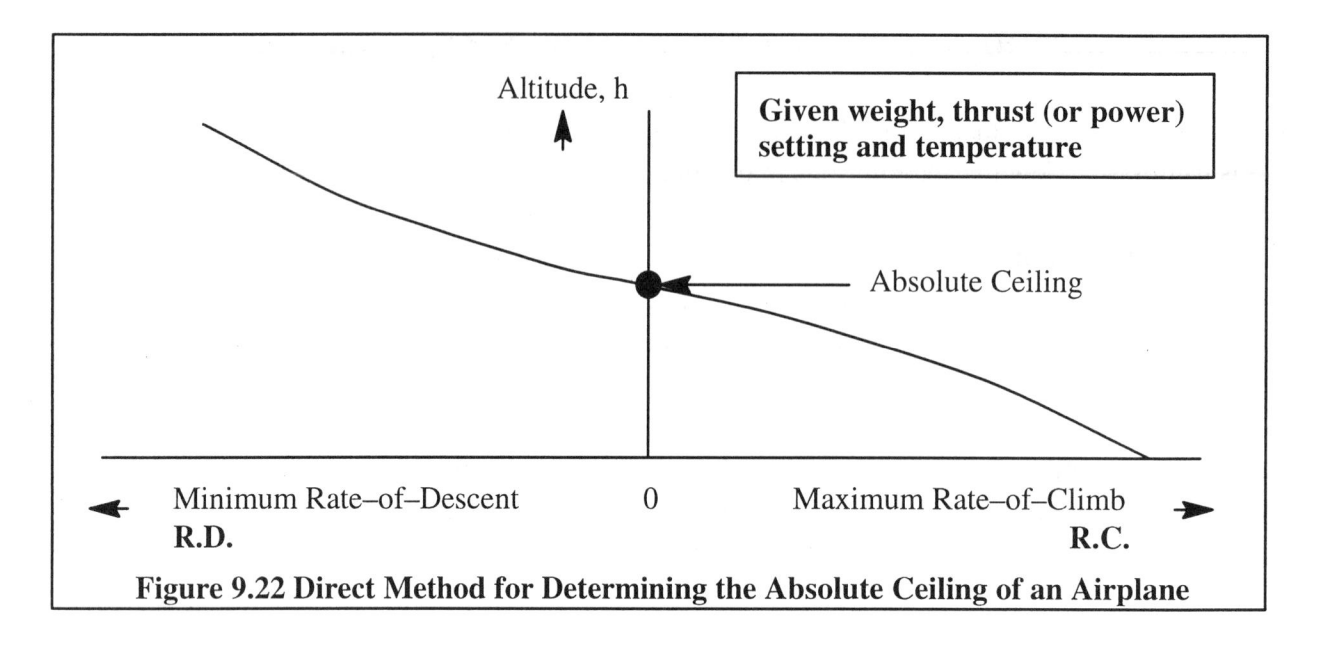

Figure 9.22 Direct Method for Determining the Absolute Ceiling of an Airplane

Service Ceiling

The service ceiling of an airplane is that altitude at which the rate-of-climb, takes on the following values:

    at **maximum continuous thrust or power**:
    for commercial piston–propeller airplanes   R.C. = 100 ft/min.
    for commercial jet airplanes   R.C. = 500 ft/min.

    at **maximum military thrust or power**:
    for military airplanes   R.C. = 100 ft/min.

The graphical method of Figure 9.23 may be used to determine the service ceiling.

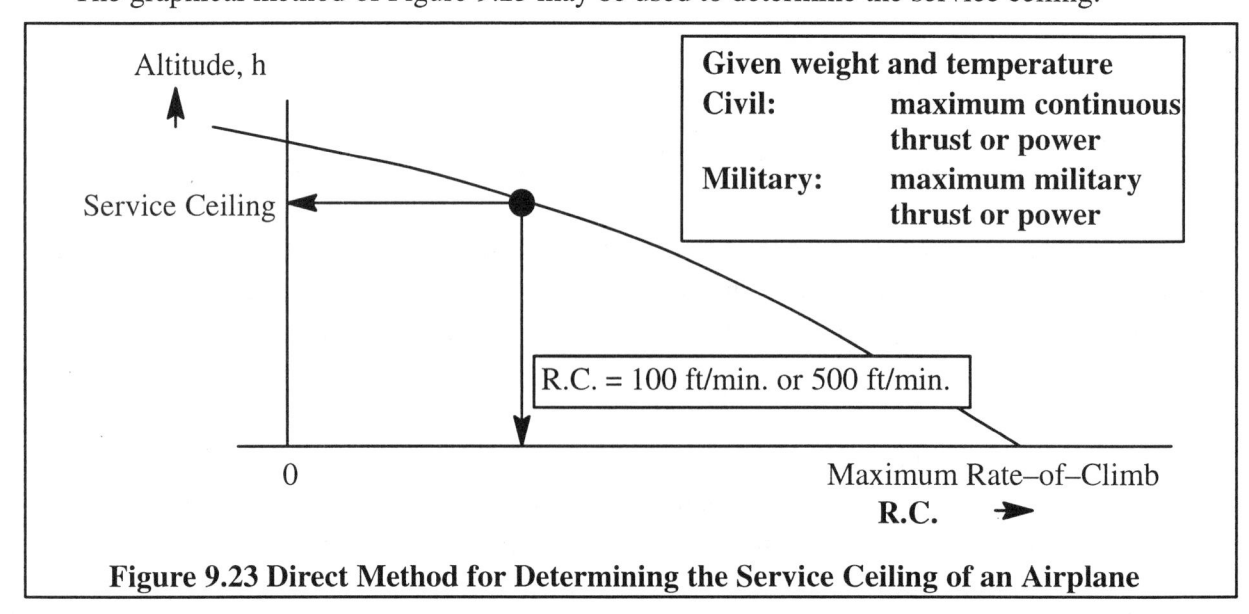

Figure 9.23 Direct Method for Determining the Service Ceiling of an Airplane

This definition of service ceiling applies to AEO and OEI operations. Obviously with OEI the available thrust (or power) will be less than that for AEO. Also, with OEI the drag of the airplane will be higher than that with AEO. Figure 9.19 provides an illustration of the difference between AEO and OEI characteristics.

Cruise and Combat Ceiling for Military Airplanes

The cruise and combat ceiling of a military airplane is that altitude at which the rate–of–climb, takes on the following values:

Cruise ceiling at maximum continuous thrust or power   for M<1.0:   300 ft/min.
                                                                                                                                                for M>1.0   1,000 ft/min.

Combat ceiling at maximum thrust or power   for M<1.0   500 ft/min.
                                                                                                                                 for M>1.0   1,000 ft/min.

The method used to determine these ceilings is similar to that suggested in Figure 9.23.

Table 9.9 shows examples of various airplane ceilings published in Jane's All the World's Aircraft of 1995–1996 (Ref. 9.7). Note the significant differences between AEO and OEI ceilings.

Table 9.9 Examples of Airplane Ceilings (Data from: Ref.9.7)				
N.A.= Not applicable Type	Weight, in lbs	Maximum R.C. at sea–level in ft/min.	Service Ceiling in feet AEO	OEI
Sukhoi Su–22M–4	42,770	45,275	49,865	Not appl.
Lockheed–Martin F–16C	27,185	N.A.	50,000	Not appl.
McDonnell–Douglas F/A–18C	36,710	N.A.	50,000 (Combat)	N.A.
Northrop–Grumman B–2A	336,500	N.A.	50,000 (Combat)	N.A.
Fairchild Metro 23	16,500	2,700	25,000	11,600
Boeing 767–200	300,000	N.A.	N.A.	21,400
Cessna Citation VI	22,000	3,700	51,000	23,500
Cessna Citationjet	10,400	3,311	41,000	26,200
Beechcraft 1900D	16,950	2,625	33,000	17,000
Piper Malibu Mirage	4,300	1,218	25,000	Not appl.

9.5 EFFECT OF FORWARD AND VERTICAL ACCELERATIONS ON CLIMB PERFORMANCE

The effects of forward and vertical acceleration as expressed by the right hand side terms in Equations (9.1) and (9.2) have been neglected in the previous sections. Therefore, only steady state climb cases could be considered. In this section, the effect of forward accelerations will be considered in Sub–section 9.5.1. A brief discussion of the effect of vertical accelerations will be given in Sub–section 9.5.2.

9.5.1 EFFECT OF FORWARD ACCELERATION

If the effect of vertical acceleration in Eqn (9.2) is neglected and small flight path angles are assumed (L = W) the resulting flight situation is referred to as quasi–steady flight. The rate–of–climb, according to Eqn (9.10) can then be expressed as:

$$\text{R.C.} = \frac{\frac{(T - D)V}{W}}{1 + \frac{V}{g}\frac{dV}{dh}} \tag{9.79}$$

This equation can be re–arranged to yield:

$$\text{R.C.}\left(1 + \frac{V}{g}\frac{dV}{dh}\right) = \frac{(T - D)V}{W} = \frac{P_{av} - P_{reqd}}{W} \tag{9.80}$$

In this form, the right hand side is referred to as the specific excess power of the airplane.

The term $(V/g)(dV/dh)$ is called the acceleration factor. The effect of this acceleration factor on climb performance will now be discussed.

To determine the numerical value of the acceleration factor, the speed–versus–altitude schedule of the airplane must be known. It has been shown that the Mach number associated with the maximum rate of climb of an airplane increases with altitude. Therefore, an airplane must accelerate along its flight path to keep flying at the maximum R.C. This would result in a fairly complicated speed–altitude schedule during a climb. In turn, this would be a difficult task for a pilot. As it turns out, for most airplanes, a constant calibrated air–speed can be identified which corresponds roughly to the speed–altitude schedule for maximum rate–of–climb.

Therefore, in practice, pilots and performance programs written for flight management systems, assume that airplanes are climbed at a constant calibrated air–speed (equivalent air–speed corrected for compressibility) until the cruise Mach number is reached. Following that, climb is continued to the initial cruise altitude at constant Mach number.

Climb and Drift-down Performance

During climbs in the troposphere (below 36,089 ft) a constant calibrated air-speed schedule is followed which implies that the correction factor, VdV/gdh can be taken at constant calibrated air-speed, V_c, which is plotted in Figure 9.24. At altitudes above 36,089 (stratosphere) most jets are climbed at constant Mach number so that dV/dh = 0.

For the reasons given before, two cases are of special interest in the troposphere:

a) Climb at constant equivalent air-speed

and

b) Climb at constant Mach number.

The correction factors which apply to these cases are derived in Sub-sub-sections 9.5.1.1 and 9.5.1.2 respectively.

9.5.1.1 Climb at Constant Equivalent Air-speed

At constant equivalent air-speed, the true air-speed will increase with increasing altitude in accordance with:

$$V = \frac{V_e}{\sqrt{\sigma}} \tag{9.81}$$

The dU/dh term in Eqn (9.80) can now be written as:

$$\frac{dV}{dh} = \frac{dV}{d\sigma}\frac{d\sigma}{dh} = -\frac{1}{2}\sigma^{-3/2}V_e\frac{d\sigma}{dh} \tag{9.82}$$

The derivative $d\sigma/dh$ depends on the altitude, h, itself. Because flight at constant equivalent air-speed is typically conducted in the troposphere (i.e. well below 36,089 ft in the standard atmosphere) it follows from Eqns (1.17) and (1.15) that:

$$\sigma = \frac{\varrho}{\varrho_0} = (1 - ah)^b \quad \text{where:} \quad a = 6.875 \times 10^{-6} \text{ ft}^{-1} \quad \text{and} \quad b = 4.2561 \tag{9.83}$$

Therefore, for climbs in the troposphere the R.C. correction factor is::

$$\frac{V}{g}\frac{dV}{dh} = \frac{V}{2g}ab\sigma^{-1/2}V_e = \frac{V^2 ab}{2g\,\theta} = \frac{V^2}{2g}\frac{ab}{V_a^2/V_{a_0}^2} = \frac{abV_{a_0}^2}{2g}M^2 = 0.567M^2 \tag{9.84}$$

Climb and Drift-down Performance

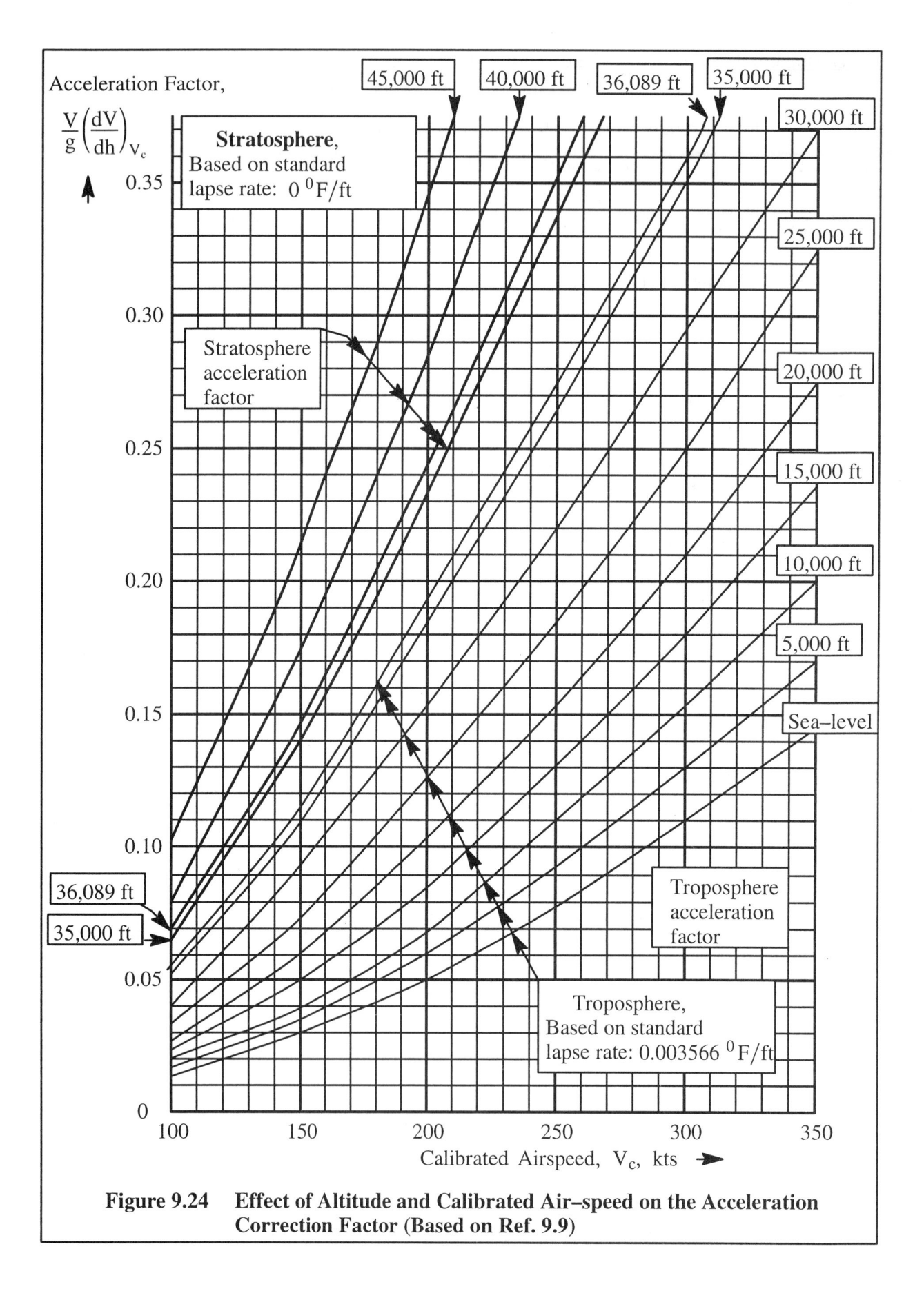

Figure 9.24 Effect of Altitude and Calibrated Air-speed on the Acceleration Correction Factor (Based on Ref. 9.9)

9.5.1.2 Climb at Constant Mach Number

In an accelerated climb profile, the cruise Mach number is normally reached well before the limit of the troposphere is reached. For flight in the troposphere the flight speed, V, can be expressed as:

$$V = M V_a = M(\gamma g R T)^{1/2} = M(\gamma g R)^{1/2}(1 - ah)^{1/2} T_0^{1/2} \qquad (9.85)$$

To arrive at Equation (9.85), use has been made of Eqns (2.31) and (1.15)

After differentiation:

$$\frac{dV}{dh} = M(\gamma g R)^{1/2} T_0^{1/2} \frac{1}{2}(-a)\left(\frac{T}{T_0}\right)^{-1/2} \qquad (9.86)$$

From this it follows that the R.C. correction factor can be written as:

$$\frac{V}{g}\frac{dV}{dh} = \frac{1}{2}\frac{\gamma g R}{g} T_0(-a)\frac{V}{V_a}M - -\frac{\gamma R T_o a}{2} M^2 = -0.133 M^2 \qquad (9.87)$$

The effect of this acceleration correction factor to R.C. is illustrated in Example 9.3.

Example 9.3: Calculate the correction factor the the rate–of–climb, R.C. for the following two cases:

    a) A typical propeller transport which climbs at a constant calibrated air–speed of $V_c = 120$ kts

    b) A typical jet transport which climbs at a constant calibrated air–speed of 280 kts (note: strictly speaking, below 10,000 ft a speed of 250 kts may not be exceeded).

Solution: The calculations for cases a) and b) are presented in Table 9.10.

The results in Table 9.10 indicate the following:

(1) To climb at a constant calibrated air–speed, the true air–speed must increase as the altitude is increased.

(2) For the propeller driven transport, if the acceleration factor is ignored, the predicted rate–of–climb will be too high by 2% at sea–level and by 10% at 40,000 ft.

(3) For the jet transport, if the acceleration factor is ignored, the predicted rate-of-climb will be too high by 9% at sea-level and by 21% at 40,000 ft.

The results obtained for the propeller transport are consistent with results obtained in Ref. 9.10.

Table 9.10 Calculation of Climb Rate Correction Factor for Two Transports

a) Propeller Transport at 120 keas

Altitude, h, ft	Density ratio, σ (Appendix A)	ΔV_c, kts (Fig. 2.9)	$V = \dfrac{V_c - \Delta V_c}{\sqrt{\sigma}}$ kts	$\dfrac{V}{g}\dfrac{dV}{dh}$ (Fig. 9.24)	$\dfrac{1}{1 + \dfrac{V}{g}\dfrac{dV}{dh}}$
0	1.0	0.0	120.0	0.018	0.982
10,000	0.73848	0.3	139.3	0.025	0.976
20,000	0.53281	0.6	163.6	0.038	0.963
30,000	0.37413	1.2	194.2	0.060	0.943
40,000	0.24617	2.0	237.8	0.115	0.897

b) Jet Transport at 280 keas

Altitude, h, ft	Density ratio, σ (Appendix A)	ΔV_c, kts (Fig. 2.9)	$V = \dfrac{V_c - \Delta V_c}{\sqrt{\sigma}}$ kts	$\dfrac{V}{g}\dfrac{dV}{dh}$ (Fig. 9.24)	$\dfrac{1}{1 + \dfrac{V}{g}\dfrac{dV}{dh}}$
0	1.0	0.0	280.0	0.096	0.912
10,000	0.73848	2.8	322.6	0.135	0.881
20,000	0.53281	6.6	374.6	0.188	0.842
30,000	0.37413	12.5	437.3	0.265	0.791
40,000	0.24617	21.0	522.0	0	1

Note: at 40,000 ft the jet is assumed to fly at constant Mach number

9.5.2 EFFECT OF VERTICAL ACCELERATION

For reasons of comfort, passenger airplanes must be flown along approximately straight lines: even small, constant vertical accelerations (perpendicular to the flight path) are not acceptable. This is the condition assumed in quasi-steady flight: only acceleration along the flight path is accepted.

If a climb is conducted while also 'pulling up', Eqn (9.2) perpendicular to the flight path shows that the airplane lift coefficient will have to be larger than in quasi-steady flight. This will affect the drag and therefore the thrust required. A brief discussion of vertical acceleration effects may be found in Chapter 12.

9.6 EFFECT OF LANDING GEAR, FLAPS, STOPPED ENGINES, TRIM REQUIREMENTS, WEIGHT AND ALTITUDE ON CLIMB PERFORMANCE

Landing gear, flaps, stopped engines, trim requirements, etc., all affect the climb capabilities of an airplane. The reason is that the drag–lift characteristics of the airplane are affected. Also, the weight of the airplane (wing–loading) and the altitude, affect climb performance. These effects are discussed as follows:

9.6.1 Effect of Landing Gear and Flaps on Climb Performance
9.6.2 Effect of Stopped Engines and Trim on Climb Performance
9.6.3 Effect of Weight and Altitude on Climb Performance

9.6.1 EFFECT OF LANDING GEAR AND FLAPS ON CLIMB PERFORMANCE

The effects of flaps and landing gear on the drag–lift characteristics of an airplane were already illustrated by the example of Figure 9.11. The rate–of–climb as well as the climb gradient decrease for increasing drag. This can be seen by inspection of Eqns (9.11), (9.12), (911a) and (9.12a).

Figure 9.25 shows the effect of the landing gear on the climb rate of a turboprop transport. Note the very large effect on climb rate at the gear down speed limit! Figure 9.26 shows the effect of flaps on the climb rate. In Figure 9.26 the landing gear is in the UP position!

9.6.2 EFFECT OF STOPPED ENGINES AND TRIM ON CLIMB PERFORMANCE

Figures 9.25 and 9.26 also show the effect on one stopped engine on the climb performance. With one engine inoperative (OEI) it is seen in Figure 9.25 that the best climb rate is down to less than 250 ft/min with the flaps up. This does not leave a pilot much margin for error. With the gear up and the flaps full down, Figure 9.26 shows that the climb rate available is very close to zero. Observe that with 40 degrees of flaps and the gear down, climb is no longer possible.

With OEI there are two effects on the drag polar of an airplane:

a) the drag on the stopped engine
b) the trim drag to overcome drag caused by control surface deflections needed to trim the airplane with OEI

These effects are illustrated in Figure 9.27 for a jet transport. For a detailed discussion of control issues during engine inoperative flight, see Ref. 9.1.

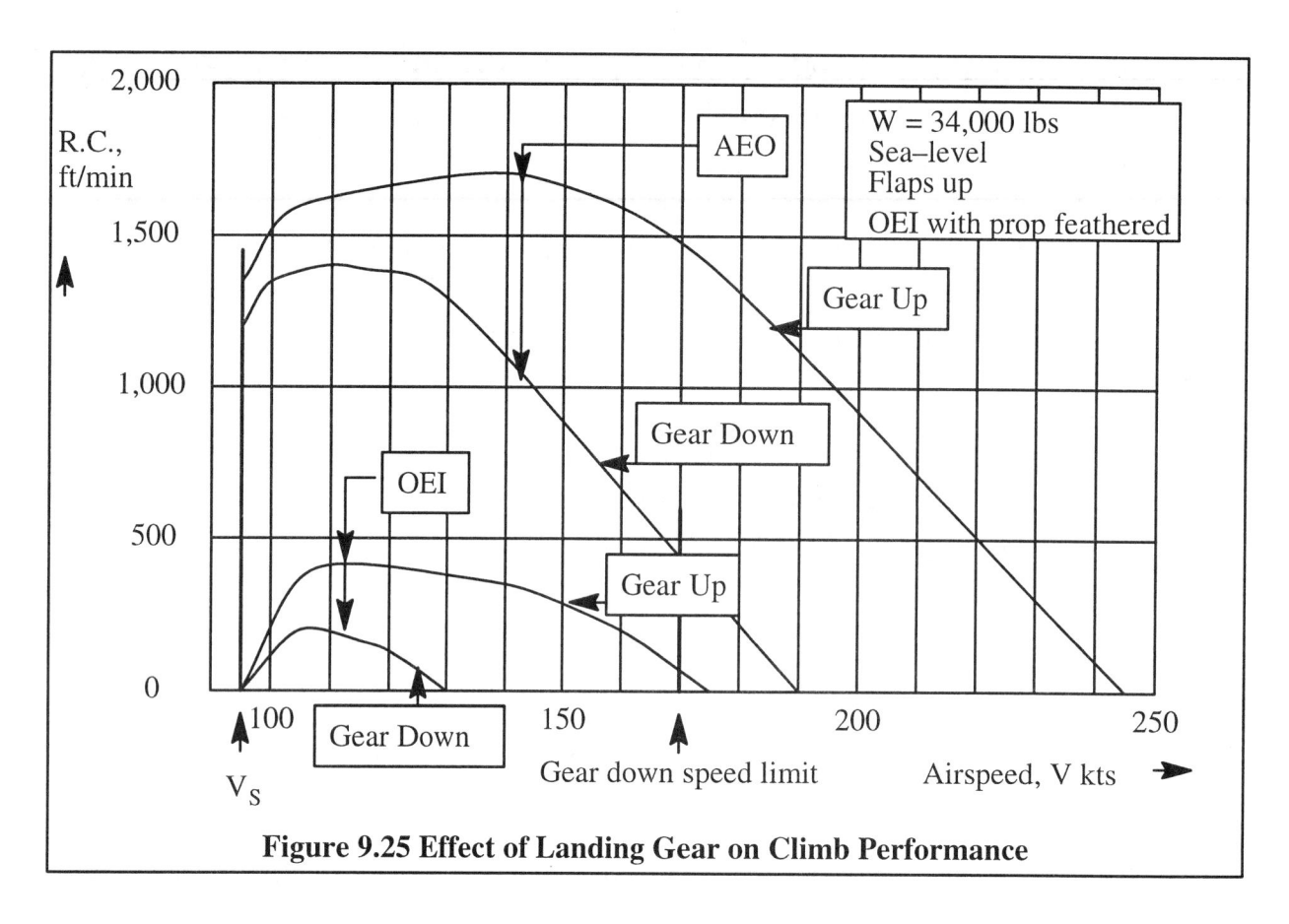

Figure 9.25 Effect of Landing Gear on Climb Performance

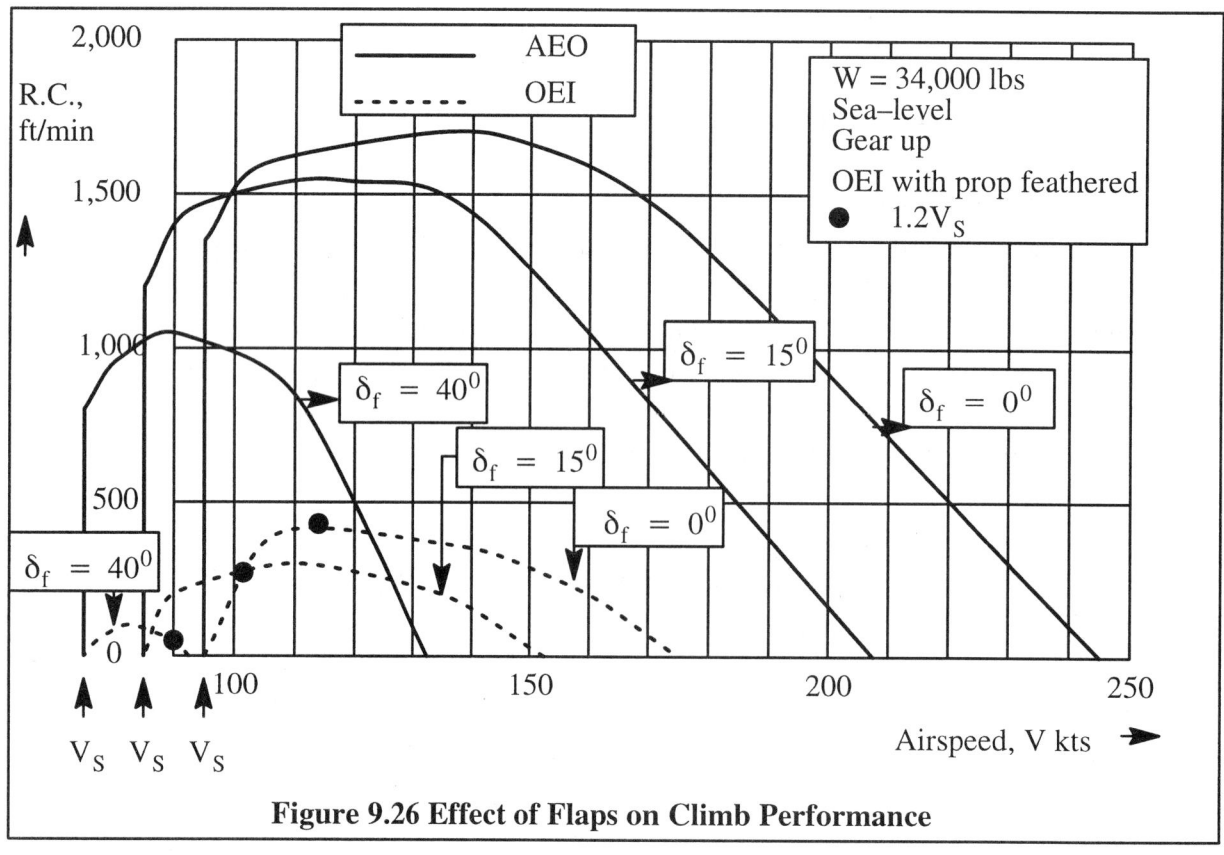

Figure 9.26 Effect of Flaps on Climb Performance

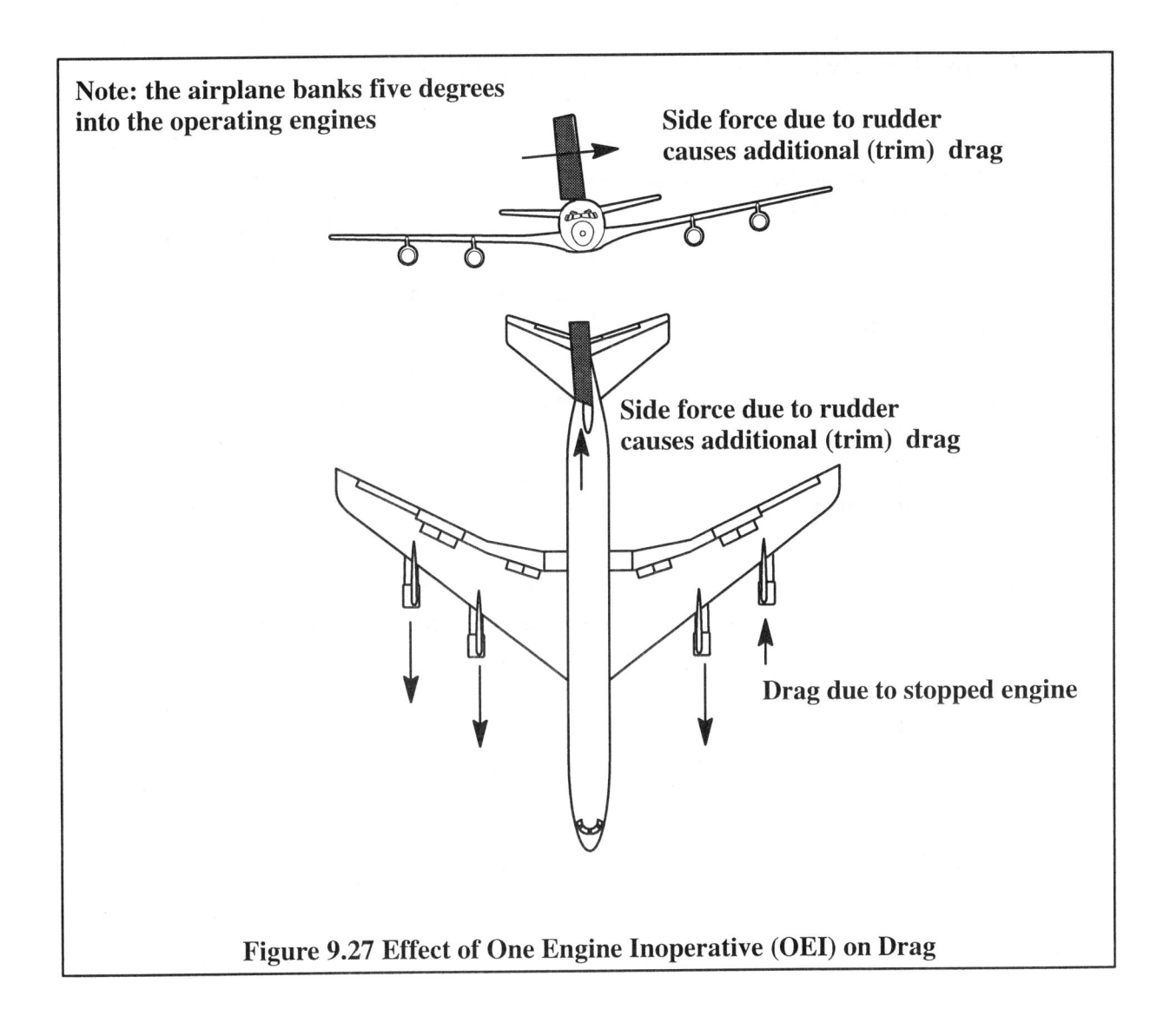

Figure 9.27 Effect of One Engine Inoperative (OEI) on Drag

(Note: the airplane banks five degrees into the operating engines. Side force due to rudder causes additional (trim) drag. Drag due to stopped engine.)

9.6.3 EFFECT OF WEIGHT AND ALTITUDE ON CLIMB PERFORMANCE

The effect of altitude on the climb performance of a jet airplane is illustrated in Figure 9.19 for a business jet. The effect of weight on achievable altitude are shown in Figure 9.18 for a large transport airplane. In both cases it is seen that the climb performance deteriorates with increasing weight and altitude. The effect of altitude on the climb performance of a turboprop airplane can be determined from Figure 9.8 by adding power available data. For speeds above 200 kts, the powers available for this example airplane are: 2,800 hp at sea–level, 2,140 hp at 10,000 ft, and 1,472 hp at 20,000 ft. The reader is asked to add these data to Figure 9.8 and plot the resulting maximum climb rates versus altitude.

The effect of weight on the climb capabilities of a turbo–prop airplane can be determined from Figure 9.9. Again, the reader is asked to compute the maximum climb rate and plot it versus weight.

9.7 CLIMB PERFORMANCE REGULATIONS

The climb performance of airplanes is subjected to a number of airworthiness regulations, all aimed at establishing adequate levels of safety with all engines operating (AEO) as well as with one engine inoperative (OEI). The climb performance requirements of airplanes (regulations) for civilian airplanes are contained in Ref. 9.2, those for military airplanes in References 9.3 and 9.4. The purpose of this Section is to summarize the most important climb performance requirements contained in these references.

For civilian airplanes, the climb performance requirements of Reference 9.2 are summarized in Tables 9.11 – 9.13 as follows:

Table 9.11 deals with Transport Category Airplanes certified according to FAR 25.

Table 9.12 deals with Normal, Utility and Acrobatic Category Airplanes certified according to FAR 23. Note that in these categories airplanes may not weigh more than 12,500 lbs nor may the number of passengers exceed 9.

Table 9.13 deals with Commuter Category Airplanes certified according to FAR 23. Note that in this category the weight may not be greater than 19,000 lbs and the number of passengers may not exceed 19

Ref. 9.2 allows a manufacturer to apply for Normal Category certification under FAR 23 rules for commuter airplanes which seat 10 or more passengers provided additional airworthiness requirements are met. These additional requirements do not affect the minimum climb requirements. The weight of such airplanes is still limited to 12,500 lbs.

The USAF and USN requirements of References 9.3 and 9.4 contain the following requirements for OEI climb:

a) a minimum gradient of 0.5% in the take–off configuration (gear down and flaps at take–off setting) with the remaining engine(s) at maximum take–off thrust or power and out of ground effect.

b) a minimum gradient of 2.5% in the take–off configuration (gear up and flaps at take–off setting) with the remaining engines at maximum take–off thrust or power and out of ground effect.

In addition, the USN has certain carrier suitability requirements which specify the minimum allowable sink following a catapult launch and the minimum acceptable vertical flight path correction during a final approach to the carrier. Detailed discussion of these requirements is beyond the scope of this text.

Table 9.11 Summary of FAR 25 Climb Performance Requirements

Transport Category								
Flight Phase/Weight FAR Paragraph	Flight and Airplane Configuration					Minimum Climb Gradient, C.G.R. in % for n engines		Minimum Steady Rate-of-Climb in ft/min.
	Number of engines stopped	Thrust or Power on Operating Engines	Flaps	Landing gear	Speed	n = 2	n = 3	n = 4	
Take-off / $W_{TO_{max}}$									
First segment Climb FAR 25.121 a)	1	Take-off	Take-off	Down	Lift-off V_{LOF}	> 0	0.3	0.5	No Requirement
Second segment Climb FAR 25.121 b)	1	Take-off	Take-off	Up	Over 35 ft Obstacle V_2	2.4	2.7	3.0	No Requirement
Third segment Climb FAR 25.121 c)	1	Maximum Continuous	Up		$\geq 1.25(V_s)$*	1.2	1.5	1.7	No Requirement
Landing / $W_{L_{max}}$									
Go-around in Approach Configuration FAR 25.121 d)	1	Take-off	Approach	Up	$\leq 1.5(V_s)$*	2.1	2.4	2.7	No Requirement
Go-around in Landing Configuration FAR 25.119	0	8 sec. after moving throttles full from flight idle	Landing	Down	$\leq 1.3(V_s)$*	3.2	3.2	3.2	No Requirement

Note 1: * means that the stall speed is that which is pertinent to the configuration for the flight phase being used

Note 2: for reciprocating powered transports, use 80% and vapor pressures defined in 25.101

Note 3: for turbine powered transports, use 80% humidity at or below standard temperatures, use 34% humidity at temperatures of 50 degrees F above standard

Table 9.12 Summary of FAR 23 Climb Performance Requirements for Normal, Utility and Acrobatic Categories

Flight Phase/Weight FAR Paragraph	Flight and Airplane Configuration					Minimum Climb Gradient, C.G.R. in % for n engines			Minimum Steady Rate-of-Climb in ft/min.
	Number of engines stopped	Thrust or Power on Operating Engines	Flaps	Landing gear	Speed	n = 2	n = 3	n = 4	
Reciprocating Powered									
FAR 23.67 b1) $W_{TO_{max}} > 6,000$ lbs $V_{s_0} > 61$ kts	1	Maximum Continuous	Most favorable	Up	$\geq 1.2(V_{s_1})^*$	1.5**	1.5**	1.5**	No Requirement
FAR 23.67 b2) $W_{TO_{max}} \leq 6,000$ lbs $V_{s_0} \leq 61$ kts	1	Maximum Continuous	Most favorable	Up	$\geq 1.25(V_s)^*$	> 0**	> 0**	> 0**	No Requirement
FAR 23.65 a)	0	Maximum Continuous	Take-off	Up	▶land-based amphibians	0.0833 / 0.0667	0.0833 / 0.0667	0.0833 / 0.0667	≥ 300 ft/min.
Turbine Powered									
FAR 23.67 c)	1	Maximum Continuous	Most favorable	Up	$\geq 1.2(V_{s_1})^*$	1.5**	1.5**	1.5**	No Requirement
FAR 23.65 a)	0	Maximum Continuous	Take-off	Up	▶land-based amphibians	0.75*** / 0.0833 / 0.0667	0.75*** / 0.0833 / 0.0667	0.75*** / 0.0833 / 0.0667	≥ 300 ft/min.
FAR 23.65 c)	0	Maximum Continuous	Take-off	Up	▶Speed for best climb	4.0***	4.0***	4.0***	No Requirement
FAR 23.77****	0	Take-off	Landing	Down	V_A	3.3***	3.3***	3.3***	No Requirement

Note: * means that the stall speed is that which is pertinent to the configuration for the flight phase being used

Note: ** These C.G.R values must be met at 5,000 ft and 41 degrees F. standard atmosphere

Note: *** These C.G.R values must be met at 5,000 ft and 81 degrees F. standard atmosphere

Note: **** This requirement also applies to reciprocating powered airplanes

Table 9.13 Summary of FAR 23 Climb Performance Requirements for the Commuter Category

Flight Phase/Weight FAR Paragraph	Flight and Airplane Configuration					Minimum Climb Gradient, C.G.R. in % for n engines			Minimum Steady Rate-of-Climb in ft/min.
$W_{TO_{max}} \leq 19{,}000$ lbs No. of pax ≤ 19	Number of engines stopped	Thrust or Power on Operating Engines	Flaps	Landing gear	Speed	n = 2	n = 3	n = 4	
Reciprocating and Turbine Powered									
FAR 23.65 a)	0	Maximum Continuous	Take-off	Up	Speed for best climb land-based amphibians	0.0833 0.0667	0.0833 0.0667	0.0833 0.0667	≥ 300 ft/min.
Take-off Climb FAR 23.67 e-1-i)	1	Maximum Take-off	Take-off	Down	Between V_{LOF} and speed at which gear is retracted	> 0	0.3	0.5	No Requirement
Take-off Climb FAR 23.67 e-1-ii)	1	Maximum Take-off	Take-off	Up	V_2 (h = 400 ft)	2.0	2.3	2.6	No Requirement
Approach Climb FAR 23.67 e-3)	1	Maximum Take-off	Up	Up	$\leq 1.5 V_{S_1}$	2.1	2.4	2.7	No Requirement
Turbine Powered FAR 23.67 c)	1	Maximum Continuous	Most favorable	Up	$\geq 1.2(V_{S_1})$*	1.5** 0.75***	1.5** 0.75***	1.5** 0.75***	No Requirement
FAR 23.77****	0	Take-off	Landing	Down	V_A	3.3***	3.3***	3.3***	No Requirement

Note: * means that the stall speed is that which is pertinent to the configuration for the flight phase being used

Note: ** These C.G.R values must be met at 5,000 ft and 41 degrees F. standard atmosphere

Note: *** These C.G.R values must be met at 5,000 ft and 81 degrees F. standard atmosphere

Note: **** This requirement also applies to reciprocating powered airplanes

9.8 SUMMARY FOR CHAPTER 9

In this chapter methods for determining the climb and drift–down performance of airplanes were presented. First, the equations of motion for a general climb situation were derived. The general climb performance characteristics of jet and propeller driven airplanes were presented for shallow flight path angles as well as for large flight path angles.

The effect of one engine inoperative on climb and drift–down characteristics of airplanes was discussed in some detail with numerical examples for jets and propeller driven airplanes.

Methods for determining times to climb to altitude, operational ceilings and drift–down times were also given.

In many types of airplanes the effect of forward accelerations during the climb cannot be neglected. Acceleration effects on climb performance were also discussed.

Landing gears, flaps, speed brakes, stopped engines, trim requirements, etc., all affect the climb capabilities of an airplane. Also, the weight and altitude of the airplane (wing–loading) affect climb performance. Where appropriate numerical examples of these effects were given.

The climb performance of airplanes is subjected to airworthiness requirements which stipulate minimum required climb rates and/or climb gradients (angles). These requirements establish adequate levels of safety with all engines operating (AEO) as well as with one engine inoperative (OEI). The climb performance requirements of airplanes (regulations) were also summarized.

9.9 PROBLEMS FOR CHAPTER 9

9.1 The drag characteristics of a light, twin engined propeller driven airplane are given by:

$$C_D = 0.0350 + 0.051 C_L^2 + 0.00138 C_L^{13.42}$$

Weight and wing area are: $W = 4,200$ lbs and $S = 155$ ft^2 respectively. If the power available at sea–level is 310 hp, determine the maximum speed at sea–level.

9.2 For the airplane of problem 9.1 determine the following quantities:
a) the best rate–of–climb
b) the speed for the best rate–of–climb
c) the best climb gradient (C.G.R.)
d) the speed for the best climb gradient

9.3 The drag characteristics of a jet driven airplane are given by:

$$C_D = 0.0150 + 0.020 C_L^2$$

Weight and wing area are: $W = 10{,}000$ lbs and $S = 200$ ft^2 respectively. Assume that the engine delivers 3,000 lbs of thrust at sea–level, independent of speed. Determine the maximum and minimum speeds by an analytical method. If the maximum, trimmed lift coefficient of the airplane is 1.5, is the answer still valid? Hint: assume L=W and T=D, set up a quadratic equation for dynamics pressure and solve for speed.

9.4 A twin jet airplane has the following Mach independent drag characteristics:

$$C_D = 0.0180 + 0.022 C_L^2 \text{ with AEO} \qquad C_D = 0.0190 + 0.023 C_L^2 \text{ with OEI}$$

Weight and wing area are: $W = 10{,}000$ lbs and $S = 200$ ft^2 respectively. Assume that each engine delivers 2,400 lbs of thrust at sea–level and 800 lbs of thrust at 40,000 ft in the standard atmosphere.

a) Use an analytical method to determine the maximum rate–of–climb at sea–level and at 40,000 ft.
b) Use an analytical method to calculate the time–to climb to 40,000 ft
c) Use an analytical method to calculate the time–to–drift–down to the OEI ceiling if the AEO cruise altitude is 40,000 ft.

9.5 The drag polars (flaps up and down) for a Cessna Cardinal are given in Figure 8.26. The available shaft horsepower at 2,500 ft is given by: $SHP_{av} = 2.463V - 0.021V^2$ with V in kts. At 7,500 ft the available shaft horsepower is 78% of that at 2,500 ft. Extrapolate this power to sea–level, to 5,000 ft and to 10,000 ft.

Calculate and plot the maximum rate–of–climb at both altitudes, with and without flaps for altitudes ranging from sea–level to 10,000 ft in increments of 2,500 ft..

9.6 For the twin business jet of Figure 9.19, determine and plot the rate–of–climb as a function of Mach number and altitude for the case of AEO and OEI.

9.7 A jet airplane is equipped with an engine, the thrust of which is independent of speed. If the airplane drag polar has the standard parabolic form, show that the dynamic pressure for the maximum rate–of–climb is:

$$\bar{q} = \frac{T}{6 C_{D_0} S} + \sqrt{\left(\frac{T}{6 C_{D_0} S}\right)^2 - \frac{W^2}{S^2 C_{D_0} \pi e A}}$$

9.8 How much power is required for the airplane of Figure 9.11 to meet the C.G.R. requirement of FAR 25.121 d) of Table 9.11 (OEI)? Use Eqn. (9.12a) and assume that the lift coefficient is equal to 1.1. The weight of the airplane is 34,000 lbs.

9.10 REFERENCES FOR CHAPTER 9

9.1 Roskam, J.; Airplane Flight Dynamics and Automatic Flight Controls, Part I; Design, Analysis and Research Corporation, 120 East Ninth Street, Suite 2, Lawrence, Kansas 66044, 1995.

9.2 Anon.; Code of Federal Regulations, Aeronautics and Space; Part 23 and Part 25, Federal Aviation Administration, Washington, D.C., January 1, 1992

9.3 Anon.; Mil–C–5011B (USAF): Military Specification, Charts: Standard Aircraft Characteristics and Performance, Piloted Aircraft (Fixed Wing), June 1977.

9.4 Anon.; AS–9263 (USNAVY): Naval Air Systems Command Specification, Guidelines for the Preparation of Standard Aircraft Characteristics Charts and Performance Data, Piloted Aircraft (Fixed Wing), October 1986.

9.5 Roskam, J.; Airplane Design, Part VI: Preliminary Calculation of Aerodynamic, Thrust and Power Characteristics; Design, Analysis and Research Corporation, 120 East Ninth Street, Suite 2, Lawrence, Kansas 66044, 1990.

9.6 Nixon, D. and Churchill, S.; Maximum Altitude Operations; Boeing Airliner Magazine, October–December 1996, Boeing Commercial Airplane Group, Seattle, WA, 1996.

9.7 Jackson, P.; Jane's All The World's Aircraft, 1995–1996; Jane's Information Group, U.K.

9.8 George, F.; Avoiding a 'Wet Footprint'; Business and Commercial Aviation, p. 90–95, November 1996

9.9 Anon.; Jet Transport Performance Methods, Boeing Document D6–1420, 1967, The Boeing Company.

9.10 Phillips, F.C.; A Kinetic Energy Correction to Predicted Rate of Climb; Journal of the Aeronautical Sciences, Vol. 9., March 1942, pp. 172–174.

CHAPTER 10: TAKE–OFF AND LANDING

In this chapter the equations for calculating the take–off and landing distances of airplanes are developed. The corresponding civil and military airworthiness rules are discussed. Applications to various airplane types are given.

The take–off process and its relation to airworthiness standards is discussed in Section 10.1. The equations of motion during the take–off process are developed in Section 10.2. Various types of solutions to these equations which can be used in the prediction of take–off distances are presented in Section 10.3.

The landing process and its relation to airworthiness standards is discussed in Section 10.4. The equations of motion during the landing process are developed in Section 10.5. Various types of solutions to these equations (which can be used in the prediction of landing distances) are presented in Section 10.6.

Examples of how take–off and landing distance information is presented in airplane flight manuals are also given.

10.1 THE TAKEOFF PROCESS

The takeoff process is normally divided into three phases:

* ground roll   * air distance   * climb–out

Figure 10.1 shows a schematic which depicts these phases in relation to the overall takeoff path and to several important speeds. There is no scale implied in this figure: the relative magnitudes of the distances covered in the take–off ground roll, the rotation and the transition vary from one airplane to another airplane.

In Figure 10.1 the airplane is assumed to accelerate until the rotation speed, V_R, has been reached. At the rotation speed the airplane is rotated to some pitch attitude angle so that the airplane will lift off at the lift–off speed, V_{LOF}.

Once lift–off has occurred the flight path angle varies gradually from zero (at lift–off) to a

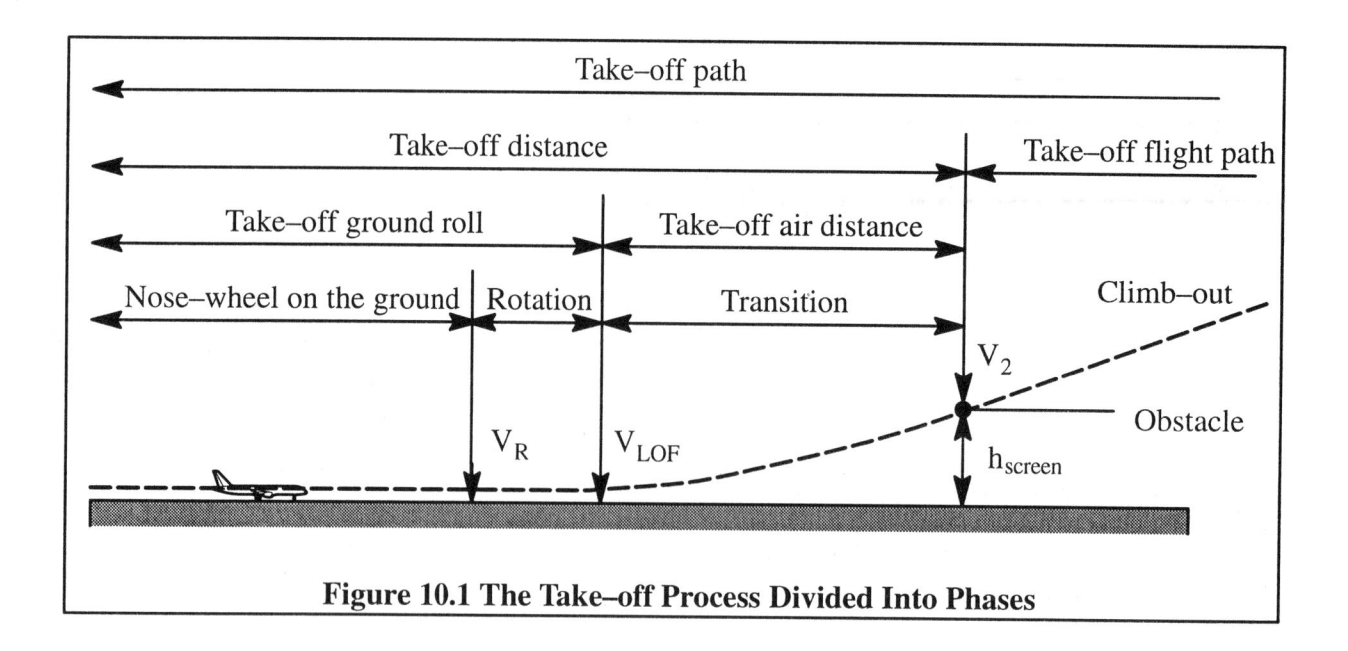

Figure 10.1 The Take–off Process Divided Into Phases

constant value at the obstacle (screen height, h_{screen}). This obstacle height varies with the type of regulation used in the certification of the airplane. For FAR 25 it is 35 ft while for FAR 23 (except commuter category) and for military airplanes it is 50 ft. The reader should consult Reference 10.1 for FAR 23 and for FAR 25. For the military regulations, see Refs 10.2 and 10.3.

Sub–sections 10.1.1 and 10.1.2 contain summaries of the civil and military rules respectively.

10.1.1 COMMERCIAL TAKE–OFF RULES

During the take–off process the configuration of the airplane (in terms of thrust or power, flap position, cooling flap position and landing gear) is kept constant. The landing gear is normally retracted soon after the airplane has lifted off. Once the airplane exceeds the obstacle (or screen) height the airplane follows the takeoff flight path until a 'safe' height of 1,500 ft above the terrain is reached. At that point the so–called climb to cruise altitude is begun.

Because of safety considerations the take–off process in a multi engine transport airplane is more complicated than that suggested in Figure 10.1. The airplane must pass through a sequence of speeds before the decision to continue the take–off may be made. This sequence is established under the assumption that an engine failure in a multi–engine airplane can occur at any time during the take–off process. Wherever such failure occurs the safety of the crew and passengers must not be compromised. Figure 10.2 illustrates the sequence of speeds through which an airplane must pass to ensure safety in a multi–engine transport airplane certified under FAR 25.

Starting from brake release at zero forward speed, V=0, the airplane is accelerated. The first reference speed is V_S, also called the calibrated stall speed or the minimum steady flight speed at

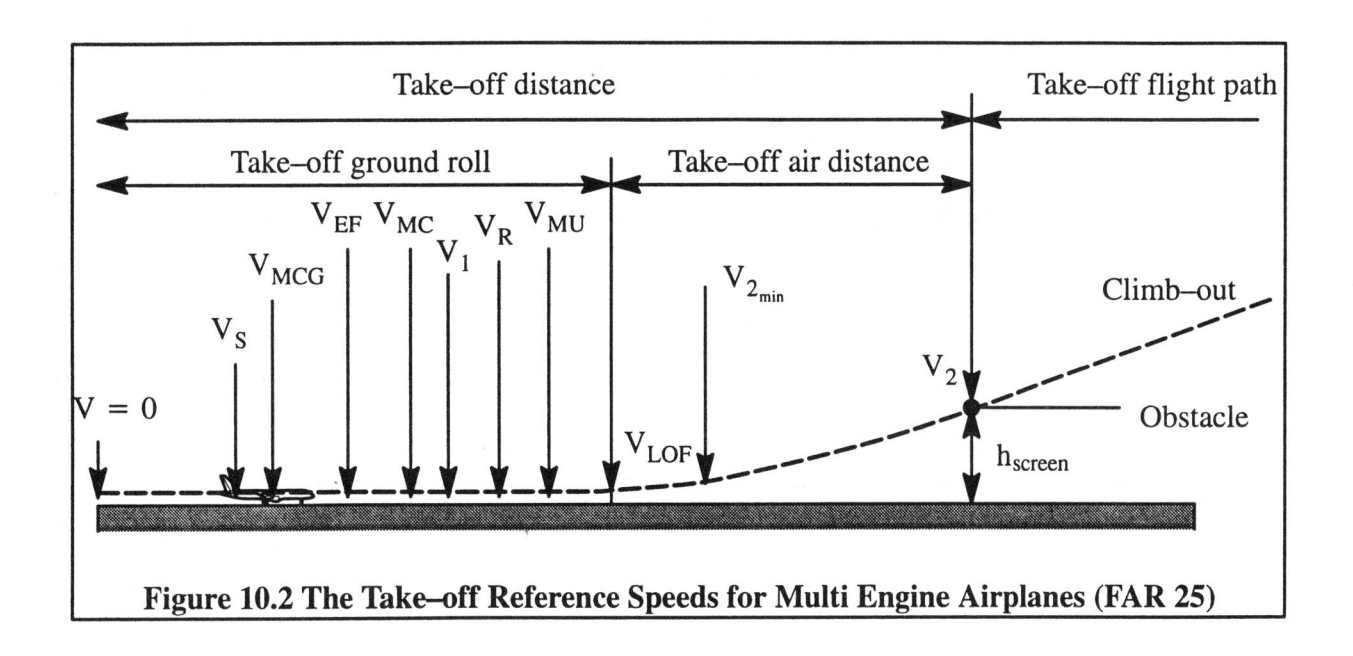

Figure 10.2 The Take–off Reference Speeds for Multi Engine Airplanes (FAR 25)

which the airplane is controllable (See FAR 25.103 for details). Definitions and determination of this stall speed are discussed in detail in Chapter 8 and in Chapter 12.

The next speed is V_{MCG}, the minimum control speed on the ground (see FAR 25.149).

The engine–failure speed, V_{EF} must be larger than V_{MCG}.

Next is the minimum control speed in the air, V_{MC} (see FAR 25.149). This is the calibrated speed at which, when the critical engine is made inoperative, it is possible to maintain control of the airplane in a straight line flight with no more than 5 degrees of bank. The minimum control speed, V_{MC} may not exceed $1.2 V_S$.

The next speed is the takeoff decision speed, V_1. When the airplane has accelerated beyond this speed the take–off must be continued even if the critical engine fails. If engine failure occurs below V_1 the take–off must be aborted. The speed V_1 must be selected by the manufacturer. However, V_1 may not be less than V_{EF}, the calibrated air–speed at which the critical engine is assumed to fail plus that speed increment gained during the interval required by the pilot to take appropriate action in terms of applying the retardation means during accelerate–stop tests. The speed, V_{EF} itself may not be less than V_{MCG}, the minimum control speed on the ground (see FAR 25.149).

The rotation speed, V_R, is that speed at which the pilot initiates rotation. The rotation speed,

V_R may not be less than V_1, nor less than 1.05 times V_{MC}, nor less than that speed which allows reaching V_2 before reaching the 35 ft obstacle height.

The minimum unstick speed, V_{MU}, is that speed at and above which the airplane can safely lift off the ground and continue the takeoff. The minimum unstick speed, V_{MU} may not be less than V_{2min} nor less than V_R plus the speed increment attained before reaching the 35 ft obstacle height.

The liftoff speed, V_{LOF}, is the calibrated speed at which the airplane becomes airborne. By inspection of the requirements summarized in Table 10.1, $V_{LOF} > V_R$ is required.

The minimum safety speed, $V_{2_{min}}$ may not be less than $1.2V_S$ for twin engined and three-engined propeller driven airplanes or for turbojet airplanes without provision for reducing the stalling speed with the most critical engine inoperative. $V_{2_{min}}$ may not be less than $1.15V_S$ for four-engined propeller driven airplanes or for turbojets with provision for reducing the stalling speed with the critical engine inoperative.

Finally, the speed, V_2, must be selected by the manufacturer to provide the climb gradient required by FAR 25.121 but may not be less than V_R plus the speed increment attained before reaching the 35 ft obstacle height.

The actual sequence of speeds is even more complicated than described in the foregoing. For details the reader should consult FAR 23.53 and FAR 25.107 in Ref. 10.1. A summary of the definitions for take-off speeds used in both FAR 23 and FAR 25 is given in Table 10.1.

The take-off distance of a FAR 25 certified transport airplane is defined by a balance between the accelerate-stop and the accelerate-go distances. This is referred to as the balanced fieldlength requirement. This requirement is the synthesis of the accelerate-stop requirement of FAR 25.109, the take-off path requirement of FAR 25.111 and the take-off distance and take-off run requirement of FAR 25.113. To calculate the stopping distance of the airplane, only brakes may be used. A graphical interpretation of the accelerate-stop requirement is given in Figure 10.3.

The accelerate-go requirement is included in FAR 25.113, called take-off distance and take-off run. A graphical interpretation of the accelerate-go requirement is given in Figure 10.4.

Operationally, the runway length must be such that the airplane can be safely operated under either of the scenarios depicted in Figures 10.3 and 10.4. This results in the definition of balanced fieldlength as depicted in Figure 10.5.

Table 10.1 Summary of Take–off Speeds			
FAR 23		**FAR 25**	
V_{S0}	Stalling speed or minimum steady speed with gear and flaps down, throttles closed.	V_S	Stalling speed or minimum steady speed with gear and flaps down, throttles closed.
V_{S1}	Stalling speed or minimum steady speed with throttles closed and airplane in the condition existing in the test for which V_{S1} is used.	V_1	Take–off decision speed $V_1 \geq V_{EF} + \Delta V_{recplusact}$
		V_{EF}	Critical engine failure speed $V_{EF} \geq V_{MCG}$
For normal, utility and acrobatic airplanes:		V_{MCG}	Minimum control speed on the ground with critical engine failed.
V_R	Rotation speed $V_R \geq V_{MC}$	$V_{2_{min}}$	Minimum take–off safety speed $V_{2_{min}} \geq 1.2 V_S$
V_{50}	Speed at the **50 ft** obstacle		For prop. driven twins and tri–s and for jets without provision for reducing the stalling speed with OEI. $V_{2_{min}} \geq 1.15 V_S$
For multi–engine airplanes: $V_{50} \geq 1.1 V_{MC}$ and $V_{50} \geq 1.2 V_{S1}$			
For single–engine airplanes: $V_{50} \geq 1.2 V_{S1}$			For prop. driven quads and for jets with provision for reducing the stalling speed with OEI. and $V_1 \geq 1.1 V_{MC}$
For commuter category airplanes:		V_2	Take–off safety speed at 35 ft $V_2 \geq V_{2_{min}}$ $V_2 \geq V_R + \Delta V_{35ft}$
V_1	Take–off decision speed $V_1 \geq 1.1 V_{S1}$ and $V_1 \geq 1.1 V_{MC}$ and $V_1 \geq V_{EF} + \Delta V_{recplusact}$	V_{MU}	Minimum un–stick speed
V_2	Take–off safety speed at **35 ft** $V_2 \geq V_1$ and $V_2 \geq 1.2 V_{S1}$	V_R	Rotation speed $V_R \geq V_1$ and $V_R \geq 1.05 V_{MC}$ and $V_R \geq V$ which allows V_2 to be reached at **35 ft** and $V_R \geq V$ which allows for a lift–off speed, V_{LOF}, of not less than $1.10 V_{MU}$ with AEO and a lift–off speed of not less than $1.05 V_{MU}$ with OEI
V_2	Must allow 23.67 climb gradient to be met		
V_{EF}	Critical engine failure speed $V_{EF} \geq V_{MC}$		
V_R	Rotation speed $V_R \geq V_1$ and $V_R \geq V$ which allows V_2 to be reached at **35 ft**		
FAR 23 and FAR 25	V_R One rotation speed must be determined which applies to AEO and OEI take–offs	$\Delta V_{recplusact}$	Speed increment gained with OEI from the time of engine failure to the time the pilots recognizes the failure and acts to slow the airplane down during accelerate–stop tests

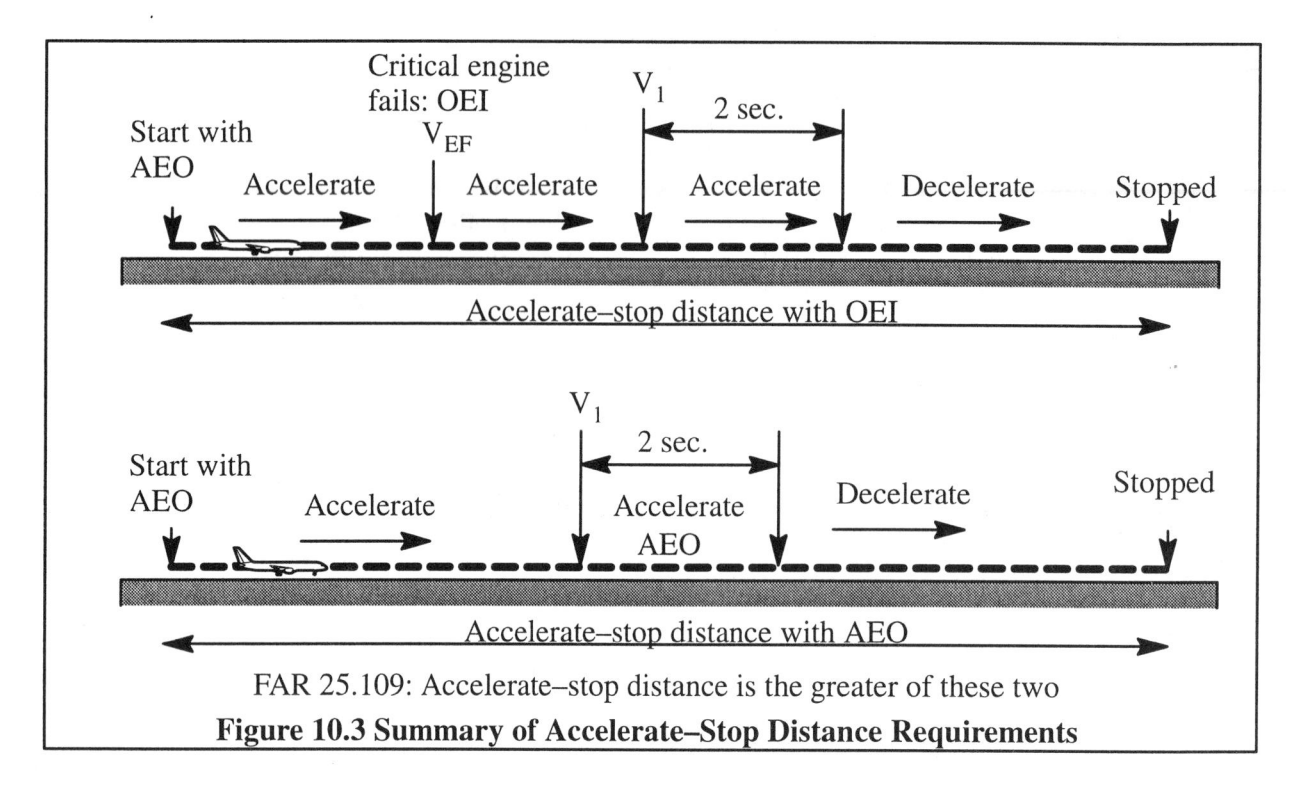

Figure 10.3 Summary of Accelerate–Stop Distance Requirements

FAR 25.109: Accelerate–stop distance is the greater of these two

FAR 25.113 is the greater of Accelerate–Go with OEI and 1.15 Accelerate–Go with AEO

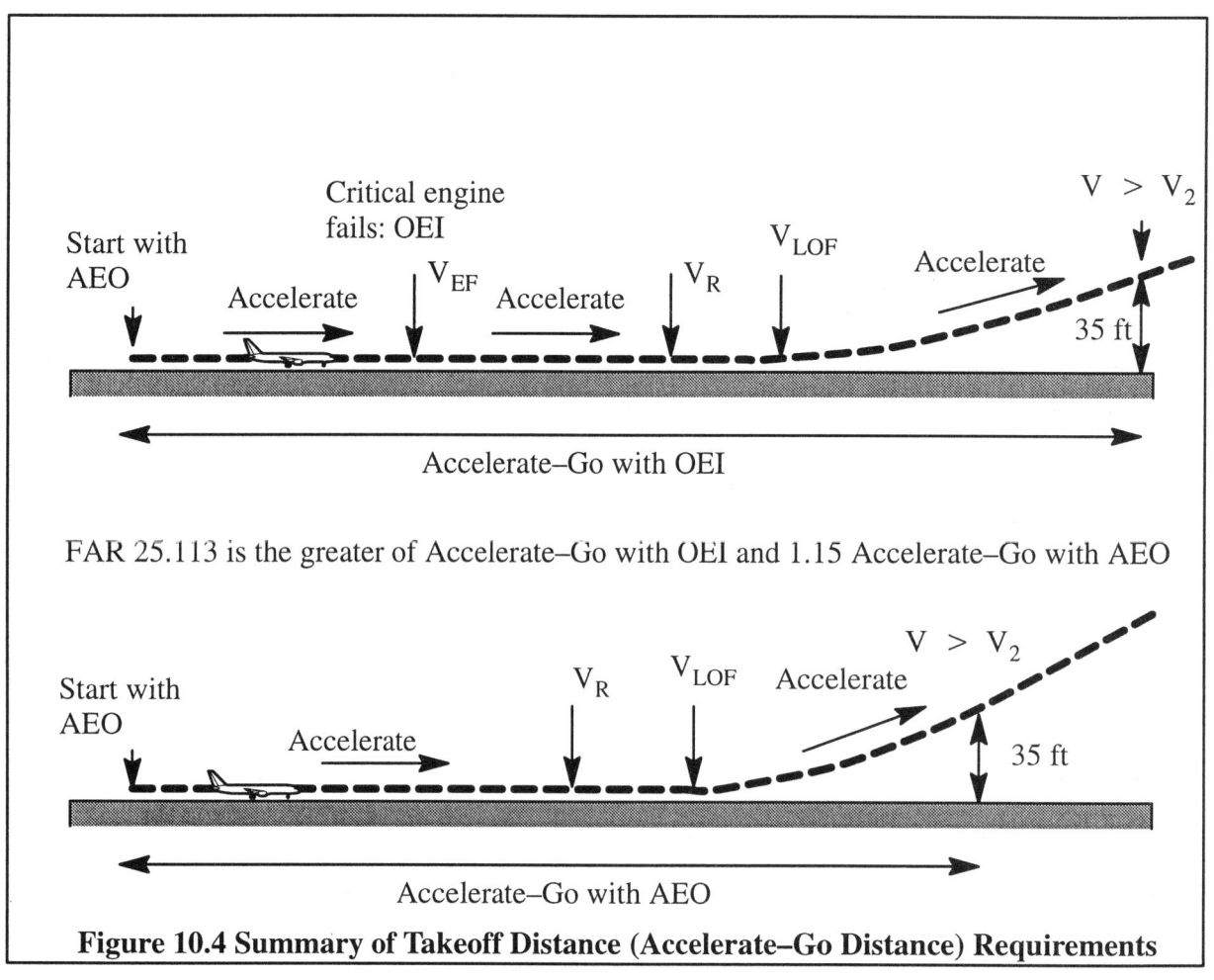

Figure 10.4 Summary of Takeoff Distance (Accelerate–Go Distance) Requirements

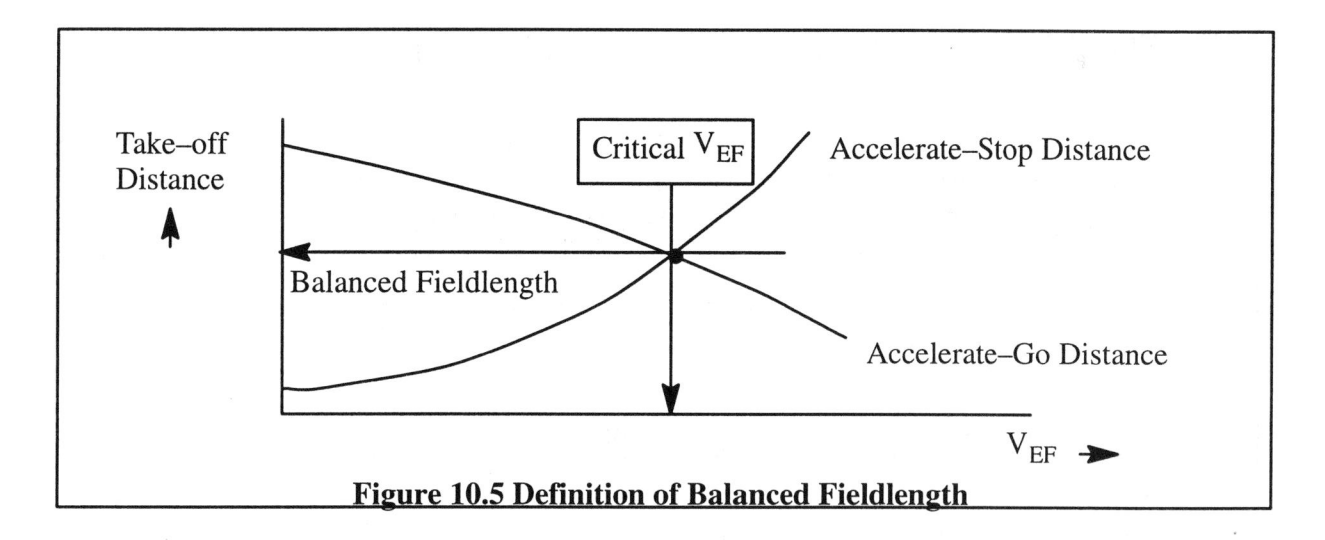

Figure 10.5 Definition of Balanced Fieldlength

To determine the take-off distances of airplanes a set of equations is needed, which describe the motion of the airplane on the ground (ground distance), the transition to a climbing flight and the climb to the obstacle height. These equations are developed in Section 10.2.

10.1.2 MILITARY TAKE-OFF RULES

According to Ref. 10.2, Par. 3.4.5.2 and Ref. 10.3, Par. 3.5.5.2 the take-off distance is defined as the sum of the ground distance and the air distance needed to arrive at a 50 ft obstacle. This is similar to the distance illustrated in Figure 10.1 for $h_{screen} = 50$ ft.

In addition, Refs 10.2 and 10.3 define a critical fieldlength. Critical fieldlength is the sum of the distance required to accelerate with AEO to critical engine failure speed plus the distance to accelerate with OEI to take-off speed or to decelerate to a stop from critical engine failure speed in the same distance. This is conceptually the same as the FAR 25 balanced fieldlength concept illustrated in Figure 10.5 with the following differences:

1) at engine failure speed the airplane must be allowed to accelerate with OEI for **3 seconds** while maintaining maximum take-off thrust on the remaining engine(s).

2) at the end of the 3 second acceleration period, all remaining engines are reduced to idle thrust, brakes are applied and any deceleration devices are deployed. The time required to achieve maximum deceleration shall be included in the calculations.

Note that in the commercial rules only brakes may be used in the calculation of balanced fieldlength! Under the military rules other means of stopping the airplane may be included in the calculations. However, these means must be shown to be reliable.

10.2 EQUATIONS OF MOTION DURING TAKE–OFF

From the discussion in Section 10.1 it follows that the take–off process consists of a take–off ground–roll and a take–off air distance. The latter in turn consists of a transition distance and a climb–to–the–obstacle distance. Figure 10.6 shows these distances.

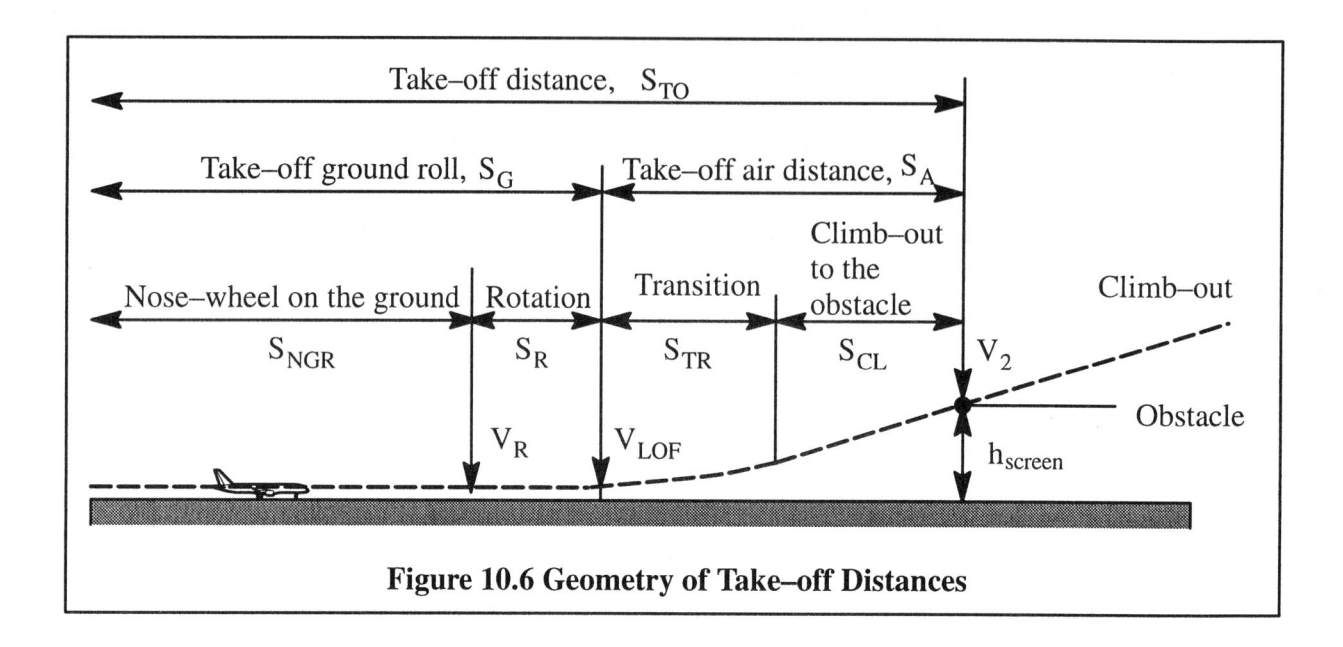

Figure 10.6 Geometry of Take–off Distances

Evidently:

$$S_{TO} = S_G + S_A = S_{NGR} + S_R + S_{TR} + S_{CL} \tag{10.1}$$

Different equations of motion apply to each of these distances. These equations are presented in Sub–sections 10.2.1 through 10.2.3 respectively:

10.2.1 Equations of motion during the take–off ground roll

10.2.2 Equations of motion during the take–off transition

10.2.3 Equations of motion during the climb–out to the obstacle

10.2.1 EQUATIONS OF MOTION DURING THE TAKE–OFF GROUND ROLL

Figure 10.7 shows the forces which act on an airplane during the ground roll with the nose–gear on the ground. Two cases are shown: level runway and runway with a constant gradient. The only difference between the two cases is the inclined force of gravity. The runway gradient is normally sufficiently small so that: $W\cos\phi = W$, and $W\sin\phi = W\phi$.

Takeoff and Landing

Figure 10.7 Forces Acting on an Airplane During the Take–off Ground Roll

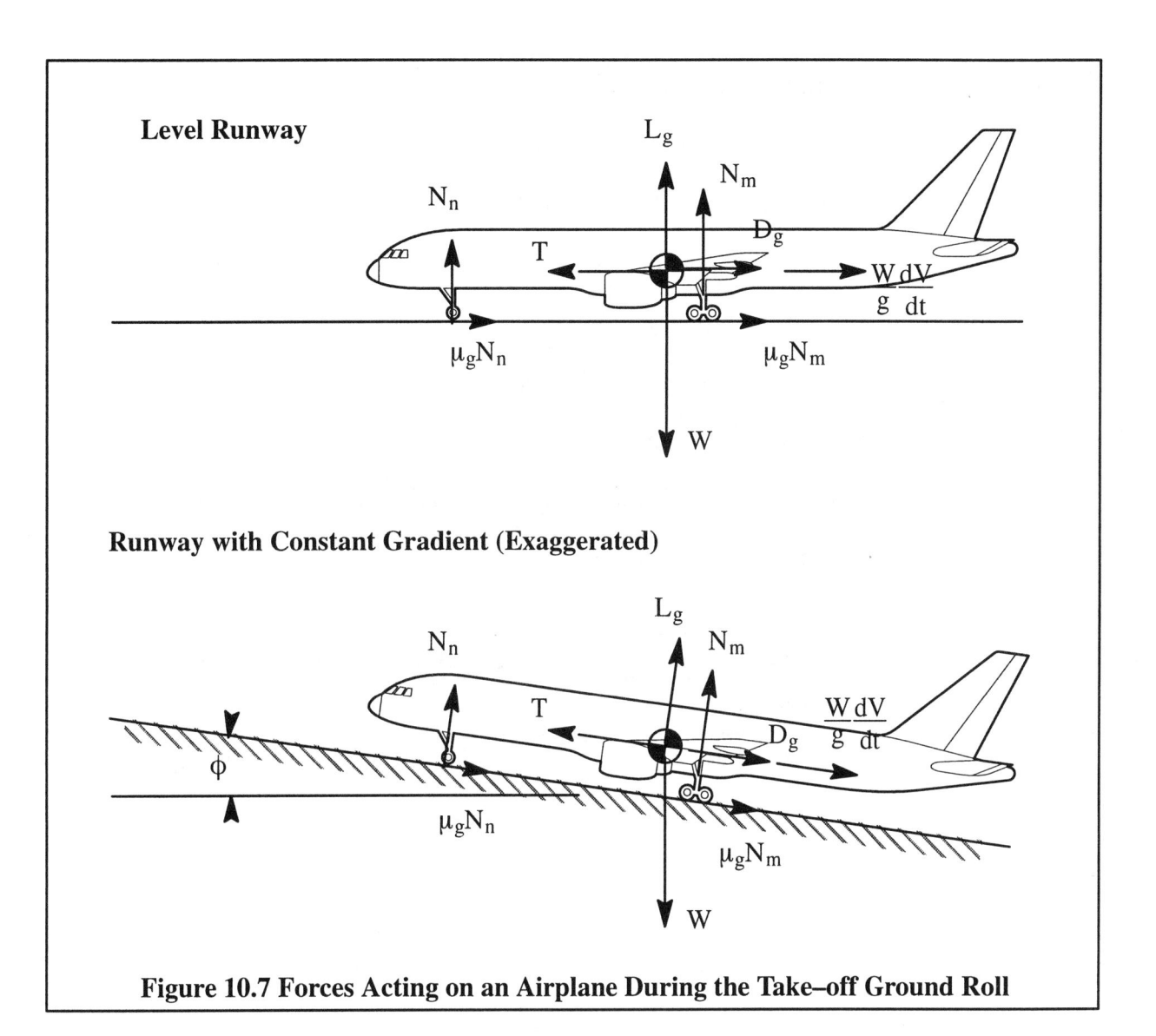

It is assumed that the airplane is in moment equilibrium and that the landing gear and tire dynamic effects are negligible. Note in Figure 10.7, that the aerodynamic forces (Lift and Drag) have been given a subscript 'g' for 'ground'. This means that during the ground roll the aerodynamic forces must be computed including the effect of ground proximity. Reference 10.4 contains methods for doing this. An explanation of how these ground effect corrections arise and how they can be predicted is also given later in this Sub–section.

The equations of motion, for the case of a runway with a gradient, ϕ are as follows:

$$L_g + N_n + N_m = W\cos\phi \approx W \tag{10.2}$$

and

$$T - D_g - \mu_g N_n - \mu_g N_m = W\sin\phi + \frac{W}{g}\frac{dV}{dt} \approx W\phi + \frac{W}{g}\frac{dV}{dt} \tag{10.3}$$

where: L_g is the airplane lift in ground effect, in lbs

N_n is the nose–gear reaction force, in lbs

N_m is the main–gear reaction force, in lbs

ϕ is the runway inclination angle (gradient) in radians: positive when unfavorable

W is the airplane weight, in lbs

T is the total installed thrust, in lbs. Note: thrust inclination is assumed to be zero.

D_g is the airplane drag in ground effect, in lbs.

μ_g is the rolling ground friction coefficient. Table 10.1b shows typical values.

V is the airplane speed relative to the ground, in ft/sec.

Table 10.1b Typical Values for the Rolling Ground Friction Coefficient	
Surface Type	μ_g
Concrete and Macadam	0.02 – 0.03 (0.025 is typically used)
Hard Turf	0.05
Short Grass	0.05
Long Grass	0.10
Soft Ground	0.10 – 0.30

By combining Eqns (10.2) and (10.3) it can be shown that the acceleration along the runway can be expressed as:

$$\frac{dV}{dt} = \frac{g}{W}(T - D_g - \mu_g W + \mu_g L_g - W\phi) \tag{10.4}$$

It is usually convenient to rewrite this as:

$$\frac{dV}{dt} = g\left\{\left(\frac{T}{W} - \mu_g\right) - \frac{(C_{D_g} - \mu_g C_{L_g})\bar{q}}{W/S} - \phi\right\} = a_g \tag{10.5}$$

where: C_{D_g} is the airplane drag coefficient in ground effect

C_{L_g} is the airplane lift coefficient in ground effect

a_g is the acceleration along the ground

Takeoff and Landing

To develop a formula for the ground roll distance, consider the general case where the airplane is taking off with a wind speed of +/− V_w, where '+' means a tail–wind and '−' means a head–wind. If V is the air–speed, the ground speed will be $V \pm V_w$. It is noted that at zero ground speed, $V = \mp V_w$. Therefore:

$$\frac{dS_G}{dt} = V \pm V_w \quad (10.6)$$

If the acceleration along the ground is called: $dV/dt = a_g$, then it is possible to write:

$$dS_G = \frac{V \pm V_w}{a_g} dV \quad (10.7)$$

The ground distance can then be obtained by combining Eqns (10.5) and (10.7):

$$S_G = \int_{\mp V_w}^{V_{LOF}} \frac{V \pm V_w}{a_g} dV = \int_{\mp V_w}^{V_{LOF}} \frac{V \pm V_w}{g\left\{\left(\frac{T}{W} - \mu_g\right) - \frac{(C_{D_g} - \mu_g C_{L_g})\bar{q}}{W/S} - \phi\right\}} dV \quad (10.8)$$

The upper sign in Eqn (10.8) is for a down–wind take–off. The lift–off speed, V_{LOF}, varies with the certification category. However, $V_{LOF} > V_R$ must be satisfied by inspection of the take–off speed requirements summary in Table 10.1.

Assuming: a) a runway with a constant gradient,
b) constant weight during the take–off run,
c) that the airplane attitude on its landing gear is constant so that C_{D_g} and C_{L_g} are constant,
d) ground friction, μ_g is also constant,
e) variation of thrust with speed is known,

it is possible to carry out the integration implied by Eqn (10.8).

Several practical solutions for Eqn (10.8) will be presented in Section 10.3.

It is of interest to understand the typical variation of the various contributions to acceleration along the runway in Eqn (10.5). Figure 10.8 shows the general trend of these accelerations.

Once the rotation speed is reached the nose–gear will leave the ground. Eqn (10.8) is still valid, but the nose–gear load should be set equal to zero and the lift and drag coefficients will increase due

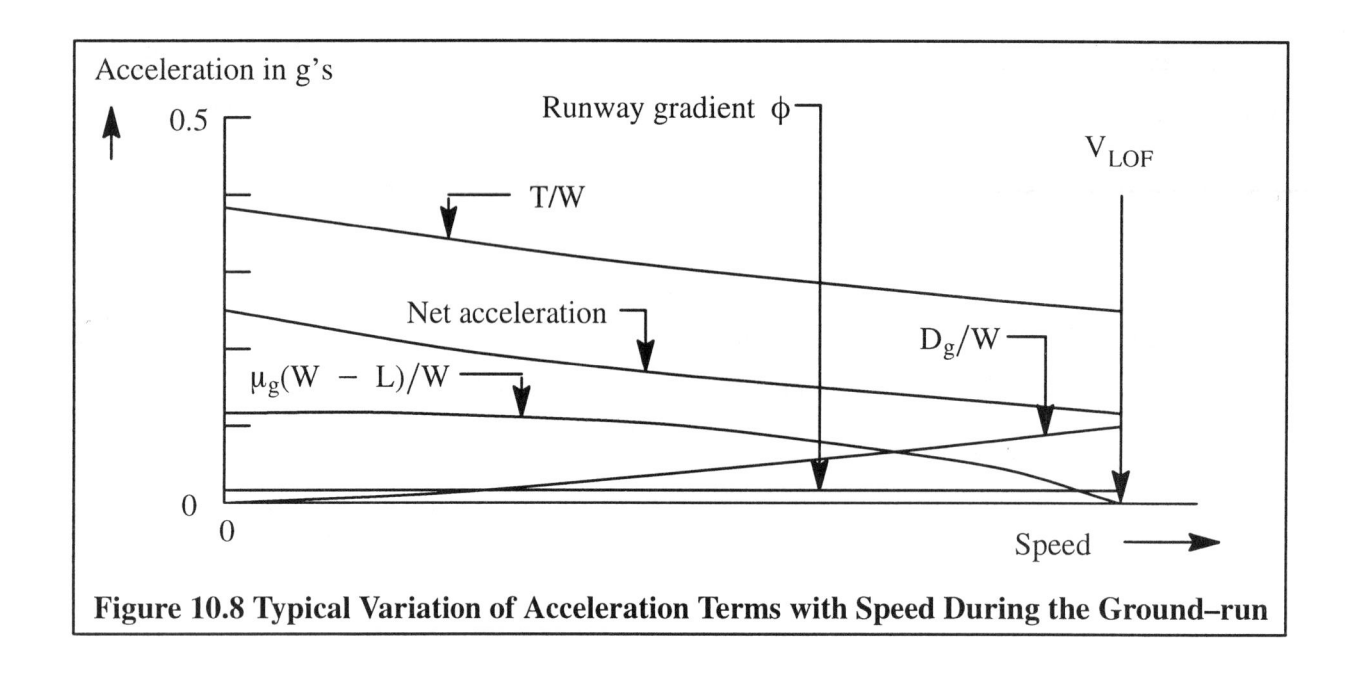

Figure 10.8 Typical Variation of Acceleration Terms with Speed During the Ground–run

to the increasing pitch attitude of the airplane relative to the runway. These effects can be included by assuming an average angular acceleration from the time of initiation of the rotation to the time of main gear lift–off. At that point the transition begins.

A brief discussion of: the effect of wind on take–off

the effect of speed on take–off thrust

the effect of the ground on lift and drag

the effect of thrust deflection on take–off

is given in Sub–sub–sections 10.2.1.1 through 10.2.1.4 respectively.

10.2.1.1 Effect of Wind on Take–off

The effect of wind, altitude and temperature on take–off performance must be accounted for by manufacturers and by pilots. The manufacturer must publish the numerical effect of wind on the take–off distance in the flight manual of a certified airplane. The pilot must account for these factors by obtaining airfield advisories on wind, altitude and temperature and consulting the flight manual before each flight, to ensure that the take–off can be conducted safely.

It must be kept in mind, that winds reported for airfields are typically measured at 50 ft above the surface. Because of shear effects in the boundary layer of air moving over the runway the following correction must be applied:

$$\frac{\text{Wind}_{\text{height 1}}}{\text{wind}_{\text{height 2}}} = \left(\frac{\text{height 1}}{\text{height 2}}\right)^{1/7} \qquad (10.9)$$

This correction factor is based on experimentally obtained values of wind–shear.

On the other hand, the FAA requires that pilots include a factor of conservatism to any take–off (and landing) distance calculations. Only 50% of the wind component measured at 50 ft may be used if the wing is favorable (i.e. head–wind). On the contrary, 150% of the wind measured at 50 ft must be accounted for if the wind is unfavorable (i.e. tail–wind). In other words:

use only half of the wind which improves performance and:
use one–and–one–half times the wind which hinders performance!

10.2.1.2 Effect of Speed on Take–off Thrust

Figure 10.9 shows actual thrust variations during take–off for a business jet and for a propeller driven twin. Note that in both cases the variation of thrust with speed is approximately linear.

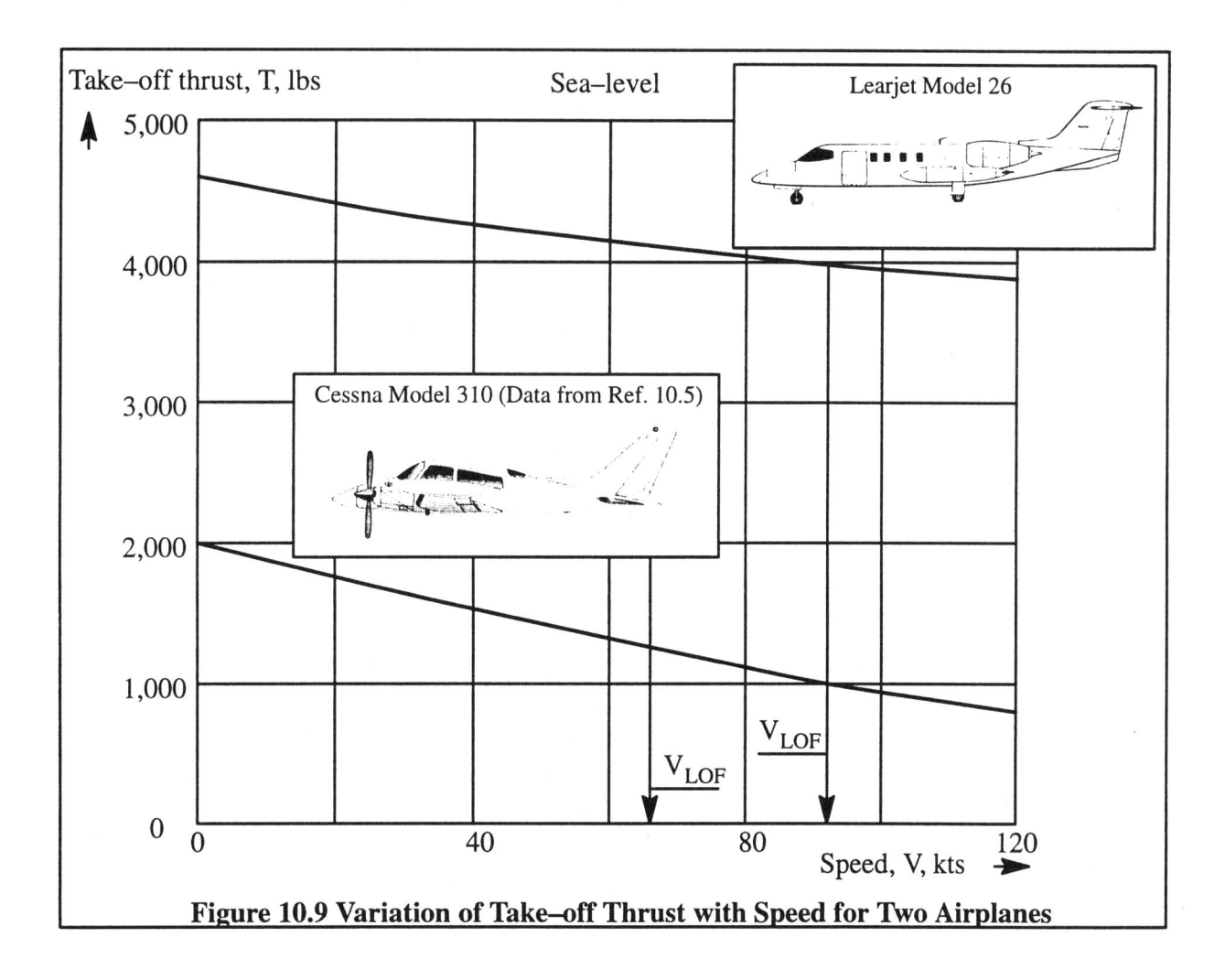

Figure 10.9 Variation of Take–off Thrust with Speed for Two Airplanes

10.2.1.3   Effect of the Ground on Lift and Drag

The lift and drag coefficients in Eqn (10.7) must be evaluated in ground effect. An easy way to visualize and estimate the effect of the ground on lift and drag is to use the image vortex system sketched in Figure 10.10.

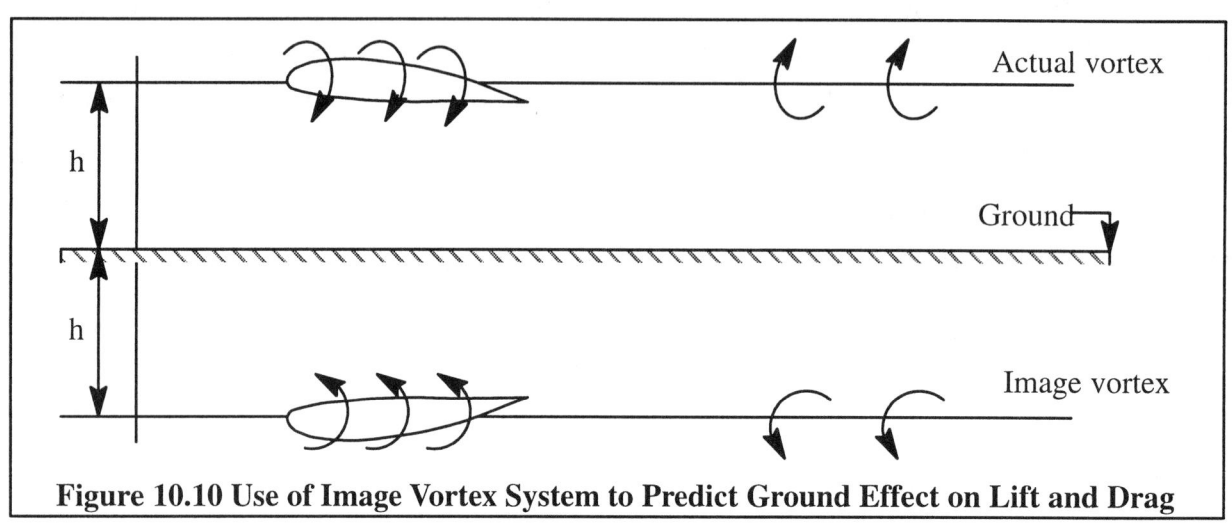

Figure 10.10 Use of Image Vortex System to Predict Ground Effect on Lift and Drag

The image vortex system takes the place of the ground in the mathematical modelling of the flow around the wing. This image vortex system has the same circulation magnitude as that of the wing but is of opposite sign. It is seen that the result of the image vortex is to increase the up–wash on the wing so that lift is increased and induced drag is decreased.

The increase in lift coefficient, C_{L_g}, at any given angle of attack, α, can be viewed as being caused by an effective increase in the lift–curve–slope, $C_{L_{\alpha_g}}$, due to an increase in the effective aspect ratio, A_e. The effective aspect ratio in ground effect can be determined from Figure 10.11.

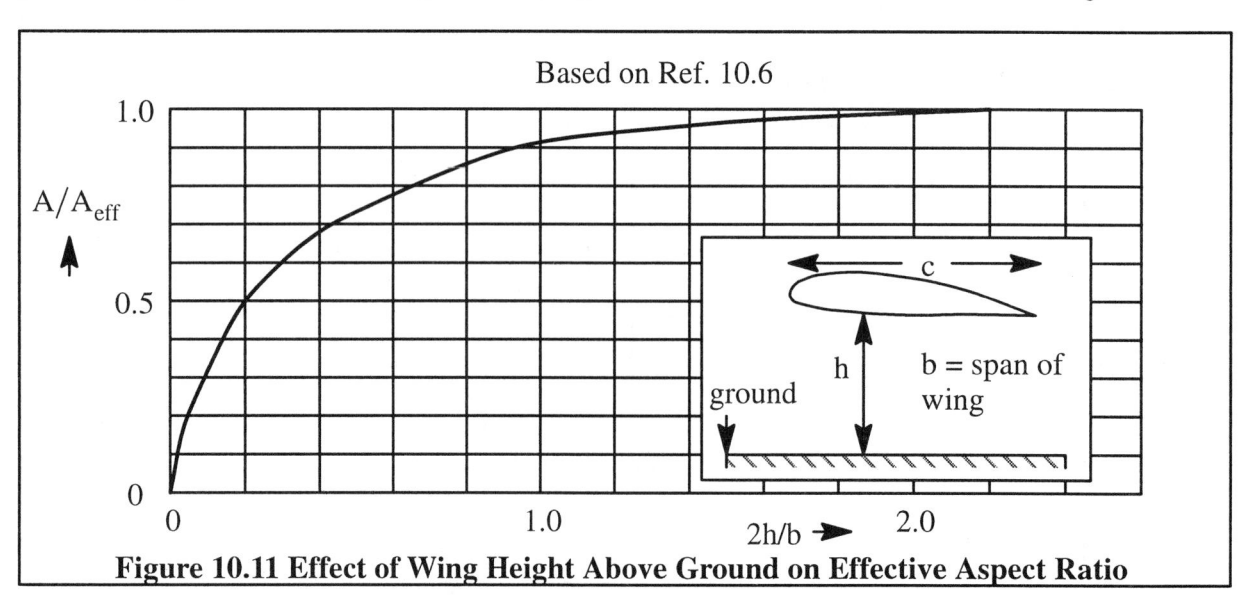

Figure 10.11 Effect of Wing Height Above Ground on Effective Aspect Ratio

The effective lift–curve–slope can then be calculated using Eqn (4.33). However, to calculate the lift coefficient in ground effect, the angle of attack for zero lift, α_{0_g}, must also be known. As a general rule, the angle of attack for zero lift is also reduced by ground effect (Ref. 10.7). According to the theory of Ref. 10.8, $\Delta\alpha_{0_g}$ is proportional to the airfoil thickness ratio, t/c. Using the least square method to curve–fit the data of Ref. 10.7 the following approximate equation for $\Delta\alpha_{0_g}$ is found:

$$\Delta\alpha_{0_g} = \left(\frac{t}{c}\right)\left\{-0.1177\frac{1}{(h/c)^2} + 3.5655\frac{1}{(h/c)}\right\} \text{ in deg.} \tag{10.10}$$

where: t/c is the thickness to chord ratio of the airfoil
       h/c is defined in Figure 10.11

By applying these ground effect corrections to the airfoil at the wing mean geometric chord, the following approximate equation may be used for calculating airplane lift in ground effect:

$$C_{L_g} = C_{L_{\alpha_g}}(\alpha - \alpha_0 - \Delta\alpha_{0_g}) = C_L\frac{C_{L_{\alpha_g}}}{C_{L_\alpha}} - C_{L_{\alpha_g}}\Delta\alpha_{0_g} \tag{10.11}$$

With large flap deflections, the ground effect on lift can be adverse (i.e. the lift in ground effect will be reduced) as shown in Ref. 10.9. A handbook method for determining airplane lift coefficient versus angle of attack curves in and out of ground effect is given in Ref. 10.4.

Ground effect on drag is primarily a change in induced drag. According to Refs 10.10 and 10.11 the decrease in induced drag due to ground effect may be expressed as:

$$\Delta C_{D_{i_g}} = -\sigma'\frac{C_L^2}{\pi A} \quad \text{Note : e = 1 in this theory} \tag{10.12}$$

where: C_L is the lift coefficient in the appropriate configuration out of ground effect

    σ' is the induced drag ground influence coefficient which may be estimated from:

$$\sigma' = \frac{1 - 1.32(h/b)}{1.05 + 7.4(h/b)} \quad \text{for} \quad 0.033 < (h/b) < 0.25 \tag{10.13}$$

Figure 10.12 shows how σ' varies with (h/b). Typical values for (h/b) are:

    (h/b) is approximately equal to 0.1 for low wing airplanes

    (h/b) is approximately equal to 0.2 for high wing configurations.

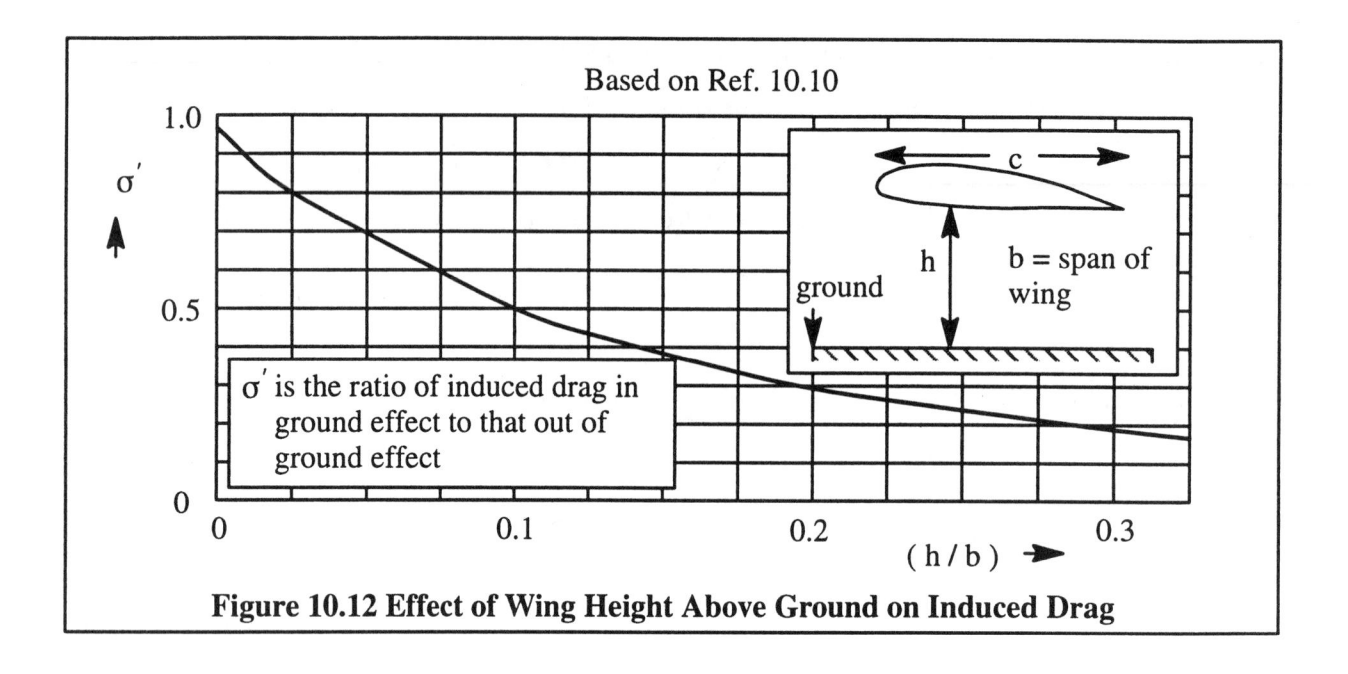

Figure 10.12 Effect of Wing Height Above Ground on Induced Drag

(Based on Ref. 10.10)

σ' is the ratio of induced drag in ground effect to that out of ground effect

The reader is cautioned not to use Figure 10.12 for large flap deflections. As a preliminary suggestion: 40 degrees or more would constitute a large flap deflection.

An example application of these ground effect corrections will now be given.

Example 10.1: The wing characteristics of the Douglas F5D–1 airplane (see Figure 10.13 for a three–view) are as follows:

$A = 2.20$ $\quad\quad$ $h/\bar{c} = 0.33$ $\quad\quad$ $2h/b = 0.36$ $\quad\quad$ $t/c = 0.05$

$S = 557 \text{ ft}^2$ $\quad\quad$ $b = 33.5 \text{ ft}$ $\quad\quad$ $\Lambda_{c/2} = 35^0$ Estimate the ground effect on airplane lift–coefficient–versus–angle–of–attack and on induced drag.

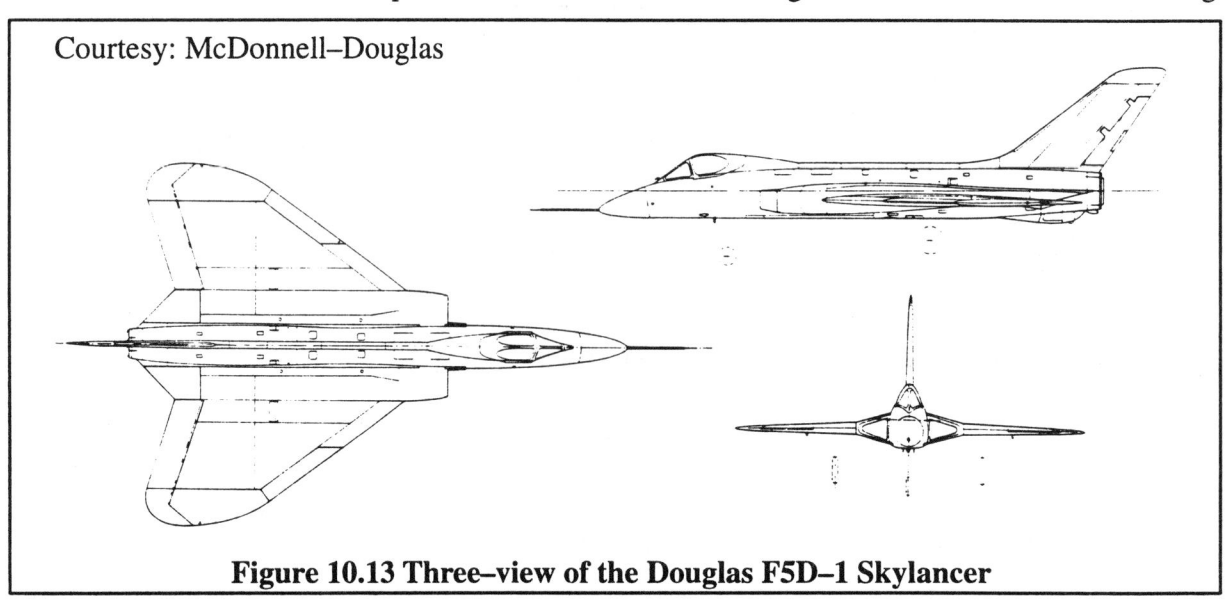

Figure 10.13 Three–view of the Douglas F5D–1 Skylancer

Courtesy: McDonnell–Douglas

Solution: From Figure 10.11 it follows that: $A/A_{eff} = 0.67$. Therefore it follows that: $A_{eff} = 2.02/0.67 = 3.01$. The lift–curve–slopes out of and in ground effect follow with Eqn. (4.32) as:

$$C_{L_\alpha} = \frac{2\pi \times 2.02}{2 + \sqrt{2.02^2(1 + \tan^2 35^0) + 4}} = 2.45 \text{ rad}^{-1} \quad \text{and}$$

$$C_{L_{\alpha_g}} = \frac{2\pi \times 3.01}{2 + \sqrt{3.01^2(1 + \tan^2 35^0) + 4}} = 3.06 \text{ rad}^{-1}$$

The change in the zero–lift–angle–of–attack follows from Eqn (10.10) as:

$$\Delta\alpha_{0_g} = 0.05\left(\frac{-0.1177}{0.329^2} + \frac{3.5655}{0.329}\right) = 0.5 \text{ deg.}$$

The relationship between the lift coefficients out of and in ground effect is found from Eqn (10.11) as:

$$C_{L_g} = C_L\frac{3.06}{2.45} - 3.06 \times \frac{\pi}{180} \times 0.5 = 1.25 C_L - 0.03$$

From Figure 10.12 or from Eqn (10.13) it is found that the ratio of induced drag in ground effect to that out of ground effect is: $\sigma' = 0.32$. The basic airplane zero–lift drag coefficient has been estimated to be: $C_{D_0} = 0.0083$, while Oswald's efficiency factor, e, is: e = 1.0 in this theory.

Figure 10.14 shows a comparison of lift and drag characteristics in and out of ground effect for the clean F5D–1 airplane.

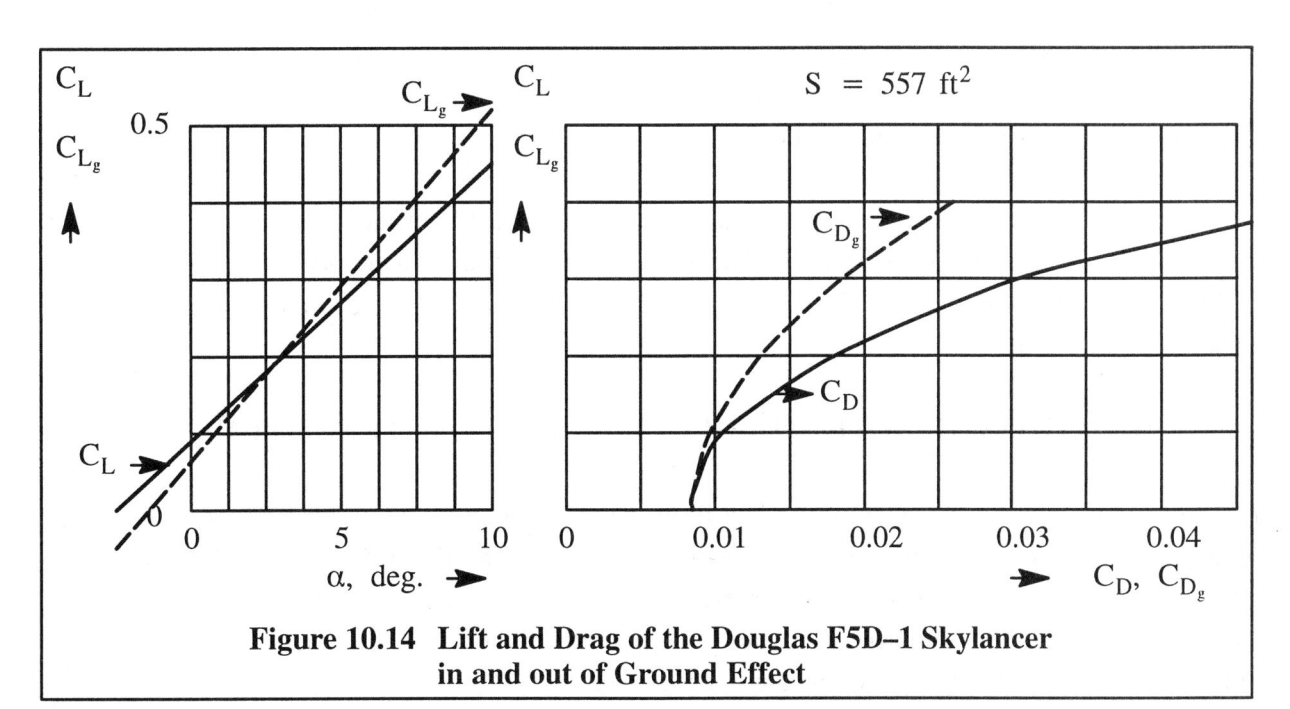

Figure 10.14 Lift and Drag of the Douglas F5D–1 Skylancer in and out of Ground Effect

10.2.1.4 Effect of Thrust Deflection on Take–off

In the case of a deflected propeller slipstream, the situation depicted in Figure 10.15 prevails.

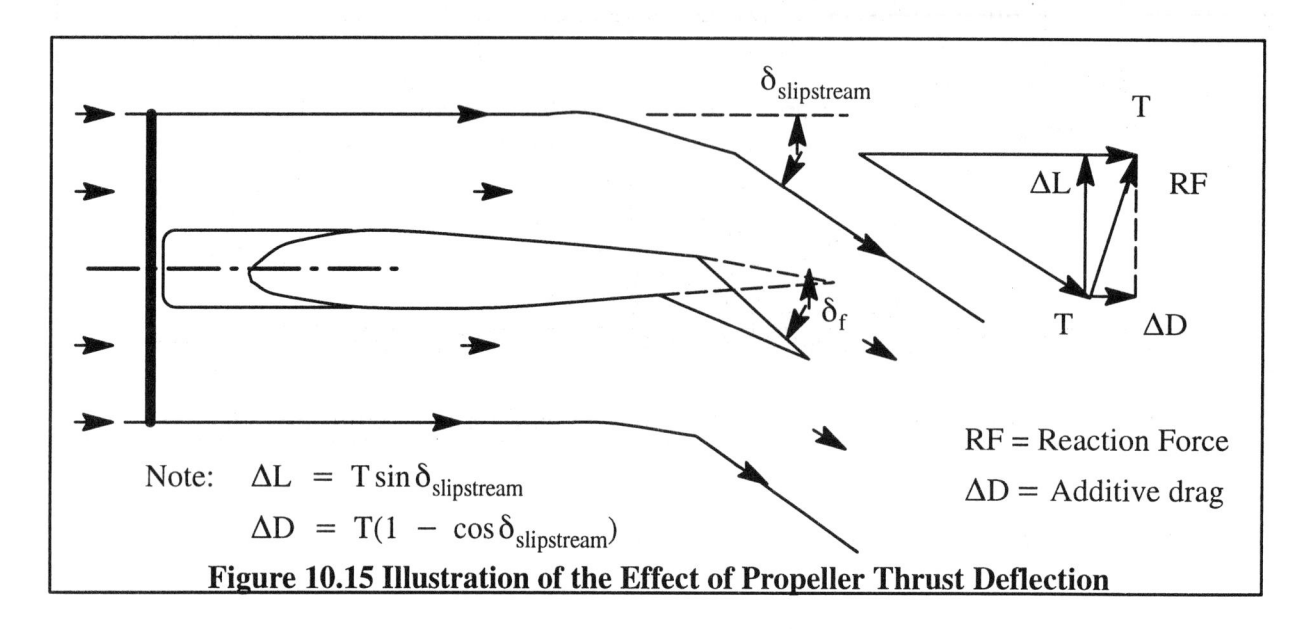

Note: $\Delta L = T \sin \delta_{slipstream}$
$\Delta D = T(1 - \cos \delta_{slipstream})$

RF = Reaction Force
ΔD = Additive drag

Figure 10.15 Illustration of the Effect of Propeller Thrust Deflection

With the flaps deflected in the take–off position, the propeller slipstream will be turned over an angle, $\delta_{slipstream}$, which may be somewhat less than the flap deflection angle, δ_f.

The slipstream deflection implies a change in the momentum of the flow. This will cause a reaction force, RF, which produces both lift (desirable) and drag (undesirable, but inevitable). As a general rule, the reaction force will also cause a pitching moment which has to be trimmed out. As a first order approximation, the magnitude of the reaction force, and the corresponding extra drag and lift can be estimated from the vector diagram shown in Figure 10.15. These extra forces must be accounted for to obtain accurate take–off predictions. FAR 23.457 also requires the reaction force to be accounted for in the design of the flap structure.

In the case of a deflected jet, when the thrust is deflected (such as in a thrust–vectored configuration), the thrust available for longitudinal acceleration along the runway in Eqn (10.8) will not be T, but: Txcos(deflection angle). From a momentum viewpoint, the same situation prevails as the one shown in Figure 10.15.

10.2.2 EQUATIONS OF MOTION DURING THE TAKE–OFF TRANSITION

Figure 10.16 shows a typical transition flight path followed by the climb–out to the obstacle. The transition starts at lift–off and continues until a straight line flight path toward the obstacle is established. During the transition the accelerations along and perpendicular to the flight path cannot be neglected. Equations (9.1) and (9.2) can be used to model the transition. These equations, modified to include ground effect, are repeated here as Eqns (10.14) and (10.15):

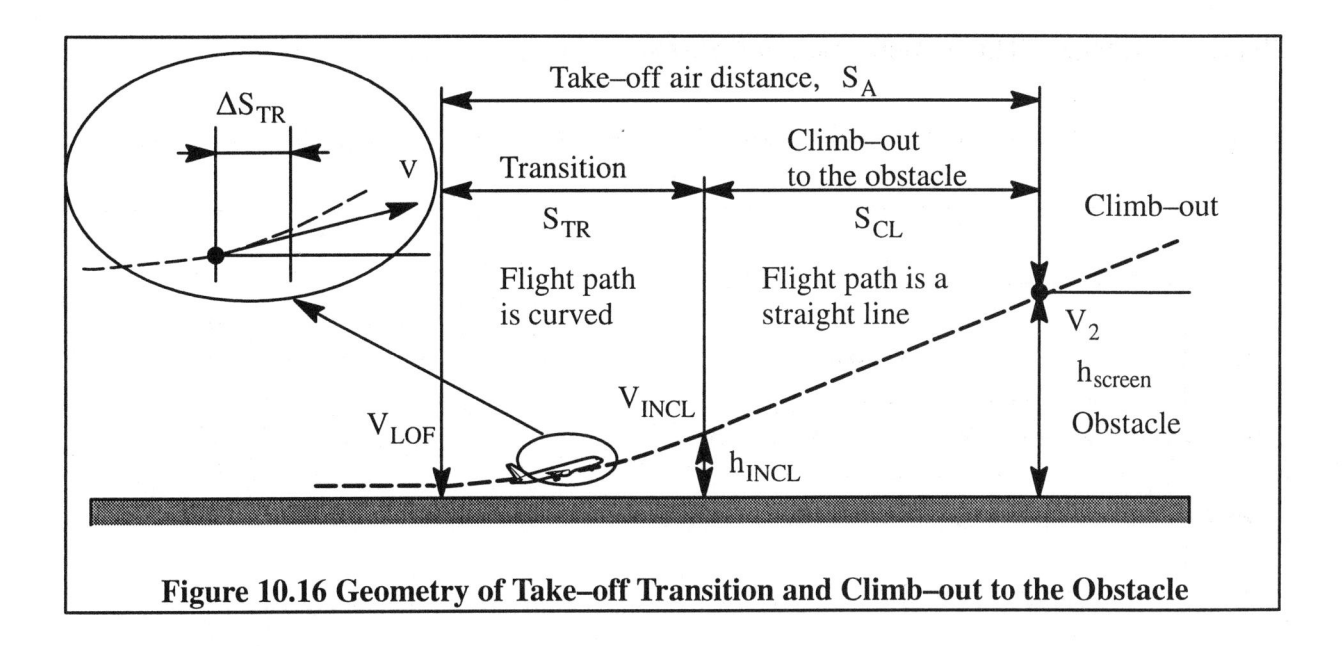

Takeoff and Landing

Figure 10.16 Geometry of Take–off Transition and Climb–out to the Obstacle

$$T\cos(\alpha + \phi_T) - D_g - W\sin\gamma = \frac{W}{g}\frac{dV}{dt} \qquad (10.14)$$

$$T\sin(\alpha + \phi_T) + L_g - W\cos\gamma = C.F. = \frac{W}{g}V\dot{\gamma} \qquad (10.15)$$

The drag and the lift coefficients in Eqns (10.14) and (10.15) must be computed in ground effect. Methods to account for ground effect were given in Sub–section 10.2.1.

10.2.3 EQUATIONS OF MOTION DURING THE CLIMB–OUT TO THE OBSTACLE

Figure 10.16 also shows the flight path during climb–out to the obstacle. During this part of the take–off the acceleration perpendicular to the flight path is zero. However, the acceleration along the flight path is not zero because the airplane must accelerate to the obstacle speed, V_2. The equations of motion are derived from Eqns (10.14) and (10.15):

$$T\cos(\alpha + \phi_T) - D_g - W\sin\gamma = \frac{W}{g}\frac{dV}{dt} \qquad (10.16)$$

$$T\sin(\alpha + \phi_T) + L_g - W\cos\gamma = 0 \qquad (10.17)$$

Depending on the size and the configuration of the airplane, ground effect may or may not be negligible. This needs to be verified for each type of airplane.

10.3 PREDICTION OF THE TAKE–OFF DISTANCE

Four methods for the calculation of the take–off distance will be presented:

10.3.1 Accurate method 10.3.2 Statistical method
10.3.3 Approximate, analytical method 10.3.4 Balanced fieldlength

Methods for presentation of take–off performance data are given in Sub–section 10.3.5.

10.3.1 ACCURATE METHOD FOR PREDICTING THE TAKE–OFF DISTANCE

10.3.1.1 Ground–distance (Accurate method)

Figure 10.6 shows the geometry of the ground–roll. Assume that the variation of the acceleration along the runway, as expressed by Eqn (10.5), is known as a function of airplane speed. The integration to find the ground–roll part of the take–off distance, S_G, according to Eqn (10.8), can now be carried out with great precision. To account for wind (assumed to be constant) during the take–off run, it is convenient to plot a curve of $(V \pm V_w)/a_g$ versus V is plotted. An example, for the case of head–wind, is shown in Figure 10.17. Simpson's Rule (shown in Figure 10.17) or any other convenient numerical integration scheme can then be used to find the take–off ground distance, S_G.

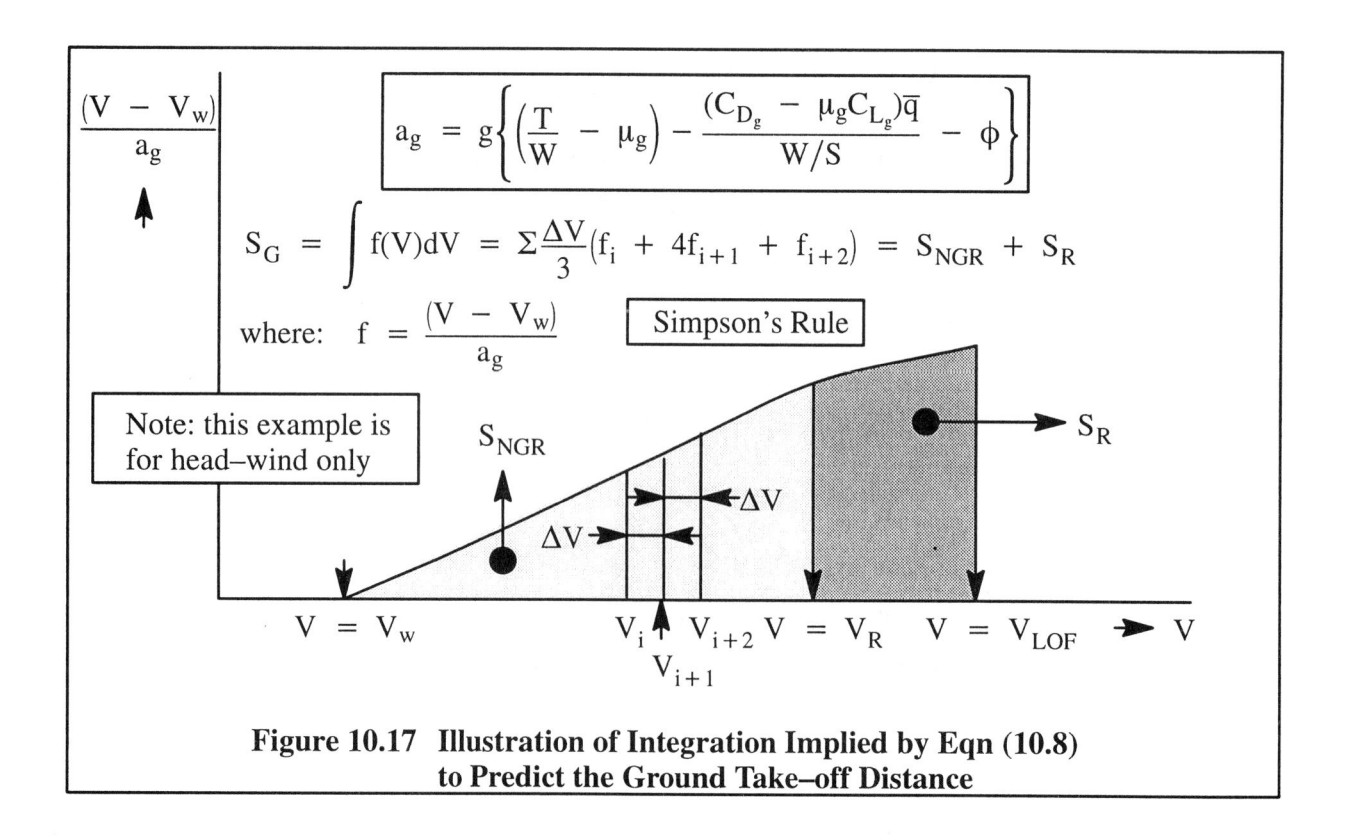

$$a_g = g\left\{\left(\frac{T}{W} - \mu_g\right) - \frac{(C_{D_g} - \mu_g C_{L_g})\bar{q}}{W/S} - \phi\right\}$$

$$S_G = \int f(V)dV = \Sigma\frac{\Delta V}{3}(f_i + 4f_{i+1} + f_{i+2}) = S_{NGR} + S_R$$

where: $f = \frac{(V - V_w)}{a_g}$

Note: this example is for head–wind only

Figure 10.17 Illustration of Integration Implied by Eqn (10.8) to Predict the Ground Take–off Distance

10.3.1.2 Transition Distance (Accurate method)

Figure 10.16 shows the geometry of the take–off transition. To calculate the transition distance, S_{TR}, the shape of the flight path and the variation of speed along the flight path must be known. The equations of motion during transition are (10.14) and (10.15). These equations can be integrated numerically between the lift–off speed, V_{LOF}, and the initial climb–out speed, V_{INCL}. The integration process should account for the fact that the airplane (at constant thrust or power) must reach the obstacle at a speed of V_2 in accordance with the speed requirements of Table 10.1. This implies some schedule of $\dot{\gamma}$ versus t. The transition distance, S_{TR}, follows from:

$$S_{TR} = \sum_i \Delta S_{TR_i} = \sum_i \left(V_i \cos\gamma_i \times \Delta t\right) \approx \sum_i (V_i \Delta t) \qquad (10.18)$$

where the angles, γ_i, are considered to be small.

10.3.1.3 Climb Distance (Accurate method)

Figure 10.16 shows the geometry of the climb–out path to the obstacle.

The climb distance to the obstacle (35 ft for FAR 25 and commuter category FAR 23, 50 ft for other airplanes in FAR 23) can be determined from knowledge of:

* the flight path angle at the height at the end of the transition, h_{INCL}, which can be assumed to be the same at the flight path angle at the obstacle height, $h_{screen} = 35$ ft or 50 ft

* the initial climb–out speed at the end of the transition, V_{INCL}

* the required speed at the obstacle, V_2

Although a numerical integration of the equations of motion (10.16) and (10.17) gives the most accurate result, an adequate estimate can be obtained from:

$$S_{CL} = \left(\frac{V_2 + V_{INCL}}{2}\right)\left[\frac{h_{screen} - h_{INCL}}{\left(\frac{V_2 + V_{INCL}}{2}\right)\gamma}\right] = \left(\frac{h_{screen} - h_{INCL}}{\gamma}\right) \qquad (10.19)$$

10.3.2 STATISTICAL METHOD FOR PREDICTING THE TAKE–OFF DISTANCE

The following statistical method for predicting the take–off distance is taken from Ref. 10.12 but is based on a method described by Loftin in Ref.10.13. The method applies to:

* Propeller driven, FAR 23 airplanes as presented in Sub–sub–section 10.3.2.1

* Jet driven FAR 25 airplanes as presented in Sub–sub–section 10.3.2.2

10.3.2.1 Statistical Method for Predicting the Take–off Distance of Propeller Driven, FAR 23 Airplanes

The method predicts the total ground distance, S_G, and the total air distance, S_A. The FAR 23 take–off distance from hard–surfaces runways may be computed from:

$$S_{TO} = 1.66\, S_G \tag{10.20}$$

with:

$$S_G = 4.9\,(TOP_{23}) + 0.009\,(TOP_{23})^2 \tag{10.21}$$

(TOP_{23}) is the so–called FAR–23 take–off parameter which is defined as follows:

$$(TOP_{23}) = \frac{(W/S)_{TO}\,(W/P)_{TO}}{\sigma\, C_{L_{max_{TO}}}} \tag{10.22}$$

where: $(W/S)_{TO}$ is the wing–loading at take–off, in lbs/ft²

$(W/P)_{TO}$ is the power–loading at takeoff, in lbs/hp

σ is the ambient density ratio at take–off

$C_{L_{max_{TO}}}$ is the trimmed maximum lift coefficient at take–off

Combining Eqns (10.20) and (10.21) yields:

$$S_{TO} = 8.134\,(TOP_{23}) + 0.0149\,(TOP_{23})^2 \tag{10.23}$$

Figure 10.18 shows a plot of Eqns (10.21) and (10.23) versus the parameter, (TOP_{23}). The data points represent actual data from Ref. 10.13. The scatter comes about largely because of differences in pilot technique. An example application is given in Figure 10.19.

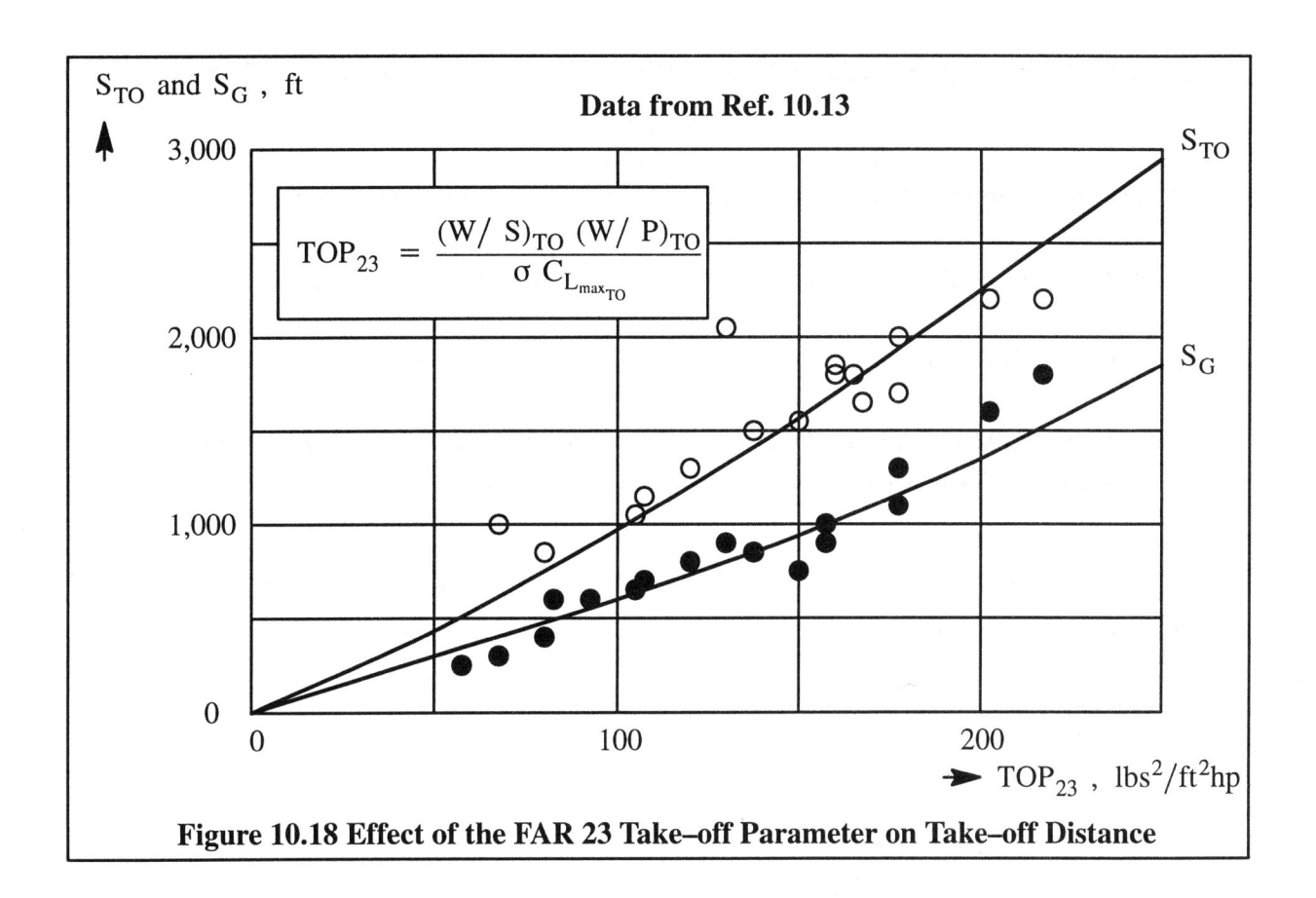

Takeoff and Landing

$$TOP_{23} = \frac{(W/S)_{TO}(W/P)_{TO}}{\sigma C_{L_{max_{TO}}}}$$

Data from Ref. 10.13

Figure 10.18 Effect of the FAR 23 Take–off Parameter on Take–off Distance

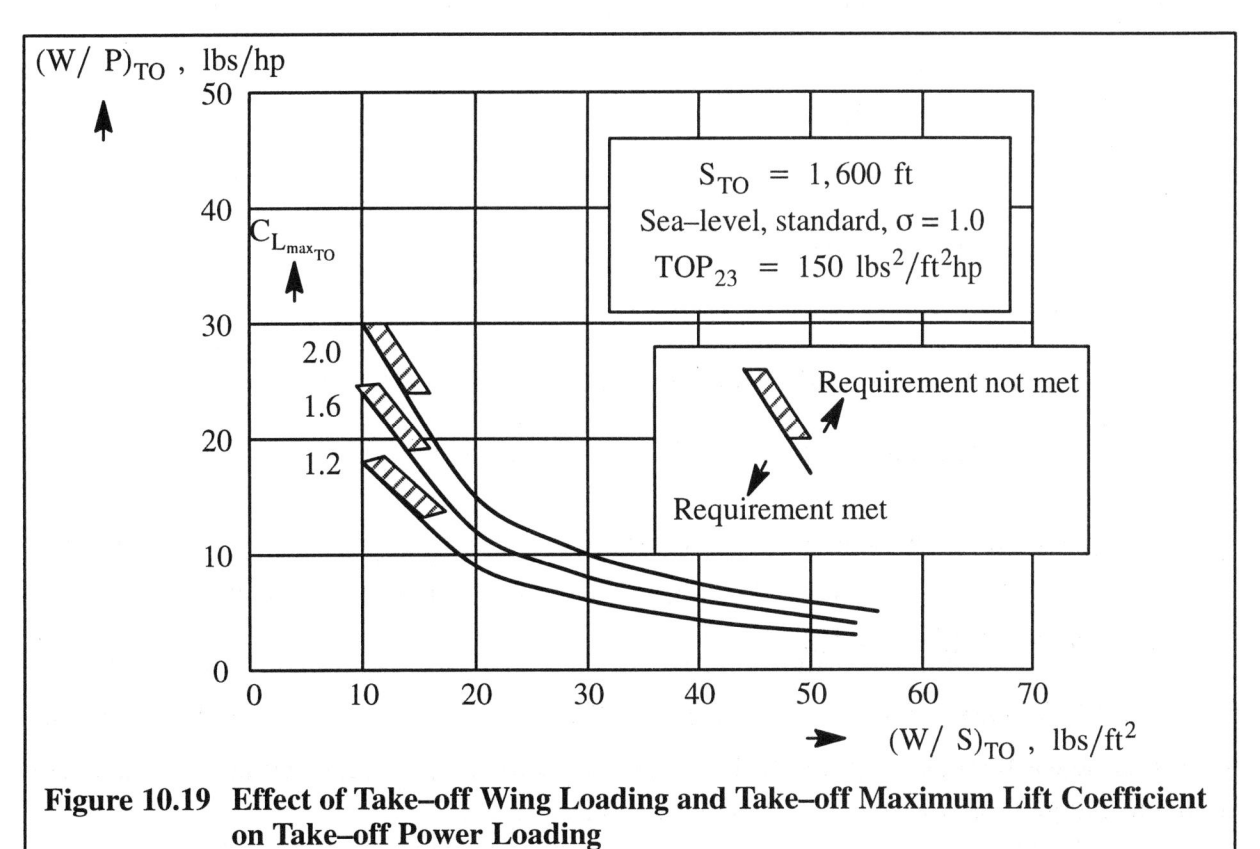

S_{TO} = 1,600 ft
Sea–level, standard, σ = 1.0
TOP_{23} = 150 lbs^2/ft^{2}hp

Requirement not met

Requirement met

Figure 10.19 Effect of Take–off Wing Loading and Take–off Maximum Lift Coefficient on Take–off Power Loading

Chapter 10

This statistical method is very useful in the early stages of preliminary design. By specifying the required take–off distance, S_{TO}, the field altitude and the temperature (this defines the density ratio, σ) and the maximum lift coefficient at take–off, $C_{L_{max_{TO}}}$, a relation between $(W/S)_{TO}$ and $(W/P)_{TO}$ is established.

Figure 10.19 shows the effect of various values of $C_{L_{max_{TO}}}$ on this relation. It is clear that increasing $C_{L_{max_{TO}}}$ will reduce the amount of take–off power required. The reader should observe that the vertical axis in Figure 10.19 is weight over power (and not power over weight).

10.3.2.2 Statistical Method for Predicting the Take–off Distance of Jet Driven, FAR 25 Airplanes

The method predicts the total FAR 25 balanced field–length distance, S_{TOBFL}, from:

$$S_{TOBFL} = 37.5 \, TOP_{25} \qquad (10.24)$$

where: (TOP_{25}) is the so–called FAR–25 take–off parameter which is defined as follows:

$$(TOP_{25}) = \frac{(W/S)_{TO}}{\sigma \, C_{L_{max_{TO}}} (T/W)_{TO}} \qquad (10.25)$$

where: $(T/W)_{TO}$ is the thrust–to–weight ratio at take–off, which is dimensionless

Figure 10.20 shows a plot of Eqn (10.24) versus the parameter, (TOP_{25}). The data points represents actual data from Ref. 10.11. The scatter comes about largely because of differences in pilot technique.

This statistical method is very useful in the early stages of preliminary design. Figure 10.21 presents a numerical example. By specifying the required take–off balanced field–length distance, S_{TOBFL}, the field altitude and temperature (this defines the density ratio, σ) and the maximum lift coefficient at take–off, $C_{L_{max_{TO}}}$, a relation between $(W/S)_{TO}$ and $(T/W)_{TO}$ is established.

The example in Figure 10.21 shows the powerful effect of $C_{L_{max_{TO}}}$ on the required thrust–to–weight ratio at take–off, $(T/W)_{TO}$, for a given wing loading at take–off, $(W/S)_{TO}$.

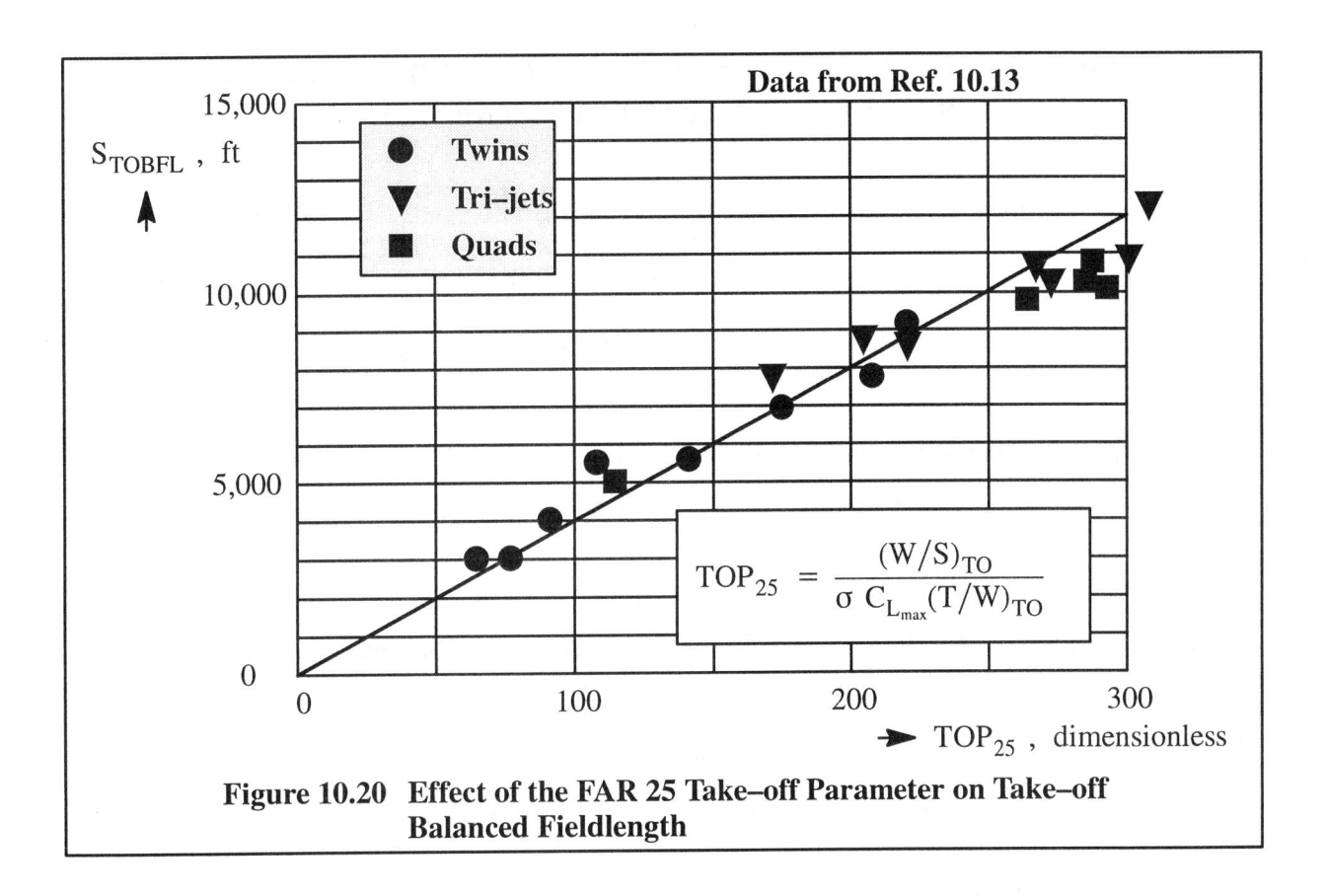

Takeoff and Landing

$$TOP_{25} = \frac{(W/S)_{TO}}{\sigma\, C_{L_{max}}(T/W)_{TO}}$$

Figure 10.20 Effect of the FAR 25 Take–off Parameter on Take–off Balanced Fieldlength

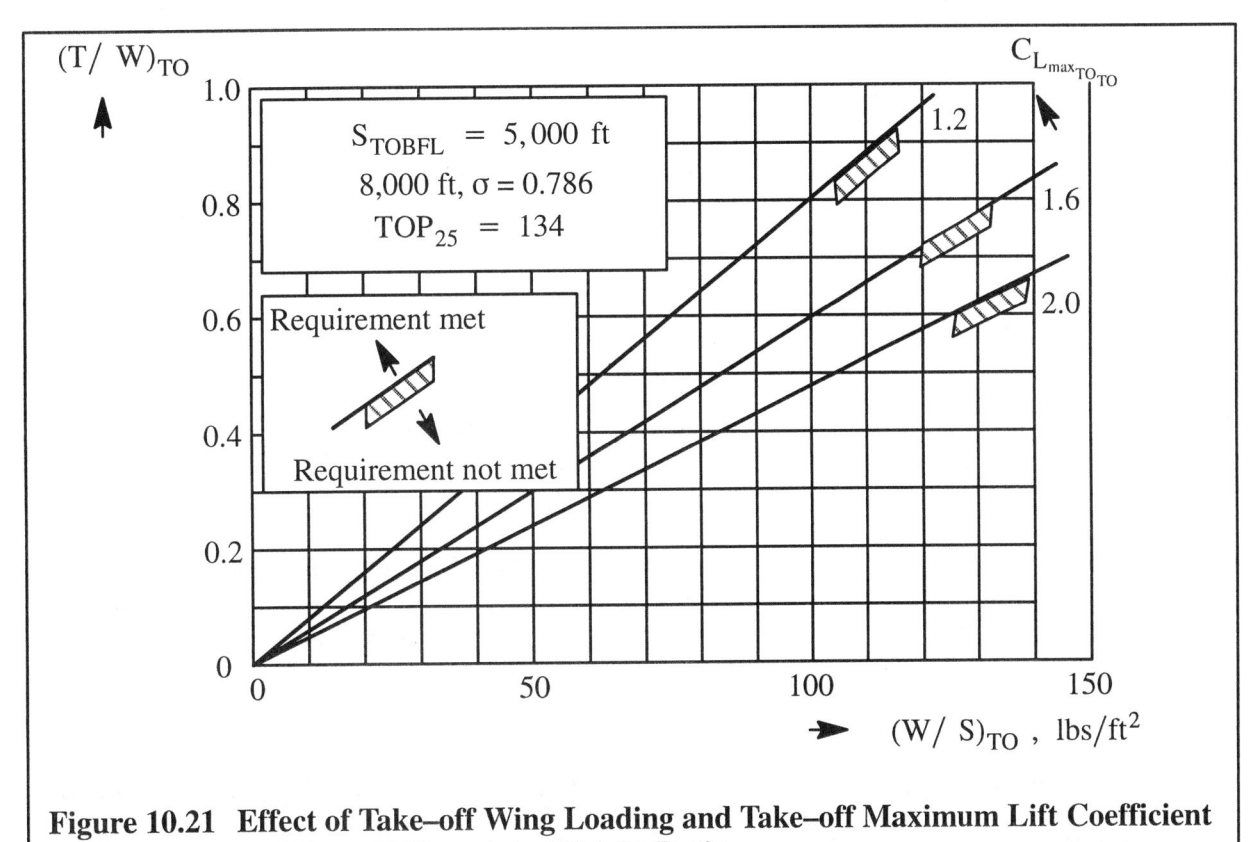

$S_{TOBFL} = 5{,}000$ ft

$8{,}000$ ft, $\sigma = 0.786$

$TOP_{25} = 134$

Figure 10.21 Effect of Take–off Wing Loading and Take–off Maximum Lift Coefficient on Take–off Thrust–to–Weight Ratio

10.3.3 APPROXIMATE, ANALYTICAL METHOD FOR PREDICTING THE TAKE–OFF DISTANCE AND THE TIME TO TAKE–OFF

In this approximate, analytical method, the total take–off distance, S_{TO} is divided into the four components shown in Figure 10.6 and expressed in Eqn (10.1). Methods for determining S_{NGR}, S_R, S_{TR} and S_{CL} are presented in Sub–sub–sections 10.3.3.1 through 10.3.3.4 respectively.

An alternate method for calculating the total ground distance, $S_G = S_{NGR} + S_R$, is given in Sub–sub–section 10.3.3.5. A method for predicting the time required to take–off is presented in Sub–sub–section 10.3.3.6.

10.3.3.1 Approximate Method for Calculating the Ground Distance Component, S_{NGR}

During this part of the ground run the airplane accelerates from $V = 0$ to $V = V_R$. Referring to Eqn (10.5) it is seen that both drag and lift depend on the square of velocity, V^2. Both drag, D, and lift, L, can thus be thought of as linear functions of V^2. It is also reasonable to approximate the low speed variation of jet engine thrust and propeller thrust with speed as a quadratic function. In such a case, the acceleration on the ground, a_g, can be written as:

$$a_g = a_{g_{V=0}} - \frac{\left(a_{g_{V=0}} - a_{g_{V=V_R}}\right) V^2}{V_R^2} \tag{10.26}$$

where:
$$a_{g_{V=0}} = g\left(\frac{T_{V=0}}{W} - \mu_g - \phi\right) \tag{10.27}$$

$$a_{g_{V=V_R}} = g\left\{\left(\frac{T_{V=V_R}}{W} - \mu_g\right) - \frac{(C_{D_g} - \mu_g C_{L_g})\varrho V_R^2}{2W/S} - \phi\right\} \tag{10.28}$$

Both accelerations, $a_{g_{V=0}}$ and $a_{g_{V=V_R}}$ are known constants for a given airplane at take–off. The lift–off speed, V_{LOF} must be selected so that: $V_{LOF} > V_R$ is satisfied in accordance with the requirements summary in Table 10.1.

For preliminary analysis purposes, $V_{LOF} \approx 1.15 V_S$ and $V_R \approx 1.10 V_S$, may be assumed.

Eqn (10.7) will now be used to find the ground distance component, S_{NGR} for two cases:

Case 1: Zero Wind and **Case 2: Non–zero Wind**.

Case 1: Zero Wind

For zero wind, Eqn (10.7) may be written as:

$$S_{NGR} = \int_0^{V_R} \frac{VdV}{a_{g_{V=0}} - \frac{\left(a_{g_{V=0}} - a_{g_{V=V_R}}\right)V^2}{V_R^2}} = \frac{1}{2}\int_0^{V_R^2} \frac{dV^2}{a_{g_{V=0}} - \frac{\left(a_{g_{V=0}} - a_{g_{V=V_R}}\right)V^2}{V_R^2}} = 0$$

$$= \frac{V_R^2}{2a_{g_{ave}}} \tag{10.29}$$

where: $a_{g_{ave}}$ may be regarded as the average acceleration during the ground roll, with:

$$a_{g_{ave}} = a_{g_{V=0}}\left\{\frac{1 - \frac{a_{g_{V=V_R}}}{a_{g_{V=0}}}}{\ln\left(\frac{a_{g_{V=0}}}{a_{g_{V=V_R}}}\right)}\right\} = k\, a_{g_{V=0}}, \quad \text{with}: k = \left\{\frac{1 - \frac{a_{g_{V=V_R}}}{a_{g_{V=0}}}}{\ln\left(\frac{a_{g_{V=0}}}{a_{g_{V=V_R}}}\right)}\right\} \tag{10.30}$$

The constant 'k' may be determined numerically from Eqn (10.30) or from Figure 10.22, once the accelerations at zero speed and at rotation speed have been calculated.

Case 2: Non–zero Wind

For the case of non–zero wind, Eqn (10.7) can be written as:

$$S_{NGR} = \int_{\mp V_w}^{V_R} \frac{VdV}{a_{g_{V=0}} - \frac{\left(a_{g_{V=0}} - a_{g_{V=V_R}}\right)V^2}{V_R^2}} \pm V_w \int_{t=0}^{t=t_R} dt \tag{10.31}$$

There are two integrals in Eqn (10.31). The first integral can be evaluated exactly. To obtain the second integral, an estimate for the time to lift–off, t_R, is required.

The reader is asked to show that the first integral follows from:

$$\frac{1}{2}\int_{\mp V_w^2}^{V_R^2} \frac{dV^2}{a_{g_{V=0}} - \frac{\left(a_{g_{V=0}} - a_{g_{V=V_R}}\right)^2 V^2}{V_R^2}} = \frac{1}{2}\left(\frac{V_R^2 - V_w^2}{a_{g_{\text{ave with wind}}}}\right) \qquad (10.32)$$

where:

$$a_{g_{\text{ave with wind}}} = k_w a_{g_{V=0}} \text{ with}: k_w = \frac{\left(1 - \frac{a_{g_{V=V_R}}}{a_{g_{V=0}}}\right)\left(1 - \frac{V_w^2}{V_R^2}\right)}{\ln\left\{\frac{a_{g_{V=0}}}{a_{g_{V=V_R}}}\left(1 - \frac{V_w^2}{V_R^2}\right) + \frac{V_w^2}{V_R^2}\right\}} \qquad (10.33)$$

The constant, k_w, may be evaluated from Eqn (10.33) or from Figure 10.22.

For the ground distance component, S_{NGR}, it therefore follows that:

$$S_{NGR} = \frac{1}{2}\left(\frac{V_R^2 - V_w^2}{a_{g_{\text{ave with wind}}}}\right) \pm V_w t_R \qquad (10.34)$$

The time elapsed during this part of the ground run, t_R, may be estimated from:

$$t_R = \left(\frac{V_R \pm V_w}{a_{g_{\text{ave with wind}}}}\right) \qquad (10.35)$$

Finally, by combining Eqns (10.34) and (10.35), the ground roll distance component, S_{NGR}, follows from:

$$S_{NGR} = \frac{1}{2}\left(\frac{V_R^2 - V_w^2}{a_{g_{\text{ave with wind}}}}\right) \pm V_w\left(\frac{V_R \pm V_w}{a_{g_{\text{ave with wind}}}}\right) = \frac{(V_R \pm V_w)^2}{2 a_{g_{\text{ave with wind}}}} \qquad (10.36)$$

The elapsed time, t_R, plays a role in determining the fuel used during take–off. Methods to determine the fuel used for a given airplane mission profile are discussed in Chapter 11.

10.3.3.2   Approximate Method for Calculating the Ground Distance Component, S_R

The process of rotating an airplane to eventually lift off typically takes 1–3 seconds. Pilot technique does have a significant effect on this time element. The following numerical values are suggested for the time to rotate to lift–off, t_{rotate}:

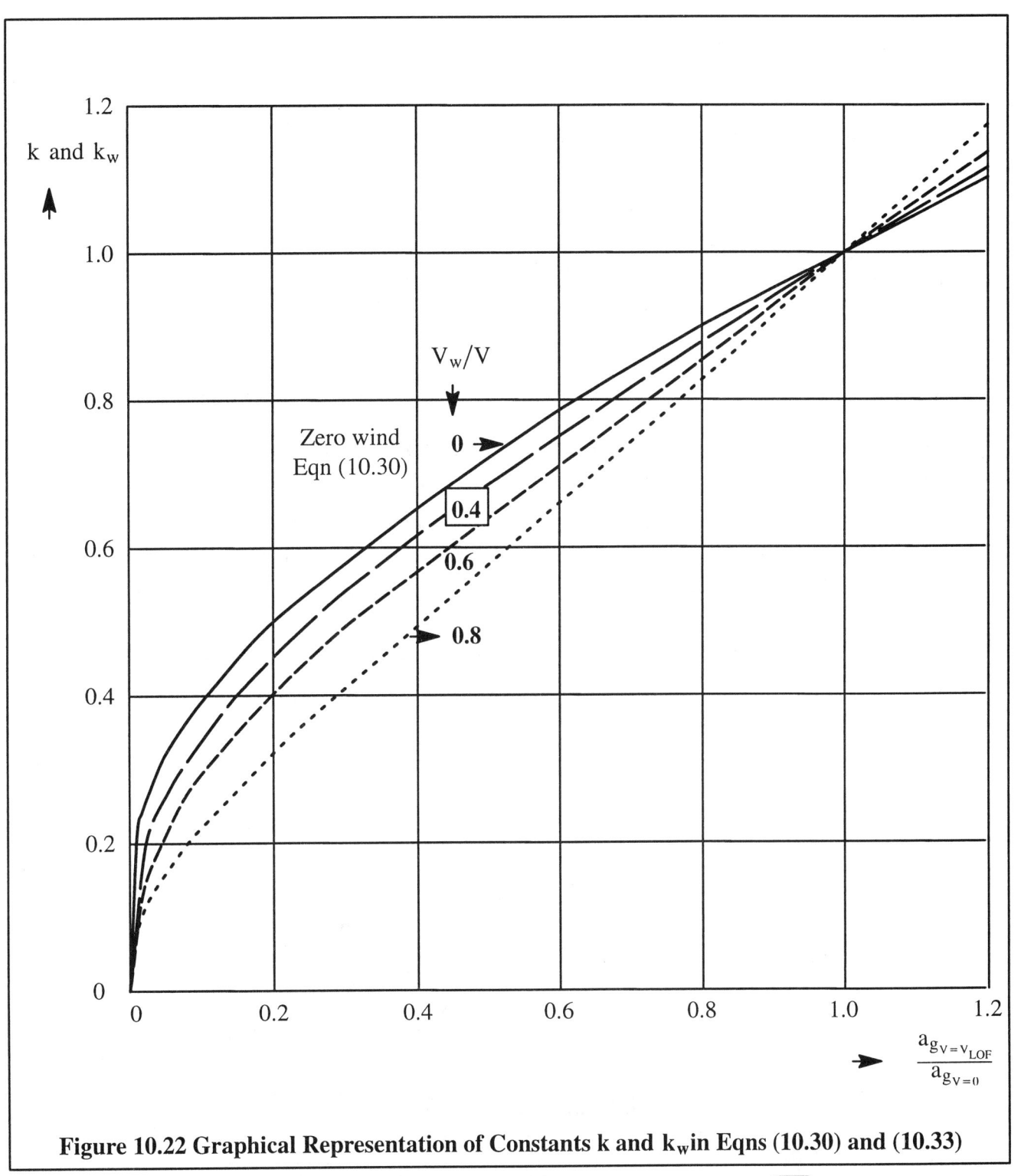

Figure 10.22 Graphical Representation of Constants k and k_w in Eqns (10.30) and (10.33)

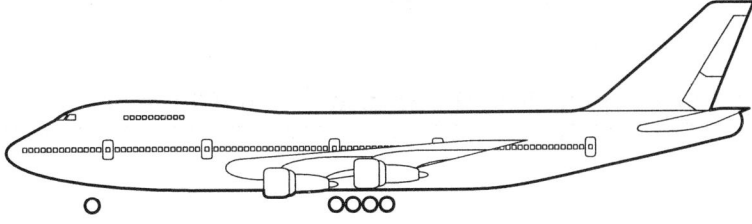

X-axis: $\dfrac{a_{g_{V=V_{LOF}}}}{a_{g_{V=0}}}$

Y-axis: k and k_w

Curves labeled by V_w/V: 0 (Zero wind Eqn (10.30)), 0.4, 0.6, 0.8

For light airplanes: $t_{rotate} = 1$ sec.

For fighters: $t_{rotate} = 2$ sec.

For transports airplanes: $t_{rotate} = 3$ sec.

It is assumed that the speed during the rotation process is the average between V_R and V_{LOF}. The ground distance component, S_R, including the effect of wind, then follows from:

$$S_R = \frac{1}{2}\{(V_R + V_{LOF}) \pm V_w\} t_{rotate} \qquad (10.37)$$

10.3.3.3 Approximate Method for Calculating the Air Distance Component, S_{TR}

Pilot technique has a significant effect on the type of flight path and the distance covered before a straight line flight path to climb out to the obstacle height is established. For preliminary analysis purposes it has been found acceptable to assume a circular flight path for this transition.

Figure 10.23 shows the geometry of the transition. It is assumed that the speed during the transition remains constant at $V = V_{LOF}$. The lift coefficient during the transition may be written as:

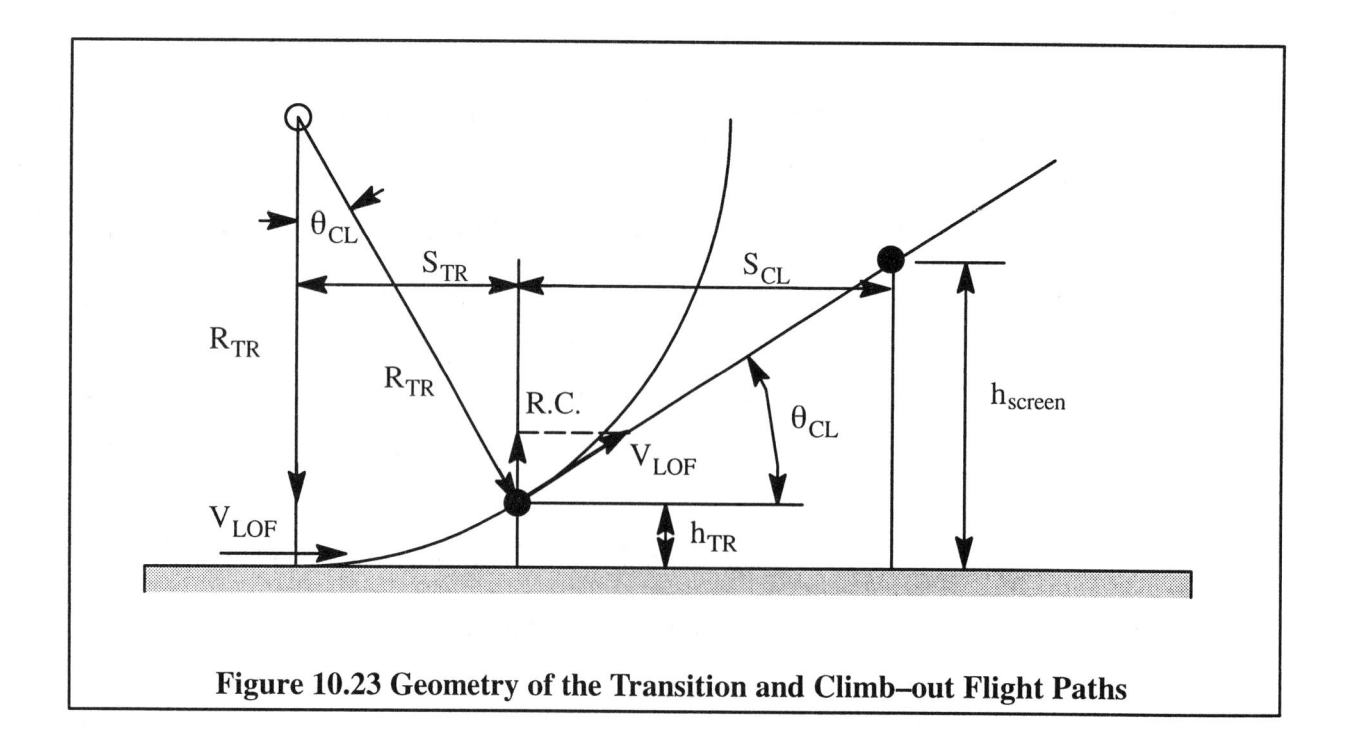

Figure 10.23 Geometry of the Transition and Climb–out Flight Paths

Takeoff and Landing

$$C_{L_{TR}} = \frac{W}{\frac{1}{2}\varrho V_{LOF}^2 S} + \Delta C_L = \frac{W}{\frac{1}{2}\varrho V_S^2 S}\left(\frac{V_S^2}{V_{LOF}^2}\right) + \Delta C_L =$$

$$= C_{L_{max}}\left(\frac{V_S^2}{V_{LOF}^2}\right) + \Delta C_L \qquad (10.38)$$

The resulting lift must be equal to the sum of the weight and the centrifugal force:

$$W + \frac{W}{g}\frac{V_{LOF}^2}{R_{TR}} = \left[C_{L_{max}}\left(\frac{V_S^2}{V_{LOF}^2}\right) + \Delta C_L\right]\frac{1}{2}\varrho V_{LOF}^2 S = W + \Delta C_L \frac{1}{2}\varrho V_{LOF}^2 S \qquad (10.39)$$

Therefore:

$$\frac{W}{g}\frac{V_{LOF}^2}{R_{TR}} = \Delta C_L \frac{1}{2}\varrho V_{LOF}^2 S \qquad (10.40)$$

The transition radius, R_{TR}, follows from:

$$R_{TR} = 2\frac{(W/S)}{\varrho g \Delta C_L} \qquad (10.41)$$

Based on data from Ref. 10.12 and assuming a "normal effort" transition, the following equation is suggested for ΔC_L:

$$\Delta C_L = \frac{1}{2}\left\{\left(\frac{V_{mean}}{V_S}\right)^2 - 1\right\}\left[C_{L_{max_{TO}}}\left\{\left(\frac{V_S}{V_{mean}}\right)^2 - 0.53\right\} + 0.38\right] \qquad (10.42)$$

where: V_{mean} is the mean speed measured in the transition path during various flight tests. For practical purposes this mean speed is assumed to be equal to the lift–off speed: $V_{mean} \approx V_{LOF}$.

As an example, for a jet transport with a maximum lift coefficient at take–off of 2.0 and a lift–off speed of: $V_{LOF} \approx 1.2 V_{S_{TO}}$ it is found that: $\Delta C_L = 0.15$.

To calculate the transition distance, S_{TR}, both the transition radius, R_{TR} and the climb angle, θ_{CL}, are required. From Figure 10.23, the rate of climb is seen to be:

$$R.C. = V_{LOF}\sin\theta_{CL} = \frac{V_{LOF}(T - D)}{W} \qquad (10.43)$$

The climb angle therefore is:

$$\theta_{CL} \approx \left\{\frac{(T-D)}{W}\right\}_{V=V_{LOF}} \qquad (10.44)$$

The transition distance, S_{TR} now may be computed from:

$$S_{TR} = R_{TR} \sin\theta_{CL} \qquad (10.45)$$

10.3.3.4 Approximate Method for Calculating the Air Distance Component, S_{CL}

From Figure 10.23 it is seen that the climb distance component of the air distance, S_{CL} may be determined from:

$$S_{CL} = \frac{h_{screen} - h_{TR}}{\tan\theta_{CL}} \qquad (10.46)$$

where: $h_{TR} \approx R_{TR} - R_{TR}\cos\theta_{CL} \approx \frac{S_{TR}}{\sin\theta_{CL}}(1 - \cos\theta_{CL}) \approx \frac{S_{TR}\theta_{CL}}{2} \qquad (10.47)$

The reader must take care to note that:

a) if $h_{TR} > h_{screen}$, then: $S_{CL} = 0$

b) if $h_{TR} < h_{screen}$, then S_{CL} follows from Eqn (10.46).

The reader must also check with Table 10.1 to determine the applicable certification basis. This determines whether the screen height is 35 ft or 50 ft.

It should also be noted, that for military airplanes the screen height is always 50 ft.

10.3.3.5 Alternate Method for Calculating the Total Ground Distance, S_G

In this alternate method, the total ground distance, $S_G = S_{NGR} + S_R$, is calculated on the assumption that the acceleration on the ground varies linearly with V^2 all the way to V_{LOF}.

In that case the average value of the ground acceleration may be computed at $V^2 = (V_{LOF}^2)/2$, or at: $V = V_{LOF}/\sqrt{2}$. Eqn (10.8) then yields:

$$S_G = \frac{1}{a_{g_{\text{ave at }V=V_{LOF}/\sqrt{2}}}} \int_{\mp V_w}^{V_{LOF}} V \pm V_w \, dV = \frac{1}{a_{g_{\text{ave at }V=V_{LOF}/\sqrt{2}}}} \left[\frac{1}{2}V^2 \pm V_w V\right]\Bigg|_{\mp V_w}^{V_{lof}} = \quad (10.48)$$

$$= \frac{1}{2}a_{g_{\text{ave at }V=V_{LOF}/\sqrt{2}}}\left[V_{LOF}^2 \pm 2V_w V_{LOF} - V_w^2 + 2V_w^2\right] = \frac{1}{2}a_{g_{\text{ave at }V=V_{LOF}/\sqrt{2}}}\left(V_{LOF} \pm V_w\right)^2 =$$

$$= \frac{(V_{LOF} \pm V_w)^2}{2g\left\{\left(\frac{T}{W} - \mu_g\right) - \frac{(C_{D_g} - \mu_g C_{L_g})\bar{q}}{W/S} - \phi\right\}_{\text{at }V=V_{LOF}/\sqrt{2}}} \quad (10.48 \text{ Cont'd})$$

10.3.3.6 Method for Calculating the Time Required to Take–off, t_{TO}

The total time required for take–off, t_{TO}, may be determined from:

$$t_{TO} = t_{NGR} + t_R + t_{TR} + t_{CL} \quad (10.49)$$

Because $a_g = dV/dt$, it is possible to find the time required to reach the rotation speed, V_R, by integration from:

$$t_{NGR} = \int_{\mp V_w}^{V_R} \frac{dV}{2g\left\{\left(\frac{T}{W} - \mu_g\right) - \frac{(C_{D_g} - \mu_g C_{L_g})\bar{q}}{W/S} - \phi\right\}} \quad (10.50)$$

If the acceleration–versus–speed relation is known, Eqn (10.49) can be integrated numerically.

To integrate Eqn (10.50) approximately, the assumption is made that the average acceleration may be computed at $V = V_R/\sqrt{2}$ in a manner similar to the method of Sub–sub–section 10.3.3.6. This yields:

$$t_{NGR} = \frac{(V_R \pm V_w)}{g\left\{\left(\frac{T}{W} - \mu_g\right) - \frac{(C_{D_g} - \mu_g C_{L_g})\bar{q}}{W/S} - \phi\right\}_{\text{at }V=V_R/\sqrt{2}}} \quad (10.51)$$

As before, the upper sign is for a tail–wind take–off.

The time to rotate, t_R, is usually taken to be 1–3 seconds. This was discussed before in Sub–sub–section 10.3.3.2, see page 463.

The transition time, t_{TR}, may be approximated from:

$$t_{TR} = \frac{S_{TR}}{V_{LOF}} \qquad (10.52)$$

where: S_{TR} is found from Eqn (10.45).

The time–to–climb to the screen height may be approximated from:

$$t_{CL} = \frac{2S_{CL}}{V_{LOF} + V_2} \quad OR: \quad t_{CL} = \frac{2S_{CL}}{V_{LOF} + V_{50}} \qquad (10.53)$$

where: V_2 and V_{50} are defined in Table 10.1 depending on the certification basis.

An application of the approximate, analytical method for calculating the take–off distance and the take–off time will be given in Example 10.2.

Example 10.2: The example concerns a propeller driven airplane with the thrust characteristics of Figure 10.9. Calculate the take–off distance and time for no wind and for zero runway gradient. Assume sea–level standard conditions. The following data are available for this airplane:

$W = 4,600$ lbs $\qquad \left(\frac{t}{c}\right)_{ave} = 0.14 \qquad V_w = 0 \qquad \phi = 0$

$\delta_f = 15^0 \qquad S = 175$ ft$^2 \qquad b = 35$ ft $\qquad A = 7$

$\mu_g = 0.03 \qquad C_{L_{max}} = 1.69 \qquad e = 0.80$ in free air

Solution: The solution is given in Table 10.2, pages 469 and 470.

Table 10.2 shows that the total take–off distance is predicted to be 1,754 ft (bottom of page 470). The take–off time is predicted to be: 30.2 seconds (bottom of Page 471).

It is of interest to compare this with the result of the statistical method of Sub–section 10.3.2. The airplane used in this example has two 260 shp engines. Assuming a 10% loss, the take–off shp available is 0.9x2x260=468 hp. The FAR 23 take–off parameter of Eqn (10.22) then is:

$$TOP_{23} = (4600/175)(4600/468)/1.69 = 153 \text{ lbs}^2/\text{ft}^2\text{hp}$$

This statistical method, with Eqn (10.23) then predicts a take–off distance of 1,593 ft. This result differs by 10% from that of the more detailed method.

Takeoff and Landing

Table 10.2 Calculation of Take–off Distance and Time for a Propeller Airplane					
Step	Description of step	Symbol	Value	Comments
Step 1	Determine the lift coefficient at lift–off: $C_{L_{LOF}} = \dfrac{C_{L_{max}}}{V_{LOF}^2 / V_S^2}$	$C_{L_{LOF}}$	1.28	$V_{LOF} = 1.15 V_S$
Step 2a	Determine the lift coefficient during the groundrun out of ground effect	$C_{L_{(out\ of\ ground\ effect)}}$	0.83	Assume this as a given
Step 2b	Determine the lift coefficient during the groundrun in ground effect	C_{L_g}	0.89	Follow example 10.1 for this airplane
Step 3	Determine the drag coefficient during the groundrun out of ground effect. Drag polar with flaps at take–off is: $C_D = 0.0620 + \dfrac{C_L^2}{\pi A e}$	$C_{D_{g(out\ of\ ground\ effect)}}$	0.1012	Note: this is out of ground effect
Step 4	Determine the height of the m.g.c. above the ground, see Figure 10.12	h	3.6 ft	From a three–view
Step 5	Calculate h/b	h/b	0.103	
Step 6	Calculate the ground effect factor from Figure 10.12	σ	0.48	
Step 7	Calculate the induced drag increment due to ground effect from Eqn (10.12) $\Delta C_{D_{i_g}} = -\sigma' \dfrac{(C_L)^2_{OGE}}{\pi A}$	$\Delta C_{D_{i_g}}$	– 0.0150	
Step 8	Determine the drag coefficient in ground effect: $C_{D_g} = $ Step 4 + Step 7	C_{D_g}	0.0862	
Step 9	Determine:	$C_{D_g} - \mu_g C_{L_g}$	0.0595	
Step 10	Determine the stall speed out of ground effect: $V_S = \sqrt{\dfrac{2W}{\rho C_{L_{max}} S}}$	V_S	114.4 ft/sec 67.8 kts	
Step 11	Determine the lift–off speed	V_{LOF}	131.6 ft/sec 77.9 kts	$V_{LOF} = 1.15 V_S$

Chapter 10

Table 10.2 (Cont'd) Calculation of Take–off Distance and Time for a Propeller Airplane				
Step	Description of step	Symbol	Value	Comments
Step 12	Determine the dynamic pressure at lift–off	$\bar{q}_{LOF}$	20.6 psf	
Step 13	Determine the rotation speed	V_R	125.8 ft/sec 74.5 kts	$V_R = 1.10 V_S$
Step 14	Determine the dynamic pressure at rotation	$\bar{q}_R$	18.8 psf	
Step 15	Determine the take–off thrust at zero speed	$T_{V=0}$	2,000 lbs	Figure 10.9
Step 16	Determine the take–off thrust at rotation speed	$T_{V=V_R}$	1,200 lbs	Figure 10.9
Step 17	Determine the take–off acceleration at zero speed:	$a_{g_{V=0}}$	12.9 ft/sec²	Eqn (10.27)
Step 18	Determine the take–off acceleration at rotation speed:	$a_{g_{V=V_R}}$	6.1 ft/sec²	Eqn (10.28)
Step 19	Determine k from Eqn (10.30): Determine $a_{g_{ave}}$ from Eqn(10.30) :	k $a_{g_{ave}}$	0.7 9.1 ft/sec²	
Step 20	Determine S_{NGR} from Eqn(10.29):	S_{NGR}	870 ft	
Step 21	Determine S_R from Eqn (10.37) :	S_R	129 ft	Use: $t_{rotate} = 1.0$ sec.
Step 22	Determine ΔC_L from Eqn(10.42)	ΔC_L	0.12	Use: $V_{mean} \approx V_{LOF}$
Step 23	Determine R_{TR} from Eqn(10.41) :	R_{TR}	5,721 ft	Transition radius
Step 24	Determine T − D at V = V_{LOF} :	$(T-D)_{V=V_{LOF}}$	591 lbs	1,150–559
Step 25	Determine θ_{CL} from Eqn(10.44) :	θ_{CL}	0.128 rad	
Step 26	Determine S_{TR} from Eqn(10.45) :	S_{TR}	732 ft	
Step 27	Determine h_{TR} from Eqn(10.47) :	h_{TR}	47 ft	Note: $h_{TR} < h_{screen}$
Step 28	Determine S_{CL} from Eqn(10.46) :	S_{CL}	23 ft	
Step 29	Determine S_{TO} from Eqn(10.1) :	S_{TO}	870 + 129 + 732 + 23 = 1,754 ft	Sum steps 20, 21, 26 and 28

Table 10.2 (Cont'd) Calculation of Take–off Distance and Time for a Propeller Airplane					
Step	Description of step	Symbol	Value	Comments
Step 30	Determine the time to reach rotation speed $$t_{NGR} = \frac{(V_R)}{g\left\{\left(\frac{T}{W} - \mu_g\right) - \frac{(C_{D_g} - \mu_g C_{L_g})\bar{q}}{W/S}\right\}_{at\ V = V_R/\sqrt{2}}}$$	t_{NGR}	23.4 sec	Eqn (10.51)
	Determine the dynamic pressure at $V = V_R/\sqrt{2} = 89$ fps	$\bar{q}_{V = 89\ fps}$	9.4 psf	
	Determine the take–off thrust at $V = V_R/\sqrt{2} = 89$ fps	$T_{V = 89\ fps}$	1,000 lbs	Figure 10.9
Step 31	Determine the time to rotate	t_R	1 sec	From p. 463
Step 31	Determine the transition time $S_{TR} = 732$ ft from Step 26 $V_{LOF} = 131.6$ fps from Step 11	t_{TR}	5.6 sec	Eqn (10.52)
Step 32	Determine the time to climb to the screen height $S_{CL} = 23$ ft from Step 28 $V_{LOF} = 131.6$ fps from Step 11 $V_{50} = 1.2V_{S_1} = 1.2 \times 114.4$ fps $= 137.3$ fps from Step 10	t_{TCL}	0.2 sec	Eqn (10.53)
Step 32	Determine the time to take–off	t_{TO}	30.2 sec	Sum steps 30, 31, 32 and 33 Eqn (10.49)

10.3.4 APPROXIMATE, ANALYTICAL METHOD FOR PREDICTING THE BALANCED FIELDLENGTH

The idea of balanced fieldlength was introduced on page 438 and a graphical interpretation was given in Figure 10.5. A schematic which shows how the various speeds and distances fit together in the balanced fieldlength philosophy is shown in Figure 10.24.

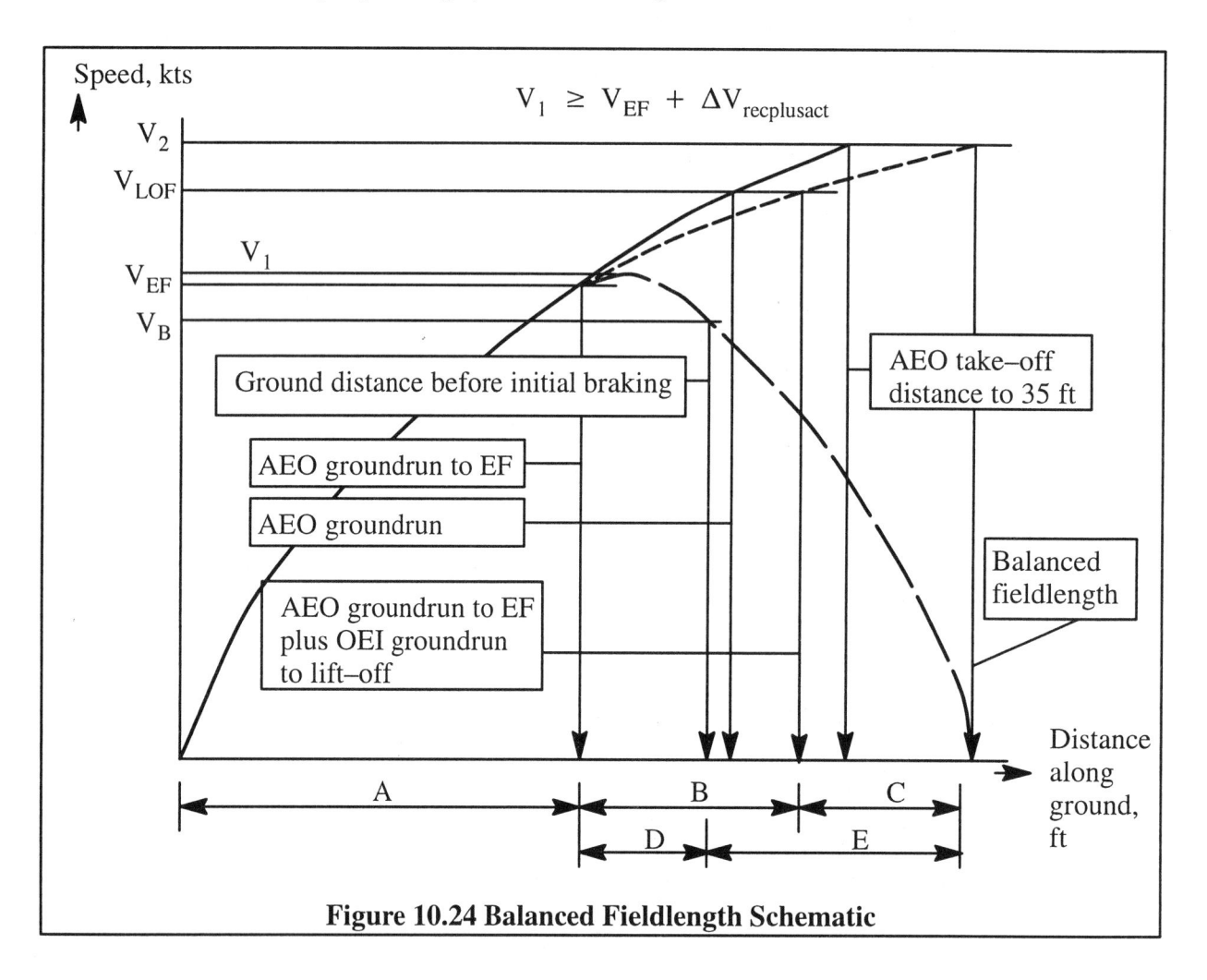

Figure 10.24 Balanced Fieldlength Schematic

The balanced fieldlength is defined in such a manner that: $B + C = D + E$, where the distances are as defined in Figure 10.24. If an engine fails below the take–off decision speed, V_1, the pilot should stop the airplane. If an engine fails above V_1 the pilot should continue the take–off with that engine failed. The calculation of segments B and C can be accomplished with the methods presented in Sub–section 10.33. However, the following modifications must be made:

1) The thrust must be reduced by that of the stopped engine
2) The drag due to the inoperative engine (including the trim drag due to the rudder) must be added to the airplane drag

To calculate segment D it is necessary to account for several human and systems factors:

a) the recognition time between engine failure and brake application
b) pilot reaction time
c) the time between brake application and power reduction on the remaining engines
   a), b) and c) are human factors
d) the time required for the brakes to act
e) the time required for the engine thrust or power to reduce to idle
   d) and e) are system delay factors

The time elapsed because of a) through e) is called: $t_{recplusact}$. During this time the speed of the airplane will in fact increase slightly by an amount called: $\Delta V_{recplusact}$. The latter quantity is noted at the bottom of Table 10.1. The 2–second time interval shown in Figure 10.3 is the minimum FAR required interval to account for these factors:

$$t_{recplusact} = 2 \text{ sec. per FAR 25 (and FAR 23 for commuters)} \qquad (10.54)$$

The authors suggest to use 3 seconds for conservatism. This is consistent with the military specifications of Ref. 10.2.

The ground distance for segment D can then be estimated from:

$$S_D = \left(V_{ave_D} \pm V_w\right) t_{recplusact} \qquad (10.55)$$

where: V_{ave_D} is the average speed during segment D. For preliminary analysis purposes it is

suggested to use: $\qquad V_{ave_D} = V_1 + 2.5 \text{ kts} \qquad (10.56)$

The calculation of segment E can be performed with the method of Sub–sub–sections 10.6.1.4 or 10.6.3.3 located in Section 10.6 for calculating landing distances.

An analytical expression for the balanced fieldlength was derived by Torenbeek in Ref. 10.14 and is presented here without proof:

$$BFL = \left(\frac{0.863}{1 + 2.3\Delta\gamma_2}\right)\left(\frac{W_{TO}/S}{\rho g C_{L_2}} + h_{screen}\right)\left(\frac{1}{\overline{T}/W_{TO} - \mu'} + 2.7\right) + \frac{\Delta S_{TO}}{\sqrt{\sigma}} \qquad (10.57)$$

where: $\Delta\gamma_2 = \gamma_2 - \gamma_{2_{min}}$

with: γ_2 is the second segment climb gradient. In turn, γ_2, may be computed from
Eqn (9.11a) and (9.12a) may be used to compute this climb gradient at the appropriate speed, weight, thrust and/or power

The following regulatory climb gradients should be observed:

$\gamma_{2_{min}} = 0.024$ for $N_e = 2$

$\gamma_{2_{min}} = 0.027$ for $N_e = 3$

$\gamma_{2_{min}} = 0.030$ for $N_e = 4$

W_{TO} is the take-off weight in lbs

S   is the wing area in ft²

C_{L_2} is the lift coefficient at V_2. Since $V_2 \approx 1.2V_S$ or $1.2V_{S_1}$ it follows that in most case: $C_{L_2} \approx 0.694 C_{L_{max_{TO}}}$ can be assumed.

h_{screen} is 35 ft under either FAR 25 rules or FAR 23 commuter category rules.

$\overline{T}$ is the mean thrust during the take-off run in lbs.

For jet airplanes it is suggested to use:

$$\overline{T} = 0.75 \left(\frac{5 + \lambda}{4 + \lambda} \right) T_{TO} \tag{10.58}$$

where: λ is the engine by-pass ratio

T_{TO} is the static take-off thrust in lbs

For propeller driven airplanes it is suggested to use:

$$\overline{T} = k_p P_{TO} \left(\frac{\sigma N_e D_p^2}{P_{TO}} \right)^{1/3} \tag{10.59}$$

where: $k_p = 5.75$ slugs$^{1/3}$/ftsec²   (conversion constant)

P_{TO} is the take-off horsepower in hp

$\sigma = \frac{\varrho}{\varrho_0}$ is the density ratio of the atmosphere

N_e is the number of engines

D_p is the propeller diameter in ft

$$\mu' = 0.010 C_{L_{max_{TO}}} + 0.02 \tag{10.60}$$

for flaps in the take-off position. For a zero flaps this quantity is zero.

ΔS_{TO} is a take-off distance increment equal to 655 ft

It is noted that Eqn (10.57) applies to jet airplanes as well as to propeller driven airplanes when certified under either FAR 25 rules or FAR 23 commuter category rules.

10.3.5 PRESENTATION OF TAKE–OFF PERFORMANCE DATA

The requirements of FAR 23 and FAR 25 stipulate several take–off distance data which must be determined and published in the flight manual of an airplane. There are many different ways to present take–off performance data. In the following, two types of data presentation will be included:

a) Manufacturers BFL data as correlated with take–off thrust–to–weight ratio
b) Fieldlength data which include the effect of payload
c) Manufacturers BFL data as presented in typical flight manuals

a) Manufacturers data as correlated with take–off thrust–to–weight ratio

Figure 10.25 shows the relationship between balanced fieldlength (BFL) and take–off thrust–to–weight ratio as obtained from manufacturer's data in the 1996–1997 issue of Ref. 10.15. The data are for standard sea–level conditions. It is seen that for jet transports (including some large business jets) the take–off thrust–to–weight ratio ranges from a low of around 0.25 (for 4–engine jets) to a high of around 0.39 for 2–engine jets. It is also seen that the BFL of jet transport ranges from around 5,000 ft to around 11,000 ft. It is noted that this data presentation does not account for the amount of payload which can be carried out of a given field and over a given range.

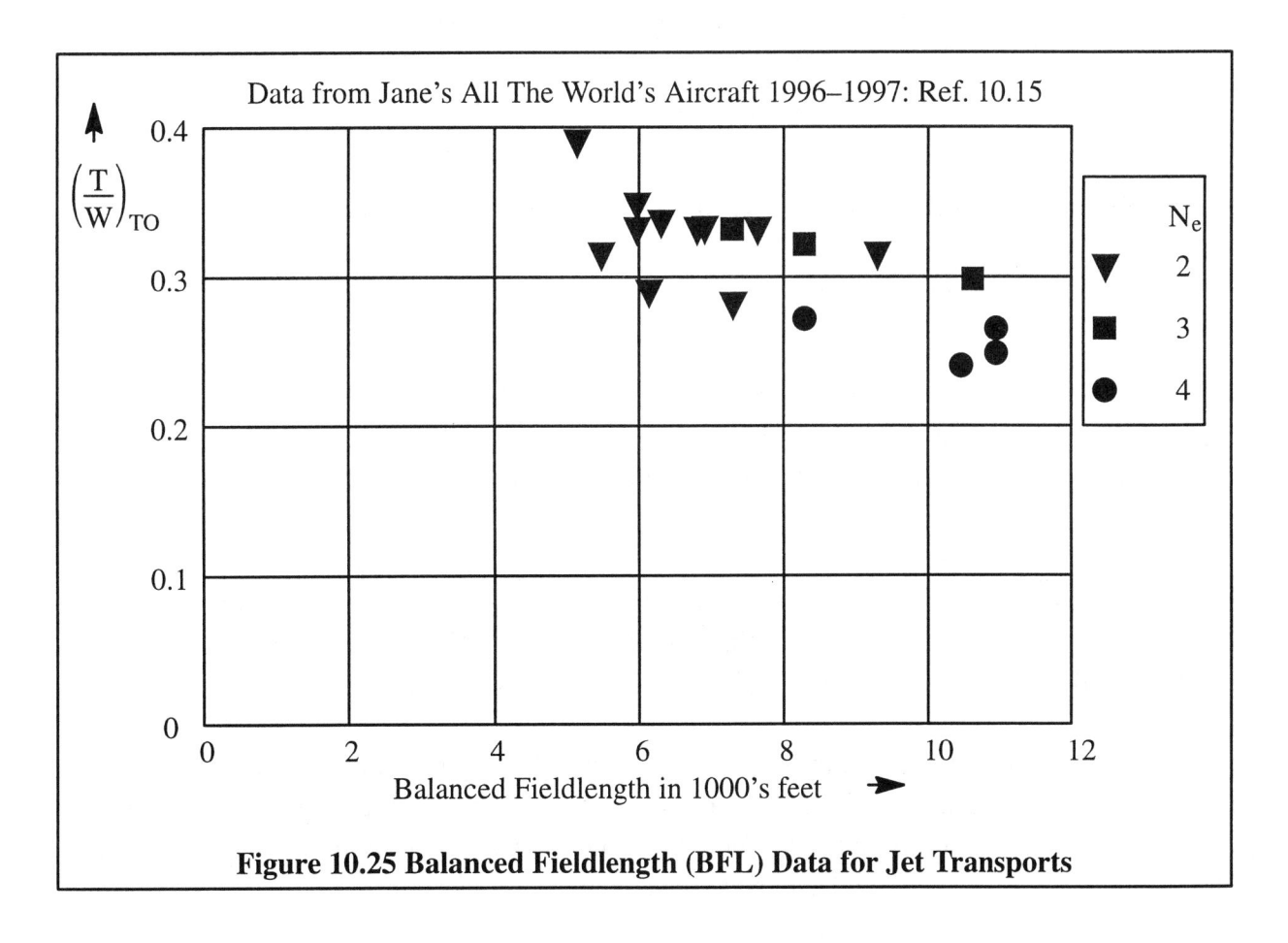

Figure 10.25 Balanced Fieldlength (BFL) Data for Jet Transports

b) Fieldlength data which include the effect of payload

The type of data provided by Figure 10.25 is very useful to a designer. It is not very useful to an operator who must decide whether a given airplane can meet certain payload–range requirements when operating from a given field. For example, consider an operator who needs to fly a large payload, W_{PL}, out of a relatively short field at high altitude and with high prevailing temperatures.

In such a situation the data presentation of Figure 10.26 is preferred.

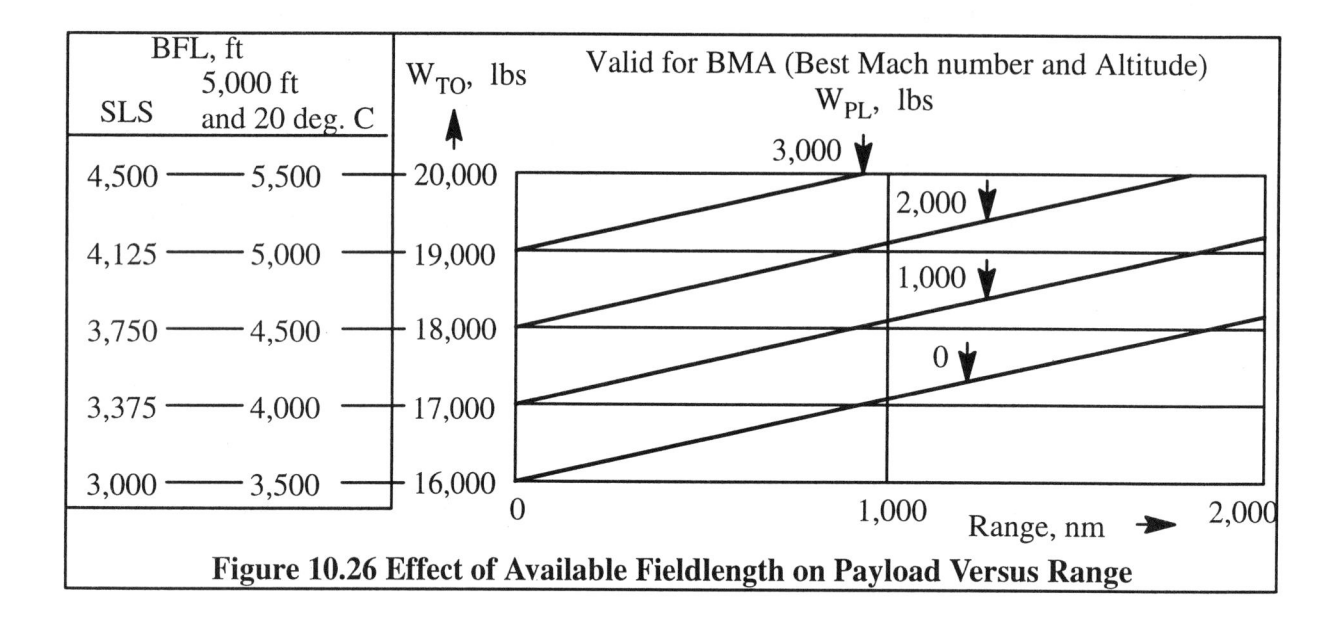

Figure 10.26 Effect of Available Fieldlength on Payload Versus Range

c) Manufacturers BFL data as presented in typical flight manuals

Figure 10.27 shows manufacturer's BFL data including the effect of: flap setting, wind, runway gradient, pressure altitude and outside air temperature (OAT). Flight manuals typically include this type of information. In addition, reference speeds such as: V_1, V_R and V_2 are also provided in flight manuals. Flight manuals also include data related to engine operation: EPR and N1 (Engine Pressure Ratio and Critical Turbine RPM) for various OAT and pressure altitude values.

In Figure 10.27 the following example can be traced through the figure: find the allowable maximum take–off weight when the available fieldlength is 10,750 ft, the head–wind is 5 kts, the runway gradient is –1.6% (favorable), the flap setting is 15 degrees, the OAT is 30 degrees C and the pressure altitude is 3,000 ft. The solution is: 720,000 lbs.

In modern transports, the allowable take–off weight for a given fieldlength, or the required fieldlength for a given take–off weight are automatically displayed in the cockpit at the request of the pilot. It is noted that this data presentation does not account for the amount of payload which can be carried out of a given field and over a given range.

Takeoff and Landing

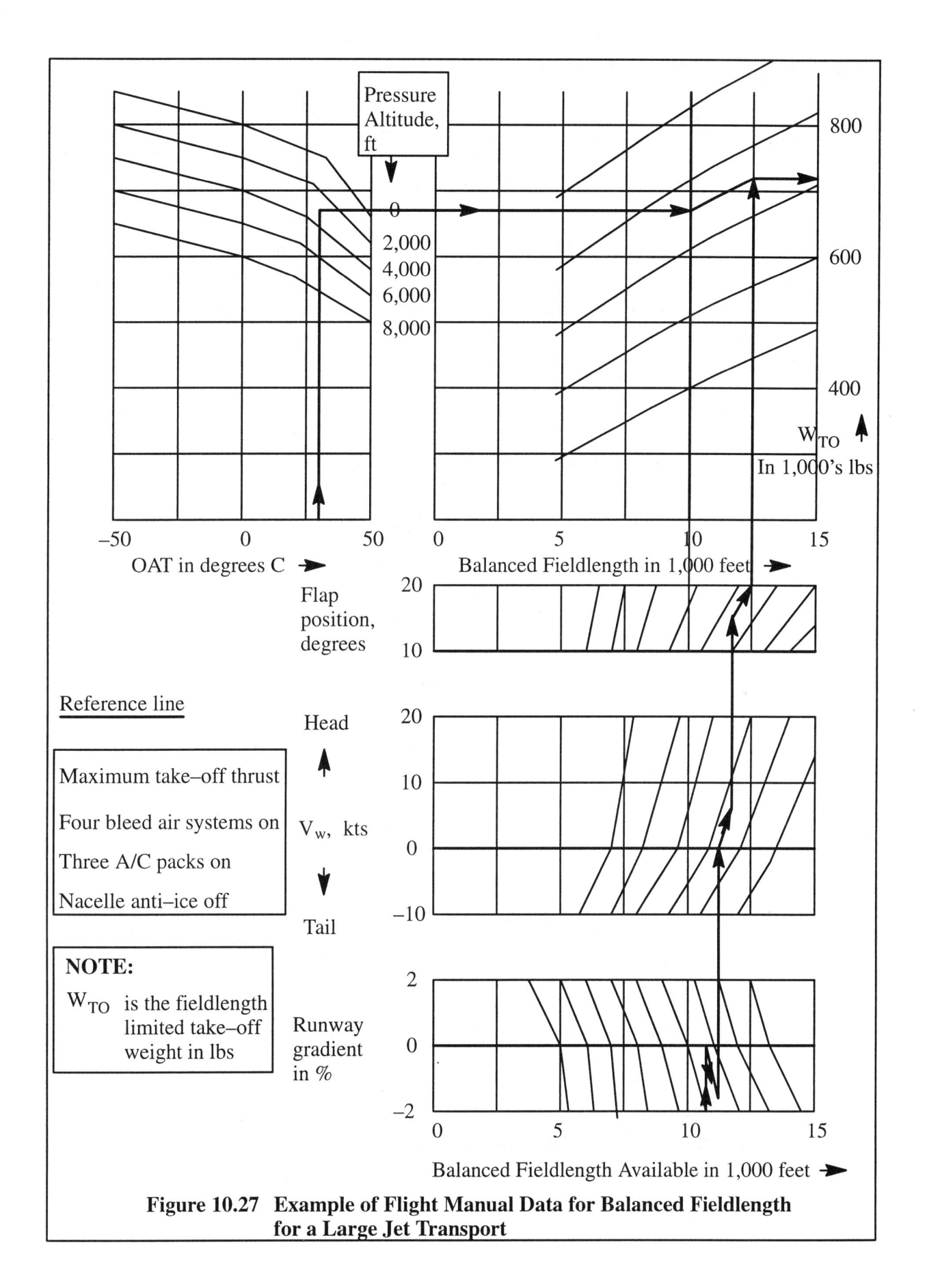

Figure 10.27 Example of Flight Manual Data for Balanced Fieldlength for a Large Jet Transport

Chapter 10   477

10.4 THE LANDING PROCESS

The landing process is normally divided into three phases:

* approach   * air distance   * ground roll

Figure 10.28 shows a schematic which depicts these phases in relation to the overall landing path and to several important speeds. There is no scale implied in this figure: the relative magnitudes of the distances covered in the approach, the air distance (transition) and the ground roll vary from one airplane to another airplane.

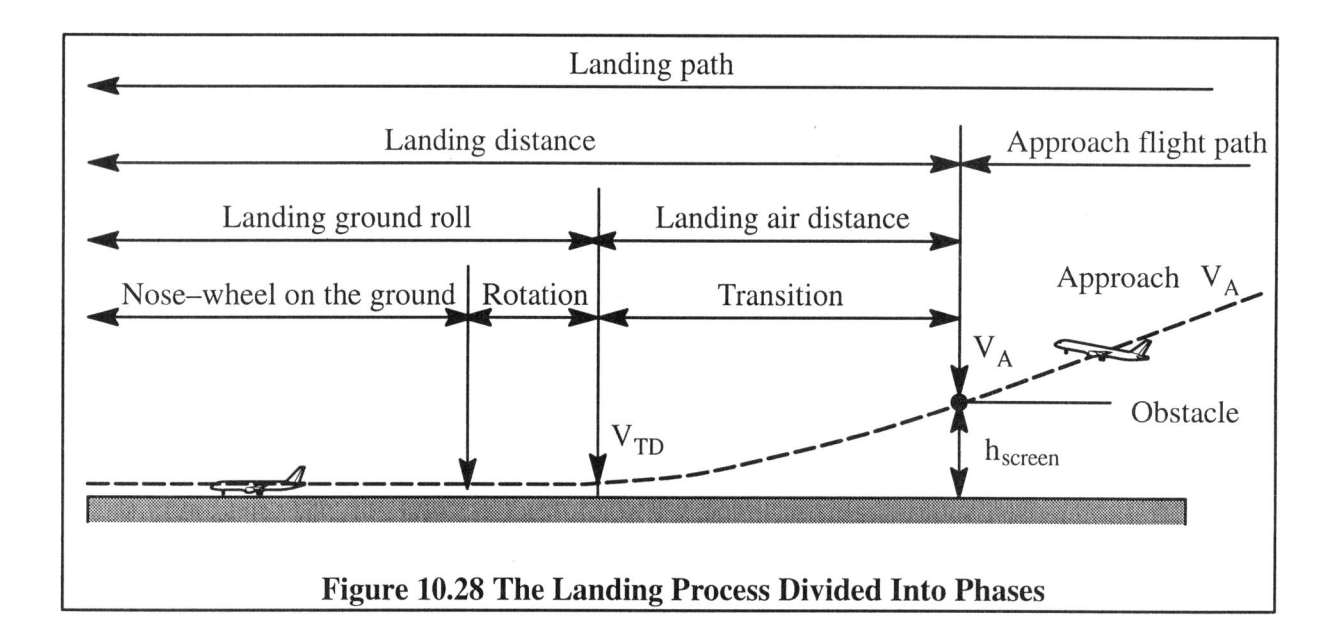

Figure 10.28 The Landing Process Divided Into Phases

In Figure 10.28 the airplane, during the approach, is flown along a straight line flight path at the approach speed, V_A, which must satisfy: $V_A \geq 1.3 V_{S_{approach}}$ until the airplane reaches the screen (or obstacle) height. The screen height is 50 ft for both FAR 23 and FAR 25 regulations (see Ref. 10.1). For military regulations, see Refs 10.2 and 10.3. Sub-sections 10.4.1 and 10.4.2 contain summaries of these regulations.

When the airplane has descended to the screen height, a transition is started until the main gear touches down at the touchdown speed, V_{TD}. At that point the nose is lowered until the nose-wheel contacts the runway. Braking may be started as soon as the wheels touch the ground. **From a landing distance certification point of view, only braking may be used.** Operationally, thrust reversers and/or other retardation devices may be used as appropriate for a given airplane.

10.4.1 COMMERCIAL LANDING RULES

During the final approach the airplane must be configured to the landing configuration. This usually means gear down and flaps in the landing position. In determining the landing distances the effect of gear and flaps on lift and drag must be accounted for. FAR 23.75, SFAR 23 as well as FAR 25.125 require that the landing distance from the 50 ft obstacle, while stabilized at a speed of $V_A = 1.3 V_{S_{approach}}$, be determined. The FAR's also specify conditions on braking, brake wear, vertical acceleration and piloting skills which must be met. All landing distance data must include correction factors for the effect of wind: 50% of the favorable effect due to head–wind but 150% of the unfavorable effect due to tail–wind must be accounted for.

The touchdown speed, V_{TD} is not specified in the FAR's. In most instances it is assumed that the touchdown speed is approximated by: $V_{TD} = 1.15 V_{S_{approach}}$. The wheels–to–ground friction coefficient while braking must be determined so that consistent braking performance during normal operations can be expected.

Balked landings initiated at the 50 ft obstacle mus be possible. The required climb performance for balked landings (go–around) is given in Chapter 9 as Tables 9.11, 9.12 and 9.13.

10.4.2 MILITARY LANDING RULES

During the final approach the airplane must be configured to the landing configuration. This usually means gear down and flaps in the landing position. In determining the landing distances the effect of gear and flaps on lift and drag must be accounted for.

For USAF airplanes (Ref. 10.2), the landing distance from the 50 ft obstacle, while stabilized at a speed of $V_A = 1.2 V_{S_{approach}}$, be determined. This assumes that the airplane is controllable at that speed and that it can develop a climb gradient of at least 0.025 in the landing configuration when dry take–off power is applied. The touchdown speed is assumed to be at least $V_{TD} = 1.1 V_{S_{approach}}$.

For USNavy airplanes, (Ref. 10.3), the landing distance from the 50 ft obstacle, while stabilized at a speed of $V_A = 1.1 V_{S_{PA}}$, be determined, where: $V_{S_{PA}} = 1.15 V_{S_L}$. In addition, conditions to ensure controllability as well as flight path correction capability must be met.

The wheels–to–ground friction coefficient while braking must be assumed to be 0.30. Without braking the wheels–to–ground friction coefficient while braking must be assumed to be 0.025. The touchdown speed must be assumed to be at least $V_{TD} = 1.1 V_{S_{approach}}$.

10.5 EQUATIONS OF MOTION DURING LANDING

It was seen in Section 10.4 that the landing process consists of an air distance and a ground roll (distance). These distances are typically sub–divided as shown in Figure 10.29.

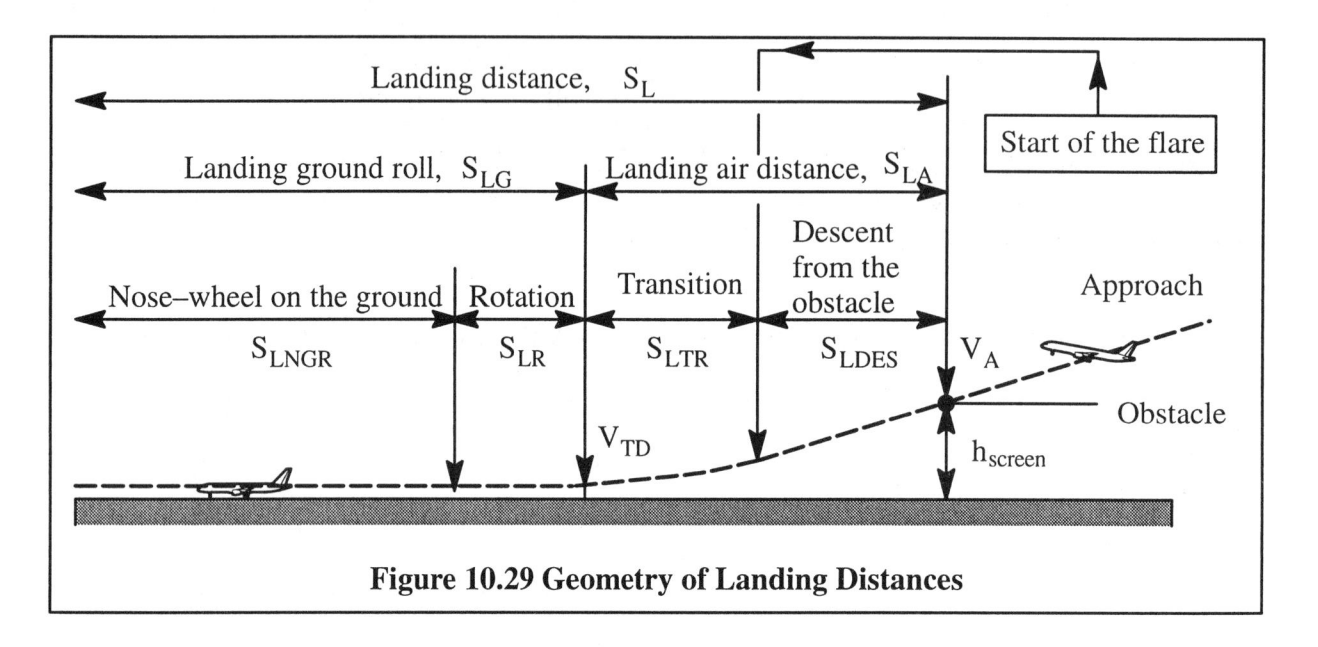

Figure 10.29 Geometry of Landing Distances

Evidently:

$$S_L = S_{LG} + S_{LA} = S_{LNGR} + S_{LR} + S_{LTR} + S_{LDES} \tag{10.61}$$

Different equations of motion apply to these distances. These equations are presented in Sub–sections 10.5.1 through 10.5.3 respectively:

10.5.1 Equations of motion during the landing approach and the descent from the obstacle
10.5.2 Equations of motion during the landing transition
10.5.3 Equations of motion during the landing ground roll

10.5.1 EQUATIONS OF MOTION DURING THE LANDING APPROACH AND THE DESCENT FROM THE OBSTACLE

Figure 10.29 also shows the flight path during the descent toward the obstacle and from the obstacle. During this part of the landing approach the acceleration perpendicular to the flight path is assumed to be zero. Ideally, if the approach is stabilized, the approach speed will be constant at $V_A = 1.3 V_{S_{approach}}$. Therefore, the deceleration along the flight path is also zero. The airplane is in a constant speed, straight line descent. The corresponding equations of motion as derived from Eqns (9.1) and (9.2) are:

$$T\cos(\alpha + \phi_T) - D_g + W\sin\bar{\gamma} = 0 \qquad (10.62)$$

$$T\sin(\alpha + \phi_T) + L_g - W\cos\bar{\gamma} = 0 \qquad (10.63)$$

Depending on the size and the configuration of the airplane, ground effect on lift and drag may or may not be negligible. This needs to be verified for each type of airplane. The approach flight path angle, $\bar{\gamma}$, is the opposite of the climb flight path angle: $\gamma : \bar{\gamma} = -\gamma$. Typical commercial transport approach flight path angles are 2.5 to 3 degrees.

10.5.2 EQUATIONS OF MOTION DURING THE LANDING TRANSITION

Figure 10.30 shows the flight path during the transition. The transition starts at the so-called flare height, h_{FL}, and continues until touchdown. Touchdown normally occurs with the main gear touching the ground before the nose-gear. During the transition the decelerations along and perpendicular to the flight path cannot be neglected. Equations (9.1) and (9.2) can be used to model the transition. These equations, modified to include ground effect, are repeated here as Eqns (10.64) and (10.65):

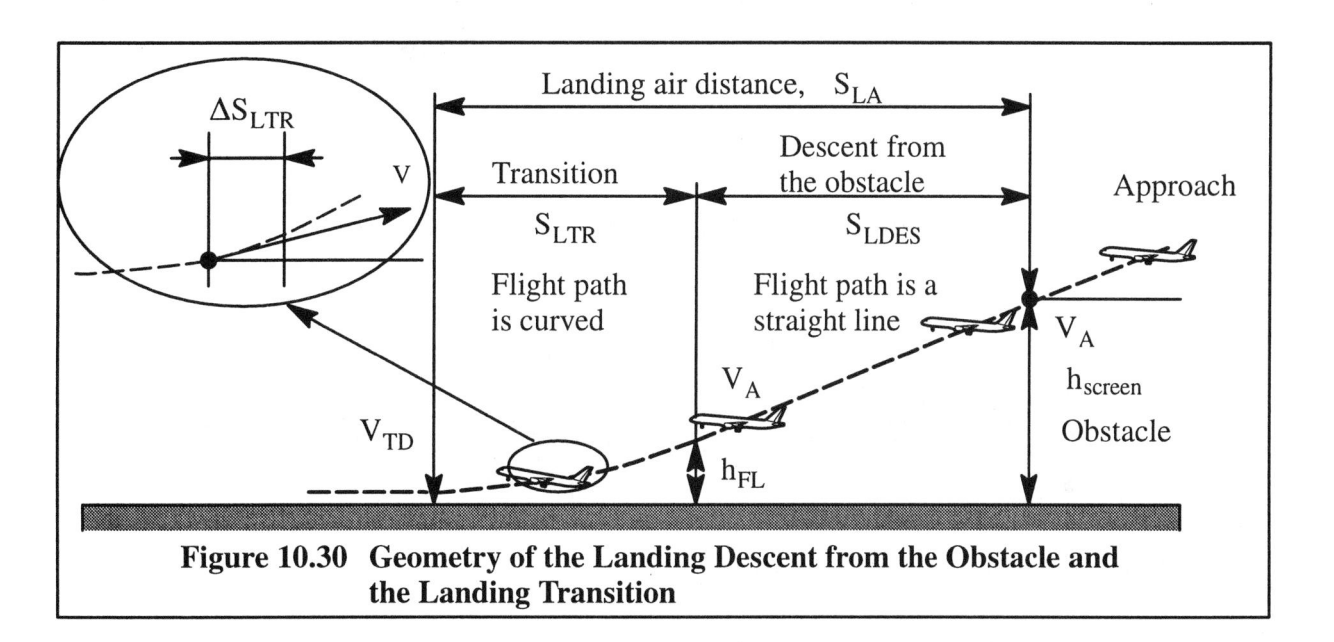

Figure 10.30 Geometry of the Landing Descent from the Obstacle and the Landing Transition

$$T\cos(\alpha + \phi_T) - D_g + W\sin\bar{\gamma} = \frac{W}{g}\frac{dV}{dt} \qquad (10.64)$$

$$T\sin(\alpha + \phi_T) + L_g - W\cos\bar{\gamma} = \text{C.F.} = \frac{W}{g}V\dot{\bar{\gamma}} \qquad (10.65)$$

The drag and the lift coefficients in Eqns (10.64) and (10.65) must be computed in ground effect. Methods to account for ground effect were given in Sub-section 10.2.1.

10.5.3 EQUATIONS OF MOTION DURING THE LANDING GROUND ROLL

Figure 10.31 shows the forces which act on the airplane during different parts of the ground roll depending on whether the nose–gear is touching the ground.

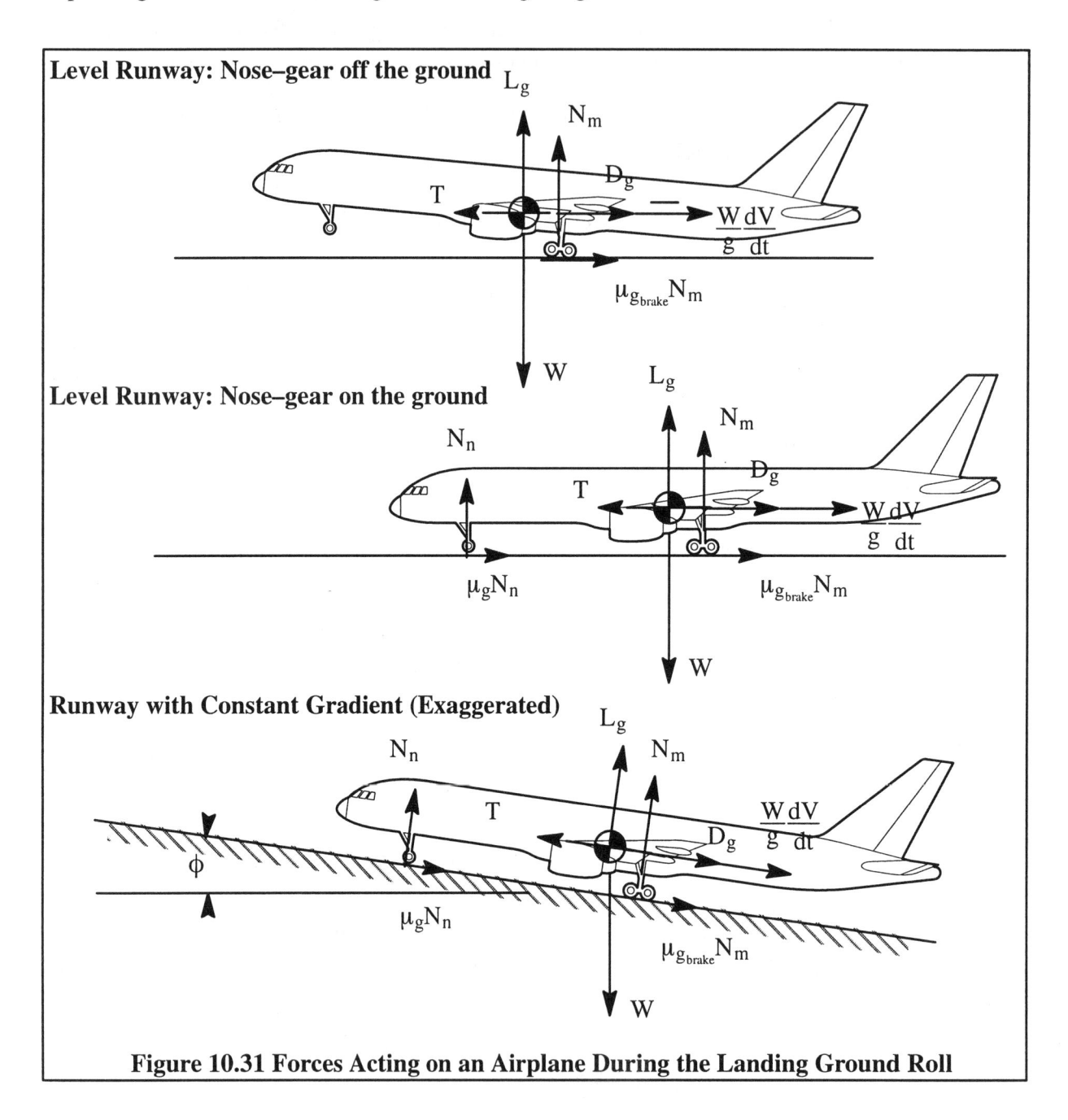

Figure 10.31 Forces Acting on an Airplane During the Landing Ground Roll

The main gear is assumed to be on the ground during the entire ground roll, the nose–gear will be on the ground only after the landing rotation (S_{LR} in Figure 10.29) has been completed. Lift and drag forces must be evaluated in ground effect. Since many airplanes employ lift dumpers (spoilers) during the ground roll, the effect of these devices on lift and drag must be included. The thrust (or power) is that which exists with the throttles retarded to the idle position. In actual operations the thrust may be reversed during part of the ground operation. However, for certification purposes, only braking may be assumed. Observe in Figure 10.31 that **only braking on the main gear** is assumed. This is typical of nearly all airplanes.

It is also assumed that the airplane is in moment equilibrium and that the landing gear and tire dynamic effects on the equations of motion are negligible.

Observe in Figure 10.31, that the aerodynamic forces (Lift and Drag) have been given a subscript 'g' for 'ground'. This means that during the ground roll the aerodynamic forces must be computed including the effect of ground proximity. Reference 10.4 contains methods for doing this. How these ground effect corrections arise is explained in Sub–section 10.2.1.

The equations of motion, for the case of a runway with a gradient, ϕ are as follows:

$$L_g + N_n + N_m = W\cos\phi \approx W \tag{10.66}$$

and

$$T - D_g - \mu_g N_n - \mu_{g_{brake}} N_m = W\sin\phi + \frac{W}{g}\frac{dV}{dt} \approx W\phi + \frac{W}{g}\frac{dV}{dt} \tag{10.67}$$

where: L_g is the airplane lift in ground effect, in lbs. The effect of spoilers, if deployed, must be accounted for.

N_n is the nose–gear reaction force, in lbs

N_m is the main–gear reaction force, in lbs

ϕ is the runway inclination angle (gradient) in radians: positive when unfavorable

W is the airplane weight, in lbs

T is the total idle thrust, in lbs. Note: thrust inclination is assumed to be zero. Reversed thrust may be used to calculate its effect on the landing ground–roll. However: for certification purposes, only braking may be assumed.

D_g is the airplane drag in ground effect, in lbs. The effect of extra drag due to lift dumpers (spoilers), if deployed, must be accounted for.

$\mu_{g_{brake}}$ is the ground braking friction coefficient. Table 10.3 shows typical values.

V is the airplane speed relative to the ground, in ft/sec.

dV/dt is the acceleration along the runway: it will be negative during the ground–roll.

Table 10.3 Typical Values for the Ground Braking Friction Coefficient	
Surface Type	$\mu_{g_{brake}}$
Dry Concrete and Macadam	0.65 – 0.80 (varies inversely with speed)
Wet Concrete and Macadam	0.25 – 0.75 (varies inversely with speed)
Slush on Concrete or Macadam	0.15 – .40 (varies inversely with speed)
Dry ice or packed snow on concrete or macadam	0.05

It is of interest to examine the behavior of the braking friction coefficient in some detail. To that end, consider Figure 10.32.

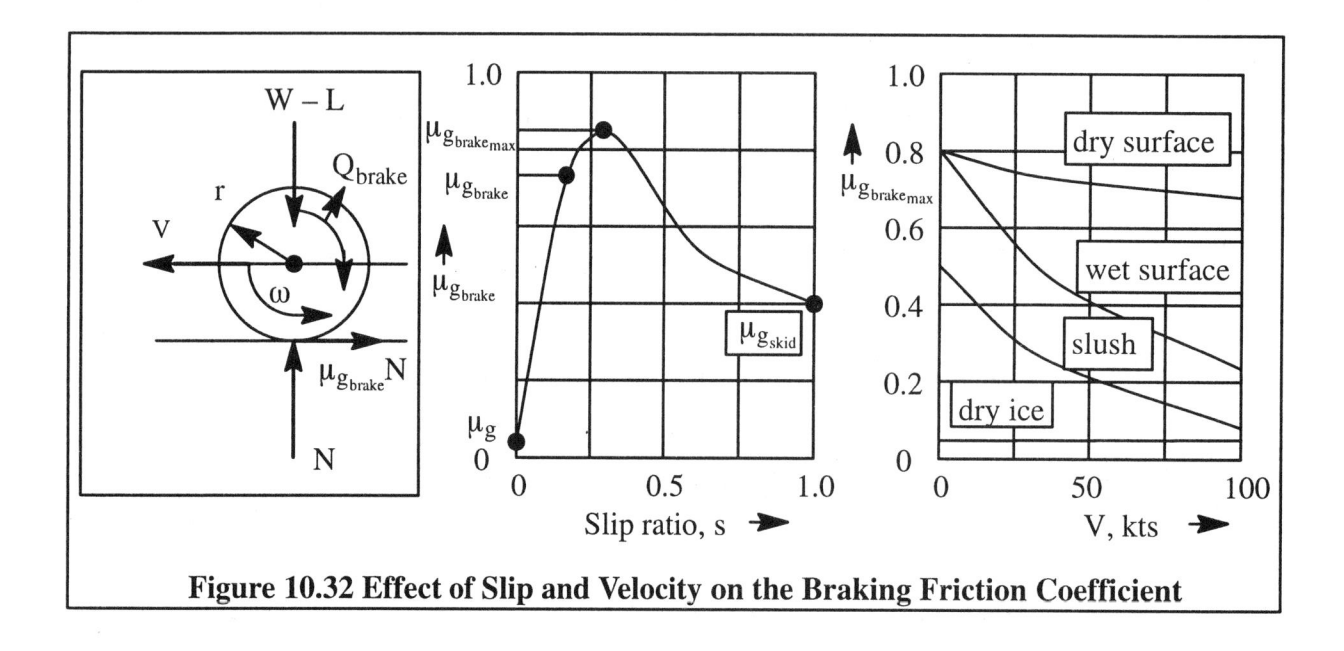

Figure 10.32 Effect of Slip and Velocity on the Braking Friction Coefficient

As soon as a braking torque, Q_{brake}, is applied to the wheel it is reacted by a counter–torque due to the increased ground friction of the braking wheel. The wheel must slow its rotational velocity, ω, so that its tangential velocity, ωr, is lowered relative to the speed, V. The following equation of motion applies to the braking wheel:

$$I_{wheel}\frac{d\omega}{dt} = \mu_{g_{brake}}Nr - Q_{brake} \qquad (10.68)$$

where: I_{brake} is the moment of inertia of the wheel about its spin axis.

For braking at a steady rotational velocity, Eqn (10.68) yields:

$$\mu_{g_{brake}}N = \frac{Q_{brake}}{r} \qquad (10.69)$$

The braking friction coefficient, $\mu_{g_{brake}}$, varies with the so–called slip ratio, s, of the wheel. This slip ratio is defined as:

$$s = \frac{V - \omega r}{V} \tag{10.70}$$

Note that the slip ratio varies from 0, at $\omega = V/r$, to 1.0, at $\omega = 0$. Observe from Figure 10.31 that at zero slip ratio: $s = 0$ (where the wheel is freely rolling) the braking friction coefficient is equal to μ_g (0.025 on dry concrete or macadam). When the wheel is completely locked: $s = 1.0$ (and therefore skidding) the braking friction coefficient equals $\mu_{g_{skid}}$. Note also, that the maximum braking friction coefficient occurs at a relatively low slip ratio. Typical braking friction coefficient values were given in Table 10.3 and are also shown in Figure 10.32 as a function of airplane speed and runway condition.

By combining Eqns (10.66) and (10.67) it can be shown that the acceleration along the runway can be expressed as:

$$\frac{dV}{dt} = \frac{g}{W}\{T - D_g - \mu_{g_{brake}}W + \mu_{g_{brake}}L_g + N_n(\mu_{g_{brake}} - \mu_g) - W\phi\} \tag{10.71}$$

It is usually convenient to rewrite this as:

$$\frac{dV}{dt} = g\left\{\left(\frac{T}{W} - \mu_{g_{brake}}\right) - \frac{(C_{D_g} - \mu_{g_{brake}}C_{L_g})\bar{q}}{W/S} + \frac{N_n}{W}(\mu_{g_{brake}} - \mu_g) - \phi\right\} = a_g \tag{10.72}$$

where: C_{D_g} is the airplane drag coefficient in ground effect

C_{L_g} is the airplane lift coefficient in ground effect

a_g is the acceleration along the ground, negative during the landing ground–roll

$\frac{N_n}{W}$ is the ratio of nose–gear ground reaction to weight. This quantity depends on the landing gear arrangement and the center of gravity location of the airplane. For preliminary analysis purposes $\frac{N_n}{W} = 0.08$ may be used during the second part of the ground roll. During the first part of the landing roll the nose gear is not on the ground and therefore its normal force is zero: $\frac{N_n}{W} = 0$. After the nose gear is on the ground, its friction coefficient is: μ_g and not $\mu_{g_{brake}}$.

To develop a formula for the ground roll distance, consider the general case where the airplane is landing with a wind speed of +/- V_w, where '+' means a tail-wind and '-' means a head-wind. If V is the air-speed, the ground speed will be $V \pm V_w$. It is noted that at zero ground speed, $V = \mp V_w$. Therefore:

$$\frac{dS_{LG}}{dt} = V \pm V_w \tag{10.73}$$

If the acceleration along the ground is called: $dV/dt = a_g$, then it is possible to write:

$$dS_{LG} = \frac{V \pm V_w}{a_g} dV \tag{10.74}$$

The reader is reminded that dV as well as a_g will be negative during the landing run. The ground distance can now be obtained by combining Eqns (10.72) and (10.74):

$$S_{LG} = \int_{V_{TD}}^{\mp V_w} \frac{V \pm V_w}{a_g} dV = \tag{10.75}$$

$$= \int_{V_{TD}}^{\mp V_w} \frac{V \pm V_w}{g\left\{\left(\frac{T}{W} - \mu_{g_{brake}}\right) - \frac{(C_{D_g} - \mu_{g_{brake}} C_{L_g})\overline{q}}{W/S} + \frac{N_n}{W}(\mu_{g_{brake}} - \mu_g) - \phi\right\}} dV$$

The upper sign in Eqn (10.75) is for a down-wind landing. The touch-down speed, V_{TD}, varies with pilot technique. Typically, $V_{TD} = 1.1 V_{S_{approach}}$, may be used for purposes of predicting the landing ground-roll. To simplify the problem a number of assumptions are introduced:

Assumptions:
  a) a runway with a constant gradient,
  b) constant weight during the ground run,
  c) that the airplane attitude on its landing gear is constant so that C_{D_g} and C_{L_g} are constant,
  d) ground braking friction coefficient, $\mu_{g_{brake}}$ is also constant,
  e) variation of thrust with speed is known.

With these assumptions it is possible to carry out the integration implied by Eqn (10.75).

Various types of solutions to Eqn (10.75) will be discussed in Section 10.6.

10.6 PREDICTION OF THE LANDING DISTANCE

Three methods for the calculation of the landing distance will be presented:

10.6.1 Accurate method   10.6.2 Statistical method

10.6.3 Approximate, analytical method

Methods for presentation of landing performance data are given in Sub–section 10.6.4.

10.6.1 ACCURATE METHOD FOR PREDICTING THE LANDING DISTANCE

10.6.1.1 Air Distance Covered During Descent from the Obstacle (Accurate method)

Figure 10.30 shows the geometry of the airborne flight path during the landing. It is seen, that the landing distance, measured along the ground, S_{LA}, is split into two components. In accordance with Eqn (10.61): $S_{LA} = S_{LDES} + S_{LTR}$.

During the landing descent from the obstacle, the speed changes from V_A to V_{FL}. It is usually assumed that $V_{FL} \approx 0.95 V_A$. The flight path is a straight line. The landing distance along the ground, S_{LDES}, is determined from:

$$S_{LDES} = \frac{(h_{screen} - h_{flare})}{\tan \overline{\gamma}_A} \tag{10.76}$$

where: h_{screen}, the screen height, is 50 ft for FAR 23, for FAR 25 and for military aircraft

h_{flare}, the flare height, varies primarily with the size of an airplane. The flare height may be estimated by assuming an instantaneous circular transition path as shown in Figure 10.33. A method for determining the flare height, h_{flare}, is given in Sub–sub–section 10.6.1.2.

$\overline{\gamma}_A = -\gamma_A$, the approach flight path angle. For a typical instrument approach this flight path angle is 2.5 to 3.0 degrees.

A geometric interpretation of these quantities is given in Figure 10.33. Next, a method for evaluating the approach flight path angle, $\overline{\gamma}_A$, will be presented.

Takeoff and Landing

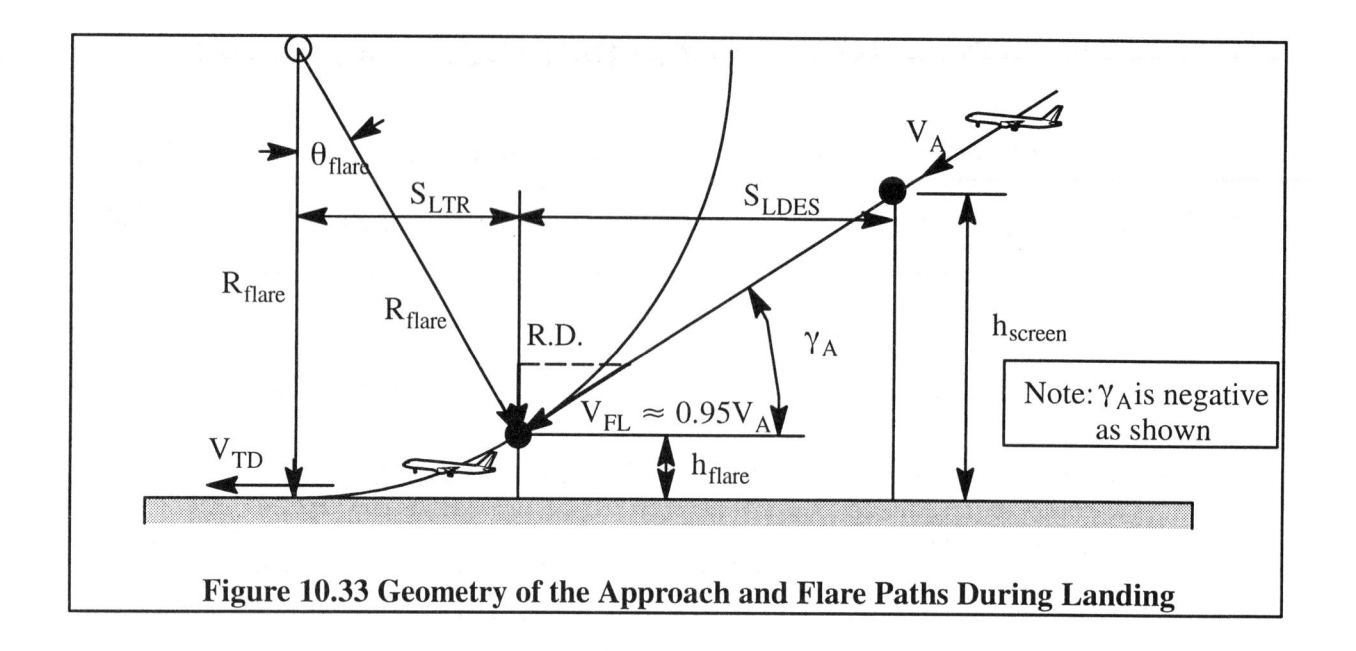

Figure 10.33 Geometry of the Approach and Flare Paths During Landing

The equations of motion during the approach flight path are obtained by introducing appropriate adjustments to Eqns (10.64) and (10.65):

$$T\cos(\alpha + \phi_T) - D + W\sin\overline{\gamma} = 0 \tag{10.77}$$

$$T\sin(\alpha + \phi_T) + L - W\cos\overline{\gamma} = \frac{W}{g}V\dot{\overline{\gamma}} \tag{10.78}$$

For the straight line component of the landing approach ($\dot{\gamma} = 0$), and for sufficiently small values of angle of attack and thrust inclination angle, $(\alpha + \phi_T) \approx 0$, these equations become:

$$T - D + W\overline{\gamma}_A = 0 \tag{10.79}$$

$$L - W = 0 \tag{10.80}$$

By combining Eqns (10.79) and (10.80) it is seen that:

$$\overline{\gamma}_A \approx -\frac{T}{W} + \frac{D}{L} \tag{10.81}$$

If the engines are at flight idle, T = 0. However, most approaches are carried out as powered approaches so that the thrust is not zero. Since the approach speed is known ($V_A = 1.3V_{S_L}$), the required lift coefficient can be computed from Eqn (10.80):

$$C_{L_A} = \frac{2W}{\rho S V_A^2} \tag{10.82}$$

This in turn allows the determination of the drag coefficient, from the drag polar in the approach flight condition. Therefore, D/L in Eqn (10.81) is now known. The thrust–to–weight ratio, T/W is selected to match the required glide slope angle. As indicated before, the glide slope angle is typically 2.5 to 3 degrees although larger angles are possible.

10.6.1.2 Air Distance Covered During the Transition or Flare (Accurate method)

During the landing transition (also called flare), the speed changes from V_{FL} to V_{TD}. The flare speed, V_{FL}, is assumed to be: $V_{FL} \approx 0.95 V_A$. This is shown in Figure 10.33. The flight path is a curved line, normally assumed to be circular. For the case of a circular flare, the transition landing distance along the ground, S_{LTR}, is determined from:

$$S_{LTR} = R_{flare} \bar{\gamma}_A \tag{10.83}$$

The flare height, h_{flare}, may be computed from:

$$h_{flare} = R_{flare}(1 - \cos\bar{\gamma}_A) = R_{flare}\frac{(\bar{\gamma}_A)^2}{2} = \frac{1}{2}S_{LDES}\bar{\gamma}_A \tag{10.84}$$

The flare radius, R_{flare}, may be determined in a manner quite analogous to the one used during the take–off transition in Equations (10.38) through (10.41). First, the lift coefficient during the landing transition may be written as:

$$C_{L_{flare}} = C_{L_{FL}} + \Delta C_{L_{FL}} = \frac{W}{\frac{1}{2}\varrho V_{FL}^2 S} + \Delta C_{L_{FL}} = \frac{W}{\frac{1}{2}\varrho V_{S_A}^2 S}\left[\frac{V_{S_A}^2}{V_{FL}^2}\right] + \Delta C_{L_{FL}} =$$

$$= C_{L_{max_A}}\left[\frac{V_{S_A}^2}{V_{FL}^2}\right] + \Delta C_{L_{FL}} \tag{10.85}$$

This assumes an instantaneous change in the lift coefficient as shown in Figure 10.34. The resulting lift must be equal to the sum of the weight and the centrifugal force:

$$W + \frac{W}{g}\frac{V_{FL}^2}{R_{flare}} = \left[C_{L_{max_A}}\left[\frac{V_{S_A}^2}{V_{FL}^2}\right] + \Delta C_{L_{FL}}\right]\frac{1}{2}\varrho V_{FL}^2 S = W + \Delta C_{L_{FL}}\frac{1}{2}\varrho V_{FL}^2 S \tag{10.86}$$

Therefore:

$$\frac{W}{g}\frac{V_{FL}^2}{R_{flare}} = \Delta C_{L_{FL}}\frac{1}{2}\varrho V_{FL}^2 S \tag{10.87}$$

The flare radius, R_{flare}, follows from:

$$R_{flare} = \frac{2(W/S)}{\varrho g \Delta C_{L_{FL}}} = \frac{V_{FL}^2}{g}\frac{C_{L_A}}{\Delta C_{L_{FL}}} = \frac{V_{FL}^2}{g(n_{FL} - 1)} \tag{10.88}$$

The load factor in the flare, n_{FL}, depends on pilot technique and is usually about 1.04 to 1.08.

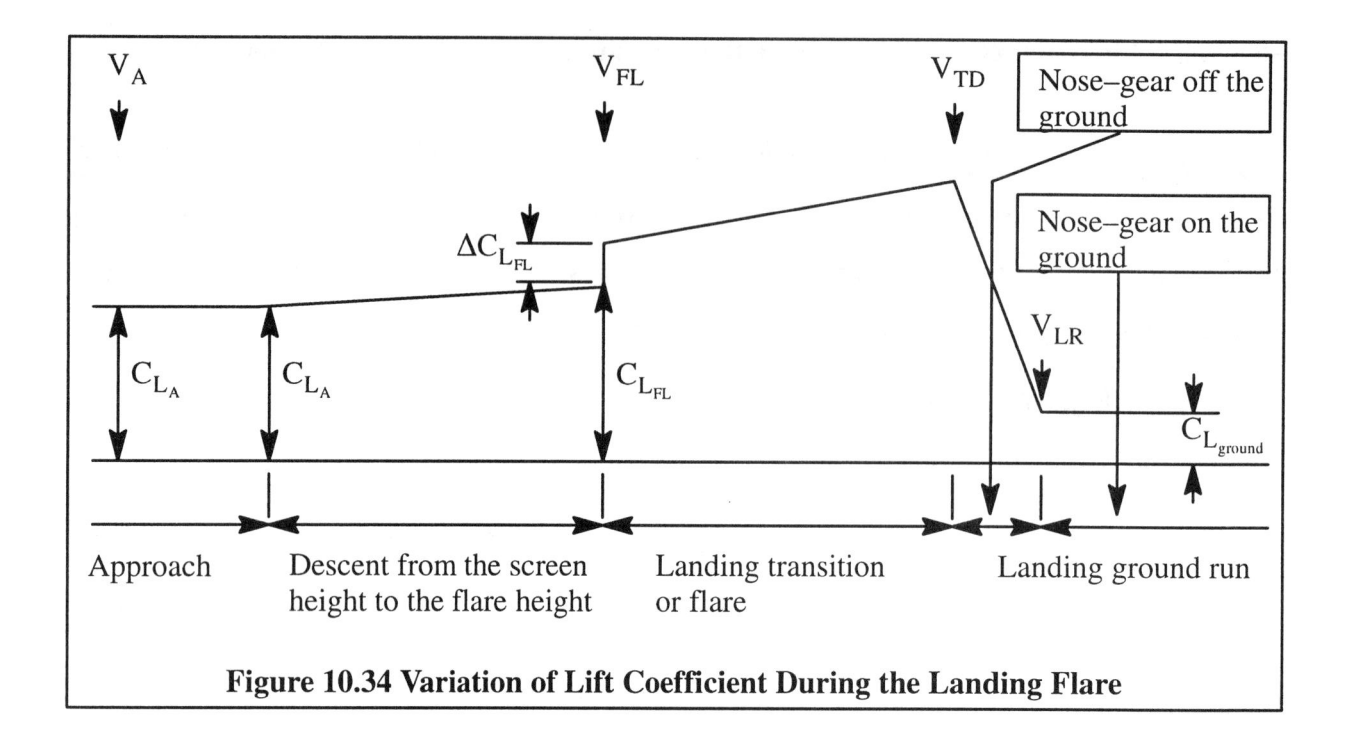

Figure 10.34 Variation of Lift Coefficient During the Landing Flare

By combining Eqns (10.83) and (10.88), the transition landing distance along the ground, S_{LTR}, is finally determined from:

$$S_{LTR} = \frac{V_{FL}^2}{g(n_{FL} - 1)} \overline{\gamma}_A \quad \left[\text{or}: S_{LTR} = \frac{V_{FL}^2 \left(-\frac{T}{W} + \frac{D}{L} \right)}{g(n_{FL} - 1)} \text{ with Eqn (10.81)} \right] \quad (10.89)$$

To determine the touchdown speed, V_{TD}, the following energy balance equation may be used:

$$\frac{W}{2g} \left(V_{FL}^2 - V_{TD}^2 \right) + W h_{flare} = \overline{(T - D)}_{TD} S_{LTR} \quad (10.90)$$

where: $\overline{(T - D)}_{TD}$ is the average retarding force during the flare.

Eqn (10.90) may be solved for the touchdown speed:

$$V_{TD} = \sqrt{\frac{W}{2g} \left(V_{FL}^2 \right) + W h_{flare} - \overline{(T - D)}_{TD} S_{LTR}} \quad (10.91)$$

The average retarding force during the flare, $\overline{(T - D)}_{TD}$, is assumed to be the average of the retarding force at V_{FL} and at V_{TD}. This average may be computed as follows:

At the start of the flare, the flight path angle is still equal to: γ_A. For that reason, the retarding force may be found by modifying Eqn (10.81) as follows:

$$(T - D)_{FL} \approx -\overline{\gamma}_A \tag{10.92}$$

At the end of the flare, that is at the touchdown speed, the retarding force may be found from:

$$(T - D)_{TD} \approx -D_{TD} \approx -W\left(\frac{C_D}{C_L}\right)_{TD} \tag{10.93}$$

The average retarding force in the flare, $\overline{(T - D)}_{TD}$ is found by averaging Eqns (10.92) and (10.93) as:

$$\overline{(T - D)}_{TD} = \frac{1}{2}\left\{-\overline{\gamma}_A - W\left(\frac{C_D}{C_L}\right)_{TD}\right\} \tag{10.94}$$

Typically, the touchdown speed, V_{TD}, is found to be about $1.15V_A$.

10.6.1.3   Ground Distance Covered with the Nose–gear Off the Ground (Accurate method)

Figure 10.29 shows the geometry of the ground–roll. Assume that the variation of the acceleration along the runway, as expressed by Eqn (10.75), is known as a function of airplane speed. The integration to find the ground–roll part of the landing distance, S_{LG}, according to Eqn (10.75), can now be carried out with great precision.

It is convenient to split the ground roll part of the landing distance into two parts, in accordance with Eqn (10.61): $S_{LG} = S_{LNGR} + S_{LR}$. During the rotation part, S_{LR}, the speed decreases from V_{TD} to V_{LR}. During the "all–wheels–on–the–ground' part, S_{LNGR}, the speed decreases from V_{LR} to zero. To estimate the ground distance during the rotation, S_{LR}, it is necessary to know the time needed to get the nose–wheel on the ground. This typically takes 1–3 seconds, depending on airplane type and pilot technique. If the average deceleration during this time is called, $a_{g_{LR}}$, the speed, V_{LR}, may be estimated from:

$$V_{LR} = V_{TD} + (a_{g_{LR}})t_{LR} \tag{10.95}$$

The reader should again realize that $a_{g_{LR}} < 0$ in Eqn (10.95). The distance covered during this rotation period, S_{LR}, now follows from:

$$S_{LR} = V_{TD} t_{LR} + (0.5 a_{g_{LR}}) t_{LR}^2 \qquad (10.96)$$

where: $a_{g_{LR}}$ is found from Eqn (10.72) by setting the nose-gear load, N_n equal to zero:

$$a_{g_{LR}} = g \left\{ \left(\frac{T}{W} - \mu_{g_{brake}} \right) - \frac{(C_{D_g} - \mu_{g_{brake}} C_{L_g}) \bar{q}}{W/S} - \phi \right\} \qquad (10.97)$$

To obtain S_{LR}, Eqn (10.75) can now be integrated from $(V_{TD} \mp V_w)$ to $(V_{LR} \mp V_w)$:

$$S_{LR} = \int_{V_{TD} \mp V_w}^{V_{LR} \mp V_w} \frac{V \pm V_w}{a_{g_{LNGR}}} dV = \qquad (10.98)$$

$$= \int_{V_{TD} \mp V_w}^{V_{LR} \mp V_w} \frac{V \pm V_w}{g \left\{ \left(\frac{T}{W} - \mu_{g_{brake}} \right) - \frac{(C_{D_g} - \mu_{g_{brake}} C_{L_g}) \bar{q}}{W/S} - \phi \right\}} dV$$

To account for wind (assumed to be constant) during the landing ground roll, it is convenient to plot a curve of $(V \pm V_w)/a_g$ versus V is plotted as shown in Figure 10.35.

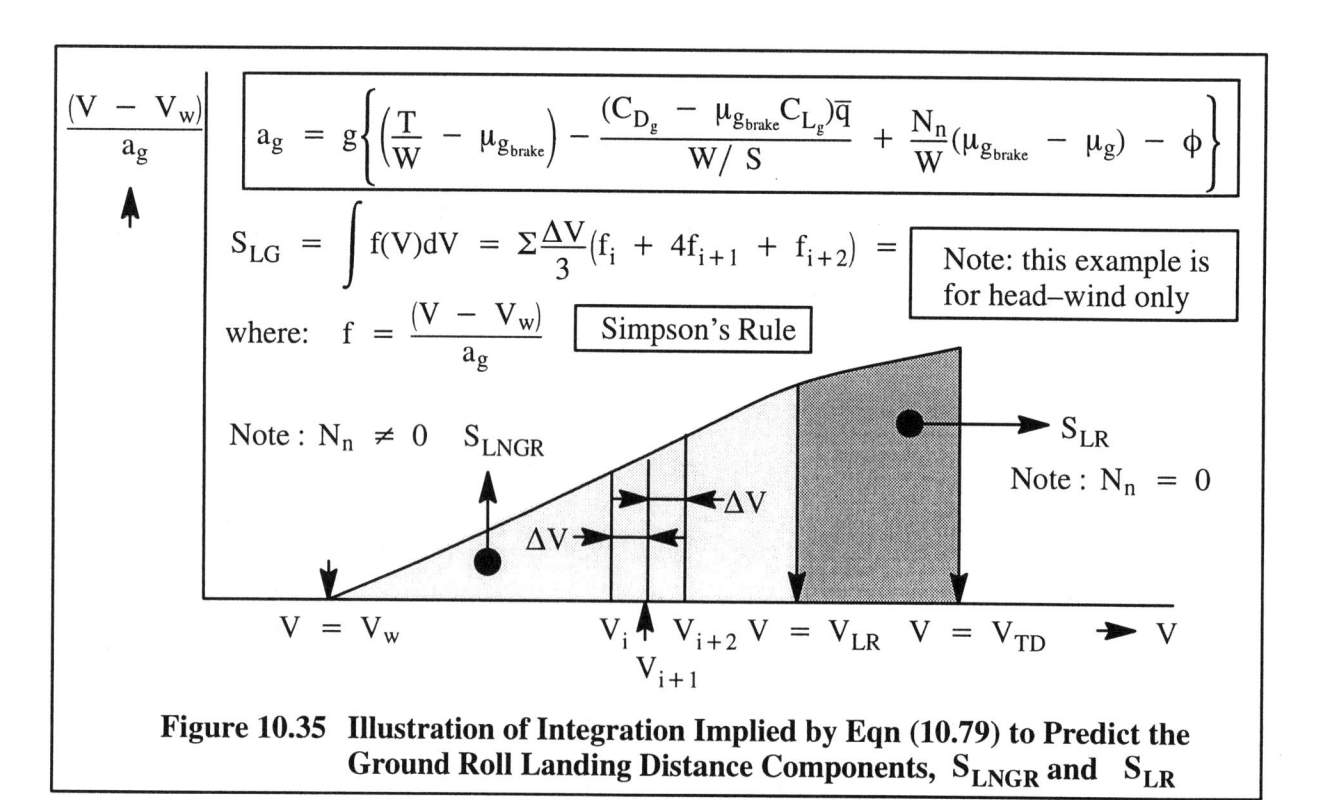

Figure 10.35 Illustration of Integration Implied by Eqn (10.79) to Predict the Ground Roll Landing Distance Components, S_{LNGR} and S_{LR}

In evaluating Eqn (10.98), two scenarios must be considered:

a) The thrust is set at flight idle (T/W close to zero, but normally slightly positive). For determining certified landing ground rolls, idle thrust should be used.

b) The thrust is reversed. In this case, T/W < 0. Note: reversed thrust may not be used to determine ground rolls for certification purposes. However, reverse thrust may be used operationally and its effect on the ground roll is then published in the flight manual.

c) With the nose–gear off the ground, thrust reversing is normally not used.

Simpson's Rule may be used in determining S_{LR}.

10.6.1.4 Ground Distance Covered with the Nose–gear On the Ground (Accurate method)

The part of the ground roll during which the nose–wheel is on the ground, S_{LNGR}, may be estimated by integrating Eqn (10.75) from $(V_{LR} \mp V_w)$ to $(\mp V_w)$:

$$S_{LNGR} = \int_{V_{LR} \mp V_w}^{\mp V_w} \frac{V \pm V_w}{a_{g_{LNGR}}} dV = \qquad (10.99)$$

$$= \int_{V_{LR} \mp V_w}^{\mp V_w} \frac{V \pm V_w}{g\left\{\left(\dfrac{T}{W} - \mu_{g_{brake}}\right) - \dfrac{(C_{D_g} - \mu_{g_{brake}} C_{L_g})\bar{q}}{W/S} + \dfrac{N_n}{W}(\mu_{g_{brake}} - \mu_g) - \phi\right\}} dV$$

To account for wind (assumed to be constant) during the landing ground roll, it is convenient to plot a curve of $(V \pm V_w)/a_g$ versus V is plotted as shown in Figure 10.35. Again, two scenarios must be considered:

a) The thrust is set at flight idle (T/W close to zero, but normally slightly positive). For determining certified landing ground rolls, idle thrust should be used.

b) The thrust is reversed. In this case, T/W < 0. Note: reversed thrust may not be used to determine ground rolls for certification purposes. However, reverse thrust may be used operationally and its effect on the ground roll is then published in the flight manual.

Simpson's Rule (shown in Figure 10.35) or any other convenient numerical integration scheme can then be used to find the take–off ground roll component, S_{LNGR}.

10.6.2 STATISTICAL METHOD FOR PREDICTING THE LANDING DISTANCE

The following statistical method for predicting the landing distance is taken from Ref. 10.10 but is based on a method described by Loftin in Ref.10.11. The method applies to:

* Propeller driven, FAR 23 airplanes   * Jet driven FAR 25 airplanes

10.6.2.1 Statistical Method for Predicting the Landing Distance of Propeller Driven, FAR 23 Airplanes

The method predicts the landing ground distance, S_{LG}, and the total landing distance, S_L, as defined in Figure 10.29. The FAR 23 landing distance from hard–surfaces runways may be computed from:

$$S_{LG} = 0.265 \, (V_{S_{approach}})^2 \qquad (10.100)$$

where: $V_{S_{approach}}$ is the stall speed of the airplane in the approach configuration.

The total landing distance, S_L (from a 50 ft obstacle), may be determined from:

$$S_L = 1.938 \, S_{LG} \qquad (10.101)$$

The stall speed in the approach (or landing) configuration follows from:

$$V_{S_{approach}} = \sqrt{\frac{2(W_L/S)}{\varrho C_{L_{max_{approach}}}}} \qquad (10.102)$$

Equation (10.100) is based on the statistical data shown in Figure 10.36. Equation (10.101) is based on the statistical data shown in Figure 10.37. Both figures apply to FAR 23 airplanes only. The scatter in the data occurs because of variations in pilot technique.

Equations (10.100) through (10.102) suggest that the landing ground roll distance depends on the following parameters:

a) Wing loading, W_L/S, in the landing configuration

b) Altitude and temperature (both of which define the air density, ϱ)

c) Maximum lift coefficient, $C_{L_{max_{approach}}}$, in the approach (=landing) configuration

10.6.2.2 Statistical Method for Predicting the Landing Distance of Jet Driven, FAR 25 Airplanes

In FAR 25, the total landing distance, S_L (from a 50 ft obstacle), is divided by a factor of safety,

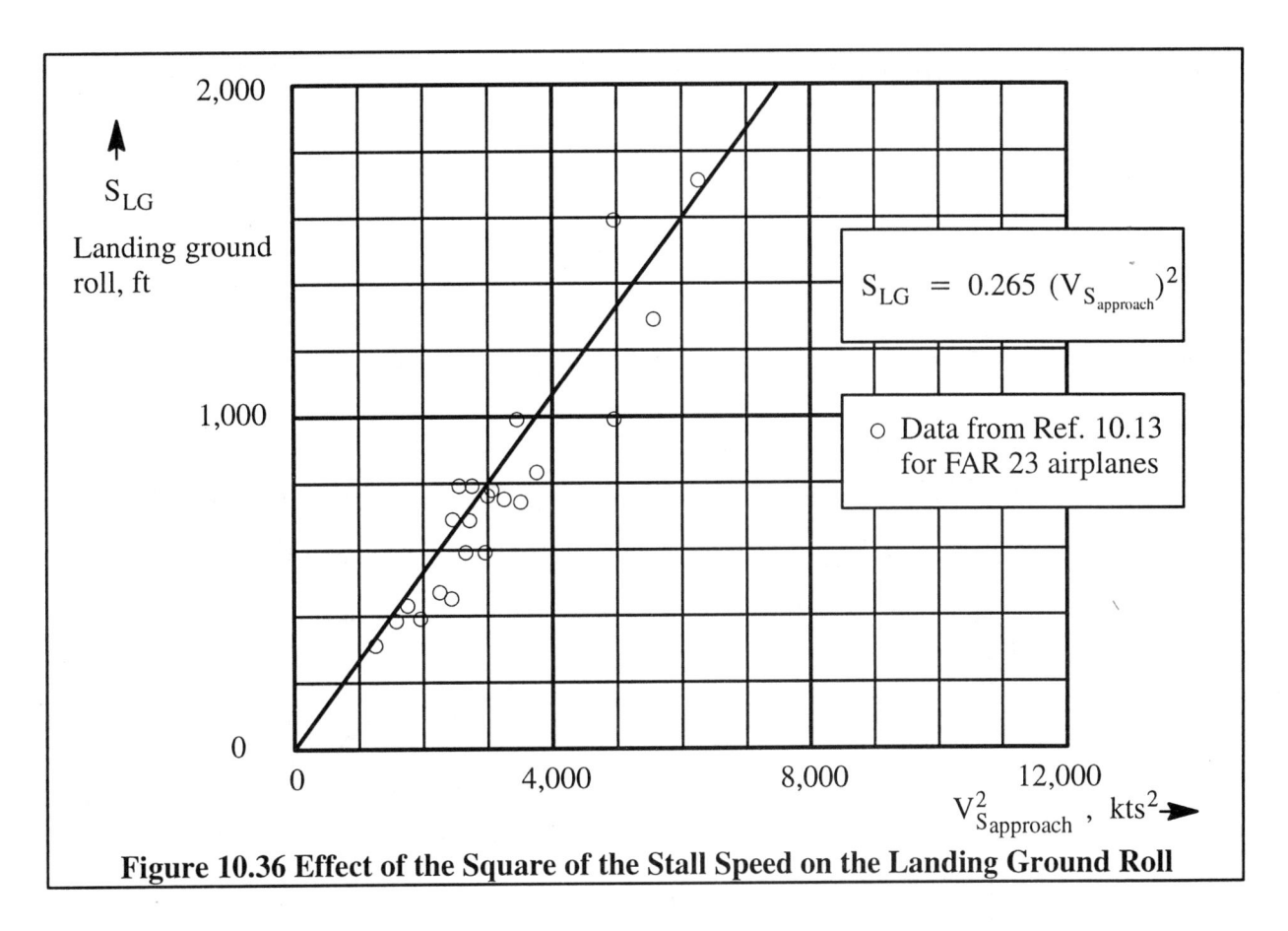

Figure 10.36 Effect of the Square of the Stall Speed on the Landing Ground Roll

S_{LG} Landing ground roll, ft

$$S_{LG} = 0.265\,(V_{S_{approach}})^2$$

○ Data from Ref. 10.13 for FAR 23 airplanes

$V^2_{S_{approach}}$, kts²

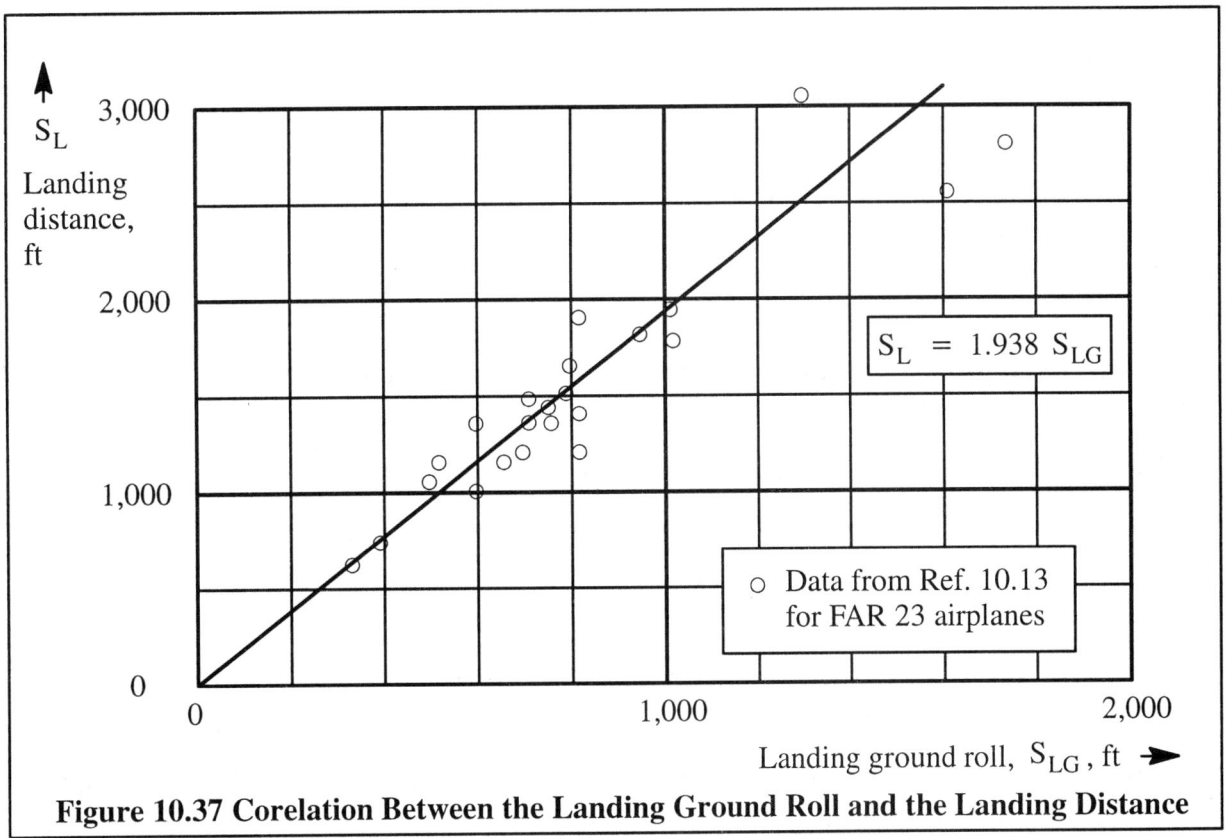

Figure 10.37 Corelation Between the Landing Ground Roll and the Landing Distance

S_L Landing distance, ft

$$S_L = 1.938\,S_{LG}$$

○ Data from Ref. 10.13 for FAR 23 airplanes

Landing ground roll, S_{LG}, ft

0.60, to produce the FAR 25 field-length, S_{FL}:

$$S_{FL} = \frac{S_L}{0.60} \tag{10.103}$$

The FAR 25 landing field-length, S_{FL}, is statistically related to the square of the approach speed, V_A^2, in kts2. The statistical data were taken from Ref. 10.13 and are shown in Figure 10.38.

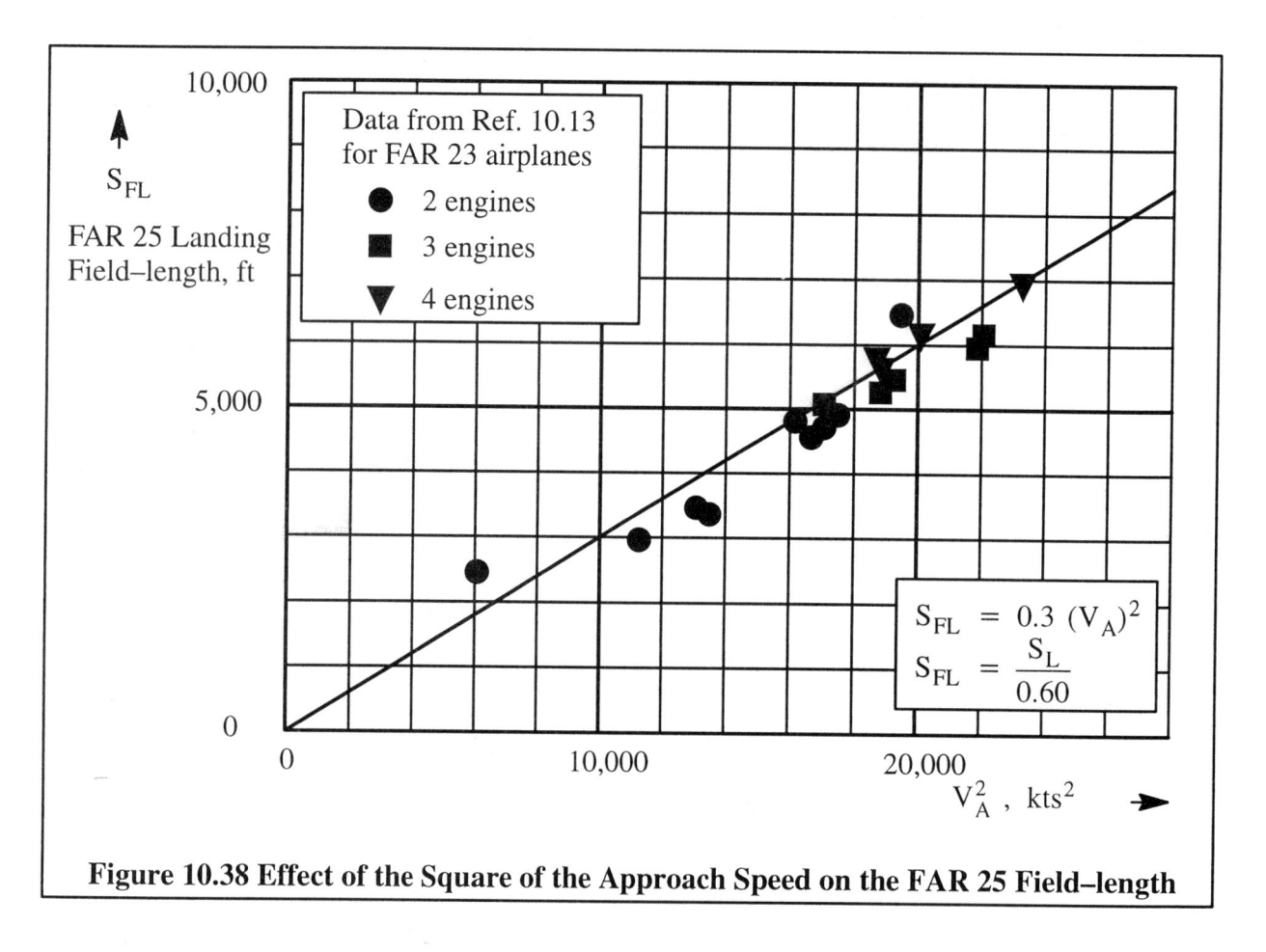

Figure 10.38 Effect of the Square of the Approach Speed on the FAR 25 Field-length

Data from Ref. 10.13 for FAR 23 airplanes
- ● 2 engines
- ■ 3 engines
- ▼ 4 engines

$S_{FL} = 0.3\,(V_A)^2$
$S_{FL} = \dfrac{S_L}{0.60}$

The scatter in the data occurs because of variations in pilot technique.

The approach speed, V_A, is related to the maximum lift coefficient in the landing configuration:

$$V_A = 1.3\sqrt{\frac{2(W_L/S)}{\varrho C_{L_{max_{approach}}}}} \tag{10.104}$$

If the field-length is known, V_A follows from Figure 10.38. The required maximum lift coefficient in the landing (approach) configuration can then be determined from Eqn (10.103), once the wing loading in the landing configuration is known.

10.6.3 APPROXIMATE, ANALYTICAL METHOD FOR PREDICTING THE LANDING DISTANCE AND THE TIME TO LAND

In this approximate, analytical method, the total landing distance, S_L is divided into three components shown in Figure 10.29 and expressed in Eqn (10.61). Methods for determining S_A, S_{LTR} and S_{LDES} are presented in Sub–sub–sections 10.6.3.1 through 10.6.3.3 respectively. A method for predicting the time required to take–off is presented in Sub–sub–section 10.6.3.4.

The reader should be aware of the fact that the landing field–length, $S_{L_{fieldlength}}$, for FAR 25 is defined as:

$$S_{L_{fieldlength}} = \frac{S_L}{0.6} \tag{10.105}$$

10.6.3.1 Approximate Method for Calculating the Air Distance Component, S_{LA}

Figure 10.39 (a modified version of Figure 10.33) depicts the geometry of the landing path.

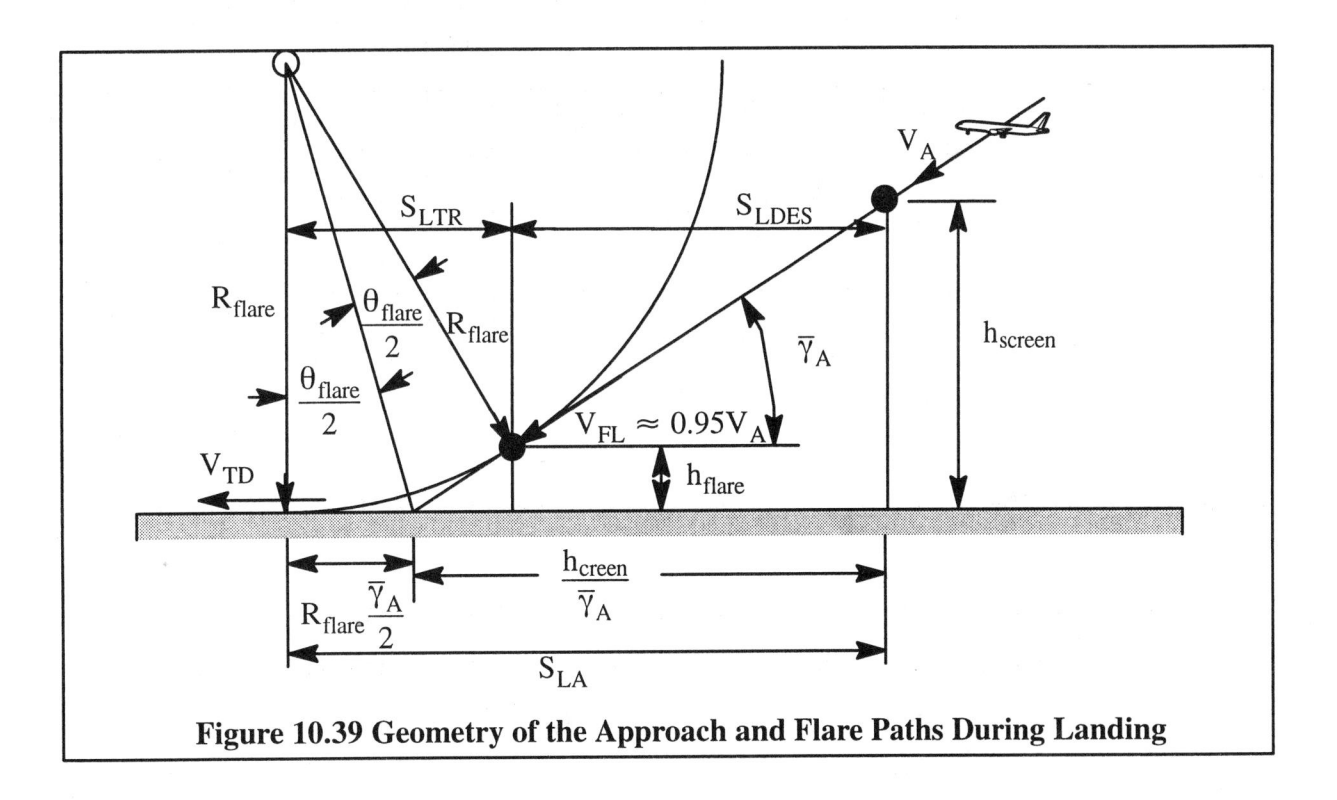

Figure 10.39 Geometry of the Approach and Flare Paths During Landing

From Figure 10.39 it is observed that:

$$S_{LA} \approx \frac{h_{screen}}{\overline{\gamma}_A} + R_{flare}\frac{\overline{\gamma}_A}{2} \tag{10.106}$$

The approach flight path angle, $\bar{\gamma}_A$, may be determined from Eqn (10.81) or a constant (2.5, 3 or more degrees) may be assumed. However, if a constant angle is assumed, the reader is cautioned to make sure that Eqn (10.81) is satisfied in terms of appropriate values for L/D and T/W in the approach flight condition.

The flare radius, R_{flare}, follows from Eqn (10.88):

$$R_{flare} = \frac{V_{FL}^2}{g(n_{FL} - 1)} \tag{10.107}$$

The load factor in the flare, n_{FL}, depends on pilot technique and is usually about 1.04 to 1.08. The flare velocity may be assumed to be:

$$V_{FL} = 0.95 \, V_A \tag{10.108}$$

10.6.3.2 Approximate Method for Calculating the Ground Distance Component, S_{LR}

After touchdown, during the rotation part, S_{LR} (see Figure 10.28), the speed decreases from V_{TD} to V_{LR}, while the nose–gear is lowered to the ground.

The time taken in this maneuver may be assumed to be:

$$t_{LR} = 1 \text{ to } 3 \text{ seconds} \tag{10.109}$$

For light airplanes it is suggested to use 1 second, for transports, use 3 seconds. The problem is in the determination of V_{LR}. It is conservative to assume that the airplane is free–rolling on the main gear while the nose–gear is being lowered to the ground during this transition maneuver. If it is assumed that the deceleration is negligible, so that: $V_{LR} = V_{TD}$, the ground distance component, S_{LR}, may be estimated from:

$$S_{LR} = 0.5 \, (V_{TD} + V_{LR}) \, t_{LR} \approx V_{TD} \, t_{LR} \tag{10.110}$$

10.6.3.3 Approximate Method for Calculating the Ground Distance Component, S_{LNGR}

During the ground distance component, S_{LNGR}, the speed decreases from V_{LR} to 0. A modification of Equation (10.75) can be used to find an approximate, closed form solution for the ground distance component, S_{LNGR}.

To simplify the problem, the following additional assumptions are made:

1) $V_{LR} = V_{TD}$ 2) Automatic braking 3) Runway inclination: $\phi = 0$

4) No wind: $V_w = 0$ 5) Constant angle of attack: C_{D_g} and C_{L_g} are independent of speed

With these assumptions, Eqn (10.75) yields:

$$S_{LNGR} = \int_{V_{TD}}^{0} \frac{V dV}{g\left\{\left(\frac{T}{W} - \mu_{g_{brake}}\right) - \frac{(C_{D_g} - \mu_{g_{brake}} C_{L_g})\bar{q}}{W/S} + \frac{N_n}{W}(\mu_{g_{brake}} - \mu_g)\right\}} \qquad (10.111)$$

This can be written as:

$$S_{LNGR} = \int_{0}^{V_{TD}} \frac{dV^2}{2g\left\{\left(\mu_{g_{brake}} - \frac{T}{W}\right) + \frac{(C_{D_g} - \mu_{g_{brake}} C_{L_g})\varrho V^2}{2W/S} - \frac{N_n}{W}(\mu_{g_{brake}} - \mu_g)\right\}} =$$

$$= \int_{0}^{V_{TD}} \frac{dV^2}{A + BV^2} = \frac{1}{B}\ln(A + BV^2)\Big|_{0}^{V_{TD}} = \frac{1}{B}\ln\left(1 + \frac{B}{A}V_{TD}^2\right) \qquad (10.112)$$

where: $A = 2g\left\{\left(\mu_{g_{brake}} - \frac{T}{W}\right) - \frac{N_n}{W}(\mu_{g_{brake}} - \mu_g)\right\}$ \hfill (10.113)

$$B = \frac{g\varrho(C_{D_g} - \mu_{g_{brake}} C_{L_g})}{W/S} \qquad (10.114)$$

Note: A will normally be positive. However, B can be negative. When that happens, the ln term will also be negative. This yields:

$$S_{LNGR} = \left\{\frac{W/S}{g\varrho(C_{D_g} - \mu_{g_{brake}} C_{L_g})}\right\} \ln\left[1 + \frac{(V_{TD})^2 \varrho(C_{D_g} - \mu_{g_{brake}} C_{L_g})}{2 W/S \left\{\left(\mu_{g_{brake}} - \frac{T}{W}\right) - \frac{N_n}{W}(\mu_{g_{brake}} - \mu_g)\right\}}\right]$$

(10.115)

The reader must decide what numerical value to use for the braking friction coefficient, $\mu_{g_{brake}}$. For numerical guidance, Table 10.3 may be used.

For a transport, the drag coefficient on the ground, C_{D_g}, will be extra high due to the effect of ground spoilers. Similarly, the lift coefficient on the ground, C_{L_g}, will be low due to the same effect.

Typical values for the ratio of nose–gear load to weight ratio during a landing, N_n/W, may range from 0.08 to 0.2. The value of T/W can be positive or negative, depending on whether idle thrust or reverse thrust is used.

10.6.3.4   Method for Calculating the Time Required to Land, t_L

The total time required for landing, t_L, may be important for purposes of calculating the fuel used during the landing part of a mission. Methods to calculate this amount of fuel are presented in Chapter 11. The total time required for landing, t_L, may be determined from:

$$t_L = t_{LA} + t_{LR} + t_{LNGR} \tag{10.116}$$

During the descent from the obstacle, the speed was assumed to be constant at V_A. The approach speed, V_A, may be determined from Eqn (10.82). The time, t_{LA}, follows from:

$$t_{LA} = \frac{S_{LA}}{V_A} \tag{10.117}$$

where: S_{LA} follows from Eqn (10.106).

For the landing ground rotation time, t_{LR}, it was suggested in Eqn (10.109) to use:

$$t_{LR} = 1 \text{ to } 3 \text{ seconds}: 1 \text{ sec for small airplanes and } 3 \text{ sec for transports} \tag{10.118}$$

The time spent during the landing groundrun, t_{LNGR}, can be computed by remembering that in general, $dt = dS/V$, and then suitably modifying Eqn (10.111):

$$t_{LNGR} = \int_{V_{TD}}^{0} \frac{dV}{g\left\{\left(\frac{T}{W} - \mu_{g_{brake}}\right) - \frac{(C_{D_g} - \mu_{g_{brake}} C_{L_g})\bar{q}}{W/S} + \frac{N_n}{W}(\mu_{g_{brake}} - \mu_g)\right\}} \tag{10.119}$$

This can be written as:

Takeoff and Landing

$$t_{LNGR} = \int_0^{V_{TD}} \frac{dV}{g\left\{\left(\mu_{g_{brake}} - \frac{T}{W}\right) + \frac{(C_{D_g} - \mu_{g_{brake}}C_{L_g})\varrho V^2}{2W/S} - \frac{N_n}{W}(\mu_{g_{brake}} - \mu_g)\right\}} =$$

$$= \int_0^{V_{TD}} \frac{dV}{C + DV^2} = \left(\frac{1}{\sqrt{CD}}\right)\arctan\left(V_{TD}\sqrt{\frac{D}{C}}\right) \quad (10.120)$$

where:
$$C = g\left\{\left(\mu_{g_{brake}} - \frac{T}{W}\right) - \frac{N_n}{W}(\mu_{g_{brake}} - \mu_g)\right\} \quad (10.121)$$

$$D = \left\{\frac{g\varrho(C_{D_g} - \mu_{g_{brake}}C_{L_g})}{2W/S}\right\} \quad (10.122)$$

This yields:

$$t_{LNGR} = \left[g\sqrt{\left\{\left(\mu_{g_{brake}} - \frac{T}{W}\right) - \frac{N_n}{W}(\mu_{g_{brake}} - \mu_g)\right\}\left\{\frac{\varrho(C_{D_g} - \mu_{g_{brake}}C_{L_g})}{2W/S}\right\}}\right] \times$$

$$\times \arctan\left\{(V_{TD})\sqrt{\frac{(C_{D_g} - \mu_{g_{brake}}C_{L_g})\varrho}{2W/S\left\{\left(\mu_{g_{brake}} - \frac{T}{W}\right) - \frac{N_n}{W}(\mu_{g_{brake}} - \mu_g)\right\}}}\right\} \quad (10.123)$$

Note: C will normally be positive. However, D can be negative. In that case the solution indicated by Eqn (10.124) must be used.

$$t_{LNGR} = \left(\frac{1}{\sqrt{-CD}}\right)\text{arctanh}\left(V_{TD}\sqrt{\frac{-D}{C}}\right) \quad (10.124)$$

An example application of these approximate methods for predicting the landing distance and landing time will be presented in Example 10.3.

Example 10.3   The example concerns a propeller driven airplane with the thrust characteristics of Figure 10.9. Calculate the landing distance for no wind and for zero runway gradient. Also calculate the time to land. Assume sea–level standard conditions. The following data are available for this airplane:

$W_L = 4,600$ lbs $\left(\frac{t}{c}\right)_{ave} = 0.14$ $V_w = 0$ $\phi = 0$

$$\delta_f = 45^0 \qquad S = 175 \text{ ft}^2 \qquad b = 35 \text{ ft} \qquad A = 7$$

$$\mu_g = 0.03 \qquad C_{L_{max_L}} = 2.12 \qquad e = 0.80 \text{ in free air}$$

$$\mu_{g_{brake}} = 0.4 \qquad T = 260 \text{ lbs in the landing flare}$$

Landing drag polar (flaps and gear down): $C_{D_A} = 0.1000 + \dfrac{(C_{L_A})^2}{\pi Ae}$

Solution: The solution is given in Table 10.4. First the landing distance.

It is seen at the end of Table 10.3 that the landing distance is predicted to be:

$S_L = 1,539$ ft. This is the actual predicted landing distance.

The corresponding landing fieldlength is given by Eqn (10.103) as:

$S_{FL} = 1,539/0.6 = 2,565$ ft.

Next is the time to land. This is also given in Table 10.4. It is seen that the total time required to land is 21 seconds.

It is of interest to compare the result for S_L with that which is predicted with the (much simpler) statistical method of Sub-section 10.6.2.

The square of the stall speed is: $(102.1/1.689)^2 = 3,654$ kts^2. With Eqn (10.100) this yields:

$S_{LG} = 0.265 \times 3,654 = 968$ ft. With Eqn (10.101) this results in:

$S_L = 1.938 \times 968 = 1,876$ ft. The difference is 22%.

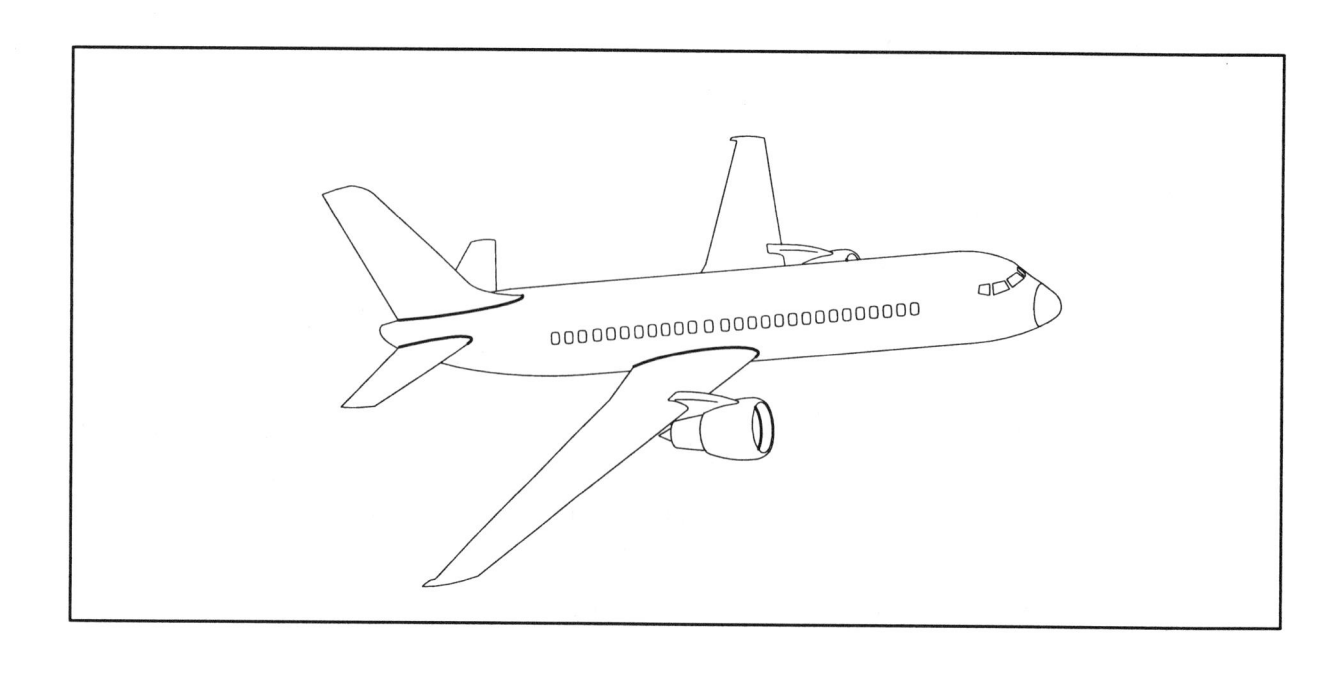

Takeoff and Landing

\multicolumn{5}{c}{Table 10.4 Calculation of Landing Distance and Time for a Propeller Airplane}
Step	Description of step	Symbol	Value	Comments
Step 1	Determine the stall speed in the landing configuration: $V_{S_L} = \sqrt{\dfrac{2W_L/S}{\varrho C_{L_{max_L}}}}$	V_S	102.1 fps	
Step 2	Determine the approach speed: $V_A = 1.3 V_{S_L}$	V_A	132.7 fps	Eqn (10.104)
Step 3	Determine the flight path angle during the descent from the screen height: $T/W = 260/4{,}600 = 0.0565$ $C_{L_A} = 2.12/1.69 = 1.25$ $C_{D_A} = 0.1000 + \dfrac{(C_{L_A})^2}{\pi A e} = 0.1889$ $\overline{\gamma}_A = -(0.0565 - 0.1889/1.25) = 0.0946 \text{ rad} = 5.4 \text{ deg}$	$\overline{\gamma}_A$	4.6 deg	Eqn (10.81)
Step 4	Determine the flare radius: $V_{FL} = 0.95\, V_A = 126.1$ fps $n_{FL} = 1.08$ assumed high because of steep flight path angle	R_{flare}	6,173 ft	Eqn (10.107) Eqn (10.108)
Step 5	Determine the landing air distance: assume $h_{screen} = 50$ ft	S_{LA}	821 ft	Eqn (10.106)
Step 6	Determine the landing ground distance component: $V_{TD} \approx 1.15 V_{S_L} = 117.4$ fps Assume: $t_{LR} \approx 1$ sec	S_{LR}	117 ft	Eqn (10.110)
Step 7	Determine the landing ground distance component: Assume: $C_{L_g} \approx 0.40$ with flaps 45^0 on the ground Assume: $C_{D_g} \approx 0.3000$ with flaps 45^0 on the ground Assume: $T/W \approx 0.0565$ on the ground Assume: $N_n/W = 0.08$ Then: $A = 20.22$ and $B = 0.0004078$ with Eqns (10.113) and (10.114)	S_{LNGR}	601 ft	Eqn (10.115)
Step 8	Determine the total landing distance $S_L = S_{LA} + S_{LR} + S_{LNGR}$	S_L	1,539 ft	Eqn (10.61) Modified

Chapter 10

Table 10.4 (Cont'd) Calculation of Landing Distance and Time for a Propeller Airplane				
Step	Description of step	Symbol	Value	Comments
Step 9	Determine the time required for the air distance: 821/132.7	t_{LA}	6 sec	Eqn (10.117)
Step 10	Determine the time required between touchdown and nose-gear on the ground:	t_{LR}	3 sec	Eqn (10.118)
Step 11	Determine the time required to stop with the nose-gear on the ground:	t_{LNGR}	12 sec	Eqn (10.123)
	Assume: $C_{L_g} \approx 0.40$ with flaps 45^0 on the ground Assume: $C_{D_g} \approx 0.3000$ with flaps 45^0 on the ground Assume: $T/W \approx 0.0565$ on the ground Assume: $N_n/W = 0.08$ Then: $C = 10.11$ and $D = 0.0002039$ with Eqns (10.121) and (10.122)			
Step 12	Determine the total time required to land:	t_L	21 sec	Eqn (10.116)

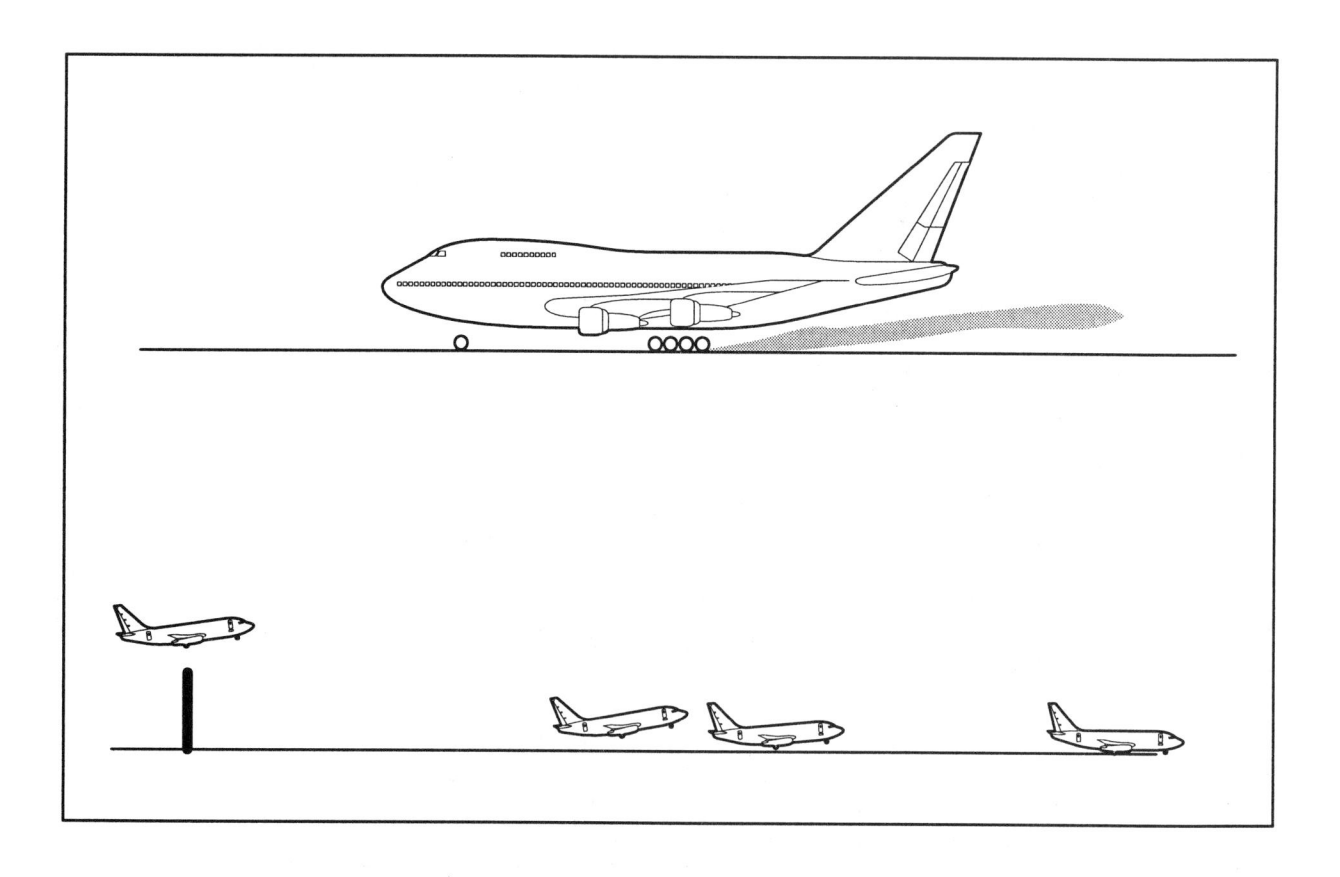

10.6.4 PRESENTATION OF LANDING DISTANCE DATA

In the following, a typical method for presenting landing distance data in airplane flight manuals are given. In determining the landing distance performance of an airplane the following factors should be accounted for:

* Outside air temperature
* Pressure altitude
* Condition of the runway surface
* Condition of the brakes and tires
* Anti–skid system operative or not

* Effect of head–wind or tail–wind
* Anti–icing system on or off
* Obstacle (screen) height
* Manual or automatic spoilers
* Ice accumulation on airplane

Figure 10.40 shows an example of the type of landing distance data which may be found in various airplane flight manuals. The dark line indicates how the data can be used.

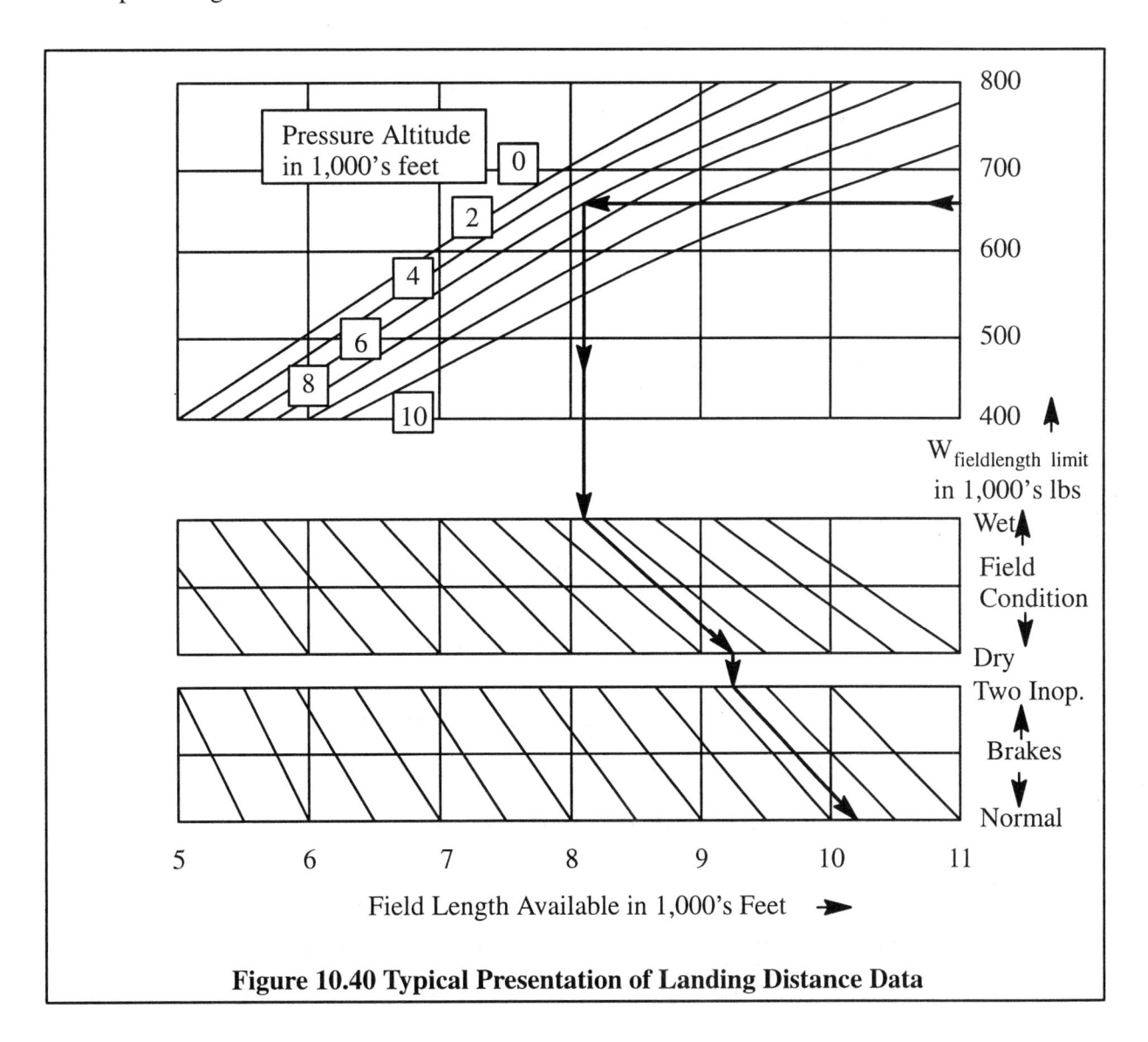

Figure 10.40 Typical Presentation of Landing Distance Data

10.7 SUMMARY FOR CHAPTER 10

In this chapter several methods for calculating the take–off and landing distances of airplanes are developed. A detailed discussion of the take–off and landing processes, in terms of important reference speeds, is provided first. The corresponding civil and military airworthiness rules are also discussed. Next, the equations of motion are developed.

Based on the equations of motion, an accurate method for their integration is presented. Then, a rapid, statistical method for predicting take–off and landing distances is given. Finally, an approximate numerical method for calculating take–off and landing distances is provided. Applications to various airplane types are given.

Examples of how take–off and landing distance data are presented in flight manuals are given.

10.8 PROBLEMS FOR CHAPTER 10

10.1 A jet airplane has the following characteristics:

$W_{TO} = 56,000$ lbs $\quad S = 900$ ft^2 $\quad V_w = 0$ $\quad \phi = 0$

The take–off drag polar, in ground effect is: $C_D = 0.0160 + 0.04\, C_L^2$

$\mu_g = 0.02$ $\quad C_{L_{max_{TO}}} = 1.80$ $\quad C_{L_g} = 1.00$

It is desired to operate this airplane out of a field with a ground–run distance of 3,000 ft. Assume that the lift–off speed is: $V_{LOF} = 1.2 V_{S_{TO}}$. Assuming a constant take–off thrust with speed, determine the required engine thrust. Work this problem by using the method of Sub–section 10.3.3. Assume standard sea–level conditions.

10.2 Use the statistical method of Sub–sub–section 10.3.2.2 to determine the thrust required of the airplane of Problem 10.1 assuming that the required balanced fieldlength is 5,000 ft at sea–level standard and at sea–level with a temperature of 95 degrees F.

10.3 A jet fighter for carrier operation has the following characteristics:

$W_L = 18,000$ lbs $\quad S = 320$ ft^2 $\quad C_{L_{max_L}} = 2.40$ $\quad C_{L_g} = 1.80$

The landing drag polar, in ground effect is: $C_D = 0.4 + 0.05\, C_L^2$. Assume that the friction coefficient while braking is: $\mu_{g_{brake}} = 0.40$. The touchdown speed, as related to the stall speed in the landing configuration, is given by: $V_{TD} = 1.15\, V_{S_L}$. The flight deck length available for the landing ground roll is 700 ft. How fast must the carrier be steaming for the landing to be completed successfully?

Assume no surface wind: $V_w = 0$ and standard sea–level conditions. Also assume that there is no effective thrust during the landing ground roll.

10.4 Assume that during a take–off ground run the angle of attack of an airplane stays constant. Also assume that the airplane thrust is independent of angle of attack. Show, that to obtain the maximum net force of forward acceleration (and hence the minimum ground run distance), the airplane lift coefficient on the ground must be equal to: $C_{L_g} = 0.5 \, (\pi A e \mu_g)$.

10.5 A company is considering the design of a new regional jet transport. Preliminary design studies have evolved the following basic aircraft characteristics:

$W_{TO} = 77,000$ lbs $\qquad$ h/b = 0.12 $\qquad$ $T_{TO} = 19,000$ lbs

$C_{L_{max_{TO}}} = 2.10$ $\qquad$ $C_{L_g} = C_{L_{LOF}}$ $\qquad$ $V_{LOF} = 1.2 \, V_{S_{TO}}$

The take–off drag polar, out of ground effect is: $C_D = 0.0342 + 0.0424 \, C_L^2$. Also assume that the following conditions apply:

$V_w = 0$ $\qquad$ $\phi = 0$ $\qquad$ $\mu_g = 0.02$ $\qquad$ $S_G = 2,500$ ft

If the take–off thrust can be assumed to be constant with speed during the take–off process, determine the required wing area.

10.6 For the airplane of example 10.2 (see Section 10.3) calculate the ground lift coefficient in ground effect if the free air lift coefficient at the same (ground) angle of attack is 0.83.

10.7 For example 10.3 perform a trade study which shows the effect of the following parameters on the landing distance:
- a) ground lift and drag coefficient
- b) friction coefficient due to braking
- c) thrust to weight ratio (include effect of reverse thrust)

10.9 REFERENCES FOR CHAPTER 10

10.1 $\qquad$ Anon.; Code of Federal Regulations (CFR), Title 14, Parts 1 to 59, January 1, 1996; U.S. Government Printing Office, Superintendent of Documents, Mail Stop SSOP, Washington, D.C. Note: FAR 23 and FAR 25 are components of this CFR.

10.2 $\qquad$ Anon.; MIL–C–005011B (USAF), Military Specification, Charts: Standard Aircraft Characteristic and Performance, Piloted Aircraft (Fixed Wing); June, 1977.

10.3 Anon.; AS–5263 (US Navy), Guidelines for the Preparation of Standard Aircraft Characteristic Charts and Performance Data, Piloted Aircraft (Fixed Wing); October, 1986.

10.4 Roskam, J.; Airplane Design, Part VI: Preliminary Calculation of Aerodynamic, Thrust and Power Characteristics; DAR Corporation, 120 East Ninth Street, Suite 2, Lawrence, Kansas, 66044; 1996.

10.5 Anon.; Cessna Aerodynamics Handbook; Cessna Research department, Wichita, Kansas, 1957

10.6 Nicolai, L. M.; Fundamentals of Aircraft Design; School of Engineering, University of Dayton, Dayton, Ohio 45469, 1975. Distributed by: METS, Inc., 6520 Kingsland Court, CA 95120.

10.7 Fink, M.P. and Lastinger, J.L.; Aerodynamic Characteristics of Low–Aspect–Ratio Wings in Close Proximity to the Ground; NASA TN D–926, 1961

10.8 Bagley, J.A.; The Pressure Distribution on Two–Dimensional Wings Near the Ground; British Aeronautical Research Council, R&M No. 3238, February 1960.

10.9 Gratzer, L.B. and Mahal, A.S.; Ground Effects in STOL Operation; Journal of Aircraft, Vol. 9, March 1972, pp. 236–242.

10.10 Wieselsberger, C.; Wing Resistance near the Ground; NACA TM 77, 1922.

10.11 Hoak, D.E.; USAF Stability and Control Datcom; Air Force Flight Dynamics Laboratory, WPAFB, Ohio, April 1978.

10.12 Roskam, J.; Airplane Design, Part I: Preliminary Sizing of Airplanes; DAR Corporation, 120 East Ninth Street, Suite 2, Lawrence, Kansas, 66044; 1996.

10.13 Loftin, Jr., L.K.; Subsonic Aircraft: Evolution and the Matching of Size to Performance; NASA Reference Publication 1060, 1980.

10.14 Torenbeek, E.; Synthesis of Subsonic Airplane Design; Delft University Press, The Netherlands, 1982

10.15 Jackson, P.; Jane's All The World's Aircraft, 1996–1997; Jane's Information Group Ltd.; Sentinel House, 163 Brighton Road, Coulsdon, Surrey CR5 2 NH, U.K.

CHAPTER 11: RANGE, ENDURANCE AND PAYLOAD–RANGE

In this chapter various methods for determining the range (cruise) and endurance (loiter) characteristics of airplanes are presented. Basic range and endurance equations for propeller driven airplanes and for jet airplanes are discussed in Sections 11.1 and 11.2 respectively. The problem of wing sizing for best range and for best endurance is also addressed in these sections.

The payload which can be carried while flying a given range and/or endurance is of prime interest to commercial and to military operators. The so–called payload–range diagram is discussed in Section 11.3. Various methods for presenting the useful range characteristics of an airplane, accounting for the ability to take–off from given field–lengths are also discussed in Section 11.3.

When flying a given route, airplanes must be operated so that there are adequate fuel reserves on board to cope with various contingencies. Typical of these are:

* Destination airport is closed and landing at an alternate airport is required
* Head–winds are stronger than expected
* An engine fails during cruise and the flight continues with increased drag and increased fuel consumption

The FAR's (Parts 135, 121) have several rules for establishing the required fuel reserves. An overview is given in Section 11.4.

Sizing an airplane to meet certain range and/or endurance requirements (while accounting for fuel used during take–off, climb, descent and landing) is one of several tasks designers must perform during early design of a new airplane. A simple weight sizing method is discussed in Section 11.5.

11.1 PROPELLER DRIVEN AIRPLANES

As an airplane progresses through its mission it burns fuel and, as a result, looses weight. The amount of fuel used during a given part of an airplane mission depends on two factors: the power required, SHP_{reqd} (measured in horsepower), and the specific fuel consumption, c_p (measured in lbs of fuel per shaft–horsepower per hour). The following expression reflects this:

$$\dot{W}_F = SHP_{reqd}\, c_p = \frac{P_{reqd}}{\eta_{p_{installed}}} c_p \qquad (11.1)$$

where: $\dot{W}_F$ is the fuel flow in lbs per hour.

c_p is the specific fuel consumption, in lbs/shp/hr

Range, Endurance and Payload–Range

$\eta_{p_{installed}}$ is the installed propeller efficiency

P_{reqd} is the power required for level flight, in hp, and given by:

$$P_{reqd} = \frac{6,076}{550 \times 60 \times 60} D V = \frac{D V}{326} \quad \text{in hp} \tag{11.2}$$

where: D   is the drag in level flight, measured in lbs

          V   is the level flight speed, measured in knots (nautical miles per hour)

The weight change of the airplane, resulting from this fuel flow is defined as:

$$dW = -\dot{W}_F \, dt = \frac{-P_{reqd}}{\eta_{p_{installed}}} c_p \, dt \tag{11.3}$$

To determine airplane endurance, a relationship between weight and time is needed. Such a relationship is obtained by re–arranging Eqn (11.3):

$$\frac{dt}{dW} = \frac{-\eta_{p_{installed}}}{P_{reqd} \, c_p} \quad \text{in hrs/lbs} \quad \text{(This is known as the endurance factor)} \tag{11.4}$$

To determine airplane range, a relationship between weight and range is needed. Since V = ds/dt and therefore, dt = ds/V, it is possible to cast Eqn (11.3) in the following format:

$$\frac{ds}{dW} = \frac{-V \eta_{p_{installed}}}{P_{reqd} \, c_p} \quad \text{in nm/lbs} \quad \text{(This is known as the range factor)} \tag{11.5}$$

With these fundamental relationships established endurance and range equations can be derived by integration of Eqns (11.4) and (11.5). In Sub–sections 11.1.1 – 11.1.4 this is done in in an approximate manner. This leads to the so–called Breguet equations. The french airplane designer, Louis Breguet, was the first to derive closed form range and endurance equations for an airplane. In general, closed form solutions to Eqns (11.4) and (11.5) cannot be obtained and numerical integration is normally used for accurate calculations. This is discussed in Sub–section 11.1.5.

11.1.1 BREGUET EQUATIONS FOR RANGE AND ENDURANCE

By combining Eqns (11.2) and (11.5) it may be shown that:

$$\frac{ds}{dW} = \frac{326 \times (-\eta_{p_{installed}})}{D c_p} = \frac{-326 \times \eta_{p_{installed}}}{c_p}\left(\frac{L}{D}\right)\left(\frac{1}{W}\right) \tag{11.6}$$

The following definitions are now introduced:

     the weight of the airplane at the start of cruise is: W_{begin}

     the weight of the airplane at the end of cruise is: W_{end}

Next, it will be assumed, that the factors c_p, $\eta_{p_{installed}}$ and L/D are approximately constant during the cruising flight*. The range of the airplane may be found by integration of Eqn (11.6) between these weight limits:

$$R = \int_{W_{begin}}^{W_{end}} ds = \frac{-326x\eta_{p_{installed}}}{c_p}\left(\frac{C_L}{C_D}\right)\int_{W_{begin}}^{W_{end}}\left(\frac{dW}{W}\right) = 326\frac{\eta_{p_{installed}}}{c_p}\frac{C_L}{C_D}\ln\left(\frac{W_{begin}}{W_{end}}\right) \text{ in n.m. (11.7)}$$

In reality, particularly for very long ranges (say, longer than 1,000 n.m.), the factors c_p, $\eta_{p_{installed}}$ and L/D will vary during the flight. Therefore, it is prudent to carry out the integration implied by Eqn (11.7) over a number of range segments. The factors c_p, $\eta_{p_{installed}}$ and L/D must then be re-evaluated for each new range segment.

This procedure is illustrated in Figure 11.1 for the case of three segments. It is a matter of judgment to decide how many segments are adequate.

For each range segment, best range occurs when the combined parameter $(\eta_{p_{installed}}/c_p)(C_L/C_D)$ is maximized.

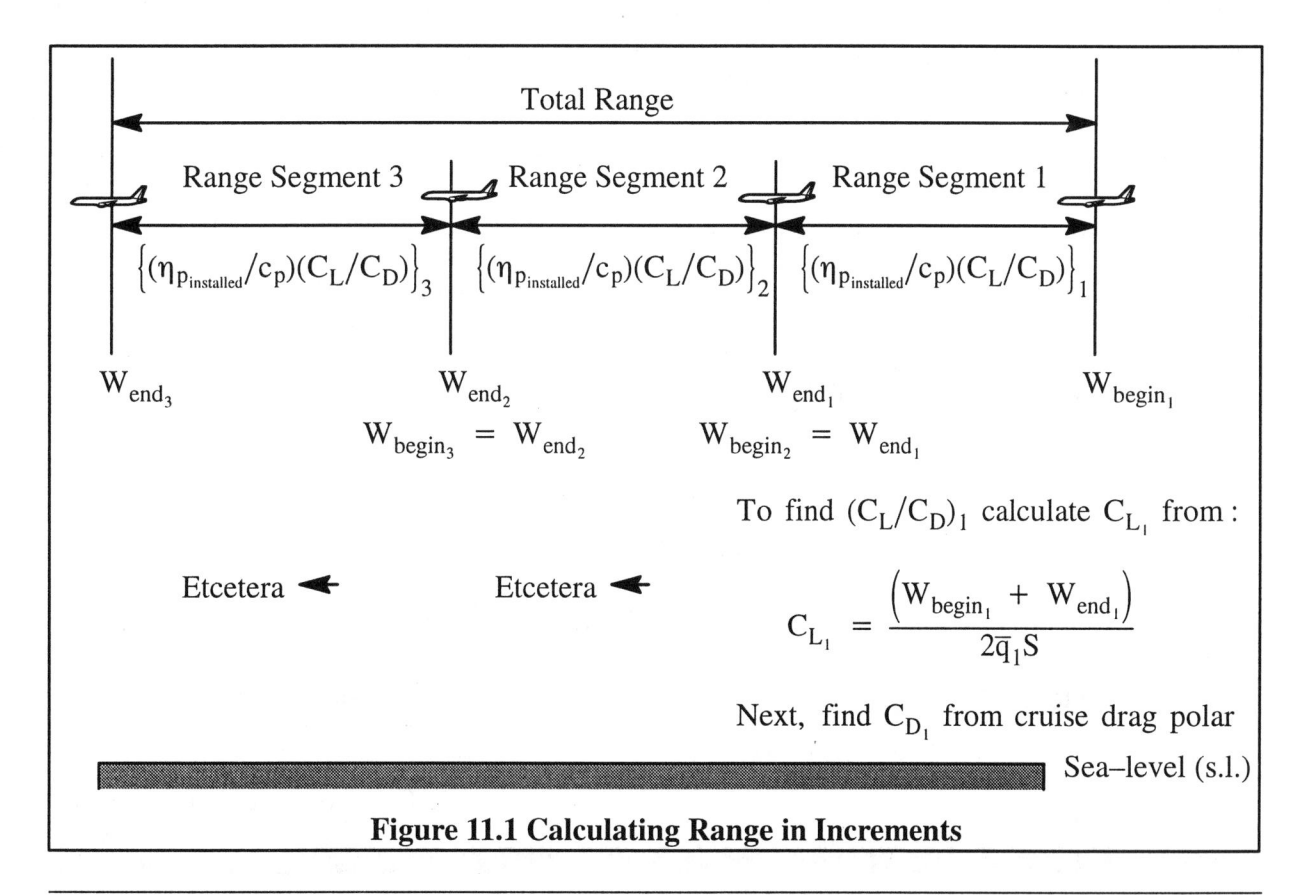

Figure 11.1 Calculating Range in Increments

---

* The reader is asked to show that the constant L/D assumption implies a climbing cruise.

The endurance of the airplane may be found by integrating Eqn (11.4) to produce:

$$E = \int_{W_{begin}}^{W_{end}} dt = -550 \frac{\eta_{p_{installed}}}{c_p} \int_{W_{begin}}^{W_{end}} \frac{dW}{D \times V} = -550 \frac{\eta_{p_{installed}}}{c_p} \int_{W_{begin}}^{W_{end}} \left(\frac{L}{D}\right) \frac{dW}{\left(W \sqrt{\frac{2W}{\varrho C_L S}}\right)} =$$

$$= -550 \frac{\eta_{p_{installed}}}{c_p} \frac{C_L^{3/2}}{C_D} \sqrt{\frac{\varrho S}{2}} \int_{W_{begin}}^{W_{end}} \frac{dW}{W^{3/2}} = 550 \frac{\eta_{p_{installed}}}{c_p} \frac{C_L^{3/2}}{C_D} \sqrt{2\varrho S} \left[\frac{1}{\sqrt{W_{end}}} - \frac{1}{\sqrt{W_{begin}}}\right] =$$

$$= 778 \frac{\eta_{p_{installed}}}{c_p} \frac{C_L^{3/2}}{C_D} \sqrt{\varrho S} \left[\frac{1}{\sqrt{W_{end}}} - \frac{1}{\sqrt{W_{begin}}}\right] \text{ in hours} \qquad (11.8)$$

The integration was carried out by also assuming constant altitude. In reality, particularly for very long endurances (say, longer than 3 hours), the factors ϱ, c_p, $\eta_{p_{installed}}$ and $C_L^{3/2}/C_D$ will vary during the flight. For that reason it is prudent to carry out the integration implied by Eqn (11.8) over a number of endurance segments. The factors c_p, $\eta_{p_{installed}}$ and $C_L^{3/2}/C_D$ must then be re–evaluated for each new endurance segment. The procedure is shown in Figure 11.2 for the case of two segments. It is a matter of judgment to decide how many segments are adequate.

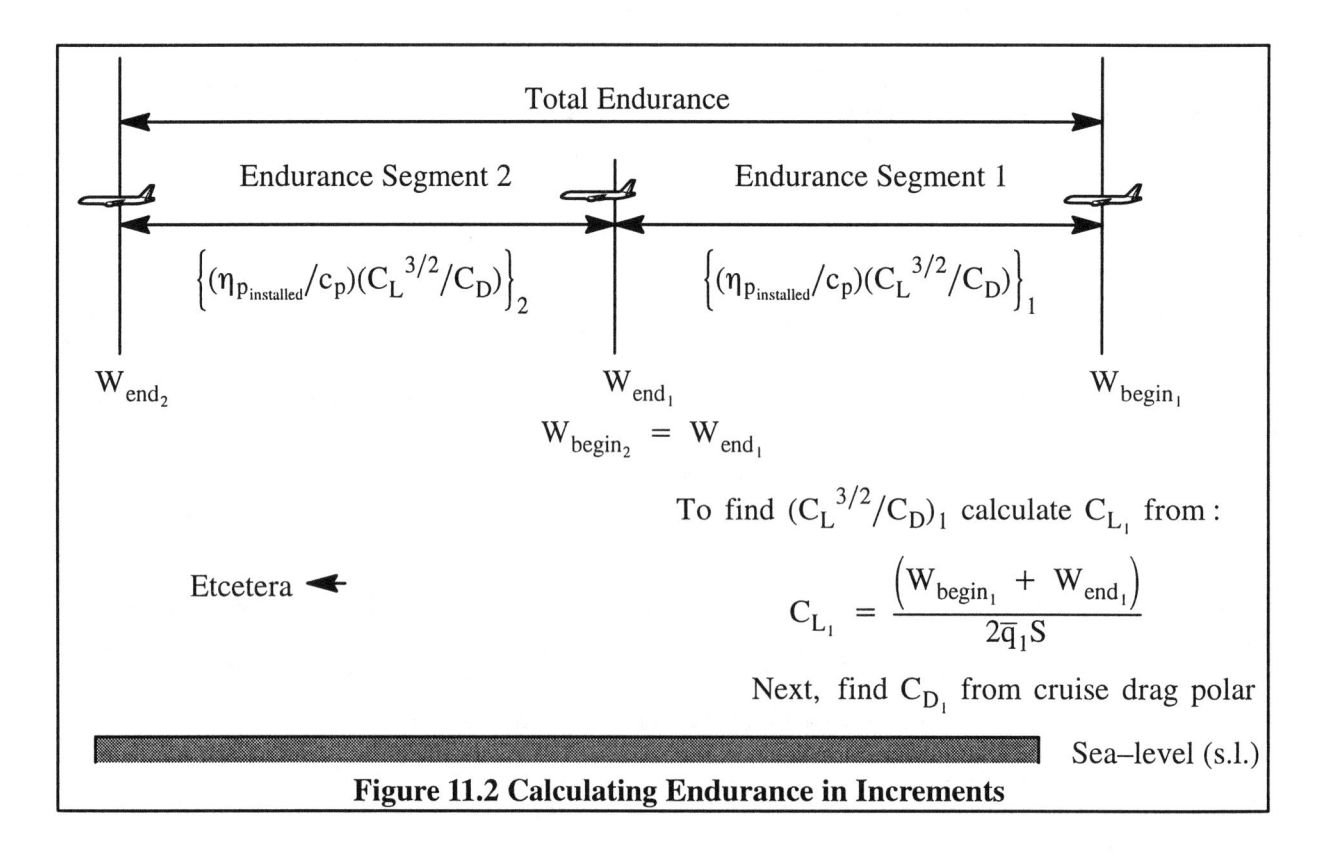

Figure 11.2 Calculating Endurance in Increments

For each endurance segment, best endurance occurs when the combined parameter $(\eta_{p_{installed}}/c_p)(C_L^{3/2}/C_D)$ is maximized, when flying at a given altitude. Note that the atmospheric density, ρ, occurs in the term: $\sqrt{\rho S}$ in the endurance equation. The reader should **not** conclude from this, that best endurance occurs at sea–level.

It should be understood that in both Equations (11.7) and (11.8) the amount of fuel consumed during the range or endurance segments of the mission is:

$$W_{F_{used}} = W_{begin} - W_{end} \tag{11.9}$$

To determine the total amount of fuel needed to complete a given mission (including fuel required to start, taxi, take–off, climb, cruise, loiter, descend, landing and shutdown) the reader is referred to Sub–sub–section 11.1.5.1.

11.1.2. MAXIMUM RANGE AND ENDURANCE FOR PARABOLIC DRAG POLARS

11.1.2.1 Maximum Range for Parabolic Drag Polars

It is noted from the range equation (11.7) that, for given values of c_p and $\eta_{p_{installed}}$ maximum range occurs when flying at the maximum value of the lift–to–drag ratio, L/D. In Chapter 8, maximum L/D and the lift and drag coefficients at which it occurs were shown to take on the following expressions for an airplane with a parabolic drag polar:

$$(C_L/C_D)_{max} = 0.5 \sqrt{\frac{\pi A e}{C_{D_0}}} \tag{11.10}$$

$$C_{L_{L/D_{max}}} = \sqrt{\pi A e C_{D_0}} \qquad\qquad C_{D_{L/D_{max}}} = 2C_{D_0} \tag{11.11}$$

It is of great interest to design an airplane so that it cruises at or close to the best possible lift–to–drag ratio. Flying at best L/D implies flying at constant angle of attack. If a cruising flight is started at the angle of attack for best L/D, to maintain this (in view of the decreasing weight due to fuel consumption) requires a climbing cruise. In the real world of Air Traffic Control, airplanes are not allowed to climb at will. Instead, airplanes are typically relegated to constant altitude cruising flight. If traffic conditions permit, pilots may get permission to change altitude to allow a more favorable L/D to be reached. During a typical trans–oceanic flight this may occur several times.

The latter is one of many reasons why sizing wing and engines of an airplane to achieve best L/D can be a very difficult task. Figures 11.3a–c illustrate three cruise scenarios. In Figure 11.3a, Point A represents the condition for best L/D. This can be achieved only, if the corresponding lift coefficient, $C_{L_{cruise}} = W_{cruise}/\bar{q}S$. In other words, weight, altitude, speed and wing area must be just right, to produce $C_{D_{cruise}} = 2C_{D_0}$. If there is a mis–match in weight, altitude, speed and/or wing

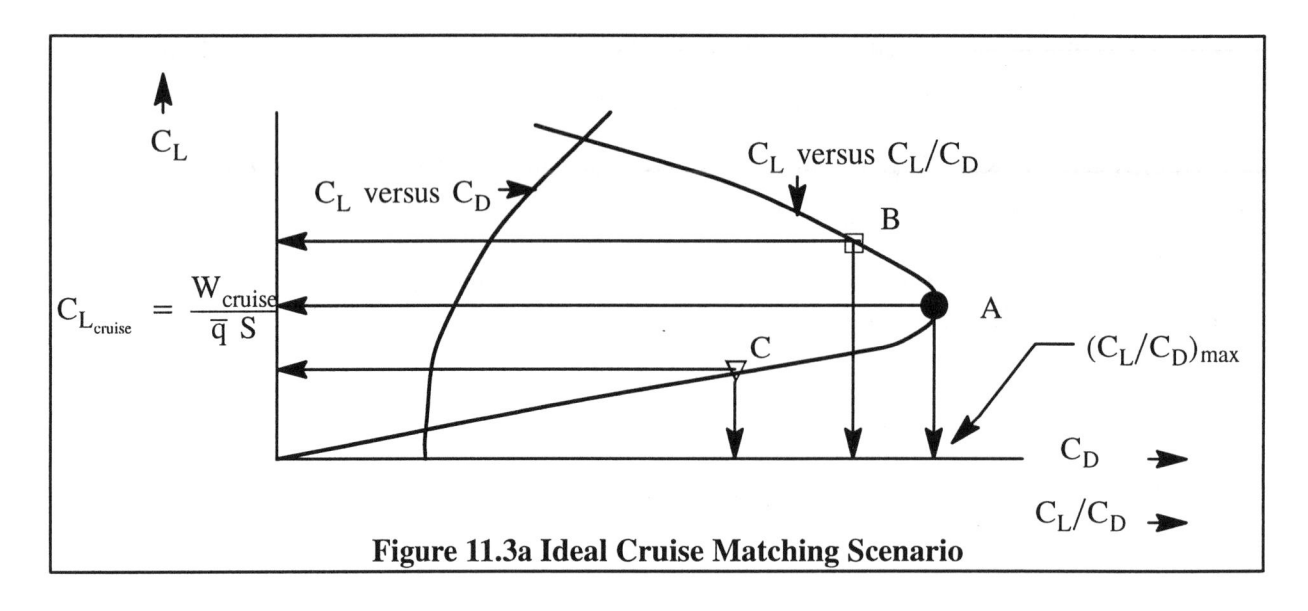

Range, Endurance and Payload–Range

Figure 11.3a Ideal Cruise Matching Scenario

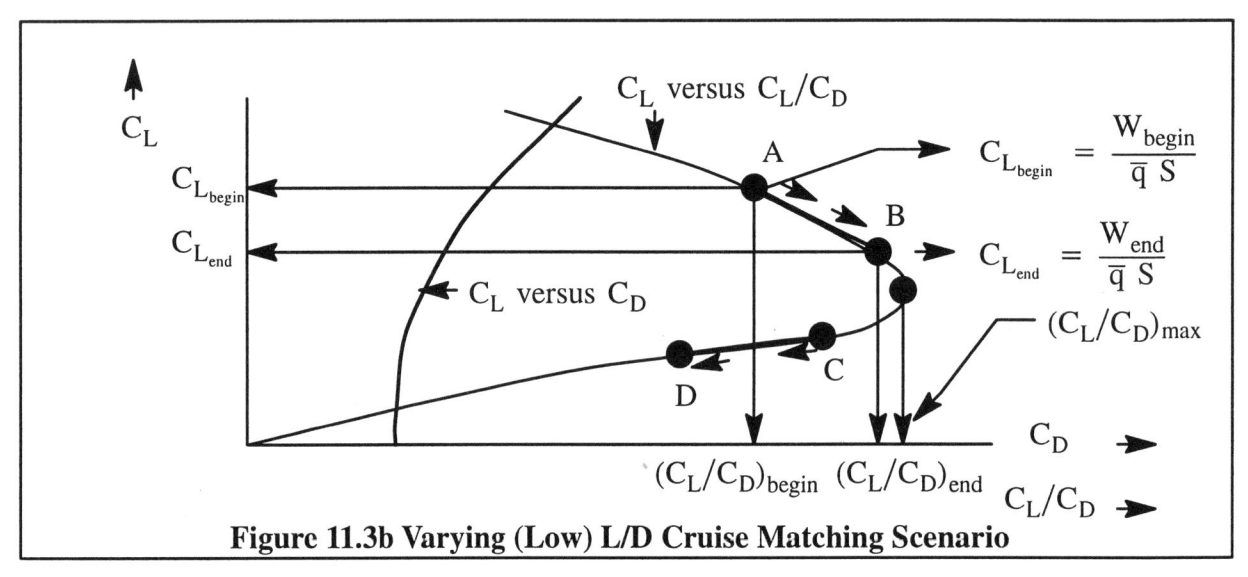

Figure 11.3b Varying (Low) L/D Cruise Matching Scenario

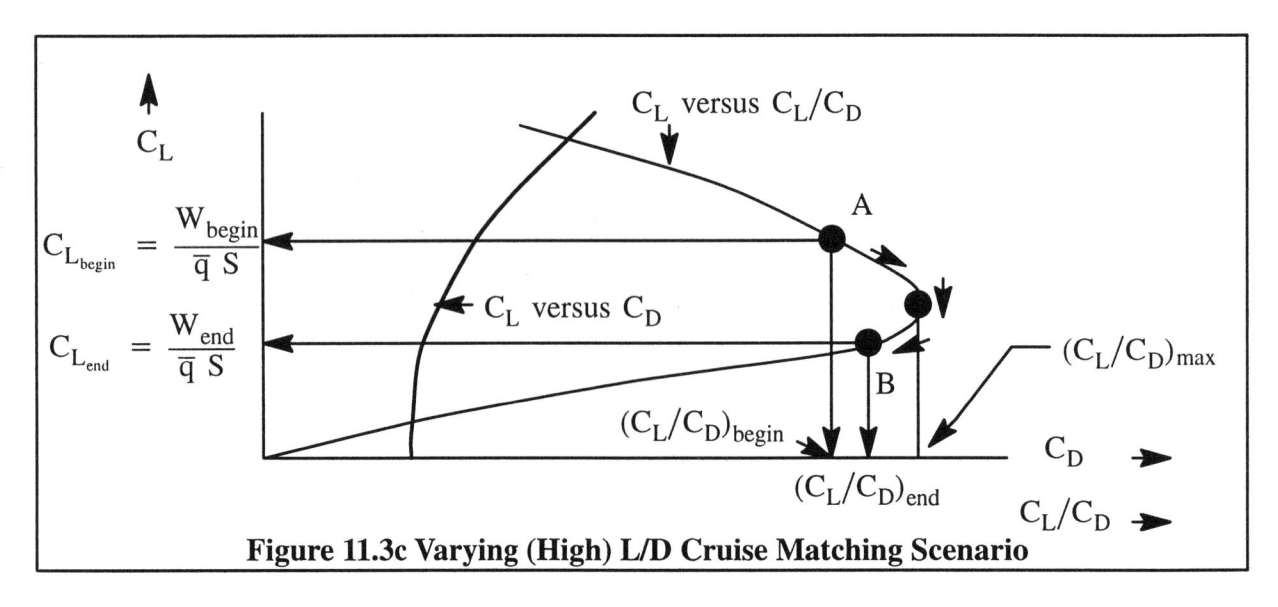

Figure 11.3c Varying (High) L/D Cruise Matching Scenario

area, point A shifts to either point B or point C. Either way, the L/D is lower than the best L/D and the range will be correspondingly less.

In Figure 11.3b two scenarios are depicted:

1) the cruise lift coefficient, which decreases with weight if altitude and speed are maintained, is above the lift coefficient for best L/D. This is the case for A to B.

2) the cruise lift coefficient, which decreases with weight if altitude and speed are maintained, is below the lift coefficient for best L/D. This is the case for C to D.

Scenario C–D is typical for fighter aircraft in the ground attack mode. Because of the very high dynamic pressures, the lift coefficient during a high speed, low altitude operation is typically much below that for best L/D. Scenario C–D is also fairly typical for low altitude general aviation airplanes: they also tend to operate at relatively low lift coefficients.

Scenario A–B is typical for very high altitude airplanes, where the lift coefficient at the beginning of cruise tends to be high due to the high weight.

Figure 11.3c depicts the ideal scenario for a long range cruise airplane which is more or less committed to cruise at constant altitude. By sizing the wing so that the airplane "rounds" the L/D corner during some part of the cruising flight, a good average L/D can be maintained.

These results bring up the important problem of how to size the wing to attain the best possible lift–to–drag ratio. For the case of an airplane with a parabolic drag polar this issue is discussed in Sub–sub–section 11.1.2.2.

11.1.2.2   Sizing the Wing for Maximum Range (Maximum C_L/C_D)

In this discussion it will be assumed that an airplane with a given weight, W, is designed to cruise at a given altitude and Mach number and that it should have a given value of maximum lift–to–drag ratio, $(L/D)_{max}$. Therefore W, $\bar{q}_{cruise}$ and $(L/D)_{max}$ are known. Using Eqn (11.11) it is seen that:

$$C_{L_{L/D_{max}}} = \sqrt{\pi A e C_{D_0}} = \frac{W}{\bar{q}_{cruise} S} \qquad (11.12)$$

The product πAe, with the help of Eqn (11.10) may be written as:

$$\pi Ae = 4 C_{D_0} \left(\frac{L}{D}\right)^2_{max} \qquad (11.13)$$

A problem is, that the zero–lift drag coefficient, C_{D_0}, itself is a function of wing area.

This is easily demonstrated by considering the total airplane equivalent parasite area, f_A, to consist of two components: the wing, f_W, and the rest of the airplane, f_{A-W}:

$$f_A = f_W + f_{A-W} \tag{11.14}$$

The equivalent parasite area, f_A, can be related to the zero–lift drag coefficient to produce:

$$C_{D_0}S = C_{D_{0_W}}S + f_{A-W} \tag{11.15}$$

From this it follows:

$$C_{D_0} = C_{D_{0_W}} + \frac{f_{A-W}}{S} \tag{11.16}$$

Substitution of Eqn (11.16) into Eqn (11.12) and solving for the required wing area yields:

$$S = \frac{\frac{W}{\bar{q}_{cruise}} - 2\left(\frac{L}{D}\right)_{max}f_{A-W}}{2\left(\frac{L}{D}\right)_{max}C_{D_{0_W}}} = \frac{\frac{W}{2\bar{q}_{cruise}} - \left(\frac{L}{D}\right)_{max}f_{A-W}}{\left(\frac{L}{D}\right)_{max}C_{D_{0_W}}} \tag{11.17}$$

Take as an example the Lockheed P–3C Orion. The following data are available for this airplane:

$W \approx 135,000$ lbs $S = 1,300$ ft^2 $f_A = 32$ ft^2 $f_{A-W} = 22$ ft^2

$(L/D)_{max} \approx 13.8$ $M_{cruise} \approx 0.54$ $h = 25,000$ ft

Substitution of these data into Eqn (11.17) produces for the following wing area for best range $S = 1,103$ ft^2, which is below the actual wing area of the airplane. The reason is that the wing of this airplane is sized for best endurance: it is a long endurance patrol airplane.

The reader is reminded of the fact that the wing size may very well be determined from other design considerations, such as take–off and landing fieldlength, fuel volume, maneuverability at altitude and various other considerations.

11.1.2.3 Effect of Wind on Maximum Range

For airplanes with a parabolic drag polar, Hale and Steiger (Ref. 11.1) derived an analytical range equation which includes the effect of wind. They showed that with a tail–wind equal to 1/2 of the best range speed without wind, the range can be improved by 2% if the airspeed is reduced by about 10% from the best range speed without wind.

11.1.2.4 Maximum Endurance for Parabolic Drag Polars

It is noted from the endurance equation (11.8) that, for given values of c_p and $\eta_{p_{installed}}$ maximum range occurs when flying at the maximum value of the parameter $\left(C_L^{3/2}/C_D\right)$. The maximum of this parameter occurs when its square, $\left(C_L^3/C_D^2\right)$, is a maximum. In Chapter 8, maximum $\left(C_L^3/C_D^2\right)$ and the lift and drag coefficients at which it occurs were shown to take on the following expressions for an airplane with a parabolic drag polar:

$$(C_L^3/C_D^2)_{max} = \frac{3}{16}\pi Ae \sqrt{\frac{3\pi Ae}{C_{D_0}}} \tag{11.18}$$

$$C_{L_{\left(C_L^3/C_D^2\right)_{max}}} = \sqrt{3\pi AeC_{D_0}} \qquad C_{D_{\left(C_L^3/C_D^2\right)_{max}}} = 4C_{D_0} \tag{11.19}$$

The wings of conventional transport airplanes are normally sized for best range operation. For fighters and for general aviation airplanes, different wing sizing criteria may apply. However, airplanes with a high altitude, on–station type mission may very well require that the wing be sized for best endurance. Sizing wing area and engines of an airplane to achieve best $\left(C_L^3/C_D^2\right)$ can be a very difficult task. Figure 11.4 shows a typical composite plot of C_D, C_L/C_D and C_L^3/C_D^2 versus the lift coefficient, C_L.

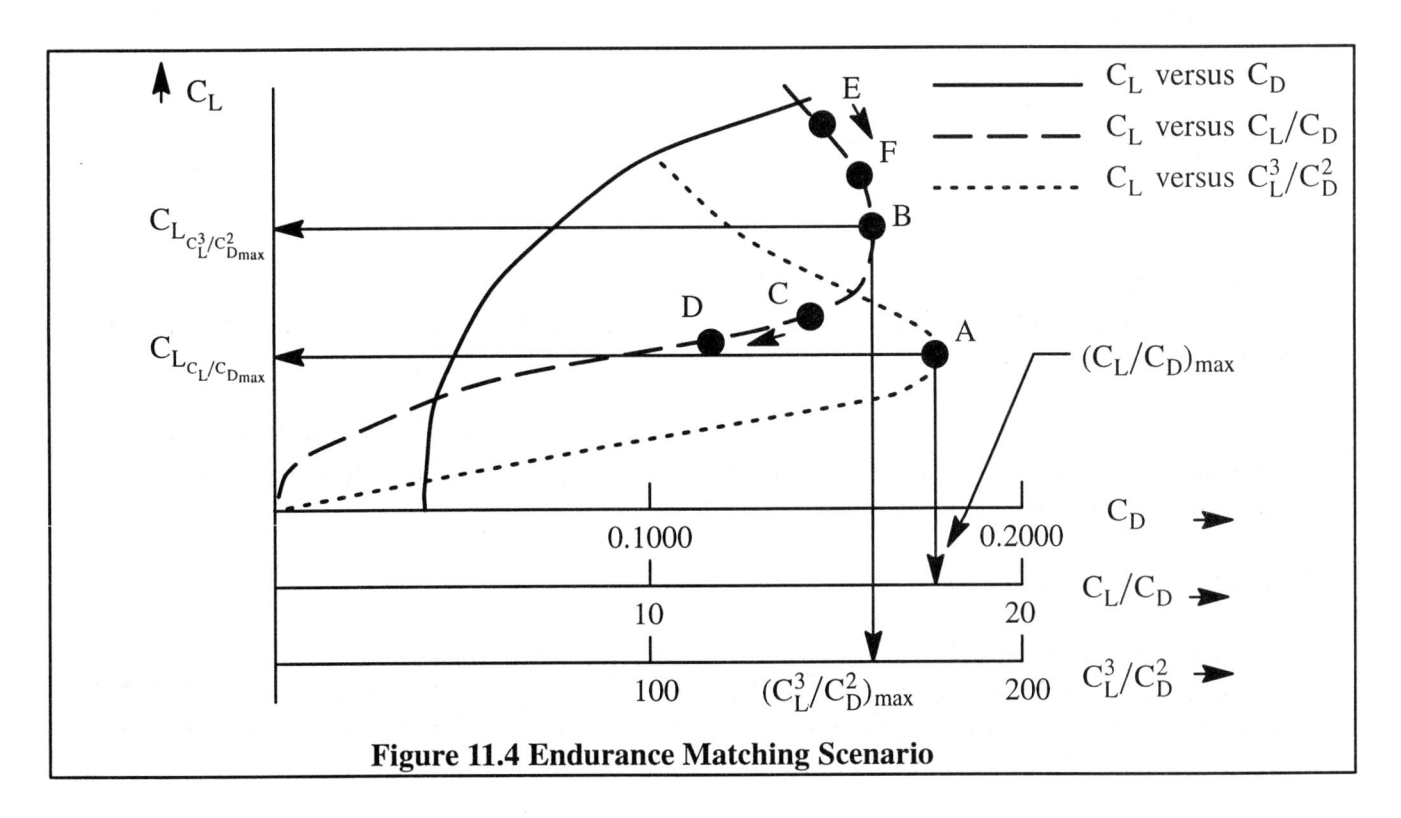

Figure 11.4 Endurance Matching Scenario

In Figure 11.4, Point A represents the condition for best L/D.

Point B in Figure 11.4 represents the condition for best (C_L^3/C_D^2). The latter can be achieved only, if the corresponding lift coefficient, $C_{L_{endurance}} = W_{endurance}/\bar{q}\,S = C_{L_{C_L^3/C_D^2 max}}$

In other words, weight, altitude, speed and wing area must be just right, to produce $C_{D_{endurance}} = 4C_{D_0}$. If there is a mis–match in weight, altitude, speed and/or wing area, point B shifts to either point C or point E. This results in two endurance scenarios: C–D and E–F. Either way, the (C_L^3/C_D^2) is lower than the best (C_L^3/C_D^2) and the endurance will be correspondingly less.

These results raise the important question of wing sizing to attain the best possible value of (C_L^3/C_D^2). The issue is addressed for the case of a parabolic drag polar in Sub–sub–section 11.1.2.5.

11.1.2.5 Sizing the Wing for Maximum Endurance (Maximum C_L^3/C_D^2)

In this discussion it will be assumed that an airplane with a given weight, W, is designed to cruise at a given altitude and Mach number and that it should have a given maximum value of the parameter (C_L^3/C_D^2). Therefore W, $\bar{q}_{cruise}$ and $(C_L^3/C_D^2)_{max}$ are known. Using Eqn (11.19) it is seen that:

$$C_{L_{(C_L^3/C_D^2)_{max}}} = \sqrt{3\pi A e C_{D_0}} = \frac{W}{\bar{q}_{endurance} S} \qquad (11.20)$$

The product πAe, with the help of Eqn (11.18) may be written as:

$$\pi Ae = \left[\frac{256}{27} C_{D_0} \left(\frac{C_L^3}{C_D^2}\right)_{max}^2\right]^{1/3} \qquad (11.21)$$

It was recognized before, that the zero–lift drag coefficient, C_{D_0}, itself is a function of wing area as seen from Eqn (11.16). By combining Eqns (11.16), (11.20) and (11.21) the following polynomial equation in the wing area, S, is obtained:

$$S^3 + \left[\frac{2f_{A-W}}{C_{D_{0_w}}}\right]S^2 + \left[\frac{f_{A-W}}{C_{D_{0_w}}}\right]^2 S - \left\{\frac{(W/\bar{q}_{cruise})^3}{16.0\,(C_L^3/C_D^2)_{max}(C_{D_{0_w}})^2}\right\} = 0 \qquad (11.22)$$

For a given airplane design, W, $\bar{q}_{cruise}$ and $(C_L^3/C_D^2)_{max}$ as well as f_{A-W} and $C_{D_{0_W}}$ will be known. The wing area, S, required for best endurance can then be solved from Eqn (11.22).

As an example, consider a high altitude antenna platform. For this airplane, the following data are assumed:

$W \approx 10,000$ lbs $\qquad f_{A-W} = 5$ ft^2 $\qquad \left(C_L^3/C_D^2\right)_{max} \approx 400$

$M_{loiter} \approx 0.25$ $\qquad h = 35,000$ ft $\qquad C_{D_{0_W}} \approx 0.0040$

Substitution into Eqn (11.22) yields the following result:

$$S^3 + 2,500 \, S^2 + 1,250^2 \, S - 944 \times 10^6 = 0 \qquad (11.23)$$

The solution for wing area for best endurance is: S = 363 ft^2.

It is of interest to study a numerical example of a plot of C_D, C_L/C_D and C_L^3/C_D^2 versus the lift coefficient, C_L. This is done for a business turbo–prop in Sub–sub–section 11.1.2.6.

11.1.2.6 Numerical Example of a Composite Plot of C_D, C_L/C_D and C_L^3/C_D^2 versus C_L

Figure 11.5 shows a composite plot of C_D, C_L/C_D and C_L^3/C_D^2 versus the lift coefficient, C_L, for an advanced, corporate turbo–prop airplane in cruise. Note the numerical value of the lift coefficient for maximum C_L/C_D: 0.94.

The airplane is assumed to cruise at an altitude of 39,000 ft and at a long range Mach number of 0.60. A typical begin weight is 11,300 lbs and a typical end weight is 9,300 lbs. The corresponding cruise lift coefficients are 0.62 and 0.51 respectively. Points A and B represent these cruise lift coefficients. It is seen in Figure 11.5 that this airplane is of the type represented by scenario C–D in Figure 11.3b.

It is also observed from Figure 11.5, that for best endurance at 39,000 ft, the airplane would have to fly at a lift coefficient of around 1.62. This would correspond to a Mach number of around 0.37. The maximum, trimmable lift coefficient for this airplane is around 1.8. A lift coefficient of 1.62 still leaves some room for maneuvering so that it would be possible to do this.

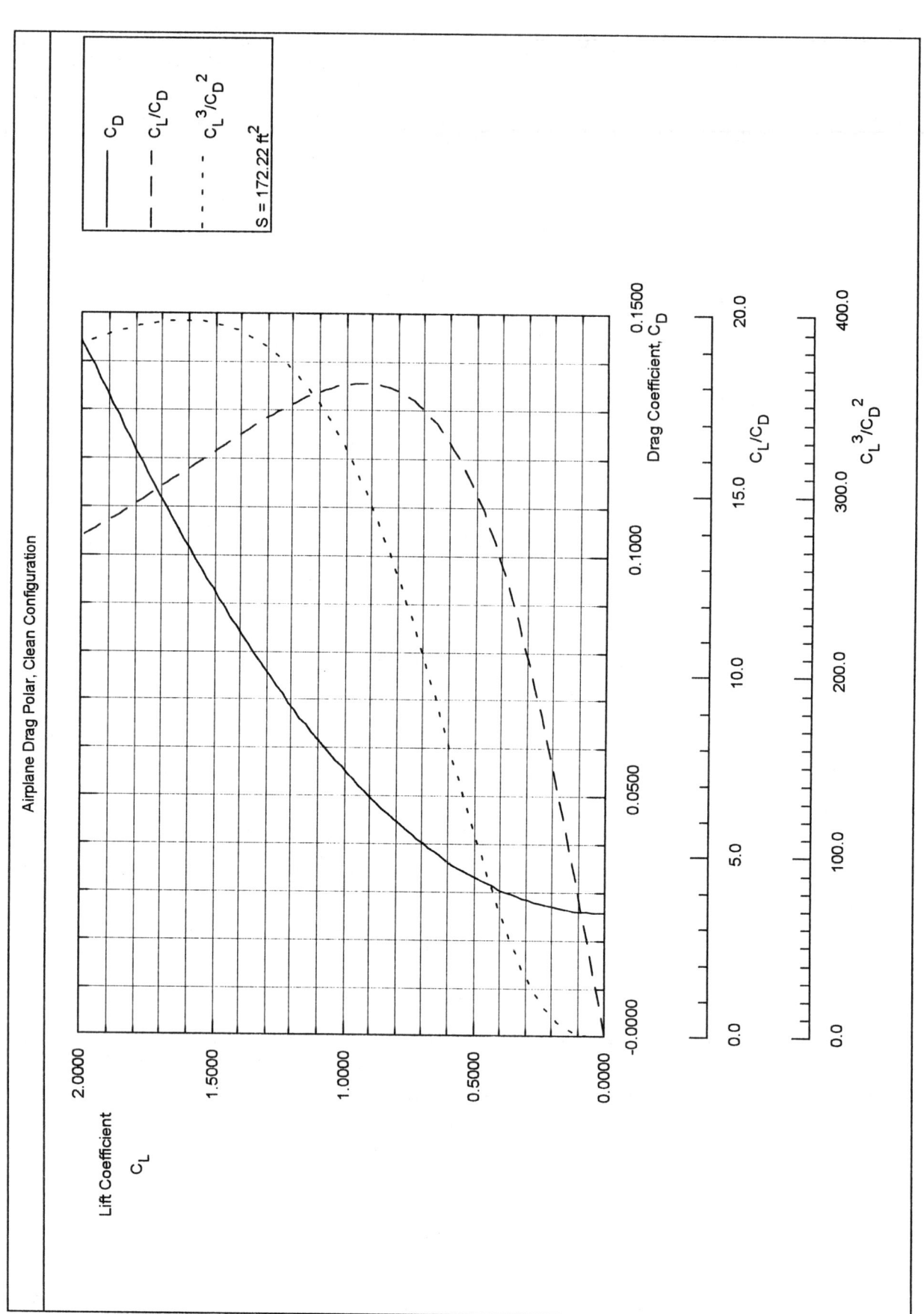

Figure 11.5 Composite Plot of C_D, C_L/C_D and C_L^3/C_D^2 versus C_L for a Business Turboprop in Cruise

11.1.3. APPLICATION OF THE RANGE AND ENDURANCE EQUATIONS TO A PROPELLER DRIVEN AIRPLANE WITH A PARABOLIC DRAG POLAR

Example 11.1: A propeller driven cargo airplane has the following characteristics:

$W_{begin} = 30,000$ lbs $\qquad c_p = 0.45$ lbs/shp/hr

$C_D = 0.0200 + 0.05\, C_L^2 \qquad \eta_{p_{installed}} = 0.87$

$S = 300$ ft$^2 \qquad h = 28,000$ ft

The airplane is supposed to carry 3,000 lbs of supplies, airdrop these supplies at a distance of 1,500 nautical miles from base and return to base without re–fuelling. Determine:

1) The total amount of cruise fuel consumed
2) the total flying time

Solution: The calculation will be performed in two parts:

a) Outbound mission b) Return mission

a) Outbound mission

The range Eqn (11.7) can be re–written as:

$$\ln\left(\frac{W_{begin}}{W_{end}}\right) = \frac{R\, c_p}{326\, \eta_{p_{installed}}} \frac{1}{C_L/C_D} \quad \text{where: R is in nautical miles}$$

For best range, C_L/C_D must be at its maximum. From the drag polar it is seen that: $C_{D_0} = 0.0200$ and $\pi Ae = 1/0.05 = 20.0$. Therefore, according to Eqn (11.10): $(C_L/C_D)_{max} = 0.5\sqrt{20.0/0.0200} = 15.8$.

Substitution of the required data in the range equation yields:

$$\ln\left(\frac{W_{begin}}{W_{end}}\right) = \frac{1,500 \times 0.45}{326 \times 0.87} \frac{1}{15.8} = 0.1506$$

Therefore: $W_{begin}/W_{end} = 1.1626$ which yields:

$W_{end} = 30,000/1.1626 = 25,805$ lbs

Amount of fuel used during the outbound mission: $30,000 - 25,805 = 4,195$ lbs.

Flying at maximum C_L/C_D, Eqn (11.11) yields:

$$C_L = \sqrt{20.0 \times 0.0200} = 0.632$$

The initial flight speed for the outbound mission is:

$$V_{begin} = \sqrt{\frac{2 \times 30{,}000}{0.0009567 \times 0.632 \times 300}} = 575 \text{ fps} = 341 \text{ kts}$$

The flying time is estimated to be: $1500/341 = 4.4$ hrs.

The reader is asked to verify this by using the endurance equation (11.8).

Hint: compute C_L^3/C_D^2 at the lift coefficient for best C_L/C_D!

b) Return mission

For the return mission: $W_{begin} = 25{,}805 - 3{,}000 = 22{,}805$ lbs

Assuming the return mission is also conducted at best C_L/C_D:

$W_{begin}/W_{end} = 1.1626$ still applies, so that:

$$W_{end} = 22{,}805/1.1626 = 19{,}616 \text{ lbs}$$

Amount of fuel used during the return mission: $22{,}805 - 19{,}616 = 3{,}189$ lbs.

The initial flight speed for the return mission is:

$$V_{begin} = \sqrt{\frac{2 \times 22{,}805}{0.0009567 \times 0.632 \times 300}} = 501 \text{ fps} = 297 \text{ kts}$$

The flying time is estimated to be: $1500/297 = 5.1$ hrs.

The reader is asked to verify this by using the endurance equation (11.8).

Hint: compute C_L^3/C_D^2 at the lift coefficient for best C_L/C_D!

The total fuel used in the mission is: $4{,}195 + 3{,}189 = 7{,}384$ lbs

The total flying time for the mission is: $4.4 + 5.1 = 9.5$ hrs.

11.1.4. MAXIMUM RANGE AND ENDURANCE FOR AIRPLANES WITH NON–PARABOLIC DRAG POLARS INCLUDING THE EFFECT OF WIND

For airplanes with non–parabolic drag polars a numerical solution is required. The drag polar can still be converted to the form illustrated in Figure 11.5.

Consider an airplane with the drag polar of Figure 11.6. Table 11.1 shows how this drag polar can be converted into lines of (C_L/C_D) and (C_L^3/C_D^2), also plotted versus lift coefficient. Points for best (C_L/C_D) and best (C_L^3/C_D^2) are now determined graphically as shown in Figure 11.6.

By computing the power required and power available data for an airplane with a given, general drag polar it is possible to determine the conditions for best range and for best endurance graphically.

From Eqns (9.13) and (9.14) it can be seen that:

$$\frac{P_{reqd}}{V} = \frac{W}{C_L/C_D} \qquad (11.24)$$

According to Eqn (11.7), maximum range occurs at the best value of (C_L/C_D). Eqn (11.24) shows that maximum range therefore occurs at the best value of P_{reqd}/V. This is convenient, because the best value of P_{reqd}/V then occurs at the tangent of a line drawn through the origin to a plot of power required, P_{reqd}, versus speed, V. This is shown graphically in Figure 11.7.

Point A represents the speed for best range if there is no wind. To determine the effect of tail wind (Point B) and head wind (Point C) the origin of the P_{reqd}, versus speed, curve is simply shifted as shown in Figure 11.7.

From Eqn (9.45) it is seen that:

$$P_{reqd} = \frac{W}{550}\sqrt{\frac{2W}{\varrho S}}\left[\frac{C_D}{C_L^{3/2}}\right] \qquad (11.25)$$

With Eqns (11.8) and (11.25) it is clear that best endurance occurs at the speed for minimum power required. That speed is indicated as Point D in Figure 11.7. Observe, that the speed for best endurance does not change with wind speed, while the speed for best range does!

Once the flight speed for best range has been determined from a graph such as Figure 11.7 the maximum range can be calculated from:

Figure 11.6 Example of a Non-Parabolic Drag Polar

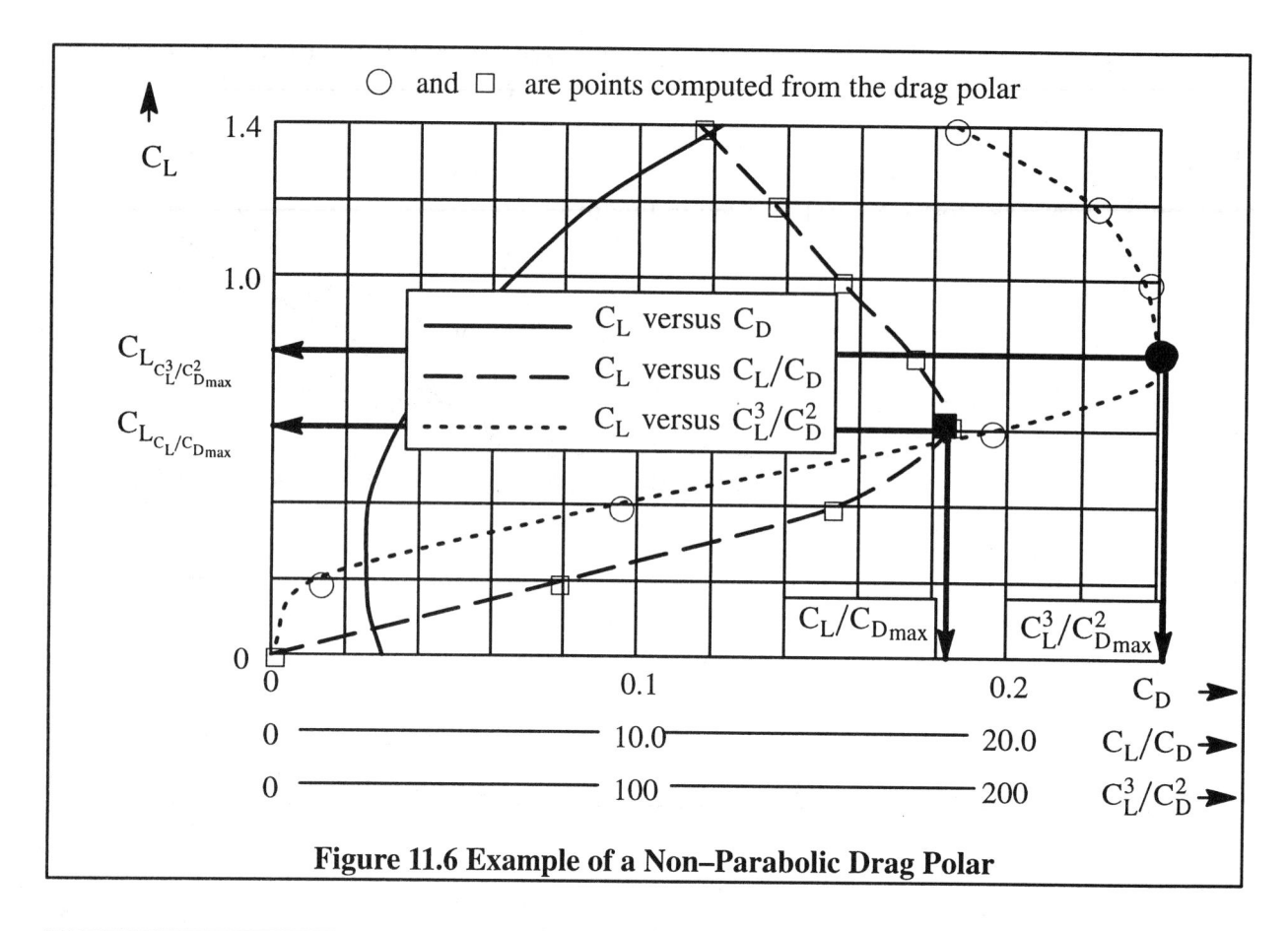

○ and □ are points computed from the drag polar

Legend:
- —— C_L versus C_D
- – – – C_L versus C_L/C_D
- ······ C_L versus C_L^3/C_D^2

Axes: C_D, C_L/C_D, C_L^3/C_D^2

Indicated points: $C_{L_{C_L/C_{D_{max}}}}$, $C_{L_{C_L^3/C_D^2_{max}}}$, $C_L/C_{D_{max}}$, $C_L^3/C_D^2_{max}$

Table 11.1 Calculation of C_L/C_D and C_L^3/C_D^2 From a Non-Parabolic Drag Polar

C_L	C_D Read From Figure 11.6	C_L/C_D – – – □ in Figure 11.6	C_L^3/C_D^2 ······ ○ in Figure 11.6
0	0.028	0	0
0.2	0.025	8.0	12.8
0.4	0.026	15.4	94.7
0.6	0.033	18.2	198.3
0.8	0.046	17.4	242.0
1.0	0.065	15.4	236.7
1.2	0.088	13.6	223.1
1.4	0.122	11.5	184.0

Range, Endurance and Payload–Range

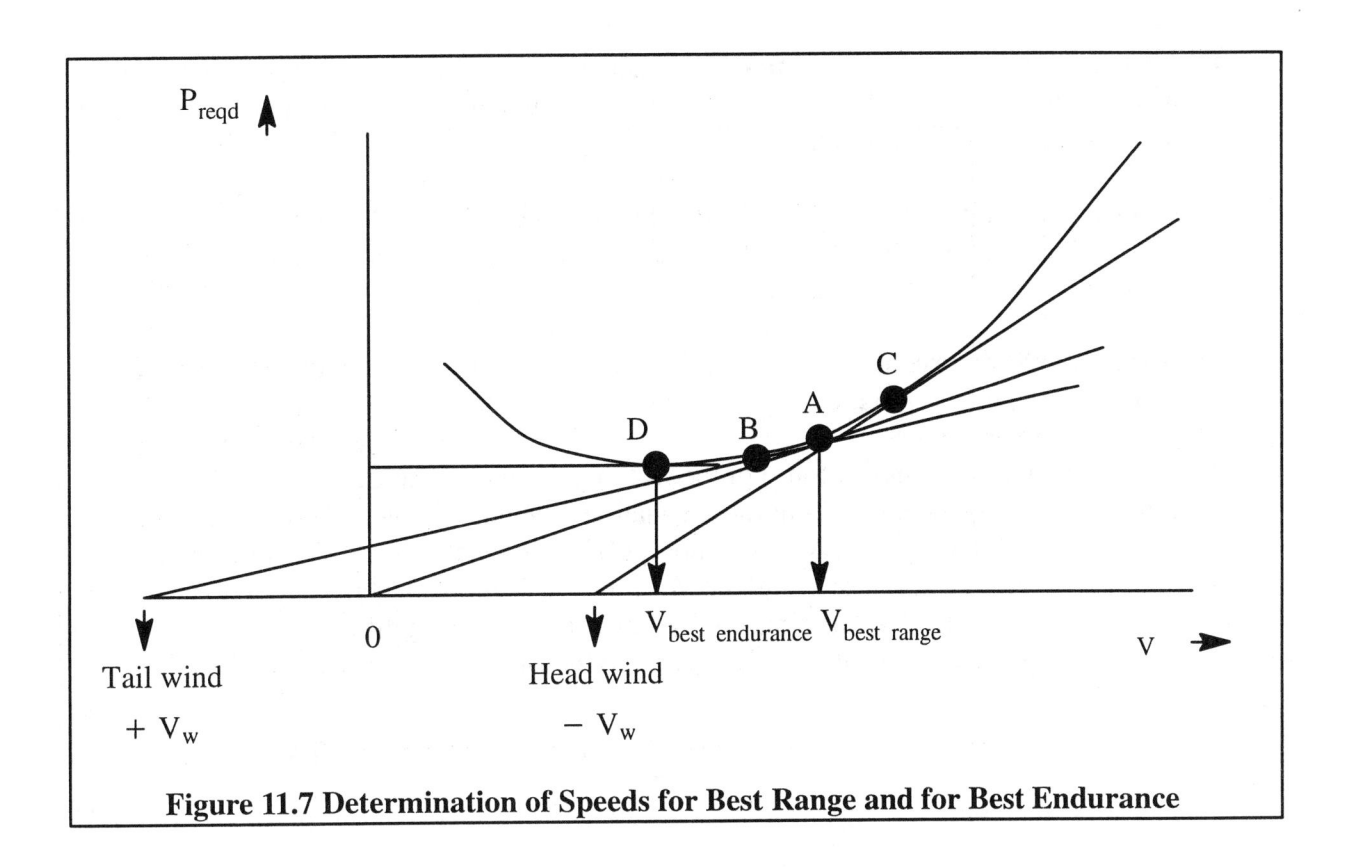

Figure 11.7 Determination of Speeds for Best Range and for Best Endurance

$$R_{max\ with\ wind} = 326\frac{\eta_p}{c_p}\frac{C_L}{C_D} \ln \frac{W_{begin}}{W_{end}} \pm 778\frac{V_w \eta_p}{c_j}\left[\frac{C_L^{3/2}}{C_D}\right]\sqrt{\varrho S}\left[\frac{1}{\sqrt{W_{end}}} - \frac{1}{\sqrt{W_{begin}}}\right] \text{ in nm} \quad (11.26)$$

where: + is for tail wind and – is for head wind, while V_w, is in knots.

The reader must ensure that the numerical values for (C_L/C_D) and for $(C_L^{3/2}/C_D)$ which are used in Eqn (11.26) are those corresponding to the proper speed for best range from Figure 11.7.

It is clear from the material presented sofar that, for a given propeller efficiency, $\eta_{p_{installed}}$, and specific fuel consumption, c_p, maximum range or maximum endurance is obtained by flying at a constant angle of attack. Only by doing that, can the parameters (C_L/C_D) and for $(C_L^{3/2}/C_D)$ be kept at their maximum values.

It is useful to consider the implications of this in view of the fact that L = W must also be satisfied at each point in the cruise or loiter part of the mission:

$$L = W = \frac{1}{2}\varrho V^2 C_L S = 1,481.3\ \delta\ M^2\ C_L\ S \quad (11.27)$$

By examining Eqns (11.6) and (11.27) side–by–side it is seen that flight for maximum range can be achieved only under the following scenarios:

1) **Constant speed or Mach number:** this requires the cruise altitude of the airplane to be steadily increased as fuel is consumed which leads to a decrease in weight, W. This is the so–called cruise–climb flight.

2) **Constant altitude:** this requires the speed (or Mach number) of the airplane to be steadily reduced as fuel is consumed which leads to a decrease in weight, W. This is the so–called constant-altitude flight.

3) **Adjust dynamic pressure:** this requires the dynamic pressure to be adjusted inversely proportional to the weight which decreases due to fuel consumption.

It should be emphasized that, because of the fact that the endurance Eqn (11.8) was derived by integration at constant altitude. This endurance equation is therefore valid only for loiter at constant altitude. This requires the speed (or Mach number) of the airplane to be steadily reduced as fuel is consumed which leads to a decrease in weight, W. Unless a flight management system is available to accomplish this automatically, loitering, while continually adjusting the speed, may not be a reasonable procedure. This does point out a limitation of the Breguet equations as derived here.

Finally, it is clear from Eqn (11.7) that to optimize range, if propeller efficiency, $\eta_{p_{installed}}$, and specific fuel consumption, c_p, are adjustable during cruise (they are in most airplanes) it is the combined factor $(\eta_{p_{installed}}/c_p)(C_L/C_D)$ which must be optimized. This was already pointed out in the discussion of range segments on page 372.

In the case of high speed, propeller driven airplanes, the reader should also realize that the lift-to–drag ratio itself becomes a function of Mach number as a result of compressibility effects.

11.1.5. ACCURATE DETERMINATION OF RANGE AND ENDURANCE BY NUMERICAL INTEGRATION OF SPECIFIC RANGE AND ENDURANCE

The specific range, S.R., of an airplane is defined as the number of nautical miles which can be flown per pound of fuel. This quantity is also known as the range factor. Remembering that dW is negative (due to fuel usage) it follows from Eqn (11.6) that:

$$\text{S.R.} = \frac{-ds}{dW} = \frac{V\eta_{p_{installed}}}{P_{reqd}c_p} = \frac{V}{\dot{W}_F} \quad \text{in nm/lbs} \tag{11.28}$$

where: V is the speed in nm/hr

$\dot{W}_F$ is the fuel flow in lbs/hr

The S.R. can be calculated for a range of airplane weights, speeds and altitudes. From these calculations it is possible to determine optimum range flight paths. The range can be determined accurately by integration:

$$R = \int_{W_{CR_{begin}}}^{W_{CR_{end}}} S.R. \, dW \tag{11.29}$$

The specific endurance, S.E., of an airplane is defined as the number of hours which can be flown per lbs of fuel. This quantity is also known as the endurance factor. Remembering that dW is negative (due to fuel usage) it follows from Eqn (11.3) that:

$$S.E. = \frac{-dt}{dW} = \frac{1}{\dot{W}_F} \quad \text{in hrs/lbs} \tag{11.30}$$

The S.E. can be calculated for a range of airplane weights, speeds and altitudes. From these calculations it is possible to determine optimum endurance flight paths. The endurance can be determined accurately by integration:

$$E = \int_{W_{LTR_{begin}}}^{W_{LTR_{end}}} S.E. \, dW \tag{11.31}$$

In principle, Eqns (11.28) and (11.30) can be used to determine range and endurance of airplanes. However, in reality, doing this for an arbitrary airplane mission is a fairly complicated process. For that reason, step–by–step procedures for determining range and endurance are given in Sub–sub–sections 11.1.5.1 and 11.1.5.2 respectively..

11.1.5.1 Step–by–step Procedure for Determining Range

Step 1: Define a mission profile for the airplane. Figure 11.8 shows two example mission profiles: one for a commercial transport airplane and the other for an attack airplane. The procedure outlined here is for a commercial transport.

Note the definition of the mission range, $R_{mission}$.

Step 2: Determine the operating empty weight of the airplane, W_{OWE}. The operating empty weight of an airplane is the sum of the empty weight, W_E, and the trapped fuel and oil weight, W_{tfo}. Note: the weight of the cabin and cockpit crew is included in the payload weight.

$$W_{OWE} = W_E + W_{tfo} \tag{11.32}$$

Step 3: Specify the sum of the payload and crew weight to be carried over the mission range, $W_{PL} + W_{crew}$. This weight would include such items as: passengers and their luggage, cockpit and cabin crew and their luggage; cargo, revenue mail etc.

Step 4: Specify the amount of mission fuel to be carried by the airplane, W_F.

Commercial Transport

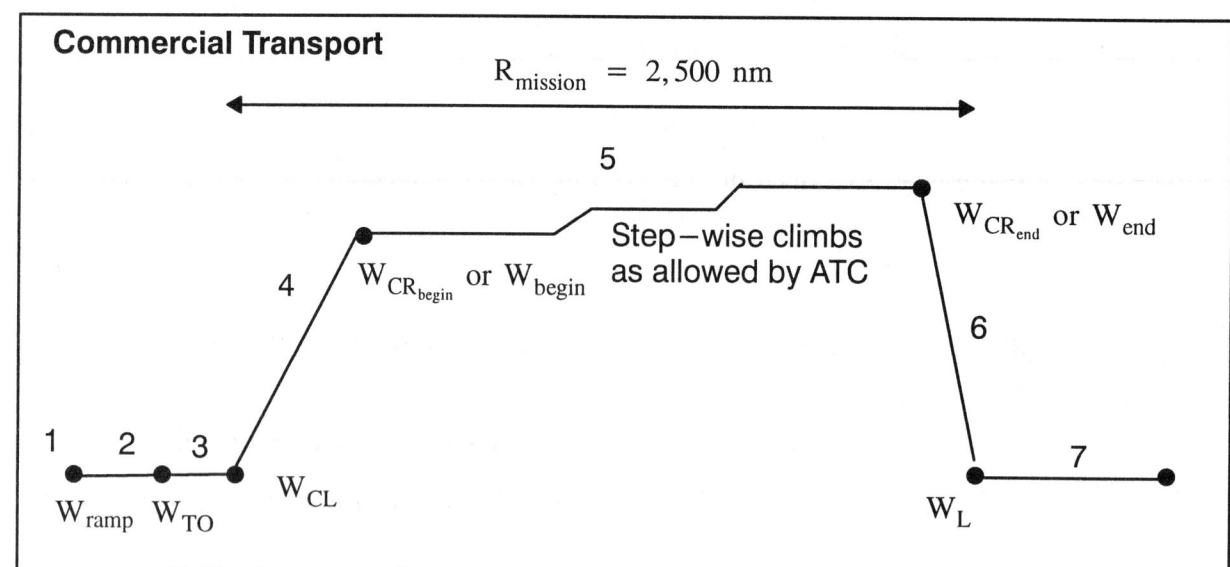

$R_{mission} = 2{,}500$ nm

$W_{CR_{begin}}$ or W_{begin}

Step-wise climbs as allowed by ATC

$W_{CR_{end}}$ or W_{end}

W_{ramp} W_{TO} W_{CL} W_L

1) Engine start and warm-up
2) Taxi
3) Takeoff
4) Climb to 45,000 ft
5) Cruise
6) Descent
7) Landing, taxi, shutdown

Note: no reserve mission shown

Attack Bomber

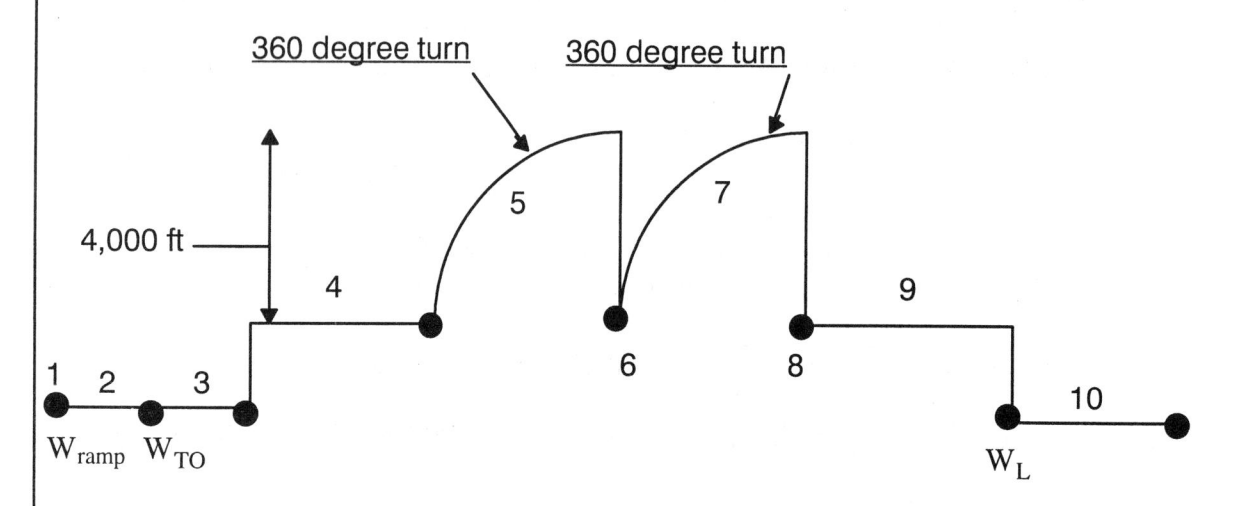

360 degree turn

360 degree turn

4,000 ft

W_{ramp} W_{TO} W_L

1) Engine start and warm-up
2) Taxi
3) Takeoff and accelerate to 350 kts
4) Dash 200 nm at 350 kts
5) 360 degree, sustained, 3g turn at 350 kts including a 4,000 ft altitude gain
6) Release 2 bombs and fire 50% ammo
7) 360 degree, sustained, 3g turn at 350 kts, at sea-level including a 4,000 ft altitude gain
8) Release 2 bombs and fire 50% ammo
9) Dash 200 nm at 350 kts
10) Landing, taxi, shutdown (no reserves)

Figure 11.8 Typical Airplane Mission Profiles

Step 5: Calculate the ramp weight of the airplane, W_{ramp}:

$$W_{ramp} = W_{OWE} + W_{PL} + W_{crew} + W_F \quad (11.33)$$

This ramp weight must satisfy the following condition:

$$W_{ramp} \leq W_{ramp_{max}} \quad (11.34)$$

where: $W_{ramp_{max}}$ is the maximum allowable ramp weight of the airplane.

Step 6: Determine the amount of fuel required to start the engines and warm–up, $W_{F_{sewu}}$, plus the amount of fuel required to taxi to the take–off position, $W_{F_{taxi}}$, and then calculate the take–off weight, W_{TO} from:

$$W_{TO} = W_{ramp} - W_{F_{sewu}} - W_{F_{taxi}} \quad (11.35)$$

This take–off weight must satisfy the following condition:

$$W_{TO} \leq W_{TO_{max}} \quad (11.36)$$

where: $W_{TO_{max}}$ is the maximum allowable take–off weight of the airplane.

Step 7: Determine the amount of fuel required for take–off, $W_{F_{TO}}$. This may be done with the following equation:

$$W_{F_{TO}} = \dot{W}_{F_{TO}} t_{TO} \quad (11.37)$$

where: $\dot{W}_{F_{TO}}$ is the fuel flow during the take–off in lbs/hr

t_{TO} is the time required for take–off according to Eqn (10.49)

Step 8: Determine the amount of fuel required to climb and accelerate to the initial cruise altitude and speed. Also, determine the range covered during the climb. Methods for determining these quantities are discussed in Chapter 9.

Eqn (9.69) can be used for the calculation of the climb fuel weight, $W_{F_{CL}}$.

Eqn (9.68) can be used for the calculation of the horizontal distance covered during the climb (climb range), R_{CL}.

It is advisable to calculate the fuel required to climb for a range of take–off weights and to a range of cruise altitudes and cruise speeds. The engine throttle and/or propeller settings used may represent additional variables to be considered.

Step 9: Determine the weight of the airplane at the start of initial cruise:

$$W_{CR_{begin}} = W_{begin} = W_{TO} - W_{F_{TO}} - W_{F_{CL}} \qquad (11.38)$$

where: W_{TO} follows from Step 6

$W_{F_{TO}}$ follows from Step 7

$W_{F_{CL}}$ follows from Step 8.

It is advisable to define a range of starting weights and cruise flight conditions.

Step 10: Determine the propeller efficiency as a function of altitude and speed. Depending on the engine type used, the propeller efficiency will also be a function of the throttle setting employed. For that reason, the propeller efficiency data are often plotted against speed for a range of throttle settings. For a generic example, see Figure 11.9.

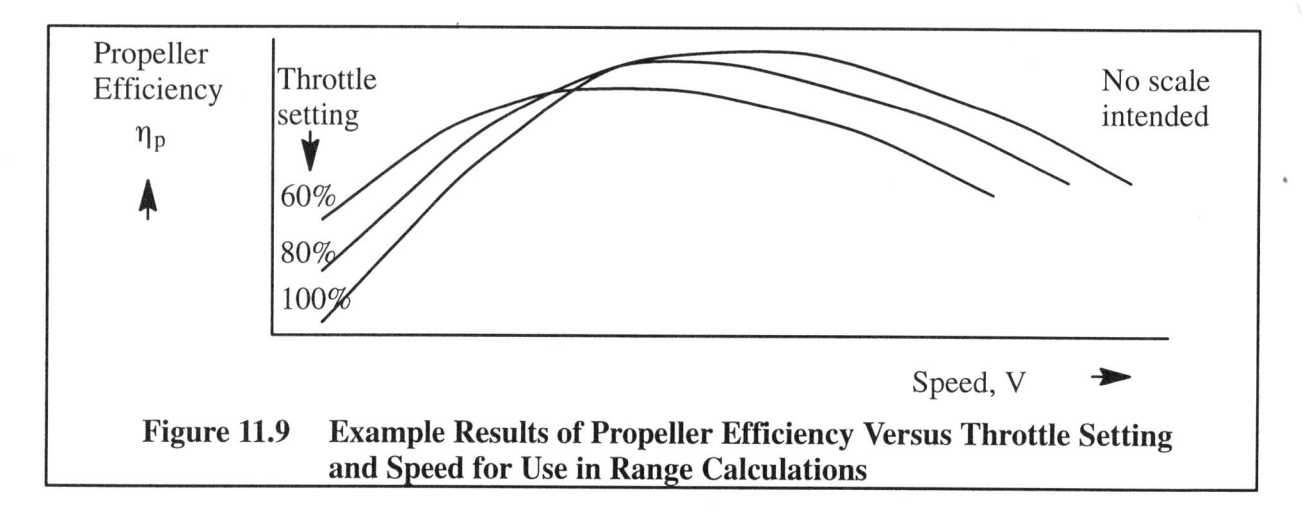

Figure 11.9 Example Results of Propeller Efficiency Versus Throttle Setting and Speed for Use in Range Calculations

Step 11: Determine the power required (P_{reqd}) for level, un-accelerated flight for a range of weights, altitudes and speeds. Chapter 8 provides methods for doing this. Eqn (8.71) may be used in these calculations.

Step 12: Determine the shaft-horsepower required from:
$$SHP_{reqd} = \frac{P_{reqd}}{\eta_p} \qquad (11.39)$$
This must be done for the same range of weights, altitudes and speeds as in Step 12.

Step 13: Determine the fuel flow from installed engine data at the level of required SHP for each engine. This defines the fuel flow, $\dot{W}_F$, in lbs/hr.

Step 14: Calculate the specific range, S.R. from Eqn (11.28) for the assumed range of cruise speeds, weights and altitudes. The results are often plotted in the manner shown in Figure 11.10.

Step 15: The specific range for a selected cruise condition can now be plotted as a function of weight as shown in Figure 11.11.

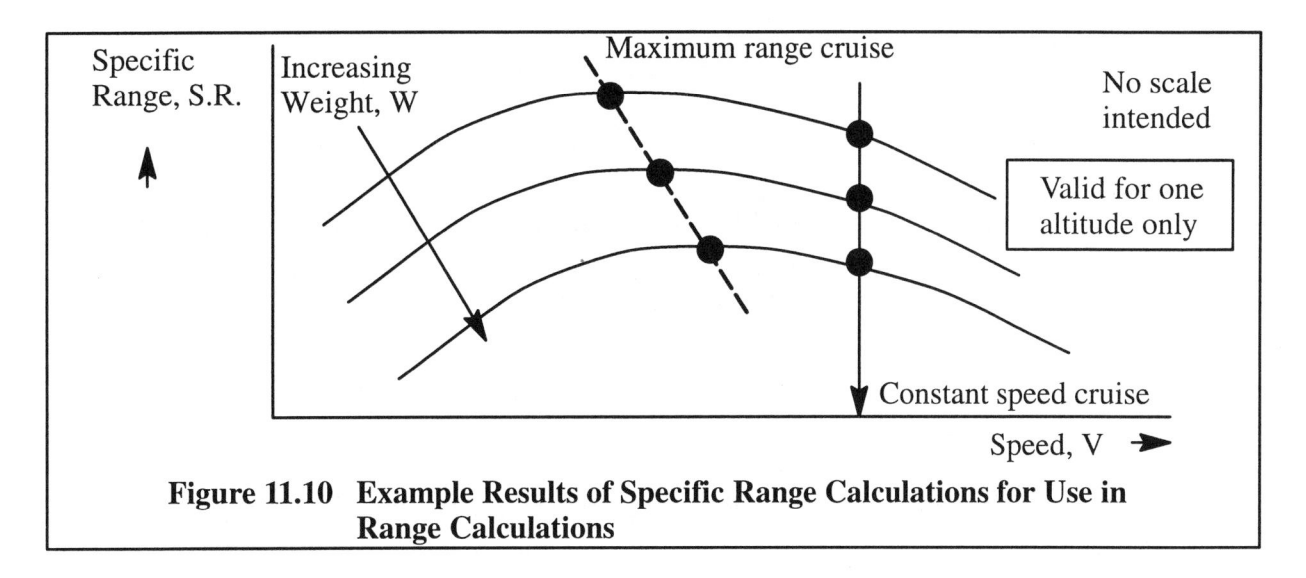

Figure 11.10 Example Results of Specific Range Calculations for Use in Range Calculations

The range can be numerically evaluated with Eqn (11.29) by using numerical integration between the limits $W_{CR_{begin}}$ and $W_{CR_{end}}$:

$$R = \Sigma (S.R.)_{ave} \Delta W \tag{11.40}$$

This can be done in tabular form or with a spreadsheet.

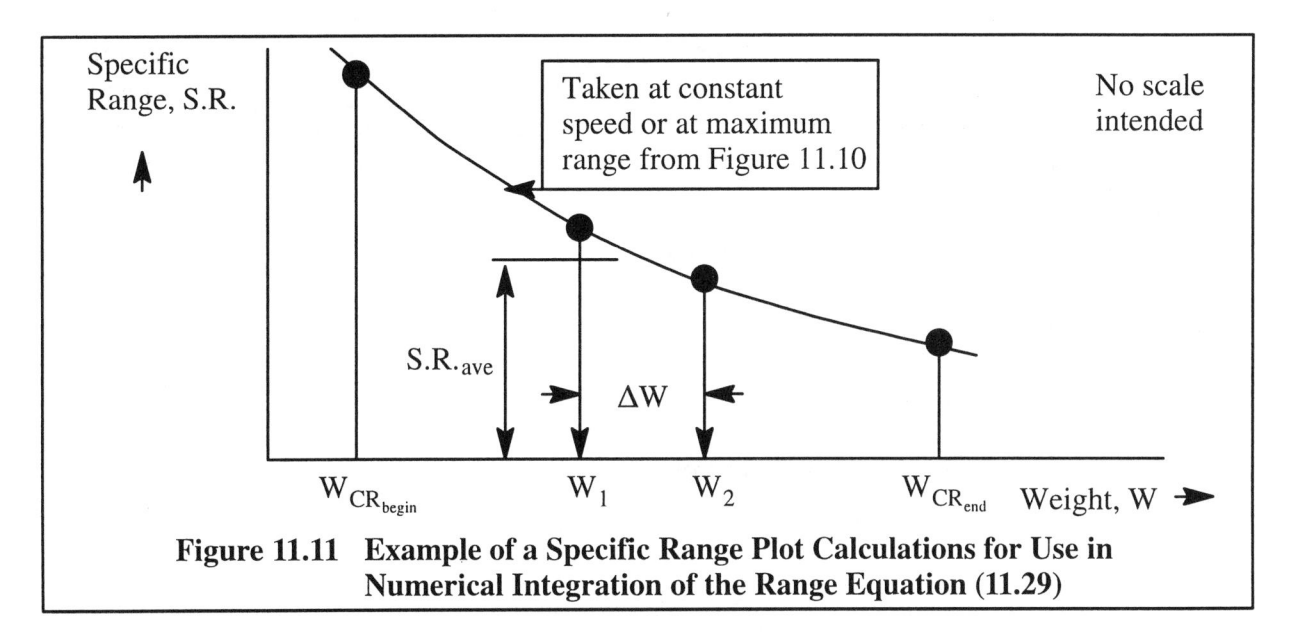

Figure 11.11 Example of a Specific Range Plot Calculations for Use in Numerical Integration of the Range Equation (11.29)

The problem now is that although $W_{CR_{begin}}$ is known, $W_{CR_{end}}$ is not. A certain amount of fuel must be set aside for the descent, for landing, and for shut–down. In addition, the required fuel reserves cannot be used for revenue range. For these reasons, the following additional steps are needed to determine the lower allowable limit, $W_{CR_{end}}$: see Steps 16 – 20.

Step 16: Determine the amount of fuel used in descent, $W_{F_{DE}}$. For a descent from one altitude to another this can be done with Eqn (9.78).

Step 17: Determine the amount of fuel required to land, W_{F_L}:

$$W_{F_L} = \dot{W}_{F_L} t_L \qquad (11.41)$$

where: $\dot{W}_{F_{DE}}$ is the fuel flow during the landing. This can be taken to be the fuel flow with the throttles at idle unless reverse thrust is used. In the latter case the fraction of the landing time during which reverse thrust is used must be estimated.

t_L is the time it takes to land. See Eqn (10.116).

Step 18: Determine the amount of fuel required to taxi and shut-down, $W_{F_{taxi/shut}}$.

Step 19: Determine the amount of fuel reserves required: $W_{F_{res}}$.

Sometimes, fuel reserves are defined as a fraction of the mission fuel required. More frequently, fuel reserve requirements are set by the customer or by the operational requirements of FAR Parts 135.97 and 121.641 or 643 (non-turbine and turbo-propeller-powered), whichever is the more severe. Section 11.4 contains an overview of these rules.

In most cases the required fuel reserves are defined by a so-called alternate mission which must be negotiated without running out of fuel. Figure 11.12 shows a typical alternate mission. The fuel needed to fly this alternate mission can be determined by retracing steps 8, 9, 11, 12, 13, 14, 15 and 16 but now for the alternate mission.

Step 20: Determine the weight at the end of cruise, $W_{CR_{end}}$:

$$W_{CR_{end}} = W_{CR_{begin}} - W_{F_{res}} - W_{F_{DE}} - W_{F_L} - W_{F_{taxi/shut}} \qquad (11.42)$$

where: $W_{CR_{begin}}$ is found from Step 9

$W_{F_{DE}}$ is found from Step 16

W_{F_L} is found from Step 17

$W_{F_{taxi/shut}}$ is found from Step 18

Step 21: Determine the total amount of fuel used during the mission, excluding the reserves: $W_{F_{used}}$:

$$W_{F_{used}} = W_{F_{sewu}} + W_{F_{taxi}} + W_{F_{TO}} + W_{F_{CL}} + W_{F_{CR}} + W_{F_{DE}} + W_{F_L} - W_{F_{taxi/shut}} \qquad (11.43)$$

The amount of fuel actually used during a mission, $W_{F_{used}}$, is important from a cost viewpoint.

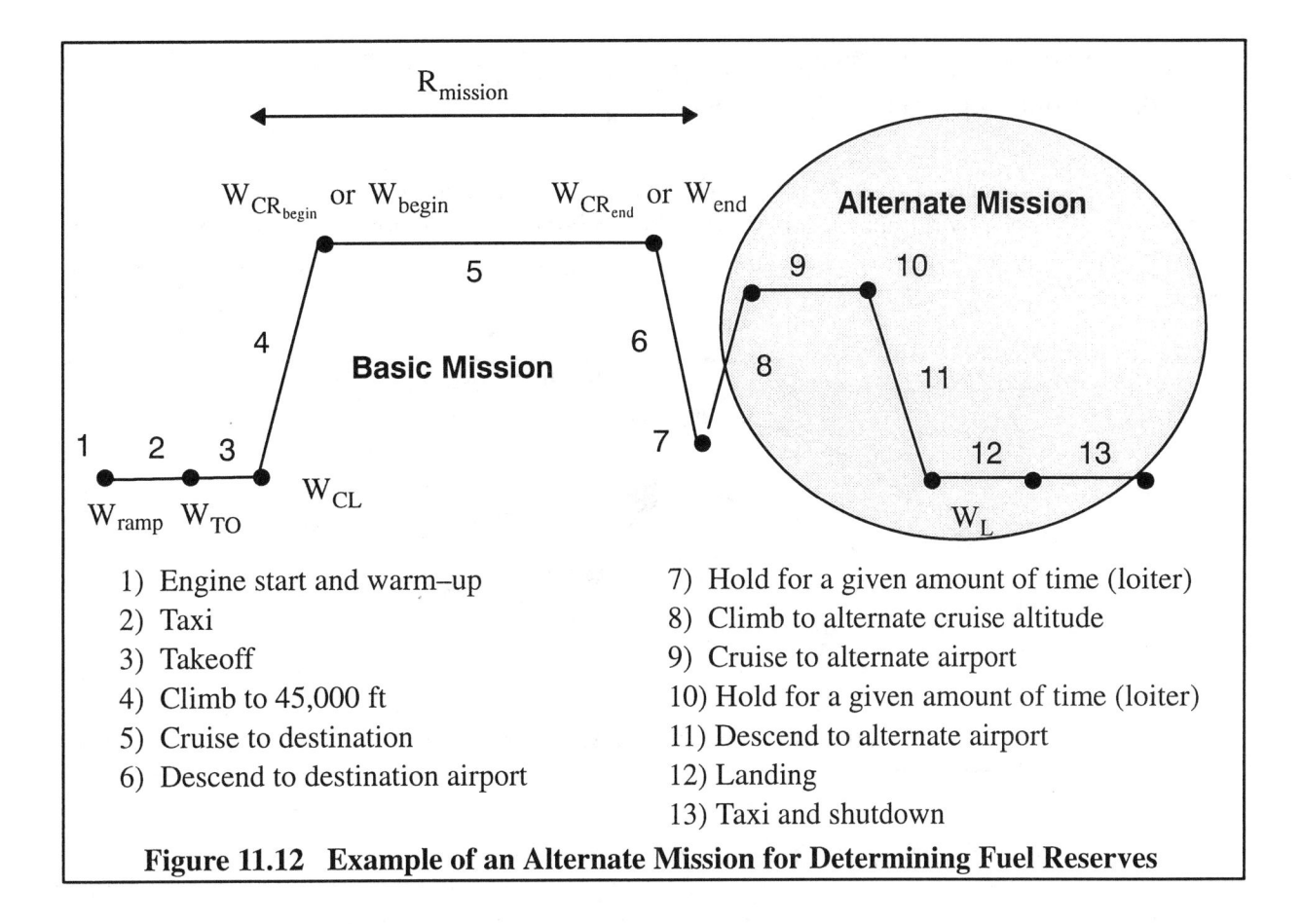

Figure 11.12 Example of an Alternate Mission for Determining Fuel Reserves

Mission phases:
1) Engine start and warm–up
2) Taxi
3) Takeoff
4) Climb to 45,000 ft
5) Cruise to destination
6) Descend to destination airport
7) Hold for a given amount of time (loiter)
8) Climb to alternate cruise altitude
9) Cruise to alternate airport
10) Hold for a given amount of time (loiter)
11) Descend to alternate airport
12) Landing
13) Taxi and shutdown

Step 22: Calculate the total amount of mission fuel required, W_F:

$$W_F = W_{F_{used}} + W_{F_{res}} \qquad (11.44)$$

This amount of fuel should equal the amount specified in Step 4.

Step 23: Verify that total mission fuel requirement, W_F, satisfies the following relationship:

$$W_F \leq W_{F_{volumetric}} \qquad (11.45)$$

11.1.5.2 Step–by–step Procedure for Determining Endurance (or Loiter)

In some airplanes there is a requirement for a certain amount of loiter time as part of the mission profile. One example was shown in the alternate mission of Figure 11.12. In certain airplanes there can be a requirement for flying to a specified altitude and then holding that position in a shallow turn for a long period of time. In both cases there is said to be a requirement for endurance.

A requirement for endurance (or loiter) is normally an integral part of a mission which also contains the mission flight phases illustrated in the mission profiles of Figures 11.8 and 11.12. A step–by–step procedure for evaluating endurance is not meaningful except as part of a more general mission. The reader should modify the previous procedure for calculating range to suit his purpose.

Critical steps in evaluating endurance are those steps which define the weight at the beginning and at the end of the endurance: $W_{LTR_{begin}}$ and $W_{LTR_{end}}$.

Once these limits are known, Eqn (11.31) can be used to numerically determine the endurance. This is typically done by retracing Steps 11 through 15 but now for endurance (loiter). Figure 11.13 shows a generic plot of specific endurance, S.E. from Eqn (11.30) for an assumed range of loiter speeds, weights and altitudes.

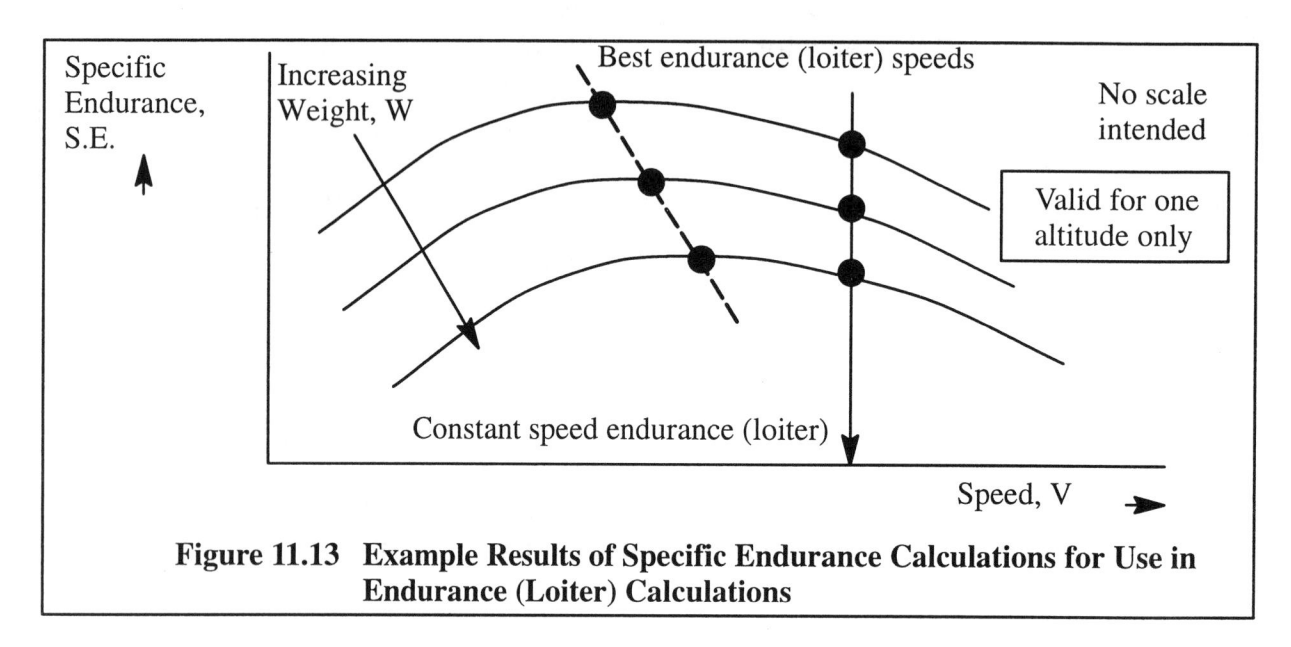

Figure 11.13 Example Results of Specific Endurance Calculations for Use in Endurance (Loiter) Calculations

The specific endurance for a selected loiter condition can now be plotted as a function of weight as shown in Figure 11.14. The endurance time (or loiter time) can be obtained by numerically evaluating Eqn (11.31):

$$E = \Sigma(S.E.)_{ave} \Delta W \qquad (11.46)$$

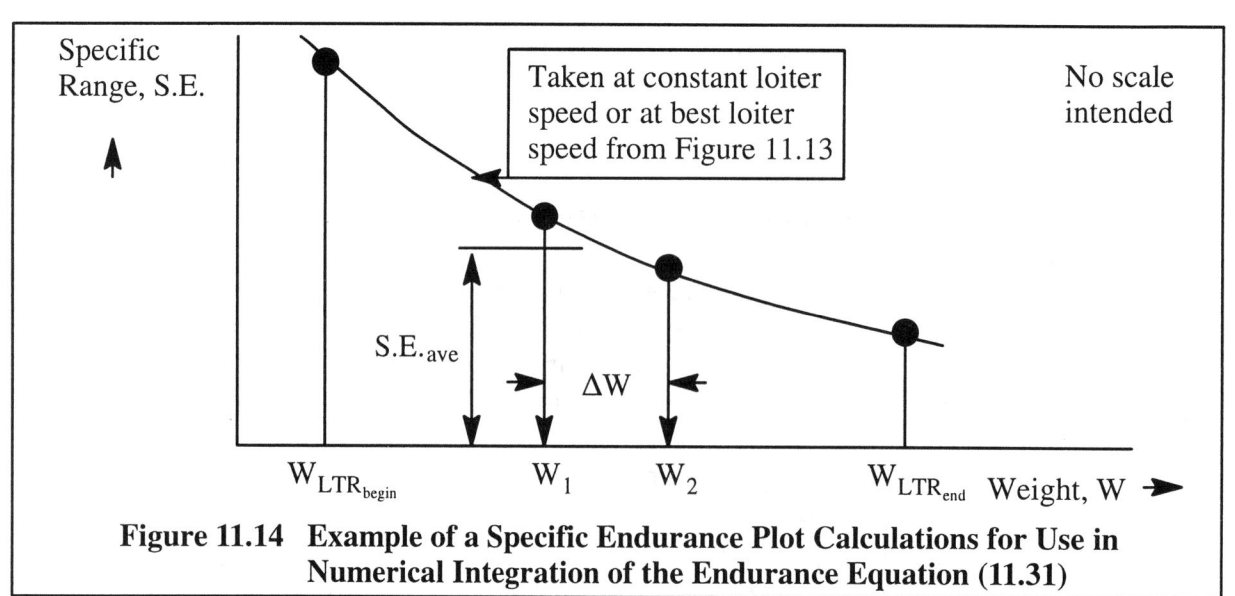

Figure 11.14 Example of a Specific Endurance Plot Calculations for Use in Numerical Integration of the Endurance Equation (11.31)

Range, Endurance and Payload–Range

11.2 JET AIRPLANES

As an airplane progresses through its mission it burns fuel and, as a result, looses weight.. The amount of fuel used during a given part of the mission of a jet airplane depends on two factors: the thrust required, T_{reqd}, (in lbs) and the specific fuel consumption, c_j, (in lbs/hr/lbs). The following expression reflects this:

$$\dot{W}_F = T_{reqd} \, c_j \tag{11.47}$$

where: $\dot{W}_F$ is the fuel flow in lbs per hour.

T_{reqd} is the thrust required for level flight, in lbs

c_j is the specific fuel consumption, in lbs/hr/lbs

The weight change of the airplane, resulting from this fuel flow is defined as:

$$dW = -\dot{W}_F \, dt = -T_{reqd} \, c_j \, dt \tag{11.48}$$

To determine airplane endurance, a relationship between weight and time is needed. Such a relationship is obtained by re–arranging Eqn (11.48):

$$\frac{dt}{dW} = \frac{-1}{T_{reqd} \, c_j} \quad \text{in hrs/lbs} \quad \text{(This is known as the endurance factor)} \tag{11.49}$$

To determine airplane range, a relationship between weight and range is needed. Since V = ds/dt and therefore, dt = ds/V, it is possible to cast Eqn (11.48) in the following format:

$$\frac{ds}{dW} = \frac{-V}{T_{reqd} \, c_j} \quad \text{in nm/lbs} \quad \text{(This is known as the range factor)} \tag{11.50}$$

With these fundamental relationships established endurance and range equations can be derived by integration of Eqns (11.49) and (11.50). In Sub–sections 11.2.1 – 11.2.3 this is done in an approximate manner. This leads to the so–called Breguet equations. The french airplane designer, Louis Breguet, was the first to derive closed from range and endurance equations for an airplane.

In the general case, closed form solutions to Eqns (11.49) and (11.50) cannot be obtained and numerical integration is normally used for accurate calculations. One approach to this, using specific range and endurance, is discussed in Sub–section 11.2.4.

11.2.1 BREGUET EQUATIONS FOR RANGE AND ENDURANCE

Two cases will be considered:

11.2.1.1 Constant Altitude Cruise and Endurance, and:

11.2.2.2 Constant Speed Cruise and Endurance

11.2.1.1 Constant Altitude Cruise and Endurance

It is observed, that the thrust required for level flight, T_{reqd}, can be written as:

$$T_{reqd} = D = \left(\frac{D}{L}\right) W = W \frac{C_D}{C_L} \tag{11.51}$$

By combining Eqns (11.51) and (11.50) it may be shown that:

$$\frac{ds}{dW} = \frac{-V(C_L/C_D)}{W\, c_j} \tag{11.52}$$

The flight speed, V, can be written as:

$$V = \sqrt{\frac{2W}{\varrho\, S\, C_L}} \quad \text{in ft/sec or}: \quad V = \frac{1}{1.689}\sqrt{\frac{2W}{\varrho\, S\, C_L}} \quad \text{in knots} \tag{11.53}$$

Substitution of Eqn (11.53) into Eqn (11.52) yields:

$$\frac{ds}{dW} = \frac{-1}{1.689\, c_j}\sqrt{\frac{2}{\varrho\, S}}\left[\frac{\sqrt{C_L}}{C_D}\right]\frac{dW}{\sqrt{W}} \quad \text{in nm/lbs} \tag{11.54}$$

This result shows, that for a given weight and altitude, the range is maximum when the factor $\sqrt{C_L}/C_D$ is a maximum.

The following definitions are now introduced:

> the weight of the airplane at the start of cruise is: W_{begin}
>
> the weight of the airplane at the end of cruise is: W_{end}

Next, it will be assumed that the factors c_j, $\sqrt{C_L}/C_D$ and the air density, ϱ, are constant during the flight. Integration of Eqn (11.54) under these assumptions yields:

$$R = -\int_{W_{CR_{begin}}}^{W_{CR_{end}}} \frac{1}{1.689 \, c_j} \sqrt{\frac{2}{\varrho S}} \left[\frac{\sqrt{C_L}}{C_D}\right] \frac{dW}{\sqrt{W}} =$$

$$= \frac{1.675}{c_j \sqrt{\varrho S}} \left[\frac{\sqrt{C_L}}{C_D}\right] \left(\sqrt{W_{CR_{begin}}} - \sqrt{W_{CR_{end}}}\right) \quad \text{in nm} \qquad (11.55)$$

In reality, particularly for very long ranges (say, longer than 1,000 n.m.), the factors c_j, $\sqrt{C_L}/C_D$ and the air density, ϱ, will vary during the flight. Therefore, it is prudent to carry out the integration implied by Eqn (11.55) over a number of range segments. The factors c_j, $\sqrt{C_L}/C_D$ and the air density, ϱ, must then be re–evaluated for each new range segment.

This procedure is illustrated in Figure 11.15 for the case of three segments. It is a matter of judgment to decide how many segments are adequate.

For each range segment at constant altitude*, best range occurs when the combined parameter $(\sqrt{C_L}/c_j C_D)$ is maximized.

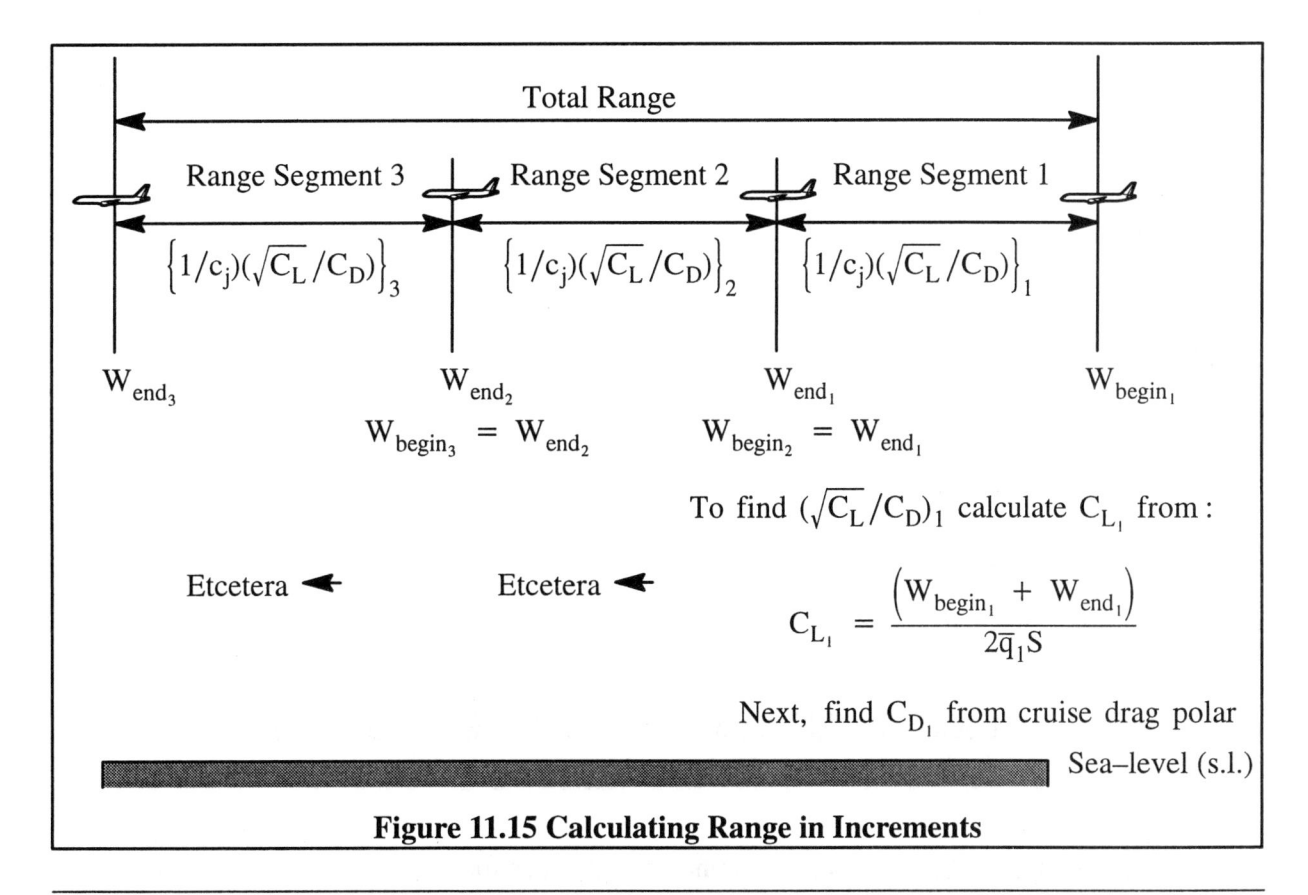

Figure 11.15 Calculating Range in Increments

———

* The reader is asked to show that the constant L/D assumption implies a climbing cruise.

The endurance equation (11.49) can be integrated in a similar manner. Combining Eqns (11.49) and (11.51) yields:

$$\frac{dt}{dW} = \frac{-(C_L/C_D)}{W\, c_j} \qquad (11.56)$$

The latter equation can be integrated to yield:

$$E = \int_{W_{LTR_{begin}}}^{W_{LTR}} \left(\frac{-(C_L/C_D)}{c_j}\right) \frac{dW}{W} = \frac{1}{c_j}\left(\frac{C_L}{C_D}\right) \ln\left(\frac{W_{LTR_{begin}}}{W_{LTR_{end}}}\right) \text{ in hrs} \qquad (11.57)$$

The integration was carried out by also assuming constant altitude. In reality, particularly for very long endurances (say, longer than 3 hours), the factors c_j and C_L/C_D will vary during the flight. For that reason it is prudent to carry out the integration implied by Eqn (11.57) over a number of endurance segments. The factors c_j and C_L/C_D must then be re–evaluated for each new endurance segment. The procedure is shown in Figure 11.16 for the case of two segments. It is a matter of judgment to decide how many segments are adequate.

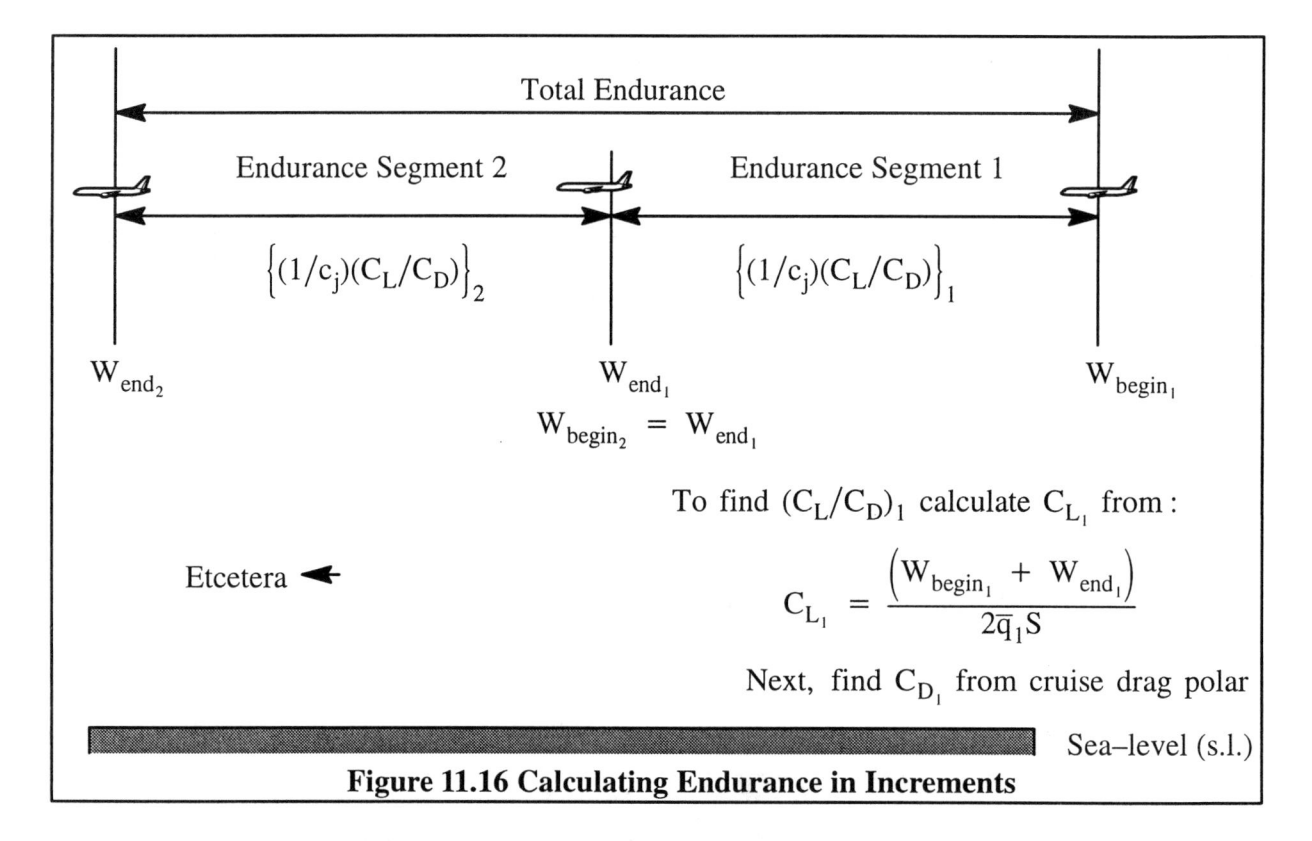

Figure 11.16 Calculating Endurance in Increments

For each endurance segment, best endurance occurs when the combined parameter $(C_L/c_j C_D)$ is maximized, when flying at a given altitude.

Range, Endurance and Payload–Range

It should be understood that in both Equations (11.55) and (11.57) the amount of fuel consumed during the range or endurance segments of the mission is:

$$W_{F_{used}} = W_{begin} - W_{end} \qquad (11.58)$$

To determine the total amount of fuel needed to complete a given mission (including fuel required to start, taxi, take–off, climb, cruise, loiter, descend, landing, shutdown and reserves) the reader is referred to Sub–sub–section 11.2.4.1

11.2.1.2 Constant Speed Cruise and Endurance

If the true airspeed, V, is kept constant, Eqn (11.52) still applies:

$$\frac{ds}{dW} = \frac{-V(C_L/C_D)}{W c_j} \qquad (11.59)$$

Next, it will be assumed that the factors c_j, C_L/C_D and the true airspeed, V, are constant during the flight. Integration of Eqn (11.54) under these assumptions yields:

$$R = \int_{W_{CR_{begin}}}^{W_{CR_{end}}} \left(\frac{-V\,C_L/C_D}{c_j}\right) \frac{dW}{W} = \left(\frac{VC_L/C_D}{c_j}\right) \ln \frac{W_{CR_{begin}}}{W_{CR_{end}}} \quad \text{in nm} \qquad (11.60)$$

It is seen that, to maximize range, the factor VC_L/C_D must be maximized. However, because of Eqn (11.53) this means that again the factor $\sqrt{C_L}/C_D$ must be maximized.

At high flight speeds, V, compressibility can have significant effects on the drag polar and therefore on the lift–to–drag ratio, C_L/C_D. For that reason it has been found convenient to re–write Eqn (11.60) in terms of the flight Mach number, M. Consider Eqn (9.26):

$$V = MV_a = M\sqrt{\theta}\,V_{a_0} \qquad (11.61)$$

where: V_a is the speed of sound at a given altitude

θ is the temperature ratio at a given altitude

V_{a_0} is the speed of sound at sea–level in kts.

Substitution of Eqn (11.61) into Eqn (11.60) yields:

$$R = = \left[\frac{V_{a_0}\sqrt{\theta}\,M\,C_L/C_D}{c_j}\right] \ln \frac{W_{CR_{begin}}}{W_{CR_{end}}} \quad \text{in nm} \qquad (11.62)$$

Range, Endurance and Payload–Range

It is seen that, to maximize range, the factor $M(C_L/C_D)$ must be maximized.

Next, consider Figure 11.17 which shows the generic effect of Mach number on the drag polar. By drawing the tangents, maximum values for C_L / C_D can be obtained. The product $M(C_L/C_D)$ is next plotted versus Mach number: see also Figure 11.17. The value of the flight Mach number which maximizes $M(C_L/C_D)$ is readily obtained from this graph.

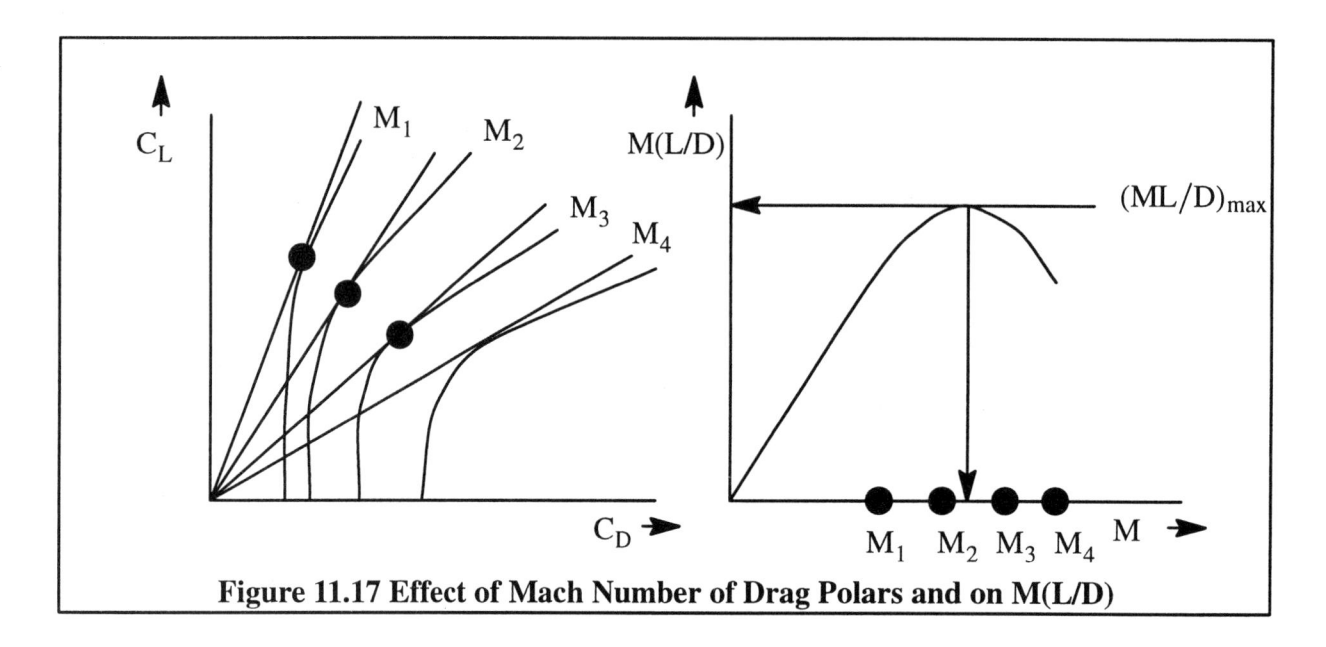

Figure 11.17 Effect of Mach Number of Drag Polars and on M(L/D)

Because both Mach number and C_L / C_D are maintained constant, the altitude has to increase has fuel is burned off. Air Traffic Control requirements may not allow this in a practical situation.

In reality, particularly for very long ranges (say, longer than 1,000 n.m.), the factors c_j, $\sqrt{C_L} / C_D$ and the air density, ϱ, will vary during the flight. Therefore, it is prudent to carry out the integration implied by Eqn (11.60) over a number of range segments. The factors c_j, $\sqrt{C_L} / C_D$ and the air density, ϱ, must then be re-evaluated for each new range segment.

This procedure is illustrated in Figure 11.18a for the case of three segments. It is a matter of judgment to decide how many segments are adequate. For each range segment at constant altitude, best range occurs when the combined parameter $(\sqrt{C_L}/c_j C_D)$ is maximized.

For constant speed endurance, Eqn (11.49) and (11.51), when combined, can be integrated to yield:

$$E = - \int_{W_{LTR_{begin}}}^{W_{LTR_{end}}} \left(\frac{1}{c_j}\frac{C_L}{C_D}\right)\frac{dW}{W} = \left(\frac{1}{c_j}\frac{C_L}{C_D}\right) \ln \frac{W_{LTR_{begin}}}{W_{LTR_{end}}} \quad \text{in hrs} \qquad (11.63)$$

Range, Endurance and Payload–Range

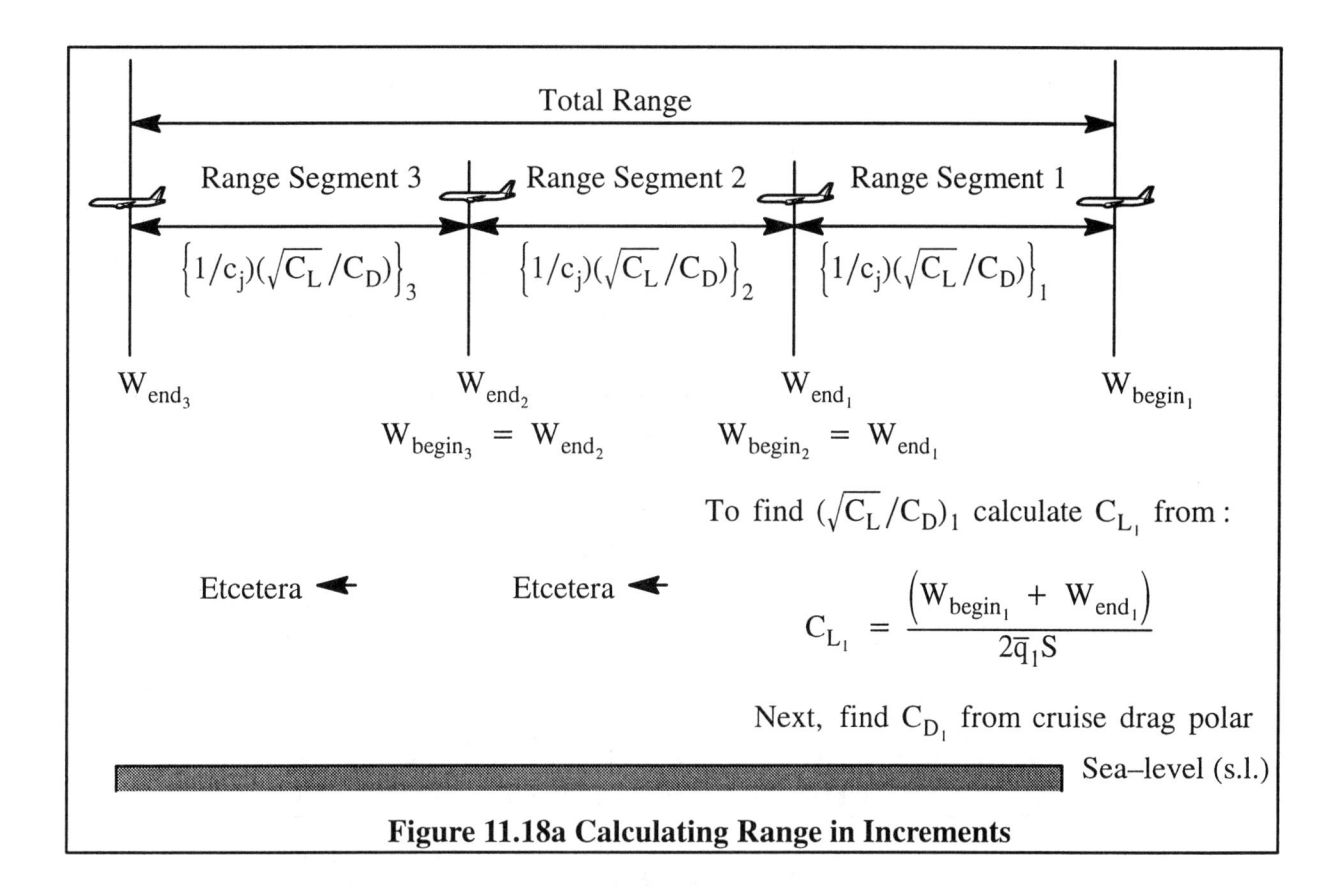

Figure 11.18a Calculating Range in Increments

The figure shows the Total Range divided into Range Segment 1, Range Segment 2, and Range Segment 3, each with the factor $\{(1/c_j)(\sqrt{C_L}/C_D)\}_i$. The weights are defined as:

- W_{end_3}, W_{end_2}, W_{end_1}, W_{begin_1}
- $W_{begin_3} = W_{end_2}$
- $W_{begin_2} = W_{end_1}$

To find $(\sqrt{C_L}/C_D)_1$ calculate C_{L_1} from:

$$C_{L_1} = \frac{(W_{begin_1} + W_{end_1})}{2\bar{q}_1 S}$$

Next, find C_{D_1} from cruise drag polar.

Sea–level (s.l.)

Etcetera ← Etcetera ←

In reality, particularly for very long endurances (say, longer than 3 hours), the factors c_j and C_L/C_D will vary during the flight. For that reason it is prudent to carry out the integration implied by Eqn (11.63) over a number of endurance segments. The factors c_j and C_L/C_D must then be re-evaluated for each new endurance segment.

The procedure is shown in Figure 11.18b for the case of two segments. It is a matter of judgment to decide how many segments are adequate.

For each endurance segment, best endurance occurs when the combined parameter $(C_L/c_j C_D)$ is maximized, when flying at a given speed and altitude.

It should be understood that in both Equations (11.55) and (11.57) the amount of fuel consumed during the range or endurance segments of the mission is:

$$W_{F_{used}} = W_{begin} - W_{end} \qquad (11.64)$$

To determine the total amount of fuel needed to complete a given mission (including fuel required to start, taxi, take–off, climb, cruise, loiter, descend, landing, shutdown and reserves) the reader is referred to Sub–sub–sections 11.1.5.1 and 11.2.4.1.

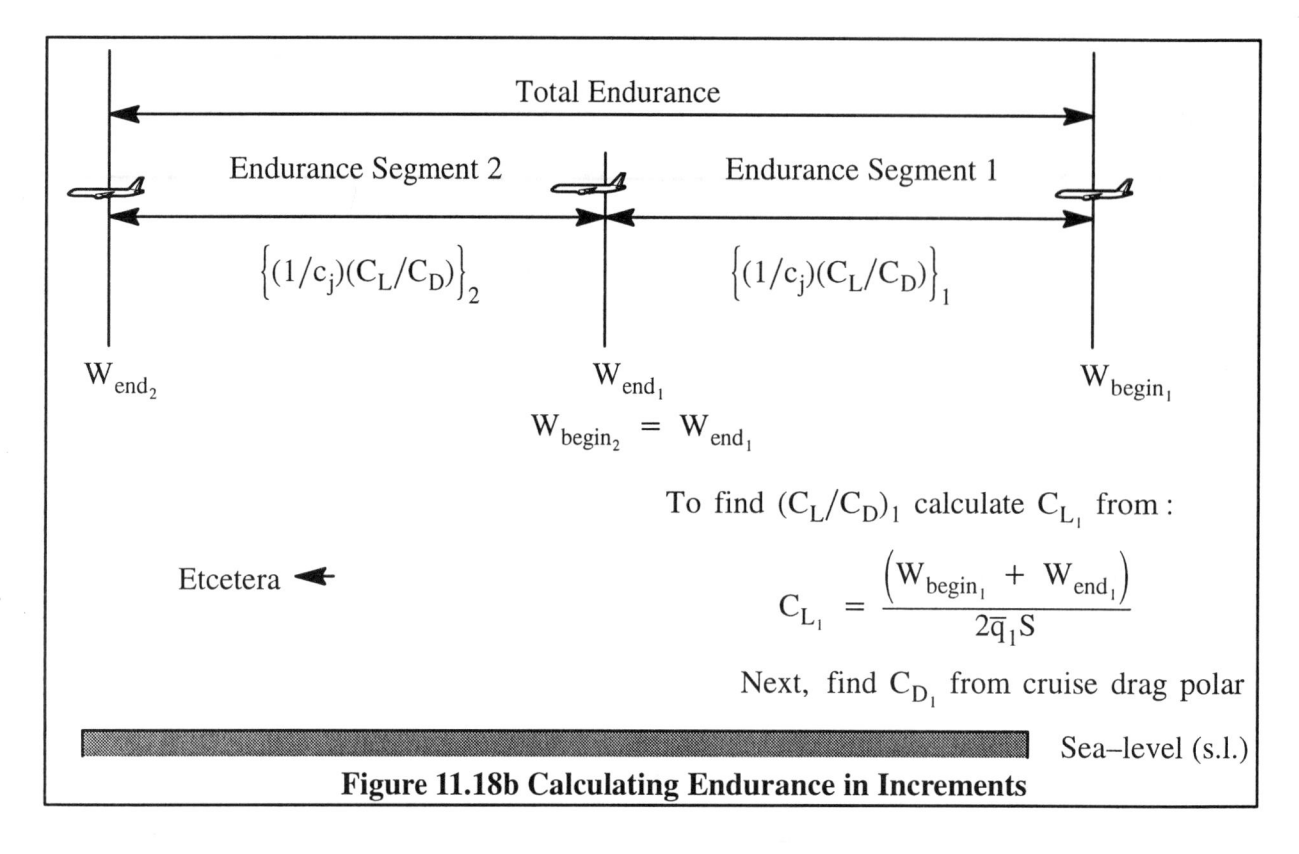

Figure 11.18b Calculating Endurance in Increments

11.2.2. MAXIMUM RANGE AND ENDURANCE FOR PARABOLIC DRAG POLARS

11.2.2.1 Maximum Range for Parabolic Drag Polars (Constant Altitude)

Observe that in the range equation (11.55) the air density, ρ, appears in the denominator. This is one of the fundamental reasons why jet airplanes must be operated at high altitude to achieve good range. For a given altitude, maximum range is obtained by flying at an angle of attack which corresponds to the maximum value of the parameter $\sqrt{C_L}/C_D$.

It is noted from the range equation (11.55) that, for given values of c_j and altitude, maximum range occurs when flying at the maximum value of the parameter $\sqrt{C_L}/C_D$. The reader is asked to show through a process of differentiation that maximum $\sqrt{C_L}/C_D$ and the lift and drag coefficients at which it occurs take on the following expressions for an airplane with a parabolic drag polar:

$$(\sqrt{C_L}/C_D)_{max} = \frac{3}{4}\left[\frac{\pi A e}{3(C_{D_0})^3}\right]^{1/4} \tag{11.65}$$

$$C_{L_{(\sqrt{C_L}/C_D)_{max}}} = \sqrt{\frac{\pi A e C_{D_0}}{3}} \qquad C_{D_{(\sqrt{C_L}/C_D)_{max}}} = \frac{4}{3}C_{D_0} \tag{11.66}$$

To fly an airplane at the maximum value of $\sqrt{C_L}/C_D$ (or at maximum C_L/C_D^2) required that the airplane be flown at the maximum value of drag divided by speed, D/V. This can be seen by considering the fact that:

$$V = \sqrt{\frac{W}{S}\frac{2}{\varrho}\frac{1}{C_L}} \qquad \text{and:} \qquad D = \frac{C_D}{C_L}W \qquad (11.67)$$

Figure 11.19 shows how the conditions for maximum $\sqrt{C_L}/C_D$ (best range) and maximum C_L/C_D (best endurance) compare with each other. Note that the speed for best range is higher than the speed for best endurance.

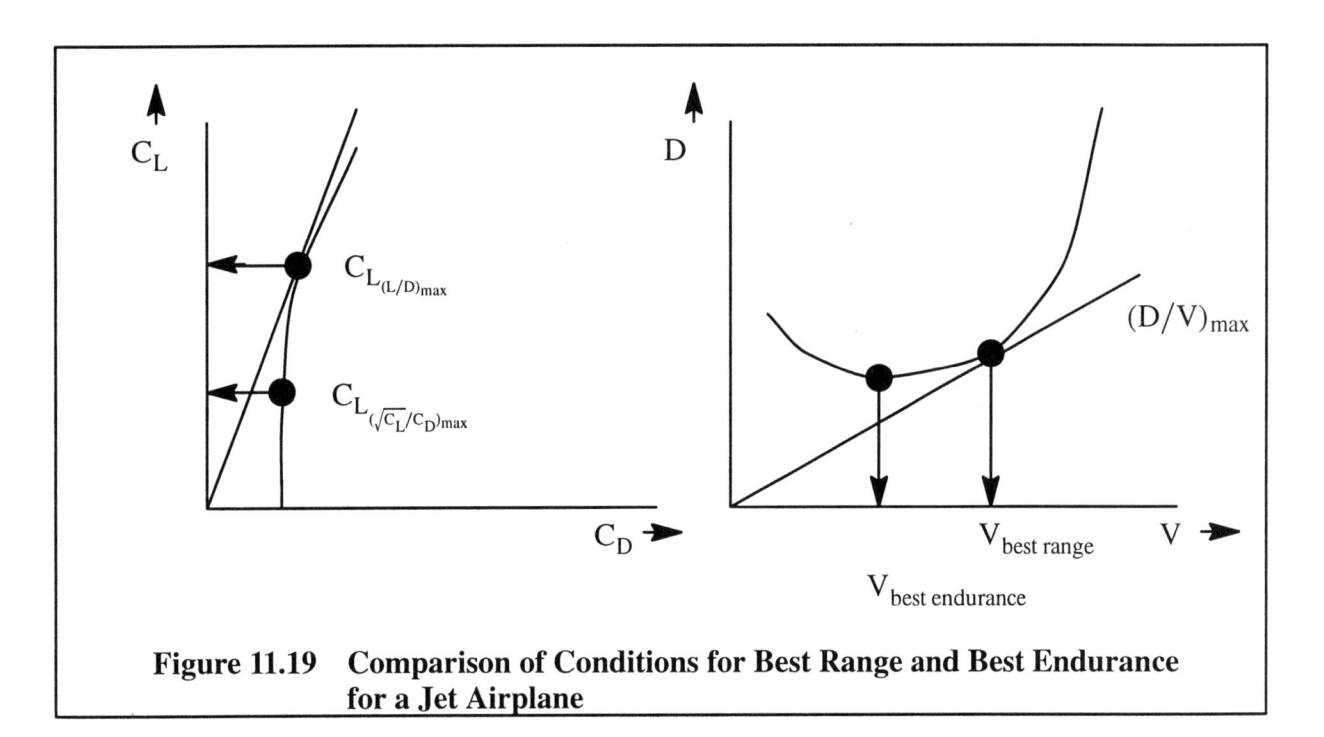

Figure 11.19 Comparison of Conditions for Best Range and Best Endurance for a Jet Airplane

11.2.2.2 Maximum Range for Parabolic Drag Polars (Constant Mach Number)

Cruise at constant Mach number with the altitude varying as ATC conditions permit, is the usual way in which jet transports are operated. It is seen from the range equation for constant Mach number (Eqn 11.62) that maximum range occurs when the parameter $M(C_L/C_D)$ is at its maximum. For a given Mach number that means the lift–to–drag ratio, (C_L/C_D), must be maximized. The conditions for which this occurs were discussed in Sub–sub–section 11.1.2.1: Eqns (11.10) and (11.11) and Figures 11.3a–c.

11.2.2.3 Sizing the Wing for Maximum Range (Maximum C_L/C_D)

The method of Sub–sub–section 11.1.2.2 will be used to see how large a typical jet transport wing should be for maximum cruise range.

Take as an example the Boeing 707–320B. The following data are available for this airplane:

$W \approx 320,000$ lbs $\quad S = 2,892$ ft^2 $\quad f_A = 42$ ft^2 $\quad f_{A-W} = 25$ ft^2

$(L/D)_{max} \approx 17$ $\quad M_{cruise} \approx 0.8$ $\quad h = 35,000$ ft

Using Eqn (11.17) it is found that: $S = 2,912$ ft^2. This is remarkably close to the actual wing area of this airplane.

11.2.2.4 Effect of Wind on Maximum Range

For airplanes with a parabolic drag polar, Hale and Steiger (Ref. 11.1) derived an analytical range equation which includes the effect of wind. They showed that with a tail–wind equal to 1/2 of the best range speed without wind, the range can be improved by 2% if the airspeed is reduced by about 10% from the best range speed without wind. However, in the case of jet transports because of the large fluctuation of wind with altitude it may sometimes be better to also change altitude. This is discussed in more detail in 11.2.4.3.

11.2.2.5 Maximum Endurance for Parabolic Drag Polars (Constant Mach Number)

It can be seen from the endurance equation (11.63) that best endurance occurs at best (C_L/C_D). The conditions for which this occurs were discussed in Sub–sub–section 11.1.2.1: Eqns (11.10) and (11.11) and Figures 11.3a–c.

11.2.2.6 Sizing the Wing for Maximum Endurance (Maximum C_L/C_D)

Because best endurance at constant Mach number also occurs at the best (C_L/C_D) the discussion in Sub–sub–section 11.2.2.3 applies here as well.

11.2.2.7 Numerical Example of a Composite Plot of C_D, C_L/C_D and $\sqrt{C_L}/C_D$ versus C_L

Figure 11.20 shows a composite plot of C_D, C_L/C_D and $\sqrt{C_L}/C_D$ versus the lift coefficient, C_L, for an advanced, regional jet transport in cruise. Note the numerical value of the lift coefficient for maximum C_L/C_D: 0.73. The lift coefficient for maximum $\sqrt{C_L}/C_D$ is 0.40. The latter is quite compatible with the operating environment for such a transport.

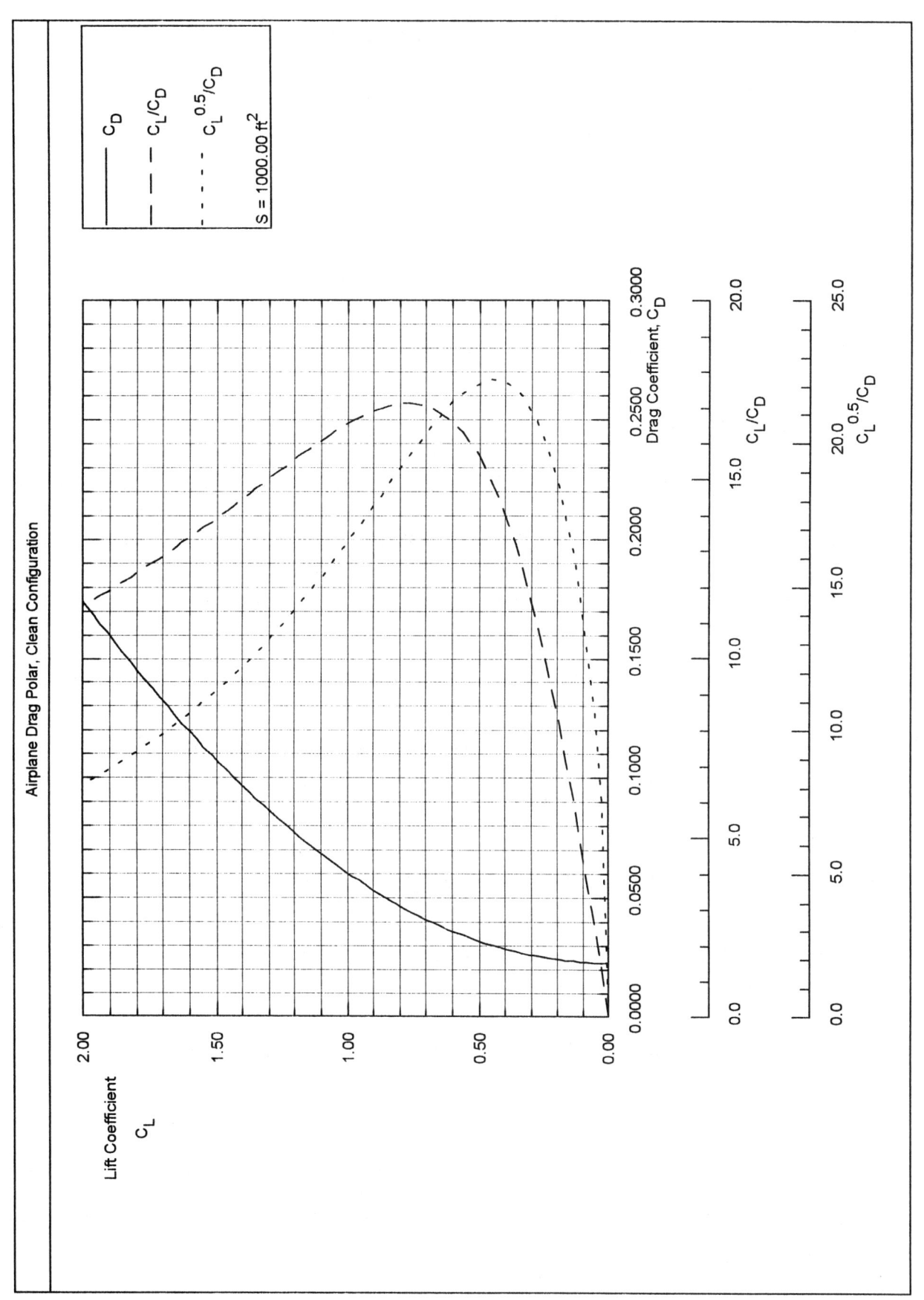

Figure 11.20 Composite Plot of C_D, C_L/C_D and $\sqrt{C_L}/C_D$ versus C_L for a Regional Jet Transport in Cruise

11.2.3. APPLICATION OF THE RANGE AND ENDURANCE EQUATIONS TO A JET TRANSPORT AIRPLANE WITH A PARABOLIC DRAG POLAR

In Example 11.2 range and endurance are determined for a jet transport cruising at constant Mach number. In Example 11.3 the maximum range and endurance are determined for the same transport under similar conditions.

Example 11.2: A small jet transport airplane has the following characteristics:

$W_{TO} = 50,000$ lbs $\qquad W_{OWE} = 28,000$ lbs

$C_D = 0.0190 + 0.055 C_L^2 \qquad h = 35,000$ ft $\qquad A = 8$

$S = 500$ ft$^2 \qquad c_j = 0.65$ lbs/hr/lbs at $M = 0.75$ and $h = 35,000$ ft

Assume, that the begin weight in the cruise phase is: $W_{CR_{begin}} = 49,000$ lbs.

Assume that the end weight in the cruise phase is; $W_{CR_{end}} = 39,000$ lbs.

Determine the cruise range and the cruise endurance at constant Mach number

Solution: The calculation will be performed by using Eqns (11.62) and (11.63).
For range, the following input data are available for Eqn (11.62):

$V_{a_0} = 661.5$ kts $\qquad \sqrt{\theta} = 0.8714$ at $35,000$ ft

In the cruise phase at M=0.75 at 35,000 ft, where: $\bar{q}/M^2 = 348.6$ psf, the dynamic pressure may be found from: $\bar{q} = (0.75)^2 348.6 = 196.1$ psf.

Since L/D will vary during the mission, an average will be used. The following lift coefficients are found for the begin and end of cruise:

$C_{L_{CR_{begin}}} = 49,000/196.1 \times 555 = 0.50$

$C_{L_{CR_{begin}}} = 39,000/196.1 \times 555 = 0.40$

With the cruise drag polar, this yields the following values for L/D:

$(C_L/C_D)_{begin} = 0.50/0.0328 = 15.2$

$(C_L/C_D)_{end} = 0.40/0.0278 = 14.4$

The average L/D value of the airplane in cruise is 14.8. The range now follows from Eqn (11.62) as:

$$R = \left(\frac{661.5 \times 0.8714 \times 0.75 \times 14.8}{0.65}\right) \ln \frac{49,000}{39,000} = 2,247 \text{ nm}$$

The endurance in this cruise condition follows from Eqn (11.63) as:

$$E = \left(\frac{14.8}{0.65}\right) \ln \frac{49,000}{39,000} = 5.2 \text{ hrs}$$

This finding should be checked with the flight speed of V=0.75x576.4=432.3 kts. Diving this into 2,247 yields for the flight time: 5.2 hrs!

It is noted with Eqns (11.10) and (11.11): $L/D_{max} = 15.5$ at $C_L = 0.59$. The wing of the airplane is therefore closely matched to this cruise condition although the airplane corresponds to scenario C–D in Figure 11.3b.

Example 11.3: For the same jet transport airplane of Example 11.2, determine the maximum range for constant altitude and for constant Mach number. Discuss the results.

Solution: The calculation will be performed by using Eqn (11.62).

Eqn (11.55) will be used for the calculation of maximum range at constant altitude. The density of air at 35,000 ft is: $\varrho = 0.0007369$ slugs/ft^3.

From Eqn (11.66) the lift and drag coefficients for maximum $\sqrt{C_L}/C_D$ are respectively:

$$C_{L_{(\sqrt{C_L}/C_D)_{max}}} = \sqrt{\frac{\pi A e C_{D_0}}{3}} = 0.34 \qquad C_{D_{(\sqrt{C_L}/C_D)_{max}}} = \frac{4}{3} C_{D_0} = 0.0253$$

The maximum value of $\sqrt{C_L}/C_D$ now is: $\sqrt{0.34}/0.0253 = 23.0$. This yields for the best range at constant altitude:

$$R = \frac{1.675}{0.65 \times \sqrt{0.0007369 \times 500}} \times 23.0 \times \left(\sqrt{49,000} - \sqrt{39,000}\right) = 2,331 \text{ nm}$$

By using $L/D_{max} = 15.5$ at $C_L = 0.59$ in Eqn (11.62) it follows that: R = 2,353 nm.

Discussion: this shows that flying at constant Mach number is marginally better than flying at constant altitude. It is of interest to check the speed for the constant altitude case. Eqn (11.67) yields for the speed at the begin of cruise :

$$V = \sqrt{\frac{W}{S} \frac{2}{\varrho} \frac{1}{C_L}} = \sqrt{\frac{49,000}{500} \frac{2}{0.0007369} \frac{1}{0.34}} = 884.5 \text{ ft/sec} = 524 \text{ kts}$$

This speed represents a Mach number of 0.91 which is way too high for such a transport. The reason for this absurd result is that the drag polar was kept constant with Mach number. For an airplane designed for a low cruise Mach number (M=0.75), the drag rise would be considerable.

11.2.4. ACCURATE DETERMINATION OF RANGE AND ENDURANCE FOR JET AIRPLANES WITH NON–PARABOLIC DRAG POLARS BY USE OF SPECIFIC RANGE AND ENDURANCE

The specific range, S.R., of an airplane is defined as the number of nautical miles which can be flown per pound of fuel. This quantity is also known as the range factor. Remembering that dW is negative (due to fuel usage) it follows from Eqn (11.50) that:

$$\text{S.R.} = \frac{-ds}{dW} = \frac{V}{T_{reqd} c_j} = \frac{V}{\dot{W}_F} \quad \text{in nm/lbs} \tag{11.68}$$

where: V is the speed in nm/hr (kts)

$\dot{W}_F$ is the fuel flow in lbs/hr

The S.R. can be calculated for a range of airplane weights, speeds and altitudes. From these calculations it is possible to determine optimum range flight paths. The range can be determined accurately by integration:

$$R = \int_{W_{CR_{begin}}}^{W_{CR_{end}}} \text{S.R. } dW \tag{11.69}$$

The specific endurance, S.E., of an airplane is defined as the number of hours which can be flown per lbs of fuel. This quantity is also known as the endurance factor. Remembering that dW is negative (due to fuel usage) it follows from Eqn (11.49) that:

$$\text{S.E.} = \frac{-dt}{dW} = \frac{1}{T_{reqd} c_j} = \frac{1}{\dot{W}_F} \quad \text{in hrs/lbs} \tag{11.70}$$

The S.E. can be calculated for a range of airplane weights, speeds and altitudes. From these calculations it is possible to determine optimum endurance flight paths. The endurance can be determined accurately by integration:

$$E = \int_{W_{LTR_{begin}}}^{W_{LTR_{end}}} \text{S.E. } dW \tag{11.71}$$

In principle, Eqns (11.69) and (11.71) can be used to determine range and endurance of airplanes. However, in reality, doing this for an arbitrary airplane mission is a fairly complicated process. For that reason, step–by–step procedures for determining range and endurance are given in Sub–sub–sections 11.2.4.1 and 11.2.4.2 respectively. These procedures partially duplicate those given for propeller driven airplanes in Sub–sub–sections 11.1.5.1 and 11.1.5.2.

Because of the large range of altitudes, Mach number and operationally encountered winds and temperatures, numerical procedures which account for these effects are recommended. One approach is discussed in Sub–sub–section 11.2.4.3.

11.2.4.1 Step–by–step Procedure for Determining Range

Step 1 through Step 9: See Sub–sub–section 11.1.5.1.

Step 10: Determine the thrust required (T_{reqd}) for level, un–accelerated flight for a range of weights, altitudes and speeds. Chapter 8 provides methods for doing this. Eqn (8.64) may be used in these calculations.

Step 11: Determine the fuel flow from installed engine data at the level of required thrust for each engine. This defines the fuel flow, $\dot{W}_F$, in lbs/hr.

Step 12: Calculate the specific range, S.R. from Eqn (11.68) for the assumed range of cruise speeds, weights and altitudes. The results are often plotted in the manner shown in Figure 11.21.

Step 13: The specific range for a selected cruise condition can now be plotted as a function of weight as shown in Figure 11.22.

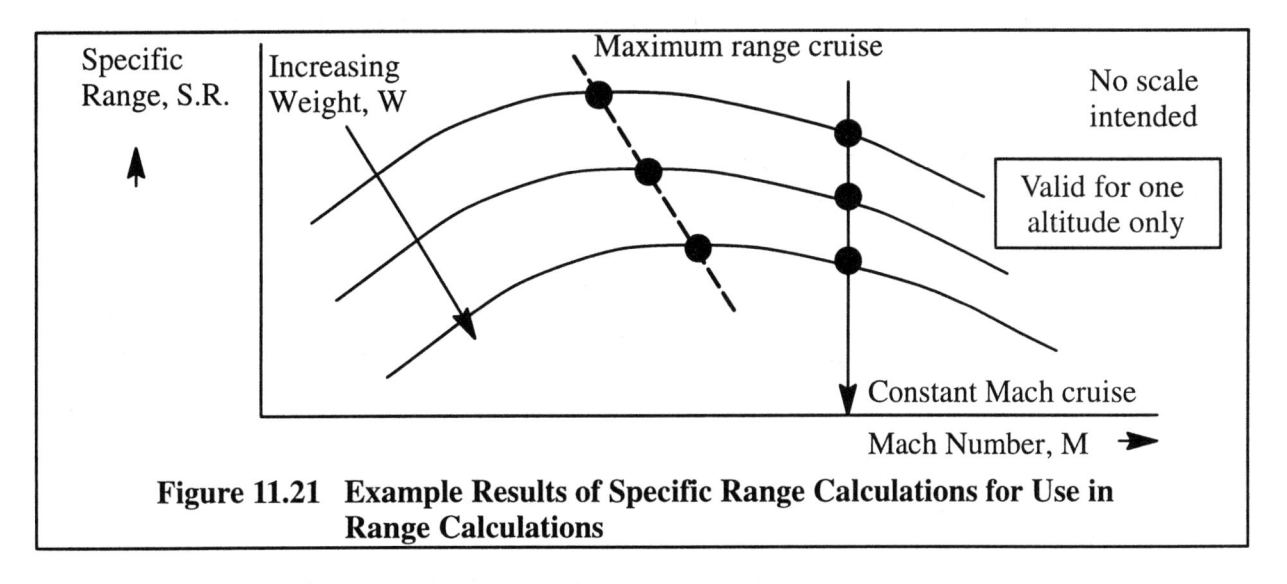

Figure 11.21 Example Results of Specific Range Calculations for Use in Range Calculations

The range can be numerically evaluated with Eqn (11.69) by using numerical integration between the limits $W_{CR_{begin}}$ and $W_{CR_{end}}$:

$$R = \Sigma(S.R.)_{ave} \Delta W \tag{11.72}$$

This can be done in tabular form or with a spreadsheet.

The problem now is that although $W_{CR_{begin}}$ is known, $W_{CR_{end}}$ is not. A certain amount of fuel must be set aside for the descent, for landing, and for shut–down. In addition, the required fuel reserves cannot be used for revenue range. For these reasons, the following additional steps are needed to determine the lower allowable limit, $W_{CR_{end}}$: see Steps 14 – 17.

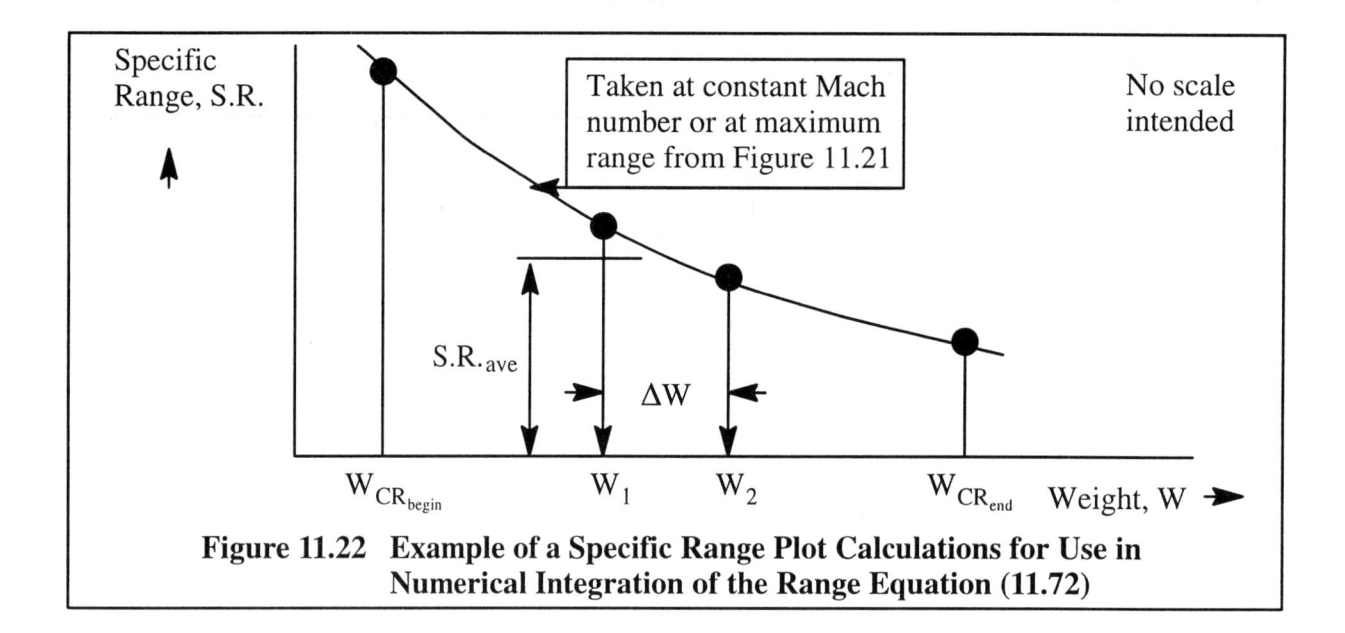

Figure 11.22 Example of a Specific Range Plot Calculations for Use in Numerical Integration of the Range Equation (11.72)

Step 14: Determine the amount of fuel used in descent, $W_{F_{DE}}$. For a descent from one altitude to another this can be done with Eqn (9.78).

Step 15: Determine the amount of fuel required to land, W_{F_L}:

$$W_{F_L} = \dot{W}_{F_L} t_L \qquad (11.73)$$

where: $\dot{W}_{F_{DE}}$ is the fuel flow during the landing. This can be taken to be the fuel flow with the throttles at idle unless reverse thrust is used. In the latter case the fraction of the landing time during which reverse thrust is used must be estimated.

t_L is the time it takes to land. See Eqn (10.116).

Step 16: Determine the amount of fuel required to taxi and shut-down, $W_{F_{taxi/shut}}$.

Step 17: Determine the amount of fuel reserves required: $W_{F_{res}}$.

Sometimes, fuel reserves are defined as a fraction of the mission fuel required. More frequently, fuel reserve requirements are set by the customer or by the operational requirements of FAR Parts 135.197 and 121.645 (turbine powered airplanes), whichever is the more severe. Section 11.4 contains an overview of these rules.

In most cases the required fuel reserves are defined by a so-called alternate mission which must be negotiated without running out of fuel. Figure 11.12 shows a typical alternate mission. The fuel needed to fly this alternate mission can be determined by retracing steps 8 through 16 but now for the alternate mission.

Step 18: Determine the weight at the end of cruise, $W_{CR_{end}}$:

$$W_{CR_{end}} = W_{CR_{begin}} - W_{F_{res}} - W_{F_{DE}} - W_{F_L} - W_{F_{taxi/shut}} \qquad (11.74)$$

where: $W_{CR_{begin}}$ is found from Step 9

$W_{F_{DE}}$ is found from Step 14

W_{F_L} is found from Step 15

$W_{F_{taxi/shut}}$ is found from Step 16

Step 19: Determine the total amount of fuel used during the mission, excluding the reserves: $W_{F_{used}}$:

$$W_{F_{used}} = W_{F_{sewu}} + W_{F_{taxi}} + W_{F_{TO}} + W_{F_{CL}} + W_{F_{CR}} + W_{F_{DE}} + W_{F_L} - W_{F_{taxi/shut}} \qquad (11.75)$$

The amount of fuel actually used during a mission, $W_{F_{used}}$, is important from a cost viewpoint.

Step 20: Calculate the total amount of mission fuel required, W_F:

$$W_F = W_{F_{used}} + W_{F_{res}} \qquad (11.76)$$

This amount of fuel should equal the amount specified in Step 4.

Step 21: Verify that total mission fuel requirement, W_F, satisfies the following relationship:

$$W_F \leq W_{F_{volumetric}} \qquad (11.77)$$

11.2.4.2 Step–by–step Procedure for Determining Endurance (Loiter)

In some airplanes there is a requirement for a certain amount of loiter time as part of the mission profile. One example was shown in the alternate mission of Figure 11.12. In certain airplanes there can be a requirement for flying to a specified altitude and then holding that position in a shallow turn for a long period of time. In both cases there is said to be a requirement for endurance.

A requirement for endurance (or loiter) is normally an integral part of a mission which also contains the mission flight phases illustrated in the mission profiles of Figures 11.8 and 11.12. A step–by–step procedure for evaluating endurance is not meaningful except as part of a more general mission. The reader should modify the previous procedure for calculating range to suit his purpose.

Critical steps in evaluating endurance are those steps which define the weight at the beginning and at the end of the endurance: $W_{LTR_{begin}}$ and $W_{LTR_{end}}$.

Once these limits are known, Eqn (11.31) can be used to numerically determine the endurance. This is typically done by retracing Steps 11 through 15 but now for endurance (loiter). Figure 11.23a shows a generic plot of specific endurance, S.E. from Eqn (11.30) for an assumed range of loiter speeds, weights and altitudes.

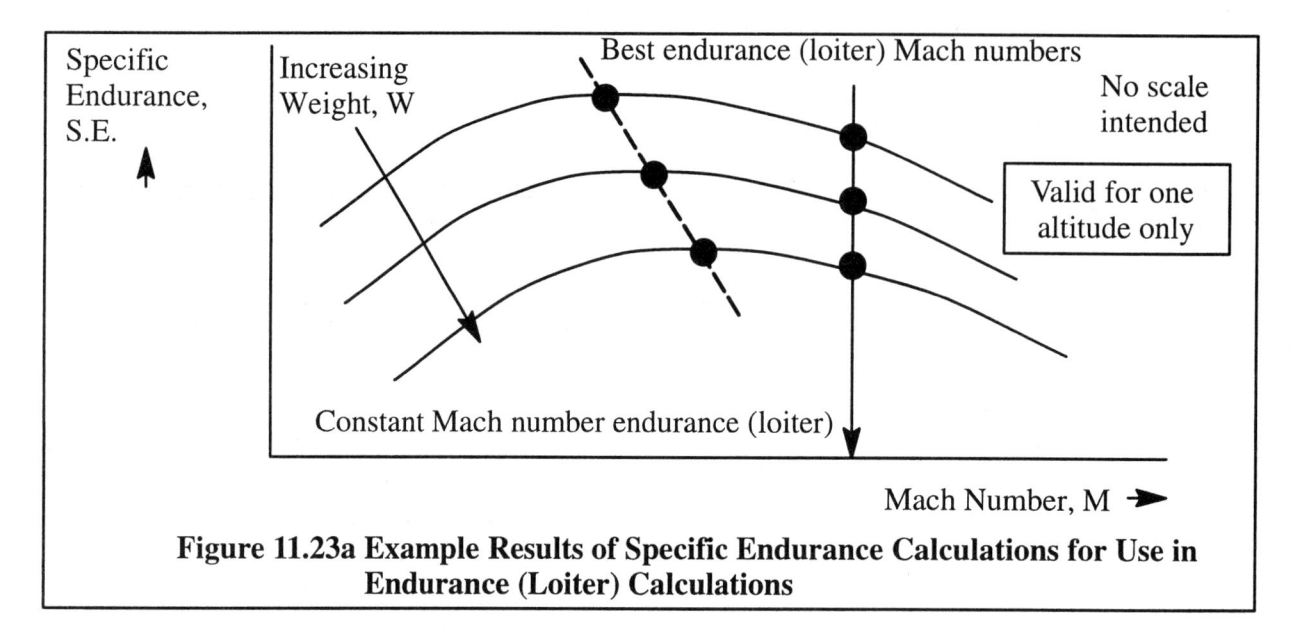

Figure 11.23a Example Results of Specific Endurance Calculations for Use in Endurance (Loiter) Calculations

The specific endurance for a selected loiter condition can now be plotted as a function of weight as shown in Figure 11.23b. The endurance time (or loiter time) can be obtained by numerically evaluating Eqn (11.31):

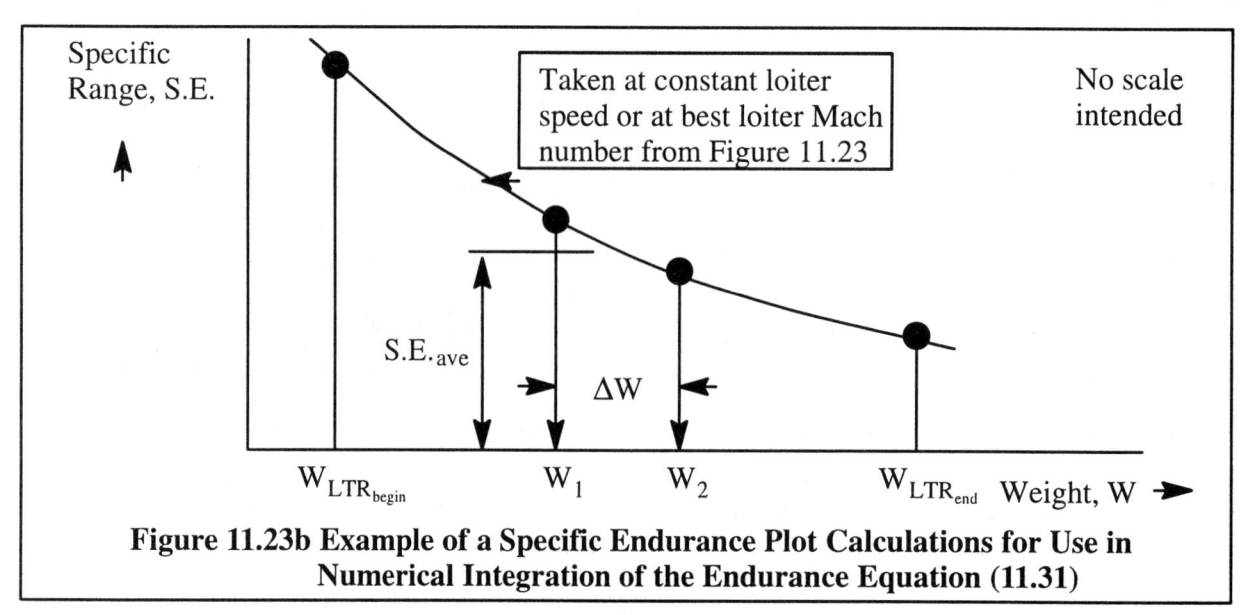

$$E = \Sigma(S.E.)_{ave} \Delta W \qquad (11.78)$$

Figure 11.23b Example of a Specific Endurance Plot Calculations for Use in Numerical Integration of the Endurance Equation (11.31)

11.2.4.3 Suggested Procedure to Account for the Effect of Altitude, Temperature and Wind on Range and Endurance

Because of the large range of altitudes, Mach number and operationally encountered winds and temperatures, numerical procedures which account for these effects are recommended. One approach is discussed in this Sub–sub–section.

First, the expression for specific range, S.R., in Eqn (11.68) can be written as:

$$\text{S.R.} = \frac{V}{T_{reqd} \, c_j} = \frac{V_{a_0} \, M \, \sqrt{\theta}}{c_j \, T_{reqd}} \quad \text{in nm/lbs} \tag{11.79}$$

Second, a table such as Table 11.2a is constructed to determine the S.R. for as many weights, Mach numbers and altitudes as are required.

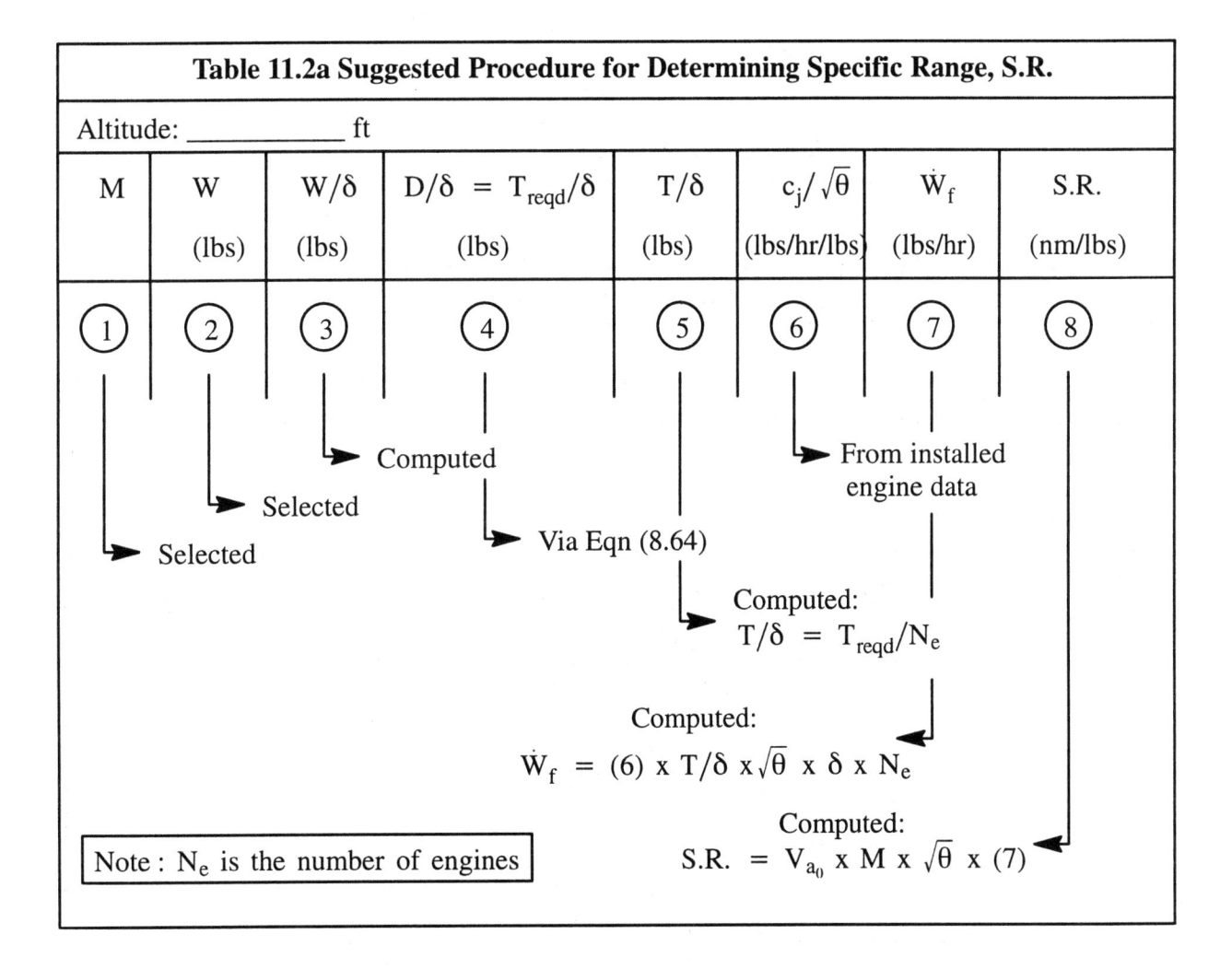

Table 11.2a Suggested Procedure for Determining Specific Range, S.R.

Altitude: _____ ft

M	W (lbs)	W/δ (lbs)	D/δ = T_{reqd}/δ (lbs)	T/δ (lbs)	$c_j/\sqrt{\theta}$ (lbs/hr/lbs)	$\dot{W}_f$ (lbs/hr)	S.R. (nm/lbs)
①	②	③	④	⑤	⑥	⑦	⑧

- ① Selected
- ② Selected
- ③ Computed
- ④ Via Eqn (8.64)
- ⑤ Computed: T/δ = T_{reqd}/N_e
- ⑥ From installed engine data
- ⑦ Computed: $\dot{W}_f$ = (6) x T/δ x $\sqrt{\theta}$ x δ x N_e
- ⑧ Computed: S.R. = V_{a_0} x M x $\sqrt{\theta}$ x (7)

Note: N_e is the number of engines

Third, the S.R. data are plotted as shown generically in Figure 11.24 for a given altitude. Plots like Figure 11.24 are then generated for a range of altitudes.

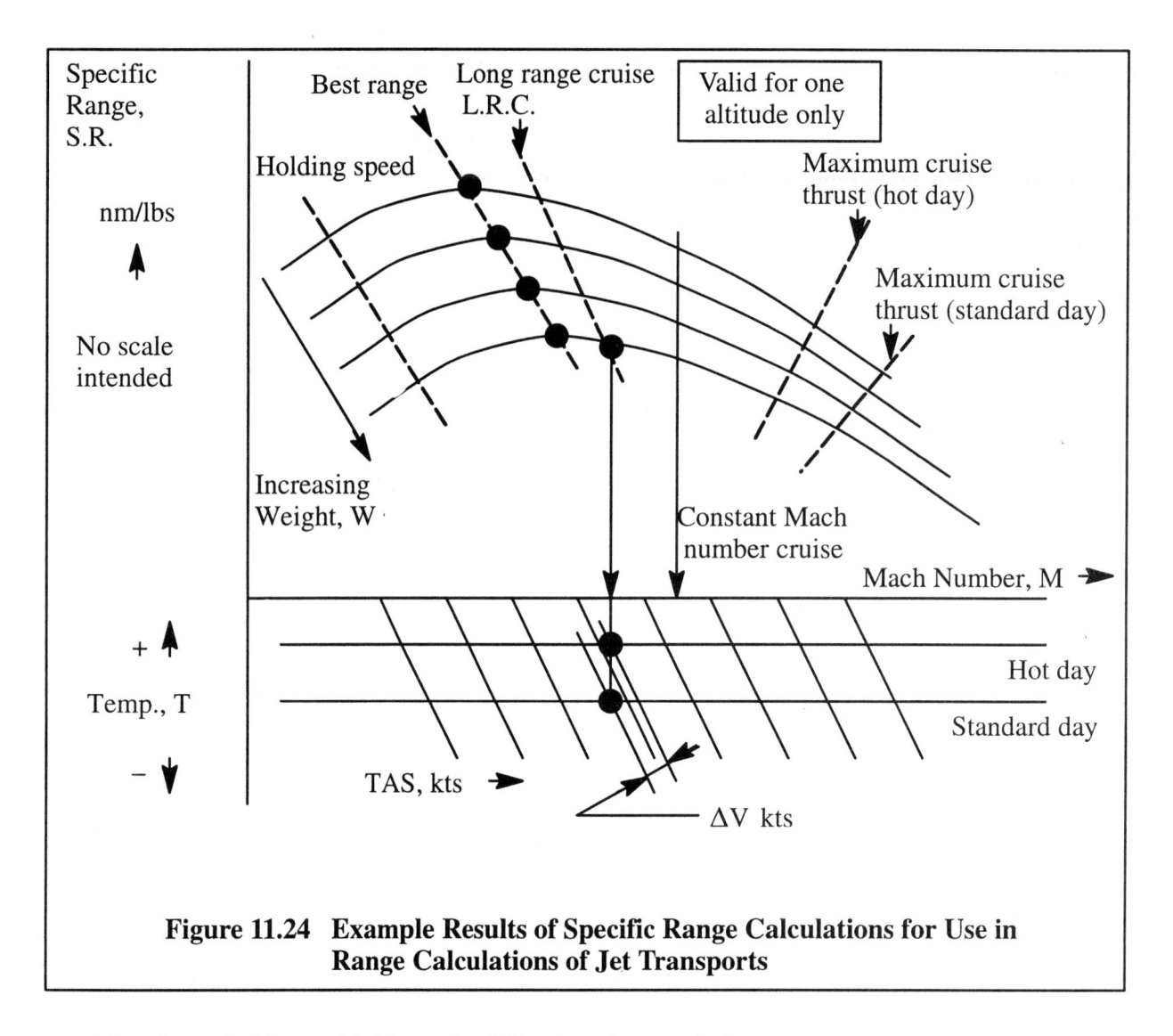

Figure 11.24 Example Results of Specific Range Calculations for Use in Range Calculations of Jet Transports

Also shown in Figure 11.24 are the following characteristics:

a) The effect of increasing temperature on the maximum available thrust

b) The effect of constant Mach number cruise

c) The holding speed is defined as the speed at minimum allowable fuel flow (flame–out). In some cases the holding speed is below the minimum drag speed. In that case, for reasons of speed stability, the minimum drag speed is used as the holding speed.

d) Long range cruise (L.R.C.) is defined as the condition for which the S.R. is at 99% of the maximum specific range.

Table 11.2b contains a summary of typical cruise options available. Once the specific range has been determined for whatever cruise option has been selected, Figure 11.22 and Eqn (11.72) can be used to compute the corresponding cruise range.

Table 11.2b Summary of Cruise Options

Cruise Option		Cruise Condition for Weight Decrease
①	Constant M and constant W/δ	Increasing altitude
②	Constant altitude for maximum range or for L.R.C. (Long Range Cruise)	Decreasing M and decreasing thrust
③	Constant M and constant altitude	Decreasing thrust
④	Rated thrust	a) Increasing altitude at constant M b) Increasing M at constant altitude

The reader must keep in mind that, to carry out the integration implied by Eqn (11.72), it is still necessary to trace through the Step–by–step procedure of Sub–sub–section 11.2.4.1 to determine the lower and upper limits of cruise weight: $W_{CR_{begin}}$ and $W_{CR_{end}}$

When an airplane is flying at a constant value of W/δ, then its weight and altitude are fixed. However, assume that weather reports predict a strong tail–wind to exist at some other altitude than the one corresponding to W and W/δ. In that case, it might be advantageous to fly at the new altitude, despite the fact that the new altitude is not optimum from a specific range viewpoint at the current value of W/δ.

The effect of wind at a given altitude is to change the ground–speed of the airplane. Therefore, the time to fly through the air is given by:

$$t = \frac{SAD}{V_{CR}} \tag{11.80}$$

where: SAD is the still air distance

V_{CR} is the cruise speed

However, to negotiate a given ground distance the time required is:

$$t = \frac{GD}{V_{CR} \pm V_w} \tag{11.81}$$

where: GD is the ground distance

V_w is the wind velocity

For a given period of time Eqns (11.80) and (11.81) can be equated to yield:

$$\text{SAD} = \text{GD} \frac{V_{CR}}{V_{CR} \pm V_w} \tag{11.82}$$

Eqn (11.82) can be used to construct a chart which shows the effect of wind on ground distance. This chart can be constructed by selecting a range of values for head-wind and tail-wind for assumed values of cruise speed and then substituting into Eqn (11.82). This establishes the relative values of ground distance and still air distance. Figure 11.25 shows a generic example of such a chart.

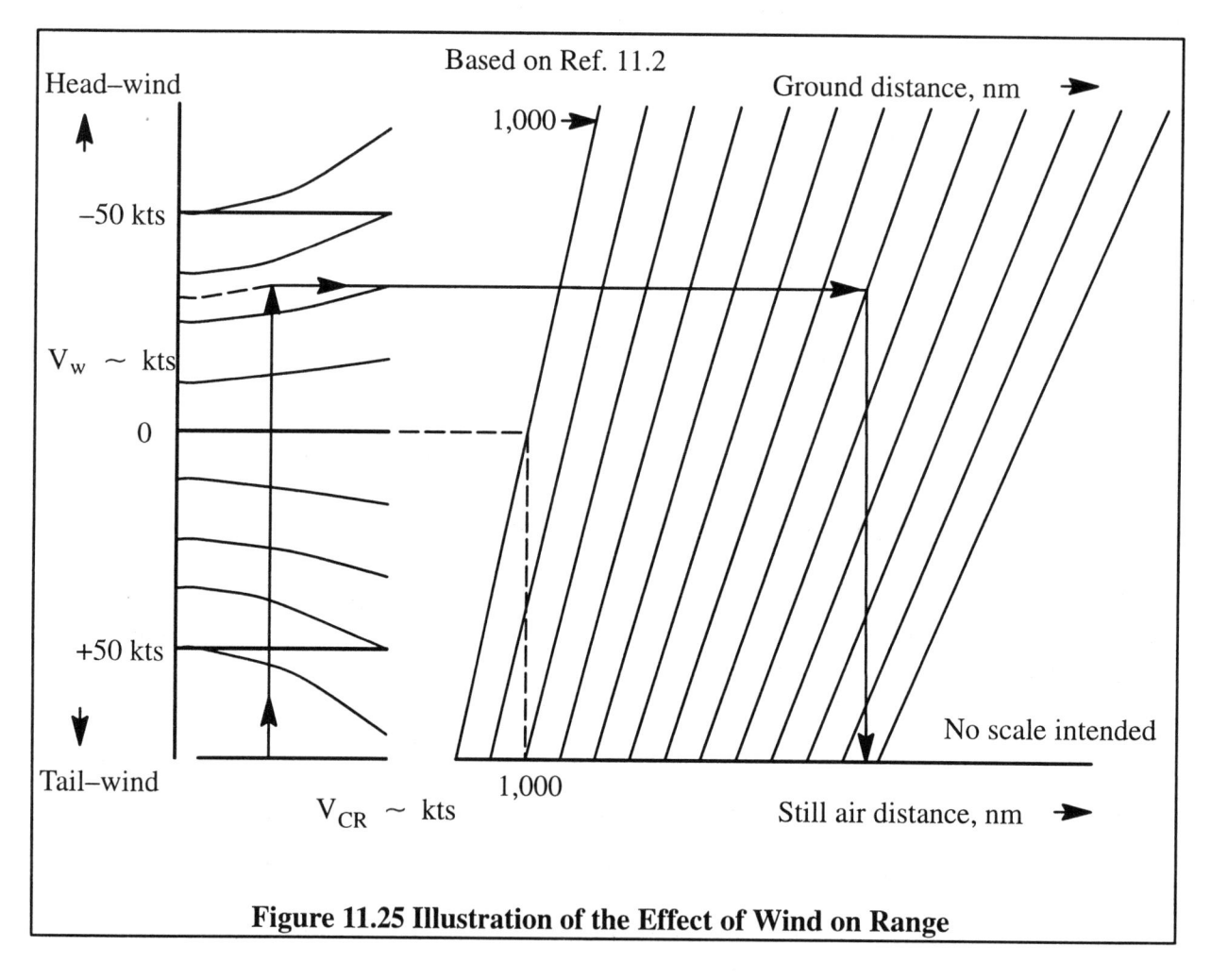

Figure 11.25 Illustration of the Effect of Wind on Range

The study of Ref. 11.1 also includes analytical expressions for best range (ground distance) with wind effect for jet airplanes. This study assumed parabolic drag polars. It was found that range improvement in the presence of head-wind is possible by increasing the cruise speed, V_{CR}. The study showed that with a head-wind equal to 0.4 times the best range speed in still air, a 6.4% increase in range can be attained by increasing the still air best range speed by 22%.

For high speed jet airplanes this finding is not practical because it would mean a cruise speed well into the dragrise. Again, the effect of Mach number on the cruise drag polar was not accounted for in this study.

In a general situation the following procedure is recommended to determine the specific range, S.R. with wind effect. Replace Eqn (11.79) by:

$$\text{S.R.} = \frac{V_g}{T_{reqd} c_j} \quad \text{in nm/lbs} \tag{11.83}$$

where: $V_g = V_{CR} \pm V_w$, the ground speed in kts

Next, for a given altitude, carry out the following step-by-step procedure:

Step 1: Select a range of values for V or M

Step 2: Calculate the ground speed: $V_g = V_{CR} \pm V_w$

Step 3: Calculate C_L/C_D with $C_L = W/\bar{q}S$ and account for the effect of M on C_D.

Step 4: Obtain the thrust required from: $T_{reqd} = D = WC_D/C_L$.

Step 5: Obtain c_j from engine data at the appropriate values of M, altitude and thrust

Step 6: Find S.R. with Eqn (11.83)

The problem of determining the best endurance performance is similar to that for obtaining the best range. Best endurance occurs when the specific endurance, S.E. is maximized. It can be obtained by superimposing fuel flow data on thrust required data as shown generically in Figure 11.26.

Lines of constant $\dot{W}_f/\delta\sqrt{\theta}$ are seen to increase in the direction of lines for higher T_{reqd}/δ. Therefore, a minimum fuel flow at a given value of W/δ will occur at the point of tangency between the two sets of curves. At the point of tangency for a given W/δ, the value of $\dot{W}_f/\delta\sqrt{\theta}$ will assume a minimum. Since S.E. is inversely proportional to $\dot{W}_f$, it will be maximized at the same point.

In most practical situations it is found that at the speed for minimum fuel flow the controllability of a jet airplane may become a problem. Therefore, with very little sacrifice in fuel maximum endurance missions are typically flown at speeds slightly above that for minimum fuel flow. This is also indicated in Figure 11.26.

Figure 11.26 Procedure for Determining Endurance of Jet Airplanes

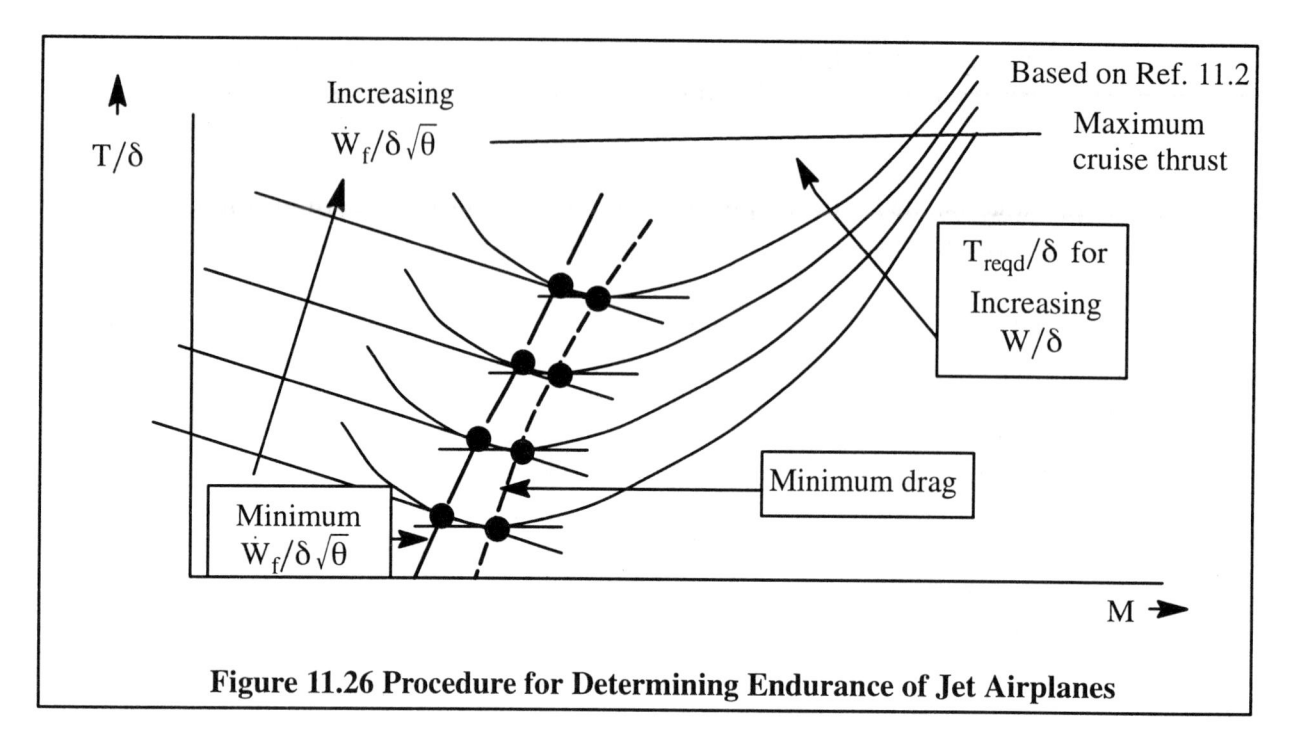

Holding at constant altitude is another aspect of best endurance which is of great practical interest. The technique used in obtaining the best endurance at constant altitude is similar to that explained previously. The speed schedule is determined from a thrust versus velocity or Mach number plot at minimum drag, for a series of selected, constant altitudes. Using this speed schedule, the fuel flow for the corresponding weight values and for the selected altitudes is determined. The results can be plotted as shown generically in Figure 11.27. It is seen, that for a given weight, the best endurance for a constant altitude operation would occur at a very specific altitude.

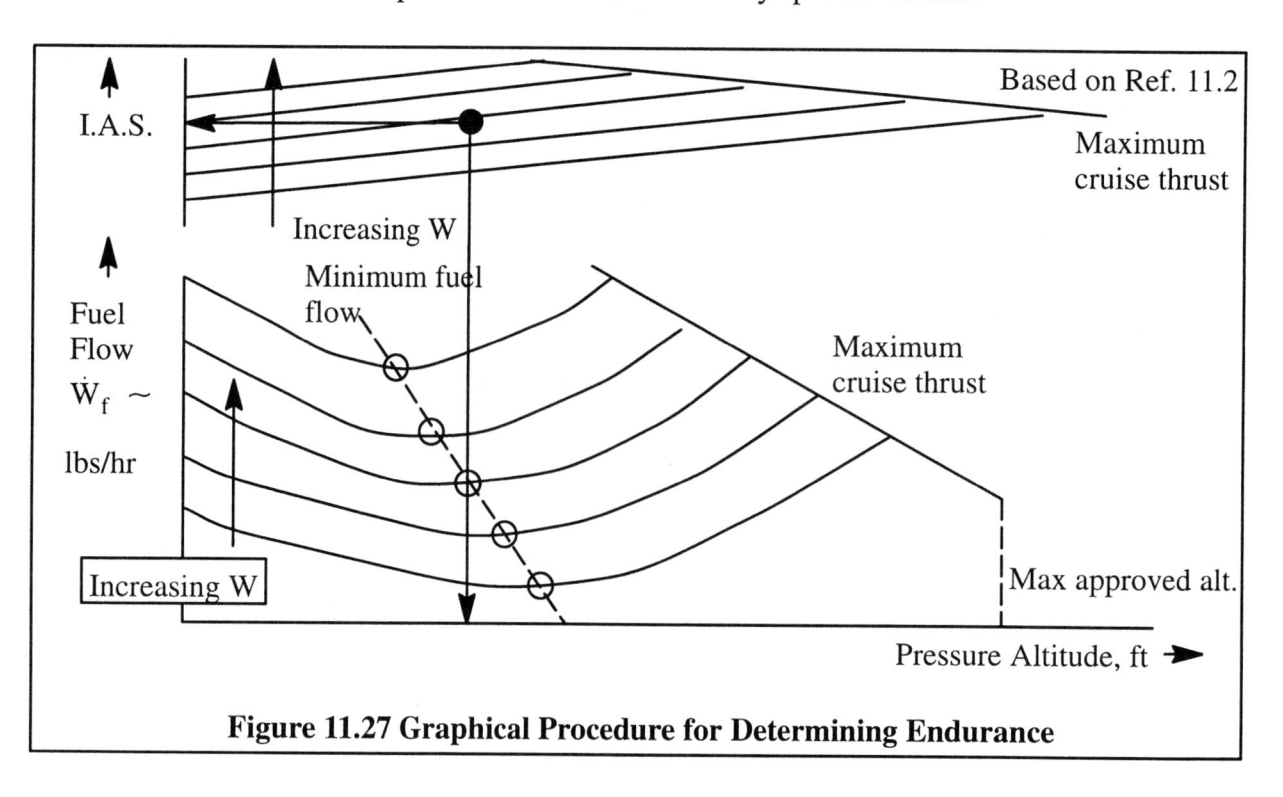

Figure 11.27 Graphical Procedure for Determining Endurance

11.3 THE PAYLOAD–RANGE DIAGRAM AND METHODS FOR PRESENTING USEFUL PAYLOAD–RANGE DATA

The payload which can be carried while flying a given range is of prime interest to commercial and military operators. Before actual payload range data can be presented in any form it is necessary to determine the amount of fuel used during all parts of an airplane mission, not just for range and endurance segments. Methods for determining the total amount of fuel used during a given mission were presented in Sections 11.1 and 11.2. An explanation of the so–called payload–range diagram is given in Sub–section 11.3.1.

A method for presenting the range characteristics of an airplane, accounting for the ability to take–off from given field–lengths is also given in Sub–section 11.3.2.

11.3.1 THE PAYLOAD–RANGE DIAGRAM

Figure 11.28 shows how a payload–range diagram is constructed. Starting at the bottom, the useful load of an airplane is defined as the difference between the maximum allowable take–off weight and the operating weight empty:

$$W_{useful} = W_{TO_{max}} - W_{OWE} \qquad (11.84)$$

The useful load consists of the fuel weight and the payload weight:

$$W_{useful} = W_F + W_{PL} \qquad (11.85)$$

Although of no practical interest, if the useful load is equal to the payload weight, the allowable fuel weight is zero and therefore the negotiable range will be zero. This situation is represented by point A in Figure 11.28. Transport airplanes are typically designed to carry a design payload weight, $W_{PL_{design}}$, over a given design range, R_{design}. The corresponding amount of fuel needed to negotiate that range is the fuel weight for the design payload, $W_{F_{design\ payload}}$. In Figure 11.28 this case is represented by point B. The design range corresponding to point B is also called the harmonic range.

If it is desired to fly a longer range, more fuel will be required. Because the useful weight defined by Eqn (11.85) cannot be exceeded, it is necessary to reduce payload while substituting fuel. This can continue until the volumetric limit of the airplane fuel tanks is reached. That case is reflected by Point C in Figure 11.28. The corresponding range is the maximum fuel volume range, $R_{max\ fuel}$.

The only way to achieve a longer range beyond point C is to eliminate payload weight. This can continue until there is no payload left. That situation is depicted by Point D in Figure 11.28. The corresponding range is called the ferry range: R_{ferry}.

It is conventional to plot payload versus range as shown in the upper part of Figure 11.28. This type of plot is referred to as the payload–range diagram. The shaded area is referred to as the profit potential of the airplane.

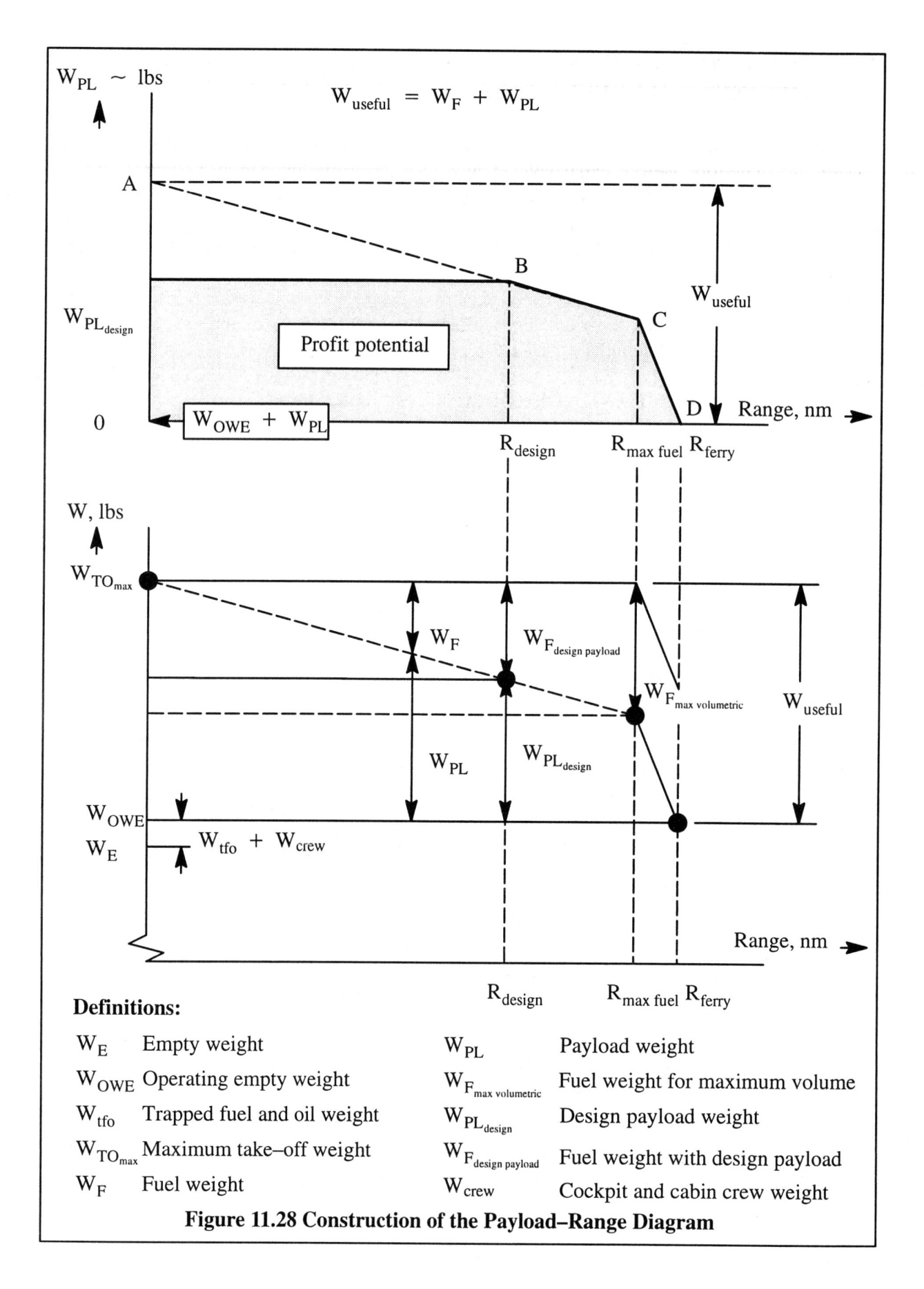

Range, Endurance and Payload–Range

Figure 11.28 Construction of the Payload–Range Diagram

Definitions:

W_E	Empty weight	W_{PL}	Payload weight
W_{OWE}	Operating empty weight	$W_{F_{max\ volumetric}}$	Fuel weight for maximum volume
W_{tfo}	Trapped fuel and oil weight	$W_{PL_{design}}$	Design payload weight
$W_{TO_{max}}$	Maximum take-off weight	$W_{F_{design\ payload}}$	Fuel weight with design payload
W_F	Fuel weight	W_{crew}	Cockpit and cabin crew weight

11.3.2 METHOD FOR PRESENTING PAYLOAD–RANGE DATA TO ACCOUNT FOR FIELDLENGTH CAPABILITY

The payload range diagram is useful for designers and for operators in comparing payload–range capabilities of various airplanes. However, it must not be forgotten that airplanes also have take–off fieldlength limitations. This was discussed in Chapter 10. It is quite possible that one airplane can carry a given payload out of a given field and negotiate some desired range, while another airplane cannot! Therefore, in comparing airplane capabilities this important feature must be considered. One way to represent the ability of an airplane to carry a given payload out of a given field, while negotiating some desired range was shown in Figure 10.26.

11.4 REGULATIONS FOR RANGE AND ENDURANCE FUEL RESERVES

When flying a given route, airplanes must be operated so that there are adequate fuel reserves on board to cope with various contingencies. Typical of these are:

* Destination airport is closed and landing at an alternate airport is required

* Head–winds are stronger than expected

* An engine fails while in cruise and the airplane must either return to the airport of origin, continue to its destination or deviate to an alternate airport.

The amount of fuel reserves which must be carried on a commercial airplane is regulated by FAR Parts 121 and 135 of Reference 11.3. For military airplanes the rules of Refs 11.4 and 11.5 apply. However, the military services reserve the right to redefine fuel reserves for any airplane depending on in–flight refuelling capabilities and other mission oriented needs.

A summary of the fuel reserve requirements of FAR Parts 121 and 135 is presented in Sub–section 11.4.1. An summary of the military requirements is given in Sub–section 11.4.2.

11.4.1 SUMMARY OF FAR PARTS 121 AND 135 REQUIREMENTS FOR FUEL RESERVES

The fuel reserves which must be carried on board civil airplanes are defined in FAR Parts 121.641, 121.643 and 121.645. A summary of these fuel reserve requirements is given in Table 11.3.

In addition, FAR 121.647 stipulates that in determining the fuel reserves required for any mission, the following factors must be considered:

a) Wind and other weather conditions forecast
b) Anticipated traffic delays
c) One instrument approach and possible missed approach at destination
d) Any other conditions that may delay landing of the airplane

Table 11.3 **Summary of Fuel Reserve Requirements According to FAR Parts 121 and 135 (Ref. 11.3)**

Fuel reserves are in addition to the fuel required to fly to and land at the airport of destination

Part 121.641: Non–turbine and turbo–propeller–powered airplanes: flag air carriers

For domestic as well as for international operations:

- a) Fuel needed to fly to and land at the most distant alternate airport specified in the dispatch release.

- b) Fuel needed to fly for 30 minutes plus 15% of the total time required to fly at normal cruise fuel consumption to the airport of destination or the airport of a) OR:
  to fly for 90 minutes at normal cruise fuel consumption, whichever is less.

The reserve requirements a) and b) are additive.

- c) Fuel needed to fly for three hours at normal cruise fuel consumption after reaching the airport of destination

Required fuel reserves are those defined by {a) + b)} or by c), whichever is greater.

Part 121.643: Non–turbine and turbo–propeller–powered airplanes: supplemental air carriers and commercial operators

For domestic operations:

- a) Fuel needed to fly to and land at the most distant alternate airport specified in the dispatch release.

- b) Fuel needed to fly for 45 minutes at normal cruise fuel consumption after reaching the airport of destination

For international operations:

- c) Fuel needed to fly for 30 minutes plus 15% of the total time required to fly at normal cruise fuel consumption to the airport of destination or the airport of a) OR:
  to fly for 90 minutes at normal cruise fuel consumption, whichever is less.

For domestic as well as for international operations:

- d) Fuel needed to fly for three hours at normal cruise fuel consumption after reaching the airport of destination

Required fuel reserves are those defined by {a) + b)} or by c) or by d) whichever is greater.

Table 11.3	(Cont'd) Summary of Fuel Reserve Requirements According to FAR Parts 121 and 135 (Ref. 11.3)

Fuel reserves are in addition to the fuel required to fly to and land at the airport of destination

Part 121.645: Turbine–engine powered airplanes, other than turbo–propeller–powered airplanes: flag and supplemental air carriers and commercial operators

For domestic and for international operations of flag air carriers and for international operations of supplemental air carriers and commercial operators:

    a) Fuel needed to fly to and land at the most distant alternate airport specified in the dispatch release.

    b) Fuel needed to fly for 10% of the total time required to fly at normal cruise fuel consumption to the airport of destination.

    c) Fuel needed to fly for 30 minutes at holding speed at 1,500 ft above the alternate airport (or at the destination airport if no alternate is required) under standard temperature conditions..

The reserve requirements a), b) and c) are additive.

    d) Fuel needed to fly for two hours at normal cruise fuel consumption after reaching the airport of destination if no alternate airport has been specified.

Required fuel reserves are those defined by {a) + b) + c)} or by d), whichever is greater.

For domestic operations of supplemental air carriers or commercial operators:

    The requirements of Part 121.643 apply.

Part 135.97: Air taxi operators and commercial operators of small airplanes. See Part 135.1 for applicability to airplane types and operations.

    a) Day time (VFR): fuel needed to fly for 30 minutes at normal cruise fuel consumption to the airport of destination.

    b) Night time (VFR): fuel needed to fly for 60 minutes at normal cruise fuel consumption to the airport of destination.

    c) (IFR): fuel needed to fly to and land at a suitable alternate airport.

11.4.2 SUMMARY OF MILITARY REQUIREMENTS FOR FUEL RESERVES

Military rules for fuel reserves are given in Refs 11.4 and 11.5. A summary of these rules is given in Table 11.4. However, the military services reserve the right to redefine fuel reserves for any airplane depending on in–flight refuelling capabilities and other mission oriented needs.

Table 11.4 Summary of Military Fuel Reserve Requirements According to Refs. 11.4 and 11.5.
Ref. 11.4: MIL–C–005011B (USAF) Fuel reserves are in addition to the fuel required to fly to and land at the airfield of destination and may be specified by one of the following requirements at the option of the procuring activity. Examples of fuel reserve specifications are: a) Fuel needed to fly a ground controlled approach; a wave–off, go–around and a second successful landing. b) Fuel as a percentage of the fuel required for the mission. c) Fuel consumed during a specified time of operation at a specified power and at a specified altitude d) The greater of the fuel required for 10% of the mission time or 20 minutes minutes at maximum endurance speed at 10,000 ft. e) Fuel required to fly to an alternate field (specify distance) plus a specified time at a specified speed at a specified altitude and account for landing.
Ref. 11.5: AS–5263 (US NAVY) Fuel reserves are in addition to the fuel required to fly to and land at the airfield of destination and may be specified by one of the following requirements at the option of the procuring activity. Examples of fuel reserve specifications are the same as those of Ref. 11.4. For carrier operations the required fuel reserves are the greater of: a) Fuel equal to 10% of the mission fuel required, including one approach to the carrier, a wave–off, a go–around, a second approach and a trap. b) Fuel equal to 20 minutes of loiter (30 minutes for cargo and transport airplanes) at sea–level at speeds for maximum endurance with all engines operating plus 5% of required mission fuel (internal plus external).

11.5 SIZING OF WEIGHT OF AIRPLANES TO GIVEN MISSION PAYLOAD, RANGE AND ENDURANCE REQUIREMENTS

Sizing an airplane to meet certain range and/or endurance requirements is one of several tasks designers must perform during early design of a new airplane. An airplane is said to have been sized for a given mission, if its take–off weight, W_{TO}, empty weight, W_E, and required mission fuel weight, W_F, have been established. A simple sizing method is discussed in this section. The method is based on that of Ref. 11.6. Software to conduct the sizing of an arbitrary airplane to an arbitrary mission requirement (civil or military) is described in Appendix B.

It is assumed that a mission specification and a mission profile are available for the airplane. An example mission specification is shown in Table 11.5.

It is shown in Ref. 11.6 that the take–off weight for an arbitrary airplane, sized to an arbitrary mission specification, can be solved from the following logarithmic equation:

$$\log_{10} W_{TO} = A + B \log_{10}(C\, W_{TO} - D) \tag{11.86}$$

where: A is a regression coefficient of known magnitude for a given airplane type

B is a regression coefficient of known magnitude for a given airplane type

A and B are coefficients which relate the airplane take–off weight to an allowable (state–of–the–art) airplane empty weight through the following equation:

$$\log_{10} W_{TO} = B \log_{10} W_E + A \tag{11.87}$$

Examples of numerical values for A and B and how they are determined are given in Sub–section 11.5.1.

C is defined as follows:

$$C = \{1 - (1 + M_{res})(1 - M_{ff}) - M_{tfo}\} \tag{11.88}$$

where: M_{ff} is defined as the mission, used fuel fraction. It is expressed as a fraction of the take–off weight, W_{TO}:

$$M_{ff} = \left(\frac{W_1}{W_{TO}}\right) \prod_{i=1}^{i=n} \left(\frac{W_{i+1}}{W_i}\right) \tag{11.89}$$

where: W_i is defined as the airplane weight at the end of mission phase i.

Table 11.5 shows a mission profile with 7 mission phases. Military missions can have many more mission phases. A method for determining the various weight fractions, W_{i+1}/W_i, will be presented in Sub-section 11.5.2.

M_{res} is defined as the reserve fuel fraction. It can be specified as a fraction of the mission fuel used, $W_{F_{used}}$, required:

$$W_{F_{res}} = M_{res}(1 - M_{ff})W_{TO} \qquad (11.90)$$

If the fuel reserves are defined in terms of an alternate mission (reserve fuel mission) the reserve fuel will be included in the mission, used fuel fraction, M_{ff}. In that case, and only in that case: $M_{res} = 0$.

M_{tfo} is defined as the trapped fuel and oil fraction. It is expressed as a fraction of the take-off weight:

$$W_{tfo} = M_{tfo} W_{TO} \qquad (11.91)$$

The trapped fuel and oil fraction is typically less than 0.5% of W_{TO}.

D is defined as follows:

$$D = W_{PL} + W_{crew} \qquad (11.92)$$

where: W_{PL} is the payload weight in lbs. This follows from the mission specification.

W_{crew} is the weight of the cockpit and cabin crew (with their luggage) needed to operate the airplane, in lbs.

Because Eqn (11.86) is logarithmic, a closed form solution cannot be given. However, iterative solutions are easily found. Whether or not solutions exist depends on the coefficients A, B, C and D. A discussion of three solution scenarios is given in Sub-section 11.5.3.

Once the coefficients A, B, C and D are determined and a solution for the take-off weight is found, the empty weight and the fuel weight are found from:

$$W_E = \text{invlog}_{10}\left\{\frac{(\log_{10} W_{TO} - A)}{B}\right\} \qquad (11.93)$$

and:

$$W_F = (1 - M_{ff})W_{TO} + W_{F_{res}} \qquad (11.94)$$

An example application is presented in Sub-section 11.5.4.

Range, Endurance and Payload–Range

Table 11.5 Mission Specification for a 50 Passenger Jet Transport

Role: Trans–continental air transportation is currently done with airplanes such as the B–737, MD–80/82 and B–757. There are many markets where passenger demand does not justify utilizing these relatively large airplanes. For such 'long, thin' markets a small passenger airplane with very long range capability might be ideal.

Payload: 50 Passengers, 175 lbs each plus 30 lbs of baggage
4 first class seats, 46 coach class seats

Crew: Cockpit, 2; Cabin, 2, 175 lbs each plus 30 lbs of baggage

Performance:
- Range: Still air range of 2,500 nm with reserves equal to 15 percent of mission fuel
- Speed: Mach 0.82 at 45,000 ft (470 kts)
- Fieldlength: 5,000 ft at sea–level, 90 deg. F. day
- Climb: Service ceiling of 50,000 ft is desired
  Direct climb to 45,000 ft in 25 minutes is desired

Powerplants: Two turbo–fans

Pressurization: 8,000 ft cabin at 45,000 ft   **Certification:** FAR 25

1) Engine start and warm–up
2) Taxi
3) Takeoff
4) Climb to 45,000 ft
5) Cruise
6) Descent
7) Landing, taxi, shutdown

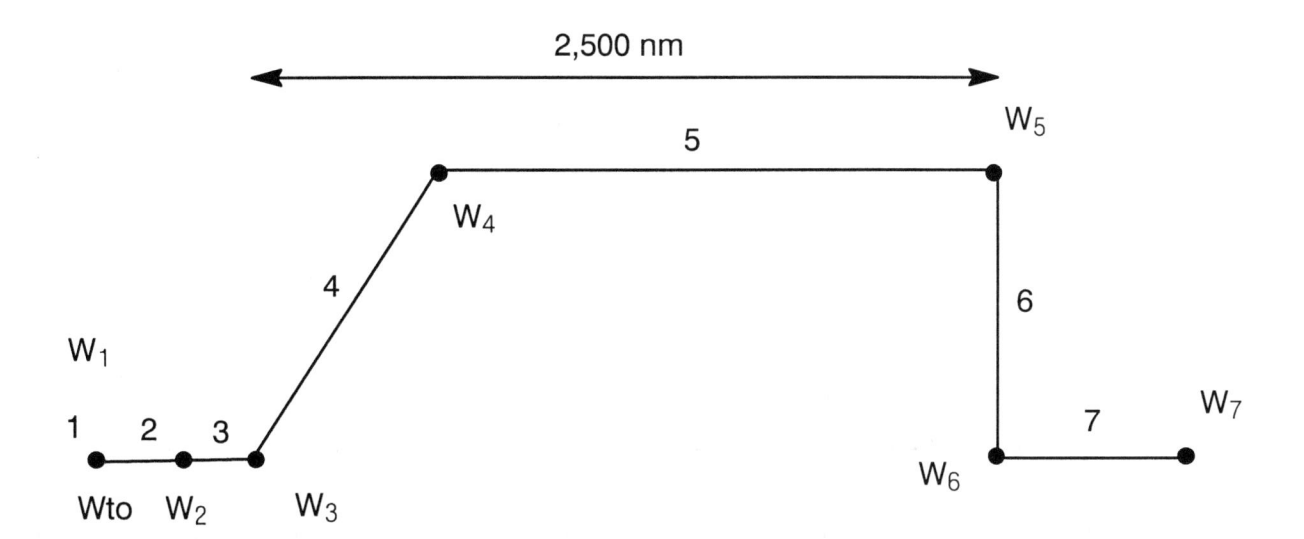

Chapter 11

567

11.5.1 DETERMINATION OF THE REGRESSION COEFFICIENTS A AND B WITH NUMERICAL EXAMPLES

For a given airplane type, the regression coefficients A and B are obtained by first constructing a logarithmic plot of actual airplane data for take–off weight versus empty weight. An example is shown in Figure 11.29. Second, a straight line is determined such that the mean square error relative to the empty weight prediction for a given take–off weight is minimized. The resulting straight line is represented by Eqn (11.87). The coefficient A can be thought of as the intercept coefficient. The coefficient B should be interpreted as the slope of the line.

Data for the construction of a plot like Figure 11.29 can be obtained from manufacturers, from Ref. 11.7 (Jane's All the World's Aircraft) and/or from various aeronautical magazines.

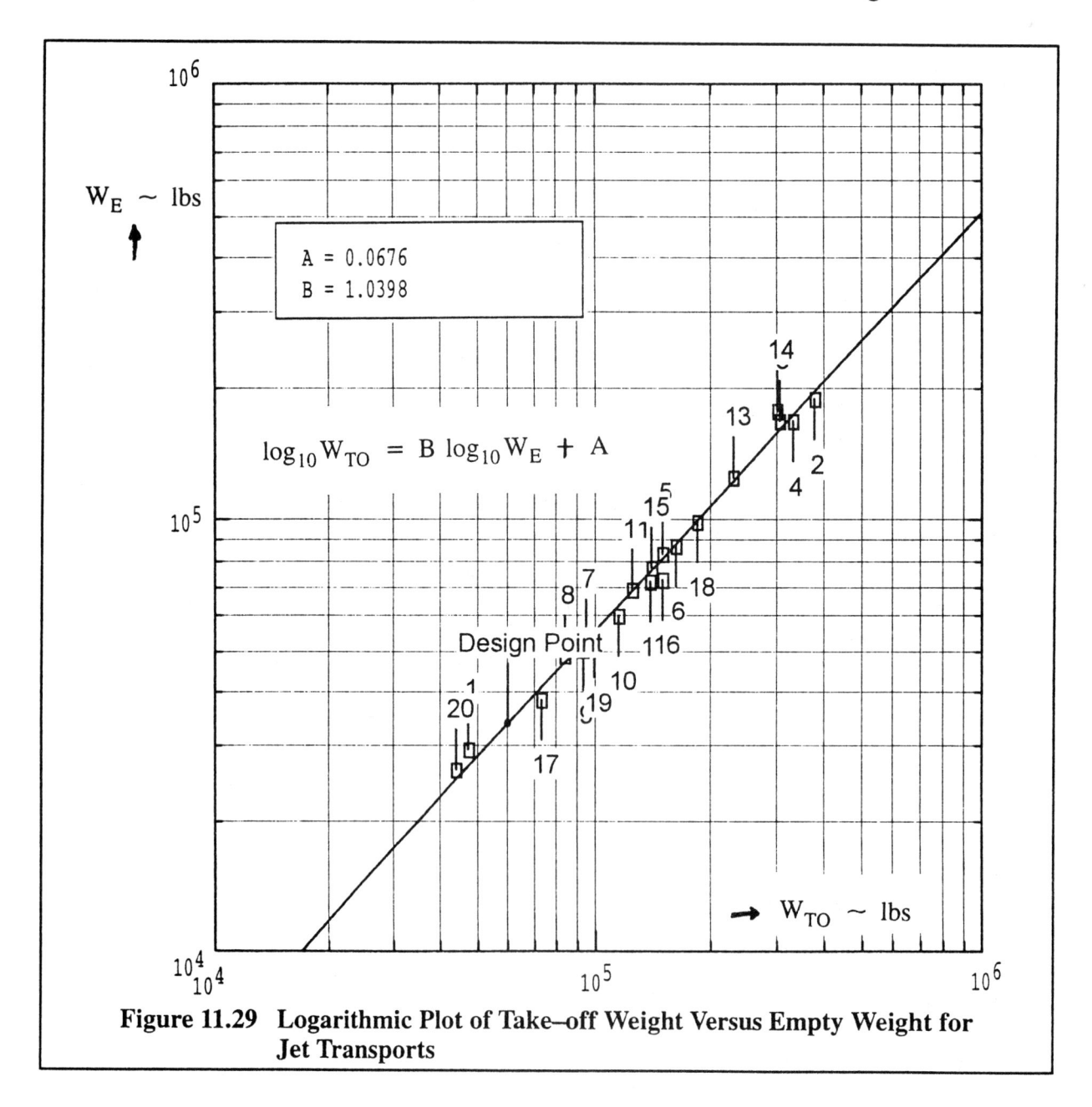

Figure 11.29 Logarithmic Plot of Take–off Weight Versus Empty Weight for Jet Transports

Numerical values for the regression coefficients A and B were determined for 12 types of airplanes. The result is given in Table 11.6. For any new project the reader is encouraged to determine his own regression coefficients from the most recent airplane weight data which may be available. The software described in Appendix B can be used to rapidly determine these coefficients and also to generate a plot such as Figure 11.29.

Table 11.6	Numerical Examples for the Regression Coefficients A and B in Eqn (11.87) for twelve Types of Airplanes					
Airplane Type	A	B	Airplane Type	A	B	
1. Homebuilts, Pers. Fun and Transp.	0.3411	0.9519	8. Military trainers Jets	0.6632	0.8640	
Scaled fighters	0.5542	0.8654	Turbo–props	0.1677	0.9978	
Composites	0.8222	0.8050	Piston–props	0.5627	0.8761	
2. Single engine propeller driven	–0.1440	1.1162	9. Fighters Jets with external load	0.5091	0.9505	
3. Twin engine propeller driven	0.0966	1.0298	Jets w/o external load	0.1362	1.0116	
Composites	0.1130	1.0403	Turbo–props with external load	0.2705	0.9830	
4. Agricultural	–0.4398	1.1946	10. Mil. Patrol, Bomb and transports Jets	–0.2009	1.1037	
5. Business jets	0.2678	0.9979	Turbo–props	–0.4179	1.1446	
6. Regional turbo–props	0.3774	0.9647	11. Flying boats, amphib. and float airplanes	0.1703	1.0083	
7. Transport jets	0.0833	1.0383	12. Supersonic cruise	0.4221	0.9876	

11.5.2 METHODS FOR DETERMINING THE WEIGHT FRACTIONS IN EQN (11.88)

In this Sub–section, methods will be presented for the numerical evaluation of each weight fraction in Eqn (11.88). Before doing this, recall the definition of the weight W_i : it is is the airplane weight at the end of mission phase i.

Each mission profile consists of mission phases which are numbered sequentially. Table 11.5 shows an example of a transport mission profile with 7 mission phases. Military airplane missions can have many more mission phases.

The fuel fraction of a mission phase is defined as the ratio of airplane weight at the end of that phase to the airplane weight at the beginning of that phase. Therefore, the weight fraction, W_{i+1}/W_i , represents the fuel fraction for phase (i+1).

All mission flight phases are split into two categories:

1.) Fuel un–intensive flight phases. These are flight phases during which a relatively small amount of fuel is used. Examples are: engine start and warm–up, taxi, take–off, short climbs, short descents, landing, taxi and shut–down.

2.) Fuel intensive flight phases. These are flight phases during which a relatively large amount of fuel is used. Examples are: prolonged climbs, cruise, loiter and prolonged descents.

During the initial weight sizing of an airplane it is acceptable to use statistically obtained data for the fuel fractions of fuel un–intensive flight phases. However, for the fuel intensive flight phases more accurate estimates must be made. The latter is done with the help of the Breguet equations for range and for endurance which were developed in Sections 11.1 and 11.2.

Fuel fractions for fuel un–intensive flight phases may be obtained from Table 11.7.

To obtain the fuel fractions for fuel intensive flight phases the Breguet equations for range and endurance are used. These equations are taken from Eqns (11.7) and (11.8) for propeller driven airplanes, and from Eqns (11.60) and (11.63) for jet airplanes. These equations are:

For propeller driven airplanes:

$$R = 326 \frac{\eta_{p_{installed}}}{c_p} \frac{C_L}{C_D} \ln\left(\frac{W_{CR_i}}{W_{CR_{i+1}}}\right) \quad \text{in n.m.} \tag{11.95}$$

$$E = 326 \frac{\eta_{p_{installed}}}{c_p V_{LTR}} \frac{C_L}{C_D} \ln\left(\frac{W_{LTR_i}}{W_{LTR_{i+1}}}\right) \quad \text{in hours} \quad \text{with } V_{LTR} \text{ in kts} \tag{11.96}$$

For jet airplanes:

$$R = \left(\frac{V_{CR} \, C_L/C_D}{c_j}\right) \ln \frac{W_{CR_i}}{W_{CR_{i+1}}} \quad \text{in nm} \quad \text{with } V_{CR} \text{ in kts} \tag{11.97}$$

$$E = \left(\frac{1}{c_j} \frac{C_L}{C_D}\right) \ln \frac{W_{LTR_i}}{W_{LTR_{i+1}}} \quad \text{in hrs} \tag{11.98}$$

For each particular cruise phase, loiter phase or prolonged climb phase, the range, R, or the endurance (loiter time), E, will be known from the mission specification. The Breguet equations (11.95) through (11.98) are then solved for the inverses of the begin to end weight ratios to produce the corresponding fuel fractions: W_{i+1}/W_i.

Values for the various efficiency parameters, $\eta_{p_{installed}}$, c_p, C_L/C_D and c_j are obtained either from actual data or from "state–of–the–art" ranges as given in Table 11.8.

Table 11.7 Suggested Fuel Fractions for Fuel Un–intensive Mission Phases

Mission Phase / Airplane Type	Start & Warm–up	Taxi	Take–off	Brief descent	Brief Climb	Landing, Taxi & Shutdown
1. Homebuilts	0.998	0.998	0.998	0.995	0.995	0.995
2. Single Engine	0.995	0.997	0.998	0.992	0.993	0.993
3. Twin Engine	0.992	0.996	0.996	0.990	0.992	0.992
4. Agricultural	0.996	0.995	0.996	0.998	0.999	0.998
5. Business Jets	0.990	0.995	0.995	0.980	0.990	0.992
6. Regional TBP's	0.990	0.995	0.995	0.985	0.985	0.995
7. Transport Jets	0.990	0.990	0.995	0.980	0.990	0.992
8. Mil. Trainers	0.990	0.990	0.990	0.980	0.990	0.995
9. Fighters	0.990	0.990	0.990	0.960–0.900	0.990	0.995
10. Mil. Patrol, Bmb & Trspt	0.990	0.990	0.995	0.980	0.990	0.992
11. Flying Boats, Amph. & Floats	0.992	0.990	0.996	0.985	0.990	0.990
12. Supersonic Cruise	0.990	0.995	0.995	0.920–0.870	0.985	0.992

Table 11.8 Suggested Values for the Efficiency Parameters in Eqns (11.95 – 11.98)

Mission Phase / Airplane Type	Cruise C_L/C_D	Cruise c_j	Cruise c_p	Cruise $\eta_{p_{installed}}$	Loiter C_L/C_D	Loiter c_j	Loiter c_p	Loiter $\eta_{p_{installed}}$
1. Homebuilts	8 – 10		0.6 – 0.8	0.70	10 – 12		0.5 – 0.7	0.65
2. Single Engine	8 – 10		0.5 – 0.7	0.80	10 – 12		0.5 – 0.7	0.70
3. Twin Engine	8 – 10		0.5 – 0.7	0.82	9 – 11		0.5 – 0.7	0.72
4. Agricultural	5 – 7		0.5 – 0.7	0.82	8 – 10		0.5 – 0.7	0.72
5. Business Jets	10 – 13	0.5 – 0.8			12 – 14	0.4 – 0.6		
6. Regional TBP's	11 – 13		0.4 – 0.6	0.85	14 – 16		0.4 – 0.6	0.77
7. Transport Jets	13 – 16	0.5 – 0.8			14 – 17	0.4 – 0.6		
8. Mil. Trainers	8 – 10	0.5 – 1.0	0.4 – 0.6	0.82	10 – 14	0.4 – 0.6	0.5 – 0.7	0.77
9. Fighters	5 – 7	0.5 – 1.2	0.5 – 0.7	0.82	6 – 9	0.6 – 0.8	0.5 – 0.7	0.77
10. Mil. Patrol, Bmb & Trspt	13 – 16	0.5 – 0.8	0.4 – 0.7	0.82	14 – 17	0.4 – 0.6	0.5 – 0.7	0.77
11. Flying Boats, Amph. & Floats	10 – 12	0.5 – 0.8	0.5 – 0.7	0.82	13 – 16	0.4 – 0.6	0.5 – 0.7	0.77
12. Supersonic Cruise	5 – 9	0.7 – 1.2			6 – 9	0.6 – 0.8		

11.5.3 THREE SOLUTION SCENARIOS FOR EQN (11.86)

As indicated before, solutions to for airplane take–off weight, W_{TO}, from Eqn (11.86) depend on the numerical values of the coefficients A, B, C and D. Solutions to Eqn (11.86) can be visualized by plotting the left hand side, $\log_{10} W_{TO}$, and the right hand side, $\{A + B \log_{10}(C W_{TO} - D)\}$, as a function of $\log_{10} W_{TO}$. A generic example is shown in Figure 11.30.

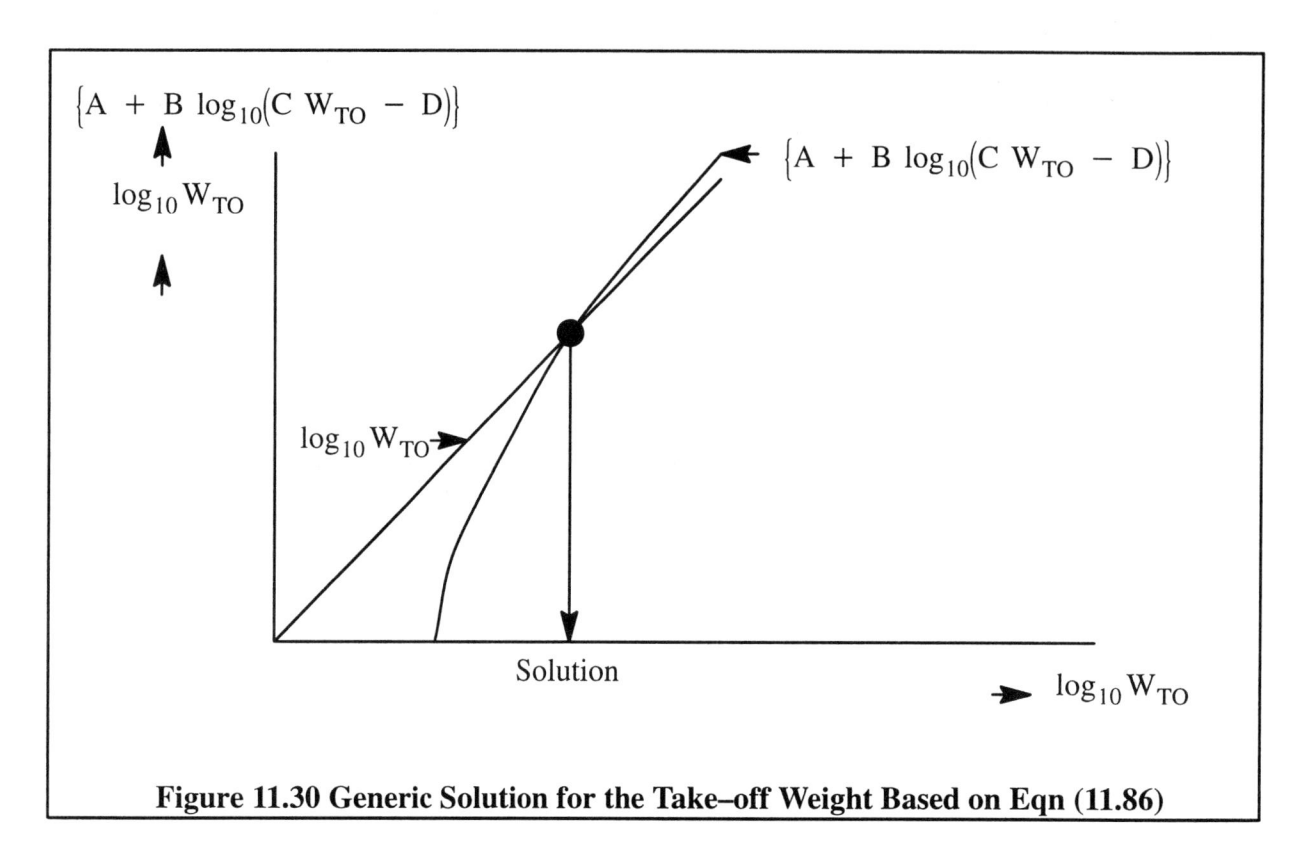

Figure 11.30 Generic Solution for the Take–off Weight Based on Eqn (11.86)

Depending on the aggressiveness of the mission requirement (extremely long range with high payload would be an example of an aggressive mission requirement), which determines the coefficients C and D, and depending on the airplane state–of–the–art, which determines the coefficients A and B, three solution scenarios are possible.

These three solution scenarios are shown in Figure 11.31.

Whenever one solution is indicated, all is well.

Whenever two solutions are indicated, the lowest weight solution should normally be used.

Whenever no solution exists, either the mission requirement is too aggressive or inappropriate values were used for the efficiency parameters $\eta_{p_{installed}}$, c_p, C_L/C_D and c_j in defining the fuel fractions for the various mission phases. An example application is presented in Sub–section 11.5.4.

Range, Endurance and Payload–Range

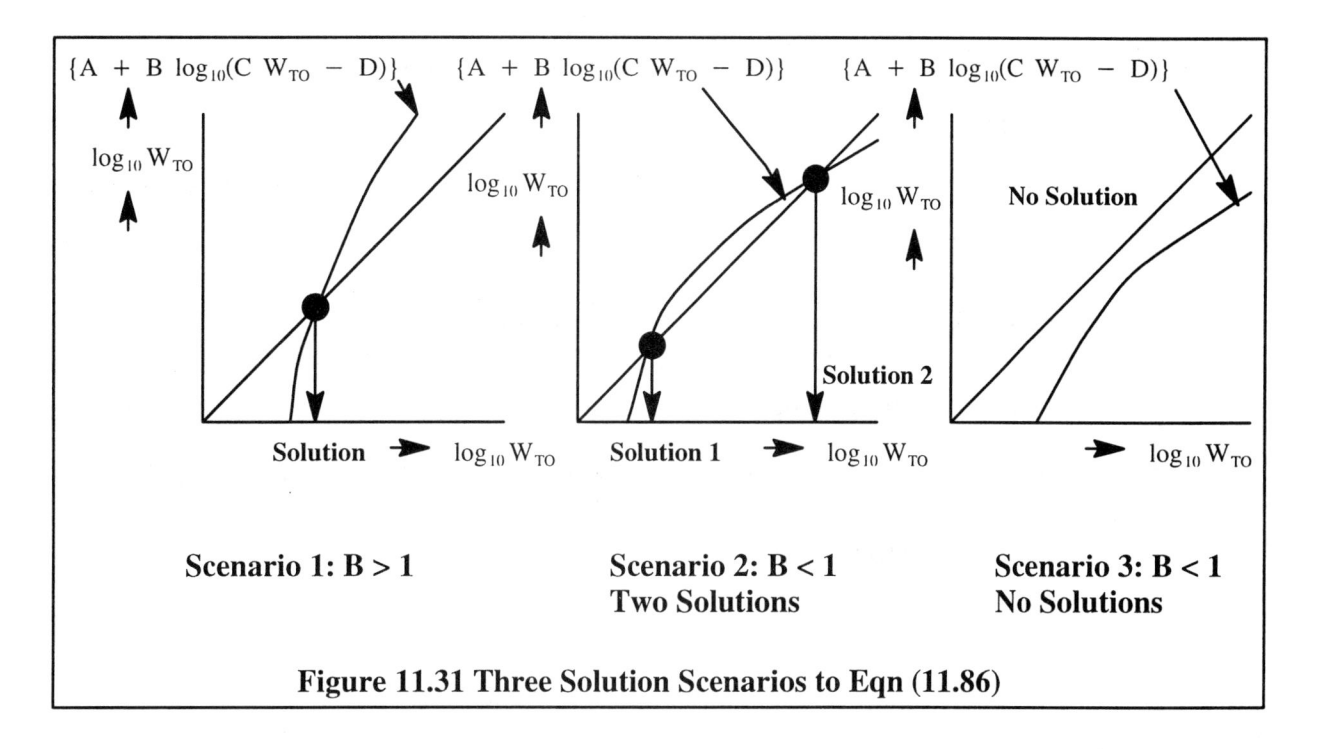

Figure 11.31 Three Solution Scenarios to Eqn (11.86)

11.5.4 EXAMPLE APPLICATION OF AIRPLANE WEIGHT SIZING

In this example the airplane with the mission specification of Table 11.5 will be used. The mission profile is seen to consist of seven phases.

A step–by–step procedure will be used to determine the following weights for this airplane:

Take–off weight, W_{TO} Empty weight, W_E Mission fuel weight, W_F

Step 1: Decide which mission phases are fuel intensive and which are not.

For this airplane it will be assumed that Phases 1, 2, 3 and 7 are fuel un–intensive. That means Table 11.7 will be used for estimating the corresponding fuel fractions.

Phases 4, 5 and 6 will be assumed to be fuel intensive. Breguet equations (11.97) and (11.98) will be used to determine the corresponding fuel fractions. Also, the data in Table 11.8 will be used to estimate the efficiencies: C_L/C_D and c_j.

Step 2: Determine the fuel fraction for Phase 1: W_1/W_{TO}. From Table 11.7 it is seen that:

$W_1/W_{TO} = 0.990$.

Step 3: Determine the fuel fraction for Phase 2: W_2/W_1. From Table 11.7 it is seen that:

$W_2/W_1 = 0.990$.

Chapter 11   573

Step 4: Determine the fuel fraction for Phase 3: W_3/W_2. From Table 11.7 it is seen that:

$W_3/W_2 = 0.995$.

Step 5: Determine the fuel fraction for Phase 4: W_4/W_3. Since this phase is considered to be fuel intensive, Breguet equation (11.98) will be used. The climb phase will be treated as an endurance phase. Table 11.5 calls for a direct climb to 45,000 ft in 25 minutes. Therefore, the endurance for this phase is: $E = 25/60 = 0.42$ hrs.

The corresponding climb s.f.c. is estimated with Table 11.8 at: $c_j = 0.6$ lbs/hr/lbs.

The corresponding climb L/D is estimated with Table 11.8 at: $C_L/C_D = 15.0$.

Substitution into Eqn (11.98) yields:

$$0.42 = \left(\frac{1}{0.60} \times 15.0\right) \ln \frac{W_3}{W_4} \quad \text{or:} \quad W_4/W_3 = 0.983$$

Assuming an average climb speed of 325 kts, the range covered during the climb is:

$R_{CL} = 0.42 \times 325 = 137$ n.m.

Step 6: Determine the fuel fraction for Phase 5: W_5/W_4. Since this phase is considered to be fuel intensive, Breguet equation (11.97) will be used. Table 11.5 calls for a cruise speed of $V_{CR} = 470$ kts. Table 11.5 also calls for a total range of 2,500 n.m. The climb range can be subtracted to yield a design cruise range of:

$R_{CR} = 2,500 - 137 = 2,363$ n.m.

The corresponding cruise s.f.c. is estimated with Table 11.8 at:
$c_j = 0.57$ lbs/hr/lbs.

The corresponding climb L/D is estimated with Table 11.8 at: $C_L/C_D = 16.0$.

Substitution into Eqn (11.97) yields:

$$2,363 = \left(\frac{470 \times 16.0}{0.57}\right) \ln \frac{W_4}{W_5} \quad \text{or:} \quad W_5/W_4 = 0.836$$

Step 7: Determine the fuel fraction for Phase 6: W_6/W_5. Since this phase is considered to be fuel intensive, Breguet equation (11.98) will be used. The descent phase will be treated as an endurance phase.

Assuming an average rate-of-descent of 2,000 ft/min, the descent time is 45,000/2,000 = 22.5 min. = 0.38 hrs.

The corresponding descent s.f.c. is estimated with Table 11.8 at:

$c_j = 0.4$ lbs/hr/lbs.

The corresponding climb L/D is estimated with Table 11.8 at: $C_L/C_D = 15.0$.

Substitution into Eqn (11.98) yields:

$$0.38 = \left(\frac{1}{0.40} \times 15.0\right) \ln\frac{W_5}{W_6} \quad \text{or:} \quad W_6/W_5 = 0.990$$

Important note:

The average forward speed during the descent is assumed to be 350 kts. This would yield a descent range of 133 n.m. However, the mission profile of Table 11.5 suggests not to take credit for the descent range (note the vertical line for Phase 6. The design range of Step 6 therefore stands at 2,363 n.m.

Step 8: Determine the fuel fraction for Phase 7: W_7/W_6. From Table 11.7 it is seen that:

$W_7/W_6 = 0.992$.

Step 9: Determine the overall mission fuel fraction with Eqn (11.89):

$M_{ff} = 0.990 \times 0.990 \times 0.995 \times 0.983 \times 0.836 \times 0.990 \times 0.992 = 0.7874$

Step 10: Determine the coefficients A, B, C and D for Eqn (11.86).

A and B are determined by plotting empty weight versus take-off weight data for 20 transports. The data are taken from Ref. 11.7 and are shown in Table 11.9. These data are plotted in Figure 11.29. The resulting regression coefficients which describe the line drawn in Figure 11.29 are: A = 0.0676 and B = 1.0398.

The coefficient C is determined from Eqn (11.88).

The reserve fuel requirements are given in Table 11.5. This results in a reserve fuel fraction of: $M_{res} = 0.15$. Assuming that the trapped fuel and oil fraction is:

$M_{tfo} = 0.005$, the following is found for C, with Eqn (11.88):

$C = \{1 - (1 + 0.15)(1 - 0.7874) - 0.005 = \} = 0.7505$

The coefficient D is determined from Eqn (11.92) and Table 11.5:

$D = W_{PL} + W_{crew} = 54 \times 205 = 11,070$ lbs

Step 11: Calculate the take-off weight from Eqn (11.86).
The result is: $W_{TO} = 59,744$ lbs.

Step 12: Calculate the empty weight from Eqn (11.93).
The result is: $W_E = 33,771$ lbs.

Step 13: Calculate the required fuel reserves from Eqn (11.90):
The results is:

$$W_{F_{res}} = 0.15\,(1 - 0.7874)\,59,744 = 1,905 \text{ lbs}$$

Step 14: Calculate the mission fuel weight, W_F, from Eqn (11.94):
The results is:

$$W_F = (1 - 0.7874)\,59,744 + 1,905 = 14,607$$

It is of interest to determine and compare the amounts of fuel used during the various mission phases. This can be done most readily with the software described in Appendix B.

The result is shown in Figure 11.32. It is clear that climb and cruise are the most fuel intensive mission phases in this case.

When comparing the weight data obtained in Steps 11, 12, 13 and 14 with the software generated data in Figure 11.32 some differences occur. These differences are due to round-offs.

11.6 SUMMARY FOR CHAPTER 11

In this chapter methods for determining range and endurance of propeller driven and jet airplanes were developed and discussed. In addition to the classical Breguet equations, numerical procedures for the accurate determination of range and endurance were presented.

Methods to account for the fuel used during the entire mission, including the fuel reserves, were also discussed. The civil and military regulations for determining fuel reserves were summarized in tabular form.

An explanation was given of how a payload-range diagram can be constructed. How much payload weight can be carried over a given range while taking off from a given airfield is important to operators. A example diagram which allows an operator to determine this was presented.

Finally, a method for sizing an airplane to an arbitrary mission was presented.

Table 11.9 Correlation of Take-off Weight and Empty Weight Data for Twenty Jet Transports and the Regression Coefficients A and B for Eqn (11.87)

Take-off and Empty Weight Regression Coefficient Calculation

Input Parameter

Number	20

Output Parameters

A	0.0676
B	1.0398

Empty Weight - Take-Off Weight Table

#	Airplane Name	W_{TO} lb	W_E lb
1	1	47250.0	29430.0
2	2	375885.0	189900.0
3	3	305560.0	169100.0
4	4	330695.0	169360.0
5	5	149915.0	84000.0
6	6	162040.0	87493.0
7	7	95000.0	53300.0
8	8	84000.0	48700.0
9	9	93000.0	50100.0
10	10	115500.0	60200.0
11	11	124500.0	69100.0
12	12	138500.0	72400.0
13	13	230000.0	125760.0
14	14	300000.0	178100.0
15	15	140000.0	78121.0
16	16	149500.0	72857.0
17	17	73000.0	38383.0
18	18	184800.0	99700.0
19	19	99650.0	53462.0
20	20	44000.0	26550.0

Range, Endurance and Payload–Range

Take-Off Weight

Input Parameters

A	0.0676		$W_{p_{expend}}$	0.0	lb	M_{tfo}	0.500	%	$\Delta W_{F_{exp}}$	0.0	lb
B	1.0398		$W_{TO_{est}}$	110000.0	lb	M_{res}	15.000	%			
$M_{ff_{uc}}$	0.7874		W_p	11070.0	lb	ΔW_e	0.1000	%			

Output Parameters

W_E	33771.5	lb	W_F	14604.6	lb	W_{tfo}	298.7	lb	M_{ff}	0.7874
$W_{F_{used}}$	12699.7	lb	$W_{F_{reserve}}$	1905.0	lb	W_{TO}	59744.9	lb		

Mission Profile Table

	Mission Profile	Weight lb	W_{f_i} lb
1	Warmup	59744.9	597.4
2	Taxi	59147.5	591.5
3	Take-off	58556.0	292.8
4	Climb	58263.2	963.1
5	Cruise	57300.1	9396.5
6	Descent	47903.6	479.0
7	Land/Taxi	47424.6	379.4

Figure 11.32 Results of Airplane Weight Sizing and Estimation of Fuel Used

11.7 PROBLEMS FOR CHAPTER 11

11.1 Determine the maximum range, maximum endurance as well as the speeds for maximum range and endurance for an airplane with the following characteristics:

$S = 200$ ft^2 $W_{TO} = 10,000$ lbs $W_{F_{max}} = 4,000$ lbs

$h = 10,000$ ft std atm $c_p = 0.52$ lbs/hr/hp $\eta_{p_{installed}} = 0.90$

The airplane has the following power characteristics:

$V \sim$ kts	$P_{reqd} \sim$ hp	$V \sim$ kts	$P_{reqd} \sim$ hp
109	240	174	250
113	220	217	400
122	205	261	600
130	200	304	925
152	215	350	1,350

11.2 Determine the maximum range for the airplane of Problem 11.1 with a tail–wind of 30 kts.

11.3 A jet airplane has the following characteristics:

$S = 400$ ft^2 $W_{TO} = 18,000$ lbs $W_{F_{max}} = 6,000$ lbs

$h = 30,000$ ft std atm $c_j = 0.85$ lbs/hr/lbs $C_D = 0.0200 + 0.06\, C_L^2$

What is the maximum range and endurance of this airplane?

Assume that the drag divergence Mach number is 0.85 and that the drag polar is not affected by Mach number below M = 0.85.

11.4 A business airplane has the following characteristics:

$S = 232$ ft^2 $W_{TO} = 14,000$ lbs $W_{F_{max}} = 6,000$ lbs

$h = 31,000$ ft std atm $c_j = 0.96$ lbs/hr/lbs $C_D = 0.0230 + 0.0936\, C_L^2$

Determine the maximum range in nm and the altitude at the end of cruise, assuming the s.f.c. remains constant during the mission. Assume that the drag divergence Mach number is 0.84 and that the drag polar is not affected by Mach number below M = 0.84.

11.5 Assume that the airplane of problem 11.4 drops a payload of 1,500 lbs at some destination and then flies back at a different altitude. The weight at the destination, where the payload was dropped, is 9,000 lbs after release of this payload. Determine:

a) The distance flown when the payload was released
b) The fuel required to return to its base.
c) Whether or not there is enough fuel to return to its base.

11.6 For a propeller driven airplane in cruise flight, show that the best range (br) speed with wind effect is given by:

$$m_{br} = \left\{ \frac{2m_{br} \pm (V_w/V_{md})}{2m_{br} \pm 3(V_w/V_{md})} \right\}^{1/4} \qquad \text{where:} \quad m = V/V_{md}$$

V_{md} is the minimum drag speed

Assume that the drag polar is parabolic. The reader should consult Ref. 11.1.

11.7 Perform a trade study showing the effect of engine s.f.c. and cruise lift–to–drag ratio on the weights estimated in Sub–section 11.5.4. Plot the results.

11.8 Use the method of Section 11.5 to determine the estimated weights for the Boeing 777. Compare your estimates with the data in Ref. 11.7.

11.9 Use the method of Section 11.5 to determine the estimated weights for the Cessna 208 Caravan. Compare your estimates with the data in Ref. 11.7.

11.8 REFERENCES FOR CHAPTER 11

11.1 Hale, F.J. and Steiger, A.R.; Effects of Wind on Aircraft Cruise Performance; Journal of Aircraft, Volume 16, June 1979, pp. 382 – 387.

11.2 Anon.; Jet Transport Performance Methods, Boeing Document D6–1420, The Boeing Company, Seattle, WA, 1967.

11.3 Anon.; Code of Federal Regulations (CFR), Title 14, Parts 1 to 59, January 1, 1996; U.S. Government Printing Office, Superintendent of Documents, Mail Stop SSOP, Washington, D.C. Note: FAR 23 and FAR 25 are components of this CFR.

11.4 Anon.; MIL–C–005011B (USAF), Military Specification, Charts: Standard Aircraft Characteristic and Performance, Piloted Aircraft (Fixed Wing); June, 1977.

11.5 Anon.; AS–5263 (US Navy), Guidelines for the Preparation of Standard Aircraft Characteristic Charts and Performance Data, Piloted Aircraft (Fixed Wing); October, 1986.

11.6 Roskam, J.; Airplane Design, Part I; Preliminary Sizing of Airplanes; DAR Corporation, 120 East Ninth Street, Suite 2, Lawrence, KS 66046, 1997.

11.7 Jackson, P.; Jane's All the World's Aircraft; Jane's Information Group Ltd, 163 Brighton Road, Coulsdon, Surrey CR5 2NH, U.K., 1996–1997

CHAPTER 12: MANEUVERING AND THE FLIGHT ENVELOPE

In this chapter the performance of the airplane during maneuvers will be discussed. Airplane maneuverability is limited by several constraints. Examples of such constraints are:

* stall speed and minimum controllable speed
* engine temperature limitations
* low speed buffet
* structural limitations
* engine inlet distortion limitations
* engine thrust limits
* high speed buffet

When put together, these constraints form the flight envelope of an airplane. Maneuverability at the edges of the flight envelope must be determined for any airplane. The material in this chapter is organized as follows:

12.1 Stall speeds and minimum speeds
12.2 Buffet limits
12.3 Level flight maximum speeds and ceilings
12.4 The V–n diagram
12.5 Pull–up: instantaneous and sustained
12.6 Steady, level and coordinated turn: instantaneous and sustained
12.7 Spin
12.8 Accelerated flight performance

12.1 STALL SPEEDS AND MINIMUM SPEEDS

As shown in Chapters 8 through 11 the take–off, landing, approach and climb performance of airplanes is greatly affected by the stall speed and (or) the minimum controllable speed. These speeds are defined as follows:

Definition 1:

The so–called 1–g stall speed of an airplane is defined by:

$$V_S = \sqrt{\frac{2W}{\varrho C_{L_{max_{trim}}} S}} \qquad (12.1)$$

where: $C_{L_{max_{trim}}}$ is the maximum trimmed lift coefficient of the airplane. This lift coefficient may be taken as the 'top' of the C_L versus α curve if the airplane is controllable at the stall angle of attack. If the airplane is not controllable at the stall angle of attack, $C_{L_{max_{trim}}}$ must be determined from that angle of attack, below the stall, where the airplane is controllable.

Note that the notation: V_S is used in FAR 25. In FAR 23 the notations V_{S0} and V_{S1} are used with the same meanings. See Ref. 12.1 for FAR 23 and FAR 25. The reader may wish to consult Tables 9.11 – 9.13 (in Chapter 9) for more information.

Definition 2:

The minimum speed at which an airplane is controllable is defined by:

$$V_S = \sqrt{\frac{2W}{\varrho C_{L_{max(controllable)}} S}} \tag{12.2}$$

where: $C_{L_{max(controllable)}}$ is the maximum trimmed lift coefficient at which the airplane is controllable.

As seen from Eqns (12.1) and (12.2), the stall speed of an airplane increases with increasing altitude. For a given true airspeed, the Mach number will increase with increasing altitude (because the speed of sound decreases with increasing altitude). This results in compressibility effects which will tend to decrease $C_{L_{max_{trim}}}$. The net result is a further increase in the stall speed with increasing altitude.

When the stall is approached, increased airflow separation results in a phenomenon called: buffeting. There are two types of buffet: low-speed and high-speed buffet. Low-speed buffet is caused by the turbulence of the separated flow shaking part of the airplane, usually the wing or the tail. High-speed buffet is usually caused by shock induced flow separation. High speed buffet usually starts somewhere on the wing. Both types of buffet can usually be felt and heard in the cockpit.

The stall speed of an airplane is also affected by flaps. When the flaps are deflected a higher value of $C_{L_{max_{trim}}}$ is produced, usually at a lower angle of attack. Therefore, the corresponding stall speed is reduced. A discussion of typical flap performance which can be obtained with mechanical flaps is given in Chapter 4.

The load factor encountered by an airplane in flight is defined by:

$$n = \frac{L}{W} \tag{12.3}$$

Clearly, in level flight, n = 1. A 1-g flight condition is hard to achieve in flight test at speeds close to the stall speed. This is because various dynamic effects which accompany flight close to stall. Most flight tests close to the stall encounter a load factor of about n = 0.9. Therefore, an effective stall lift coefficient may be computed from:

$$C_{L_{max(effective)}} = \frac{C_{L_{max}}(n=1)}{n}, \text{ with } n = 0.9 \tag{12.4}$$

The numerical magnitude of $C_{L_{max}}(n=1)$ can be determined from windtunnel tests, corrected for Reynolds number effects. Using this effective maximum lift coefficient, the airplane stall speeds with the flaps up or down may be obtained. Eqn (12.4) implies that the airplane stall speed is approximately 94% of its 1–g stall speed.

FAR 23.49, FAR 25.103 (and SFAR 23) define the stalling speed as the flight speed in knots calibrated airspeed (KCAS) at the most unfavorable c.g. location. The manufacturer must identify that c.g. location. For a conventional (tail–aft) airplane this will normally be the most forward c.g. location. The stall speed must be obtained from flight test data. During the flight test the airplane must be trimmed at a speed slightly above the expected stall speed (FAR 23 and SFAR 23) or between $1.2V_S$ and $1.4V_S$ (FAR 25). Elevator control is then applied in such a manner that the speed reduction of the airplane <u>does not exceed one–knot–per–second</u>. Therefore, the variation of airspeed with time must be recorded during the flight test and plotted as in Figure 12.1.

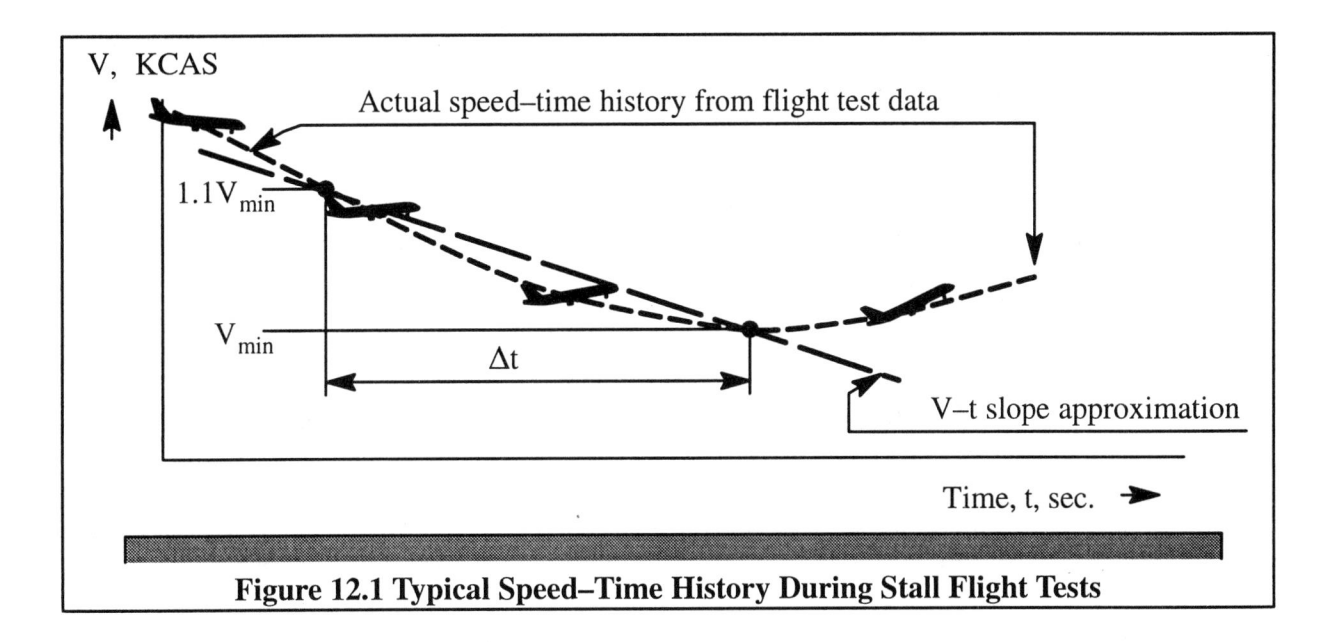

Figure 12.1 Typical Speed–Time History During Stall Flight Tests

The entry rate into the stall–speed demonstration maneuver may be computed from:

$$\frac{dV}{dt} = \frac{V_{min} - 1.1 V_{min}}{\Delta t} \tag{12.5}$$

At the minimum speed, V_{min}, the corresponding lift coefficient, $C_{L_{max_{trim}}}$, may be determined from Eqn (12.1). However, this lift coefficient must be corrected for the most critical c.g. location*. To do this, consider Figure 12.2.

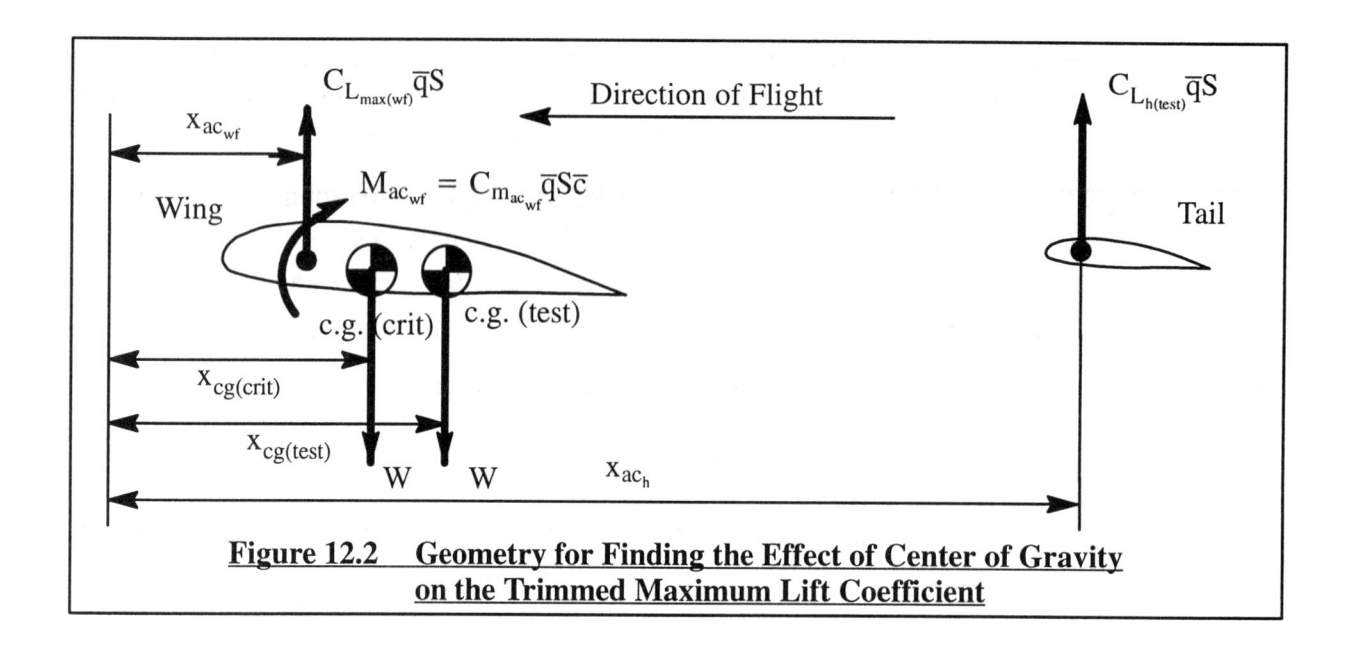

Figure 12.2 Geometry for Finding the Effect of Center of Gravity on the Trimmed Maximum Lift Coefficient

As the c.g. is moved forward, the tail load to trim the airplane is decreased. The tail load to trim is shown in Figure 12.2 as an up–load. Since the coefficient, $C_{m_{ac_{wf}}}$ is normally negative (particularly with the flaps down), the tail load to trim will normally also be negative (i.e. down). Therefore, the tail load to trim will increase in the negative direction, as the c.g. is moved forward.

When the airplane is trimmed, the airplane lift coefficient is the sum of the wing/fuselage lift coefficient and the tail lift coefficient (the latter corrected for S_h/S and $\bar{q}_h/\bar{q}$). Therefore:

At the test c.g.:
$$C_{L_{max(test)}} = C_{L_{max(wf)}} + C_{L_{h(test)}} \tag{12.6}$$

At the most critical c.g.:
$$C_{L_{max(crit)}} = C_{L_{max(wf)}} + C_{L_{h(crit)}} \tag{12.7}$$

The corresponding moment (i.e. trim) equilibrium conditions are:

Test c.g.:
$$C_{m_{ac_{wf}}}\bar{c} - C_{L_{max(wf)}}(x_{ac_{wf}} - x_{c.g.(test)}) = C_{L_{h(test)}}(x_{ac_h} - x_{c.g.(test)}) \tag{12.8}$$

Most critical c.g.:
$$C_{m_{ac_{wf}}}\bar{c} - C_{L_{max(wf)}}(x_{ac_{wf}} - x_{c.g.(crit)}) = C_{L_{h(crit)}}(x_{ac_h} - x_{c.g.(crit)}) \tag{12.9}$$

By defining that: $(x_{ac_h} - x_{c.g.(crit)}) = l_h$, it can be shown that:

$$C_{L_{max_{c.g.(crit)}}} = C_{L_{max_{c.g.(test)}}}\left(1 + \frac{x_{c.g.(crit)} - x_{c.g.(test)}}{l_h}\right) \tag{12.10}$$

---

* The manufacturer must identify what constitutes the most critical c.g. location. This location depends on the configuration of the airplane and on the way it is trimmed. For a conventional configuration the most critical c.g. location is typically the most forward c.g. location.

It is seen that the more forward the critical c.g. is, the lower the trimmed maximum lift coefficient is. It should be recognized that this derivation also assumes that the horizontal tail itself is not stalled. The effect of tail stall on trimmability is discussed in References 12.2 and 12.3.

The reader is asked in problems 12.4 and 12.5 to re–derive Eqn (12.10) for a canard and for a three–surface configuration. Reference 12.2 is recommended for detail discussions of airplane trim.

The maximum lift coefficient, $C_{L_{max_{c.g.}}}$, is determined for each individual flap setting by plotting the flight test measured values of $C_{L_{max_{c.g.}}}$ as a function of the entry rate produced during the test. Figure 12.3 shows how the numerical value of $C_{L_{max_{c.g.}}}$ may be determined.

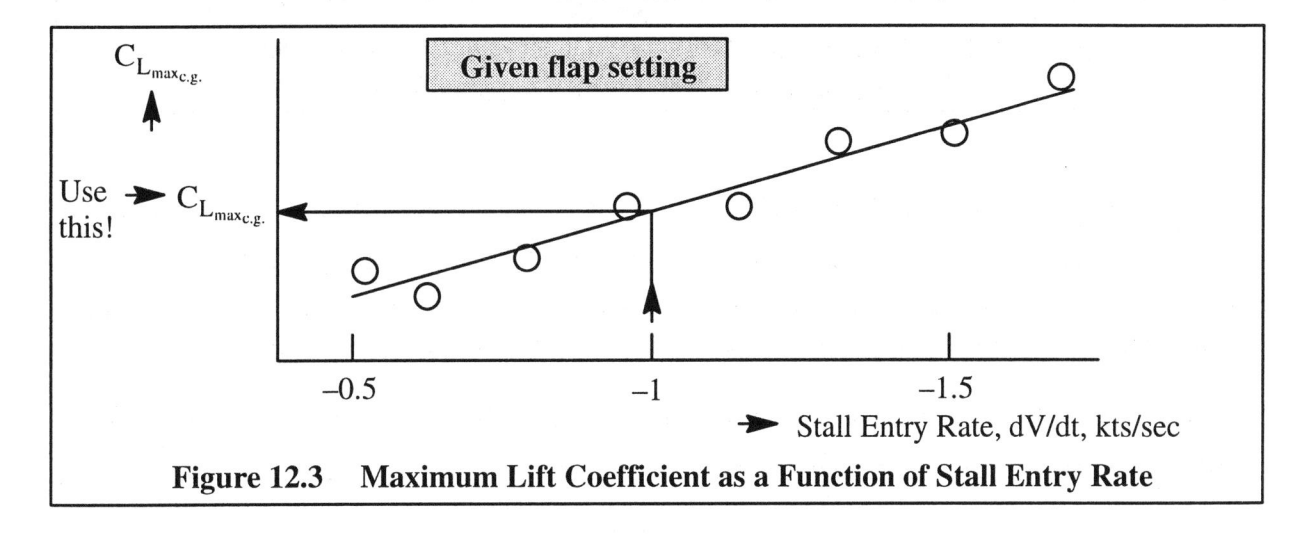

Figure 12.3   Maximum Lift Coefficient as a Function of Stall Entry Rate

If the stall flight tests are not conducted at zero thrust, corrections for thrust must be applied to the data. The tests are repeated for those flap settings for which the airplane is to be certified. The results of these flight tests are normally presented in the form of a plot such as shown in Figure 12.4.

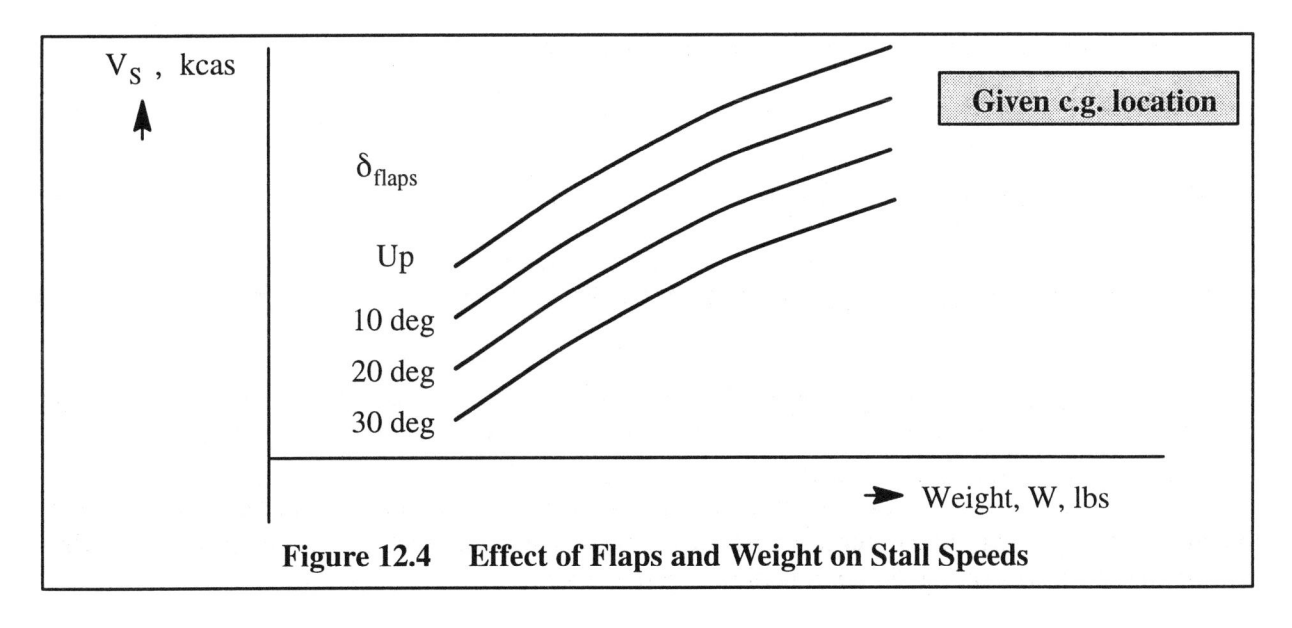

Figure 12.4   Effect of Flaps and Weight on Stall Speeds

12.2 BUFFET LIMITS

Buffeting is defined as the structural response of an airframe in response to aerodynamic excitation which in turn is produced when the flow separates. Examples of buffeting are:

a) wing flow separation exciting the wing structure
b) separated wing flow hitting another airplane component (such as the horizontal tail)
c) separated flow from spoilers hitting the horizontal tail (this happens in many jet transports when a pilot deploys the spoilers to slow the airplane down from cruise flight)
d) air intake flow breakdown (also called inlet buzz)
e) flow interference between external stores

Buffeting can be very annoying to pilots and passengers. In military aircraft certain magnitudes of buffet can interfere with the pilot's ability to control the airplane. For that reason, Ref. 10.3 contains another definition for minimum speed: that at which buffet becomes intolerable. whenever that speed is higher than that according to Definitions 1 or 2 (See Section 10.1) the minimum speed based on some buffeting criterion must be used in setting the airplane reference speeds.

Flow separation from the wing is the most frequently occurring cause of buffet. Wing flow separation may be characterized by several types, four of which are given in Figure 12.5.

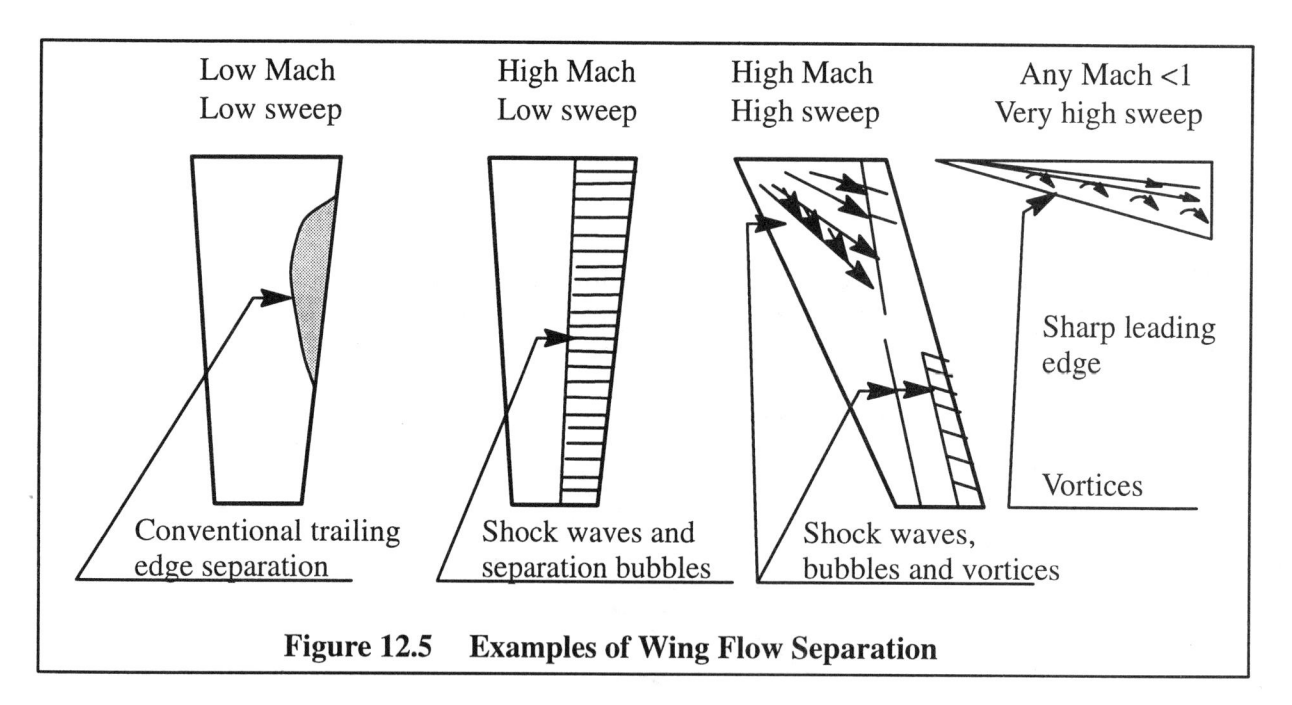

Figure 12.5 Examples of Wing Flow Separation

Wings with low sweep angles are characterized by leading–edge or trailing–edge separations. These separations form bubbles on the wing which usually excite strong buffeting. At transonic speeds (which for an unswept wing can occur at relatively low Mach numbers), strong shock waves can induce flow separation. For swept wings, buffeting may also be induced by separated vortices.

Tail buffeting is normally caused by separated flow from the wing. However, it can also be caused by separated flow on the tail itself.

A wide variety of methods is used to predict the onset of buffet. Fifteen of these are summarized in Figure 12.6. The most widely used method employs the measurement of the wing root bending moment: method 1 in Figure 12.6. With method 1, the onset of buffet is defined as occurring at that lift coefficient for which the root mean square (r.m.s.) of the root bending moment exceeds some numerical value.

In method 2 the divergence in pressure at the trailing edge of the wing is measured. This measurement should be taken at roughly 80% of the wing span to give reliable results at high subsonic and transonic speeds.

In the case of method 3 it is assumed that buffet onset coincides with flow separation from the wing which results in departure of the lift–coefficient–versus–angle–of–attack curve from linearity.

In some airplanes the change in lift–coefficient–versus–angle–of–attack triggers an unstable pitch break. In that case the angle of attack where the pitch break occurs can be thought of as coinciding with buffet onset. That is the thought behind method 4.

Sudden changes in the width of the wake emanating from the wing trailing edge can also be associated with flow separation and therefore buffet. That is reflected by method 5.

Whenever shocks trigger flow separation, the position of the shock in terms of the wing chord can be used as an indicator of buffet onset. Method 6 is an example of this.

In method 7 the nature of a shock wave is examined. Shocks are often characterized by a sharp change in pressure. Whenever the Mach number downstream of a shock itself is sonic, separation is triggered and again buffet begins. The angle of attack where this happens is that associated with buffet onset.

Since buffet is often detectable by sudden, sharp fluctuations in airframe load–factor or local accelerations, the reading of a properly placed accelerometer can be used to determine when buffet begins. In method 8 accelerometer readings are used to predict buffet onset.

A sudden deviation in axial force (See method 9) with angle of attack can be used to predict the onset of buffet. This method is quite reliable.

If the flow separation of a wing occurs on one wing before the other, buffet onset can be correlated to a sudden change in rolling moment behavior. This is used in method 10.

Various other experimental methods (11–15) are also listed in Figure 12.6. Ultimately, pilot opinion on what constitutes intolerable buffet must be used to help set the limits in angle of attack and Mach number where buffet becomes unacceptable. It should be noted that the boundary between acceptable and unacceptable buffet is not rigorously defined. The upper limit is a function of airplane type and mission. In a transport airplane, buffet limits may be imposed because of structural fatigue and passenger comfort considerations. In a fighter, the ability of the pilot to fly the airplane (handling qualities) in the presence of turbulence may be more important.

Maneuvering and the Flight Envelope

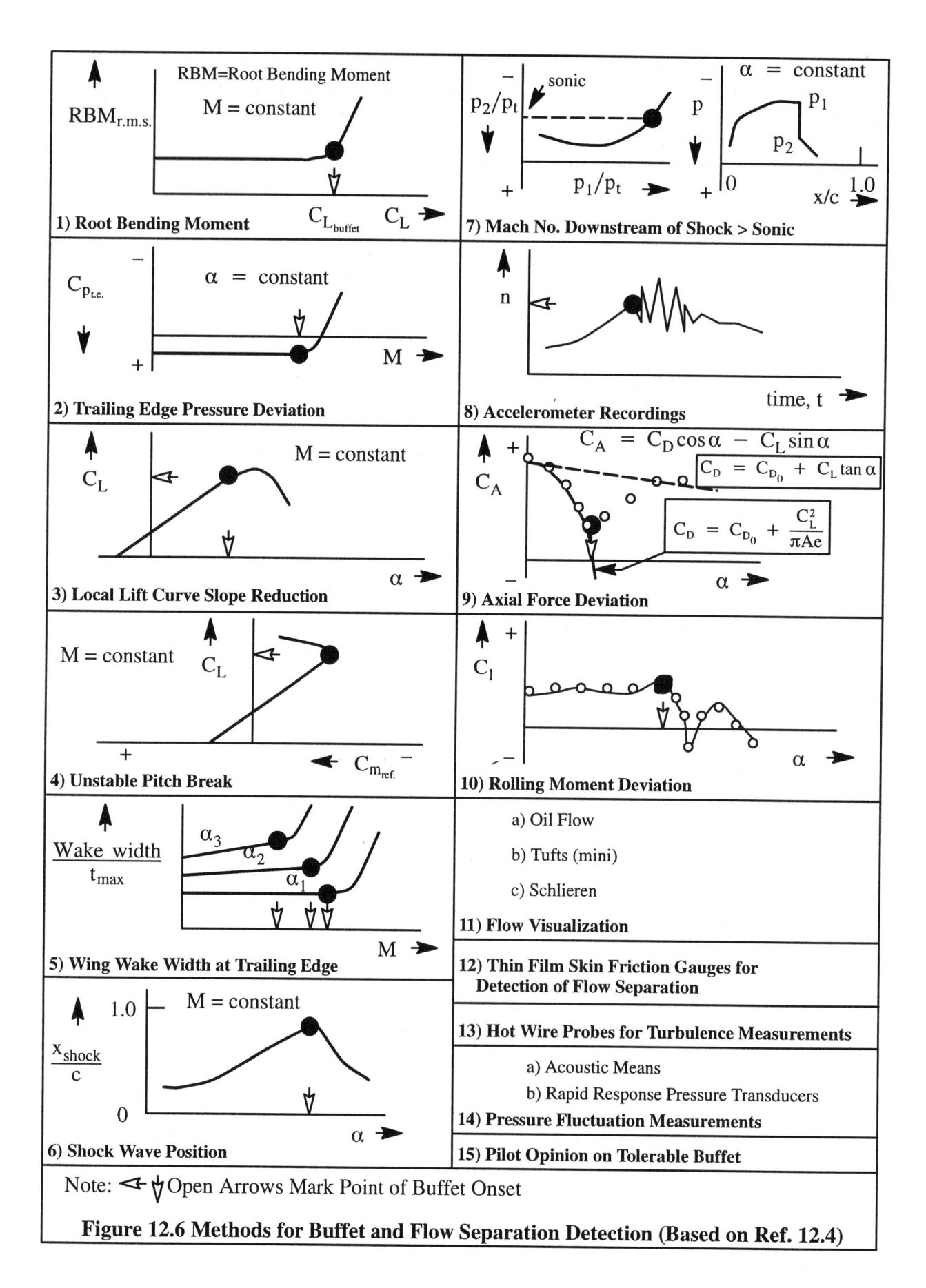

1) **Root Bending Moment** — $RBM_{r.m.s.}$ vs C_L, $M = $ constant, buffet onset at $C_{L_{buffet}}$

2) **Trailing Edge Pressure Deviation** — $C_{p_{t.e.}}$ vs M, $\alpha = $ constant

3) **Local Lift Curve Slope Reduction** — C_L vs α, $M = $ constant

4) **Unstable Pitch Break** — C_L vs $C_{m_{ref.}}$, $M = $ constant

5) **Wing Wake Width at Trailing Edge** — $\dfrac{\text{Wake width}}{t_{max}}$ vs M, at $\alpha_1, \alpha_2, \alpha_3$

6) **Shock Wave Position** — $\dfrac{x_{shock}}{c}$ vs α, $M = $ constant

7) **Mach No. Downstream of Shock > Sonic** — p_2/p_t vs p_1/p_t (sonic); p vs x/c, $\alpha = $ constant

8) **Accelerometer Recordings** — n vs time, t

9) **Axial Force Deviation** — C_A vs α
$$C_A = C_D \cos\alpha - C_L \sin\alpha$$
$$C_D = C_{D_0} + C_L \tan\alpha$$
$$C_D = C_{D_0} + \frac{C_L^2}{\pi A e}$$

10) **Rolling Moment Deviation** — C_l vs α

11) **Flow Visualization**
 a) Oil Flow
 b) Tufts (mini)
 c) Schlieren

12) **Thin Film Skin Friction Gauges for Detection of Flow Separation**

13) **Hot Wire Probes for Turbulence Measurements**

14) **Pressure Fluctuation Measurements**
 a) Acoustic Means
 b) Rapid Response Pressure Transducers

15) **Pilot Opinion on Tolerable Buffet**

Note: ⇐ ⇓ Open Arrows Mark Point of Buffet Onset

Figure 12.6 Methods for Buffet and Flow Separation Detection (Based on Ref. 12.4)

Maneuvering and the Flight Envelope

A typical curve showing the buffet limit on lift coefficient is shown in Figure 12.7. For a given Mach number, the lift coefficient corresponding to buffet onset can be found. Note the large decrease in maximum lift coefficient with increasing Mach number. The trend shown in Figure 12.7 is typical for transport type airplanes.

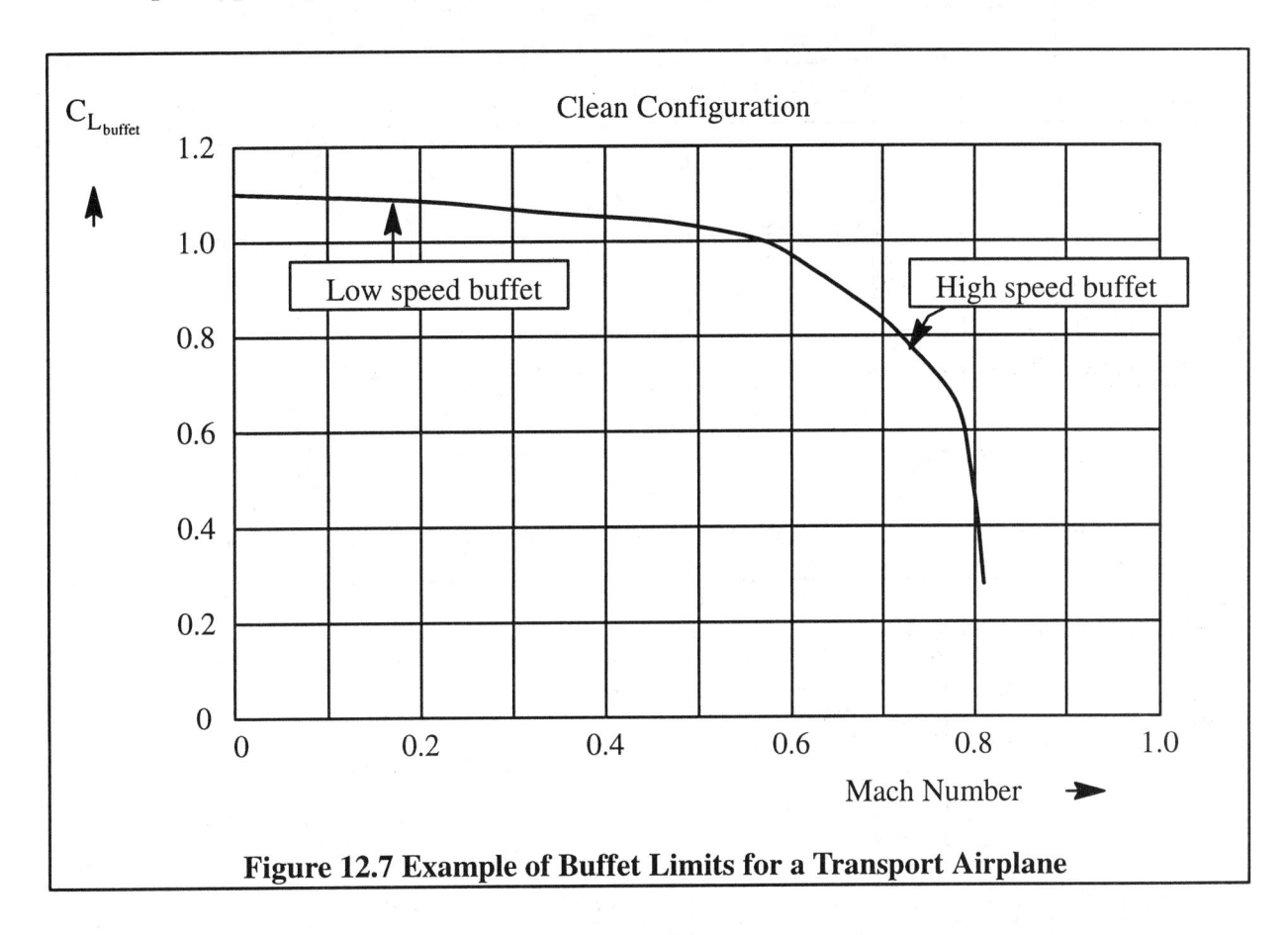

Figure 12.7 Example of Buffet Limits for a Transport Airplane

For any given weight, the altitude where, at that value of lift coefficient, buffet would begin, can be calculated from:

$$\delta = \frac{W}{1,481.3 M^2 C_{L_{buffet}} S} \tag{12.11}$$

The results can be represented graphically as shown in Figure 12.8. The information presented in Figures 12.7 and 12.8 can be combined as in Figure 12.9 so that the instantaneous maneuver load factor capability of an airplane can be determined. By arranging the data in the form of Figure 12.10 it is possible to determine the instantaneous maneuver load factor capability of an airplane more conveniently. The data presentation of Figure 12.10 is used in flight manuals of transport airplanes.

In the example of Figure 12.10, at a weight of 104,000 lbs, a Mach number of 0.72 and a pressure altitude of 31,000 ft, a load factor of 1.68 g's (corresponding to a bank angle of 53 degrees in a level turn), is possible before buffet occurs. The 1–g stall speed at that altitude is seen to occur at a Mach number of 0.49.

Other applications of the material presented here are discussed in Sections 12.5 and 12.6.

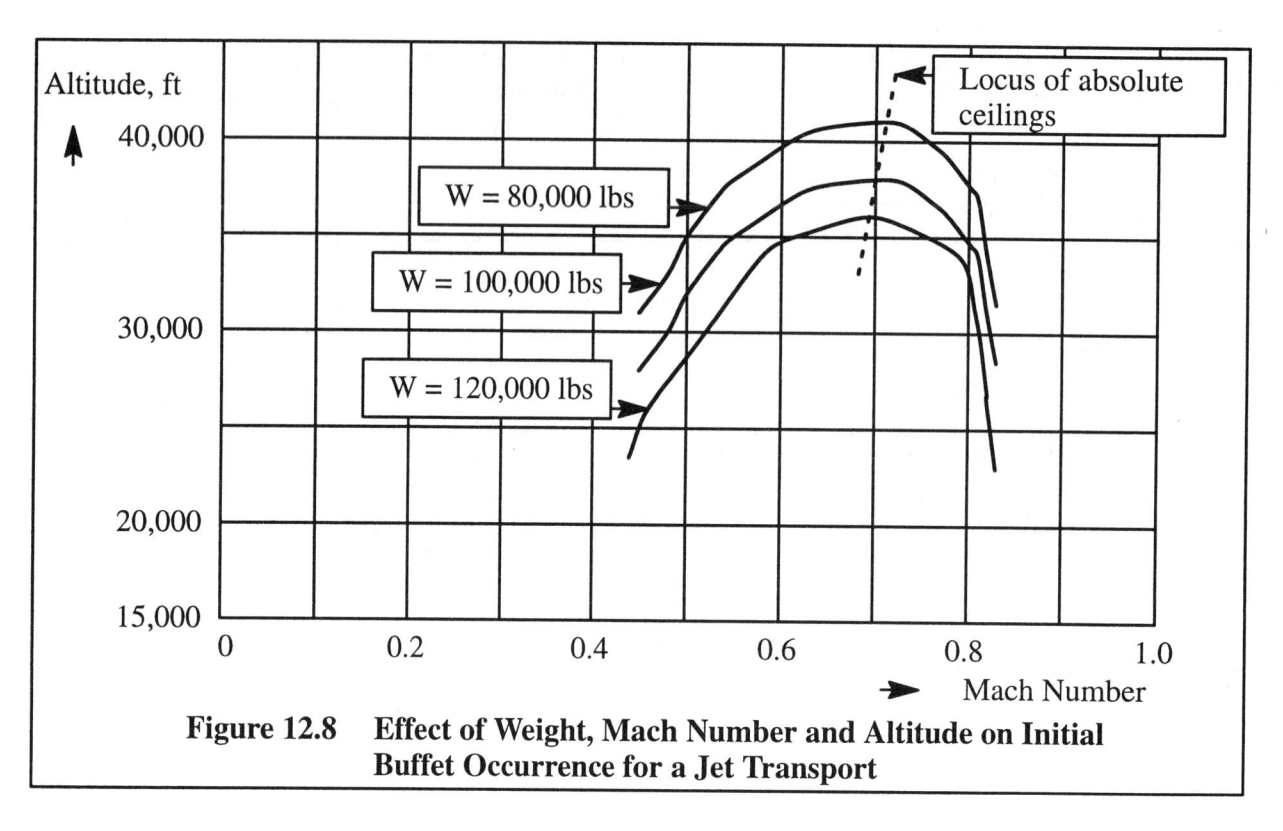

Figure 12.8 Effect of Weight, Mach Number and Altitude on Initial Buffet Occurrence for a Jet Transport

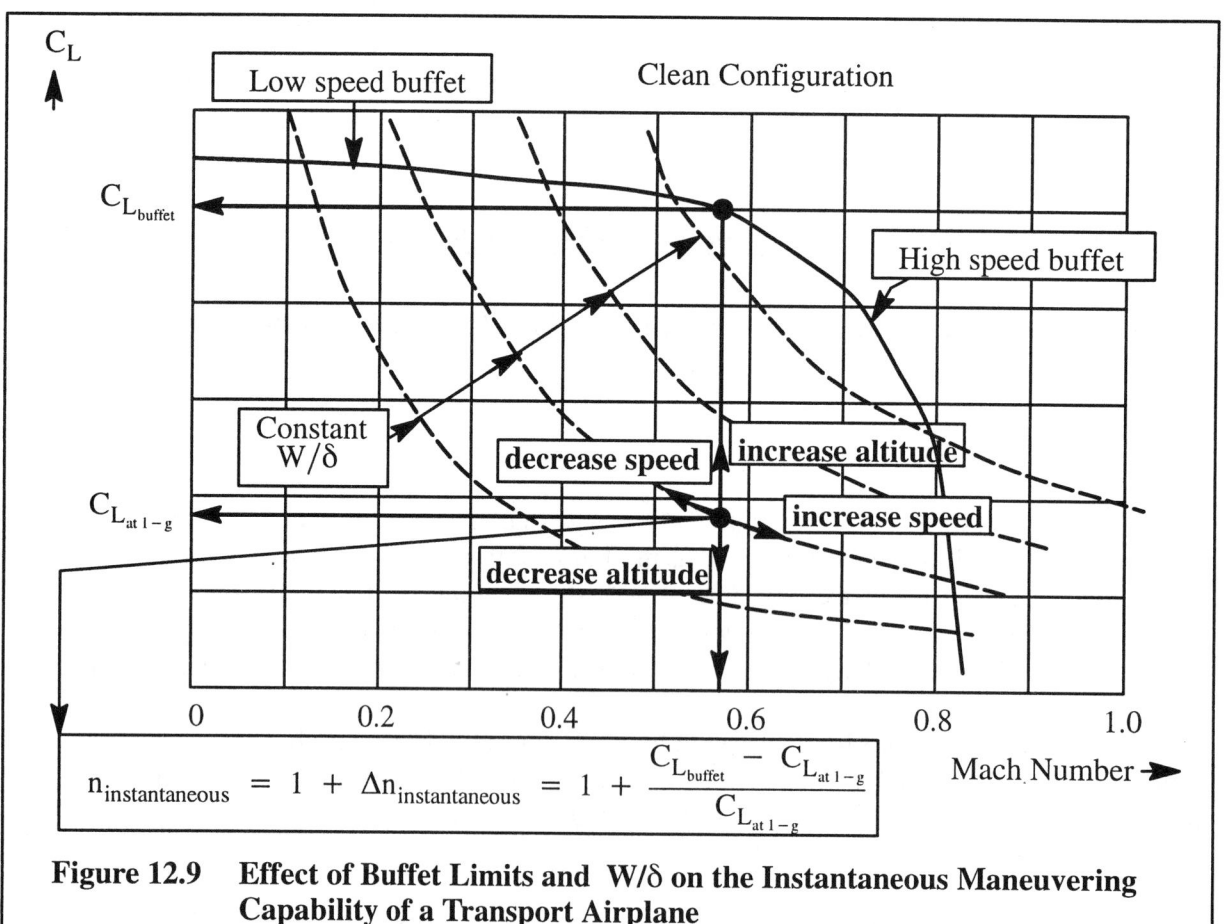

$$n_{instantaneous} = 1 + \Delta n_{instantaneous} = 1 + \frac{C_{L_{buffet}} - C_{L_{at\,1-g}}}{C_{L_{at\,1-g}}}$$

Figure 12.9 Effect of Buffet Limits and W/δ on the Instantaneous Maneuvering Capability of a Transport Airplane

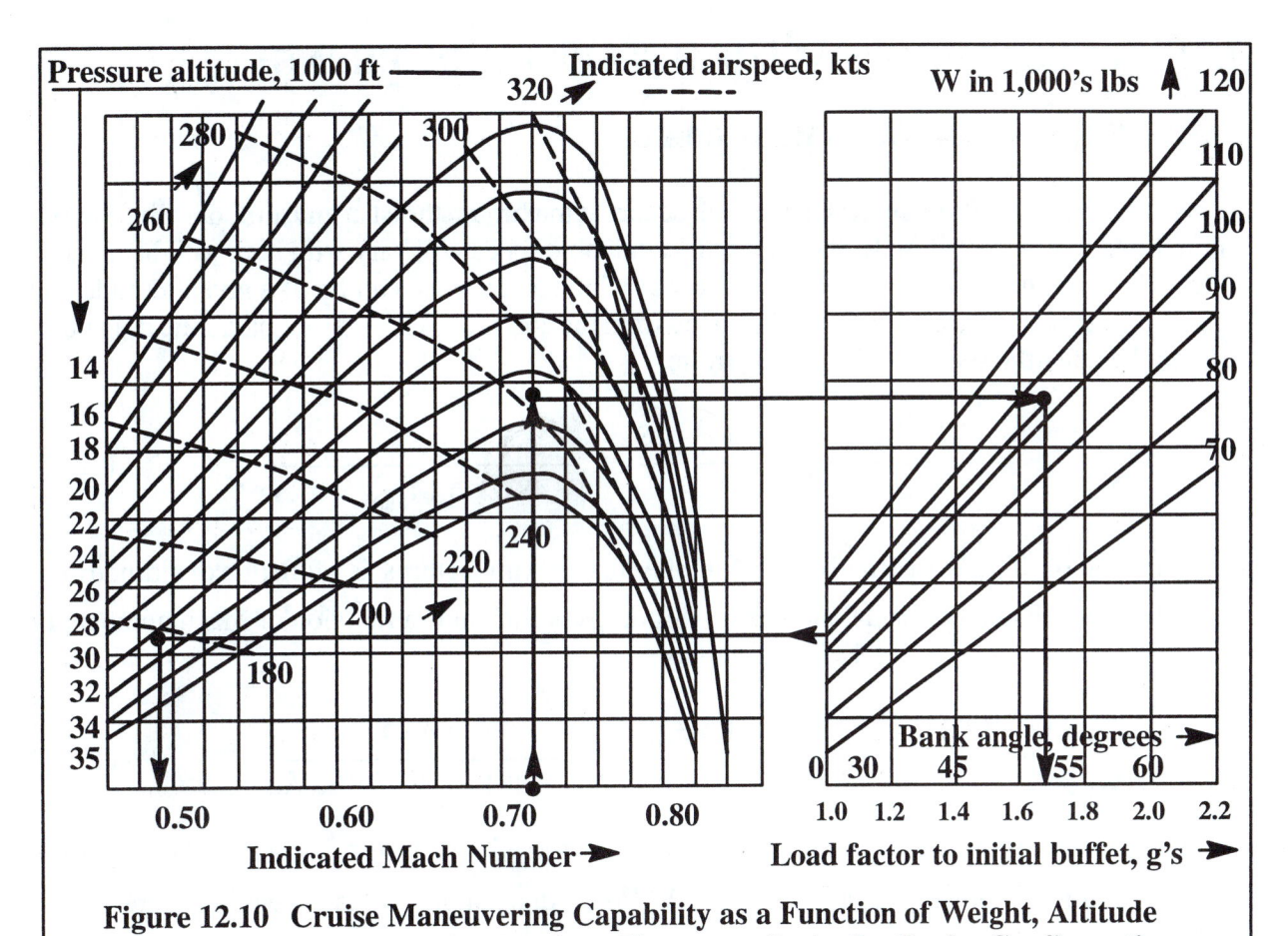

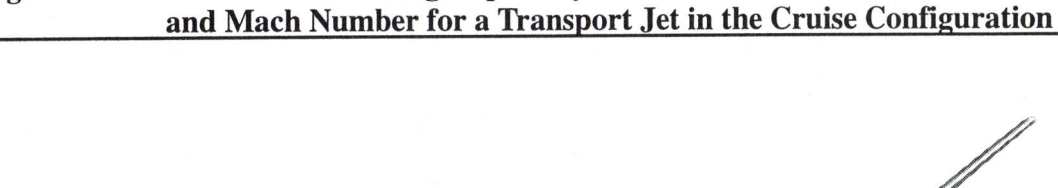

Figure 12.10 Cruise Maneuvering Capability as a Function of Weight, Altitude and Mach Number for a Transport Jet in the Cruise Configuration

12.3 LEVEL FLIGHT MAXIMUM SPEEDS AND CEILINGS

12.3.1 LEVEL FLIGHT MAXIMUM SPEEDS

A very important flight limit is that which occurs at maximum thrust at any altitude. Since maximum thrust varies with altitude, the maximum attainable speed will vary with both weight and altitude. The maximum Mach number (or speed) at which an airplane can fly at a given altitude and at maximum available thrust can be determined from the condition T = D. For a parabolic drag polar this implies that the speed can be solved from:

$$\frac{T_{max}(\delta, M)}{W} = \frac{C_{D_0}(M)\, 1{,}481.3\, \delta\, M_{max}^2\, S}{W} + \frac{W}{1{,}481.3\, \delta\, M_{max}^2\, S\, \pi\, Ae} \qquad (12.12)$$

The notation: $T_{max}(\delta, M)$ and $C_{D_0}(M)$ implies that maximum thrust is a function of altitude (and possibly of Mach number) and that the zero–lift drag coefficient is a function of Mach number. In many practical cases, Eqn (12.12) is written as:

$$\frac{(T_{max}(\delta, M)/\delta)}{(W/\delta)} = \frac{C_{D_0}(M)\, 1{,}481.3\, M_{max}^2\, S}{(W/\delta)} + \frac{(W/\delta)}{1{,}481.3\, M_{max}^2\, S\, \pi\, Ae} \qquad (12.13)$$

For a given altitude, at a given Mach number and at maximum thrust, Eqn (12.13) can be used to solve iteratively for M_{max} as a function of (W/δ). An example plot is shown in Figure 12.11.

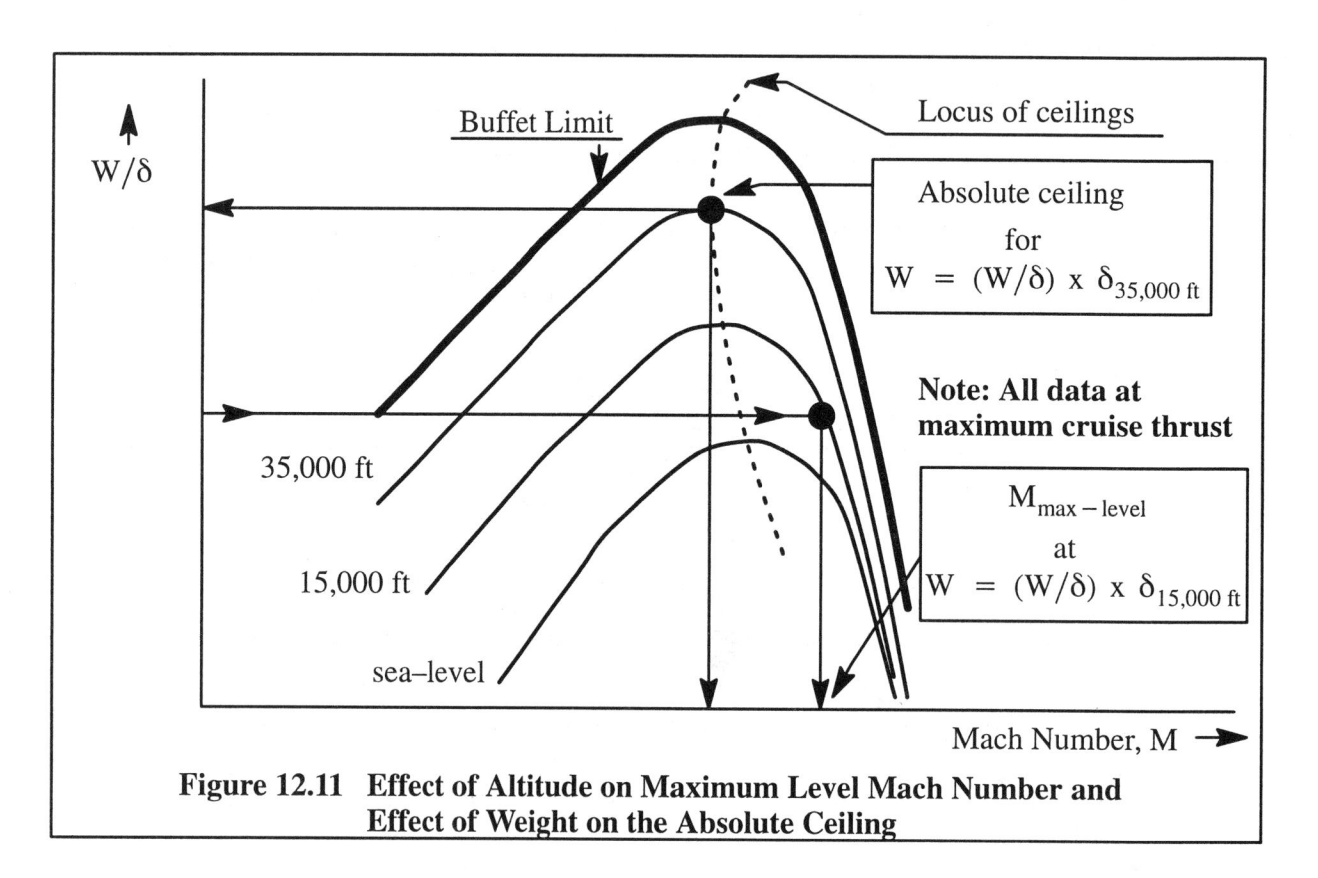

Figure 12.11 Effect of Altitude on Maximum Level Mach Number and Effect of Weight on the Absolute Ceiling

If the maximum thrust and the zero–lift drag coefficient are independent of Mach number it is possible to solve for the maximum attainable Mach number as shown in Equation (12.14):

$$M_{max} = \sqrt{\frac{T_{max}}{2 C_{D_0} S (1,481.3 \delta)} + \sqrt{\frac{T_{max}^2}{4\{C_{D_0} S (1,481.3 \delta)\}^2} - \frac{(W/S)^2}{C_{D_0} \pi Ae (1,481.3 \delta)^2}}}$$

(12.14)

For a detailed discussion of methods for determining the maximum speed of jet–driven and propeller–driven airplanes the reader is referred to Chapter 8.

12.3.2 LEVEL FLIGHT CEILINGS

Methods to determine the ceiling capability of airplanes were also discussed in Chapter 9. Assuming that there is sufficient thrust available to fly level at a given altitude, the altitude at which n=1 flight is possible at the stall or buffet limit, is called the absolute ceiling. Figure 12.8 shows how absolute ceilings can be represented graphically. Another way of showing this type of information is given in Figure 12.11.

12.4 THE V–n DIAGRAM

The purpose of this section is to familiarize the reader with the construction and use of V–n diagrams. These diagrams are used primarily in the determination of combinations of flight conditions and load factors to which the airplane structure must be designed. However, V–n diagrams are also useful in determining the maneuvering capability of an airplane.

The particular V–n diagram to which an airplane must be designed depends on the certification basis selected by the manufacturer and/or the customer. V–n diagrams for FAR 23, FAR 25 and Military aircraft are discussed in Sub–sections 12.4.1 through 12.4.3 respectively.

12.4.1 FAR–23

The basic V–n diagram for FAR 23 certified airplanes is defined in FAR 23, Part 23.335 of Reference 12.1. A graphical presentation is given in Figure 12.12.

There are four important speeds used in the V–n diagram: V_S, the 1–g stall speed; V_A, the design maneuvering speed; V_C, the design cruise speed, and V_D, the design diving speed.

Rules for determining these speeds are presented in Sub–sub–sections 12.4.1.1 through 12.4.1.4.

Rules for determining the various envelope lines are discussed in Sub–sub–sections 12.4.1.5 through 12.4.1.7.

Figure 12.12 V–n Diagram According to FAR 23

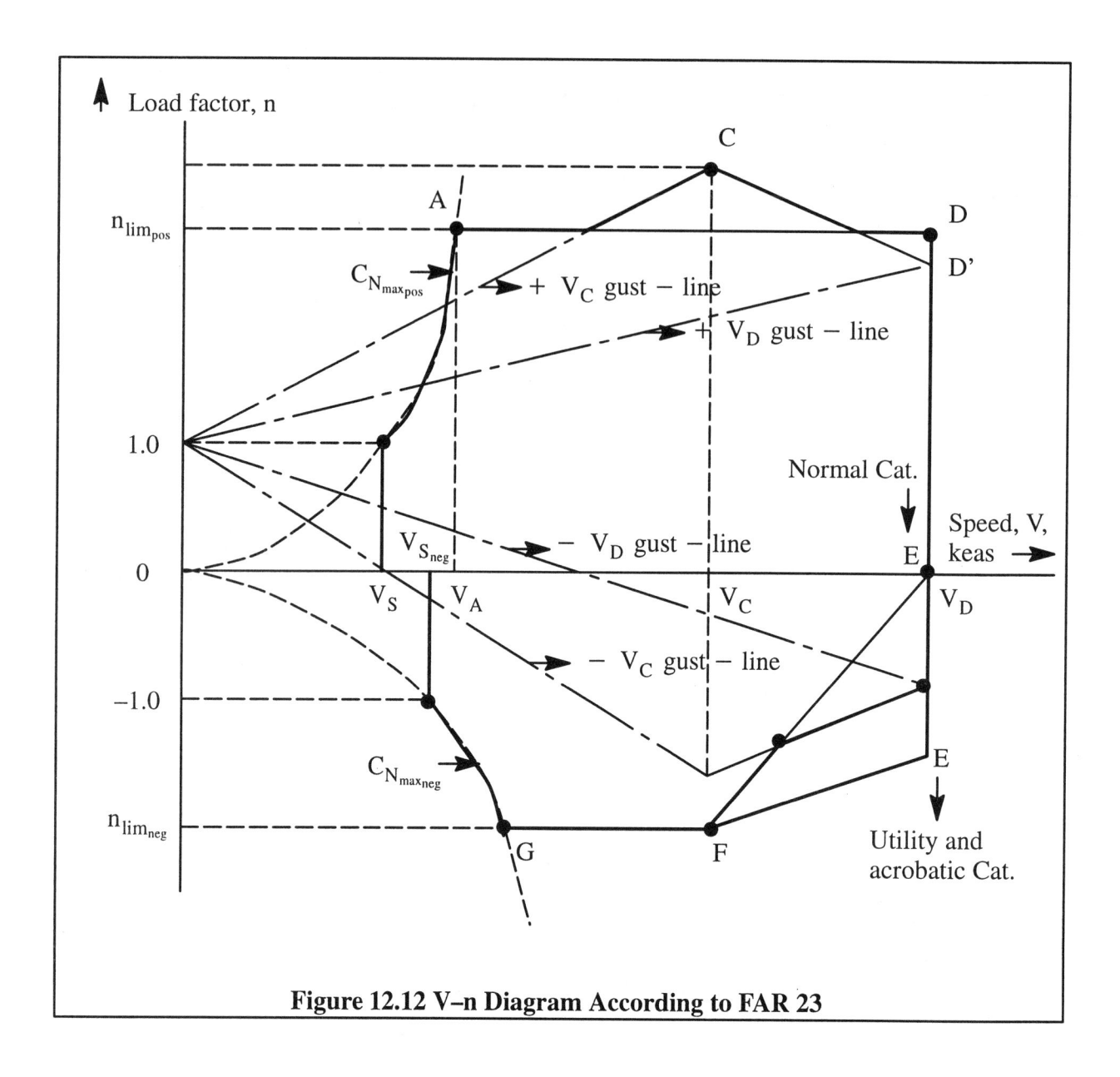

12.4.1.1 The 1–g Stall Speed, V_S

For purposes of constructing a V–n diagram, the 1–g stall speed, V_S may be computed from:

$$V_S = \sqrt{\frac{2W_{FDWG}}{\rho C_{N_{max(controllable)}} S}} \tag{12.15}$$

where: W_{FDGW} is the flight design gross weight. For most civil airplanes: W_{FDGW} equals the maximum design take–off weight.

Observe that the maximum normal force coefficient is used instead of the maximum lift coefficient such as in Eqn (12.2). The difference between these two coefficients is usually not very large.

The maximum normal force coefficient follows from:

$$C_{N_{max(controllable)}} = \sqrt{\left(C_{L_{max(controllable)}}\right)^2 + \left(C_{D_{at\,C_{L_{max(controllable)}}}}\right)^2} \tag{12.16}$$

Note: C_N is the same as C_R in Chapter 8.

For preliminary analysis purposes it is acceptable to use:

$$C_{N_{max(controllable)}} = 1.1 C_{L_{max(controllable)}} \tag{12.17}$$

The parabolic line 0–A (in Figure 12.12) follows from:

$$V_{S_{n=n}} = \sqrt{\frac{2nW_{FDGW}}{\varrho C_{N_{max(controllable)}} S}} \tag{12.18}$$

12.4.1.2 The Design Maneuvering Speed, V_A

The design maneuvering speed, V_A, must be selected by the designer, but must satisfy the following relationship:

$$V_A \geq V_S \sqrt{(n_{lim_{pos}})} \tag{12.19}$$

where: $n_{lim_{pos}}$ is the positive design limit load factor as defined in Sub–sub–section 12.4.1.6

However, V_A need not exceed V_C as defined next.

12.4.1.3 The Design Cruising Speed, V_C

The design cruising speed, V_C (in keas), must be selected by the designer, but must satisfy the following relationship:

$$V_C \geq k_c \sqrt{(W_{FDGW})/S} \tag{12.20}$$

where: $k_c = 33$ for normal and utility category airplanes with wing loadings up to $W_{FDGW}/S = 20$ psf.

k_c varies linearly from 33 to 28.6 as the wing loading varies from 20 to 100 psf, for normal and utility category airplanes

$k_c = 36$ for acrobatic category airplanes

However, V_C need not exceed $0.9V_H$, where V_H is the maximum level speed obtained with maximum thrust or power.

12.4.1.4 The Design Diving Speed, V_D

The design diving speed, V_D, must satisfy the following relationship:

$$V_D \text{ (or } M_D) \geq 1.25V_c \text{ (or } 1.25M_C) \tag{12.21}$$

where: V_C follows from Eqn (12.20).

12.4.1.5 The Negative 1–g Stall Speed, $V_{S_{neg}}$

For purposes of constructing a V–n diagram, the negative 1–g stall speed, $V_{S_{neg}}$ may be computed from:

$$V_{S_{neg}} = \sqrt{\frac{2W_{FDGW}}{\varrho C_{N_{max(controllable)_{neg}}} S}} \tag{12.22}$$

The maximum negative normal force coefficient follows from:

$$C_{N_{max(controllable)_{neg}}} = \sqrt{\left(C_{L_{max(controllable)_{neg}}}\right)^2 + \left(C_{D_{at\ C_{L_{max(controllable)_{neg}}}}}\right)^2} \tag{12.23}$$

For preliminary analysis purposes it is acceptable to use:

$$C_{N_{max(controllable)_{neg}}} = 1.1 C_{L_{max(controllable)_{neg}}} \tag{12.24}$$

The parabolic line 0–G (in Figure 12.12) follows from:

$$V_{S_{n=n_{neg}}} = \sqrt{2n_{neg} \frac{W_{FDGW}}{\varrho C_{N_{max(controllable)_{neg}}} S}} \tag{12.25}$$

12.4.1.6 The Design Limit Load Factors, $n_{lim_{pos}}$ and $n_{lim_{neg}}$

The positive design limit load factor, $n_{lim_{pos}}$ must be selected by the designer, but must meet the following condition:

$$n_{lim_{pos}} \geq 2.1 + \frac{24,000}{W_{FDGW} + 10,000} \tag{12.26}$$

However, $n_{lim_{pos}}$ need not be greater than 3.8

$$n_{lim_{pos}} = 4.4 \text{ for utility category airplanes}$$

$$n_{lim_{pos}} = 6.0 \text{ for acrobatic category airplanes}$$

The negative design limit load factor, $n_{lim_{neg}}$ must be selected by the designer, but must meet the following condition:

$$n_{lim_{neg}} \geq 0.4 n_{lim_{pos}} \text{ for normal and for utility category airplanes} \tag{12.27}$$

$$n_{lim_{neg}} \geq 0.5 n_{lim_{pos}} \text{ for acrobatic category airplanes} \tag{12.28}$$

12.4.1.7 The Gust Load Factor Lines in Figure 12.12

The two positive gust load factor lines in Figure 12.12 are defined by the following equation:

$$n_{lim} = 1 \pm \frac{K_g U_{de} V C_{L_\alpha}}{498(W_{FDGW}/S)} \tag{12.29}$$

where: K_g is the gust alleviation factor, defined by:

$$K_g = \frac{0.88 \mu_g}{5.3 + \mu_g} \quad \text{For subsonic airplanes} \tag{12.30}$$

$$\text{and}: K_g = \frac{\mu_g^{1.03}}{6.9 + \mu_g^{1.03}} \quad \text{For supersonic airplanes}$$

with:

$$\mu_g = \frac{2(W_{FDGW}/S)}{\varrho \, \bar{c} \, g \, C_{L_\alpha}} \tag{12.31}$$

C_{L_α} is the airplane lift–curve–slope in 1/rad

U_{de} is the derived gust velocity, which is defined as follows:

For V_C gust lines :

$$U_{de} = 50 \text{ fps} \quad \text{for altitudes from sea-level to 20,000 ft} \quad (12.32)$$

$$U_{de} = 66.67 - 0.000833 \, h \quad \text{for altitudes from 20,000 ft to 50,000 ft} \quad (12.33)$$

For V_D gust lines:

$$U_{de} = 25 \text{ fps} \quad \text{for altitudes from sea-level to 20,000 ft} \quad (12.34)$$

$$U_{de} = 33.34 - 0.000417 \, h \quad \text{for altitudes from 20,000 ft to 50,000 ft} \quad (12.35)$$

If the slope of the V_C gust line is such that point C is above line AD in Figure 12.12, the design limit load factor at the speed, V_C, is given by the projection of point C onto the load-factor axis.

A similar situation can arise with regard to point D': if point D' is above point D in Figure 12.12, then the design limit load factor at the speed, V_D is given by the projection of point D' onto the load-factor axis. If any one of these conditions is satisfied the airplane is said to be gust-critical.

Clearly, the slope of both gust lines is related to the lift-curve slope and the wing loading of the airplane. Observe, that airplanes with a low wing loading and a high lift-curve slope may become gust critical.

Whenever the V-n diagram indicates a higher design limit load factor to be required, the structural weight of the airplane (and thus the empty weight) tends to be higher.

The two negative gust load factor lines in Figure 12.12 are defined also by Eqn (12.29) except that their slopes are negative.

12.4.2 FAR-25

There are two types of V-n diagram for FAR 25 certified airplanes: 1) V-n the maneuver diagram and 2) the V-n gust diagram. Both types are defined in FAR 25, Part 25.335 of Reference 12.1.

Graphical presentations for both diagrams are given in Figures 12.13 and 12.14.

12.4.2.1 The 1-g Stall Speed, V_{S_1}

For purposes of constructing a V-n diagram, the 1-g stall speed, V_{S_1} may be computed from:

$$V_{S_1} = \sqrt{\frac{2W_{FDWG}}{\varrho C_{N_{max(controllable)}} S}} \quad (12.36)$$

The normal force coefficient may be determined with Eqn (12.16) or (12.17).

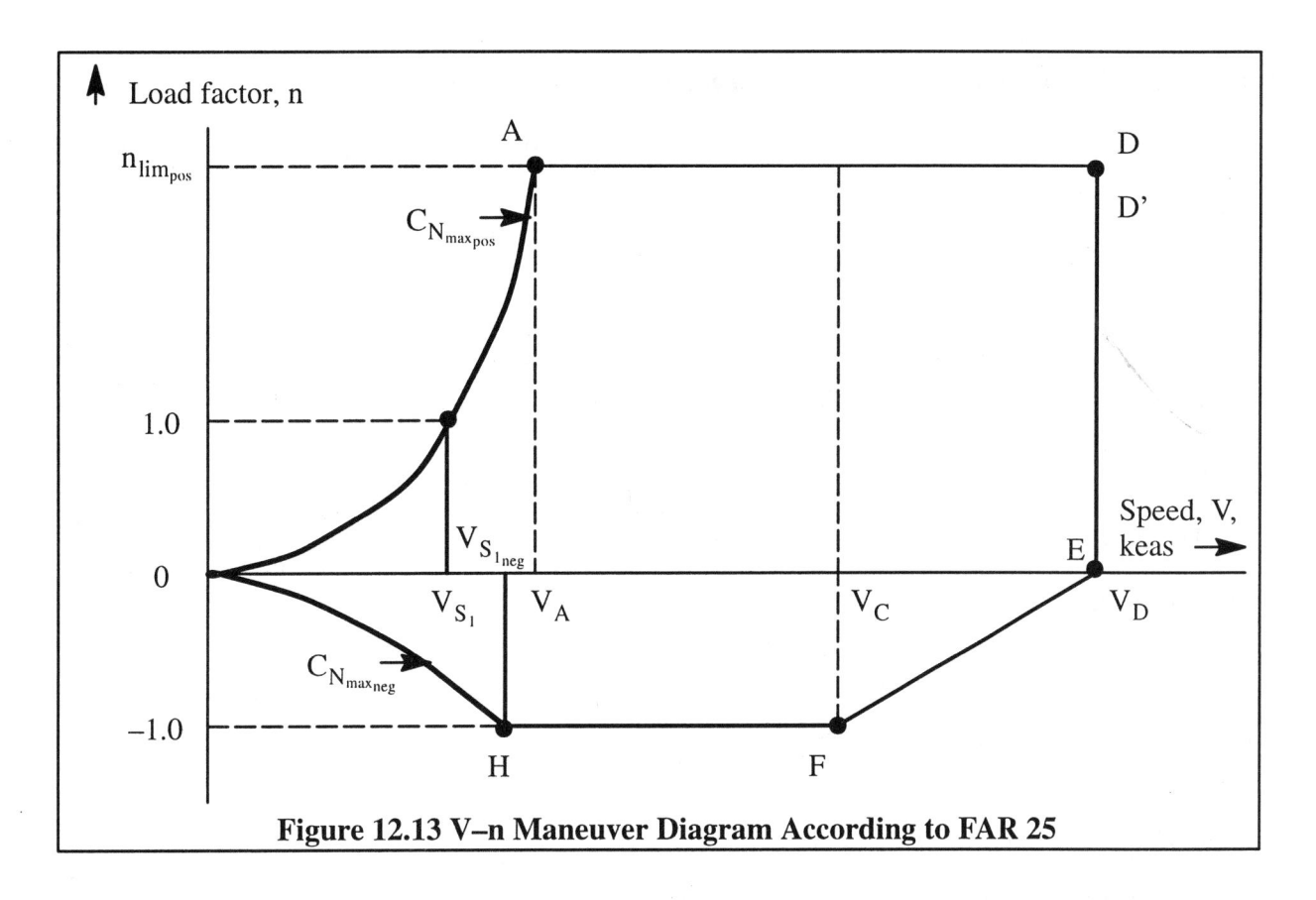

Maneuvering and the Flight Envelope

Figure 12.13 V–n Maneuver Diagram According to FAR 25

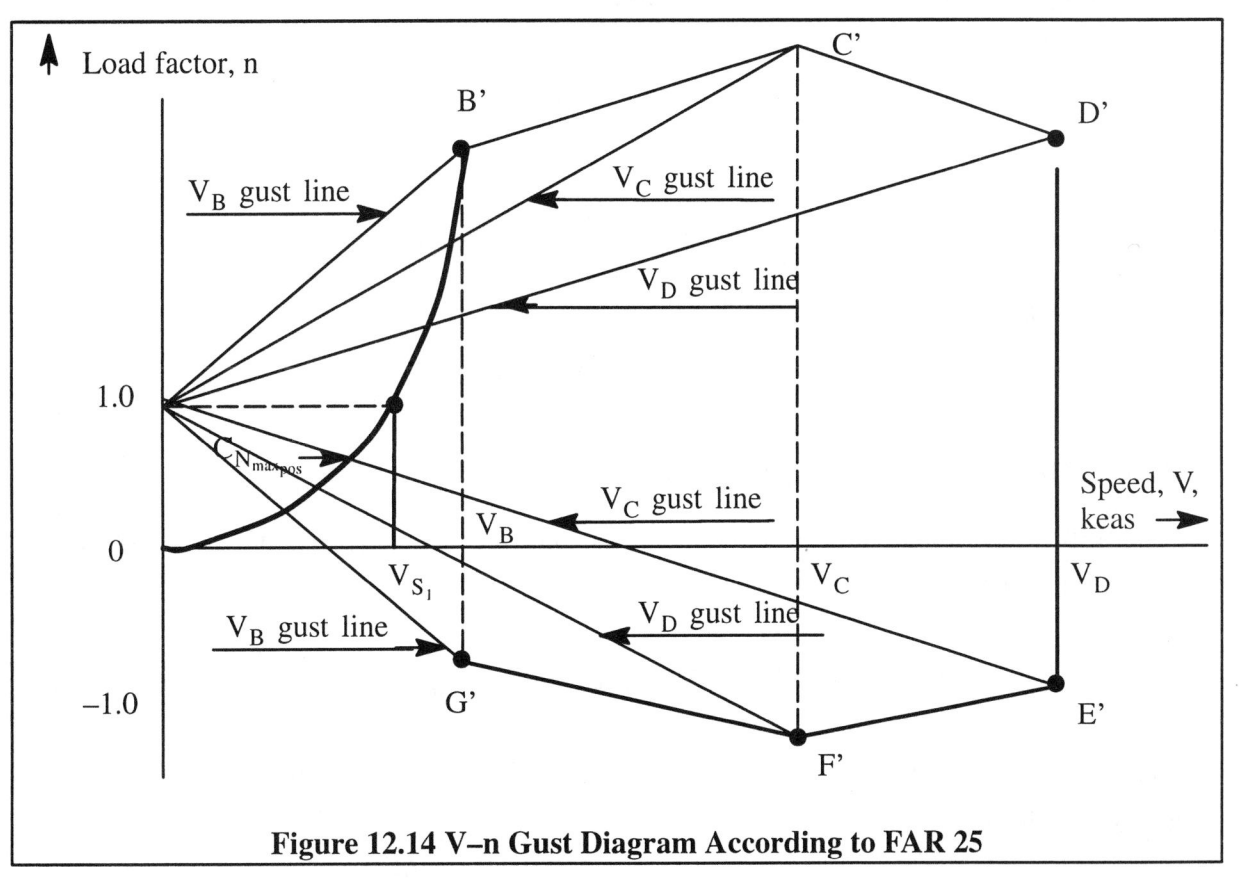

Figure 12.14 V–n Gust Diagram According to FAR 25

12.4.2.2 The Design Maneuvering Speed, V_A

The design maneuvering speed, V_A, must be selected by the designer, but must satisfy the following relationship:

$$V_A \geq V_{S_1}\sqrt{(n_{lim_{pos}})} \tag{12.37}$$

where: $n_{lim_{pos}}$ is the positive design limit load factor as defined in Sub–sub–section 12.4.2.6

However, V_A need not exceed V_C as defined in Sub–sub–section 12.4.2.?

12.4.2.3 The Design Speed for Maximum Gust Intensity, V_B

The design speed for maximum gust intensity, V_B (see Figure 12.14), need not be greater than the cruise speed, V_C.

However, it may also not be less than the speed determined from the intersection of the $C_{N_{max_{pos}}}$ line and the gust line marked, V_B.

12.4.2.4 The Design Cruising Speed, V_C

The design cruising speed, V_C (in keas), must be selected by the designer, but must satisfy the following relationship:

$$V_C \geq V_B + 43 \text{ keas} \tag{12.38}$$

12.4.2.5 The Design Diving Speed, V_D

The design diving speed, V_D, must satisfy the following relationship:

$$V_D \text{ (or } M_D) \geq 1.25 V_c \text{ (or } 1.25 M_C) \tag{12.39}$$

where: V_C follows from Eqn (12.38).

12.4.2.6 The Negative 1–g Stall Speed, $V_{S_{1_{neg}}}$

For purposes of constructing a V–n diagram, the negative 1–g stall speed, $V_{S_{1_{neg}}}$ may be computed with the method from Sub–sub–section 12.4.1.5.

12.4.2.7 The Design Limit Load Factors, $n_{lim_{pos}}$ and $n_{lim_{neg}}$

The positive design limit load factor, $n_{lim_{pos}}$ must be selected by the designer, but must meet the following condition:

Maneuvering and the Flight Envelope

$$n_{lim_{pos}} \geq 2.1 + \frac{24,000}{W_{FDGW} + 10,000} \tag{12.40}$$

However, $n_{lim_{pos}}$ must be greater than 2.5 at all times

$n_{lim_{pos}}$ need not be greater than 3.8 at $W = W_{FDGW}$

The negative design limit load factor, $n_{lim_{neg}}$ must be selected by the designer, but must meet the following condition:

$$n_{lim_{neg}} \geq -1.0 \text{ for speeds } \leq V_C \tag{12.41}$$

$n_{lim_{neg}}$ varies linearly between V_C and V_D (12.42)

12.4.2.8 The Gust Load Factor Lines in Figure 12.14

The three positive gust load factor lines in Figure 12.14 are defined by Equations (12.29) through (12.31). The gust line marked, V_B, is defined as follows:

For V_B gust lines:

$$U_{de} = 66 \text{ fps} \quad \text{for altitudes from sea–level to 20,000 ft} \tag{12.43}$$

$$U_{de} = 84.67 - 0.000933 \text{ h} \quad \text{for altitudes from 20,000 ft to 50,000 ft} \tag{12.44}$$

The V_C and V_D gust lines are as defined by Eqns (12.32) through (12.35).

12.4.3 MILITARY AIRPLANES

There are two types of V–n diagram for military airplanes: 1) the V–n maneuver diagram and 2) the V–n gust diagram.

The military gust diagrams are the same as those of FAR 25. The maneuver diagram is depicted in Figure 12.15. The design maximum load factors are given in Table 12.1. The various military procurement authorities reserve the right to modify design load factors as required by the mission of any new airplane.

The speeds V_H and V_L are defined as follows:

V_H is the maximum level flight speed which can be attained at a particular combination of weight and altitude. This speed may be determined with the methods of Sub–section 12.3.1.

V_L is the maximum design dive speed. Typically: $V_L = 1.25 V_H$.

Figure 12.15 V–n Maneuver Diagram for Military Airplanes

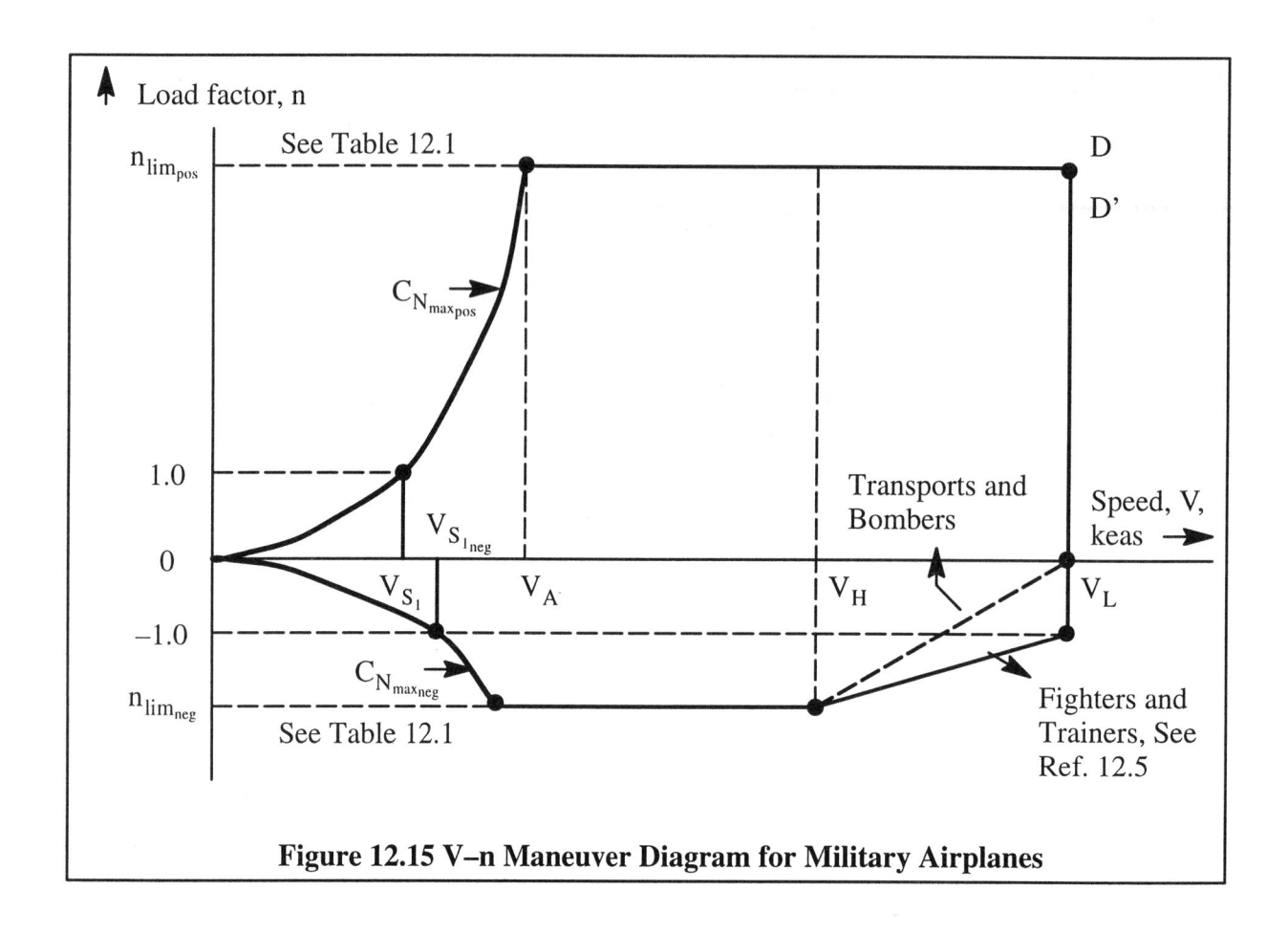

Table 12.1 Design Limit Load Factors for Military Airplanes {MIL–A 8861 (ASG)}

Airplane Type		Design Limit Load Factor at Flight Design Gross Weight: W_{FDGW}	
USAF	USN	$n_{lim_{pos}}$	$n_{lim_{neg}}$
Fighter		8.67	−3.00
Attack	Fighter, Attack, Trainer	7.33	−3.00
	Observation	6.00	−3.00
Trainer		5.67	−2.33
Utility	Utility	4.00	−2.00
Small Bomber		3.67	−1.67
Medium Bomber, Assault Transport	Patrol, Weather, Anti–submarine, Reconnaissance	3.00	−1.00
Medium Transport		2.50	−1.00
Heavy Bomber, Heavy Transport		2.00	−1.00

12.5 SYMMETRICAL PULL–UP: INSTANTANEOUS AND SUSTAINED

12.5.1 INSTANTANEOUS PULL–UP

Consider an airplane diving along a path in a vertical plane (vertical to the horizon). Figure 12.16 depicts a typical situation when the airplane is being recovered back to level flight. The flight path is assumed to be a circle with radius, R_{loop}. This maneuver is referred to as a steady symmetrical pull–up. A push–over maneuver would be just the opposite.

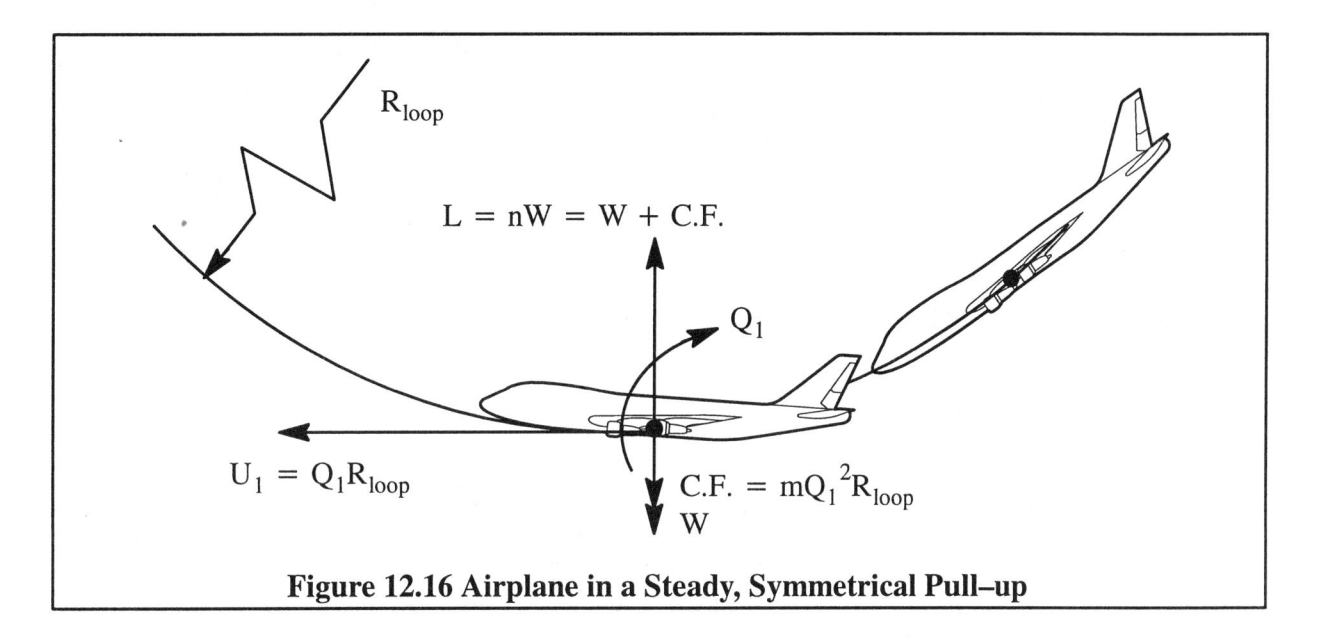

Figure 12.16 Airplane in a Steady, Symmetrical Pull–up

To pull n 'g'–s in such a pull–up requires the lift to equal: $L = nW$. The maximum load factor which can be developed from an aerodynamic viewpoint is given by:

$$n_{max} = \frac{L_{max}}{W\,(=L)} = \frac{0.5\varrho V^2 C_{L_{max}} S}{0.5\varrho V_S^2 C_{L_{max}} S} = \frac{V^2}{V_S^2} = \frac{C_{L_{max}}}{C_L} \qquad (12.45)$$

The reader should note that this result is consistent with that depicted in Figure 12.9. Observe, that at a speed equal to twice the stall speed, a n=4 load factor can be pulled. The problem is that this applies only in a steady state. If the airplane is pitched up rapidly, the build–up of the normal force coefficient with angle of attack becomes a function of the rate of build–up of angle of attack with time. Figure 12.17 shows the results of test data on an airfoil which indicate the dependence on the rate of change of angle of attack with time. Note that the maximum available normal force coefficient is substantially greater than it is according to static test data. This is referred to as a dynamic stall.

In a push–over maneuver the load factor will be negative. The magnitude of the maximum negative normal force coefficient tends to be less than that of the maximum positive normal force coefficient. The reason is that most wings are built with positive cambered airfoils.

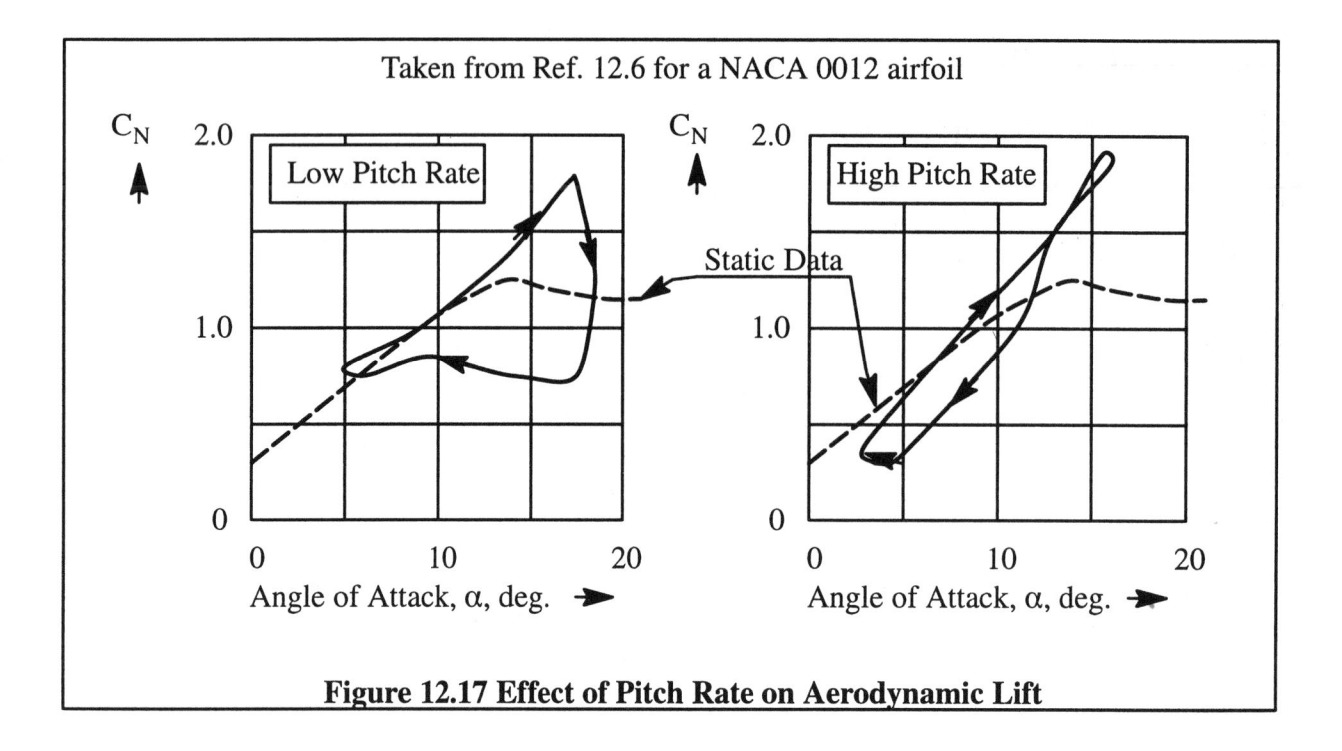

Figure 12.17 Effect of Pitch Rate on Aerodynamic Lift

12.5.2 SUSTAINED PULL–UP

In the discussion of the instantaneous pull–up, only the lift equilibrium was considered. It is quite possible that the thrust is insufficient to overcome the extra drag due to the higher required lift coefficient. If speed can be maintained, the maneuver is called: sustained. This implies sufficient power or thrust is available to overcome the extra induced drag associated with the higher lift in the pull–up. In the following analysis it is assumed that the pitch rate is sufficiently small, so that static data can be used.

For jet–propelled airplanes, the thrust required to sustain speed in a n–g pull–up can be written with Eqn (8.59) as::

$$T_{reqd_{n-g}} = D_{n-g} = C_D \bar{q} S = \left(C_{D_0} + \frac{(n\,C_L)^2}{\pi A e}\right)\bar{q}S \tag{12.46}$$

where: C_L is the lift coefficient in level flight at the same speed.

For propeller driven airplanes, the power required to sustain speed in a n–g pull–up can be written with En (8.69) as:

$$P_{reqd_{n-g}} = D_{n-g}V = C_D \bar{q} S V = \left(C_{D_0} + \frac{(nC_L)^2}{\pi A e}\right)\bar{q}SV \tag{12.47}$$

Figure 12.18 illustrates the dramatic difference between instantaneous and sustained load–factor capability of the F–16 fighter throughout its flight envelope.

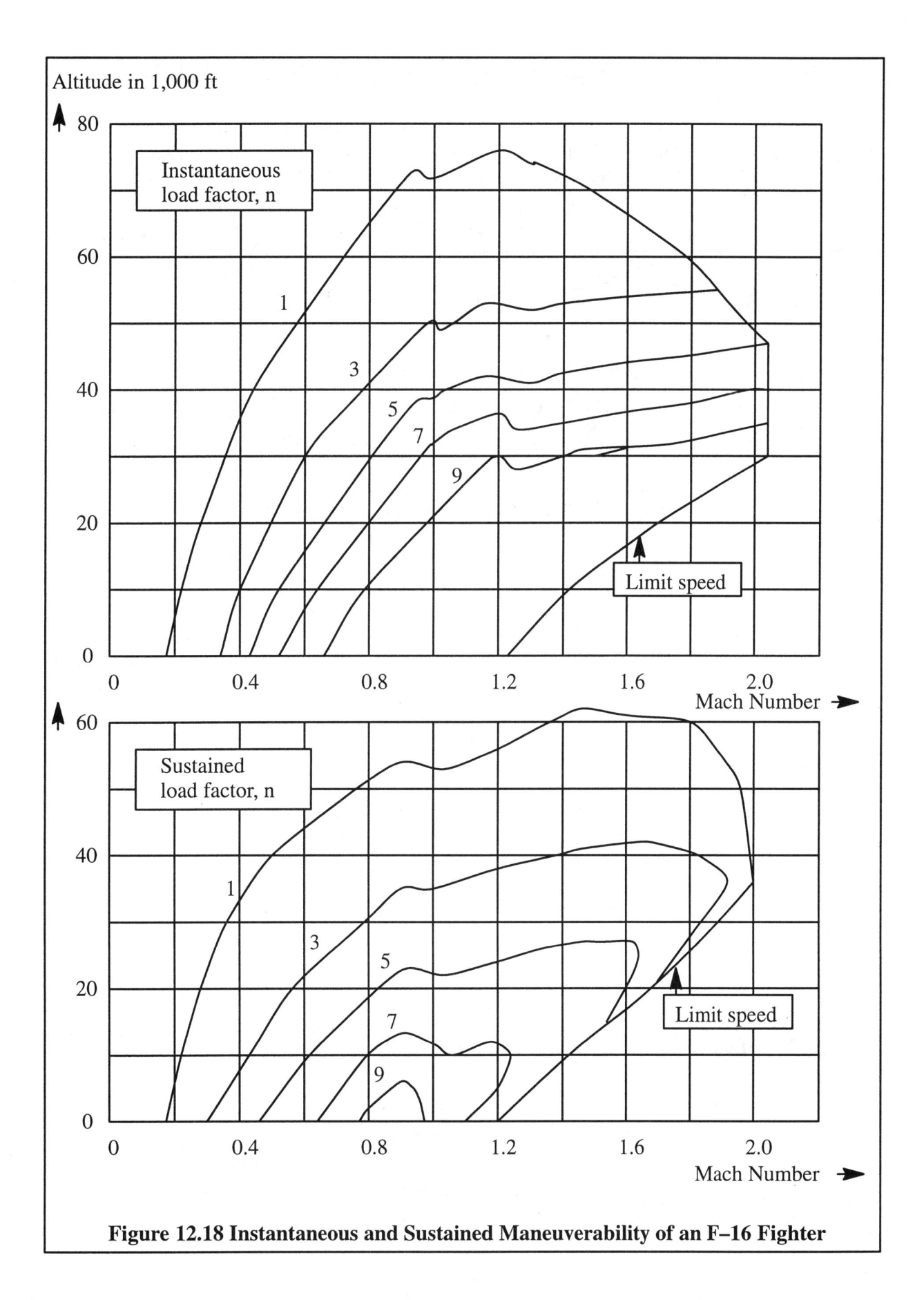

Figure 12.18 Instantaneous and Sustained Maneuverability of an F–16 Fighter

12.6 STEADY, LEVEL AND COORDINATED TURNS: INSTANTANEOUS AND SUSTAINED

Consider an airplane in a steady, level, coordinated turn. A coordinated turn is one in which the net acceleration in the airplane Y–axis direction is zero (ball in center). Figure 12.19 depicts the geometry and the forces in such a turn.

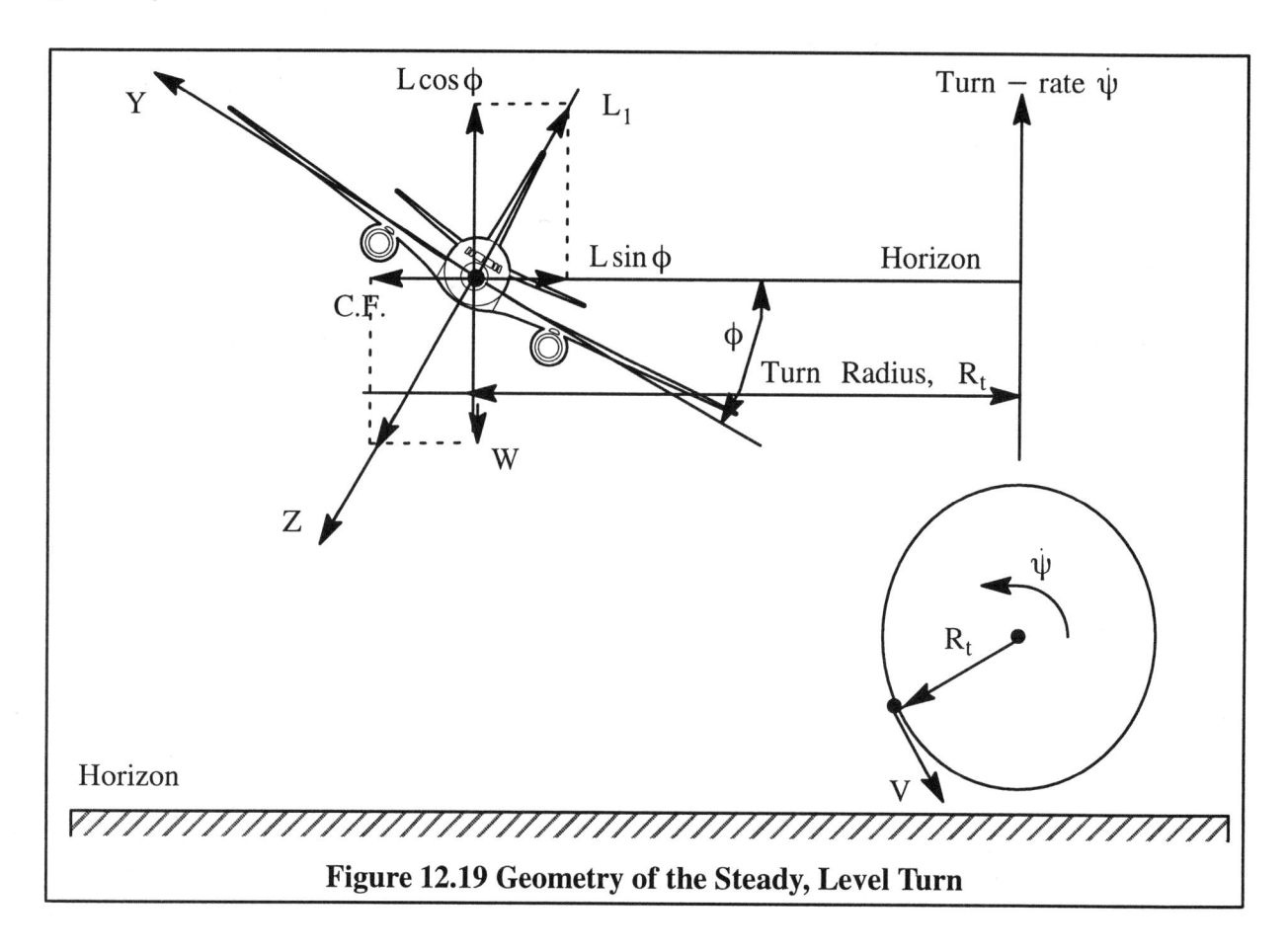

Figure 12.19 Geometry of the Steady, Level Turn

Evidently:

$$L \cos\phi = W \tag{12.48}$$

$$L \sin\phi = C.F. = \frac{W\,V^2}{g\,R_t} \tag{12.49}$$

From the definition of load factor it is also seen that:

$$n = \frac{L}{W} = \frac{1}{\cos\phi} \tag{12.50}$$

Upon dividing Eqn (12.49) by Eqn (12.48) it can be shown that the turn radius, R_t, becomes:

$$R_t = \frac{V^2}{g\,\tan\phi} = \frac{V^2}{g\sqrt{n^2 - 1}} \tag{12.51}$$

The turn radius is an important performance parameter for any airplane which must turn to avoid terrain (such as might happen when flying up a box canyon) or which must turn inside a potential adversary. The latter is particularly of military significance. In that case, the turn rate is also important. The corresponding turn rate, $\dot{\psi}$, follows from:

$$\dot{\psi} = \frac{V}{R_t} \tag{12.52}$$

The situation depicted in Figure 12.19 applies to a sustained turn since the speed V is supposed to be constant in a steady, level turn. The geometry and the forces can however be applied to the instantaneous turn, with the understanding that in reality the speed and/or the altitude will bleed off. Sub–section 12.6.1 deals with the instantaneous case while 12.6.2 deals with the sustained case.

12.6.1 INSTANTANEOUS TURN

An interesting observation from Eqn (12.51) is that the minimum turn radius at any given speed depends only on the maximum load factor which the airplane can develop. This maximum load factor may be determined on an instantaneous basis from Eqn (12.45) or from the insert in Figure 12.9:

$$n_{instantaneous} = 1 + \Delta n_{instantaneous} = 1 + \frac{C_{L_{buffet}} - C_{L_{at\,1-g}}}{C_{L_{at\,1-g}}} \tag{12.53}$$

This load factor can be used to estimate the instantaneous turn radius and the corresponding instantaneous turn rate.

It is important to recognize the fact that in a turn the stall speed becomes a function of the load factor and thus of the bank angle. The reader is asked to show that the following holds for the instantaneous stall speed at load factors larger than unity:

$$V_{S_{turn}} = V_{S_{1-g}} \sqrt{n_{instantaneous}} \tag{12.54}$$

This is also called the stall speed in turning flight.

12.6.2 SUSTAINED TURN

To sustain a turn requires that sufficient thrust (or power) is available to over come the extra induced drag generated in a coordinated turn. If the turn is not coordinated, the drag can even be higher because of the non–zero sideslip. Since in a turn the lift is n times that in level flight, the induced drag will be n^2 times larger! Therefore, in a 2–g turn (corresponding to a 60 degrees bank angle), the induced drag is 4 times larger than in level flight!

Equations (12.46) and (12.47) can be used to estimate the thrust (or power) required in a maximum effort, sustained, turn.

Figure 12. 20 shows the relationship between speed, turn radius and turn rate in a steady (= sustained) level turn. Note the logarithmic nature of the graph.

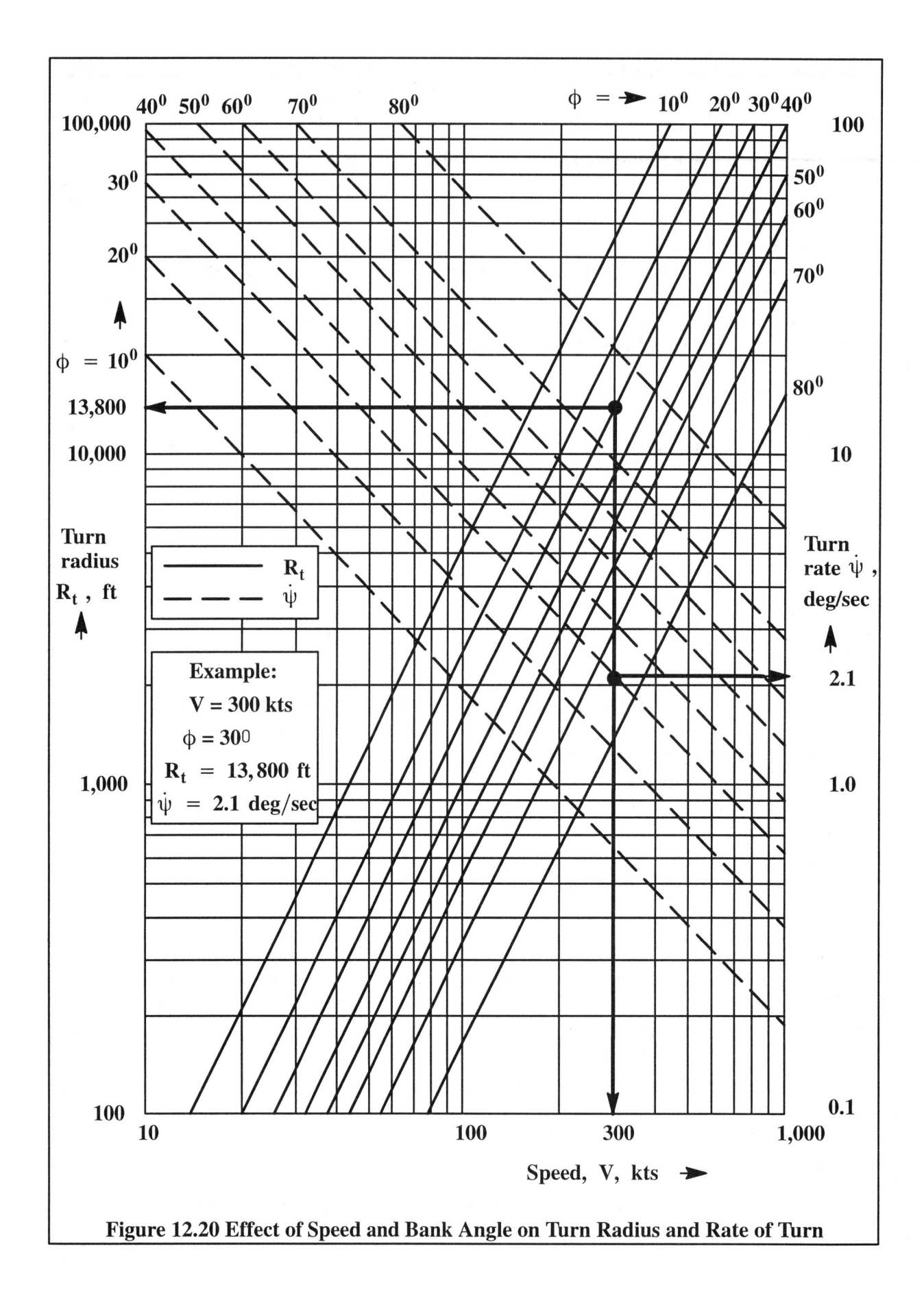

Figure 12.20 Effect of Speed and Bank Angle on Turn Radius and Rate of Turn

Example:
V = 300 kts
$\phi = 30°$
R_t = 13,800 ft
$\dot{\psi}$ = 2.1 deg/sec

As a general rule, the turning performance of an airplane may be limited by four factors:

1) Aerodynamic: stall and/or buffet
2) Structural: positive or negative design limit load factor
3) Thrust or power: maximum available thrust or power, engine temperature limit
4) Inlet: inlet distortions at high angles of attack in high load–factor turns

The absolute minimum turn radius which can be achieved in a turn, occurs at the stall speed in the turn. By combining Eqns (12.51) and (12.54) it is found that:

$$R_{t_{min.}} = \frac{V_{S_t}^2}{g\sqrt{n^2-1}} = \frac{V_S^2}{g}\frac{n}{\sqrt{n^2-1}} \qquad (12.55)$$

As the load factor, n, increases, the factor: $n/\sqrt{n^2-1}$ will decrease and approach unity, when $n \to \infty$. The absolute minimum turn radius is given by:

$$R_{t_{absolute\ min.}} = \frac{V_S^2}{g} \qquad (12.56)$$

The reader should keep in mind the fact that V_S is a constant for a given airplane, at a given weight and altitude. A smaller radius of turn than that given by Eqn (12.56) **cannot** be achieved in a sustained (steady) level turn.

Observe also from Eqn (12.55) that when n = 1 (steady level flight) the turn radius is infinite.

These properties are illustrated in Figure 12.21a).

Increasing the speed above the turning stall speed will eventually produce a load factor which is equal to or exceeds the design limit load factor from a structural viewpoint. Increasing speed beyond this point (assuming there is sufficient thrust or power to do this) will make the airplane exceed its structural limit. Design limit load factors were discussed in Section 12.5.

When the bank angle and the load factor are held constant to satisfy a given structural limit, the turn radius varies with V^2. The intersection of the aerodynamic limit and the structural limit is referred to as the maneuver speed, V_A, as shown in Figure 12.21. This is also the maneuver speed used in the V–n diagrams of Figures 12.13, 12.14 and 12.16.

Therefore, the maneuver speed, V_A, is the minimum speed necessary to develop aerodynamically the design limit load factor, $n_{lim_{pos}}$, and it produces the minimum turn radius at the aerodynamic and structural limits simultaneously.

Figure 12.21 Limits on Maneuvering Performance of Airplanes

a) Aerodynamic and Structural Limits on Turning Performance

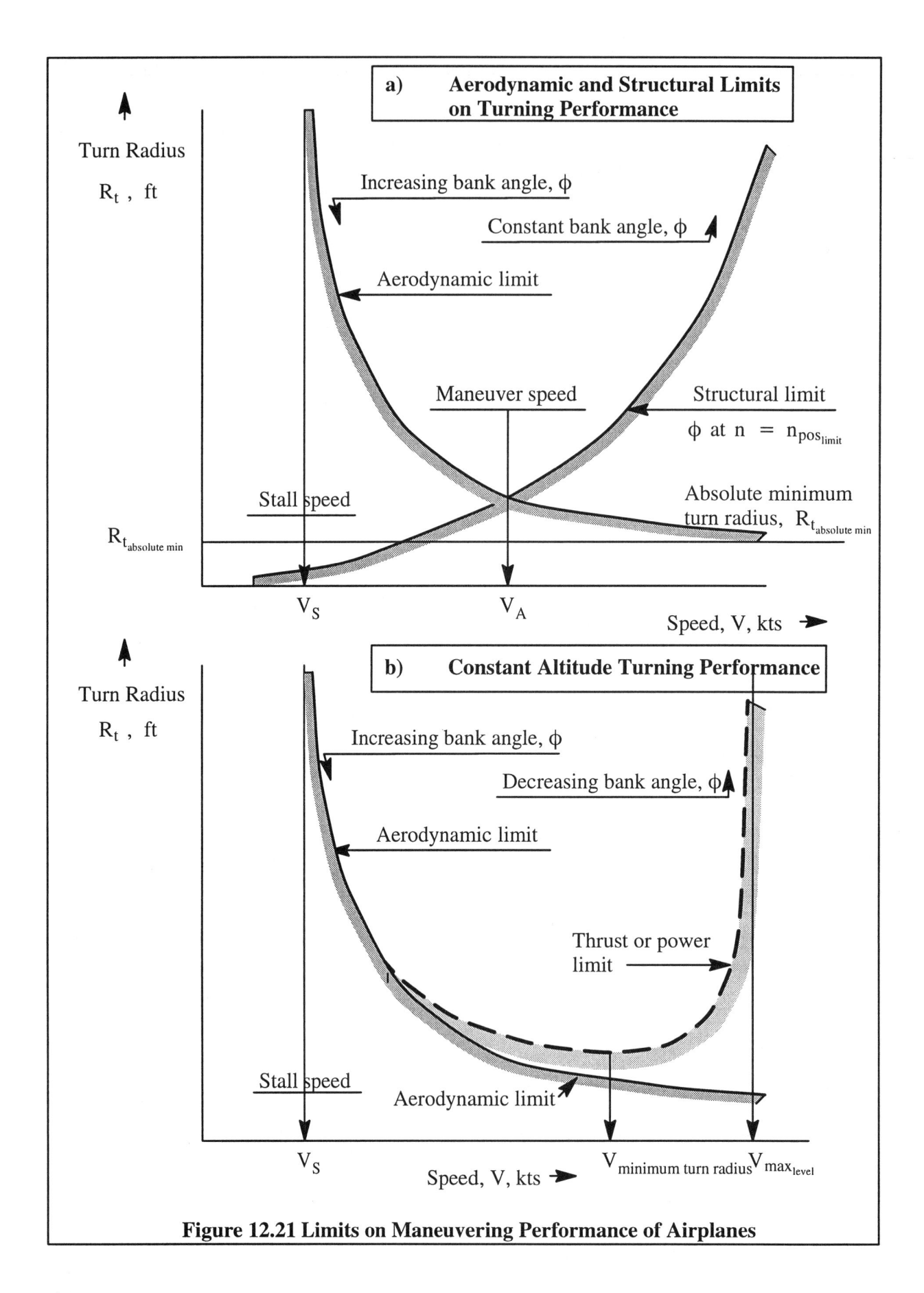

- Increasing bank angle, ϕ
- Constant bank angle, ϕ
- Aerodynamic limit
- Maneuver speed
- Structural limit, ϕ at $n = n_{pos_{limit}}$
- Stall speed
- Absolute minimum turn radius, $R_{t_{absolute\,min}}$

Axes: Turn Radius R_t, ft vs. Speed, V, kts; markers at V_S and V_A.

b) Constant Altitude Turning Performance

- Increasing bank angle, ϕ
- Decreasing bank angle, ϕ
- Aerodynamic limit
- Thrust or power limit
- Stall speed
- Aerodynamic limit

Axes: Turn Radius R_t, ft vs. Speed, V, kts; markers at V_S, $V_{minimum\,turn\,radius}$, $V_{max_{level}}$.

Maneuvering and the Flight Envelope

In these considerations, the maximum lift coefficient was assumed to be independent of the flight speed (or Mach number). In reality this is not the case as the reader may see from Figure 12.7. Because of the drop–off in maximum lift coefficient at high Mach numbers, the turn radius at high Mach numbers tends to be thrust limited, rather than structures limited.

At the maximum achievable level flight speed, where maximum thrust equals the drag, the turn radius becomes infinite.

As the speed is reduced below that for maximum level speed, the parasite drag is reduced while the induced drag is increased. Minimum turn radius turns can now be achieved as limited by thrust or power available. The reader is cautioned again about the fact that the maximum lift coefficient may decrease significantly with Mach number as shown in Figure 12.7. In such a case an iterative solution is indicated.

These properties are illustrated in Figure 12.21b). Most generally, aerodynamic and structural limit pre–dominate at low altitude while aerodynamic and thrust(or power) limits pre–dominate at high altitudes. An example application will now be discussed.

Example 12.1: An airplane is flying straight and level at sea–level and a speed of 300 ft/sec. The pilot puts the airplane in a level, coordinated turn with a radius of 2,850 ft, while maintaining the same angle of attack as the one the airplane had in the straight and level flight condition. The pilot adjusts the engine thrust as required to maintain the speed at 300 ft/sec (sustained turn). Without changing the angle of attack or the engine thrust, the pilot next brings the airplane out of the turn, to a wings level climb condition. Estimate the rate of climb, if the lift–to–drag ratio is 9.0. Assume no acceleration along or perpendicular to the flight path in the climb.

Solution: In straight and level flight: $\quad L_{level} = W \quad$ (1)

In level, coordinated turning flight: $\quad L_{turn} \cos \phi = W = L_{level} \quad$ (2)

As long as the angle of attack remains the same, the lift–to–drag ratio will remain the same. Therefore: $\quad (L/D)_{turn} = (L/D)_{level} \quad$ (3)

Because of (2): $\quad V_{turn}^2 \cos \phi = V_{level}^2 \quad$ (4)

In a steady level turn, according to Eqn (12.51) and using (2):

$$\tan \phi = \frac{V_{turn}^2}{gR_t} = \frac{V_{level}^2}{gR_t \cos \phi} \quad (5)$$

Therefore: $\quad \sin \phi = \dfrac{V_{level}^2}{gR_t} = \dfrac{300^2}{32.2 \times 2,850} = 0.9807 \quad$ (6)

The ratio of the thrust required to overcome the drag in the turn to that in level flight is found from:

$$\frac{T_{turn}}{T_{level}} = \frac{D_{turn}}{D_{level}} = \frac{L_{turn}}{L_{level}} = \frac{1}{\cos\phi} = \frac{1}{\sqrt{1 - 0.9807^2}} = 5.12 \quad (7)$$

Once the airplane is out of the turn and in the un-accelerated climb condition, the following equations of motion apply:

$$L_{climb} - W\cos\gamma = 0 \quad (8)$$

$$T_{climb} - W\sin\gamma - D_{climb} = 0 \quad (9)$$

Elimination of W yields:

$$T_{climb} = L_{climb}\tan\gamma + D_{climb} = D_{climb}\left(\frac{L_{climb}}{D_{climb}}\tan\gamma + 1\right) \quad (10)$$

Division by: $T_{level} = D_{level}$ yields:

$$\frac{T_{climb}}{T_{level}} = \frac{D_{climb}}{D_{level}}\left(\frac{L_{climb}}{D_{climb}}\tan\gamma + 1\right) \quad (11)$$

However, the following is also correct (using (3) and (8)):

$$\frac{D_{climb}}{D_{level}} = \frac{D_{climb}}{L_{climb}}\frac{L_{climb}}{L_{level}} = \frac{D_{climb}}{L_{climb}}\frac{W\cos\gamma}{D_{level}} = \frac{D_{climb}}{L_{climb}}\frac{L_{level}}{D_{level}}\cos\gamma = \cos\gamma \quad (12)$$

Therefore, using (11):

$$\frac{T_{climb}}{T_{level}} = \frac{T_{turn}}{T_{level}} = \cos\gamma\left(\frac{L_{climb}}{D_{climb}}\tan\gamma + 1\right) = \frac{L_{climb}}{D_{climb}}\sin\gamma + \cos\gamma \quad (13)$$

Next, using (7) and the fact that the lift-to-drag ratio is 9.0 due to the constant angle of attack, it follows that:

$$5.12 = \cos\gamma + 9\sin\gamma \quad (14)$$

By quadrature it may be shown that this cab be re-written as:

$$82\cos^2\gamma - 10.23\cos\gamma - 54.79 = 0 \quad (15)$$

From this it follows that: $\cos\gamma = 0.881$ and $\sin\gamma = 0.474$
From (8) it follows that:

$$0.5\varrho V_{climv}^2 C_{L_{climb}} S - 0.5\varrho V_{level}^2 C_{L_{level}} S\cos\gamma = 0 \quad (16)$$

Because the angle of attack is presumed to be constant, from level flight to turning flight and then to climbing flight, the lift coefficients are identical.
Thus: $C_{L_{climb}} = C_{L_{turn}} = C_{L_{climb}}$. Therefore, (16) yields:

$$V_{climb} = V_{level}\sqrt{\cos\gamma} = 300\sqrt{0.881} = 282 \text{ft/sec} \quad (17)$$

The rate-of-climb, R.C. is thus given by:
R.C. = $V\sin\gamma$ = 282 x 0.474 = 134 ft/sec = 8,040 ft/min.

12.7 SPIN

Spin entry is usually caused by the stalling of the wing of an airplane. For that reason, the stall characteristics and the associated stability behavior of the airplane at high angles of attack (i.e. high lift coefficients) are important in defining the initial tendencies of an airplane following wing stall.

During spins the balance between aerodynamic forces and moments on the one hand and the inertial properties of the airplane on the other hand dominate the picture. Because airplanes have large differences in external aerodynamic configuration and internal mass distribution it has not been possible to arrive at a spin theory which covers all airplanes and which can be used during early design to produce an airplane with "good" spin characteristics. Reference 12.7 contains detailed discussions of the spin behavior of various types of airplanes.

To understand how an airplane may enter a spin, after a wing has stalled and after disturbances in yaw and roll have occurred, consider Figure 12.22 and focus on a wing section on the right wing. The yaw rate disturbance, r, is seen by that section as a local increase in forward velocity, u_{yaw}. The roll rate disturbance, p, is seen as a change in vertical velocity, w_{roll}. If the reference angle of attack of the airplane is: α, the combined effect of the two disturbances is to increase the local section angle of attack from α to $\alpha_{right} = \alpha + \Delta\alpha_{right}$. The reader should refer to the vector diagram in the upper left corner of Figure 12.22.

A similar vector diagram on the right of Figure 12.22 shows that for a wing section on the left wing the combined effect of the disturbances is to decrease the local section angle of attack from α to $\alpha_{left} = \alpha - \Delta\alpha_{left}$.

Next, refer to the $C_L - C_D - \alpha$ plot (Figure 12.22a) for a conventional, low performance airplane. Such airplanes typically have unswept, moderate aspect ratio wings. Note, that the right (or down–going) wing section experiences a decrease in lift and an increase in drag relative to the reference angle of attack. The left wing section is seen to experience an increase in lift and a decrease in drag relative to the reference angle of attack. Together these forces create moments which will enforce the rolling and yawing (i.e. spinning) of the airplane. The airplane is said to auto–rotate.

In conventional, low performance airplane the spinning motion is dominated by rolling with only moderate yawing. This is because the lift changes are stronger than the drag changes.

Next, refer to the $C_L - C_D - \alpha$ plot (Figure 12.22b) for a highly swept, high performance airplane. The maximum lift coefficient is not very well defined because of the shallow (flat) nature of the lift–curve at high angles of attack. The difference in lift between the right wing and the left wing is very small so that the rolling motion is weak. On the other hand, the drag changes are very large which tends to produce a strong yawing motion.

In highly swept, high performance airplane the spinning motion is dominated by yawing with only moderate rolling. This is because the lift changes are weaker than the drag changes.

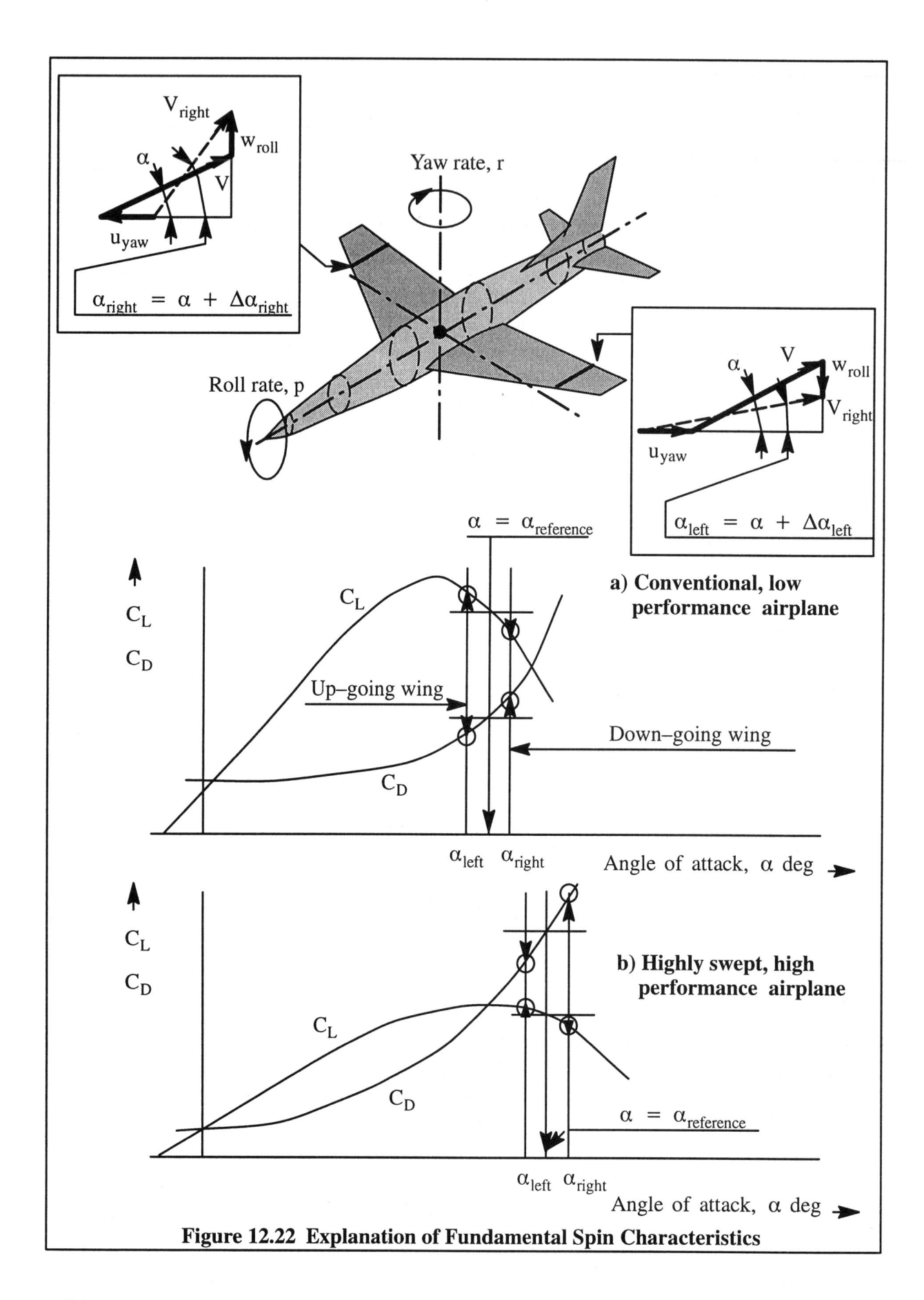

Figure 12.22 Explanation of Fundamental Spin Characteristics

After the spin is entered, different spin phases may be identified:

(1) The incipient phase: this is the non–steady state portion of the spin. It is also referred to as "post–stall gyration".

(2) The developed phase: this is the phase involving a balance between aerodynamic and inertial moments and forces. The situation depicted in Figure 12.22 deals with this phase.

(3) The recovery phase: this is another non–steady phase during which the controls are moved to break the steady state, auto–rotation phase.

Whether or not the controls are effective in bringing about a recovery from the spin depends on the ability of these controls to generate moments which are sufficient to halt the auto–rotation. Practical experience has shown that the use of the rudder to oppose the spin rotation is often effective in arresting a spin. However, whether or not this works is very strongly dependent on details of the airplane configuration. For more detailed discussions of the effect of configuration on spin behavior the reader is referred to References 12.8, 12.9 and 12.10.

In all instances, if the airplane angle of attack can be reduced below that of wing stall, the spin can be arrested. This requires the stick to be moved forward. However, the longitudinal controls are not always sufficiently effective to do this.

In the previous discussion, the words inertial moments were used several times. It is useful to examine the general airplane equations of motion of Ref. 12.2, Chapter 1. It is seen that several inertial terms appear in the rolling and yawing moment equations (1.52a) and (1.52c) of Ref. 12.2.

In the yawing moment equation (1.52c) of Ref.12.2, the term $\{(I_{xx} - I_{yy})/I_{zz}\}pq$ is of great interest. For high performance airplanes it is generally true that $I_{yy} \gg I_{xx}$. By exciting the airplane in pitch and roll (to generate positive p and q) a significant inertial yawing acceleration can be generated which can help in arresting the spin. This requires that the stick be moved back.

In the rolling moment equation (1.52a) of Ref. 12.2, the term $\{(I_{zz} - I_{yy})/I_{xx}\}rq$ is not usually very significant, because in many airplanes $I_{zz} \approx I_{yy}$.

In the pitching moment equation (1.52b) of Ref.12.2, the term $\{(I_{xx} - I_{zz})/I_{yy}\}pr$ can also be of great interest. For high performance airplanes it is generally true that $I_{zz} \gg I_{xx}$. By exciting the airplane in roll and yaw (p and r) a significant inertial pitching acceleration can be generated which can help in arresting the spin. Whenever the rudder is not very effective in exciting yaw rate, a pumping motion of the stick may be tried to induce pitching oscillations, which, once large enough can be used to reduce the angle of attack.

References 12.9 and 12.10 should be consulted for further information on spin behavior and recovery. Aerodynamic design for spin prevention was the focus of research reported in Ref. 12.11.

12.8 ACCELERATED FLIGHT PERFORMANCE

In the discussions of this chapter, the accelerations along the flight path have been neglected, and only accelerations perpendicular to the flight path were considered. For low performance airplanes this is acceptable. For high performance airplanes and in particular for supersonic airplanes this is not acceptable. The airplane equations of motion in this case are given by Eqns (9.1) and (9.2) which are re–numbered here:

$$T\cos(\alpha + \phi_T) - D - W\sin\gamma = \frac{W}{g}\frac{dV}{dt} \qquad (12.57)$$

$$T\sin(\alpha + \phi_T) + L - W\cos\gamma = \text{C.F.} = \frac{W}{g}V\dot{\gamma} \qquad (12.58)$$

The forces and angles in these equations were defined in Figure 9.1. For the convenience of the reader, this figure is repeated here as Figure 12.23.

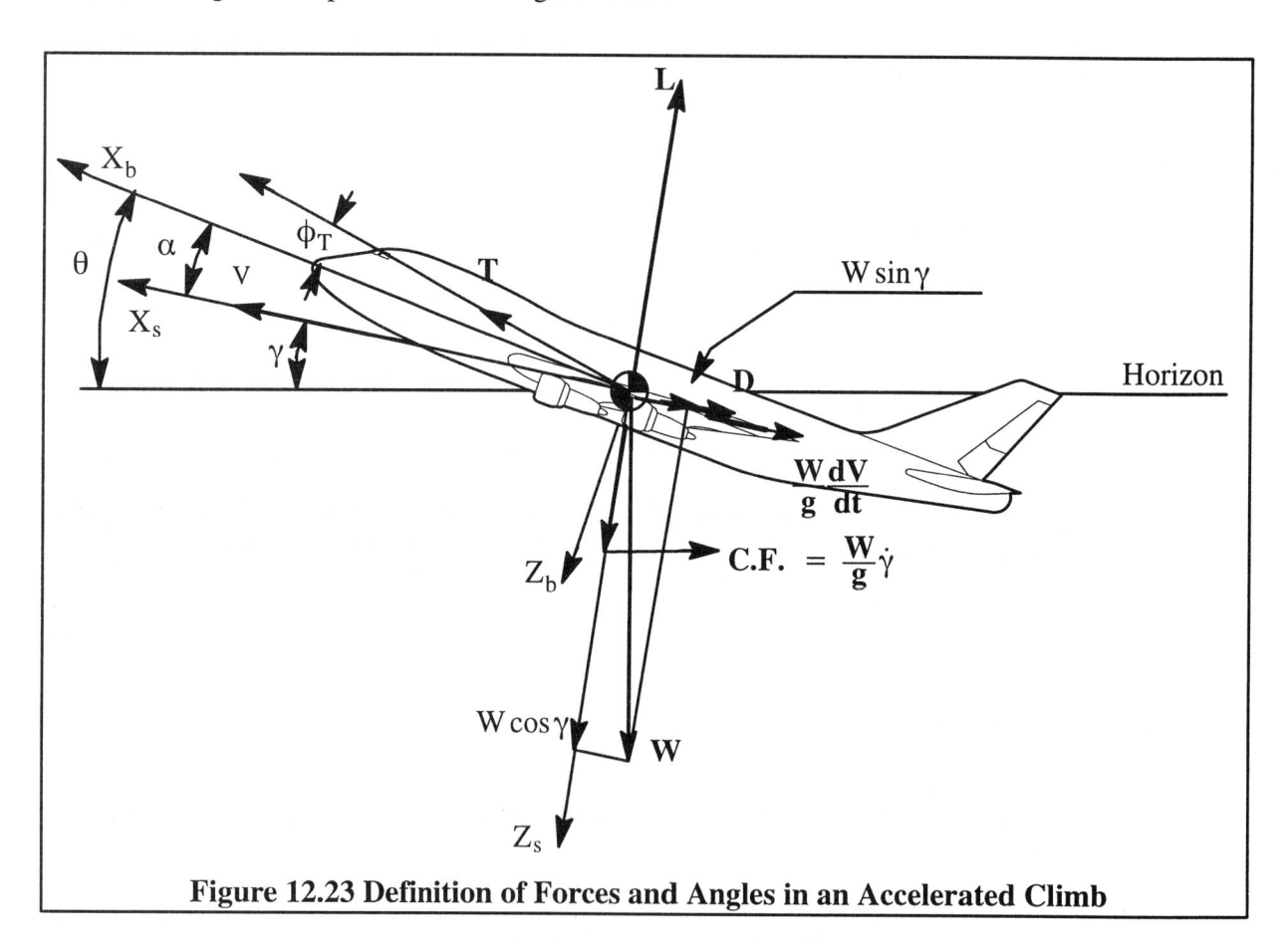

Figure 12.23 Definition of Forces and Angles in an Accelerated Climb

These equations must be augmented by the following additional equations:

$$\text{R.C.} = \dot{h} = V\sin\gamma \qquad (12.59)$$

$$\dot{x} = V\cos\gamma \qquad (12.60)$$

$$\dot{W} = -\dot{W}_f \qquad (12.61)$$

The variables in these equations were already identified in Chapter 9. Additional variables introduced by equations (12.59) through (12.61) are: altitude, h, horizontal distance, x and fuel flow rate, $\dot{W}_f$. Observe, that the following functional relations exist:

$$L = L(V, h, \alpha), \quad D = D(V, h, \alpha), \quad T = T(V, h) \text{ and } \dot{W}_f = \dot{W}_f(V, h) \tag{12.62}$$

In general, the only practical way to solve these five equations for V, γ, h, x and W is through numerical integration. For quick estimation it has been found that the so–called "energy–state approximation" can be used. References 12.12 through 12.14 contain examples.

In Sub–section 9.4.1 an application of the energy–state approximation to accelerated flight performance will be presented.

12.8.1 APPLICATION OF ENERGY–STATE APPROXIMATION TO ACCELERATED FLIGHT PERFORMANCE

Equation (12.57) can be re–written as:

$$\frac{1}{g}\frac{dV}{dt} + \sin\gamma = \frac{T\cos(\alpha + \phi_T) - D}{W} \tag{12.63}$$

Multiplying both sides by V and replacing (Vsin γ) by dh/dt yields:

$$\frac{d}{dt}\left(\frac{V^2}{2g}\right) + \frac{dh}{dt} = \frac{\{T\cos(\alpha + \phi_T) - D\}V}{W} \tag{12.64}$$

The left hand side of Eqn (12.64) represents the time rate of change of the total energy (kinetic plus potential) per unit weight. It is referred to as the specific energy and defined as:

$$h_e = \frac{V^2}{2g} + h \tag{12.65}$$

It therefore follows that:

$$\frac{dh_e}{dt} = \frac{d}{dt}\left(\frac{V^2}{2g}\right) + \frac{dh}{dt} = \frac{\{T\cos(\alpha + \phi_T) - D\}V}{W} \tag{12.66}$$

The time derivative of specific energy, dh_e/dt, is referred to as the "specific excess power" of the airplane (S.E.P.). Eqn (12.66) can be interpreted to say that, if the S.E.P of an airplane is known, its rate of climb, R.C.=dh/dt and its acceleration, dU/dt can be determined.

As an example, consider the F–104G flying at M=0.8 and 20,000 ft altitude (V = 829 ft/sec) with a weight of W=18,000 lbs. Assume that the maximum thrust available in that condition is T=10,000 lbs while the drag in level (n=1) flight is D=2,086 lbs. It can be assumed that $(\alpha + \phi_T) = 0$.

The specific excess power in this case is:

$$\frac{dh_e}{dt} = \frac{(10,000 - 2,086)\,829}{18,000} = 365 \text{ ft/sec} \tag{12.67}$$

Because $dh_e/dt > 0$ at the maximum thrust, it implies that the F–104G will accelerate and/or climb, depending on what the pilot wants to do.

If the true airspeed is held constant ($dV/dt = 0$) the climb rate in this flight condition is given by:

$$\text{R.C.} = \frac{dh}{dt} = \frac{dh_e}{dt} = 365 \text{ ft/sec} = 21,900 \text{ ft/min} \tag{12.68}$$

If, on the other hand, the flight path is held level so that the R.C. is zero, the airplane will accelerate as follows:

$$\frac{dV}{dt} = \frac{g}{U}\frac{dh_e}{dt} = \frac{32.2}{829} \times 365 = 14.2 \text{ ft/sec}^2 \tag{12.69}$$

If the pilot wants to fly at a constant speed (say 829 ft/sec) he must either reduce thrust or perform a level turn to pull g's and thereby increase the drag at the same thrust.

A useful way in which to indicate the performance potential of an airplane is to plot constant specific energy curves and constant specific excess power (S.E.P.) curves on a plot of altitude versus Mach number. This is done in Figure 12.24 for the F–104G. An interpretation of the various points and lines in Figure 12.24 is given next.

(1) At any point, A, on the $dh_e/dt = 0$ line, the airplane is in a steady state, level flight condition. It cannot climb at the speed it is flying at any point A. At any point B or C, where $dh_e/dt = 200$ ft/sec, the airplane will climb, accelerate or both. It should be noted that the actual thrust and drag levels at points A, B and C are different.

Point D represents the maximum energy level that can be reached by this airplane. Point E is above the $dh_e/dt = 0$ line. The airplane can reach point E only by zooming up there with an energy level of 9.5×10^4 ft. The airplane is not capable of staying at point E because $dh_e/dt < 0$.

(2) The R.C. of an airplane can be determined from the specific energy plot if the climb schedule of the airplane (desired Mach number versus altitude) is known. This may be seen from Eqn (9.9) by re-writing it in the following form:

$$\frac{dh}{dt} = \frac{dh_e/dt}{1 + \frac{V}{g}\frac{dV}{dh}} \tag{12.70}$$

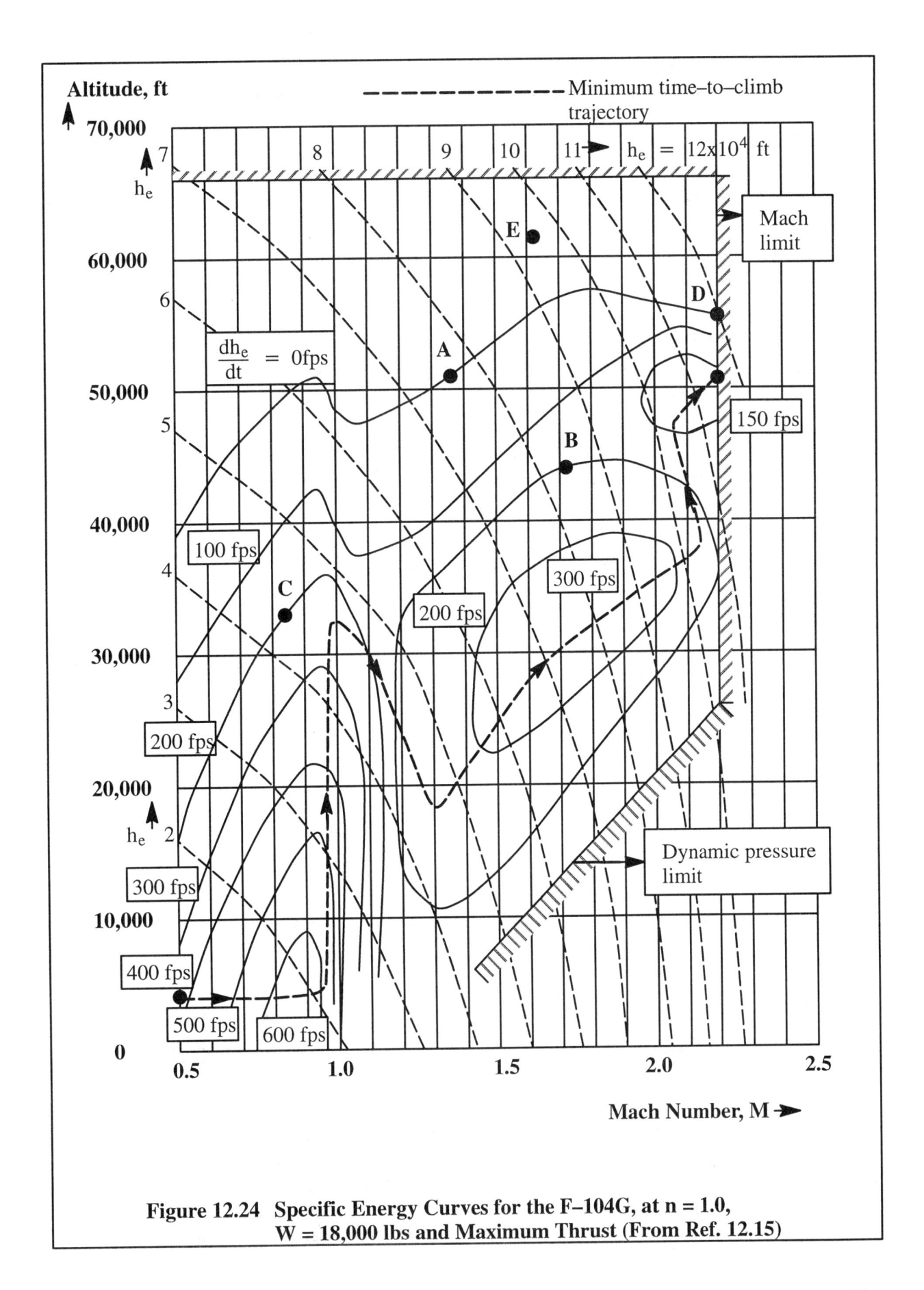

Figure 12.24 Specific Energy Curves for the F-104G, at n = 1.0, W = 18,000 lbs and Maximum Thrust (From Ref. 12.15)

By first plotting the desired climb schedule or trajectory on the h–V diagram (Figure 12.24) the R.C. can be obtained by taking the slope dV/dh at each point on the h–V diagram.

(3) Determination of a minimum time–to–climb schedule. The time required to climb from one energy level to another is given by:

$$t = \int_{h_{e_1}}^{h_{e_2}} \frac{1}{dh_e/dt} dh_e \qquad (12.71)$$

where: dh_e/dt must be regarded as a function of h_e and Mach number, M or speed, V. The condition for minimizing the integral in Eqn (12.71) can be obtained from the Calculus of Variations (See for example, Ref. 12.14) as:

$$\frac{\partial}{\partial V}\left\{\frac{1}{(dh_e/dt)_{h_e = \text{constant}}}\right\} = \frac{-\partial/\partial V\left\{(dh_e/dt)_{h_e = \text{constant}}\right\}}{\left\{(dh_e/dt)_{h_e = \text{constant}}\right\}^2} = 0 \qquad (12.72)$$

However, because $dh_e/dt \neq 0$, this condition can be written as:

$$\frac{\partial}{\partial V}\left\{(dh_e/dt)_{h_e = \text{constant}}\right\} = 0 \qquad (12.73)$$

In the conventional method for determining the best climb performance of airplanes the condition which corresponds to Eqn (12.73) is:

$$\frac{\partial}{\partial V}\left\{(dh_e/dt)_{h = \text{constant}}\right\} = 0 \qquad (12.74)$$

Eqn (12.74) implies that the speed schedule for minimum time–to–climb is chosen such that <u>at a given altitude</u> the specific excess power, dh_e/dt is a maximum.

On the other hand, Eqn (12.74) implies that the corresponding speed schedule should be selected where the contours of $dh_e/dt = $ constant are tangent to contours of constant specific energy, h_e.

The difference between the conventional and the energy method is illustrated in Figure 12.25 for a subsonic fighter airplane. It is observed, that the difference is very small. Having said this, it must be also observed, that the conventional method is useless for a supersonic airplane because of the large speed change involved if in fact the speed at the equivalent point E is to be supersonic. The results of applying Eqn (12.73) to a supersonic fighter are shown in Figure 12.26 for the F–104G. The discontinuity in the region of M = 1.0 can be determined by assuming a subsonic climb

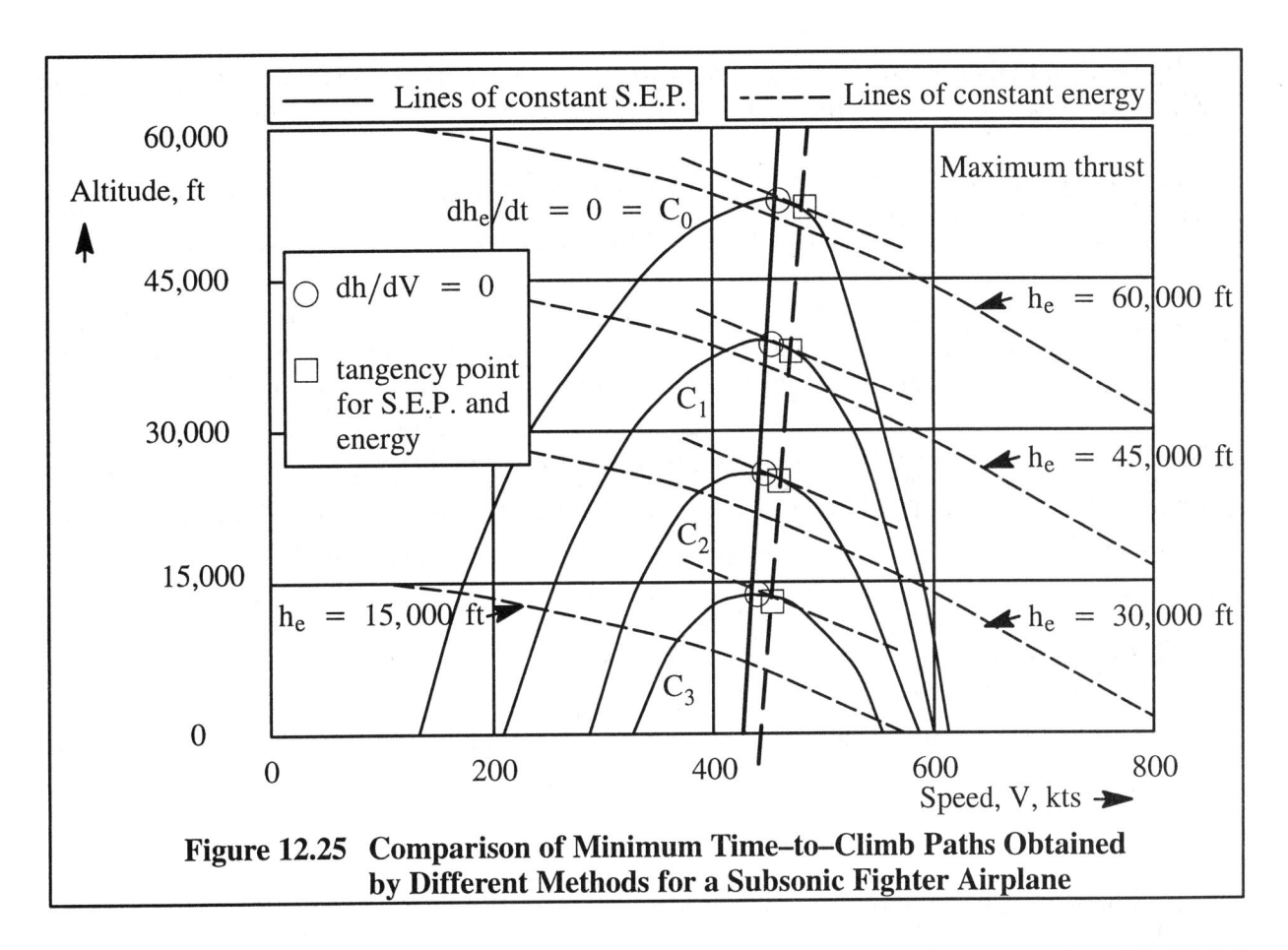

Figure 12.25 Comparison of Minimum Time–to–Climb Paths Obtained by Different Methods for a Subsonic Fighter Airplane

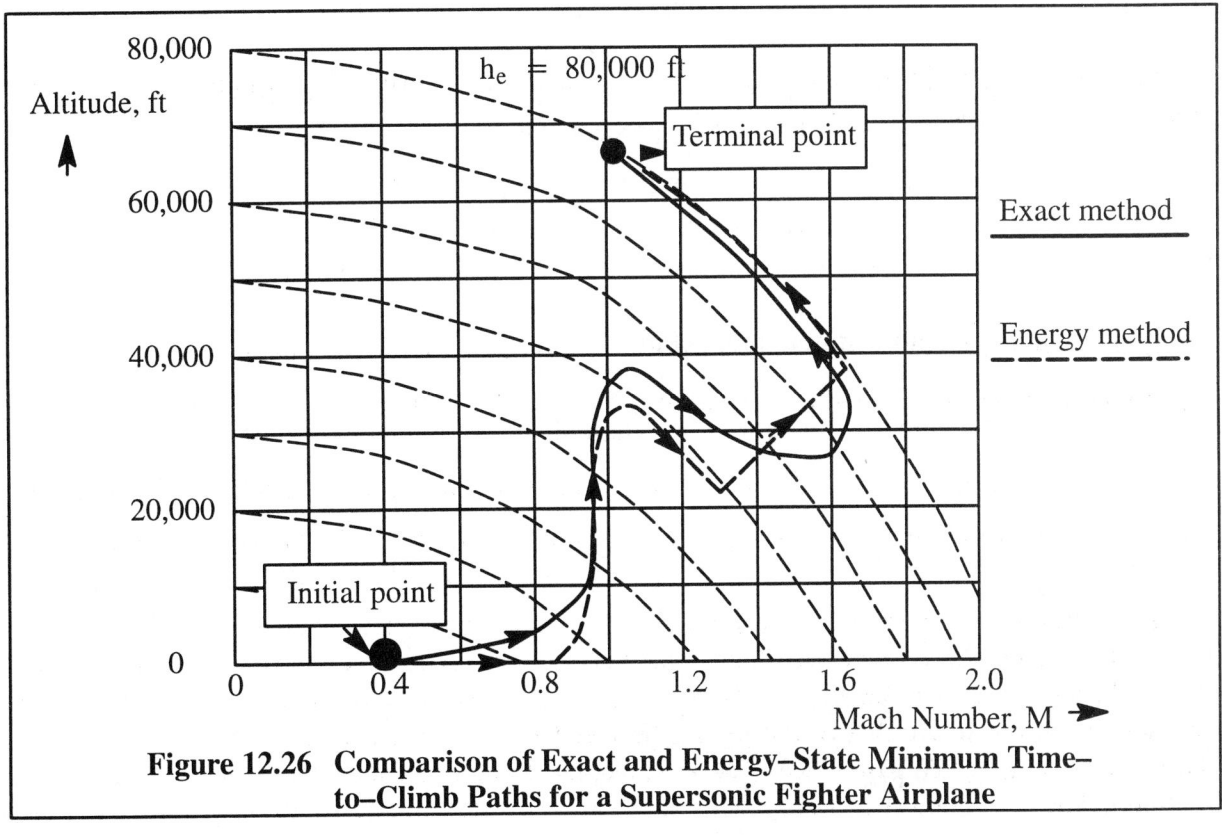

Figure 12.26 Comparison of Exact and Energy-State Minimum Time–to–Climb Paths for a Supersonic Fighter Airplane

until intersecting a line of constant h_e which is tangent to two equal valued dh_e/dt lines, one subsonic, the other supersonic.

Returning to Figure 12.24, for the F–104G, to climb from 3,500 ft altitude at M = 0.5 to 51,000 ft altitude at M = 2.2 in a minimum amount of time, the following trajectory has to be followed.

First, accelerate form M = 0.5 to about M = 0.9 at 3,500 ft altitude. Then, climb at constant Mach number (0.9) to an altitude of 32,000 ft. Next, dive at constant energy to an altitude of 19,000 ft and a Mach number of 1.3. This is followed by an accelerating climb, at constant calibrated airspeed, to 38,000 ft and a Mach number of 2.15. Then, zoom–climb to 47,000 ft at constant h_e, allowing the Mach number to decrease to about 2.02. Finally, an accelerating climb to 50,000 ft and M = 2.2.

Figure 12.26 shows the reasonableness of this procedure by comparing the energy climb path with the exact solution obtained by numerical optimization. The energy–climb path is much easier to follow (control) by a pilot. The exact solution flight path can be flown with the help of a flight management computer (Ref. 12.17).

To estimate the total time in the climb, Eqn (12.71) is approximated by the following summation:

$$t \approx \sum \frac{1}{dh_e/dt} \Delta h_e \qquad (12.75)$$

A problem with this approximation is that in a zoom or dive along a curve of constant specific energy, h_e, $\Delta h_e = 0$ would have to be satisfied. That would mean that the motion takes place in zero time, according to Eqn (12.75). This is a deficiency of the energy–state approximation. As a consequence, this method must be supplemented with other methods, such as exact integration, to determine the total time. Another possible procedure is to use the quasi–steady method discussed in Chapter 9 incrementally, to find the flight path angle, γ. This is done by setting $\dot{\gamma}$ and $\dot{V}$ equal to zero and solving Eqns (12.63) and (12.64) for the flight path angle, γ. With the flight path angle known, the time to climb to a new altitude can be computed incrementally.

(4) Determination of a minimum fuel climb schedule.

The amount of fuel required to climb from altitude h_1 to altitude h_2 can be computed from:

$$W_F = \int_{h_{e_1}}^{h_{e_2}} \frac{1}{dh_e/dW_F} dh_e \qquad (12.76)$$

where:

$$\frac{dh_e}{dW_F} = \left(\frac{dh_e}{dt}\right)\left(\frac{dt}{dW_F}\right) = \frac{dh_e/dt}{dW_F/dt} = \frac{dh_e/dt}{Tc_j} \qquad (12.77)$$

where: c_j is the engine specific fuel consumption in lbs/hr/lbs.

Observe that, in accordance with Eqn (12.66):

$$\frac{dh_e}{dt} = \frac{\{T\cos(\alpha + \phi_T) - D\}V}{W} \tag{12.78}$$

By combining Eqn (12.77) with Eqn (12.78) it is seen that:

$$\frac{dh_e}{dW_F} = \frac{dh_e/dt}{Tc_j} = \frac{\{T\cos(\alpha + \phi_T) - D\}V}{WTc_j} \tag{12.79}$$

The idea used in minimum time–to–climb trajectories was to use curves for constant dh_e/dt. In this case, these curves are replaced by curves for constant dh_e/dW_f in the same h–V diagram.

There are many more performance optimization problems which can be solved with the energy-state approximation. Examples are:

a) find the maximum range at fixed throttle for a given amount of fuel

b) find the maximum range negotiable in a given amount of time

c) etc. The reader may wish to consult Refs 12.13 through 12.16.

Example 12.2: Construct a specific energy plot for a supersonic fighter with $S = 530$ ft² and $W = 42,000$ lbs. The drag and thrust characteristics for this airplane are given in Figures 12.27 and 12.28 respectively. Assume that the thrust inclination angle, $\phi_T = 0$, and that the load factor $n = 1.0$. This example airplane is the same one considered in Ref. 12.13.

Solution: Curves for constant h_e can be constructed by plotting equation (12.33) on h – V or h – M diagrams. In the following solution, only curves for constant dh_e/dt will be discussed. Their determination is most easily accomplished with a table as shown in Table 12.2.

The results of the calculations in Table 12.2 are plotted in Figure 12.29. By constructing a family of curves over a wide range of altitudes, these curves are re–plotted for constant values of dh_e/dt similar to those shown in Figure 12.24. To save space in the book, the calculations in Table 12.2 are performed only for sea–level and for 20,000 ft. The remaining calculations are left as an exercise for the reader.

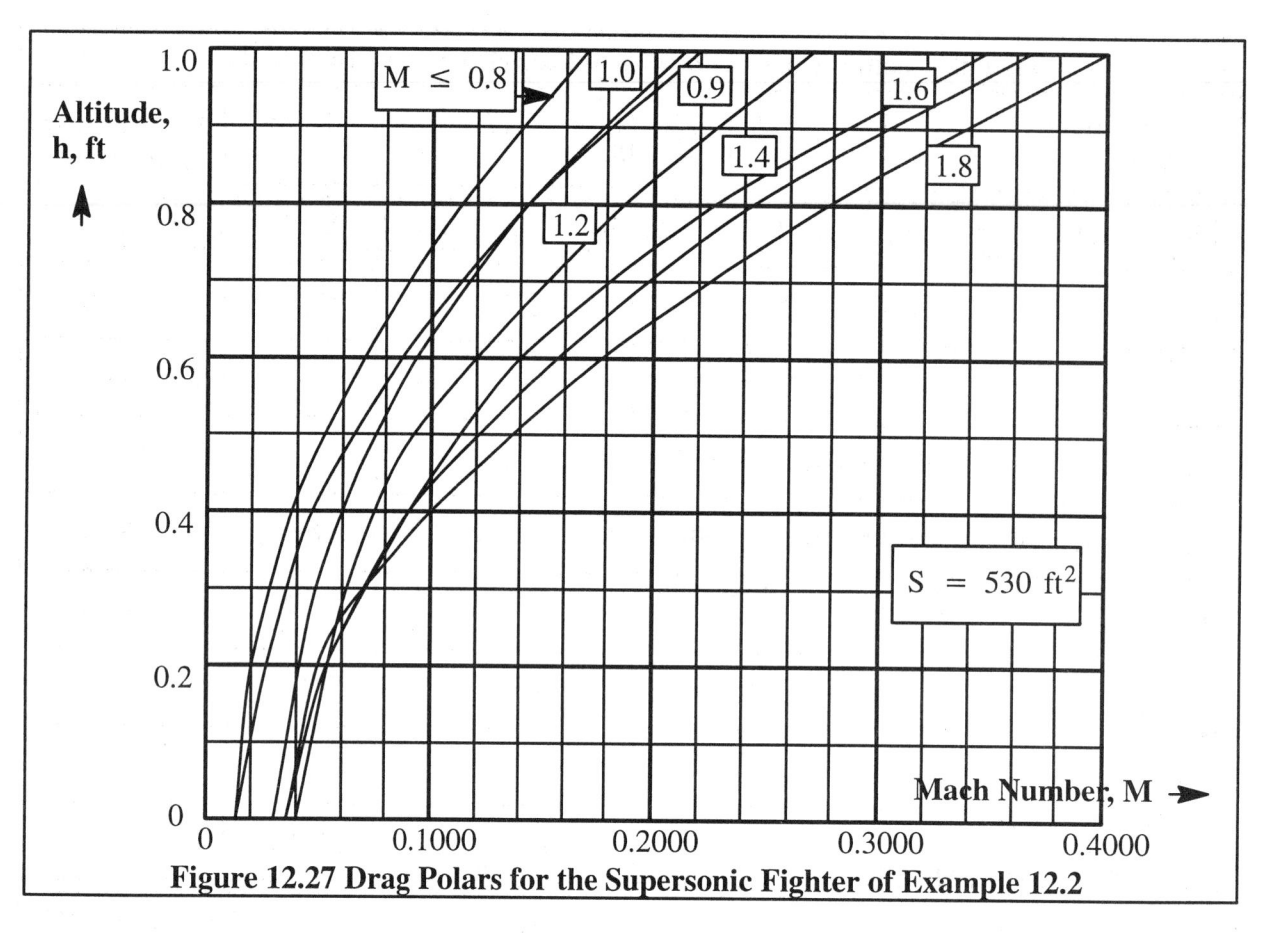

Figure 12.27 Drag Polars for the Supersonic Fighter of Example 12.2

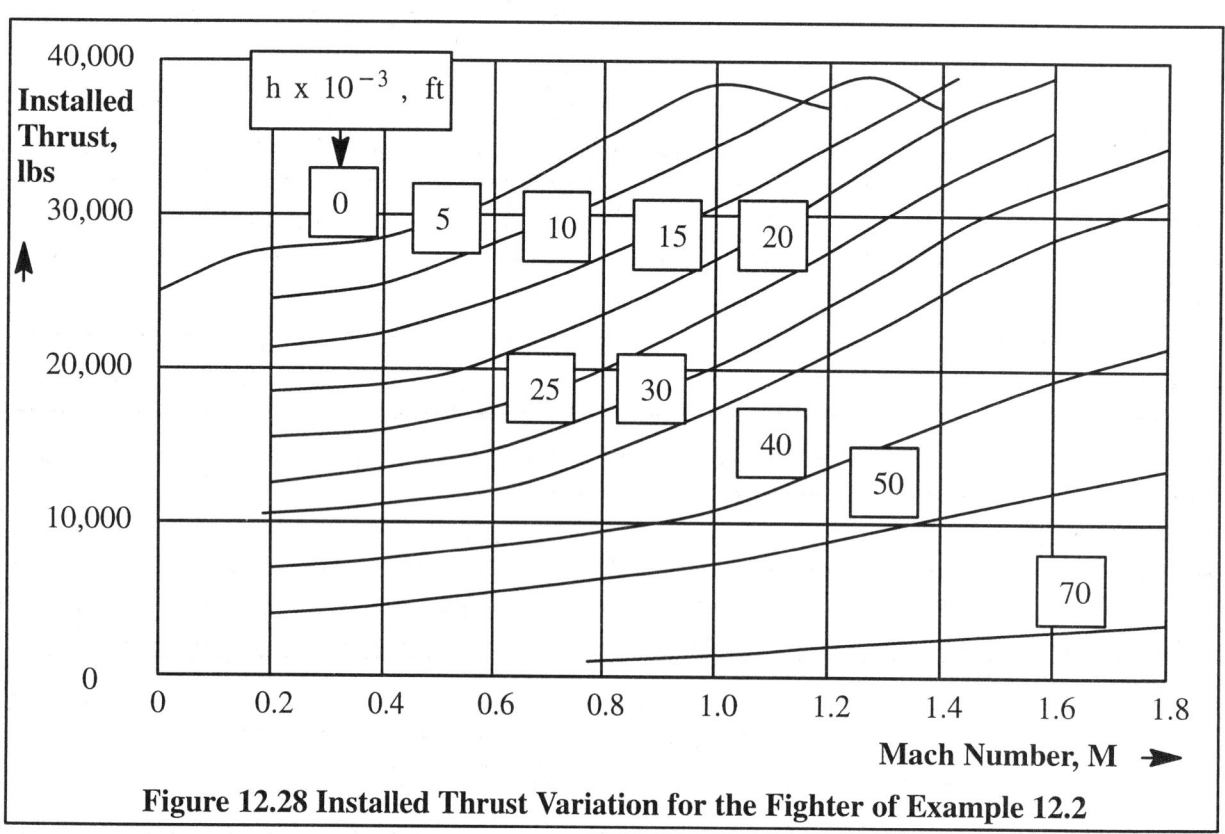

Figure 12.28 Installed Thrust Variation for the Fighter of Example 12.2

Table 12.2 Example Calculation of Lines for Constant dh_e/dt

M	V, ft/sec	C_L	C_D	C_L/C_D	$D = C_D \bar{q} S$, lbs	T, lbs	dh_e/dt, ft/sec Eqn (12.66)
h = 0 ft, δ = 1.0, C_L = W/$\bar{q}$S, lbs, $\bar{q}$ = 1,481.3 δ M^2 lbs/ft²							
0.3	334.9	0.594	0.0680	8.74	4,805	28,200	186.6
0.4	446.6	0.334	0.0310	10.77	3,894	28,300	259.5
0.6	669.8	0.149	0.0160	9.31	4,522	30,800	419.1
0.7	781.5	0.109	0.0149	7.32	5,732	32,600	499.9
0.8	893.1	0.084	0.0140	6.00	7,034	34,500	584.0
0.9	1,004.8	0.066	0.0149	4.43	9,475	36,200	639.3
1.0	1,116.4	0.053	0.0310	1.71	24,338	37,900	360.5
1.2	1,339.7	0.037	0.0410	0.90	46,352	36,100	−327.0
h = 20,000 ft, δ = 0.4595, C_L = W/$\bar{q}$S, lbs, $\bar{q}$ = 1,481.3 δ M^2 lbs/ft²							
0.4	414.7	0.728	0.0960	7.58	5,541	15,900	102.3
0.6	622.1	0.323	0.0290	11.14	3,766	17,300	200.5
0.8	829.4	0.182	0.0180	10.11	4,156	19,800	308.9
0.9	933.1	0.144	0.0183	7.87	5,347	21,600	361.1
1.0	1,036.8	0.116	0.0330	3.52	11,905	23,300	281.3
1.2	1,244.2	0.081	0.0420	1.93	21,818	27,300	162.4
1.4	1,451.5	0.059	0.0400	1.48	28.283	31,600	114.6
1.6	1,658.9	0.045	0.0370	1.22	34,170	35,700	60.4

Courtesy: Lockheed–Martin

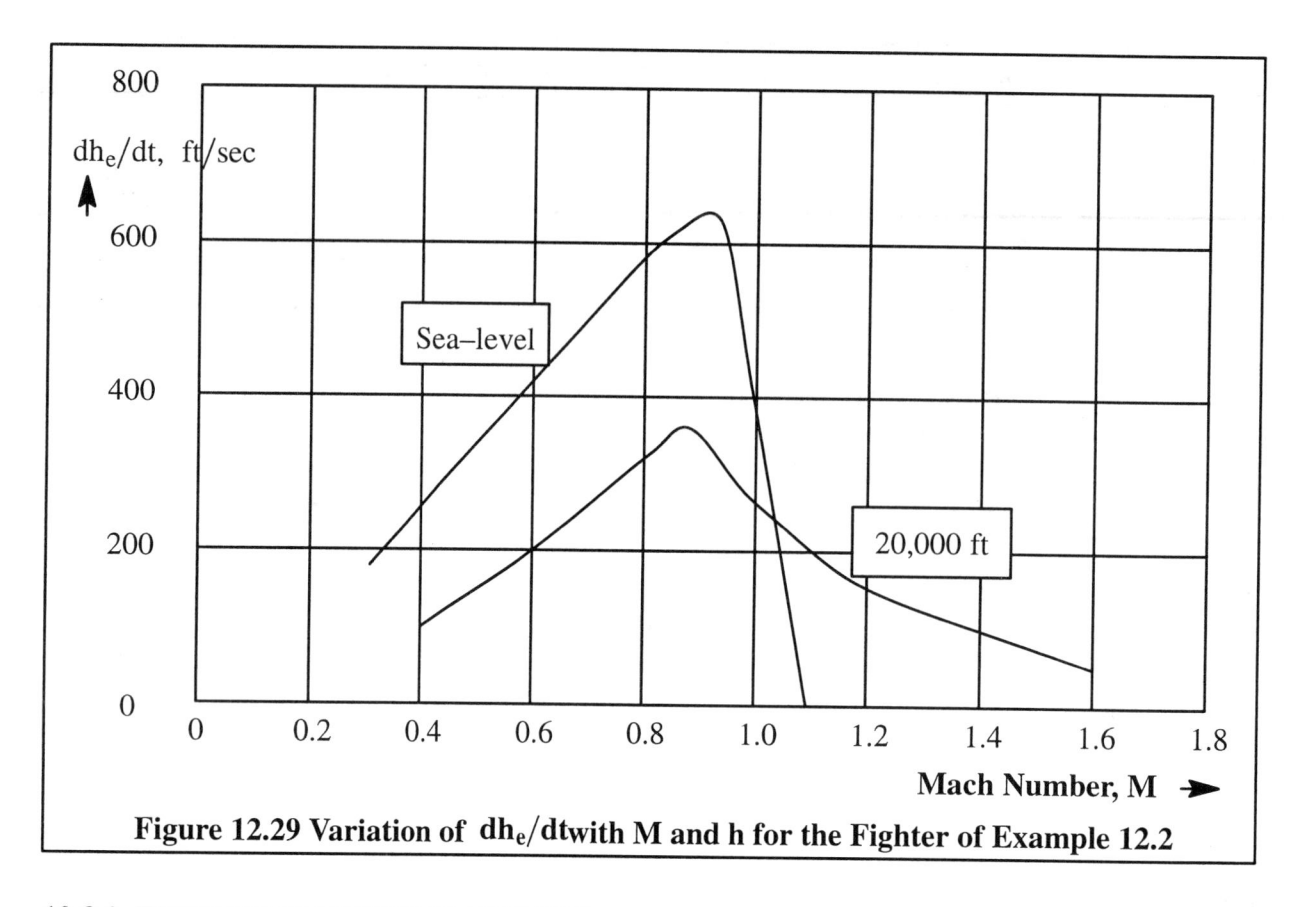

Figure 12.29 Variation of dh_e/dt with M and h for the Fighter of Example 12.2

12.8.2 ENERGY MANEUVERABILITY

The V–n diagram discussed in Section 12.4 can be used to indicate whether one airplane has a turn advantage over the other. However, the maneuvering capability which can be deduced from the V–n diagram is "instantaneous" and not sustained. To investigate the difference between the sustained maneuverability of two airplanes, plots of constant specific energy can be used.

A fighter will have an advantage over an adversary if:

1) an engagement can be entered at a higher energy level than that of the adversary
2) if the higher energy level can be maintained
3) if an engagement can be entered at a lower energy level, but, due to superior acceleration potential a higher energy level can be reached very quickly.

To determine whether one fighter has an advantage over another, its specific energy curves can be superimposed on a similar plot for the adversary. An example is shown in Figure 12.30. From this figure it is seen that airplane B is superior only at high subsonic speeds, above 10,000 ft. In all other areas of the flight envelope, airplane A would be superior. It should be observed, that to construct specific energy curves for load factors other than 1.0, Example 12.2 can still be used, with one difference: the level flight lift coefficient, C_L, should be replaced by: nC_L.

To indicate whether advantage 3) exists, the optimum energy path of Sub–section 12.8.1 may be used to find whether one airplane can reach a certain flight condition sooner than the other.

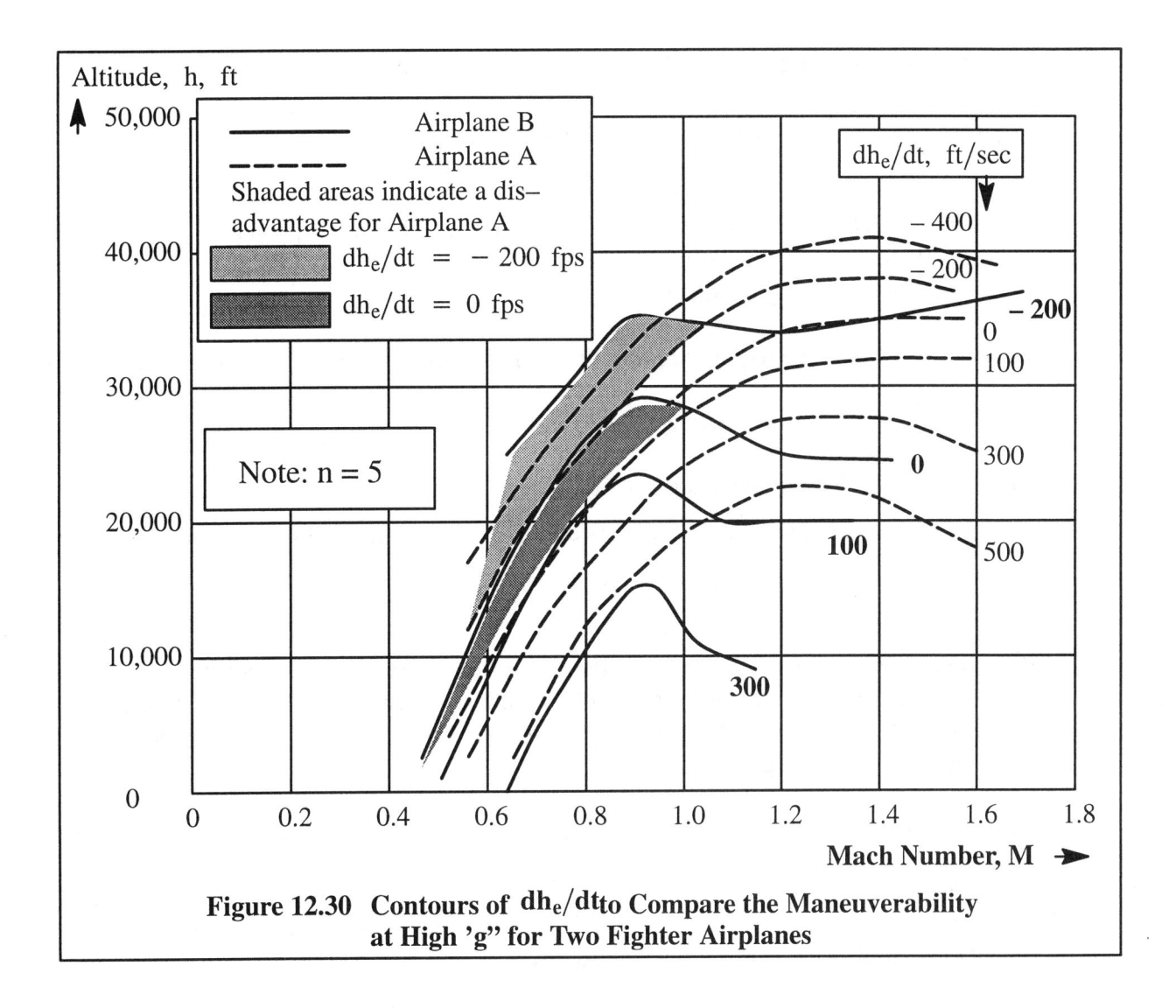

Figure 12.30 Contours of dh_e/dt to Compare the Maneuverability at High 'g' for Two Fighter Airplanes

12.9 SUMMARY FOR CHAPTER 12

In this chapter various effects of the flight envelope on the maneuverability of airplanes were discussed. In addition to stall speeds, airplane buffet limits, level flight maximum speeds, instantaneous and sustained maneuvering capabilities were analyzed. Specifically, pull–ups and turns were discussed in some detail.

Because airplane maneuverability may also be limited by structural considerations the construction of V–n diagrams was also illustrated.

Spin entry and spin recovery were analyzed from an aerodynamic viewpoint. An explanation of what constitutes a spin was given.

Accelerated flight performance was analyzed from a viewpoint of the energy state approximation. The concepts of energy contours and energy maneuverability were introduced and applied to fighter type airplanes.

12.10 PROBLEMS FOR CHAPTER 12

12.1 A jet trainer has the following characteristics:

$$W = 5,000 \text{ lbs} \qquad S = 86.1 \text{ ft}^2 \qquad C_{L_{max}} = 1.75$$

The lift–drag relationship is defined in Figure 5.4. The airplane is powered by a single jet engine with a maximum thrust of 2,000 lbs at sea–level.
Determine the following characteristics:

a) The minimum time required to turn through 180 degrees at sea–level at constant speed

b) The corresponding load factor

c) The airplane lift (in lbs) in such a turn

Hint: Assume a range of level flight speeds, V_0, and use the following equations:

$$n = T/T_0 \qquad t = \pi/\dot{\psi} \qquad V = V_0 \sqrt{n}$$

12.2 An airplane, which weighs 2,000 lbs, is flying at an equivalent airspeed of 80 kts at an altitude of 10,000 ft. It makes a 90 degree turn in 15 seconds while maintaining a constant altitude and a constant angle of attack. The wing loading of the airplane is 20 psf. At the given airspeed, the lift–to–drag ratio is 10.0.

Calculate the radius of the turn, the load factor in the turn and the thrust horsepower required.

12.3 Complete the calculation of Example 12.2 over the range of altitudes shown in Figure 12.27 and construct the corresponding specific energy plot. Next, sketch in this plot the minimum–time climb path from M=0.85 at sea–level to M–1.8 at 37,000 ft.

12.4 Re–derive Eqn (12.10) for a canard configuration.

12.5 Re–derive Eqn (12.10) for a three–surface configuration.

12.11 REFERENCES FOR CHAPTER 12

12.1 Anon.; Code of Federal Regulations (CFR), Title 14, Parts 1 to 59, January 1, 1996; U.S. Government Printing Office, Superintendent of Documents, Mail Stop SSOP, Washington, D.C. Note: FAR 23 and FAR 25 are components of this CFR.

12.2 Roskam, J.; Airplane Flight Dynamics and Automatic Flight Controls, Part I; DAR Corporation, 120 East Ninth Street, Suite 2, Lawrence, Kansas, 66044; 1996.

12.3 Roskam, J.; Ice on Your Tail Can Be Deadly; Professional Pilot Magazine, p.74–76 December 1994.

12.4   Anon.; The Effects of Buffeting and Other Transonic Phenomena on Maneuvering Combat Aircraft; AGARD–AR–82, 1975.

12.5   Lang, J.D. and Carling, J.C.; Aircraft Performance, Stability and Control; USAF Academy, Volume II, 1978.

12.6   Hancock, G.J.; Problems of Aircraft Behavior at High Angles of Attack; AGARDograph 136, April 1969.

12.7   Stinton, Darrol; Flying Qualities and Flight Testing of the Airplane; AIAA Education Series, AIAA, Washington, D.C. and Blackwell Scientific Publications, Oxford, U.K., 1996.

12.8   Roskam, J.; Airplane Design, Part III, Layout Design of Cockpit, Fuselage, Wing and Empennage: Cutaways and Inboard Profiles; DAR Corporation, 120 East Ninth Street, Suite 2, Lawrence, Kansas, 66044; 1989.

12.9   Burk, S.M. Jr., Bowman, J.S. Jr. and While, W.L.; Spin–Tunnel Investigation of the Spinning Characteristics of a Typical Single–Engine General Aviation Airplane Design; NASA TP 1009, 1977.

12.10 Bihrle, W. Jr.; Effects of Several Factors on the Theoretical predictions of Airplane Spin Characteristics; NASA CR 132521, 1974.

12.11 Staff of NASA Langley Research Center; Exploratory Study of the Effects of Wing–Leading–Edge Modifications on the Stall/Spin Behavior of a Light General Aviation Airplane; NASA TP–1589, Dec. 1979.

12.12 Rutowski, E.S.; Energy Approach to the General Aircraft Performance Problem; Journal of the Aeronautical Sciences, Volume 21, March 1954, pp. 187–195.

12.13 Bryson, A.E. Jr., Desai, M.N. and Hoffman, W.C.; Energy–State Approximation on Performance Optimization of Supersonic Aircraft; Journal of Aircraft, Vol. 6, Nov.–Dec. 1969, pp. 481–488.

12.14 Lang, J.D.; Aircraft Performance, Stability and Control; USAF Academy, Colorado, August 1971. Revised by J.C. Carling in May 1974: Chapter 5 of Volume I.

12.15 Nicolai, L.M.; Fundamentals of Aircraft Design; University of Dayton, Ohio, 1975.

12.16 Leimann, G. (Ed.); Optimization Techniques with Applications to Aerospace Systems; Academic Press, 1962, Chapters 1 and 4.

12.17 Zagalsky, N.R.; Aircraft Energy Management; AIAA Paper No. 73–228, January 1973.

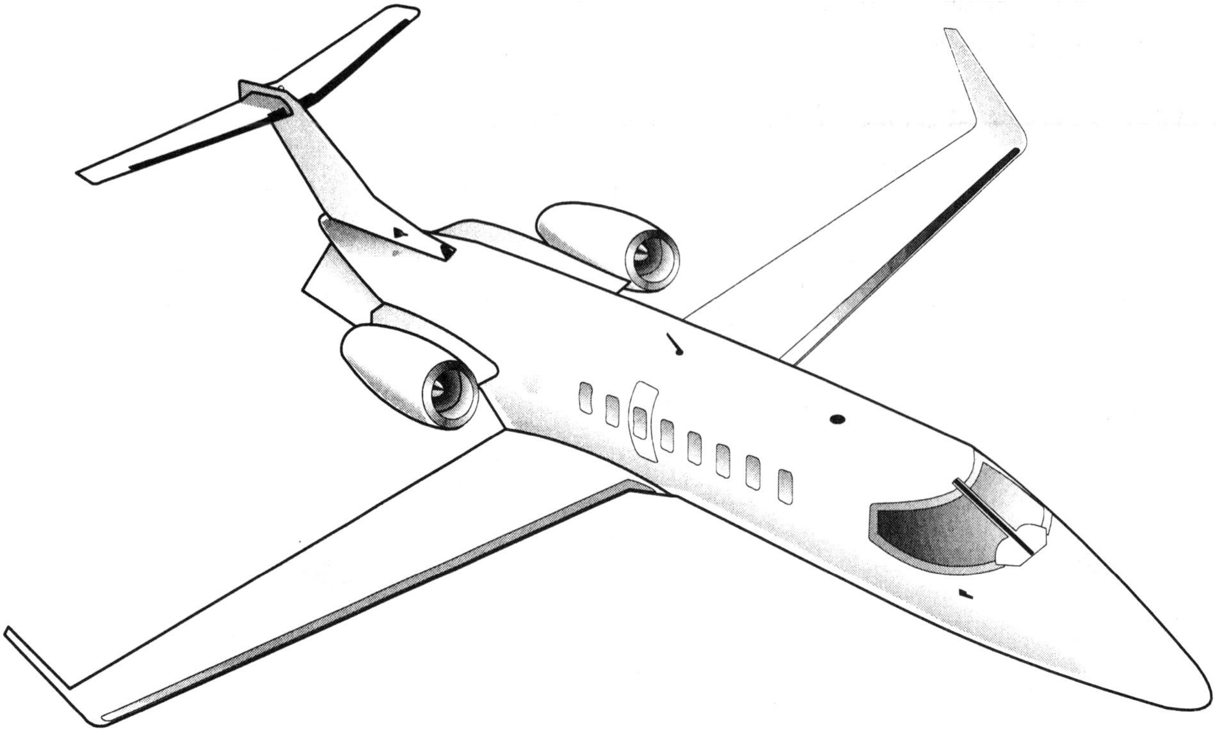

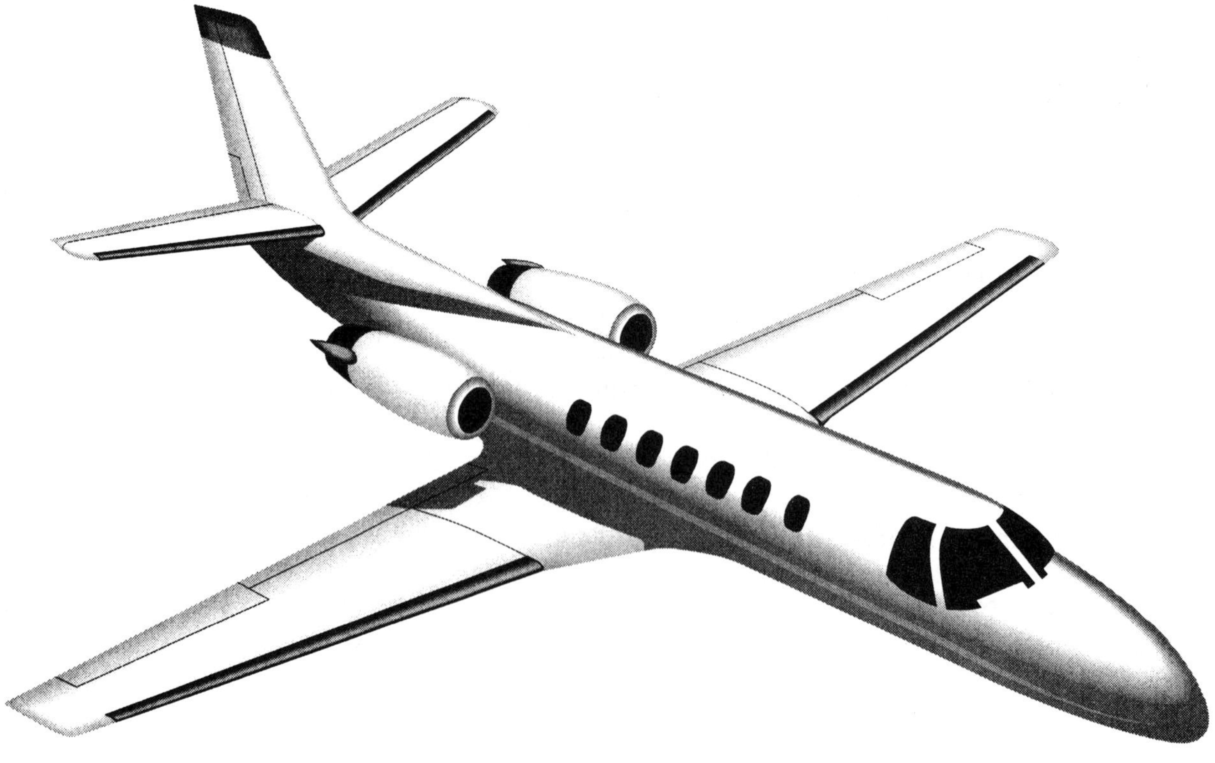

APPENDIX A: PROPERTIES OF THE U.S. STANDARD ATMOSPHERE AND CONVERSION FACTORS

In this appendix the properties of the 1962 U.S. Standard Atmosphere are tabulated in accordance with Eqns (1.15), (1.16), (1.17), (1.21) and (1.22) of Chapter 1. Two tables are given: Table A1 in English units and Table A2 in Metric units. Conversion factors are given in Table A3.

TABLE A1 U.S. STANDARD ATMOSPHERE IN ENGLISH UNITS

Alt.	Temp.	Temp. Ratio	Press.	Press. Ratio	Density	Density Ratio	Coeff. of Viscosity	Speed of Sound
h (ft) Geopotential	T (^0R)	θ	p (psi)	δ	ϱ (slugs/ft^3) ($\times 10^{-3}$)	σ	μ (lbs $-$ sec/ft^2) ($\times 10^{-7}$)	V_a (ft/sec)
0	518.7	1.0000	14.70	1.0000	2.377	1.0000	3.737	1,116.4
1,000	515.1	0.9932	14.17	0.9644	2.3081	0.97106	3.717	1,112.6
2,000	511.5	0.9863	13.66	0.9298	2.2409	0.94277	3.697	1,108.7
3,000	508.0	0.9794	13.17	0.8962	2.1751	0.91512	3.677	1,104.9
4,000	504.4	0.9725	12.69	0.8637	2.1109	0.88809	3.657	1,101.0
5,000	500.8	0.9657	12.23	0.8320	2.0481	0.86167	3.636	1,097.1
6,000	497.3	0.9588	11.78	0.8014	1.9868	0.83586	3.616	1,093.2
7,000	493.7	0.9519	11.34	0.7716	1.9268	0.81064	3.596	1,089.2
8,000	490.1	0.9450	10.92	0.7428	1.8683	0.78602	3.575	1,085.3
9,000	486.6	0.9382	10.50	0.7148	1.8111	0.76196	3.555	1,081.4
10,000	483.0	0.9313	10.11	0.6877	1.7553	0.73848	3.534	1,077.4
11,000	479.4	0.9244	9.720	0.6614	1.7008	0.71555	3.513	1,073.4
12,000	475.9	0.9175	9.346	0.6360	1.6476	0.69317	3.493	1,069.4
13,000	472.3	0.9107	8.984	0.6113	1.5957	0.67133	3.472	1,065.4
14,000	468.7	0.9038	8.633	0.5857	1.5451	0.65003	3.451	1,061.4
15,000	465.2	0.8969	8.294	0.5643	1.4956	0.62924	3.430	1,057.3
16,000	461.6	0.8900	7.965	0.5420	1.4474	0.60896	3.409	1,053.2
17,000	458.0	0.8831	7.647	0.5203	1.4004	0.58919	3.388	1,049.2
18,000	454.5	0.8763	7.339	0.4994	1.3546	0.56991	3.366	1,045.1
19,000	450.9	0.8694	7.041	0.4791	1.3100	0.55112	3.345	1,041.0
20,000	447.3	0.8625	6.754	0.4595	1.2664	0.53281	3.324	1,036.8

Appendix A

Properties of the U.S. Standard Atmosphere and Conversion Factors

TABLE A1 (CONT'D) U.S. STANDARD ATMOSPHERE IN ENGLISH UNITS

Alt.	Temp.	Temp. Ratio	Press.	Press. Ratio	Density	Density Ratio	Coeff. of Viscosity	Speed of Sound
h (ft) Geopotential	T (^0R)	θ	p (psi)	δ	ϱ (slugs/ft^3) ($\times 10^{-3}$)	σ	μ (lbs $-$ sec/ft^2) ($\times 10^{-7}$)	V_a (ft/sec)
21,000	443.8	0.8556	6.475	0.4406	1.2240	0.51497	3.302	1,032.7
22,000	440.2	0.8488	6.207	0.4223	1.1827	0.49758	3.281	1,028.5
23,000	436.6	0.8419	5.947	0.4046	1.1425	0.48065	3.259	1,024.4
24,000	433.1	0.8350	5.696	0.3876	1.1033	0.46417	3.238	1,020.2
25,000	429.5	0.8281	5.454	0.3711	1.0651	0.44812	3.216	1,016.0
26,000	426.0	0.8213	5.220	0.3552	1.0280	0.43250	3.194	1,011.7
27,000	422.4	0.8144	4.994	0.3398	0.9919	0.41730	3.172	1,007.5
28,000	418.8	0.8075	4.777	0.3250	0.9567	0.40251	3.150	1,003.2
29,000	415.3	0.8006	4.567	0.3107	0.9225	0.38812	3.128	999.0
30,000	411.7	0.7938	4.364	0.2970	0.8893	0.37413	3.106	994.7
31,000	408.1	0.7869	4.169	0.2837	0.8569	0.36053	3.084	990.3
32,000	404.6	0.7800	3.981	0.2709	0.8255	0.34731	3.061	986.0
33,000	401.0	0.7731	3.800	0.2586	0.7950	0.33447	3.039	981.6
34,000	397.4	0.7663	3.626	0.2467	0.7653	0.32199	3.016	977.3
35,000	393.9	0.7594	3.458	0.2353	0.7365	0.30987	2.994	972.9
36,000	390.3	0.7525	3.297	0.2243	0.7086	0.29811	2.971	968.5
36,089	390.0	0.7519	3.282	0.2234	0.7062	0.29710	2.969	968.1
37,000	390.0	0.7519	3.142	0.2138	0.6759	0.28435	2.969	968.1
38,000	390.0	0.7519	2.994	0.2038	0.6442	0.27101	2.969	968.1
39,000	390.0	0.7519	2.854	0.1942	0.6139	0.25829	2.969	968.1
40,000	390.0	0.7519	2.720	0.1851	0.5851	0.24617	2.969	968.1
41,000	390.0	0.7519	2.592	0.1764	0.5577	0.23462	2.969	968.1
42,000	390.0	0.7519	2.471	0.1681	0.5315	0.22361	2.969	968.1
43,000	390.0	0.7519	2.355	0.1602	0.5065	0.21311	2.969	968.1
44,000	390.0	0.7519	2.244	0.1527	0.4828	0.20311	2.969	968.1
45,000	390.0	0.7519	2.139	0.1455	0.4601	0.19358	2.969	968.1
46,000	390.0	0.7519	2.039	0.1387	0.4385	0.18450	2.969	968.1
47,000	390.0	0.7519	1.943	0.1322	0.4180	0.17584	2.969	968.1
48,000	390.0	0.7519	1.852	0.1260	0.3983	0.16759	2.969	968.1
49,000	390.0	0.7519	1.765	0.1201	0.3796	0.15972	2.969	968.1
50,000	390.0	0.7519	1.682	0.1145	0.3618	0.15223	2.969	968.1

TABLE A1 (CONT'D) U.S. STANDARD ATMOSPHERE IN ENGLISH UNITS

Alt.	Temp.	Temp. Ratio	Press.	Press. Ratio	Density	Density Ratio	Coeff. of Viscosity	Speed of Sound
h (ft) Geopotential	T (^0R)	θ	p (psi)	δ	ϱ (slugs/ft^3) ($\times 10^{-3}$)	σ	μ (lbs $-$ sec/ft^2) ($\times 10^{-7}$)	V_a (ft/sec)
51,000	390.0	0.7519	1.603	0.1091	0.3449	0.14509	2.969	968.1
52,000	390.0	0.7519	1.528	0.1040	0.3287	0.13828	2.969	968.1
53,000	390.0	0.7519	1.456	0.09909	0.3133	0.13179	2.969	968.1
54,000	390.0	0.7519	1.388	0.09444	0.2985	0.12560	2.969	968.1
55,000	390.0	0.7519	1.323	0.09001	0.2845	0.11971	2.969	968.1
56,000	390.0	0.7519	1.261	0.08578	0.2712	0.11409	2.969	968.1
57,000	390.0	0.7519	1.201	0.08176	0.2585	0.10874	2.969	968.1
58,000	390.0	0.7519	1.145	0.07792	0.2463	0.10364	2.969	968.1
59,000	390.0	0.7519	1.091	0.07426	0.2348	0.098772	2.969	968.1
60,000	390.0	0.7519	1.040	0.07078	0.2238	0.094137	2.969	968.1
61,000	390.0	0.7519	0.9913	0.06746	0.2133	0.089720	2.969	968.1
62,000	390.0	0.7519	0.9448	0.06429	0.2032	0.085509	2.969	968.1
63,000	390.0	0.7519	0.9005	0.06127	0.1937	0.081497	2.969	968.1
64,000	390.0	0.7519	0.8582	0.05840	0.1846	0.077672	2.969	968.1
65,000	390.0	0.7519	0.8179	0.05566	0.1760	0.074027	2.969	968.1

Properties of the U.S. Standard Atmosphere and Conversion Factors

TABLE A2 U.S. STANDARD ATMOSPHERIC IN METRIC UNITS

Alt.	Temp.	Temp. Ratio	Press.	Press. Ratio	Density	Density Ratio	Coeff. of Viscosity	Speed of Sound
h (m) Geopotential	T (^{0}K)	θ	p (N/m^2)	δ	ϱ (Kg/m^3)	σ	μ (N − sec/m^2) (x10^{-5})	V_a (m/sec)
0	288.2	1.0000	101,325	1.0000	1.2250	1.0000	1.789	340.3
500	284.9	0.9888	95,460	0.9421	1.1673	0.9529	1.774	338.4
1,000	281.7	0.9775	89,874	0.8870	1.1116	0.9075	1.758	336.4
1,500	278.4	0.9662	84555	0.8345	1.0581	0.8637	1.742	334.5
2,000	275.2	0.9549	79495	0.7846	1.0065	0.8216	1.726	332.5
2,500	271.9	0.9436	74682	0.7371	0.95686	0.7811	1.710	330.6
3,000	268.7	0.9324	70108	0.6919	0.90912	0.7421	1.694	328.6
3,500	265.4	0.9211	65764	0.6490	0.86323	0.7047	1.678	326.6
4,000	262.2	0.9098	61640	0.6083	0.81913	0.6687	1.661	324.6
4,500	258.9	0.8985	57728	0.5697	0.77677	0.6341	1.645	332.6
5,000	255.7	0.8872	54019	0.5331	0.73612	0.6009	1.628	320.5
5,500	252.4	0.8760	50506	0.4985	0.69711	0.5691	1.612	318.5
6,000	249.2	0.8647	47181	0.4656	0.65970	0.5385	1.595	316.4
6,500	245.9	0.8534	44034	0.4346	0.62384	0.5093	1.578	314.4
7,000	242.7	0.8421	41060	0.4052	0.58950	0.4812	1.561	312.4
7,500	239.4	0.8309	38251	0.3775	0.55662	0.4544	1.544	310.2
8,000	236.2	0.8196	35599	0.3513	0.52517	0.4287	1.527	308.1
8,500	232.9	0.8083	33099	0.3267	0.49509	0.4042	1.510	305.9
9,000	229.7	0.7970	30742	0.3034	0.46635	0.3807	1.492	303.8
9,500	226.4	0.7857	28523	0.2815	0.43890	0.3583	1.475	301.6
10,000	223.2	0.7745	26436	0.2609	0.41271	0.3369	1.457	299.5
10,500	219.9	0.7632	24474	0.2415	0.38773	0.3165	1.439	297.3
11,000	216.7	0.7519	22632	0.2234	0.36392	0.2971	1.422	295.1
11,500	216.7	0.7519	20916	0.2064	0.33633	0.2746	1.422	295.1
12,000	216.7	0.7519	19330	0.1908	0.31083	0.2537	1.422	295.1

TABLE A2 (CONT'D) U.S. STANDARD ATMOSPHERIC IN METRIC UNITS

Alt.	Temp.	Temp. Ratio	Press.	Press. Ratio	Density	Density Ratio	Coeff. of Viscosity	Speed of Sound
h (m) Geopotential	T (^0K)	θ	p (N/m^2)	δ	ϱ (Kg/m^3)	σ	μ ($N-sec/m^2$) ($\times 10^{-5}$)	V_a (m/sec)
12,500	216.7	0.7519	17864	0.1763	0.28726	0.2345	1.422	295.1
13,000	216.7	0.7519	16510	0.1629	0.26548	0.2167	1.422	295.1
13,500	216.7	0.7519	15258	0.1506	0.24536	0.2003	1.422	295.1
14,000	216.7	0.7519	14101	0.1392	0.22675	0.1851	1.422	295.1
14,500	216.7	0.7519	13032	0.1286	0.20956	0.1711	1.422	295.1
15,000	216.7	0.7519	12044	0.1189	0.19367	0.1581	1.422	295.1
15,500	216.7	0.7519	11131	0.1099	0.17899	0.1461	1.422	295.1
16,000	216.7	0.7519	10287	0.1015	0.16542	0.1350	1.422	295.1
16,500	216.7	0.7519	9507	0.09383	0.15288	0.1248	1.422	295.1
17,000	216.7	0.7519	8787	0.08672	0.14129	0.1153	1.422	295.1
17,500	216.7	0.7519	8121	0.08014	0.13058	0.1066	1.411	295.1
18,000	216.7	0.7519	7505	0.07407	0.12068	0.09851	1.422	295.1
18,500	216.7	0.7519	6936	0.06845	0.11153	0.09104	1.422	295.1
19,000	216.7	0.7519	6410	0.06326	0.10307	0.08414	1.422	295.1
19,500	216.7	0.7519	5924	0.05847	0.095257	0.07776	1.422	295.1
20,000	216.7	0.7519	5475	0.05403	0.088035	0.07187	1.422	295.1

TABLE A3 CONVERSION FACTORS

Multiply	by	To obtain
feet (ft)	0.3048	meters (m)
meters (m)	3.281	feet (ft)
statute miles (s.m.)	5,280	feet
	1.609	kilometers
	0.8690	nautical miles (n.m.)
nautical miles (n.m.)	6,076	feet
	1.852	kilometers (km)
	1.1508	statute miles (s.m.)

TABLE A3 (CONT'D) CONVERSION FACTORS

Multiply	by	To obtain
feet/sec	0.5921	n.m./hr (kts)
	0.6818	s.m./hr
	1.097	km/hr
n.m./hr or knots (kts)	1.689	ft/sec
	1.151	s.m./hr
	1.852	km/hr
s.m./hr (mph)	1.467	ft/sec
	1.609	km/hr
	0.8684	kts
rad/sec	0.1592	revolutions/sec
	9.549	revolutions/min (rpm)
	57.296	deg/sec
revolutions/min (rpm)	0.10472	rad/sec
slug	14.59	kilograms
kilogram	0.06854	slugs
pound (lbs)	4.448	newtons
pounds per square feet (psf)	47.88	newtons per square meter
pounds per square inch (psi)	6.895	newtons per square meter
British Thermal Unit	0.0003927	horsepower–hours
foot–pound (ft–lbs)	1.356	newton–meters
horsepower	550	foot–lbs/sec

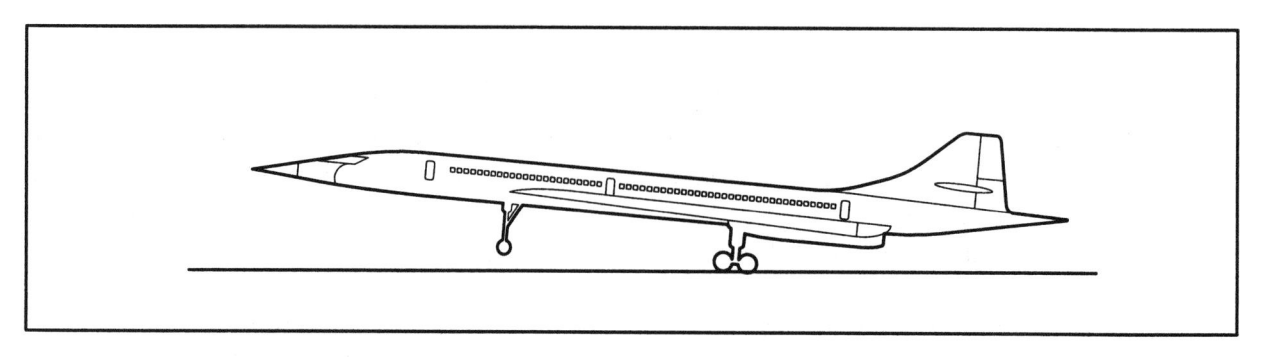

APPENDIX B:
ADVANCED AIRCRAFT DESIGN AND ANALYSIS SOFTWARE

In this appendix brief descriptions are given of advanced aircraft design and analysis software packages which are available from DARcorporation, publisher of this textbook. These descriptions include the following:

B1 AAA: ADVANCED AIRCRAFT ANALYSIS

B2 G.A.–CAD: GENERAL AVIATION COMPUTER AIDED DESIGN

B3. OTHER SOFTWARE PACKAGES

B4 REFERENCES

B1 AAA: ADVANCED AIRCRAFT ANALYSIS

This program is available on UNIX platforms (SUN, IBM, SGI and HP) and on IBM Compatible PC (Windows 3.1, 95 and NT).

B1.1 GENERAL CAPABILITIES OF THE AAA PROGRAM

The Advanced Aircraft Analysis (AAA) program described in this appendix was developed to reduce the time and cost required to design and analyze new and existing, fixed wing airplane configurations. In the process of designing a new airplane or analyzing an already existing airplane, engineers must determine the following characteristics of that airplane:

* drag   * installed thrust (or power)   * performance

* weight   * weight breakdown   * inertias

* stability   * control (open and closed loop)   * cost

The AAA program was developed to allow engineers to do this very rapidly and in a truly user–friendly manner. The AAA program is arranged in a modular fashion. Each module addresses a specific phase of design and/or analysis of the preliminary design decision making process. The various modules of the program are briefly described in Sub–section B1.2.

The AAA program is based on material presented in References B1, B2 and B3.

The AAA Program is available on a variety of platforms under several licensing options. The AAA program as well as References B1–B3 are marketed by:

Design, Analysis and Research Corporation

120 East Ninth Street, Suite 2

Lawrence, Kansas 66044, USA

Tel. 785–832–0434   Fax: 785–832–0524

e–mail: info@darcorp.com   http://www.darcorp.com

B1.2   BRIEF DESCRIPTION OF AAA PROGRAM MODULES

The AAA program consists of 15 independent modules. Each module is designed to perform tasks which need to be completed in the evaluation of a given airplane at each stage of its preliminary design development. New capabilities are constantly added to the program. The description given here applies to Version 1.7 which was released in January of 1996.

The program applies to civil as well as to military fixed–wing airplanes. **Conventional, pure canard and three–surface airplanes** with jet, turboprop or piston–prop propulsive installations can be handled by the program. At any stage of the design process, a report quality screen–dump and/or parameter print–out can be commanded.

The user can, at any time, ask for a definition of a particular design parameter. Where needed, a graphical or tabular definition is also available. In many instances the program suggests ranges of values for certain parameters.

B1.2.1 Weight Sizing Module

This module allows determination of mission segment fuel fractions as well as estimates of take–off weight, empty weight and fuel weight for an arbitrary mission specification. For military airplanes, the effect of dropping weapons or stores can be accounted for.

Sensitivity of take–off weight with respect to various mission, aerodynamic and propulsion parameters can be determined.

B1.2.2 Performance Sizing Module

With this module, the user can determine the relation between take–off thrust–to–weight ratio (or weight–to–power ratio) and take–off wing–loading for any airplane, based on mission and airworthiness performance requirements. All pertinent civil and military airworthiness regulations of FAR 23, FAR 25, MIL–C–005011B (USAF) and AS–5263 (USNAVY) can be considered. Plots of $(W/S)_{TO}$ versus $(T/W)_{TO}$ {or versus $(W/P)_{TO}$} can be generated on any scale.

B1.2.3 Geometry Module

In this module, the planform geometry of a straight tapered and cranked lifting surface can be determined. All estimated geometry parameters (such as area, mean geometric chord, aspect ratio, taper ratio and sweep angle) can be displayed as 2–D drawings. This module can be used for wings, canards, horizontal tails, vertical tails and fuselages.

B1.2.4 High Lift Module

This module allows the user to calculate the maximum lift coefficients of wings with and without trailing edge flaps. The effect of airfoil type and Reynolds number are accounted for. The program will calculate and display the flap size needed to achieve a given maximum lift coefficient.

B1.2.5 Drag Polar Module

Two methods for estimating airplane drag polars are available in this module: a simplified (Class I) method and a detailed (Class II) method. In the Class I method, the program can calculate drag polars based on statistical relations between airplane type, take–off weight and wetted area.

Plots of: C_L versus C_D,

C_L versus C_L/C_D,

C_L versus $(C_L)^3/(C_D)^2$ and

C_L versus $\sqrt{C_L}/C_D$ can be generated on any scale desired by the user. In the Class II method, the program determines the drag polar of all individual components of an airplane. Plots of component drag polar contributions and total drag polar can be generated. Corrections for laminar flow can be included at the option of the user. Drag can also be plotted as a function of Mach number and flap angle. The Class II method applies to subsonic, transonic and supersonic flight.

B1.2.6 Stability and Control Module

This module can be used for sizing of horizontal and vertical tail surfaces to given stability or volume coefficient requirements. Longitudinal trim diagrams, which account for the effect of control surface stall, can be generated. The effect of stick, wheel and/or rudder pedal forces on longitudinal and lateral–directional trim can be determined. The program also calculates all required gradients, such as: $\frac{\partial F_s}{\partial n}, \frac{\partial F_s}{\partial V}$, etc. Neutral points, stick–free and stick–fixed can be determined. Effects of bob–weight and down–spring can be included at the option of the user.

B1.2.7 Weight and Balance Module

Two methods for estimating airplane weight breakdown are available in this module: a simplified (Class I) method and a detailed (Class II) method. In the Class I method, the program can calculate the component weight breakdown of an airplane as well as its inertias, based on statistical data. In the Class II method, the program estimates the weight of airplanes with the help of various formulas which relate component weight to significant design parameters. Plots of center of gravity versus weight can be generated for arbitrary load and un–load scenarios. V–n diagrams for FAR 23, FAR 25 and MIL–A–8861(ASG) specifications can be determined.

B1.2.8 Installed Thrust Module

With this module the installed thrust (or power) as well as specific fuel consumption can be estimated from given engine manufacturers data for piston, turboprop, prop–fan and turbojet/fan engines. Installed data can be plotted versus altitude and Mach number for a range of engine ratings. Propeller characteristics can be estimated from manufacturer's data. The user can also estimate effects of power extractions (mechanical, electrical and hydraulic). Inlets and nozzles can be sized.

B1.2.9 Performance Analysis Module

In this module the field length, stall, climb, range, endurance, dive, glide and maneuvering characteristics of an airplane can be evaluated. Payload–weight diagrams can be generated. This module uses more sophisticated methods than the Performance Sizing Module of B1.2.2.

B1.2.10   Stability and Control Derivatives Module

In this module the subsonic stability and control derivatives of conventional, canard and three–surface configurations can be analyzed. In addition, elevator control surface hinge–moment derivatives for various nose shapes and horn configurations can be estimated.

B1.2.11   Dynamics Module

With this module the open–loop airplane transfer functions can be determined, including the standard flying quality parameters of the short period, phugoid, roll, spiral and dutch–roll modes. The sensitivity of airplane undamped natural frequency, damping ratio and time–constant to arbitrary changes in stability derivatives and inertial characteristics can be plotted to any scale desired by the user. The program can determine whether the airplane meets the Level I, II or III flying quality requirements of MIL–F–8785C. Roll coupling stability diagrams can be generated. The open–loop airplane transfer functions are automatically transferred to the Control Module.

B1.2.12   Control Module

In this module, Root–Locus plots and Bode plots can be generated for open–loop and closed–loop control systems. The transfer functions of sensors, actuators and other loop components can be entered at the option of the user. The effect of tilt angles on yaw damper and roll damper performance can be evaluated.

B1.2.13   Cost Analysis Module

Estimates for airplane RDTE (Research, Development, Technology and Evaluation), manufacturing, prototype, direct and indirect operating cost and life cycle cost can be made for military as well as for civil airplanes. Quick estimates of airplane and engine prices can be made.

B1.2.14 Database Module

This module manages the data for all AAA modules. Design information can be stored or retrieved and parameters in the data–base can be viewed, edited or printed. This module also allows the user to determine atmospheric characteristics for standard and off–standard conditions.

B1.2.15 Help/Setup Module

This module displays help in using the AAA program. Setup allows the user to select between S.I. and British units. Project and company information can be entered here.

The AAA program has been found to be very useful in an instructional environment: students can rapidly discover which design parameters have the greatest effect on items B1.2.1–B1.2.13. Students can also be given real world homework assignments. AAA takes away the need for hours of tedious preparation of input data.

The program provides on–line written and graphical definitions for input parameters. This is done by selecting INFO or HELP on the calculator pad. The calculator pad pops onto the screen any time an input variable or output variable is selected by the user. Selection is done with a mouse/cursor combination. By selecting THEORY the program also identifies the theoretical models and assumptions made in various modules. This helps the user understand the underlying theories and equations.

B2   G.A.–CAD: GENERAL AVIATION COMPUTER AIDED DESIGN

This is a software package for the preliminary design of general aviation airplanes. The CAD component of the software allows for a complete 3D definition of an airplane and all of its components. The package contains a very extensive airfoil data base which can be used in the design of wings, tails, fins and pylons. The geometric data resulting from drawing an airplane in CAD are automatically read into the analysis part of the program. In the analysis part, drag, stability and control, loads and structural analyses can be performed. Most of the modules of the AAA software described in B.1 are also part of G.A.–CAD.

G.A.–CAD is available on IBM Compatible PC (Windows 3.1, 95 and NT).

B3.   OTHER SOFTWARE PACKAGES

Depending on the individual needs of a customer, DARcorporation can package down–scaled, more economical versions of its software.

B4 REFERENCES

B1 Roskam, J.; Airplane Flight Dynamics and Automatic Flight Controls, Parts I and II; 1994 and 1995 respectively.
(Part I contains approximately 576 pages and Part II contains approximately 383 pages)

B2 Lan, C. E. and Roskam, J.; Airplane Aerodynamics and Performance; 1997.
(This text contains 700 pages)

B3 Roskam, J.; Airplane Design, Parts I – VIII; 1989.
(This eight–volume text contains 202, 310, 454, 416, 209, 550, 351 and 368 pages respectively)

Sales and Marketing for References B1–B3 and all software packages described here:

Design, Analysis and Research Corporation, 120 East Ninth Street, Suite 2, Lawrence, Kansas, 66044, USA.
Tel. 785–832–0434   Fax: 785–832–0524

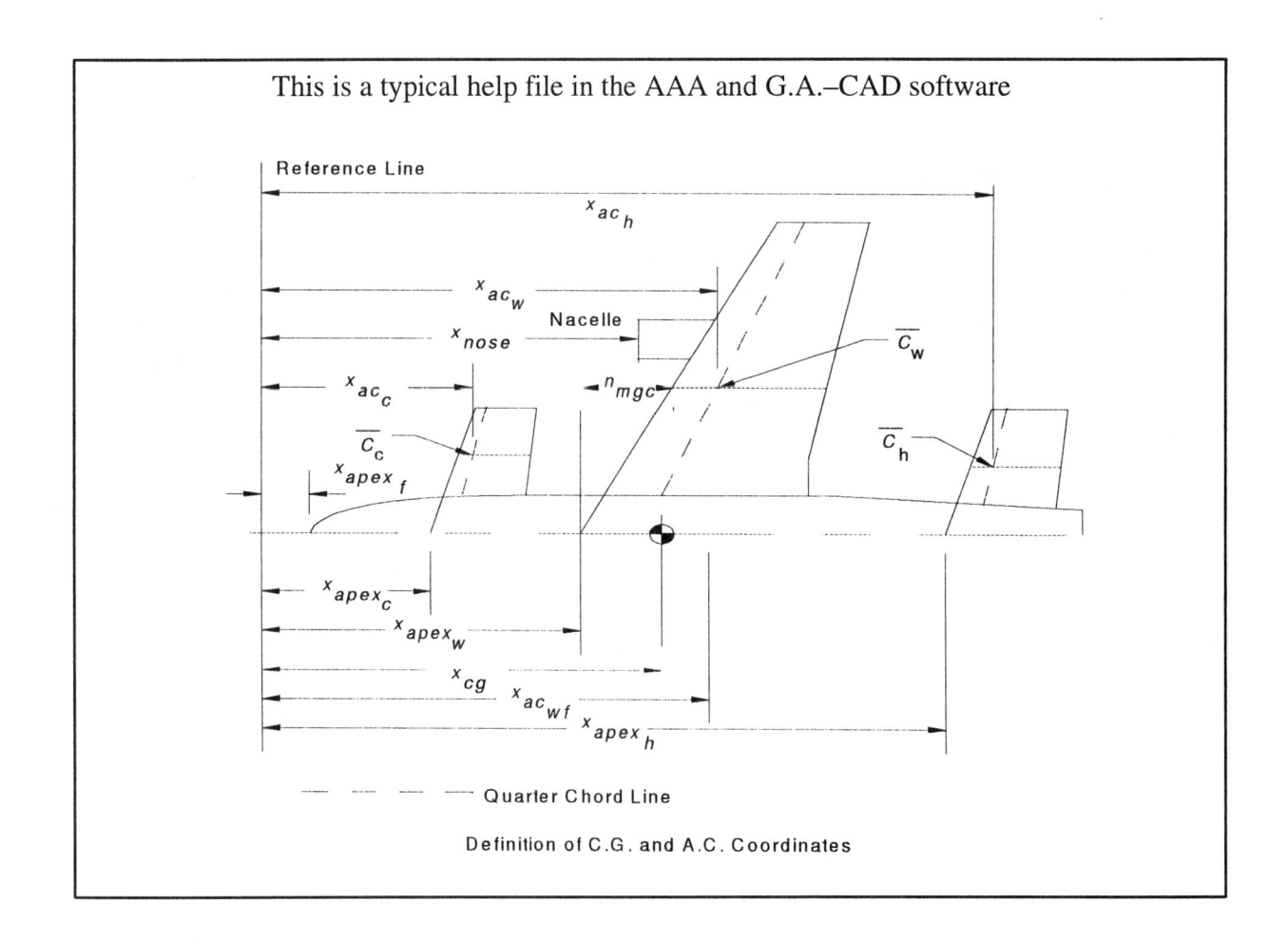

This is a typical help file in the AAA and G.A.–CAD software

Definition of C.G. and A.C. Coordinates

APPENDIX C: PROPELLER PERFORMANCE CHARTS

In this appendix, propeller performance charts are presented for the following cases:

1) **Three–bladed propellers**

   a) Ratio of thrust coefficient to power coefficient, C_T/C_P, versus the power coefficient, C_P, for a range of integrated design lift coefficient, C_{L_i}, in Charts C1 through C4.

   b) Free propeller efficiency, $\eta_{P_{free}}$, as a function of the power coefficient, C_P, and the advance ratio, J, in Charts C9 through C24.

2) **Four–bladed propellers**

   a) Ratio of thrust coefficient to power coefficient, C_T/C_P, versus the power coefficient, C_P, for a range of integrated design lift coefficient, C_{L_i} in Charts C5 through C8.

   b) Free propeller efficiency, $\eta_{P_{free}}$, as a function of the power coefficient, C_P, and the advance ratio, J, in Charts C25 through C40.

Note well: For propellers with 2 blades or with more than 4 blades, an extrapolation procedure is described in Section 7.6 of Chapter 7 in this text.

The data in this appendix are presented through the courtesy of Hamilton–Standard.

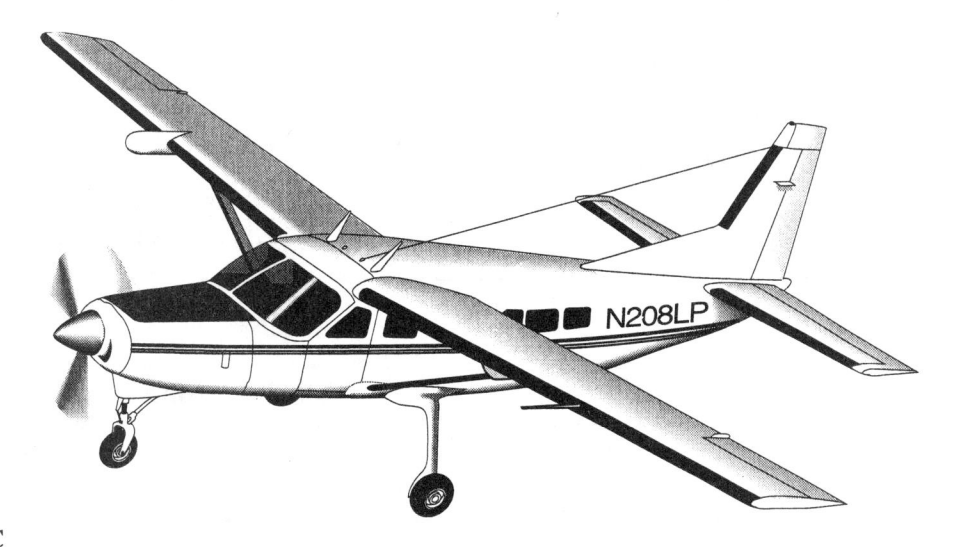

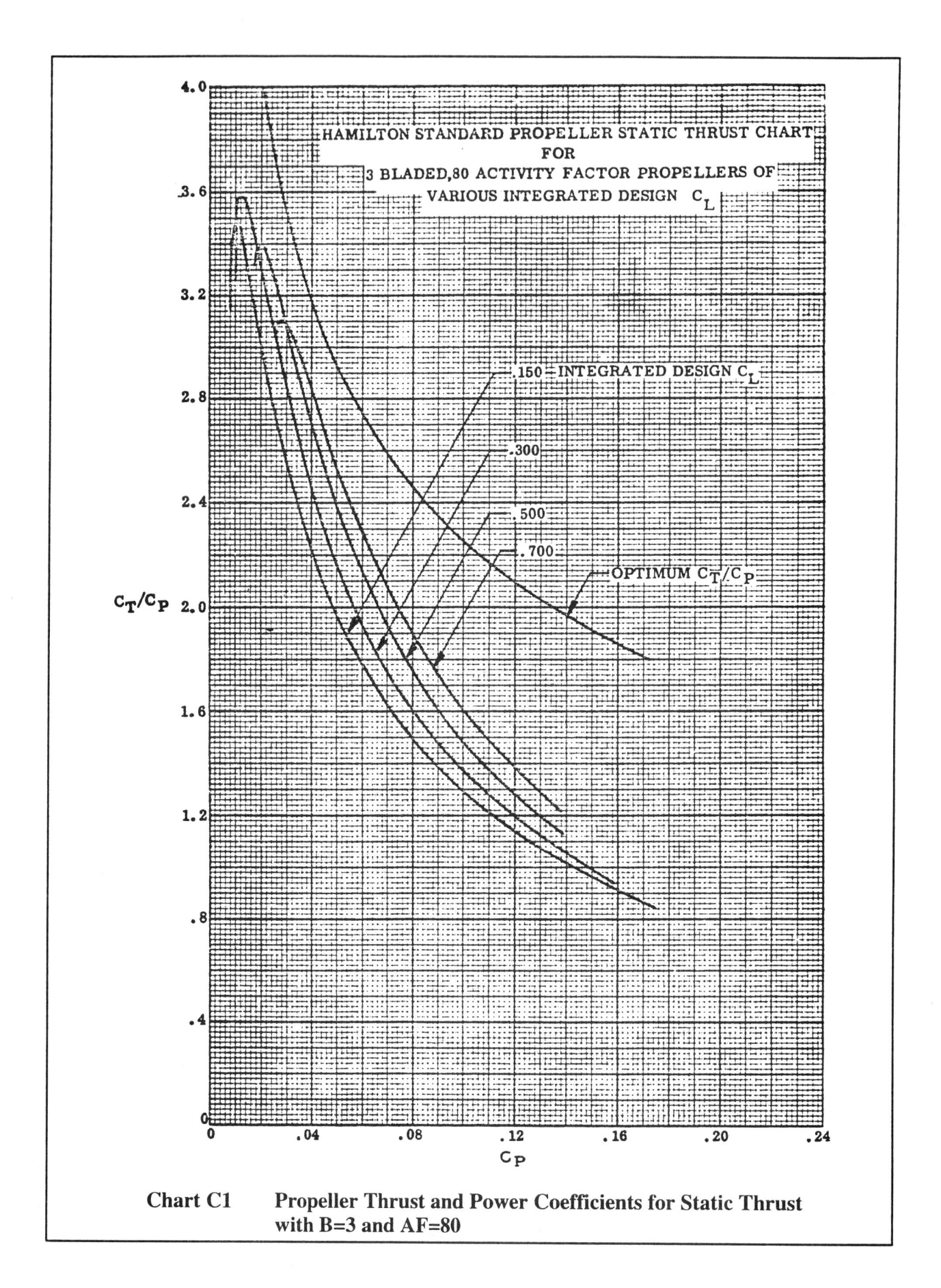

Chart C1   **Propeller Thrust and Power Coefficients for Static Thrust with B=3 and AF=80**

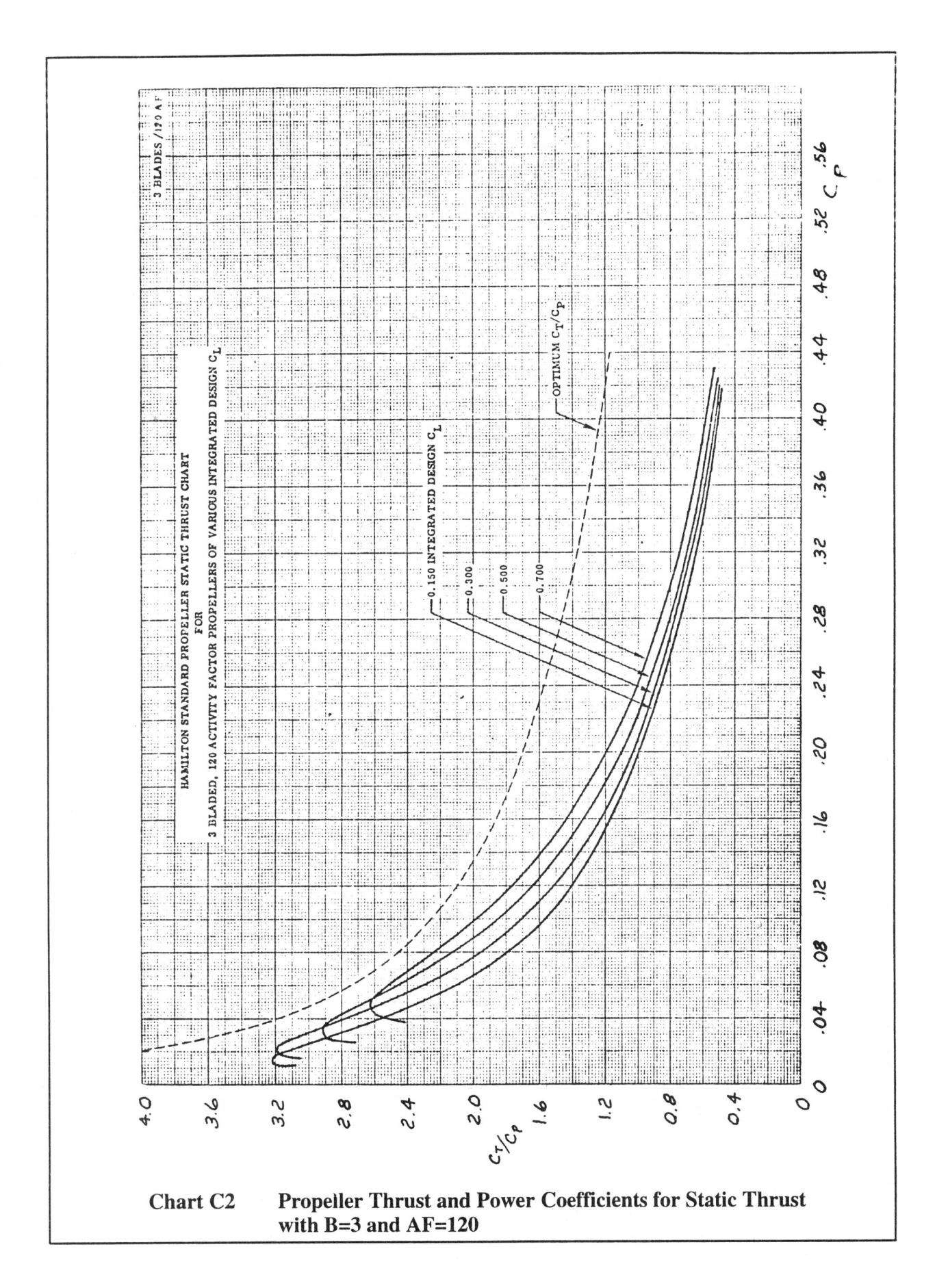

Chart C2   **Propeller Thrust and Power Coefficients for Static Thrust with B=3 and AF=120**

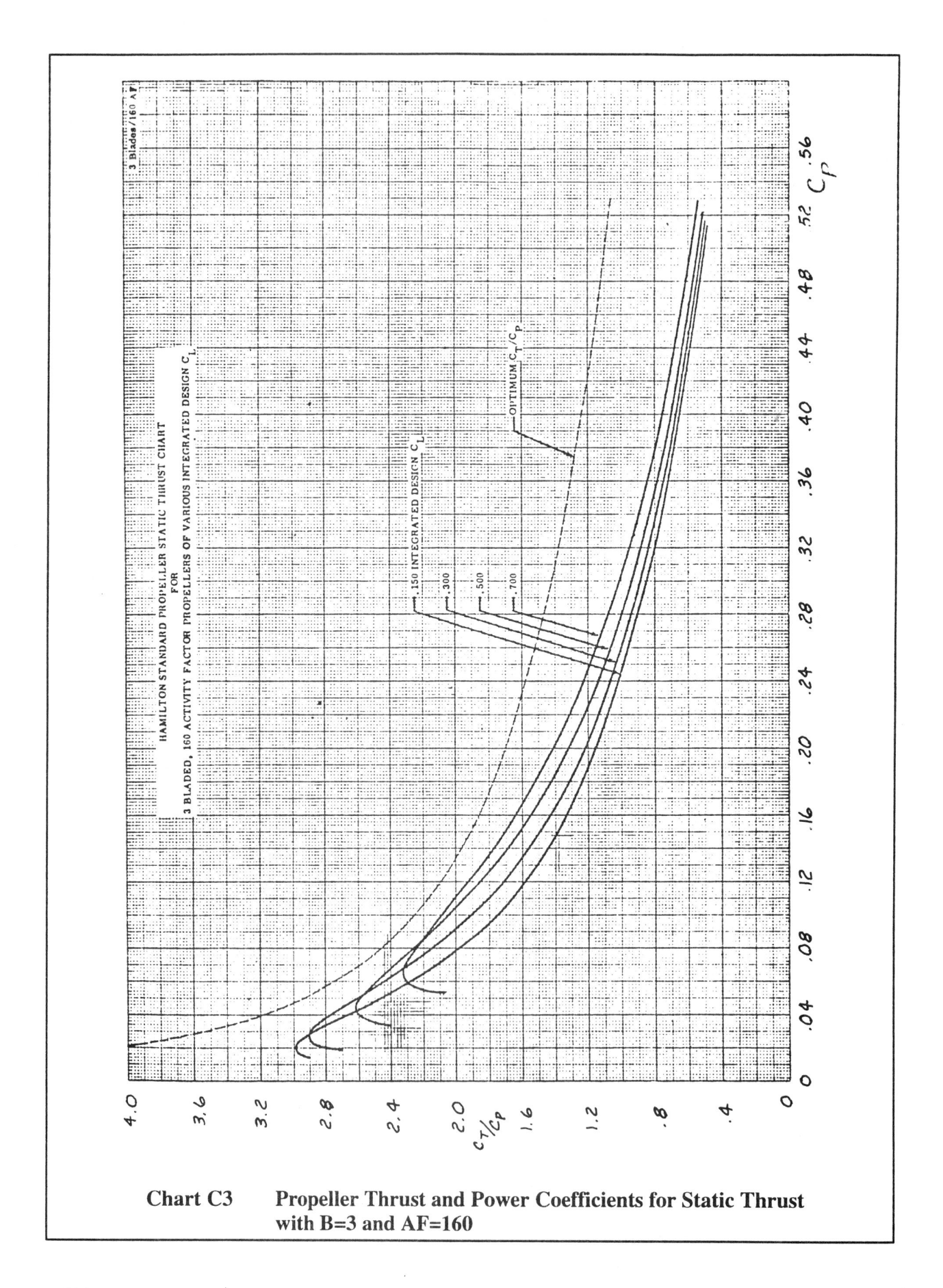

Chart C3 Propeller Thrust and Power Coefficients for Static Thrust with B=3 and AF=160

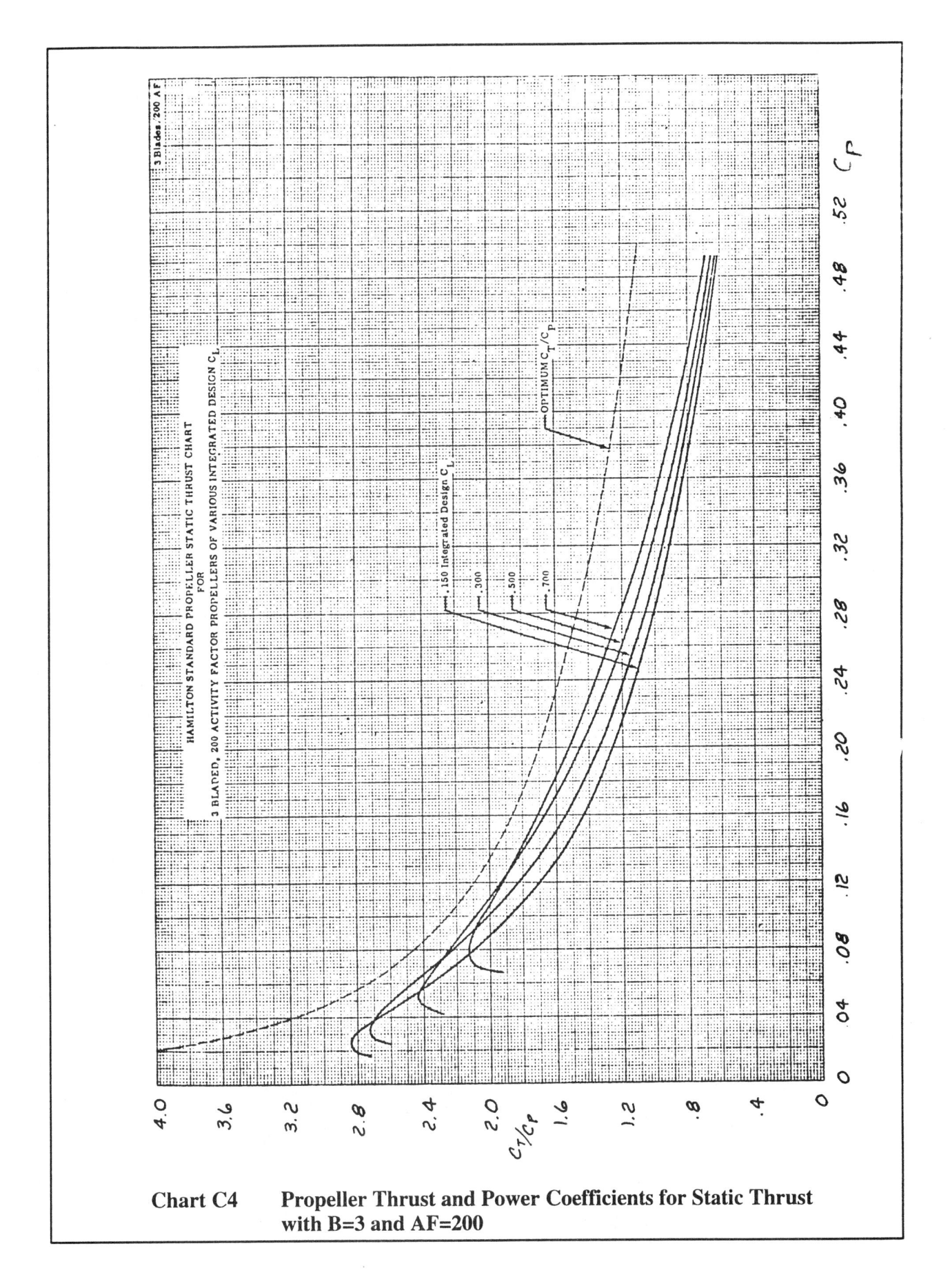

Propeller Performance Charts

Chart C4 Propeller Thrust and Power Coefficients for Static Thrust with B=3 and AF=200

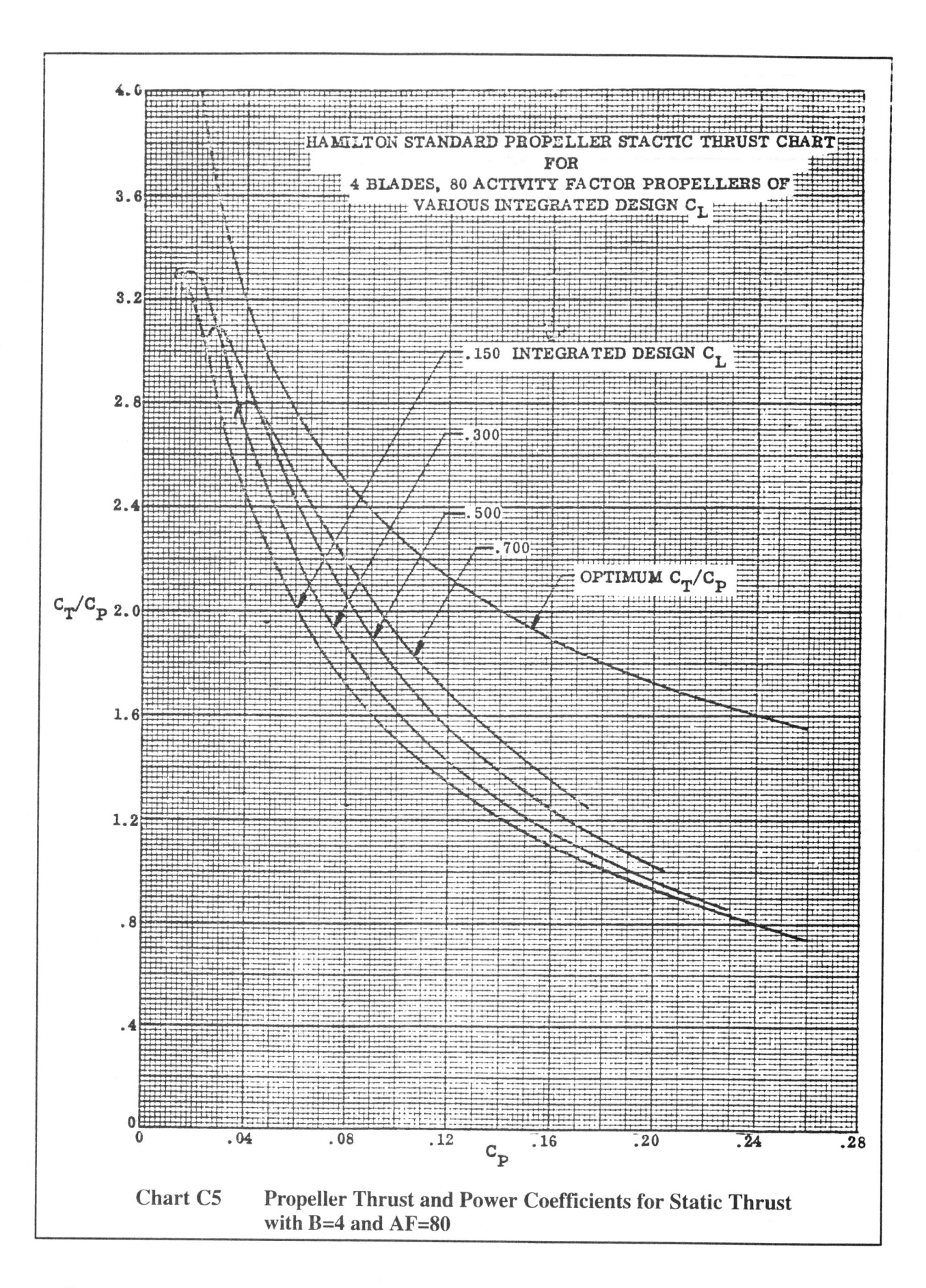

Chart C5 Propeller Thrust and Power Coefficients for Static Thrust with B=4 and AF=80

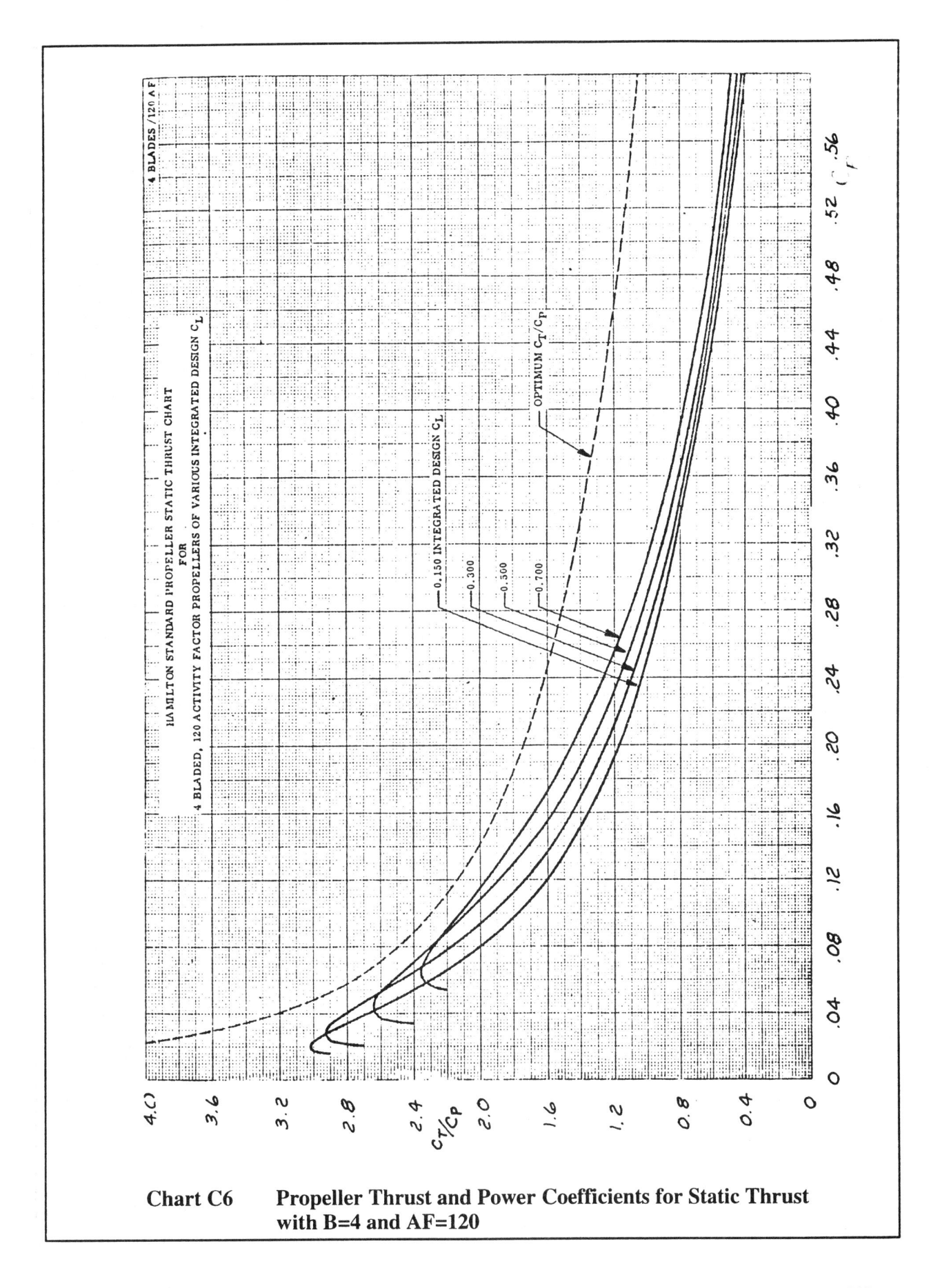

Propeller Performance Charts

Chart C6 Propeller Thrust and Power Coefficients for Static Thrust with B=4 and AF=120

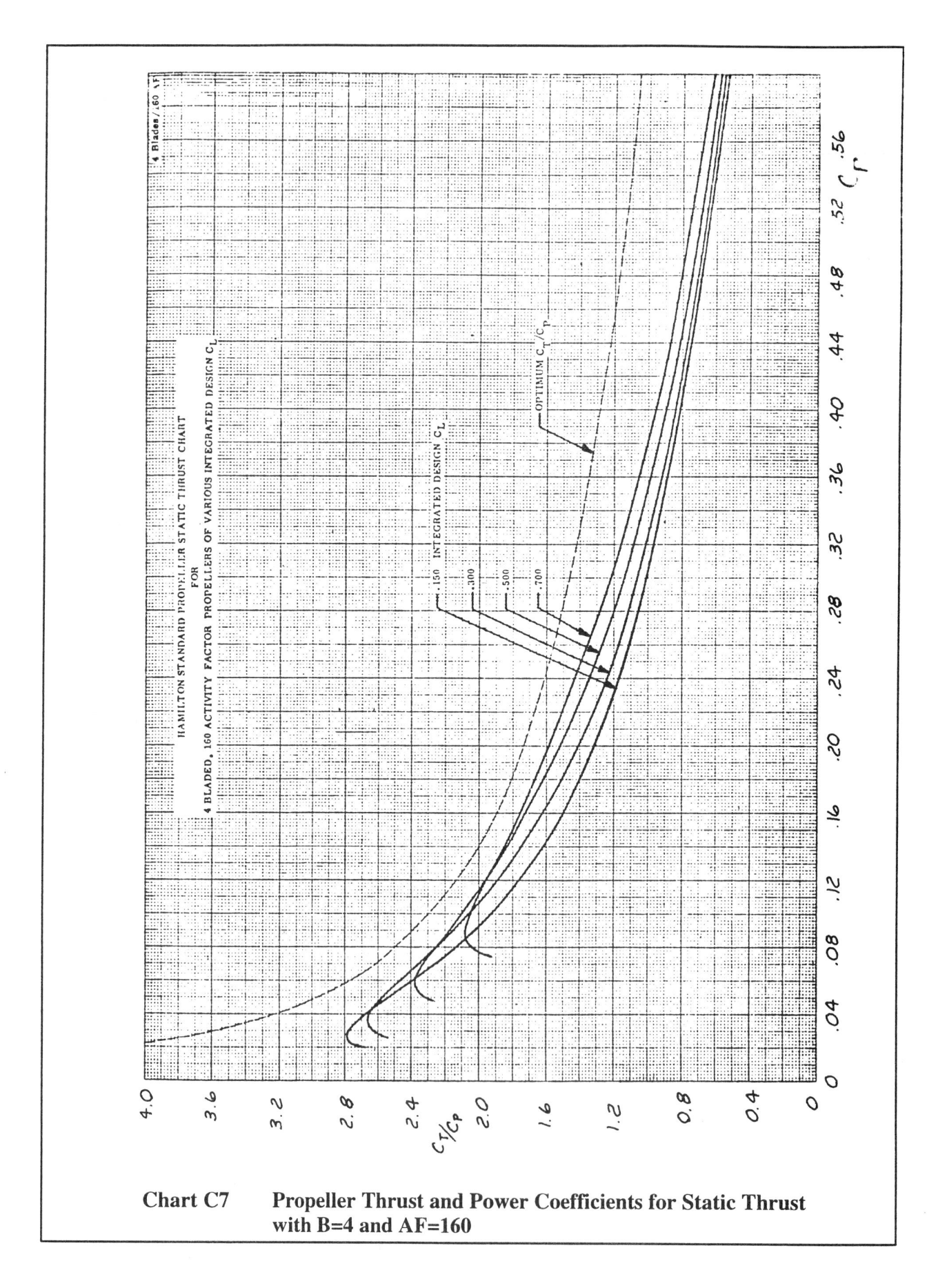

Chart C7 Propeller Thrust and Power Coefficients for Static Thrust with B=4 and AF=160

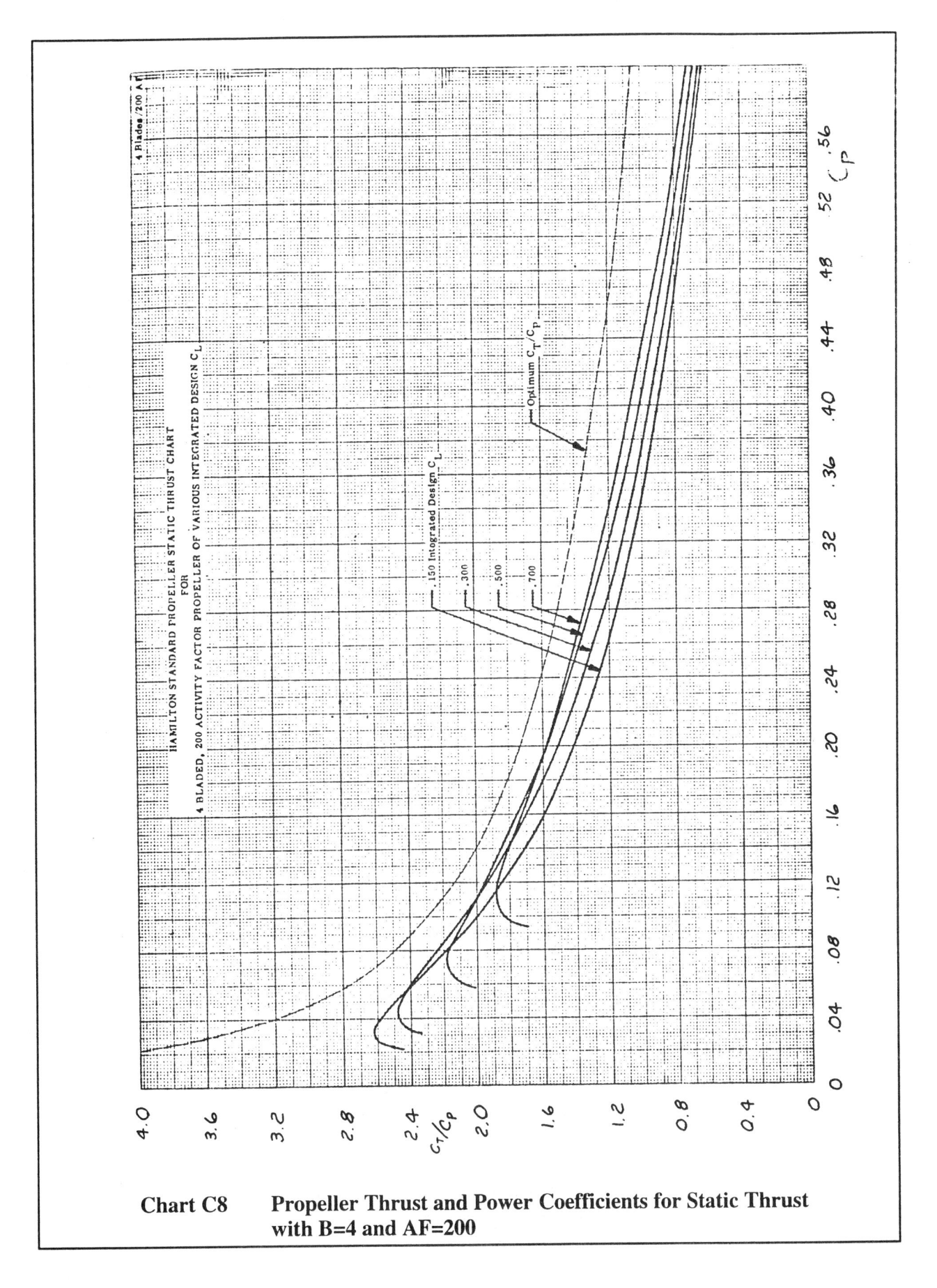

Chart C8 Propeller Thrust and Power Coefficients for Static Thrust with B=4 and AF=200

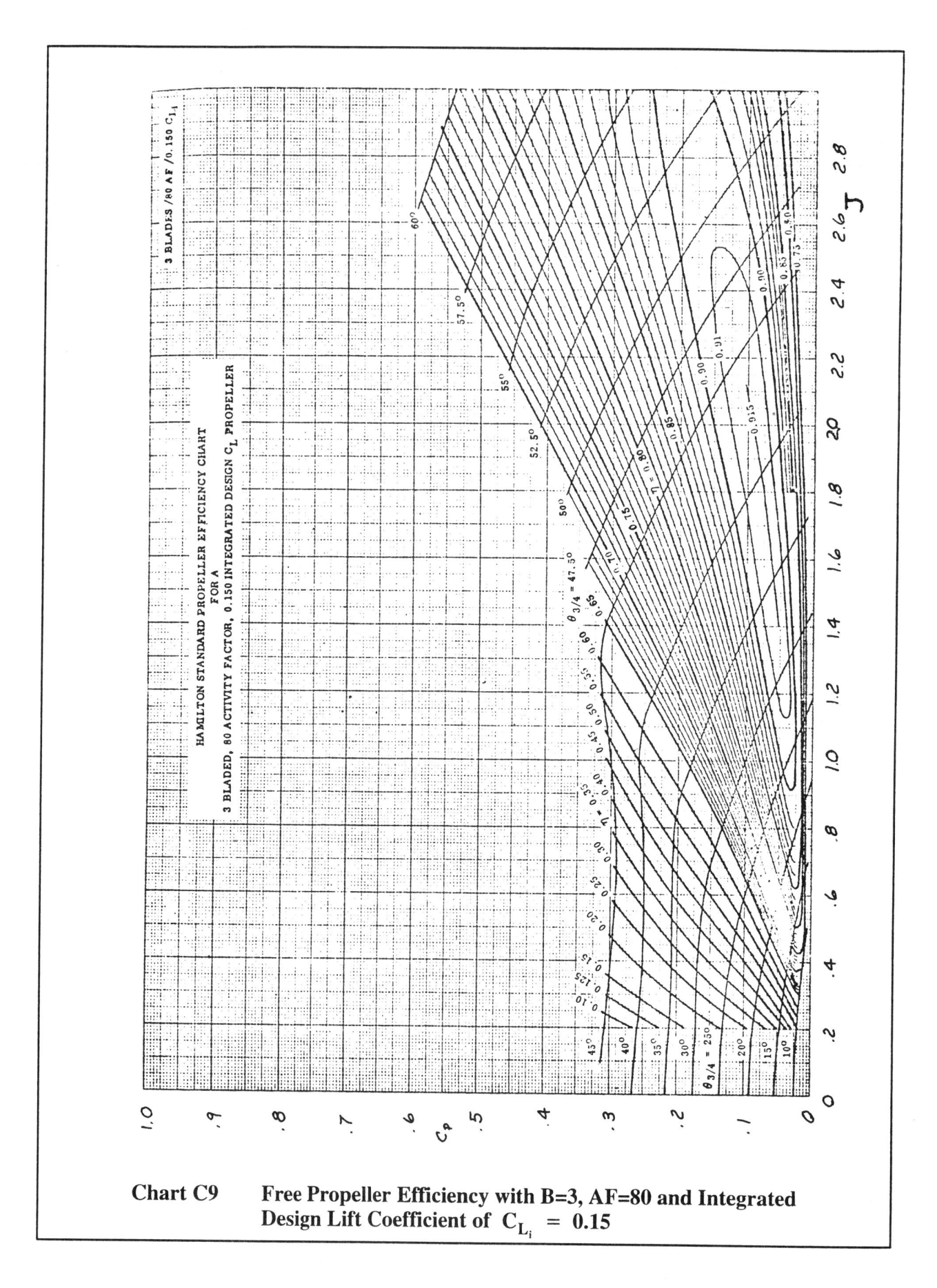

Chart C9   Free Propeller Efficiency with B=3, AF=80 and Integrated Design Lift Coefficient of $C_{L_i} = 0.15$

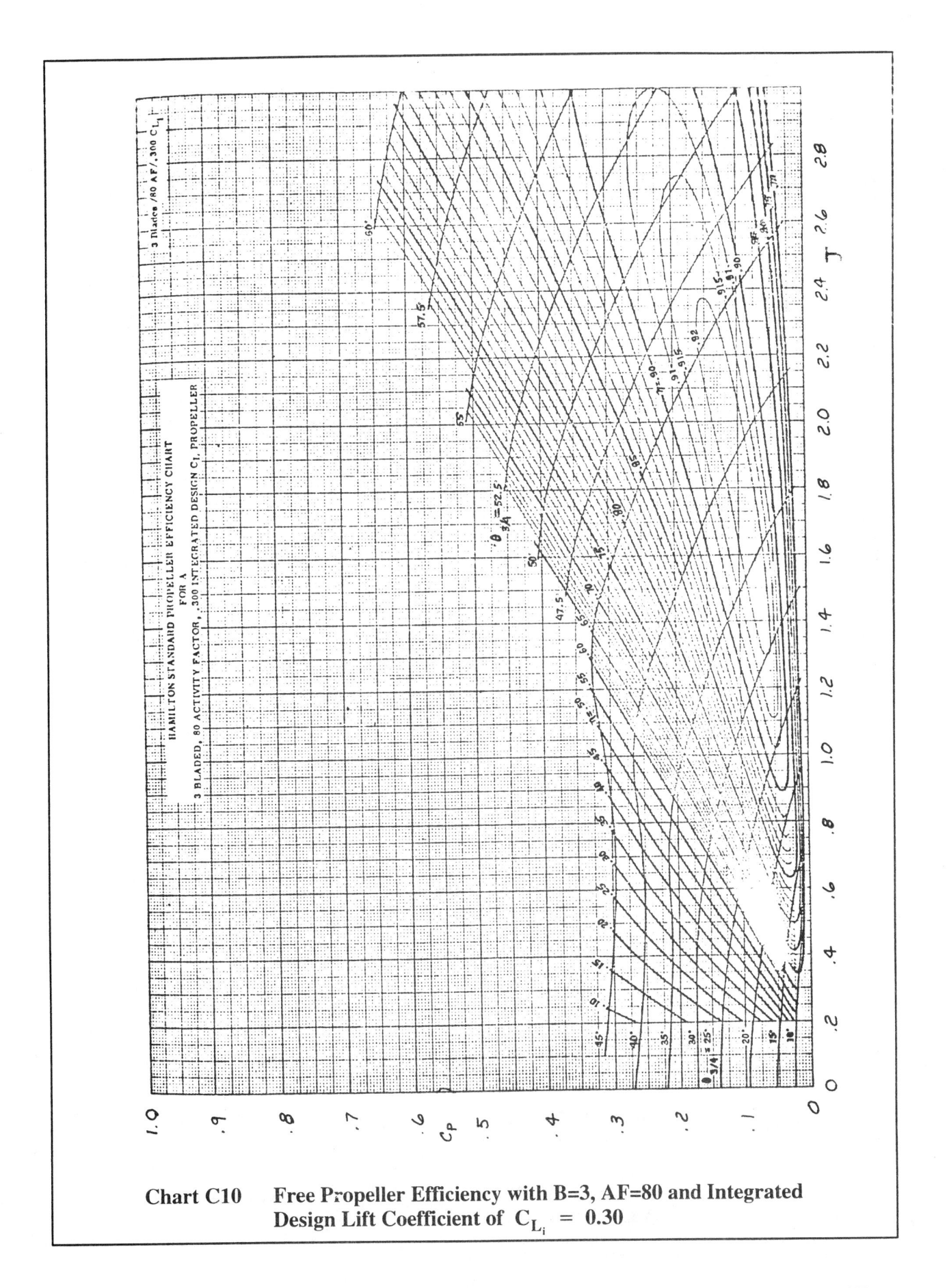

Chart C10 Free Propeller Efficiency with B=3, AF=80 and Integrated Design Lift Coefficient of $C_{L_i} = 0.30$

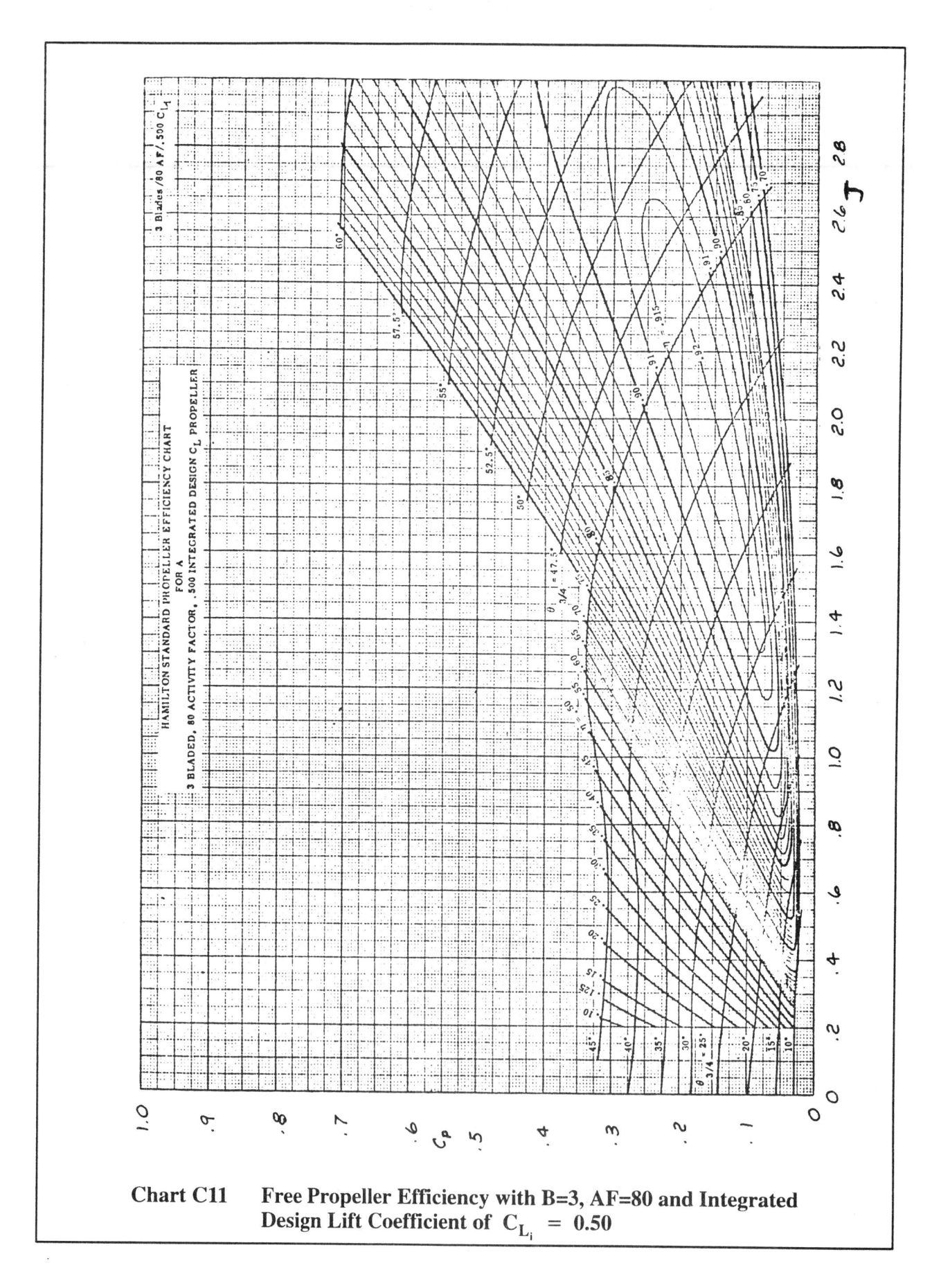

Chart C11   Free Propeller Efficiency with B=3, AF=80 and Integrated Design Lift Coefficient of $C_{L_i} = 0.50$

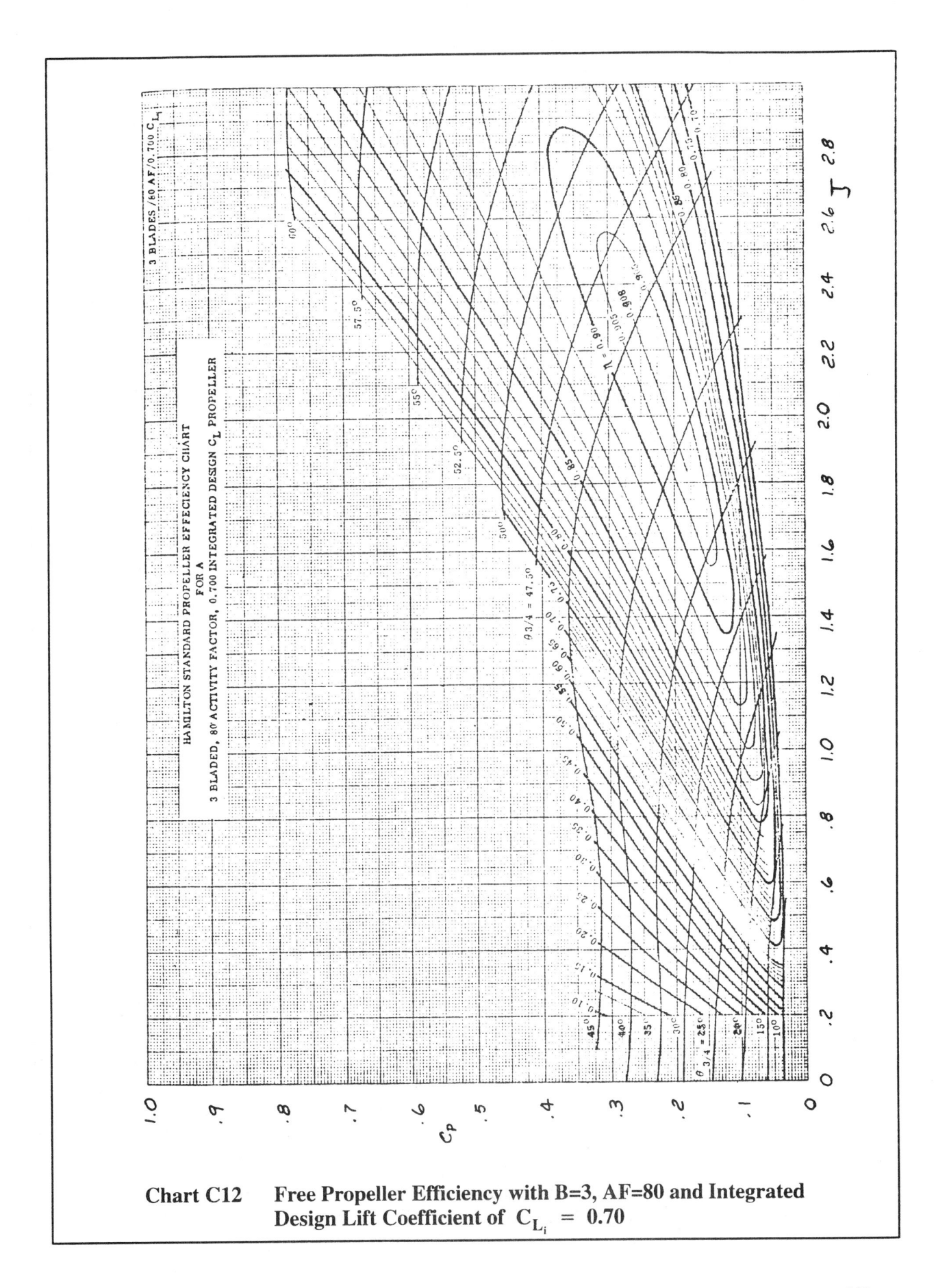

Propeller Performance Charts

Chart C12 Free Propeller Efficiency with B=3, AF=80 and Integrated Design Lift Coefficient of $C_{L_i} = 0.70$

Appendix C

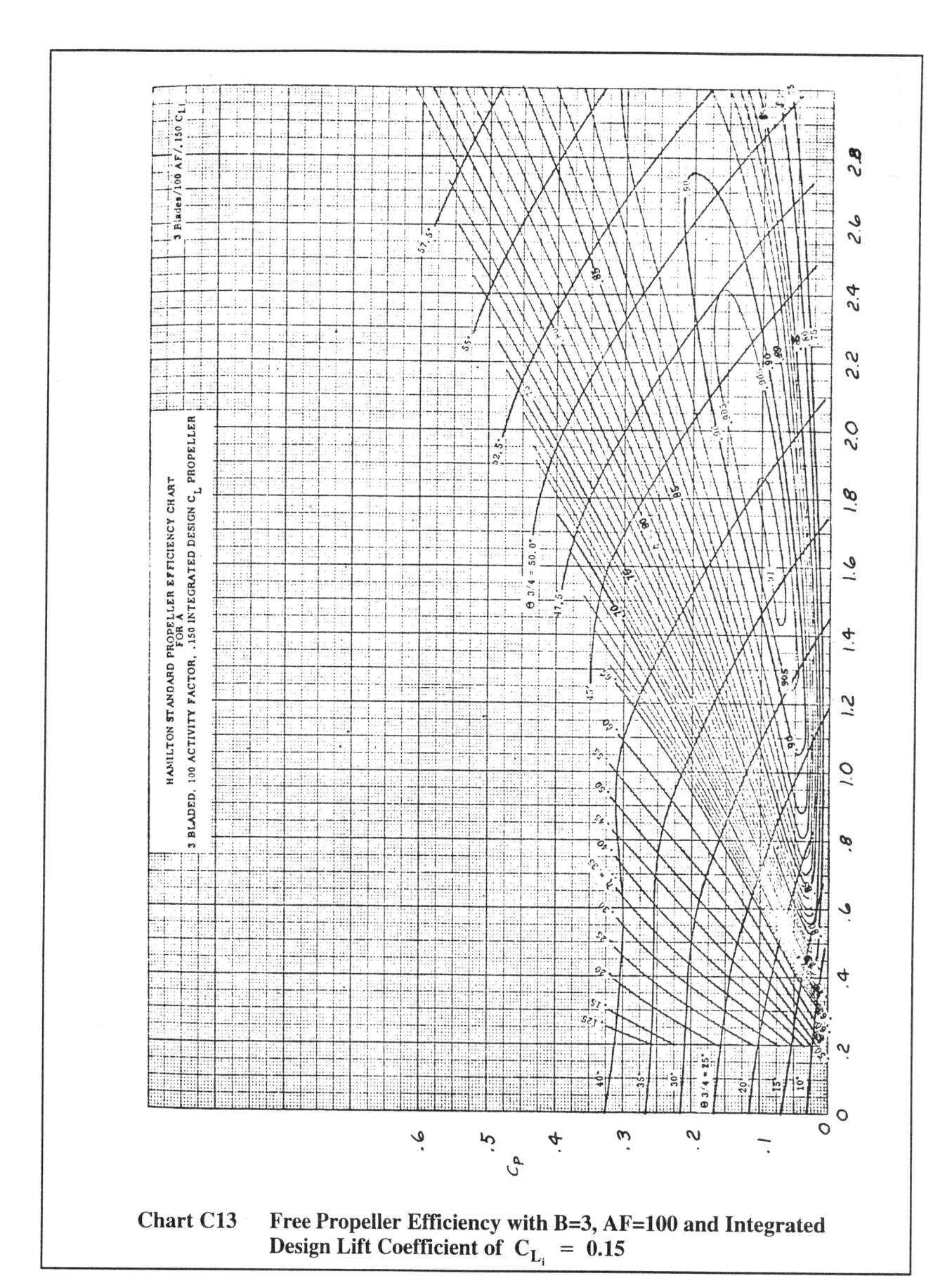

Chart C13 Free Propeller Efficiency with B=3, AF=100 and Integrated Design Lift Coefficient of $C_{L_i} = 0.15$

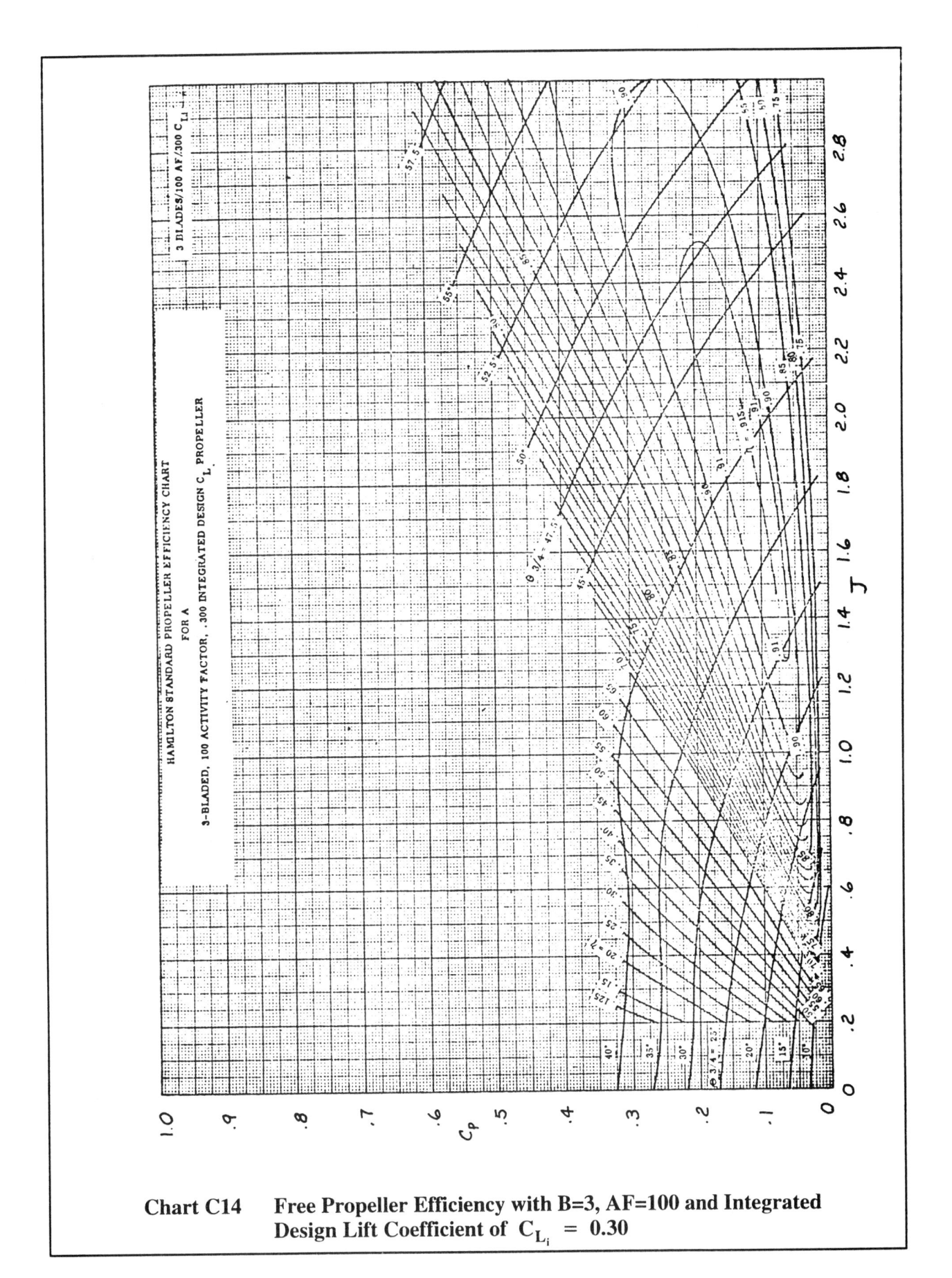

Chart C14 Free Propeller Efficiency with B=3, AF=100 and Integrated Design Lift Coefficient of $C_{L_i} = 0.30$

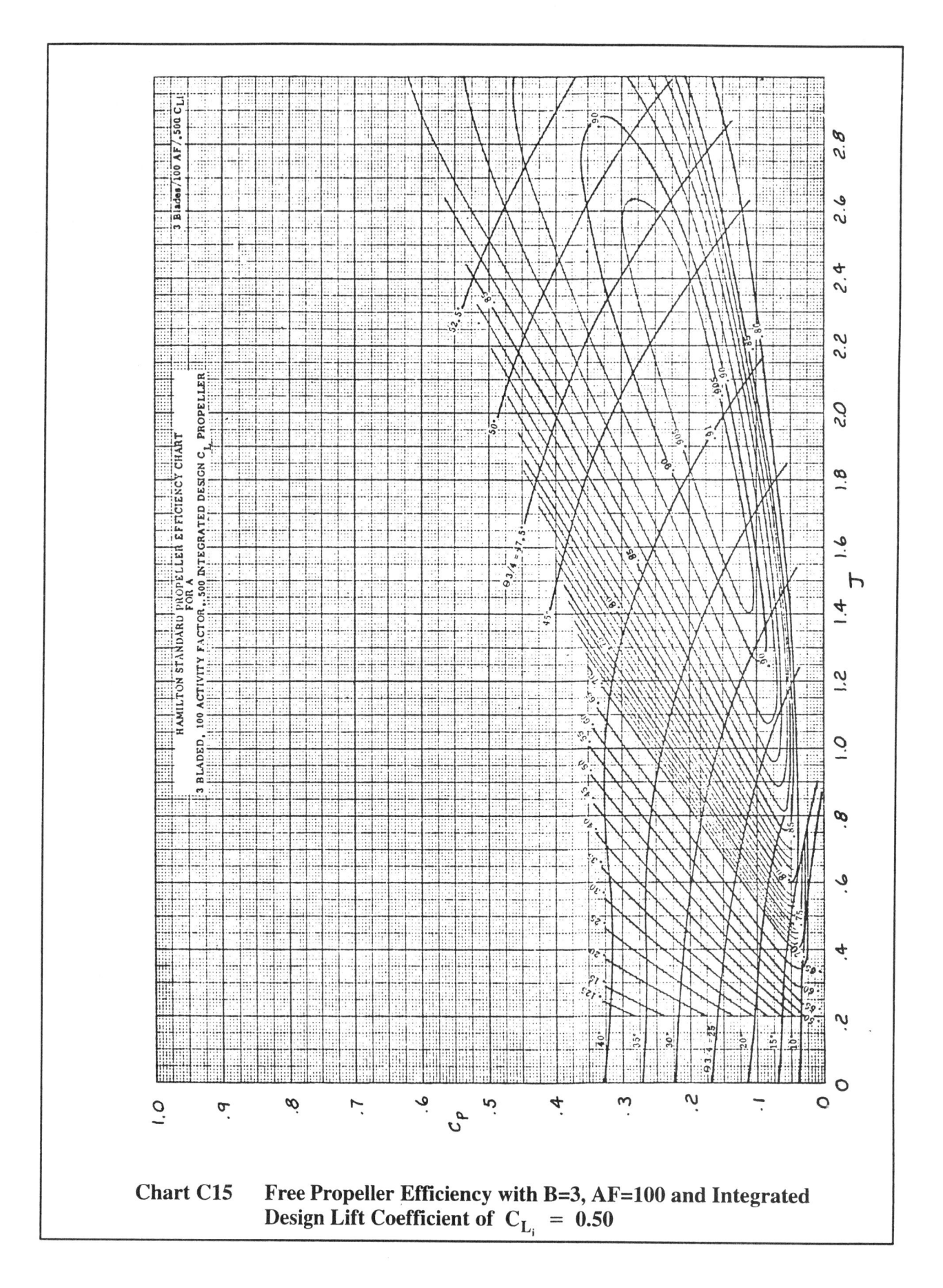

Chart C15 Free Propeller Efficiency with B=3, AF=100 and Integrated Design Lift Coefficient of $C_{L_i} = 0.50$

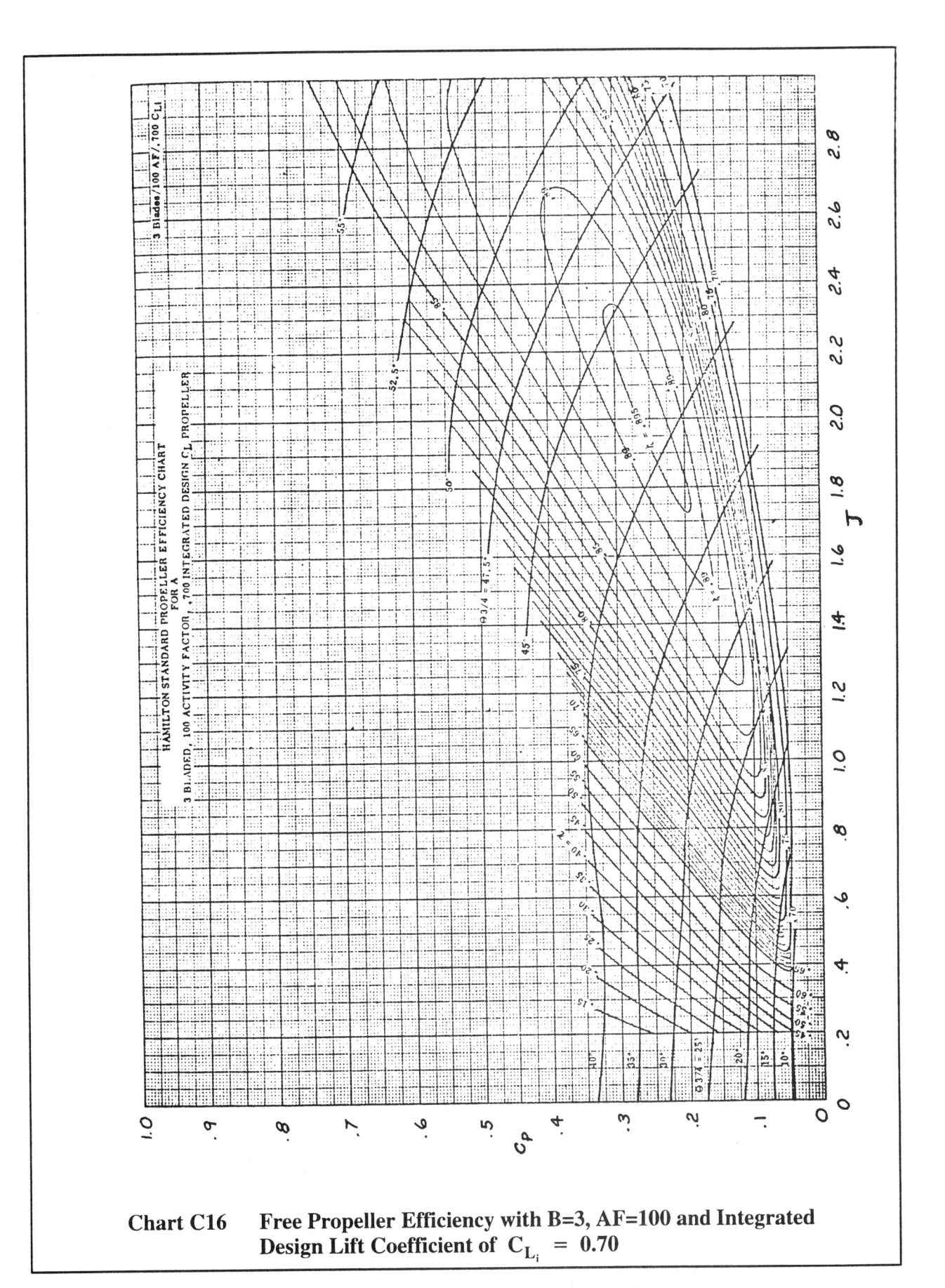

Chart C16 Free Propeller Efficiency with B=3, AF=100 and Integrated Design Lift Coefficient of $C_{L_i} = 0.70$

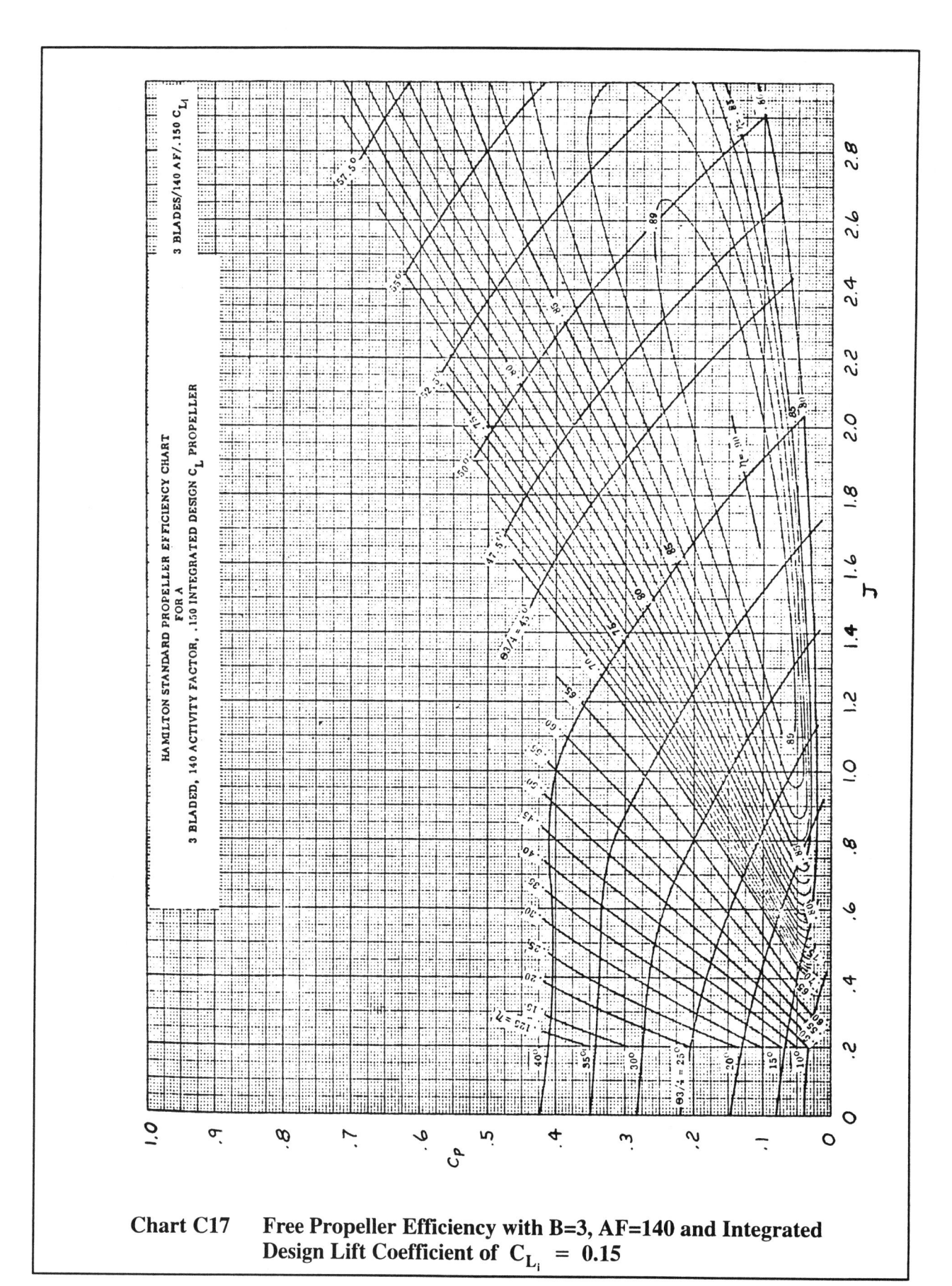

Chart C17 Free Propeller Efficiency with B=3, AF=140 and Integrated Design Lift Coefficient of $C_{L_i} = 0.15$

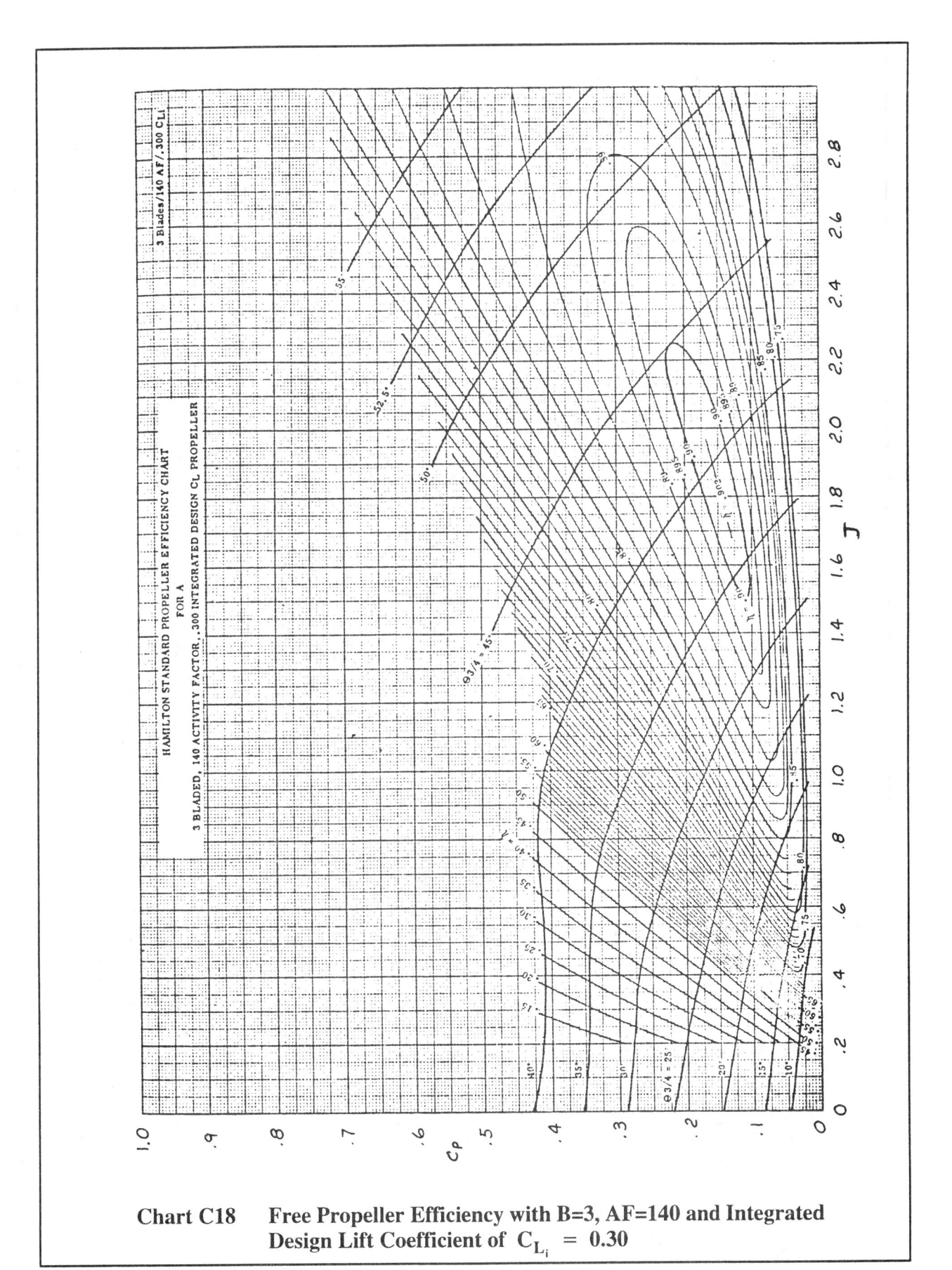

Propeller Performance Charts

Chart C18 Free Propeller Efficiency with B=3, AF=140 and Integrated Design Lift Coefficient of $C_{L_i} = 0.30$

Appendix C

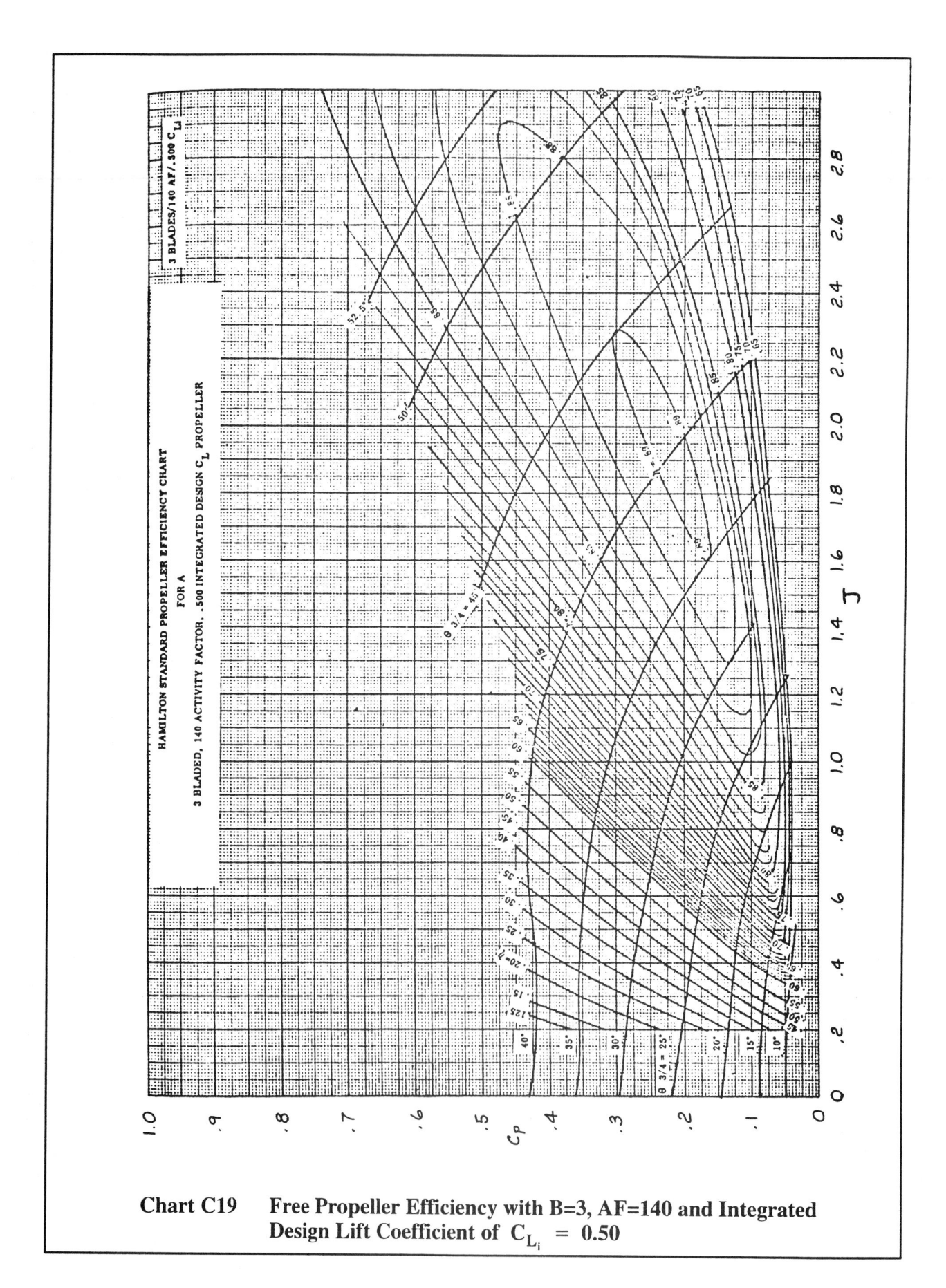

Chart C19 Free Propeller Efficiency with B=3, AF=140 and Integrated Design Lift Coefficient of $C_{L_i} = 0.50$

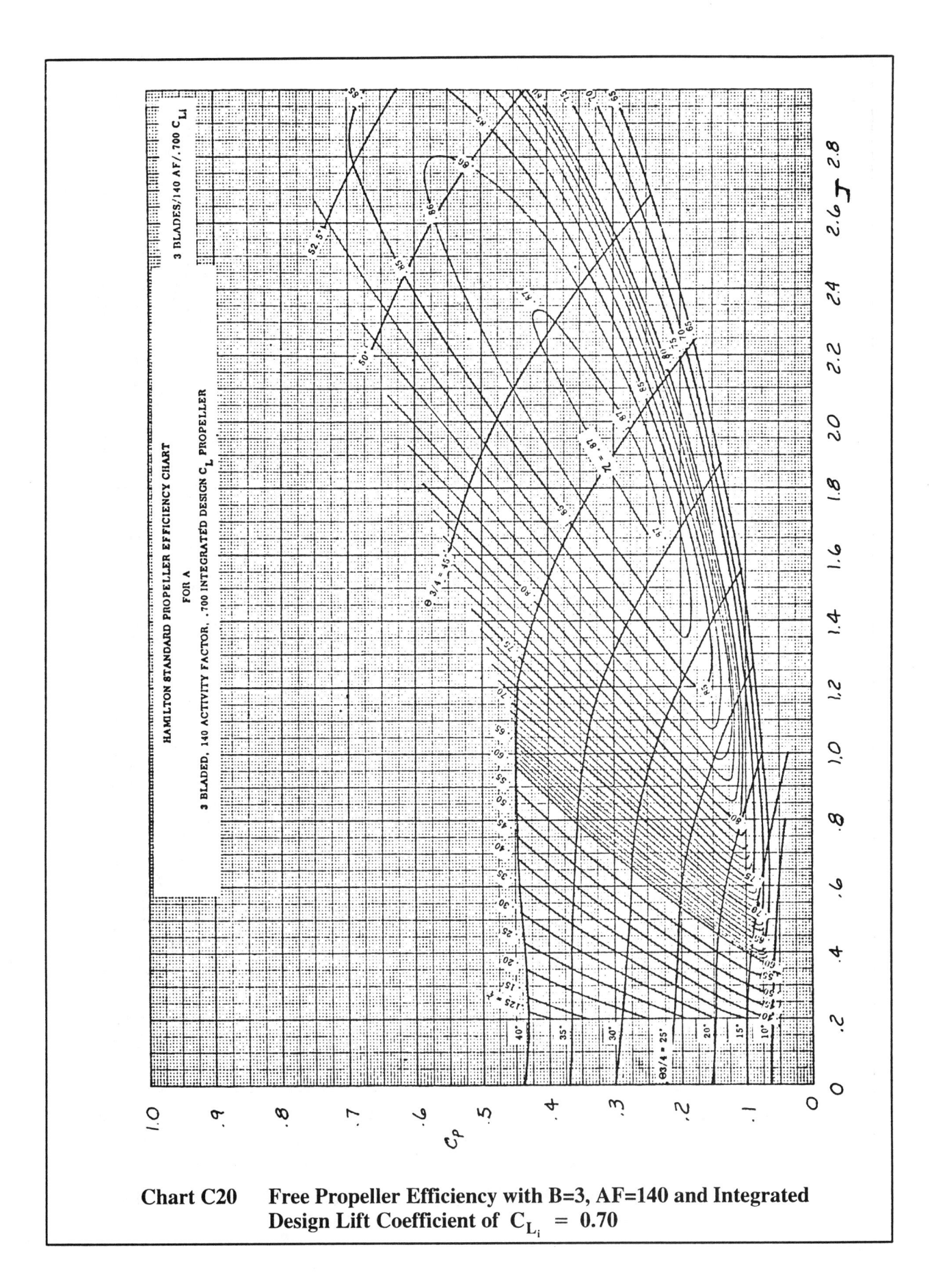

Chart C20 Free Propeller Efficiency with B=3, AF=140 and Integrated Design Lift Coefficient of $C_{L_i} = 0.70$

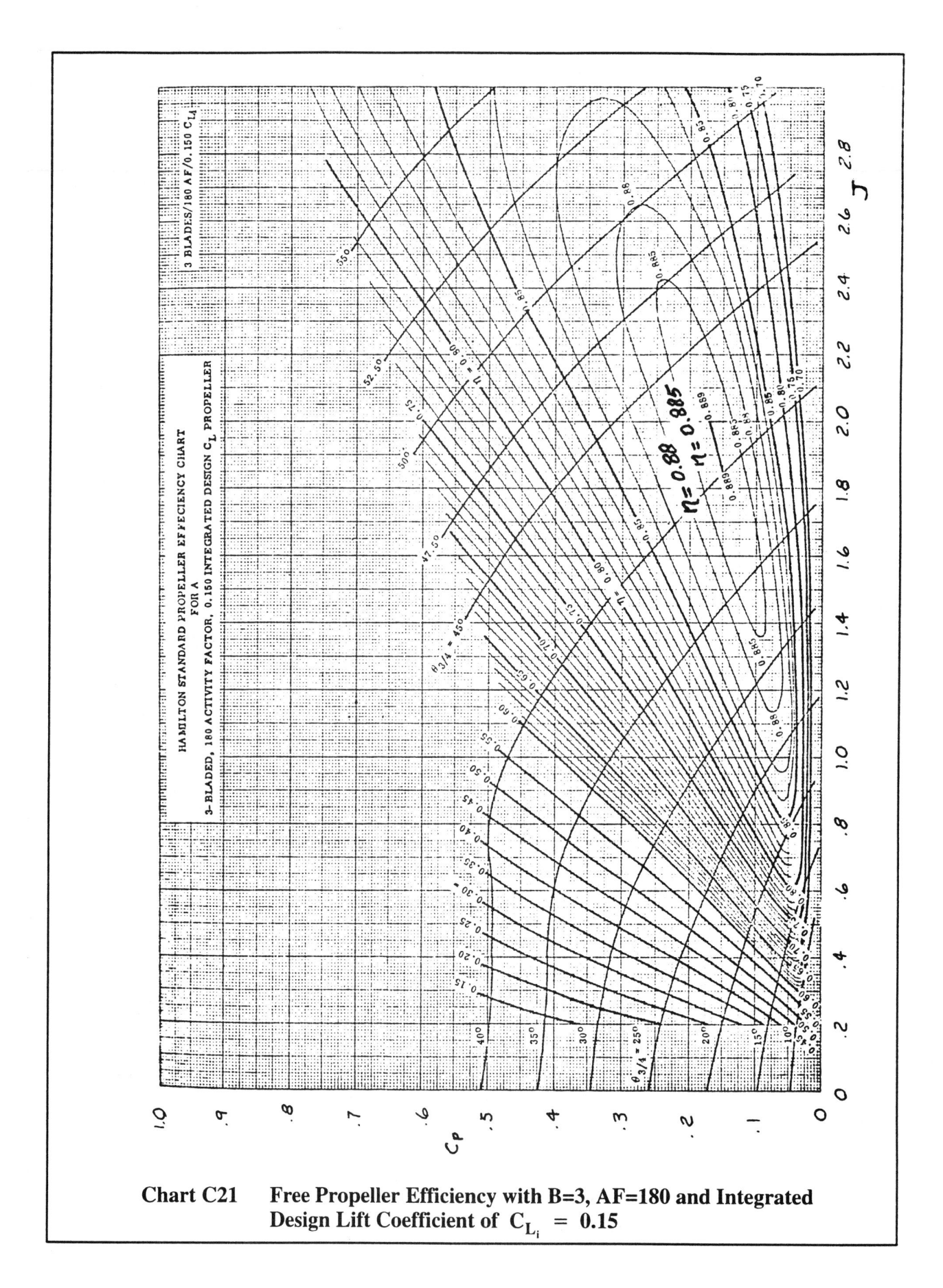

Chart C21 Free Propeller Efficiency with B=3, AF=180 and Integrated Design Lift Coefficient of $C_{L_i} = 0.15$

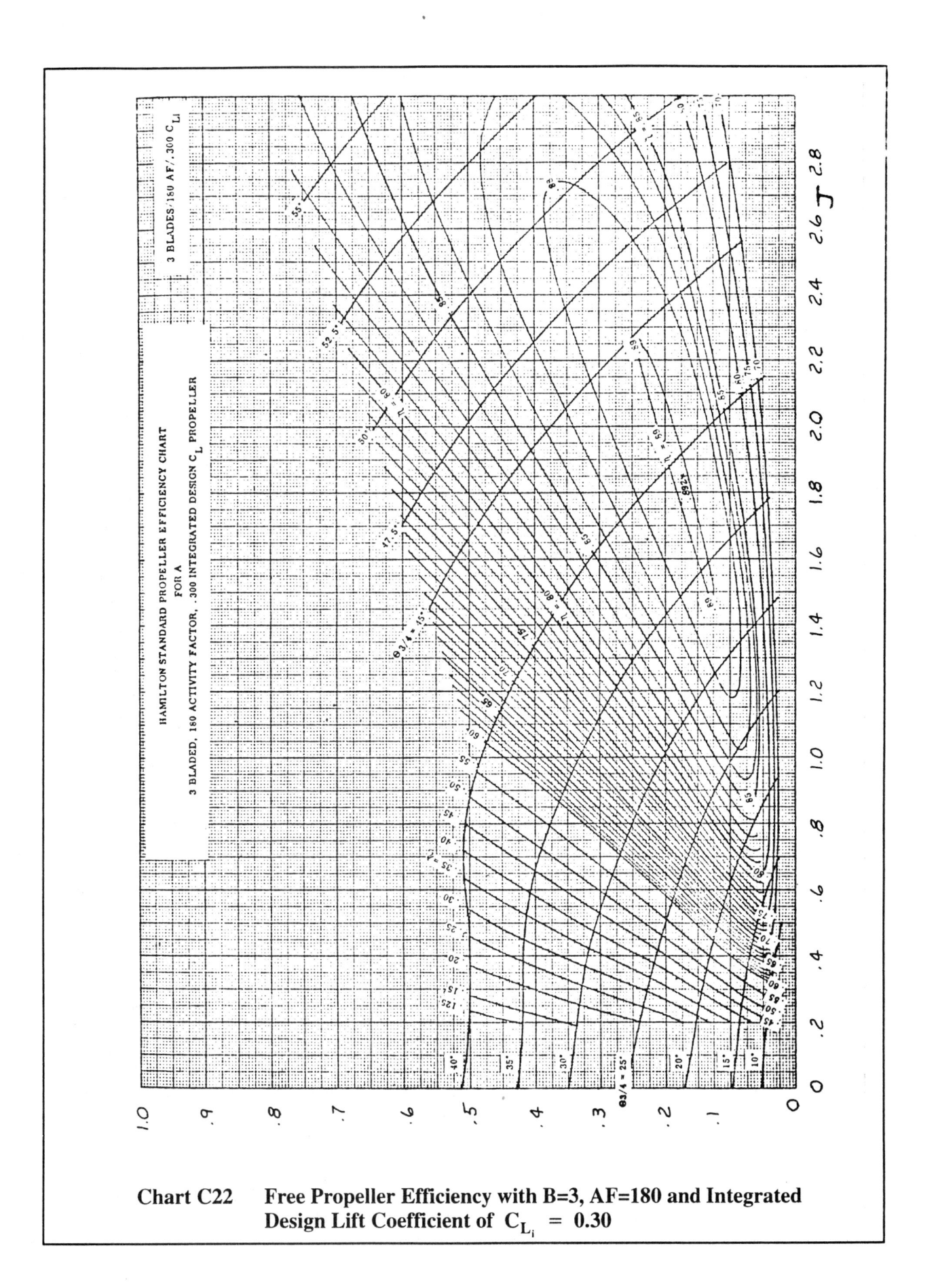

Chart C22 Free Propeller Efficiency with B=3, AF=180 and Integrated Design Lift Coefficient of $C_{L_i} = 0.30$

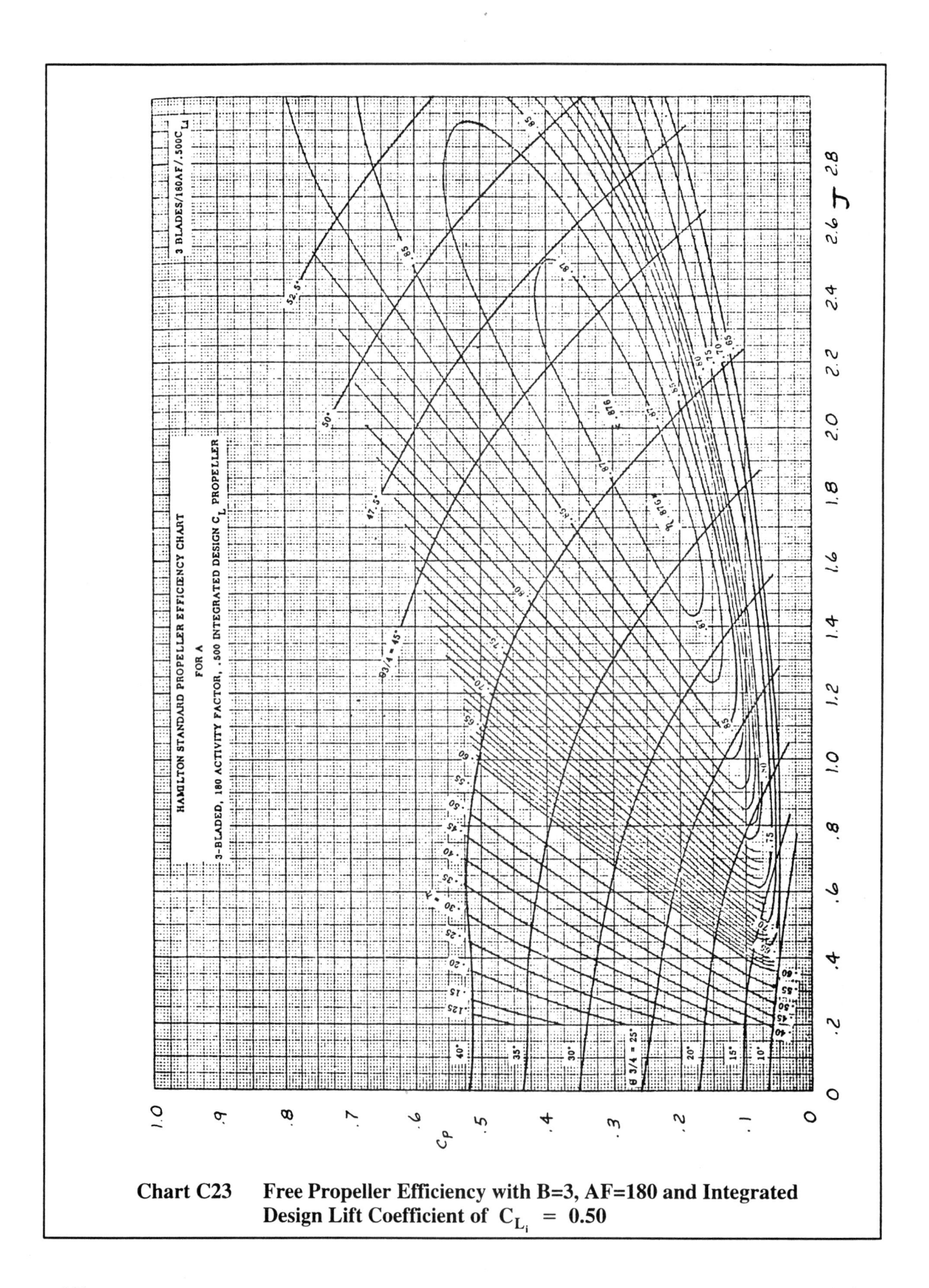

Chart C23 Free Propeller Efficiency with B=3, AF=180 and Integrated Design Lift Coefficient of $C_{L_i} = 0.50$

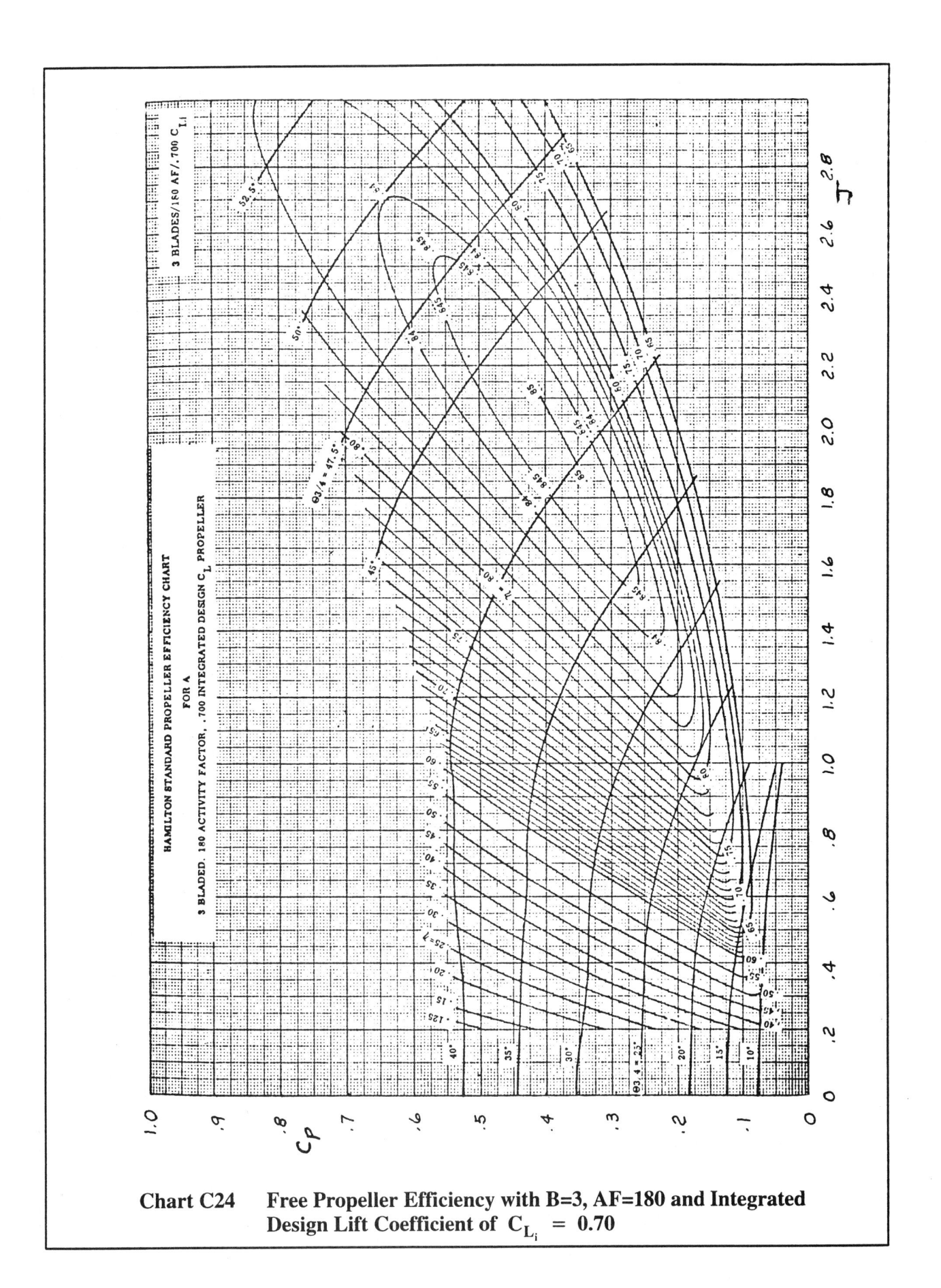

Chart C24 Free Propeller Efficiency with B=3, AF=180 and Integrated Design Lift Coefficient of $C_{L_i} = 0.70$

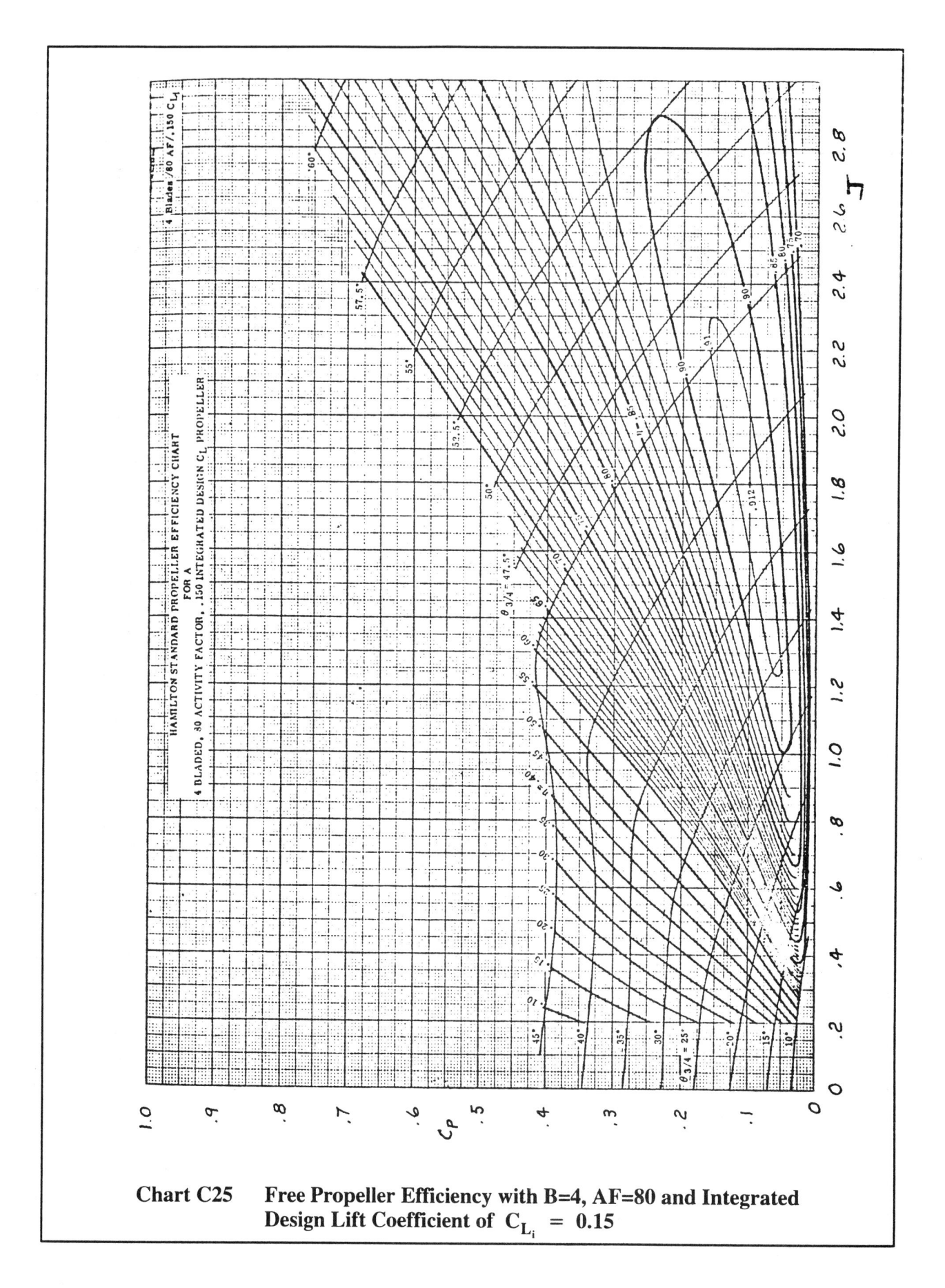

Chart C25   Free Propeller Efficiency with B=4, AF=80 and Integrated Design Lift Coefficient of $C_{L_i} = 0.15$

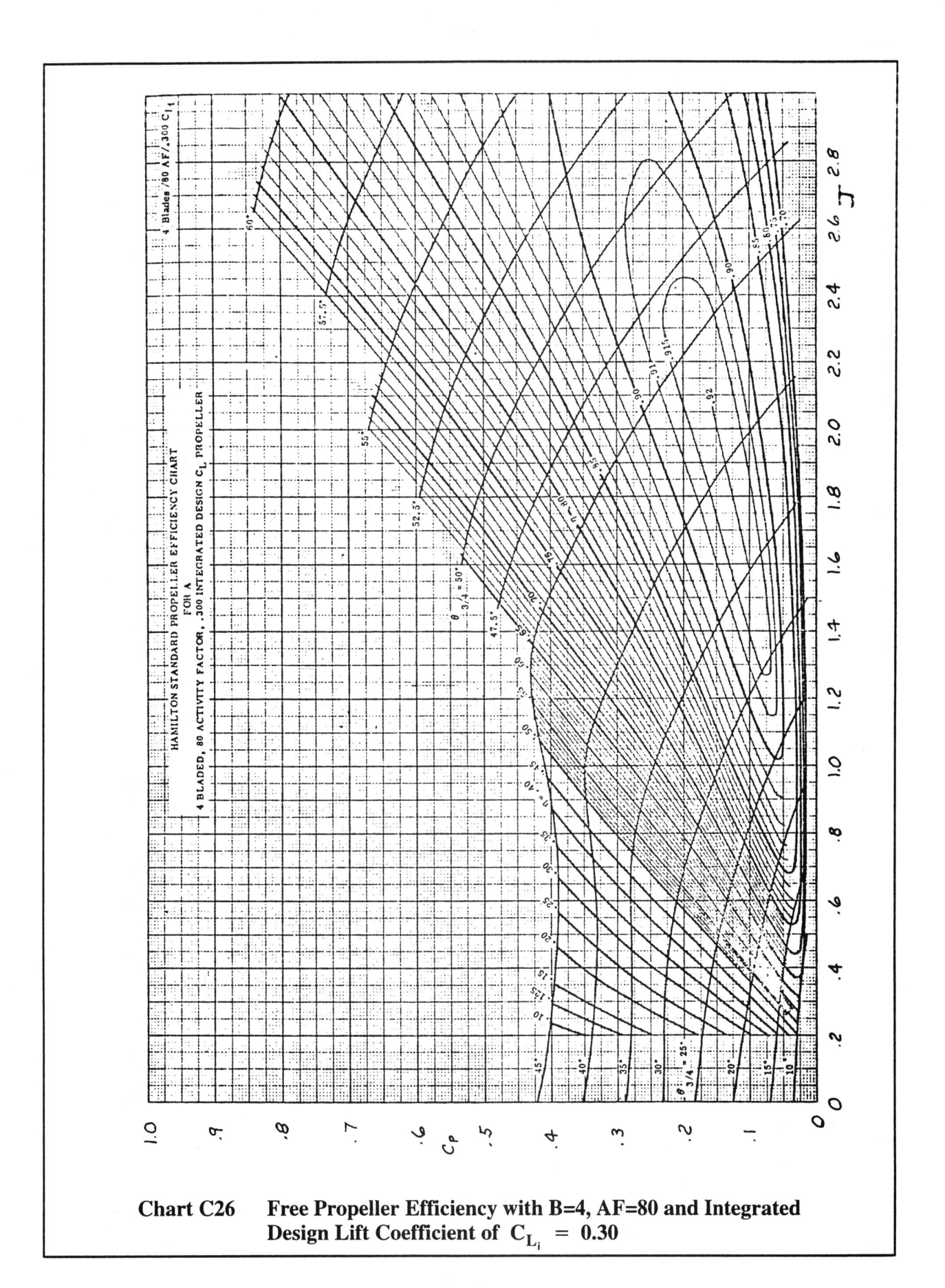

Chart C26 Free Propeller Efficiency with B=4, AF=80 and Integrated Design Lift Coefficient of $C_{L_i} = 0.30$

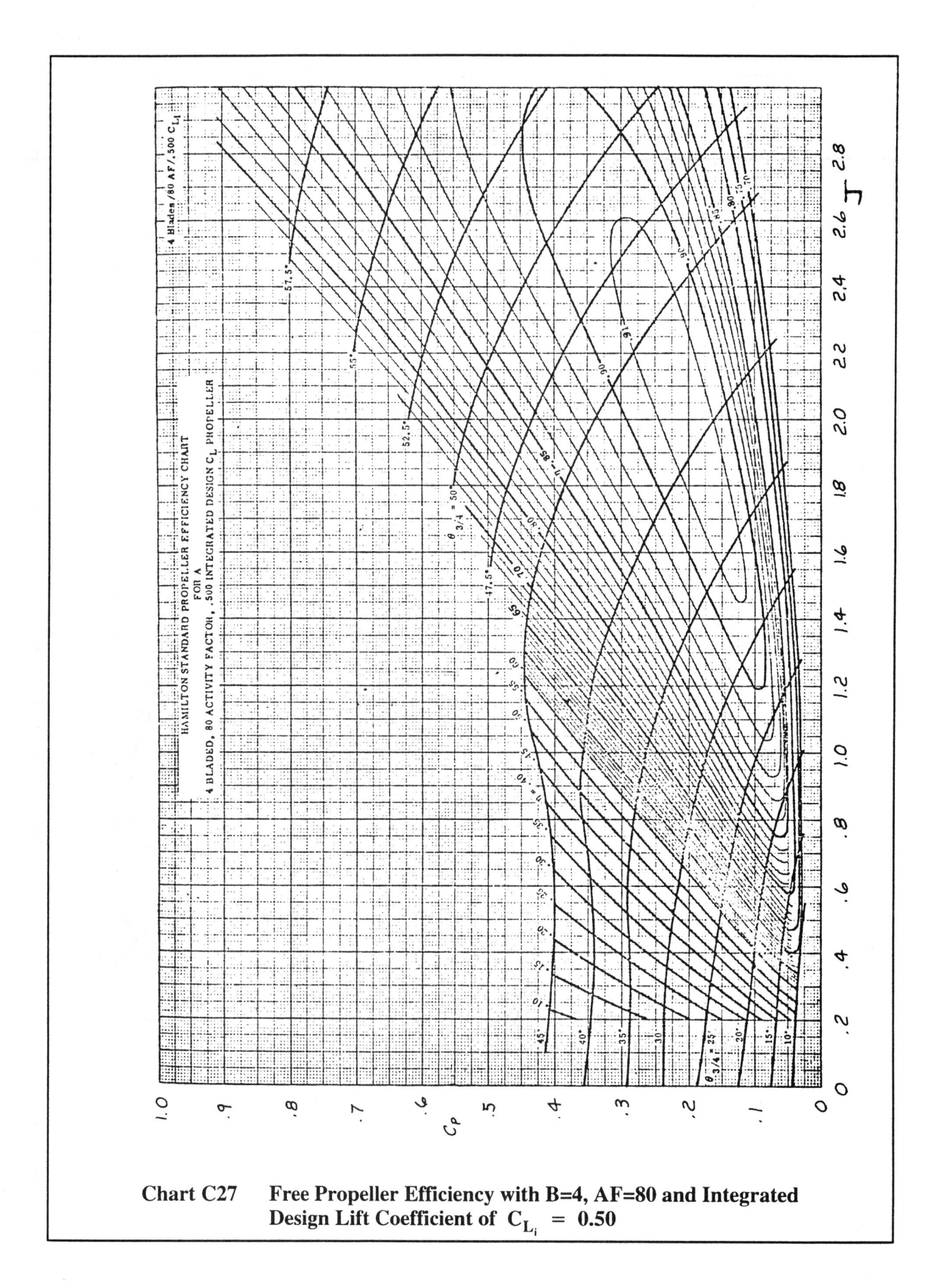

Chart C27 Free Propeller Efficiency with B=4, AF=80 and Integrated Design Lift Coefficient of $C_{L_i} = 0.50$

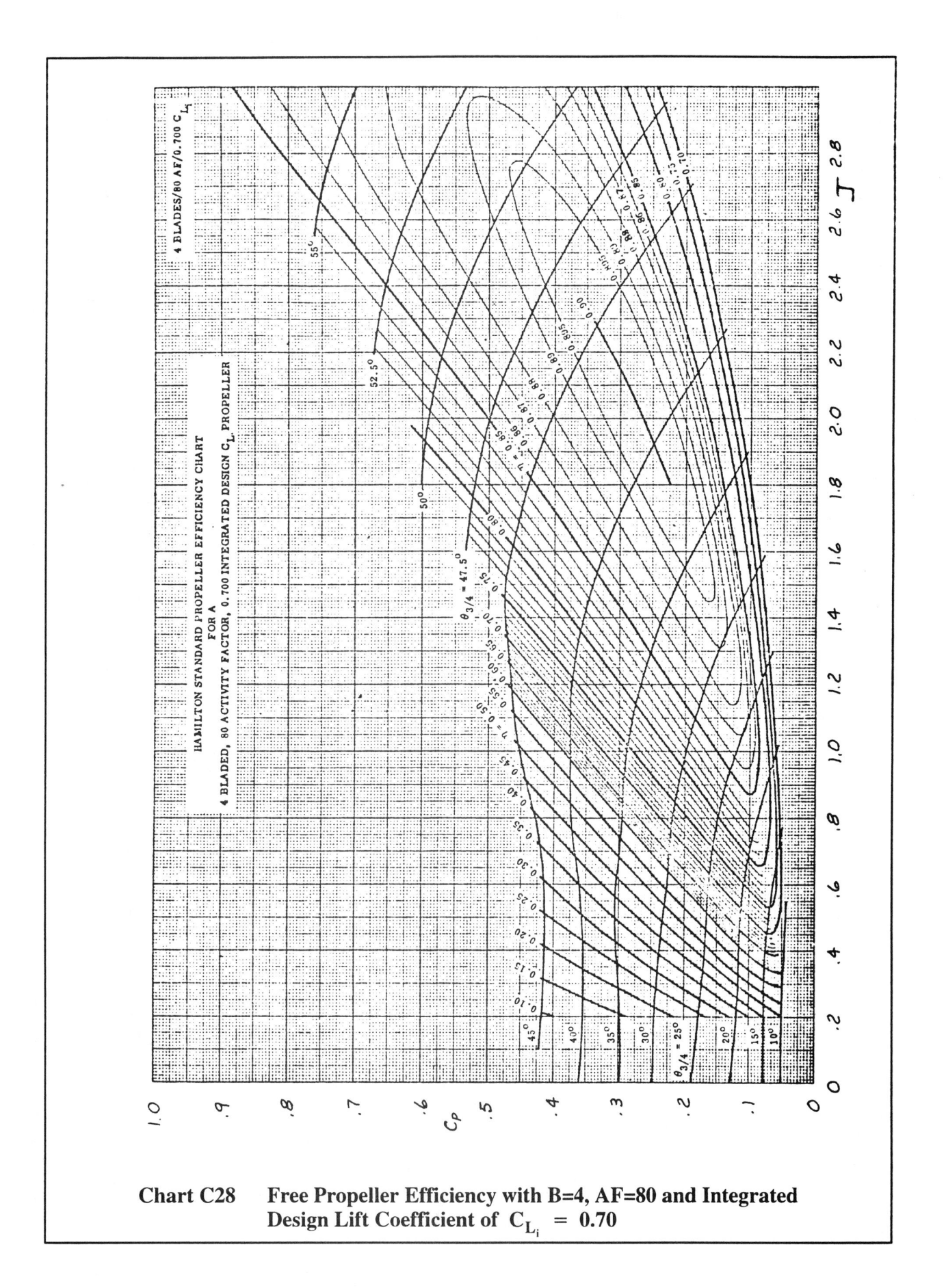

Chart C28 Free Propeller Efficiency with B=4, AF=80 and Integrated Design Lift Coefficient of $C_{L_i} = 0.70$

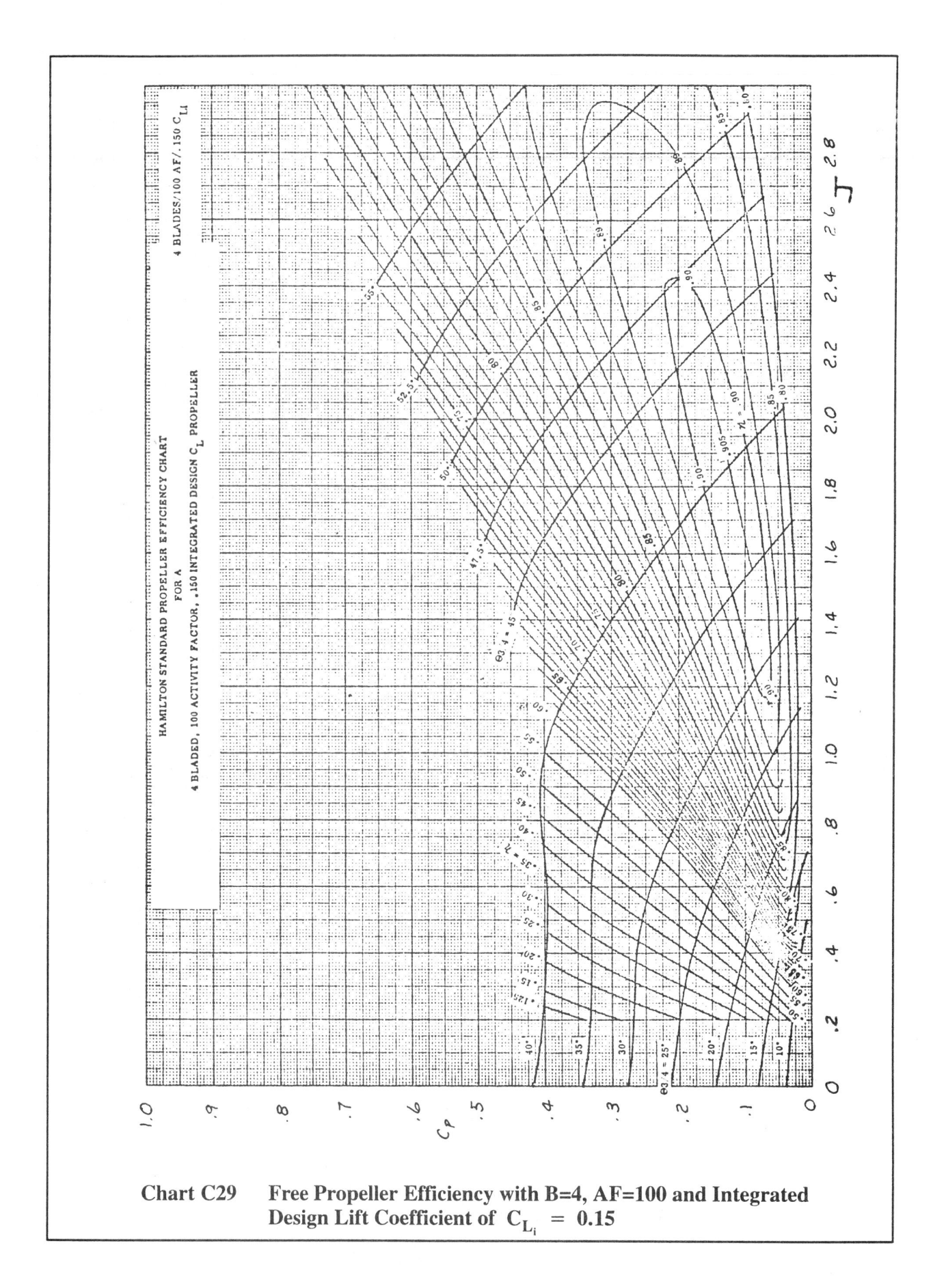

Chart C29 Free Propeller Efficiency with B=4, AF=100 and Integrated Design Lift Coefficient of $C_{L_i} = 0.15$

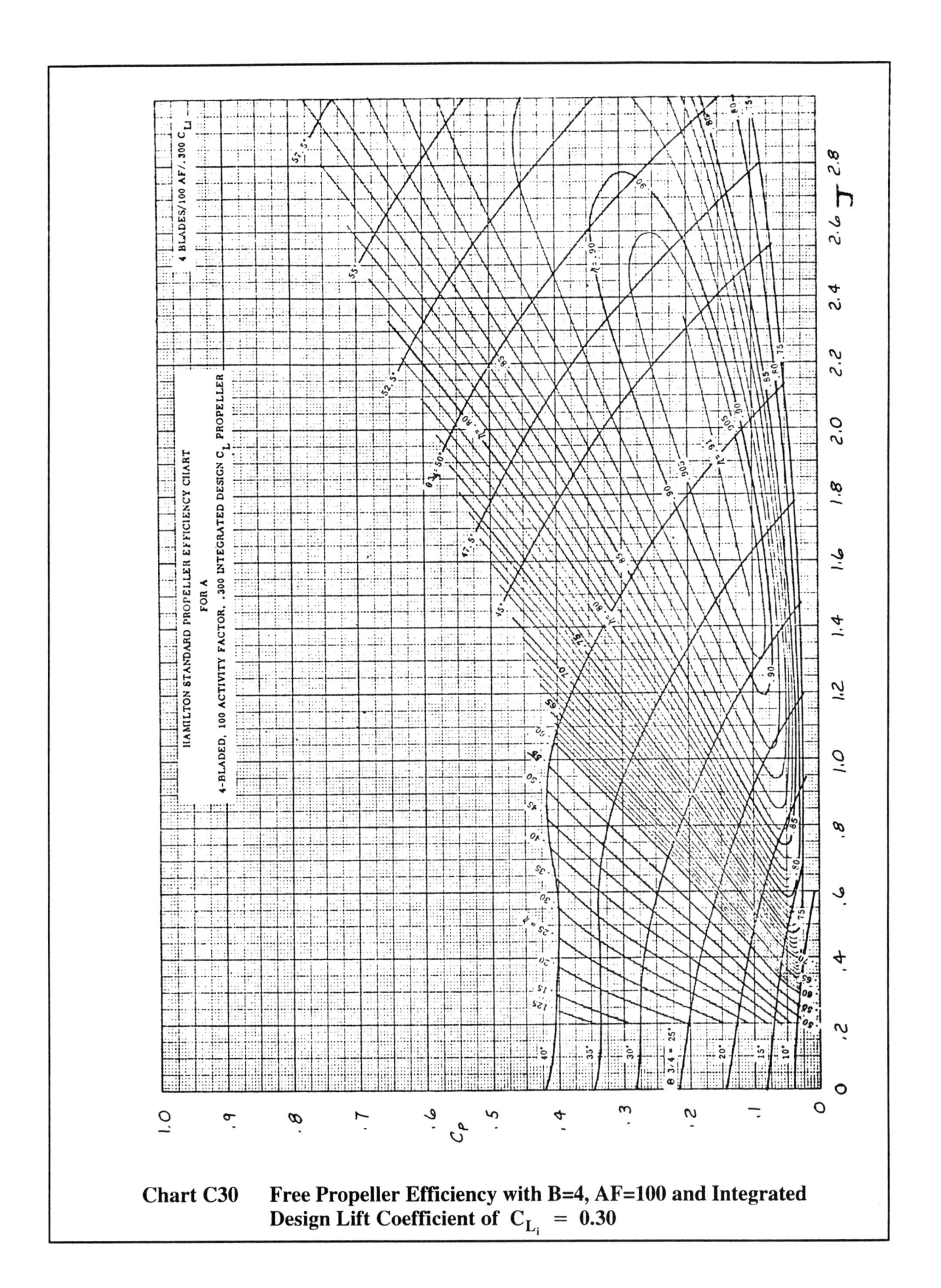

Chart C30   Free Propeller Efficiency with B=4, AF=100 and Integrated Design Lift Coefficient of $C_{L_i} = 0.30$

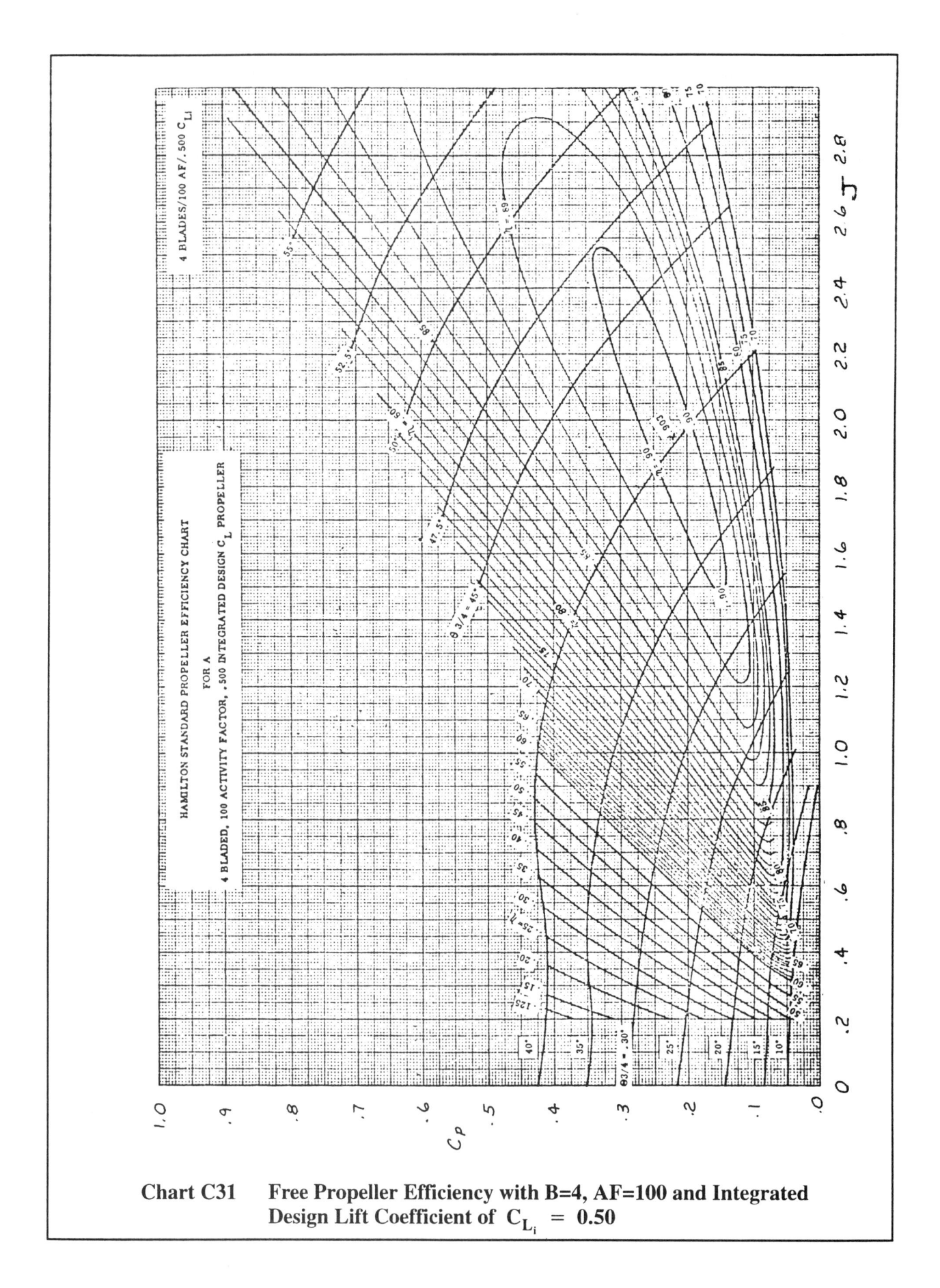

Propeller Performance Charts

Chart C31 Free Propeller Efficiency with B=4, AF=100 and Integrated Design Lift Coefficient of $C_{L_i} = 0.50$

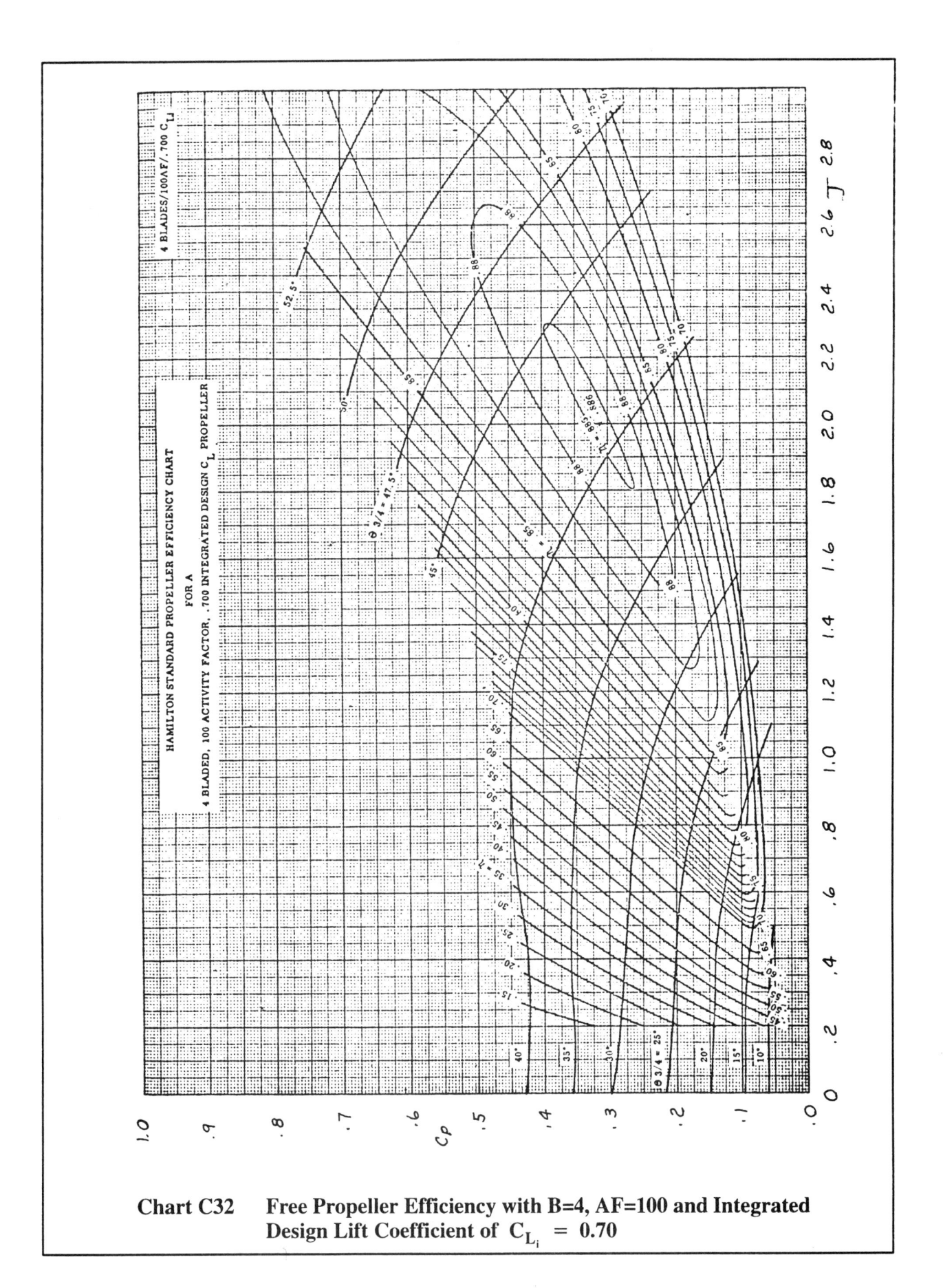

Chart C32 Free Propeller Efficiency with B=4, AF=100 and Integrated Design Lift Coefficient of $C_{L_i} = 0.70$

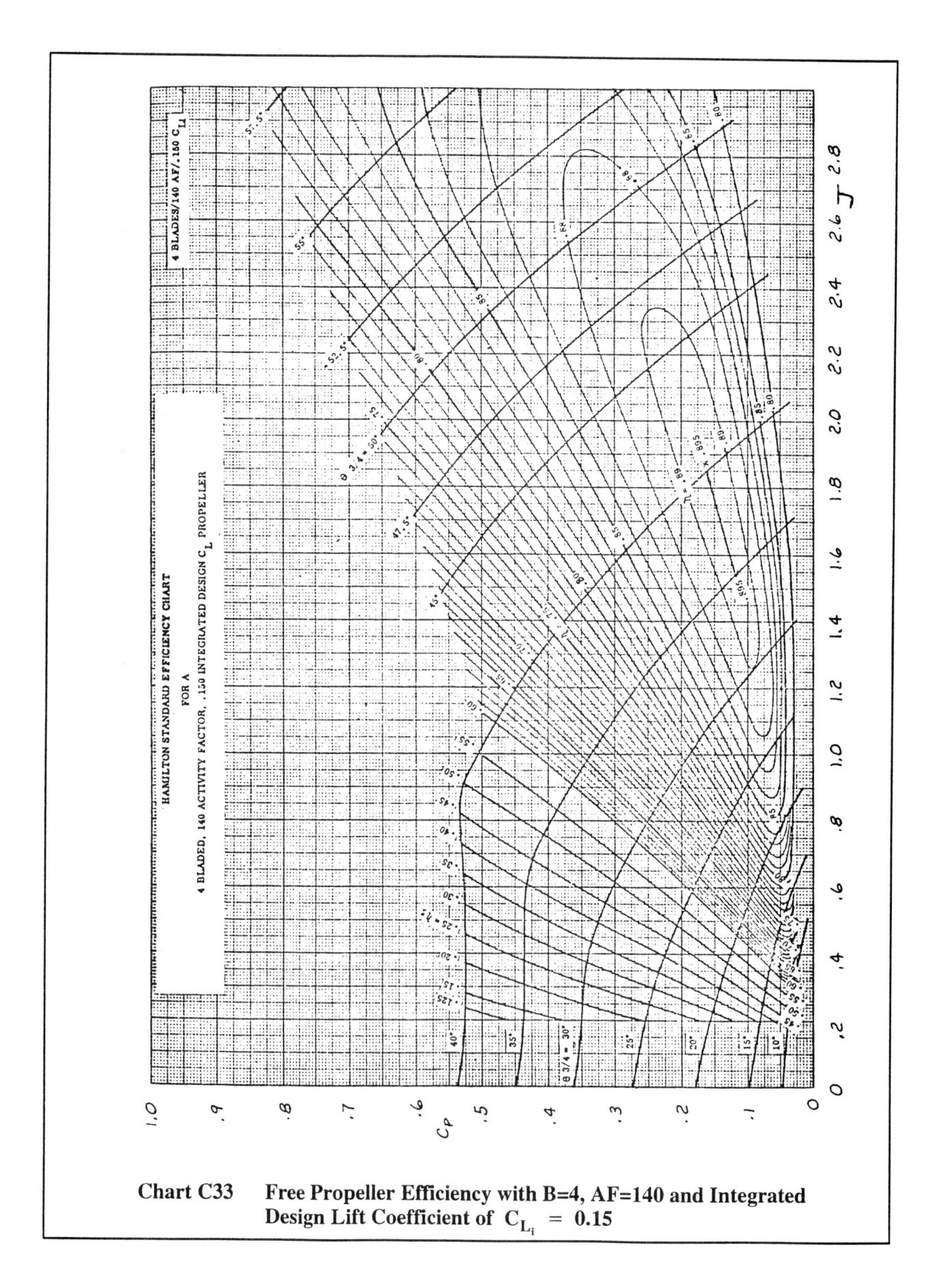

Chart C33 Free Propeller Efficiency with B=4, AF=140 and Integrated Design Lift Coefficient of $C_{L_i} = 0.15$

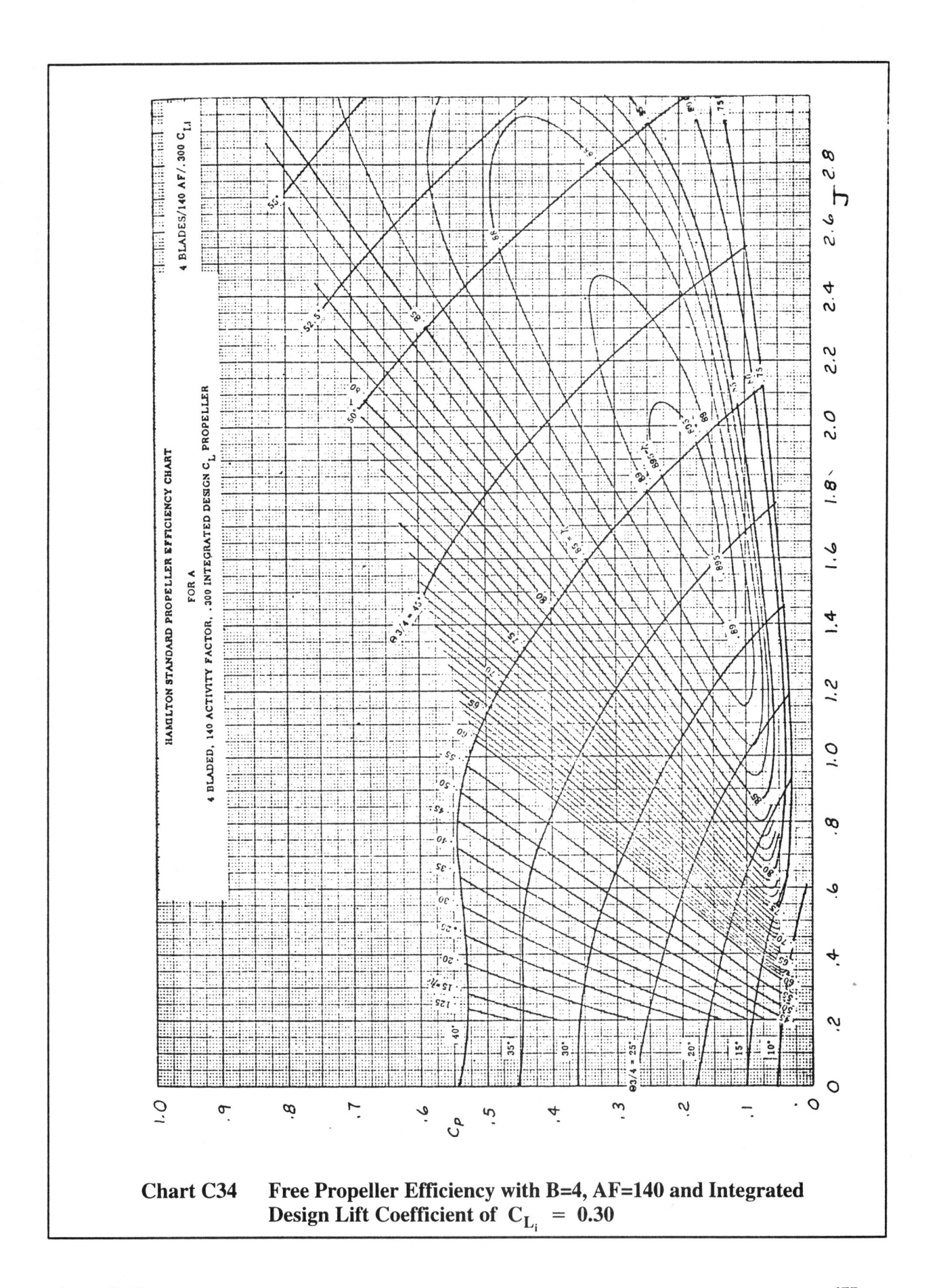

Propeller Performance Charts

Chart C34 Free Propeller Efficiency with B=4, AF=140 and Integrated Design Lift Coefficient of $C_{L_i} = 0.30$

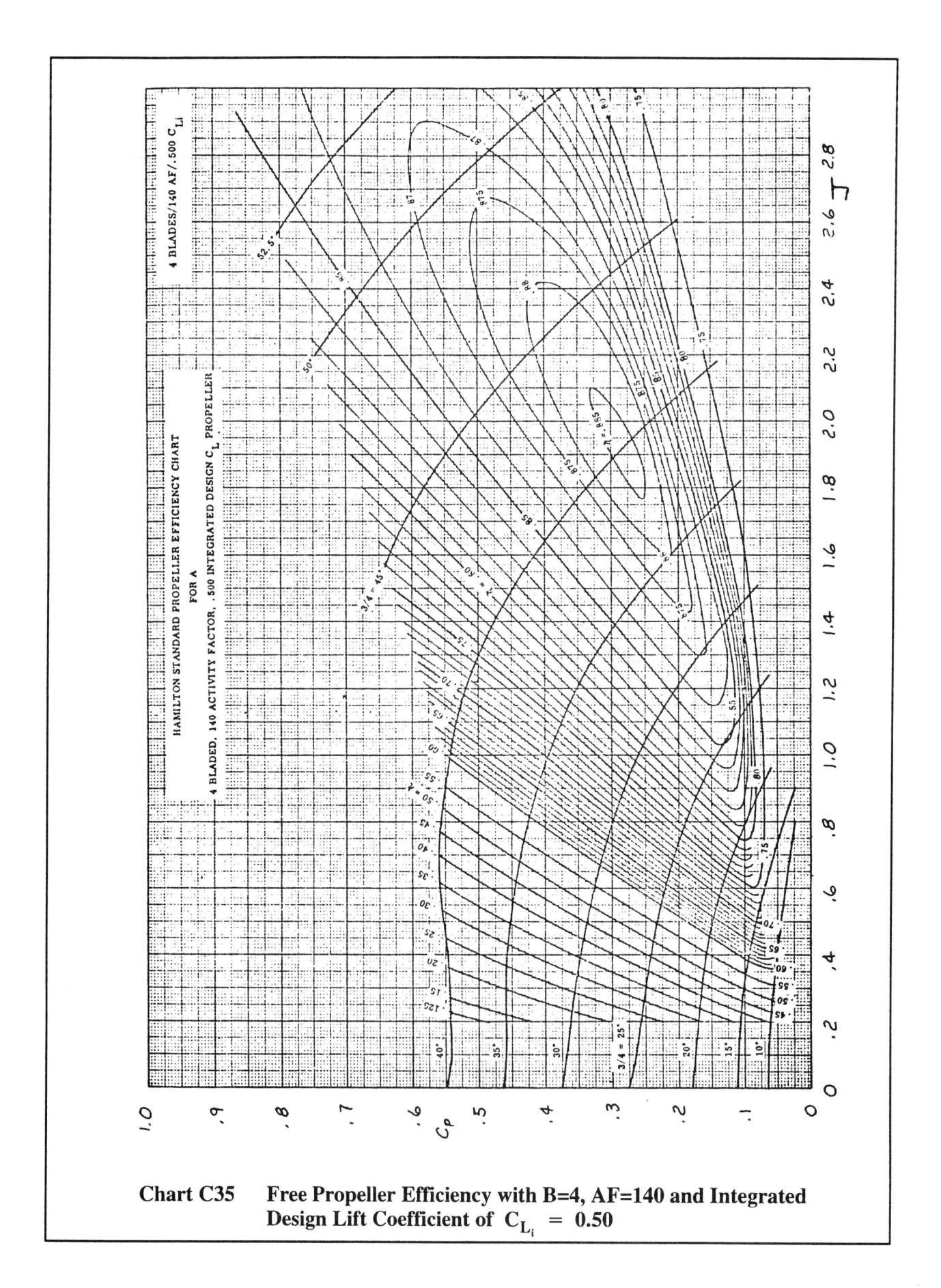

Chart C35   Free Propeller Efficiency with B=4, AF=140 and Integrated Design Lift Coefficient of $C_{L_i} = 0.50$

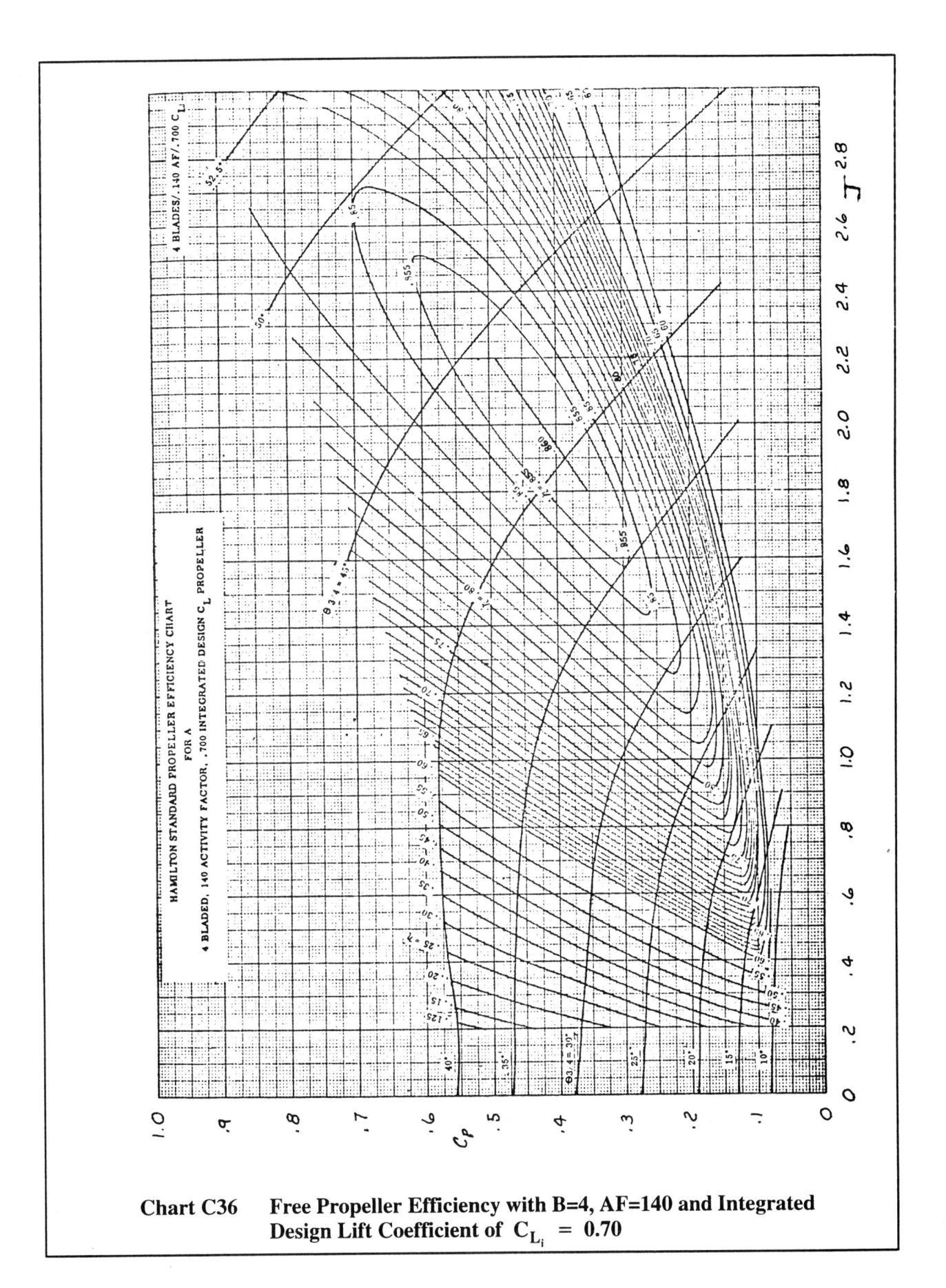

Chart C36 Free Propeller Efficiency with B=4, AF=140 and Integrated Design Lift Coefficient of $C_{L_i} = 0.70$

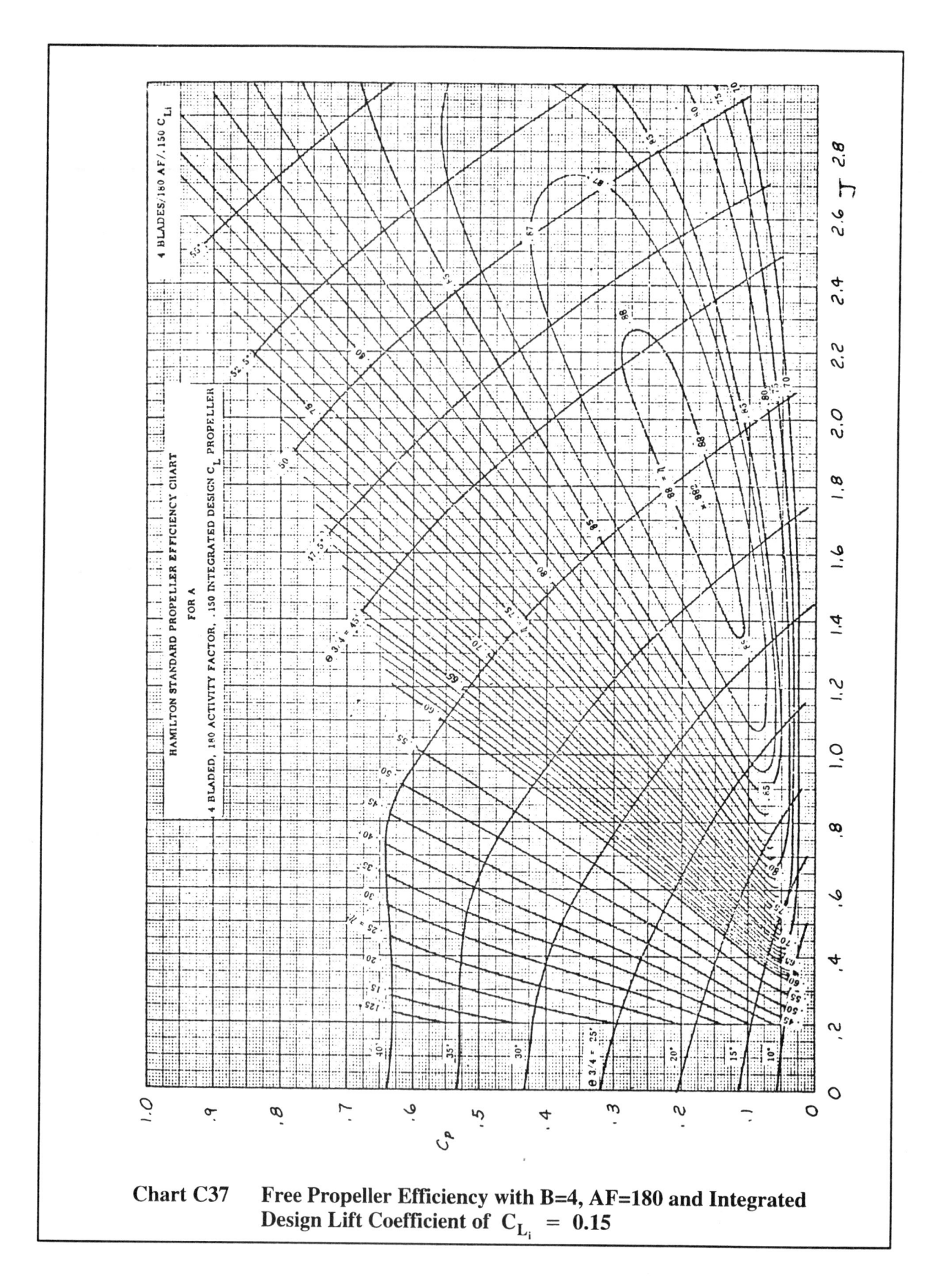

Chart C37 Free Propeller Efficiency with B=4, AF=180 and Integrated Design Lift Coefficient of $C_{L_i} = 0.15$

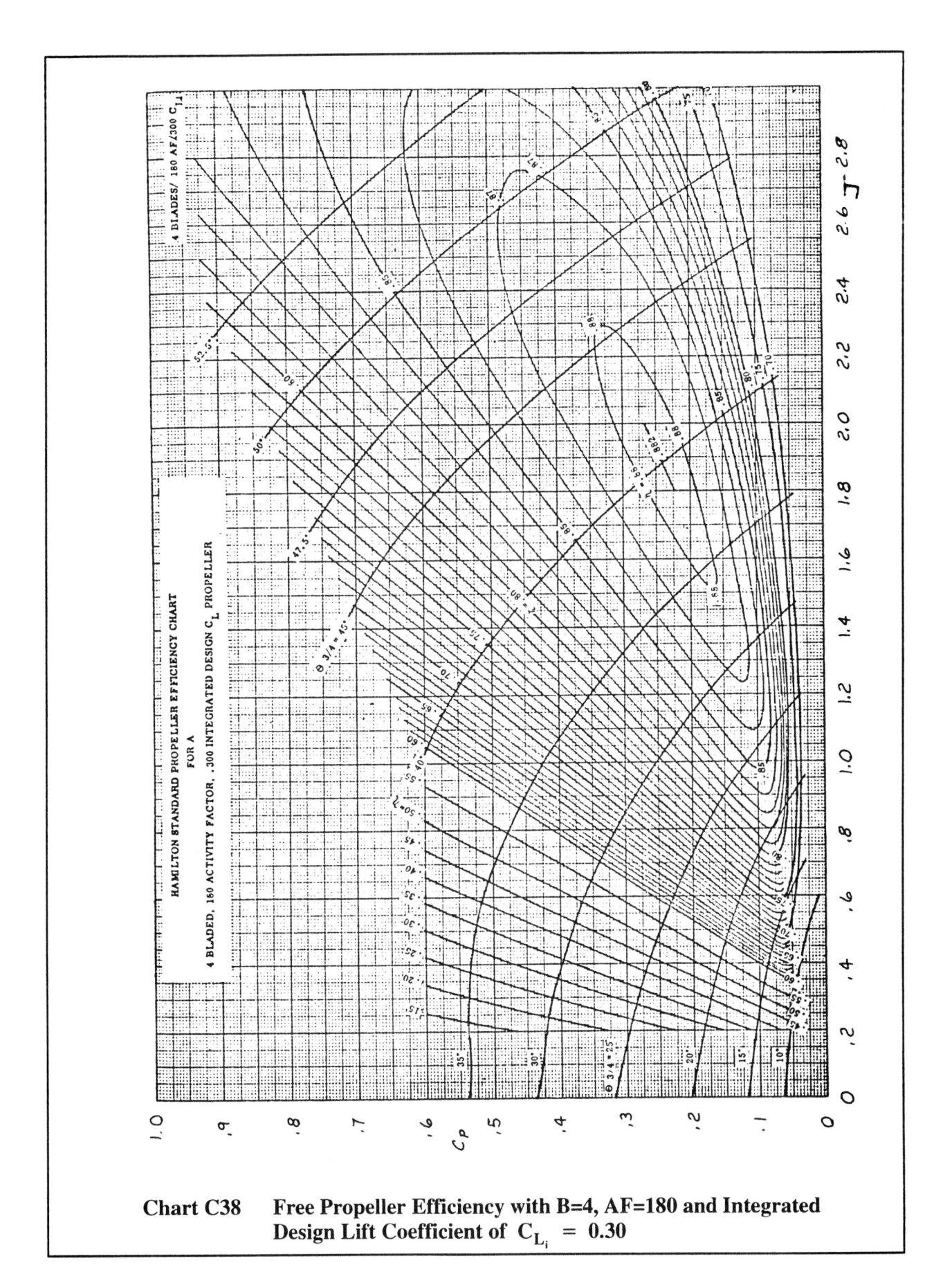

Chart C38 Free Propeller Efficiency with B=4, AF=180 and Integrated Design Lift Coefficient of $C_{L_i} = 0.30$

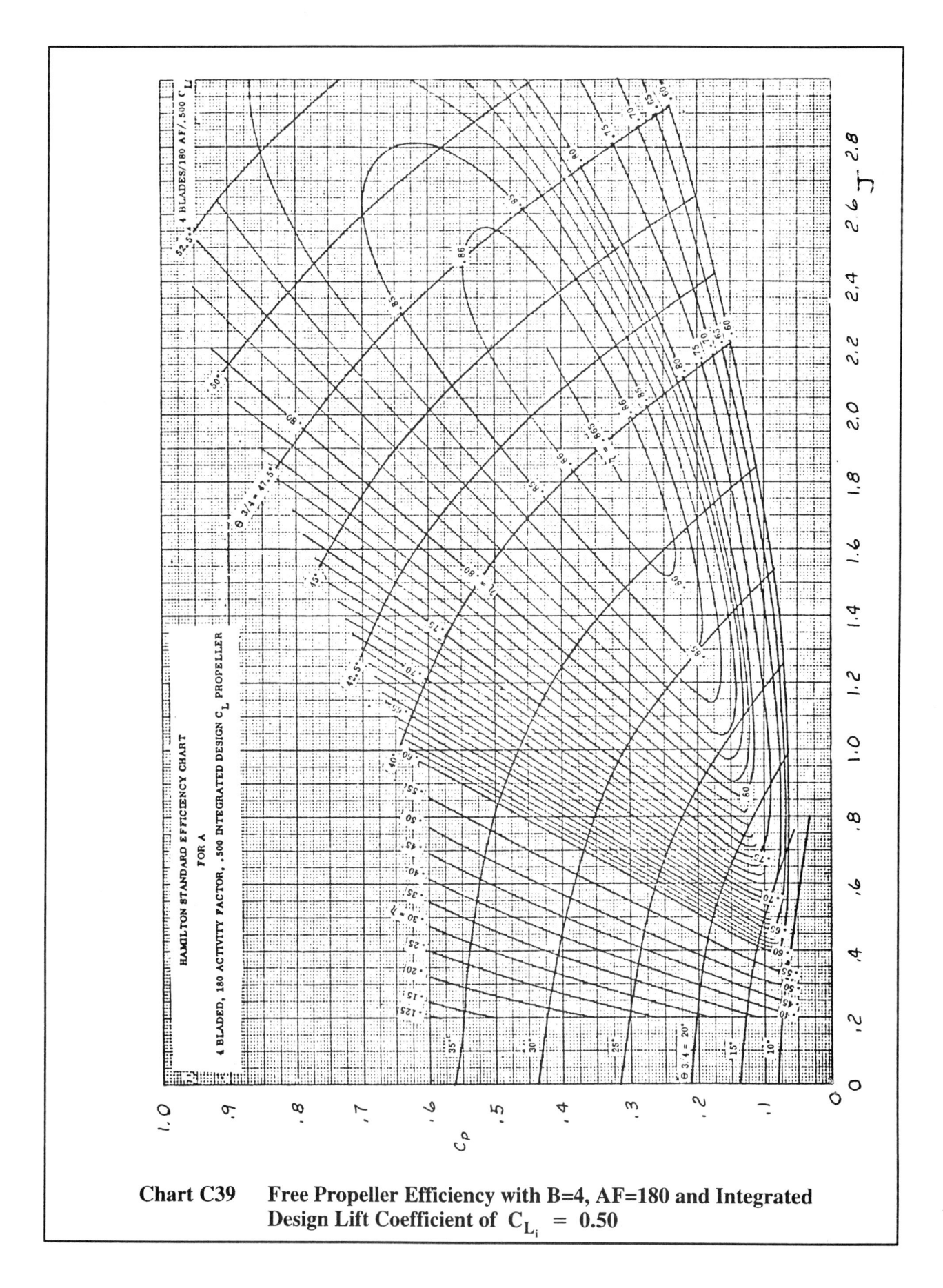

Chart C39 Free Propeller Efficiency with B=4, AF=180 and Integrated Design Lift Coefficient of $C_{L_i} = 0.50$

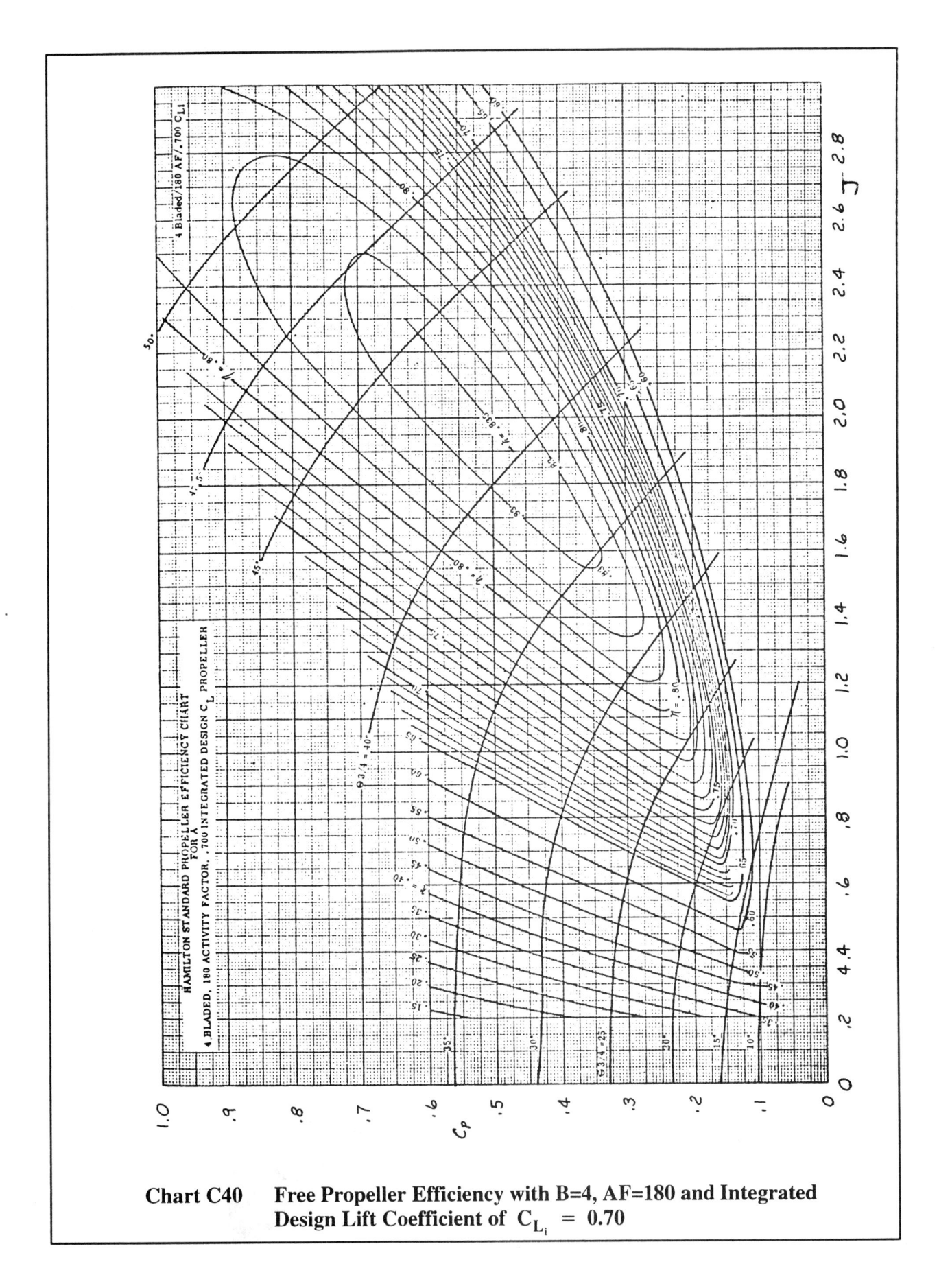

Propeller Performance Charts

Chart C40 Free Propeller Efficiency with B=4, AF=180 and Integrated Design Lift Coefficient of $C_{L_i} = 0.70$

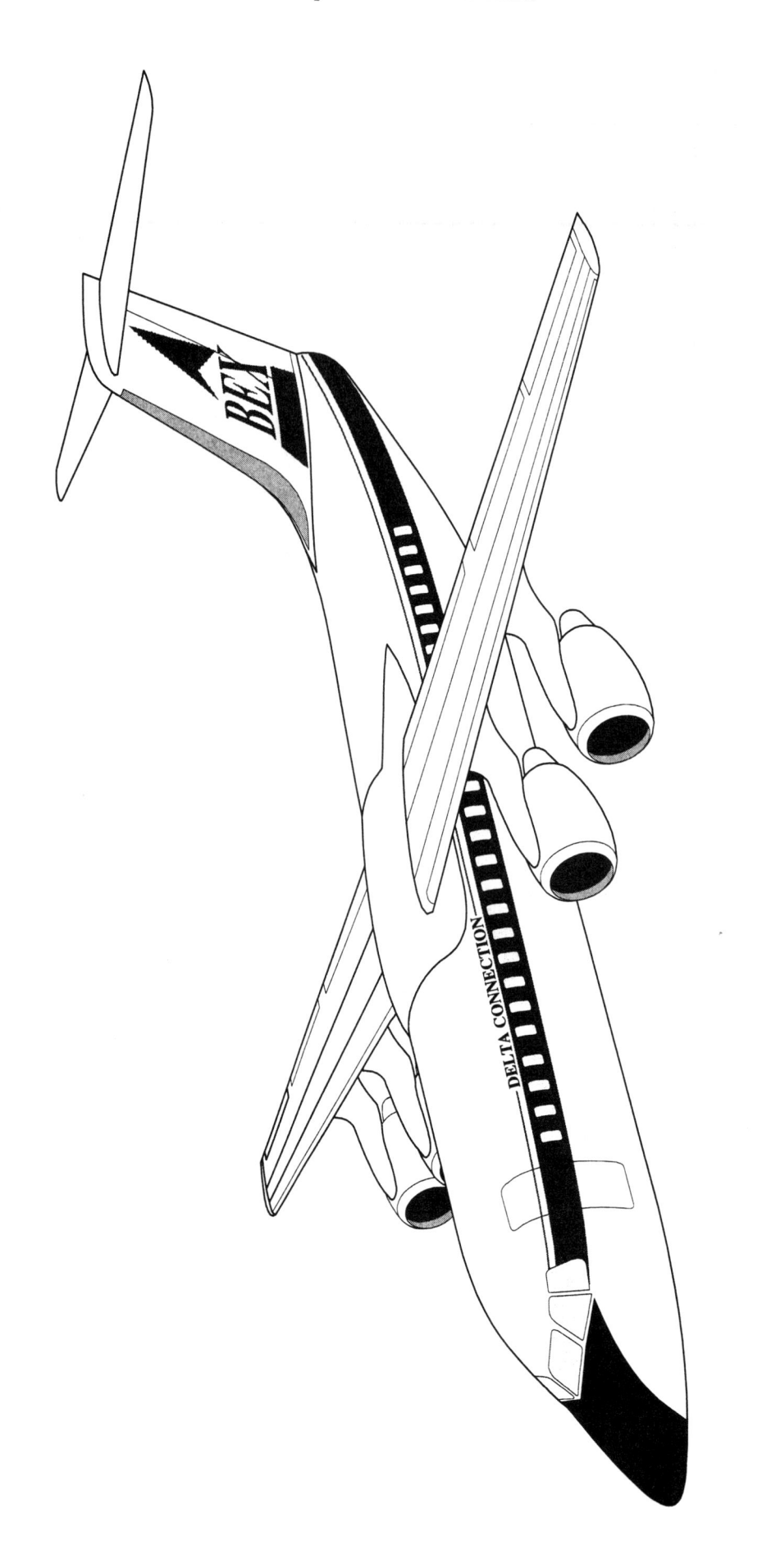

Propeller Performance Charts

APPENDIX D: PROPELLER NOISE CHARTS

In this appendix, propeller noise charts are presented as follows:

Chart D1   Far Field Partial Noise Level Based on Power and Tip Speed

Chart D2   Far Field Partial Noise Level Based on Blade Count and Propeller Diameter

Chart D3   Atmospheric Absorption and Spherical Spreading of Sound

Chart D4   Directivity Index

Chart D5   Perceived Noise Level Adjustment – 2 Bladed Propellers

Chart D6   Perceived Noise Level Adjustment – 3 Bladed Propellers

Chart D7   Perceived Noise Level Adjustment – 4 Bladed Propellers

Chart D8   Perceived Noise Level Adjustment – 6 and 8 Bladed Propellers

Chart D9   Near Field Partial Noise Level Based on Power and Propeller Diameter

Chart D10 Near Field Partial Noise Level Based on Tip Speed and Tip Clearance

Chart D11 Variation of Overall, Free–Space Propeller Noise Levels with Axial Position, x/D Fore and Aft of the Propeller Plane

Chart D12 Effect of Reflecting Surfaces in the Pressure Field

Chart D13 Near Field Harmonic Noise Level Distribution

The data in this appendix are presented through the courtesy of Hamilton–Standard.

Propeller Noise Charts

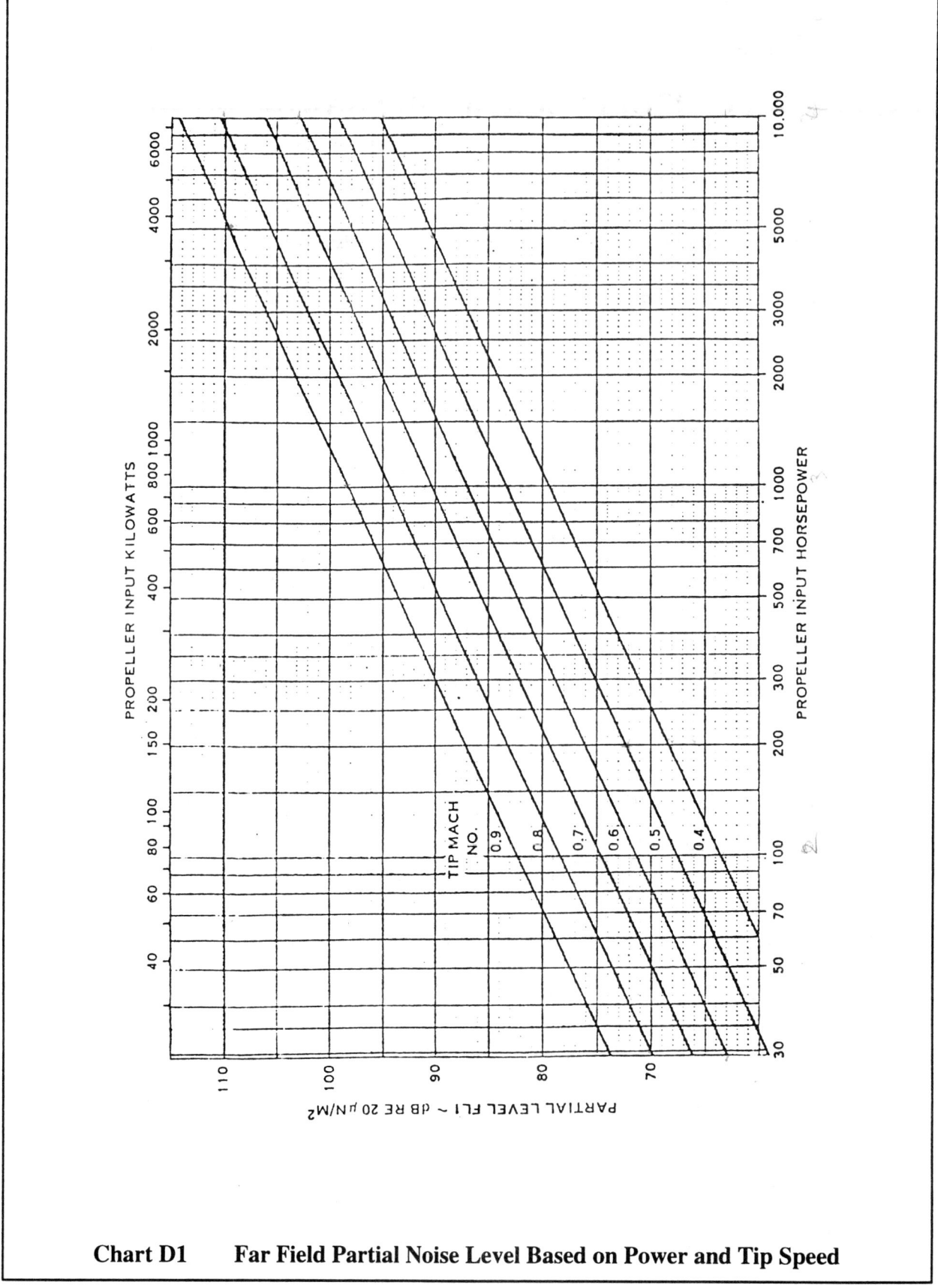

Chart D1   Far Field Partial Noise Level Based on Power and Tip Speed

Propeller Noise Charts

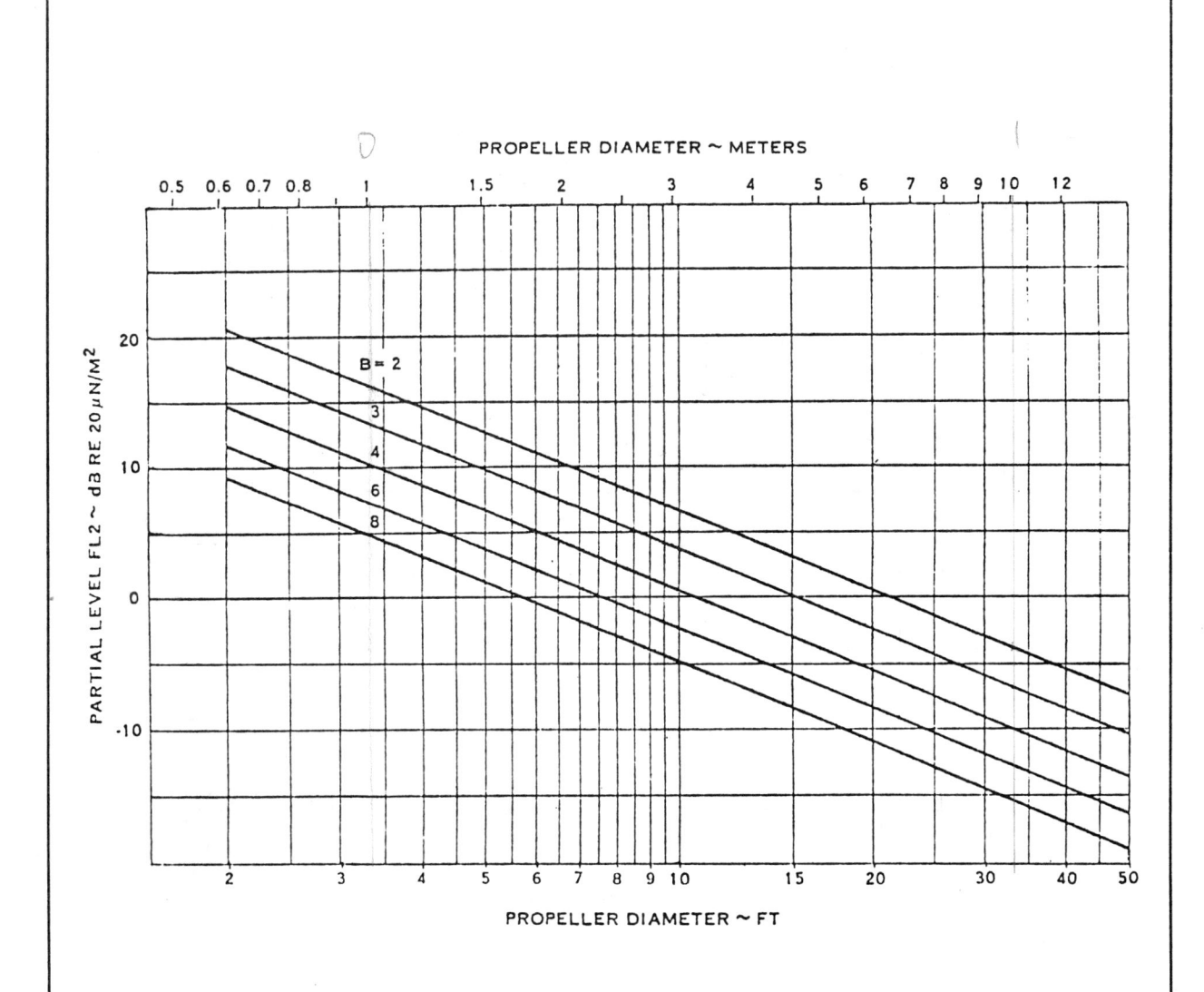

Chart D2 Far Field Partial Noise Level Based on Blade Count and Propeller Diameter

Propeller Noise Charts

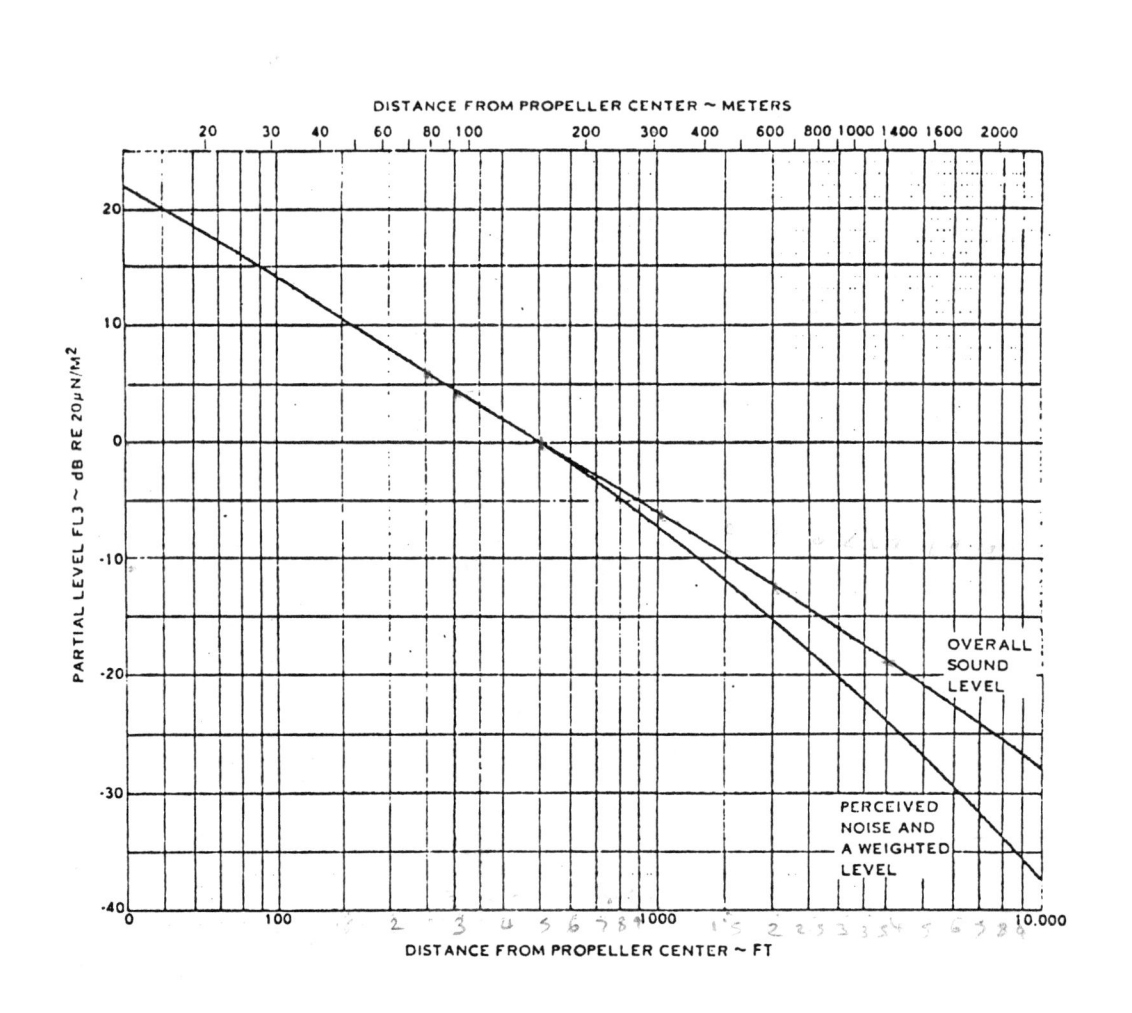

Chart D3   Atmospheric Absorption and Spherical Spreading of Sound

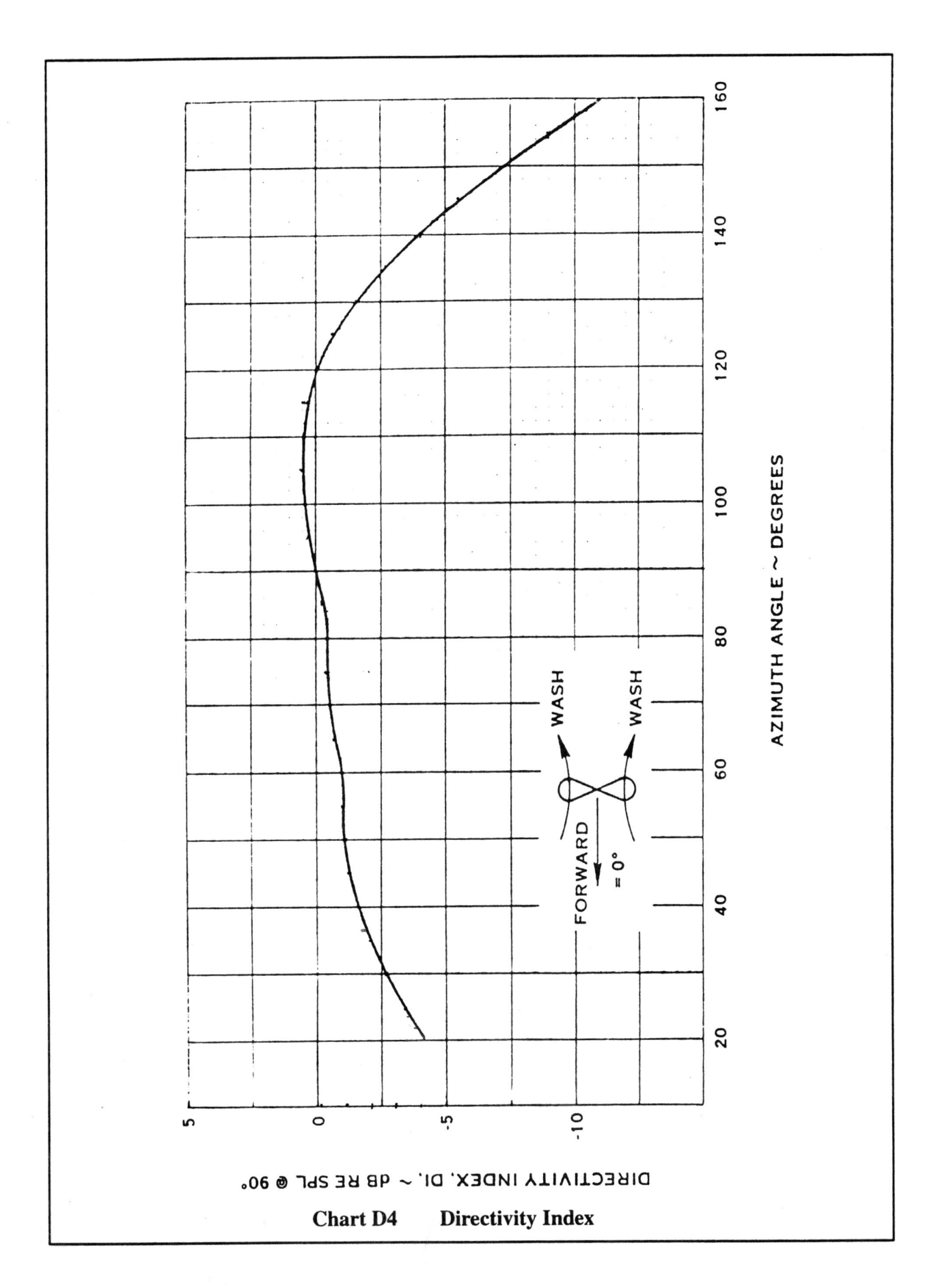

Propeller Noise Charts

Chart D4   Directivity Index

Appendix D

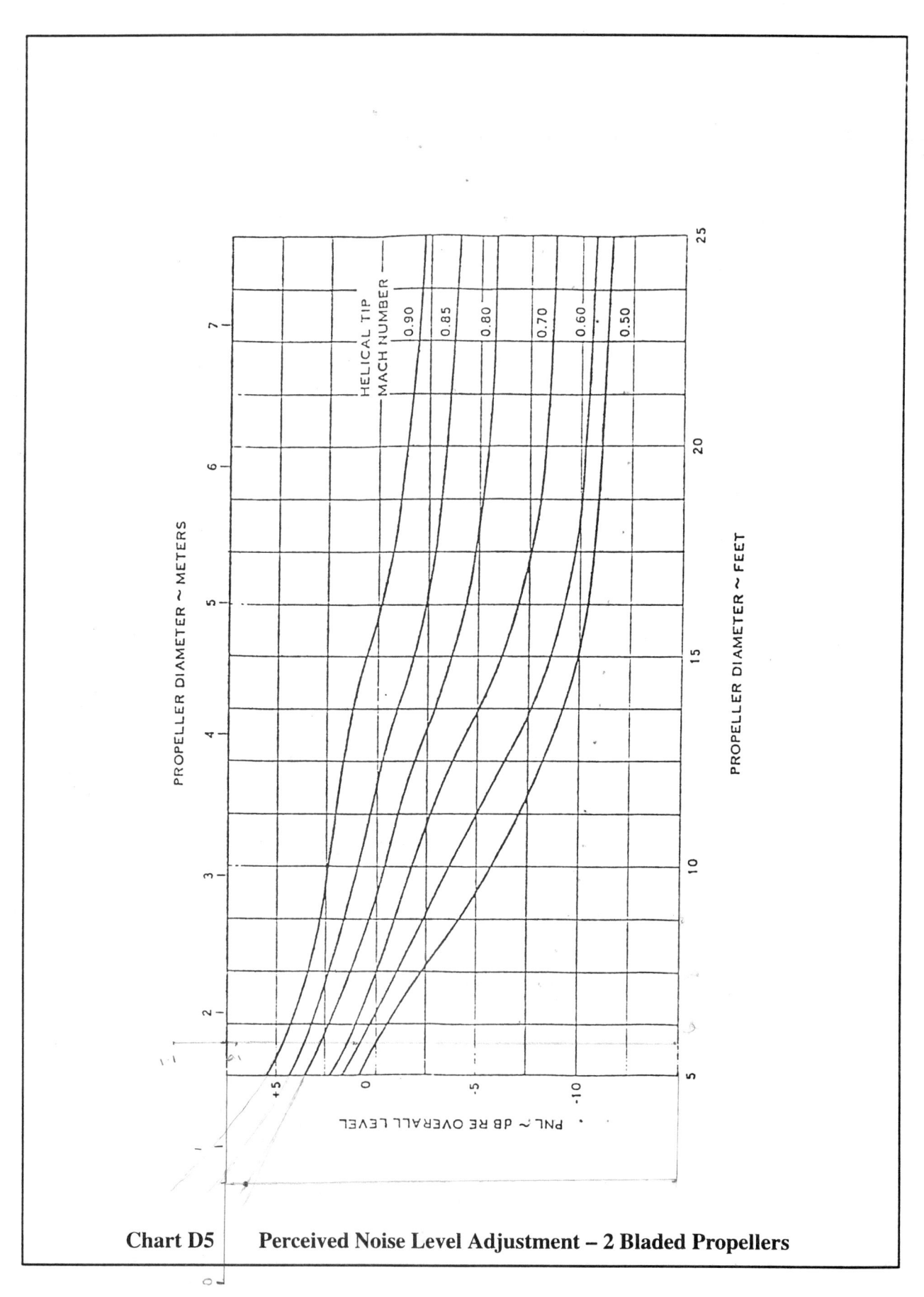

Chart D5 Perceived Noise Level Adjustment – 2 Bladed Propellers

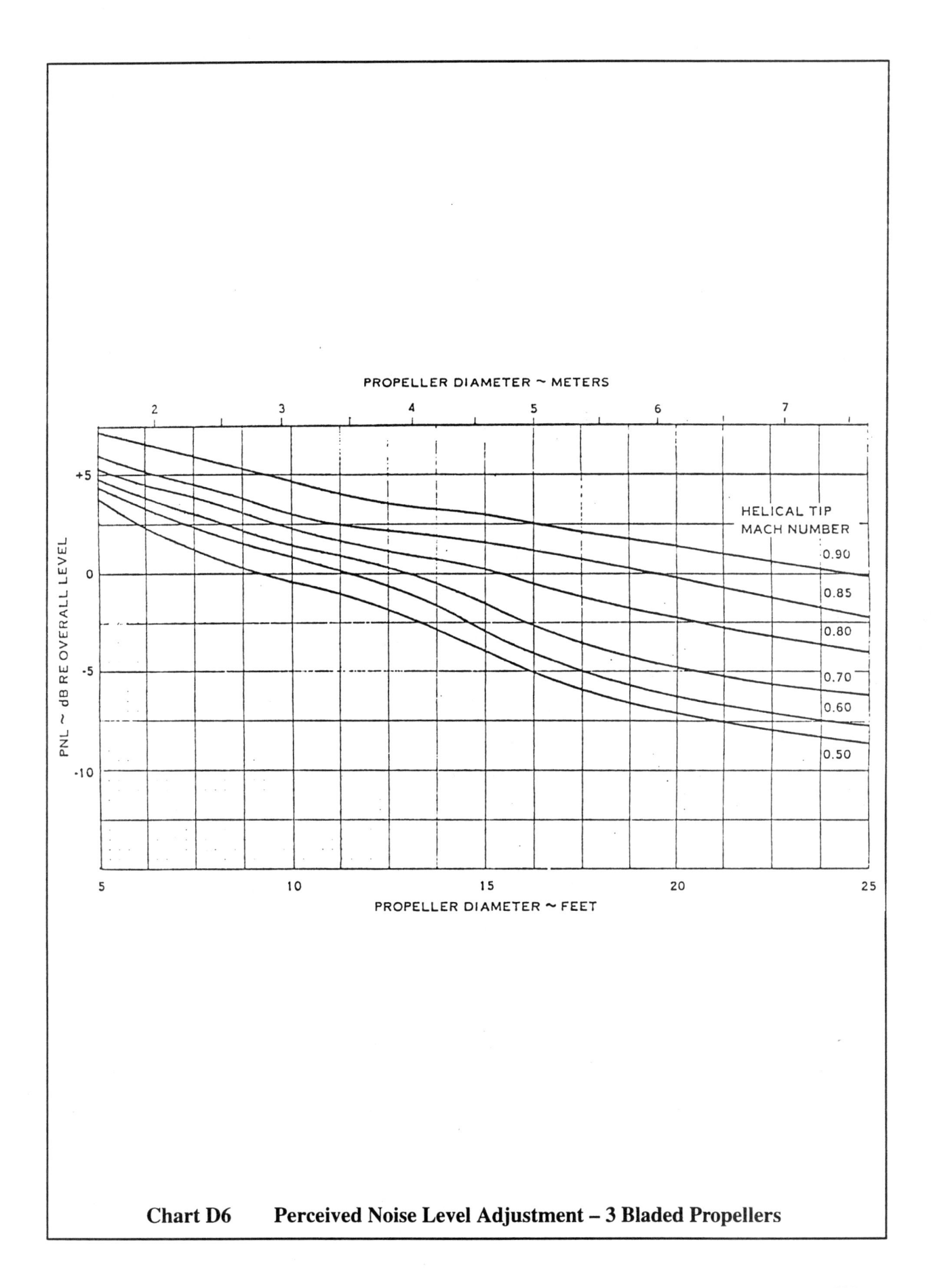

Chart D6   **Perceived Noise Level Adjustment – 3 Bladed Propellers**

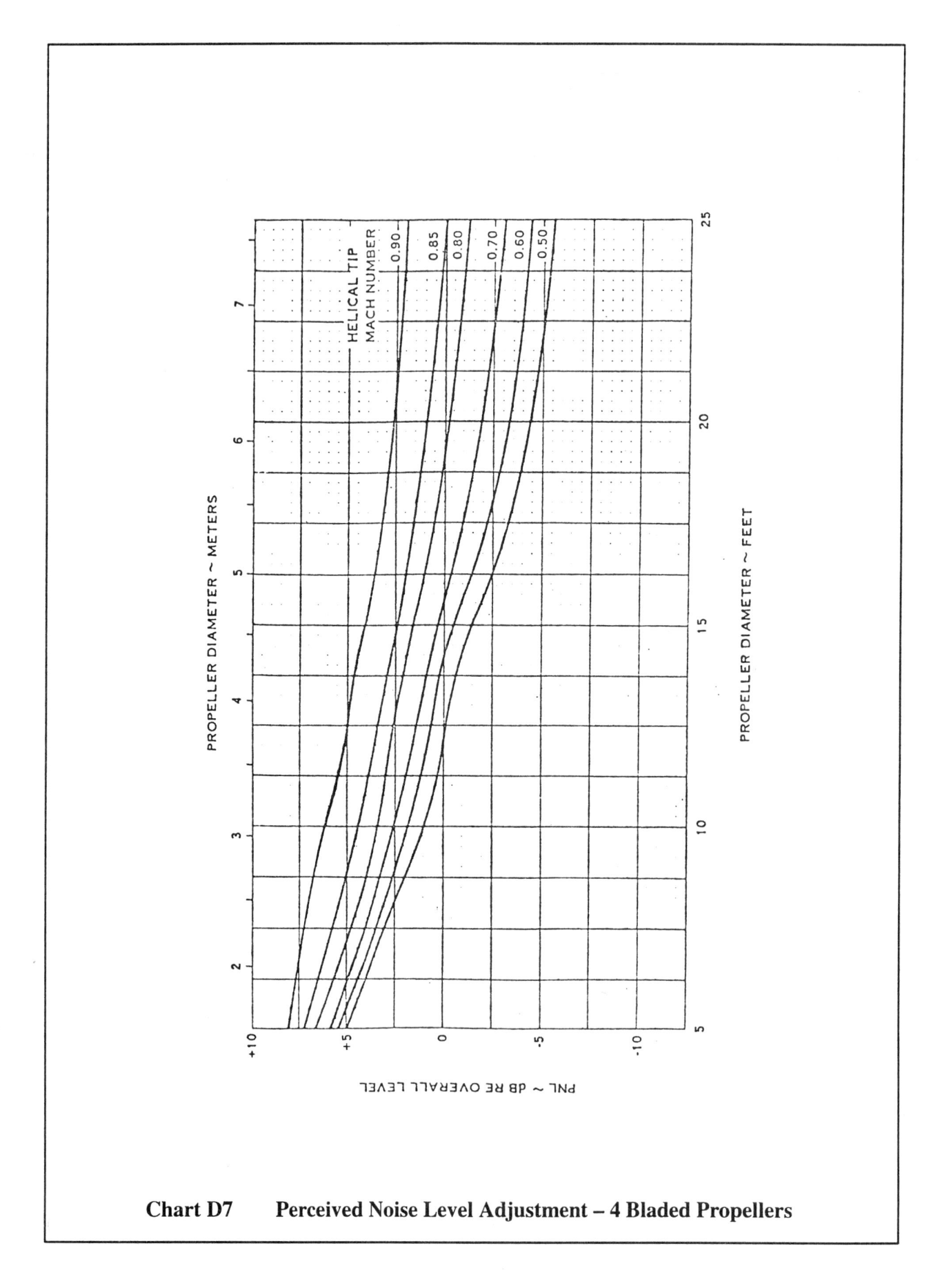

Chart D7   Perceived Noise Level Adjustment – 4 Bladed Propellers

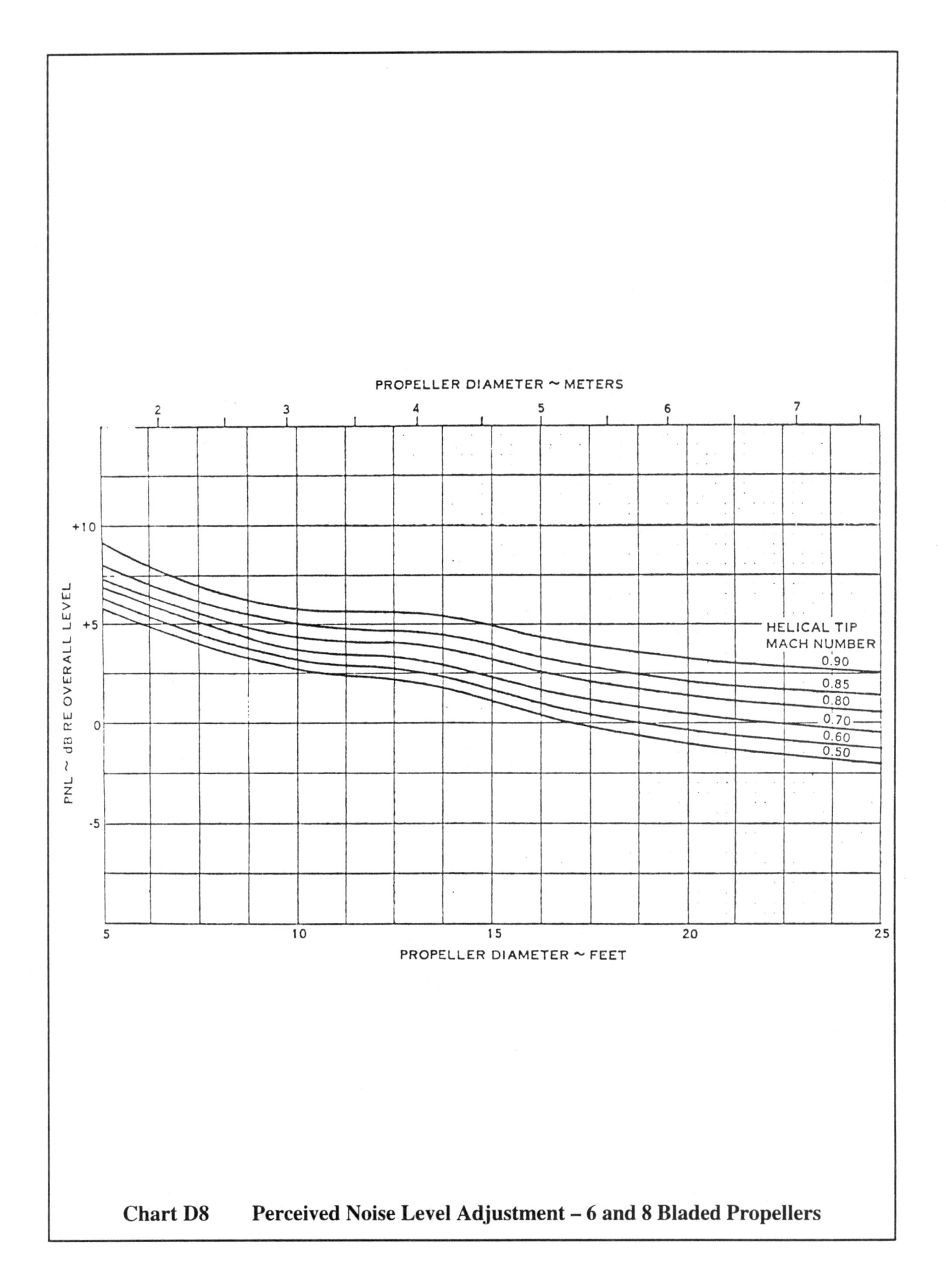

Chart D8   Perceived Noise Level Adjustment – 6 and 8 Bladed Propellers

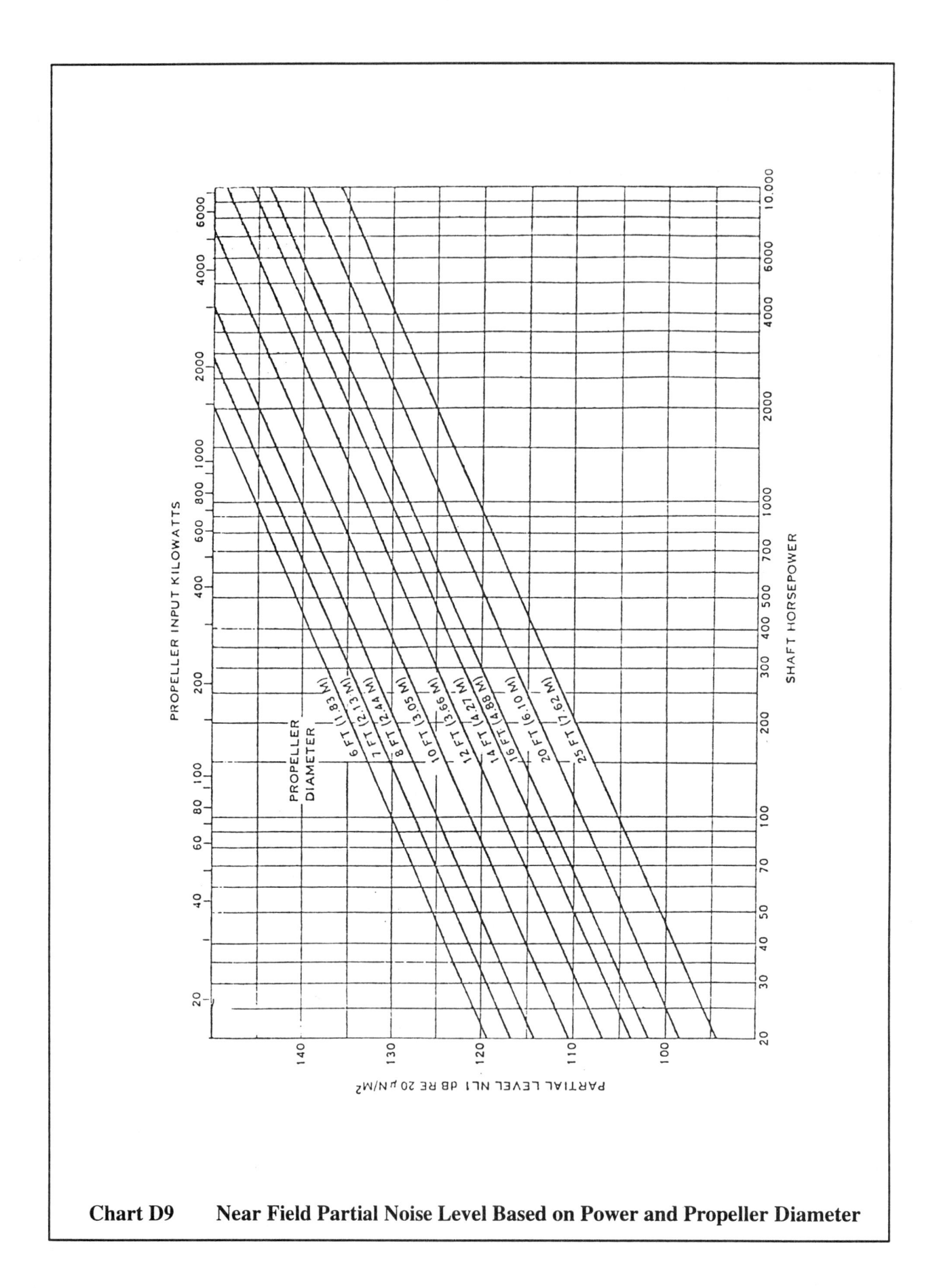

Chart D9   Near Field Partial Noise Level Based on Power and Propeller Diameter

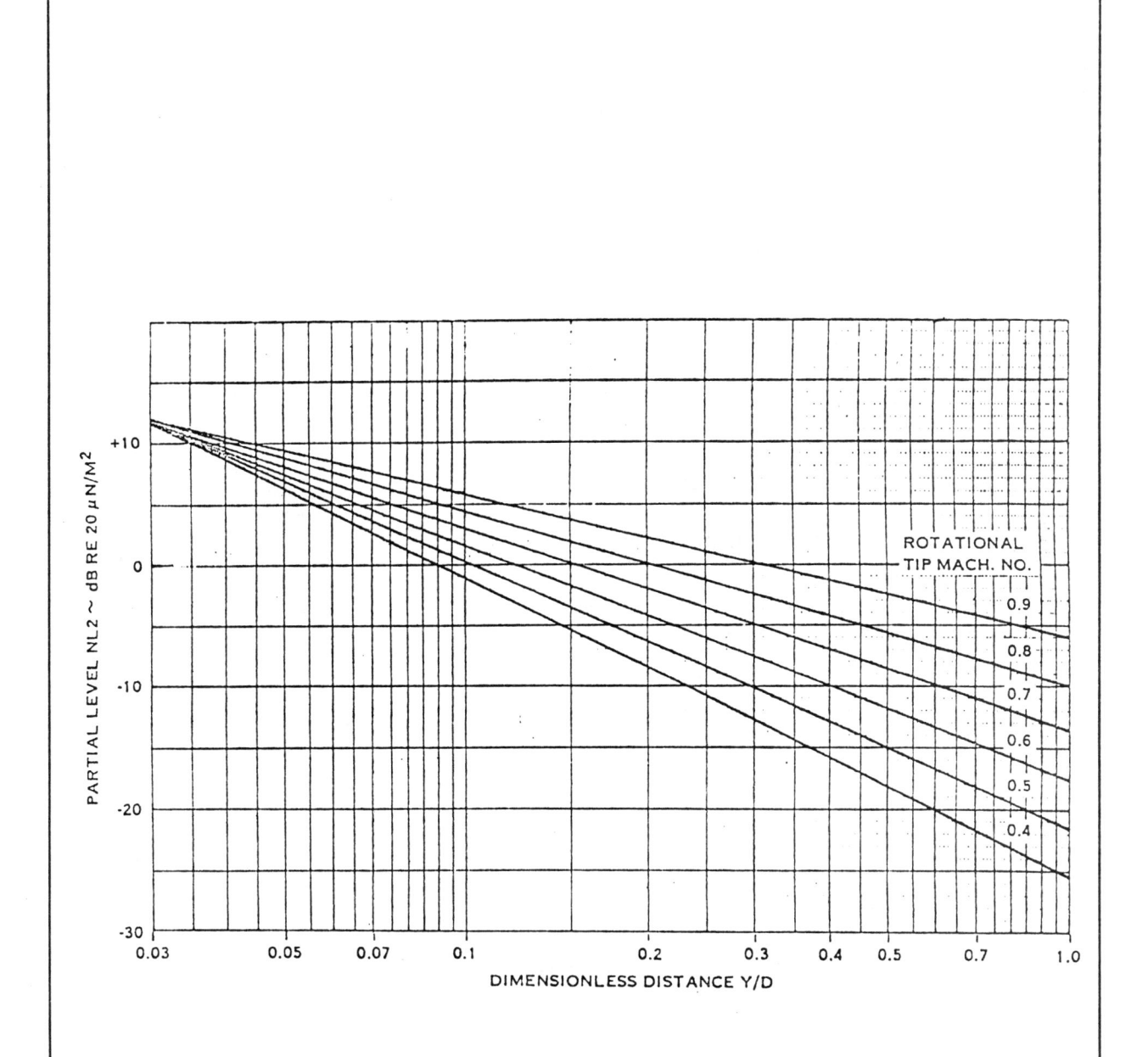

Chart D10 Near Field Partial Noise Level Based on Tip Speed and Tip Clearance

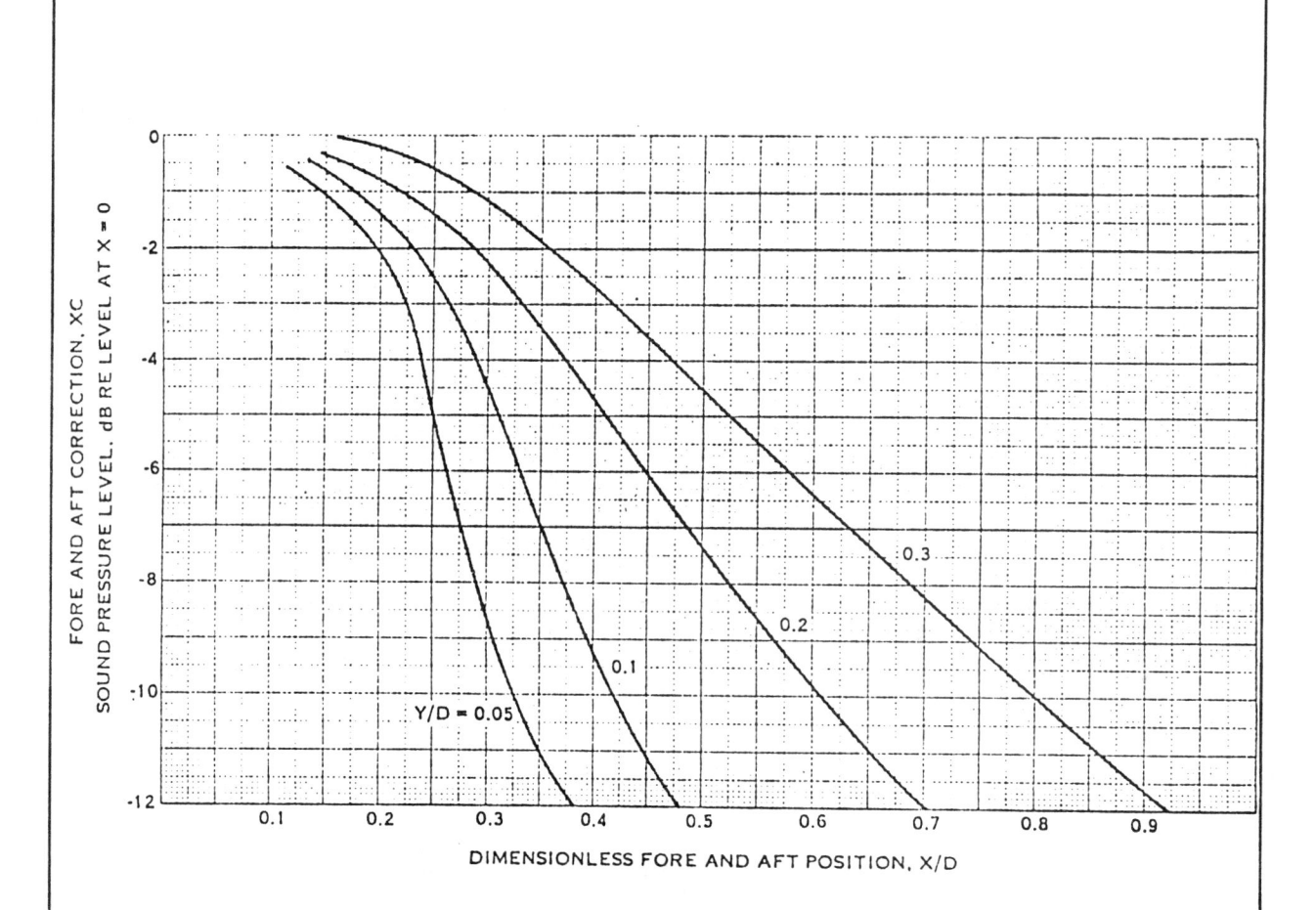

Chart D11 Variation of Overall, Free–Space Propeller Noise Levels with Axial Position x/D Fore and Aft of the Propeller Plane

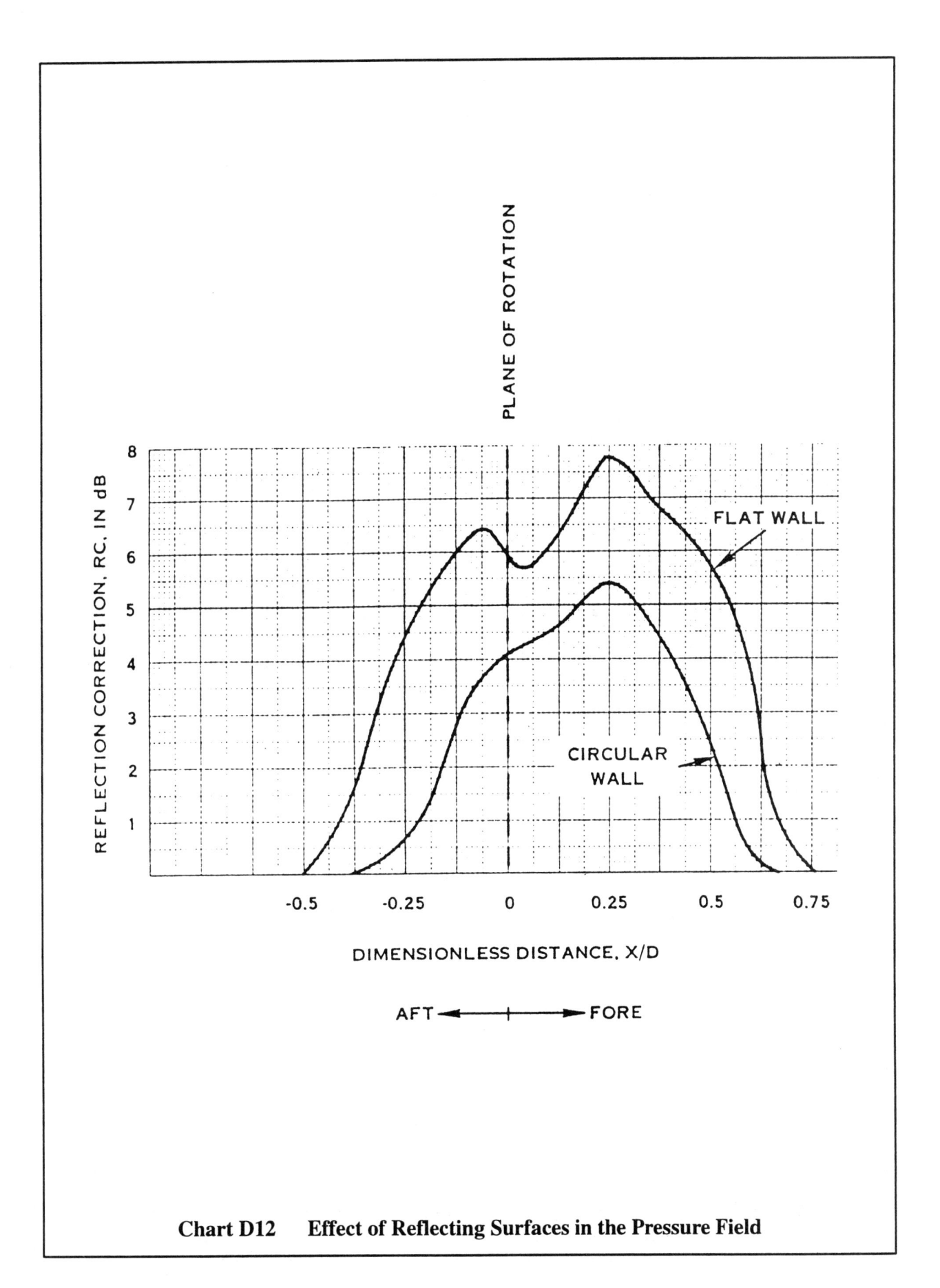

Chart D12   **Effect of Reflecting Surfaces in the Pressure Field**

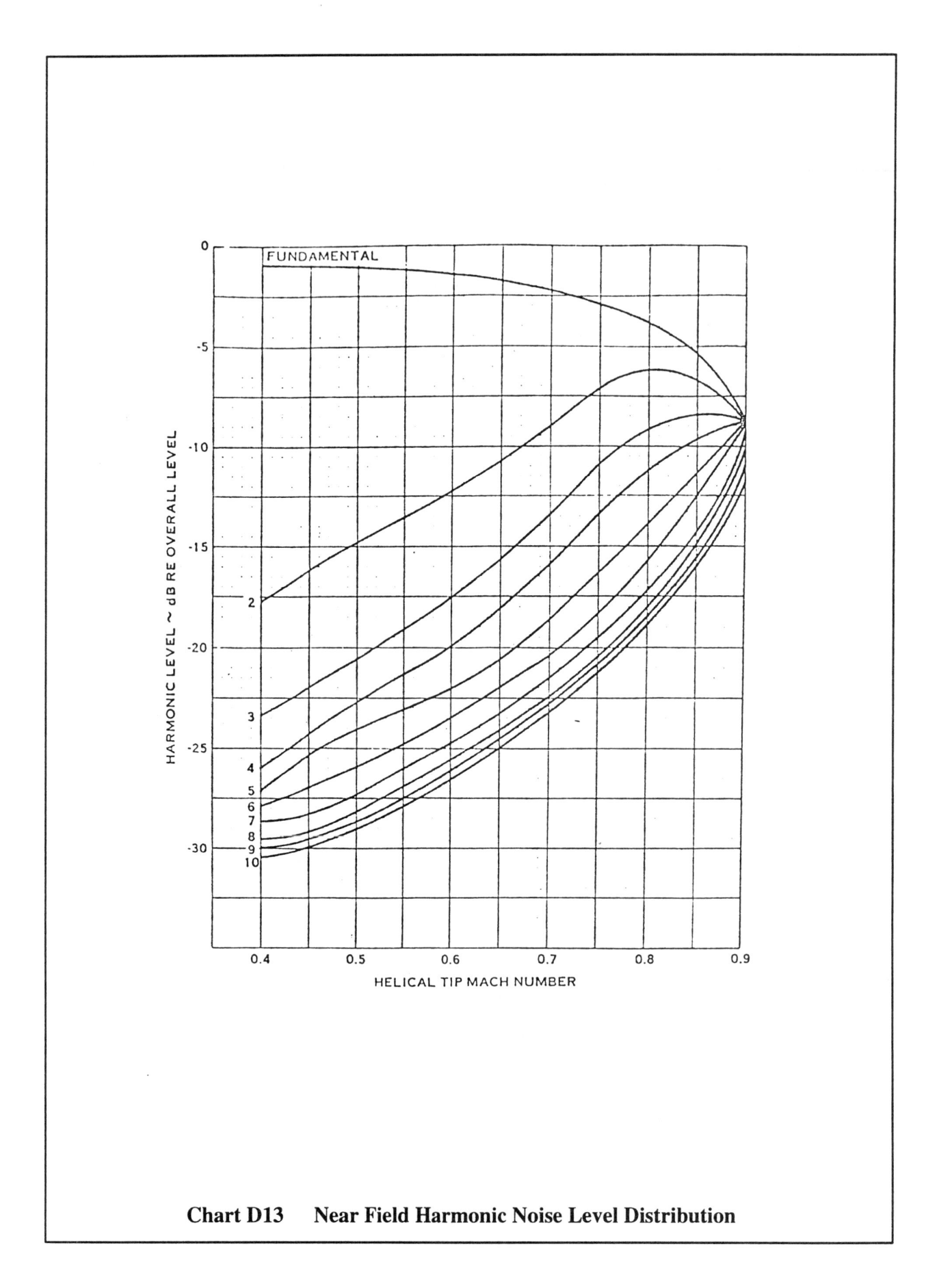

Chart D13 Near Field Harmonic Noise Level Distribution

INDEX

Abrupt stall	77
Absolute ceiling	592, 417
Absolute temperature	3
Accelerated flight performance	616
Accelerate–go distance	440, 438
Accelerate–stop distance	440, 438
Acceleration effect on climb performance	420
Acceleration factor	420, 375
Acceleration of gravity	3
Activity factor	322
Adiabatic compressible flow	26
Aerodynamic center	108, 60, 59, 1
Aerodynamic principles	13
Aerodynamics	1
Afterburner	231, 230
Air density	3
Air distance (landing)	497, 489, 487, 478
Air distance (take–off)	464, 435
Airfoil applications	76, 74
Airfoil design	70
Airfoil designations	72, 70
Airfoil geometry	51
Airfoil maximum lift	80, 77
Airfoil pressure distribution	60
Airfoil stall	110
Airfoil theory	51, 1
Airfoil thickness location parameter	153
Airspeed corrections	26
Airspeed indicator	22, 21
Airspeed measurement	21, 13
Airworthiness regulations	1
All–engines–operating (AEO) ceilings	417, 404
Altimeters	8
Altitude effect on glide performance	347
Altitude effect on power	215
Altitude effect on range and endurance	553
Anti–skid system	505
Approach	480, 478
Area rule	154
Aspect ratio	111, 110, 96
Atmosphere	631, 3, 1

Atmospheric fundamentals	3
Atmospheric pressure	3
Augmentor wing	88
Available thrust	376, 354
Available thrust horsepower	376, 307
Axial flow compressor	235
Axis systems	331

Baffle plates	222
Balanced fieldlength	472, 441, 438
Barometric altimeters	9, 8
Base drag	160
Bernoulli	33, 13
Bernoulli equation	266, 46
Best endurance	513
Best range	511
Betz	35, 34
Binomial expansion	25
Biot and Savart law	32
Blade element theory	275
Boeing definition of drag divergence	141, 65
Boundary layer	13
Boundary layer control	122, 87
Boundary layer splitter (bleed)	44, 43
Bound vortex	97
Brake–horse–power	213
Braking friction coefficient	483
Braking torque	484
Brayton–cycle	225
Breguet equations	536, 535, 510
British thermal unit	214
Buckingham's π–theorem	238, 53
Buffet	581, 410, 142
Buffet boundary (limit)	586, 581, 142
By–pass ratio	231

Calculus of variations	620
Calibrated airspeed	421, 26, 22
Calibrated stall speed	436
Camber	110, 79, 77, 52
Canard configuration	131
Canard drag	161
Canopy/windshield effect on drag	169
Center of pressure	108

Centrifugal flow compressor	235
Chord line	52
Circulation	97, 34, 32, 1
Clean airplane drag polar	137
Climb at constant equivalent airspeed	421
Climb at constant Mach number	423
Climb distance (take–off)	455
Climb gradient	429, 376
Climb–out (take–off)	435
Climb performance	420, 401, 389, 387, 379, 373, 1
Climb performance regulations	428
Combat ceiling	419
Combustion	214, 201
Combustion chamber	225
Compounding	217
Compressibility	1
Compressibility correction	30
Compressibility effects	296, 290, 286, 154, 119, 62, 17, 13
Compressible Bernoulli equation	22, 20, 13
Compressible continuity equation	14
Compressible propeller theory	273
Compression ratio	213
Conservation of mass	13
Constant altitude endurance	536
Constant altitude range	536
Constant speed endurance	539
Constant speed range	539
Continuity equation	46, 19, 13
Conventional configuration	131
Conversion factors	635, 1
Cooling of piston engines	221
Cooling of turbine engines	249
Crankshaft	213
Critical Mach number	119, 65, 62
Critical pressure coefficient	63, 62
Cruise ceiling	419

Dalton's Law	3
Density altitude	8
Density of mixture of dry air and water vapor	4
Density variation with altitude	6
Descent performance	480, 1
Design cruising speed	600, 595
Design diving speed	600, 596
Design lift coefficient	323, 57

Design limit load factor	600, 597
Design maneuvering speed	600, 594, 593
Design speed for maximum gust intensity	600
Diameter (propeller)	322
Diesel	201
Diesel engines	208
Dive brakes	132, 121
Double slotted flaps	85, 80
Douglas definition for drag divergence	141, 65
Downwash	97, 1
Drag	137, 19, 1
Drag breakdown	175, 174
Drag build–up	146
Drag coefficient	97
Drag contributions	146
Drag determination in the wind–tunnel	178
Drag divergence	65
Drag divergence Mach number	141, 120, 67, 66
Drag due to lift	154
Drag on stopped engines	311
Drag polar	137, 57, 1
Drag reduction	177, 176, 174
Drift–down performance	403, 389, 388, 379, 373, 1
Ducted propellers	313, 311
Dynamic pressure	22, 21

Effective aspect ratio	100
Electric motors	261, 202, 201
Empennage drag	161
Endurance	533, 526, 509, 1
Endurance equation	510
Endurance factor	510
Energy maneuverability	626
Energy method	99, 98
Energy state approximation	617
English units	4
Equation of state	3
Equivalent airspeed	421, 26, 21
Equivalent body drag	156
Equivalent flat plate drag	188
Equivalent parasite area	188
Equivalent shaft–horsepower	234
Euler's equation	19, 16
Exosphere	5

Far–field noise	317
First law of thermodynamics	17
Fixed pitch propeller	286
Flap effect on climb performance	425
Flap effect on drag	167, 145
Flight envelope	581, 207, 202, 1
Flight mechanics	1
Flight path angle	332
Flow–field corrections	178
Flow separation	586, 42, 37, 13
Forward sweep	120
Four–stroke piston engine	209
Fowler flaps	82, 80
Free propeller efficiency	298
Free turbine	235
Friction coefficient	444
Fuel consumed in climb	404
Fuel consumed in drift–down	412
Fuel flow	535, 509
Fuel reserves	1
Fuselage camber	171
Fuselage drag	157

Gas constant	3
Geometric pitch angle (propeller)	275
Glauert	63
Glide angle	340, 336
Global Positioning System (GPS)	8
Gradual stall	77
Ground effect on lift and drag	448
Ground distance or roll (landing)	498, 493, 491, 482, 478
Ground distance or roll (take–off)	466, 460, 454, 442, 435
Ground Proximity Warning System (GPWS)	8
Gust load factor line	601, 597

High lift devices	128, 127, 121, 80, 1
Hodograph	344
Hoerner	68
Horizontal tail drag	161

Incompressible Bernoulli equation	266, 16, 15, 13
Incompressible continuity equation	14
Incompressible propeller theory	269, 266

Indicated airspeed	26
Indicated shaft–horse–power	213
Induced drag	146, 100, 98, 97, 1
Induced drag coefficient	137
Induced velocity (propeller)	267
Installation of piston engines	221
Installation of turbine engines	251, 249
Installed propeller efficiency	298
Instantaneous pull–up	603
Instantaneous turn	607, 606
Instrument error	27, 26
Instrument indicated airspeed	26
Integrated design lift coefficient	323
Integration drag	163
Interference	148
Interference drag	172
International standard atmosphere	4
International Civil Aviation Organization (ICAO)	4
Internal energy	17
Ionosphere	5
Isentropic equation of state	22, 19, 17, 13

Jet engines	225, 202, 201

Karman–Tsien	65, 64
Kinematic viscosity	38
Kinetic energy	16
Krueger flap	87
Kutta–Joukowski theorem	47, 34, 32, 13

Laminar boundary layer	39, 38
Laminar skin friction	39
Landing distance	497, 494, 487
Landing obstacle	478
Landing performance	435, 1
Landing performance data	505
Landing process	478
Landing gear effect on climb performance	425
Landing gear effect on drag	168, 145
Leading edge devices	116, 86
Leading edge radius	78, 77, 52

Index

Leading edge shape	78
Leading edge vortex	121
Level flight ceiling	593
Lift	1
Lift coefficient	97
Lift curve slope	56
Lifting line method	100
Lifting surface correction factor	151
Lifting surface method	126
Lifting surfaces	1
Lift–off speed	435
Lift–to–drag ratio	337, 57
Linear momentum principle	47, 36, 13
Location of maximum thickness	79, 77
Lowry	123, 104

Mach number	40, 22
Mach indicator	30
Maneuvering	581, 1
Manifold absolute pressure	215
Mass flow rate	226, 14
Maximum lift	343, 127, 80
Maximum endurance	548, 542, 523, 517, 513
Maximum range	548, 542, 523, 513, 511
Maximum speed	592
Mean (camber) line	52
Mechanical arrangements of turbojets, turbofans and turboprops	235
Mechanical efficiency	213
Mechanical output	214
Metric units	4
Minimum control speed	582, 581, 437
Minimum control speed on the ground	437
Minimum safety speed	438
Minimum speed	581, 342
Minimum time–to–climb	620
Minimum unstick speed	438
Momentum losses	283
Momentum method	266, 98
Most critical center of gravity (c.g.)	584

NACA (National Advisory Committee on Aeronautics)	70, 51
NASA (National Aeronautics and Space Administration)	88, 70, 51
Nacelle–pylon drag	162
Near–field noise	320

Negative–1–g–stall speed	600, 596
Negative propeller thrust	308, 269
Net thrust	234
Noise considerations for propellers	315
Noise considerations for turbine engines	252
Northern hemisphere	4

One–engine–inoperative (OEI) ceilings	417, 404
One–g–stall speed	598, 594, 593
Oswald's efficiency factor	137, 100
Otto, Otto cycle	201
Overall propulsion efficiency	214

Pacer airplane	27
Parabolic drag polar	137, 104
Parasite drag	146
Partial span flaps	122
Payload–range	559, 509, 1
Payload–range data	561
Payload–range diagram	559
Peak pressure	61
Perfect gas law	19, 4, 3
Performance	331, 1
Piston engine performance charts	217
Piston engines	208, 201
Pitch attitude angle	332
Pitching moment coefficient	109, 97
Pitching moment curve	59
Pitch break	113
Pitch–up	112
Pitot–static tube	21
Plain flaps	80
Polhamus	123, 104
Position error	27, 26
Positive propeller thrust	304, 300, 273, 266
Power available	376, 213
Power required	509, 376, 307
Power disk loading coefficient	284
Prandtl–Glauert transformation	104, 65, 64, 63
Prandtl lifting line theory	100
Prandtl–Schlichting skin friction formula	40
Pressure altitude	8
Pressure gauge	21

Pressure gradient	42
Pressure variation with altitude	6
Pressure wave	19
Profile drag	150
Propeller advance velocity	275
Propeller advance ratio	293, 288, 282
Propeller blade angle	286
Propeller blade geometry	287
Propeller blade loading	288
Propeller blockage	292
Propeller design variables	322
Propeller disk	266
Propeller efficiency	298, 292, 282, 268, 267, 265, 213
Propeller in–flight thrust	304
Propeller installation effects	292
Propeller noise	316, 315, 1
Propeller noise charts	685
Propeller performance	300, 1
Propeller performance charts	643
Propeller power coefficient	282
Propeller power losses	282
Propeller selection	322
Propeller shaft extension	295
Propeller shank	289, 286
Propeller static thrust	300
Propeller theory	273, 269, 265, 1
Propeller thrust coefficient	282, 280
Propeller thrust reversing	308
Propeller torque	276
Propeller torque coefficient	282, 280
Prop–fan	311, 235
Propulsion	1
Propulsion systems	201
Propulsive efficiency	233, 227
Pull–up	603, 1
Pulse–jets	258, 225
Pusher	299, 221

Radar altimeters	9, 8
Ram–jets	259, 230, 225
Range	527, 526, 509, 1
Range equation	510
Range factor	510
Range–payload	1
Rate-of-climb	404, 375, 332

Rate–of–descent	378, 341, 338, 336, 332
Reciprocating engines	1
Reynolds number	178, 107, 40, 38
Reynolds number effects	186, 80, 68
Rocket engines	260, 201
Rotation speed	435
Regulations: climb performance	428
Regulations: fuel reserves	561
Regulations: landing performance	479
Regulations: take–off performance	436
Regulations: V–n diagram	593
Required thrust	356, 334
Rolling ground friction coefficient	444
Rotary engines	202, 1
Roughness grit	68
Runway gradient	443

Scalar	37
Scaling laws for piston engines	222
Scaling laws for turbine engines	256
Scavenging	210
Schlichting's formula	68, 40
Screen height	478, 436
Scrubbing effects	296, 292
Sectional drag coefficient	55
Sectional lift coefficient	55
Sectional pitching moment coefficient	55
Service ceiling	418
Shaft–horse–power	234, 213
Shape factor	68
Sizing of weight of airplanes	565
Shock wave	20, 19
Skin friction drag	39
Slats	116, 87, 86
Slip ratio	485
Slots	116, 87, 86
Slotted flaps	82, 80
Specific excess power	1
Specific fuel consumption (s.f.c.)	509, 225
Specific heat	18
Specific endurance	525
Specific excess power	617
Specific range	526
Speed	1
Speed brakes	145, 132, 121, 1

Speed of sound	19, 13
Speed polar	349, 344
Span efficiency factor	104, 100
Spin	613, 581
Split flaps	82, 80
Splitter	43
Spoilers	132, 121, 1
Stagnation pressure	21
Stall characteristics	109, 1
Stall control devices	115
Stall fences	116
Stall snags	116
Stall speed	585, 581, 436, 342
Stall strips	·117
Static port	28, 27
Steady, powered flight	357, 354, 331
Steady, un–powered flight	335, 331
Steep angle climb performance	401, 387
Sting type model installation	178
Stopped engine effect on climb performance	425
Store drag	169
Strain gauge balance	184, 178
Streamline	13
Streamtube	15
Stratosphere	7, 6, 5
Strut type model installation	178
Subsonic	22
Supercharging	216
Supercritical airfoil	73, 71
Supersonic	22
Supersonic aerodynamics	4
Sustained pull–up	604
Sustained turn	607, 606
Sweep angle	165, 112, 110, 96

Take–off climb–out to the obstacle	453
Take–off distance	460, 456, 454
Take–off obstacle	436
Take–off performance	435, 1
Take–off performance data	475
Take–off process	435
Take–off time	467, 460
Take–off transition	452
Tandem configuration	165
Taper ratio	110, 96

Tare corrections	178
Temperature altitude	8
Temperature effect on range and endurance	553
Temperature variation with altitude	5
Thermal efficiency	214
Thickness	52
Thickness ratio	78, 77
Thrust available	376, 354
Thrust required	354
Thrust reversal for propellers	308
Thrust reversal for turbine engines	255
Thrust specific fuel consumption (t.s.f.c.)	229
Time–to–climb	410, 404
Time–to–drift–down	411, 404
Time–to–land	500, 497
Tip Mach number	291
Total pressure	21
Tractor	299, 221
Trailing edge flaps	80
Transition distance (take–off)	455
Transition distance (landing)	481
Transition point	38
Transition Reynolds number	38
Trim	127
Trim drag	170
Trip strip	186, 68
Tripping wire	43
Tropopause	5
Troposphere	6, 5
Turbine engines	1
Turbofans	231, 225
Turbojets	225
Turbo–props	233, 225
Turbulence factor	185
Turbulent boundary layer	40, 39, 38
Turbulent skin friction	152, 150, 40
Turn	606, 1
Twist	165, 115, 114, 110
Two–stroke piston engine	210

Un–powered glide	352
U.S. Standard Atmosphere	4

Variable pitch propeller	286
Vertical acceleration effect on climb	424
Vertical tail drag	161
Viscosity	10
Viscous effects	47, 37, 13
V–n diagram	593
Vortex	32
Vortex generator	117

Wash–out	115
Water vapor	3
Weight effect on glide performance	348
Weight sizing	1
Wet footprint	416
Wetted area	150
Whitcomb	71
Wind effect on glide performance	350, 347
Wind effect on range	516
Windmilling	311
Wind tunnel	15, 1
Wing area	95
Wing aspect ratio	96
Wing drag	148
Winglets	166
Wing mean geometric chord	96
Wing planform effect on drag	163
Wing properties	95
Wing sizing for maximum endurance	518
Wing sizing for maximum range	515
Wing stall	110
Wing sweep angle	96
Wing taper ratio	96
Wing theory	95, 1

Zero lift drag coefficient	148, 137, 100

Notes

Notes

Notes

Airplane Aerodynamics and Performance
Dr. Jan Roskam and Dr. C.T. Lan

Nearly all aerospace engineering curricula include as required material a course on airplane aerodynamics and airplane performance. This textbook delivers a comprehensive account of airplane aerodynamics and performance.

The atmosphere • basic aerodynamic principles and applications • airfoil theory • wing theory • airplane drag • airplane propulsion systems • propeller theory • fundamentals of flight mechanics for steady symmetrical flight • climb performance and speed • take-off and landing performance • range and endurance • maneuvers and flight envelope

Item 110 • ISBN 1-884885-44-6 • softcover • 752 pages

Airplane Flight Dynamics and Automatic Flight Controls, Parts I and II
Dr. Jan Roskam

Airplane Flight Dynamics & Automatic Flight Controls Part I

Exhaustive coverage is provided of the methods for analysis and synthesis of the steady and perturbed state stability and control of fixed wing airplanes. This widely used book has been updated with modern flying quality criteria.

General steady and perturbed state equations of motion for a rigid airplane • concepts and use of stability & control derivatives • physical and mathematical explanations of stability & control derivatives • solutions and applications of the steady state equations of motion from a viewpoint of airplane analysis and design • emphasis on airplane trim, take-off rotation and engine-out control • open loop transfer functions • analysis of fundamental dynamic modes: phugoid, short period, roll, spiral and dutch roll • equivalent stability derivatives and the relation to automatic control of unstable airplanes • flying qualities and the Cooper-Harper scale: civil and military regulations • extensive numerical data on stability, control and hingemoment derivatives

Item 210 • ISBN 1-884885-17-9 • softcover • pages 576

Airplane Flight Dynamics & Automatic Flight Controls Part II

Exhaustive coverage is provided of the methods for analysis and synthesis of automatic flight control systems using classical control theory. This widely used book has been updated with the latest software methods.

Elastic airplane stability and control coefficients and derivatives • method for determining the equilibrium and manufacturing shape of an elastic airplane • subsonic and supersonic numerical examples of aeroelasticity effects on stability & control derivatives • bode and root-locus plots with open and closed loop airplane applications, and coverage of inverse applications • stability augmentation systems: pitch dampers, yaw dampers and roll dampers • synthesis concepts of automatic flight control modes: control-stick steering, auto-pilot hold, speed control, navigation and automatic landing • digital control systems using classical control theory applications with Z-transforms • applications of classical control theory • human pilot transfer functions

Item 220 • ISBN 1-884885-18-7 • softcover • 381 pages

DARcorporation, 120 East Ninth Street, Suite 2, Lawrence, Kansas 66044, USA
Telephone: (785) 832-0434   Fax: (785) 832-0524   e-mail: info@darcorp.com   http://www.darcorp.com

1-800-327-7144

Airplane Design, Parts I - VIII
Dr. Jan Roskam

Airplane Design Part I
Estimating take-off gross weight, empty weight and mission fuel weight • sensitivity studies and growth factors • estimating wing area • take-off thrust and maximum clean, take-off and landing lift • sizing to stall speed, take-off distance, landing distance, climb, maneuvering and cruise speed requirements • matching of all performance requirements via performance matching diagrams
Item 310 • ISBN 1-884885-42-X • softcover **• 226 pages**

Airplane Design Part II
Selection of the overall configuration • design of cockpit and fuselage layouts • selection and integration of the propulsion system • Class I method for wing planform design • Class I method for verifying clean airplane maximum lift coefficient and for sizing high lift devices • Class I method for empennage sizing and disposition, control surface sizing and disposition, landing gear sizing and disposition, weight and balance analysis, stability and control analysis and drag polar determination
Item 320 • ISBN 1-884885-43-8 • softcover **• 336 pages**

Airplane Design Part III
Cockpit (or flight deck) layout design • aerodynamic design considerations for the fuselage layout • interior layout design of the fuselage • fuselage structural design considerations • wing aerodynamic and operational design considerations • wing structural design considerations • empennage aerodynamic and operational design considerations • empennage structural and integration design consideration • integration of propulsion system • preliminary structural arrangement, material selection and manufacturing breakdown
Item 330 • ISBN 1-884885-06-3 • hardcover **• 454 pages**

Airplane Design Part IV
Landing gear layout design • weapons integration and weapons data • flight control system layout data • fuel system layout design • hydraulic system design • electrical system layout design • environmental control system layout design • cockpit instrumentation, flight management and avionics system layout design • de-icing and anti-icing system layout design • escape system layout design • water and waste systems layout design • safety and survivability considerations
Item 340 • ISBN 1-884885-07-1 • hardcover **• 416 pages**

Airplane Design Part V
Class I methods for estimating airplane component weights and airplane inertias • Class II methods for estimating airplane component weights, structure weight, powerplant weight, fixed equipment weight and airplane inertias • methods for constructing v-n diagrams • Class II weight and balance analysis • locating component centers of gravity •
Item 350 • ISBN 1-884885-08-X • hardcover **• 209 pages**

Airplane Design Part VI
Summary of drag causes and drag modeling • Class II drag polar prediction methods •airplane drag data • installed power and thrust prediction methods • installed power and thrust data • lift and pitching moment prediction methods • airplane high lift data • methods for estimating stability, control and hingemoment derivatives • stability and control derivative data
Item 360 • ISBN 1-884885-09-8 • hardcover **• 550 pages**

Airplane Design Part VII
Controllability, maneuverability and trim • static and dynamic stability • ride and comfort characteristics • performance prediction methods • civil and military airworthiness regulations for airplane performance and stability and control • the airworthiness code and the relationship between failure states, levels of performance and levels of flying qualities
Item 370 • ISBN 1-884885-10-1 • hardcover **• 351 pages**

Airplane Design Part VIII
Cost definitions and concepts • method for estimating research, development, test and evaluation cost • method for estimating prototyping cost • method for estimating manufacturing and acquisition cost • method for estimating operating cost • example of life cycle cost calculation for a military airplane • airplane design optimization and design-to-cost considerations • factors in airplane program decision making
• Item 380 • ISBN 1-884885-11-X • hardcover **• 368 pages**

DARcorporation, 120 East Ninth Street, Suite 2, Lawrence, Kansas 66044, USA
Telephone: (785) 832-0434   Fax: (785) 832-0524   e-mail: info@darcorp.com   http://www.darcorp.com

1-800-327-7144

Airplane Design and Analysis Software for Windows

Advanced Aircraft Analysis (AAA) is an integrated airplane design and analysis system, taking users from weights to stability and control within the same environment and used by Universities and Airplane Manufacturers in 24 countries. The full AAA program is available on UNIX platforms (SUN, IBM, SGI, HP) and on IBM Compatible PC (Windows 3.1, 95, NT).

G.A.-CAD (General Aviation Computer Aided Design) is a software package for the preliminary design of general aviation aircraft. G.A.-CAD will reduce design and development time by using computer aided design methods and eliminating tedious hand calculations. G.A.-CAD is designed after Advanced Aircraft Analysis (AAA), but for general aviation airplane types only.

G.A.-CAD Lite is based on G.A.-CAD and contains Preliminary Sizing of Airplanes, Drag, Lift, Geometry, Stability & Control Sizing, Weights, Performance and CAD.

G.A.-HomeBuilt is based on G.A.-CAD and contains Preliminary Sizing of Airplanes, Drag, Lift, Geometry, Stability & Control Sizing, Weights and Performance.

Stability & Control PROfessional contains Stability & Control Sizing, Stability & Control Derivatives & Hingemoment Derivatives plus calculation of airplane aerodynamic center shift due to fuselage, nacelles & stores, and supports multiple flight conditions. S&C PROfessional uses the same methods as AAA, and is based on Airplane Flight Dynamics I & II, Airplane Design Part II and Airplane Design Part VI, by Dr. Jan Roskam.

AeroDynamics PROfessional contains Drag, Clean Wing Lift, Lifting Line, Flap Sizing & Verification, CLo/Cmo. AeroDynamics PROfessional uses the same methods as AAA, and is based on Airplane Design Part I, Airplane Design Part II and Airplane Design Part VI, Aerodynamics & Performance, by Dr. Jan Roskam.

AeroXpress contains Stability & Control PROfessional, AeroDynamics PROfessional, Trim Diagram & Geometry. AeroXpress uses the same methods as AAA, and is based on Airplane Flight Dynamics I & II, Airplane Design Part I, Airplane Design Part II and Airplane Design Part VI, Aerodynamics & Performance, by Dr. Jan Roskam.

FlightDynamics PROfessional calculates longitudinal & lateral-directional transfer functions. The open loop flying qualities of the airplane are checked against FAR, JAR, VLA & MIL-F-8785C requirements. The transfer functions can be used to generate Bode & Root-Locus plots for open & closed loop control systems. FlightDynamics uses the same methods as AAA, and is based on Airplane Flight Dynamics I & II, and Airplane Design Part VI, by Dr. Jan Roskam.

Preliminary Sizing of Airplanes performs preliminary weight & performance sizing for new & existing airplanes, determines preliminary estimates of airplane maximum take-off weight, fuel weight & payload weight. Estimates parabolic drag polar, and determines wing area. Preliminary Sizing of Airplanes uses the same methods as AAA, and is based on Airplane Design Part I, by Dr. Jan Roskam.

DARcorporation, 120 East Ninth Street, Suite 2, Lawrence, Kansas 66044, USA

Telephone: (785) 832-0434   Fax: (785) 832-0524   e-mail: info@darcorp.com   http://www.darcorp.com

1-800-327-7144

Airplane Design and Analysis Software for Windows

Drag offers drag estimation methods for wing, horizontal tail, vertical tail, canard, fuselage, nacelle, flap (leading and trailing edge), landing gear (fixed and retractable), canopy/windshield, stores, trim, spoiler and engine inlets and outlets. Airplane examples are provided. Drag uses the same methods as AAA, and is based on Airplane Design Part I and Airplane Design Part VI, by Dr. Jan Roskam.

Stability & Control Sizing performs preliminary sizing of airplane stability surfaces & complete airplane trim analysis. S&C Sizing will size canard, horizontal tail & vertical tail surface areas to stability requirements. S&C Sizing uses the same methods as AAA, and is based on Airplane Flight Dynamics I and Airplane Design Part II, by Dr. Jan Roskam.

Stability & Control Derivatives provides complete longitudinal & lateral-directional S&C derivative calculations for a variety of airplane configurations. Stability & control derivatives are calculated for tail aft, canard & three surface configurations with single or twin vertical tails. S&C Derivatives uses the same methods as AAA, and is based on Airplane Flight Dynamics I & II and Airplane Design Part VI, by Dr. Jan Roskam.

Hingemoment Derivatives uses the geometry of the control surface & the airplane flight condition to determine hingemoment derivatives for longitudinal & lateral-directional control surfaces. Hingemoment derivatives are calculated for aileron, rudder, elevator & canardvator. The hingemoment due to angle of attack derivative & hingemoment due to control surface deflection derivative are calculated for a given flight condition. Hingemoment Derivatives uses the same methods as AAA, and is based on Airplane Design Part VI, by Dr. Jan Roskam.

Cost performs preliminary cost estimation for newly designed airplanes. Inflation effects are accounted for using cost escalation factors based on the consumer price index history. The estimation of research, development, test and evaluation (RDTE) cost is broken down into airframe engineering and design cost, development support and testing cost, flight test airplanes cost, flight test operations cost, test and simulation facilities cost, RDTE profit, and cost to finance the RDTE phase. Cost estimates are represented in 'then year' dollars. Cost uses the same methods as AAA, and is based on Airplane Design Part VIII, by Dr. Jan Roskam.

Atmosphere will estimate properties of the atmosphere. The properties of atmosphere established by the International Civil Aviation Organization (ICAO) are used as standard atmospheric properties. Calculates: Speed of sound - Acceleration of gravity - Air kinematic viscosity - Air pressure - Air temperature - Air density - Air pressure ratio - Air temperature ratio - Air density ratio. Atmosphere uses the same methods as AAA, and is based on Aerodynamics & Performance, by Dr. Jan Roskam and Dr. C.T. Lan.

DARcorporation, 120 East Ninth Street, Suite 2, Lawrence, Kansas 66044, USA
Telephone: (785) 832-0434   Fax: (785) 832-0524   e-mail: info@darcorp.com   http://www.darcorp.com

1-800-327-7144

Design, Analysis and Research Corporation

Design, Analysis and Research Corporation (DARcorporation) was founded by Dr. Jan Roskam in 1991 with the following objectives:

- Market, support and develop the Advanced Aircraft Analysis (AAA) computer program. This user-friendly program runs under UNIX and the X Window System environment on workstations and MS Windows on PC's.

- Market, support and develop the following PC/Windows based computer programs:
    - Stability & Control PROfessional
    - AeroDynamics PROfessional
    - AeroXpress
    - FlightDynamics PROfessional
    - Atmosphere
    - Hingemoment Derivatives
    - Stability & Control Derivatives
    - Stability & Control Sizing
    - Cost
    - Preliminary Sizing of Airplanes
    - Drag

- Provide consulting services in the general area of aircraft configuration design and analysis, stability and control and estimation of airplane models for flight simulators and aircraft technical analysis.

- Market and distribute the Aircraft Design, Aircraft Performance and Flight Dynamics textbooks written by Dr. Jan Roskam.

- Develop and market the General Aviation Computer Aided Design (G.A.-CAD) software. A PC based preliminary design and analysis tool for the General Aviation industry.

- Market AERO-CAD a new Computer Aided Drafting program for PC's with specific Aeronautical Engineering tools built-in.

AIRPLANE DESIGN AND ANALYSIS

DARcorporation, 120 East Ninth Street, Suite 2, Lawrence, Kansas 66044, USA
Telephone: (785) 832-0434   Fax: (785) 832-0524   e-mail: info@darcorp.com   http://www.darcorp.com

1-800-327-7144

Notes

DARcorporation
Design, Analysis and Research Corporation

Information Request Form

FOR FASTEST SERVICE, CALL	1-800-327-7144	(U.S and Canada only)
	(785) 832-0434	
FAX	1-888-224-6139	(U.S and Canada only)
	(785) 832-0524	
e-mail	info@darcorp.com	

Date _____

Phone _____

Fax _____

Name _____

Title _____

Company/University _____

_____

Address _____

_____

_____

E-mail _____

❑ I would like to receive your *free* Newsletter

❑ Please add my name to your mailing list

❑ Please send me a book order form

Consulting/Contracting Services

Please have a representative contact me. I am interested in the following areas:

❑ New Design

❑ Modification Analysis

❑ Verification Analysis

G.A.-CAD

❑ Please send me your information packet

Advanced Aircraft Analysis

❑ Please send me a detailed product description of AAA

❑ I would like to arrange for a free 30-day evaluation copy of AAA

❑ Please have a representative contact me

Please check all that apply:

❑ Professional   ❑ Professor   ❑ Homebuilder   ❑ Student

no. employees: _____   no. students in class: _____

AIRPLANE DESIGN AND ANALYSIS

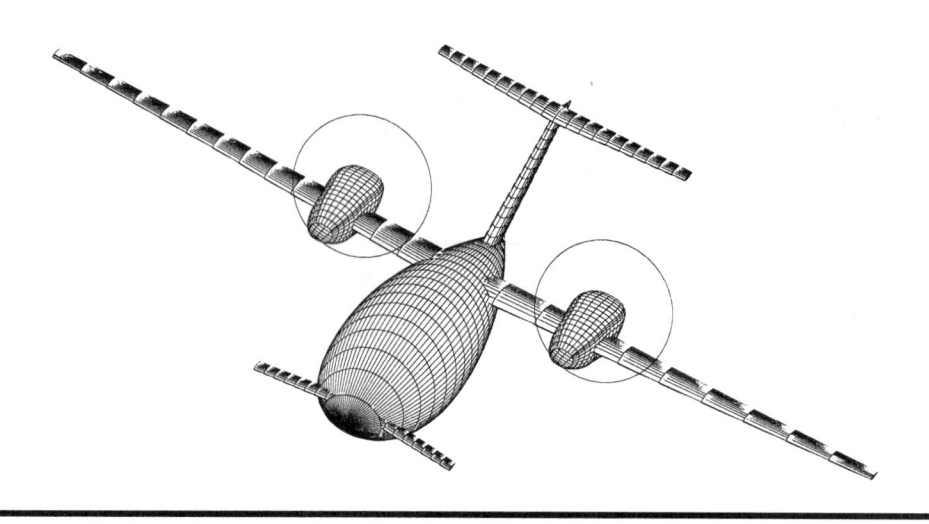

--FIRST FOLD--

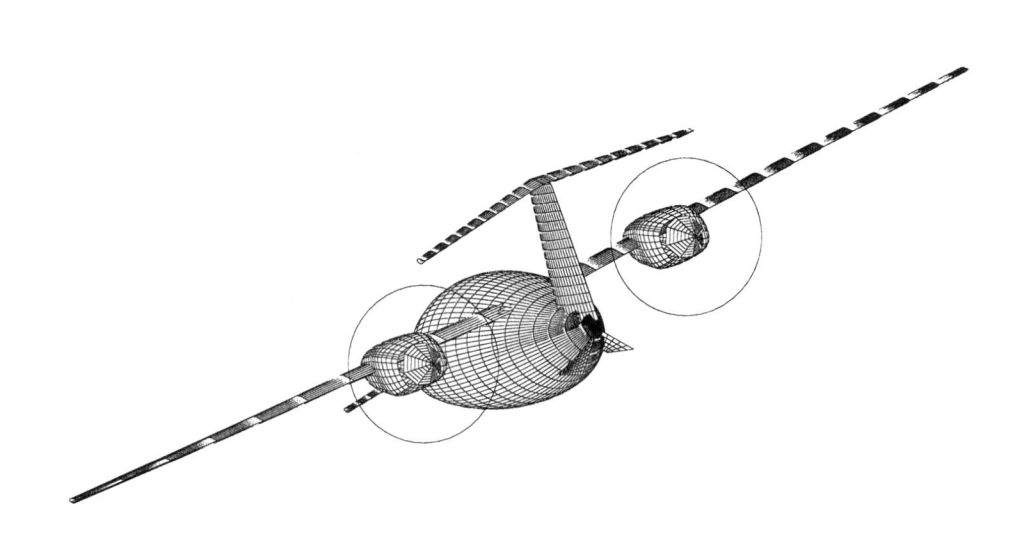

--SECOND FOLD THEN TAPE--

BUSINESS REPLY MAIL
FIRST CLASS MAIL   PERMIT NO.697   LAWRENCE, KS

POSTAGE WILL BE PAID BY ADDRESSEE

NO POSTAGE
NECESSARY
IF MAILED
IN THE
UNITED STATES

DESIGN, ANALYSIS AND RESEARCH CORPORATION
120 E 9TH ST   STE 2
LAWRENCE KS 66044-9799